Prefixes and Suffixes in Anatomical and Medical Terminology

Element	Definition and Example
mito-	thread: *mitochondrion*
mono-	alone, one, single: *monocyte*
morph-	form, shape: *morphology*
multi-	many, much: *multinuclear*
myo-	muscle: *myology*
narc-	numbness, stupor: *narcotic*
necro-	corpse, dead: *necrosis*
neo-	new, young: *neonatal*
nephr-	kidney: *nephritis*
neuro-	nerve: *neurolemma*
noto-	back: *notochord*
ob-	against, toward, in front of: *obturator*
oc-	against: *occlusion*
-oid	resembling, likeness: *sigmoid*
oligo-	few, small: *oligodendrocyte*
-oma	tumor: *lymphoma*
oo-	egg: *oocyte*
or-	mouth: *oral*
orchi-	testis: *orchiectomy*
ortho-	straight, normal: *orthopnea*
-ory	pertaining to: *sensory*
-ose	full of: *adipose*
osteo-	bone: *osteoblast*
oto-	ear: *otolith*
ovo-	egg: *ovum*
par-	give birth to, bear: *parturition*
para-	near, beyond, beside: *paranasal*
path-	disease, that which undergoes sickness: *pathology*
-pathy	abnormality, disease: *neuropathy*
ped-	children: *pediatrician*
pen-	need, lack: *penicillin*
-penia	deficiency: *thrombocytopenia*
per-	through: *percutaneous*
peri-	near, around: *pericardium*
phag-	to eat: *phagocyte*
-phil	have an affinity for: *neutrophil*
phleb-	vein: *phlebitis*
-phobia	abnormal fear, dread: *hydrophobia*
-plasty	reconstruction of: *rhinoplasty*
platy-	flat, side: *platysma*
-plegia	stroke, paralysis: *paraplegia*
-pnea	to breathe: *apnea*
pneumo(n)-	lung: *pneumonia*
pod-	foot: *podiatry*
-poiesis	formation of: *hemopoiesis*
poly-	many, much: *polyploid*
post-	after, behind: *postnatal*
pre-	before in time or place: *prenatal*
pro-	before in time or place: *prophase*

Element	Definition and Example
proct-	anus: [illegible]
pseudo-	false: *pseudostratified*
psycho-	mental: *psychology*
pyo-	pus: *pyorrhea*
quad-	fourfold: *quadriceps femoris*
re-	back, again: *repolarization*
rect-	straight: *rectus abdominis*
ren-	kidney: *renal*
rete-	network: *rete testis*
retro-	backward, behind: *retroperitoneal*
rhin-	nose: *rhinitis*
-rrhagia	excessive flow: *menorrhagia*
-rrhea	flow or discharge: *diarrhea*
sanguin-	blood: *sanguineous*
sarc-	flesh: *sarcoma*
-scope	instrument for examining a part: *stethoscope*
-sect	cut: *dissect*
semi-	half: *semilunar*
-sis	process or action: *dialysis*
steno-	narrow: *stenosis*
-stomy	surgical opening: *tracheostomy*
sub-	under, beneath, below: *subcutaneous*
super-	above, beyond, upper: *superficial*
supra-	above, over: *suprarenal*
syn- (sym-)	together, joined, with: *synapse*
tachy-	swift, rapid: *tachycardia*
tele-	far: *telencephalon*
tens-	stretch: *tensor tympani*
tetra-	four: *tetrad*
therm-	heat: *thermogram*
thorac-	chest: *thoracic cavity*
thrombo-	lump, clot: *thrombocyte*
-tomy	cut: *appendectomy*
tox-	poison: *toxemia*
tract-	draw, drag: *traction*
trans-	across, over: *transfuse*
tri-	three: *trigone*
trich-	hair: *trichology*
-trophy	a state relating to nutrition: *hypertrophy*
-tropic	turning toward, changing: *gonadotropic*
ultra-	beyond, excess: *ultrasonic*
uni-	one: *unicellular*
-uria	urine: *polyuria*
uro-	urine, urinary organs or tract: *uroscope*
vas-	vessel: *vasoconstriction*
viscer-	organ: *visceral*
vit-	life: *vitamin*
zoo-	animal: *zoology*
zygo-	union, join: *zygote*

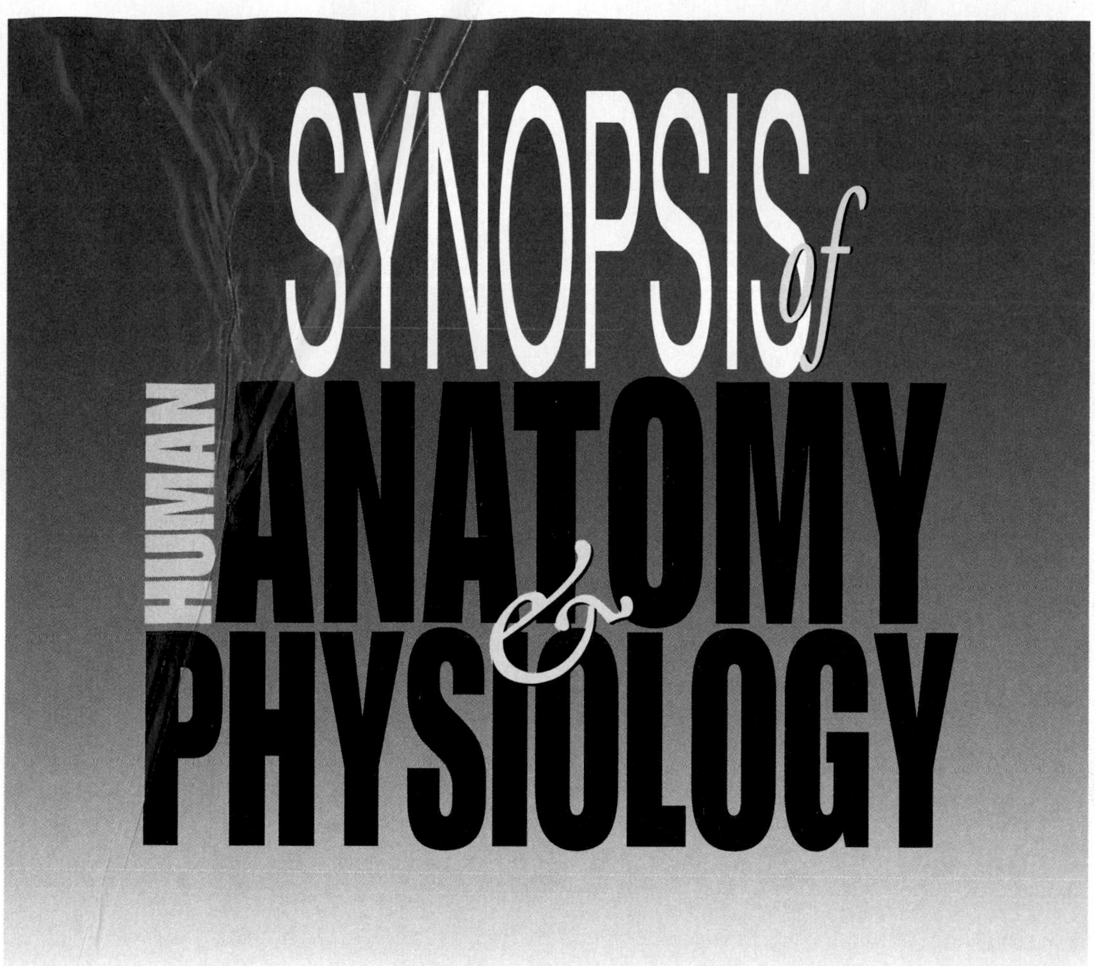

Kent M. Van De Graaff
Weber State University

Stuart Ira Fox
Pierce College

Karen M. LaFleur
Greenville Technical College

WCB Wm. C. Brown Publishers

Dubuque, IA Bogotá Buenos Aires Caracas Chicago Guilford, CT London
Madrid Mexico City Seoul Singapore Sydney Taipei Tokyo Toronto

Project Team

Editor *Kristine Noel*
Developmental Editor *Kelly Drapeau*
Production Editor *Jane Matthews*
Marketing Manager *Tom Lyon*
Designer *K. Wayne Harms*
Art Editor *Jennifer L. Osmanski, Kathleen Timp*
Photo Editor *John C. Leland*
Advertising Coordinator *Heather Wagner*
Permissions Coordinator *Karen L. Storlie*

Wm. C. Brown Publishers

President and Chief Executive Officer *Beverly Kolz*
Vice President, Director of Editorial *Kevin Kane*
Vice President, Sales and Market Expansion *Virginia S. Moffat*
Vice President, Director of Production *Colleen A. Yonda*
Director of Marketing *Craig S. Marty*
National Sales Manager *Douglas J. DiNardo*
Executive Editor *Michael D. Lange*
Advertising Manager *Janelle Keeffer*
Production Editorial Manager *Renée Menne*
Publishing Services Manager *Karen J. Slaght*
Royalty/Permissions Manager *Connie Allendorf*

Copyedited by Ann Mirels

Cover illustration by Lisa Gravunder

The credits section for this book begins on page 675 and is considered an extension of the copyright page.

Library of Congress Catalog Card Number: 95–83328

ISBN 0–697–04296–0

Printed in the United States of America

10 9 8 7 6 5 4 3 2 1

The Dynamic Human

*E*xperience anatomy and physiology in an entirely new dimension. **The Dynamic Human** CD-ROM interactively illustrates the complex relationships between anatomical structures and their functions in the human body. Realistic, three-dimensional visuals are the premier features of this exciting learning tool. After a brief introduction, **The Dynamic Human** covers each body system—demonstrating to the viewer the anatomy, physiology, histology, and clinical applications of each system.

Synopsis of Human Anatomy and Physiology, by Van De Graaff et al., has been correlated to **The Dynamic Human.** Throughout the text, a "dancing man" icon appears in many figure legends, signaling the reader that information relating to this figure can be found on **The Dynamic Human** CD-ROM. A complete correlation guide is found in the preface of the book.

Windows Version 0-697-37910-8
Macintosh Version 0-697-37909-4

Contents:

Anatomical Orientation
Skeletal System
Muscular System
Nervous System
Endocrine System
Cardiovascular System
Lymphatic System
Digestive System
Respiratory System
Urinary System
Reproductive System

To order any of these products, contact your bookstore manager or call our Customer Service Department at 800-338-5578.

Look at the great anatomy and physiology study tools Wm. C. Brown Publishers has to offer!

Student Study Guide for Synopsis of Human Anatomy and Physiology

by Kent Van De Graaff
ISBN: 0-697-03397-X

For each chapter in this text, there is a corresponding study guide chapter that contains focus questions, mastery quizzes, study activities, and answer keys with explanations.

Study Cards for Human Anatomy and Physiology, 3/e

by Kent Van De Graaff et al.
ISBN: 0-697-26447-5

Make studying a breeze with this boxed set of 300, two-sided study cards. Each card provides a complete description of terms, clearly labeled drawings, pronunciation guides, and clinical information on diseases. The *Study Cards* offer a quick and effective way for students to review human anatomy and physiology.

Coloring Guide to Anatomy and Physiology

by Judith Stone and Robert Stone
ISBN: 0-697-17109-4

This helpful manual emphasizes learning through the process of color association. The *Coloring Guide* provides a thorough review of anatomical and physiological concepts. By labeling and coloring each drawing, you will easily learn key anatomical and physiological structures and functions.

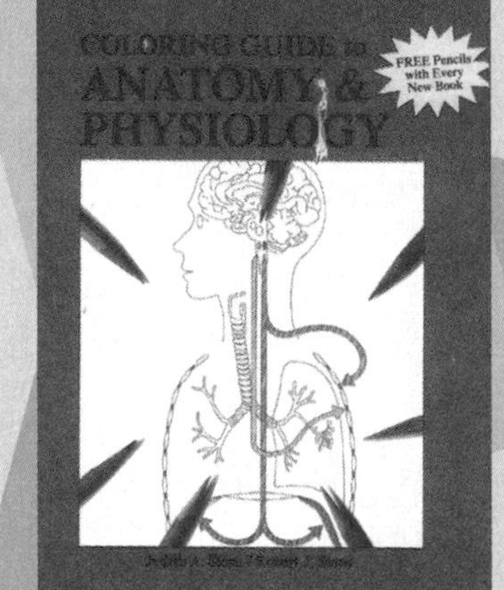

Atlas of Skeletal Muscles

by Judith Stone and Robert Stone
ISBN: 0-697-13790-2

This inexpensive atlas depicts all of the human skeletal muscles in precise drawings that show their origin, insertion, action, and innervation. The illustrations will help you locate muscles and understand their actions.

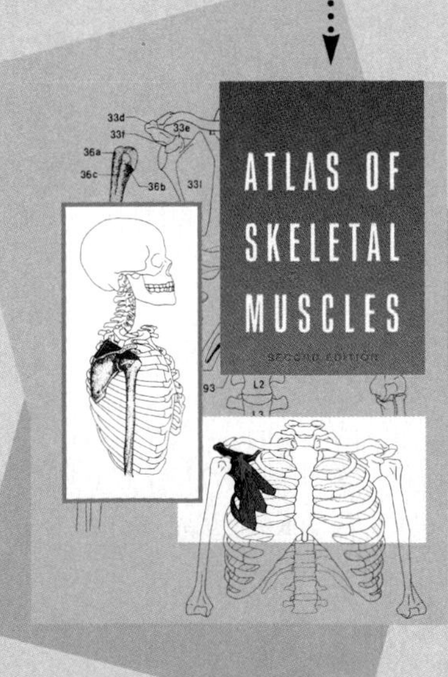

To order any of these products, contact your bookstore manager or call our Customer Service Department at 800-338-5578.

QuickStudy:

Computerized Study Guide for *Synopsis of Human Anatomy and Physiology* by Van De Graaff

Available on diskette:
IBM: 0-697-28958-3
Macintosh: 0-697-29859-1

Focus study time only in the areas where you need the most help.

How does it work?

- Take the computerized test for each chapter.
- The program presents you with feedback for each answer.
- A page-referenced study plan is then created for all incorrect answers.
- Look up the answers to the questions you missed.
- This easy-to-use *QuickStudy* is not a duplicate of the printed study guide that accompanies your text. It's full of new study aids.

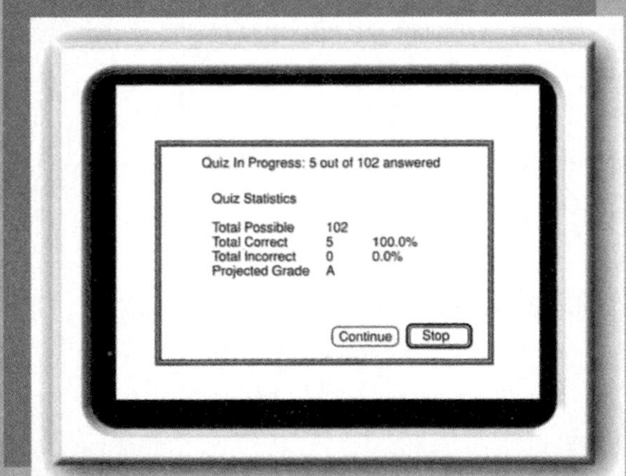

Provides instant feedback on quiz results.

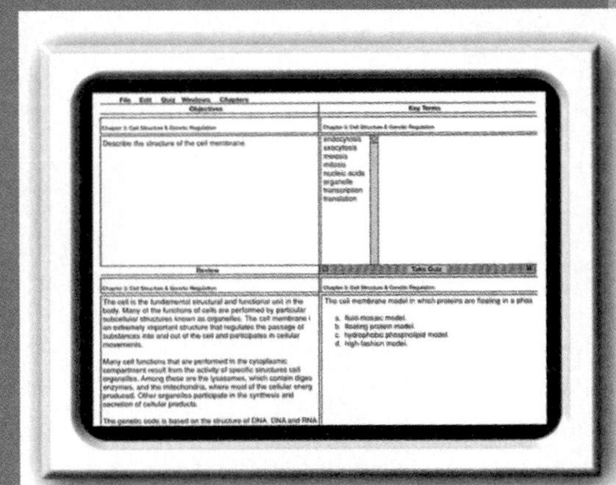

Four different review methods available.

- **Learning Objectives** outline the major points covered in each text chapter.
- **Key Terms** and their definitions are listed for each chapter.
- **Review** presents important concepts and facts in detail.
- **Take Quiz** allows testing on any combination of—or all—chapters.

To order any of these products, contact your bookstore manager or call our Customer Service Department at 800-338-5578.

BRIEF CONTENTS

CONTENTS

3 Cell Structure and Function 41

4 Histology 74

Unit II

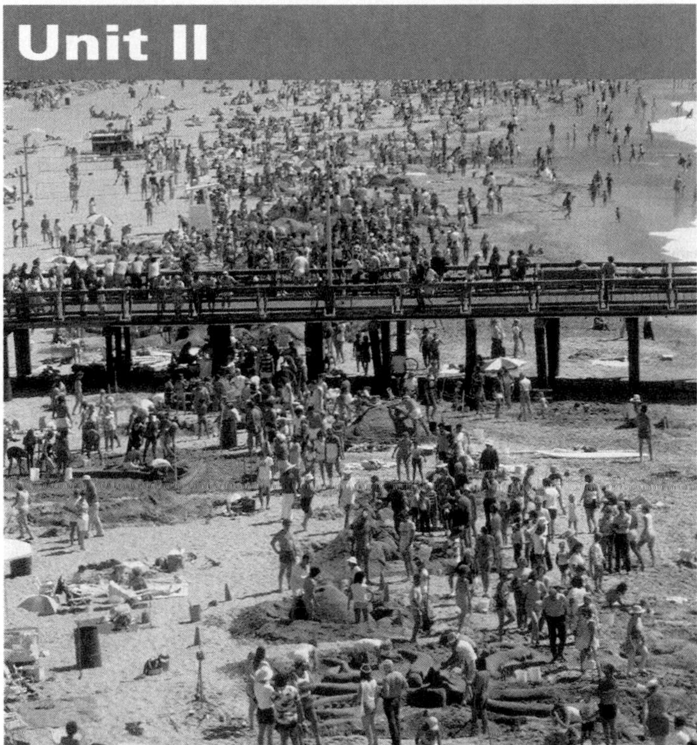

Support and Movement

5 Integumentary System 95

6 Skeletal System 114

7 Articulations 154

8 Muscular System 174

Unit III

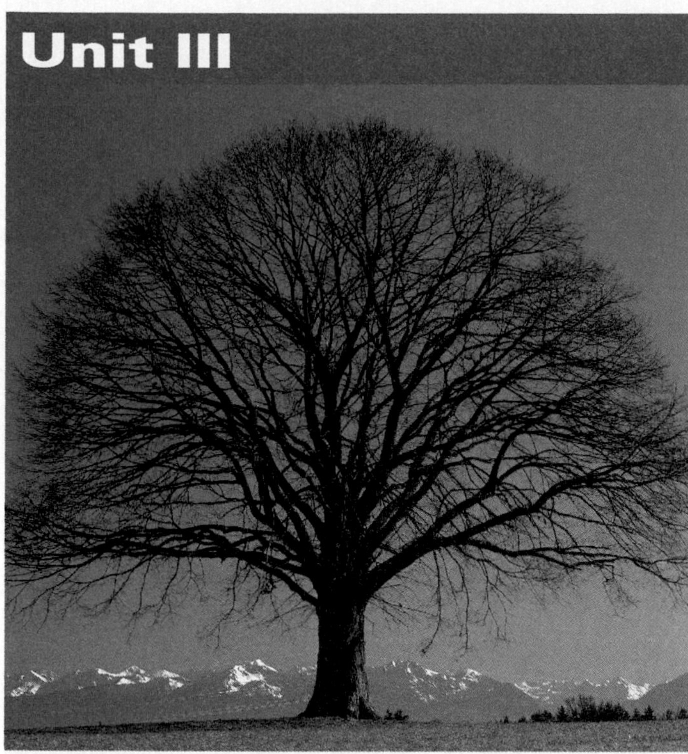

Integration and Control

9 Functional Organization of Nervous Tissue 230

10 Nervous System 247

11 Autonomic Nervous System 287

12 The Senses 300

Unit IV

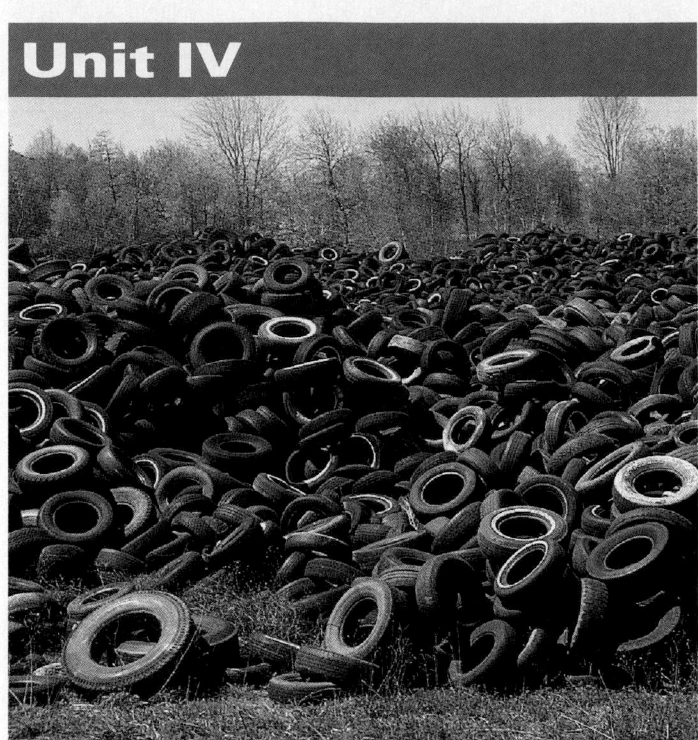

Regulation and Maintenance

Unit V

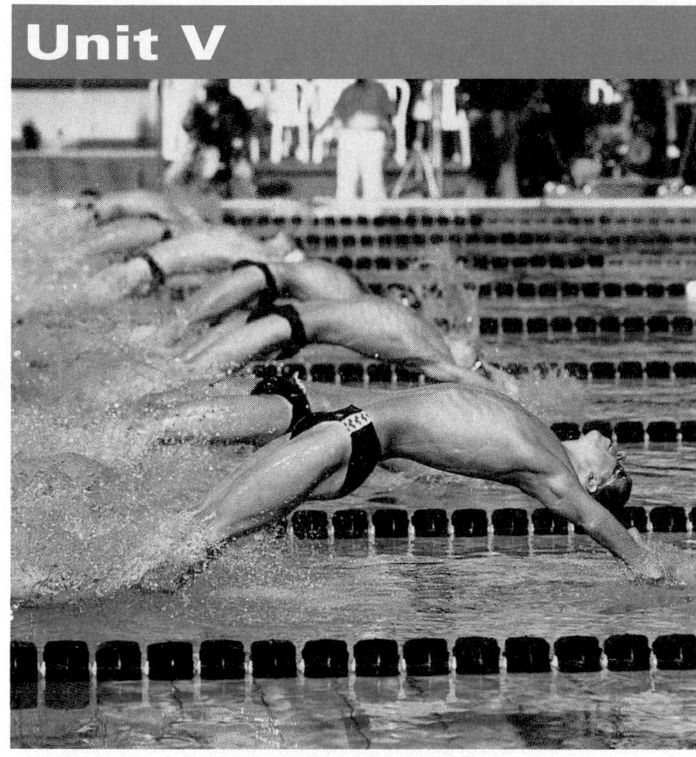

Reproduction and Development

22 Male Reproductive System 564

23 Female Reproductive System 591

24 Prenatal Development and Inheritance 615

BUILDING SYNOPSIS

The Authors

Kent M. Van De Graaff received his Ph.D. in zoology from Northern Arizona State University. For more than two decades, he has been a teacher and researcher of anatomy and physiology at the University of Minnesota, Brigham Young University, and Weber State University. He is the author of the nationally acclaimed textbook *Human Anatomy,* currently in its fourth edition, and co-author, with Dr. Fox, of *Concepts of Human Anatomy and Physiology,* also in its fourth edition. A prolific writer, Kent has authored and co-authored a wealth of other learning materials, including atlases, manuals, and guides. As an educator, he has been honored many times with awards of excellence for teaching, student advisement, and academic service.

Kent is the father of six children. Two of his sons are physicians and a third is currently in medical school. Outside the classroom, he enjoys jogging and painting wildlife.

Stuart I. Fox earned his Ph.D. in medical physiology from the University of Southern California (USC) Medical School while teaching full-time at Los Angeles City College. He is the author of several widely used laboratory manuals and co-author of the textbook *Concepts of Human Anatomy and Physiology.* He is also the author of *Human Physiology,* now in its fifth edition and the leading undergraduate text in its field.

Stuart has taught at California State University, Northridge, and is now a full-time faculty member at Pierce College, where his daughter, Laura, is currently a student. The numerous awards he has received reflect his love for teaching. Among his favorite pastimes are amateur astronomy, mountain biking, fly fishing, hiking, and cross-country skiing in the Sierras.

Karen M. LaFleur earned her Ed.D. from Clemson University after having taught anatomy and physiology to nursing students for more than 15 years. She is currently a full-time faculty member at Greenville Technical College in South Carolina.

After coordinating the 1991 Conference of the Human Anatomy and Physiology Society (HAPS) as a volunteer, Karen served as an elected board member and first chair of the Grants and Awards Committee. Her commitment to HAPS culminated in her election as president in 1996. Karen is active in the South Carolina Academy of Science and the National Science Teachers Association. She is also a board member of the Foothills Trail Conference, which protects and preserves a 100-mile trail system in the escarpment of the Blue Ridge Mountains.

The Contributors

Ann Mirels (contributing editor) earned a B.A. in English and a B.F.A. in advertising design from Syracuse University. In her more than 25 years in college textbook publishing, she has held the positions of production manager, text designer, and managing editor. She has received several national design awards, but her first love is working with words: "This work constantly demands new solutions in order to best serve the reader. I never seem to tire of it."

Ann is currently working as a freelancer from her home in Columbus, Ohio. In the rare moments when she is not editing manuscripts, she likes to paint in acrylics.

Christopher H. Creek (illustrator) received his B.F.A. in illustration from Brigham Young University. He has been a freelance medical illustrator for more than 10 years and has illustrated numerous college-level textbooks and atlases. At BYU, he was a part-time instructor in the Design Department, where he taught undergraduate foundations and figure drawing classes.

Chris in now the president of Zygote Media Group, Inc., a company that builds 3-D computer-generated anatomy and figure models for multimedia titles. In what spare time he can find, he paints portraits and landscapes. He also enjoys running marathons.

PREFACE

Synopsis of Human Anatomy and Physiology is expressly designed for introductory level students who are interested in pursuing a career in the allied fields or who simply want to gain some basic knowledge about the structure and function of the human body. Because these students may have only a limited science background, the narrative of *Synopsis* is deliberately conversational. Its accessibility encourages students to read for meaning, to think for themselves, and to solve problems creatively.

Throughout the writing of *Synopsis* our goal was to have students actively participate in achieving their learning objectives. To this end, we sought feedback from a large contingent of vigilant reviewers, most of whom are currently teaching anatomy and physiology courses to nonscience majors. Both the text and illustration programs went through numerous drafts to ensure a level of presentation appropriate for these students. Only if the content is within students' grasp can we expect them to be active learners.

The information in *Synopsis* is organized into modules, each of which can be read as a separate entity. Although numerous cross-references tie together and reinforce important ideas, each module can be understood without having read any of the others. This arrangement allows the instructor maximum flexibility in topic presentation. Some modules may be emphasized; others omitted entirely. Furthermore, the modules may be taught in any number of sequences, again at the discretion of the instructor.

One of the significant attributes of *Synopsis* is its conceptual approach. Each module is prefaced by a concept statement—a succinct expression of the main idea, or organizing principle, of the information that follows. These concept statements help students see the "big picture" before delving into the details. Mastering concepts allows the students to apply information to real-life situations, whereas memorizing tidbits of information does not.

Although a large number of new terms must be learned in any anatomy and physiology course, we kept the technical terminology in *Synopsis* within manageable limits. In addition, the reviewers were particularly valuable in helping us streamline the narrative so as not to burden students with a lot of extraneous detail. This "weeding out" process wasn't always easy, but it strengthened the text enormously; it kept it on target in meeting students' needs.

Central to *Synopsis* are three organizing themes: (1) the complementary nature of structure and function in the human body; (2) the dynamic constancy of the body's internal environment, or homeostasis, which is generally ensured by intricate regulating mechanisms; and (3) the interrelatedness of the systems of the body, each of which depends on the others in maintaining normal body functioning. These themes are the unifying threads that make up the fabric of the text. They not only underscore the beauty and complexity of the human body, but also the need to care for it properly.

Learning Aids: A Guide for the Student

The pedagogical devices in *Synopsis* are designed to help you learn anatomy and physiology. Don't just read this text as you would a novel. Interact with it and make it your own. Use the pedagogical devices as tools to gain the most benefit from the text.

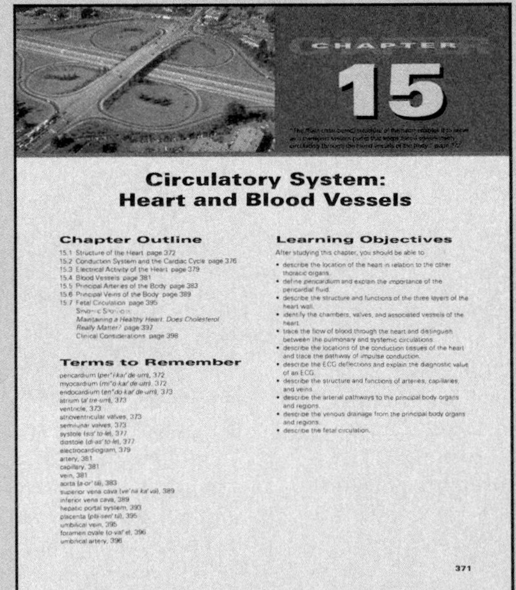
CHAPTER 15

Circulatory System: Heart and Blood Vessels

Chapter Outline

15.1 Structure of the Heart page 372
15.2 Conduction System and the Cardiac Cycle page 376
15.3 Electrical Activity of the Heart page 379
15.4 Blood Vessels page 381
15.5 Principal Arteries of the Body page 383
15.6 Principal Veins of the Body page 389
15.7 Fetal Circulation page 395
Maintaining a Healthy Heart: Does Cholesterol Really Matter? page 397
Clinical Considerations page 398

Terms to Remember

pericardium (*per"i-kar' de-um*), 372
myocardium (*mi"o-kar' de-um*), 372
endocardium (*en"do-kar' de-um*), 373
atrium (*a' tre-um*), 373
ventricle, 373
atrioventricular valves, 373
semilunar valves, 373
systole (*sis' to-le*), 377
diastole (*di-as' to-le*), 377
electrocardiogram, 379
artery, 381
capillary, 381
vein, 381
aorta (*a-or' ta*), 383
superior vena cava (*ve' na ka' va*), 389
inferior vena cava, 389
hepatic portal system, 393
placenta (*pla-sen' ta*), 395
umbilical vein, 395
foramen ovale (*o-val' e*), 396
umbilical artery, 396

Learning Objectives

After studying this chapter, you should be able to

- describe the location of the heart in relation to the other thoracic organs.
- define *pericardium* and explain the importance of the pericardial fluid.
- describe the structure and functions of the three layers of the heart wall.
- identify the chambers, valves, and associated vessels of the heart.
- trace the flow of blood through the heart and distinguish between the pulmonary and systemic circulations.
- describe the locations of the conduction tissues of the heart and trace the pathway of impulse conduction.
- describe the ECG deflections and explain the diagnostic value of an ECG.
- describe the structure and functions of arteries, capillaries, and veins.
- describe the arterial pathways to the principal body organs and regions.
- describe the venous drainage from the principal body organs and regions.
- describe the fetal circulation.

371

Chapter Introductions Each chapter begins with an overview of the contents of the chapter in outline form. The overview is followed by a list of page-referenced key terms, each of which is defined in the chapter and also in the glossary at the end of the text. Learning objectives are also included in each chapter introduction to provide you with a focus for study and to facilitate your review of the material.

Terminology Aids Where each technical term first appears in the narrative, it is set off by boldface or italic type and is often followed by a phonetic pronunciation in parentheses.

Pause in your reading to learn the correct pronunciation of a term so that you will be better able to remember it. The phonetic spellings are easy to interpret using the guide that precedes the glossary.

Word Derivations The derivations of some of the terms in the text are provided in footnotes at the bottom of the page on which the term is introduced. Don't skip over these footnotes; they are often interesting in themselves. Furthermore, if you know how a word was derived, it becomes more meaningful and is easier to remember. You can identify the roots of each term by referring to the glossary of prefixes and suffixes on the inside front cover of the text.

Illustration Program *Synopsis* is much more than just words. Because anatomy and physiology are visually oriented sciences, the preparation of an outstanding art and photo program was a top priority from the start. An abundance of carefully rendered full-color figures supplements the text to maximize your learning. Every one of them was checked and rechecked to ensure the conceptual clarity, precision, and accuracy of the linework, color usage, labels, and caption. Following chapter 8, seven full-page illustrations of the male and female trunk are included for your reference.

In addition to the anatomical renderings, flowcharts and other color graphics are used throughout *Synopsis* to clarify complex physiological processes. Light and scanning electron micrographs are also used where appropriate, and a comprehensive set of anatomical reference plates, along with photographs of dissected human cadavers, is included as Appendix B.

Although many of the figures can be admired for their beauty alone, keep in mind that they were created for one primary purpose—to illustrate concepts presented in the text. "Reading" *Synopsis* also means analyzing the figures. Each one is placed as close as possible to its text reference to spare you the trouble of flipping through pages.

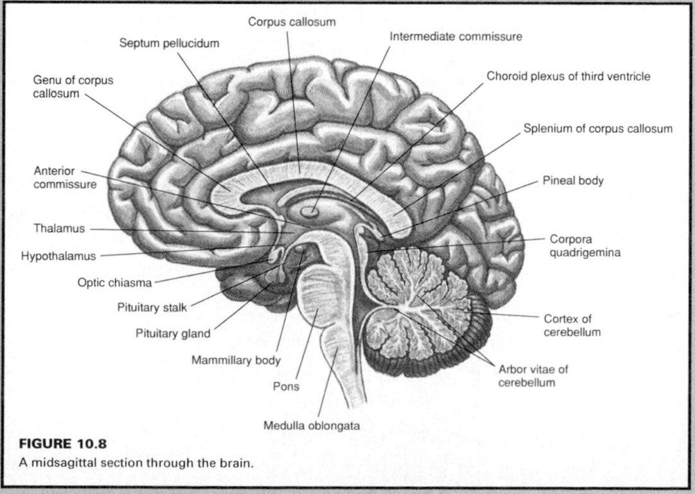

FIGURE 10.8
A midsagittal section through the brain.

Clinical Commentaries Set off from the text narrative are short paragraphs on a color background containing information of a clinical nature. This interesting information is relevant to the discussion that precedes it, but more importantly, it demonstrates how basic scientific knowledge is applied. In addition to these in-text commentaries, many chapters contain a separate "Clinical Considerations" section describing selected disorders, diseases, or dysfunctions of specific organ systems.

The stimulatory effect of TSH on the thyroid is dramatically revealed in people who develop an iodine-deficiency (endemic) goiter. *In the absence of sufficient dietary iodine, the thyroid cannot produce adequate amounts of T_4 and T_3. The resulting lack of negative feedback inhibition causes abnormally high levels of TSH secretion. High TSH cannot stimulate increased thyroxine secretion under these circumstances, but it does stimulate abnormal growth of the thyroid gland (fig. 13.15). Such abnormal growth is known as a* goiter.

Inside Information—In certain parts of the world, including central Europe and the Great Lakes region of North America, iodine is essentially absent from the soil. Before iodine was added to table salt, the Great Lakes region was sometimes referred to as the "goiter belt."

"Inside Information" These brief but noteworthy items, set off by bars above and below, let you know just how amazing your body really is. They also offer surprising facts about ancient medicine, major scientific breakthroughs, diet and health, and people who have made history. You might even find yourself scanning the text for these items between the bars. This is the kind of information you will want to share with your friends!

Synoptic Spotlight Many of the chapters in *Synopsis* include a special "Synoptic Spotlight"—an in-depth discussion on a topic of general interest. Some of the Spotlights contain practical information on such health-related concerns as skin cancer, cholesterol and the heart, exercise training, and blood transfusions. A number of them focus on issues that have received a good deal of media attention in recent years—the development and testing of new drugs, DNA fingerprinting, tissue and organ transplants, anabolic steroids, AIDS, and air pollution, for example. Still others are expanded discussions of material covered in the text. These include "More Than Meets the Eye," "The Inimitable Liver," and "Control of Reproduction." All of the Spotlights relate to the chapter content, but they are meant to be enjoyed in and of themselves.

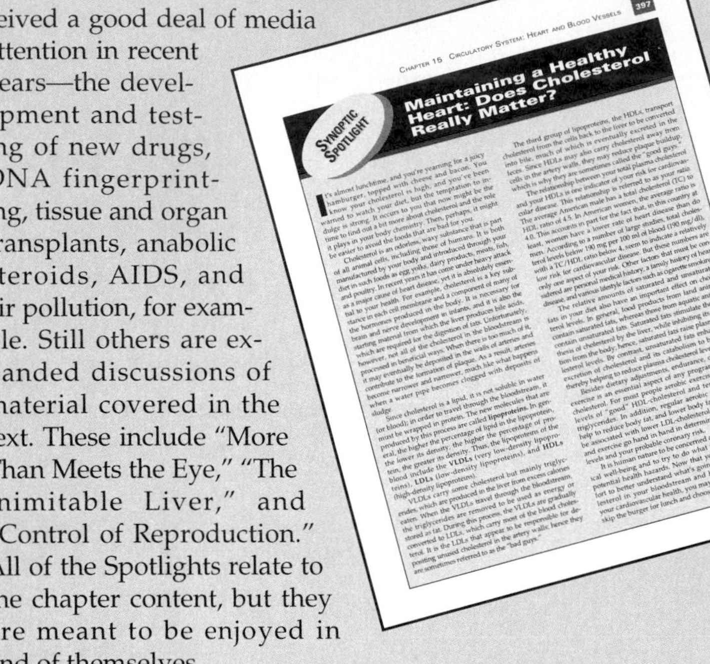
CHAPTER 15 CIRCULATORY SYSTEM: HEART AND BLOOD VESSELS 397

SYNOPTIC SPOTLIGHT

Maintaining a Healthy Heart: Does Cholesterol Really Matter?

Chapter Summaries At the end of each chapter the material is summarized for you in outline form, following the sequence of the modules. Review each summary after studying the chapter to be sure that you have not missed any points. In addition, use the chapter summaries in preparing for exams.

Review Activities Sets of objective, essay, and critical thinking questions follow each chapter summary, along with a labeling exercise that employs all or part of a key illustration in the chapter. Complete these review activities to obtain feedback as to the depth of your understanding and learning. The critical thinking questions challenge you to use the chapter information in novel ways toward the solution of practical problems. Answers to the objective questions are provided in Appendix A.

Glossary The glossary of terms at the end of the text is particularly noteworthy for its comprehensiveness. The definitions for almost all of the terms are accompanied by pronunciation keys, and synonyms are indicated as appropriate. The majority of the glossary terms are accompanied by a page number indicating where the term is discussed in the text narrative. Look to the glossary as you review to check your understanding of the technical terminology.

Multimedia Tie-ins

A videotape icon appears in many of the figure legends throughout the chapters. This icon alerts the reader that an animation related to the figure is available from the WCB Life Science Animations videotape series. Students can purchase the videotapes, and the tapes are available to qualified adopters. A correlation guide is included in this preface. The five tape titles are:

1. Chemistry, The Cell, and Energetics (1–11)
2. Cell Division, Heredity, Genetics, Reproduction, and Development (12–21)
3. Animal Biology #1 (22–35)
4. Animal Biology #2 (26–45)
5. Plant Biology, Evolution, and Ecology (46–53)

A running man icon also appears in many of the figure legends throughout the chapters. This icon represents the Dynamic Human CD-ROM and signals the reader that information relating to this figure can be found on the CD-ROM. The Dynamic Human CD-ROM can be packaged with *Synopsis of Human Anatomy and Physiology* for minimal fee or can be bought separately. A correlation guide can be found at the end of the preface.

Supplementary Materials

The supplement package that accompanies *Synopsis of Human Anatomy and Physiology* is designed to help students in their learning activities and to guide instructors in planning course work and presentations. All of the ancillaries are available to qualified adopters. Following are brief descriptions of these support materials.

1. An *Instructor's Manual and Test Item File* prepared by Jeffrey and Karianne Prince provides instructional support in the use of the textbook. It also contains a test item file with approximately 70 items for each chapter to aid instructors in constructing examinations.
2. A *Student Study Guide* written by Kent M. Van De Graaff features focus questions, mastery quizzes, study activities, and answer keys with explanations.
3. *Transparencies* consist of 200 acetate sheets that can be used to enrich and clarify classroom lectures. They can also be used for short quizzes. The transparencies were made from text illustrations that were chosen for their value in reinforcing lecture presentations.
4. *Slides* that accompany the textbook include:
   a. 100 histology slides showing basic tissue types representative of all the human body systems;
   b. 70 clinical application slides depicting important pathological conditions; and
   c. 25 radiographic slides.

5. *Microtest* is a computerized testing service for generating examinations and quizzes. A complete test item file is available on computer diskette for use with DOS, Windows, and Macintosh computers.
6. *QuickStudy* is a computerized study guide that contains true-false, multiple-choice, and fill-in-the-blank questions on a diskette. It is available for use with IBM and Macintosh computers.
7. *Study Cards for Anatomy and Physiology* prepared by Kent Van De Graaff et al. is a boxed set of 300 3-by-5-inch cards. It serves as an illustrated summary of the key information presented in the text. The Study Cards offer students a quick and effective way to review human anatomy and physiology.
8. *The Dynamic Human* CD-ROM illustrates the important relationships between anatomical structures and their functions in the human body. Realistic three-dimensional visualizations make this CD-ROM particularly useful.
9. *The Dynamic Human Videodisc* contains all the animations (200+) from the CD-ROM. A bar code directory is also available.
10. *WCB Life Science Animations Videotape Series* is a series of five videotapes containing 53 animations that cover many key physiological processes. Another videotape, *Physiological Concepts of Life Science*, contains similar animations.
11. *Explorations in Human Biology* CD-ROM consists of 15 interactive modules that stress human physiology. Students can actively investigate vital processes as they explore these modules, each of which is filled with color, sound, and movement. The CD-ROM is available for use with Macintosh and IBM Windows computers.
12. *Explorations in Cell Biology and Genetics* CD-ROM features interactive concepts related to key topics covered in an anatomy and physiology course. It can be used by an instructor in the classroom or placed in a lab or resource center for students. The CD-ROM is available for use with Macintosh and IBM Windows computers.
13. *Life-Science Living Lexicon* CD-ROM is a comprehensive collection of life science terms, including definitions of their roots, prefixes, and suffixes. Audio pronunciations and illustrations are included as well. The Lexicon is student-interactive, featuring quizzing and note-taking capabilities.
14. *Anatomy of Physiology Videodisc* is a four-sided videodisc containing more than 30 animations of physiological processes, as well as line art and micrographs. A bar code directory is also available.
15. *WCB Anatomy and Physiology Video Series* consists of the following exceptional videotapes:
    a. Introduction to the Human Cadaver and Prosection;
    b. Introduction to Cat Dissection: Musculature;
    c. Blood Cell Counting, Identification, & Grouping;
    d. Internal Organs and the Circulatory System of the Cat.
16. *Coloring Guide to Anatomy and Physiology* prepared by Robert and Judith Stone emphasizes learning through color association. The Coloring Guide provides a thorough review of anatomical and physiological concepts.
17. *Atlas of the Skeletal Muscles* prepared by Robert and Judith Stone is a guide to the structure and function of human skeletal muscles. The illustrations help students locate muscles and understand their actions.
18. *Case Histories in Human Physiology* prepared by Donna Van Wynsberghe and Gregory Cooley, M.D., provides students with opportunities to integrate their thinking toward the solution of problems. An answer key is also available.

Acknowledgments

When several years are devoted to the preparation of a textbook, there are many people to thank. We are most appreciative of our students first of all; they provide us with our professional challenges and rewards. A special thanks is extended to John M. Adams and Steven T. Parker, students who critically examined portions of the manuscript as part of their classroom activities. Their perspective as potential users of the text was invaluable in helping us shape final drafts. We are also grateful to our colleagues, many of whom offered excellent suggestions for improvement.

Ann Mirels is the person to whom we are most indebted in the writing of *Synopsis*. In large measure, the success of this text is due to her talent as an editor and her support of us as authors. Ann understands all aspects of textbook development. She is a creative person who has a keen sense for how best to reach students in helping them learn. Thanks Ann.

Two other professionals also contributed immensely. One of them, Chris Creek, rendered much of the art program. His scrupulousness in achieving anatomical accuracy in his illustrations benefited virtually every aspect of the text. We are also grateful to Eric J. Van De Graaff, M.D., whose clinical input so greatly enhanced the text's practical value.

The editors at Wm. C. Brown were totally supportive and committed to this project from the start. Our thanks to Acquisitions Editor Colin H. Wheatley for his friendship and encouragement. Many thanks are also due to Kris Noel, who orchestrated the complex process of development, and Jane Matthews, who shepherded the project through the production stages. The superb organizational skills of these editors went far in bringing *Synopsis* to fruition. In addition, we thank all of the individuals who reviewed the manuscript in its numerous drafts. They played a major role in guiding our efforts to achieve a text of the highest quality.

Finally, we are indebted to our families. For their love and understanding, and for their many sacrifices over the long course of writing and revising, we are grateful beyond measure.

Reviewers

Jerri Adler
Lane Community College

Karen Bajinski
St. Lawrence College
St. Laurent–Brockville Campus
St. Vincent De Paul Hospital

Dr. Reuben E. Barrett
Prairie State College

Karen L. Borg
Midlands Technical College

Monica L. Cauley
Mac Arthur State Technical College

Steve B. Chandler
Florida A & M University

Steven D. Collins
Niagara College

Janet Colwell
St. Francis School of Practical Nursing

Redding I. Corbett III
Midlands Technical College

Brian Curry
Grand Valley State University

Darrell Davies
Kalamazoo Valley Community College

Connie S. Dempsey
Stark Technical College

William J. Despain
North Central Technical College

Gurmeet K. Dhaliwal
Wilbur Wright College

Kathleen Tatum Flickinger
Iowa State University

Robert E. Friar
Ferris State University

Patrick Galliart
North Iowa Area Community College

Luanne Gogolin
Ferris State University

Harold J. Grau
Christopher Newport University

Bruce E. Grayson
University of Miami

Charles J. Grossman
Xavier University
VA Medical Center
University of Cincinnati, College of Medicine

Charles E. Harmon
Sarasota, Florida High School

Vickie Hennessy
Sinclair Community College

Ann Henninger
Wartburg College

John F. Hertner
University of Nebraska–Kearney

Gloria Hillert
Triton College

Nelda W. Hinckley
John A. Logan College

Judith Hoffman
Alexandria Technical College

Dawn Holtzmeier
Hocking College

Rita Hoots
Yuba College

Taraneh Javad
Aiken Technical College

Linda Kaeding
Institute of Medical/Dental Technology

Gary Kwiecinski
University of Scranton

Richelle S. Laipply
Stark Technical College

Richard Landesman
University of Vermont

Pamela M. Langley
New Hampshire Technical Institute

Lynne J. Lauer
Lakeland College

Stephen G. Lebsack
Linn-Benton Community College

Charlene L. Newby
Lakeshore Technical College

Steven J. Person
Lake Superior State University

Michael Postula
Parkland College

Lyn Prendergast
Mount Ida College

Elaine T. Princevalli
Eli Whitney Regional Vocational Technical School

Mary E. Pryor
Daytona Beach Community College

Tom J. Reeves
Midlands Technical College

Mary K. Ringsven
Dakota County Technical College

Donna Ritch
University of Wisconsin, Green Bay

Karl J. Roberts
Prince George's Community College

Barbra A. Roller
Florida International University

Arleen L. Sawitzke
Salt Lake Community College

M. Isabelle Seburn
Niagara College

Laurie Shearer
Ivy Tech State College

Cathleen Shrier
Azusa Pacific University

Barbara M. Stout
Stark Technical College

James G. Timmons
Lee College

Tuan A. Tran
Los Angeles College of Chiropractic

Margaret Vogel
Riverland Technical College

Karen Wagner
Wascana Institute

Russell F. Wells
St. Lawrence University

Shaun Willson
Pamlico Community College

Carolyn S. Zeiger
Hannibal Public School of Practical Nursing

Jean A. Zorko
Stark Technical College

It is our sincere hope that students and instructors will enjoy using this text and that it will serve them well. We invite all readers to send us comments and suggestions for future editions.

Life Science Animations (LSA)

Figure 2.2 LSA 1	Formation of an Ionic Bond
Figure 3.1 LSA 2	Journey into a Cell
Figure 3.5 LSA 3	Endocytosis
Figure 3.13 LSA 16	Transcription of a Gene
Figure 3.14 LSA 17	Protein Synthesis
Figure 3.15 LSA 17	Protein Synthesis
Figure 3.16 LSA 15	DNA Replication
Figure 3.18 LSA 12	Mitosis
Figure 8.9 LSA 30	Sliding Filament Model of Muscle Contraction
Figure 9.6 LSA 22	Formation of Myelin Sheath
Figure 9.7 LSA 22	Formation of Myelin Sheath
Figure 9.12 LSA 23	Saltatory Nerve Conduction
Figure 10.27 LSA 25	Reflex Arcs
Figure 12.24 LSA 27	The Organ of Corti
Figure 14.5 LSA 40	A, B, O Blood Types
Figure 15.8 LSA 38	Production of Electrocardiogram
Figure 15.25 LSA 39	Common Congenital Defects of the Heart
Figure 17.9 LSA 42	Structure and Function of Antibodies
Figure 17.10 LSA 41	B-Cell Immune Response
Figure 17.12 LSA 43	Types of T-cells
Figure 20.2 LSA 34	Digestion of Carbohydrates
Figure 20.35 LSA 36	Digestion of Lipids
Figure 20.36 LSA 36	Digestion of Lipids
Figure 22.10 LSA 19	Spermatogenesis
Figure 23.8 LSA 20	Oogenesis
Figure 24.9 LSA 21	Human Embryonic Development

Dynamic Human CD-ROM

Chapter 1

1.9 Anatomical Orientation/Planes
1.14 Anatomical Orientation/Visible Human/Thorax
Anatomical Orientation/Visible Human/Abdomen

Chapter 3

3.4 Immune and Lymphatic Systems/Explorations/Nonspecific Immunity/Phagocytosis

Chapter 4

4.1 Digestive System/Histology/Duodenal Villi
Digestive System/Histology/Duodenum
Digestive System/Histology/Esophagus
Urinary System/Histology/Renal Cortex
4.2 Urinary System/Histology/Bladder
4.3 Digestive System/Histology/Submandibular Gland
4.4 Digestive System/Histology/Submandibular Gland
4.7 Skeletal System/Histology/Elastic Cartilage
Skeletal System/Histology/Fibrocartilage
Skeletal System/Histology/Hyaline Cartilage
4.8 Skeletal System/Histology/Compact Bone
Skeletal System/Histology/Spongy Bone
Skeletal System/Explorations/Cross Section of a Long Bone
4.10 Muscular System/Histology/Cardiac Muscle
Muscular System/Histology/Smooth Muscle
Muscular System/Histology/Skeletal Muscle longitudinal
Muscular System/Histology/Skeletal Muscle cross section
Muscular System/Anatomy/Cardiac Muscle
Muscular System/Anatomy/Smooth Muscle
Muscular System/Anatomy/Skeletal Muscle
Cardiovascular System/Histology/Cardiac Muscle
4.11 Nervous System/Histology/Dorsal Root Ganglion
Nervous System/Histology/Spinal Neurons
Nervous System/Explorations/Motor and Sensory Pathways

Chapter 6

6.1 Skeletal System/Anatomy/Gross Anatomy
Skeletal System/Explorations/Walking Skeleton Animation
6.2 Muscular System/Explorations/Muscle Action around Joints
6.3 Skeletal System/Histology/Compact Bone
Skeletal System/Histology/Spongy Bone
Skeletal System/Explorations/Cross Section of a Long Bone
6.7 Skeletal System/Explorations/Cross Section of a Long Bone
6.8 Skeletal System/Histology/Spongy Bone
Skeletal System/Histology/Compact Bone
6.9 Skeletal System/Histology/Compact Bone
6.10 Skeletal System/Histology/Compact Bone
6.14 Skeletal System/Anatomy/3D Viewer: Cranial Anatomy
Skeletal System/Anatomy/Gross Anatomy
6.15 Skeletal System/Anatomy/3D Viewer: Cranial Anatomy
Skeletal System/Anatomy/Gross Anatomy
6.16 Skeletal System/Anatomy/3D Viewer: Cranial Anatomy
Skeletal System/Anatomy/Gross Anatomy
6.17 Skeletal System/Anatomy/3D Viewer: Cranial Anatomy

continued

	Skeletal System/Anatomy/Gross Anatomy
6.21	Skeletal System/Anatomy/3D Viewer: Cranial Anatomy
	Skeletal System/Anatomy/Gross Anatomy
6.22	Skeletal System/Anatomy/3D Viewer: Cranial Anatomy
	Skeletal System/Anatomy/Gross Anatomy
6.23	Skeletal System/Anatomy/Gross Anatomy
6.24	Skeletal System/Anatomy/Gross Anatomy
6.25	Skeletal System/Anatomy/Gross Anatomy
6.26	Skeletal System/Anatomy/Gross Anatomy
6.27	Skeletal System/Anatomy/Gross Anatomy
6.28	Skeletal System/Anatomy/3D Viewer: Thoracic Anatomy
	Skeletal System/Anatomy/Gross Anatomy
6.29	Skeletal System/Anatomy/Gross Anatomy
6.30	Skeletal System/Anatomy/Gross Anatomy
6.31	Skeletal System/Anatomy/Gross Anatomy
	Skeletal System/Clinical Concepts/ Dislocated Shoulder
6.32	Skeletal System/Anatomy/Gross Anatomy
6.33	Skeletal System/Anatomy/Gross Anatomy
6.34	Skeletal System/Anatomy/Gross Anatomy
6.35	Skeletal System/Anatomy/Gross Anatomy
6.36	Skeletal System/Anatomy/Gross Anatomy
6.37	Skeletal System/Anatomy/Gross Anatomy
6.39	Skeletal System/Anatomy/Gross Anatomy
6.40	Skeletal System/Anatomy/Gross Anatomy
6.41	Skeletal System/Anatomy/Gross Anatomy
6.43	Skeletal System/Clinical Concepts/ Fractured Femur

Chapter 7

7.3	Skeletal System/Explorations/Types of Joints/Synovial/Generic Synovial
7.4	Skeletal System/Explorations/Types of Joints/Fibrous
7.5	Skeletal System/Explorations/Types of Joints/Fibrous
7.7	Skeletal System/Explorations/Types of Joints/Cartilaginous
7.8	Skeletal System/Explorations/Types of Joints/Cartilaginous
7.9	Skeletal System/Explorations/Types of Joints/Synovial/Generic Synovial
	Skeletal System/Histology/Synovial Joint
7.10	Skeletal System/Explorations/Types of Joints/Synovial/Types of Synovial Joints/Gliding
7.11	Skeletal System/Explorations/Types of Joints/Synovial/Synovial Joint Motion/Hinge Joint
	Skeletal System/Explorations/Types of Joints/Synovial/Synovial Joint Motion/ Elbow (Hinge) Joint
7.12	Skeletal System/Explorations/Types of Joints/Synovial/Types of Synovial Joints/Pivot
7.13	Skeletal System/Explorations/Types of Joints/Synovial/Types of Synovial Joints/Condyloid
7.14	Skeletal System/Explorations/Types of Joints/Synovial/Types of Synovial Joints/ Saddle
7.15	Skeletal System/Explorations/Types of Joints/Synovial/Synovial Joint Motion/ Ball and Socket
	Skeletal System/Explorations/Types of Joints/Synovial/Synovial Joint Motion/Shoulder
7.23	Skeletal System/Clinical Concepts/ Arthroscopy

Chapter 8

8.2	Muscular System/Anatomy/Skeletal Muscle
	Muscular System/Explorations/Sliding Filament Theory
	Muscular System/Histology/Skeletal Muscle longitudinal
	Muscular System/Histology/Skeletal Muscle cross section
8.3	Muscular System/Anatomy/Skeletal Muscle
	Muscular System/Explorations/Sliding Filament Theory
	Muscular System/Histology/Skeletal Muscle longitudinal
	Muscular System/Histology/Skeletal Muscle cross section
8.4	Muscular System/Anatomy/Skeletal Muscle
	Muscular System/Explorations/Sliding Filament Theory
	Muscular System/Histology/Skeletal Muscle longitudinal
	Muscular System/Histology/Skeletal Muscle cross section
8.5	Muscular System/Anatomy/Skeletal Muscle
	Muscular System/Explorations/Sliding Filament Theory
	Muscular System/Histology/Skeletal Muscle longitudinal
	Muscular System/Histology/Skeletal Muscle cross section
8.7	Nervous System/Explorations/Motor and Sensory Pathways
8.8	Muscular System/Clinical Concepts/ Isometric vs. Isotonic Contractions
8.9	Muscular System/Explorations/Sliding Filament Theory
8.10	Muscular System/Explorations/Sliding Filament Theory
8.11	Muscular System/Explorations/Sliding Filament Theory
8.12	Muscular System/Explorations/Sliding Filament Theory
8.13	Muscular System/Explorations/Sliding Filament Theory

Chapter 9

9.2	Nervous System/Histology/Spinal Neurons
9.3	Nervous System/Explorations/Motor and Sensory Pathways
	Nervous System/Histology/Spinal Neurons
9.13	Muscular System/Explorations/ Neuromuscular Junction

Chapter 10

10.3	Nervous System/Anatomy/Gross Anatomy
	Nervous System/Anatomy/3D Viewer: Cranial Anatomy
10.4	Nervous System/Anatomy/Gross Anatomy
	Nervous System/Anatomy/3D Viewer: Cranial Anatomy
10.6	Nervous System/Explorations/Motor and Sensory Pathways
10.10	Endocrine/Anatomy/Gross Anatomy
10.17	Nervous System/Anatomy/Spinal Cord Anatomy
10.21	Nervous System/Anatomy/Gross Anatomy
10.23	Nervous System/Anatomy/Spinal Cord Anatomy
	Nervous System/Anatomy/3D Viewer: Cranial Anatomy
10.26	Nervous System/Anatomy/Spinal Cord Anatomy
	Nervous System/Anatomy/3D Viewer: Cranial Anatomy
10.27	Nervous System/Explorations/Motor and Sensory Pathways
	Nervous System/Explorations/Reflex Arc
10.28	Nervous System/Explorations/Motor and Sensory Pathways
	Nervous System/Explorations/Reflex Arc

Chapter 11

11.1	Nervous System/Explorations/Motor and Sensory Pathways
	Nervous System/Explorations/Reflex Arc

Chapter 12

12.3	Nervous System/Explorations/Olfaction
12.4	Nervous System/Explorations/Taste
	Nervous System/Histology/Vallate Papillae
12.5	Nervous System/Explorations/Taste
	Nervous System/Explorations/ Innervation of Tongue
12.7	Nervous System/Histology/Eye
12.9	Nervous System/Histology/Eye
12.10	Nervous System/Explorations/Vision
12.11	Nervous System/Explorations/Vision
12.12	Nervous System/Histology/Retina
12.13	Nervous System/Histology/Retina
12.15	Nervous System/Explorations/Vision
12.19	Nervous System/Explorations/Hearing
12.20	Nervous System/Explorations/Hearing
12.21	Nervous System/Explorations/Dynamic Equilibrium
12.22	Nervous System/Explorations/Hearing
12.23	Nervous System/Explorations/Hearing
12.24	Nervous System/Explorations/Hearing

12.25 Nervous System/Explorations/Dynamic Equilibrium
12.26 Nervous System/Explorations/Dynamic Equilibrium

Chapter 13
13.1 Endocrine System/Anatomy/Gross Anatomy
13.4 Endocrine System/Anatomy/Gross Anatomy/Hypothalamus and Pituitary
13.6 Endocrine System/Anatomy/Gross Anatomy/Hypothalamus and Pituitary/Posterior Lobe
13.7 Endocrine System/Anatomy/Gross Anatomy/Hypothalamus and Pituitary/ Hypothalamo-Hypophyseal Portal System
13.8 Endocrine System/Anatomy/Gross Anatomy/Thyroid
Endocrine System/Explorations/ Hypothalamo-Pituitary-Thyroid Axis
13.9 Endocrine System/Anatomy/Gross Anatomy/Gonads
13.10 Endocrine System/Anatomy/Gross Anatomy/Adrenal Gland
Endocrine System/Histology/Adrenal Medulla
Endocrine System/Histology/Adrenal Cortex
13.13 Endocrine System/Anatomy/Gross Anatomy/Thyroid
13.14 Endocrine System/Histology/Thyroid
13.15 Endocrine System/Explorations/ Hypothalamo-Pituitary-Thyroid Axis
13.16 Endocrine System/Anatomy/Gross Anatomy/Thyroid/ Parathyroid
13.17 Endocrine System/Anatomy/Gross Anatomy/Pancreas
Endocrine System/Clinical Concepts/ Diabetes
13.19 Immune and Lymphatic Systems/ Anatomy/Gross Anatomy/Thymus
13.20 Endocrine System/Explorations/ Endocrine Function
13.21 Endocrine System/Explorations/ Endocrine Function
13.22 Endocrine System/Explorations/ Endocrine Function

Chapter 14
14.5 Immune and Lymphatic Systems/ Clinical Concepts/Blood Type

Chapter 15
15.1 Cardiovascular System/Anatomy/Gross Anatomy
Cardiovascular System/Anatomy/ 3D Viewer: Thoracic Anatomy
15.2 Cardiovascular System/Anatomy/Gross Anatomy
Cardiovascular System/Anatomy/ 3D Viewer: Thoracic Anatomy
15.3 Cardiovascular System/Explorations/ Heart Dynamics/Conduction Cycle
15.5 Cardiovascular System/Explorations/ Heart Dynamics/Cardiac Cycle
Cardiovascular System/Explorations/ Heart Dynamics/Heart Sounds
15.6 Cardiovascular System/Explorations/ Heart Dynamics/Heart Sounds
15.7 Cardiovascular System/Explorations/ Heart Dynamics/Conduction Cycle
Cardiovascular System/Explorations/ Heart Dynamics/ECG
15.8 Cardiovascular System/Explorations/ Heart Dynamics/ECG
Cardiovascular System/Explorations/ Heart Dynamics/Blood Flow
15.9 Cardiovascular System/Explorations/ Generic Vasculature
15.11 Cardiovascular System/Anatomy/Gross Anatomy
15.22 Cardiovascular System/Explorations/ Generic Portal System

Chapter 16
16.8 Immune and Lymphatic Systems/ Explorations/Lymph Formation and Movement
16.12 Cardiovascular System/Explorations/ Generic Vasculature

Chapter 17
17.1 Immune and Lymphatic Systems/ Anatomy/Gross Anatomy
17.2 Immune and Lymphatic Systems/ Anatomy/Gross Anatomy
Immune and Lymphatic Systems/ Anatomy/Microscopic Contents
Immune and Lymphatic Systems/ Histology/Lymph Node
17.3 Immune and Lymphatic Systems/ Anatomy/Gross Anatomy
Immune and Lymphatic Systems/ Histology/Spleen
17.4 Immune and Lymphatic Systems/ Anatomy/Gross Anatomy
Immune and Lymphatic Systems/ Histology/Spleen
17.6 Immune and Lymphatic Systems/ Anatomy/Microscopic Contents
Immune and Lymphatic Systems/ Explorations/Nonspecific Immunity/ Phagocytosis
17.8 Immune and Lymphatic Systems/ Explorations/Specific Immunity/ Antibody-mediated Immunity
17.9 Immune and Lymphatic Systems/ Explorations/Specific Immunity/ Antibody-mediated Immunity
17.11 Immune and Lymphatic Systems/ Explorations/Specific Immunity/ Antibody-mediated Immunity
17.12 Immune and Lymphatic Systems/ Explorations/Specific Immunity
17.13 Immune and Lymphatic Systems/ Explorations/Specific Immunity/ T Helper Cell Function
17.14 Immune and Lymphatic Systems/ Explorations/Specific Immunity/ Cytotoxic T Cell Function
17.15 Immune and Lymphatic Systems/ Explorations/Specific Immunity
17.16 Immune and Lymphatic Systems/ Explorations/Specific Immunity

Chapter 18
18.1 Respiratory System/Anatomy/Gross Anatomy
Respiratory System/Anatomy/3D Viewer: Thoracic Anatomy
18.2 Respiratory System/Anatomy/Gross Anatomy
18.3 Respiratory System/Anatomy/Gross Anatomy
18.4 Respiratory System/Anatomy/Gross Anatomy
Respiratory System/Clinical Concepts/ Bronchoscopy 1&2
18.6 Respiratory System/Histology/Alveoli
Respiratory System/Histology/Lung
18.7 Respiratory System/Histology/Alveoli
Respiratory System/Histology/Lung
18.8 Respiratory System/Histology/Alveoli
Respiratory System/Histology/Lung
18.9 Anatomical Orientation/Visible Human/ Thorax
18.13 Respiratory System/Explorations/ Mechanics of Breathing
18.14 Respiratory System/Explorations/ Mechanics of Breathing
18.16 Respiratory System/Explorations/Gas Exchange
18.20 Respiratory System/Explorations/ Oxygen Transport
18.21 Respiratory System/Explorations/ Oxygen Transport
18.22 Respiratory System/Explorations/ Oxygen Transport
18.23 Respiratory System/Explorations/ Oxygen Transport

Chapter 19
19.1 Urinary System/Anatomy/Gross Anatomy
19.2 Urinary System/Anatomy/Gross Anatomy
19.3 Urinary System/Anatomy/Kidney Anatomy
Urinary System/Anatomy/Nephron Anatomy
Urinary System/Anatomy/3D Viewer: Nephron
19.4 Urinary System/Anatomy/Kidney Anatomy
19.5 Urinary System/Anatomy/Nephron Anatomy
Urinary System/Anatomy/3D Viewer: Nephron
19.6 Urinary System/Histology/Renal Cortex
Urinary System/Histology/Renal Medulla

continued

19.7 Urinary System/Explorations/Urine Formation
19.8 Urinary System/Explorations/Urine Formation
19.10 Urinary System/Explorations/Urine Formation
19.11 Urinary System/Explorations/Urine Formation
19.12 Urinary System/Explorations/Urine Formation
19.13 Urinary System/Explorations/Urine Formation
19.14 Urinary System/Explorations/Urine Formation
19.20 Urinary System/Anatomy/Gross Anatomy
19.22 Urinary System/Histology/Bladder

Chapter 20

20.1 Digestive System/Anatomy/Gross Anatomy
20.4 Digestive System/Anatomy/Gross Anatomy
20.5 Digestive System/Histology/Duodenal Villi
Digestive System/Histology/Duodenum
20.6 Digestive System/Histology/Duodenal Villi
Digestive System/Histology/Duodenum
20.8 Digestive System/Explorations/Oral Cavity
20.9 Digestive System/Explorations/Oral Cavity
Digestive System/Anatomy/Gross Anatomy
20.12 Digestive System/Histology/Tooth
20.13 Digestive System/Anatomy/Gross Anatomy
Digestive System/Histology/Submandibular Gland
Digestive System/Explorations/Oral Cavity
20.14 Digestive System/Anatomy/Gross Anatomy
Digestive System/Explorations/Stomach
Digestive System/Histology/Fundic Stomach
20.15 Digestive System/Explorations/Stomach
Digestive System/Histology/Fundic Stomach
20.16 Digestive System/Explorations/Stomach
Digestive System/Histology/Fundic Stomach
20.17 Digestive System/Anatomy/Gross Anatomy
Digestive System/Explorations/Duodenum
Digestive System/Histology/Duodenal Villi
Digestive System/Histology/Duodenum
20.18 Digestive System/Histology/Duodenal Villi
Digestive System/Histology/Duodenum
20.19 Digestive System/Histology/Duodenal Villi
20.20 Digestive System/Histology/Duodenal Villi
Digestive System/Histology/Duodenum
20.23 Digestive System/Anatomy/Gross Anatomy
20.25 Digestive System/Clinical Concepts/Barium Radiograph
20.26 Digestive System/Anatomy/Gross Anatomy
Digestive System/Histology/Liver
Digestive System/Anatomy/3D Viewer: Digestive Anatomy
Digestive System/Histology/Gallbladder
Digestive System/Explorations/Gallbladder & Liver
20.27 Digestive System/Histology/Liver
Digestive System/Explorations/Gallbladder & Liver
20.28 Digestive System/Histology/Liver
Digestive System/Explorations/Gallbladder & Liver
20.29 Digestive System/Explorations/Gallbladder & Liver
20.30 Digestive System/Anatomy/Gross Anatomy
Digestive System/Anatomy/3D Viewer: Digestive Anatomy
Digestive System/Explorations/Gallbladder & Liver
20.31 Digestive System/Explorations/Pancreas
Endocrine System/Anatomy/Gross Anatomy/Pancreas
Digestive System/Anatomy/Gross Anatomy
Digestive System/Anatomy/3D Viewer: Digestive Anatomy
20.39 Digestive System/Clinical Concepts/Gallstones
20.40 Digestive System/Clinical Concepts/Ulcer

Chapter 22

22.1 Reproductive System/Anatomy/3D Viewer: Male Reproductive System
Reproductive System/Anatomy/Male Reproductive System
22.7 Reproductive System/Anatomy/3D Viewer: Male Reproductive System
Reproductive System/Anatomy/Male Reproductive System
Reproductive System/Histology/Testis
22.8 Reproductive System/Histology/Testis
22.10 Reproductive System/Explorations/Male/Spermatogenesis
22.11 Reproductive System/Explorations/Male/Spermatogenesis
Reproductive System/Histology/Testis
22.13 Reproductive System/Anatomy/Male Reproductive System
Reproductive System/Anatomy/3D Viewer: Male Reproductive System
22.14 Reproductive System/Explorations/Male/Erection
22.16 Reproductive System/Clinical Concepts/Vasectomy

Chapter 23

23.1 Reproductive System/Anatomy/Female Reproductive System
Reproductive System/Anatomy/3D Viewer: Female Reproductive System
23.4 Reproductive System/Anatomy/Female Reproductive System
Reproductive System/Anatomy/3D Viewer: Female Reproductive System
23.6 Reproductive System/Anatomy/Female Reproductive System
Reproductive System/Anatomy/3D Viewer: Female Reproductive System
23.7 Reproductive System/Explorations/Female/Oogenesis
Reproductive System/Histology/Ovary
Reproductive System/Histology/Ovarian Follicle
Reproductive System/Explorations/Female/Ovarian Cycle
23.8 Reproductive System/Explorations/Female/Oogenesis
Reproductive System/Histology/Ovary
Reproductive System/Histology/Ovarian Follicle
Reproductive System/Explorations/Female/Ovarian Cycle
23.9 Reproductive System/Explorations/Female/Ovarian Cycle
23.10 Reproductive System/Explorations/Female/Ovarian Cycle
23.11 Reproductive System/Explorations/Female/Ovarian Cycle
Reproductive System/Histology/Ovarian Follicle
Reproductive System/Histology/Ovary
23.12 Reproductive System/Explorations/Female/Ovarian Cycle
Reproductive System/Explorations/Female/Menstrual Cycle
23.13 Reproductive System/Histology/Ovary
23.14 Reproductive System/Histology/Ovary
23.19 Reproductive System/Clinical Concepts/Tubal Ligation

Chapter 24

24.3 Reproductive System/Explorations/Female/Ovarian Cycle

CHAPTER

1

"The human body exhibits a number of levels of structural and functional complexity." *page 6*

The Human Body: Organizing Principles

Chapter Outline

Terms to Remember

anatomy (*ă-nat' ŏ-me*), 2
physiology (*fiz"e-ol' ŏ-je*), 2
scientific method, 4
cell, 6
tissue, 6
organ, 6
system, 7
organism, 7
thorax (*thor' aks*), 12
abdomen (*ab' dŏ-men*), 13
viscera (*vis' er-ă*), 14
pleura (*ploor' ă*), 16
pericardium (*per"ĭ-kar' de-um*), 16
peritoneum (*per"ĭ-tŏ-ne' um*), 16
homeostasis (*ho" me-o-sta' sis*), 17

Learning Objectives

After studying this chapter, you should be able to

- list some of the historic events that helped to define the sciences of anatomy and physiology.
- explain what is meant by the scientific method and discuss its significance in furthering our understanding of anatomy and physiology.
- describe the essential life functions.
- list the physical conditions required to sustain life.
- describe the different levels of organization in the human body.
- describe the general function of each body system.
- identify the planes of reference used to depict the structural arrangement of the human body.
- describe the anatomical position and use the descriptive and directional terms that refer to body structures, surfaces, and regions.
- list the regions of the body and the localized areas within each region.
- identify the body cavities and list the organs found within each.
- define *homeostasis* and explain how negative feedback helps to maintain homeostasis.

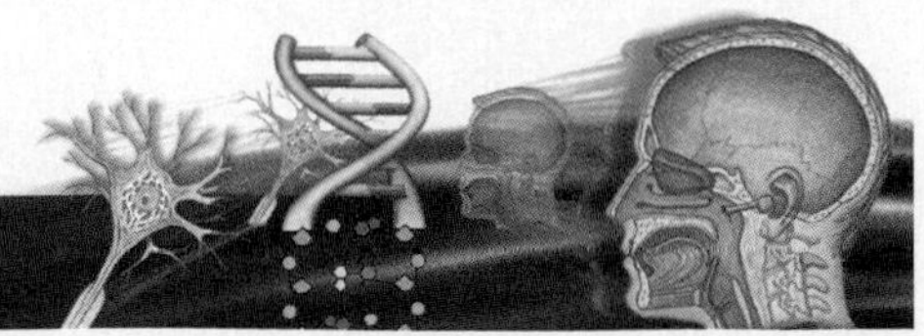

1.1 The Sciences of Anatomy and Physiology

Anatomy and physiology are integrated, dynamic sciences with an exciting heritage. These sciences provide the foundation for personal health and clinical applications.

As soon as we begin to explore our body parts as infants we become students of anatomy and physiology, and in a sense we remain students for life. Besides being naturally curious about our bodies and how they work, we can't help but take notice when, at times, they fail to function as we want them to. All of us are actively engaged to some degree in maintaining a healthy body, and our knowledge about its form and function is constantly expanding.

Anatomy and Physiology Defined

The science of **anatomy** (*ă-nat'ŏ-me*) deals with the *structure* of the human body and the relationships between body parts. Dissection of human cadavers is basic to understanding body structure; indeed, the term *anatomy* is derived from a Greek work meaning "to cut up." Knowledge of anatomy is gained primarily through observation.

The science of **physiology** (*fiz"e-ol'ŏ-je*) is concerned with the *functions* of the body. It attempts to explain the physical and chemical processes that direct the body's activities. *Physiology* is derived from another Greek word—this one meaning "the study of nature." The nature of an organism is, in effect, its function. Knowledge of physiology is gained through scientific experimentation.

The structure and function of the human body are intimately bound. Each body structure is designed to perform a specific function, and structure often determines what functions can take place. This relationship between structure and function will become more meaningful to you as you study the body systems in later chapters.

Pioneering Minds

The history of anatomy and physiology parallels that of medicine, since knowledge of body structure and function was often pursued out of a need to treat ailments. *Hippocrates* (ca. 460–ca. 377 BCE), a Greek physician, holds the honor of being considered the father of medicine. It was he who separated medicine from religion and philosophy and established the rational system of medicine as a science. In Hippocrates' view, diseases developed from natural causes; they were not "the wrath of the gods." His creed is memorialized as the Hippocratic Oath (table 1.1), which many graduating medical students repeat as a promise of professional stewardship and duty to humankind.

Medicine at the time of Hippocrates advocated that an individual's health depended on the balance of four body fluids, or humors, *which represented the four elements (earth, air, fire, and water). This* humoral theory *maintained that if blood, phlegm, choler, and black bile were present in the body in proper proportions, the person would be healthy with a good disposition. Bloodletting, practiced for centuries after the time of Hippocrates, was an attempt to establish a balance of the body humors.*

TABLE 1.1 The Hippocratic Oath

I swear by Apollo Physician and Aesculapius and Hygeia and Panacea and all the gods and goddesses, making them my witnesses, that I will fulfill according to my ability and judgment this oath and this convenant:

To hold him who has taught me this art as equal to my parents and to live my life in partnership with him, and if he is in need of money to give him a share of mine, and to regard his offspring as equal to my brothers in male lineage and to teach them this art—if they desire to learn it—without fee and convenant; to give a share of precepts and oral instruction and all the other learning to my sons and to the sons of him who has instructed me and to pupils who have signed the convenant and have taken an oath according to the medical law, but to no one else.

I will apply dietetic measures for the benefit of the sick according to my ability and judgment; I will keep them from harm and injustice.

I will neither give a deadly drug to anybody if asked for it, nor will I make a suggestion to this effect. Similarly I will not give to a woman an abortive remedy. In purity and holiness I will guard my life and my art.

I will not use the knife, not even on sufferers from stone, but will withdraw in favor of such men as are engaged in this work.

Whatever houses I may visit, I will come for the benefit of the sick, remaining free of all intentional injustice, of all mischief, and in particular of sexual relations with both female and male persons, be they free or slaves.

What I may see or hear in the course of the treatment or even outside of the treatment in regard to the life of men, which on no account one must spread abroad, I will keep to myself, holding such things shameful to be spoken about.

If I fulfill this oath and do not violate it, may it be granted to me to enjoy life and art, being honored with fame among all men for all time to come; if I transgress it and swear falsely, may the opposite of all this be my lot.

---
anatomy: Gk. *ana,* up; *tome,* a cutting
physiology: Gk. *physis,* nature; *logos,* study
humor: L. *humor,* fluid
phlegm: Gk. *phlegm,* inflammation
choler: Gk. *chole,* bile

Claudius Galen (A.D. 130–201), a Greek living under Roman domination, was the most influential writer of all time on medical subjects. For nearly 1500 years (during the Middle Ages), the writings of Galen went unchallenged as the unquestionable authority on anatomical and medical matters. Unfortunately, many of Galen's works contained errors, primarily because his religious beliefs biased his scientific endeavors. In addition, the culture of his day prevented him from dissecting humans, forcing him to draw conclusions about the human body based on dissections of other animals.

Andreas Vesalius (1514–64), a Flemish physician, repudiated Galenic tradition in a magnificently illustrated treatise, *De Humani Corporis Fabrica* (*On the Structure of the Human Body*) (fig. 1.1). With so bold a challenge to entrenched ideas, he stirred up a host of controversy and incurred the wrath of many of his colleagues. Vesalius, who was among the first to perform legal dissections of human cadavers, is regarded as the father of anatomy.

William Harvey (1578–1657), an Englishman who served as physician to King James I, not only challenged Galen's anatomical teachings, but his beliefs about physiology as well. In 1628, Harvey published *On the Movement of the Heart and Blood in Animals.* This research established proof that blood flows through vessels in one direction, refuting the two-way flow through the same vessel that Galen had assumed. Controversy over the circulation of the blood continued to rage for the next 20 years, however, until Harvey's experiments were duplicated by other scientists. Harvey (fig. 1.2) is regarded as the father of physiology, and his work serves as a classic example of the scientific method (to be discussed shortly).

The microscope added an entirely new dimension to anatomy and physiology. Once the earliest microscopes had been constructed, it was only a matter of time before cells were discovered and described. This was accomplished by the Englishman *Robert Hooke* (1635–1703), although the importance of cellular structure did not become apparent until about 150 years after his discovery.

Inside Information—The microscope was accidentally invented when a Dutch optician got two lenses stuck in a tube. *Antony van Leeuwenhoek* (1632–1723) owned a fabric shop in Holland, but he quickly capitalized on this happy circumstance. Leeuwenhoek became one of the early masters at constructing homemade microscopes and is reported to have owned 247 of them. It was the microscope of *Marcello Malpighi* (1628–94), however, that first displayed blood moving through capillaries, thus putting the finishing touches on Harvey's theory of blood circulation.

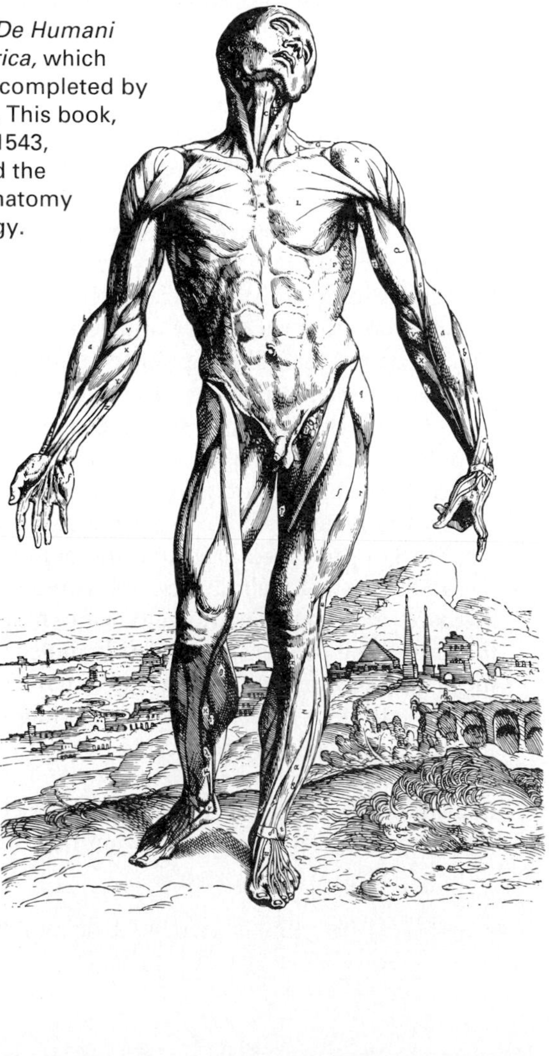

FIGURE 1.1

A plate from *De Humani Corporis Fabrica,* which Vesalius had completed by the age of 28. This book, published in 1543, revolutionized the sciences of anatomy and physiology.

FIGURE 1.2

In the early seventeenth century, the English physician William Harvey demonstrated that blood circulates, and that it does not flow back and forth through the same vessels.

Improvements in the microscope during the eighteenth century paved the way for formulation of the cell theory in the mid-nineteenth century by two German scientists, *Matthias Schleiden* and *Theodor Schwann.* In addition, the improved microscope was invaluable for understanding the causes of diseases, and thus for discovering cures for many of them.

The Scientific Method

The **scientific method** is a disciplined approach to gaining information (facts or data) about the world. Simply stated, the scientific method depends on a systematic search for information and a continual checking and rechecking to see whether previous ideas still hold up in the light of new evidence.

The scientific method often begins with an *observation* that an event has occurred repeatedly. This is followed by question formulation and exploration of other sources of knowledge. The culmination of this first step is the *formulation of a hypothesis*—a statement that proposes a relationship. You might hypothesize, for example, that people who exercise regularly have a lower resting pulse rate than people who do not. You would then perform experiments or make other observations to *test the hypothesis,* and then you would *analyze the results.* In so doing, you would use a *control,* or data base for comparison. Finally, you would *draw conclusions* as to whether your experiments or observations supported or refuted your hypothesis. Once a hypothesis survives testing, it may be incorporated into a more general *theory.* Scientific theories are statements about the natural world that incorporate a number of proven hypotheses. They serve as a logical framework by which these hypotheses can be interrelated.

The hypothesis in the preceding example is scientific because it is *testable;* the pulse rates of 100 athletes and 100 "couch potatoes" can be measured to see whether there are statistically significant differences. If there are, the statement that athletes, on the average, have lower resting pulse rates than inactive people is justified *based on these data.* In order for a discovery to become generally accepted and included in textbooks, however, other scientists must be able to duplicate the results. Thus, theories are based on *reproducible data.*

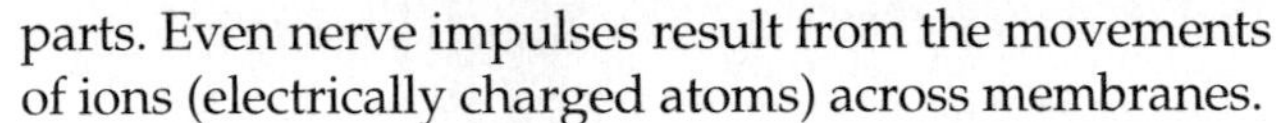

1.2 Maintenance of Life

Staying alive depends on the ability to carry on vital operations within the body.

Requirements for Life

All living organisms display a number of structural and functional characteristics that are vital to their continued existence. Taken together, these characteristics broadly define what we mean by the process of life.

1. **Organization.** Each cell and organ within our body is programmed by genetics (and adapted) to carry out specific activities.
2. **Responsiveness.** Responsiveness is the ability of an organism (or cell) to monitor internal or external conditions and react, or respond, to changes in accordance with the needs of the body. A white blood cell, for example, will respond to toxins (poisons) from bacteria and move toward the pathogens in an attempt to dispose of them. Pulling a hand away from a hot object demonstrates responsiveness, as does drinking water to satisfy thirst.
3. **Movement.** Movement results in the pumping of blood, breathing, and changes in the position of our body parts. Even nerve impulses result from the movements of ions (electrically charged atoms) across membranes.
4. **Growth.** Growth refers both to an increase in numbers of cells and to the development of an organism (fig. 1.3).
5. **Differentiation.** It is during our embryonic development that undifferentiated (unspecialized) tissues undergo cellular differentiation, resulting in organs that are structurally and functionally distinct.
6. **Metabolism.** Metabolism includes all of the chemical processes carried out in each living body cell. A cell is like a factory that requires materials to be brought in from the outside so that products can be manufactured and dispersed. The bloodstream connects each cell to a supply of nutrients and oxygen; it also serves for dispersal of produced products and elimination of metabolic wastes. Cellular metabolism includes the breakdown of complex molecules into more usable forms and the processing of simple molecules to form more complex molecules. Metabolism makes energy available to cells, creates body heat, and enables cells to produce (synthesize) and secrete useful molecules (such as regulatory hormones).
7. **Digestion.** Digestion is the process by which complex food substances are converted to simpler forms,

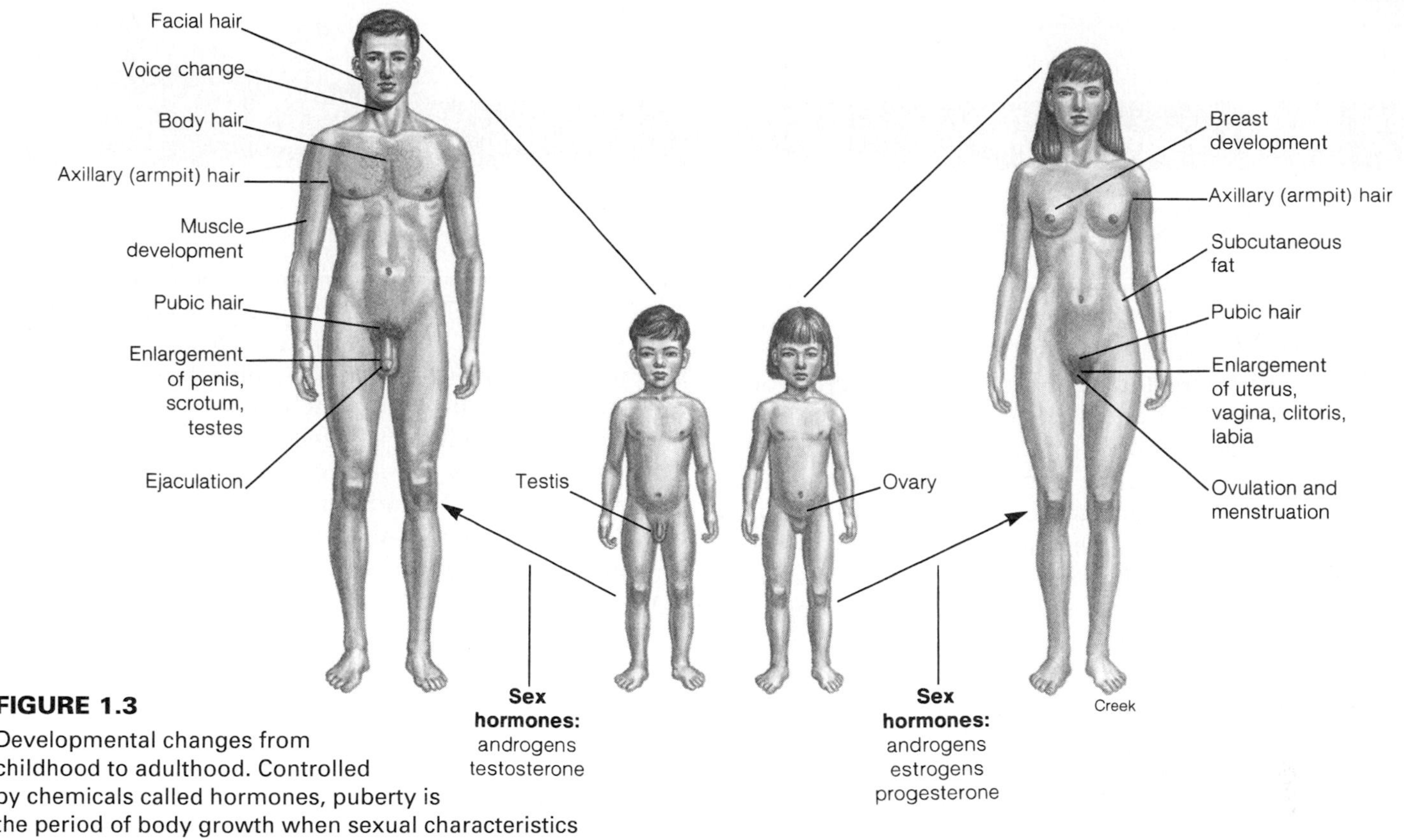

FIGURE 1.3
Developmental changes from childhood to adulthood. Controlled by chemicals called hormones, puberty is the period of body growth when sexual characteristics are expressed and the sex organs become functional.

allowing them to be absorbed through the intestinal wall and transported to the cells by the blood.

8. **Excretion.** Excretion is the removal of wastes produced by metabolism. These wastes are frequently toxic to cells and must be eliminated from the body.
9. **Reproduction.** Reproduction refers to both cellular replication, important for our body's growth and repair, and to sexual reproduction, the process by which parents produce offspring.
10. **Survival instinct.** An inborn instinct to survive is essential for an organism to avoid death long enough to produce offspring.

TABLE 1.2 Physical Needs for Life

Physical Needs	Examples of Physiological Effects
Water	Accounts for about 60% of body weight and is the medium in which most chemical reactions in the body take place; essential for homeostasis (metabolic balance) in the body
Oxygen	Accounts for 20% of normal air and is used in the body in releasing energy from food substances
Food	Nutrient-containing substances that can be broken down and used for energy, building new compounds or regulating metabolic processes in the body
Heat	A form of energy expressed as temperature: the body can tolerate a wide range in environmental temperature but maintains a narrow range in body temperature; many physiological mechanisms ensure a consistent internal body temperature
Pressure	External, such as atmospheric pressure, important for respiration; internal, such as blood pressure, important for filtration of blood in kidneys
Protection	Clothing and shelter create artificial environments for our bodies

Physical Conditions Required for Life

The world provides us with an environment suitable for life. Not only are our bodies adapted to take advantage of environmental parameters, but we have become dependent upon the consistency of these factors for survival. Human life is fragile and is strictly dependent on the presence of the particular environmental conditions summarized in table 1.2.

1.3 Body Organization

The human body exhibits a number of levels of structural and functional complexity.

Cellular Level

The **cell** is the basic structural and functional component of life. We are multicellular organisms composed of 60 to 100 trillion cells. It is at the microscopic cellular level (fig. 1.4) that such vital functions of life as metabolism, growth, response to stimuli, repair, and reproduction are carried on. All cells contain a selectively permeable **cell (plasma) membrane** that encloses a semifluid substance called **cytoplasm.** Certain cytoplasmic structures and molecules are arranged in small functional units called **organelles** (*or"gă-nelz*). Each organelle carries out a specific function within the cell.

Your body contains many distinct kinds of cells, each specialized to perform specific functions. Examples are bone cells, muscle cells, fat cells, and nerve cells. Each of these cell types has a unique structure directly related to its function.

Tissue and Organ Levels

Tissues are layers or groups of similar cells that perform a common function. The entire human body is composed of only four principal kinds of tissues: *epithelial, connective, muscular,* and *nervous tissues.* The outer layer of our skin, for example, is a tissue (epithelium) because it is composed of similar cells that together serve as a protective shield for the body. *Histology* is the science concerned with the microscopic study of tissues. The characteristic role of each tissue type is discussed fully in chapter 4.

An **organ** is a group of two or more tissue types that performs a specific function. Organs occur throughout your body and vary greatly in size and function. Examples of organs are the heart, pancreas, ovary, skin, and

cell: L. *cella,* small room
tissue: Fr. *tissu,* woven; from L. *texo,* to weave
organ: Gk. *organon,* instrument

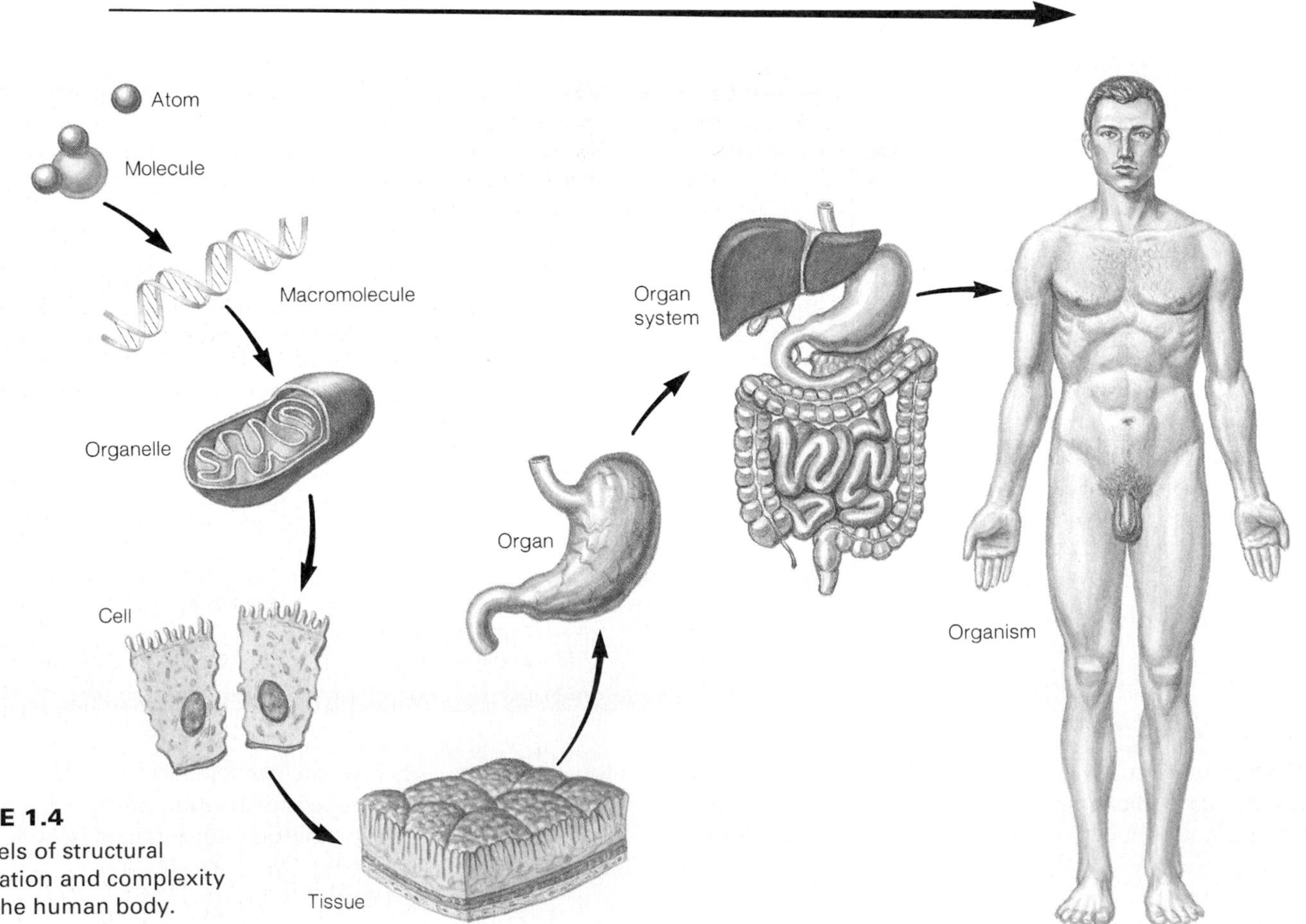

FIGURE 1.4
The levels of structural organization and complexity within the human body.

even any one of the bones. Most organs contain all four principal tissues. In the stomach, for example, the inside epithelial lining performs the functions of secretion and absorption. The wall of the stomach, however, also contains muscle tissue (for contractions of the stomach), nervous tissue (for regulation), and connective tissue, which binds the other tissues together.

System Level

The **systems** of the body constitute the next level of organization. A body system consists of various organs that have similar or related functions. Examples of systems are the circulatory system, nervous system, digestive system, and endocrine system. Certain organs may serve several systems. The pancreas, for example, functions with both the endocrine and digestive systems. All of the systems of the body are interrelated and function together, making up the **organism.**

A *systemic approach* to studying anatomy and physiology emphasizes the purposes of various organs within a system. For example, the functional role of the digestive system can be best understood if all of the organs within that system are studied together. In a *regional approach,* all of the organs and structures in one particular region of the body are examined at the same time. The regional approach has merit in graduate professional schools (medical, dental, etc.) because the structural relationships of portions of several systems can be observed simultaneously. This is important for surgeons, who must be familiar with all of the structures within a particular region. Dissections of cadavers are usually conducted on a regional basis. Radiographs (X rays) and new medical technologies, including computerized tomography (CT), magnetic resonance imaging (MRI), and positron emission tomography (PET) scans, allow the organs within different body regions to be safely visualized in a patient (fig. 1.5).

X rays were discovered in 1895 by Wilhelm Konrad Roentgen—an accomplishment for which he was awarded the first Nobel prize in physics. The X-ray image that is produced on film is frequently referred to as a roentgenograph. *The recent development of the computer tomographic technique has been hailed as the greatest advancement in diagnostic medicine since the discovery of X rays themselves.*

system: Gk. *systema,* being together

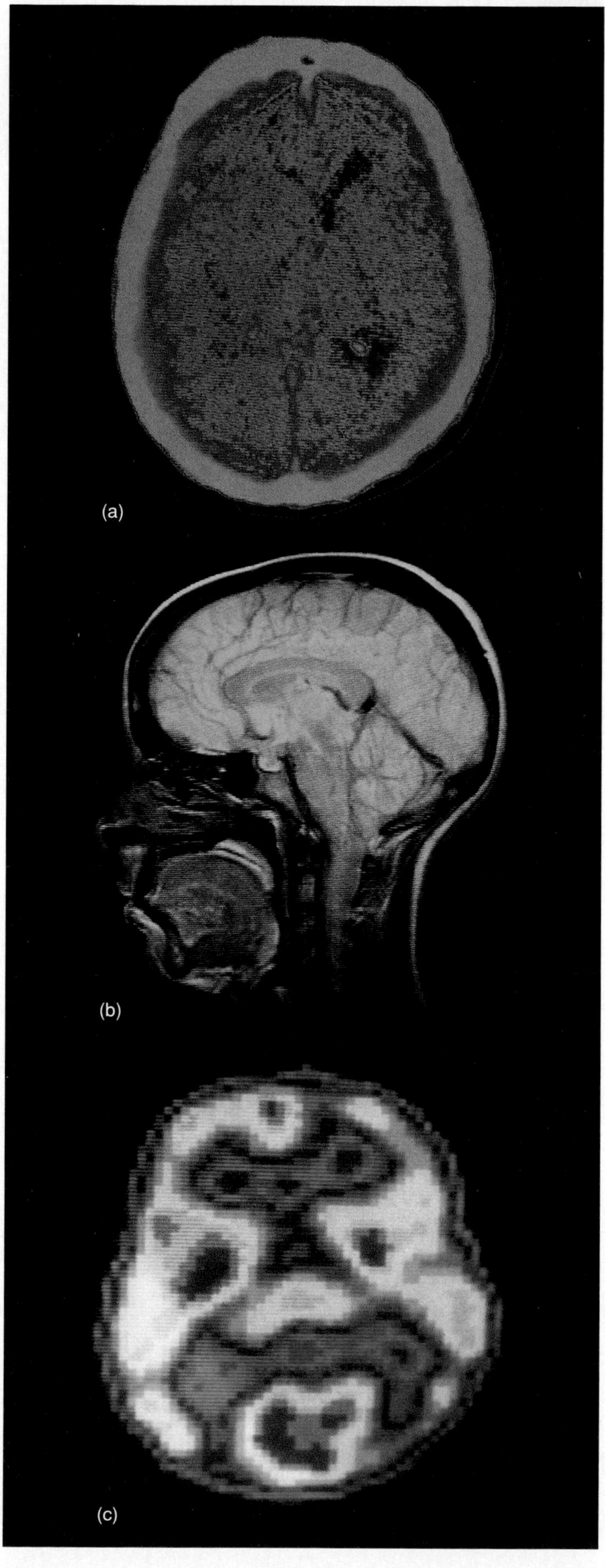

FIGURE 1.5

Different techniques in radiological anatomy provide unique perspectives of the anatomy of the human head. (*a*) A CT (computerized tomography) scan, (*b*) an MR (magnetic resonance) image (MRI), and (*c*) a PET (positron emission tomography) scan.

Integumentary system
Function: external support and protection of body

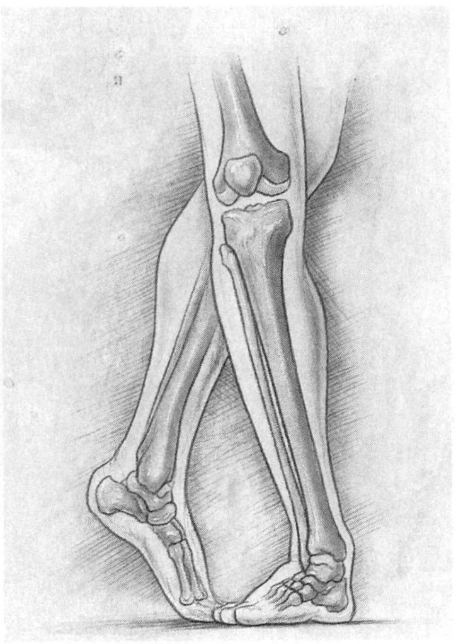

Skeletal system
Function: internal support and flexible framework for body movement; production of blood cells

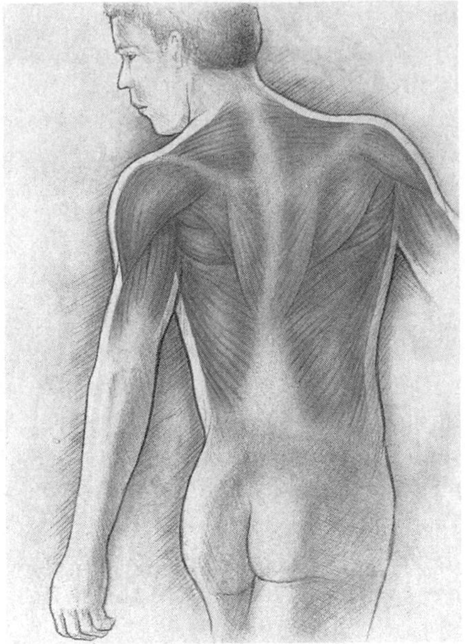

Muscular system
Function: body movement; production of body heat

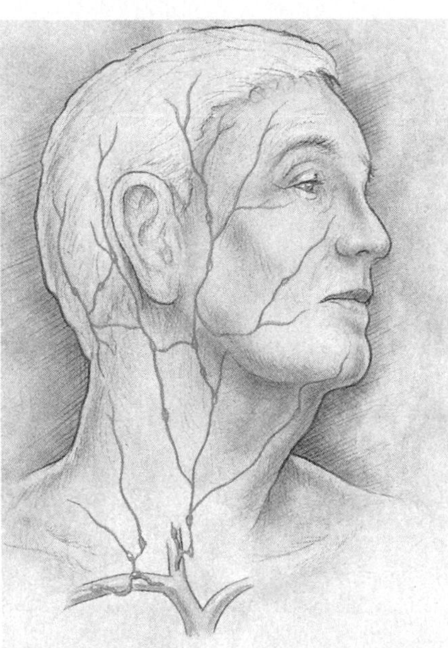

Lymphatic system
Function: body immunity; absorption of fats; drainage of tissue fluid

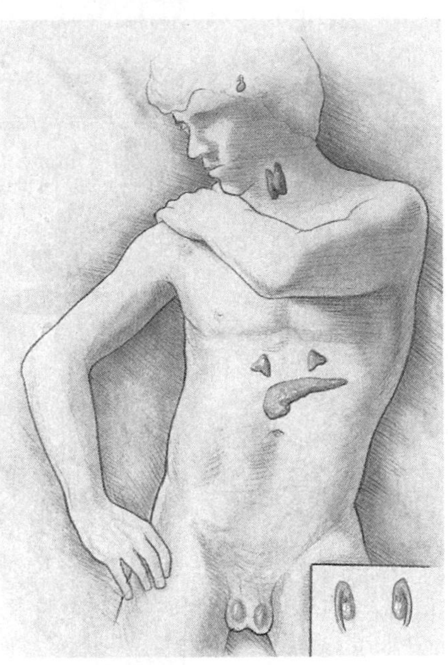

Endocrine system
Function: secretion of hormones for chemical regulation

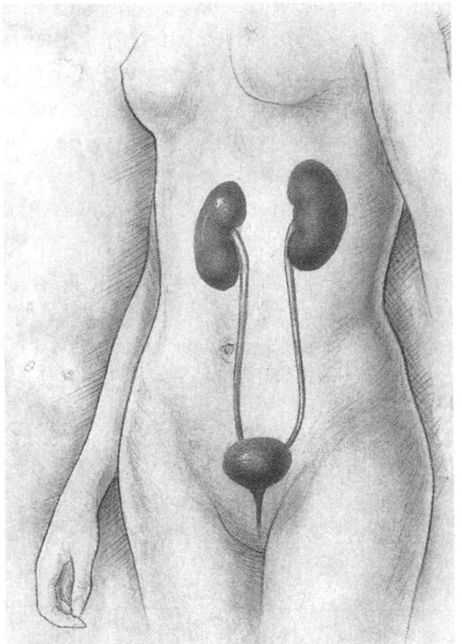

Urinary system
Function: filtration of blood; maintenance of volume and chemical composition of blood; removal of metabolic wastes from body

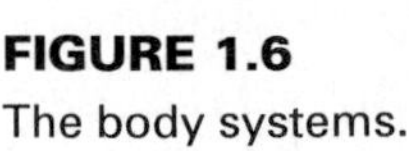

FIGURE 1.6
The body systems.

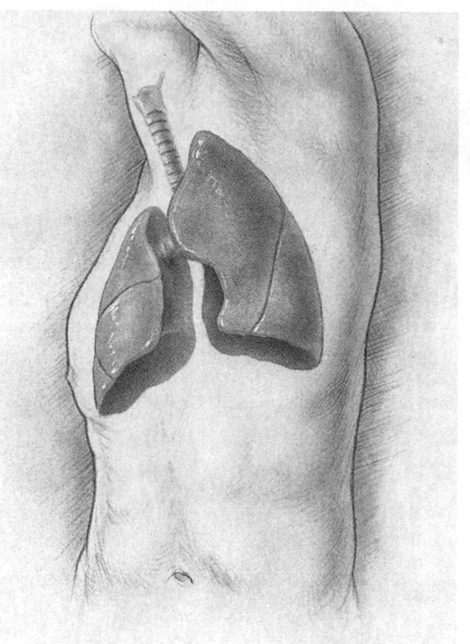

Respiratory system
Function: gaseous exchange between external environment and blood

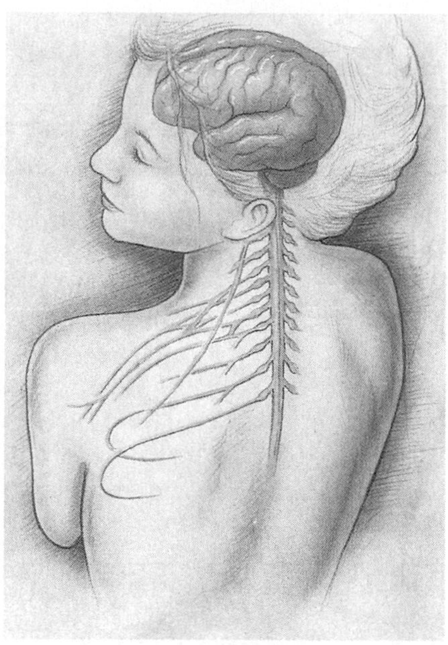

Nervous system
Function: control and regulation of all other systems of the body

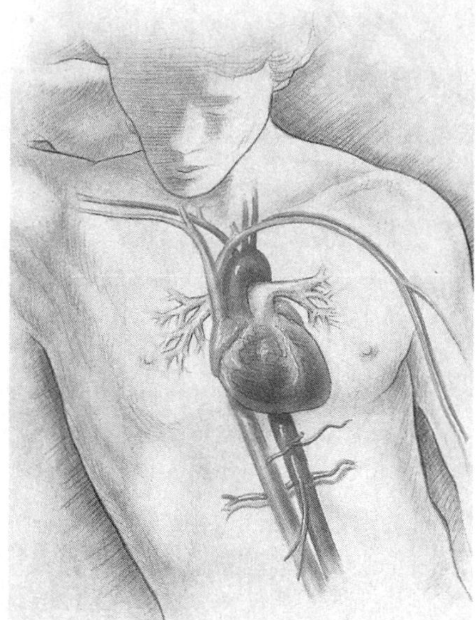

Circulatory system
Function: transport of life-sustaining materials to body cells; removal of metabolic wastes from cells

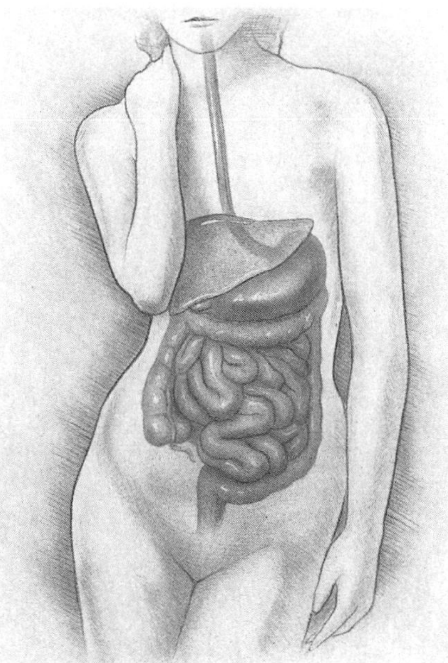

Digestive system
Function: breakdown and absorption of food materials

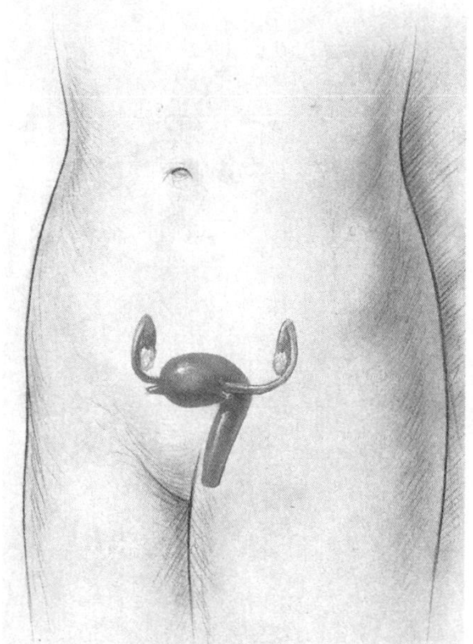

Female reproductive system
Function: production of female sex cells (ova); receptacle for sperm from male; site for fertilization of ovum, implantation, and development of embryo and fetus; delivery of fetus

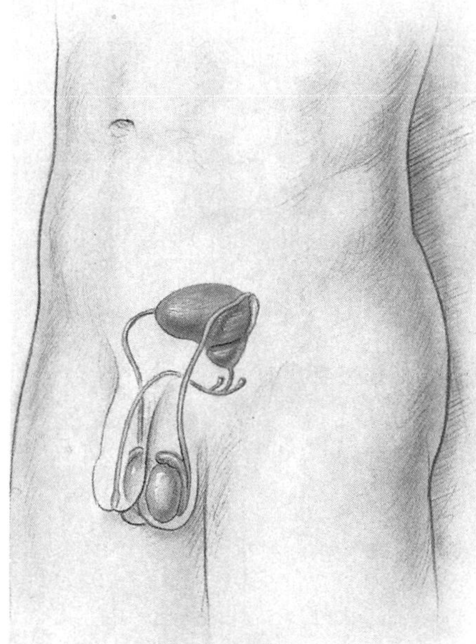

Male reproductive system
Function: production of male sex cells (sperm); transfer of sperm to reproductive system of female

1.4 Planes of Reference and Descriptive Terminology

All of the planes of reference and terms indicating direction are made with reference to the body in a specific orientation, known as the anatomical position. Most anatomical terms are derived from Greek and Latin.

Anatomical Position and Directional Terms

All terms of direction that describe the relationship of one body part to another are made in reference to the **anatomical position.** In this position, a person is standing with feet parallel and flat on the floor, eyes directed forward, and arms at the sides of the body with palms forward and fingers pointing downward (see table 1.3, photo at left).

Directional terms are used to locate structures and regions of the body relative to the anatomical position (table 1.3). For example, when a person is in the anatomical position, the thumb is always lateral to the little finger, and the nose is anterior to the ears.

TABLE 1.3 Directional Terms for the Human Body

Term	Definition	Example
Superior (cranial, cephalic)	Toward the head; toward the top	The thorax is superior to the abdomen.
Inferior (caudal)	Away from the head; toward the bottom	The legs are inferior to the trunk.
Anterior (ventral)	Toward the front	The navel is on the anterior side of the body.
Posterior (dorsal)	Toward the back	The kidneys are posterior to the intestine.
Medial	Toward the midline of the body	The heart is medial to the lungs.
Lateral	Toward the side of the body	The ears are lateral to the nose.
Internal (deep)	Away from the surface of the body	The brain is internal to the cranium.
External (superficial)	Toward the surface of the body	The skin is external to the muscles.
Proximal	Toward the main mass of the body	The knee is proximal to the foot.
Distal	Away from the main mass of the body	The hand is distal to the elbow.

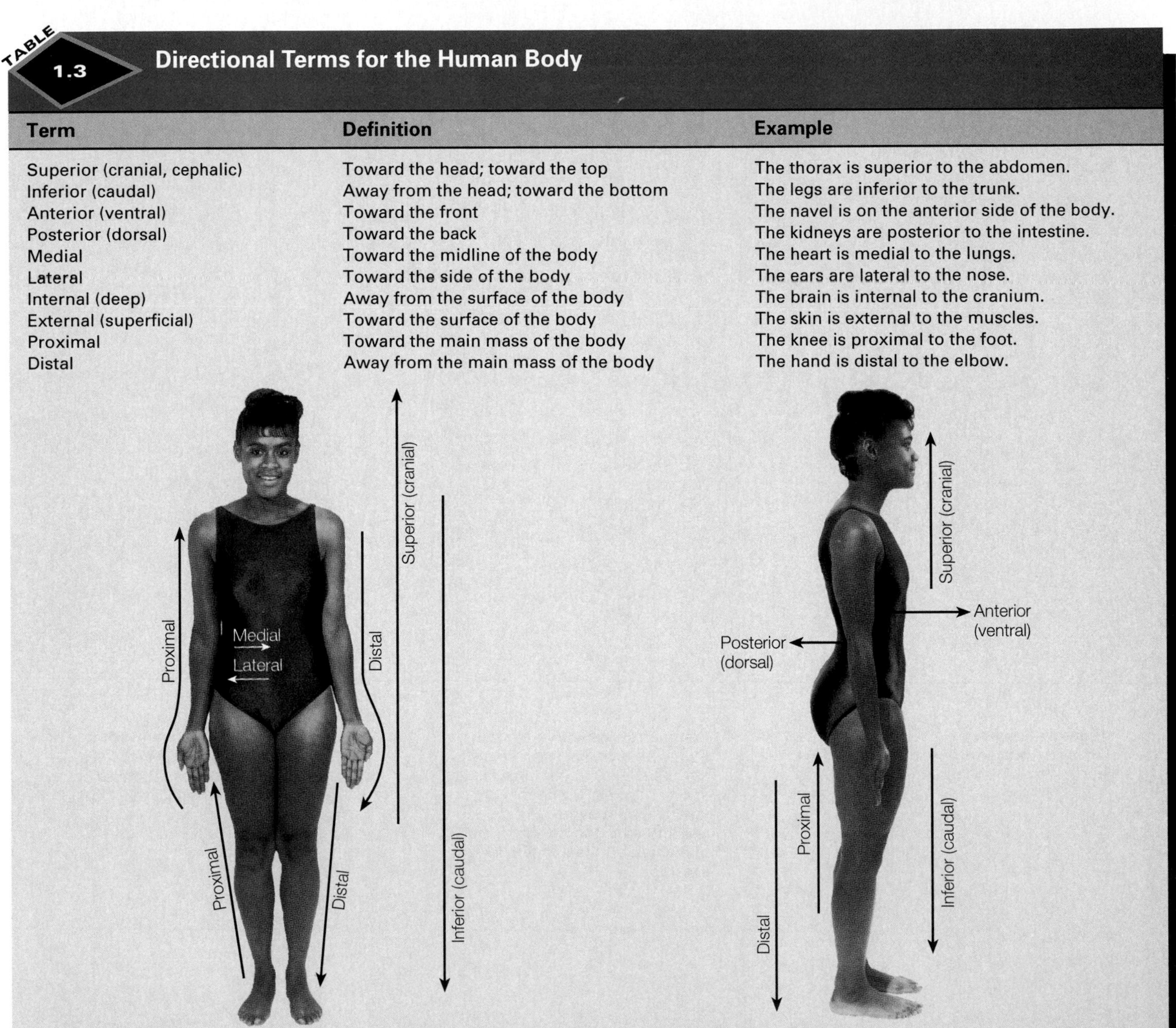

Planes of Reference

In order to see and study the structural arrangements of various organs, the body may be sectioned (cut) according to three planes of reference: a **midsagittal plane,** a **coronal plane,** and a **transverse plane** (figs. 1.7 and 1.8).

coronal: L. *corona,* crown.

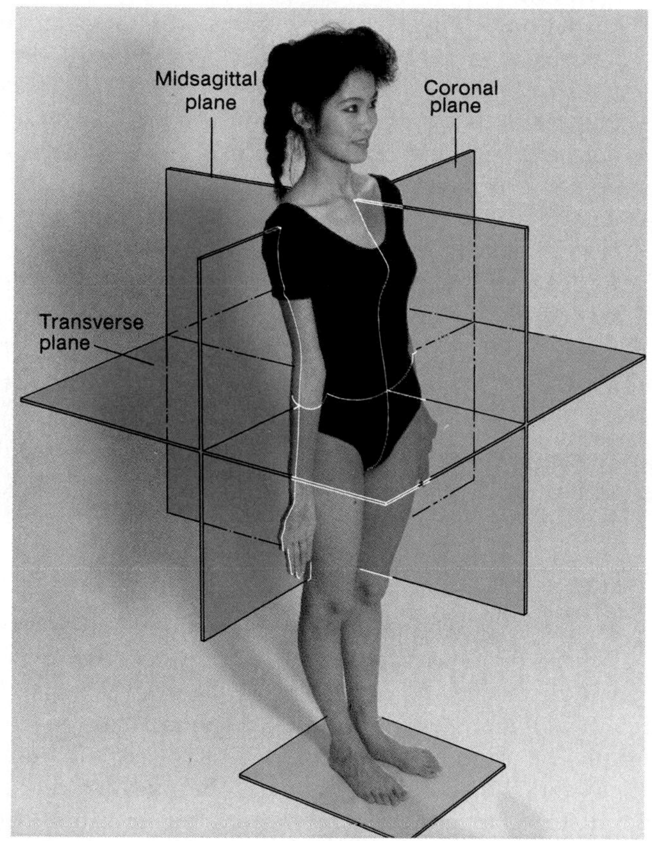

FIGURE 1.7
Planes of reference through the body in anatomical position.

A midsagittal plane passes lengthwise through the body, dividing it into equal right and left halves. Other *sagittal planes* extend parallel to the midsagittal plane, but off-center, and divide the body into unequal right and left portions. Coronal, or *frontal, planes,* also pass lengthwise and divide the body into anterior (front) and posterior (back) portions. Transverse planes, also called *horizontal,* or *cross-sectional, planes,* divide the body into superior (upper) and inferior (lower) portions.

The value of the computerized tomographic (CT) scan is that it displays an image along a transverse plane (see fig. 1.5) similar to that which could otherwise be obtained only in an actual section through the body. Prior to the development of this X-ray technique, conventional X-ray images were on a vertical plane, and the dimensions of the body irregularities were difficult, if not impossible, to ascertain.

Word Derivations

Studying anatomical and physiological terminology can be fun. Not only is it a way to understand how words are formed, but it reinforces your learning. The majority of scientific terms are of Greek or Latin derivation, but some are German, French, and Arabic derivatives. Some anatomical structures bear the names of the people who discovered or described them; unfortunately, these names, or *eponyms,* simply must be memorized.

Many Greek and Latin terms were coined more than 2500 years ago. It is exciting to decipher the meanings of these terms and gain a glimpse into our medical history. Many terms referred to common plants or animals. The term *vermis* means worm; *cochlea,* snail shell; *cancer,* crab; and *uvula,* grape. Other terms reveal the warlike environment of the Greek and Latin era. *Thyroid,* for

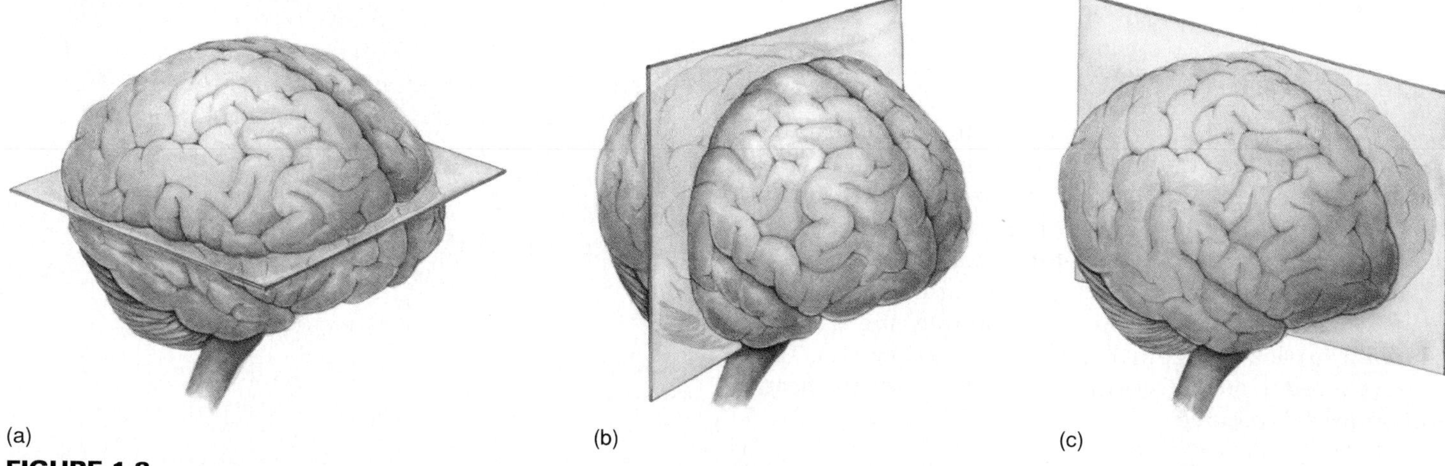

FIGURE 1.8
The human brain sectioned along (*a*) a transverse plane, (*b*) a coronal plane, and (*c*) a midsagittal plane.

example, means shield and *thorax,* breastplate. *Sella* means saddle and *stapes* means stirrup. In a reference to the god who held up the world in Greek mythology, the *atlas* is the bone that supports the skull.

You will encounter many new terms throughout your study of anatomy and physiology. You can learn these terms more easily if you understand their prefixes and suffixes. Use the glossary of prefixes and suffixes (on the inside front cover) as an aid in learning new terms. Saying these terms out loud will also help you learn them.

Inside Information—The tradition of using surnames for body parts (he who finds it names it) has complicated the task of learning anatomical terminology; frequently there are equivalent, more descriptive, terms also in use. The current trend is to standardize terms and do away with "tombstone names," however familiar they may be.

Clinical Procedures

Certain clinical (medical) procedures are important in determining anatomical structure and function in a living person. The most common of these are as follows:

1. **Inspection.** Visually observing the body to note symptoms such as abnormal skin color, swelling, or rashes. Other observations may include needle marks on the skin, irregular breathing, or abnormal behavior.
2. **Palpation.** Applying the fingers with firm pressure to the surface of the body to feel surface landmarks, lumps, tender spots, or pulsations.
3. **Percussion.** Tapping sharply on the body walls to detect vibrations as an aid in locating excess fluids or organ abnormalities.
4. **Auscultation.** Listening to the sounds that various organs make (breathing, heartbeat, and so forth).
5. **Reflex testing.** Observing a person's automatic response to a stimulus.

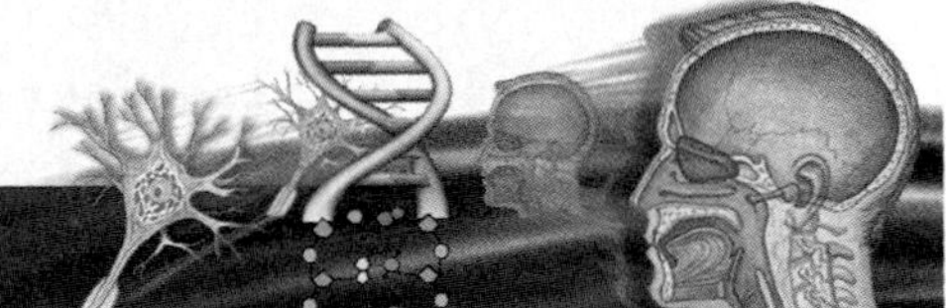

1.5 Body Regions and Body Cavities

The human body is divided into regions and specific areas that can be identified on the surface. The head and trunk contain internal organs housed in distinct body cavities.

Body Regions

The body is divided into several surface regions. Learning the terms that refer to these regions now will help you learn the names of underlying structures later. The major body regions are the *head, neck, trunk* (*torso*), *upper extremity,* and *lower extremity.* The trunk is divided into the thorax and abdomen. The body regions are illustrated in figure 1.9.

Head and Neck

The head is divided into a **facial region,** which includes the eyes, nose, and mouth, and a **cranial region,** or **cranium,** which covers and supports the brain. The identifying names for specific surface regions of the face are based on associated organs—for example, the orbital (eye), nasal (nose), oral (mouth), and auricular (ear) regions—or underlying bones—for example, the frontal, temporal, parietal, zygomatic, and occipital regions. The neck, referred to as the **cervical region,** supports the head and permits it to move.

Thorax

The thorax (*thor'aks*), or **thoracic region,** is often referred to as the chest. The **mammary region** of the thorax surrounds the nipple and in mature females develops as the breast. Between the mammary regions is the **sternal region.** The armpit is called the **axillary fossa,** or simply **axilla,** and the surrounding area, the **axillary region.** Paired **scapular regions** can be identified from the back of the thorax. The **vertebral region,** following the vertebral column, extends the length of the back.

The heart and lungs are located within the thoracic cavity. Surface landmarks are helpful in physically examining them. A physician must know, for example, where the valves of the heart can best be detected and where to listen for respiratory sounds. The axilla is important in examining for infected lymph nodes. When fitting a patient for crutches, a physician will instruct the patient to avoid supporting the weight of the body on the axillary areas because of the possibility of damaging the underlying nerves and vessels.

thorax: L. *thorax,* chest
mammary: L. *mamma,* breast
axillary: L. *axilla,* armpit

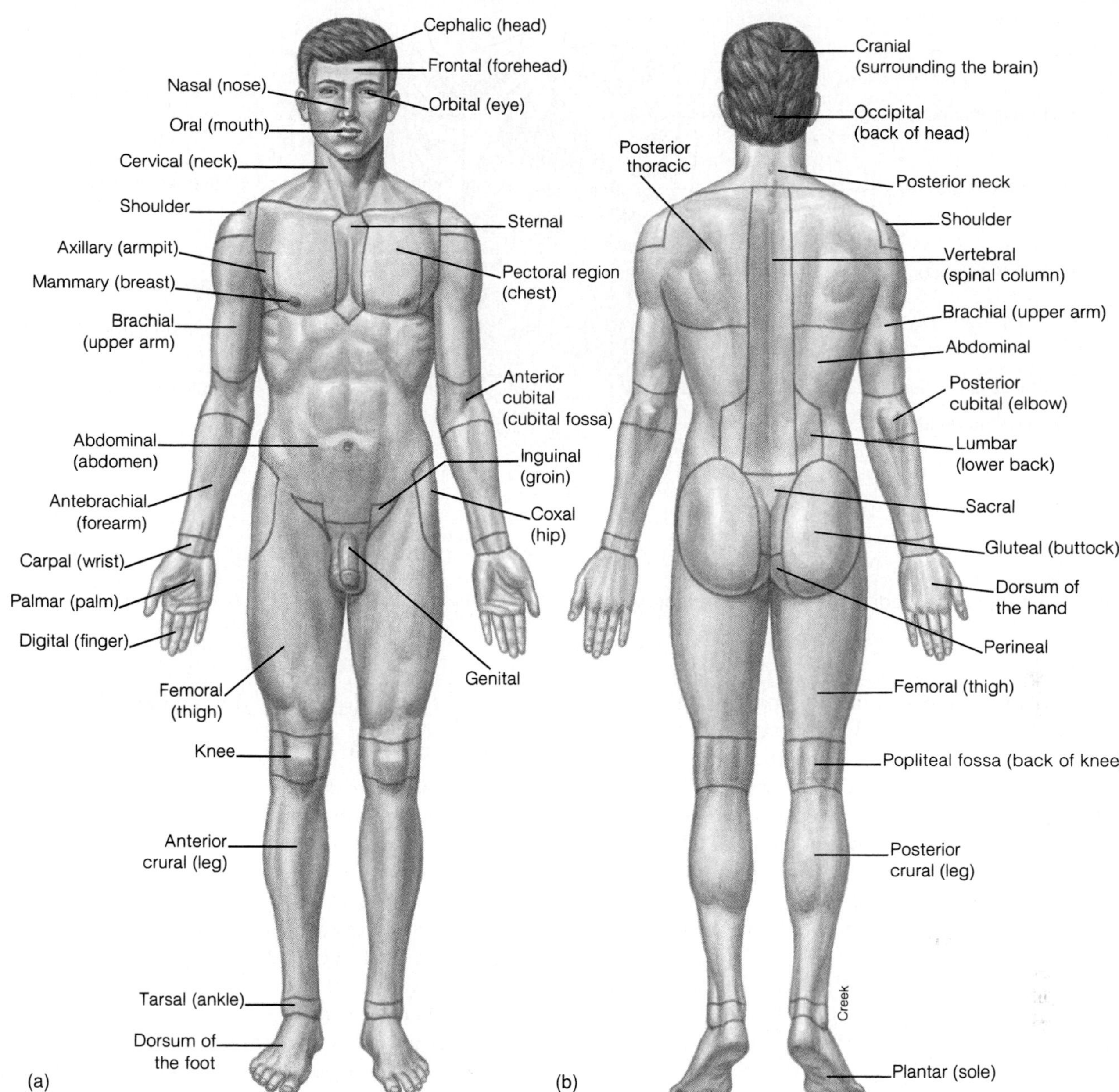

FIGURE 1.9
Body regions. (*a*) Anterior view and (*b*) posterior view.

Abdomen

The abdomen (*ab'dŏ-men*) is located below the thorax. The **umbilicus,** or **navel,** is an obvious landmark on the front and center of the abdomen.

The **pelvic region** forms the lower portion of the abdomen. Within the pelvic region is the **pubic area,** which is covered with pubic hair in sexually mature individuals. The **perineum** (*per"ĭ-ne'um*) is the region containing the external sex organs and the anal opening. The center of the back side of the abdomen is the **lumbar region.** The **sacral region** is located farther down, at the point where the vertebral column terminates. Large hip muscles form the **buttock,** or **gluteal region.** A specific part of this region is a common injection site for hypodermic needles.

A common clinical practice is to subdivide the abdomen into nine regions (fig. 1.10) for ease of reference in describing abnormal conditions in the abdominal area.

Upper and Lower Limbs

The upper limb is anatomically divided into the **shoulder, brachium** (arm), **antebrachium** (forearm), and **manus** (hand). The shoulder is the region between the

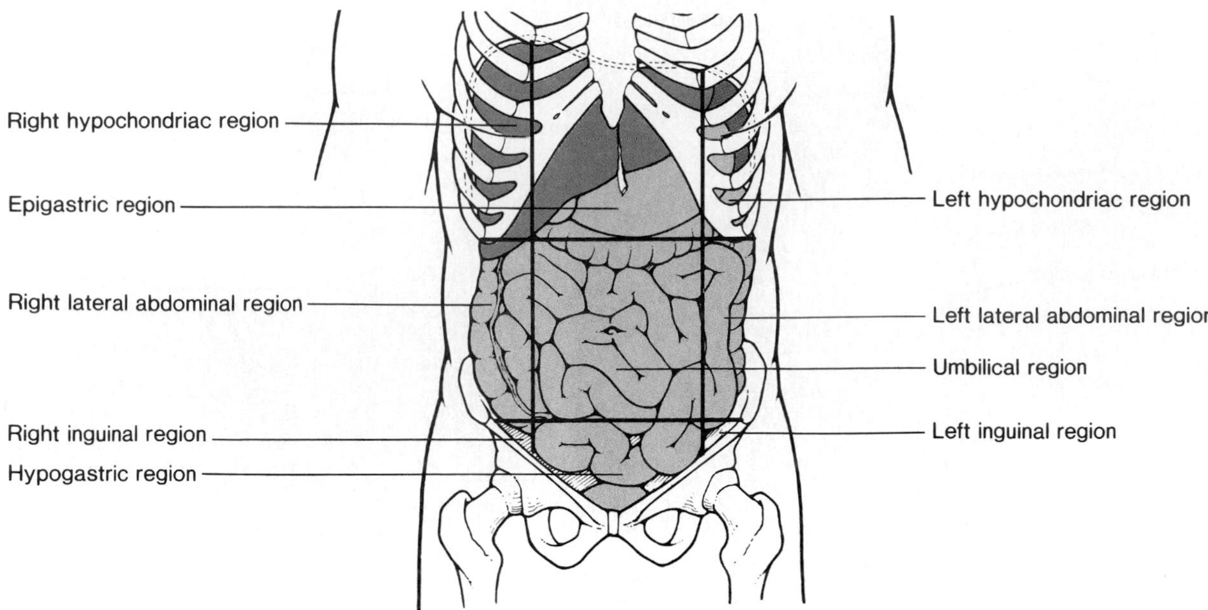

FIGURE 1.10
The abdomen is frequently subdivided into nine regions.

pectoral girdle and the brachium that contains the shoulder joint. Between the arm and forearm is a flexible joint called the *elbow*. The area of the elbow is known as the **cubital region.** The front of the elbow is also known as the **cubital fossa,** an important site for intravenous injections and for the withdrawal of blood. The *wrist* is the flexible junction between the forearm and the hand. The front of the hand is referred to as the **palm,** or **palmar surface,** and the back of the hand is called the **dorsum of the hand.**

The lower limb consists of the **hip, thigh, knee, leg,** and **pes** (foot). The hip is the region between the pelvic girdle and the thigh that contains the hip joint. The knee joint has two surfaces: the front surface is the **patellar region,** or *kneecap;* the back of the knee is called the **popliteal** (*pop″lĭ-te′al*) **fossa.** The *shin* is a prominent bony ridge extending longitudinally along the leg. The *ankle* is the junction between the leg and the foot. The *heel* is the back of the foot, and the *sole* is referred to as the **plantar surface.** The **dorsum of the foot** is the top surface.

The popliteal fossa is checked in elderly patients with degenerative conditions. Bulging, weakened areas, called aneurysms, *within the walls of the popliteal blood vessels are common, as are infected popliteal lymph nodes. A superficial vein, called the small saphenous vein, may become* varicose *(greatly distended) in elderly people.*

Body Cavities and Associated Membranes

Body Cavities

Body cavities are spaces within the body. They contain organs that are protected, divided into compartments, and supported by thin membranes. There are two principal body cavities: the **posterior** (dorsal) **body cavity** containing the brain and spinal cord, and the larger **anterior** (ventral) **body cavity** containing the internal organs of the trunk (figs. 1.11 and 1.12).

Organs within the anterior body cavity are collectively called **viscera** (*vis′er-ă*), or **visceral organs.** The anterior body cavity is partitioned by the muscular *diaphragm* into an upper **thoracic cavity,** or chest cavity, and a lower **abdominopelvic cavity.** Within the thoracic cavity are two **pleural cavities** surrounding the right and left lungs and a **pericardial cavity** surrounding the heart. The **mediastinum** is the area between the lungs. It contains the heart, trachea, esophagus, and all other structures of the thoracic cavity except the lungs themselves.

The abdominopelvic cavity consists of an upper **abdominal cavity** and a lower **pelvic cavity.** The abdominal cavity contains the stomach, small intestine, most of the large intestine, liver, gallbladder, pancreas, spleen, and kidneys. The pelvic cavity is occupied by part of the large intestine, the urinary bladder, and the internal reproductive organs.

cubital: L. *cubitis,* elbow
popliteal: L. *poples,* ham (hamstring muscles) of the knee
visceral: L. *viscus,* internal organ of trunk

Body cavities confine organs and systems that have related functions. The major portion of the nervous system occupies the posterior cavity; the principal organs of the respiratory and circulatory systems, the thoracic cavity; the primary organs of digestion, the abdominal cavity; and the reproductive organs, the pelvic cavity. Not only do these cavities house and support various body organs, they also effectively compartmentalize them. Infections and diseases rarely spread from one compartment to another, and trauma to the body is minimized by this compartmentalization. For example, a serious injury to the thoracic cavity may cause one lung to collapse, but usually not both. If both lungs collapse, a person will soon die.

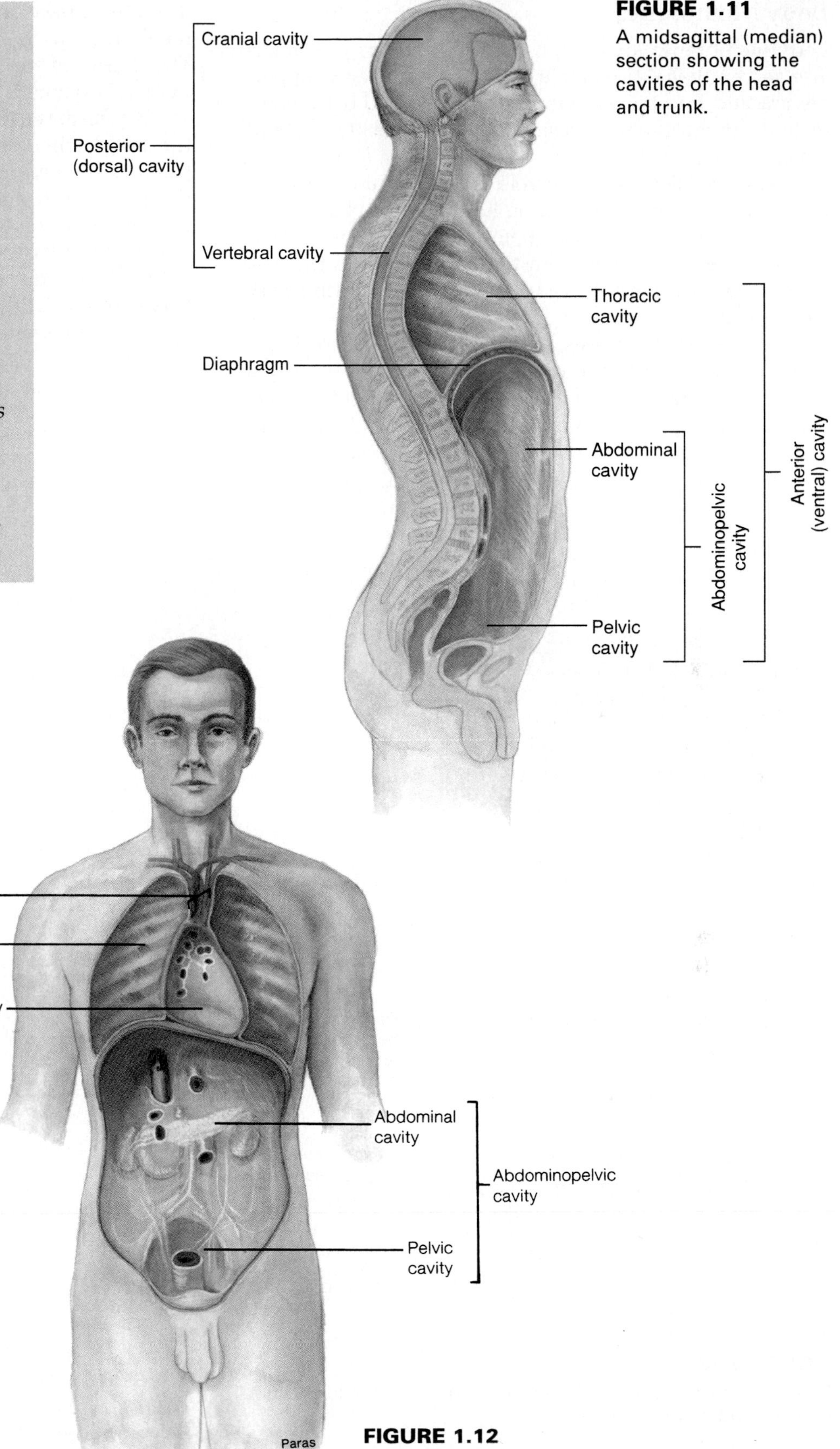

FIGURE 1.11
A midsagittal (median) section showing the cavities of the head and trunk.

FIGURE 1.12
An anterior view showing the cavities of the trunk.

Body Membranes

Body membranes are composed of thin layers of connective and epithelial tissue that cover, separate, and support viscera and line body cavities. There are two basic types of body membranes: *mucous* (*myoo'kus*) *membranes* and *serous* (*ser'us*) *membranes.*

Mucous membranes secrete a thick liquid substance called *mucus.* Generally, mucus lubricates or protects the associated organs. Mucous membranes line cavities and tubes that enter or exit from the body, such as the oral (mouth) and nasal cavities and the tubes of the respiratory and digestive systems.

Serous membranes line the thoracic and abdominopelvic cavities and cover visceral organs, secreting a watery lubricant called *serous fluid.* **Pleurae** (singular, *pleura* [*ploor'ă*]) are serous membranes associated with the lungs (fig. 1.13*a*). Each pleura (pleura of right lung and pleura of left lung) has two parts. The **visceral pleura** is attached to the outer surface of the lung, while the **parietal pleura** lines the thoracic walls and the thoracic side of the diaphragm. The moistened space between the two pleurae is the **pleural cavity.** Each lung is surrounded by its own pleural cavity.

Pericardial membranes are the serous membranes of the heart (fig. 1.13*b*). The thin **visceral pericardium** (*per"ĭ-kar'de-um*) is the outer layer of tissue attached to the heart, and a thicker **parietal pericardium** is the durable covering that surrounds the heart. The space between these two membranes is the **pericardial cavity.** Like the pleural cavity, the pericardial cavity contains a small amount of fluid.

Serous membranes of the abdominal cavity are called **peritoneal membranes** (fig. 1.14). The **parietal peritoneum** (*per"ĭ-tŏ-ne'um*) attaches to the abdominal wall, and the **vis-**

peritoneum: Gk. *peritonaion,* stretched over

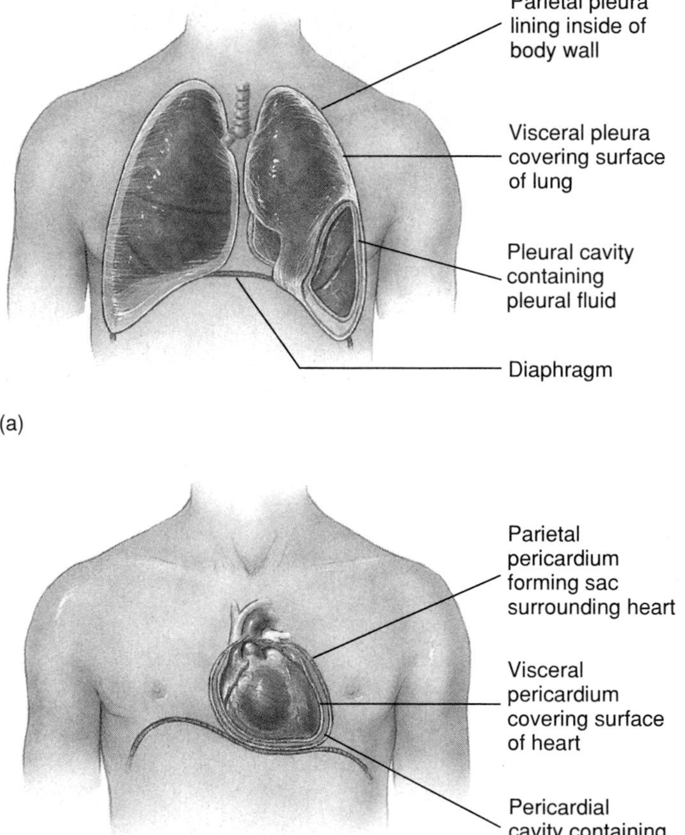

FIGURE 1.13
Serous membranes of the thorax (*a*) surrounding the lungs and (*b*) surrounding the heart.

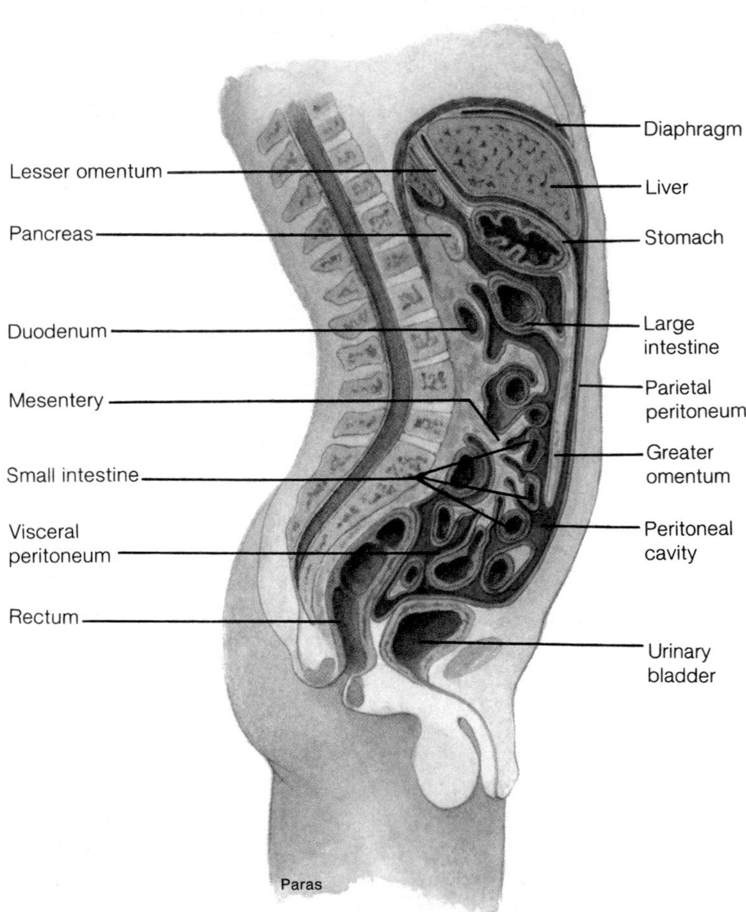

FIGURE 1.14
Visceral organs of the abdominopelvic cavity and the supporting serous membranes.

ceral peritoneum attaches to the visceral organs. The **peritoneal cavity** is the fluid-filled space within the abdominopelvic cavity between the parietal and visceral peritoneal membranes. A **mesentery** (*mes'en-ter"e*) is a fused double layer of parietal peritoneum that connects the parietal peritoneum to the visceral peritoneum. Mesenteries help to hold many of the visceral organs in place, at the same time permitting involuntary digestive movement.

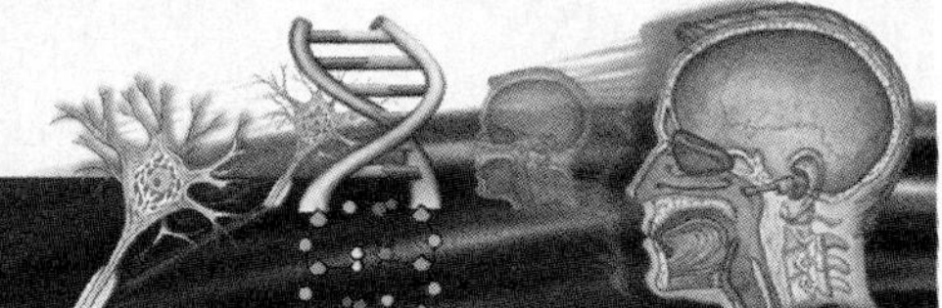

1.6 Homeostasis

Homeostasis refers to the ability of an organism to maintain stability of its internal environment by adjusting its physiological processes. Each cell, tissue, and body organ functions within a range of values suitable to support life.

Over a century ago, the French physiologist Claude Bernard observed that the *milieu intérieur* (internal environment) remains remarkably constant despite changing conditions in the external environment. In a book entitled *The Wisdom of the Body*, published in 1932, the American physiologist Walter Cannon coined the term **homeostasis** (*ho"me-o-sta'sis*) to describe this internal constancy. Cannon suggested that mechanisms of physiological regulation exist for one purpose—the maintenance of internal constancy.

The concept of homeostasis has been of immense value in the study of anatomy and physiology because it allows diverse regulatory mechanisms to be understood in terms of their "why" as well as their "how." The concept of homeostasis also provides a basis for medical diagnostic procedures. When a particular measurement of the internal environment deviates significantly from the normal range, we can conclude that homeostasis of a specific body function is not being maintained. A number of such measurements, together with clinical observations, may allow a particular defective mechanism, or a disease, to be identified.

The systems of the body are regulated by negative feedback mechanisms that operate to maintain homeostasis. Positive feedback mechanisms are rare in the human body and tend to lead away from homeostasis. We will now consider these types of feedback mechanisms in turn.

Negative Feedback Mechanisms

In order for internal constancy to be maintained, your body must have *sensors* that are able to detect deviations from a *set point*. The set point is like the temperature that is set on a house thermostat (fig. 1.15). In a similar manner, there is a set point for your body temperature, blood glucose (sugar) level, the tension on a tendon, and so on. When a sensor detects a deviation from a particular set point, it must relay this information to an *integrating center* (fig. 1.16), which usually receives information from many different sensors. The integrating center is often a particular region of the brain or spinal cord, but in some cases it can also be cells of endocrine glands. The relative strengths of different sensory inputs are weighed in the integrating center, and, in response, the integrating center either increases or decreases the activity of particular *effectors*, which are generally muscles or glands.

If, for example, your body temperature exceeds the set point of 37°C (98.6°F), sensors in a part of the brain detect this deviation and, acting via an integrating center (also in the brain), stimulate activities of effectors (including sweat glands) that lower the temperature. As another example, if your blood glucose concentration falls below

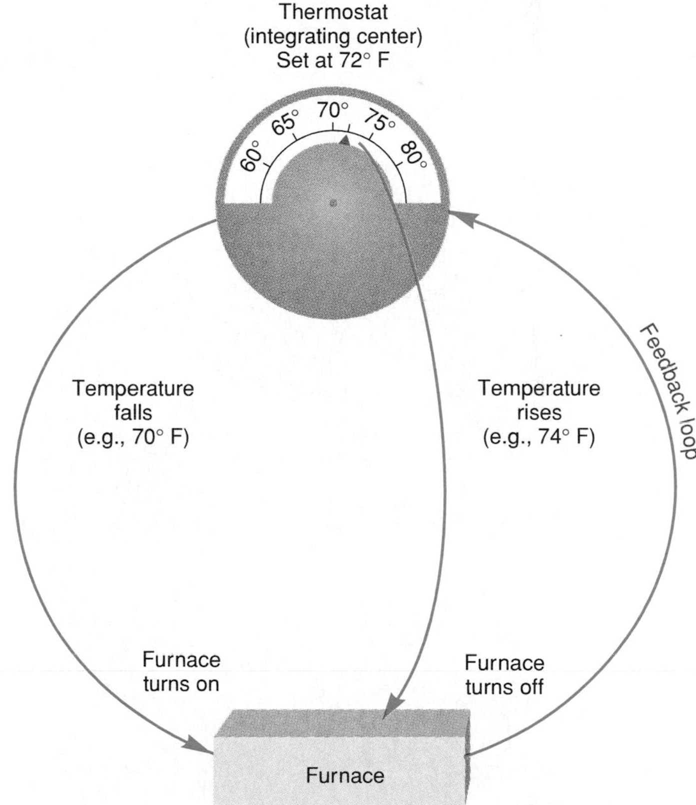

FIGURE 1.15
An example of a negative feedback mechanism. A thermostat in a house is sensitive to temperature changes. When the temperature drops, the thermostat electronically signals the furnace to start producing heat. As the room warms, the thermostat signals the furnace to shut off.

"Where Do We Fit In?"—A Classification Scheme

Zoologists have established a classification scheme to organize the structural and evolutionary relationships among living organisms. The highest taxonomic level is the kingdom, and the most specific level is the species. As humans, we are a species belonging to the kingdom Animalia.

We also belong to the phylum Chordata and subphylum Vertebrata, along with fishes, amphibians, reptiles, birds, and other mammals. All chordates have three structures in common: a *notochord* (*no'tŏ-kord*), a *dorsal hollow nerve cord*, and *pharyngeal* (*fă-rin'je-al*) *pouches* (see the figure below). These chordate characteristics are well expressed during our embryonic stage of development and persist, to a certain extent, into adulthood. The **notochord** is a flexible rod of tissue that extends along the back of an embryo. Traces of the notochord are found in the gelatinous center within each intervertebral disc of the vertebral column. The **dorsal hollow nerve cord** is positioned above the notochord and develops into the brain and spinal cord. **Pharyngeal pouches** form gill openings in fish and some amphibians. In other chordates, such as humans, embryonic pharyngeal pouches develop, but persist as chambers within the ear region.

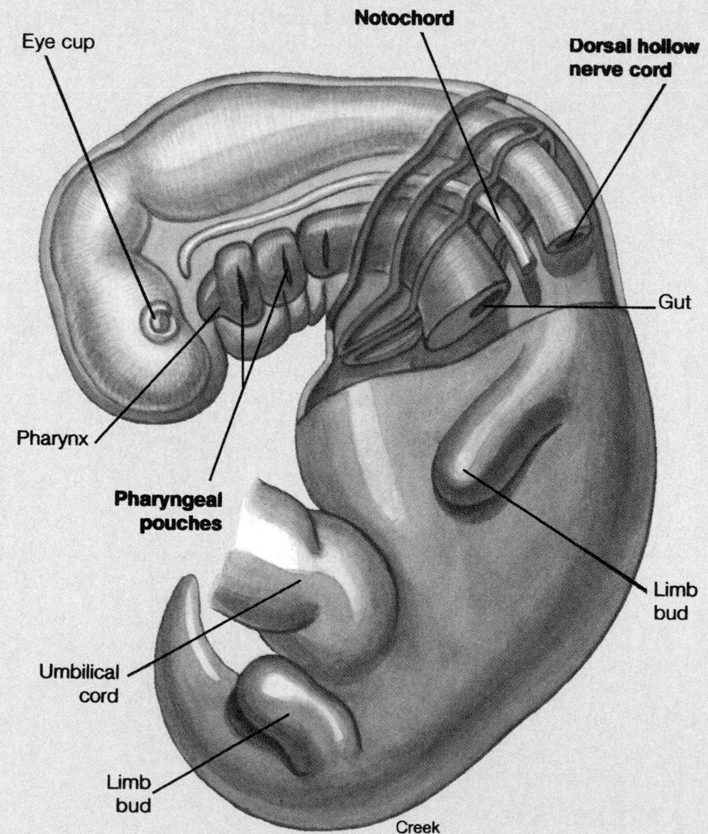

A schematic diagram of a chordate embryo. The three diagnostic chordate characteristics are indicated in bold type.

Being members of the class Mammalia, we are vertebrates with hair and mammary glands. Hair is a protective covering for most mammals, and mammary glands serve for suckling the young. Other characteristics of mammals include three small middle-ear bones, a fleshy outer ear, teeth of various shapes (such as incisors and molars), a lower jaw formed by a single bone, usually seven cervical vertebrae (neck bones), an attached placenta, well-developed facial muscles, a muscular diaphragm in the trunk, and a four-chambered heart with a left aortic arch.

The class Mammalia is subdivided into 19 groups called orders. Along with monkeys and apes, we belong to the order Primates. Members of this order have grasping hands and a relatively large, well-developed brain.

Humans are the only living members of the family Hominidae. *Homo sapiens* is included within this family, to which all the ethnic groups of humans belong. Our classification is presented in table 1.4.

A few of our anatomical characteristics are so specialized that they are diagnostic in separating us from other animals—even from other closely related mammals. Other of our characteristics are equally well developed in other animals, but when these function with the human brain, they provide us with remarkable and unique capabilities. Some of our distinguishing characteristics are as follows:

1. **A large, well-developed brain.** The adult human brain weighs about 3 pounds (1350 grams). This gives us a large brain-to-body-weight ratio. But more important, the brain has specialized structures that account for emotion, thought, reasoning, and memory.

2. **Bipedal locomotion.** Because we walk on two limbs, our style of locomotion is referred to as *bipedal.* Our upright posture imposes other structural features, such as the *sigmoid* (S-shaped) *curvature* of the spine and arched feet.
3. **An opposable thumb.** Our thumb joint is adapted for versatility in grasping. Most primates have opposable thumbs.
4. **Articulated speech.** The structure of our vocal organs and our well-developed brain account for our ability to speak.

We also differ from other mammals in the number and arrangement of our vertebrae, the kinds and number of our teeth, the extent of our facial muscle development, and in certain features of our body organs that are uniquely human.

TABLE 1.4 Classification Scheme of Human Beings

Taxon	Designated Grouping	Characteristics
Kingdom	Animalia	Eucaryotic cells that lack walls, plastids, and photosynthetic pigments
Phylum	Chordata	Dorsal hollow nerve cord; notochord; pharyngeal pouches
Subphylum	Vertebrata	Vertebral column
Class	Mammalia	Mammary glands; hair
Order	Primates	Well-developed brain; prehensile hands
Family	Hominidae	Large cerebrum; bipedal locomotion
Genus	*Homo*	Flattened face; prominent chin and nose with inferiorly positioned nostrils
Species	*sapiens*	Largest cerebrum

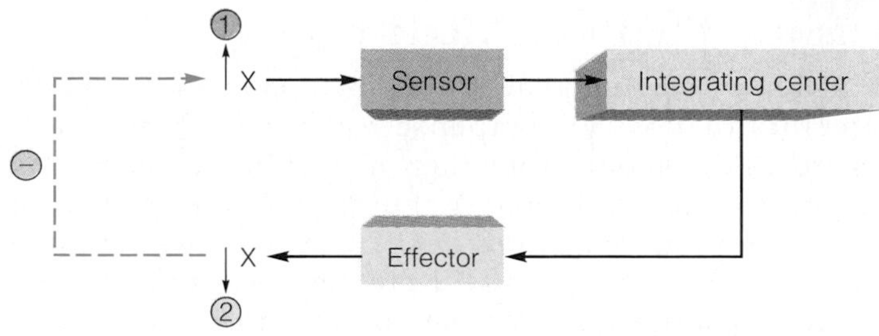

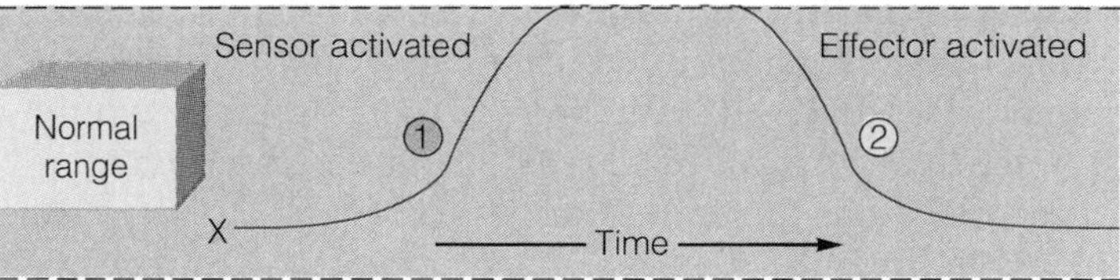

FIGURE 1.16

A rise in some factor of the internal environment (↑X) is detected by a sensor. Acting through an integrating center, this causes an effector to produce a change in the opposite direction (↓X). The initial rise is thus reversed, completing a negative feedback loop (shown by dashed arrow and negative sign). The numbers indicate the sequence of changes.

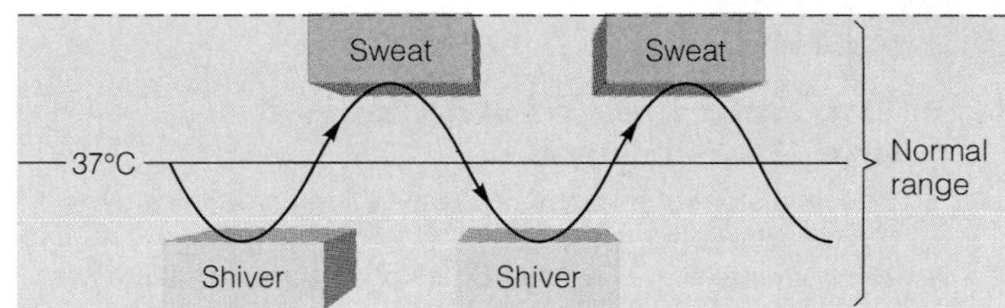

FIGURE 1.17

A simplified scheme by which body temperature is maintained within the normal range (with a set point of 37°C) by two antagonistic mechanisms—shivering and sweating. Shivering is induced when the body temperature falls too low; it gradually subsides as the temperature rises. Sweating occurs when the body temperature is too high; it diminishes as the temperature falls. Most aspects of the internal environment are regulated by the antagonistic actions of different effector mechanisms.

normal, the effectors act to increase the blood glucose. One can think of the effectors as "defending" the set points against deviations. Since the activity of the effectors is influenced by the effects they produce, and since this regulation is in a negative, or reverse, direction, this type of control system is known as a **negative feedback loop.** (Notice that in fig. 1.16 negative feedback is illustrated by a dashed line and a negative sign.)

Most factors in the internal environment are controlled by several effectors that often have *antagonistic,* or opposite, *actions.* This allows a finer degree of control than could be achieved by simply switching one effector on or off. Normal body temperature, for example, is maintained at a set point of about 37°C by the antagonistic effects of sweating, shivering, and other mechanisms (fig. 1.17).

Positive Feedback Mechanisms

Constancy of the internal environment is maintained by effectors that act to compensate for the change that served as the stimulus for their activation; in short, by negative feedback loops. A thermostat, for example, maintains a constant temperature by increasing heat production when it is cold and decreasing heat production when it is warm. The opposite occurs during **positive feedback**—in this case, the action of effectors *amplifies* those changes that stimulated the effectors. A thermostat that worked by positive feedback, for example, would increase heat production in response to a rise in temperature. Although there are a few positive feedback mechanisms in the body, it is clear that homeostasis must ultimately be maintained by negative rather than by positive feedback mechanisms.

The secretion of the hormone oxytocin during the labor of childbirth is an example of a positive feedback mechanism. As the baby is forced toward the birth canal (vagina) through uterine contractions, the increased pressure on the mother's cervix (the muscular outlet of the uterus) stimulates nerve cells in the wall of the cervix. Nerve impulses travel to the pituitary gland of the brain causing the release of oxytocin. This hormone is then transported in the blood to the uterus, where it stimulates the uterine muscle to contract even more vigorously and frequently, forcing the baby farther into the birth canal. This cyclic series of events results in increasingly powerful contractions that occur closer together. Once the baby is born, the pressure stimulus for oxytocin release is eliminated, shutting off the positive feedback mechanism.

SUMMARY

1.1 The Sciences of Anatomy and Physiology

1. The history of anatomy and physiology parallels that of medicine.
2. Hippocrates, the famous Greek physician, is regarded as the father of medicine. The Roman physician Galen was the ultimate authority on anatomy and medical treatment for nearly 1500 years. The father of anatomy was Vesalius, a Flemish physician who challenged Galen's teachings and accelerated research in anatomy during the Renaissance. The English physician, Harvey, is regarded as the father of physiology because of his work on the circulation of blood.
3. The scientific method is a disciplined approach for gaining information about the world. It includes observation, gathering data, and formulating a hypothesis that is testable and capable of being refuted by further observations.

1.2 Maintenance of Life

1. Requisites for life include organization, responsiveness, movement, growth, differentiation, metabolism, digestion, excretion, reproduction, and a survival instinct.
2. Our physical needs include water, oxygen, food, heat, pressure, and protection.

1.3 Body Organization

1. The cell is the basic structural and functional component of life. Tissues are aggregations of similar cells that perform a common function.
2. An organ is an aggregate of two or more tissues that performs specific functions.
3. A body system consists of various organs that have similar or related functions.
4. The interrelated systems of the body, working together, make up the organism.

1.4 Planes of Reference and Descriptive Terminology

1. In the anatomical position, the person is erect, facing forward, with arms at the sides and palms turned forward.
2. Directional terms are used to describe the location of one body part with respect to another part, with the body in the anatomical position.
3. The majority of anatomical and physiological terms are of Greek or Latin derivation.

1.5 Body Regions and Body Cavities

1. The body is divided into regions, which can be identified on the surface. Each region contains internal organs, the location of which are anatomically, physiologically, and clinically important.
2. For functional and protective purposes, the viscera (internal organs of the trunk) are divided into compartments and supported in specific body cavities by serous membranes.

1.6 Homeostasis

1. Homeostasis is the dynamic constancy of the internal environment. It is maintained by mechanisms that act through negative feedback loops.
2. In a negative feedback loop, an effector acts to cause changes in the internal environment that compensate for the initial deviations that are detected by sensors.
3. Positive feedback loops serve to amplify those changes that stimulated the effectors; thus, they tend to lead away from homeostasis.

Review Activities

Objective Questions

1. The publication of *De Humani Corporis Fabrica* earned this person the right to be called the father of human anatomy. He was (*module 1.1*)
   a. Galen.
   b. Harvey.
   c. Hippocrates.
   d. Vesalius.

2. Chemical interactions and reactions that occur continuously in each body cell are referred to as (*module 1.2*)
   a. metabolism.
   b. differentiation.
   c. adaptation.
   d. synthesis.

3. Which of the following is *not* one of the four principal kinds of tissue? (*module 1.3*)
   a. epithelial tissue
   b. nervous tissue
   c. connective tissue
   d. integumentary tissue
   e. muscle tissue

4. In the anatomical position, (*module 1.4*)
   a. the arms are extended away from the body.
   b. the body is horizontal, facing upward.
   c. the body is erect and the palms face anteriorly.
   d. the body is in fetal position.

5. Listening to sounds that an organ makes is (*module 1.4*)
   a. percussion.
   b. palpation.
   c. audiotation.
   d. auscultation.

6. The cubital fossa is located in (*module 1.5*)
   a. the thorax.
   b. the abdomen.
   c. the upper limb.
   d. the lower limb.

7. Regarding serous membranes, which of the following word pairs is *incorrect?* (*module 1.5*)
   a. visceral pleura/heart
   b. parietal pleura/body wall
   c. visceral peritoneum/small intestine
   d. parietal peritoneum/body wall

8. Which of the following statements about homeostasis is *true?* (*module 1.6*)
   a. The internal environment is maintained relatively constant.
   b. Positive feedback mechanisms act to correct changes in the internal environment.
   c. Homeostasis is maintained by a continuous interaction of positive and negative feedback mechanisms.
   d. All of the above are true.

9. The taxonomic scheme from specific to general is (*Synoptic Spotlight*)
   a. species, class, order, and phylum.
   b. genus, family, kingdom, and phylum.
   c. species, family, class, and kingdom.
   d. genus, phylum, class, and kingdom.

Essay Questions

1. Discuss the impact of the microscope on the advancement of anatomy, physiology, and medicine. (*module 1.1*)
2. Describe the scientific method and explain why it serves as a basis for anatomical and physiological research. (*module 1.1*)
3. Describe the levels of organization of the human body. (*module 1.3*)
4. What is meant by the anatomical position? Why is it important? (*module 1.4*)
5. Define *serous membrane* and explain how this type of membrane aids the functioning of visceral organs. (*module 1.5*)
6. Explain why an understanding of homeostasis is important in diagnosing and treating diseases. (*module 1.6*)

Labeling Exercise

Label the serous membranes indicated on the figure to the right.

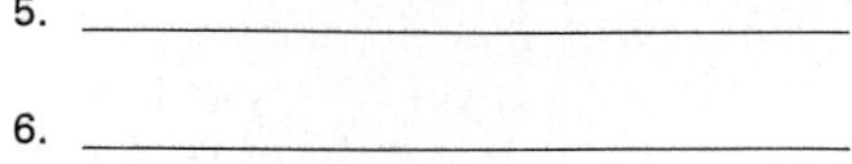

1. ______________________________
2. ______________________________
3. ______________________________
4. ______________________________
5. ______________________________
6. ______________________________

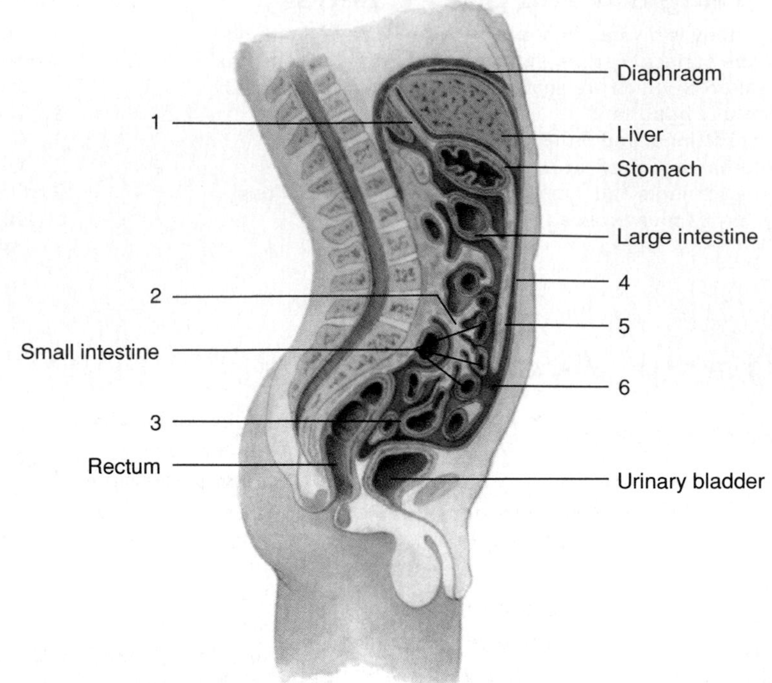

Critical Thinking Questions

1. Smooth muscle is a tissue type. Using examples, discuss the level of body organization "below" smooth muscle tissue and the level of body organization "above" smooth muscle tissue.
2. Arteries transport blood away from the heart and veins transport blood toward the heart. Using the terms *proximal, distal, medial,* and *lateral,* describe blood flow from the heart to the right thumb and back to the heart. In answering this question, keep anatomical position in mind. Also, be aware that adding an "ly" to these directional terms changes them from nouns to adjectives.
3. Using an example other than the ones described in this chapter, explain how a feedback mechanism works. Why is homeostasis a state of dynamic constancy rather than a steady state?
4. During your next physical examination, the physician may assess the condition of some of the organs and structures in your body through the techniques of palpation, percussion, and auscultation. Describe how each of these techniques may be used for a particular body organ or structure.
5. Physicians used to believe that following childbirth a woman should be confined to bed for a minimum of 5 days. Design an experiment consistent with the principles of the scientific method to determine the effect of inactivity on recovery time of postpartum women.

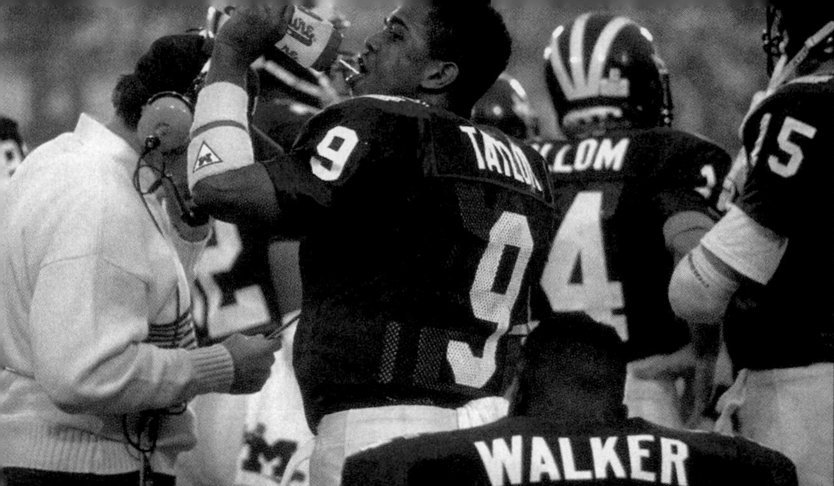

CHAPTER 2

"Water is the most abundant molecule in the body and is the solvent that transports nutrients into and wastes out of body cells." *page 26*

Chemical Composition of the Body

Chapter Outline

Terms to Remember

atom, 24
isotope (*i'sŏ-tōp*), 25
covalent bond (ko-va'lent), 25
ion (*i'on*), 26
acid, 27
base, 27
pH scale, 27
buffer, 28
carbohydrate, 29
hydrolysis (*hi-drol'ĭ-sis*), 30
lipid (*lip'id*), 31
steroid (*ster'oid*), 32
protein, 33
enzyme (*en'zīm*), 35
ATP, 37

Learning Objectives

After studying this chapter, you should be able to

- list the components of an atom and explain how they are organized.
- explain how chemical bonds are formed and state what is meant by an ionic compound.
- define *acid* and *base,* and explain the pH scale.
- state the general formula for a carbohydrate molecule and describe the subcategories of carbohydrates.
- identify the subclasses of lipids and explain why they are all classified as lipids.
- describe the structure and functions of proteins.
- describe the mechanism of enzyme action and explain why ATP is called the universal energy carrier.

2.1 Atoms and Elements

Each chemical element is composed of fundamental structures called atoms. It is the electrons in the outermost shell of an atom that participate in chemical reactions.

To understand cellular structure and function, you must first become familiar with the basic principles of chemistry. All of the normal activities that occur in the body have a chemical basis, and many diseases result from abnormal chemical processes.

Elements are the simplest chemical substances. Four elements compose over 95% of the body's mass. These elements and the percentages of body weight they account for are oxygen (O) 65%, carbon (C) 18%, hydrogen (H) 10%, and nitrogen (N) 3%. Other common elements within the body include calcium (Ca), potassium (K), sodium (Na), phosphorus (P), magnesium (Mg), and sulfur (S).

Each element is composed of yet more fundamental structures called **atoms.** The atoms of a particular element share the same characteristics but differ from the atoms of any other element. An atom consists of a central *nucleus* surrounded by a characteristic number of *electrons.* The nucleus contains two types of particles: *protons,* which are positively charged, and *neutrons,* which are uncharged. The mass of a proton is approximately equal to the mass of a neutron. The **atomic mass** of an atom is equal to the sum of the protons and neutrons in its nucleus. An atom of carbon, for example, contains six protons and six neutrons, and therefore has an atomic mass of 12 (table 2.1).

The **atomic number** of an atom is equal to its number of protons. Because carbon has six protons, it has an atomic number of 6. All atoms have a net charge of zero, however, because they have the same number of negatively charged electrons as positively charged protons. Since a carbon atom has six protons, for example, it must also have six electrons outside its nucleus (fig. 2.1).

atom: Gk. *atmos,* indivisible

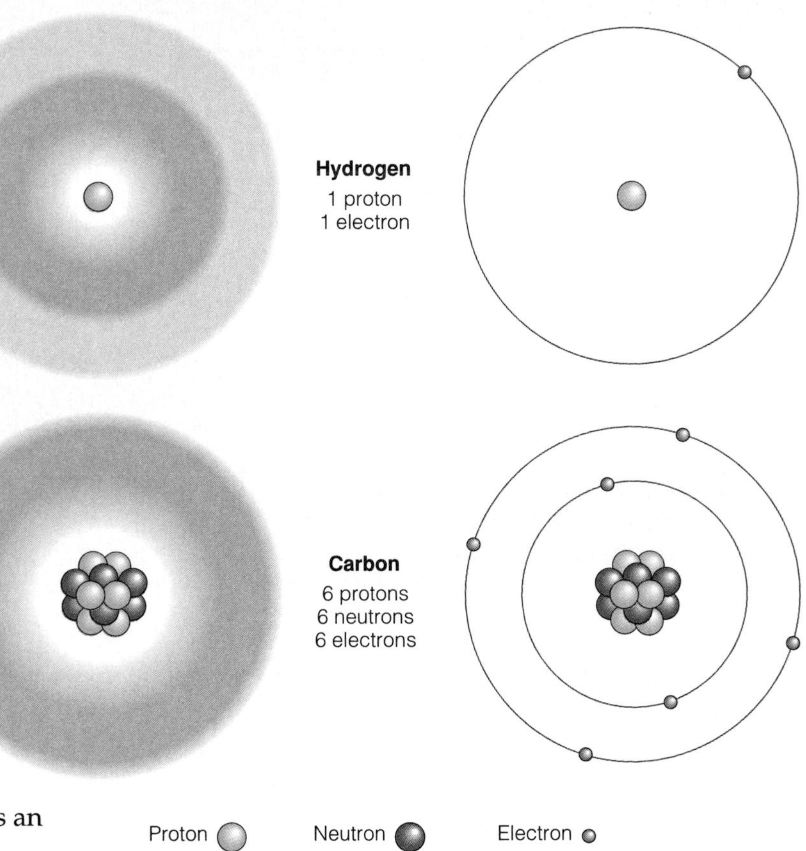

FIGURE 2.1
Diagrams of the hydrogen and carbon atoms. The electron shells on the left are represented by shaded spheres that indicate probable positions of the electrons. The shells on the right are represented by concentric circles.

TABLE 2.1 Selected Atoms of Importance in the Human Body

Atom	Symbol	Atomic Number	Atomic Mass	Orbital 1	Orbital 2	Orbital 3	Number of Chemical Bonds
Hydrogen	H	1	1	1	0	0	1
Carbon	C	6	12	2	4	0	4
Nitrogen	N	7	14	2	5	0	3
Oxygen	O	8	16	2	6	0	2
Sulfur	S	16	32	2	8	6	2

The electrons that move around the nucleus of an atom occupy volumes of space known as *electron shells.* These are often diagrammed as a series of concentric circles, or orbitals, indicating the electrons' probable location. The first shell, closest to the nucleus, can contain only two electrons. Any additional electrons must occupy other shells. The second and third shells can contain a maximum of eight electrons each, and the fourth shell can contain a maximum of 18 electrons. The shells are filled from the innermost outward. Carbon, with six electrons, has two electrons in its first shell and four electrons in its second (fig. 2.1). *Only the electrons in the outermost shell participate in chemical reactions and form chemical bonds.* These outermost electrons are known as the **valence electrons** of the atom.

Isotopes

A particular atom with a given number of protons in its nucleus may exist in several forms. These forms, called **isotopes** (*i'sŏ-tōps*), differ in their number of neutrons. Their atomic number is the same, but their atomic mass is different. The term *element* refers to all of the different isotopes of a particular atom. The element hydrogen, for example, has three isotopes. The most common isotope of hydrogen has a nucleus consisting of just one proton. Another isotope, *deuterium,* has one proton and one neutron in its nucleus; a third isotope, *tritium,* has one proton and two neutrons. Tritium is radioactive in that it emits electromagnetic radiations; it is used in many laboratory procedures.

Inside Information—Hydrogen, with its three isotopes, has the least of any element. The elements with the most isotopes (36 each) are xenon (Xe) and cesium (Cs).

2.2 Chemical Bonds

Molecules are formed when two or more atoms are joined together by interactions between the electrons of their outer electron shells.

A **molecule** is produced by the chemical bonding of atoms. *Chemical bonds* involve the valence electrons (those in the outer shell) of each atom of the molecule. These outer electrons may be shared equally between the atoms, thus producing the strongest of chemical bonds. The valence electrons also may be unequally shared, spending more time closer to one than to the other atom. The less equal the sharing, the weaker the bond. At the extreme, one or more electrons may be completely transferred from one atom to another. This produces charged atoms known as *ions,* as discussed below.

Covalent Bonds

Covalent (*ko-va'lent*) **bonds** are those formed by the sharing of electrons between atoms. The number of covalent bonds that each atom can form with other atoms is determined by the number of electrons needed to complete the outermost electron shell. Hydrogen, for example, must obtain only one more electron—and can thus form only one chemical bond—to complete the first shell of two electrons (fig. 2.2). Carbon, by contrast, must obtain four more electrons—and can thus form four chemical bonds—to complete the second shell of eight electrons (fig. 2.3, *left*).

Covalent bonds formed between identical atoms, as in oxygen gas (O_2) and hydrogen gas (H_2), are the strongest because their electrons are equally shared. Since the electrons are equally distributed between the two atoms, these molecules are said to be **nonpolar,** and they are bonded by *nonpolar covalent bonds.* When covalent bonds are formed between atoms of different elements, however, the electrons may be pulled more toward one atom

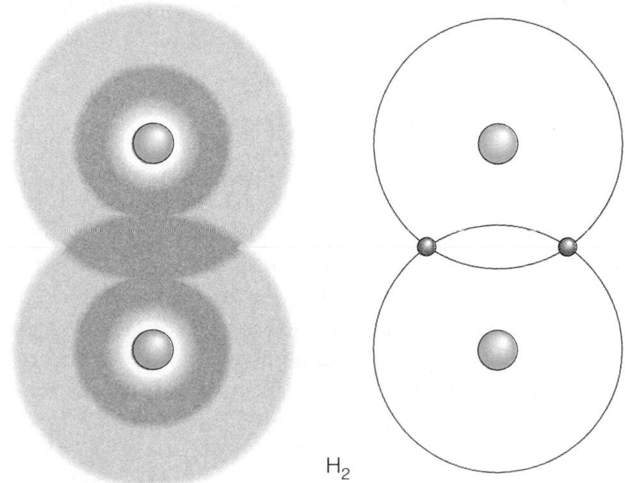

FIGURE 2.2
The hydrogen molecule, showing the covalent bonds between hydrogen atoms formed by the equal sharing of electrons.

than the other. The end of the molecule toward which the electrons are pulled is electrically negative in comparison to the other end. Such a molecule is said to be **polar** (it has a positive and negative "pole"). Atoms of oxygen, nitrogen, and phosphorus have a particularly strong tendency to pull electrons toward themselves when they bond with other atoms. These atoms thus form *polar covalent bonds* when they share electrons with other atoms.

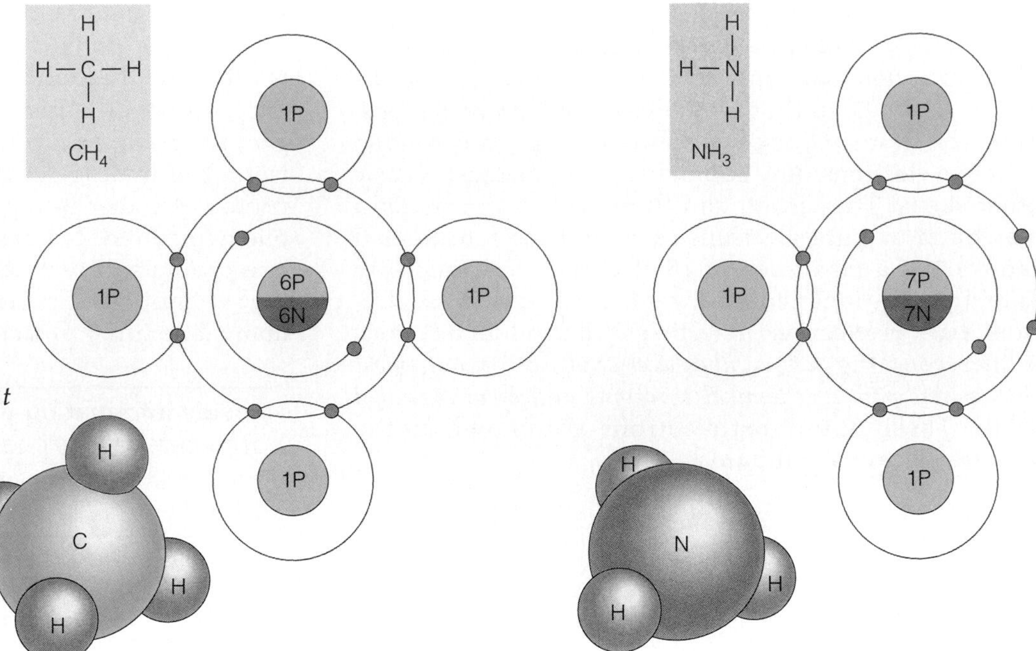

FIGURE 2.3
The molecules methane and ammonia represented in three different ways. Notice that a bond between two atoms consists of a pair of shared electrons from the outer shell of each atom.

Water

Water is the most abundant molecule in the body and is the solvent that transports nutrients into and wastes out of body cells. Water is a good solvent because it is polar; the oxygen atom pulls electrons from the two hydrogens toward its side of the water molecule, so that the oxygen side is negatively charged compared to the hydrogen side of the molecule (see fig. 2.6).

Charged atoms (ions), such as Na^+ and Cl^-, attract the polar water molecules. This allows ions, as well as polar molecules (those with polar covalent bonds), to dissolve in water. Ions and polar molecules are thus said to be **hydrophilic** (*hi"drŏ-fil'ik*); literally "water loving." Molecules that are primarily formed with nonpolar covalent bonds, such as the hydrocarbon chains of fat molecules, lack charges and cannot dissolve in water. Such molecules are said to be **hydrophobic** (*hi"drŏ-fo'bik*)—"water fearing."

Ions and Ionic Compounds

In particular cases, one or more of the outer electrons of an atom are completely transferred from that atom to another; thus, they are not shared at all. This produces an imbalance in the numbers of protons and electrons in the atom. As previously mentioned, atoms with a net charge obtained in this way are known as **ions** (*i'onz*). An atom that has lost electrons becomes a positively charged ion called a *cation;* one that has gained electrons becomes a negatively charged ion called an *anion.* The cation and anion can attract each other to form an **ionic compound**—a *compound* being a molecule composed of two or more different elements.

Common table salt, or sodium chloride (NaCl), is an example of an ionic compound. Sodium, with a total of 11 electrons, has 2 in its first electron shell, 8 in its second, and only 1 in its third shell. Chlorine, conversely, has 7 electrons in its third shell and is only one electron short of completing its outermost shell of 8 electrons. The lone electron in sodium's outer electron shell is attracted to chlorine's outer shell. This creates a chloride ion (represented as Cl^-) and a sodium ion (Na^+). Although table salt is shown as NaCl, it is actually composed of Na^+Cl^- (fig. 2.4).

Hydrogen Bonds

When hydrogen forms a polar covalent bond with an atom of oxygen or nitrogen, the hydrogen gains a slight positive charge as the electron is pulled toward the other atom. This other atom is thus described as being *electronegative.* Since this hydrogen has a slight positive charge, it will have a weak attraction for a second electronegative atom (oxygen or nitrogen) that may be located near it. This weak attraction is called a **hydrogen bond.** Hydrogen bonds are usually indicated by dotted or dashed lines in structural formulas (fig. 2.5) to distinguish them from strong covalent bonds, which are indicated by solid lines.

hydrophilic: Gk. *hydor,* water; *philos,* fond
hydrophobic: Gk. *hydor,* water; *phobos,* fear

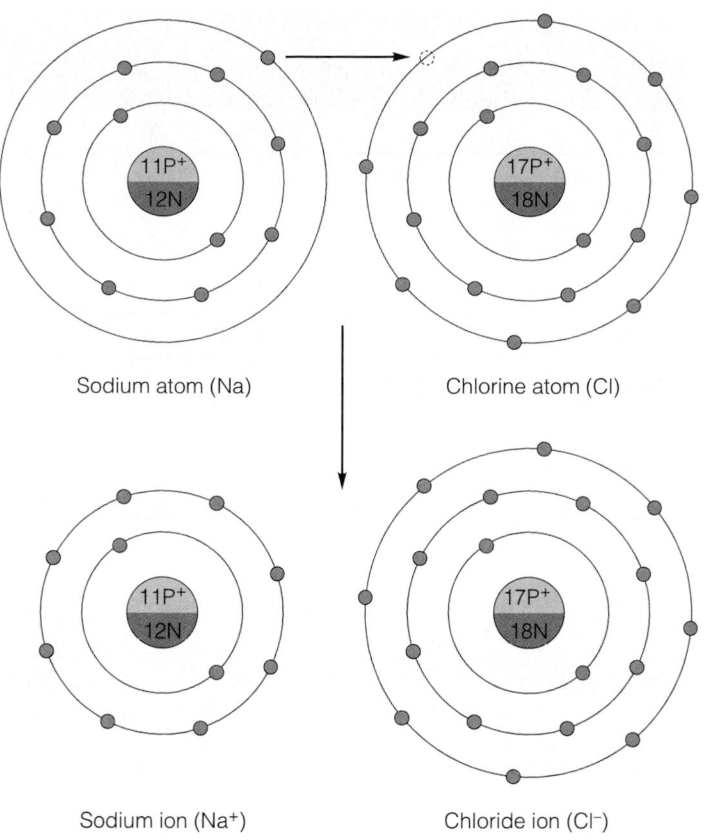

FIGURE 2.4
The dissociation of sodium and chlorine to produce sodium and chloride ions. The positive sodium ions and the negative chloride ions attract each other to produce the ionic compound sodium chloride (NaCl).

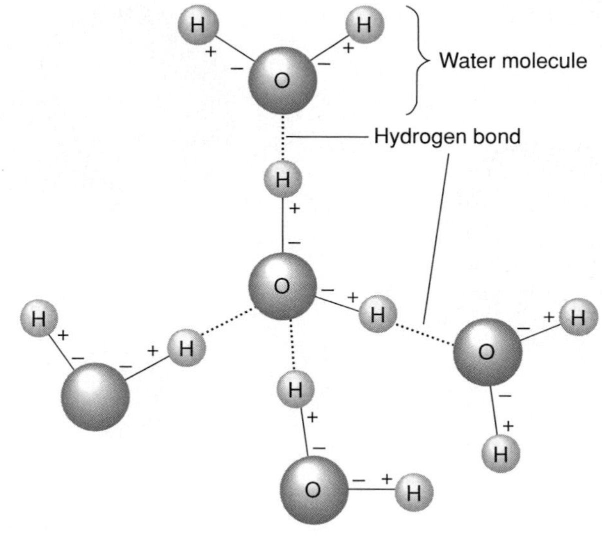

FIGURE 2.5
Water molecules are weakly joined by the attraction of the electronegative oxygen for the positively charged hydrogen atoms of adjacent water molecules. These weak bonds are called *hydrogen bonds.*

Hydrogen bonds are formed between adjacent water molecules (fig. 2.5). The hydrogen bonding between water molecules is responsible for many of the physical properties of water, including the property of its surface to behave like a stretched film (*surface tension*). Hydrogen bonds also occur within proteins, where they help to stabilize the delicate and complex structures of these large organic molecules.

2.3 Acids, Bases, and the pH Scale

Acids donate H^+ (protons) to a solution. The pH value of a solution is a measure of its H^+ concentration.

Acids and Bases

The bonds in water molecules joining hydrogen and oxygen atoms together are polar covalent bonds. Although these bonds are strong, a relatively small number will break as the electron from the hydrogen atom is completely transferred to oxygen. When this occurs, the water molecule ionizes to form a *hydroxyl ion* (OH^-) and a hydrogen ion (H^+), which is simply a free proton (fig. 2.6).

Ionization of water molecules produces equal amounts of OH^- and H^+. A solution in which the OH^- and H^+ concentrations are equal is said to be *neutral.* A solution that has a higher H^+ concentration than that of water is called *acidic;* one with a lower H^+ concentration is called *basic.* An **acid** is a molecule that can release protons (H^+) into a solution; it is a "proton donor." A **base** is a "proton acceptor." Most strong bases release OH^- into a solution; the OH^- combines with H^+ to form water, thus lowering the H^+ concentration. Examples of common acids and bases are listed in table 2.2.

pH Scale

The H^+ concentration of a solution is usually indicated in pH units on a **pH scale** that runs from 0 to 14. The pH value is equal to the negative logarithm of the H^+ concentration:

$$pH = -\log_{10}[H^+]$$

where $[H^+]$ = molar H^+ concentration. (*Molarity*, indicated by *M*, is one way of expressing the concentration of a solution.)

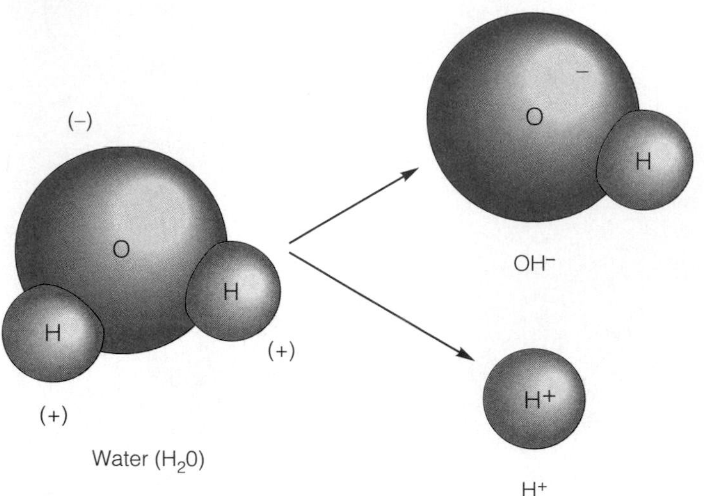

FIGURE 2.6
A model of a water molecule showing its polar nature. Notice that the oxygen side of the molecule is negative, whereas the hydrogen side is positive. Polar covalent bonds are weaker than nonpolar covalent bonds. As a result, some water molecules ionize to form a hydroxyl ion (OH^-) and a hydrogen ion (H^+). The H^+ combines with water molecules to form hydronium ions (H_3O^+) (not shown).

Pure water has a H^+ concentration of 10^{-7} *M,* and thus has a pH of 7 (neutral). Because of the logarithmic relationship, a solution with 10 times the hydrogen ion concentration (10^{-6} *M*) has a pH of 6, whereas a solution with one-tenth the H^+ concentration (10^{-8} *M*) has a pH of 8. A change of one whole number on the pH scale thus represents a tenfold change in the H^+ concentration.

Notice that the pH value is *inversely related* to the H^+ concentration. A solution with a higher H^+ concentration has a lower pH; one with a lower H^+ concentration has a higher pH. A strong acid with a high H^+ concentration of 10^{-2} *M,* for example, has a pH of 2, whereas a solution with an H^+ concentration of only 10^{-10} *M* has a pH of 10. **Acidic solutions,** therefore, have a pH of less than 7 (that of pure water), whereas **basic solutions** have a pH between 7 and 14 (table 2.3).

Buffers

The concentrations of acids and bases in our body fluids must be kept stable in order to avoid potentially serious damage to the body. A **buffer** is a system of molecules and ions that acts to prevent drastic changes in H^+ concentration and thus serves to stabilize the pH of a solution. In blood plasma, for example, the pH is stabilized by the following reversible reaction involving the bicarbonate ion (HCO_3^-) and carbonic acid (H_2CO_3):

$$HCO_3^- + H^+ \rightleftharpoons H_2CO_3$$

The double arrows indicate that the reaction could go either to the right or to the left; the net direction depends on the concentration of molecules and ions on each side. If an acid (such as lactic acid) should release more H^+ into the solution, for example, the increased concentration of H^+ would drive the equilibrium to the right. Under the opposite condition, when the concentration of free H^+ in the solution was falling, the reaction would be reversed and additional H^+ would be released.

TABLE 2.2 Common Acids and Bases

Acid	Symbol
Hydrochloric acid	HCl
Phosphoric acid	H_3PO_4
Nitric acid	HNO_3
Sulfuric acid	H_2SO_4
Carbonic acid	H_2CO_3
Base	**Symbol**
Sodium hydroxide	NaOH
Potassium hydroxide	KOH
Calcium hydroxide	$Ca(OH)_2$
Ammonium hydroxide	NH_4OH

TABLE 2.3 The pH Scale

	H^+ Concentration (*molar*)	pH	OH^- Concentration (*molar*)
Acids	1.0	0	10^{-14}
	0.1	1	10^{-13}
	0.01	2	10^{-12}
	0.001	3	10^{-11}
	0.0001	4	10^{-10}
	10^{-5}	5	10^{-9}
	10^{-6}	6	10^{-8}
Neutral	10^{-7}	7	10^{-7}
Bases	10^{-8}	8	10^{-6}
	10^{-9}	9	10^{-5}
	10^{-10}	10	0.0001
	10^{-11}	11	0.001
	10^{-12}	12	0.01
	10^{-13}	13	0.1
	10^{-14}	14	1.0

Inside Information—Arterial blood has a normal pH of 7.35 to 7.45. If the arterial blood pH falls below 7.35, the condition is called *acidosis.* An increase in blood pH above 7.45 is known as *alkalosis.* Normal urine pH averages about 6.0, and saliva has a pH between 6.0 and 7.4. The pH of gastric juice is less than 2.0.

2.4 Carbohydrates

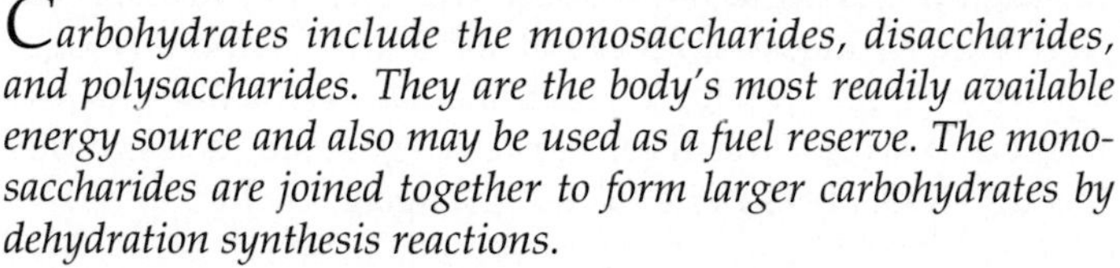

Carbohydrates include the monosaccharides, disaccharides, and polysaccharides. They are the body's most readily available energy source and also may be used as a fuel reserve. The monosaccharides are joined together to form larger carbohydrates by dehydration synthesis reactions.

Molecules may be **organic** or **inorganic.** Organic molecules are those that contain carbon, hydrogen, and oxygen atoms. All other molecules are considered to be inorganic. Note that although the term *organic* means "life," organic molecules are not necessarily the product of living organisms. Some are found in meteorites, and many are manufactured in chemistry laboratories.

Carbohydrates are organic molecules that contain carbon, hydrogen, and oxygen in the ratio described by their name—*carbo* (carbon) and *hydrate* (water, H_2O). The general formula for a carbohydrate molecule is thus CH_2O; the molecule contains twice the number of hydrogen atoms as it contains carbon or oxygen atoms.

Monosaccharides, Disaccharides, and Polysaccharides

Carbohydrates include **monosaccharides** (*mon"ŏ-sak'ă-rīdz*), or simple sugars, and longer molecules that contain a number of monosaccharides joined together. The suffix *-ose* denotes a sugar molecule; *hexose,* for example, refers to a six-carbon monosaccharide with the formula $C_6H_{12}O_6$. This formula is adequate for some purposes, but it does not distinguish between related hexose sugars that are *structural isomers* of each other. The structural isomers glucose, galactose, and fructose, for example, are monosaccharides that have the same ratio of atoms arranged in slightly different ways (fig. 2.7).

Two monosaccharides can be joined covalently to form a **disaccharide,** or double sugar. Common disaccharides include table sugar, or *sucrose* (composed of glucose and fructose); milk sugar, or *lactose* (composed of glucose and galactose); and malt sugar, or *maltose* (composed of two glucose molecules).

When many monosaccharides are joined together in long chains, the resulting molecule is called a **polysaccharide.** *Starch,* for example, is a polysaccharide found in many plants, and is formed by the bonding together of thousands of glucose subunits. **Glycogen** is animal starch that is found in the liver and muscles. As a polysaccharide, glycogen consists of repeating glucose molecules. Glycogen differs from plant starch, however, in that it is more highly branched (fig. 2.8).

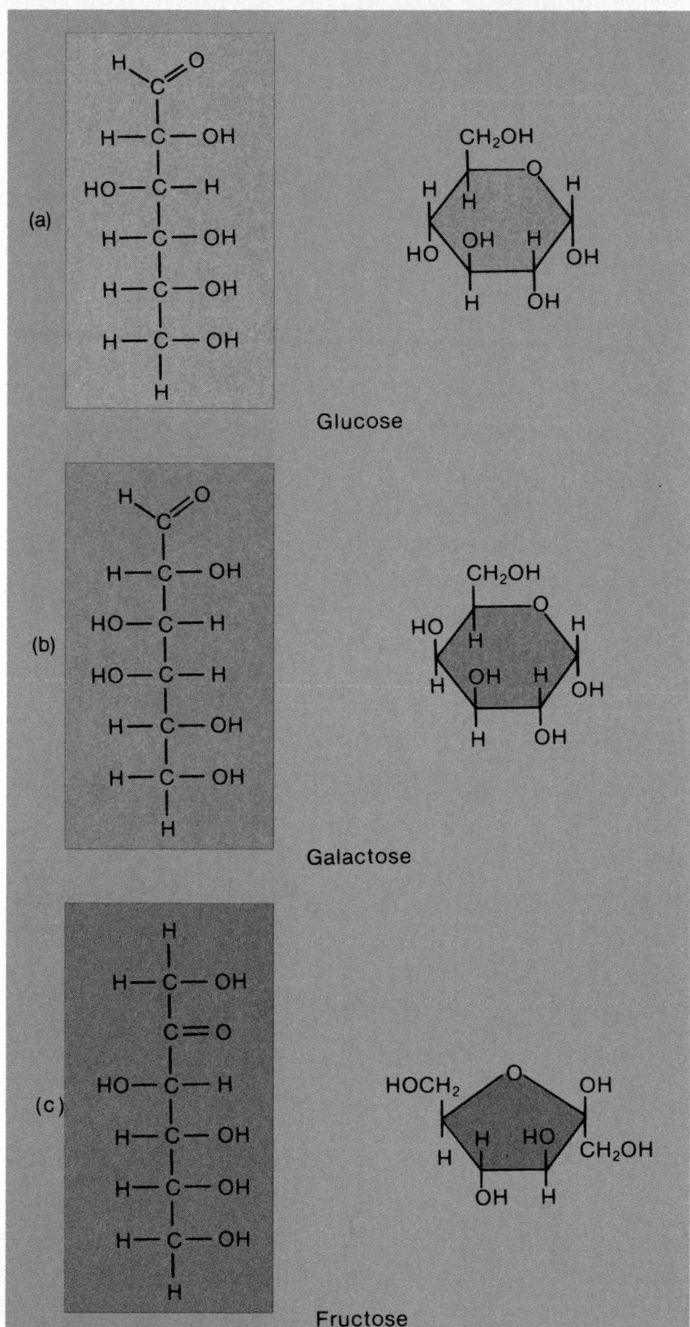

FIGURE 2.7
The structural formulas of three hexose (six-carbon) sugars: (*a*) glucose, (*b*) galactose, and (*c*) fructose. All three have the same ratio of atoms—$C_6H_{12}O_6$—and therefore are isomers of each other.

monosaccharide: Gk. *monos,* single; *sakcharon,* sugar

Dehydration Synthesis and Hydrolysis

In the formation of disaccharides and polysaccharides, the separate subunits (monosaccharides) are bonded covalently by a type of reaction called **dehydration synthesis,** or **condensation.** This reaction requires the participation of specific enzymes, or biological catalysts (discussed in module 2.7). In dehydration synthesis, a hydrogen is removed from one monosaccharide and a hydroxyl group (OH) is removed from another. As a covalent bond is formed between the two monosaccharides, water (H_2O) is produced. A dehydration synthesis reaction is illustrated in figure 2.9*a*.

When you consume disaccharides and polysaccharides, or when the stored glycogen in your liver and muscles is to be used by tissue cells, the covalent bonds joining the monosaccharide subunits must be broken. These *digestion reactions* occur by means of **hydrolysis** (*hi-drol'ĭ-sis*). This is the reverse of dehydration synthesis—in hydrolysis, a water molecule is split. The resulting hydrogen atom is added to one of the free glucose molecules and the hydroxyl group is added to the other (fig. 2.9*b*).

Hydrolysis reactions occur when food is digested, and dehydration synthesis reactions occur when glycogen and other energy-storage molecules are produced in the body tissues. When you eat a potato, for example, the starch within it is hydrolyzed into separate glucose molecules within the small intestine. This glucose is absorbed into the blood and carried to the tissues. The liver and muscles can store excess glucose from the blood in the form of glycogen by dehydration synthesis reactions. During fasting or prolonged exercise, the liver can add glucose to the blood through hydrolysis of its stored glycogen.

hydrolysis: Gk. *hydor,* water; *lysis,* break

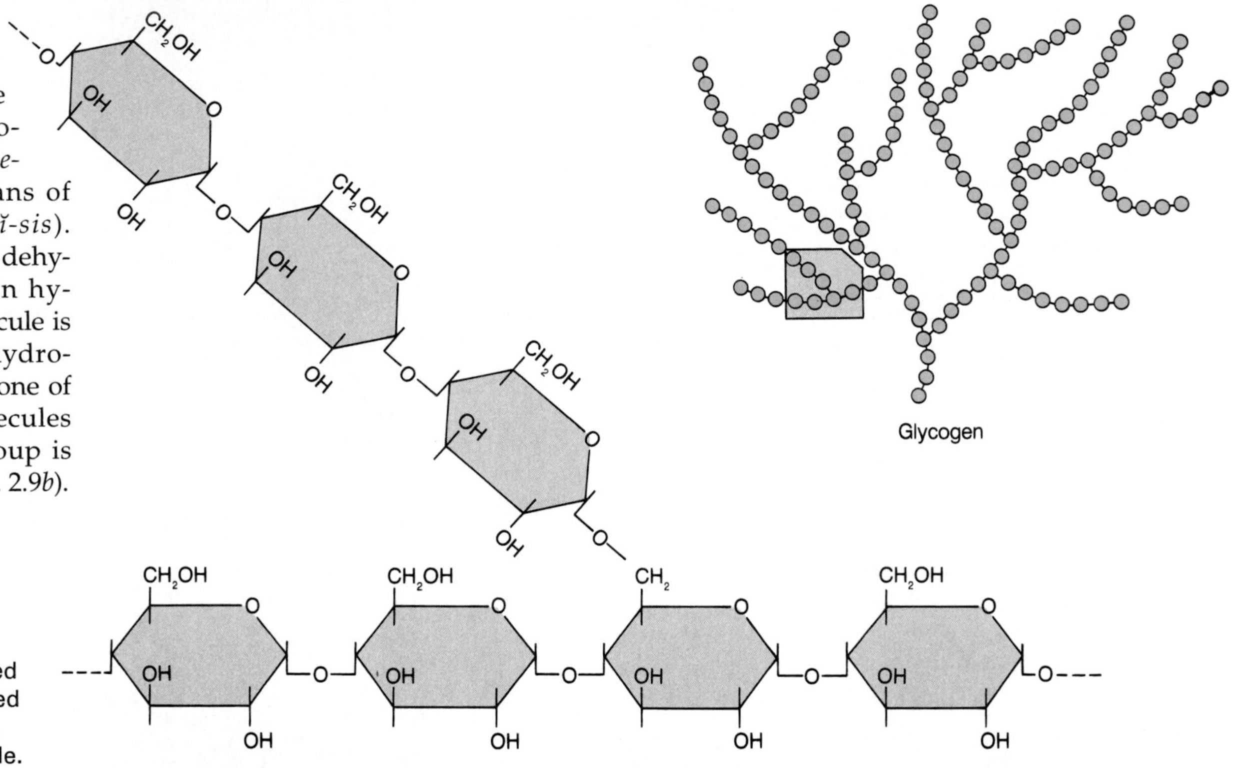

FIGURE 2.8
Glycogen is a polysaccharide composed of glucose subunits joined together to form a large, highly branched molecule.

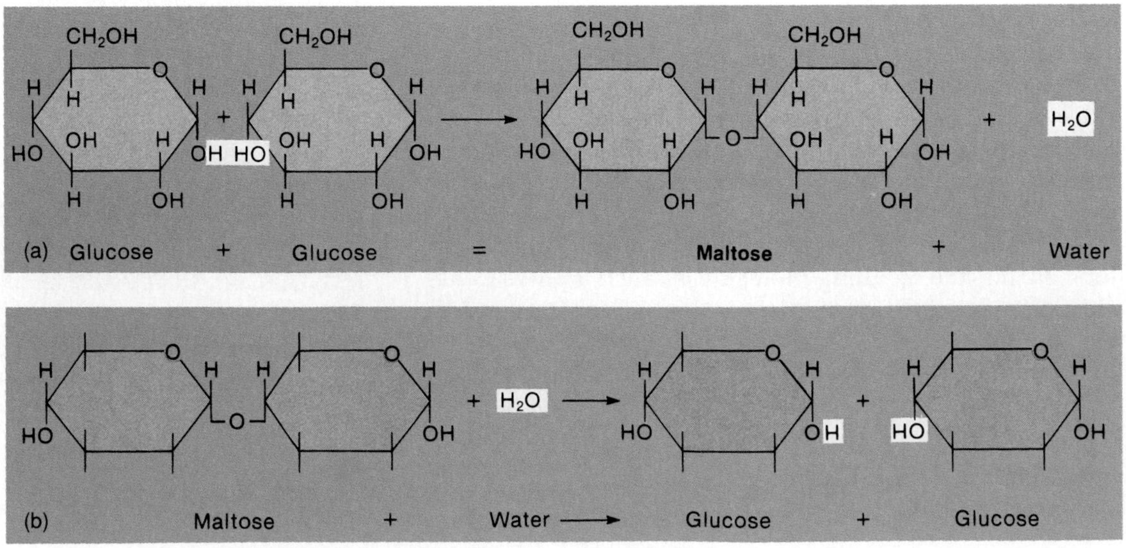

FIGURE 2.9
Dehydration synthesis (*a*) of two glucose molecules produces a molecule of maltose. Hydrolysis (*b*) of maltose yields two free glucose molecules.

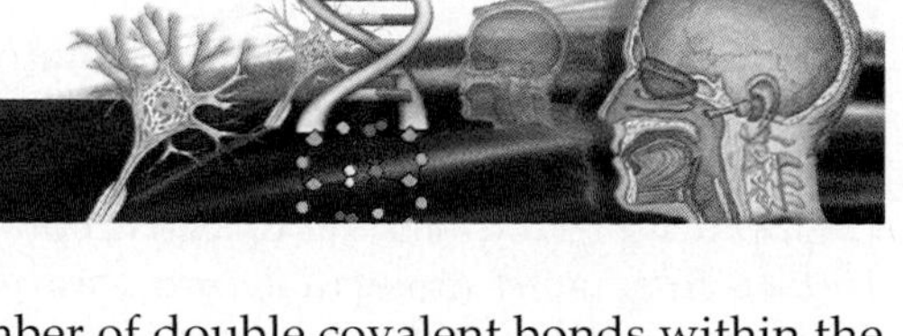

2.5 Lipids

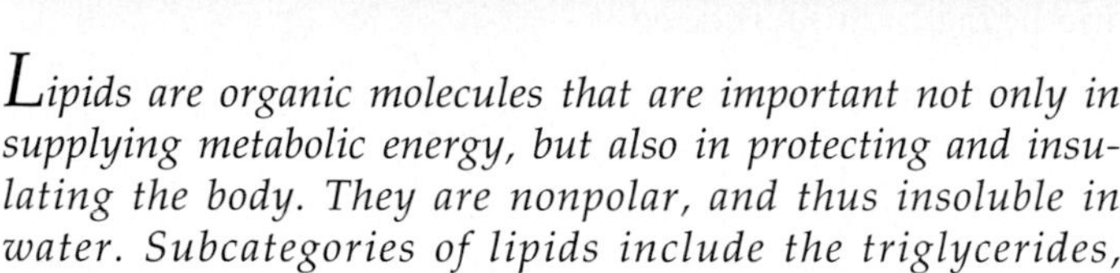

Lipids are organic molecules that are important not only in supplying metabolic energy, but also in protecting and insulating the body. They are nonpolar, and thus insoluble in water. Subcategories of lipids include the triglycerides, steroids, and phospholipids.

The diverse subcategories of **lipids** (*lip'idz*) have a physical property in common—they are all *insoluble in water.* This is because lipids consist primarily of hydrocarbon chains and rings, which are nonpolar, and thus hydrophobic.

Triglycerides

Triglycerides are a subcategory of lipids that includes fats and oils. These molecules are formed by the condensation of one molecule of *glycerol* (a three-carbon alcohol) with three molecules of *fatty acids.* Each fatty acid molecule (fig. 2.10) consists of a nonpolar hydrocarbon chain with a carboxylic acid group on one end. If the carbon atoms within the hydrocarbon chain are joined by single covalent bonds so that each carbon atom can also bond with two hydrogen atoms, the fatty acid is said to be *saturated.* If there are a number of double covalent bonds within the hydrocarbon chain, so that each carbon atom can bond with only one hydrogen atom, the fatty acid is said to be *unsaturated.* Within the adipose cells of the body, triglycerides are formed as the carboxylic acid ends of fatty acid molecules condense with the hydroxyl groups of a glycerol molecule (fig. 2.10).

The saturated fat content (expressed as a percentage of total fat) for selected food items is as follows: canola oil (6%), olive oil (14%), margarine (17%), chicken fat (31%), palm oil (51%), beef fat (52%), butter fat (66%), and coconut oil (77%). People with high cholesterol levels are encouraged to reduce their intake of saturated fats, since this type of dietary fat may contribute to the development of atherosclerosis, *a disease of the blood vessels. In general, health authorities recommend (1) that total fat intake not exceed 30% of the total energy intake per day and (2) that saturated fat contribute less than 10% of total fat intake for good cardiovascular health.*

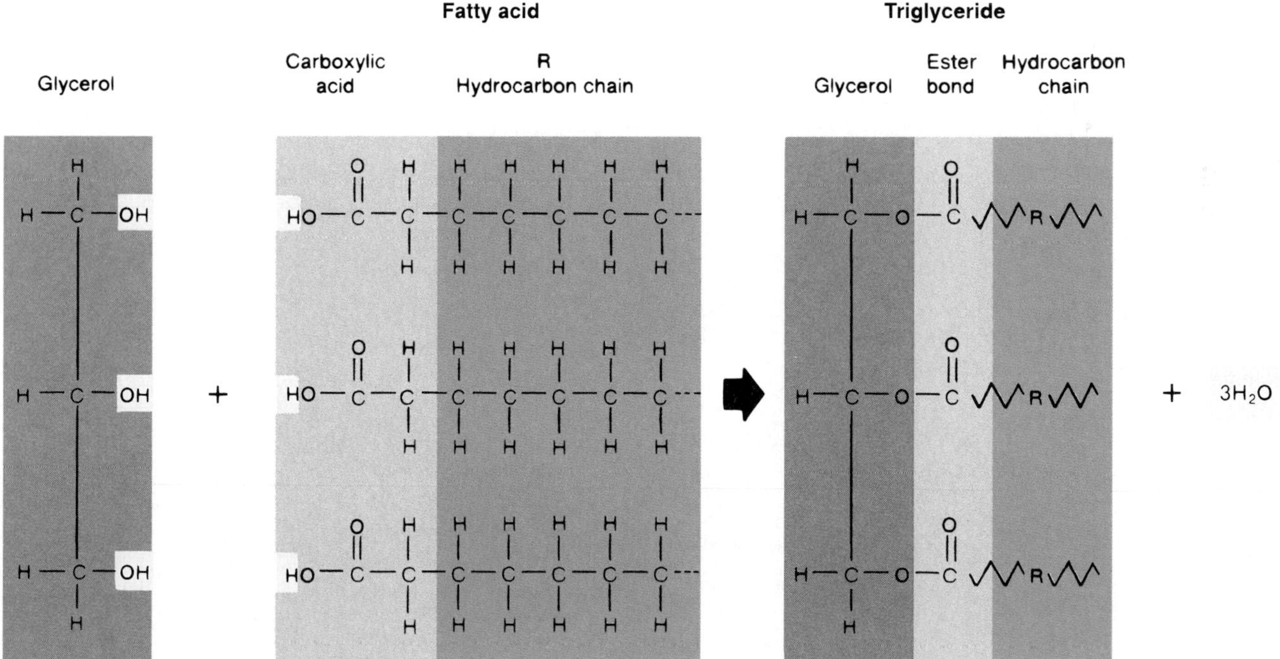

FIGURE 2.10
Dehydration synthesis of a triglyceride molecule from a glycerol and three fatty acids. Sawtooth lines represent carbon chains which are symbolized by an R.

Phospholipids

The class of lipids known as **phospholipid** molecules includes a number of categories of lipids, all of which contain a phosphate group. The most common type of phospholipid molecule has the three-carbon glycerol attached to two fatty acid molecules and to a phosphate group. This phosphate group is, in turn, bonded to another molecule. If the phosphate group is attached to a nitrogen-containing *choline* molecule, the phospholipid is known as **lecithin** (fig. 2.11). Figure 2.11 illustrates a simple way of showing the structure of a phospholipid—the parts of the molecule capable of ionizing (and thus becoming polar) are shown as a circle, whereas the nonpolar hydrocarbon chains are represented by lines.

Because of their dual nature (part polar, part nonpolar), phospholipid molecules have important physiological functions. Besides forming the major component of the cell membrane, they also decrease the surface tension of water. This latter function—as *surfactants* (surface-active agents)—is important in preventing the collapse of the lungs (see chapter 18, module 18.6).

Steroids

Steroid (*ster'oid*) molecules are quite different from triglyceride or phospholipid molecules structurally. Yet steroids are still included in the lipid category of molecules because they are nonpolar and insoluble in water. All steroid molecules have the same basic structure; three six-carbon rings are joined to one five-carbon ring (fig. 2.12). Different kinds of steroids, however, have different functional groups attached to this basic structure. They also vary in the number and position of the double covalent bonds between the carbon atoms in the rings.

Cholesterol molecules are important in the body because they are the precursors (parent molecules) for steroid hormones produced by the testes, ovaries, and adrenal cortex. The testes and ovaries (collectively called the *gonads*) secrete **sex steroids,** which include estradiol and progesterone from the ovaries and testosterone from the testes. The adrenal cortex secretes the **corticosteroids,** including hydrocortisone.

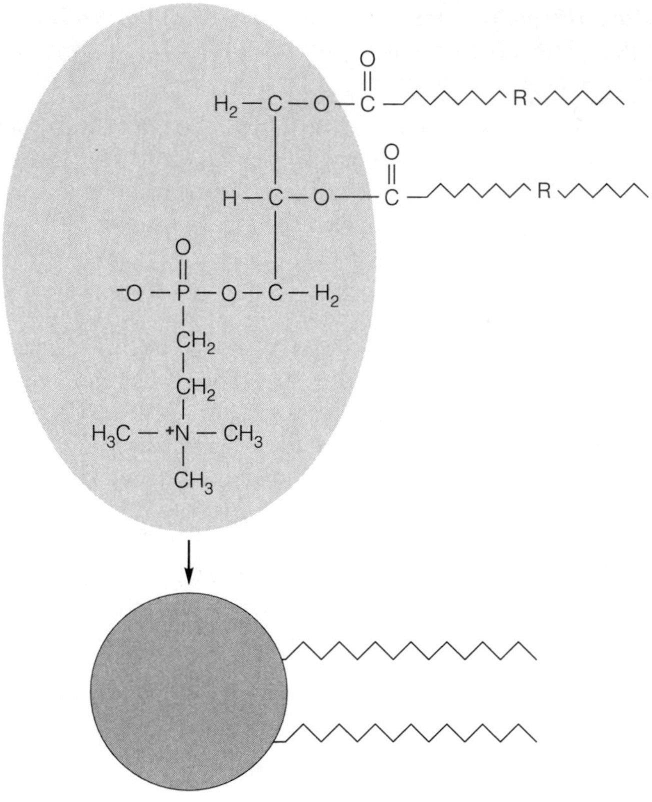

FIGURE 2.11
The structure of lecithin, a typical phospholipid (*top*) and its more simplified representation (*bottom*).

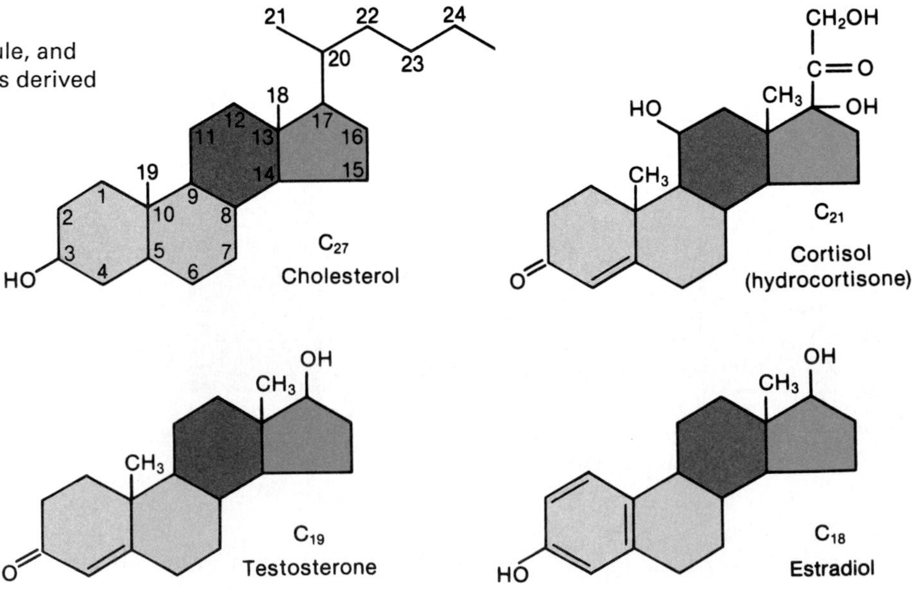

FIGURE 2.12
The cholesterol molecule, and some steroid hormones derived from cholesterol.

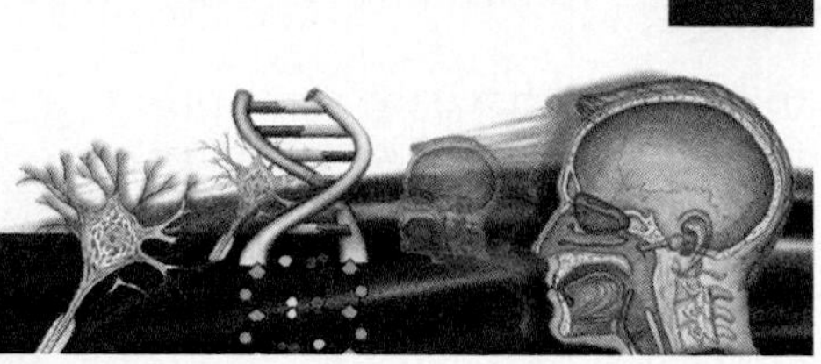

2.6 Proteins

Proteins are large molecules composed of amino acid subunits. There are 20 different types of amino acids that can contribute to a given protein. This variety allows each type of protein to be constructed to function in very specific ways.

The enormous diversity of protein structure results from the fact that 20 different building blocks—the **amino** (*ă-me'no*) **acids**—can be used to form a protein. The specific sequence of amino acids in a protein, and thus the specific functional structure of the protein, is determined by genetic information. This genetic information for protein synthesis is contained in another category of organic molecules—the *nucleic acids.* The nucleic acids are described in chapter 3 (see module 3.11).

Structure of Proteins

Proteins are very large molecules, or *macromolecules,* composed of *amino acids.* As the name implies, each amino acid contains an *amino group* (NH_2) on one end of the molecule. A *carboxyl group* (COOH) is on the other end of the molecule. Each of the 20 different amino acids, however, has a unique structure and chemical properties. These differences are due to differences in the *functional groups* of these amino acids. The abbreviation for the functional group in the general formula for an amino acid is "R" (fig. 2.13). The R symbol actually stands for the word *residue,* but can be more easily remembered as indicating the "*rest* of the molecule."

When amino acids are joined together by dehydration synthesis, the hydrogen from the amino end of one amino acid combines with the hydroxyl group of the carboxylic acid end of another amino acid to produce water. A covalent bond is formed between the two amino acids as the water is split off (fig. 2.14). The bond formed in this way between adjacent amino acids is called a **peptide bond.** When many amino acids are joined by many peptide bonds, a chain of amino acids, a **polypeptide,** is produced. When the length of a polypeptide chain becomes very long (greater than about 100 amino acids), the molecule is called a **protein.**

The structure of a protein can be described at four different levels. At the first level, the sequence of amino acids in the protein is described. This is the **primary structure** of the protein. Each different type of protein has a different primary structure, which is coded by a different gene. Each gene, however, makes billions of *copies* of its protein, and all of

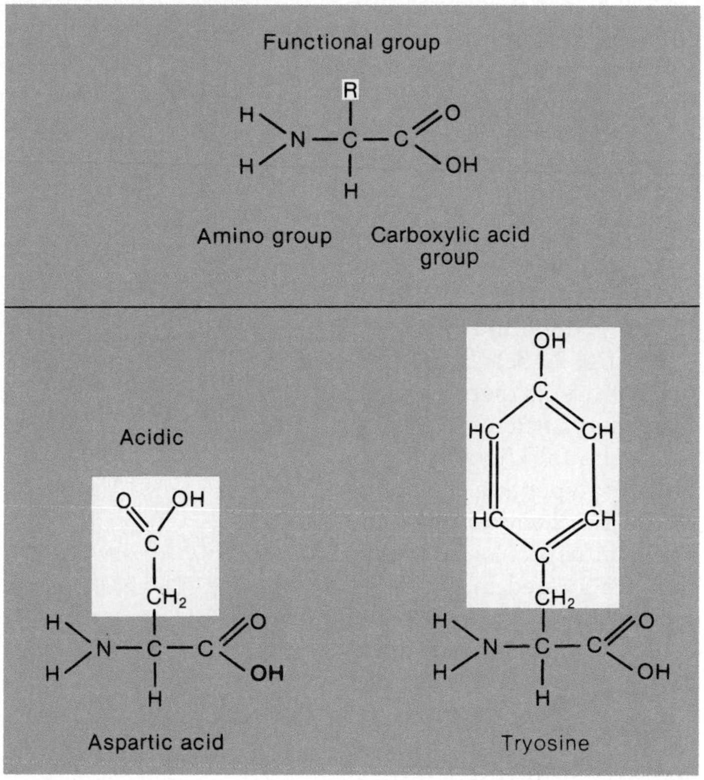

FIGURE 2.13
Representative amino acids, showing different types of functional (R) groups.

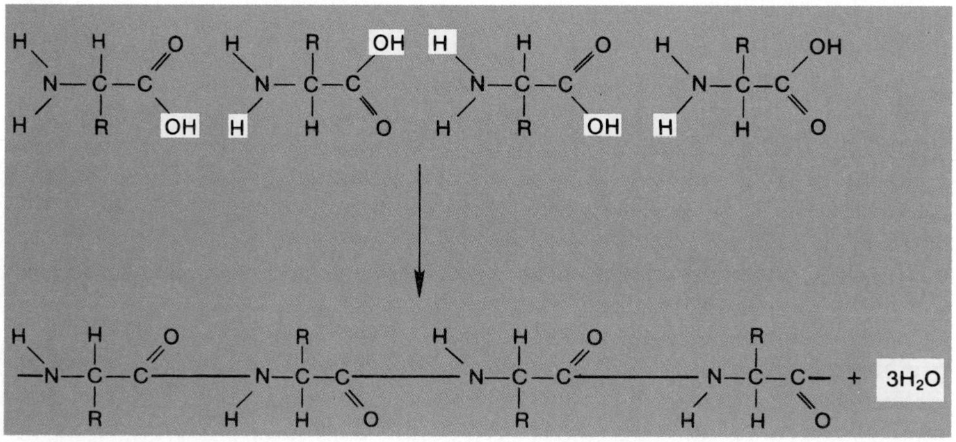

FIGURE 2.14
The formation of peptide bonds by dehydration synthesis reactions between amino acids.

those copies have the same primary structure. The primary structure of a protein is illustrated in figure 2.15*a*.

Weak interactions (such as hydrogen bonds) between amino acids in neighboring positions in the polypeptide chain cause the chain to twist into a *helix*. The extent and location of the helical structure is different for each protein because of differences in amino acid composition. A description of the helical structure of a protein is termed its **secondary structure** (fig. 2.15*b*).

Most polypeptide chains bend and fold upon themselves to produce complex three-dimensional shapes, called the **tertiary structure** of the proteins (fig. 2.15*c*). Each type of protein has its own characteristic tertiary structure. This is because the folding and bending of the polypeptide chain is produced by chemical interactions between particular amino acids that are located in different regions of the chain. Some proteins (such as hemoglobin and insulin) are composed of a number of polypeptide chains covalently bonded together. This is the **quaternary structure** of these proteins.

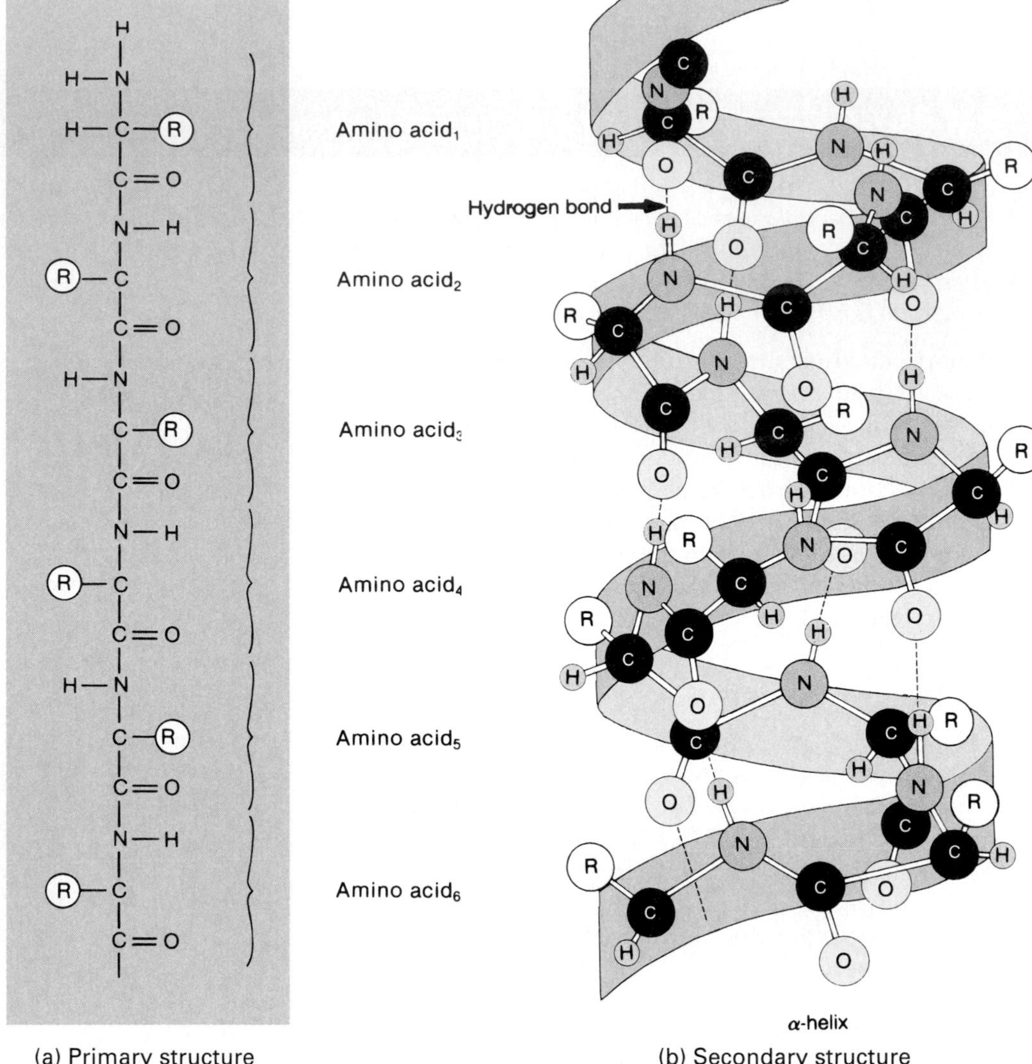

(a) Primary structure
(b) Secondary structure

(c) Tertiary structure

FIGURE 2.15
Three levels of protein structure.

Functions of Proteins

Because of their tremendous diversity of structure, proteins can serve a wider variety of functions than any other type of molecule in the body. The roles they assume are both passive and active. Many proteins, for example, contribute significantly to the structure of different tissues; in this way, they assume a passive role in the functions of these tissues. Examples of **structural proteins** include *collagen* in the connective tissues and *keratin* in the epidermis of the skin. Other proteins assume a more active role in the body, exerting some form of control of metabolism. *Enzymes* and *antibodies* are good examples of **functional proteins.** So too are proteins in cell membranes that serve as *receptors* for specific regulatory molecules (such as hormones) and as *carriers* that transport specific molecules across the membrane.

2.7 Enzymes and ATP

Cellular metabolism depends on the action of enzymes, which are biological catalysts. Metabolic energy is captured in the bonds of ATP.

Function of Catalysts

Most of the chemical reactions within a cell depend upon the action of *enzymes* (*en'zīmz*). Biochemists have demonstrated that enzymes are proteins that act as *biological catalysts.* A **catalyst** is a chemical that increases the rate of a reaction without changing the nature of the reaction or its final result, and without itself being altered as a result of the reaction. Catalysts speed up a chemical reaction; the same reaction would occur to the same degree in the absence of the catalyst, but it would progress at a much slower rate.

In order for a given reaction to occur, the reactants must have sufficient energy. The amount of energy required for a reaction to proceed is called the **activation energy.** By analogy, a match will not burn and release heat energy unless it is first "activated" by the heat of friction while being struck, or by the heat of a flame.

Catalysts make a reaction go faster at lower temperatures by *lowering the activation energy* required for the reaction. A lowered activation energy means that a larger percentage of the population of reactant molecules will have sufficient energy to participate in the reaction (fig. 2.16).

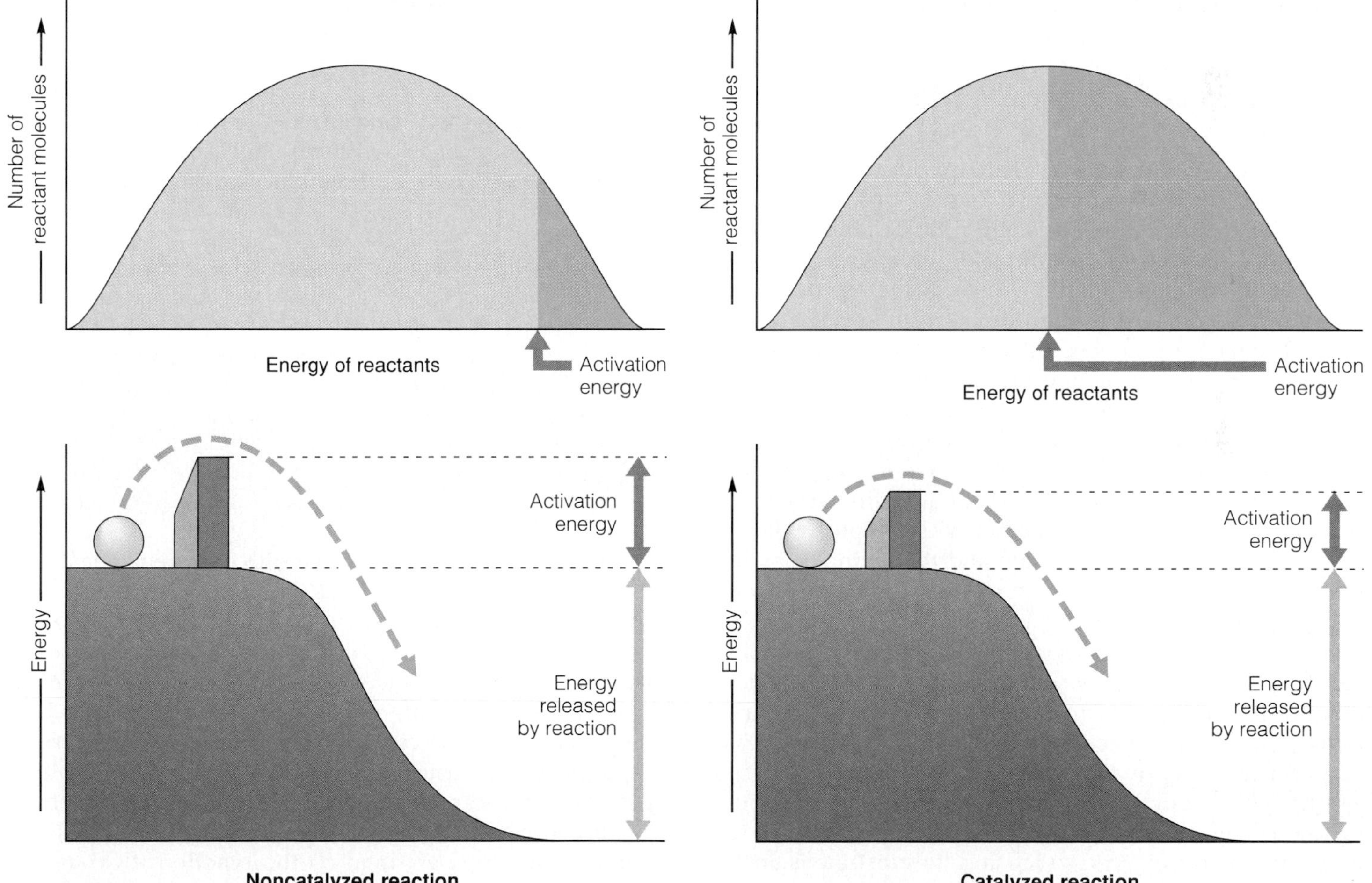

FIGURE 2.16

A comparison of a noncatalyzed reaction with a catalyzed reaction. The upper figures compare the proportion of reactant molecules that have sufficient activation energy to participate in the reaction (green). This proportion is increased in the enzyme-catalyzed reaction because enzymes lower the activation energy required for the reaction (shown as a barrier on top of an energy "hill" in the lower figures). Reactants that can overcome this barrier are able to participate in the reaction, as shown by arrows pointing to the bottom of the energy hill.

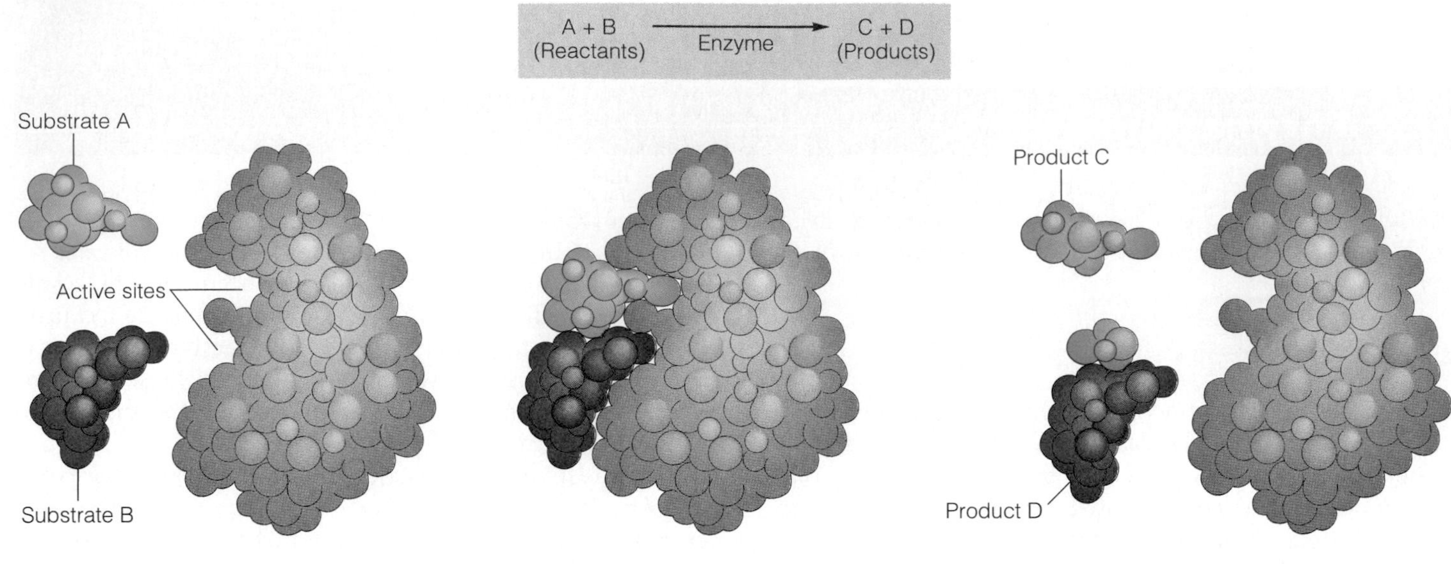

FIGURE 2.17

The lock-and-key model of enzyme action. (*a*) Substrates A and B fit into active sites in the enzyme, forming an enzyme-substrate complex (*b*). This then dissociates (*c*), releasing the products of the reaction and the free enzyme.

Mechanism of Enzyme Action

The ability of enzymes to lower the activation energy for a reaction derives from their structure. Enzymes are very large proteins with complex, highly ordered, three-dimensional shapes produced by physical and chemical interactions between their amino acids. Each type of enzyme has a characteristic shape, or *conformation,* with ridges, grooves, and pockets that are lined with specific amino acids. The particular pockets that are active in catalyzing a reaction are called the **active sites** of the enzyme.

The reactant molecules, called the **substrates** of the enzyme, have shapes that allow them to fit into the active sites. The fit may not be perfect at first, but a perfect fit may be induced as a substrate gradually slips into an active site. This induced fit, together with temporary bonds that form between the substrate and the amino acids lining the active sites of the enzyme, weakens the existing bonds within the substrate molecules so that they may be more easily broken. New bonds are more easily formed as substrates are brought close together in the proper orientation. The *enzyme-substrate complex,* formed temporarily in the course of the reaction, then dissociates to yield *products* and the free, unaltered enzyme (which can be used again). This model of enzyme activity is known as the **lock-and-key model** (fig. 2.17) because the nature of the interaction between an enzyme and its substrate is analogous to the specificity of a key for a lock.

Notice in figure 2.17 that the term *enzyme* is written below the arrow. This is because an enzyme is neither a reactant nor a product of the reaction. Most enzyme names describe their function and end in the suffix *-ase.* The enzyme glycogen synthetase, for example, catalyzes the synthesis of glycogen from glucose.

When organs are damaged by disease, their cells liberate specific enzymes into the blood. These enzymes can be measured and used to diagnose the disease. Creatine phosphokinase (CPK), *for example, may be released by damaged heart muscle, skeletal muscle, or brain tissue. Different forms of this enzyme are released by the different tissues and can be identified by clinical laboratory procedures.*

Since the shape of the active sites of the enzyme determines how the enzyme works, anything that changes the shape of the enzyme protein could reduce the action of the enzyme. In particular, changes in pH and increases in temperature alter protein shape, and thereby influence enzyme activity. For this reason, each enzyme works best only under specific conditions of pH and temperature. Fever, as well as acidosis and alkalosis (changes in blood pH), therefore, can affect the activity of tissue enzymes.

Because of the specific shape of its active site, each enzyme is highly specific as to the reaction it catalyzes. There is a different enzyme for almost every different chemical reaction involved in the body's metabolism. Since the information for the construction of each enzyme

SYNOPTIC SPOTLIGHT

From Lab to Medicine Cabinet: The Development of Pharmaceuticals

The drugs we have come to rely on in treating and preventing disease usually have their beginnings in basic physiological research, often at cellular and molecular levels. It may be that a new family of drugs is developed using cells in tissue culture (in vitro, or outside the body). For example, cell physiologists, studying membrane transport, may discover that a particular drug blocks membrane channels for Ca^{++}. Based on their knowledge of physiology, other scientists may predict that a drug of this nature could be useful in the treatment of hypertension (high blood pressure). This drug may then be tried in experimental animals.

If a drug is effective at extremely low concentrations in vitro, there is a chance that it may work in vivo (in the body) at concentrations low enough not to be toxic (poisonous). This possibility must be thoroughly tested utilizing experimental animals, primarily rats and mice. More than 90% of drugs tested in experimental animals are too toxic for further development. Only in those rare cases when the toxicity is low enough in experimental animals may development progress to human/clinical trials.

In **phase I clinical trials,** the drug is tested on healthy human volunteers. This is done to test its toxicity in humans and to study how the drug is "handled" by the body—how it is metabolized, how rapidly it is removed from the blood by the liver and kidneys, how it can be most effectively administered, and so on. If no toxic effects are observed, the drug can proceed to the next stage. In **phase II clinical trials,** the drug is tested on the target human population (for example, those with hypertension). Only in those exceptional cases where the drug seems to be effective but has minimal toxicity does testing move to the next phase. **Phase III trials** are conducted in many research centers across the country to maximize the number of test subjects involved. At this point, the test population must include large numbers of subjects of both sexes, as well as people of different ethnic groups. In addition, people are tested who have other health problems besides the one the drug is intended to benefit. For example, those who have diabetes in addition to hypertension would be included in this phase. If the drug passes phase III trials, it goes to the *Federal Drug Administration* (*FDA*) for approval. **Phase IV trials** test other potential uses of the drug.

A very low percentage of drugs make it all the way through these trials to eventually become approved and marketed—and, as you might imagine, the process is slow. While rigorous standards must be maintained to protect people from drugs that do more harm than good, the FDA has come under attack for not making certain drugs (for AIDS and cancer, for example) available to the public more quickly.

is coded by a different gene (region of DNA), genes can direct the pathways of metabolism by causing different enzymes to be produced at different times. The subject of genetic control of protein synthesis is covered in chapter 3 (see module 3.13).

ATP

In order to carry out their basic life activities, cells require the input of energy. This energy comes ultimately from the combustion of carbohydrates, lipids, and proteins into carbon dioxide and water, as described in chapter 21. These reactions are not used to power the cell directly, however. Instead, they are used to derive energy for the synthesis of **adenosine triphosphate** (*ă-den'o-sēn tri-fos'fāt*) **(ATP)** from adenosine diphosphate (ADP) and inorganic phosphate (abbreviated P_i).

Later, when enzymes reverse this reaction and convert ATP to ADP and P_i, a large amount of energy is released. Energy released from the breakdown of ATP is used to power the energy-requiring processes in all cells (fig. 2.18). Because it is the immediate source of energy for such activities as muscle contraction, active transport, and synthesis reactions, ATP is called the **universal energy carrier.**

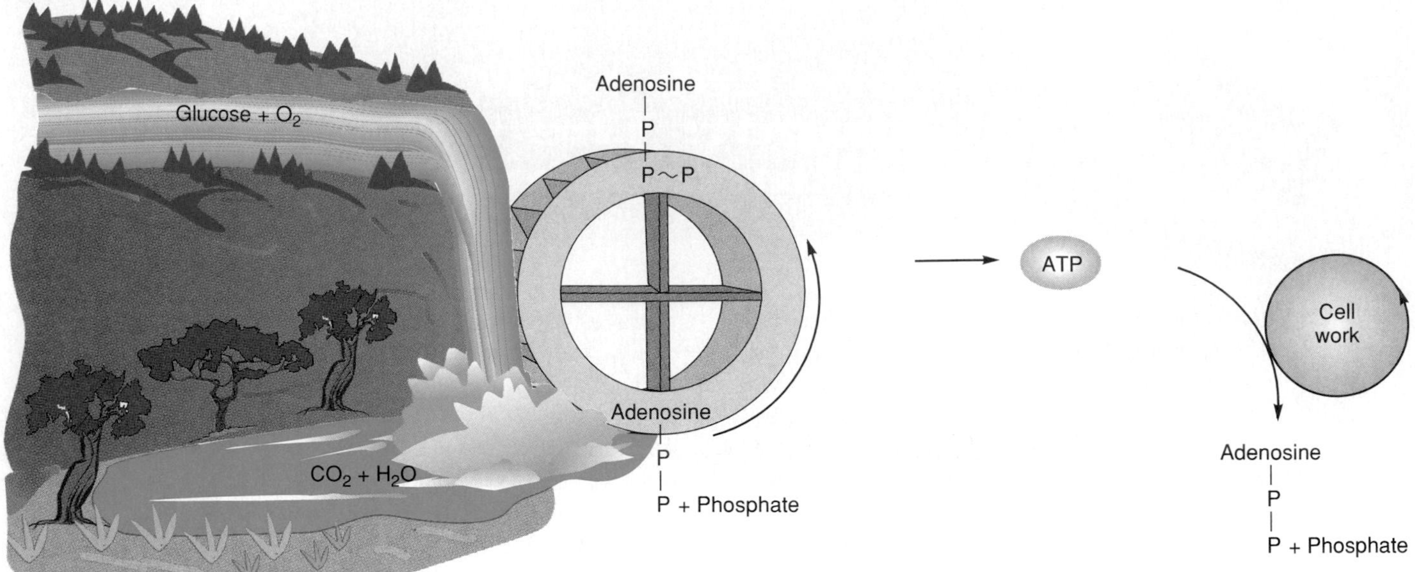

FIGURE 2.18
The role of ATP as a universal energy carrier. The energy needed to make ATP is derived from the breakdown of particular molecules, such as glucose. When ATP is itself broken down, the energy released is used to power the work of the cell.

SUMMARY

2.1 Atoms and Elements

1. An element is a chemical substance composed of atoms of only one kind.
2. An atom consists of a nucleus, which contains protons and neutrons, and electrons that move around the nucleus in volumes of space called electron shells.
3. The number of electrons in an atom is equal to the number of protons.
4. Isotopes are different atomic forms of the same element.

2.2 Chemical Bonds

1. Covalent bonds are produced when two or more atoms share pairs of electrons in their outermost shells.
2. An ion is an atom that has either gained or lost electrons; it has a net charge, either negative or positive.
3. Hydrogen bonds are weak attractive forces that develop between the negative oxygen atoms and the positive hydrogen atoms of two polar molecules; for example, between two water molecules.

2.3 Acids, Bases, and the pH Scale

1. An acid is a molecule that releases H^+ into a solution, thus raising the H^+ concentration.
2. A base is a molecule that functions as a H^+ acceptor. Many bases release OH^- into a solution and lower the H^+ concentration of that solution by forming H_2O.
3. A neutral solution has a pH of 7; an acidic solution has a pH of less than 7; and a basic solution has a pH greater than 7.

2.4 Carbohydrates

1. Carbohydrates contain carbon, hydrogen, and oxygen in the ratio $C_nH_{2n}O_n$.
2. Monosaccharides, or simple sugars, are the smallest subclass of carbohydrates; an example is glucose.
3. Monosaccharides can be bonded together by dehydration synthesis reactions to form disaccharides, such as maltose and sucrose.
4. A polysaccharide is a molecule consisting of many monosaccharides bonded together.
5. Hundreds of glucose subunits bonded together covalently produce a molecule of glycogen. Glycogen is stored in the liver and skeletal muscles.

2.5 Lipids

1. Lipids are organic molecules that are nonpolar, and thus insoluble in water.
2. Triglycerides are a subclass of lipids that includes fats and oils.
3. Steroids are a class of lipids that includes cholesterol and its steroid hormone derivatives.
4. Phospholipids contain a polar group with phosphate; the rest of the molecule is nonpolar.

2.6 Proteins

1. Proteins are very large molecules composed of amino acid subunits.
2. Twenty different amino acids serve as protein building blocks. The sequence of amino acids in a particular protein is the primary structure of the protein.
3. The secondary structure of a protein refers to the helical nature of the polypeptide chain. The tertiary structure refers to the three-dimensional shape of the protein.
4. Some proteins have a quaternary structure in that they are composed of more than one polypeptide chain.

2.7 Enzymes and ATP

1. Enzymes are protein catalysts that increase the rate of reactions without being changed by the reaction.
2. Enzymes are specific in their ability to convert particular substrates into products.
3. ATP is the universal energy donor of the cell.

Review Activities

Objective Questions

1. Which of the following statements about the valence electrons of an atom is *true?* (*module 2.1*)
   a. They are the outermost electrons.
   b. They participate in chemical reactions.
   c. They can be shared between atoms.
   d. In some cases, they can be transferred from one atom to another.
   e. All of the above are true.

2. The bond between oxygen and hydrogen in a water molecule is (*module 2.2*)
   a. a hydrogen bond.
   b. a polar covalent bond.
   c. a nonpolar covalent bond.
   d. an ionic bond.

3. The type of bond formed between two molecules of water is (*module 2.2*)
   a. a hydrolytic bond.
   b. a polar covalent bond.
   c. a nonpolar covalent bond.
   d. a hydrogen bond.

4. If solution A has a pH of 2 and solution B has a pH of 10, which of the following statements is *true?* (*module 2.3*)
   a. Solution A has a higher H^+ concentration than solution B.
   b. Solution B is basic.
   c. Solution A is acidic.
   d. All of the above are true.

5. Glucose is (*module 2.4*)
   a. a disaccharide.
   b. a polysaccharide.
   c. a monosaccharide.
   d. a steroid.

6. Carbohydrates are stored in the liver and muscles in the form of (*module 2.4*)
   a. glucose.
   b. triglycerides.
   c. glycogen.
   d. cholesterol.

7. Lecithin is (*module 2.5*)
   a. a carbohydrate.
   b. a protein.
   c. a steroid.
   d. a phospholipid.

8. The tertiary structure of a protein is determined *directly* by (*module 2.6*)
   a. genes.
   b. peptide bonds.
   c. the primary structure of the protein.
   d. enzymes that mold the protein.

9. The carbon-to-nitrogen bond that joins amino acids together is (*module 2.6*)
   a. a glycosidic bond.
   b. a peptide bond.
   c. a hydrogen bond.
   d. a double bond.

10. The energy needed for cellular work comes *directly* from (*module 2.7*)
   a. the breakdown of food molecules.
   b. the formation of ATP.
   c. the breakdown of ATP.
   d. none of the above.

Essay Questions

1. Draw a model of an atom and label the components. Distinguish between atomic mass and atomic number. (*module 2.1*)
2. Compare and contrast polar and nonpolar covalent bonds. (*module 2.2*)
3. Explain what happens to the valence electrons of sodium and chlorine when they form the ionic compound of sodium chloride. (*module 2.2*)
4. Define *acid* and *base* and explain how acids and bases influence the pH of a solution. (*module 2.3*)
5. With reference to dehydration synthesis and hydrolysis reactions, explain the relationship between the starch in your lunch, liver glycogen, and blood glucose. (*module 2.4*)
6. All fats are lipids, but not all lipids are fats. Explain why this is so. (*module 2.5*)
7. Explain the relationship between the primary structure of a protein and its secondary and tertiary structure. (*module 2.6*)
8. Explain the significance of ATP in the body. (*module 2.7*)
9. Describe the lock-and-key model of enzyme action. (*module 2.7*)

Labeling Exercise

Label the molecules shown in the figure to the right.

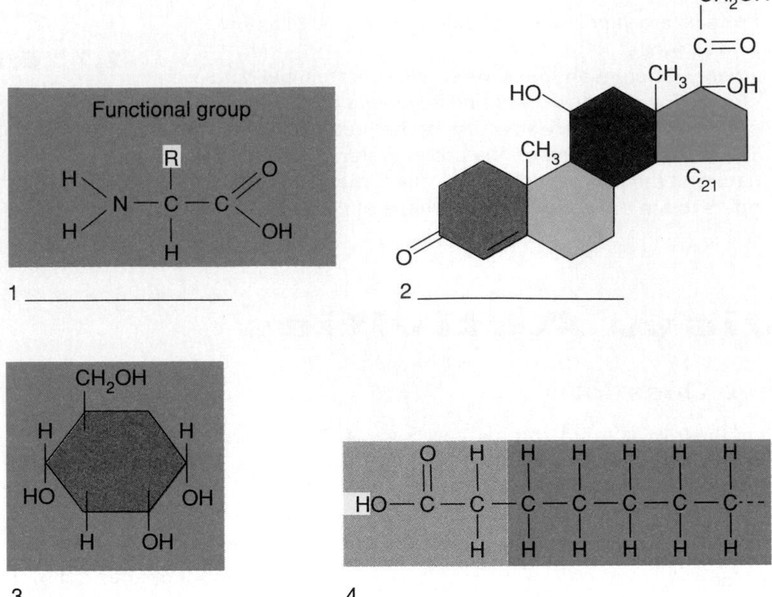

Critical Thinking Questions

1. When you make a pot roast, do the hydrogen bonds or the covalent bonds of the meat proteins break? How do you know? (Hint: What happens to the meat when you cook it?)
2. Insulin, a very important hormone, is a polypeptide. Why can't a diabetic take insulin pills? What would the person get from such pills?
3. Suppose the pH of your child's wading pool was too high. What could you add to lower it? What would happen to the pH if you added plain table salt? Explain.
4. An advertisement promotes a line of costly natural vitamins, including one for vitamin C (ascorbic acid). Would the natural vitamin C be any different from a cheap bottle of artificially synthesized ascorbic acid? Would your body be able to tell the difference?
5. A chemist would describe a solution with a pH of 7.2 as being slightly basic. A health professional would describe a blood pH of 7.2 as acidemic. Explain how these words are used by the two professionals.

CHAPTER

3

"Complexity of function demands that cells demonstrate complexity of structure at the subcellular level." *page 42*

Cell Structure and Function

Chapter Outline

Terms to Remember

nucleus (*noo'kle-us*), 43
cytoplasm (*si'tŏ-plaz"em*), 43
organelle (*or"gă-nel'*), 43
cilia (*sil'e-ă*), 45
flagellum (*flă'jel'um*), 45
phagocytosis (*fag"ŏ-si-to'sis*), 45
pinocytosis (*pin"ŏ-si-to'sis*), 45
endocytosis (*en"do-si-to'sis*), 46
exocytosis (*ek"so-si-to'sis*), 47
diffusion, 47
osmosis (*oz-mo'sis*), 50
facilitated diffusion, 53
active transport, 53
lysosome (*li'sŏ-sōm*), 56
mitochondrion (*mi"tŏ-kon'dre-on*), 57
endoplasmic reticulum (*en'do-plaz'mik rĕ-tik'yŭ-lum*), 57
ribosome (*ri'bŏ-sōm*), 57
Golgi apparatus (*gol'je*), 58
DNA, 60
RNA, 61
mitosis (*mi-to'sis*), 67

Learning Objectives

After studying this chapter, you should be able to

- describe the structure of the cell membrane and the structure and functions of cilia and flagella.
- describe amoeboid movements and explain the significance of endocytosis and exocytosis.
- explain what is meant by passive transport and describe the conditions under which osmosis occurs.
- describe the characteristics of carrier-mediated transport and explain how facilitated diffusion occurs.
- explain what is meant by active transport and describe how the Na^+/K^+ pumps work.
- describe the structure and functions of lysosomes, mitochondria, the Golgi apparatus, and the endoplasmic reticulum.
- describe the structure of nucleotides and explain the law of complementary base pairing.
- explain how DNA directs the synthesis of RNA, and how RNA directs the synthesis of proteins.
- explain how DNA replication occurs and list the stages of mitosis.

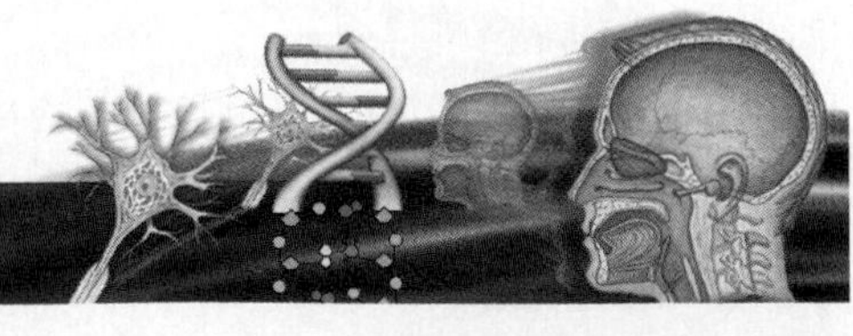

3.1 Basic Cell Structure

The cell is the most basic structural and functional unit of the human body and of all living organisms. Many cell functions are performed by cell membranes and particular subcellular structures known as organelles.

Because the cell looks so small and simple when viewed with the ordinary (light) microscope, it is difficult to imaging that each cell is a living entity unto itself. Equally amazing is the fact that the physiology of our organs and systems derive from the complex functions of the cells of which they are composed. Complexity of function demands that cells demonstrate complexity of structure at the subcellular level.

As the basic functional unit of the body, each cell is a highly organized molecular factory. The estimated 60 trillion to 100 trillion cells that make up your body come in a great variety of shapes and sizes. This great diversity, which is also apparent in the subcellular structures within different cells, reflects the diversity of function of different cells in the body. All cells, however, have certain features in common. They are all surrounded by a cell membrane and most of them contain the structures listed in table 3.1. Although no single cell can be considered "typical," a model of a generalized cell is shown in figure 3.1.

For descriptive purposes, a cell can be divided into three principal parts:

1. **Cell (plasma) membrane.** The selectively permeable cell membrane surrounds the cell. It controls the passage of molecules into and out of the cell and separates the cell's internal structures from the extracellular environment.

TABLE 3.1 Structure and Function of Cellular Components

Component	Structure	Function
Cell (plasma) membrane	Membrane composed of phospholipid and protein molecules	Gives form to cell and controls passage of materials in and out of cell
Cytoplasm	Fluid, jellylike substance in which organelles are suspended	Serves as matrix in which chemical reactions occur
Endoplasmic reticulum	System of interconnected membrane-forming canals and tubules	Smooth endoplasmic reticulum metabolizes nonpolar compounds and stores Ca^{++} in striated muscle cells; rough endoplasmic reticulum assists in protein synthesis
Ribosomes	Granular particles composed of protein and RNA	Synthesize proteins
Golgi apparatus	Cluster of flattened, membranous sacs	Synthesizes carbohydrates and packages molecules for secretion; secretes lipids and glycoproteins
Mitochondria	Double membrane with folded inner partitions	Release energy from food molecules and transform energy into usable ATP
Lysosomes	Membranous sacs	Digest foreign molecules and worn and damaged cells
Peroxisomes	Spherical membranous vesicles	Contain enzymes that produce hydrogen peroxide and use this for various oxidation reactions
Centrosome	Nonmembranous mass of two rodlike centrioles	Helps organize spindle fibers and distribute chromosomes during mitosis
Vacuoles	Membranous sacs	Store and excrete various substances within the cytoplasm
Fibrils and microtubules	Thin, hollow tubes that form cytoskeleton and other structures	Support cytoplasm and transport materials within the cytoplasm
Cilia and flagella	Minute cytoplasmic extensions from cell, composed of microtubules	Move particles along surface of cell or move cell
Nuclear membrane	Membrane surrounding nucleus, composed of protein and lipid molecules	Supports nucleus and controls passage of materials between nucleus and cytoplasm
Nucleolus	Dense, nonmembranous mass composed of protein and RNA molecules	Forms ribosomes
Chromatin	Fibrous strands composed of protein and DNA molecules	Controls cellular activity for carrying on life processes

FIGURE 3.1
A generalized cell and the principal organelles. Keep in mind that this illustration is not of an actual living cell; rather, it is a composite of features common to most cells.

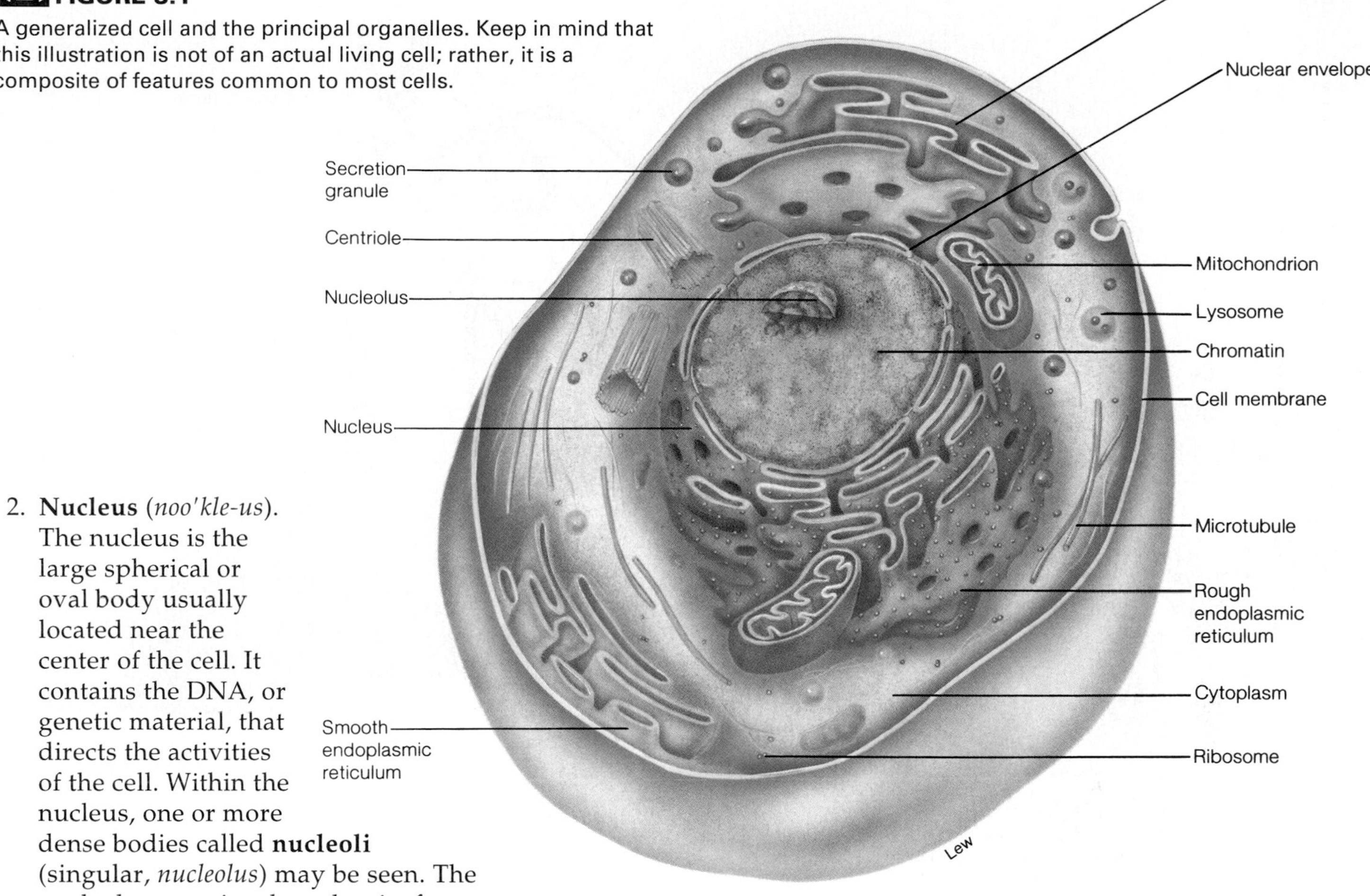

2. **Nucleus** (*noo'kle-us*). The nucleus is the large spherical or oval body usually located near the center of the cell. It contains the DNA, or genetic material, that directs the activities of the cell. Within the nucleus, one or more dense bodies called **nucleoli** (singular, *nucleolus*) may be seen. The nucleolus contains the subunits for ribosomes, the structures that serve as sites for protein synthesis.
3. **Cytoplasm** (*si'tŏ-plaz"em*) and **organelles** (*or"gă-nelz'*). Cytoplasm is the cellular material between the nucleus and the cell membrane. The term *cytosol* is frequently used to describe the thick, semifluid portion of the cytoplasm; that is, the part that cannot be removed by centrifugation. Organelles are specialized structures within the cell that perform specific functions.

Inside Information—Cells were first named and described by Robert Hooke in 1665. Actually, what he saw through his compound microscope were slices of dead cork, and what he called "cells" were merely the cell walls of plant cells (animal cells do not have cell walls). The notion that cells could be alive did not occur to Hooke at the time.

3.2 Cell Membrane

The cell membrane allows selective communication between the intracellular and extracellular environments and participates in cellular movements.

Structure of the Cell Membrane

The **cell membrane** (also called the **plasma membrane**), and indeed all of the membranes surrounding organelles within the cell, are composed primarily of phospholipids and proteins. As described in the previous chapter (see module 2.5), phospholipids are polar on the end that contains the phosphate group and nonpolar (and hydrophobic) throughout the rest of the molecule. Since there is an aqueous (watery) environment on each side of the membrane, the hydrophobic parts of the molecules

aqueous: L. *aqua,* water

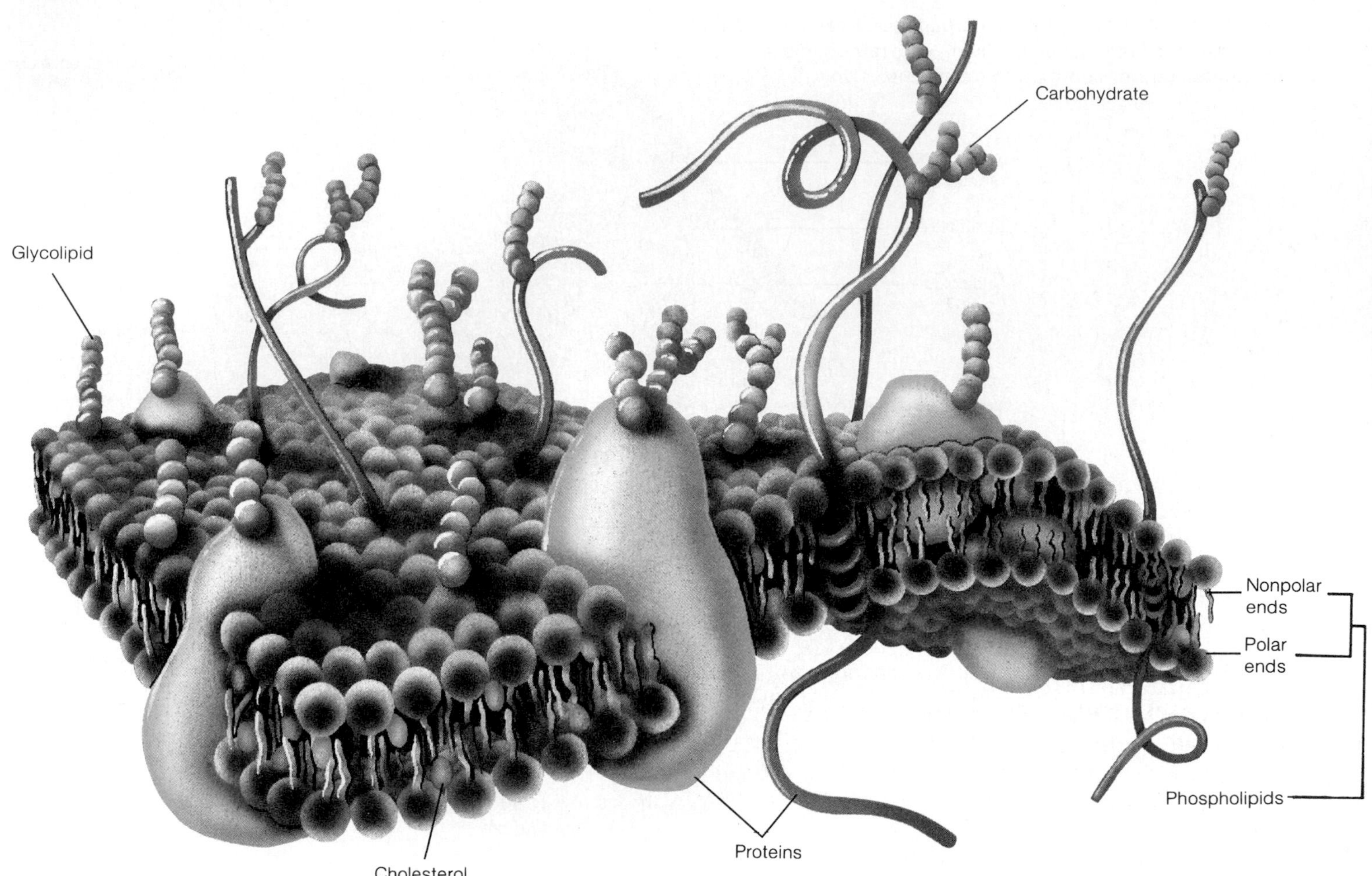

FIGURE 3.2
The fluid-mosaic model of the cell membrane. The membrane consists of a double layer of phospholipids, with the phosphates (shown by spheres) oriented outward and the hydrophobic hydrocarbons (wavy lines) oriented toward the center. Proteins may completely or partially span the membrane. Carbohydrates are attached to the outer surface.

"huddle" in the center of the membrane, leaving the polar ends exposed to water on both surfaces. This results in the formation of a double layer (bilayer) of phospholipids in the cell membrane.

The hydrophobic core of the membrane restricts the passage of water-soluble molecules and ions. Certain of these polar compounds, however, do pass through the membrane. The specialized functions and selective transport properties of the membrane are believed to be due to its protein content. These proteins are stuck between the phospholipids, and most span the membrane completely from one side to the other. Like icebergs floating in a "sea" of phospholipids, the proteins are free to move within the membrane. As a result, they are not uniformly distributed, but rather form a constantly changing mosaic. This structural arrangement of the membrane is known as **fluid-mosaic model** (fig. 3.2).

Functionally, the proteins found in the cell membrane

1. provide structural support;
2. transport molecules across the cell membrane;
3. control chemical reactions at the cell surface by enzymatic action;
4. serve as receptors for hormones and other regulatory molecules that arrive at the outer surface of the cell membrane; and
5. serve as cellular "markers" (antigens) that induce the production of antibodies.

In addition to phospholipids and proteins, the cell membrane also contains carbohydrates and cholesterol that serve specific functions. Moreover, in order for the cell membrane to carry out its functions more efficiently, its surface area is sometimes increased through numerous foldings. These foldings produce fingerlike projections called **microvilli.** In the small intestine, for example, microvilli increase the surface area of membrane exposed to food. This improves the efficiency with which the food is digested and transported through the cells into the blood (see chapter 20, module 20.6).

Cilia and Flagella

Cilia (*sil'e-ă*) are tiny hairlike structures that project from the cell. Like rowers in a boat, they stroke in unison. Cilia in the human body are found on the *apical surface* (the surface facing the lumen, or cavity) of stationary epithelial cells in the respiratory and female reproductive tracts. In the respiratory system, the cilia transport strands of mucus to the throat (pharynx), where the mucus can either be swallowed or expectorated (spit out). In the female reproductive tract, ciliary movements in the epithelial lining draw the egg (ovum) into the uterine tube and move it toward the uterus.

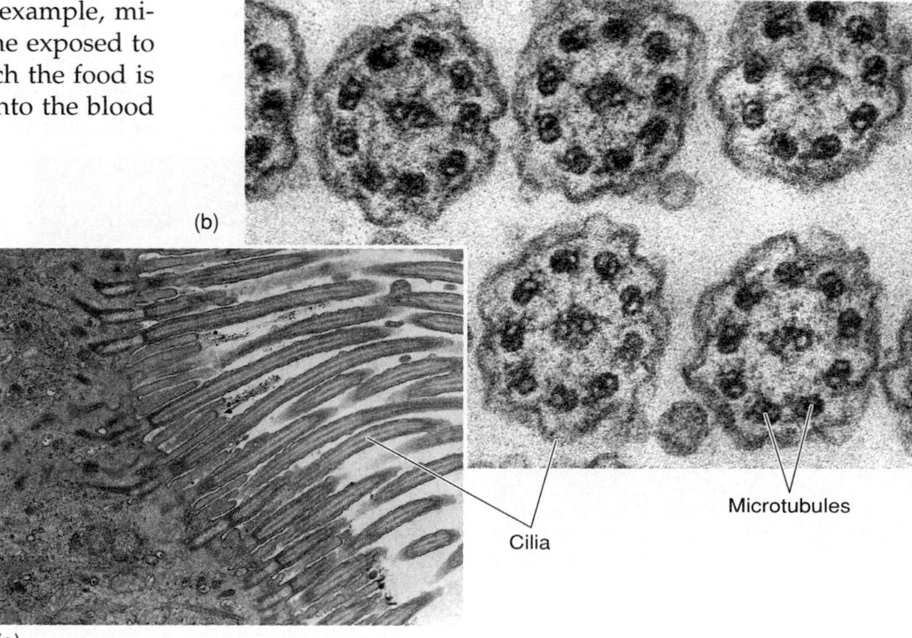

FIGURE 3.3
Electron micrographs of cilia showing (*a*) longitudinal and (*b*) cross sections. Notice the characteristic "9 + 2" arrangement of microtubules in the cross sections.

Sperm are the only cells in the human body that have **flagella** (*flă-jel'ă*). The flagellum is a single whiplike structure that propels the sperm cell through its environment. Both cilia and flagella are composed of *microtubules* (formed from proteins) arranged in a characteristic way. One pair of microtubules in the center of a cilium or flagellum is surrounded by nine other pairs of microtubules to produce what is often described as a "9 + 2" arrangement (fig. 3.3).

3.3 Cell Movements

Movements of the cell membrane and cytoplasm may result in amoeboid movement of the entire cell. "Cell eating" and "drinking" also involve movements of the membrane and cytoplasm.

Amoeboid Movement and Phagocytosis

Some body cells—including certain white blood cells and macrophages in connective tissues—are able to move like an amoeba (a single-celled organism). This **amoeboid movement** is accomplished by projections of the cytoplasm called **pseudopods** (*soo'dŏ-podz*), which attach to a substrate and pull the cell along.

Cells that exhibit amoeboid movement (and also certain liver cells that are not mobile) use pseudopods to surround and engulf particles of organic matter, such as bacteria. This process—a type of cellular "eating" called **phagocytosis** (*fag"ŏ-si-to'sis*)—protects the body from invading microorganisms and removes extracellular debris.

Phagocytic cells surround their victim with pseudopods, which join together and fuse (fig. 3.4). After the inner membrane of the pseudopods becomes continuous around the ingested particle, it pinches off from the inner surface of the cell membrane. The ingested particle is now contained in an organelle called a *food vacuole* within the cell. The particle will subsequently be digested by enzymes contained in a different organelle—the lysosome, described in module 3.9.

Pinocytosis (*pin"ŏ-si-to'sis*) is a type of cellular "drinking." In this process, the engulfed material is liquid rather than solid. Instead of forming pseudopods, the cell membrane invaginates (becomes infolded) to produce a deep, narrow furrow. The membrane near the surface of this

cilia: L. *cili,* small hair
flagellum: L. *flagrum,* whip
pseudopod: Gk. *pseudo,* false; *pod,* foot

phagocytosis: Gk. *phagein,* to eat; *kytos,* hollow body
pinocytosis: Gk. *pinein,* to drink; *kytos,* hollow body

furrow then fuses, and a small vacuole containing the extracellular fluid is pinched off and enters the cell. In this way, a cell can take in proteins and other large molecules that may be present in the extracellular fluid.

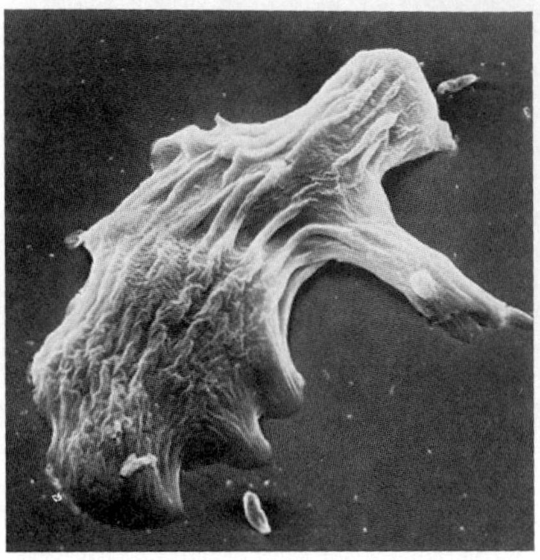

(a)

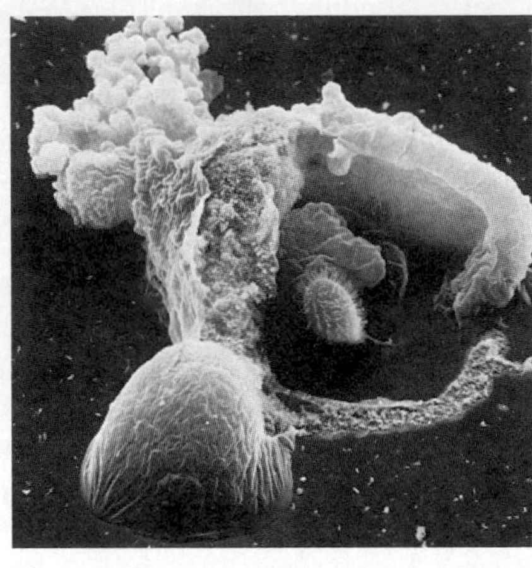

(b)

FIGURE 3.4

Scanning electron micrographs of phagocytosis showing (*a*) the formation of pseudopods and (*b*) the entrapment of the prey within a food vacuole.

Endocytosis and Exocytosis

Phagocytosis and pinocytosis, in which part of the cell membrane invaginates to form first a pouch and then a vacuole, are two means by which extracellular materials can be engulfed by a cell. The general term for this process is **endocytosis** (*en″do-si-to′sis*). Another type of endocytosis involves the smallest area of cell membrane, and it occurs only in response to specific molecules in the extracellular environment. Since the extracellular molecules must bond with very specific *receptor proteins* in the cell membrane, this process is known as *receptor-mediated endocytosis.*

In receptor-mediated endocytosis, the interaction of specific molecules in the extracellular fluid with specific membrane receptor proteins causes the membrane to invaginate, fuse, and pinch off as a *vesicle* (a small vacuole), within the cell (fig. 3.5). Vesicles formed in this way contain extracellular fluid

endocytosis: Gk. *endo,* within; *kytos,* hollow body

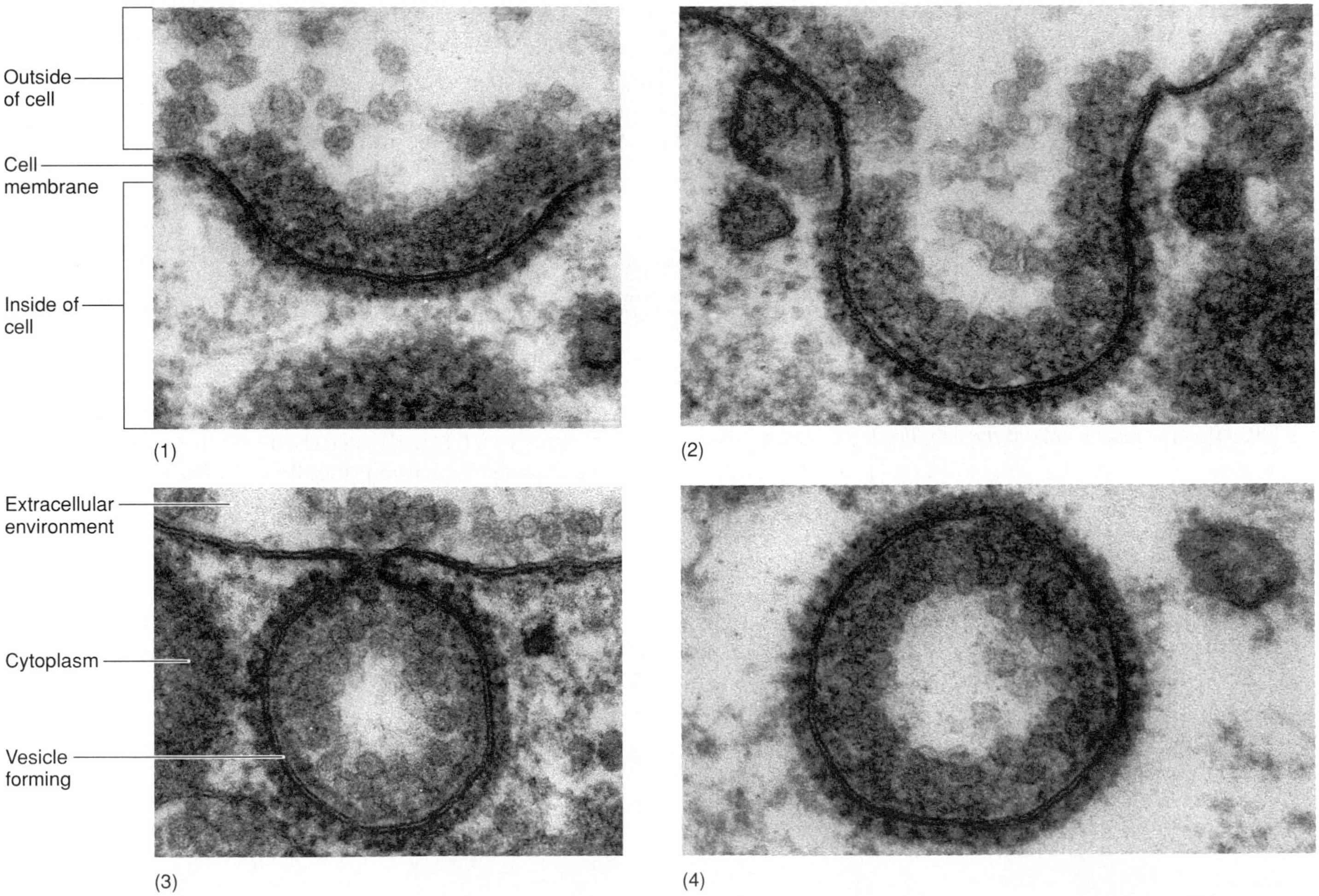

Outside of cell
Cell membrane
Inside of cell

(1)

(2)

Extracellular environment
Cytoplasm
Vesicle forming

(3)

(4)

FIGURE 3.5

Stages 1–4 of endocytosis, in which specific bonding of extracellular particles to membrane receptor proteins is believed to occur.

and molecules that could not have passed by other means into the cell. Cholesterol, for example, may be taken up into liver and artery cells by receptor-mediated endocytosis.

Molecules produced within the cell that are destined for secretion are packaged inside of the cell within vesicles. In the process of **exocytosis** (*ek"so-si-to'sis*), these secretory vesicles fuse with the cell membrane and release their contents into the extracellular environment. Exocytosis is essentially the reverse of endocytosis; it is the process by which endocrine cells secrete hormones and neurons release their neurotransmitter molecules.

Endocytosis and exocytosis account for only part of the two-way traffic between the intracellular and extracellular compartments. Most of this traffic is due to membrane transport processes—the movement of molecules and ions through the cell membrane—to be discussed next.

3.4 Membrane Transport Processes

The net diffusion of molecules and ions across the cell membrane constitutes passive transport. Active transport is involved when movement occurs against a concentration gradient.

The cell (plasma) membrane separates the intracellular environment from the extracellular environment. Proteins, nucleotides, and other molecules needed for the structure and function of the cell cannot penetrate, or "permeate," the membrane, while other molecules and many ions can. The cell membrane is thus **selectively permeable;** that is, it is permeable to some molecules and not to others. This selective permeability is one of the most important functional features of the cell.

Categories of Membrane Transport

The mechanisms involved in the transport of molecules and ions through the cell membrane may be divided into two categories:

1. transport that requires the action of specific *carrier proteins* in the membrane (*carrier-mediated transport*); and
2. transport through the membrane that is not carrier mediated.

Carrier-mediated transport may further be subdivided into *facilitated diffusion* and *active transport.* Membrane transport that does not use carrier proteins consists of the *simple diffusion* of ions, lipid-soluble molecules, and water through the membrane. The diffusion of (water) solvent through a membrane is called *osmosis.*

Membrane transport processes may also be categorized based on their energy requirements. Transport in which the net movement is from higher to lower concentration (down a concentration gradient) does not require metabolic energy; this is **passive transport.** Passive transport includes simple diffusion, osmosis, and facilitated diffusion. Transport that occurs against a concentration gradient (through a membrane to the region of higher concentration) is **active transport.** Active transport requires the expenditure of metabolic energy (ATP) and involves specific carrier proteins.

Diffusion

Molecules in a gas and molecules and ions dissolved in a solution are in constant motion, moving about randomly as a result of their thermal (heat) energy. The tendency of the molecules to be scattered evenly, or diffusely, within a given volume of gas or solution as a result of this random motion is called **diffusion.** Whenever a *concentration difference,* or *concentration gradient,* exists between two regions of a solution, therefore, random molecular motion tends to eliminate the gradient and to distribute the molecules uniformly (fig. 3.6).

As a result of random molecular motion, molecules in the region of the solution where there is a higher concentration will enter the region of lower concentration. Molecules will also move in the opposite direction, but not as frequently. As a result, there will be a *net movement* from the region of higher to the region of lower concentration until the concentration difference is eliminated. This net movement is called **net diffusion** (fig. 3.7). Net diffusion is a physical process that occurs whenever there is a concentration difference in a solution. When the concentration difference exists across a membrane, diffusion becomes a type of membrane transport.

permeable: L. *per,* through; *meare,* to pass
diffusion: L. *diffundo,* to pour in different directions

FIGURE 3.6
Diffusion is the random movement of molecules from a region of higher concentration toward one of lesser concentration. Diffusion is occurring as a sugar cube dissolves in water and the sugar molecules become evenly distributed.

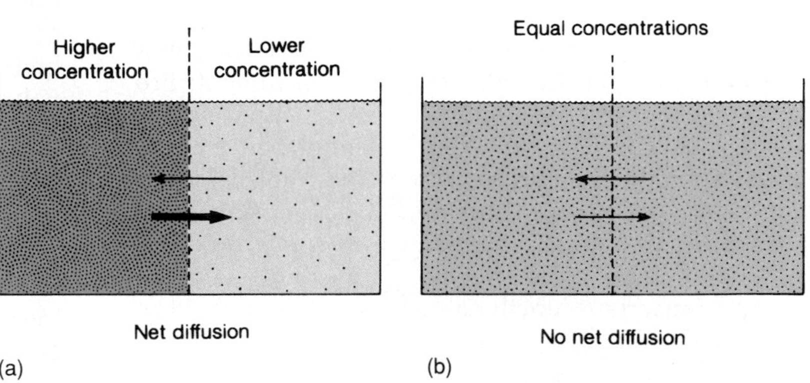

FIGURE 3.7
(*a*) Net diffusion occurs when there is a concentration difference (or concentration gradient) between two regions of a solution, provided that the membrane separating these regions is permeable to the diffusing substance. (*b*) Diffusion tends to equalize the concentration of these solutions, and thus to eliminate the concentration differences.

3.5 Passive Transport

Nonpolar molecules can penetrate the phospholipid barrier, and small inorganic ions can pass through channels in the membrane. The transport of polar molecules through the membrane is carrier mediated.

Diffusion through the Cell Membrane

Since the cell membrane consists primarily of a double layer of phospholipids, molecules that are nonpolar, and thus lipid soluble, can easily pass from one side of the membrane to the other. The cell membrane, in other words, does not present a barrier to the diffusion of nonpolar molecules such as oxygen gas (O_2) or steroid hormones. Small organic molecules that have polar covalent bonds but are uncharged, such as CO_2 (as well as ethanol and urea), are also able to penetrate the double-lipid layer. Net diffusion of these molecules can thus easily occur between the intracellular and extracellular compartments when concentration gradients are present (fig. 3.8).

Although water is not lipid soluble, water molecules can diffuse through the cell membrane because of their small size and lack of net charge. In certain membranes, however, the passage of water is restricted to specific channels that can open or close in response to physiological regulation.

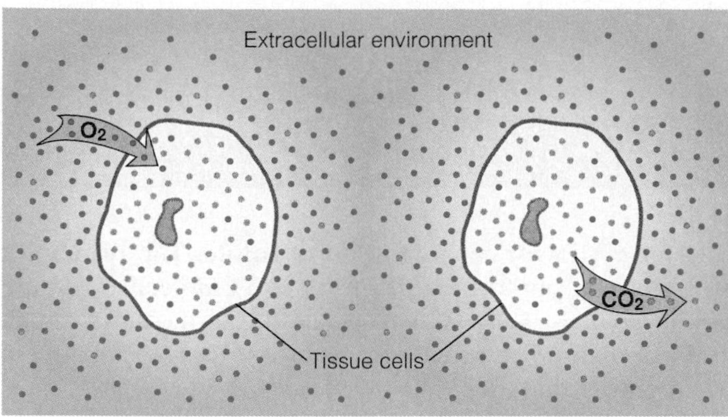

FIGURE 3.8

Gas exchange between the intracellular and extracellular compartments occurs by diffusion. The colored dots, which represent oxygen and carbon dioxide molecules, indicate relative concentrations inside the cell and in the extracellular environment.

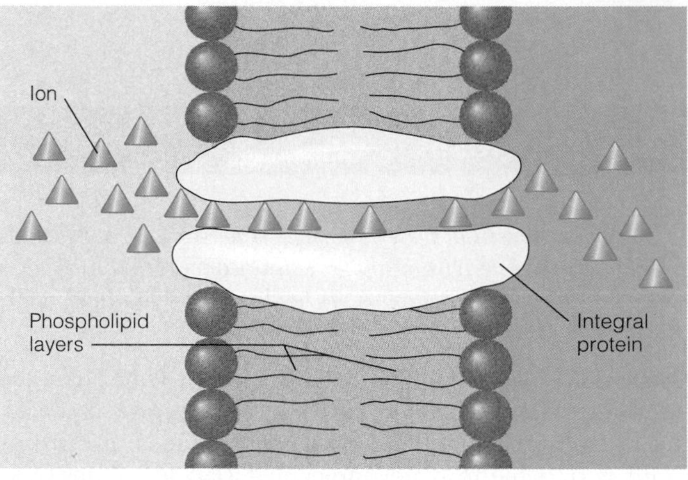

FIGURE 3.9

Inorganic ions (such as Na^+ and K^+) may be able to penetrate the membrane through pores within integral proteins that span the thickness of the double phospholipid layers.

The phospholipid portion of the membrane is impermeable to charged inorganic ions, such as Na^+ and K^+. These ions may pass through the cell membrane, however, via tiny **ion channels** through the membrane that are too small to be seen even with an electron microscope. These channels are provided by some of the *integral proteins* that span the thickness of the membrane (fig. 3.9).

Cystic fibrosis is an inherited disorder of the exocrine glands that occurs most frequently in Caucasians. As a result of this disorder, there is abnormal NaCl and water movement across wet epithelial membranes. Where such membranes line the pancreatic ductules and small airways, the secretions of these membranes become more viscous and cannot be properly cleared, leading to pancreatic and pulmonary disorders. This genetic defect involves a particular glycoprotein that forms chloride (Cl^-) channels in the apical cell membrane of the epithelial cells.

Rate of Diffusion

The rate of diffusion—a measure of the number of diffusing molecules that pass through the membrane per unit time—is influenced by

1. the magnitude of the concentration difference across the membrane (the "steepness" of the concentration gradient);
2. the permeability of the membrane to the diffusing substances; and
3. the surface area of the membrane through which the substances are diffusing.

The magnitude of the concentration difference across the membrane serves as the driving force for diffusion. Regardless of this concentration difference, however, the diffusion of a substance across a membrane will not occur if the membrane is not permeable to that substance. With a given concentration difference, the rate of diffusion through a membrane will vary directly with the degree of permeability. In a resting neuron, for example, the membrane is about 20 times more permeable to potassium (K^+) than to sodium (Na^+); consequently, K^+ diffuses much more rapidly than does Na^+. Changes in the protein structure of the membrane channels, however, can change the permeability of the membrane. This occurs during the production of a nerve impulse, when specific stimulation opens Na^+ channels temporarily and allows a faster diffusion rate for Na^+ than for K^+ (see chapter 9, module 9.4).

In areas of the body that are specialized for rapid diffusion, the surface area of the cell membranes may be increased by numerous folds. As mentioned in module 3.2, the rapid passage of the products of digestion across the epithelial membranes in the small intestine is aided by many tiny folds that form the finger-like projections called microvilli. Similar microvilli are found in the renal (kidney) tubule epithelium, which must reabsorb various molecules that are filtered out of the blood.

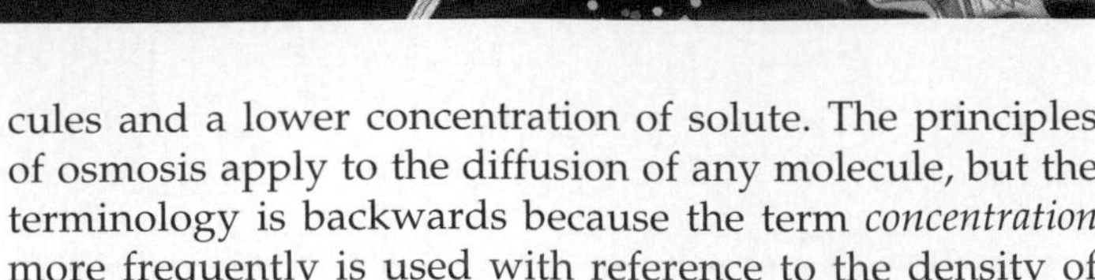

3.6 Osmosis

The net movement of water by osmosis is from the more dilute solution to the one with a higher solute concentration. The osmotic pressure is a measure of the tendency of a solution to take in water by osmosis.

Osmosis (*oz-mo'sis*) is the net diffusion of water (the solvent) across a membrane. In order for osmosis to occur, the membrane must be *semipermeable;* that is, it must be more permeable to water molecules than to solutes. Like the diffusion of solute molecules, the diffusion of water occurs when the water is more concentrated on one side of the membrane than on the other side; that is, when one solution is more dilute than the other (fig. 3.10). The more dilute solution has a higher concentration of water molecules and a lower concentration of solute. The principles of osmosis apply to the diffusion of any molecule, but the terminology is backwards because the term *concentration* more frequently is used with reference to the density of solute rather than of solvent molecules.

osmosis: Gk. *osmos,* a thrust

Imagine a cylinder divided into two equal compartments by a membrane partition that can freely move. One compartment initially contains 180 g/L (grams per liter) of glucose and the other compartment contains 360 g/L of glucose. If the membrane is permeable to glucose, glucose will diffuse from the 360-g/L compartment to the 180-g/L compartment until both compartments contain 270 g/L of glucose. If the membrane is not permeable to glucose but is permeable to water, the same result (270-g/L solutions on both sides of the membrane) will be achieved by the diffusion of water. As water diffuses from the 180-g/L compartment to the 360-g/L compartment, the former solution becomes more concentrated while the latter becomes more dilute. This is accompanied by volume changes, as illustrated in figure 3.11.

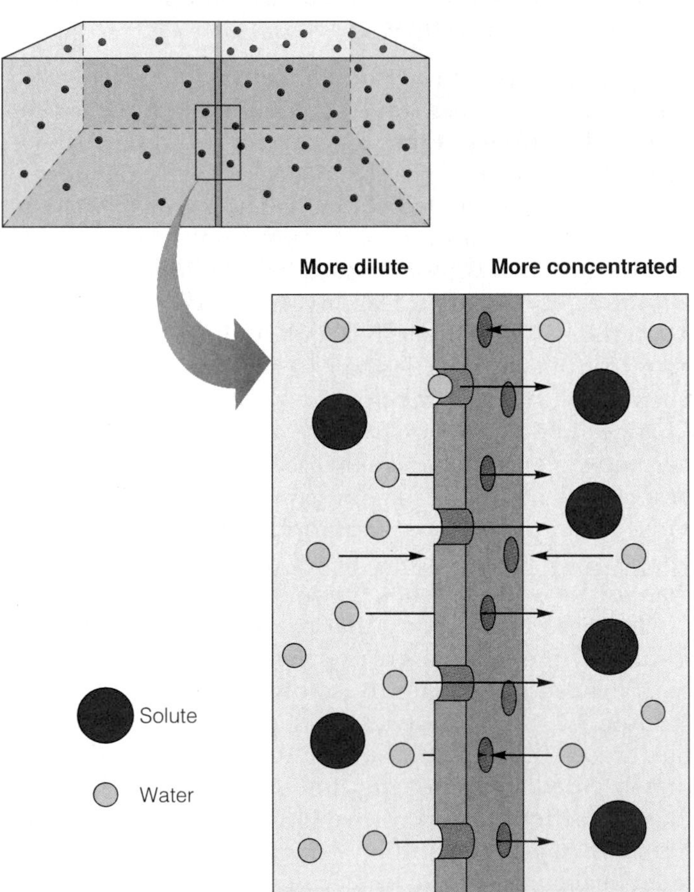

FIGURE 3.10
A model of osmosis, or the net movement of water from the solution of lesser solute concentration to the solution of greater solute concentration.

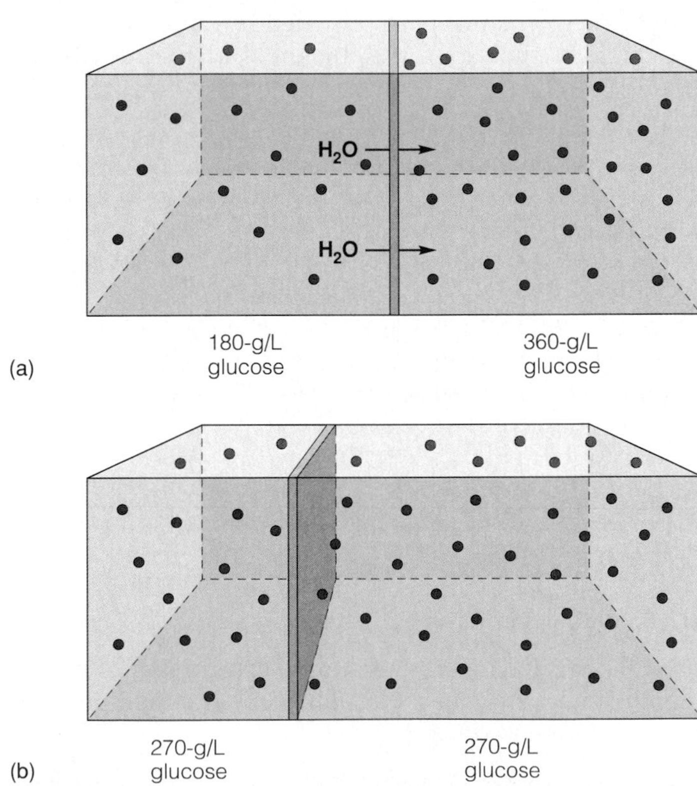

FIGURE 3.11
(*a*) A moveable semipermeable membrane (permeable to water but not glucose) separates two solutions of different glucose concentration. As a result, water moves by osmosis into the solution of greater concentration until (*b*) the volume changes equalize the concentrations on both sides of the membrane.

Osmotic Pressure and Solute Concentration

Osmosis and the movement of the membrane partition could be prevented by an opposing force. If one compartment contained 180 g/L of glucose and the other compartment contained pure water, the osmosis of water into the glucose solution could be prevented by pushing against the membrane with a certain force. This concept is illustrated in figure 3.12.

The force that would have to be exerted to prevent osmosis in this situation is the **osmotic pressure** of the solution. This backwards measurement indicates how strongly the solution "draws" water into it by osmosis. The greater the solute concentration of a solution, the greater its osmotic pressure. Pure water, thus, has an osmotic pressure of zero, and a 360-g/L glucose solution has twice the osmotic pressure of a 180-g/L glucose solution.

In summary, the osmotic pressure measures the *tendency of a solution to take in water by osmosis,* and this tendency is proportional to the concentration of solute in the solution. More accurately, the tendency of a solution to take in water depends on its concentration of nondiffusible solute; that is, solute that cannot pass through the membrane.

Scientists frequently express the solute concentration of fluids in units of **osmolality.** As discussed in chapter 19, these units of measurement are used in describing the formation of urine by the kidneys (see module 19.2). The method of calculating osmolality units need not concern us here; it is enough to understand that osmolality units measure the total number of solute molecules in a given volume of solution. If one solution has a higher osmolality than a second solution, it has a greater total concentration of solute molecules, and therefore a greater osmotic pressure. If a membrane that was impermeable to the solute separated the two solutions, water would flow from the second solution into the first.

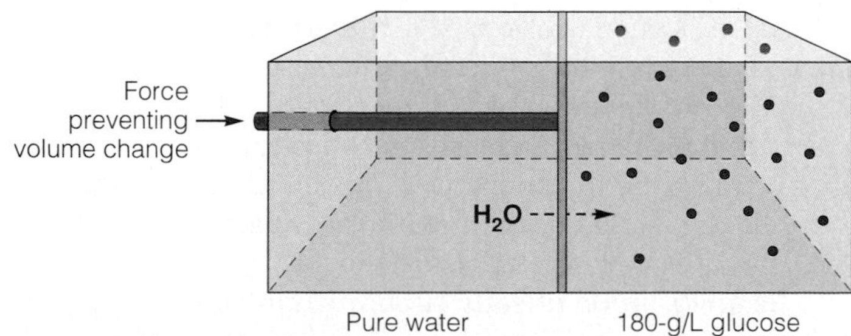

FIGURE 3.12

If a semipermeable membrane separates pure water from a 180-g/L glucose solution, water tends to move by osmosis into the glucose solution, thus creating a hydrostatic pressure that pushes the membrane to the left and expands the volume of the glucose solution. The amount of pressure that must be applied to just counteract this volume change is equal to the osmotic pressure of the glucose solution.

Regulation of Blood Osmolality

The osmolality of the blood plasma is normally maintained within very narrow limits by a variety of regulatory mechanisms. When a person becomes dehydrated, for example, the blood becomes more concentrated as the total blood volume is reduced. The increased blood osmolality and osmotic pressure stimulates *osmoreceptors,* which are neurons located in a part of the brain called the hypothalamus.

As a result of increased osmoreceptor stimulation, the person becomes thirsty and, if water is available, drinks. Along with increased water intake, a person who is dehydrated excretes a lower volume of urine. This occurs as a result of the following sequence of events: (1) increased plasma osmolality stimulates osmoreceptors in the hypothalamus of the brain; (2) the osmoreceptors stimulate the posterior pituitary to secrete **antidiuretic hormone (ADH);** and (3) ADH acts on the kidneys to promote water retention so that a lower volume of urine is excreted.

A person who is dehydrated, therefore, drinks more and urinates less. These responses represent a negative feedback loop (fig. 3.13) that acts to maintain homeostasis of the plasma concentration (osmolality) and, in the process, helps to maintain a proper blood volume.

Tonicity

Acting through the negative feedback mechanisms just described, the blood plasma is maintained at a particular concentration, or osmolality. Solutions of glucose or sodium chloride that have the same osmolality as plasma can

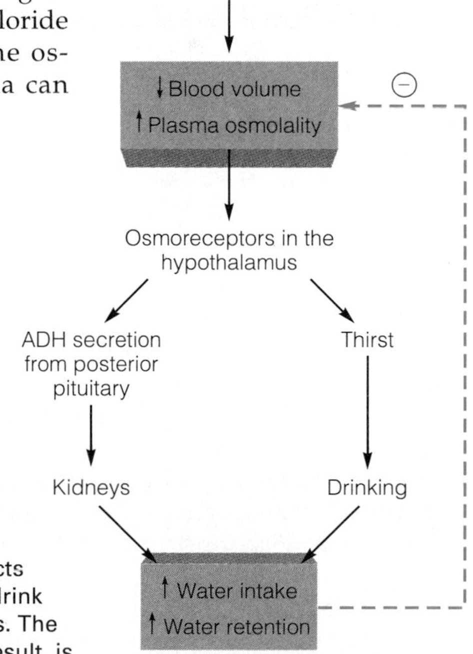

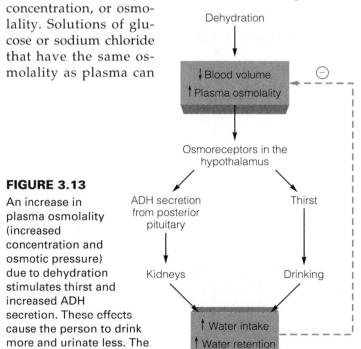

FIGURE 3.13

An increase in plasma osmolality (increased concentration and osmotic pressure) due to dehydration stimulates thirst and increased ADH secretion. These effects cause the person to drink more and urinate less. The blood volume, as a result, is increased while the plasma osmolality is decreased. These effects help bring the blood volume back to the normal range and complete the negative feedback loop (indicated by a negative sign).

easily be produced in a laboratory. Such solutions—labeled *5% dextrose* (5 g of glucose per 100 ml) and *normal saline* (0.9 g of NaCl per 100 ml)—are used clinically as intravenous fluids. Since 5% dextrose and normal saline have the same osmolality as plasma, they are said to be *isosmotic* to plasma.

The term *tonicity* is used to describe the effect of a solution on the osmotic movement of water. For example, if an isosmotic glucose or saline solution is separated from plasma by a membrane that is permeable to water but not to glucose or NaCl, osmosis will not occur. In this case the solution is said to be **isotonic** to plasma. Red blood cells placed in an isotonic solution will neither gain nor lose water.

Solutions that have a lower total concentration of osmotically active solutes and a lower osmotic pressure than plasma are said to be **hypotonic** to plasma. Red blood cells placed in hypotonic solutions gain water and may burst (a process called *hemolysis*). When red blood cells are placed in a **hypertonic** solution (such as seawater), which has a higher osmolality and osmotic pressure than plasma, they shrink due to the osmosis of water out of the cells. In this process, called *crenation*, the cell surface takes on a scalloped appearance (fig. 3.14).

Fluids delivered intravenously must be isotonic to blood in order to maintain the correct osmotic pressure and prevent cells from either expanding or shrinking due to the gain or loss of water. Common fluids used for this purpose are normal saline and 5% dextrose, which, as previously described, have about the same osmolality as normal plasma (approximately 300 mOsM). Another isotonic solution frequently used in hospitals is Ringer's lactate, *which contains glucose and lactic acid in addition to a number of different salts.*

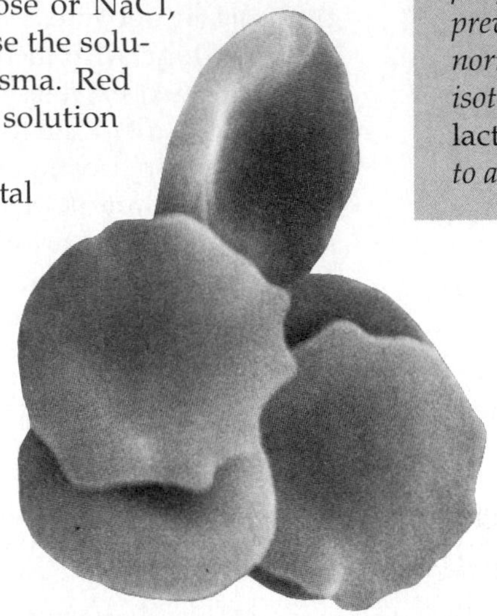

FIGURE 3.14
A scanning electron micrograph of normal and crenated red blood cells.

3.7 Carrier-Mediated Passive Transport

Carrier-mediated transport in which the net movement is down a concentration gradient, and which is therefore passive, is called facilitated diffusion.

Characteristics of Membrane Carriers

In order to sustain metabolism, cells must be able to take up glucose, amino acids, and other organic molecules from the extracellular environment. Molecules such as these, however, are too large and polar to pass through the lipid barrier of the cell membrane by simple diffusion. To move across the membrane, they need the help of **carrier proteins** within the membrane. This mediated transport has characteristics in common with enzyme activity; namely, *specificity* and *saturation.*

Like enzyme proteins, carrier proteins interact only with specific molecules. Glucose carriers, for example, interact with glucose only, to the exclusion of even closely related monosaccharides. As a further example of specificity, particular carriers for amino acids transport some types of amino acids but not others.

As the concentration of a transported molecule increases, its rate of transport also increases—but only up to a maximum. Beyond this rate, called the *transport maximum* (or T_m), further increases in concentration do not further increase the transport rate. This indicates that the carriers have reached their maximum capacity—they have become saturated (fig. 3.15).

As an example of saturation, imagine a bus stop that is serviced once per hour by a bus that can hold a maximum of 40 people (its "transport maximum"). If 10 people are waiting at the bus stop, 10 will be transported per hour. If 20 people are waiting, 20 will be transported per hour. This linear relationship will hold up to a maximum of 40 people; if there are 80 people at the bus stop, the transport rate will still be 40 per hour.

isotonic: Gk. *isos,* equal; *tonos,* tension
hypotonic: Gk. *hypo,* under; *tonos,* tension
hypertonic: Gk. *hyper,* over; *tonos,* tension
crenation: L. *crena,* a notch

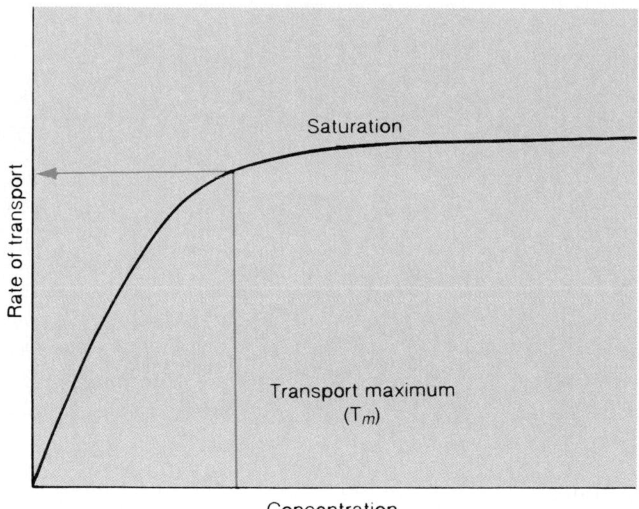

FIGURE 3.15
Carrier-mediated transport displays the characteristics of saturation, which is shown by the presence of a transport maximum (T_m).

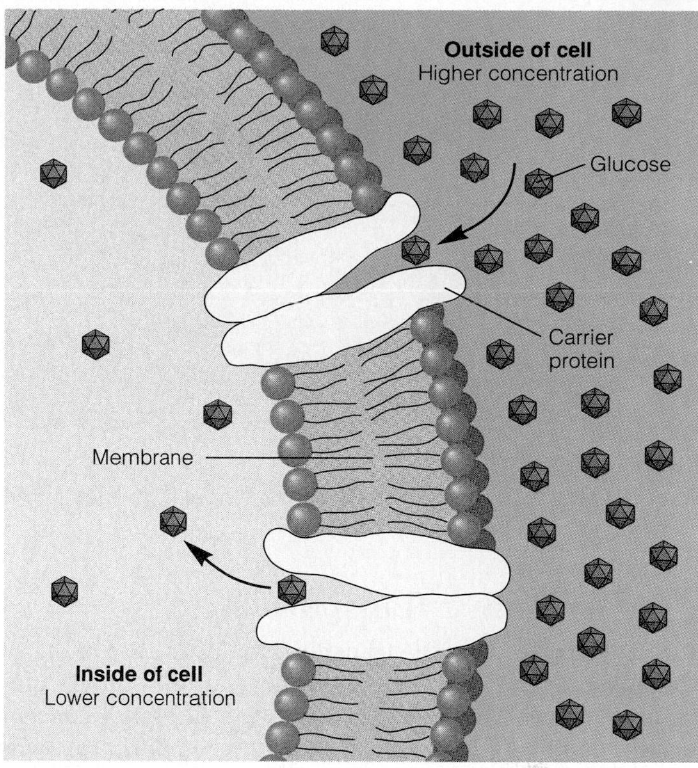

FIGURE 3.16
A model of facilitated diffusion. In this process, a molecule is transported across the cell membrane by a carrier protein.

The kidneys transport a number of molecules from the blood filtrate (which will become urine) back into the blood. Glucose, for example, is normally completely reabsorbed so that urine is normally free of glucose. However, if the glucose concentration of the blood and filtrate is too high (a condition called hyperglycemia*), the transport maximum will be exceeded. In this case, glucose will be found in the urine (a condition called* glycosuria*). This may result from eating too many sweets or from the inadequate action of the hormone insulin (in the disease* diabetes mellitus*).*

Facilitated Diffusion

The transport of glucose from the blood across the cell membranes of tissue cells occurs by a carrier-mediated process called **facilitated diffusion.** Facilitated diffusion, like simple diffusion, is powered by the thermal energy of the diffusing molecules and involves the net transport of substances through a cell membrane from the side of higher to the side of lower concentration. Active cellular metabolism is not required for either facilitated or simple diffusion.

Unlike simple diffusion, the facilitated diffusion of glucose through the cell membrane displays the properties of specificity and saturation. These are the characteristics of carrier-mediated transport. One conceptual model of the transport carriers is that each may be composed of two protein subunits that interact with glucose so as to create a channel through the membrane (fig. 3.16). By means of this channel, the glucose can move from the side of higher to the side of lower concentration.

3.8 Carrier-Mediated Active Transport

Carrier-mediated transport of molecules or ions against their concentration gradients requires metabolic energy, and thus is called active transport.

The transport of a molecule or ion "uphill" across a cell membrane, from lower to higher concentration, requires the expenditure of metabolic energy in the form of ATP. This is **active transport.** Such transport allows a cell to build up a high concentration of a particular molecule, or to eliminate molecules or ions that "leak" into the cell by diffusion. There are two subcategories of active transport, designated *primary* and *secondary.*

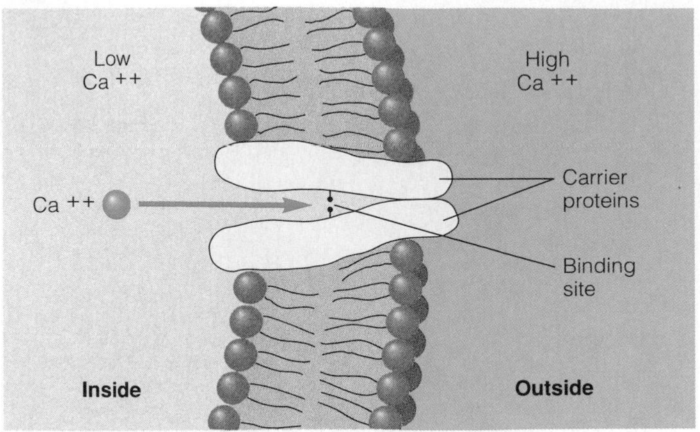

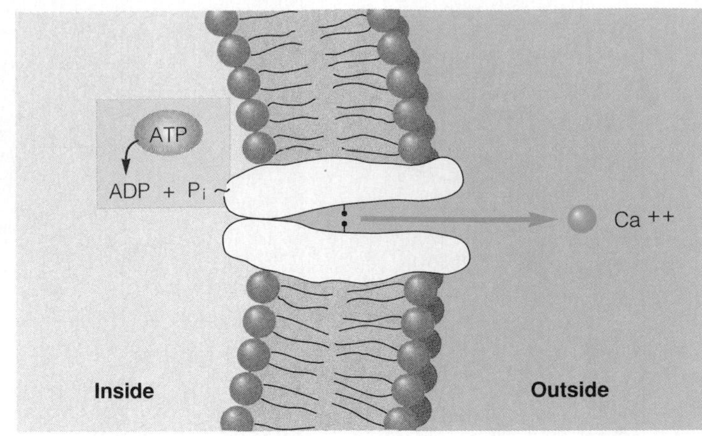

FIGURE 3.17

A model of active transport showing the hingelike motion of the protein subunits.

Primary Active Transport

As an example of active transport, the epithelial lining of the intestine and of the renal (kidney) tubules moves glucose from the side of lower to the side of higher concentration (from the lumen to the blood). Similarly, all cells extrude Ca^{++} into the extracellular environment and, by this means, maintain an intracellular Ca^{++} concentration that is up to 10,000 times lower than the extracellular Ca^{++} concentration. The movement of molecules and ions against their concentration gradients, from lower to higher concentrations, requires the expenditure of cellular energy that is obtained from ATP.

Primary active transport occurs when the hydrolysis of ATP is directly required for the function of the protein carrier. The following events are believed to occur: (1) the molecule or ion to be transported binds to a specific "recognition site" on one side of the carrier protein; (2) this bonding stimulates the breakdown of ATP, which in turn results in phosphorylation of the carrier protein; (3) as a result of phosphorylation, the carrier protein undergoes a conformational (shape) change; and (4) a hingelike motion of the carrier protein releases the transported molecule or ion on the other side of the membrane. This model of active transport is illustrated in figure 3.17.

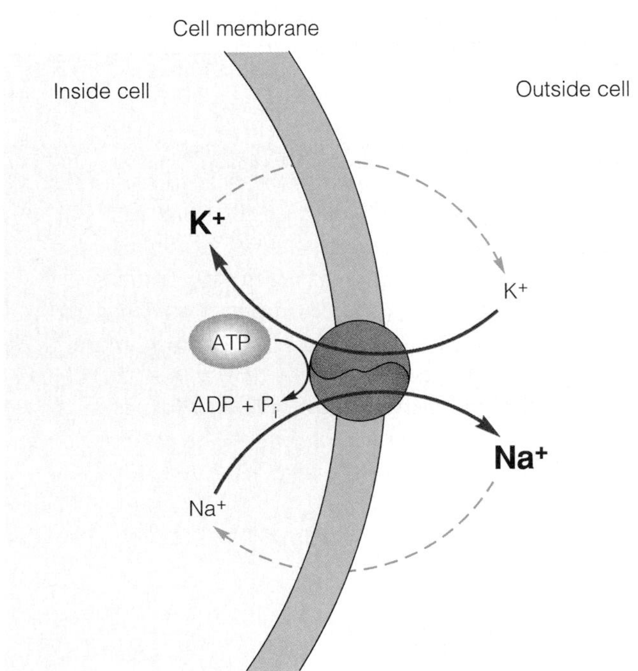

FIGURE 3.18

The Na^+/K^+ pump actively exchanges intracellular Na^+ for K^+. The carrier itself is an ATPase that breaks down ATP for energy. Dashed arrows indicate the direction of passive transport (diffusion); solid arrows indicate the direction of active transport.

The Sodium/Potassium Pump

Primary active transport carriers are often referred to as "pumps." Although some of these carriers transport only one molecule or ion at a time, other carriers exchange one molecule or ion for another. The most important of the latter type of carriers is the **Na^+/K^+pump.** This carrier protein, which is also an ATPase enzyme that converts ATP to ADP and P_i, actively extrudes three sodium ions from the cell as it transports two potassium ions into the cell. This transport is energy-dependent because Na^+ is more highly concentrated outside the cell and K^+ is more concentrated within the cell. Both ions, in other words, are moved against their concentration gradients (fig. 3.18).

Secondary Active Transport (Cotransport)

In **secondary active transport,** or **cotransport,** the energy needed for the "uphill" movement of a molecule or ion is obtained from the "downhill" transport of Na^+ into the

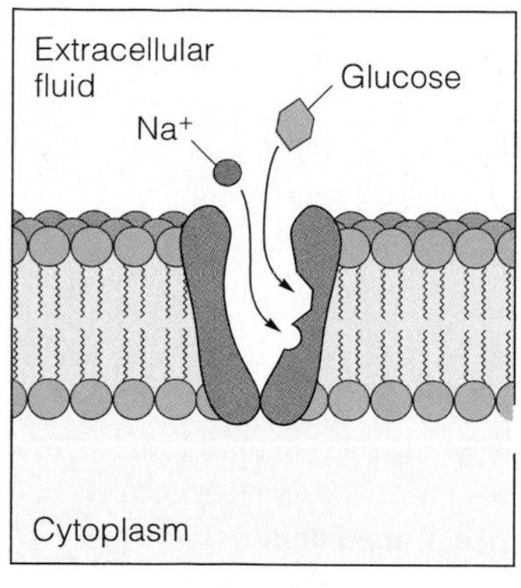

(a)

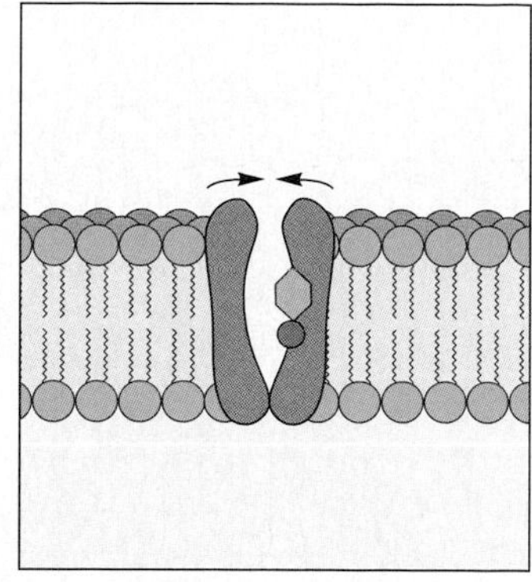

(b)

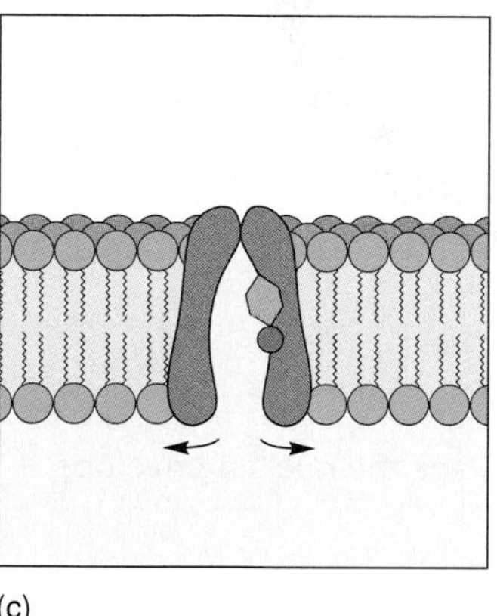

(c)

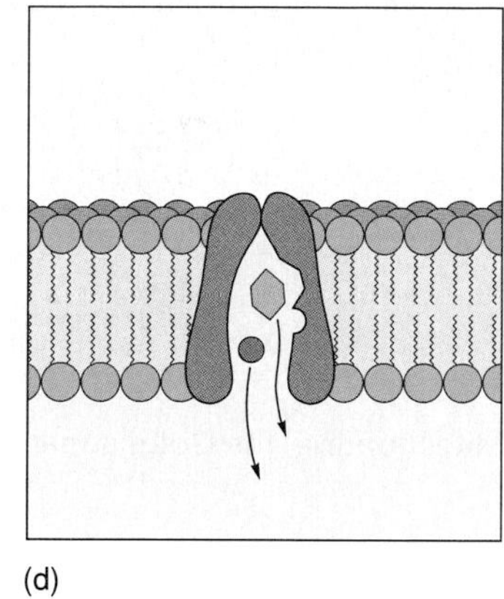

(d)

FIGURE 3.19

A model (*a–d*) for the cotransport of Na^+ and glucose into a cell. This is a secondary active transport because it is dependent upon the diffusion for Na^+ created by the Na^+/K^+ pumps.

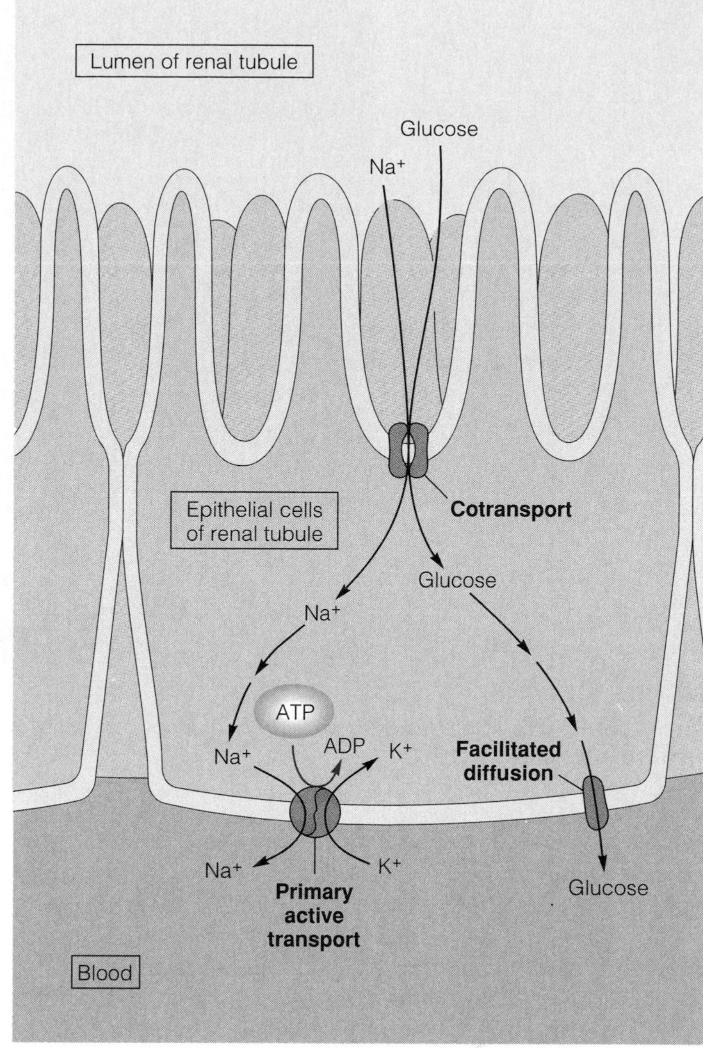

FIGURE 3.20

The transport of glucose from the fluid in the renal (kidney) tubules through the epithelial cells of the tubule and into the blood. All three types of carrier-mediated transport are used in this process.

cell. Hydrolysis of ATP by the action of the Na^+/K^+ pumps is required indirectly, in order to maintain low intracellular Na^+ concentrations. The diffusion of Na^+ into the cell then powers the uphill movement of a different ion or molecule across the cell membrane, either into or out of the cell.

In the epithelial cells of the renal tubules, for example, glucose is transported against its concentration gradient by a carrier that requires the simultaneous bonding of Na^+ (fig. 3.19). Glucose and Na^+ move in the same direction (into the cell) as a result of the Na^+ gradient created by the Na^+/K^+ pumps. Because of the distribution of specific carriers in the epithelial cell membrane, the glucose is moved from the lumen of the renal tubules into the blood (fig. 3.20).

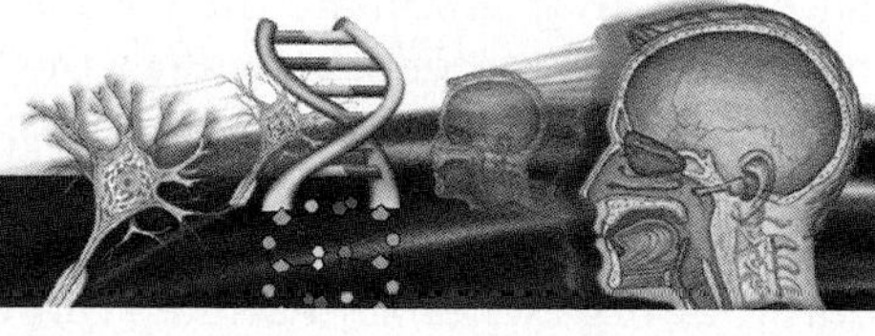

3.9 Lysosomes and Mitochondria

Lysosomes and mitochondria are two organelles that contain powerful enzymes. Lysosomes are involved in digestive reactions, and mitochondria serve as the sites for most of the cell's energy-yielding reactions.

Lysosomes

After a phagocytic cell has engulfed the proteins, polysaccharides, and lipids in a particle of "food" (such as a bacterium), these molecules are still kept isolated from the

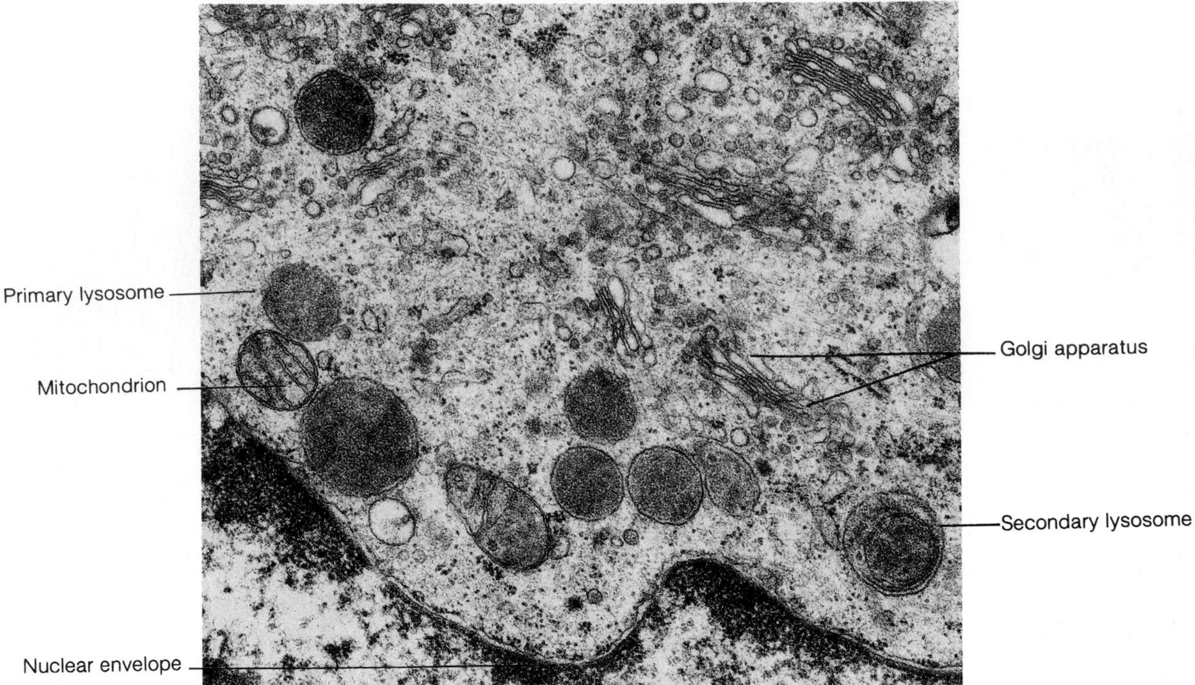

FIGURE 3.21
An electron micrograph showing primary and secondary lysosomes. Mitochondria, the Golgi apparatus, and the nuclear envelope are also indicated.

cytoplasm by the membranes surrounding the food vacuole. The large molecules of proteins, polysaccharides, and lipids must first be digested into their smaller subunits (amino acids, monosaccharides, and so on) before they can cross the vacuole membrane and enter the cytoplasm.

The digestive enzymes of a cell are isolated from the cytoplasm and concentrated within membrane-bound organelles called **lysosomes** (*li'sŏ-sōmz*) (fig. 3.21). A *primary lysosome* is one that contains only digestive enzymes (about 40 different kinds) within an environment that is considerably more acidic than the surrounding cytoplasm. A primary lysosome may fuse with a food vacuole (or with another cellular organelle) to form a *secondary lysosome* in which worn-out organelles and the products of phagocytosis can be digested. Thus, a secondary lysosome contains partially digested remnants of other organelles and ingested organic material. A lysosome that contains undigested wastes is called a *residual body*. Residual bodies may eliminate their wastes by exocytosis, or the wastes may accumulate within the cell as the cell ages.

The partly digested membranes of various organelles and other cellular debris often observed within secondary lysosomes result from **autophagy,** a process that destroys worn-out organelles so that they can be continuously replaced. Thus, lysosomes are said to constitute the "digestive system" of the cell.

lysosome: Gk. *lysis,* a loosening; *somo,* body
autophagy: Gk. *auto,* self; *phagein,* to eat

Inside Information—Lysosomes were not discovered until the early 1950s, but their existence and functions had been predicted before these organelles were actually observed in cells. Such was not the case with other organelles, whose structures generally were observed and described before their functional roles in the cell were understood.

Lysosomes are also referred to as "suicide bags," since a break in their membranes would release their digestive enzymes and thus destroy the cell. This happens normally as part of *programmed cell death,* in which tissues are destroyed in the process of embryological development. It also occurs in white blood cells during an inflammation reaction.

Most, if not all, molecules in the cell have a limited life span. They are continuously destroyed and must be continuously replaced. Glycogen and some complex lipids in the brain, for example, are digested normally at a particular rate by lysosomes. If a person, because of some genetic defect, does not have the proper amount of these lysosomal enzymes, the resulting abnormal accumulation of glycogen and lipids could destroy the tissues. Examples of such diseases include Tay–Sachs disease *and* Gaucher's disease.

FIGURE 3.22
(*a*) An electron micrograph of a mitochondrion. The outer membrane and the infoldings of the inner membrane, called cristae, are clearly seen. The fluid in the center is the matrix. (*b*) A diagrammatic representation of the structure of a mitochondrion.

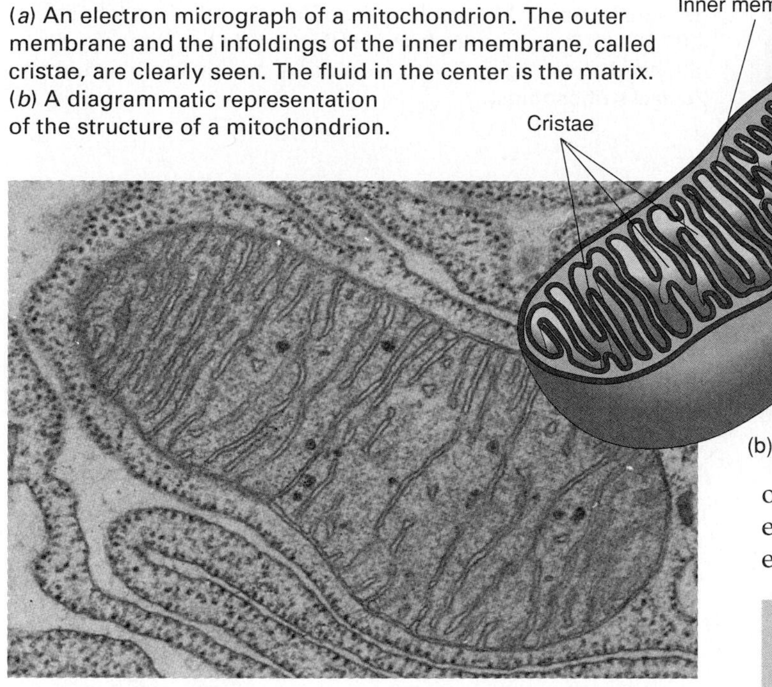

Mitochondria

All cells in the body, with the exception of mature red blood cells, have a hundred to a few thousand organelles called **mitochondria** (*mi"tŏ-kon'dre-ă*). Mitochondria serve as sites for the production of most of the energy in cells. For this reason, mitochondria are sometimes called the "powerhouses" of the cell.

Although mitochondria vary in size and shape, they all have the same basic structure (fig. 3.22). Each is surrounded by an *outer membrane* that is separated by a narrow space from an *inner membrane*. The inner membrane has many folds, called *cristae*, which extend into the central area (or *matrix*) of the mitochondrion. The cristae and the matrix compartmentalize the space within the mitochondrion, and they each have a distinct role in the generation of cellular energy.

Mitochondria are able to migrate through the cytoplasm of a cell, and it is believed that they are able to reproduce themselves. Indeed, mitochondria contain their own DNA! This is a more primitive form of DNA than that found within the cell nucleus. For these and other reasons, many scientists believe that mitochondria evolved from separate organisms, related to bacteria, that entered animal cells and remained in a state of symbiosis.

The cytoplasm of an ovum (egg cell) contains mitochondria, whereas the part of the sperm cell that fertilizes the ovum contains none. Therefore, all of the mitochondria in a fertilized egg are derived from the mother. The mitochondrial DNA replicates itself and the mitochondria divide, so that all of the mitochondria in the fertilized ovum and the cells derived from it during embryonic and fetal development are genetically identical to those in the original ovum. This provides a unique form of inheritance that is passed only from mother to child. A rare cause of blindness known as Leber's hereditary optic neuropathy, *as well as several other disorders that are inherited only along the maternal lineage, are known to be caused by defective mitochondrial DNA.*

3.10 Endoplasmic Reticulum and the Golgi Apparatus

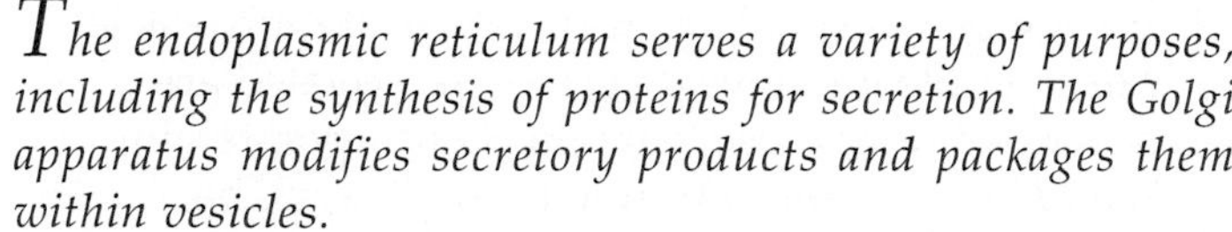

The endoplasmic reticulum serves a variety of purposes, including the synthesis of proteins for secretion. The Golgi apparatus modifies secretory products and packages them within vesicles.

Endoplasmic Reticulum

Most cells contain a system of internal membranes known as the **endoplasmic reticulum** (*en'do-plaz'mik rĕ-tik'yŭ-lum*), often abbreviated **ER.** Although the name sounds complicated, *endoplasmic* simply means "within the plasm" (cytoplasm of the cell) and *reticulum* means "network." There are two types of endoplasmic reticulum:

1. a **rough,** or **granular, endoplasmic reticulum** (fig. 3.23*b*), characterized by *ribosomes* on its surface; and
2. a **smooth endoplasmic reticulum** (fig. 3.23*c*), which lacks ribosomes.

Ribosomes (*ri'bŏ-somz*) are tiny structures composed of protein and RNA that serve as sites of protein synthesis. Ribosomes that are free in the cytoplasm of the cell serve as sites for the synthesis of cellular proteins. Ribosomes attached to the membranes of the rough endoplasmic

mitochondrion: Gk. *mitos*, thread; *chondros*, grain

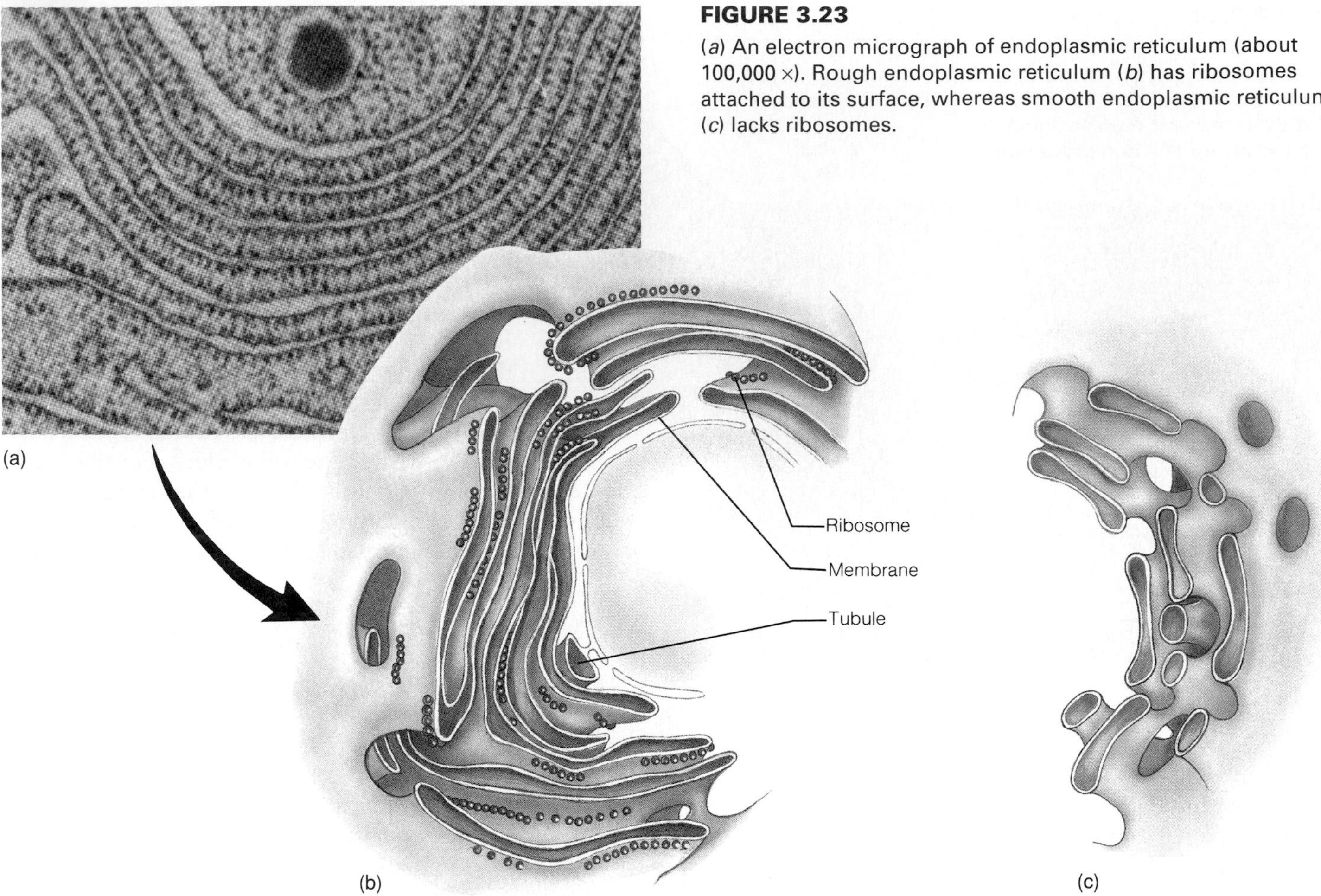

FIGURE 3.23

(*a*) An electron micrograph of endoplasmic reticulum (about 100,000 ×). Rough endoplasmic reticulum (*b*) has ribosomes attached to its surface, whereas smooth endoplasmic reticulum (*c*) lacks ribosomes.

reticulum serve as sites for the synthesis of proteins that will be secreted by the cell. The rough endoplasmic reticulum is thus found in the cells that are active in protein synthesis and secretion, such as those of many exocrine and endocrine glands.

The role of the smooth endoplasmic reticulum varies according to the type of cell in which it is found. It serves as a site for enzyme reactions in steroid hormone production and inactivation, for example, and as a site for the storage of Ca^{++} in skeletal muscle cells.

The smooth endoplasmic reticulum in liver cells contains enzymes used for the inactivation of steroid hormones and many drugs. This inactivation is generally achieved by reactions that convert these compounds to more water-soluble and less active forms that can be more easily excreted by the kidneys. When people take certain drugs (such as alcohol or phenobarbital) for a long period of time, larger and larger doses of these compounds are required to produce a given effect. This phenomenon, called *tolerance,* is accompanied by an increase in the smooth endoplasmic reticulum, and thus an increase in the enzymes charged with inactivation of these drugs.

Golgi Apparatus

Secretory proteins do not remain trapped within the rough endoplasmic reticulum; they are transported to another organelle within the cell—the **Golgi** (*gol'je*) **apparatus.** This organelle serves three interrelated functions.

1. It further alters or changes proteins (for example, by adding carbohydrates to form *glycoproteins*).
2. It separates different types of proteins according to their function and destination.
3. It packages the final products into secretory vesicles. For example, proteins that are to be secreted are separated from those that will be incorporated into the cell membrane and from those that will be introduced into lysosomes by becoming packaged within separate membrane-enclosed vesicles.

The Golgi apparatus consists of several flattened, hollow sacs that enclose *cisternae* (spaces). Proteins produced

Golgi apparatus: from Camillo Golgi, Italian histologist, 1843–1926

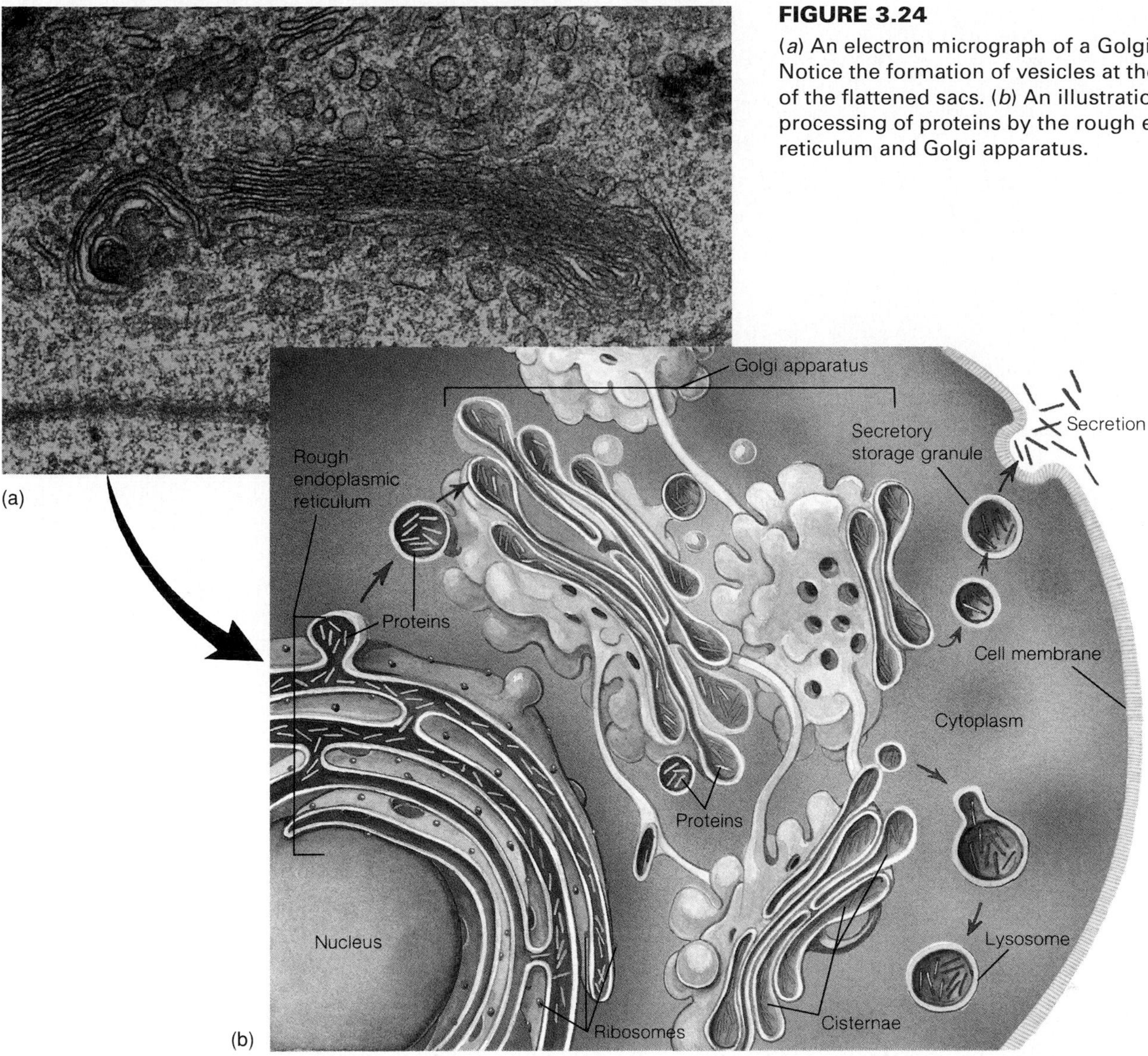

FIGURE 3.24
(*a*) An electron micrograph of a Golgi apparatus. Notice the formation of vesicles at the ends of some of the flattened sacs. (*b*) An illustration of the processing of proteins by the rough endoplasmic reticulum and Golgi apparatus.

by the rough endoplasmic reticulum are believed to travel in membrane-enclosed vesicles to the sac on one end of the Golgi apparatus. After specialized modifications of the proteins are made within one sac, the modified proteins are passed by means of vesicles to the next sac. The finished products finally leave the Golgi apparatus in vesicles that fuse with the cell membrane in the process of exocytosis (fig. 3.24).

3.11 Cell Nucleus and DNA

The genetic code is contained in the structure of DNA, which is found within the cell nucleus. DNA and a related molecule, RNA, are known as nucleic acids.

The largest, most conspicuous of the cell's organelles is the spherical or oval-shaped nucleus. Most cells in the body have a single nucleus, although some—such as skeletal muscle cells—have many nuclei. The internal region of the nucleus is filled with a gel-like fluid called *nucleoplasm,* similar in consistency to the cytosol within the cytoplasm of the cell. Surrounding this internal region is a nuclear envelope composed of inner and outer membranes. At numerous points on the surface of the nucleus, these membranes fuse to form small openings called *nuclear*

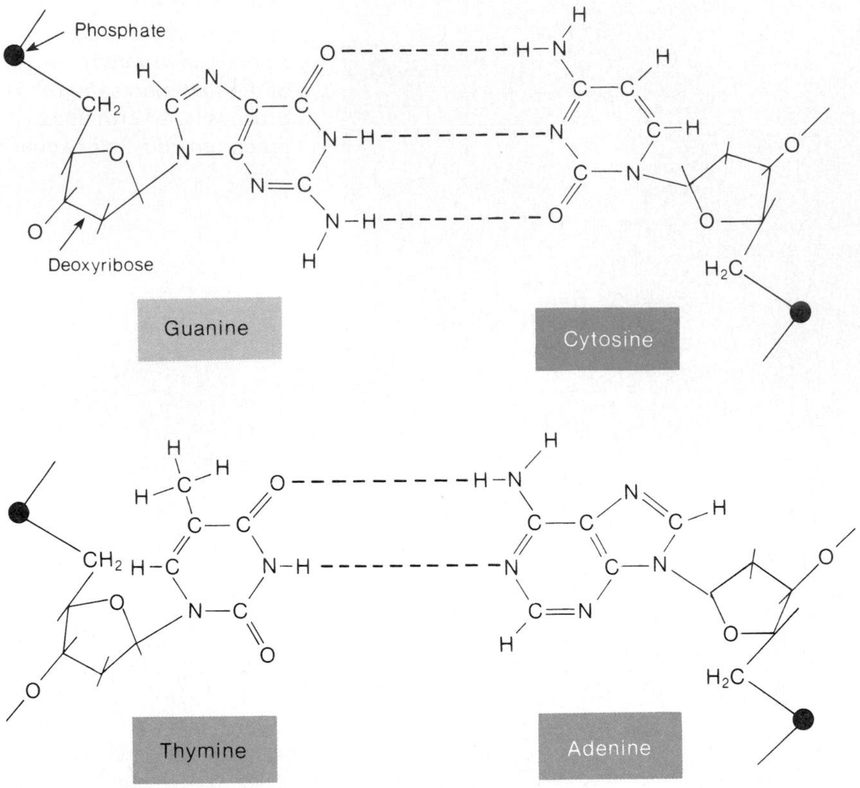

FIGURE 3.25
The four nitrogenous bases in deoxyribonucleic acid (DNA). Notice that hydrogen bonds can form between guanine and cytosine and between thymine and adenine.

pores. These small, cylindrical channels permit the movement of large molecules between the nucleoplasm and cytoplasm during protein synthesis.

Nucleic Acids

The nucleus contains molecules of DNA, and it is the structure of DNA that serves as the basis for the genetic code. The nucleus thus functions as the "control center" of the cell, directing its activities through the actions of its *nucleic acids.*

Nucleic acids include the macromolecules of **DNA** and **RNA,** and the subunits from which these molecules are formed. These subunits are known as **nucleotides.** Each nucleotide is composed of three smaller subunits: a five-carbon sugar, a phosphate group bonded to one end of the sugar, and a nitrogenous base bonded to the other end of the sugar (fig. 3.25). The nitrogenous bases are cyclic nitrogen-containing molecules of two kinds: pyrimidines and purines. The *pyrimidines* (*pĭ-rim'ĭ dēnz*) contain a single ring of carbon and nitrogen, whereas the *purines* have two such rings.

Deoxyribonucleic Acid (DNA)

The structure of DNA serves as a basis for the genetic code, but actually it has a simpler structure than that of most proteins. The model for this structure was devised by the English physicist-turned-biologist Francis Crick and the young American chemist James Watson. Building on key experiments of others, they solved the puzzle of how such a seemingly simple molecule could carry out such a complex function.

Sugar molecules in the nucleotides of DNA are a type of pentose (five-carbon) sugar called **deoxyribose** (*de-ok"se-ri'bōs*). Each deoxyribose can be covalently bonded to one of four possible bases. These bases include the two purines (*adenine* and *guanine*) and the two pyrimidines (*cytosine* and *thymine*). Thus, there are four different types of nucleotides that can be used to produce the long DNA chains.

When nucleotides combine to form a chain, the phosphate group of one nucleotide condenses with the deoxyribose sugar of another. This forms a sugar-phosphate chain as water is removed in dehydration synthesis. Since the nitrogenous bases are attached to the sugar molecules, the sugar-phosphate chain looks like a backbone from which the bases project. Each of these bases can form hydrogen bonds with other bases, which are part of a different chain of nucleotides. Such hydrogen bonding between bases produces a *double-stranded* DNA molecule; the two strands are like a staircase, with the paired bases as steps.

Actually, the two chains of DNA twist about each other to form a **double helix,** so that the molecule resembles a spiral staircase (fig. 3.26). It has been shown that the number of purine bases in DNA is equal to the number of pyrimidine bases. The reason for this is explained by the **law of complementary base pairing:** adenine can pair only with thymine (through two hydrogen bonds), whereas guanine can pair only with cytosine (through three hydrogen bonds). Applying this rule, we can predict the base sequence of one DNA strand if we know the sequence of bases in the complementary strand.

Although we can be certain of which base is opposite a given base in DNA, we cannot predict which bases will be above or below that particular pair within a single polynucleotide chain. Although there are only four bases, the number of possible base sequences along a stretch of several thousand nucleotides (the length of a gene) is almost infinite. Despite the almost infinite variety of possible sequences, almost all of the billions of copies of a particular gene in a person are identical. The mechanisms by which this is achieved will be described shortly.

Inside Information—Watson and Crick first described their vision of DNA in 1953, in the journal *Nature.* The closing sentence of their brief article (a mere 900 words) is a marvel of restraint: "It has not escaped our notice that the specific pairing we have postulated . . . immediately suggests a possible copying mechanism for the genetic material."

The Human Genome Project *was launched by Congress in 1988 with the ambitious goal of completely mapping the human genome by September 30, 2005. That allowed just 15 years to determine the exact sequences of bases with which the 3 billion base pairs are arranged to form the 50,000 to 100,000 genes in the haploid human genome (the genome of a sperm cell or ovum). Is spending the billions of dollars required for this enormous project justified? Consider that such a detailed map will provide the ultimate reference for diagnosis and treatment of the 4,000 genetic diseases that are known to be directly caused by particular abnormal genes.*

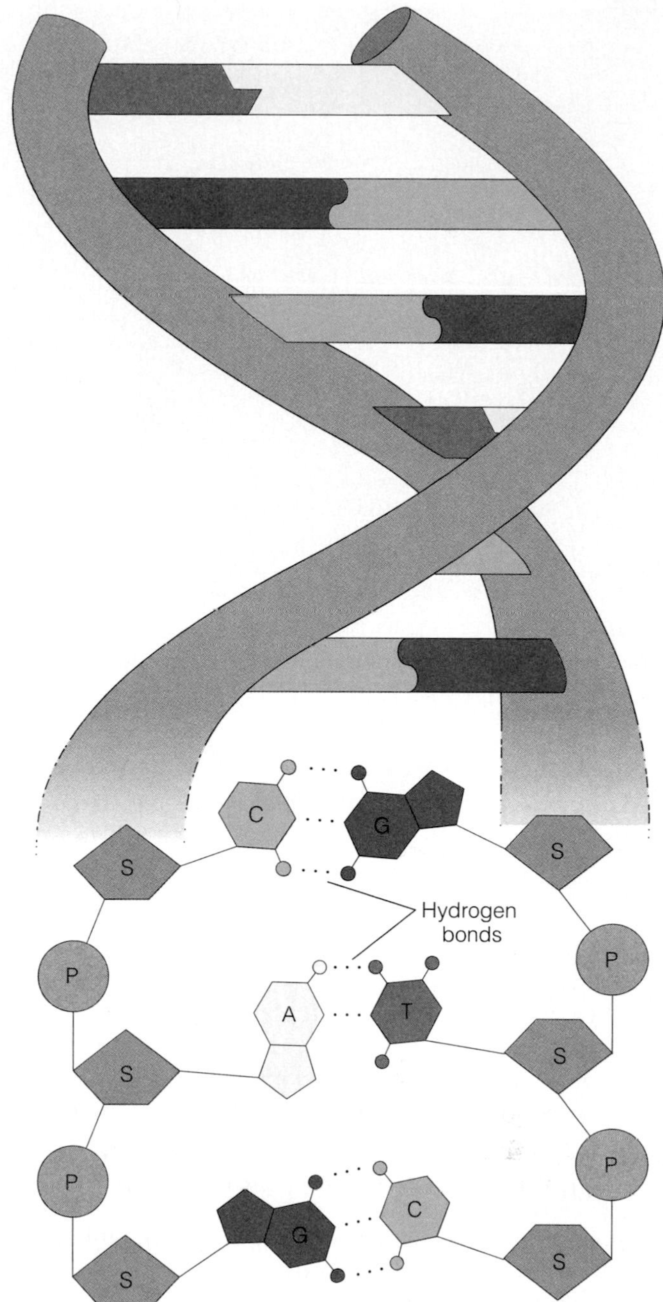

FIGURE 3.26
The double-helix structure of DNA.

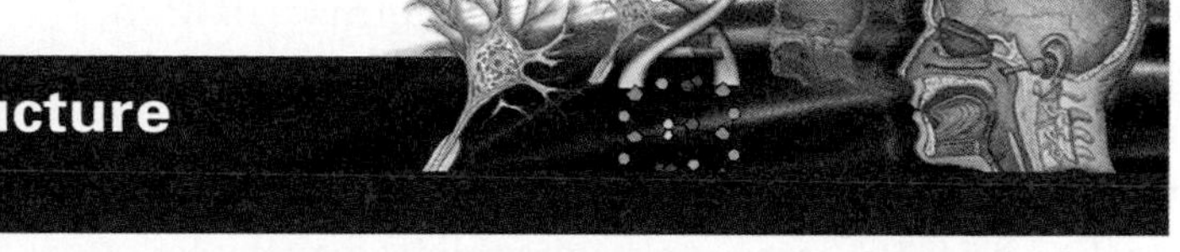

3.12 Encoding Information—RNA Structure and Synthesis

RNA is a single-stranded polynucleotide chain with a base sequence that is complementary to a region of the DNA. The genetic code of DNA is expressed through its direction of RNA synthesis.

Structure of Ribonucleic Acid (RNA)

The job of creating proteins is not performed by DNA directly. Rather, the DNA creates a messenger molecule of complementary structure to carry the genetic information

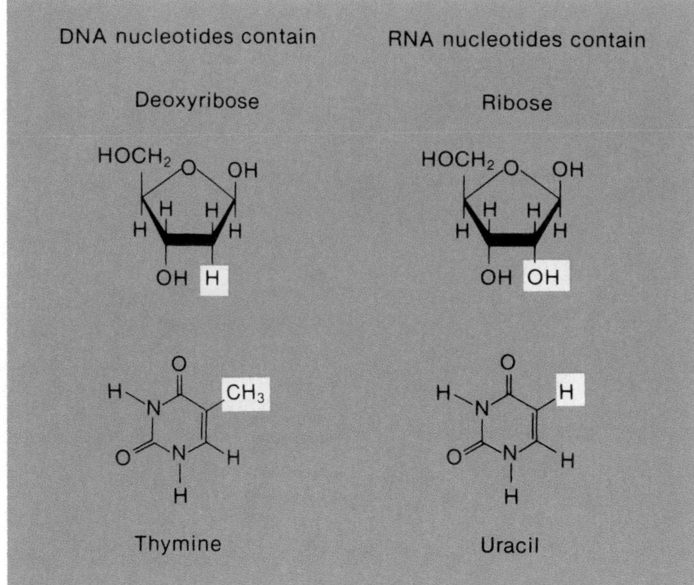

FIGURE 3.27
Differences between the nitrogenous bases and sugars in DNA and RNA.

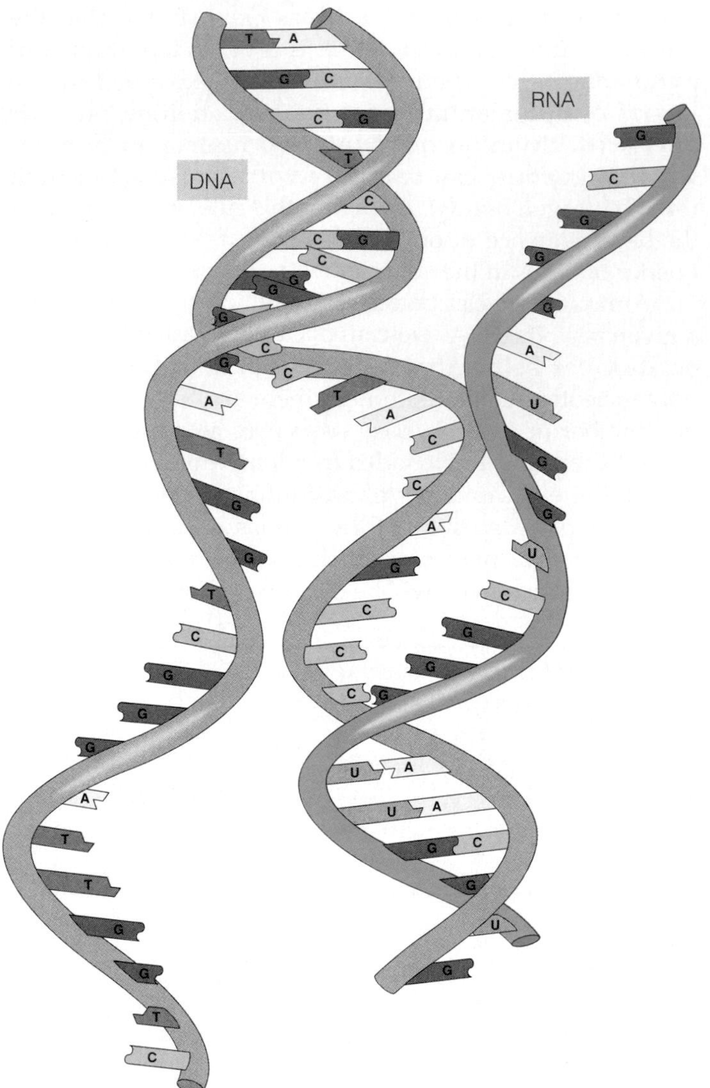

FIGURE 3.28
RNA synthesis (genetic transcription). Notice that only one of the two DNA strands is used to form a single-stranded molecule of RNA.

used in protein synthesis. Like DNA, RNA consists of long chains of nucleotides joined together by sugar-phosphate bonds. However, as shown in figure 3.27, nucleotides in RNA differ from those in DNA in the following ways:

1. A **ribonucleotide** contains the sugar *ribose* (instead of deoxyribose).
2. The base *uracil* is present in place of thymine.
3. RNA is composed of a single polynucleotide strand (it is not double-stranded like DNA).
4. RNA is considerably shorter than DNA.

Four types of RNA are produced within the nucleus, each with a different composition and function:

1. **precursor messenger RNA (pre-mRNA),** which is altered within the nucleus (through cutting and splicing) to form mRNA;
2. **messenger RNA (mRNA),** which contains the code for the synthesis of specific proteins;
3. **transfer RNA (tRNA),** which transfers amino acids and which is needed for decoding the genetic message contained in mRNA; and
4. **ribosomal RNA (rRNA),** which forms part of the structure of ribosomes.

The DNA that codes for rRNA synthesis is located in the nucleolus. Pre-mRNA and tRNA synthesis is controlled by DNA located elsewhere in the nucleus.

Genetic Transcription—RNA Synthesis

The chromosomes seen during cell division (discussed in module 3.13) are inactive packages of DNA. The genes do not become active until the chromosomes unravel. Active DNA directs the metabolism of the cell indirectly through its regulation of RNA and protein synthesis.

One gene codes for one polypeptide chain. Each gene is a stretch of DNA that is several thousand nucleotide pairs long. In order for the genetic code to be translated for the synthesis of specific proteins, the DNA code must first be transcribed into an RNA code (fig. 3.28). This is accomplished by DNA-directed RNA synthesis, or **genetic transcription.**

During RNA synthesis, the enzyme *RNA polymerase* breaks the weak hydrogen bonds between paired DNA bases. This does not occur throughout the length of DNA, but only in the regions that are to be transcribed (there are base sequences that code for "start" and "stop"). Double-stranded DNA, therefore, separates in these regions so that the freed bases can pair with the complementary RNA nucleotide bases that are freely available in the nucleus.

This pairing of bases follows the law of complementary base pairing: guanine bonds with cytosine (and vice versa), and adenine bonds with uracil (because uracil in RNA is equivalent to thymine in DNA). In RNA synthesis, only *one* of the two freed strands of DNA serves as a guide (fig. 3.28). Once an RNA molecule has been produced, it detaches from the DNA strand on which it was formed. This process can continue indefinitely, producing many thousands of RNA copies of the DNA strand being transcribed. When the gene is no longer to be transcribed, the separated DNA strands can recoil into their helical form.

In the case of pre-mRNA, the finished molecule is altered after synthesis. Within the pre-mRNA are noncoding regions known as *introns.* The introns are removed through the action of enzymes, and the coding regions are then spliced together so that they can direct the synthesis of a specific protein.

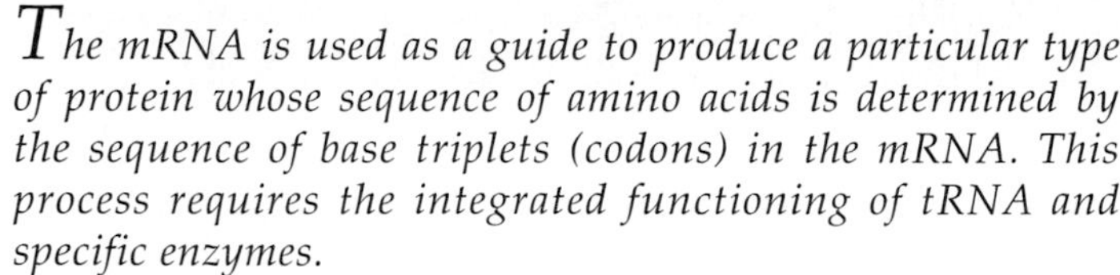

3.13 Processing the Information—Protein Synthesis

The mRNA is used as a guide to produce a particular type of protein whose sequence of amino acids is determined by the sequence of base triplets (codons) in the mRNA. This process requires the integrated functioning of tRNA and specific enzymes.

Once produced, mRNA leaves the nucleus and enters the cytoplasm, where it attaches to ribosomes. The mRNA passes through a number of ribosomes to form a "string-of-pearls" structure called a *polyribosome,* or *polysome* for short. The association of mRNA with ribosomes is needed for **genetic translation**—the production of specific proteins according to the code contained in the mRNA base sequence.

Functions of Codons and Anticodons

Each mRNA molecule contains several hundred or more nucleotides, arranged in the sequence determined by complementary base pairing with DNA during genetic transcription (RNA synthesis). Every three bases, or *base triplet,* is a "code word"—called a **codon**—for a specific amino acid. Sample codons and their amino acid "translations" are shown in table 3.2 and in figure 3.29. As mRNA moves through the ribosome, the sequence of codons is translated into a sequence of specific amino acids within a growing polypeptide chain.

Translation of the codons is accomplished by transfer RNA (tRNA) and particular enzymes. One end of each tRNA contains the **anticodon.** The anticodon consists of three nucleotides that are complementary to a specific codon in mRNA. Enzymes in the cell cytoplasm called *aminoacyl-tRNA synthetase enzymes* join specific amino acids to the ends of tRNA, so that a tRNA with a given anticodon is always bonded to one specific amino acid. There are 20 different varieties of synthetase enzymes—one for each type of amino acid. Each synthetase must not only recognize its specific amino acid, it must also be able to attach this amino acid to the particular tRNA that has the correct anticodon for that amino acid. Each

TABLE 3.2 Selected DNA Base Triplets and mRNA Codons

DNA Triplet	RNA Codon	Amino Acid
TAC	AUG	"Start"
ATC	UAG	"Stop"
AAA	UUU	Phenylalanine
AGG	UCC	Serine
ACA	UGU	Cysteine
GGG	CCC	Proline
GAA	CUU	Leucine
GCT	CGA	Arginine
TTT	AAA	Lysine
TGC	ACG	Tyrosine
CCG	GGC	Glycine
CTC	GAG	Aspartic acid

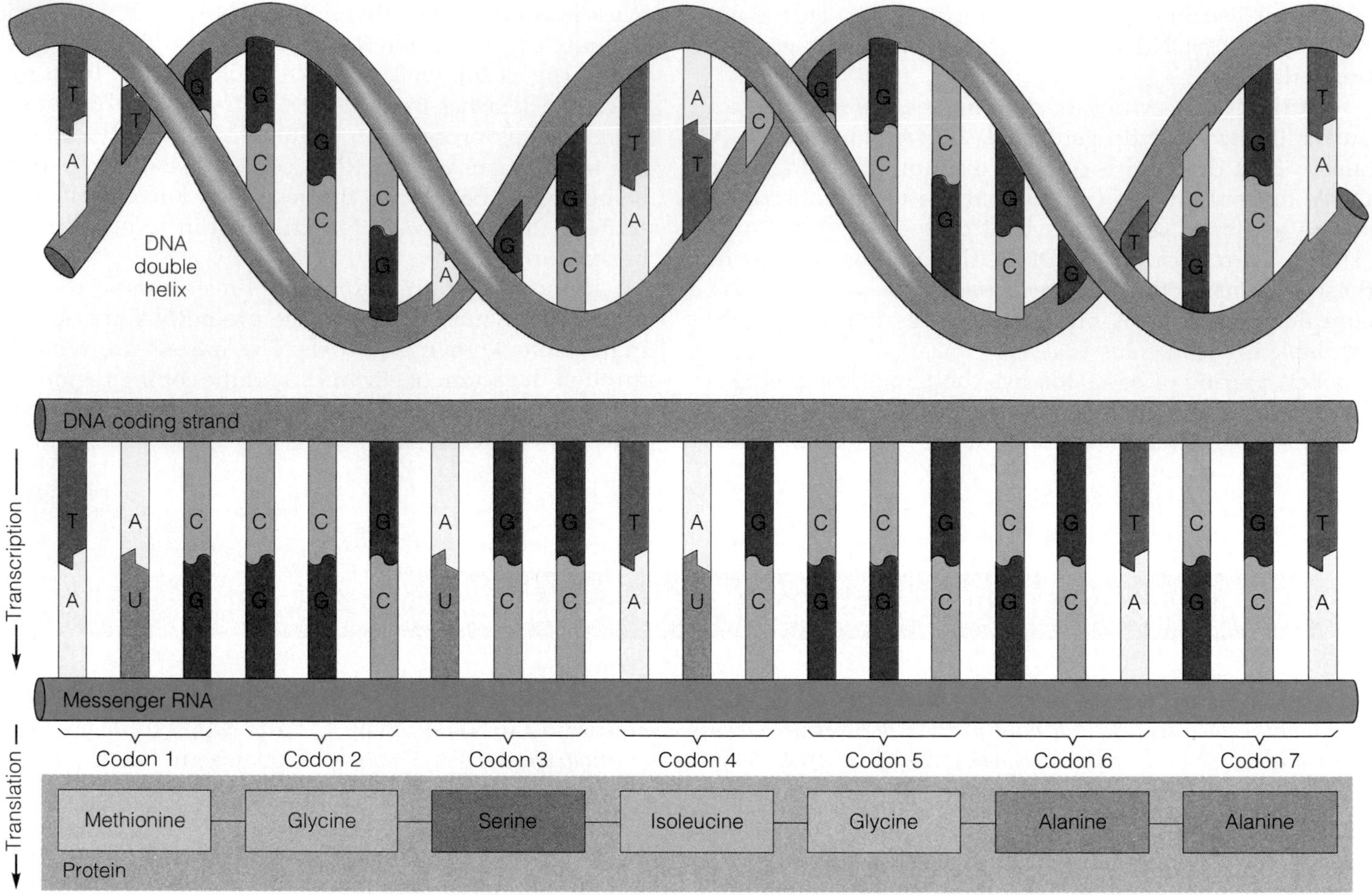

FIGURE 3.29
The genetic code is first transcribed into base triplets (codons) in mRNA and then translated into a specific sequence of amino acids in a protein.

of the tRNA molecules in the cytoplasm of a cell is thus bonded to a specific amino acid and is capable of bonding by its anticodon base triplet with a specific codon in mRNA.

Formation of a Polypeptide

The anticodons of tRNA bind to the codons of mRNA as the mRNA moves through the ribosome. Since each tRNA molecule carries a specific amino acid, the joining together of these amino acids by peptide bonds forms a polypeptide whose amino acid sequence has been determined by the sequence of codons in mRNA.

The first and second tRNA bring the first and second amino acids together, and a peptide bond forms between them. The first amino acid then detaches from its tRNA, so that a dipeptide is linked by the second amino acid to the second tRNA. When the third tRNA binds to the third codon, the amino acid it brings forms a peptide bond with the second amino acid (which detaches from its tRNA). A tripeptide is thus attached by the third amino acid to the third tRNA. The polypeptide chain thus grows as new amino acids are added to its growing tip (fig. 3.30). This growing polypeptide chain is always attached by means of only one tRNA to the strand of mRNA, and this tRNA molecule is always the one that has added the latest amino acid to the growing polypeptide.

As the polypeptide chain becomes longer, interactions between its amino acids cause the chain to twist into a helix (secondary structure) and to fold and bend upon itself (tertiary structure). At the end of this process, the new protein detaches from the tRNA as the last amino acid is added.

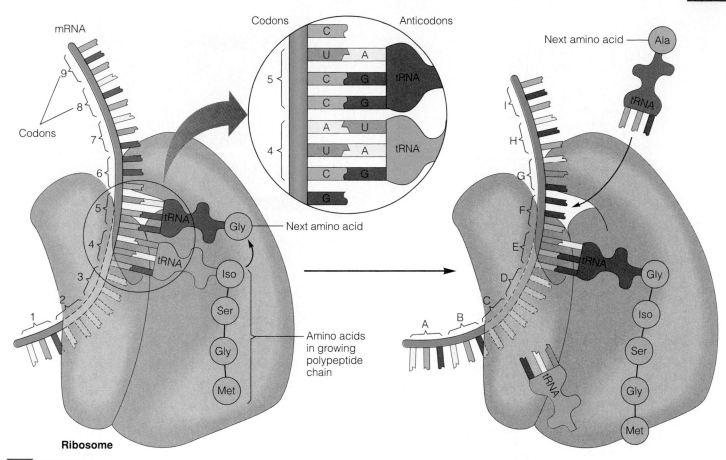

FIGURE 3.30
The actions of mRNA and tRNA in genetic translation. The three-letter abbreviations for the amino acids in the growing polypeptide chain stand for the amino acids indicated in figure 3.29.

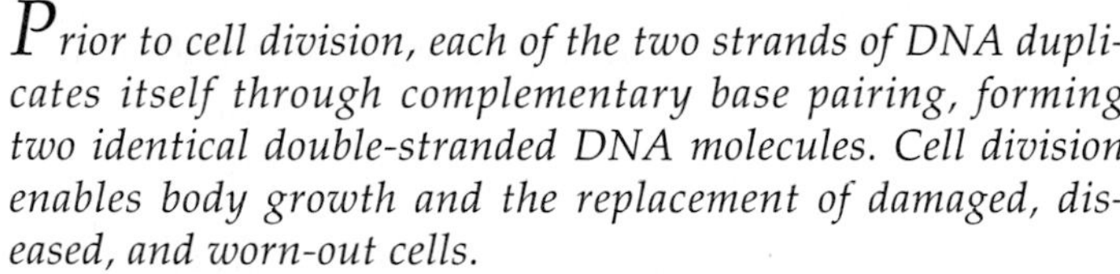

3.14 Reproduction of Information—DNA Synthesis and Mitosis

Prior to cell division, each of the two strands of DNA duplicates itself through complementary base pairing, forming two identical double-stranded DNA molecules. Cell division enables body growth and the replacement of damaged, diseased, and worn-out cells.

DNA Replication

When a cell is going to divide, each DNA molecule replicates itself, and each of the two daughter cells receives one of the identical DNA copies thus produced. Replication of DNA requires the action of a specific enzyme known as *DNA polymerase.* This enzyme moves along the DNA molecule, breaking the weak hydrogen bonds between complementary bases as it travels. As a result, the bases of each of the two DNA strands become free to bind to new complementary bases.

Because of the rules of complementary base pairing, the bases of each original strand will bind to the appropriate free nucleotides: adenine bases pair with thymine-containing nucleotides; guanine bases pair with cytosine-containing nucleotides, and so on. This

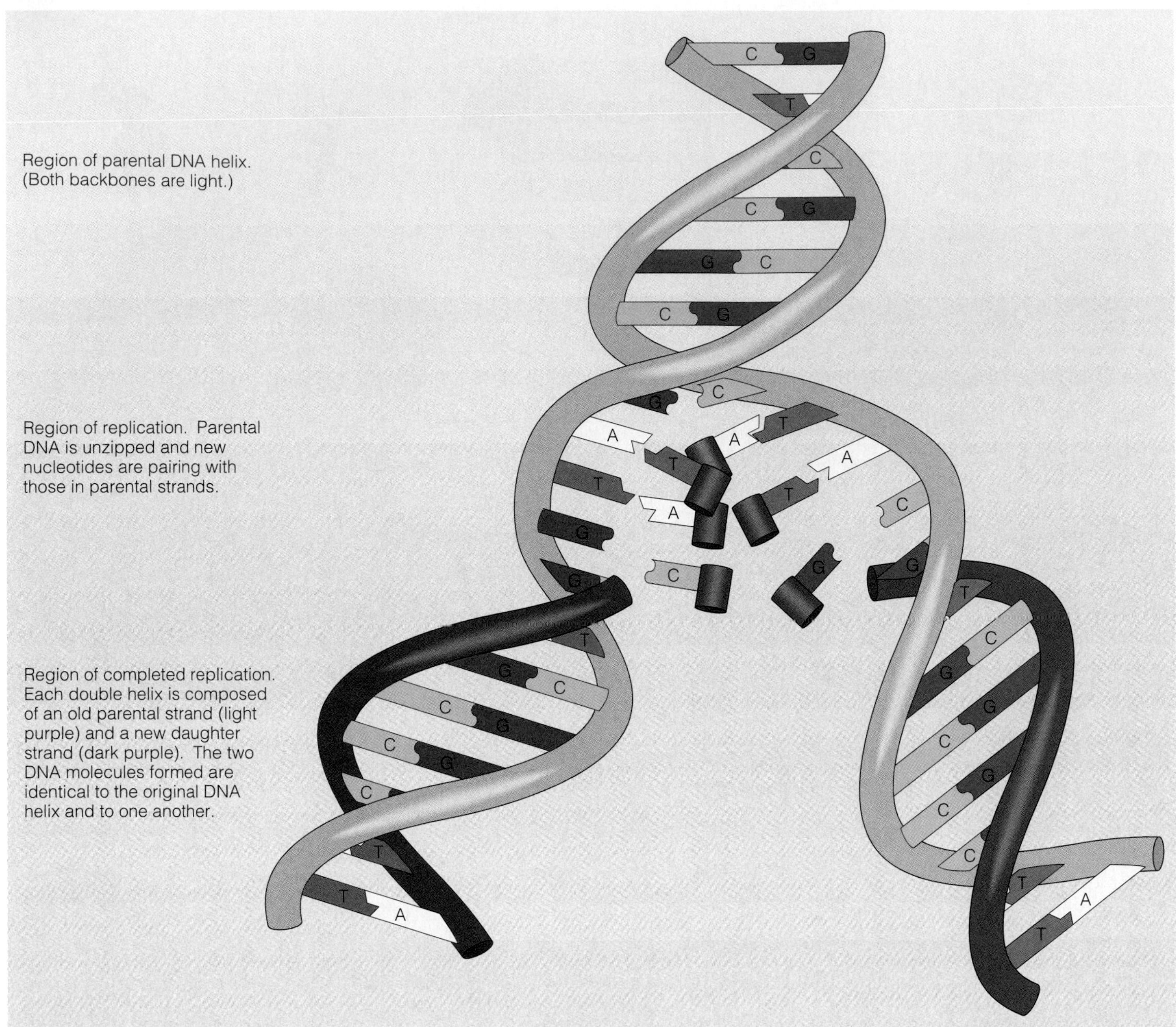

Region of parental DNA helix. (Both backbones are light.)
Region of replication. Parental DNA is unzipped and new nucleotides are pairing with those in parental strands.
Region of completed replication. Each double helix is composed of an old parental strand (light purple) and a new daughter strand (dark purple). The two DNA molecules formed are identical to the original DNA helix and to one another.

FIGURE 3.31
The replication of DNA. Each new double helix is composed of one old and one new strand. The base sequences of each of the new molecules are identical to those of the parent DNA because of complementary base pairing.

ensures that the two new double-stranded molecules of DNA will contain the same base sequence as the parent molecule (fig. 3.31). Replication is said to be *semiconservative* because one strand is "old" (it is "conserved") and the other strand is "new." Through this mechanism, the sequence of bases in DNA—which is the basis of the genetic code—is preserved from one cell generation to the next.

The nondividing cell is in a part of its life cycle known as **interphase.** During interphase, the DNA is dispersed throughout the nucleus as a network of fibers called *chromatin;* it is not organized in the short, thick, rodlike structures familiar to us as *chromosomes.* If a cell is going to divide, it replicates its DNA, which condenses to eventually form chromosomes that are easily seen in the ordinary (light) microscope.

chromatin: Gk. *chroma,* color

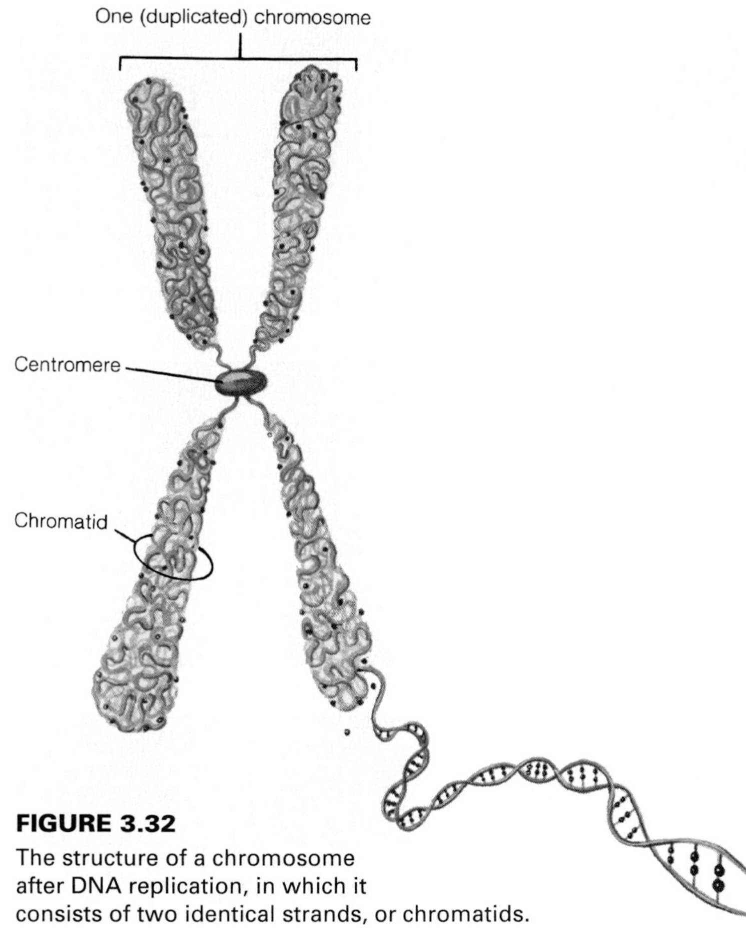

FIGURE 3.32
The structure of a chromosome after DNA replication, in which it consists of two identical strands, or chromatids.

At the end of this process, each chromosome consists of two strands called **chromatids** (*kro'mă-tidz*) that are joined together by a **centromere** (fig. 3.32). The two chromatids within a chromosome contain identical DNA base sequences because each is produced by the semiconservative replication of DNA. Each chromatid, therefore, contains a complete double-helix DNA molecule that is a copy of the DNA molecule existing prior to replication. Each chromatid will become a separate chromosome once cell division has been completed.

Mitosis

Mitosis (*mi-to'sis*) is the process of cell division that results in two identical *daughter cells,* containing the same number of chromosomes. It is through mitosis that an organism grows and replaces worn-out cells and tissues. When a cell is going to divide, an organelle called the **centrosome** replicates itself, and each centrosome takes a position on opposite sides of the nucleus. The centrosomes then produce **spindle fibers,** which are composed of microtubules. The spindle fibers will eventually pull the duplicated chromosomes to opposite poles of the cell during cell division. Cells that lack a centrosome, such as mature muscle cells and nerve cells (neurons), cannot divide.

The cell next proceeds through the various stages of mitosis: *prophase, metaphase, anaphase,* and *telophase* (fig. 3.33). The nuclear membrane disappears during prophase, and the centrosomes take up positions at opposite poles of the cell and produce spindle fibers. In metaphase of mitosis, the chromosomes line up single file along the equator of the cell. This aligning of chromosomes at the equator is believed to result from the action of the spindle fibers that are attached to the centromere of each chromosome.

Anaphase begins when the centromeres split apart and the spindle fibers shorten, pulling the two chromatids in each chromosome to opposite poles. Each pole therefore gets one copy of each of the 46 chromosomes. Division of the cytoplasm (*cytokinesis*) during telophase results in the production of two daughter cells that are genetically identical to each other and to the original parent cell.

Adult skeletal and cardiac (heart) muscle cells cannot divide by mitosis. Consequently, the growth of these organs can be achieved only through an increase in the size of their cells. This type of growth is called hypertrophy. *When it occurs in skeletal muscles in response to an increased workload—during weight training, for example—it is called* compensatory hypertrophy. *The heart muscle may also demonstrate compensatory hypertrophy when its workload increases in response, for example, to hypertension (high blood pressure). The opposite of hypertrophy is* atrophy, *where the cells become smaller than normal. This may result from the disuse of skeletal muscles associated with prolonged bed rest, various diseases, advanced age, and malnourishment.*

Inside Information—Human cells that are growing continuously may undergo mitosis once or twice a day. However, once we attain our adult state, divisions of cells in any of our tissues becomes much less frequent, which is believed to be a contributing factor in aging.

mitosis: Gk. *mitos,* thread

FIGURE 3.33
The stages of mitosis.

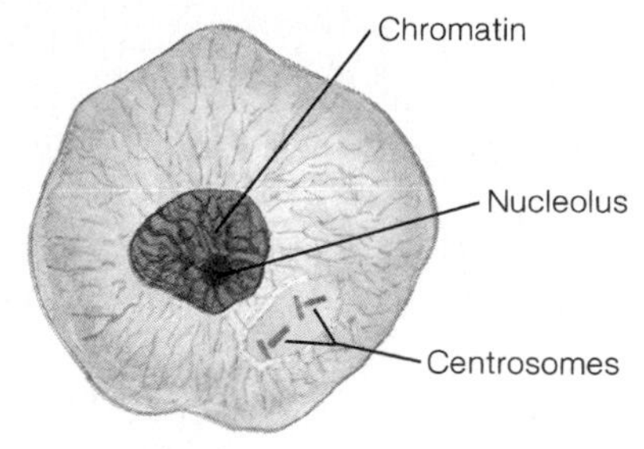

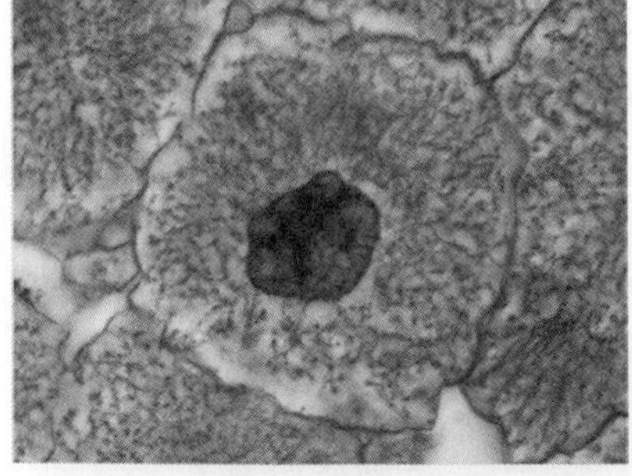

(a) Interphase
- The chromosomes are in an extended form and seen as chromatin in the electron microscope.
- The nucleus is visible.

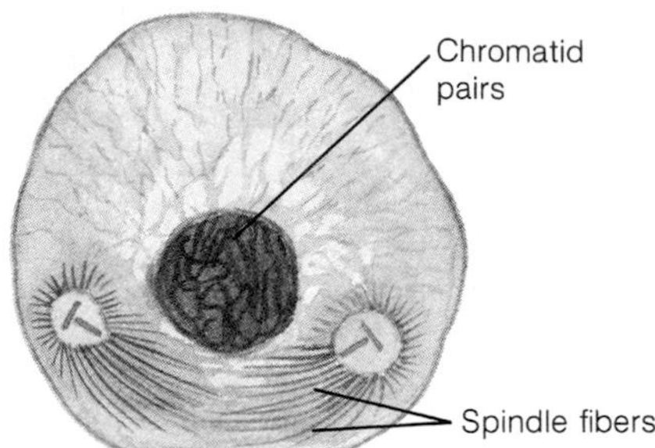

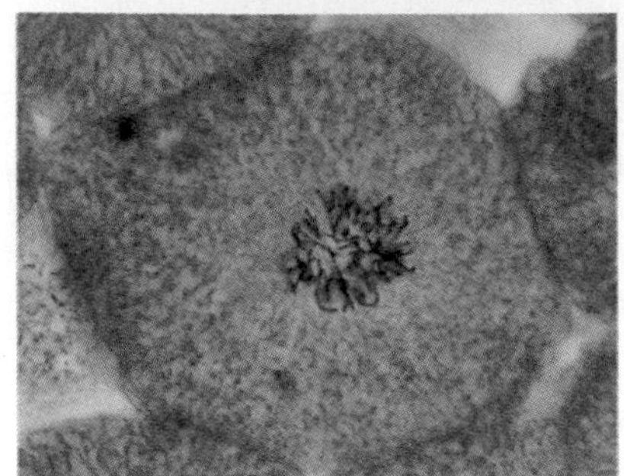

(b) Prophase
- The chromosomes are seen to consist of two chromatids joined by a centromere.
- The centrioles move apart toward opposite poles of the cell.
- Spindle fibers are produced and extended from each centrosome.
- The nuclear membrane starts to disappear.
- The nucleolus is no longer visible.

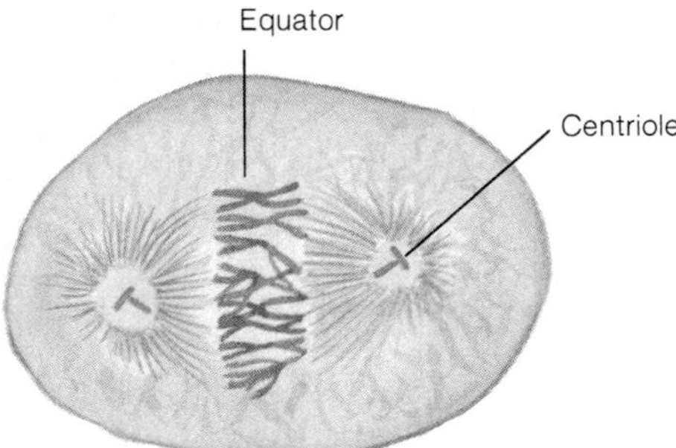

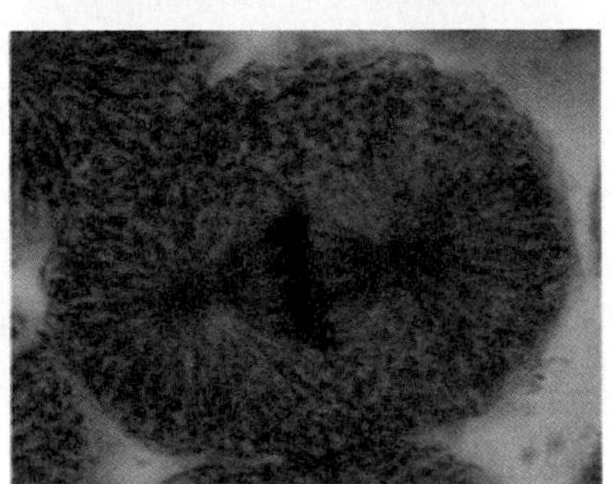

(c) Metaphase
- The chromosomes are lined up at the equator of the cell.
- The spindle fibers from each centriole are attached to the centromeres of the chromosomes.
- The nuclear membrane has disappeared.

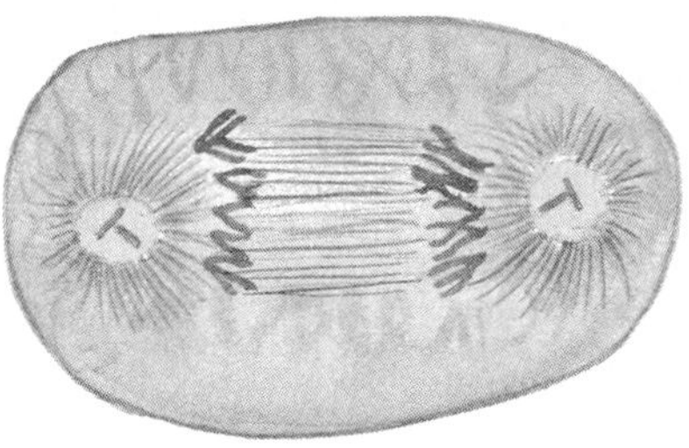

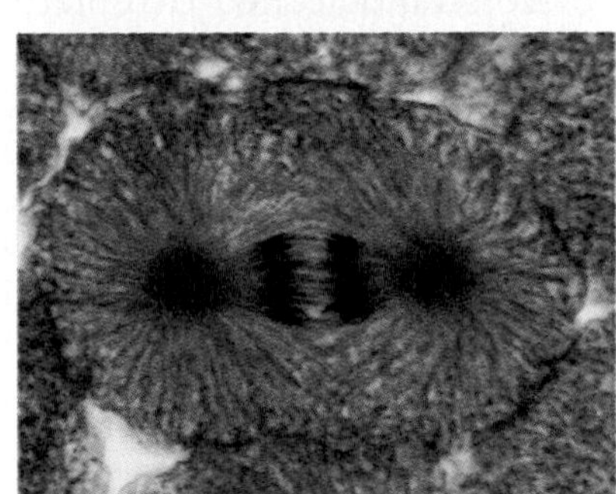

(d) Anaphase
- The centromeres split, and the sister chromatids separate as each is pulled to an opposite pole.

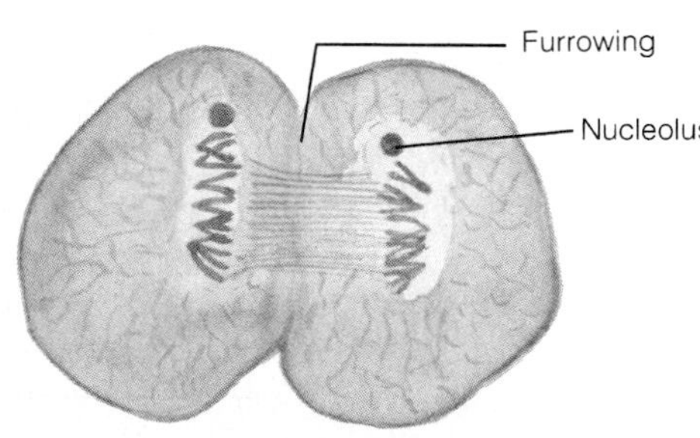

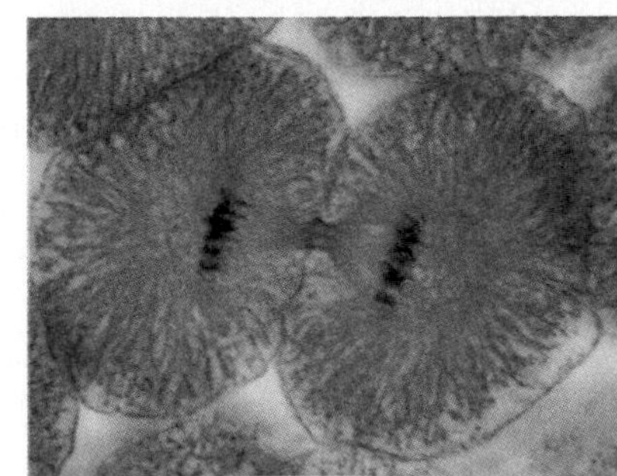

(e) Telophase
- The chromosomes become longer, thinner and less distance.
- New nuclear membranes form.
- The nucleolus reappears.
- Cell division is nearly complete.

Synoptic Spotlight

DNA in Criminal Investigations

In 1984, Alec Jeffreys, a professor of genetics at the University of Leicester in England, was studying the evolution of genes. He was particularly intrigued by the **intron**—the series of bases in chromosomes sometimes referred to as "junk DNA," whose functions are not clearly understood. When the genetic code is transferred to the complementary mRNA molecules prior to being converted into proteins, these intron sequences are snipped out by special "editor" enzymes.

While analyzing fragments of DNA, Jeffreys noticed that introns were often made up of the same sequence of bases repeated over and over again, and that these sequences varied dramatically from one individual to the next. Indeed, it is because of these highly variable regions that the DNA from two people never breaks down into precisely the same pattern.

Although Jeffreys had no special interest in murder cases at the time, it occurred to him that the individual patterns, displayed as a "DNA fingerprint," could become a powerful forensic tool. Biological evidence left behind at a crime scene could be matched to the person who committed the crime. Blood, semen, saliva, hair roots, or bone—any material containing nucleated cells—could provide a tissue sample. However, a method for obtaining the "fingerprint" needed to be found.

The DNA fingerprinting technique Jeffreys devised depends on **DNA probes**—small fragments of single-stranded DNA produced by machines called DNA synthesizers. The probes bind (by complementary base pairing) to specific base sequences of DNA from samples taken at the scene of a crime and from suspected perpetrators. Several probes are used, each labeled with a radioactive tag. This allows fragment patterns from the samples to be visualized on film and compared.

A 1987 Florida sexual assault case marked the beginning of widespread acceptance of DNA evidence in criminal trials. In fact, there have been over 20,000 cases in which attorneys have framed arguments using data derived from DNA tests on human tissues. In addition to attorneys, there are others who have welcomed the chance to use the results gleaned from DNA testing. Incarcerated criminals, claiming to have been wrongly convicted, are attempting to reopen their cases, encouraged by a well-publicized case involving a Canadian who had his decade-old rape conviction overturned.

As with comparisons of blood types in paternity suits, the DNA test works best when it can be argued that a defendant's DNA does *not* match the DNA at the crime scene. In other words, a defendant can usually be ruled out easily as the perpetrator, but it is much more difficult to prove that a particular defendant, and no other person, could possibly have committed the crime. However, it is often the case in criminal trials that the probability of a DNA sample having come from any person other than the defendant is so infinitesimally small that a jury is readily sold on the defendant's guilt.

Two procedures for DNA fingerprinting, both used on Jeffreys' original technique, are currently in use. The more time-consuming analysis, called **Restriction Fragment Length Polymorphism (RFLP),** may take several weeks to complete, and it must be based on "good" DNA samples; that is, the samples must be relatively large and uncontaminated by DNA from other organisms. Furthermore, extreme environmental conditions can damage DNA and make standard RFLP analysis uninformative.

In RFLP analysis, minute DNA fragments are removed from the entire molecule by means of specific nucleic enzymes. Radioactively tagged DNA probes bind to the DNA areas that are used to establish identity, and X-ray film detects the radioactive pattern. The X-ray film is then developed, and the DNA "fingerprint" is compared to those made from other samples. Multiple samples are compared, seeking a minimum of four matches. Once a convincing number of matches has been found, statistical predictions are made as to how likely it is that two people at random in the whole population could have that degree of a pattern match.

The second test—the **Polymerase Chain Reaction (PCR)**—is quicker to complete, taking only a few days, and can be done with much smaller tissue samples. When time is of the essence in forensic evaluations, this is the test that is used. In some respects, it can be thought of as a molecular photocopying procedure. It uses repeated cyles to reproduce a target area of DNA until enough copies are available for analysis. Thus, PCR itself is not an analytic tool; rather, it facilitates forensic applications by allowing a scientist to take a sample of DNA, which would generally be insufficient to detect the characteristics of the DNA, and amplify it. Because the products generated by one sample can serve as templates in the next cycle, the number of amplified copies doubles with each cycle. Thus, 20 to 25 cycles of PCR potentially yield about a millionfold reproduction.

Because PCR is not an evaluation of strictly random portions of the DNA molecule, statistical projections based on this test are not as convincing as those based on RFLP. For example, assessment of your DNA and mine using PCR might indicate that there was only one in a thousand chances that those two samples could have come from the same person. Using our samples in the RFLP test, the statistics would say that there was only one chance in a billion that the samples were from the same individual.

SUMMARY

3.1 Basic Cell Structure

1. The cell is surrounded by a cell, or plasma, membrane.
2. Organelles are minute structures within the cell that carry out specialized functions.

3.2 Cell Membrane

1. The cell membrane is composed of two phospholipid layers and proteins. Their arrangement is described in the fluid-mosaic model of the membrane.
2. On the surface of some membranes are oarlike structures called cilia. Sperm cells have a whiplike structure called a flagellum.

3.3 Cell Movements

1. Some cells display amoeboid movement, using pseudopods for locomotion. Pseudopods are also used in phagocytosis.
2. Materials from the extracellular environment are brought into the cell by endocytosis.
3. Materials produced by the cell can be secreted into the extracellular environment by exocytosis.

3.4 Membrane Transport Processes

1. Cell membranes are selectively permeable.
2. Transport across the cell membrane may or may not be mediated by carriers, and may or may not require the expenditure of ATP.
3. Net diffusion occurs from the region of higher to that of lower concentration as a result of the thermal energy of the molecules in solution.

3.5 Passive Transport

1. Nonpolar molecules, such as O_2, CO_2, and steroid hormones, can diffuse through the double phospholipid layer of the cell membrane.
2. Inorganic ions, such as Na^+ and K^+, can diffuse through channels within the integral proteins of the cell membrane.
3. The rate of diffusion through the membrane depends, in part, on the concentration gradient across the membrane.

3.6 Osmosis

1. Osmosis is the net diffusion of water across a membrane from a more dilute to a more concentrated solution.
2. The higher the concentration of a solution (and the greater its tendency to draw water by osmosis), the higher will be its osmotic pressure.
3. The total solute concentration of a solution is measured by its osmolality.
4. A hypotonic solution is more dilute, and a hypertonic solution is more concentrated, than a reference solution (usually normal plasma). Isotonic solutions have the same concentration and osmotic pressure as the reference solution.
5. Plasma osmolality is sensed by osmoreceptors in the hypothalamus of the brain. An increased blood osmolality stimulates thirst and the secretion of antidiuretic hormone (ADH).
6. A red blood cell will swell with water in a hypotonic solution and will shrink (crenate) in a hypertonic solution.

3.7 Carrier-Mediated Passive Transport

1. Membrane carriers are proteins that display specificity for the molecules they transport.
2. Carriers can become saturated and thus display a transport maximum.
3. Carrier-mediated transport that is passive, in which the transported molecules are moved across the membrane from the side higher to that of lower concentration, is called facilitated diffusion.

3.8 Carrier-Mediated Active Transport

1. Primary active transport moves the transported molecule across the membrane from the side of lower to the side of higher concentration. This type of transport requires the hydrolysis of ATP.
2. Primary active transport carriers are called pumps. The Na^+/K^+ pump is an example.
3. Secondary active transport (cotransport) occurs when a molecule is moved against its concentration gradient by a carrier powered by the diffusion of Na^+. ATP is indirectly involved because of the role of the Na^+/K^+ pumps in previously establishing the steep concentration gradient for Na^+.

3.9 Lysosomes and Mitochondria

1. Lysosomes are organelles that contain digestive enzymes.
2. Mitochondria are the organelles that provide most of the energy for the cell.

3.10 Endoplasmic Reticulum and the Golgi Apparatus

1. The rough endoplasmic reticulum contains ribosomes, which serve as the sites for protein synthesis.
2. The smooth endoplasmic reticulum contains specific enzymes and may function in the storage of Ca^{++}.
3. The Golgi apparatus is a system of flattened sacs that forms membranous vesicles to package secretory products.

3.11 Cell Nucleus and DNA

1. The nucleotides in DNA consist of the sugar deoxyribose, phosphate, and one of four nitrogenous bases: adenine, guanine, cytosine, or thymine.
2. The DNA molecule is in the form of a double helix.
3. According to the law of complementary base pairing, bases are specific in their bonding: adenine bonds with thymine and guanine bonds with cytosine.

3.12 Encoding Information—RNA Structure and Synthesis

1. RNA contains the sugar ribose (instead of deoxyribose) and the base uracil (in place of thymine).
2. The three major forms of RNA are mRNA, tRNA, and rRNA.
3. Messenger RNA (mRNA) is a complementary copy of a gene, which is a region of DNA that codes for one polypeptide chain.
4. RNA synthesis is called genetic transcription.

3.13 Processing the Information—Protein Synthesis

1. The genetic code in mRNA consists of three bases called codons.
2. Codons bond to anticodons, which are three bases in tRNA.
3. Each type of tRNA is bonded to a specific type of amino acid, which the tRNA brings to the growing polypeptide chain.
4. Protein synthesis is called genetic translation.

3.14 Reproduction of Information—DNA Synthesis and Mitosis

1. Each parent DNA strand acts as a template for the production of a new, complementary DNA strand.
2. This produces two daughter DNA molecules that each contain one strand from the old DNA and one new strand.
3. When a cell is going to divide, spindle fibers form from each pole of the cell and chromatids are pulled to opposite sides of the cell.
4. In this way, two daughter cells, each containing the same DNA as the parent cell, are produced.

Review Activities

Objective Questions

1. The organelle that packages proteins within vesicles so that they can be secreted is (*module 3.1*)
   a. the Golgi apparatus.
   b. the rough endoplasmic reticulum.
   c. the smooth endoplasmic reticulum.
   d. the ribosome.

2. In the cell membrane, (*module 3.2*)
   a. protein and phospholipids form a regular, repeating structure.
   b. phospholipids form a double layer, with the polar parts facing each other.
   c. proteins are free to move within a double layer of phospholipids.
   d. a double layer of proteins surrounds a core of phospholipids.

3. When a secretory vesicle fuses with the cell membrane, which of the following processes is occurring? (*module 3.3*)
   a. pinocytosis
   b. endocytosis
   c. phagocytosis
   d. exocytosis

4. If a poison (such as cyanide) stops the production of ATP, which of the following transport processes will stop? (*module 3.4*)
   a. the movement of Na^+ out of a cell
   b. osmosis
   c. the movement of K^+ out of a cell
   d. all of the above

5. Which of the following substances is able to pass by diffusion through the phospholipid portion of the cell membrane? (*module 3.5*)
   a. steroids
   b. Na^+
   c. glucose
   d. proteins

6. Which of the following statements regarding an increase in blood osmolality is *true?* (*module 3.6*)
   a. It can occur as a result of dehydration.
   b. It causes a decrease in blood osmotic pressure.
   c. It is accompanied by a decrease in ADH.
   d. All of the above are true.

7. Carrier-mediated facilitated diffusion (*module 3.7*)
   a. uses cellular ATP.
   b. is used for cellular uptake of blood glucose.
   c. is a form of active transport.
   d. transports molecules against their concentration gradients.

8. Which of the following statements about the Na^+/K^+ pump is *true?* (*module 3.8*)
   a. Na^+ is actively transported into the cell.
   b. K^+ is actively transported out of the cell.
   c. An equal number of sodium and potassium ions are transported with each cycle of the pump.
   d. The pumps are constantly active in all cells.

9. The organelle that contains digestive enzymes is (*module 3.9*)
   a. the mitochondrion.
   b. the lysosome.
   c. the endoplasmic reticulum.
   d. the Golgi apparatus.

10. Which of the following organelles serves as a site for the intracellular storage of Ca^{++}? (*module 3.10*)
   a. the ribosome
   b. the Golgi apparatus
   c. the lysosome
   d. the smooth endoplasmic reticulum
   e. the rough endoplasmic reticulum

11. Which of the following statements about DNA is *false?* (*module 3.11*)
   a. It is located in the nucleus.
   b. It is double-stranded.
   c. The bases adenine and thymine can bond together.
   d. The bases guanine and adenine can bond together.

12. The RNA nucleotide that pairs with adenine in DNA is (*module 3.12*)
   a. thymine.
   b. uracil.
   c. guanine.
   d. cytosine.

13. Which of the following statements about RNA is *true?* (*module 3.12*)
   a. It is made in the nucleus.
   b. It is double-stranded.
   c. It contains the sugar deoxyribose.
   d. It is a complementary copy of the entire DNA molecule.

14. A codon (*module 3.13*)

a. consists of three bases on mRNA.
b. binds to three bases on tRNA.
c. codes for a specific amino acid.
d. is complementary to three bases in a gene.
e. is described by all of the above.

15. After a DNA molecule has replicated, the duplicate strands are called (*module 3.14*)

a. chromosomes.
b. chromatids.
c. centromeres.
d. spindle fibers.

16. The phase of mitosis in which the chromosomes line up at the equator of the cell is called (*module 3.14*)

a. interphase.
b. prophase.
c. metaphase.
d. anaphase.
e. telophase.

Essay Questions

1. Compare the structure and function of cilia and flagella. (*module 3.2*)
2. Describe the processes of endocytosis and exocytosis and explain their significance. (*module 3.3*)
3. Explain how simple diffusion can be distinguished from facilitated diffusion and how active transport can be distinguished from passive transport. (*module 3.4*)
4. What factors influence the rate of diffusion of molecules through the cell membrane? (*module 3.5*)
5. Describe the conditions required to produce osmosis. Why does osmosis occur under these conditions? (*module 3.6*)
6. Describe the characteristics of membrane carriers. (*module 3.7*)
7. Describe the action of the Na^+/K^+ pumps and explain their significance. (*module 3.8*)
8. Why would you expect phagocytic cells to be rich in lysosomes? (*module 3.9*)
9. Distinguish between the structure and functions of a smooth and a rough endoplasmic reticulum. (*module 3.10*)
10. Explain the law of complementary base pairing. (*module 3.11*)
11. Describe how RNA is produced and list the different forms of RNA. (*module 3.12*)
12. Explain how mRNA and tRNA cooperate in the formation of a protein. (*module 3.13*)
13. Explain how one DNA strand can serve as a template for the synthesis of another DNA strand. Why is DNA replication called "semiconservative"? (*module 3.14*)

Labeling Exercise

Label the structures indicated on the figure to the right.

1. ______________________________
2. ______________________________
3. ______________________________
4. ______________________________
5. ______________________________
6. ______________________________

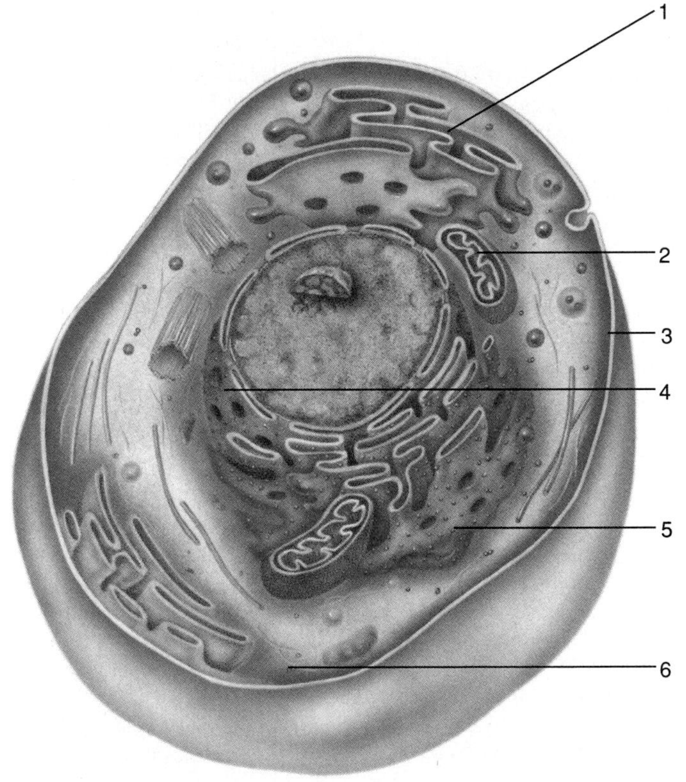

Critical Thinking Questions

1. Alcohol is hydrophobic and lipid soluble. What do these terms mean? How could alcohol affect the structure of cell membranes? Explain.
2. People born with a particular genetic defect have low numbers of receptor proteins for cholesterol in their liver cells. How could this affect their blood cholesterol concentrations? What significance do you think this could have on their health?
3. Considering the three factors that affect the rate of diffusion through a membrane, explain how surgical shortening of the small intestine may help a person suffering from morbid obesity. (Hint: Food is digested and absorbed primarily in the small intestine.)
4. What kind of solution, in terms of its concentration, should be used to wet contact lenses? Could pure distilled water be used instead? Why or why not?
5. Cyanide is a drug that blocks the production of ATP. What effect would cyanide have on osmosis? On the movement of Na^+ into a cell? On the movement of K^+ out of a cell? Explain.
6. Antibiotics can have different mechanisms of action. An antibiotic called puromycin blocks genetic translation. One called actinomycin D blocks genetic transcription. These drugs can be used to determine how regulatory molecules, such as hormones, work. For example, if a hormone's effects on a tissue were blocked immediately by actinomycin D but not by puromycin, what would that tell you about the mechanism of action of the hormone?

CHAPTER

4

"Epithelial tissue may be one layer or several layers thick. The upper surface may be exposed to the environment." *page 75*

Histology

Chapter Outline

Terms to Remember

histology, 75
epithelium (*ep"ĭ-the' le-um*), 75
mucosa (*myoo-ko' să*), 77
exocrine (*ek' sŏ-krin*), 78
mesenchyme (*mez' en-kīm*), 81
adipocyte (*ad' ĭ-po-sīt*), 83
cartilage (*kar' tl-ij*), 84
chondrocyte (*kon' dro-sīt*), 84
osteocyte (*os' te-ŏ-sīt*), 85
osteon (*os' te-on*), 86
erythrocyte (*ĕ-rith' rŏ-sīt*), 86
leukocyte (*loo' kŏ-sīt*), 86
thrombocyte (*throm' bŏ-sīt*), 87
neuron, 88
neuroglia (*noo-rog' le-ă*), 89

Learning Objectives

After studying this chapter, you should be able to

- define *tissue* and explain why it is important to understand body structure and function at the microscopic level.
- describe how the interrelationship of tissues forms the structural basis of organs.
- list the four basic tissue types and give the distinguishing characteristics of each type.
- describe the general characteristics, locations, and functions of epithelial tissue.
- define *gland* and distinguish between the various types of glands in the body.
- describe the general characteristics, locations, and functions of connective tissue.
- identify the various types of cells and fibers that constitute connective tissue and explain their functions.
- describe the characteristics, locations, and functions of muscle tissue.
- describe the structure and function of nervous tissue.
- discuss the regenerative abilities of the various tissues of the body.

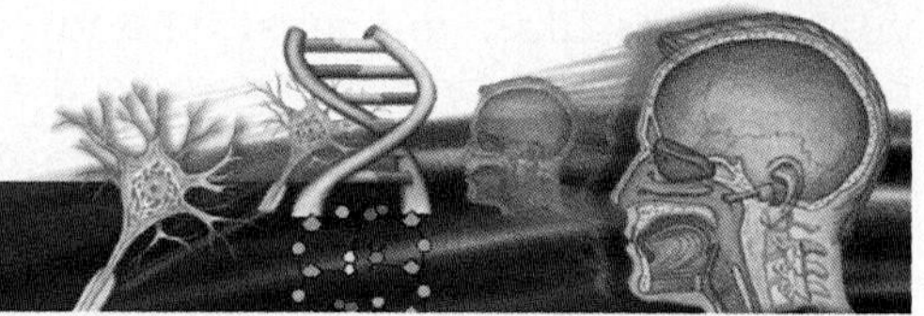

4.1 Introduction to Tissues

Histology is the science that deals specifically with the study of tissues. The four basic tissue types in the body are epithelial, connective, muscle, and nervous tissues.

Histology, the microscopic study of tissues, provides a foundation for understanding how cells are organized to form tissues, organs, organ systems, and the body as a whole. The intimate relationship between structure and function in tissues will quickly become apparent as you study the major tissue types.

Tissues are collections of similar cells that perform a common function. The various types of tissues are established during early embryonic development. As the embryo grows, organs form from specific arrangements of tissues. Many adult organs, such as the heart and skeletal muscles, contain the original cells and tissues that were formed prenatally, although some changes in function occur as the tissues are acted upon by hormones or as their effectiveness declines with age. Many diseases, in fact, drastically alter the tissues within an affected organ. *Pathology* is the study of abnormal tissues in diseased organs. By knowing the normal tissue structure, a medical practitioner can recognize the abnormal. Histology in most medical schools is followed by a course in pathology.

Tissue cells are surrounded and bound together by a nonliving extracellular material called **matrix.** Matrix varies in composition from one tissue to another and may be liquid, semisolid, or solid. Blood, for example, has a liquid matrix, permitting this tissue to flow through vessels. By contrast, bone cells are separated by a solid matrix, permitting bone tissue to support the body.

Inside Information—Blood and bone, so strikingly different in form, are each a type of connective tissue. Blood is the only fluid tissue in the body; bone is the most rigid tissue.

Based on their structure and function, the tissues of the body may be classified into four basic categories:

1. **Epithelial tissue** covers body and organ surfaces, lines the inside walls of body cavities and organs, and forms various glands.
2. **Connective tissue** joins, supports, and protects body parts.
3. **Muscle tissue** contracts to produce movement.
4. **Nervous tissue** produces nerve impulses and transmits them from one body part to another.

4.2 Epithelial Tissue

Epithelial tissue is divided into subtypes according to the physical features of the tightly packed cells that form the tissue. Epithelial tissue covers body surfaces or forms glands that are derived developmentally from the body surfaces. This tissue is adapted for protection, absorption, and secretion.

Characteristics of Membranous Epithelial Tissue

Epithelial tissue, or *epithelium* (*ep"ĭ-the-le-um*), is located throughout the body and forms such structures as the outer layer of the skin, the lining of body cavities and vessels, the covering of viscera, and the secretory part of glands. Epithelium always has one free surface (the *apical surface*) exposed to a body cavity, a lumen (hollow part of a body tube or duct), or to the skin surface.

Epithelial tissue may be one layer or several layers thick. The upper surface may be exposed to the environment, as in the case of epithelium in the integumentary and respiratory systems; to fluids, as in the circulatory and urinary systems; or to semisolids, as in the digestive tract. The deep surface of most epithelial tissues is bound to underlying tissue by a **basement membrane** consisting of glycoprotein from the epithelial cells and a meshwork of protein fibers from the underlying connective tissue. With few exceptions, epithelial tissue is without blood

histology: Gk. *histos,* web (tissue); *logos,* study
pathology: Gk. *pathos,* suffering, disease; *logos,* study
matrix: L. *matris,* mother
epithelium: Gk. *epi,* upon; *thelium,* to cover

vessels (avascular) and must be nourished by diffusion from blood vessels in nearby connective tissue. Cells that make up epithelial tissue are tightly packed together, with little extracellular matrix in the narrow spaces between them. Named on the basis of the appearance of their cells, epithelial tissue is broadly classified as *simple* (nonlayered) or *stratified* (layered).

Simple Epithelial Tissue

Simple epithelial tissue is a mere single cell layer thick! This tissue type is located where diffusion (random movement of molecules from a region of higher concentration toward one of lower concentration), filtration (movement of a substance through a membrane as a result of hydrostatic pressure), and secretion (discharge of a substance produced in a cell) are principal functions. The cells of simple epithelial tissue range in size and shape from thin, flattened cells to tall, columnar cells. Some of these cells have cilia that create currents for the movement of materials across cell surfaces. Others have microvilli that increase the surface area for absorption.

Simple Squamous Epithelium

Simple squamous epithelium is composed of flattened, irregularly shaped cells that are tightly bound together (fig. 4.1*a*). This tissue is adapted for diffusion and filtration. It occurs in the lining of air sacs within the lungs (where gaseous exchange occurs), in the kidney (where blood is filtered), on the inside walls of blood vessels, the lining of body cavities, and the covering of the viscera. The simple squamous epithelium lining the walls of blood and lymphatic vessels is referred to as **endothelium.**

Simple Cuboidal Epithelium

This type of epithelium is composed of a single layer of tightly fitted cube-shaped cells (fig. 4.1*b*). It is found lining small ducts and tubules that have excretory, secretory, or absorptive functions. Simple cuboidal epithelium occurs on the surface of the ovaries, forms a portion of the kidneys, and lines the ducts of the salivary glands and pancreas.

Simple Columnar Epithelium

Simple columnar epithelium is composed of tall, narrow cells (fig. 4.1*c*). The height of the cells varies depending on the site and function of the tissue. Specialized **goblet cells** are scattered through this tissue. Goblet cells secrete a lubricative and protective mucus along the free surfaces of the tissue. Simple columnar epithelium is found lining the inside walls of the stomach and small intestine, where it forms a highly absorptive surface and also secretes certain digestive chemicals.

Inside Information—Within the stomach, simple columnar epithelium has a tremendous mitotic rate. It replaces itself every 2 to 3 days.

Simple Ciliated Columnar Epithelium

Simple ciliated columnar epithelium is characterized by the presence of cilia along the free surface (fig. 4.1*d*). By contrast, the more usual simple columnar type lacks cilia. Cilia produce wavelike movements that transport materials through tubes or passageways. This type of epithelium occurs in the female uterine tubes to move the egg cell into the uterus.

Pseudostratified Ciliated Columnar Epithelium

As its name implies, this type of epithelium has a layered appearance (*strata* = layers). Actually, it is not multilayered (*pseudo* = false), since each cell is in contact with the basement membrane; however, not all cells are exposed to the surface (fig. 4.1*e*). This type of epithelium appears to be stratified because the nuclei of the cells are located at different levels. Numerous goblet cells and a ciliated exposed surface are characteristic of this epithelium. The trachea and the bronchial tubes are lined with this tissue; hence, it is frequently called *respiratory epithelium.* Its function is to remove dust and bacteria trapped in mucus. Coughing, sneezing, or simply clearing the throat are protective reflex mechanisms for helping to clear the respiratory passages of inhaled particles that have been trapped in the mucus.

Stratified Epithelial Tissue

Stratified epithelial tissue consists of two or more layers of cells. In contrast to simple epithelial tissue, stratified epithelial tissue is too thick to absorb or secrete. Stratified epithelial tissue generally has a protective function. The cells of this tissue also divide rapidly. The various types of stratified epithelial tissue are classified according to the shape of the surface layer of cells.

Stratified Squamous Epithelium

Stratified squamous epithelium is composed of a number of cell layers that are flattest at the surface (fig 4.2*a*). Cell divisions occur only within the deepest layer, called the *stratum basale.* As the newly produced cells grow in size, they are pushed toward the surface where they will replace the cells that are sloughed off. Movement of the epithelial cells away from the supportive basement membrane

squamous: L. *squamosus,* scaly
endothelium: Gk. *endon,* within; *thelium,* to cover

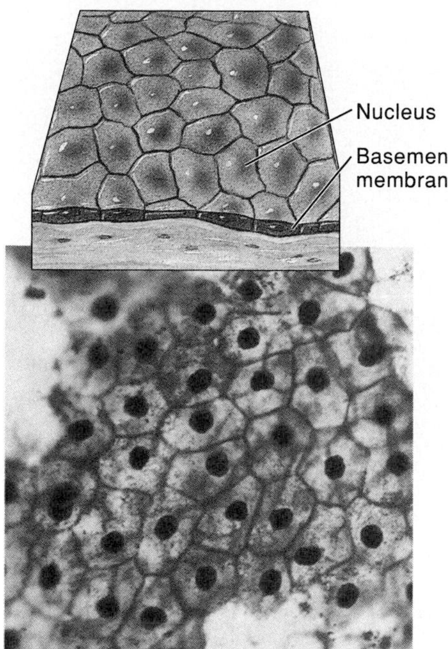

(a) Simple squamous epithelium

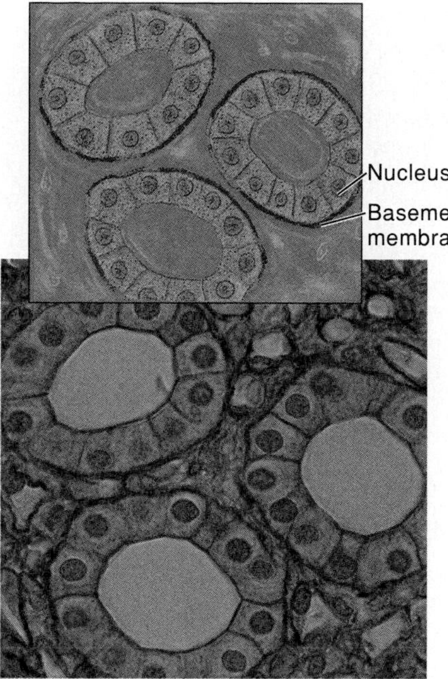

(b) Simple cuboidal epithelium

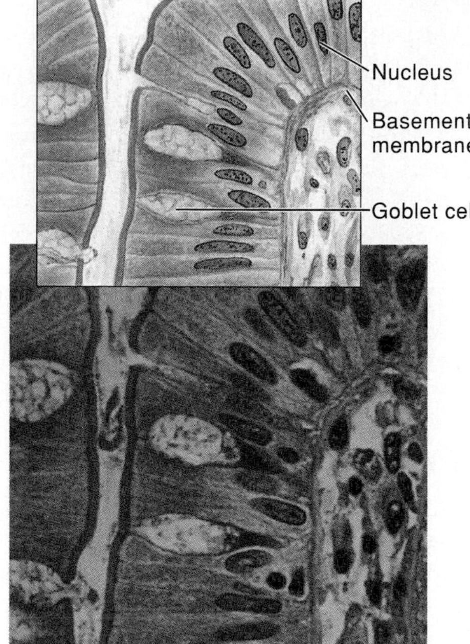

(c) Simple columnar epithelium

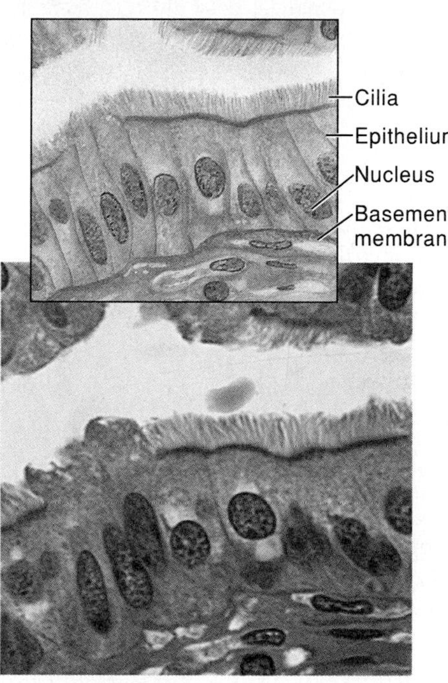

(d) Simple ciliated columnar epithelium

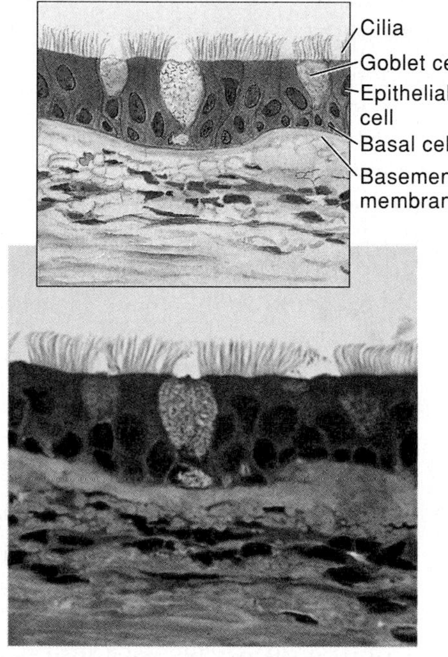

(e) Pseudostratified ciliated columnar epithelium

FIGURE 4.1
Simple epithelial tissue types.

is accompanied by the production of *keratin* (described below), progressive dehydration, and flattening.

There are two types of stratified squamous epithelial tissues: *keratinized* and *nonkeratinized.*

1. **Keratinized stratified squamous epithelium** contains *keratin,* a protein that strengthens the tissue. Keratin makes the epidermis (outer layer) of the skin somewhat waterproof and protects it from bacterial invasion. The outer layers of the skin are dead, but glandular secretions keep them soft (see chapter 5, module 5.2).
2. **Nonkeratinized stratified squamous epithelium** lines the mouth and throat, nasal cavity, vagina, and anal canal. This type of epithelium, called **mucosa** (myoo-kó-să), is well adapted to withstand moderate abrasion but not fluid loss. The cells on the exposed surface of this tissue are alive and are always moistened.

keratin: Gk. *keras,* horn

Stratified squamous epithelium is the first line of defense against the entry of disease-causing organisms. Stratification, rapid cell division, and keratinization within the epidermis of the skin are important protective features. An acidic pH of the surface of this tissue also helps prevent disease. The pH of the skin is between 4 and 6.8. The pH in the oral cavity ranges between 5.8 and 7.1, which tends to prevent the growth of microorganisms. The pH of the anal region is about 6, and the pH along the surface of the vagina is 4 or lower.

Stratified Cuboidal Epithelium

Stratified cuboidal epithelium usually consists of two or three layers of cuboidal cells (fig. 4.2*b*). This type of epithelium has a limited distribution in the body. It is found lining the large ducts of sweat glands, salivary glands, and the pancreas. The stratification of this tissue probably provides a stronger lining than would simple epithelium.

Transitional Epithelium

Transitional epithelium is similar to nonkeratinized stratified squamous epithelium except that the surface cells of the former are large and round rather than flat, and some may have two nuclei (fig. 4.2*c*). Transitional epithelium is located only within the urinary system (in the urinary bladder, ureters, and part of the urethra)—particularly in the inner surface of the urinary bladder, where it is specialized to permit stretching as the bladder fills with urine.

A summary of membranous epithelial tissue is presented in table 4.1.

Glandular Epithelial Tissue

As tissues develop in the embryo, certain epithelial cells migrate into the underlying connective tissue, forming specialized secretory structures called **exocrine** (*ek'sŏ-krin*) **glands.** The secretions from exocrine glands pass through ducts onto body surfaces or into body cavities. These glands should not be confused with endocrine glands, which are ductless, and which secrete their products (hormones) into the blood or surrounding extracellular fluid. Exocrine glands within the skin include oil (sebaceous) glands, sweat glands, and mammary glands. Exocrine glands within the digestive system include the salivary and pancreatic glands.

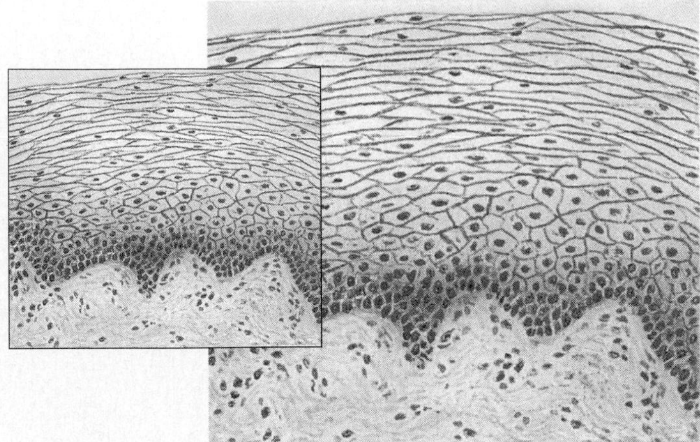

(a) Stratified squamous epithelium

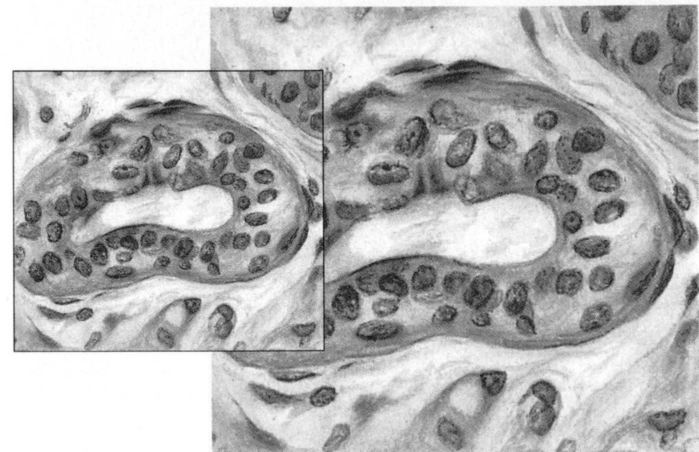

(b) Stratified cuboidal epithelium

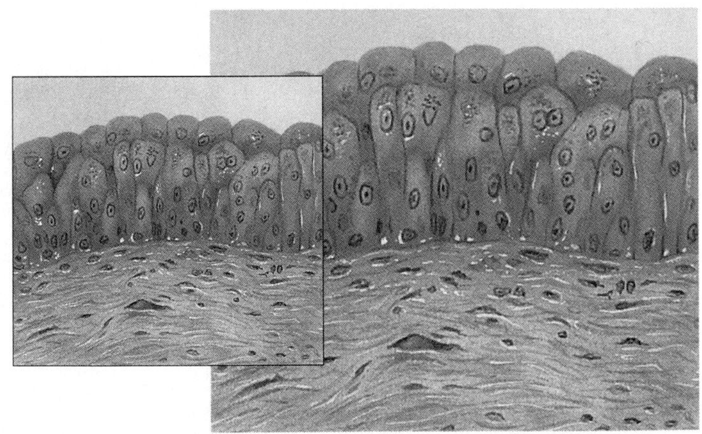

(c) Transitional epithelium

FIGURE 4.2
Stratified epithelial tissue types.

exocrine: Gk. *exo,* outside; *krinein,* to separate

TABLE 4.1 **Summary of Membranous Epithelial Tissue**

Type	Structure and Function	Location
Simple Epithelial Tissues	Single layer of cells; function varies with type	Covering visceral organs; linings of body cavities, tubes, and ducts
Simple squamous epithelium	Single layer of flattened, tightly bound cells; diffusion and filtration	Capillary walls; alveoli of lungs; covering visceral organs; linings of body cavities
Simple cuboidal epithelium	Single layer of cube-shaped cells; excretion, secretion, or absorption	Surface of ovaries; linings of renal (kidney) tubules, salivary ducts, and pancreatic ducts
Simple columnar epithelium	Single layer of nonciliated, tall, column-shaped cells; protection, secretion, and absorption	Lining of most of gastrointestinal (GI) tract
Simple ciliated columnar epithelium	Single layer of ciliated, column-shaped cells; transport role through ciliary motion	Lining of uterine tubes
Pseudostratified ciliated columnar epithelium	Single layer of ciliated, irregularly shaped cells; many goblet cells; protection, secretion, ciliary movement	Lining of respiratory passageways
Stratified Epithelial Tissues	Two or more layers of cells; function varies with type	Epidermal layer of skin; linings of body openings, ducts, and urinary bladder
Stratified squamous epithelium (keratinized)	Numerous layers containing keratin, with outer layers flattened and dead; protection	Epidermis of skin
Stratified squamous epithelium (nonkeratinized)	Numerous layers lacking keratin, with outer layers moistened and alive; protection and pliability	Linings of oral and nasal cavities, vagina, and anal canal
Stratified cuboidal epithelium	Usually two layers of cube-shaped cells; strengthening of luminal walls	Larger ducts of sweat glands, salivary glands, and pancreas
Transitional epithelium	Numerous layers of rounded, nonkeratinized cells; expansion	Walls of ureters and urinary bladder

Exocrine glands are classified according to their structure and how they discharge their products. Classified according to structure, there are two types of exocrine glands, *unicellular* and *multicellular glands.*

1. **Unicellular glands** are single-celled glands, such as *goblet cells.* They are interspersed within most columnar epithelial tissues. Goblet cells are found in the epithelial linings of the respiratory, digestive, urinary, and reproductive systems. The mucous secretion of these cells lubricates and protects the surface linings.
2. **Multicellular glands,** as their name implies, are composed of both secretory cells and cells that form the walls of the ducts. Multicellular glands are classified as *simple* or *compound glands.* The ducts of the simple glands do not branch, whereas those of the compound type do (fig. 4.3). Multicellular glands are also classified according to the shape of their secretory portion. They are identified as *tubular glands* if the secretory portion resembles a tube and as *acinar glands* if the secretory portion resembles a flask. Multicellular glands with a secretory portion that resembles both a tube and a flask are termed *tubuloacinar glands.*

Multicellular glands are also classified according to the means by which they release their product (fig. 4.4).

1. **Merocrine glands** are those that secrete a watery substance through the cell membrane of the secretory cells. Salivary glands, pancreatic glands, and certain sweat glands are of this type.
2. **Apocrine glands** are those in which the secretion accumulates on the surface of the secretory cell; then, a portion of the cell and the secretion is pinched off and discharged.
3. **Holocrine glands** are those in which the entire secretory cell and its product are discharged. An example of a holocrine gland is an oil-secreting (sebaceous) gland of the skin (see chapter 5, module 5.4).

Inside Information— Mammary glands are generally classified as apocrine glands, although very little cellular material is discharged along with the milk secretion during lactation.

A summary of glandular epithelial tissue is presented in table 4.2.

merocrine: Gk. *meros,* part; *krinein,* to separate
apocrine: Gk. *apo,* off; *krinein,* to separate
holocrine: Gk. *holos,* whole; *krinein,* to separate

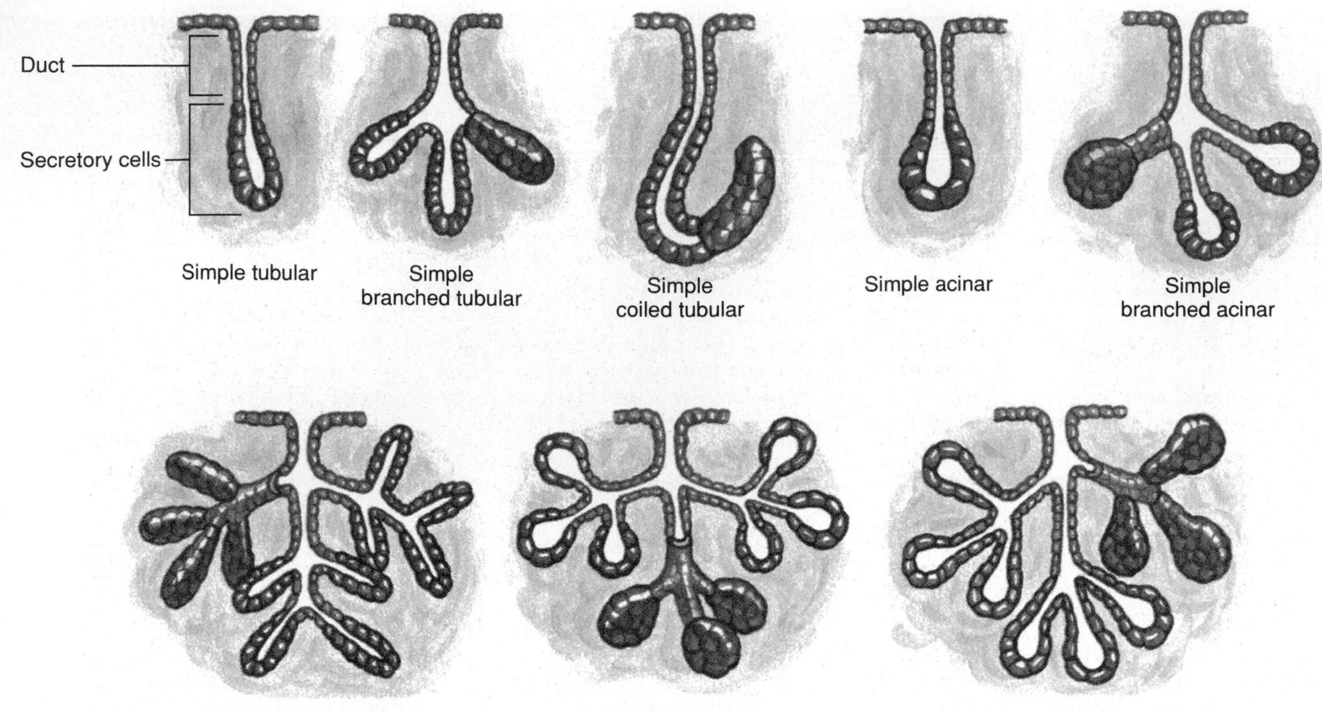

FIGURE 4.3
Structural classification of multicellular exocrine glands. The ducts of the simple glands either do not branch or have few branches, whereas those of the compound glands have multiple branches.

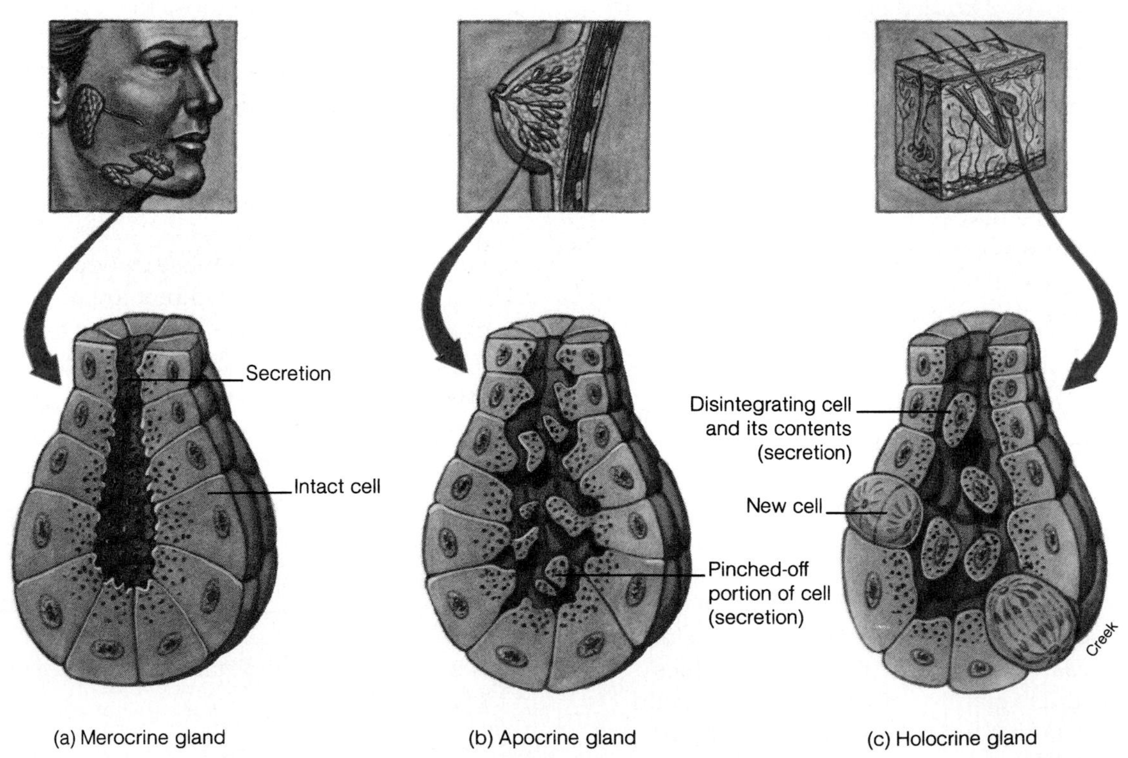

FIGURE 4.4
Examples of multicellular exocrine glands.

Table 4.2 Summary of Glandular Epithelial Tissue

Classification of Exocrine Glands by Structure		
Type	*Function*	*Example*
I. Unicellular	Lubricate and protect	Goblet cells of digestive, respiratory, urinary, and reproductive systems
II. Multicellular	Protect, cool body, lubricate, aid in digestion, maintain body homeostasis	Sweat glands, digestive glands, mammary glands, sebaceous glands
A. Simple		
1. Tubular	Aid in digestion	Intestinal glands
2. Branched tubular	Protect, aid in digestion	Uterine glands, gastric glands
3. Coiled tubular	Regulate temperature	Certain sweat glands
4. Acinar	Provide additive for spermatozoa	Seminal vesicle of male reproductive system
5. Branched acinar	Condition skin	Sebaceous skin glands
B. Compound		
1. Tubular	Lubricate urethra of male, assist body digestion	Bulbourethral gland of male reproductive system, liver
2. Acinar	Provide nourishment for infant, aid in digestion	Mammary gland, salivary glands (sublingual and submandibular)
3. Tubuloacinar	Aid in digestion	Salivary gland (parotid), pancreas

Classification of Exocrine Glands by Mode of Secretion		
Type	*Description of Secretion*	*Example*
Merocrine glands	Watery secretion for regulating temperature or enzymes that promote digestion	Salivary and pancreatic glands, certain sweat glands
Apocrine glands	Portion of secretory cell and secretion are discharged; provides milk for infant, assists in regulating temperature	Mammary glands, certain sweat glands
Holocrine glands	Entire secretory cell with enclosed secretion is discharged; conditions skin	Sebaceous glands of the skin

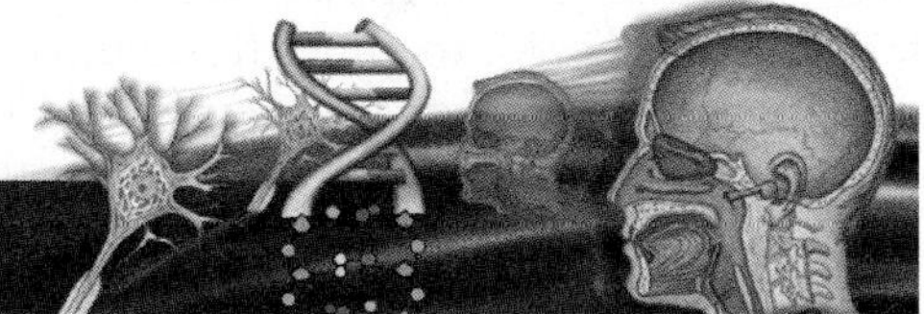

4.3 Connective Tissue

Connective tissue is divided into subtypes according to the characteristics of the matrix that binds the cells. Connective tissue provides structural and metabolic support for other tissues and organs of the body.

Characteristics of Connective Tissue

Connective tissue is the most abundant tissue in the body. It supports or binds other tissues and provides for the metabolic needs of all body organs. Certain types of connective tissue store nutritional substances; other types manufacture protective and regulatory materials.

With the exception of cartilage, connective tissue is highly vascular and well nourished. It is able to regenerate, and thus functions in the repair of body organs. Unlike epithelial tissue, which is composed of tightly fitted cells, connective tissue contains considerably more matrix (nonliving extracellular material) than cells; in fact, the cells of connective tissue rarely touch one another at all.

Embryonic Connective Tissue

During the 6-week embryonic period of development (from the beginning of the third to the end of the eighth week), all of the body organs form. At the beginning of the embryonic period, all connective tissue looks the same and is referred to as **mesenchyme** (*mez'en-kīm*). Mesenchyme is unspecialized embryonic connective tissue that consists of irregularly shaped cells positioned in a uniform, jellylike matrix (fig. 4.5). Before the end of

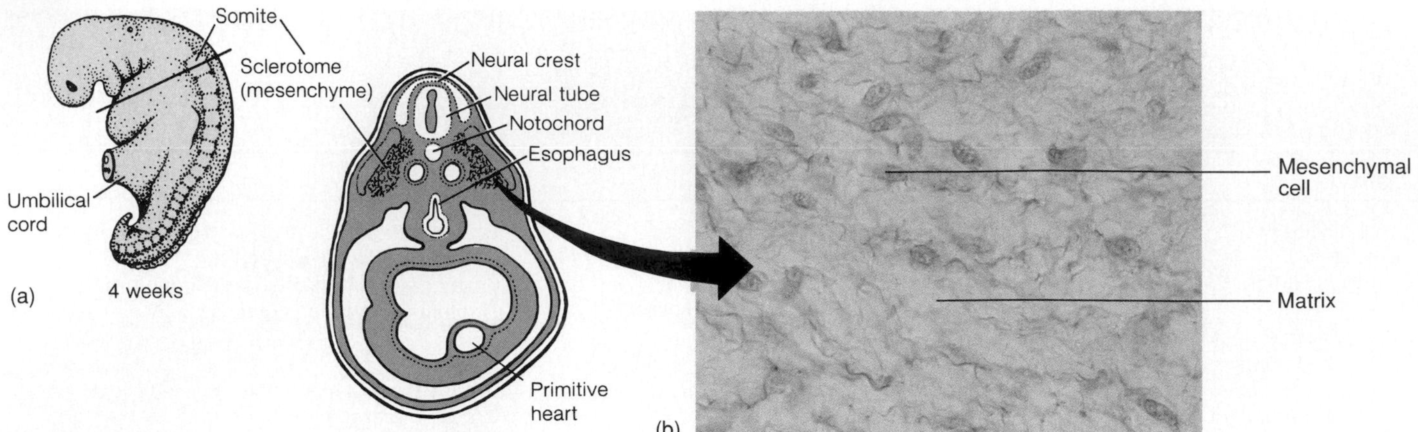

FIGURE 4.5
Mesenchyme is a type of embryonic connective tissue that can migrate and give rise to all other kinds of connective tissue. (*a*) It is found within an early developing embryo and (*b*) consists of irregularly shaped cells lying in a homogenous, jellylike matrix.

the embryonic period, mesenchyme migrates to specific sites and differentiates to form the various kinds of connective tissue that contribute to the structure of the body organs. The causes of tissue differentiation are not fully understood.

Connective Tissue Proper

Connective tissue proper has a loose, flexible matrix, frequently called *ground substance.* The most common cell within connective tissue proper is called a **fibroblast.** Fibroblasts are star-shaped cells that produce collagenous, elastic, and reticular fibers. Although extracellular, these long protein strands are a key structural element of connective tissue. **Collagenous fibers** are composed of a protein called *collagen;* they are flexible, yet have tremendous strength. **Elastic fibers** are composed of a protein called *elastin,* which provides some tissues with elasticity and extensibility. **Reticular fibers** are composed of a protein *reticulin,* which is similar in composition to collagen; they form a latticelike framework.

There are six basic kinds of connective tissue proper (fig. 4.6). These tissues are distinguished by the consistency of the ground substance and the type of reinforcement fibers.

Loose Connective (Areolar) Tissue

Loose connective tissue is characterized by randomly arranged elastic fibers, collagenous fibers, and fibroblasts scattered through the matrix. This connective tissue is distributed throughout the body, functioning as a binding and packing material. It binds the skin to the underlying muscles and is highly vascular, providing nutrients to the skin. Loose connective tissue that surrounds muscle fibers and muscle groups is known as *fascia.* It even surrounds blood vessels and nerves, providing both protection and support.

Dense Regular Connective Tissue

Dense regular connective tissue is characterized by large amounts of closely packed collagenous fibers that run parallel to the direction of force placed on this tissue during body movement. It is found where strong, flexible support is needed. *Tendons,* which attach muscles to bones, and *ligaments,* which connect bone to bone across joints, are composed of this type of tissue.

Dense Irregular Connective Tissue

Dense irregular connective tissue is characterized by large amounts of closely packed collagenous fibers that are interwoven to provide strength in any direction. This tissue is found in the dermis (deep layer) of the skin and submucosa (vascular layer) of the digestive tract. It also forms the fibrous coverings of organs and joints.

Elastic Connective Tissue

Elastic connective tissue is composed primarily of elastic fibers that are irregularly arranged. This tissue is found in the walls of large arteries, in the trachea, and in the bronchial tubes leading to the lungs. It is also present between the arches of the vertebrae that make up the vertebral column.

collagen: Gk. *kolla,* glue
reticular: L. *rete,* net or netlike

fascia: L. *fascia,* band or girdle
tendon: L. *tendere,* to stretch
ligament: L. *ligare,* bind

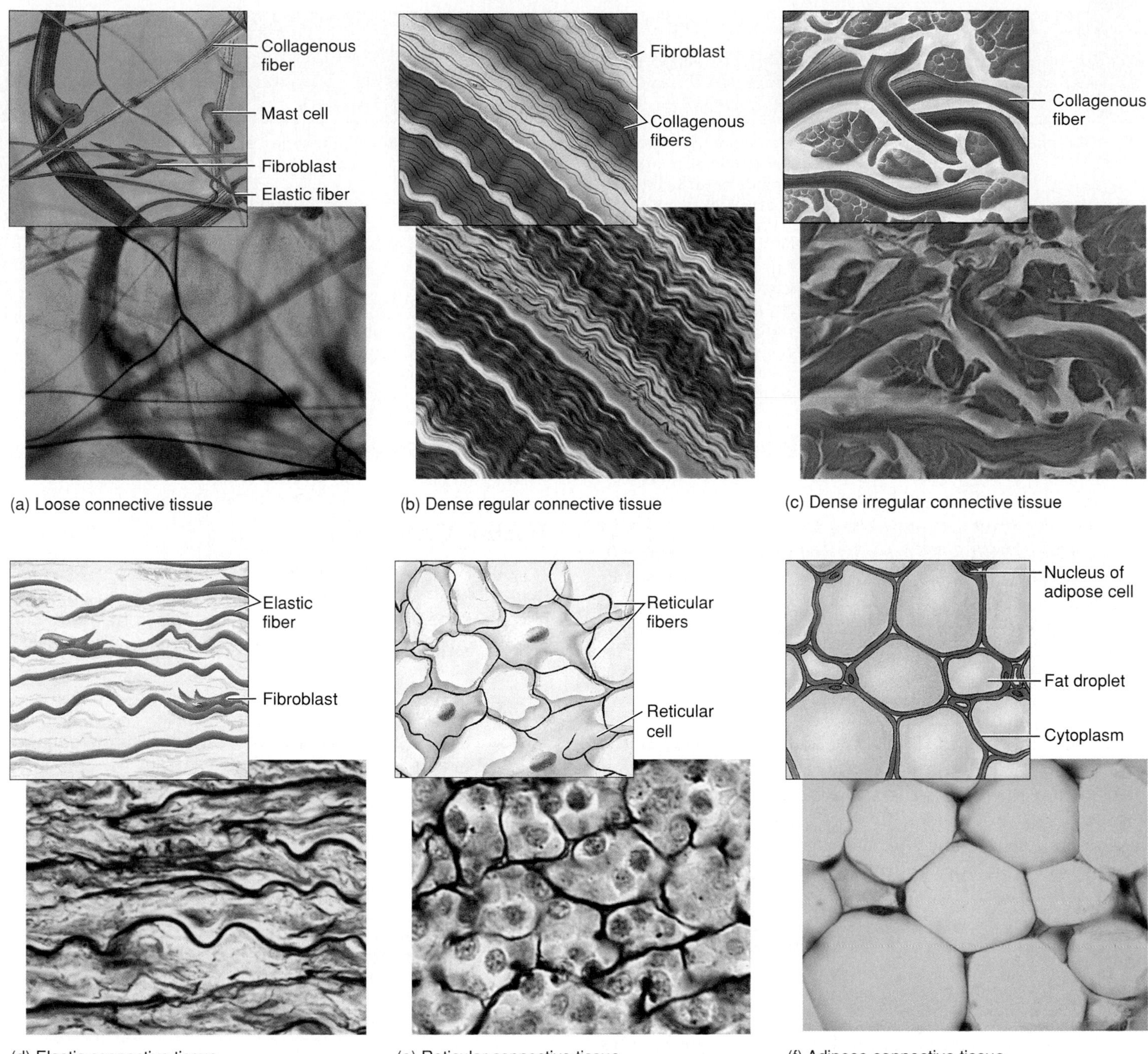

(a) Loose connective tissue (b) Dense regular connective tissue (c) Dense irregular connective tissue

(d) Elastic connective tissue (e) Reticular connective tissue (f) Adipose connective tissue

FIGURE 4.6
Connective tissue proper.

Reticular Connective Tissue

Reticular connective tissue is characterized by a network of reticular fibers woven through a jellylike matrix. Some of the cells within reticular tissue are *phagocytic,* and therefore can destroy foreign materials. The liver, spleen, lymph nodes, and bone marrow contain reticular connective tissue.

Adipose Tissue

Adipose tissue contains large numbers of **adipose cells,** or **adipocytes** (*ad'ĭ-po-sīts*). Adipose cells form prenatally and during the first year of life. These cells store fat within their cytoplasm, causing them to swell and forcing their nuclei to one side (see fig. 4.6*f*).

---

adipose: L. *adiposus,* fat

TABLE 4.3 **Summary of Connective Tissue Proper**

Type	Structure and Function	Location
Loose connective (areolar) tissue	Predominantly fibroblast cells with lesser amounts of collagen and elastin proteins; binds organs, holds tissue fluids	Surrounding nerves and vessels, between muscles, beneath the skin
Dense regular connective tissue	Densely packed collagenous fibers; provides strong, flexible support	Tendons, ligaments
Dense irregular connective tissue	Densely packed collagenous fibers arranged in a tight interwoven pattern; provides tensile strength in any direction	Dermis of skin, fibrous capsules of organs and joints
Elastic connective tissue	Irregularly arranged elastic fibers; supports, provides framework	Large arteries, lower respiratory tract, between the arches of vertebrae
Reticular connective tissue	Reticular fibers forming supportive network; stores, performs phagocytic function	Lymph nodes, liver, spleen, thymus, bone marrow
Adipose connective tissue	Adipose cells; protects, stores fat, insulates	Below the skin, surface of heart, around kidneys, back of eyeball, surrounding joints

Overfeeding an infant during its first year, when adipose cells are forming, may cause an excessive amount of adipose tissue to develop. A person with a large amount of adipose tissue is more likely to develop obesity later in life than a person with a lesser amount. Dieting eliminates the fat stored within the tissue but not the tissue itself.

Adipose tissue is found throughout the body but is concentrated just beneath the skin, around the kidneys, on the surface of the heart, surrounding joints, and in the breasts of mature females. Fat functions not only as a food reserve, but also to support and protect various organs. It helps to keep the body warm because it is a poor conductor of radiant heat.

Connective tissue proper is summarized in table 4.3.

Cartilage

Cartilage (*kar'tl-ij*) consists of cartilage cells, or **chondrocytes** (*kon'dro-sīts*). The chondrocytes occupy tiny spaces called **lacunae** (*lă-kyoo'ne*—singular, *lacuna*) within an elastic semisolid matrix. Cartilage provides support and protection. Because it has few blood vessels, cartilage must receive nutrients through diffusion from surrounding tissue. For this reason, it has a slow rate of cell division and, if damaged, heals slowly.

There are three types of cartilage (fig. 4.7), distinguished from one another by the type and amount of fibers embedded within the matrix.

lacuna: L. *lacuna,* hole or pit
hyaline: Gk. *hyalos,* glass

Hyaline Cartilage

Hyaline cartilage, commonly called *gristle,* has a homogeneous, bluish-staining matrix that gives it a glassy appearance. It is the most abundant cartilage within the body and is located in the respiratory tract, rib cage, and developing bone.

Fibrocartilage

Fibrocartilage has a matrix reinforced with many collagenous fibers. It is a durable tissue adapted to withstand tension and compression. Fibrocartilage is found at the symphysis pubis, where the two pelvic bones articulate, and between the vertebrae as intervertebral discs. It also forms the cartilaginous wedges, called *menisci,* within the knee joint.

Inside Information—As you go about your daily activities, you are actually shrinking. Pressure on the discs between the vertebrae of your spine causes them to become compacted, thus reducing your height slightly. After a night's sleep, pressure is reduced and you regain the fraction of height you have lost. Aging, however, brings with it a gradual compression of the intervertebral discs that is irreversible.

Elastic Cartilage

Elastic cartilage is similar to hyaline cartilage except for the presence of abundant elastic fibers, which makes elastic cartilage very flexible and strong. This cartilage is found in the outer ear, portions of the larynx, and in the auditory canal.

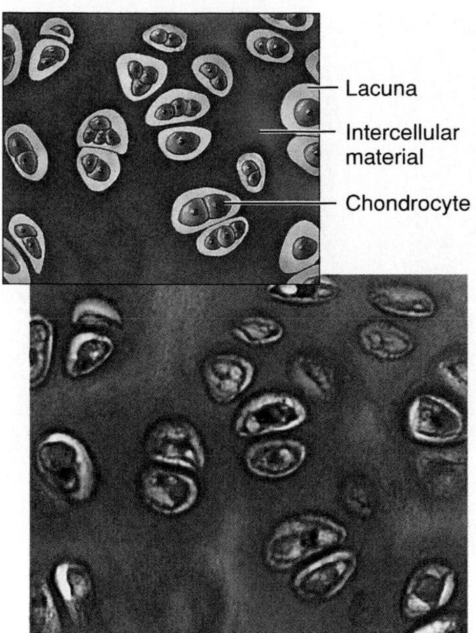

(a) Hyaline cartilage

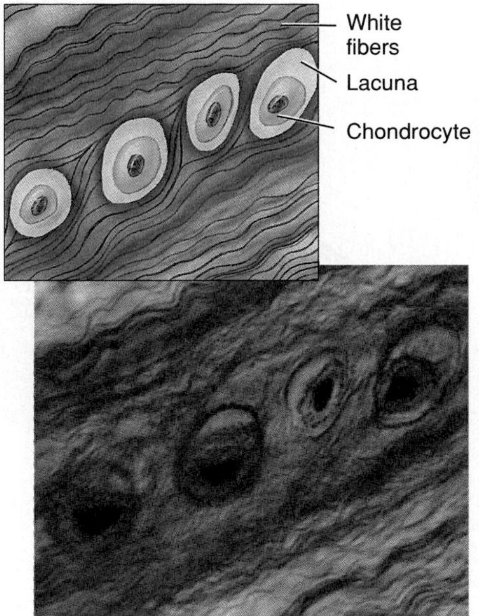

(b) Fibrocartilage

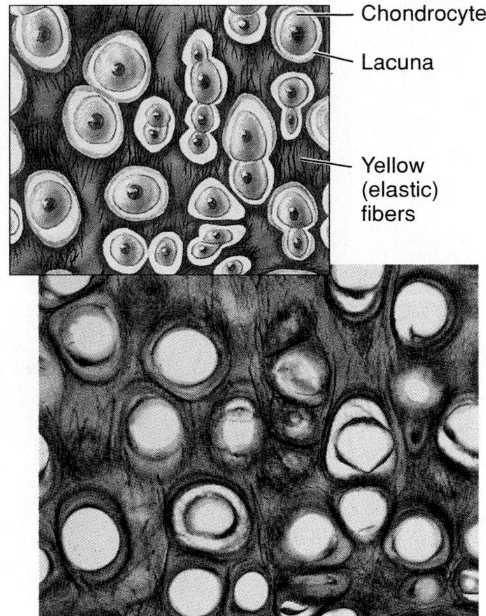

(c) Elastic cartilage

FIGURE 4.7
The principal types of cartilage.

TABLE 4.4 Cartilage

Type	Structure and Function	Location
Hyaline cartilage	Homogeneous matrix with extremely fine collagenous fibers; provides flexible support, protects, is precursor to bone	Articular surfaces of bones, nose, walls of respiratory passages, fetal skeleton
Fibrocartilage	Abundant collagenous fibers within matrix; supports, withstands compression	Symphysis pubis, intervertebral discs, knee joint
Elastic cartilage	Abundant elastic fibers within matrix; supports, provides flexibility	Framework of outer ear, auditory canal, portions of larynx

The three types of cartilage are summarized in table 4.4.

Bone (Osseous) Tissue

Bone, or **osseous, tissue** is the most rigid of all the connective tissues. Unlike cartilage, bone has a rich blood supply and is metabolically very active. The hardness of bone is due to the calcium phosphate located within the matrix. Numerous collagenous fibers, also found within the matrix, give bone some flexibility.

Based on porosity, bone tissue is classified as either compact or spongy, and most bones have both types (fig. 4.8). *Compact bone tissue* constitutes the hard outer portion of a bone, and *spongy bone tissue* constitutes the porous, highly vascular inner portion. The outer surface of a bone is covered by a connective tissue layer called the *periosteum* that serves as a site of attachment for ligaments and tendons, provides protection, and gives durable strength to the bone. Spongy bone tissue makes the bone lighter and provides a space for the bone marrow, where blood cells are produced.

When compact bone tissue is viewed through a microscope, a uniform structural arrangement can be seen. The bone cells, called **osteocytes** (*os'te-ŏ-sīts*), are arranged in rings around a **central** (haversian) **canal,** which contains blood vessels and a nerve. Each osteocyte occupies a space called a **lacuna.** Radiating from each lacuna are tiny canals, called **canaliculi,** that cross the dense matrix of the bone tissue to nearby lacunae. Nutrients diffuse through the canaliculi to reach each osteocyte. The matrix layers of bone tissue are called **lamellae.**

haversian canal: from Clopton Havers, English anatomist, 1650–1702

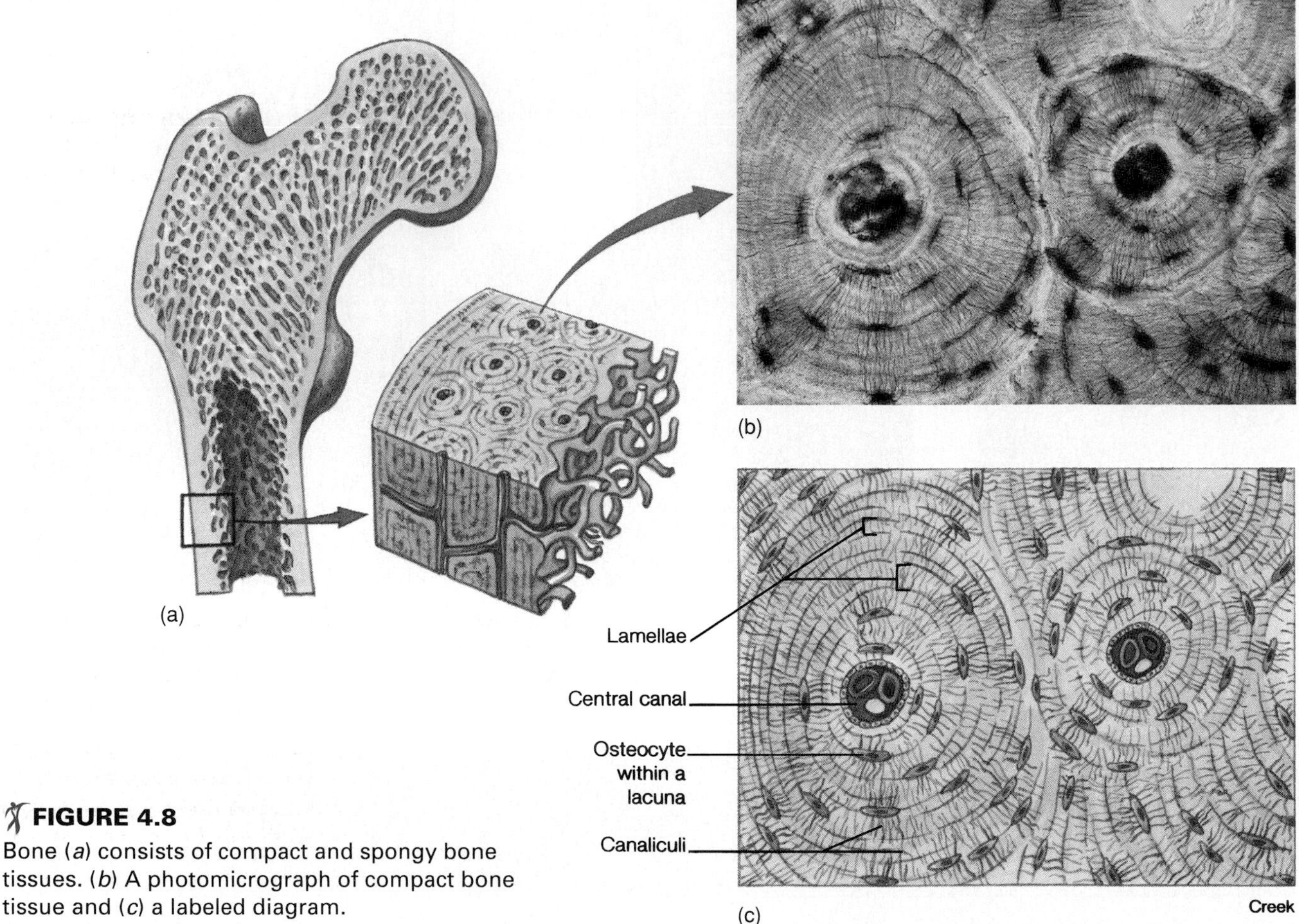

FIGURE 4.8
Bone (*a*) consists of compact and spongy bone tissues. (*b*) A photomicrograph of compact bone tissue and (*c*) a labeled diagram.

A central canal, with its surrounding osteocytes, lacunae, canaliculi, and concentric lamellae, constitutes an **osteon** (*os'te-on*), or *haversian system* (see fig. 6.10). Metabolic activity within bone tissue takes place in the osteons. Areas between the osteons contain **interstitial lamellae.** These areas have osteocytes within lacunae and associated canaliculi, but in an irregular arrangement. Bone tissue is further described in chapter 6 (see module 6.5).

Blood (Vascular) Tissue

Blood, or **vascular tissue,** is a highly specialized fluid connective tissue that plays a vital role in maintaining internal body homeostasis. **Formed elements** (red blood cells, white blood cells, and platelets) are suspended in the liquid plasma matrix. The technical names of the three types of formed elements are *erythrocytes, leukocytes,* and *thrombocytes* (fig. 4.9).

1. **Erythrocytes** (*ĕ-rith'rŏ-sīts*), or **red blood cells (RBCs),** are tiny biconcave discs that lack nuclei. Their red color is due to the protein *hemoglobin.* Oxygen attaches to and is transported on the hemoglobin molecules. In an infant, erythrocytes are sometimes produced in the spleen as well as in the bone marrow, but normally the bone marrow is the only production site after birth. The life span of erythrocytes is between 90 and 120 days.
2. **Leukocytes** (*loo'kŏ-sīts*), or **white blood cells (WBCs),** are slightly larger than erythrocytes and are nucleated. These cells exhibit amoeboid movement by forming cytoplasmic extensions that move them along, and they serve as a mobile defense force to protect the body against invasions by microorganisms. They are produced in bone marrow and lymphatic tissue and have a life span that ranges from 3 to 300 days. There are five kinds of leukocytes: *neutrophils, eosinophils, basophils, lymphocytes,* and *monocytes.*

erythrocyte: Gk. *erythros,* red; *kytos,* hollow (cell)
leukocyte: Gk. *leukos,* white; *kytos,* hollow (cell)

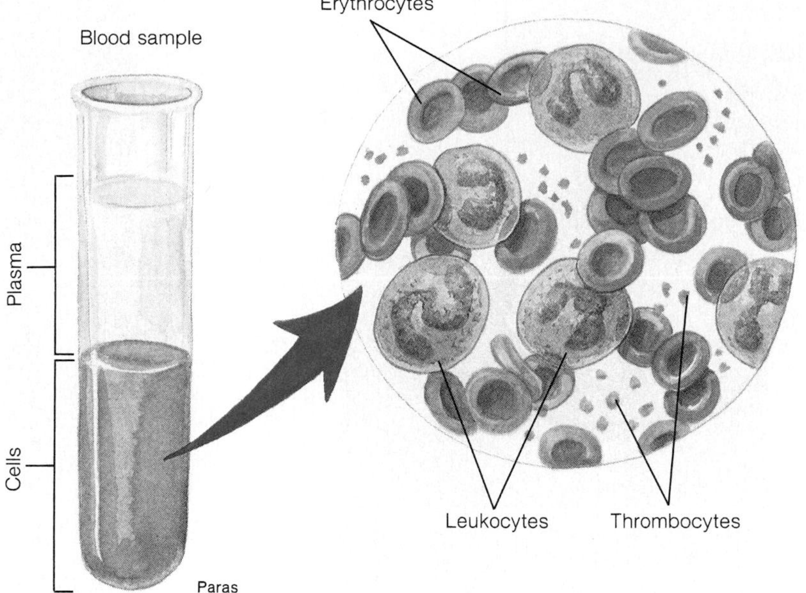

FIGURE 4.9
Blood is a type of connective tissue. Its cells, or formed elements—erythrocytes (red blood cells), thrombocytes (platelets), and leukocytes (white blood cells)—are suspended in a liquid plasma matrix.

3. **Thrombocytes** (*throm'bŏ-sīts*) are the **platelets** of the blood. Platelets, along with the protein *fibrinogen* found in the plasma, play a role in blood clotting. Platelets form from large cells in the bone marrow called **megakaryocytes.** The megakaryocytes break up into fragments, and each fragment becomes enclosed by a piece of cell membrane. Platelets have a life span between 5 and 7 days. Like leukocytes, they exhibit amoeboid movement.

Blood is discussed fully in chapter 14.

An injury may stimulate tissue repair activity, usually involving formation of connective tissue. The healing process differs in minor and major wounds. A minor scrape or cut results in platelet activity of the exposed blood and the formation of a scab. The upper layer of the skin regenerates beneath the scab. A major open wound brings about connective tissue formation, in which collagenous fibers form from surrounding fibroblasts to strengthen the traumatized area. The healed area is known as a scar.

thrombocyte: Gk. *thrombos,* a clot; *kytos,* hollow (cell)

4.4 Muscle Tissue and Nervous Tissue

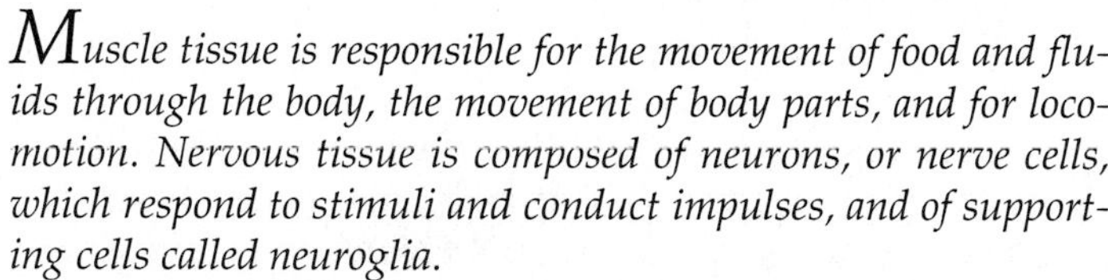

Muscle tissue is responsible for the movement of food and fluids through the body, the movement of body parts, and for locomotion. Nervous tissue is composed of neurons, or nerve cells, which respond to stimuli and conduct impulses, and of supporting cells called neuroglia.

Muscle Tissue

Muscle tissue is unique because it is able to contract, thus making movement possible. The muscle cells, or *fibers,* are long and cylindrical. There are three types of muscle tissue in the body: *smooth, cardiac,* and *skeletal muscle tissue* (fig. 4.10).

Smooth Muscle

Smooth muscle tissue is common throughout the body. In the wall of the digestive tract, wavelike contractions of smooth muscle (*peristaltic contractions*) mix food and move it along the length of the tract. Smooth muscle is also

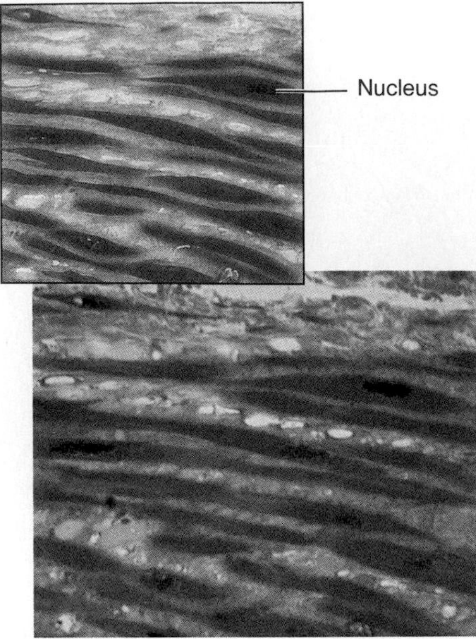

(a) Smooth muscle tissue

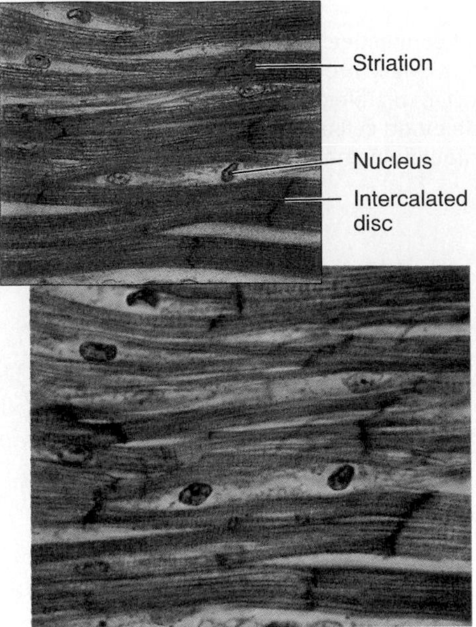

(b) Cardiac muscle tissue

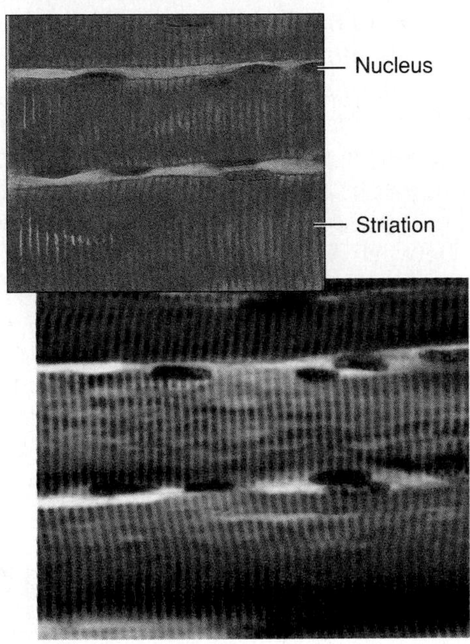

(c) Skeletal muscle tissue

FIGURE 4.10
Muscle tissue.

found in the walls of blood vessels, the walls of respiratory passages, and in the urinary and reproductive ducts. The contraction of smooth muscle is under involuntary (unconscious) nervous control.

Smooth muscle fibers are long, spindle-shaped cells that contain a single nucleus. These cells are usually grouped together in flattened sheets, forming the muscular portion of the wall around a lumen.

Cardiac Muscle

Cardiac muscle tissue makes up most of the wall of the heart. This tissue is characterized by branching fibers, a central nucleus, and banding patterns called *striations.* The cardiac muscle fibers are joined by **intercalated discs.** Intercalated discs help to hold neighboring cells together and to spread the contraction from cell to cell. Cardiac muscle tissue also contracts involuntarily.

Skeletal Muscle

Skeletal muscle tissue makes up the skeletal muscles that are attached to the bones of the skeleton. Contraction of this tissue results in voluntary or involuntary body movements. Skeletal muscle fibers are long and multinucleate (have many nuclei). The striations of this tissue are easily seen through a microscope. Fibers are grouped into bundles, which are visible without a microscope in fresh muscle. Both cardiac and skeletal muscle fibers cannot divide once tissue formation is complete. Skeletal muscle is further discussed in chapter 8.

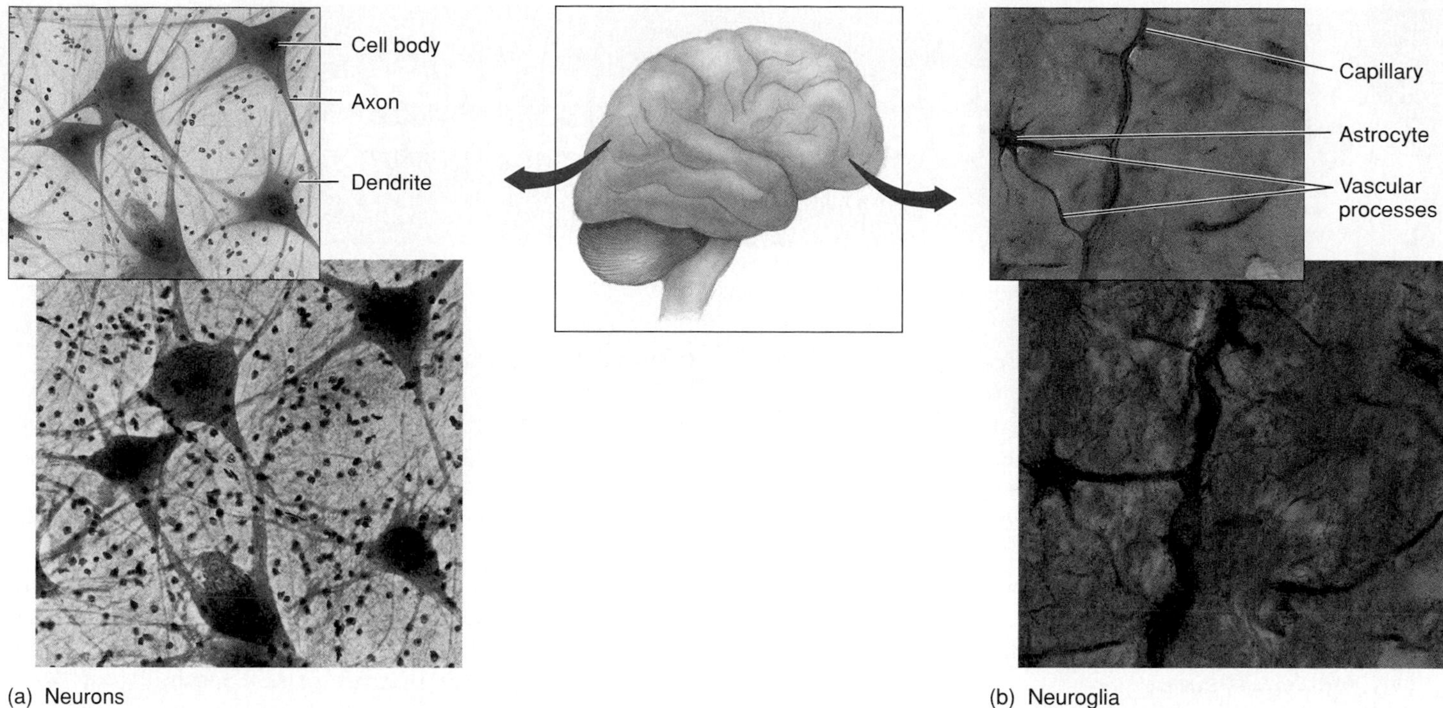

FIGURE 4.11
Nervous tissue is found within the brain and consists of two principal kinds of cells: (*a*) neurons and (*b*) neuroglia.

Nervous Tissue

Nervous tissue, contained within the brain, spinal cord, and nerves, is composed of two kinds of cells—*neurons* and *neuroglial cells.* **Neurons,** or **nerve cells,** are the basic structural and functional units of the nervous system. They are specialized to respond to physical and chemical stimuli and to generate and conduct impulses to and from the various body organs. Of all the body's cells, neurons are probably the most specialized. As with striated muscle fibers, the number of neurons is established shortly after birth, and thereafter they lack the ability to undergo mitosis. Under certain circumstances, however, a neuron that has been severed can repair itself (see the Synoptic Spotlight in chapter 9, p. 242).

A neuron has three principal components (fig. 4.11*a*):

1. The **cell body** contains the nucleus and specialized organelles and microtubules.
2. The **dendrites** function to receive a stimulus and conduct the impulse toward the cell body.
3. The **axon** is a long extension that conducts an impulse away from the cell body. The term *nerve fiber* usually refers to an axon and the myelin sheath that surrounds it.

In addition to neurons, nervous tissue is composed of supporting cells, collectively called **neuroglia** (*noo-rog'le-ă*) (fig. 4.11*b*). Neuroglial cells, sometimes called glial cells, are about five times as abundant as neurons and have a limited ability to divide. They do not transmit impulses, but rather support and bind neurons together. Certain neuroglial cells are phagocytic; others assist in providing nourishment for the neurons.

Neurons and neuroglia are discussed in detail in chapter 9.

neuron: Gk. *neuron,* sinew or nerve
neuroglia: Gk. *neuron,* nerve; *glia,* glue

Synoptic Spotlight

Moving and Matching: Tissue Transplants

The bionic man (or woman), not so long ago a figment of the imagination, is quickly becoming a reality. Largely due to recent advances in biomedical engineering and microsurgery, our ability to introduce artificial materials into the human body to replace worn or damaged parts has improved dramatically. Transplants of human tissue and organs are also common, and in many cases promise a new lease on life for the fortunate patients who receive them. In this brief space, we will consider just a few types of transplantations—those involving skin, hearts, bone marrow, and fetal tissue.

Skin transplants, more often called *grafts,* have a long history of success. In this procedure, tissues from a person's own body are "harvested" and introduced into areas where skin has sustained devastating damage to the regenerative layers. Transfer of one's own tissue is referred to as an **autograft.** Since the patient's own skin is being used, there are no complications of incompatibility and rejection.

In the earliest skin grafts, a nearby limb was temporarily attached surgically to the damaged area. The skin from a healthy arm or finger, for example, was loosened and stretched to cover the damaged tissue. When the new skin grew in to replenish the damaged cells, the limb was separated and released. An alternative technique that is more common today involves removal of sheets of skin. During the harvesting surgery, a sophisticated version of our kitchen carrot scraper is used to carefully lift sheer layers of epidermis. Part (or all) of the dermis is also removed; the depth of the sheet depends on the severity and depth of the injury. These sheer bits of skin are cut into small pieces that are not much larger than postage stamps. These fragments are carefully distributed on top of the exposed fascia. Once they establish a blood supply, the fragments begin to grow and replenish the damaged skin.

Recent advances in cell culture techniques have contributed further to the success of skin transplants. It is now possible to remove a small section of skin from a burn victim, and grow it under controlled laboratory conditions. From initial stamp-sized samples, large sheets of epidermis have been grown and used to cover burn areas. Cultured epidermis offers the added benefit of potentially scar-free healing.

Homotransplants, or **allografts,** are transplantations between individuals of the same species who are not genetically identical. These transplants are more complicated than autografts because of the possibility of tissue rejection. The key to success is the tissue compatibility of the donor and the recipient. Medically, it is far easier to prevent rejection than it is to halt an immune system that has been activated. That is why meticulous matching of tissue is critical. The first step in matching involves the blood and the ABO and Rh systems, since the marker proteins (or antigens) on these cells are widespread throughout body cells. Once the blood groups have been matched, then other tissue matches are determined.

The role of the major histocompatibility complex (MHC) proteins is particularly important in transplantations. MHC-Encoded proteins, located on the cell membranes of all body cells, make it possible for the immune system to know who the "good guys" are; that is, to recognize one's own cells as "self" and distinguish them from "nonself" cells. In mismatched transplantations, certain cells in the recipient, called *cytotoxic T cells,* "see" the transplanted cells as intruders, and act vigorously to destroy them. Hence, the fewer cell-marker differences (and the more matches) between the two individuals the better. (This is certainly a good time to have an identical twin handy.) Twin-to-twin, or **isograft, transplants**—with identical genetic material—have the greatest chance of success. Since few of us have twins, however, it is far more likely that cells with membrane markers like ours will be found in a close relative. If this is not possible, the hunt for compatible MHCs may be very long.

Heart transplants have probably received more public attention than any other organ transplant, perhaps because of the drama associated with the heart. When our hearts are failing, options are limited. The most superficially simple solution is to find a compatible donor heart that can be substituted for the diseased organ. First performed in 1987 in Maryland, the surgical technique has been vastly improved. What has *not* changed are the complicated and risk-filled procedures that precede and follow the operation. In order to be considered a candidate for transplant surgery, a patient must be evaluated by a team of physicians. They must determine whether the individual is first of all capable of surviving the actual surgery, and then of adopting a postsurgical lifestyle that would justify such an extreme measure. Once a patient passes the screening process, he or she must also be able to survive until a willing donor is found—a donor with compatible tissues. (The specific selection of candidates to receive heart transplants varies from institution to institution, but all hospital systems face similar difficult issues.)

Following the surgery, doses of immune-suppressant drugs are administered in an attempt to keep the patient's body from rejecting the new organ. One such drug, *cyclosporine,* an extract of soil fungi, has proven particularly useful in this regard. Use of the immunosuppressants must be balanced against the risk of allowing the recipient to be vulnerable to pathogens, which could take full advantage of a compromised immune system.

Bone marrow transplants offer hope to a different type of patient. When disease (e.g., aplastic anemia and some types of leukemias) or radiation exposure destroys bone marrow, then transplanting healthy bone marrow into the malfunctioning bones is called for. As with heart transplants, there must be a match of blood type. Since the most likely source of compatible tissues is from individuals with similar DNA, close relatives are asked to become donors. Once the lengthy screening process has been completed, the patient's own bone marrow is destroyed prior to the surgery by means of massive doses of radiation or chemotherapy. During surgery, small amounts of healthy marrow tissue are removed from the donor. The sample is then mixed with anticoagulants and filtered to remove some types of lymphocytes before being introduced into the marrow recipient. Once inside the recipient, it is hoped that the new tissue will infiltrate the bone marrow with healthy cells—cells that will resume the functions of the ones that had to be destroyed.

The circumstances involving fetal tissue are quite different from skin, heart, or bone marrow transplants. The anguish over a miscarriage certainly cannot be mitigated. However, grieving parents may gain a small bit of comfort by knowing that tissue from the conceptus could possibly provide hope and relief for others. Parkinson's disease patients, diabetics, elderly individuals, and women with fertility problems may all be the grateful recipients of such fetal tissue. Fetal tissue is excellent for transplantation for several reasons. The fetal tissues lack most membrane markers, making it unlikely that the recipient's immune system will recognize them as foreign. Thus, the complications of rejection are avoided. Also, because the metabolic rate of fetal tissues is very high, the rapidly growing cells quickly merge with a recipient's cells. Since fetal cells are largely undifferentiated, they are "plastic" and capable of developing into the types of cells that the recipient needs.

On the other hand, there are those who argue that using fetal tissue for therapeutic purposes will increase the number of abortions. The question of whether this type of transplantation poses an ethical dilemma is still being hotly debated.

Clinical Considerations

In addition to its importance in determining organ structure and function, the science of histology is of immense clinical importance. Many diseases are diagnosed through microscopic examination of tissue sections. Even in performing an autopsy, an examination of various tissues is vital for establishing the cause of death.

Several sciences are concerned with specific aspects of tissues. *Histopathology* is the study of diseased tissues. *Histochemistry* is concerned with the biochemical physiology of tissues as they function to maintain body homeostasis. *Histotechnology* studies the various techniques by which tissues can be better stained and observed. In all of these disciplines, a thorough understanding of normal, or healthy, tissue is imperative for recognizing altered, or abnormal, tissue.

In diagnosing a disease, it is frequently important to examine tissues from a living person histologically. The procedure of removing a section of living tissue is termed a *biopsy*—a term also used to refer to the sample of tissue removed. There are several techniques for obtaining a biopsy. *Surgical removal* is usually done on large masses or tumors. *Curettage* involves cutting and scraping tissue, as may be done in examining for uterine cancer. In a *percutaneous needle biopsy*, a biopsy needle is inserted through a small skin incision and tissue samples are withdrawn (fig. 4.12). Both normal and diseased tissues are removed for purposes of comparison.

Changes in Tissue Composition

Most diseases change tissue structure *locally*, where the disease is prevalent. Some diseases, however, called *general conditions*, cause changes that are remote from the locus of the disease. **Atrophy** (wasting of body tissue), for example, may be limited to an organ where the disease interferes with the metabolism of that organ, but it also may involve an entire limb when nourishment or nerve impulses are impaired. *Muscle atrophy* can be caused by a disease of the nervous system such as polio, or it can be the result of a diminished blood supply to a muscle. *Senescence* (*sĕ-nĕ'sens*) *atrophy*, or simply *senescence*, is the natural aging of tissues and organs within the body. *Disuse atrophy* is a local atrophy that results from the inactivity of a tissue or organ.

Necrosis is death of cells or tissues within the body. It can be recognized by physical changes in the dead tissues. Necrosis can be caused by severe injury; physical agents (trauma, heat, radiant energy, chemical poisons); or poor nutrition of tissues. When examined through a microscope, the necrotic tissue usually appears opaque, with a whitish or yellowish cast. **Gangrene** occurs when large amounts of necrotic tissue are invaded by microorganisms that live on decaying flesh.

Somatic death is the death of the whole body. Following somatic death, tissues undergo irreversible changes such as **rigor mortis** (muscular rigidity), clotting of the blood, and cooling. Postmortem (after death) changes occur under varying conditions at predictable rates of time, which is useful in estimating an approximate time of death.

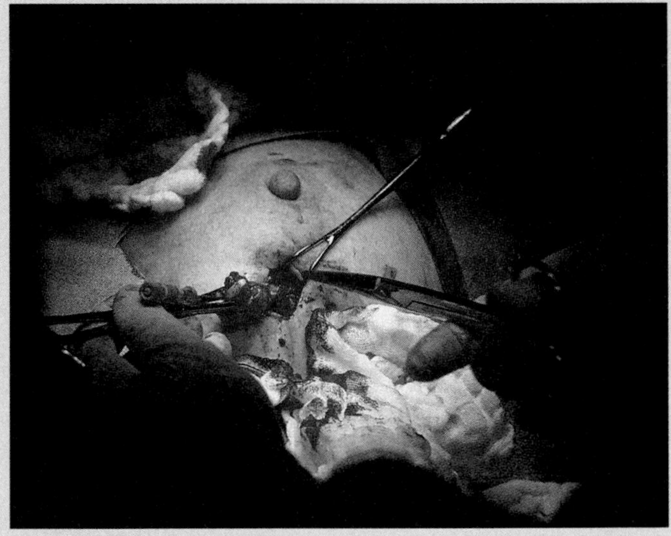

FIGURE 4.12
A biopsy permits examination of live tissue for diagnostic purposes.

SUMMARY

4.1 Introduction to Tissues

1. Tissues are collections of similar cells that perform a common function. The study of tissues is called histology.
2. Cells are separated and bound together by an extracellular matrix whose composition varies from solid to liquid.
3. The four basic tissue types are epithelial tissue, connective tissue, muscle tissue, and nervous tissue.

4.2 Epithelial Tissue

1. Epithelial tissue is adapted for protection, absorption, and secretion. It covers body surfaces, has little extracellular matrix, and lacks blood vessels. A basement membrane supports most kinds of epithelial tissue.
2. Simple epithelial tissue consists of a single layer of cells. It is located where diffusion, filtration, and secretion occur.
3. Stratified epithelial tissue consists of two or more layers of cells and is adapted for protection.
4. Transitional epithelium lines the cavity of the urinary bladder and is adapted for stretching.
5. Glands formed from developing epithelial tissue function as secretory exocrine glands.

4.3 Connective Tissue

1. Connective tissue provides metabolic and structural support for other tissues. It has abundant matrix with relatively few cells. With the exception of cartilage, it contains many blood vessels.
2. Connective tissue proper has a flexible matrix called ground substance that contains cells called fibroblasts. Fibroblasts produce collagenous, elastic, and reticular fibers, which are key structural elements of connective tissue.
3. Cartilage consists of a semisolid matrix of chondrocytes and various fibers. It provides a flexible framework for many organs.
4. Bone tissue consists of osteocytes, collagenous fibers, and a durable matrix of mineral salts.
5. Blood tissue consists of formed cellular elements (erythrocytes, leukocytes, and thrombocytes [platelets]) suspended in a fluid plasma matrix.

4.4 Muscle Tissue and Nervous Tissue

1. Muscle tissue (smooth, cardiac, and skeletal) is responsible for the movement of materials through the body, the movement of one part of the body with respect to another, and for locomotion.
2. Fibers in the three kinds of muscle tissue are adapted to contract in response to stimuli.
3. Neurons, or nerve cells, respond to stimuli and conduct impulses to and from all body organs.
4. Neuroglial cells support and bind neurons. Some are phagocytic; others provide nourishment.

Review Activities

Objective Questions

1. Which of the following is *not* a principal type of body tissue? (*module 4.1*)
   a. nervous tissue
   b. integumentary tissue
   c. connective tissue
   d. muscle tissue
   e. epithelial tissue

2. Which statement regarding epithelial tissues is *false*? (*module 4.2*)
   a. Most epithelial tissues lack a direct blood supply.
   b. They are strengthened by elastic and collagenous fibers.
   c. One side is exposed to a lumen or cavity, or to the external environment.
   d. They have cells that are tightly packed together.

3. A gastric ulcer of the stomach would involve (*module 4.2*)
   a. simple cuboidal epithelium.
   b. transitional epithelium.
   c. simple ciliated columnar epithelium.
   d. simple columnar epithelium.

4. Which structural and secretory designation describes mammary glands? (*module 4.2*)
   a. acinar, apocrine
   b. tubular, holocrine
   c. tubular, merocrine
   d. acinar, holocrine

5. Dense regular connective tissue is found in (*module 4.3*)
   a. blood vessels.
   b. the spleen.
   c. tendons.
   d. the uterine wall.

6. The phagocytic connective tissue found in the lymph nodes, liver, spleen, and bone marrow is (*module 4.3*)
   a. reticular tissue.
   b. loose fibrous tissue.
   c. mesenchyme.
   d. elastic tissue.

7. Cartilage is slow in healing following an injury because (*module 4.3*)
   a. it is located in body areas that are under constant physical strain.
   b. it is avascular.
   c. its chondrocytes cannot divide.
   d. it has a semisolid matrix.

8. Cardiac muscle tissue is characterized by (*module 4.4*)
   a. striations.
   b. intercalated discs.
   c. rhythmical involuntary contractions.
   d. all of the above.

9. Which of the following statements concerning nervous tissue is *false?* (*module 4.4*)
   a. It consists of neurons and neuroglial cells.
   b. Neurons require constant nourishment because of their rapid rate of cell division.
   c. The brain is composed of nervous tissue.
   d. Neurons conduct impulses and neuroglial cells support neurons.

Essay Questions

1. Define *tissue.* What are the differences between cells, tissues, glands, and organs? (*module 4.1*)
2. Describe the four principal tissue types. What are the general functions of each? (*module 4.1*)
3. What physiological functions are epithelial tissues adapted to perform? (*module 4.2*)
4. Identify the epithelial tissue found in each of the following structures or organs and state the function of the tissue in each case. (*module 4.2*)
   a. in the alveoli of the lungs
   b. lining the inside wall of the GI tract
   c. in the outer layer of the skin
   d. lining the inside wall of the urinary bladder
   e. lining the inside wall of the uterine tube
   f. lining the inside wall of the lower respiratory tract
5. Compare and contrast the three types of cartilage in terms of structure, location, and function. (*module 4.3*)
6. Distinguish between neurons and neuroglia. (*module 4.4*)

Labeling Exercise

Label the tissue types and the structures shown in the figure to the right.

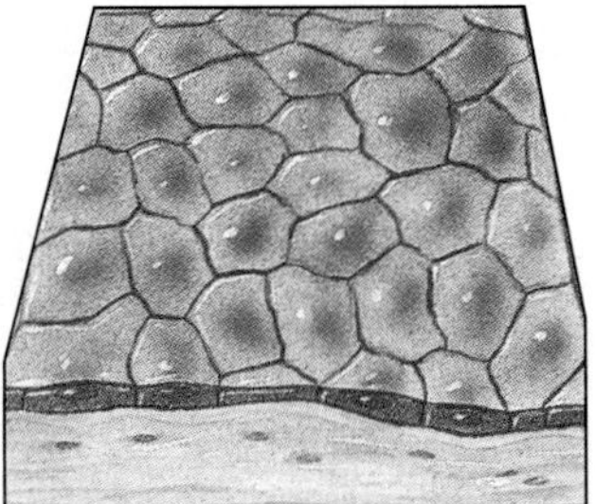
(1)

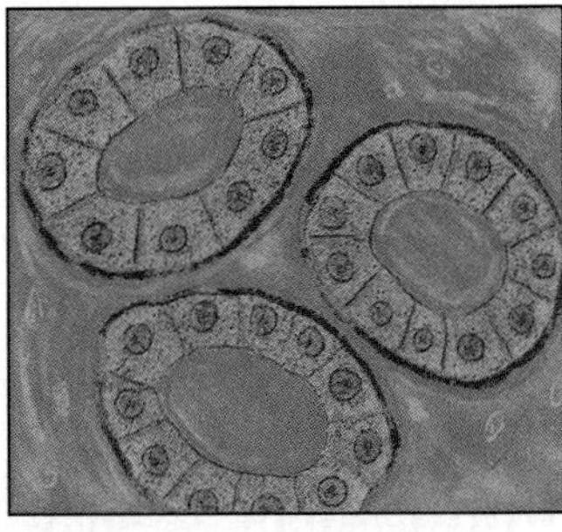
(2)

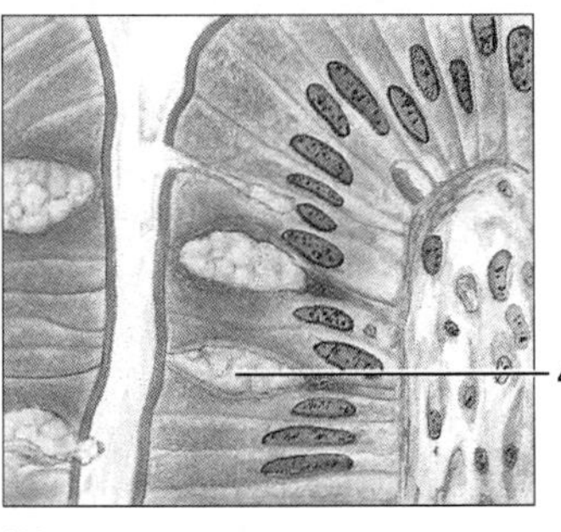

(3)

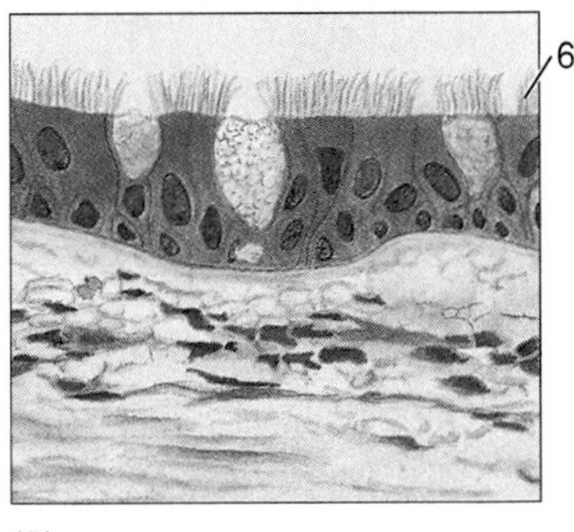

(5)

Critical Thinking Questions

1. The aorta (a principal blood vessel) has three layers surrounding its lumen. What is the predominant tissue in each of the three layers and what is the adaptive function of each?
2. Is the skin a tissue or an organ? What statements can you make to justify your response? Are there any good counterarguments?
3. The function of a tissue is actually a function of its cells. And the function of a cell is a function of its organelles. Knowing this, what type of organelles would be particularly abundant in cardiac muscle tissue that requires a lot of energy; in reticular tissue within the liver, where cellular debris and toxins are ingested; and in fibrous connective tissue that consists of tough protein strands?
4. Explain the relationship between adipose cells and fat. Discuss the function of fat and explain the potential danger of excess fat.
5. Why does cancer rarely occur in muscle fibers (for example, in the cardiac muscle fibers of the heart)? Why are most brain tumors (gliomas) composed of neuroglial cells?

CHAPTER

5

"The skin is a barrier to microorganisms, water, and excessive sunlight." *page 101*

Integumentary System

Chapter Outline

Terms to Remember

integument (*in-teg'yoo-ment*), 96
epidermis (ep"ĭ-der'mis), 97
keratin (*ker'ă-tin*), 97
cornification, 98
melanin (*mel'ă-nin*), 98
melanocyte (*mel'ă-no-sīt*), 98
dermis, 99
hypodermis (*hi"pŏ-der'mis*), 100
hair, 103
hair follicle, 104
sebaceous (*sĕ-ba'shus*) gland, 106
sudoriferous (*soo"dor-if'er-us*) gland, 106
eccrine (*ek'rin*) gland, 107
apocrine (*ap'ŏ-krin*) gland, 107
mammary gland, 107
ceruminous (*sĕ-roo'mĭ-nus*) gland, 107

Learning Objectives

After studying this chapter, you should be able to

- define the basic anatomical terms pertaining to the integumentary system.
- describe the histological characteristics of the epidermis, dermis, and hypodermis.
- summarize the process of cellular replacement within the skin and explain the transitional events that occur within each of the epidermal layers.
- discuss the factors that determine skin color.
- discuss the protective role of the skin and list the specific features that prevent desiccation (drying out) of the body.
- explain the role of the skin in warming and cooling the body.
- describe the structure and functions of hair, integumentary glands, and nails.
- explain, in general, how the skin responds to burns or other injuries.

5.1 The Skin As an Organ

The skin (integument) is the largest organ of the body, and together with its accessory organs (hair, glands, and nails), it constitutes the integumentary system. In its role as a dynamic, protective interface between the continually changing external environment and the body's internal environment, the skin functions to maintain homeostasis.

We are more aware of and concerned with our integumentary system than perhaps any other system of our body. One of the first things we do in the morning is to look in the mirror and see what we have to do to make our skin and hair presentable. Periodically, we examine our skin for wrinkles (fig. 5.1) and our scalp for gray hairs. We recognize others to a large degree by features of their skin.

The appearance of our skin frequently determines the initial impression we make on others. Unfortunately, it may also determine whether or not we succeed in gaining social acceptance. For example, social rejection as a teenager, imagined or real, can be directly associated with skin problems such as acne. A person's self-image and consequent social behavior may be closely linked to his or her physical appearance.

The skin, or **integument** (*in-teg'yoo-ment*), is much more than a showpiece, however. It helps to regulate certain body functions and protect certain body structures. Together with its accessory organs (hair, glands, and nails), it constitutes a body system of its own. Included in this integumentary system are the millions of sensory receptors of the skin and its rich supply of blood vessels.

The skin is a dynamic interface between the body and the external environment. It protects us from the environment, yet permits us to communicate with it. It is tough enough to resist more environmental assaults, yet soft enough to respond to the gentlest touch.

Inside Information—The average human body is covered with about 2 square meters (22 square feet) of skin, making skin the largest organ in surface area.

The skin is considered an organ since it consists of several kinds of tissues that are structurally arranged to function together. It is a heavy organ, accounting for approximately 7% of our adult body weight. The skin is of variable thickness, averaging 1.5 mm. It is thickest on the parts of the body exposed to wear and abrasion, such as the soles of the feet and palms of the hand. In these areas, it is about 6 mm thick. It is thinnest on the eyelids, external genitalia, and on the tympanic membrane (eardrum), where it is approximately 0.5 mm thick. Even its appearance and texture varies from the rough, callous skin covering the elbow and knuckle joints to the soft, sensitive areas of the eyelids, nipples, and genitalia.

integument: L. *integumentum*, covering

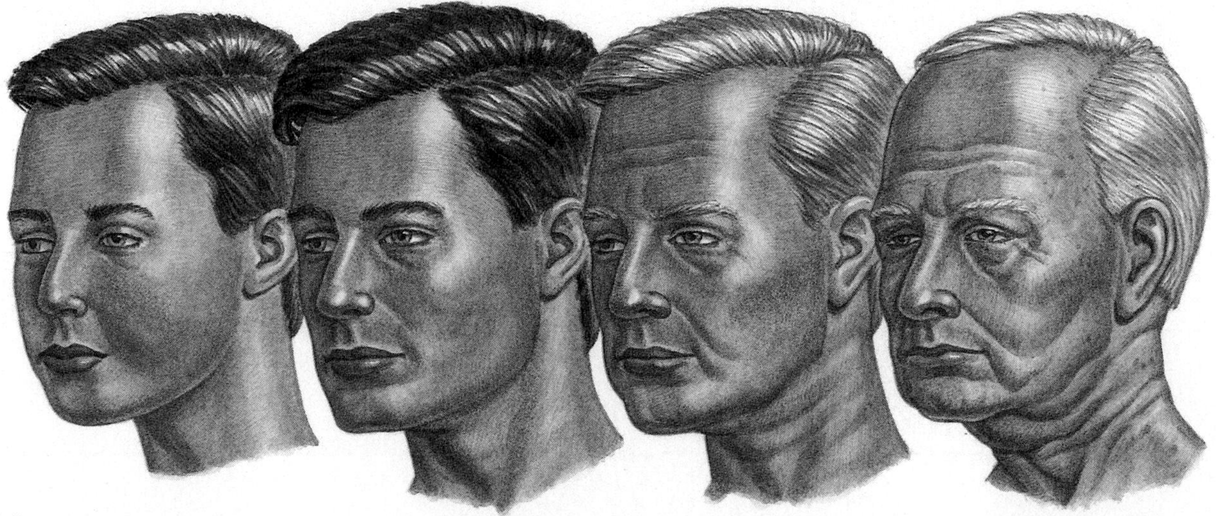

FIGURE 5.1
The aging process is most obvious in the changing integumentary system.

The general appearance and condition of the skin are clinically important because they provide clues to certain body conditions or dysfunctions. Pale skin may indicate shock, whereas red, flushed, overly warm skin may indicate fever and infection. A rash suggests allergies or local infections. Abnormal textures of the skin may be the result of glandular or nutritional problems. Even chewed fingernails may be a clue to emotional problems.

5.2 Layers of the Skin

The skin consists of two principal layers. The outer epidermis is stratified into four or five structural and functional layers, and the thick and deeper dermis consists of two layers. The hypodermis (subcutaneous tissue) connects the skin to underlying organs.

Epidermis

The **epidermis** (*ep″ĭ-der′mis*) is the superficial protective layer of the skin, composed of stratified squamous epithelium. Over most of your body, this layer is between 30 and 50 cell layers thick (00.7 to 0.12 mm). As described in chapter 4 (see module 4.2), all but the deepest layers of the epidermis are dead cells that contain a protein called **keratin** (*ker′ă-tin*). Keratin, which toughens and waterproofs the skin, is formed in a process called *keratinization.* The epidermis is composed of either four or five layers, or *strata,* depending on its location (fig. 5.2). The epidermis of

epidermis: Gk. *epi,* on; *derma,* skin
stratum: L. *stratum,* something spread out

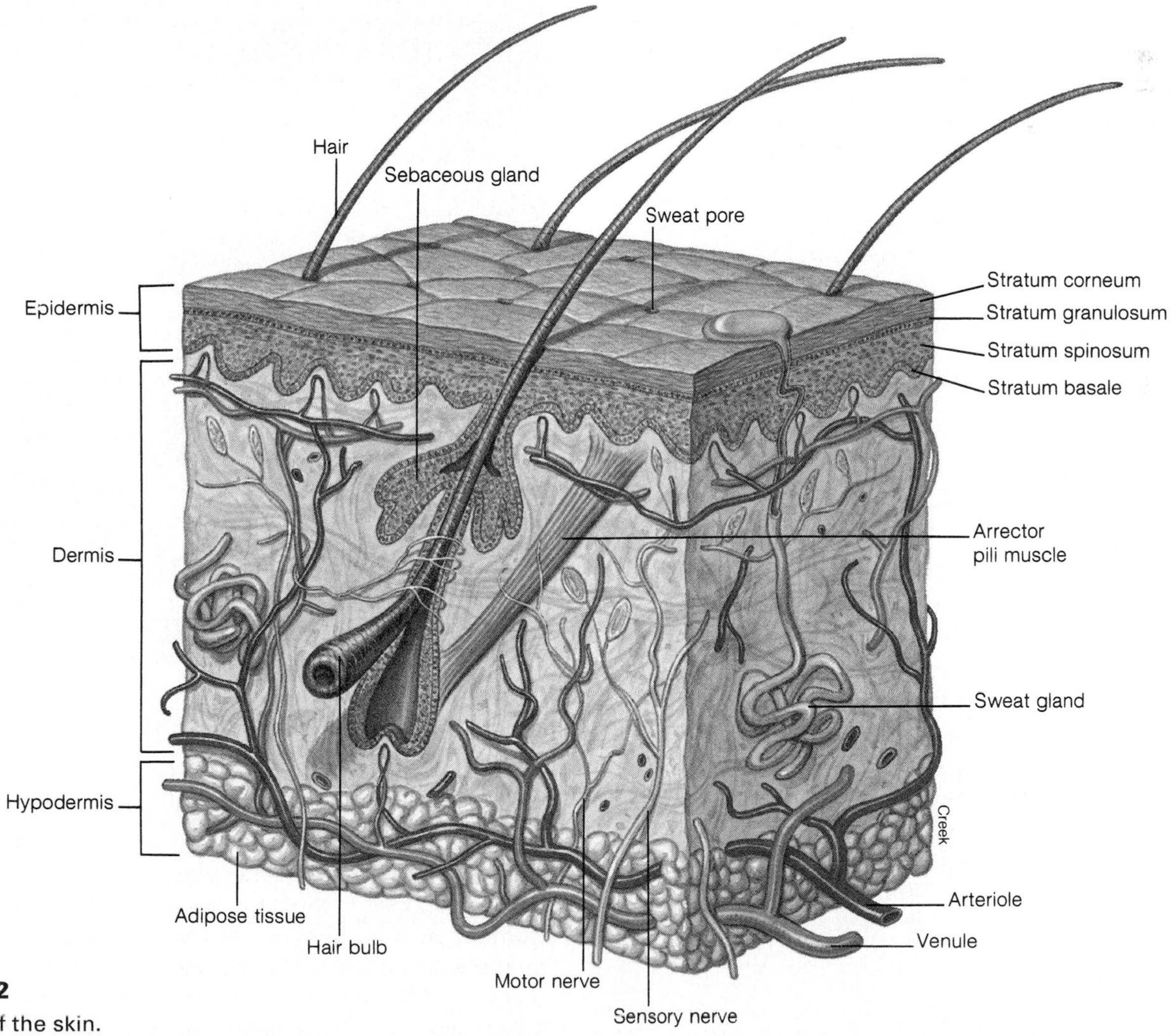

FIGURE 5.2
A diagram of the skin.

TABLE 5.1 **Layers of the Epidermis**

Layer	Location	Characteristics
Stratum corneum	Outermost layer	Composed of many layers of keratinized and cornified dead cells that are flattened; nuclei are absent
Stratum lucidum	Below the stratum corneum	A thin, clear layer found only in the epidermis of the palms and soles
Stratum granulosum	Below the stratum lucidum	Composed of several layers of granular cells that contain keratin and shriveled nuclei
Stratum spinosum	Below the stratum granulosum	Composed of several layers of cells with darkened nuclei and spinelike projections; limited mitosis
Stratum basale	Deepest layer	Composed of a single layer of cells that are constantly undergoing mitosis; contains pigment-producing melanocytes

the palms and soles has five layers because these areas are exposed to most friction. The epidermis of all other areas of the body has only four layers. The epidermal layers are listed in table 5.1, along with their locations and characteristics.

Cornification is the flattening and drying of the outer stratum corneum of the epidermis and is an important protective adaptation of the skin. Friction at the surface of the skin stimulates mitotic activity in the deepest epidermal layer—the stratum basale (*bă-sal'e*). This may lead to the formation of a *callus* for additional protection.

Tattooing colors the skin permanently (fig. 5.3) because dyes are injected below the dividing stratum basale into the dermis. Small tattoos can be removed by skin grafting; for large tattoos, mechanical abrasion of the skin is preferred.

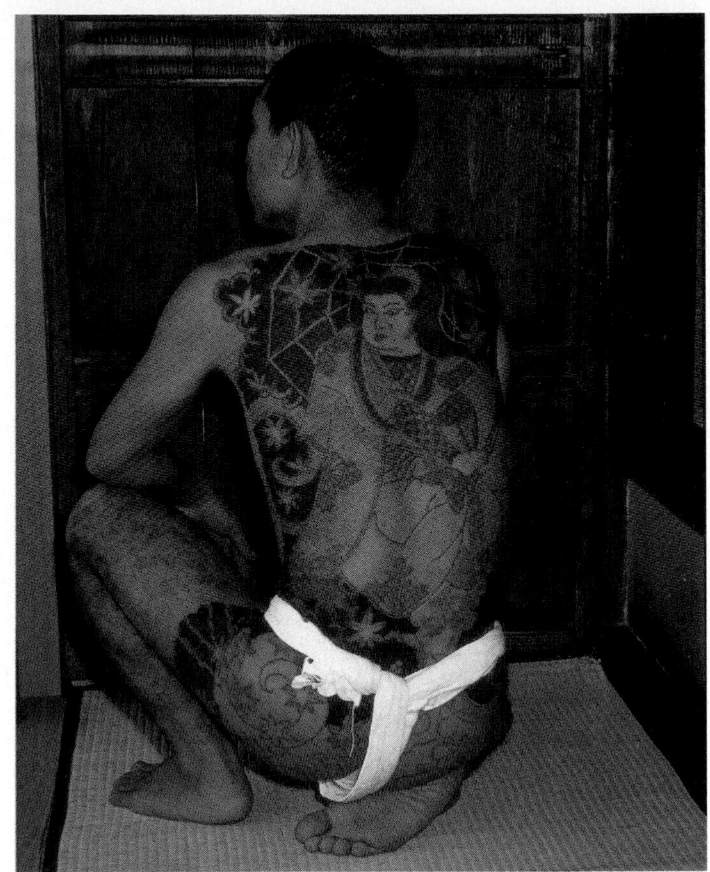

FIGURE 5.3
For centuries, and in many cultures, tattoos have been created to adorn the skin.

Coloration of the Skin

A person's normal skin color is a combination of three pigments: **melanin** (*mel'ă-nin*), **carotene,** and **hemoglobin.** Melanin is a brown-black pigment formed in cells called **melanocytes** (*mel'ă-no-sīts*). Melanocytes are found throughout the stratum basale and stratum spinosum. All individuals of similar size have approximately the same number of melanocytes, but the amount of melanin produced and the distribution of the melanin determine racial variations in skin color, such as black, brown, yellow, and white. Melanin protects the basal layer against the damaging effect of the ultraviolet (UV) rays of the sun (see the Synoptic Spotlight on page 108). A gradual exposure to the sunlight promotes the increased production of melanin within the melanocytes, and hence tanning of the skin. The skin of a person with *albinism* has the normal number of melanocytes in the epidermis but lacks the enzyme *tyrosinase* that converts the amino acid *tyrosine* to melanin. Albinism is a hereditary condition. *"Liver spots"* on elderly people, which appear as pigmented patches, are benign growths of pigment-producing melanocytes.

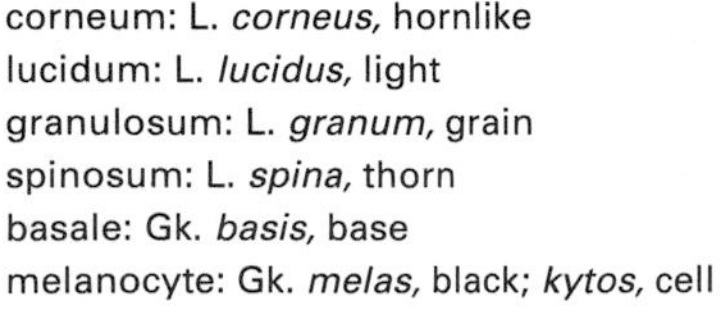
corneum: L. *corneus,* hornlike
lucidum: L. *lucidus,* light
granulosum: L. *granum,* grain
spinosum: L. *spina,* thorn
basale: Gk. *basis,* base
melanocyte: Gk. *melas,* black; *kytos,* cell

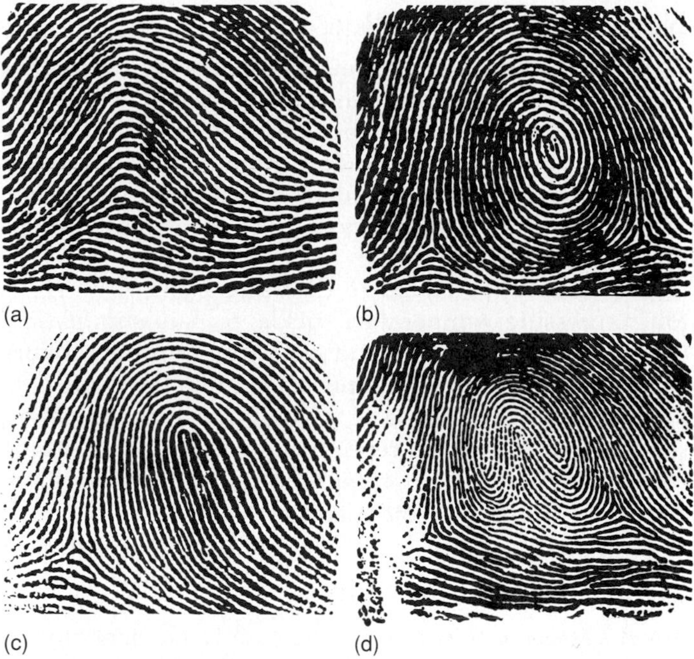

FIGURE 5.4
The four basic fingerprint patterns: (*a*) arch, (*b*) whorl, (*c*) loop, and (*d*) combination.

Carotene is a yellowish pigment found in the epidermal cells and fatty parts of the dermis. Carotene is abundant in the skin of people of Asian descent. Together with melanin, carotene accounts for their yellow-tan skin. Hemoglobin is not a pigment of the skin, but rather the oxygen-binding pigment found in red blood cells. Oxygenated blood flowing through the dermis gives the skin its pinkish tones.

The color of a person's skin may be indicative of certain conditions or diseases. Cyanosis *is a bluish hue of the skin symptomatic of certain cardiovascular or respiratory diseases. In* jaundice, *the skin becomes yellowish because of excess bile pigment in the blood. Jaundice is usually caused by liver abnormalities.*

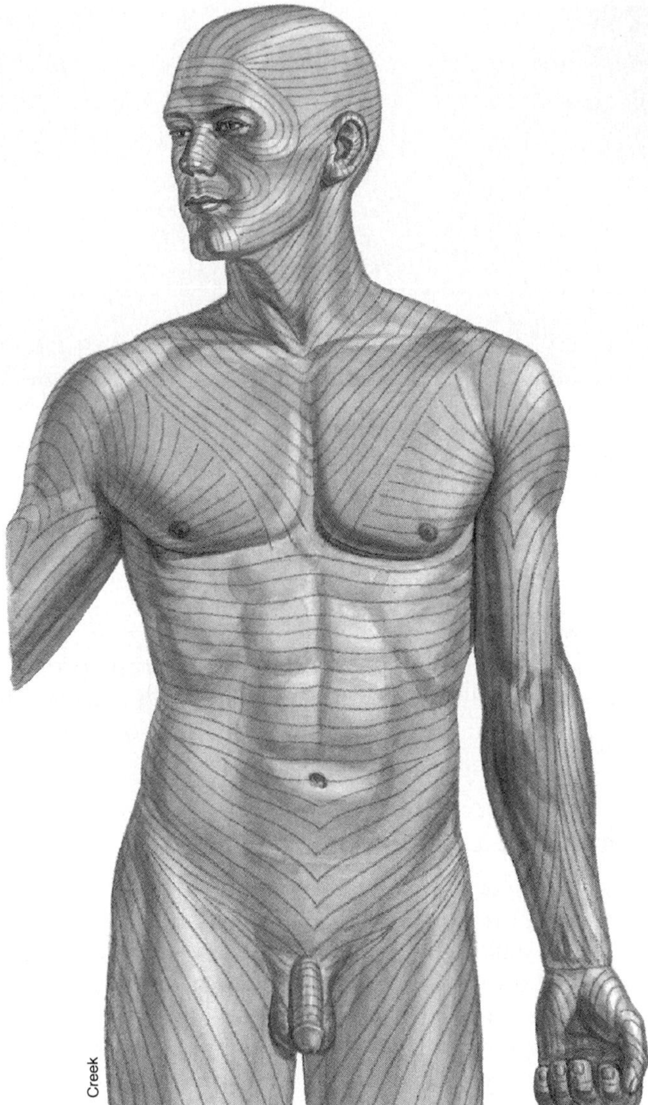

FIGURE 5.5
Lines of tension are caused by the pull of elastic and collagenous fibers within the dermis of the skin. Surgical incisions made parallel to the lines of tension heal more rapidly and create less scar tissue than incisions across the lines of tension.

Surface Patterns

The exposed surface of the skin has recognizable patterns that are either present at birth or develop later. **Fingerprints** (fig. 5.4) are congenital patterns that are present on the fingers and toes. These patterns have basic similarities but are not identical in any two individuals. They are formed by the pull of elastic fibers within the dermis and are well established prenatally. Fingerprints function to prevent slippage when grasping objects.

Acquired surface lines include the deep **flexion creases** on the palms and the shallow **flexion lines** on the knuckles.

Dermis

The **dermis,** the second principal skin layer, is deeper and thicker than the epidermis (see fig. 5.2). Elastic and collagenous fibers within the dermis are arranged in definite patterns, producing lines of tension in the skin (fig. 5.5). There are many more elastic fibers in the dermis of a young person than in an elderly one, and a decreasing number of

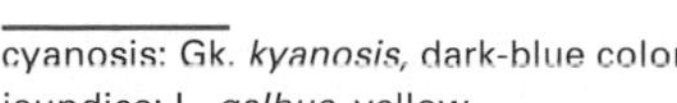

cyanosis: Gk. *kyanosis,* dark-blue color
jaundice: L. *galbus,* yellow

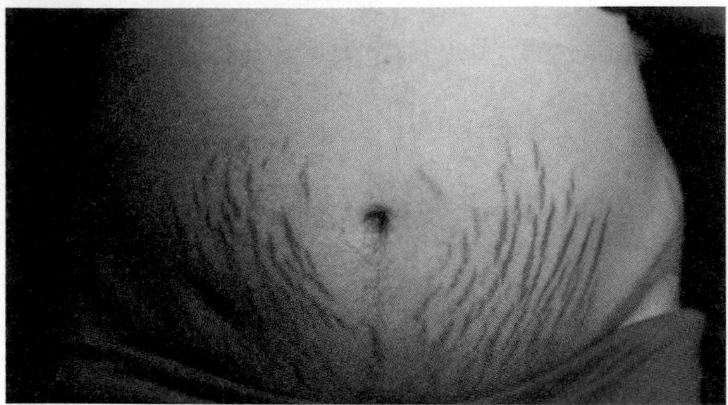

FIGURE 5.6
Stretch marks on the abdomen of a pregnant woman. Stretch marks generally fade with time but may leave permanent markings.

elastic fibers is apparently associated with aging. The extensive network of blood vessels in the dermis provides nourishment to the living portion of the epidermis. The dermis also contains many sweat glands, oil-secreting glands, nerve endings, and hair follicles.

Layers of the Dermis

The dermis is composed of two layers. The upper **papillary layer** is in contact with the epidermis. Numerous projections, called *papillae,* extend from the upper portion of the dermis into the epidermis. Papillae form the base for the friction ridges on the fingers and toes.

The deeper and thicker layer of the dermis is called the **reticular layer.** Fibers here are more dense and regularly arranged to form a tough, flexible meshwork. It is quite stretchable, as is evident in pregnant women, but it can be pulled too far, causing "tearing" of the dermis. The repair of a strained dermal area leaves a white streak called a stretch mark, or *linea albicans.* Stretch marks are usually found on the buttocks, thighs, abdomen, and breasts (fig. 5.6).

Inside Information—Your leather shoes, gloves, belts, and other items are made from the strong, pliable reticular layer of the hide of domestic mammals. In the tanning process, the animal's hide is treated with chemicals that separate this layer from the hairy epidermis and papillary layer of the dermis. The reticular layer is then softened and treated with protective chemicals before being cut and assembled into consumer goods.

Nerve Supply of the Skin

The dermis of the skin has an extensive nerve supply (innervation). Specialized glands and muscles within the dermis respond to nerve impulses transmitted from the brain and spinal cord to the skin. Stimulation of glands or muscles may result in glandular secretion, as in sweating, or the contraction of muscles that produces goose bumps (described in module 5.4).

Several types of nerve fibers respond to various touch, pressure, temperature, tickle, or pain stimulation. Some are exposed nerve endings, some form a network around hair follicles, and some extend into the papillae of the dermis. Certain areas of the body, such as the palms, soles, lips, and external reproductive organs, have a high concentration of sensory receptors and are therefore more sensitive to touch than other areas.

Blood Supply of the Skin

Blood vessels within the dermis supply nutrients to the mitotically active stratum basale and to the glands and hair follicles. Dermal blood vessels also play an important role in regulating body temperature and blood pressure. Narrowing (constriction) or widening (dilation) either directs the blood away from the superficial dermal arterioles (minute blood vessels) or permits it to flow freely throughout dermal vessels. Fever or shock can be detected by changes in the color and temperature of the skin. Blushing is the result of the involuntary vasodilation of dermal blood vessels.

It is important to maintain good blood circulation in people who are bedridden to prevent bedsores, or decubitus ulcers. *When a person lies in one position for an extended period, the dermal blood flow is restricted where the body presses against the bed. As a consequence, cells die and open wounds may develop (fig. 5.7). Changing the position of the patient frequently and periodically massaging the skin are good preventive measures against decubitus ulcers.*

Hypodermis

The hypodermis (*hi"pŏ-der'mis*), or *subcutaneous tissue,* is not actually a part of the skin, but it binds the dermis to underlying organs. The hypodermis is composed

decubitus: L. *decumbere,* lie down
ulcer: L. *ulcus,* sore
hypodermis: Gk. *hypo,* under; *derma,* skin

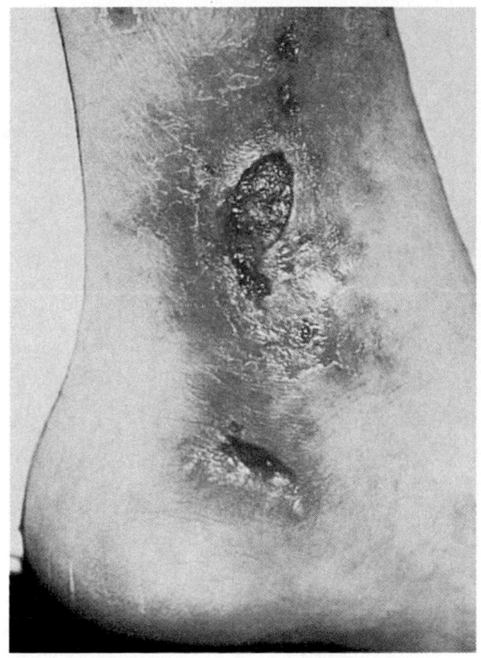

FIGURE 5.7
A bedsore (decubitus ulcer) on the medial surface of the ankle. Bedsores are most common on skin overlying a bony projection, such as at the hip, ankle, heel, shoulder, or elbow.

primarily of loose connective tissue and adipose cells interlaced with blood vessels (see fig. 5.2). The amount of adipose tissue in the hypodermis varies with the region of the body and the sex, age, and nutritional state of the individual. Females generally have about an 8% thicker hypodermis than males. This layer functions to store lipids, insulate and cushion the body, and regulate temperature.

The hypodermis is the site for subcutaneous injections. Using a hypodermic needle, medicine can be administered to patients who are unconscious or uncooperative, or when oral medications are not practical. Subcutaneous devices to administer slow-release, low-dosage medications are now available. For example, insulin may be administered in this way to treat some forms of diabetes. Even a subcutaneous birth-control device (Norplant) *is currently being marketed (see the Synoptic Spotlight in chapter 23, p. 608).*

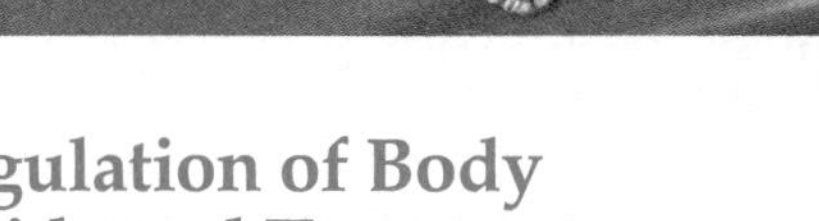

5.3 Physiology of the Skin

The skin not only protects the body from disease-causing organisms and external injury, it is a highly dynamic organ that plays a key role in maintaining homeostasis.

Protection of the Body

The skin is a barrier to microorganisms, water, and excessive sunlight. Oily secretions onto the surface of the skin form an acidic protective film (pH 4.0–6.8) that waterproofs the body and slows the growth of most disease-causing organisms. The protein *keratin* in the epidermis also waterproofs the skin, and the cornified outer layer (stratum corneum) resists scraping and keeps out microorganisms. Upon exposure to sunlight (UV light), the melanocytes in the stratum basale and stratum spinosum are stimulated to synthesize *melanin,* which absorbs sunlight. Surface friction causes the epidermis to thicken by increasing the rate of mitosis in the cells of the stratum basale, resulting in the formation of a protective *callus.*

Regulation of Body Fluids and Temperature

Fluid Loss

The layered, keratinized, cornified epidermis is adapted for continuous exposure to the air. In addition, the outer layers are dead and scalelike. Human skin is virtually waterproof, protecting the body from dehydration (desiccation) on dry land and even from water absorption when immersed in water.

Temperature Regulation

The skin plays a crucial role in regulating body temperature (fig. 5.8). Body heat comes from cellular metabolism, particularly in muscle cells. A normal body temperature of 37°C (98.6°F) is maintained in three ways, all involving the skin:

1. through radiant heat loss from dilated blood vessels;
2. through the evaporation of perspiration; and
3. through retention of heat from constricted blood vessels.

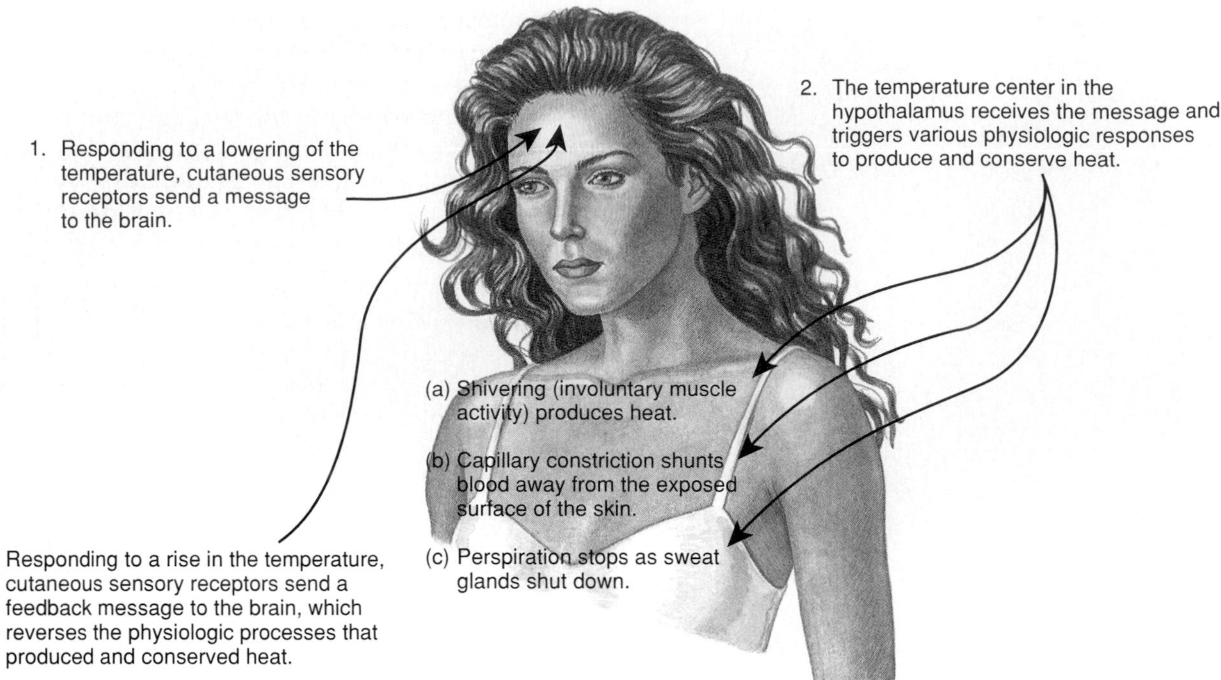

1. Responding to a lowering of the temperature, cutaneous sensory receptors send a message to the brain.
2. The temperature center in the hypothalamus receives the message and triggers various physiologic responses to produce and conserve heat.
(a) Shivering (involuntary muscle activity) produces heat.
(b) Capillary constriction shunts blood away from the exposed surface of the skin.
(c) Perspiration stops as sweat glands shut down.
3. Responding to a rise in the temperature, cutaneous sensory receptors send a feedback message to the brain, which reverses the physiologic processes that produced and conserved heat.

FIGURE 5.8
Temperature regulation involves feedback mechanisms. Messages of decreased body temperature sent to the brain trigger a response that can quickly generate up to five times the normal rate of body heat production.

The volume of perspiration produced is largely a function of how much the body is overheated. This volume increases approximately 100 to 150 mL per day for each 1°C elevation in body temperature. For each hour of hard physical work out-of-doors in the summertime, a person may produce 1 to 2 L of perspiration. Emotional stimulation also causes us to sweat, as, for example, in the "cold sweat" of fear.

A serious danger of continued exposure to heat and excessive water and salt loss is heat exhaustion, *characterized by nausea, weakness, dizziness, headache, and a decreased blood pressure.* Heat stroke *is similar to heat exhaustion, except that in heat stroke sweating is prevented (for reasons that are not clear) and body temperature rises. Convulsions, brain damage, and death may follow.*

Excessive heat loss triggers a shivering response in muscles that increases heat production. Not only do skeletal muscles contract, but tiny smooth muscles attached to hair follicles contract, causing goose bumps.

When the body's heat-producing mechanisms cannot keep pace with heat loss, hypothermia *results. A lengthy exposure to temperatures below 20°C and dampness may lead to this condition. This is why it is so important that a hiker, for example, dress appropriately for the weather conditions, especially on cold days. The initial symptoms of hypothermia are numbness, paleness, delirium, and uncontrolled shivering. If the body core temperature falls below 32°C (90°F), the heart loses its ability to pump blood and will go into fibrillation (erratic contractions). If the victim is not warmed, extreme drowsiness, coma, and death follow.*

Cutaneous Absorption

Because of the effective protective barriers of the skin already described, cutaneous absorption (absorption through the skin) is limited. Some gases, such as oxygen and carbon dioxide, may pass through the skin and enter the blood. Only small amounts of UV light, needed to produce vitamin D, penetrate readily. Unfortunately, harmful chemicals such as lipid-soluble toxins and pesticides can easily enter the body through the skin.

Synthesis

The integumentary system synthesizes melanin and keratin, which remain in the skin, and vitamin D, which is used elsewhere in the body. The integumentary cells contain a compound called *dehydrocholesterol,* from which they synthesize vitamin D in the presence of UV light. Synthesized vitamin D enters the blood and helps to regulate the metabolism of calcium and phosphorus, which are important in the development of strong, healthy bones. *Rickets* (fig. 5.9) is a disease caused by vitamin D deficiency.

Sensory Reception

Highly specialized sensory receptors that respond to heat, cold, pressure, touch, vibration, and pain are located throughout the dermis. Called *cutaneous receptors,* these sensory nerve cells are especially abundant in the skin of the face and palms, the fingers, the soles of the feet, and the external reproductive organs. As a general rule, *the thinner the skin, the greater the sensitivity.*

rickets: E. *wrick,* to twist

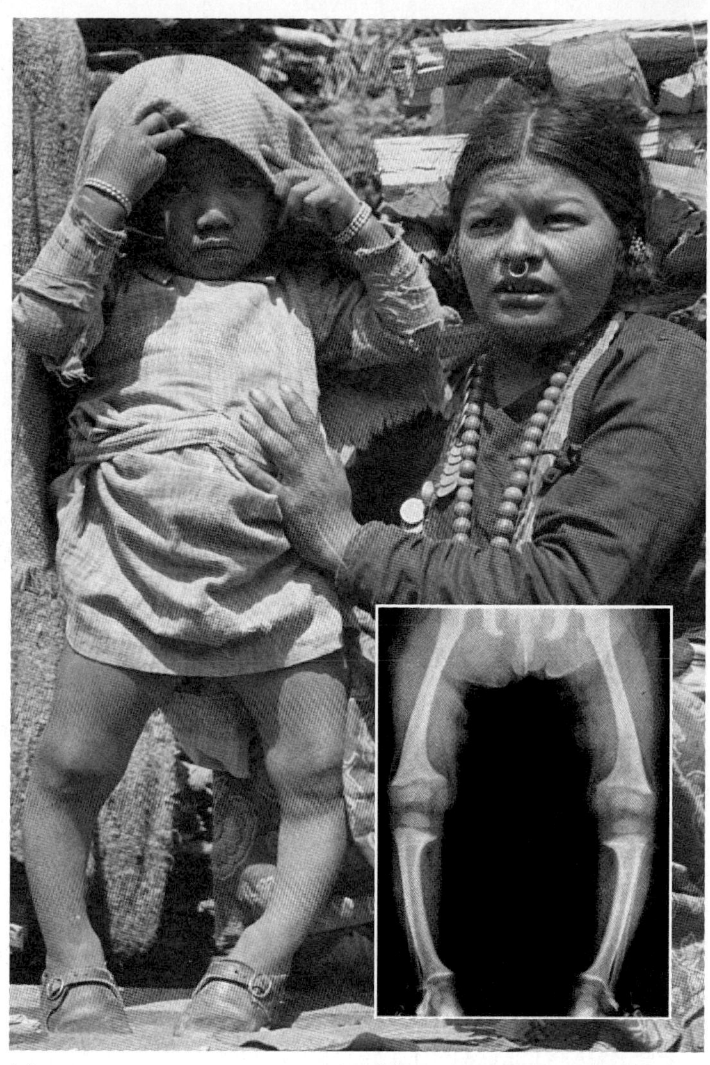

(a) (b)

FIGURE 5.9

(*a*) Rickets in a child from a Nepalese village whose inhabitants live in windowless huts. During the rainy season, which may last 5 to 6 months, the children are kept indoors. (*b*) A radiograph (X ray) of a 10-month-old child with rickets. Rickets develops from an improper diet and also from lack of the sunlight needed to synthesize vitamin D.

5.4 Skin Accessories

The hair, nails, and glands of the skin are formed in the epidermis and nourished by blood vessels in the dermis. Hair and nails have a limited functional role, whereas integumentary glands are extremely important in protecting the body and maintaining homeostasis.

Hair

The presence of **hair** on the body is one of the distinguishing features of mammals, but its distribution, function, density, and texture varies among different mammalian species. In humans, hair is found everywhere on the body, but only the scalp, face, pubic area, and underarms are densely haired. Men and women have about the same density of hair on their bodies, but hair is generally more obvious on men (fig. 5.10) due to male hormones. Certain structures and regions of the body are hairless, such as the palms, soles, lips, penis, and parts of the female genitalia.

The primary function of hair is protection, even though its effectiveness is limited. Hair on the scalp and eyebrows protects against sunlight. The eyelashes and the

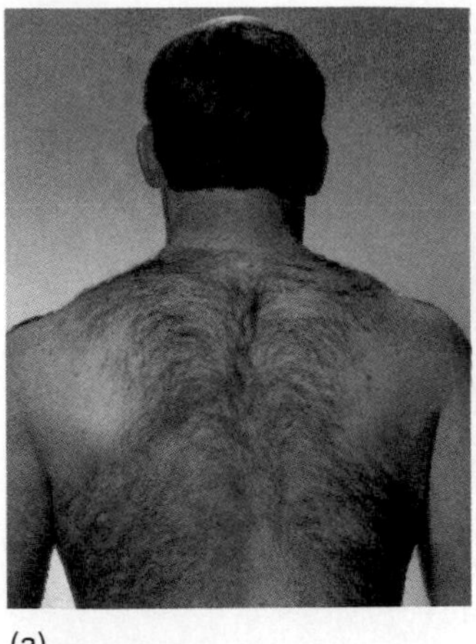
(a)

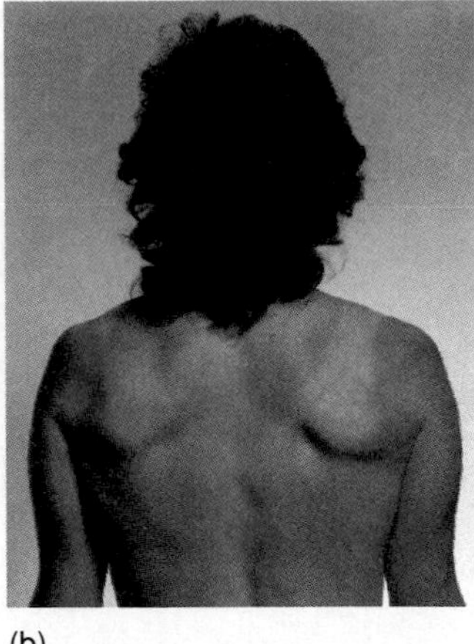
(b)

FIGURE 5.10
Comparison of male and female expression of body hair.

hair in the nostrils protect against airborne particles. Hair on the scalp may also protect against mechanical injury. Secondary functions of hair are to distinguish individuals and to serve as a sexual attractant.

Each hair consists of a diagonally positioned **shaft, root,** and **bulb** (fig. 5.11). The shaft is the visible, but dead, portion of the hair above the surface of the skin. The bulb is the enlarged base of the root within the **hair follicle.** Each hair develops from stratum basale cells within the bulb of the hair, where nutrients are received from dermal blood vessels. As the cells divide, they are pushed away from the nutrients toward the surface, and cellular death and keratinization occur. In a healthy person, hair grows at the rate of about 1 mm every 3 days.

The life span of a hair varies from 3 to 4 months for an eyelash to 3 to 4 years for a scalp hair. Each hair lost is replaced by a new hair that grows from the base of the follicle and pushes the old hair out. Between 10 and 100 hairs are lost daily. Baldness, or *alopecia,* results when hair is lost and not replaced. This condition may be disease related, but it is generally inherited and most frequently occurs in males. Genetically determined, testosterone-dependent loss of scalp hair is called *male pattern baldness.*

Three layers can be observed in a hair. An inner **medulla** (*me-dul'a*) is composed of loosely arranged cells separated by many air cells. The **cortex** surrounding the medulla consists of hardened, tightly packed cells. A **cuticle** covers the cortex and forms the toughened outer layer of the hair. Cells of the cuticle have jagged edges that give a hair a scaly appearance when observed under a dissecting scope.

Heavy metals such as lead, mercury, or arsenic will accumulate in the hair cells of people exposed to these metals. The concentration in the hair will be 10 times as great as that found in their blood or urine. For this reason, hair samples are important diagnostic clues.

Even certain metabolic diseases or nutritional deficiencies may be detected in hair samples. For example, the hair of children with cystic fibrosis *will be deficient in calcium and show an excess of sodium. The hair of malnourished individuals will show a deficiency of zinc.*

Hair color is determined by the type and amount of pigment produced in the stratum basale at the base of the follicle. Varying amounts of melanin produce hair color from blond to brunette to black; the more abundant the melanin, the darker the hair. The pigment *trichosiderin* has an iron base and is responsible for red hair. Gray and white hair is caused by lack of pigment production and by air spaces within the layers of the shaft of the hair. The texture of hair is determined by the cross-sectional shape; straight hair is round in cross section, wavy hair is oval, and kinky hair is flat.

Sebaceous glands and specialized smooth muscles, called **arrector pili** (*ă-rek'tor pī'le*), are attached to the hair follicle (fig. 5.11). Arrector pili muscles are involuntary, responding to temperature or psychological stimuli. When these muscles contract, the hair stands up, causing goose bumps.

Humans have three distinct kinds of hair.

1. **Lanugo.** Lanugo (*la-noo'go*) is a fine, silky fetal hair that appears during the last trimester of development. Lanugo is usually seen only on premature babies.
2. **Angora.** Angora hair grows continuously. It is found on the scalp and on the faces of mature males.
3. **Definitive.** Definitive hair grows to a certain length and then stops. It is the most common type of hair. Eyelashes, eyebrows, and pubic and axillary hair are examples.

medulla: L. *medulla,* marrow
cortex: L. *cortex,* bark
cuticle: L. *cuticula,* small skin

pili: L. *pilus,* hair
lanugo: L. *lana,* wool

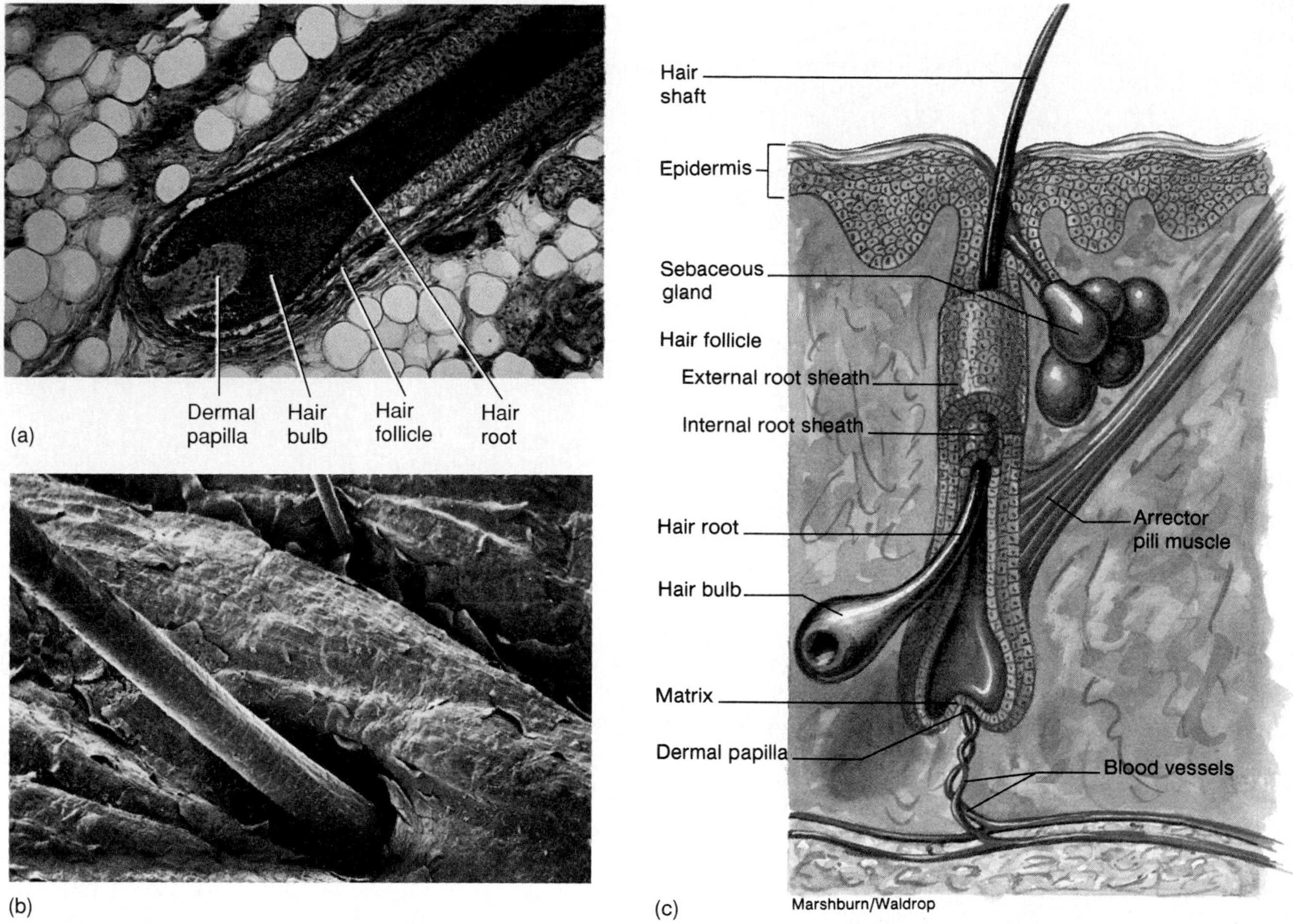

FIGURE 5.11
The structure of hair and the hair follicle. (*a*) A photomicrograph (63×) of the bulb and root of a hair within a hair follicle. (*b*) A scanning electron micrograph (280×) of a hair as it extends from a follicle. (*c*) A diagram of hair, a hair follicle, a sebaceous gland, and an arrector pili muscle.

Inside Information—Recently, hairs from the head of Napoleon (preserved by his servants) were analyzed and found to have a high arsenic content. However, it could not be concluded from this evidence that his death was caused by arsenic poisoning, as some historians have suggested.

Nails

The **nails** on the ends of the fingers and toes are formed from the compressed outer layer (stratum corneum) of the epidermis. Both fingernails and toenails protect the digits, and fingernails also aid in grasping and picking up small objects.

Each nail consists of a **body, free edge,** and a **root** (fig. 5.12). The platelike body of the nail rests on a **nail bed,** which is actually the stratum spinosum of the epidermis. The body and nail bed have a pinkish tone because of the underlying blood vessels. The sides of the nail body are protected by a **nail fold.** The free edge of the nail extends over a thickened region of the stratum corneum called the **hyponychium** (*hi"pŏ-nik'e-um*) (quick). The root of the nail is attached at the base.

An **eponychium** (cuticle) covers the root of the nail. The eponychium frequently splits, causing a hangnail. The growth area of the nail is the **nail matrix.** A small part of the nail matrix, the **lunula** (*loo'nyoo-lă*), can be seen as a half-moon-shaped area near the root of the nail.

hyponychium: Gk. *hypo,* under; *onyx,* nail
lunula: L. *lunula,* small moon

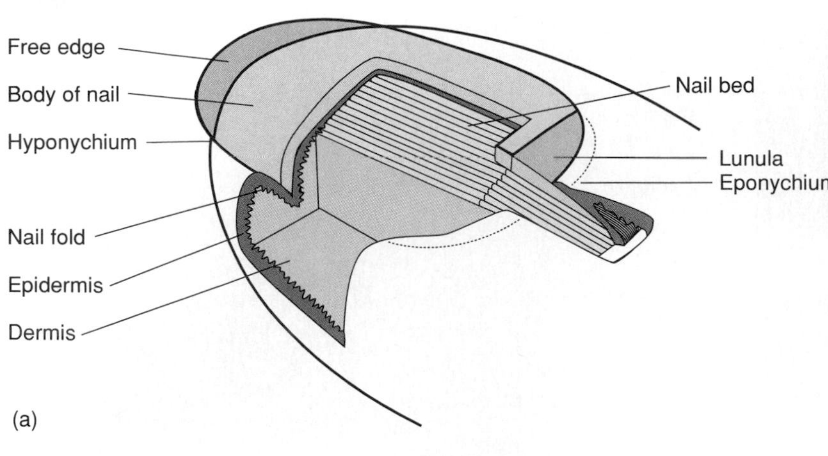

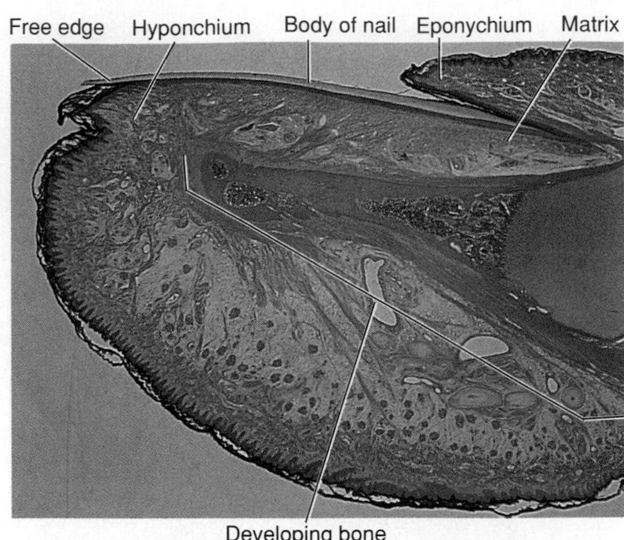

FIGURE 5.12
A fingertip showing the associated structures of the nail: (*a*) A diagram of a nail and (*b*) a photomicrograph of a nail from a fetus.

The condition of nails may be indicative of a person's general health and well-being. Nails should appear pinkish, showing the rich vascular capillaries beneath the nail. A yellowish hue may indicate glandular dysfunctions or nutritional deficiencies. Split nails may also be caused by nutritional deficiencies. A prominent bluish tint may indicate low blood oxygen levels, and spoon nails (a concave nail body) may be the result of iron-deficiency anemia. Dirty or ragged nails may indicate poor personal hygiene, and chewed nails may suggest emotional problems.

Glands

Although they originate in the epidermal layer, all of the glands of the skin are located in the dermis, where they are physically supported and receive nutrients. Glands of the skin are referred to as *exocrine* because they are externally secreting glands that either release their secretions directly or through ducts. The glands of the skin are of three basic types: *sebaceous* (*sĕ-ba'shus*), *sudoriferous* (*soo"dor-if'er-us*), and *ceruminous* (*sĕ-roo'mĭ-nus*).

Sebaceous Glands

Sebaceous, or oil, glands are branched glands attached to a hair follicle (fig. 5.13). These glands secrete a lipid mixture called **sebum** onto the shaft of the hair. The sebum then spreads along the shaft of the hair to the surface of the skin, where it lubricates and waterproofs the skin and prevents the hair from becoming brittle. If the ducts of sebaceous glands become blocked, the glands may become infected, resulting in *acne.* Sex hormones regulate the production and secretion of sebum, and hyperactivity of sebaceous glands can produce serious acne problems, particularly during the teenage years.

Good routine hygiene is very important for health and social reasons. Washing away the dried residue of perspiration and sebum eliminates dirt. Excessive bathing, however, can wash off the natural sebum and dry the skin, causing it to itch or crack. The commercial lotions used for dry skin are, for the most part, refined and perfumed lanolin, which is sebum from sheep.

Sudoriferous Glands

Sudoriferous glands, or **sweat glands,** excrete perspiration, or sweat, onto the surface of the skin. Perspiration is composed of water, salts, urea, and uric acid. It is therefore valuable not only for evaporative cooling, but also for

sebum: L. *sebum,* tallow or grease
sudoriferous: L. *sudorifer,* sweat; *ferre,* to bear

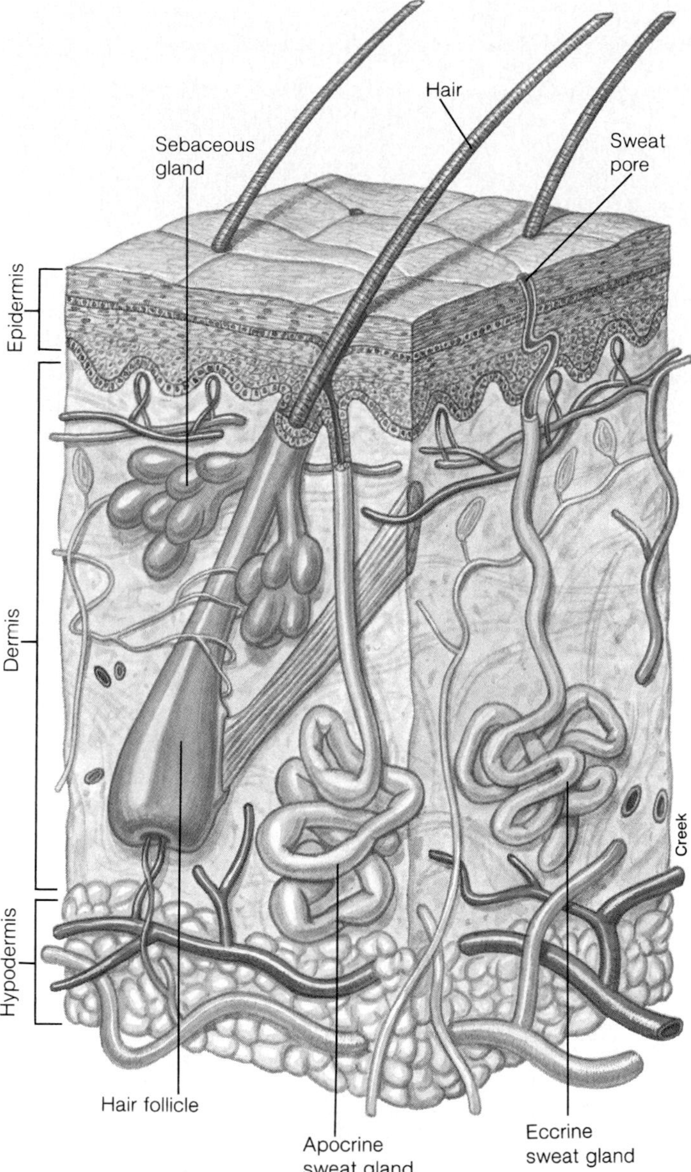

FIGURE 5.13
Types of skin glands.

the excretion of wastes. Sweat glands are coiled and tubular shaped (fig. 5.13) and are of two types: *eccrine* (*ek'rin*) and *apocrine* (*ap'ŏ-krin*) *sweat glands.*

1. **Eccrine sweat glands** are widely distributed over the body, especially on the forehead, back, palms, and soles. These glands are formed before birth and function in cooling.

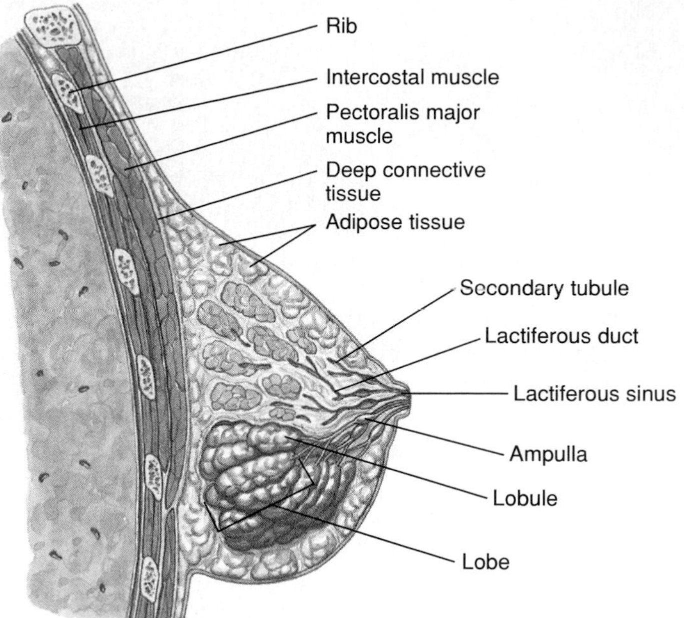

FIGURE 5.14
A mammary gland within the female breast.

2. **Apocrine sweat glands** are much larger than the eccrine glands. Apocrine glands are found in the axillary and pubic regions and secrete into hair follicles. They do not function until puberty, and their odoriferous secretion is thought to act as a sexual attractant.

Mammary glands, found within the breasts, are specialized sudoriferous glands that secrete milk during lactation (fig. 5.14). The breasts of the female are most fully developed during the childbearing years, under the stimulus of pituitary and ovarian hormones.

Ceruminous Glands

These specialized glands are found only in the external auditory canal, where they secrete **cerumen,** or earwax. Cerumen is an insect repellent and also keeps the tympanic membrane (eardrum) from drying out. Excessive amounts of cerumen may interfere with hearing.

cerumen: L. *cera,* wax

Cell Division Gone Awry: Skin Cancer

Skin cancer is the most common kind of cancer, affecting some 500,000 Americans each year. Fortunately, it is also the most treatable kind—95% of those who develop skin cancer are cured by surgery, X-ray therapy, or drug therapy. In addition, medical researchers are fairly certain that they have identified the cause of most types of skin cancers, so that preventative measures can be taken.

In 90% of skin cancer cases, sunlight is regarded as a causative factor. Sunlight radiates through the fragile atmosphere of our planet, particularly in areas where the ozone layer has been depleted. *Ultraviolet (UV) light*—the component of sunlight that is potentially damaging to the skin—is emitted in different wavelengths that vary in their ability to penetrate the ozone layer. **UVC rays,** those with the shortest wavelength, do not penetrate the ozone layer. **UVB rays,** intermediate in wavelength, are largely blocked by ozone, but enough get through to cause about 80% of sunburn damage. **UVA rays,** with the longest wavelength, are not blocked by ozone. Until the 1980s, UVA rays were thought to be relatively safe because they take longer than UVB rays to produce a burn. Now, however, it is recognized that skin damage can occur even without a burn. UVA rays penetrate the skin deeply, breaking down connective tissue and causing the skin to sag and wrinkle. They also are believed to potentiate the effect of cancer-causing UVB rays, and they themselves may lead to the development of cancer.

People who stay out in the sun too long can get a burn just as severe as that caused by fire or boiling water. Both UVA and UVB rays stimulate the pigment-producing melanocytes to produce more melanin, and it is the buildup of this melanin that produces your tan. The tan may look attractive, but it is by no means "healthy"—tanned skin is damaged skin. The consequences of repeated exposures may not appear for 15 to 20 years, but when they do, they range from wrinkles, blotches, and unattractive growths to various forms of skin cancer.

As shown in the accompanying photos, the most frequently encountered skin cancers are of three types:

- **Basal cell carcinoma,** the most common skin cancer, accounts for about 70% of total cases. It usually occurs where the sun strikes hardest—on the face and arms. This type of cancer arises from cells in the stratum basale. It appears first on the surface of the skin as a small, shiny bump. As the bump enlarges, it often develops a central crater that erodes, crusts, and bleeds. Fortunately, there is little danger that it will spread (metastasize) to other areas of the body. These carcinomas are usually treated by excision (surgical removal).
- **Squamous cell carcinoma** arises from cells immediately superficial to the stratum basale. Normally, these cells undergo very little division. But in squamous cell carcinoma they continue to divide as they produce keratin. The result is usually a firm, red keratinized tumor, confined to the epidermis. If untreated, however, it may invade the dermis and metastasize. Treatment usually consists of excision and X-ray therapy.
- **Malignant melanoma,** the most life-threatening form of skin cancer, arises from the melanocytes located in the stratum basale. Often, it begins as a small molelike growth, which enlarges,

changes color, becomes ulcerated, and bleeds easily. Metastasis occurs quickly, and unless treated early—usually by widespread excision and X-ray therapy—this cancer is often fatal.

A variety of measures are recommended to protect yourself against all three varieties of skin cancer. First, use a sunblock to protect your skin. The currently available products at SPF 15 ratings will filter out about 90% of the damaging light. If you prefer not to slather yourself with lotion, wear protective clothing. Long-sleeved, dark fabrics with tight weaves are best. Baseball caps offer some protection, but wide-brimmed hats are better. In addition, wear sunglasses designed to block both UVA and UVB radiation; they not only protect the skin around your eyes, but may also help to prevent cataracts. And, finally, stay out of tanning booths. Despite merchants' claims that they have blocked out all damaging wavelengths, patrons are typically exposed to 2 to 20 times the amount of UVA that they get in natural sunlight.

It is also important to be viligant in monitoring your body. Do you have warts or moles? Check these for changes in size, shape, color, or differences in sensation. Check also for any new, suspicious growths. If a mole has ragged edges, is multicolored, is asymmetrical, or has a diameter greater than that of a pencil eraser, consult your physician promptly.

If you take preventative measures, your chances of developing skin cancer will be greatly reduced, and you will probably enter old age looking better than your less cautious contemporaries. For those of you who are diehard "sun-worshippers," the American Cancer Society has a succinct warning: Fry now, pay later.

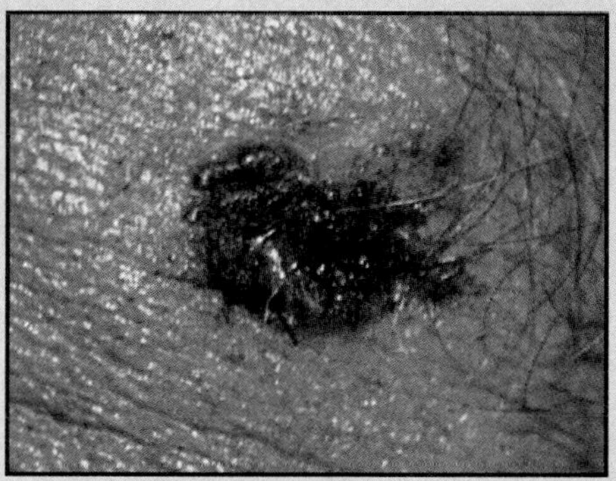

Basal cell carcinoma

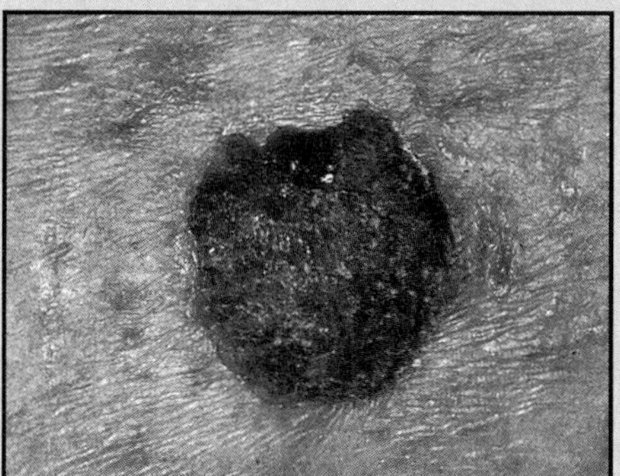

Squamous cell carcinoma

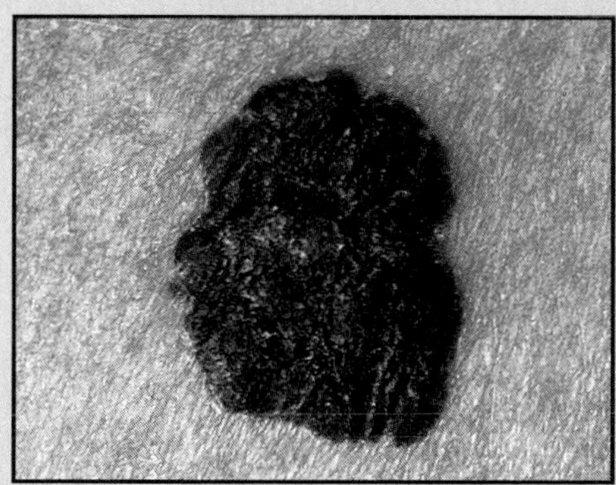

Malignant melanoma

Clinical Considerations

The skin is a buffer against the external environment and is therefore subjected to a variety of disease-causing microorganisms and physical assaults. There are a number of *infectious diseases* of the skin, which is not surprising considering how social we are. Effective prevention and treatment are available for most of these diseases, but too frequently we ignore skin problems. Infectious diseases of the skin include childhood *viral infections* (measles, chicken pox); *bacterial* ("staph") *infections* (boils, impetigo); *sexually transmitted diseases; fungi* (ringworm, athlete's foot); and *mites* (scabies).

Warts are virally caused abnormal growths of tissue that occur primarily on the hands and feet. A different type of wart, called a *venereal*, or *genital, wart*, occurs in the genital region and is sexually transmitted. Venereal warts (also called *condyloma acuminatum*) are caused by a slow-growing DNA virus. They can usually be treated effectively with chemicals, cryosurgery, cautery, or laser therapy. However, because risk factors for cervical cancer may be linked to venereal warts, it is important that they be treated aggressively.

Burns

A **burn** is an epithelial injury caused by contact with thermal, radioactive, chemical, or electrical agents. Burns generally occur on the skin, but they can involve the linings of the respiratory and digestive tracts. Burns that have a *local effect* (causing local tissue destruction) are not as serious as those that have a *systemic effect* (involving the entire body). Possible systemic effects include body dehydration, shock, reduced blood circulation and urine production, and bacterial infections.

Burns are classified as first degree, second degree, and third degree based on their severity. In **first-degree burns,** the epidermal layers of the skin are damaged and symptoms are restricted to local effects such as redness, pain, and edema (swelling). A shedding of the surface layers generally follows in a few days. A sunburn is an example. **Second-degree burns** involve both the epidermis and dermis. Blisters appear and recovery is usually complete, although slow. **Third-degree burns** (fig. 5.15) destroy the entire thickness of the skin and even underlying muscle. The skin appears charred and is insensitive to touch. As a result, ulcerating wounds develop and the body attempts to heal itself by forming scar tissue. Skin grafts are frequently used to assist recovery.

Wound Healing

The skin usually protects itself against most abrasions. If a wound occurs, a chain of events promotes rapid healing. Trauma to the epidermal layers stimulates increased cell division in the stratum basale. When there is damage to the dermis, general body responses include a temporary rise in temperature and pulse rate.

In an open wound (fig. 5.16), blood vessels are broken and bleeding occurs. Through the action of *blood platelets* and protein molecules called *fibrinogen* (see chapter 14, module 14.5), a clot forms and blocks the flow of blood. A *scab* forms from dried blood and covers and protects the damaged area. *Inflammation* destroys bacteria and helps to dispose of dead or injured cells. It also promotes healing.

The next step in healing is the appearance of *fibroblasts* at the wound margins. Together with new branches from nearby blood vessels, *granulation tissue* is formed. Phagocytic cells migrate into the wound and ingest dead cells and foreign debris. Eventually, the damaged area is repaired and the protective scab is sloughed off.

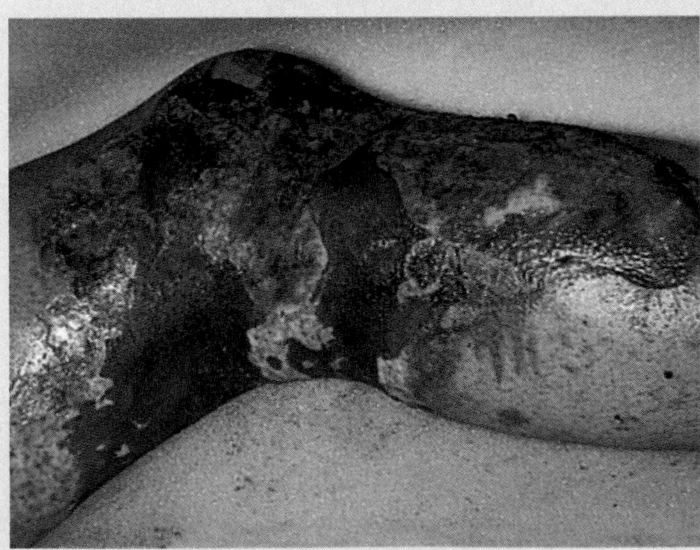

FIGURE 5.15
A third-degree burn.

If the wound is severe, the granulation tissue may develop into *scar tissue*. The collagenous fibers of scar tissue are denser than those of normal skin, and scar tissue has no epidermal layer. Scar tissue also has fewer blood vessels than normal skin and may lack hair, glands, and sensory receptors. The closer together the edges of a wound, the less granulation tissue develops and the less obvious a scar. This is one reason taping or suturing a large break in the skin reduces scarring.

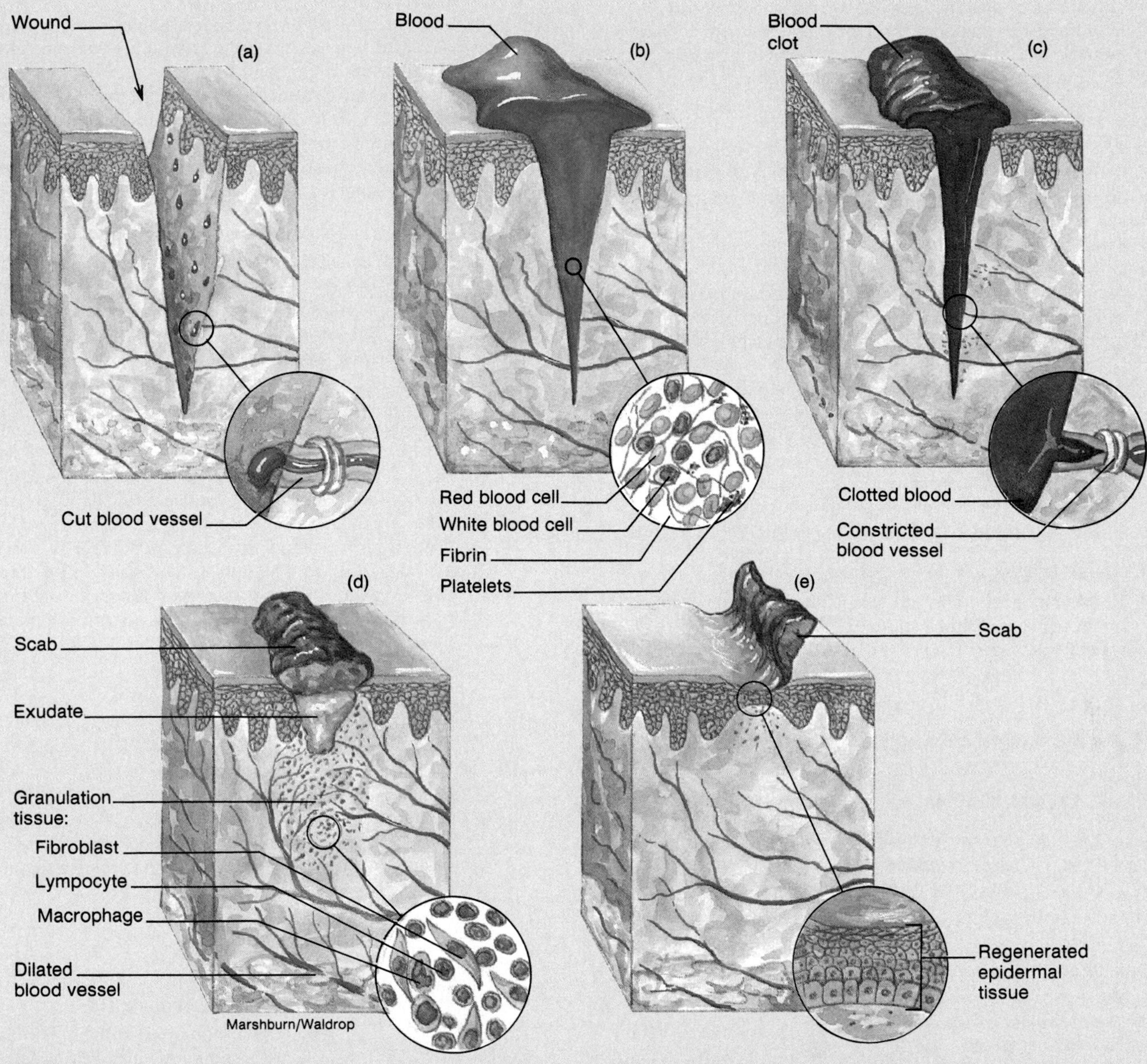

FIGURE 5.16

The process of wound healing. (*a*) A penetrating wound into the dermis ruptures blood vessels. (*b*) Blood cells, fibrinogen, and fibrin flow out of the wound. (*c*) Vessels constrict and a clot blocks the flow of blood. (*d*) A protective scab is formed from the clot, and granulation tissue forms within the site of the wound. (*e*) The scab sloughs off as the epidermal layers are regenerated.

SUMMARY

5.1 The Skin As an Organ

1. The skin is an organ because it consists of several kinds of tissues structurally organized to function together.
2. The appearance of the skin provides clues to certain body conditions or dysfunctions. Abnormal skin coloration may be due to inadequate oxygen, poisons in the blood, infections, or metabolic disorders. Certain nutritional deficiencies also cause abnormal skin coloration.

5.2 Layers of the Skin

1. The stratified squamous epithelium of the epidermis is divisible into five layers: the stratum basale, stratum spinosum, stratum granulosum, stratum lucidum, and stratum corneum.
2. Keratinization and cornification are protective changes within the epidermal cells.
3. A combination of melanin, carotene, and hemoglobin account for normal skin color. Melanin, produced by melanocytes, protects the basal layer against damaging UV rays of sunlight.
4. Surface patterns, including fingerprints, are formed by the pull of elastic fibers within the dermis.
5. The dermis is composed of fibrous connective tissue interlaced with elastic fibers. The two layers of the dermis are the upper papillary layer and the deeper reticular layer.
6. The hypodermis, composed of adipose and fibrous connective tissue, binds the dermis to underlying organs.

5.3 Physiology of the Skin

1. The thickened, keratinized, cornified epidermis protects the body from disease and external injury. Melanin provides a barrier to UV light.
2. Keratinization and cornification of the skin help to minimize fluid loss.
3. The skin controls heat loss from the body through constriction and dilation of blood vessels. Evaporation of perspiration cools the body.
4. The skin permits limited absorption of some UV light, respiratory gases, steroids, fat-soluble vitamins, and certain toxins.
5. Melanin and keratin are produced in the skin and remain in the skin; vitamin D is synthesized in the skin and used elsewhere in the body.
6. Sensory reception in the skin is provided through cutaneous receptors that respond to precise sensory stimuli.

5.4 Skin Accessories

1. Each hair consists of a shaft, root, and bulb. The bulb is the enlarged base of the root within the hair follicle. Arrector pili are specialized smooth muscles attached to hair follicles that automatically respond to cold temperatures. The three hair layers are the medulla, cortex, and cuticle.
2. Lanugo is fine, silky fetal hair; angora is continuously growing hair, such as on the scalp; definitive hair, such as eyelashes and pubic hair, grows to a certain length and then stops.
3. Hardened, keratinized nails are found on the end of each digit, where they have a protective function. Fingernails aid in grasping and picking up small objects.
4. Integumentary glands are exocrine since they secrete their products onto an epithelial surface, either directly or through ducts. Sebaceous glands secrete sebum, sudoriferous glands excrete sweat, mammary glands secrete milk during lactation, and ceruminous glands secrete cerumen.

Review Activities

Objective Questions

1. Which of the following statements concerning skin is *false?* (*module 5.1*)
   a. The skin accounts for about 7% of the adult body weight.
   b. Red and overly warm skin may indicate fever and infection.
   c. The skin is thinnest on the eyelids, tympanum, and external reproductive organs.
   d. The skin is the largest tissue of the body.

2. The epidermal layer absent in the thin skin of the face is the stratum (*module 5.2*)
   a. granulosum.
   b. lucidum.
   c. spinosum.
   d. corneum.

3. Which of the following does *not* contribute to skin color? (*module 5.2*)
   a. vitamin D
   b. melanin
   c. carotene
   d. hemoglobin

4. The epidermal layer characterized by ongoing mitosis is the stratum (*module 5.2*)
   a. spinosum.
   b. corneum.
   c. basale.
   d. granulosum.

5. Integumentary cells synthesize vitamin D in the presence of sunlight and (*module 5.3*)
   a. keratin.
   b. cortisol.
   c. trichosiderin.
   d. dehydrocholesterol.

6. Integumentary glands that empty their secretions into hair follicles are (*module 5.4*)
   a. sebaceous glands.
   b. sudoriferous glands.
   c. eccrine glands.
   d. ceruminous glands.

7. Fetal hair that is present during the last trimester of development is called (*module 5.4*)
a. angora.
b. replacement.
c. lanugo.
d. seborrhea.

8. Mammary glands are modified (*module 5.4*)
a. sweat glands.
b. oil glands.
c. sebaceous glands.
d. endocrine glands.

9. The skin of a burn victim has been severely damaged through the epidermis and into the dermis. Integumentary regeneration will be slow with some scarring, but it will be complete. Which kind of burn is this? (*Clinical Considerations*)
a. first degree
b. second degree
c. third degree

Essay Questions

1. Why is the skin considered an organ and a part of the integumentary system? (*module 5.1*)
2. Discuss the growth process and regeneration of the epidermis. (*module 5.2*)
3. What are some physical and chemical features of the skin that make it an effective protective organ? (*module 5.3*)
4. Explain how skin color and hair color and texture are determined. What three kinds of hair do we have? (*module 5.4*)
5. Distinguish between sebaceous, sudoriferous, mammary, and ceruminous glands in terms of structure and function. (*module 5.4*)

Labeling Exercise

Label the structures indicated on the figure to the right.

1. ______________________
2. ______________________
3. ______________________
4. ______________________
5. ______________________
6. ______________________

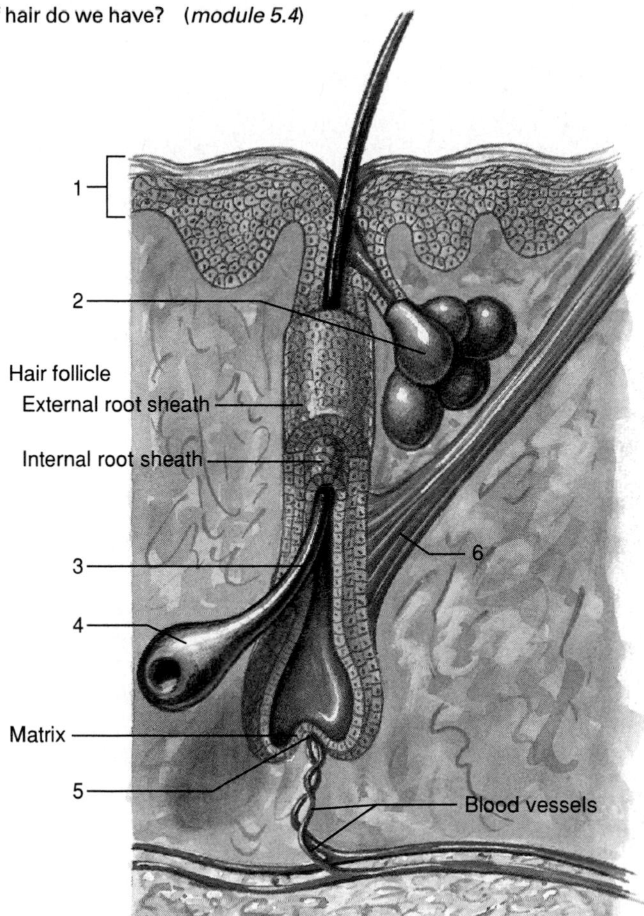

Critical Thinking Questions

1. Although your friend meant well with a gift certificate for 10 sessions at a local tanning salon, your better judgment tells you to decline the gift. Write a paragraph explaining why it can be dangerous to expose your skin to the rays emitted in tanning booths.
2. Review the anatomy and physiology of the skin by explaining (a) the mechanisms involved in sweating and shivering; (b) variations in individual skin color; (c) abnormal coloration of the skin (for example, cyanosis, jaundice, and pallor); and (d) the occurrence of acne.
3. During the aging process, the skin becomes drier, wrinkled, and slower to heal. Knowing that these are normal structural changes, how would you advise a middle-aged person to safeguard his or her skin as a protective organ?
4. A criminal wants to have his fingerprints obscured. What would he have to alter to make such a permanent change?
5. Why is scalp cancer more common in men than in women?

CHAPTER

6

"The skeleton forms a framework that supports the softer tissues and organs of the body." *page 117*

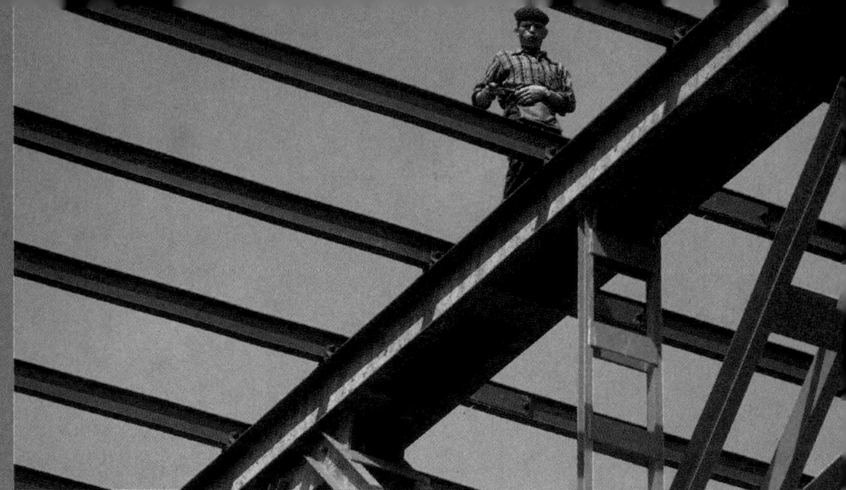

Skeletal System

Chapter Outline

Terms to Remember

hemopoiesis (*hem"ō-poi-e'sis*), 117
ossification (*os"ĭ-fĭ-ka'shun*), 118
fontanel (*fon"tă-nel'*), 119
condyle (*kon'dīl*), 121
facet (*fas'et*), 121
tubercle (*too'ber-k'l*), 121
foramen (*fō-ra'men*), 121
fossa (*fos'ă*), 121
diaphysis (*di-af'ĭ-sis*), 121
epiphysis (*ĕ-pif'ĭ-sis*), 121
articular cartilage, 122
periosteum (*per"e-os'te-um*), 122
spongy bone, 122
compact bone, 122
osteon (*os'te-on*), 123

Learning Objectives

After studying this chapter, you should be able to

- describe the division of the skeletal system into axial and appendicular components.
- discuss the principal functions of the skeleton and identify the body systems served by these functions.
- describe the features of a typical long bone and list the functions of each structure.
- list the different kinds of bone cells, describe their functions, and discuss the organization of bone tissue.
- describe the process of endochondral ossification as related to bone growth.
- identify the cranial and facial bones of the skull and describe their locations and structural features.
- identify the bones of the five regions of the vertebral column and describe the structure of a typical vertebra.
- describe the rib cage.
- list the bones of the pectoral and pelvic girdles and compare the functional roles of the two girdles.
- describe the location and structural features of each long bone in the upper and lower limbs.

6.1 Organization of the Skeletal System

The axial and appendicular parts of the adult human skeleton normally consist of 206 individual bones arranged to form a strong, yet flexible, body framework.

The adult skeletal system consists of a framework of approximately 206 bones and their associated connective tissues, including cartilage, tendons, and ligaments. The number of bones varies from person to person in accordance with age and genetic factors. At birth, the skeleton consists of about 270 bones. As further bone development (ossification) occurs during infancy, the number increases. During adolescence, however, the number of bones actually decreases, as separate bones gradually fuse. Each bone is actually an organ that plays a part in the total functioning of our skeletal system. The science concerned with the study of bones is called *osteology.*

For convenience of study, the skeleton is divided into *axial* and *appendicular portions,* as shown in figure 6.1 and summarized in table 6.1. The **axial skeleton** consists of the bones that form the vertical axis of the body and that support and protect the organs of the head, neck, and trunk. The components of the axial skeleton are as follows:

1. **Skull.** The skull consists of two sets of bones: the *cranial bones* that form the cranium, or braincase, and the *facial bones* that support the eyes, nose, and jaws.
2. **Auditory ossicles.** Three auditory ossicles (bones of the ear) are located in the middle-ear chamber of each ear and serve to transmit sound impulses.
3. **Hyoid bone.** The hyoid is located above the larynx (voice box) and below the mandible (lower jaw). It supports the tongue and assists in swallowing.
4. **Vertebral column.** The vertebral column (backbone) consists of 26 individual vertebrae separated by cartilaginous discs. In the pelvic region, several vertebrae are fused to form the *sacrum,* which provides attachment for the pelvic girdle. The lower vertebrae are fused to form the *coccyx* (tailbone).
5. **Rib cage.** The rib cage forms the bony and cartilaginous framework of the thorax. It articulates posteriorly with the thoracic vertebrae and includes the 12 pairs of *ribs,* the flattened *sternum* (breastbone), and the *costal cartilages* that connect the ribs to the sternum.

The **appendicular skeleton** is composed of the bones of the upper and lower limbs and the bony girdles that anchor the limbs to the axial skeleton. The components of the appendicular skeleton are as follows:

1. **Pectoral girdle.** The paired *scapulae* (shoulder blades) and *clavicles* (collarbones) constitute the pectoral girdle. It is not a complete girdle, having only an anterior attachment to the skeleton at the sternum via the clavicles. The primary function of the pectoral girdle is to provide attachment for the muscles that move the arm.

osteology: L. *ossis,* bone; Gk. *logos,* study
ossicle: L. *ossiculum,* little bone

TABLE 6.1 Classification of the Bones of the Adult Skeleton

Axial Skeleton			Appendicular Skeleton	
Skull—22 Bones		**Auditory Ossicles**—6 Bones	**Pectoral Girdle**—4 Bones	
14 Facial Bones	*8 Cranial Bones*	malleus (2)	scapula (2)	
maxilla (2)	frontal (1)	incus (2)	clavicle (2)	
palatine (2)	parietal (2)	stapes (2)	**Upper Limbs**—60 Bones	
zygomatic (2)	occipital (1)	**Hyoid**—1 Bone	humerus (2)	carpal bones (16)
lacrimal (2)	temporal (2)	**Vertebral Column**—26 Bones	radius (2)	metacarpal bones (10)
nasal (2)	sphenoid (1)	cervical vertebra (7)	ulna (2)	phalanges (28)
vomer (1)	ethmoid (1)	thoracic vertebra (12)	**Pelvic Girdle**—2 Bones	
inferior nasal concha (2)		lumbar vertebra (5)	os coxa (2) (each contains 3 fused bones)	
mandible (1)		sacrum (1) (5 fused bones)	**Lower Limbs**—60 Bones	
		coccyx (1) (3–5 fused bones)	femur (2)	tarsal bones (14)
		Rib Cage—25 Bones	tibia (2)	metatarsal bones (10)
		rib (24)	fibula (2)	phalanges (28)
		sternum (1)	patella (2)	

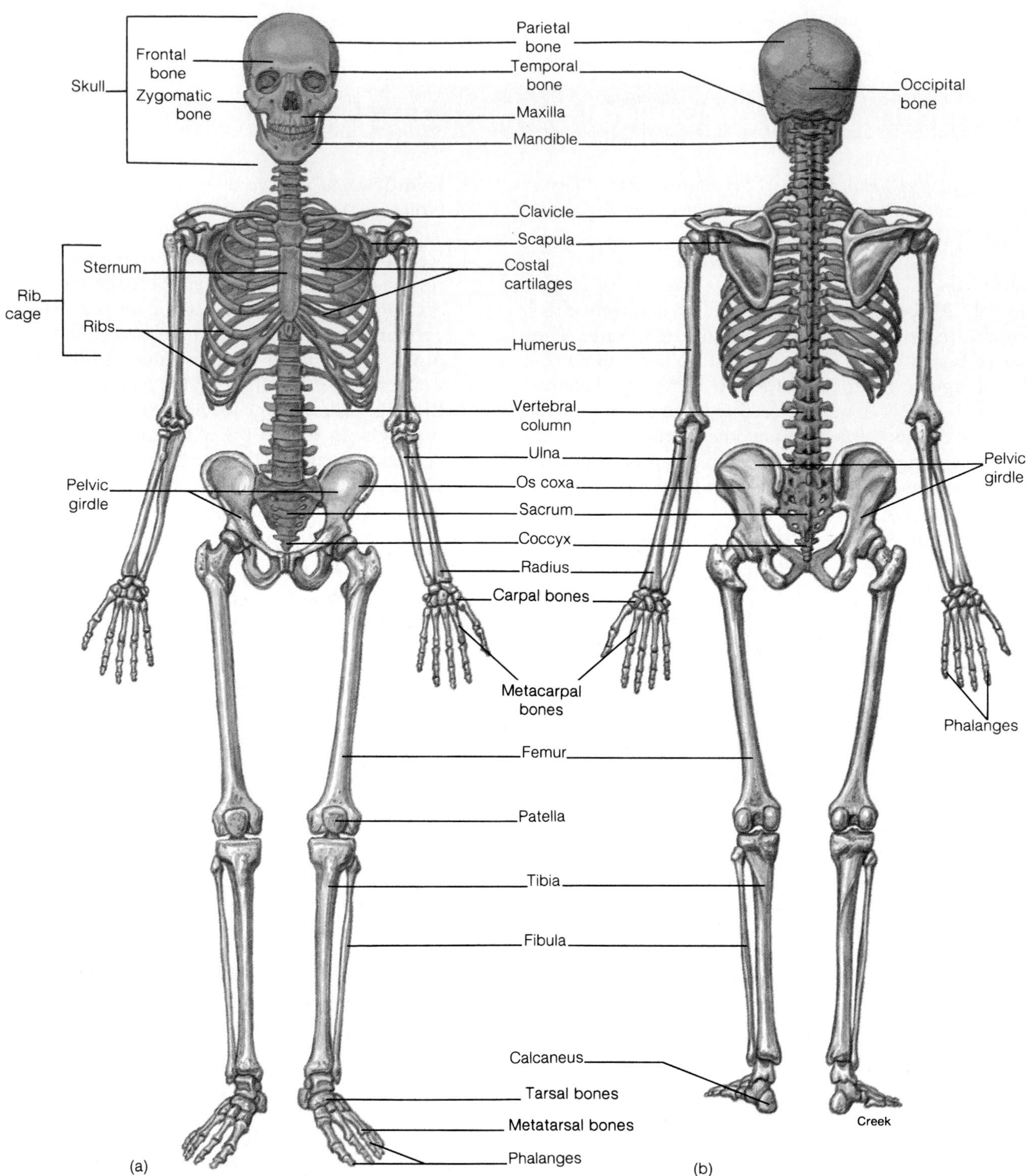

FIGURE 6.1
The human skeleton. (*a*) An anterior view and (*b*) a posterior view. The appendicular portion is colored light blue.

2. **Upper limbs.** Each upper limb consists of a proximal *humerus* within the arm (brachium), an *ulna* and *radius* within the forearm (antebrachium), the *carpal bones* of the wrist, and the *metacarpal bones* and *phalanges* (finger bones) of the hand.
3. **Pelvic girdle.** The pelvic girdle is formed by two *ossa coxae* (hipbones), united anteriorly by the *symphysis* (*sim'fĭ-sis*) *pubis* and posteriorly by the sacrum of the vertebral column. The pelvic girdle supports the weight of the body through the vertebral column and protects the lower viscera (internal organs) within the pelvic cavity. It also provides attachment for the muscles that move the thigh.
4. **Lower limbs.** Each lower limb consists of a proximal *femur* (thighbone) within the thigh, a *tibia* (shinbone) and *fibula* within the leg, the *tarsal bones* of the ankle, and the *metatarsal bones* and *phalanges* (toe bones) of the foot. In addition, the *patella* (kneecap) is located on the anterior surface of the knee joint, between the thigh and leg regions.

Inside Information—Of the 206 bones in your body, 106 are in your hands and feet.

6.2 Functions of the Skeleton

The bones of the skeleton serve the mechanical functions of supporting and protecting the body and permitting body movement. They function metabolically as sites of blood cell production and mineral storage.

The strength of bone comes from its inorganic components, of such durability that they resist decomposition even after death. Much of what we know about prehistoric animals, including humans, has been determined from preserved skeletal remains. When we think of bone, we tend to think of something hard, dry, and dead. The term *skeleton* even comes from a Greek word meaning "dried up." Living bone, however, is not inert material; it is dynamic and adaptable. It performs many body functions, including *support, protection, body movement, hemopoiesis,* and *mineral storage.*

1. **Support.** The skeleton forms a framework that supports the softer tissues and organs of the body.
2. **Protection.** The skull and vertebral column enclose the brain and spinal cord; the rib cage protects the heart, lungs, great vessels, liver, and spleen; and the pelvic girdle supports and protects the pelvic viscera. Even the site where blood cells are produced is protected within the central hollow portion of certain bones.
3. **Body movement.** Bones serve as anchoring attachments for skeletal muscles. In this capacity, the bones act as levers (with the joints functioning as pivots) when muscles contract to cause body movement (fig. 6.2).
4. **Hemopoiesis.** The process of blood cell formation is called hemopoiesis (*hem"ŏ-poi-e'sis*). It takes place in tissue called red bone marrow located in the center of some bones (fig. 6.3). In an infant, the spleen and liver produce red blood cells, but as the bones mature, the bone marrow takes over this function.

hemopoiesis: Gk. *haima,* blood; *poiesis,* a making

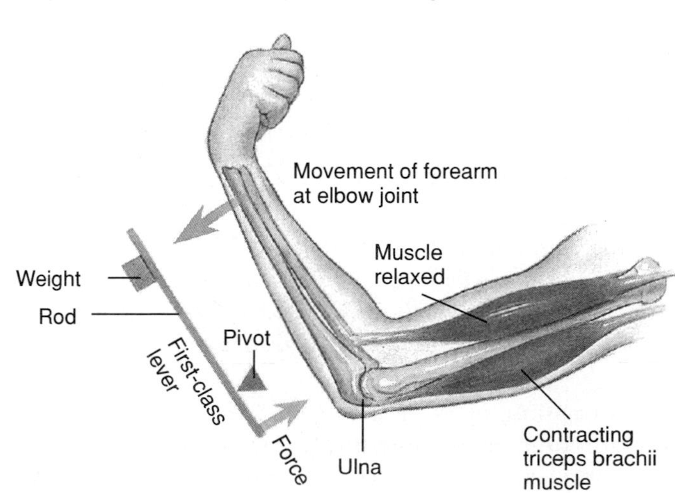

FIGURE 6.2
Movement of the skeleton is accomplished by contraction of a muscle (the force), causing a bone (the lever) to be moved at the joint (the pivot).

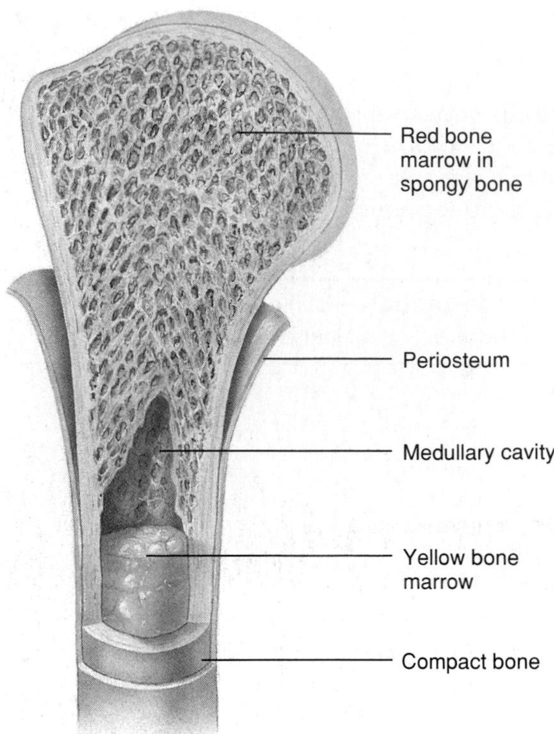

FIGURE 6.3
Hemopoiesis is the process by which blood cells are formed. In an adult, blood cells are formed in the red bone marrow.

5. **Mineral storage.** The inorganic matrix of bone is composed primarily of the minerals calcium and phosphorus. These minerals, which account for approximately two-thirds of the weight of bone, give bone its firmness and strength. About 95% of the calcium and 90% of the phosphorus within the body are located in the bones and teeth. Although blood levels of calcium and phosphorus are kept within narrow limits, both of these mineral salts are essential for other body functions. For example, calcium is needed for muscle contraction, blood clotting, and the movement of ions and nutrients across cell membranes. Phosphorus is needed for the activities of the nucleic acids DNA and RNA, as well as for ATP utilization. If mineral salts are not obtained in the diet in sufficient amounts, they may be withdrawn from the bones, weakening the bony architecture. In addition to calcium and phosphorus, small amounts of magnesium and sodium salts are stored in bone tissue.

Calcium and phosphorus are absorbed from the small intestine into the blood with the help of vitamin D. As the bones develop in a child, it is critical that diet contain an adequate amount of calcium and phosphorus (for normal ossification) and vitamin D (for absorption of these minerals). If the diet is deficient in any of these essentials, a condition known as rickets *develops (see fig. 5.9). Rickets is characterized by soft bones that may result in bowlegs and malformation of the head, chest, and pelvic girdle.*

In summary, the skeletal system is not an isolated unit. It works to assist several other body systems. It is associated with the muscular system by storing the calcium needed for muscular contraction; in addition, it provides attachment for muscles as they span the movable joints. The skeletal system serves the circulatory system by producing blood cells in the bone marrow. Directly or indirectly, the skeletal system supports and protects all of the systems of the body.

Inside Information—The skeleton's 206 bones support a mass of muscles and organs that may weigh five times as much as the bones themselves.

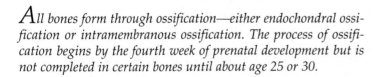

6.3 Development of the Skeletal System

All bones form through ossification—either endochondral ossification or intramembranous ossification. The process of ossification begins by the fourth week of prenatal development but is not completed in certain bones until about age 25 or 30.

Bone formation, or *ossification* (*os"ĭ-fĭ-ca'shun*), begins during embryonic development (fig. 6.4). Bone tissue derives from specialized migratory cells known as *mesenchyme.* Some of these embryonic cells will develop into primitive cartilage cells called *chondroblasts* (*kon'dro-blasts*). These cells produce a cartilage matrix that is eventually replaced by bone in a process known as **endochondral ossification.** Most of the skeleton is formed in this fashion—first it goes through a hyaline cartilage stage, which is then replaced by mature bone. The humerus, radius, ulna, femur, tibia, and fibula are examples of bones that develop in this way.

A lesser number of mesenchymal cells develop directly into bone without first going through a cartilage

chondroblast: Gk. *chondros,* cartilage; *blastos,* offspring or germ

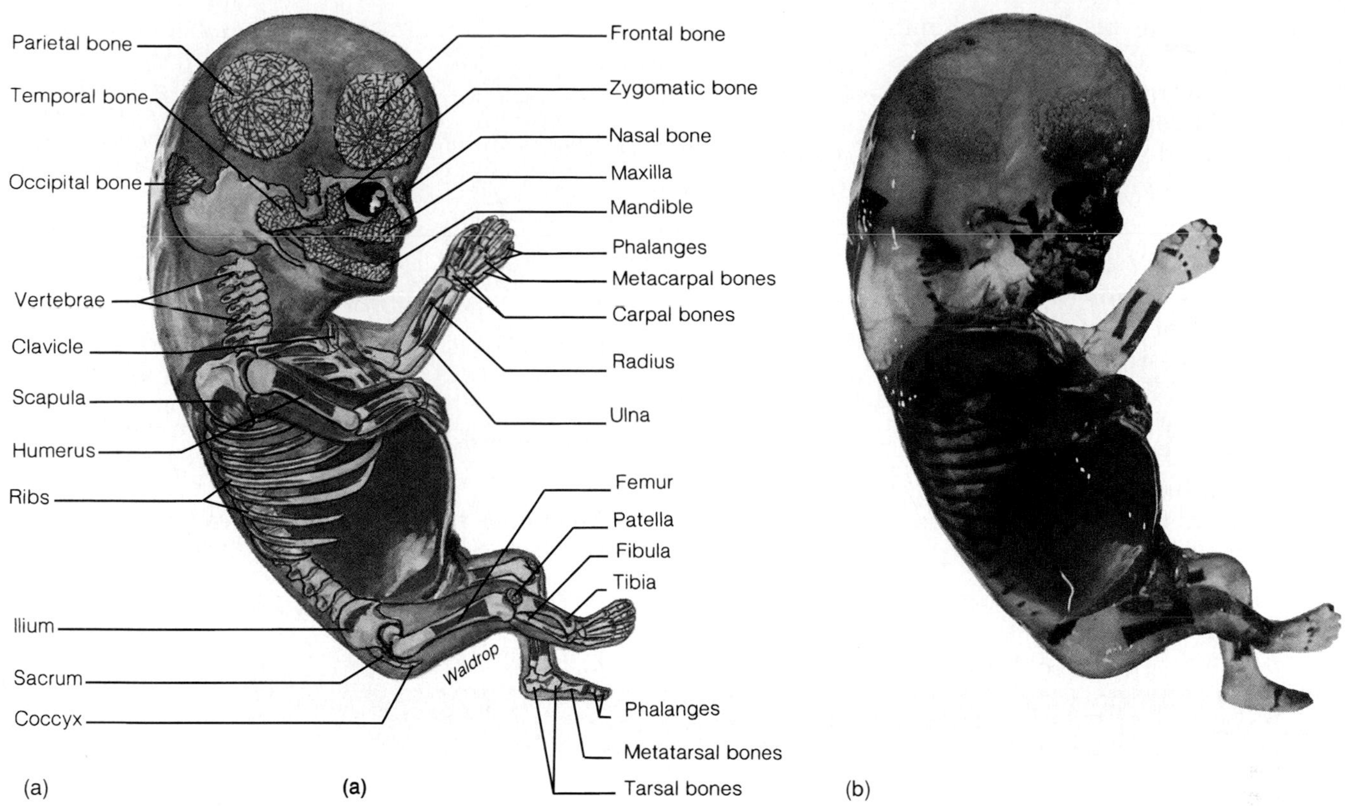

FIGURE 6.4

Ossification centers of the skeleton of a 10-week-old fetus. (*a*) The diagram depicts endochondral ossification in red and intramembranous ossification in a stippled pattern; the cartilaginous portions of the skeleton are shown in gray. (*b*) The photograph shows the ossification centers stained with a red indicator dye.

stage. This type of bone formation is referred to as **intramembranous ossification.** Facial bones and certain bones of the cranium are formed in this fashion. *Sesamoid bones* are specialized intramembranous bones that develop in tendons. The patellae (kneecaps) are sesamoid bones.

During fetal development and infancy, the bones of the skull are separated by pliable fibrous unions. There are also six large areas of connective tissue membrane that unite the gaps between the developing bones. These membranous sheets are called **fontanels** (*fon"tă-nelz'*), meaning "little fountains." The name derives from the fact that a baby's pulse can be felt surging in these "soft spots" on the skull. The fontanels permit the skull to undergo changes in shape, called *molding,* during childbirth, and they accommodate the rapid growth of the brain during infancy. Ossification of the fontanels is normally complete by 20 to 24 months after birth. The fontanels are illustrated in figure 6.5 and briefly described below.

1. **Anterior (frontal) fontanel.** The anterior fontanel is diamond-shaped and is the most prominent. It is located on the anteromedian portion of the skull.
2. **Posterior (occipital) fontanel.** The posterior fontanel is positioned at the back of the skull on the median line. It is also diamond-shaped, but smaller than the anterior fontanel.
3. **Anterolateral (sphenoid) fontanels.** The paired anterolateral fontanels are found on both sides of the skull, directly below the anterior fontanel. They are relatively small and irregularly shaped.
4. **Posterolateral (mastoid) fontanels.** The paired posterolateral fontanels, also irregularly shaped, are located on the posterolateral sides of the skull.

During normal childbirth, the fetal skull comes under tremendous pressure. Bones may even shift to lie one under another, altering the shape of the skull. This makes delivery easier for the mother. If a baby is born breech (buttocks first), these shifts do not occur. Delivery becomes much more difficult, often requiring the use of forceps.

fontanel: Fr. *fontaine,* little fountain

Sutures are unions between the bones of the skull (fig. 6.5). At birth the sutures are still in the process of formation, but as the infant's head continues to grow they gradually replace the pliable fontanels. Four of the most obvious sutures are the following:

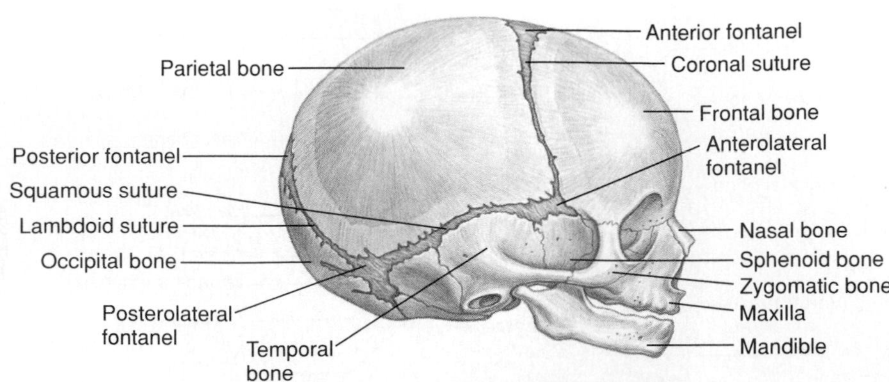

FIGURE 6.5
The fetal skull showing the fontanels and sutures.

1. **Sagittal suture.** The prominent sagittal suture extends the anteroposterior median length of the skull between the anterior and posterior fontanels.
2. **Coronal suture.** The coronal suture extends from the anterior fontanel to the anterolateral fontanel.
3. **Lambdoid suture.** The lambdoid suture extends from the posterior fontanel to the posterolateral fontanel.
4. **Squamous suture.** The squamous suture connects the posterolateral fontanel to the anterolateral fontanel.

6.4 Bone Structure

Each bone has a characteristic shape and surface features that suggest its relationship to other bones, muscles, and the body as a whole.

Each bone of the skeleton is an individual organ consisting of several types of tissue. Bone tissue is the principal tissue, but nervous, vascular, and cartilaginous tissues also contribute to the structure and function of bone.

The shape and surface features of each bone indicate its functional role. Bones that are long, for example, function as levers during body movement. Bones that support the body are massive and have large articular surfaces and processes for muscle attachment. Roughened areas on these bones may serve for the attachment of ligaments, tendons, or muscles. A flattened surface provides placement for a large muscle or may serve for protection. Grooves around the articular end of a bone indicate where a tendon or nerve passes, and openings through a bone permit the passage of nerves or vessels. The various descriptive terms used to identify the surface features of bones are summarized in table 6.2.

Shapes of Bones

The bones of the skeleton are grouped on the basis of shape into four principal categories (fig. 6.6).

1. **Long bones.** Long bones are longer than they are wide and function as levers. Most of the bones of the upper and lower limbs are of this type (e.g., the humerus, tibia, metacarpal bones, and phalanges).
2. **Short bones.** Short bones are cube-shaped and are found in tight spaces where they transfer forces of movement (e.g., the carpal and tarsal bones).
3. **Flat bones.** Flat bones have a broad, dense surface for muscle attachment or protection of underlying organs (e.g., bones of the cranium, the ribs, and the scapula).

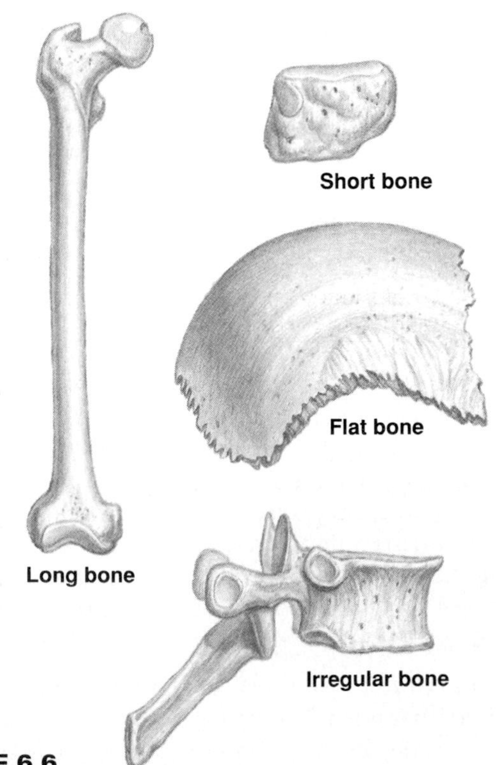

FIGURE 6.6
Examples of the four shapes of bones.

lambdoid: Gk. *lambda,* letter λ in Greek alphabet

TABLE 6.2 **Surface Features of Bone**

Structure	Description and Example
Articulating Surfaces	
condyle (*kon' dīl*)	A large, rounded articulating knob (the occipital condyle of the occipital bone)
facet (*fas' et*)	A flattened or shallow articulating surface (the costal facet of a thoracic vertebra)
head	A prominent, rounded articulating end of a bone (the head of the femur)
Nonarticulating Prominences	
crest	A narrow, ridgelike projection (the iliac crest of the os coxa)
epicondyle	A projection above a condyle (the medial epicondyle of the femur)
process	Any marked bony prominence (the mastoid process of the temporal bone)
spine	A sharp, slender process (the spine of the scapula)
trochanter	A massive process found only on the femur (the greater trochanter of the femur)
tubercle (*too' berk'l*)	A small, rounded process (the greater tubercle of the humerus)
tuberosity	A large, roughened process (the radial tuberosity of the radius)
Depressions and Openings	
alveolus	A deep pit or socket (the alveoli for teeth in the maxilla)
fissure	A narrow, slitlike opening (the superior orbital fissure of the sphenoid bone)
foramen (*fŏ-ra' men*)	A rounded opening through a bone (the foramen magnum of the occipital bone)
fossa (*fos' ă*)	A flattened or shallow surface (the mandibular fossa of the temporal bone)
sinus	A cavity or hollow space in a bone (the frontal sinus of the frontal bone)
sulcus	A groove that accommodates a vessel, nerve, or tendon (the intertubercular sulcus of the humerus)

4. **Irregular bones.** Irregular bones are varied in shape and have many surface markings for muscle attachment or articulation (e.g., the vertebrae and certain bones of the skull).

Structure of a Typical Long Bone

Bone tissue is organized as *compact bone* or *spongy bone,* and most bones have both types. Compact bone is hard and dense, and is the protective exterior portion of all bones. The spongy bone, when it occurs, is deep to the compact bone and is quite porous.

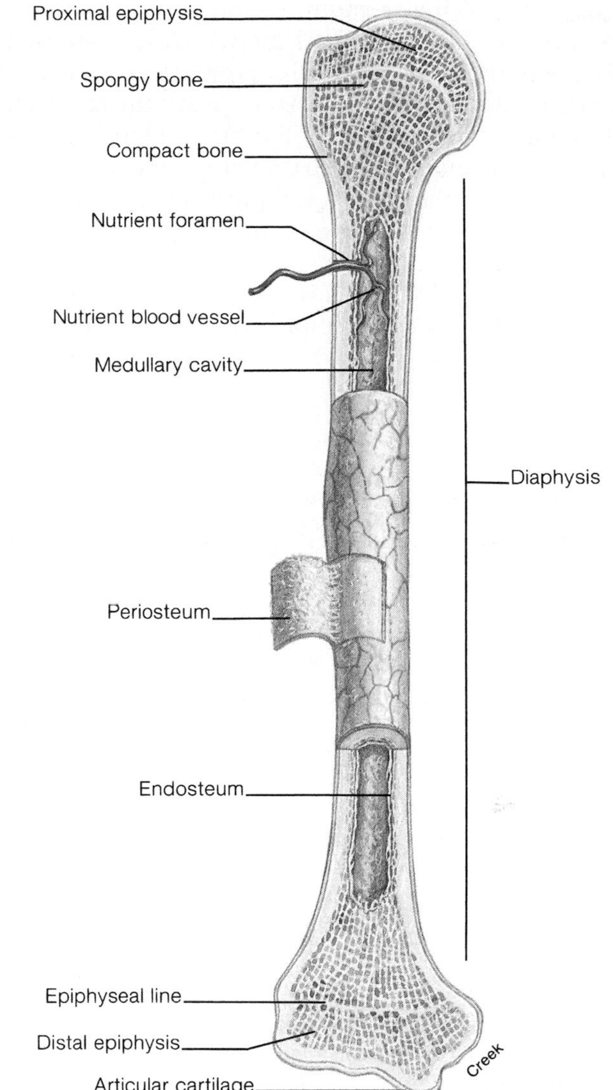

FIGURE 6.7
Diagram of a long bone shown in longitudinal section.

The long bones of the skeleton have a descriptive terminology all their own. In a long bone, the bone **shaft,** or **diaphysis** (*di-af' ĭ-sis*), consists of a cylinder of compact bone that surrounds a central cavity called the **medullary cavity** (fig. 6.7). The medullary cavity is lined with a thin layer of connective tissue called the **endosteum** (*en-dos' te-um*). The cavity in the adult contains **yellow bone marrow,** so named because it contains large amounts of yellow fat. On each end of the shaft is an **epiphysis** (*ĕ-pif' -ĭ-sis*), consisting of spongy bone surrounded by a layer of compact bone. **Red bone marrow** is found within the spongy bone,

facet: Fr. *facette,* little face
trochanter: Gk. *trochanter,* runner
tuberosity: L. *tuberosus,* lump
diaphysis: Gk. *dia,* throughout; *physis,* growth
epiphysis: Gk. *epi,* upon; *physis,* growth

especially that of the sternum, vertebrae, and portions of the ossa coxae. **Articular cartilage,** which is composed of thin hyaline cartilage, caps each epiphysis and eases joint movement. Along the diaphysis are **nutrient foramina**—small openings into the bone that allow blood vessels to pass through the bone for nourishment of the living tissue.

Between the diaphysis and epiphysis of a bone is an **epiphyseal** (*ep"ĭ-fiz'e-al*) **plate**—a region of mitotic activity responsible for *internal bone growth.* As bone growth is completed, an *epiphyseal line* replaces the plate, and final ossification occurs between the epiphysis and the diaphysis. A **periosteum** (*per"e-os'te-um*) of dense regular connective tissue covers the surface of the bone, except over the articular cartilage. The periosteum is a place of attachment for tendons and ligaments and is responsible for *external bone growth.*

Fracture of a long bone in a young person may be serious if it damages an epiphyseal plate. If such an injury is untreated, or treated improperly, growth of the bone may be arrested or slowed, resulting in a permanent shortening of the affected limb.

6.5 Bone Tissue

Bone tissue is composed of several types of bone cells embedded in a matrix of inorganic salts (calcium and phosphorus) and collagenous fibers. Bone cells and ground substance give bone flexibility and strength; the inorganic salts give it hardness.

Bone Cells

There are three principal types of bone cells. **Osteogenic cells** are found in the bone tissue in contact with the endosteum and the periosteum. These cells respond to trauma, such as fracture, by giving rise to bone-forming cells (*osteoblasts*) and bone-destroying cells (*osteoclasts*). **Osteoblasts** are bone-forming cells that synthesize and secrete unmineralized ground substance. They are abundant in areas of high metabolism; for example, under the periosteum and bordering the medullary cavity. **Osteocytes** are mature bone cells (fig. 6.8) derived from osteoblasts that deposit minerals around themselves. Osteocytes maintain healthy bone tissue by secreting enzymes and influencing bone mineral content. They also regulate the calcium release from bone tissue to blood. **Osteoclasts** are large multinuclear cells that use enzymes to break down bone tissue. These cells are important in bone growth, remodeling, and healing.

In healthy bone tissue, there is a balance between the breakdown of old bone and the production of new bone. In a disorder called *osteoporosis* (literally, "porous bones"), the breakdown occurs at a faster rate than the replacement, causing the bones to become weak and brittle. Osteoporosis is discussed further under "Clinical Considerations," p. 148.

Spongy and Compact Bone Tissues

Both spongy and compact bone tissues are present in most bones. **Spongy bone tissue** is located deep within the bone, and is quite porous. Minute spikes or plates of bone tissue, called **trabeculae,** give spongy bone a lattice-like appearance. The cavities of spongy bone are filled with **red bone marrow,** which produces red blood cells, platelets, and certain white blood cells. Spongy bone tissue is light but strong.

Compact bone tissue is superficial to the spongy bone tissue and is very hard and dense. It consists of microscopic cylindrical structures running lengthwise in

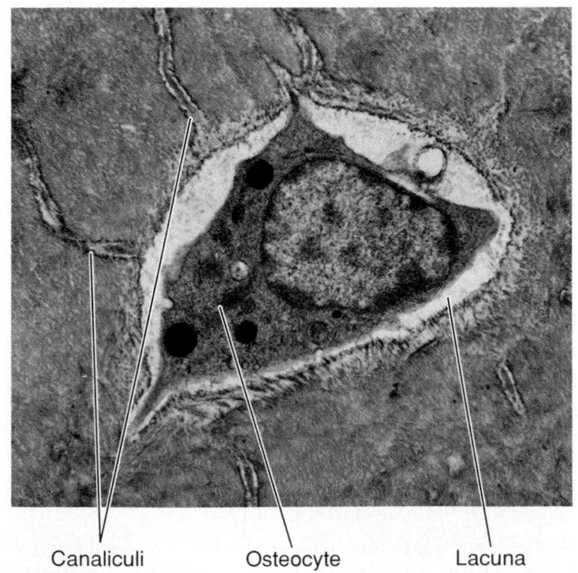

FIGURE 6.8
A photomicrograph of an osteocyte within a lacuna.

periosteum: Gk. *peri,* around; *osteon,* bone
osteoblast: Gk. *osteon,* bone; *blastos,* offspring or germ
osteoclast: Gk. *osteon,* bone; *klastos,* broken

the bone (figs. 6.9 and 6.10). These columnlike structures are the **osteons** (*os'te-onz*) of the bone tissue, also called *haversian systems.* The matrix of an osteon is laid down in concentric rings called **lamellae.** The lamellae surround a **central** (haversian) **canal** that contains tiny vessels and a nerve. Osteocytes within spaces called **lacunae** are regularly arranged between the lamellae. The lacunae are connected by **canaliculi,** through which nutrients diffuse. **Perforating** (Volkmann's) **canals** penetrate compact bone, connecting osteons with blood vessels and nerves.

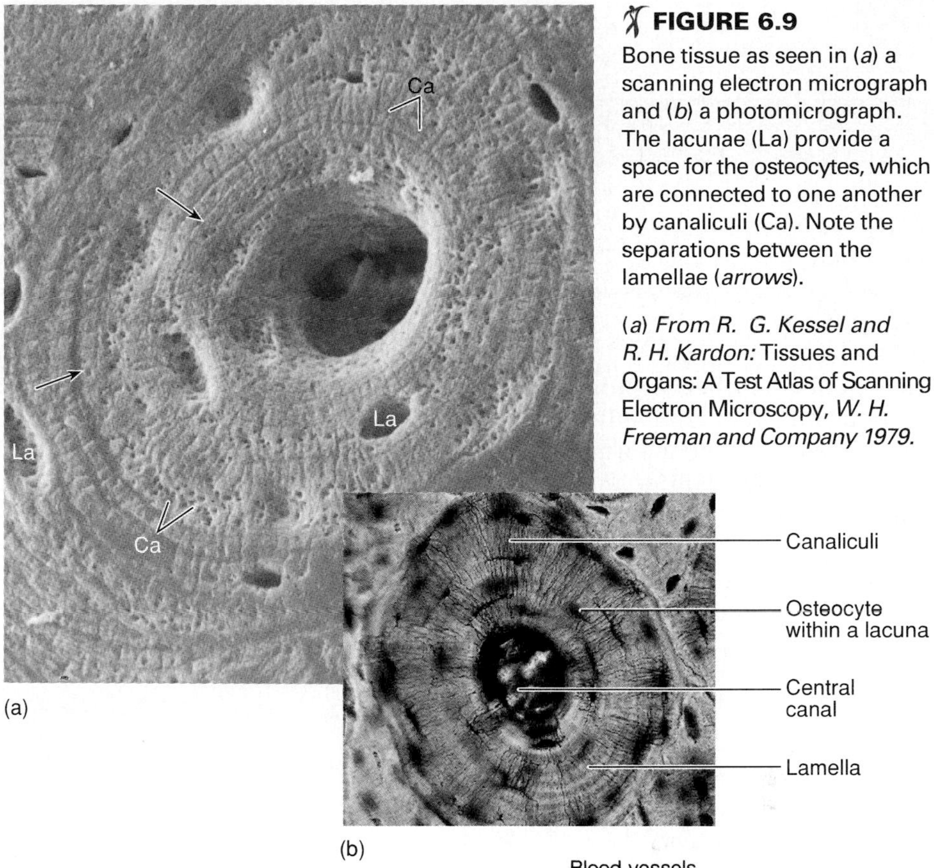

FIGURE 6.9

Bone tissue as seen in (*a*) a scanning electron micrograph and (*b*) a photomicrograph. The lacunae (La) provide a space for the osteocytes, which are connected to one another by canaliculi (Ca). Note the separations between the lamellae (*arrows*).

(*a*) *From R. G. Kessel and R. H. Kardon:* Tissues and Organs: A Test Atlas of Scanning Electron Microscopy, *W. H. Freeman and Company 1979.*

haversian system: from Clopton Havers, English anatomist, 1650–1702

Volkmann's canal: from Alfred Volkmann, German physiologist, 1800–1877

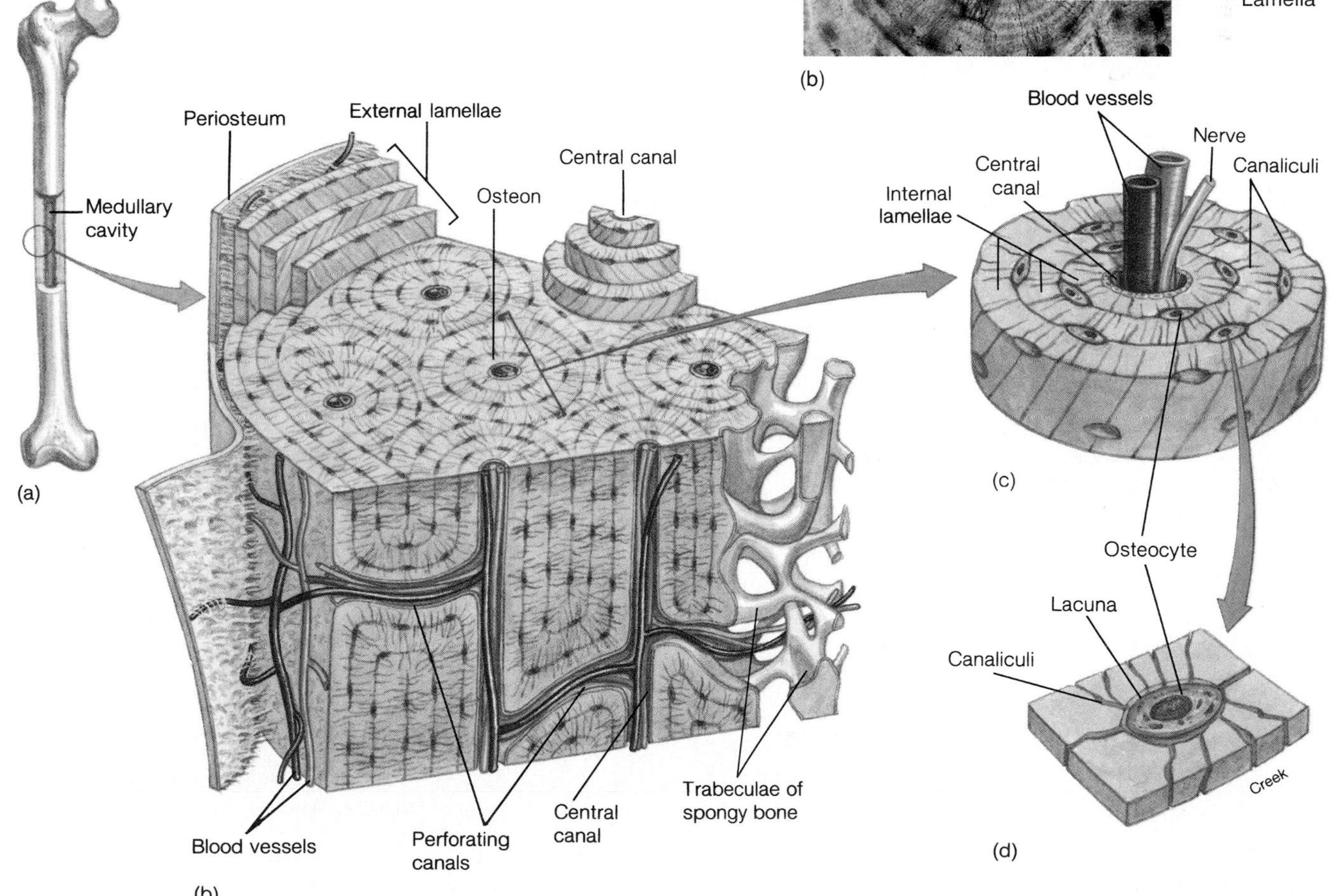

FIGURE 6.10

Compact bone tissue. (*a*) A diagram of the femur showing a cut through the compact bone into the medullary cavity. (*b*) The arrangement of the osteons within the diaphysis of the bone. (*c*) An enlarged view of an osteon showing the osteocytes within lacunae and the concentric lamellae. (*d*) An osteocyte within a lacuna.

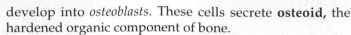

6.6 Bone Growth

The development of bone from embryo to adult depends on the orderly processes of cell division, growth, and ongoing remodeling. Bone growth is influenced by genetics, hormones, and nutrition.

As described earlier, in most bone development a cartilaginous model is gradually replaced by bone tissue during endochondral bone formation (see fig. 6.4). As the cartilage model grows, the *chondrocytes* in the center of the shaft enlarge, and minerals are deposited within the matrix in a process called *calcification* (fig. 6.11). Calcification reduces the passage of nutrients to the chondrocytes, causing them to die. At the same time, some cells of the perichondrium (connective tissue surrounding cartilage) develop into *osteoblasts*. These cells secrete **osteoid,** the hardened organic component of bone.

A **periosteal bud,** consisting of osteoblasts and blood vessels, invades the disintegrating center of the cartilage model from the periosteum. Once in the center, the osteoblasts secrete osteoid, and a **primary ossification center** is established. Ossification then expands into the deteriorating cartilage. This process is repeated in both the proximal and distal epiphyses, forming **secondary ossification centers** where spongy bone develops.

Once the secondary ossification centers have been formed, bone tissue totally replaces cartilage tissue, except at the articular ends of the bone and at the epiphyseal plates. An **epiphyseal plate** consists of five histological

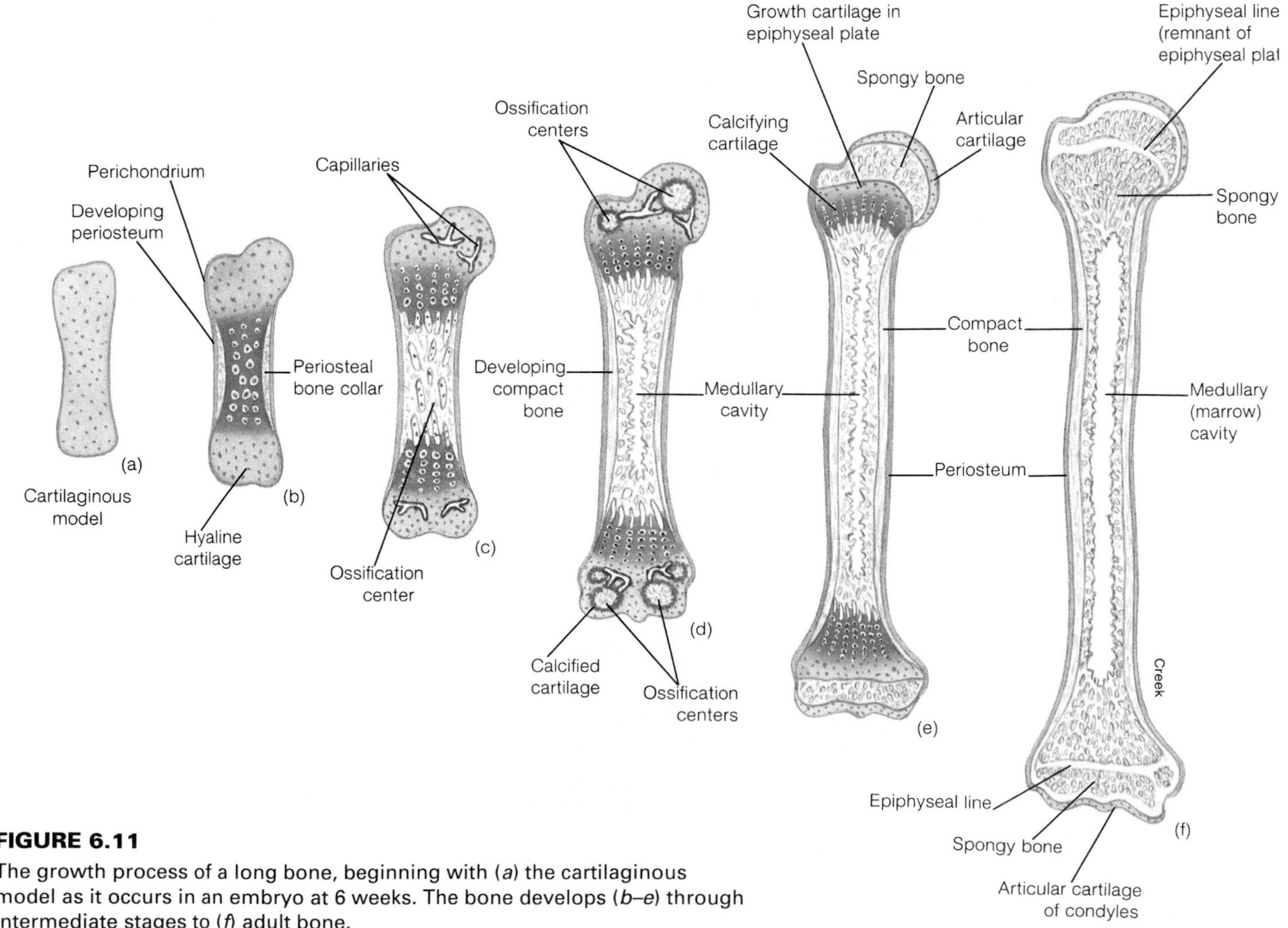

FIGURE 6.11
The growth process of a long bone, beginning with (*a*) the cartilaginous model as it occurs in an embryo at 6 weeks. The bone develops (*b–e*) through intermediate stages to (*f*) adult bone.

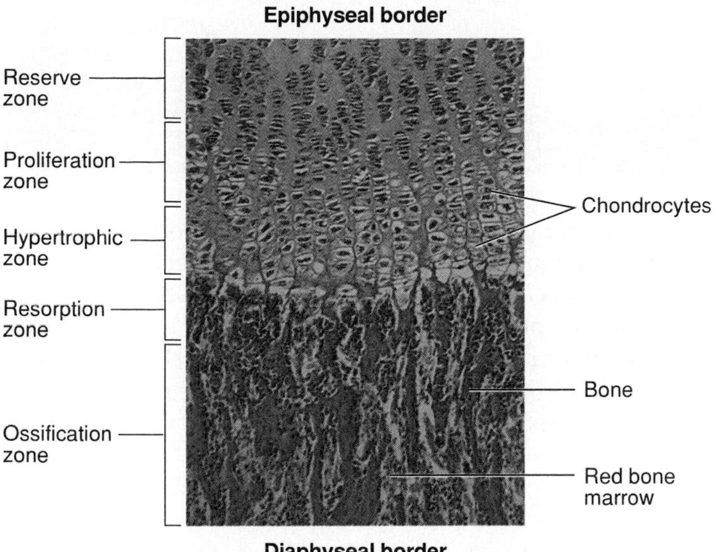

FIGURE 6.12
A photomicrograph from an epiphyseal plate (63×).

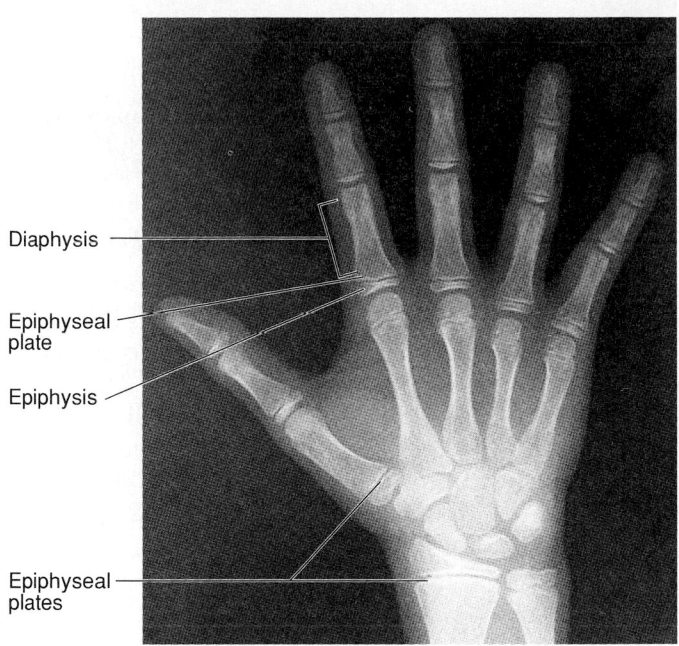

FIGURE 6.13
The presence of epiphyseal plates as seen in an X ray of a child's hand. The plates indicate that the bones are still growing in length.

zones (fig. 6.12). The *reserve zone* borders the epiphysis and consists of small chondrocytes irregularly dispersed throughout the intercellular matrix. The chondrocytes in this zone anchor the epiphyseal plate to the bony epiphysis. The *proliferation zone* consists of larger, regularly arranged chondrocytes that are constantly dividing. The *hypertrophic zone* consists of very large chondrocytes arranged in columns. The growth of length in long bones is due to the cellular proliferation at the zone of proliferating cartilage and the growth and maturation of these new cells within the hypertrophic zone. The *resorption zone* is the area where a change in mineral content is occurring. The *ossification zone* is a region of transformation from cartilage tissue to bone tissue. Chondrocytes within this zone die because the matrix surrounding them becomes calcified. Osteoclasts then break down the calcified matrix and the area is invaded by osteoblasts and capillaries. As the osteoblasts mature, more osteoids are secreted and bone tissue is formed. The result of this process is a gradual increase in the length of the bone at the epiphyseal plates.

The time at which epiphyseal plates ossify varies greatly from bone to bone, but it generally occurs between the ages of 18 and 20 within the long bones of most people. Because ossification of the epiphyseal plates within each bone occurs at predictable periods of development, radiologists can determine the ages of people who are still growing by examining X rays of their bones (fig. 6.13). Large discrepancies between bone age and chronological age may indicate a genetic or endocrine abnormality.

As you grow older, your bones are continually being remodeled. Bony prominences develop as stress is applied to the periosteum, causing the osteoblasts to secrete osteoid and form new bone tissue. The greater trochanter of the femur, for example, develops in response to forces of stress applied to the periosteum of the bone where the tendons of muscles attach. Even when you stop growing in height, your bones continue to strengthen if you remain physically active.

As new bone layers are deposited on the outside surface of the bone, osteoclasts dissolve bone tissue adjacent to the medullary cavity. In this way, the size of the cavity keeps pace with the increased growth of the bone.

Bone is highly dynamic and is continually being remodeled in response to stress or even the lack of stress. The effects of the absence of stress can best be seen in the bones of bedridden or paralyzed individuals. X rays of their bones will reveal a marked loss of bone tissue.

The movement of teeth in orthodontics involves bone remodeling. The teeth sockets (alveoli) are reshaped through the activity of osteoclast and osteoblast cells as stress is applied with braces. The use of traction in treating certain skeletal disorders has a similar effect.

6.7 Cranial Bones

Tight sutures join the eight cranial bones of the skull to enclose and protect the brain.

The skull consists of cranial bones and facial bones. The eight bones of the cranium form tight unions to enclose and protect the brain and sensory organs. The cranial bones consist of one frontal bone, two parietal bones, two temporal bones, one occipital bone, one sphenoid bone, and one ethmoid bone. Diagrams of the skull and an X ray are presented in figures 6.14 through 6.22.

Frontal Bone

The frontal bone (figs. 6.14 and 6.15) forms the anterior roof of the cranium, the forehead, the roof of the nasal cavity, and the superior arch of the *orbits* (eye sockets), which contain the eyeballs. The **supraorbital margin** is a bony ridge over the orbit. Openings along this ridge, called **supraorbital foramina,** allow passage of small nerves and vessels.

Parietal Bones

The two parietal bones form the upper sides and roof of the cranium. The frontal bone and the parietal bones meet at the **coronal suture,** and the right and left parietal bones meet at the **sagittal suture** along the cranial midline.

Temporal Bones

The two temporal bones form the lower sides of the cranium (figs. 6.14, 6.15, 6.16, and 6.21). (The term *temporal* refers to time. Gray hairs, usually indicating the passage of time, frequently appear first on the temples.) Each temporal bone is joined to its adjacent parietal bone by the **squamous** (*skwa'mus*) **suture.** Structurally, each temporal bone has four parts.

1. **Squamous part.** The squamous part is the flattened plate of bone on the sides of the skull. Projecting forward is a **zygomatic process** that forms the posterior portion of the **zygomatic arch.** On the inferior surface is the cuplike **mandibular fossa,** which forms a joint with the condyle of the mandible. This articulation is the *temporomandibular joint.*
2. **Tympanic part.** The tympanic part contains the **external acoustic meatus,** or ear canal, which is posterior to the mandibular fossa. A thin, pointed **styloid process** (figs. 6.14 and 6.15) projects downward from the tympanic part.
3. **Mastoid part.** The **mastoid process,** a rounded projection posterior to the external acoustic meatus, accounts for the mass of the mastoid portion.
4. **Petrous part.** The petrous (*pet'rus*) part can be seen in the floor of the cranium (fig. 6.22). The structures of the middle ear (fig. 6.19) and inner ear are housed in this dense part of the temporal bone.

The mastoid process of the temporal bone can be easily felt as a bony knob (fig. 6.14) behind the earlobe. This process contains small sinuses that can become infected in mastoiditis, *as a result, for example, of a prolonged middle-ear infection.*

Occipital Bone

The occipital bone forms the back and most of the base of the skull. It is fastened to the parietal bones by the **lambdoid suture.** The **foramen magnum** is the large hole in the occipital bone through which the spinal cord connects to the brain stem. On each side of the foramen magnum are the **occipital condyles** (fig. 6.21), which articulate with the first vertebra (the atlas). The **external occipital protuberance** is a prominent posterior projection on the occipital bone that can be felt as a bump under the skin.

Sphenoid Bone

The sphenoid (*sfe'noid*) bone forms the anterior base of the cranium and can be seen in several views of the skull (figs. 6.14, 6.15, 6.16., and 6.22). It consists of a **body** with laterally projecting **greater** and **lesser wings,** which form part of the bony orbit. The wedgelike body contains a saddlelike depression, the **sella turcica** (*sel'ă tur'sĭ-kă*), commonly called the "Turk's saddle," that houses the pituitary gland. A pair of **pterygoid** (*ter'ĭ-goid*) **processes** project inferiorly from the body of the sphenoid bone to help form the lateral walls of the nasal cavity.

---

zygomatic: Gk. *zygoma,* yolk
styloid: Gk. *stylos,* pillar
mastoid: Gk. *mastos,* breast
petrous: Gk. *petra,* rock
magnum: L. *magnum,* great
sphenoid: Gk. *sphenoeides,* wedgelike

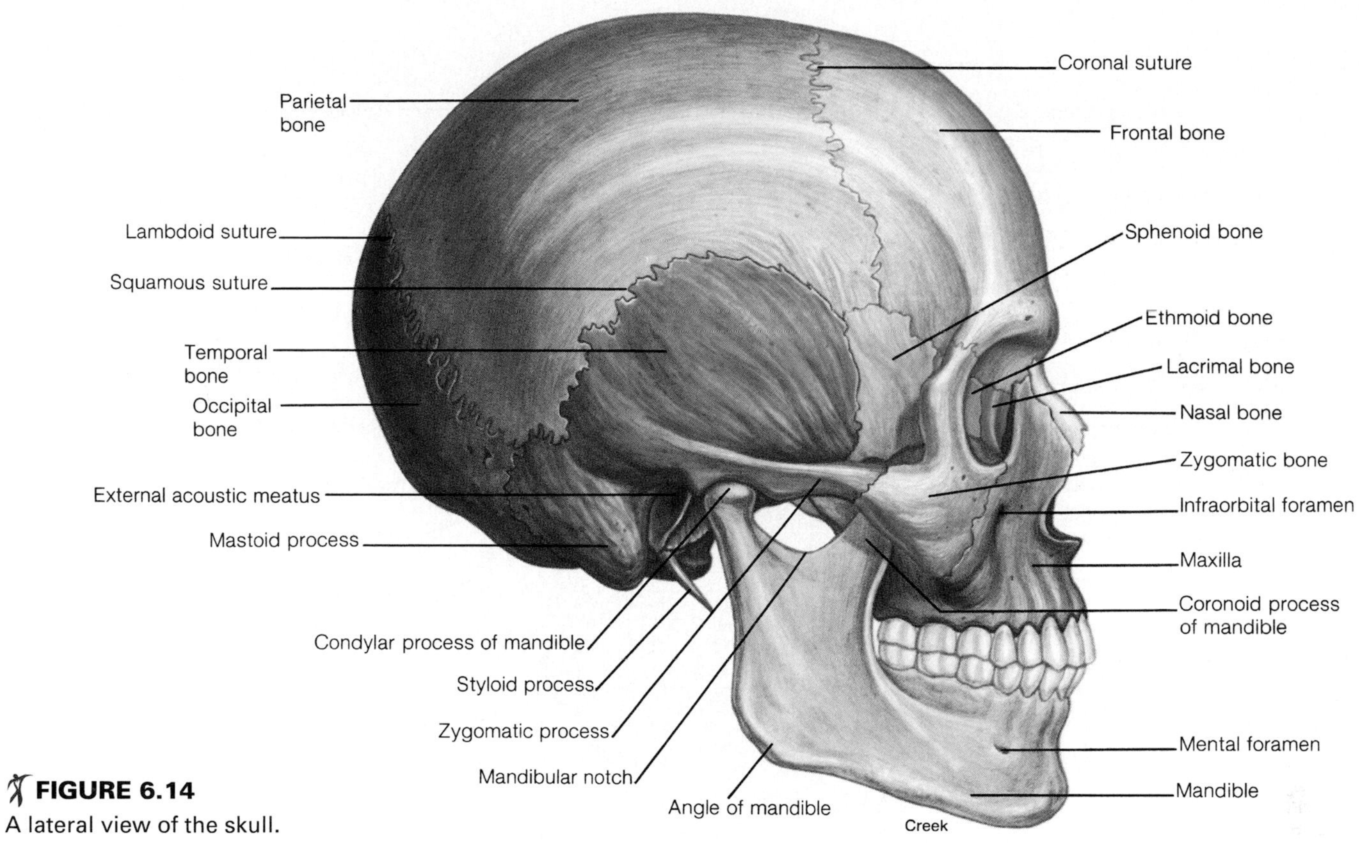

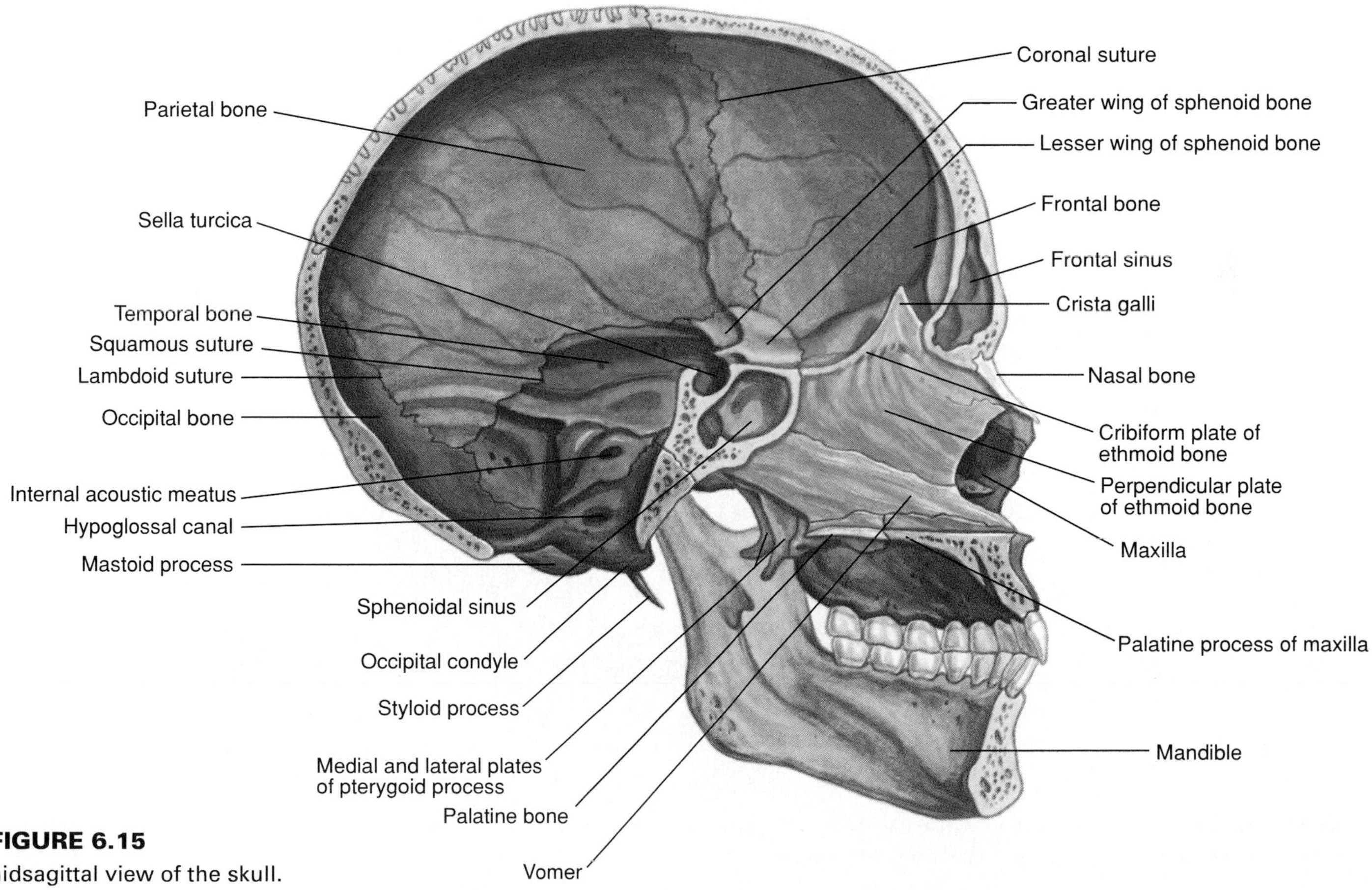

Coronal suture
Parietal bone
Frontal bone
Lambdoid suture
Squamous suture
Sphenoid bone
Ethmoid bone
Temporal bone
Lacrimal bone
Occipital bone
Nasal bone
Zygomatic bone
External acoustic meatus
Infraorbital foramen
Mastoid process
Maxilla
Coronoid process of mandible
Condylar process of mandible
Styloid process
Zygomatic process
Mental foramen
Mandibular notch
Mandible
Angle of mandible
Creek

FIGURE 6.14
A lateral view of the skull.

Coronal suture
Greater wing of sphenoid bone
Parietal bone
Lesser wing of sphenoid bone
Frontal bone
Sella turcica
Frontal sinus
Crista galli
Temporal bone
Squamous suture
Lambdoid suture
Nasal bone
Occipital bone
Cribiform plate of ethmoid bone
Perpendicular plate of ethmoid bone
Internal acoustic meatus
Hypoglossal canal
Maxilla
Mastoid process
Sphenoidal sinus
Palatine process of maxilla
Occipital condyle
Styloid process
Mandible
Medial and lateral plates of pterygoid process
Palatine bone
Vomer

FIGURE 6.15
A midsagittal view of the skull.

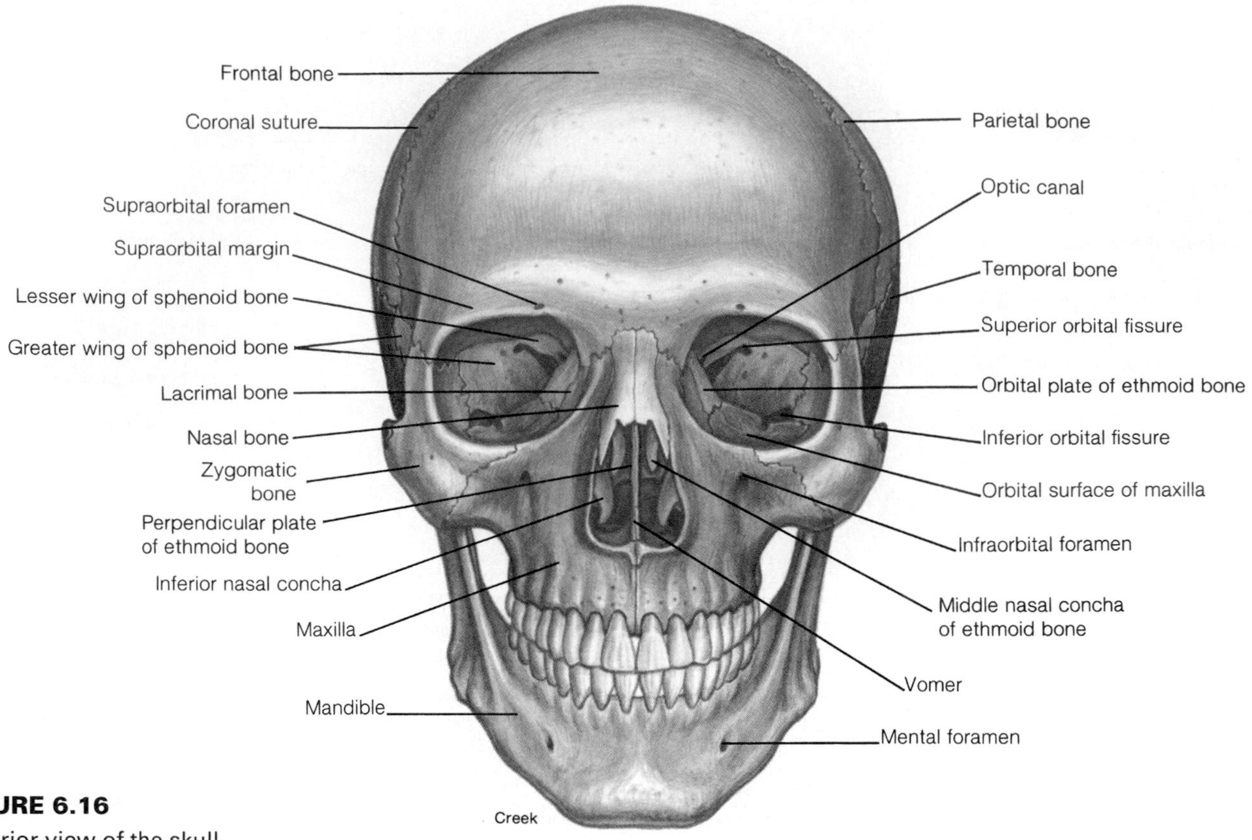

FIGURE 6.16
An anterior view of the skull.

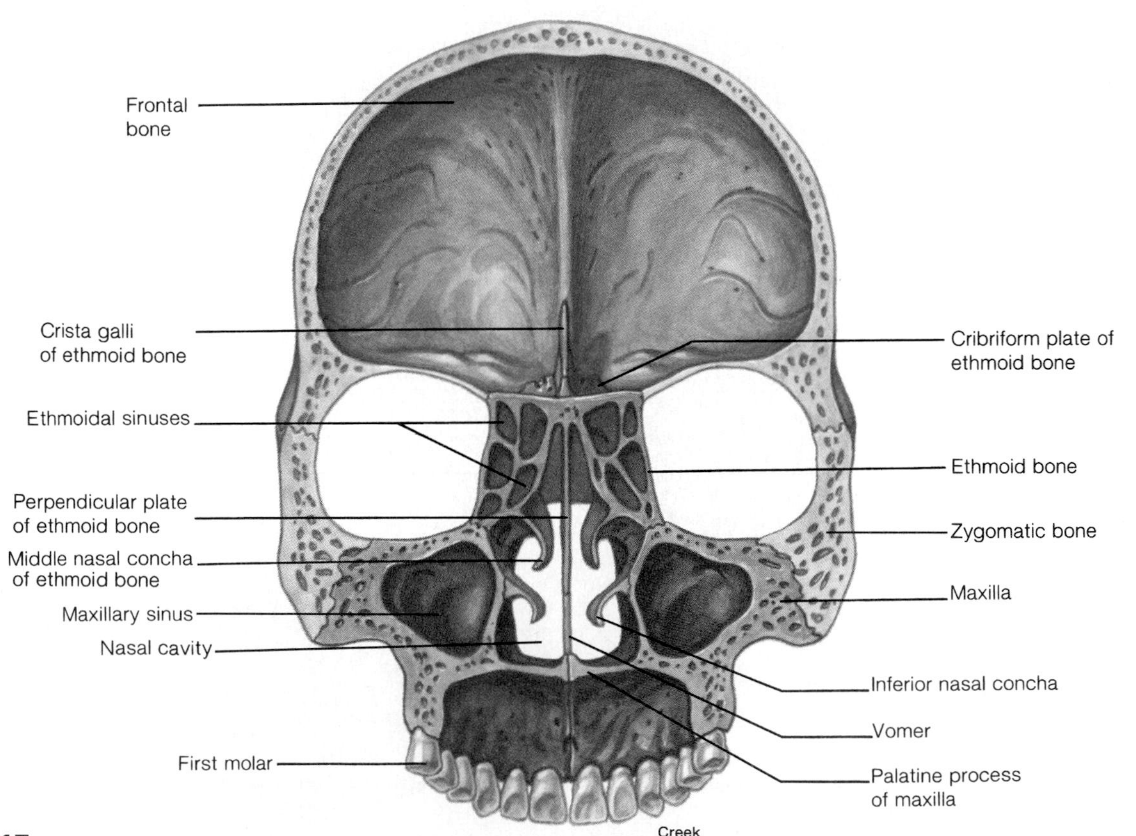

FIGURE 6.17
A posterior view of a frontal (coronal) section of the skull.

Ethmoid Bone

The ethmoid bone is located in the anterior portion of the floor of the cranium, between the orbits, where it forms the roof of the nasal cavity (figs. 6.14, 6.15, and 6.17). An inferior projection of the ethmoid bone, called the **perpendicular plate,** forms the upper part of the *nasal septum* that separates the nasal cavity into two chambers. A spinelike **crista galli** projects into the cranial cavity and provides an attachment site for the membranes covering the brain. At right angles to the perpendicular plate, within the floor of the cranium, is the **cribriform plate,** which has numerous small openings (fig. 6.22). These openings allow the olfactory nerves to enter the cranial cavity. On both lateral walls of the nasal cavity are two scroll-shaped plates of the ethmoid bone, the **superior** and **middle nasal conchae,** also known as **turbinates.**

ethmoid: Gk. *ethmos,* sieve
crista galli; L. *crista,* crest; *galli,* cock's comb
cribriform: L. *cribrum,* sieve; *forma,* like
conchae: L. *conchae,* shells

6.8 Facial Bones

The 14 facial bones shape the face, support the teeth, and provide attachment sites for the muscles that move the jaw and produce facial expression.

With the exceptions of the vomer and the mandible, all of the facial bones are paired. The articulated facial bones are illustrated in figures 6.14 through 6.22.

Maxilla

The two maxillae join at the midline to form the upper jaw. The upper teeth are contained in sockets within the maxilla (fig. 6.21). The **palatine process,** a horizontal plate of the maxilla, forms the front part of the *hard palate,* or roof of the mouth.

If the two palatine processes fail to join during prenatal development (at about 12 weeks), a cleft palate *results. A cleft palate may be accompanied by a* cleft lip *(harelip) lateral to the midline. These conditions can be surgically treated with excellent cosmetic results. An immediate problem, however, is that a baby with a cleft palate may have a difficult time nursing because it is unable to swallow effectively.*

Palatine Bones

The L-shaped palatine bones form the posterior third of the hard palate, a part of the orbits, and a part of the nasal cavity. The **horizontal plates** of the palatines contribute to the formation of the hard palate (fig. 6.21).

Zygomatic Bones

The two zygomatic bones form the cheekbones of the face. A posteriorly extending **temporal process** of this bone unites with the zygomatic process of the temporal bone (fig. 6.14) to form the **zygomatic arch.** The zygomatic bone also forms the lateral margin of the orbit.

Lacrimal Bones

The thin lacrimal bones form the anterior part of the medial wall of each orbit (figs. 6.14 and 6.16). These are the smallest of the facial bones. Each one has a **nasolacrimal canal** that permits tears to drain into the nasal cavity.

Nasal Bones

The small, rectangular nasal bones join at the midline to form the bridge of the nose (fig. 6.16). Fractures of the nasal bones or fragmentation of the associated cartilage are common facial injuries.

Inferior Nasal Conchae

The two inferior nasal conchae (*kong'ke*—singular, *concha*) are fragile, scroll-like bones that project horizontally and medially from the lateral walls of the nasal cavity (fig. 6.17). They extend into the nasal cavity just below the superior and middle nasal conchae, which are part of the ethmoid bone. The inferior nasal conchae are the largest of the three paired conchae, and, like the other two, are covered with a mucous membrane to warm, moisten, and cleanse inhaled air.

Vomer

The vomer is a thin, flattened bone that forms the lower part of the nasal septum (figs. 6.15 and 6.17). The vomer, along with the perpendicular plate of the ethmoid bone, forms the nasal septum.

Mandible

The mandible (fig. 6.18), or lower jawbone, is attached to the skull at the temporomandibular joint and is the only movable bone of the skull. The horseshoe-shaped front and sides of this bone are referred to as the **body.** Extending vertically from the posterior portion of the body are two **rami** (singular, *ramus*). Each ramus has a knoblike **condylar process,** which articulates with the mandibular fossa of the temporal bone, and a pointed **coronoid process** for muscle attachment. The depressed rim between these two processes is called the **mandibular notch.** The **angle of the mandible** is where the horizontal body and vertical ramus meet at the corner of the jaw. Like the maxillae, the mandible contains sockets for teeth.

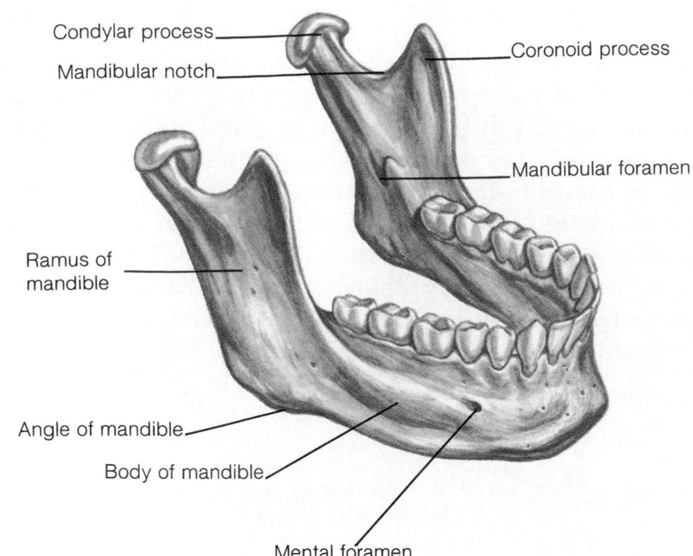

FIGURE 6.18
The mandible.

Hyoid Bone

The single hyoid bone is a unique part of the skeleton in that it does not attach directly to any other bone. It is located in the neck region, below the mandible, where it is suspended from the styloid process of the temporal bone by ligaments and muscles. You can feel this U-shaped bone by placing a thumb and a finger on either side of the upper neck under the lateral portions of the mandible and firmly squeezing medially.

The hyoid bone supports the tongue and provides attachment for some of its muscles. This bone is carefully examined in an autopsy when strangulation is suspected, since during strangulation it is frequently fractured.

6.9 Cavities and Sinuses of the Skull

The cavities of the skull surround and protect the brain and sensory organs. The sinuses provide structural support and reduce the weight of the skull.

The cavities and sinuses of the skull are formed from a complex arrangement of bones, all of which have already been described. A brief summary of these areas follows.

Cranial Cavity

The large cranial cavity (figs. 6.15, 6.17, and 6.22) supports the brain and is divided into an *anterior cranial fossa,* a *middle cranial fossa,* and a *posterior cranial fossa.*

Oral Cavity

Because of its fleshy and muscular floor, the *oral,* or *buccal, cavity* (commonly called the mouth) is a cavity of the head rather than a cavity of the skull. The oral cavity functions with the digestive system and with the respiratory system. The framework of the oral cavity is the hard palate and soft palate superiorly, the teeth and mandible anteriorly and laterally, and the fleshy, muscular floor inferiorly. The tongue is anchored to the floor of the oral cavity and to the hyoid bone.

Nasal Cavity

The nasal cavity (fig. 6.17) is part of the respiratory system. It is divided by the nasal septum into two lateral halves, each referred to as a *nasal fossa.* The nasal fossae open anteriorly through the *nostrils* (*external nares*) and posteriorly through the *choanae* (*internal nares*).

---

vomer: L. *vomer,* plowshare
mandible: L. *mandere,* to chew
ramus: L. *ramus,* branch
condylar: L. *concylus,* knucklelike

Orbital Cavities

Each of the two orbital cavities, or *bony orbits,* supports and protects an eyeball and its associated muscles, vessels, and nerves.

Middle-Ear Cavities

Three small paired bones, called **auditory ossicles,** are located within the middle-ear cavities (fig. 6.19) in the petrous part of the temporal bones (fig. 6.22). From outer to inner, these bones are the **malleus** (*mal'e-us*) (hammer), **incus** (*ing'kus*) (anvil), and **stapes** (*sta'pēz*) (stirrup). Their movements transmit sound impulses through the middle-ear cavity.

Inside Information—The tiny stapes of the middle ear is the smallest bone in the human body. Which is the largest?—see page 145.

Mastoid Sinuses

The mastoid process of each temporal bone (fig. 6.14) contains a number of small cavities, collectively called the *mastoid sinus.* There is a connection between the middle-ear chamber and the mastoid sinus. Infection of the mastoid sinus, called *mastoiditis,* may be difficult to treat with antibiotics because of the spongy bone of which the mastoid sinus is composed.

Paranasal Sinuses

Paired air spaces in certain bones of the skull are called *paranasal sinuses.* These cavities are named according to the bones in which they are found; thus, there are the **maxillary, frontal, sphenoidal,** and **ethmoidal sinuses** (figs. 6.15, 6.17, and 6.20). Each sinus drains through ducts into the nasal cavity. Paranasal sinuses may help to warm and moisten the inspired air. These sinuses are responsible for some sound resonance, but most importantly, they decrease the weight of the skull.

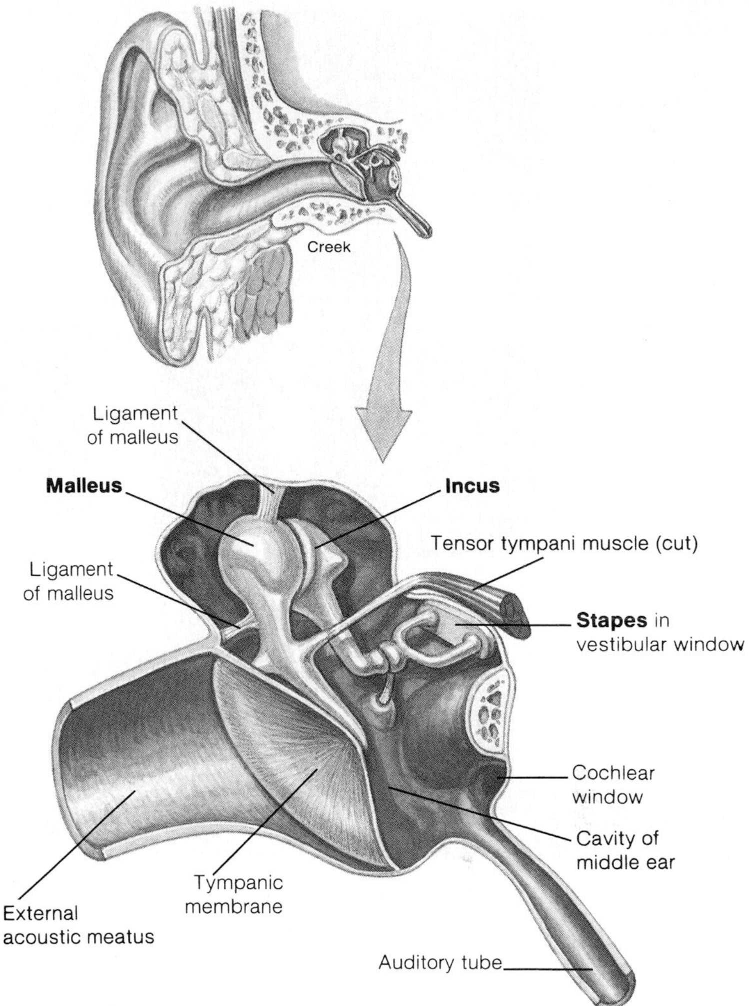

FIGURE 6.19
The three auditory ossicles (bones of the middle ear).

malleus: L. *malleus,* hammer
incus: L. *incus,* anvil
stapes: L. *stapes,* stirrup

The moist, warm vascular lining within the nasal cavity is susceptible to infections, particularly if a person is not in good health. Infections of the nasal cavity easily spread to surrounding spaces. The paranasal sinuses connect to the nasal cavity, and because microorganisms and allergies are often airborne, these sinuses frequently become infected. The eyes become reddened and swollen during a nasal infection because of the connection of the nasolacrimal duct, through which tears drain from the orbit to the nasal cavity. The paranasal ducts drain mucus from the paranasal sinuses. Blowing the nose too forcefully may force contaminated mucus back into the sinuses, causing them to become inflamed—a condition called sinusitis. *Organisms may spread via the auditory tube from the throat to the middle ear. If the nasal infection persists, organisms may even enter the cranial cavity through the cribriform plate and infect the meninges (membranes) covering the brain, causing* meningitis.

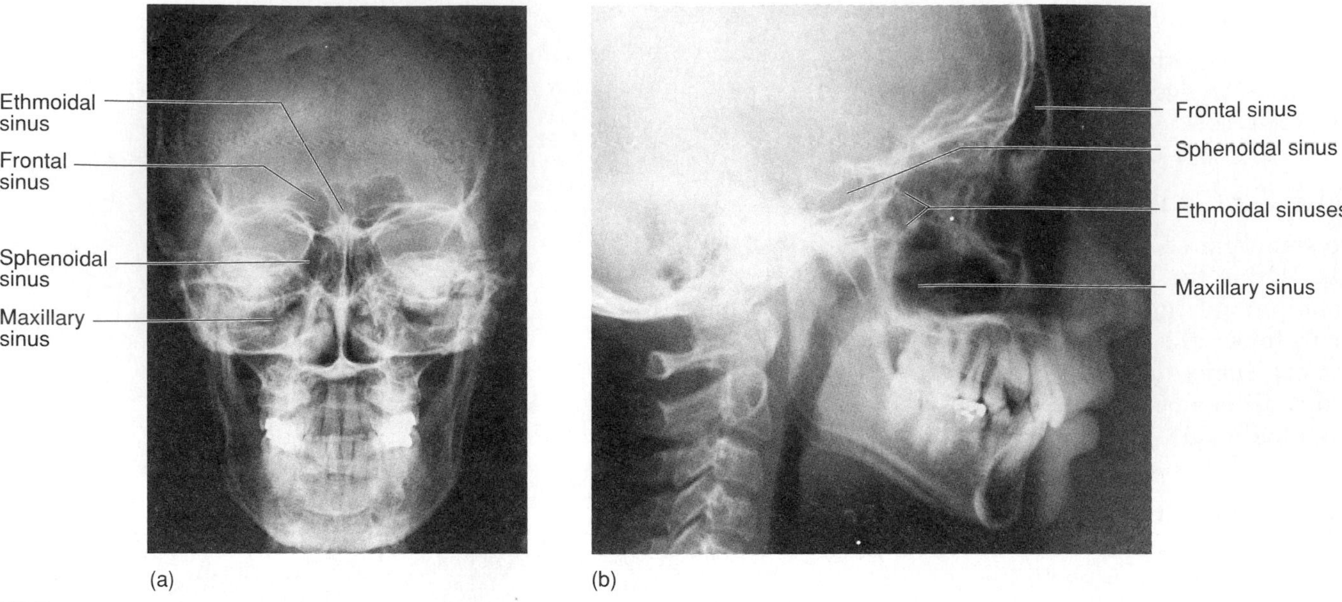

FIGURE 6.20
X rays of the skull showing the paranasal sinuses. (*a*) An anteroposterior view and (*b*) a right lateral view.

6.10 Foramina of the Skull

The foramina of the skull permit passage of nerves, vessels, and other structures to and from the brain and the sensory organs of the head. These small holes are important features in identifying the bones of the skull.

The cranial cavity is not a solid vault enclosing the brain; rather, it is pierced by small holes, called **foramina** (singular, *foramen*) (figs. 6.16, 6.21, and 6.22). Foramina permit nerves, vessels, and other structures to pass to and from the brain. The major foramina of the skull are listed in table 6.3.

The passage of cranial nerves (see chapter 10, module 10.8) or their branches accounts for many of the foramina of the skull. Vessels transporting blood to and from the brain pass through other foramina. The **foramen magnum** (figs. 6.21 and 6.22) is the largest opening in the skull and accommodates the spinal cord.

Dentists use bony landmarks of the facial region to find nerves so that anesthetics can be injected. For example, the trigeminal nerve is composed of three smaller nerves, the lower two of which send sensations from the teeth, gums, and jaws. The teeth in the lower jaw can be desensitized by an injection near the mandibular foramen. An injection near the foramen rotundum of the skull numbs the teeth on one side of the upper jaw.

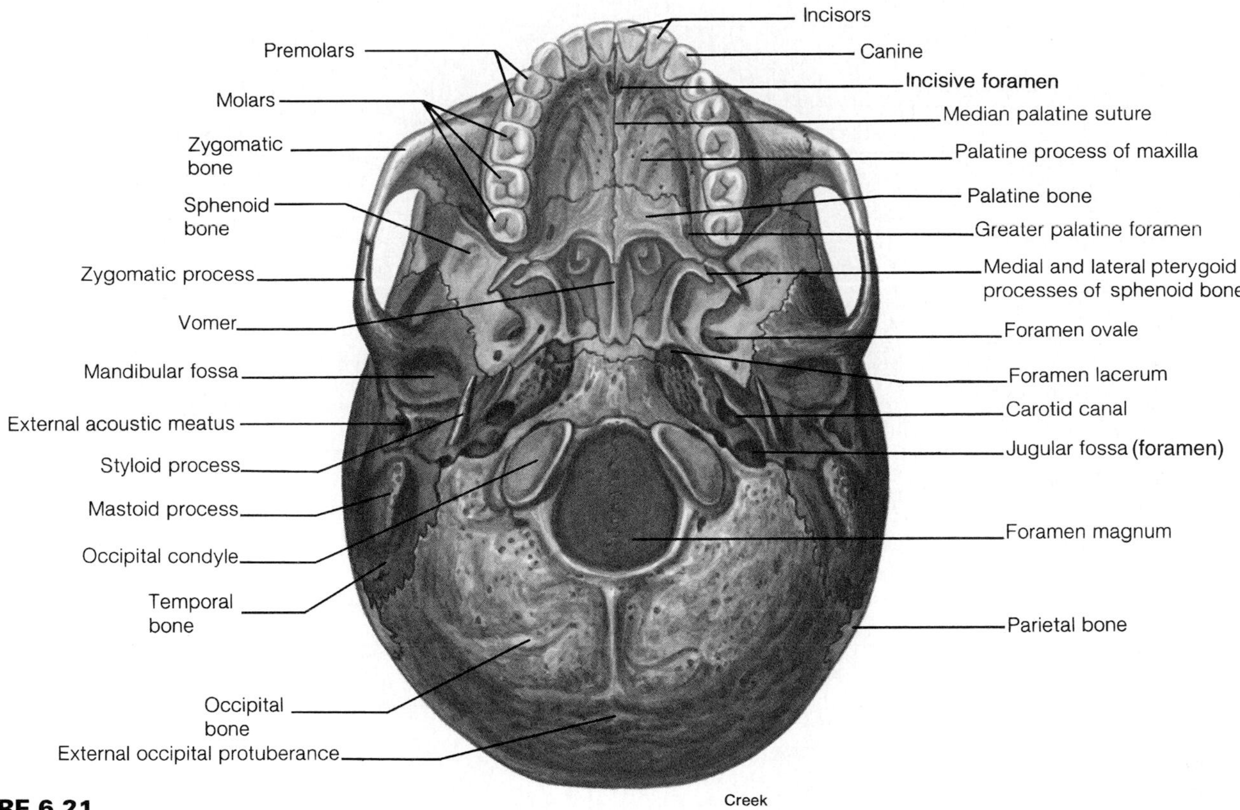

FIGURE 6.21
An inferior view of the skull.

TABLE 6.3 Summary of Major Foramina of the Skull

Foramen	Location	Structures Transmitted
Carotid canal	Petrous part of temporal bone	Internal carotid artery and sympathetic nerves
Greater palatine	Palatine bone of hard palate	Greater palatine nerve and blood vessels
Hypoglossal foramen/canal	Anterolateral edge of occipital condyle	Hypoglossal nerve
Incisive	Anterior region of hard palate, posterior to incisors	Branches of palatine vessels and nerve
Inferior orbital fissure	Between maxilla and greater wing of sphenoid bone	Maxillary nerve of trigeminal cranial nerve, zygomatic nerve, and blood vessels
Infraorbital	Inferior to orbit in maxilla	Infraorbital nerve and artery
Jugular	Between petrous part of temporal and occipital bones, posterior to carotid canal	Internal jugular vein; vagus, glossopharyngeal, and accessory nerves
Lacerum	Between petrous portion of temporal and sphenoid bones	Internal carotid artery
Magnum	Occipital bone	Union of brain and spinal cord, meningeal membranes, and accessory nerves; blood vessels
Mandibular	Medial surface of ramus of mandible	Alveolar nerve and vessels
Mental	Below second premolar on lateral side of mandible	Mental nerve and vessels
Nasolacrimal canal	Lacrimal bone	Nasolacrimal (tear) duct
Olfactory	Cribriform plate of ethmoid bone	Olfactory nerves
Optic canal	Back of orbit in lesser wing of sphenoid bone	Optic nerve
Ovale	Greater wing of sphenoid bone	Mandibular nerve of trigeminal cranial nerve
Rotundum	Within body of sphenoid bone	Maxillary nerve of trigeminal cranial nerve
Spinosum	Posterior angle of sphenoid bone	Meningeal vessels
Superior orbital fissure	Between greater and lesser wings of sphenoid bone	Four cranial nerves (oculomotor, trochlear, ophthalmic branch of trigeminal, and abducens)
Supraorbital	Supraorbital ridge of orbit	Supraorbital nerve and artery

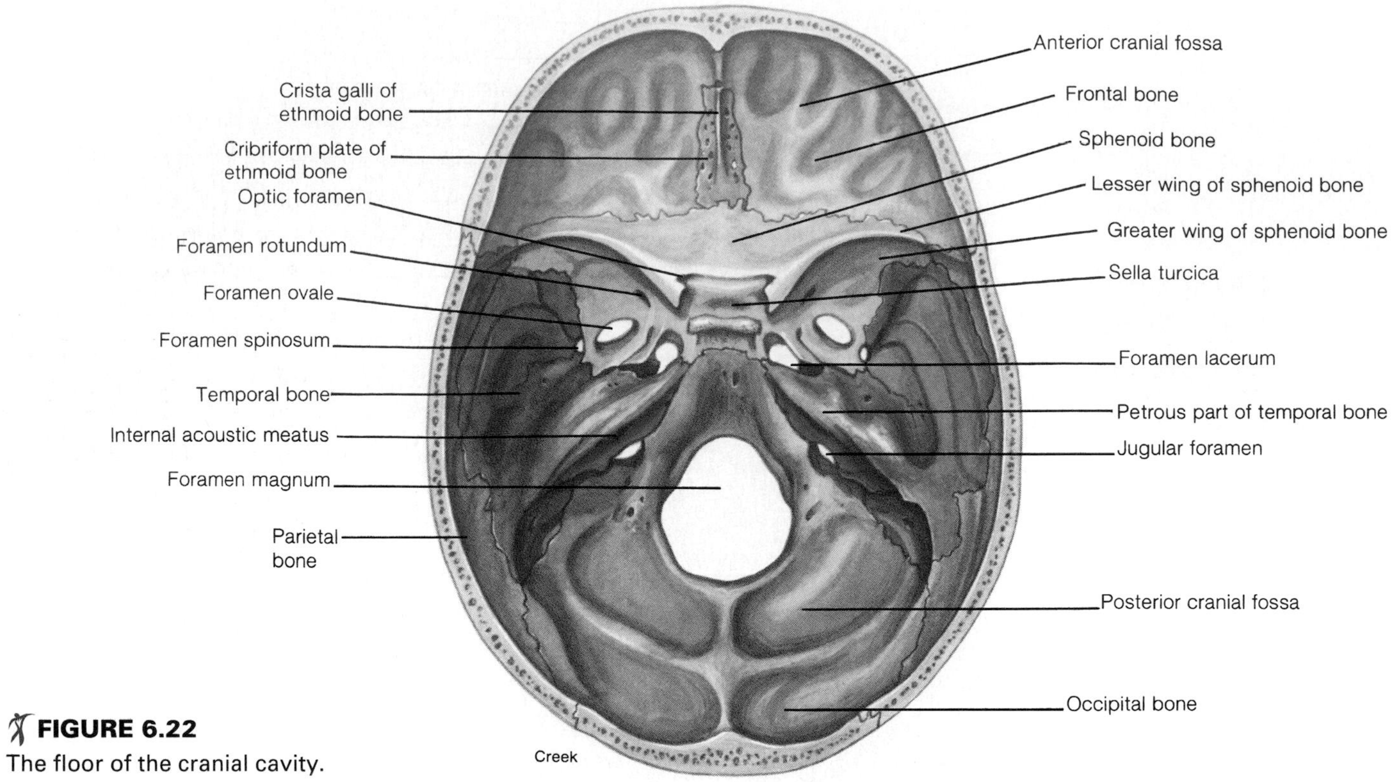

FIGURE 6.22
The floor of the cranial cavity.

6.11 Vertebral Column

The supporting vertebral column consists of a series of bones called vertebrae. Between the bodies of adjacent vertebrae are intervertebral discs that lend flexibility and absorb the stress of movement. Vertebrae enclose and protect the spinal cord, support and permit movement of the skull, articulate with the rib cage, and provide for the attachment of trunk muscles.

The **vertebral column** (backbone) and the *spinal cord* of the nervous system together constitute the *spinal column.* The vertebral column has three functions:

1. to support the head and upper limbs while permitting freedom of movement;
2. to provide attachment for various muscles, ribs, and visceral organs; and
3. to protect the spinal cord and permit passage of the spinal nerves.

The vertebral column is composed of 33 individual vertebrae, some of which are fused. There are seven **cervical,** twelve **thoracic,** five **lumbar,** five fused (in adults) **sacral,** and three to five fused **coccygeal** (*kok-sij'e-al*) vertebrae; thus, the adult vertebral column is composed of a total of 26 movable parts. Vertebrae are separated by fibrocartilaginous intervertebral discs and are secured to one another by interlocking processes and binding ligaments. This structural arrangement provides only limited movement between vertebrae but extensive movement for the entire vertebral column. Between the vertebrae are openings called **intervertebral foramina** that allow passage of spinal nerves.

When viewed from the side, four curvatures of the vertebral column can be identified (fig. 6.23). The **cervical, thoracic,** and **lumbar curves** are identified by the type of vertebrae they include. The **pelvic curve (sacral curve)** is formed by the shape of the sacrum and coccyx (*kok'siks*). The curves of the vertebral column make possible our bipedal stance.

The four vertebral curves are not present in an infant. The cervical curve begins to develop at about 3 months as the baby begins holding up its head, and it becomes more pronounced as the baby learns to sit up. The lumbar curve develops as a child begins to walk. The thoracic and pelvic curves are called *primary curves* because they retain the shape of the fetus. The cervical and lumbar curves are called *secondary curves* because they are modifications of the fetal shape.

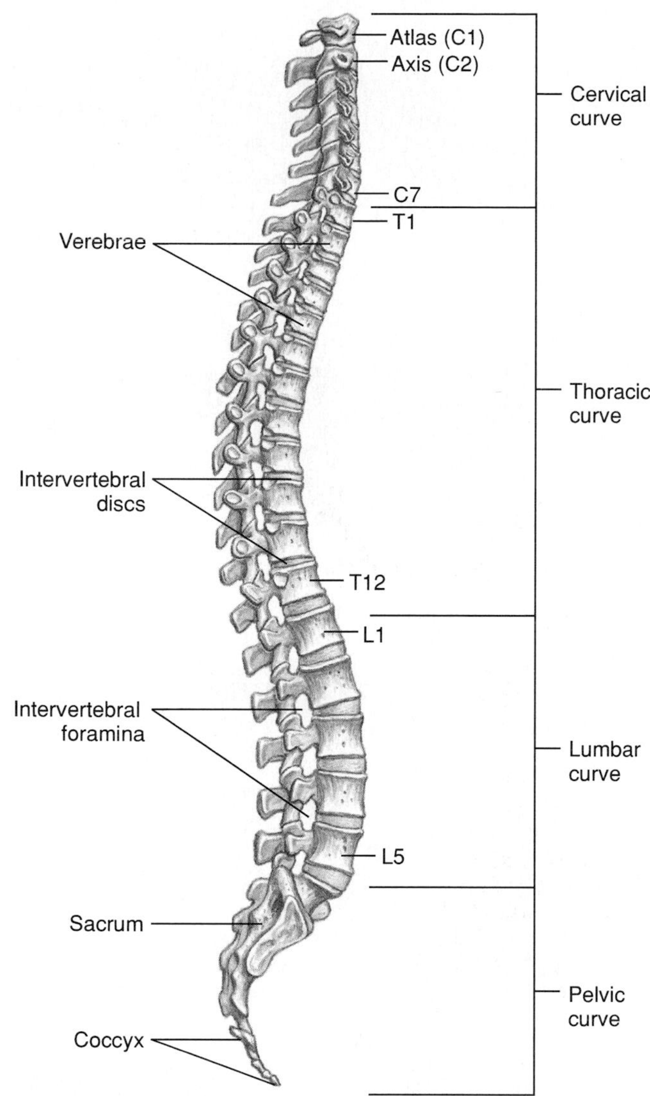

FIGURE 6.23
The vertebral column of an adult has four curves named according to the region in which they occur. The vertebrae are separated by intervertebral discs, which allow flexibility.

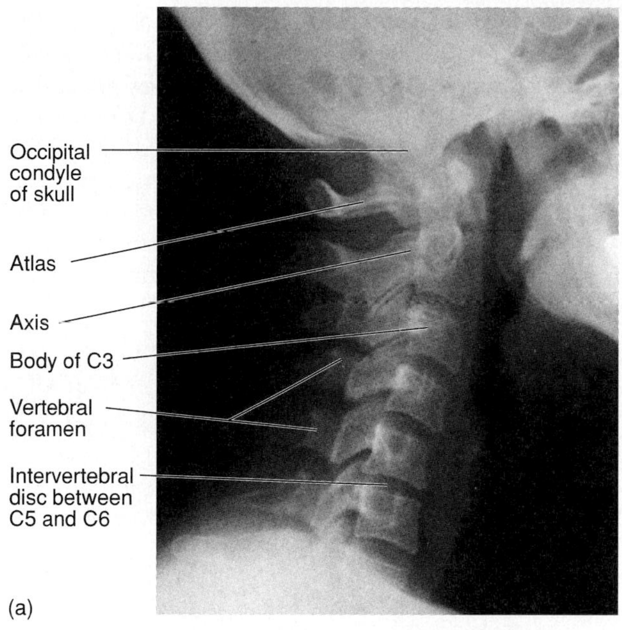

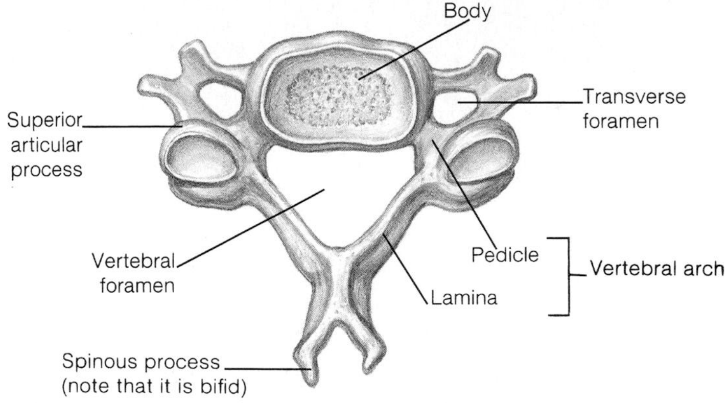

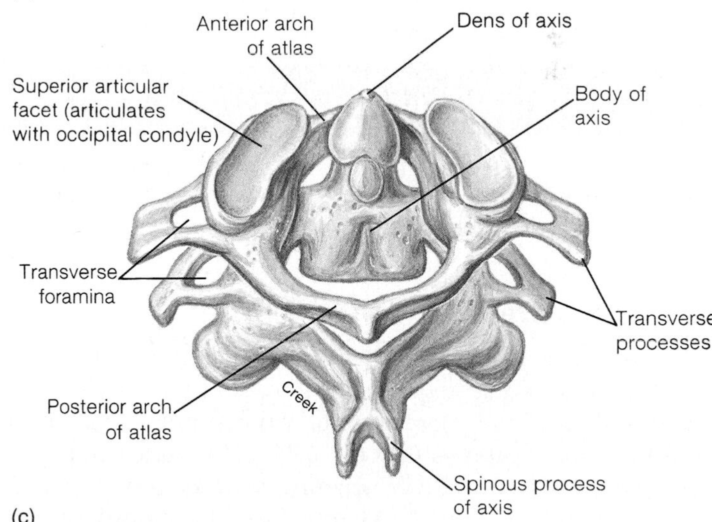

FIGURE 6.24
Cervical vertebrae. (*a*) An X ray of the cervical region, (*b*) a superior view of a typical cervical vertebra, and (*c*) the atlas and axis as they articulate.

General Structure of Vertebrae

Vertebrae are similar in their general structure from one region to another. A vertebra consists of a drum-shaped **body** (figs. 6.24 and 6.25), which is in contact with **intervertebral discs** above and below. The **vertebral arch** is attached to the posterior surface of the body and is composed of two supporting **pedicles** and two arched **laminae.** The space formed by the vertebral arch and body is the **vertebral foramen,** through which the spinal cord passes. Between the pedicles of adjacent vertebrae are the **intervertebral foramina** (see fig. 6.23), through which spinal nerves emerge as they branch from the spinal cord.

pedicle: L. *pediculus,* small foot
lamina: L. *lamina,* thin layer

Seven processes rise from the vertebral arch: the **spinous process,** two **transverse processes,** two **superior articular processes,** and two **inferior articular processes** (fig. 6.25). The first two processes serve for muscle attachment, and the last two pairs limit twisting of the vertebral column. The spinous process extends posteriorly and inferiorly. The transverse process extends laterally from each side of a vertebra. The superior articular processes of a vertebra interlock with the inferior articular processes of the bone above.

A laminectomy *is the surgical removal of the spinous processes and their supporting vertebral laminae in a particular region of the vertebral column. A laminectomy may be performed to relieve pressure on the spinal cord caused by a blood clot, a tumor, or a herniated (ruptured) disc.*

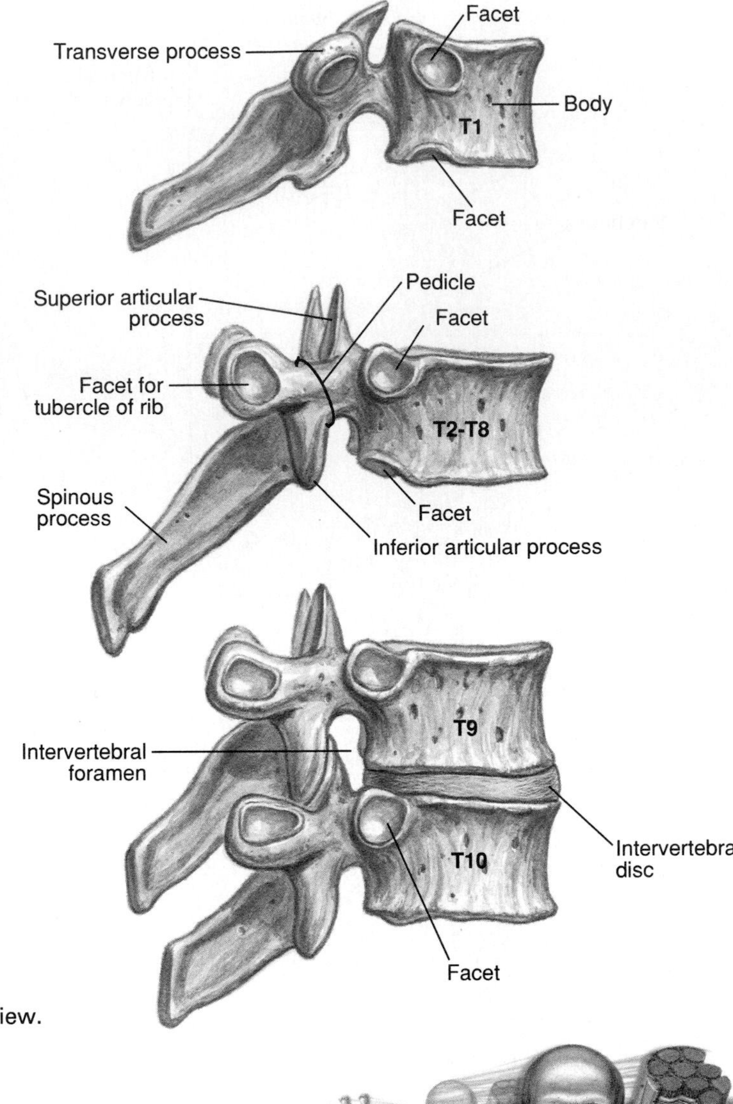

FIGURE 6.25
Thoracic vertebrae in a lateral view.

6.12 Vertebral Regions

Vertebrae within each region of the vertebral column have structural features that are adaptive for particular functions.

Cervical Vertebrae

The seven cervical vertebrae form a flexible framework for the neck and support the head. Cervical vertebrae are distinguished by the presence of a **transverse foramen** in the transverse process (see fig. 6.24). The vertebral blood vessels pass through this opening as they travel to and from the brain. Cervical vertebrae C2–C6 generally have a **bifid,** or notched, spinous process. The first cervical vertebra has no spinous process, and the process of C7 is not bifid and is larger than those of the other vertebrae.

The **atlas** is the first cervical vertebra (sometimes called cervical 1 or C1). It has cupped **superior articular surfaces** that articulate with the oval occipital condyles of the skull. This joint supports the skull and permits the nodding of the head in a "yes" movement.

The **axis** is the second cervical vertebra (C2). It has a peglike **dens** (*odontoid process*) for rotation with the atlas in turning the head from side to side, as in a "no" movement.

Whiplash *is a common term for any injury to the neck. Muscle, bone, or ligament injury in this portion of the spinal column is relatively common in individuals involved in automobile accidents and sports injuries. Joint dislocation occurs commonly between the fourth and fifth or fifth and sixth cervical vertebrae, where neck movement is greatest.*

atlas: from Gk. mythology, Atlas—the Titan who supported the heavens
axis: L. *axis,* axle
odontoid: Gk. *odontos,* tooth

Thoracic Vertebrae

Twelve thoracic vertebrae articulate with the ribs to form the posterior anchor of the rib cage. Thoracic vertebrae are larger than the cervical and increase in size from superior (T1) to inferior (T12). Each thoracic vertebra has a long spinous process, which slopes downward, and **facets** for articulation with the ribs (fig. 6.25).

Lumbar Vertebrae

The five lumbar vertebrae are easily identified by their heavy bodies and thick, blunt spinous process (fig. 6.26) for attachment of powerful back muscles. They are the largest vertebrae of the vertebral column.

Sacrum

The wedge-shaped sacrum (fig. 6.27) provides a strong foundation for the pelvic girdle. It consists of four or five sacral vertebrae that become fused after age 26. The sacrum has an extensive **auricular surface** on each side. This surface forms a junction, called the **sacroiliac joint,** with the hip. A **median sacral crest** is formed along the posterior surface where the spinous processes fuse. **Posterior sacral foramina** on both sides of the crest allow the passage of nerves from the spinal cord. The *sacral canal* is continuous with the vertebral canal. Paired **superior articular processes** join with the fifth lumbar vertebra.

The smooth anterior surface of the sacrum forms the posterior wall of the pelvic cavity. It is characterized by four **transverse lines,** indicating the fusion of the vertebral bodies, and four paired **pelvic foramina,** which allow passage of pelvic nerves. The superior border of the anterior surface of the sacrum, called the **sacral promontory,** is an important obstetric landmark for pelvic measurements.

Coccyx

The triangular coccyx, our vestigial tailbone, is composed of three to five (most often four) fused coccygeal vertebrae (fig. 6.27).

The vertebral regions are summarized in table 6.4.

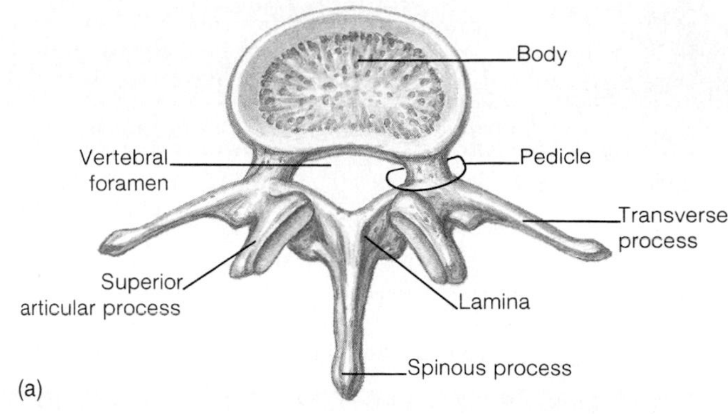

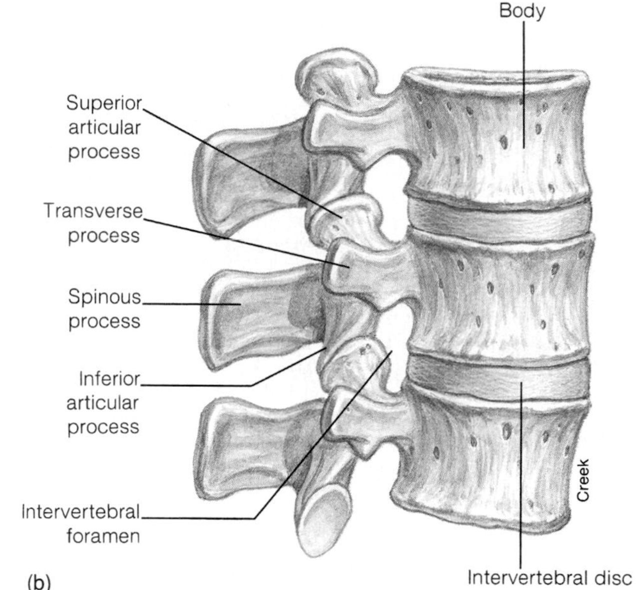

FIGURE 6.26
Lumbar vertebrae. (*a*) A superior view and (*b*) a lateral view.

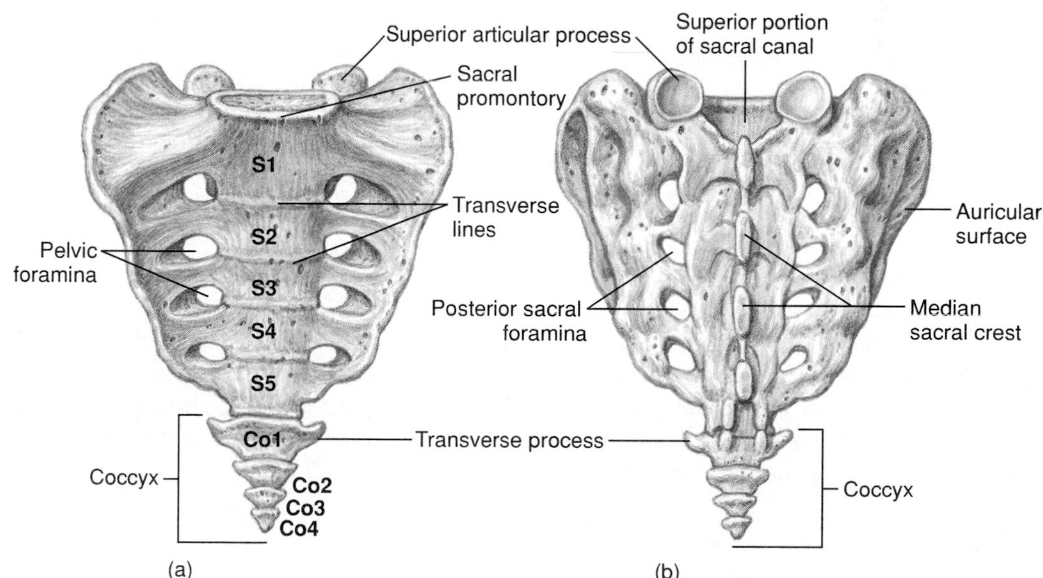

FIGURE 6.27
The sacrum and coccyx. (*a*) An anterior view and (*b*) a posterior view.

lumbar: L. *lumbus,* loin
sacrum: L. *sacris,* sacred
coccyx: Gk. *kokkyx,* like a cuckoo's beak

TABLE 6.4 Regions of the Vertebral Column

Region	Number of Bones	Diagnostic Features
Cervical	7	Transverse foramina; superior facets of atlas articulate with occipital condyle; dens of axis; spinous processes of third through sixth vertebrae are bifid
Thoracic	12	Long spinous processes; facets for articulation with ribs
Lumbar	5	Large bodies; prominent transverse processes; short, thick spinous processes
Sacrum	5 fused vertebrae	Extensive auricular surface; median sacral crest; posterior sacral foramina; sacral promontory; sacral canal
Coccyx	3 to 5 fused vertebrae	Small and triangular

6.13 Rib Cage

The cone-shaped, flexible rib cage consists of 12 thoracic vertebrae, 12 paired ribs, costal cartilages, and the sternum. It encloses and protects the thoracic viscera and is involved in breathing.

The **sternum, ribs, costal cartilages,** and the previously described thoracic vertebrae form the **rib cage** (fig. 6.28). The rib cage is anteroposteriorly compressed and narrower superiorly than inferiorly. It supports the pectoral girdle and upper limbs, protects and supports the thoracic and upper abdominal viscera, and plays a major role in breathing. In addition, the sternum of the rib cage contains red bone marrow for blood cell production.

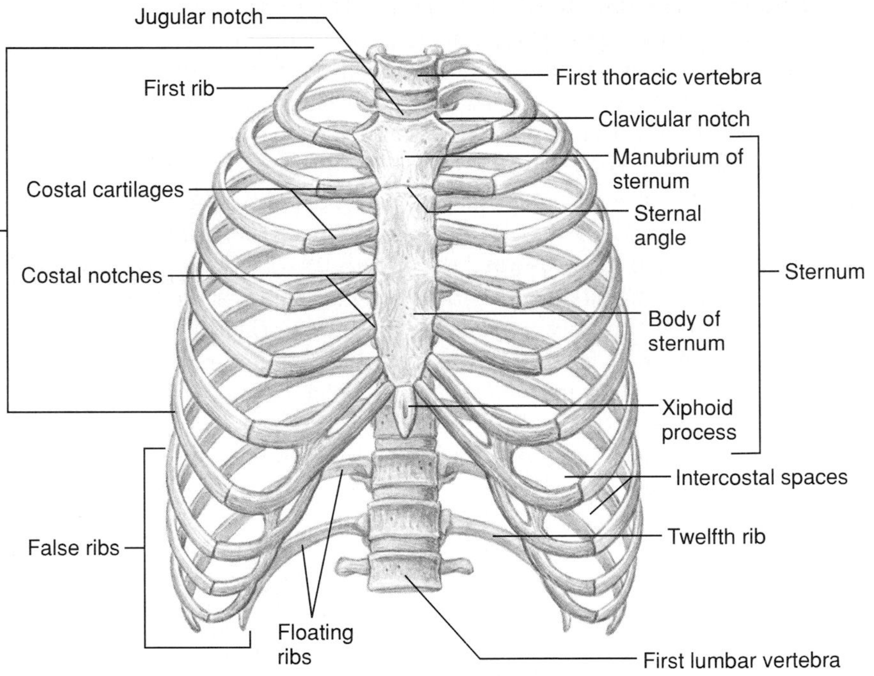

Jugular notch
First rib
First thoracic vertebra
Clavicular notch
Manubrium of sternum
Costal cartilages
Sternal angle
True ribs
Costal notches
Sternum
Body of sternum
Xiphoid process
Intercostal spaces
Twelfth rib
False ribs
Floating ribs
First lumbar vertebra

FIGURE 6.28
The rib cage.

Sternum

The sternum (breastbone) is an elongated, flattened bony plate consisting of three separate bones: the upper **manubrium,** the central **body,** and the lower **xiphoid** (*zif'oid; zi'foid*) **process.** On the lateral sides of the sternum are *costal notches,* where the **costal cartilages** attach. A **jugular notch** is formed at the superior end of the manubrium, and a **clavicular notch** for articulation with the clavicle is present on both sides of the sternal notch. The manubrium articulates with the costal cartilages of the first and second ribs. The body of the sternum attaches to the costal cartilages of the second through the seventh ribs. The xiphoid process does not attach to ribs but is an attachment for abdominal muscles.

Ribs

There are twelve pairs of ribs, each pair being attached posteriorly to a thoracic vertebra. Anteriorly, the first seven pairs are anchored to the sternum by individual

sternum: Gk. *sternon,* chest
manubrium: L. *manubrium,* a handle
xiphoid: Gk. *xiphos,* sword
costal: L. *costa,* rib

costal cartilages; these ribs are called **true ribs.** The remaining five pairs (ribs 8, 9, 10, 11, and 12) are termed **false ribs.** Because the last two pairs of false ribs do not attach to the sternum, they are referred to as **floating ribs.** The four floating ribs are embedded in the muscles of the body wall.

Although the ribs vary structurally, each of the first ten pairs has a **head** and a knoblike **tubercle** for articulation with a vertebra. The last two have a head but no tubercle. In addition, each of the twelve pairs has a **neck, angle,** and **shaft** (fig. 6.29). The head of a rib projects posteriorly and articulates with the body of a thoracic vertebra. The tubercle is a process just lateral to the head. It articulates with the facet of the transverse process. The neck is the narrow area between the head and the tubercle. The **shaft,** or **body,** is the curved main part of the rib. Along the inner surface of the shaft is a depressed canal called the **costal groove** that protects the costal vessels and nerve. Spaces between the ribs, called **intercostal spaces,** are occupied by the intercostal muscles.

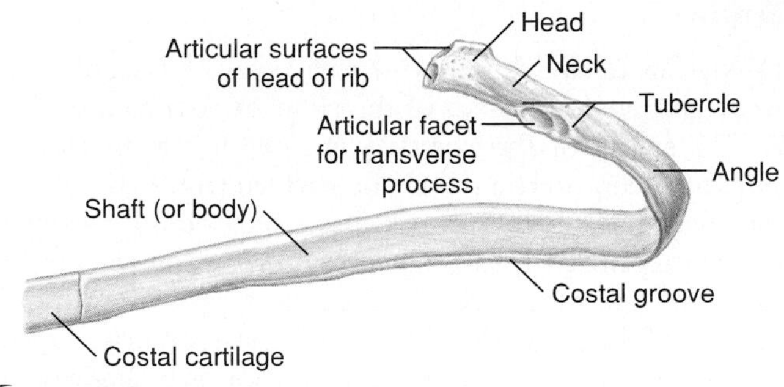

FIGURE 6.29
The structure of a rib.

Fractures of the ribs are relatively common, and most frequently occur between ribs 3 and 10. The first two pairs of ribs are protected by the clavicles; the last two pairs move freely and will give on impact. Little can be done to assist the healing of broken ribs other than binding them tightly to limit movement.

6.14 Pectoral Girdle

The structure of the pectoral girdle lends support to the upper limbs, while permitting freedom of movement. It also provides extensive surfaces for attachments of muscles.

The two scapulae and two clavicles make up the **pectoral** (shoulder) **girdle.** It is not a complete girdle, having only an anterior attachment to the axial skeleton at the sternum. The primary function of the pectoral girdle is to provide attachment for the numerous muscles that move the upper limb. The pectoral girdle is not weight bearing, and is therefore more delicate than the pelvic girdle.

Clavicle

The slender S-shaped clavicle (collarbone) binds the shoulder to the axial skeleton, at the same time allowing freedom of movement. The articulation of the **sternal extremity** (fig. 6.30) of the clavicle to the manubrium is referred to as the *sternoclavicular joint.* The **acromial extremity** of the clavicle articulates with the acromion of the scapula (fig. 6.31). This articulation is referred to as the *acromioclavicular joint.* A **conoid tubercle** is present on the inferior surface of the acromial extremity, and a **costal tuberosity** is present on the inner surface of the sternal extremity. Both processes serve as attachment for ligaments.

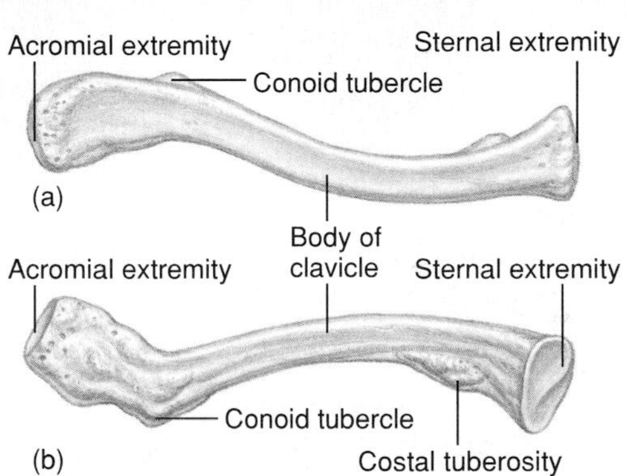

FIGURE 6.30
The right clavicle. (*a*) A superior view and (*b*) an inferior view.

The long, delicate clavicle is the most commonly broken bone in the body. When a person receives a blow to the shoulder, or attempts to break a fall with an outstretched hand, the force is placed on the clavicle, which may cause it to fracture. The most vulnerable part of this bone is through its center, to the side of the conoid tubercle. Because the clavicle is directly beneath the skin and not covered with muscle, a fracture can easily be palpated, and frequently seen.

---

clavicle: L. *clavicula,* a small key
acromial: Gk. *akros,* peak; *omos,* shoulder
conoid tubercle: Gk. *konos,* cone; L. *tuberculum,* a small swelling
costal tuberosity: L. *costa,* rib; *tuberosus,* a knob

Scapula

The scapula (shoulder blade) is a large, triangular flat bone on the posterior side of the rib cage, overlying ribs 2 to 7. The **spine** of the scapula is a prominent diagonal bony ridge seen on the posterior surface (fig. 6.32). Above the spine is the **supraspinous fossa,** and below the spine is the **infraspinous fossa.** The spine broadens toward the shoulder as the **acromion.** This process serves for the attachment of several muscles, as well as for articulation with the clavicle. Inferior to the acromion is a shallow depression, the **glenoid cavity,** into which the head of the humerus fits. The **coracoid process** is a thick upward projection that lies superior and anterior to the glenoid cavity. On the anterior surface of the scapula is a slightly concave area known as the **subscapular fossa.**

The scapula has three borders separated by three angles. The superior edge is called the **superior border.** The **medial border** is nearest the vertebral column. The **lateral border** is directed toward the arm. The **superior angle** is located at the junction of the superior and medial borders. The **inferior angle** is located at the junction of the medial and lateral borders. The **lateral angle** is located at the junction of the superior and lateral borders. It is at the lateral angle that the scapula articulates with the head of the humerus. Along the superior border is a distinct depression called the **scapular notch.**

The scapula has numerous surface features because 15 muscles attach to it. Clinically, the pectoral girdle is significant because it includes the bones most frequently broken while trying to break a fall. The acromion is used as a landmark for identifying the site for an injection in the arm.

Because the socket of the shoulder joint (the glenoid cavity) is shallow and only weakly reinforced by ligaments, dislocations of the shoulder are relatively common. In contact sports, for example, this type of injury occurs frequently. Sudden jerks to the arm can also dislocate the shoulder, especially in children, whose muscles are immature.

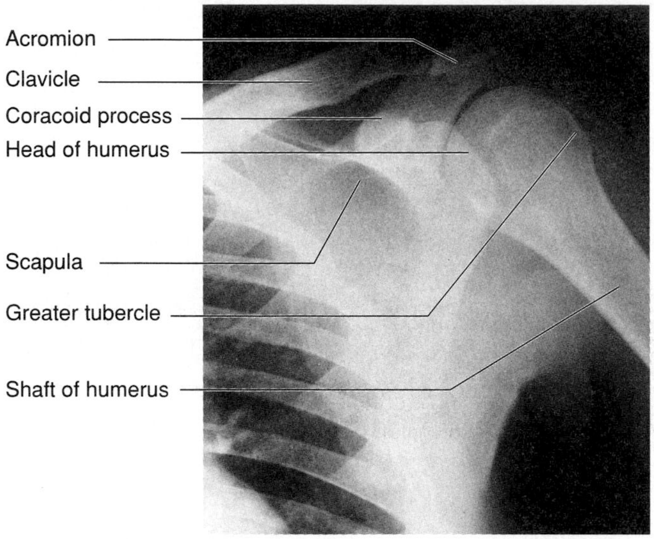

Acromion
Clavicle
Coracoid process
Head of humerus
Scapula
Greater tubercle
Shaft of humerus

FIGURE 6.31
This X ray of the right shoulder shows the positions of the clavicle, scapula, and humerus.

scapula: L. *scapula,* shoulder
glenoid: Gk. *glenoeides,* shallow form
coracoid process: Gk. *korakodes,* crow's beak

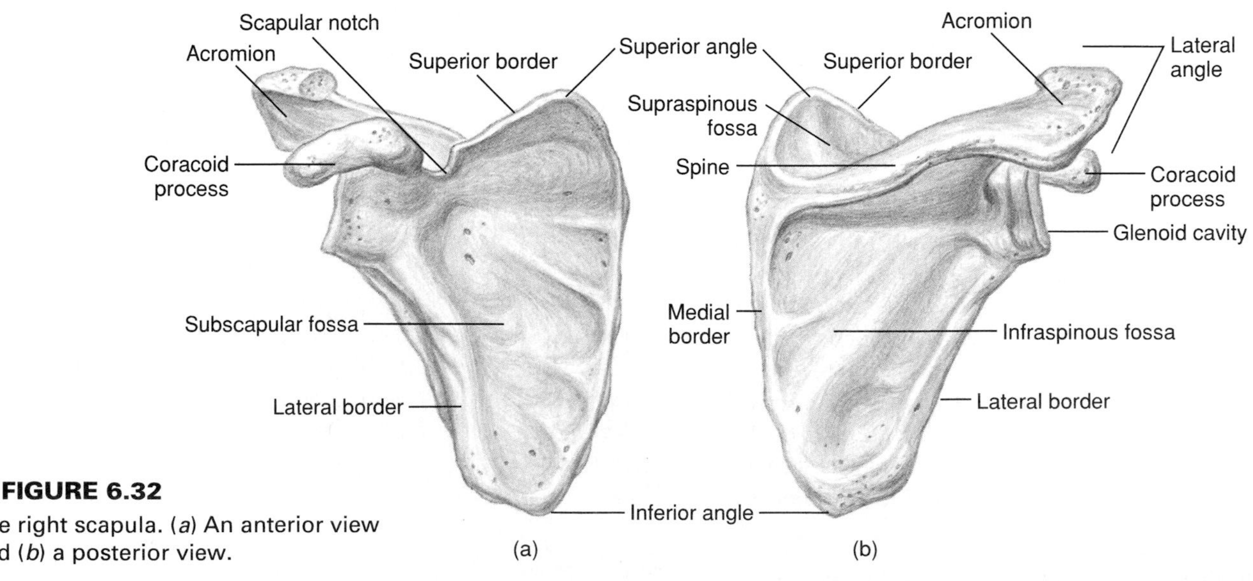

FIGURE 6.32
The right scapula. (*a*) An anterior view and (*b*) a posterior view.

6.15 Upper Limb

The upper limb consists of the arm, forearm, and hand. The bones of the upper limb serve as sites for muscle attachment and permit a wide range of movement, especially in the joints of the hand.

Arm (Brachium)

Technically speaking, the arm (brachium) refers only to that portion of the upper limb between the shoulder and elbow. The arm contains a single bone, the humerus.

Humerus

The humerus (fig. 6.33) is the longest bone of the upper limb. It consists of a proximal **head,** which articulates with the glenoid cavity of the scapula; a **shaft;** and a distal end. The **surgical neck,** the constriction just below the head, is a frequent site of fractures. The **greater tubercle** is a large knob on the lateral upper portion of the humerus. The **lesser tubercle** is slightly anterior to the greater tubercle and is separated from the greater by an **intertubercular groove.** The tendon of the biceps brachii muscle passes through this groove. Along the middle of the shaft is a roughened area, the **deltoid tuberosity,** for the attachment of the deltoid muscle. Small openings in the shaft are called **nutrient foramina.**

On the distal end of the humerus are two rounded surfaces. The **capitulum** is the lateral rounded condyle that articulates with the radius. The **trochlea** is the pulleylike medial surface that articulates with the ulna. On either side above the condyles are the **lateral** and **medial epicondyles.** The **coronoid fossa** is a depression above the trochlea on the anterior surface. The **olecranon fossa** is a depression on the distal posterior surface. Both fossae are adapted to work with the ulna during movement of the forearm.

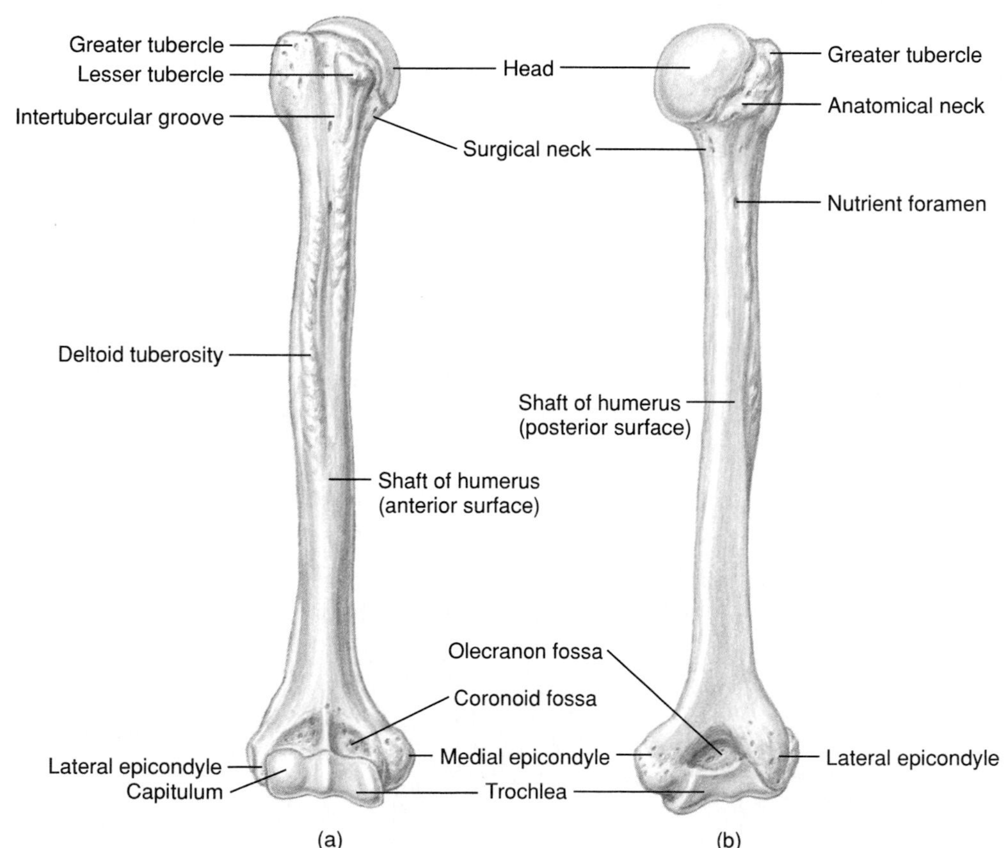

FIGURE 6.33
The right humerus. (*a*) An anterior view and (*b*) a posterior view.

Forearm (Antebrachium)

The forearm (antebrachium) contains the ulna on the medial side and radius on the lateral side (fig. 6.34).

Ulna

The proximal end of the ulna articulates with the humerus and radius. A distinct depression, the **trochlear notch,** articulates with the trochlea of the humerus. The **coronoid process** forms the anterior lip of the trochlear notch. The **olecranon** (*o-lek'ră-non*) forms the posterior portion, or point of the elbow. Lateral and inferior to the coronoid process is the **radial notch,** which articulates with the radius.

deltoid tuberosity: Gk. *deltoeides,* shaped like the letter Δ
capitulum: L. *caput,* little head
trochlea: Gk. *trochilia,* a pulley
olecranon: Gk. *olene,* ulna; *kranion,* head

On the tapered distal end of the ulna is a knobbed portion, the **head,** and a knoblike projection, the **styloid process.** The ulna articulates at both ends with the radius.

Radius

The radius consists of a **shaft** with a small proximal end and a large distal end. A proximal disc-shaped **head** articulates with the capitulum of the humerus and the radial notch of the ulna. The prominent **tuberosity of radius** for attachment of the biceps brachii muscle is on the medial side of the shaft, just below the head. On the distal end of the radius is a **styloid process** on the lateral tip and the **ulnar notch** on the medial side for receiving the distal end of the ulna.

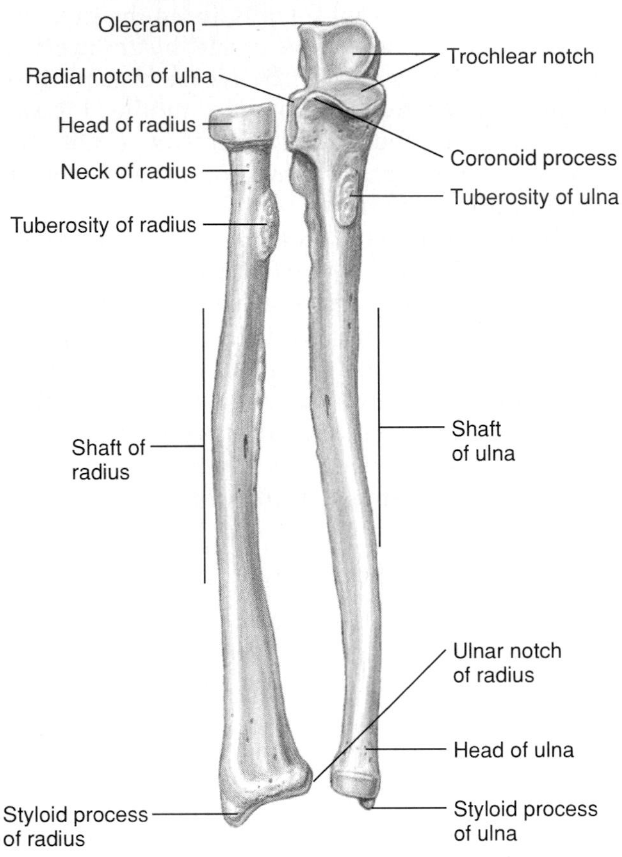

FIGURE 6.34
An anterior view of the right radius and ulna.

Hand

Each hand contains 27 bones grouped into the carpus, metacarpus, and phalanges (fig. 6.35).

Carpus

The carpus, or wrist, consists of eight carpal bones arranged in two rows of four bones each. The proximal row, naming from the lateral (thumb) to medial side, consists of the **scaphoid, lunate, triquetral,** and **pisiform.** The distal row, from lateral to medial, consists of the **trapezium, trapezoid, capitate,** and **hamate.** The scaphoid and lunate of the proximal row articulate with the distal end of the radius.

---
styloid process: Gk. *stylos,* pillar; *eidos,* resemblance
carpus: Gk. *karpos,* wrist
scaphoid: Gk. *skaphe,* boat; *eidos,* resemblance
lunate: L. *lunare,* crescent or moon-shaped
triquetral: L. *trequetrus,* three-cornered
pisiform: Gk. *pisos,* pea
trapezium: Gk. *trapesion,* small table
capitate: L. *capitatus,* head
hamate: L. *hamatus,* hook

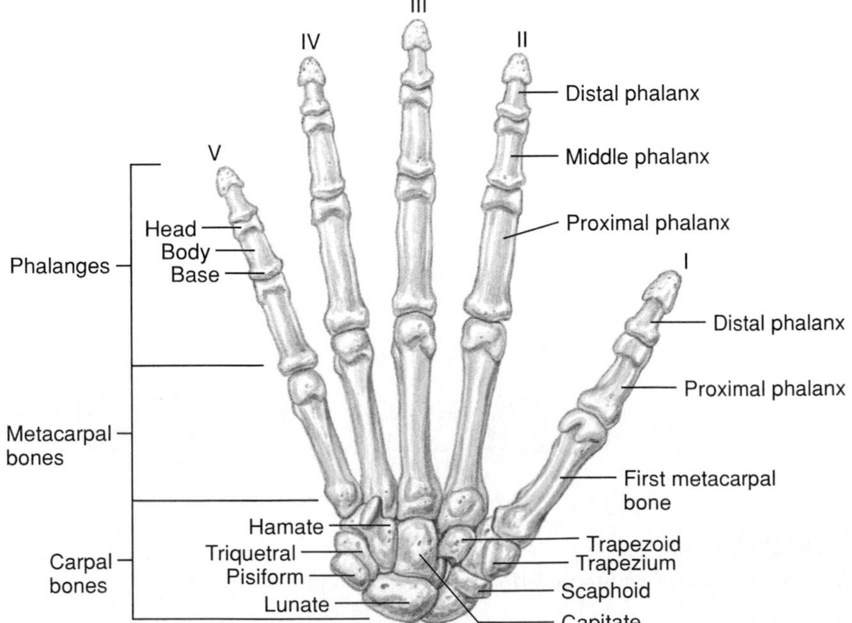

FIGURE 6.35
An anterior view of the bones of the right hand. Each digit (finger) is indicated by a Roman numeral, the first digit, or thumb, being Roman numeral I.

Metacarpus

The metacarpus, or palm of the hand, contains five metacarpal bones. Each metacarpal bone consists of a proximal **base,** a **shaft,** and a distal **head,** which is rounded for articulation with a phalanx.

Phalanges

The 14 phalanges are the bones of the digits (fingers). A single finger bone is called a **phalanx** (*fa'langks*). The phalanges of the fingers are arranged in a proximal row, a middle row, and a distal row. The thumb has only a proximal and a distal phalanx. The digits are numbered I to V, starting with the thumb.

6.16 Pelvic Girdle

The structure of the pelvic girdle is adaptive for support and locomotion. Extensive processes and other surface features of the pelvic girdle accommodate the massive muscles used in maintaining posture and in body movement.

The **pelvic girdle,** or **pelvis,** is formed by two **ossa coxae** (hipbones) united anteriorly by the **symphysis pubis** (fig. 6.36). It is attached posteriorly to the sacrum. The pelvic girdle and its binding ligaments support the weight of the body from the vertebral column. The pelvic girdle also supports and protects the lower abdominopelvic organs.

The basinlike pelvis is divided into a **greater** (false) **pelvis** and a **lesser** (true) **pelvis** (see fig. 6.38). These two components are divided by the **pelvic brim,** a curved bony rim passing inferiorly from the sacral promontory to the upper margin of the symphysis pubis. The greater pelvis is the expanded portion of the pelvis, superior to the pelvic brim. The pelvic brim not only divides the two portions but surrounds the **pelvic inlet** of the lesser pelvis.

Each os coxae (hipbone) actually consists of three separate bones: the *ilium,* the *ischium,* and the *pubis* (fig. 6.37). These bones are fused in the adult. On the lateral surface of the os coxae, where the three bones ossify, is a large circular depression, the **acetabulum** (*as"ĕ-tab'yŭ-lum*), which receives the head of the femur.

phalanx: L. *phalanx,* digital bone; from Gk., battle line (of soldiers)
coxae: L. *coxae,* hips
acetabulum: L. *acetabulum,* vinegar cup

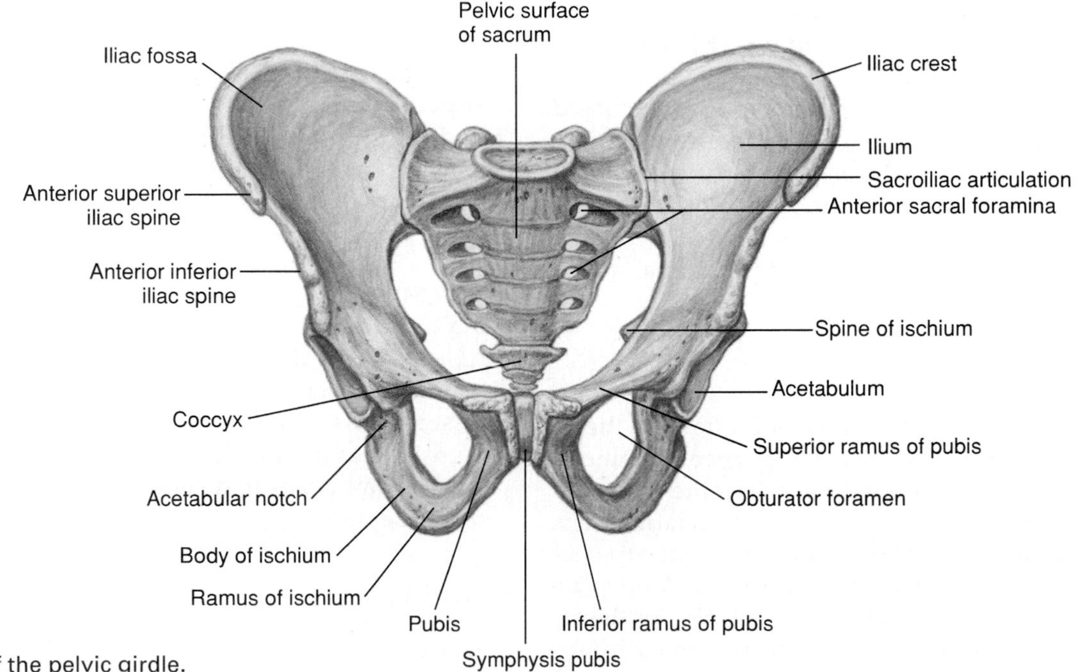

FIGURE 6.36
An anterior view of the pelvic girdle.

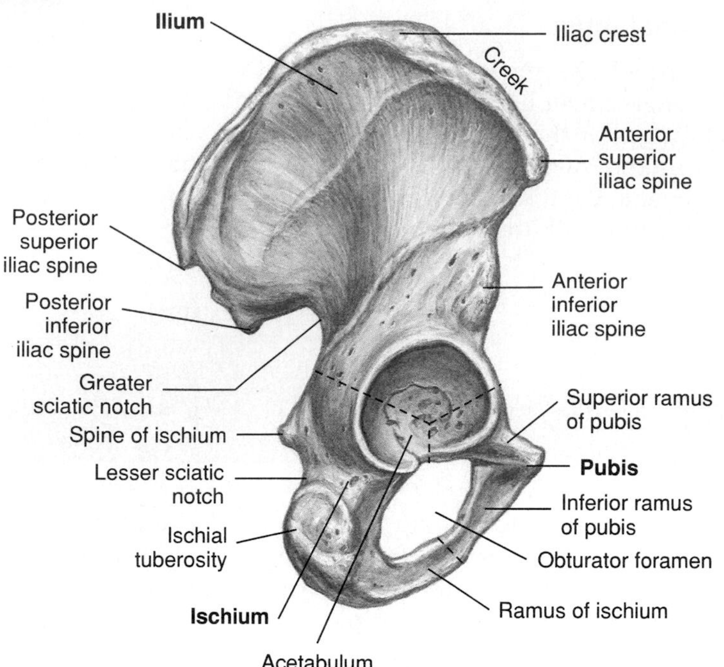

FIGURE 6.37
The lateral aspect of the right os coxae.

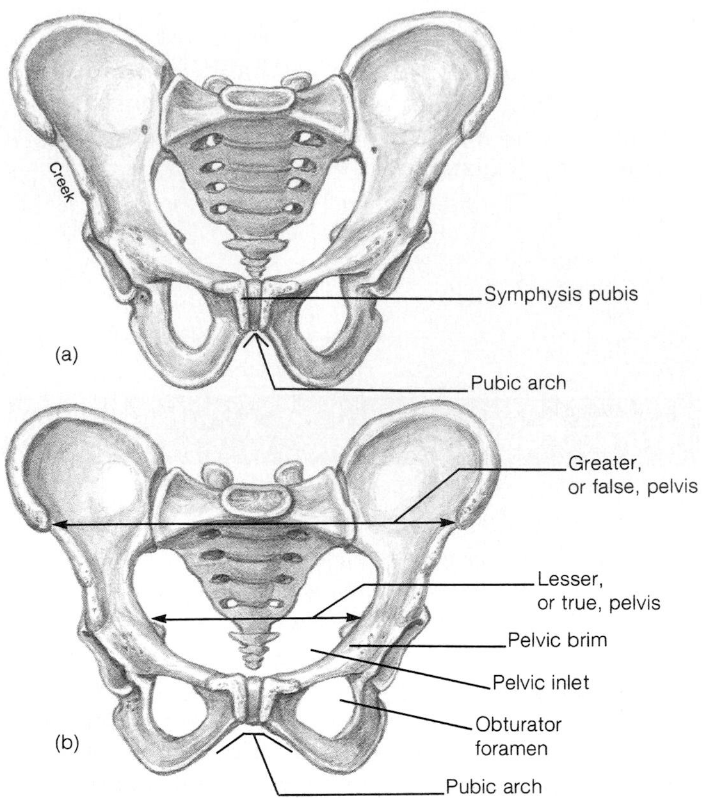

FIGURE 6.38
A comparison of (*a*) the male and (*b*) female pelves.

Ilium

The ilium is the largest and uppermost of the three pelvic bones. The ilium has a crest and four angles, or spines. These important surface landmarks serve for muscle attachment. The **iliac crest** forms the prominence of the hip. This crest terminates anteriorly as the **anterior superior iliac spine.** Just below this spine is the **anterior inferior iliac spine.** The posterior end of the iliac crest is the **posterior superior iliac spine,** and just below is the **posterior inferior iliac spine.**

Below the posterior inferior iliac spine is the curved **greater sciatic notch.** On the medial surface of the ilium is the roughened **auricular surface,** which articulates with the sacrum. The **iliac fossa** is the smooth, concave surface on the anterior portion of the ilium (see fig. 6.36). The **iliac tuberosity,** for the attachment of the sacroiliac ligament, is posterior to the iliac fossa.

Ischium

The ischium is the posteroinferior component of the os coxae. The **spine of the ischium** is the projection immediately posterior to the greater sciatic notch of the ilium and ischium. Inferior to this spine is the **lesser sciatic notch.** The **ischial tuberosity** is the bony projection that supports the weight of the body in the sitting position. A deep **acetabular notch** is present on the inferior portion of the acetabulum. The large **obturator foramen** is formed by the **ramus** of the ischium, together with the pubis. The obturator foramen is covered by the obturator membrane, to which several muscles attach.

Pubis

The pubis is the anterior bone of the os coxae. It consists of a **superior ramus** and an **inferior ramus** that support the **body** of the pubis. The body contributes to the formation of the symphysis pubis—the joint between the two ossa coxae.

Sex-Related Differences in the Pelvis

Structural differences between the pelvis of an adult male and that of an adult female (fig. 6.38 and table 6.5) reflect the female's role in pregnancy and childbirth. In a vaginal delivery, a baby must pass through its mother's lesser pelvis. *Pelvimetry* is the measurement of the dimensions and capacity of the pelvis—especially of the adult female pelvis to determine whether a cesarean section might be necessary.

ilium: L. *ilia,* loin
ischium: Gk. *ischion,* hip joint
pubis: L. *pubis,* genital area

Table 6.5 Comparison of the Male and Female Pelves

Characteristics	Male Pelvis	Female Pelvis
General structure	More massive; prominent processes	More delicate; processes not so prominent
Pelvic inlet	Heart-shaped	Round or oval
Pelvic outlet	Narrower	Wider
Anterior superior iliac spines	Not as wide apart	Wider apart
Obturator foramen	Oval	Triangular
Acetabulum	Faces laterally	Faces more anteriorly
Symphysis pubis	Deeper, longer	Shallower, shorter
Pubic arch	Angle less than 90°	Angle greater than 90°

6.17 Lower Limb

The lower limb consists of the thigh, leg, ankle, and foot. The bones of the lower limb are adapted for body support and movement.

Inside Information—In most people, the femur is stronger, pound for pound, than reinforced concrete.

Thigh

The patella will be considered along with the femur in the discussion of the thigh, even though the femur is the only bone of the thigh.

Femur

The femur (thighbone) is the largest, longest, and heaviest bone in the body (fig. 6.39). The proximal, rounded **head** of the femur articulates with the acetabulum of the os coxae. The constricted area supporting the head is called the **neck,** and is a common site for fractures in the elderly.

The **shaft** of the femur has a slight medial curve to bring the knee joint in line with the body's plane of gravity. On the lateral side of the shaft is the **greater trochanter,** and on the medial side is the **lesser trochanter.** Between the trochanters, on the anterior side, is the **intertrochanteric line;** on the posterior side is the **intertrochanteric crest.** The **linea aspera** is a roughened vertical ridge on the posterior surface of the shaft.

The distal end of the femur is expanded for articulation with the tibia. The **medial** and **lateral condyles** are the articular processes for this joint. The shallow depression between the condyles on the posterior aspect is called the **intercondylar fossa.** The **patellar surface** is located between the condyles on the anterior side. Above the condyles on the lateral and medial sides are the **epicondyles,** which serve for ligament and tendon attachment.

Patella

The patella (kneecap) is a sesamoid (specialized intramembranous) bone located on the anterior side of the knee joint (fig. 6.40). It develops in response to stress in the tendon of the large anterior thigh muscle. The patella is a triangular bone with a broad **base** and an inferiorly pointed **apex.** It protects the knee joint and strengthens the tendon of the anterior thigh muscle.

Leg

Technically speaking, the leg refers only to that portion of the lower limb between the knee and foot. The tibia and fibula are the bones of the leg. The tibia is the larger and more medial of the two bones.

Tibia

The tibia (shinbone) articulates proximally with the femur and distally with the talus. The **medial** and **lateral condyles** of the tibia (fig. 6.40) articulate with the condyles of the femur. Between the condyles is a slight upward projection called the **intercondylar eminence.** The **tibial tuberosity,** for attachment of the patellar ligament,

femur: L. *femur,* thigh
linea aspera: L. *linea,* line: *asperare,* rough
patella: L. *patina,* small plate
tibia: L. *tibia,* shinbone, pipe, flute

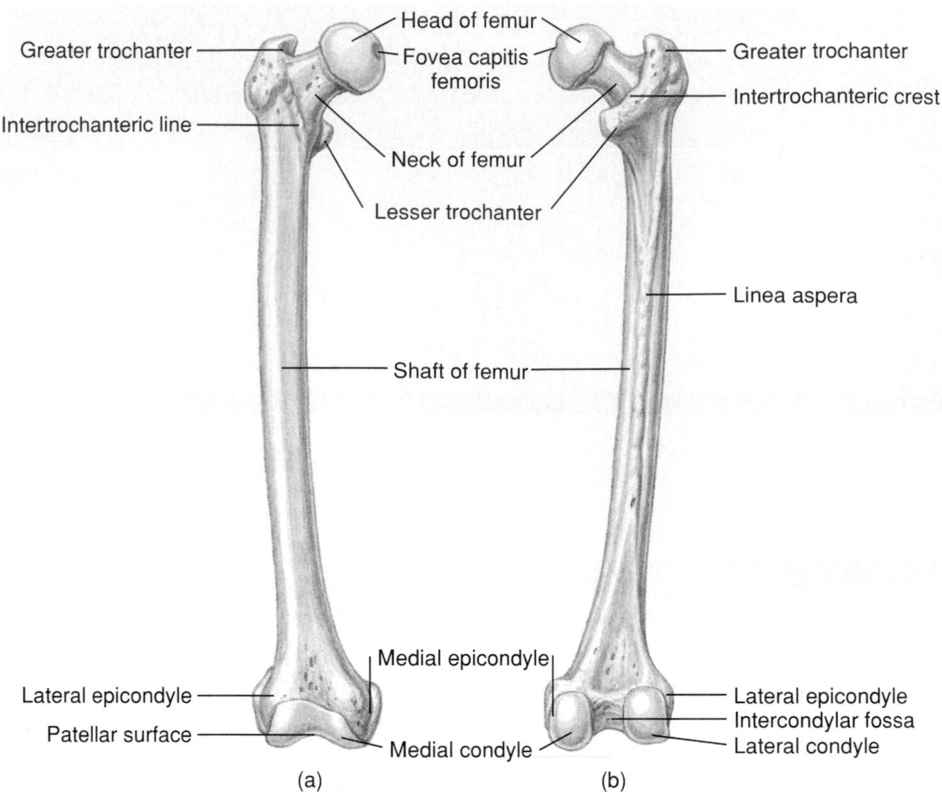

FIGURE 6.39
The right femur. (*a*) An anterior view and (*b*) a posterior view.

is located on the upper anterior portion of the shaft. The **anterior crest** is a sharp ridge along the anterior surface of the shaft. The **medial malleolus** (*mă-le'ŏ-lus*) is a prominent medial knob located on the distal end of the tibia. It can be palpated on the medial side of the ankle.

Fibula

The fibula is a long, narrow bone that is important in muscle attachment. The **head** of the fibula (fig. 6.40) articulates with the proximal end of the tibia. On the distal end is a prominent knob, the **lateral malleolus,** which can be palpated on the lateral side of the ankle.

Foot

The foot contains 26 bones, grouped into the tarsus, metatarsus, and phalanges (fig. 6.41).

Tarsus

There are seven tarsal bones. The **talus** is the tarsal bone that articulates with the tibia to form the ankle joint. The **calcaneus** is the largest of the tarsal bones and forms the heel of the foot. Anterior to the talus is the block-shaped **navicular bone.** The remaining four tarsal bones form a distal series that articulate with the metatarsal bones. They are, from the medial to lateral side, the **medial, intermediate,** and **lateral cuneiform bones** and the **cuboid bone.**

Metatarsus and Phalanges

The metatarsal bones and phalanges are similar in name and number to those of the hand. They differ in shape, however, to support the body and provide leverage during walking. The numbering of the digits is another difference. In reference to anatomical position, the thumb is the first digit of the hand and is on the lateral side—the great toe is the first digit of the foot and is on the medial side.

Arches of the Foot

The foot has two arches that are formed by the structure and arrangement of the bones held in place by ligaments and tendons (fig. 6.42). The arches are not rigid; they "give" when weight is placed on the foot, and they spring back when the weight is lifted.

---

malleolus: L. *malleolus,* small hammer
fibula: *fibula,* clasp or brooch
tarsus: Gk. *tarsos,* flat of the foot
talus: L. *talus,* ankle
calcaneus: L. *calcis,* heel

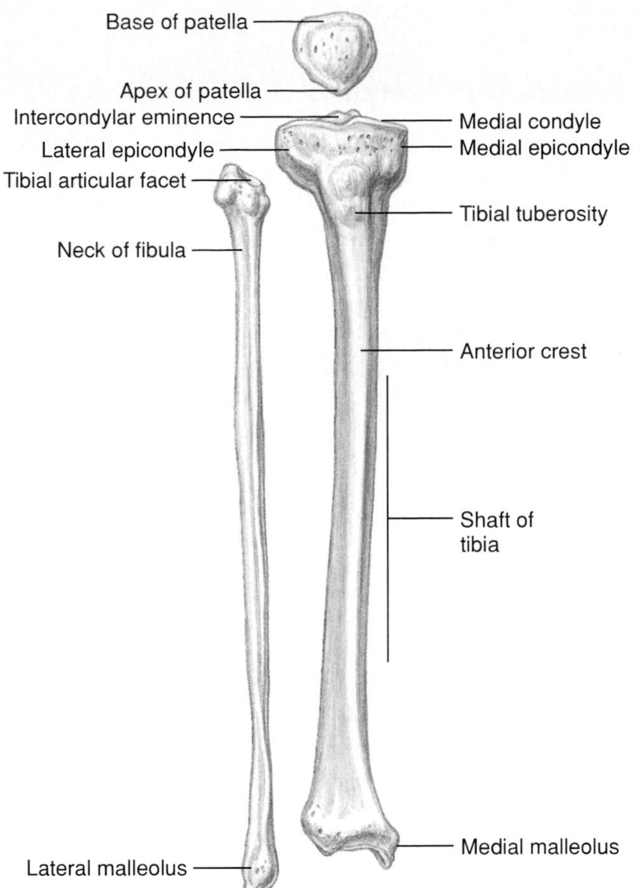

FIGURE 6.40
An anterior view of the right tibia, fibula, and patella.

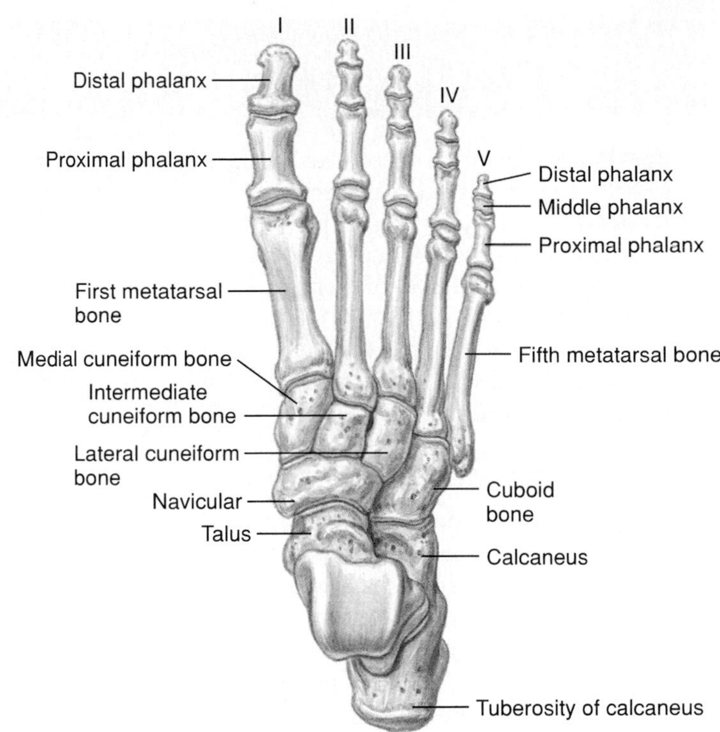

FIGURE 6.41
A superior view of the bones of the right foot. Each digit (toe) is indicated by a Roman numeral, the first digit, or great (big) toe, being Roman numeral I.

The **longitudinal arch** is divided into medial and lateral portions. The medial, or inner, part rises well above the ground. The talus is the keystone of the medial part, which originates at the calcaneus, rises at the talus, and descends to the first three metatarsal bones. The shallower lateral part consists of the calcaneus, cuboid, and fourth and fifth metatarsal bones. The cuboid is the keystone of this part.

The **transverse arch** extends across the width of the foot and is formed by the cuboid and cuneiform bones posteriorly and the bases of all five metatarsal bones anteriorly.

A weakening of the ligaments and tendons of the foot may cause the arches to "fall"—a condition known as *pes planus,* or, more commonly, flatfoot.

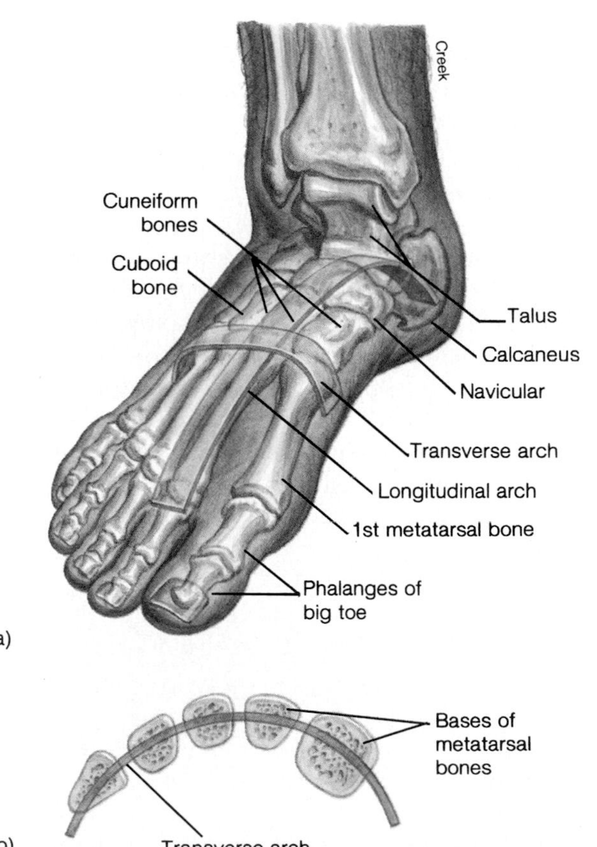

FIGURE 6.42
The arches of the foot. (*a*) A medial view of the right foot showing both arches. (*b*) A transverse view through the bases of the metatarsal bones showing a portion of the transverse arch.

Clinical Considerations

Each bone is a dynamic living organ that is influenced by hormones, diet, aging, and disease. Since the development of bones is genetically controlled, congenital abnormalities are possible. The hardness of bones gives them strength, yet they lack the resiliency to avoid fracture when they undergo excessive trauma. All of these aspects of bones make for some important and interesting clinical considerations.

Developmental Disorders

Congenital malformations account for several types of skeletal deformities. Certain bones may fail to form during osteogenesis, or they may form abnormally. **Cleft palate** and **cleft lip** are malformations of the palate and face. They vary in severity and seem to involve both genetic and environmental factors. **Spina bifida** is a congenital defect of the vertebral column resulting from a failure of the laminae of the vertebrae to fuse, leaving the spinal cord exposed.

Extra digits, called **polydactyly,** is the most common limb deformity. Usually an extra digit is incompletely formed and is not functional. **Syndactyly,** or webbed digits, is also common.

Nutritional and Hormonal Disorders

Vitamin D has a tremendous influence on bone. When there is a deficiency of this vitamin, the body is unable to properly metabolize calcium and phosphorus. Vitamin D deficiency in children causes **rickets.** As mentioned in module 6.2, the bones of a child with rickets remain soft and become deformed from the weight of the body. A deficiency of vitamin D in the adult causes bone resorption, resulting in **osteomalacia.** This condition is common in women who have repeated pregnancies and poor diets, and who do not get enough sunlight.

Osteoporosis is a weakening of the bones, primarily as a result of calcium loss. The causes of osteoporosis include aging, inactivity, poor diet, and an imbalance in hormones or other chemicals in the blood. It is most common in older women because low levels of estrogens after menopause lead to increased bone resorption, and the formation of new bone is not sufficient to keep pace.

People with osteoporosis are prone to bone fracture, particularly at the pelvic girdle and vertebrae, as the bones become too brittle to support the weight of the body. Complications of hip fractures often lead to permanent disability, and vertebral compression fractures may produce a permanent curved deformity of the spine.

Although there is no known cure for osteoporosis, good eating habits and a regular program of exercise, established at an early age and continued throughout adulthood, can minimize its effects. Treatment in women through dietary calcium, exercise, and estrogens has had limited positive results. In addition, a drug called *alendronate* (Fosamax), approved by the FDA in 1995, has been shown to be effective in managing osteoporosis. This drug works without hormones to block osteoclast activity, making it useful for women who choose not to be treated with estrogen replacement therapy.

Paget's disease, another bone disorder that affects mainly older adults, is characterized by disorganized metabolic processes within bone tissue. The activity of osteoblasts and osteoclasts becomes irregular, resulting in thick bony deposits in some areas of the skeleton and fragile, thin bone in other areas. The spine, pelvis, femur, and skull are most often involved, and become increasingly painful and deformed. Bowed leg bones, abnormal curvature of the spine, and enlargement of the skull may develop. The cause of Paget's disease is currently not known.

polydactyly: Gk. *polys,* many; *daktylos,* finger
syndactyly: Gk. *syn,* together; *daktylos,* finger
Paget's disease: from Sir James Paget, English surgeon, 1814–99

Trauma and Injury

The most common type of bone injury is a **fracture.** A fracture is the cracking or breaking of a bone. In *radiology,* X rays are used to diagnose the position and extent of a fracture. Several kinds of fractures are illustrated in figure 6.43.

When a bone fractures, treatment involves realigning the broken ends and then setting them until new bone tissue is formed and the fracture is healed. The location and severity of the fracture and the age of the patient will determine the type of immobilization. Methods of immobilization include tape, splints, casts, straps, wires, and steel pins. It has also been found

A *greenstick fracture* is incomplete, and the break occurs on the convex surface of the bend in the bone.

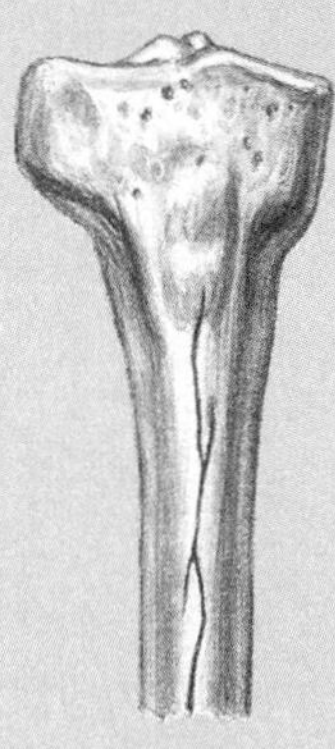

A *partial (fissured) fracture* involves an incomplete longitudinal break.

A *comminuted fracture* is complete and results in several bony fragments.

A *transverse fracture* is complete, and the break occurs horizontally.

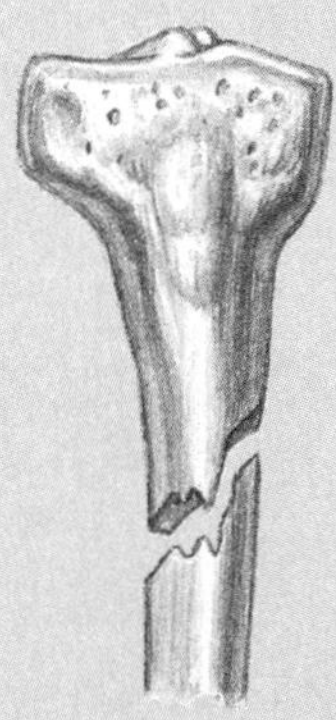

An *oblique fracture* occurs at an angle other than a right angle to the long axis of the bone.

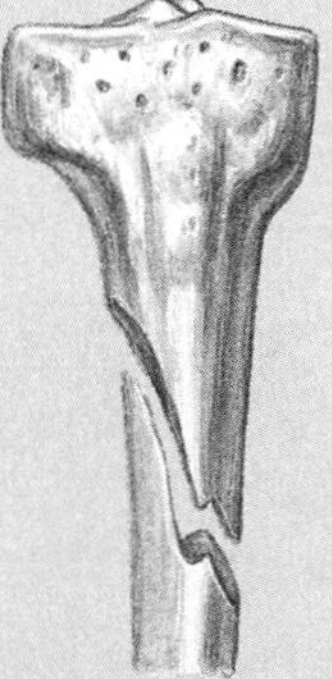

A *spiral fracture* is caused by twisting the bone.

FIGURE 6.43
Examples of fractures.

continued

that applying weak electrical currents to certain fractured bones promotes healing and reduces the time of immobilization. Physicians can realign, immobilize, and treat a fracture, but the ultimate repair must occur within the bone itself (fig. 6.44).

Bone Tumors

Malignant bone tumors are three times more common than benign tumors. Pain is the usual symptom of either type of bone tissue neoplasm.

Two types of benign bone tumors are **osteomas,** which often involve the skull, and **osteoid osteomas,** which are painful neoplasms of the long bones, usually in children. Of the two types, osteomas occur more frequently.

Osteogenic sarcoma is the most virulent type of bone cancer. It frequently metastasizes through the blood to the lungs. This disease usually originates in the long bones and is accompanied by aching and persistent pain.

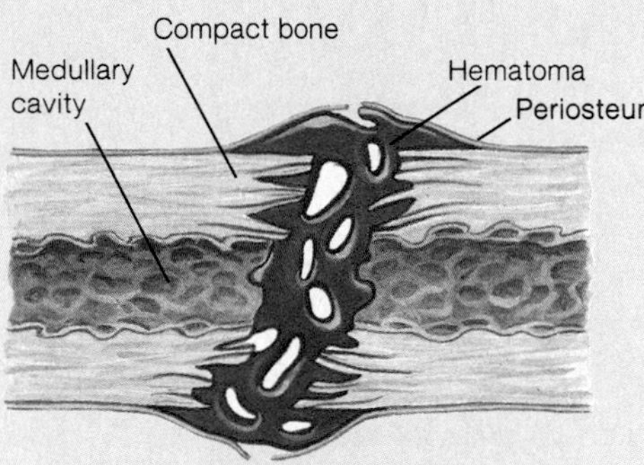

(a) Blood escapes from ruptured blood vessels and forms a hematoma.

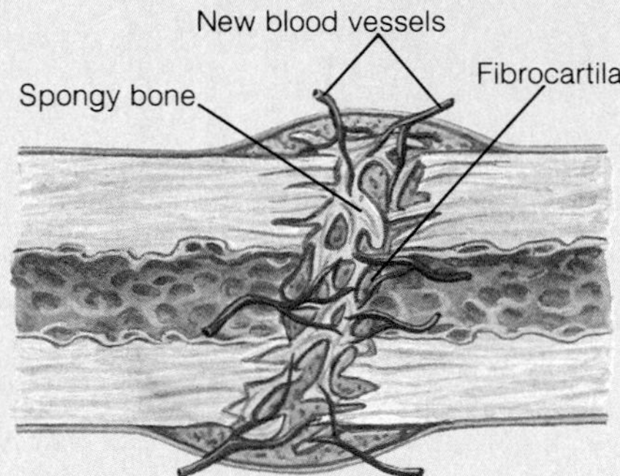

(b) Spongy bone forms in regions close to developing blood vessels; fibrocartilage forms in more distant regions.

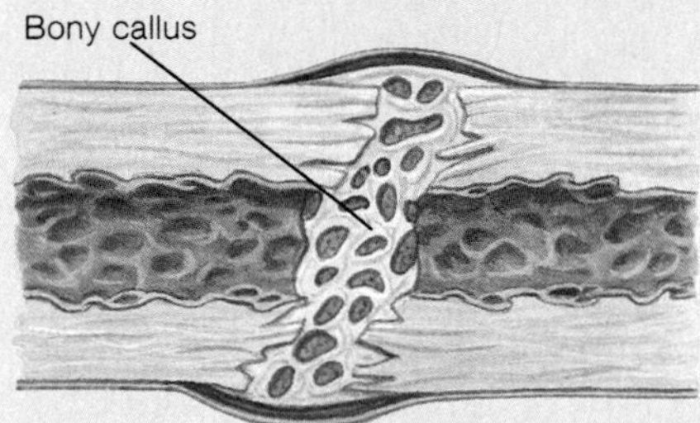

(c) Fibrocartilage is replaced by a bony callus.

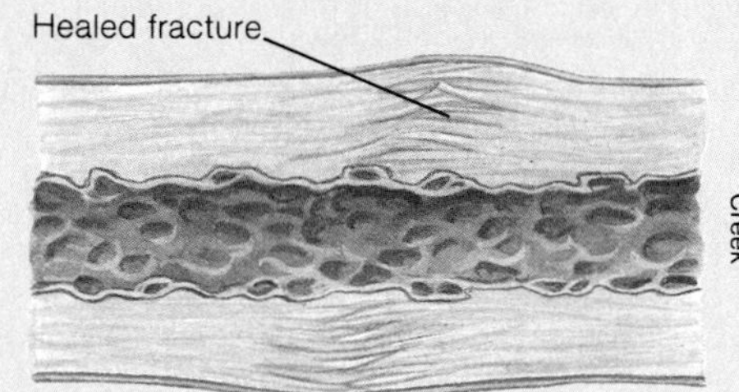

(d) Osteoclasts remove excess bony tissue, making new bone structure much like the original.

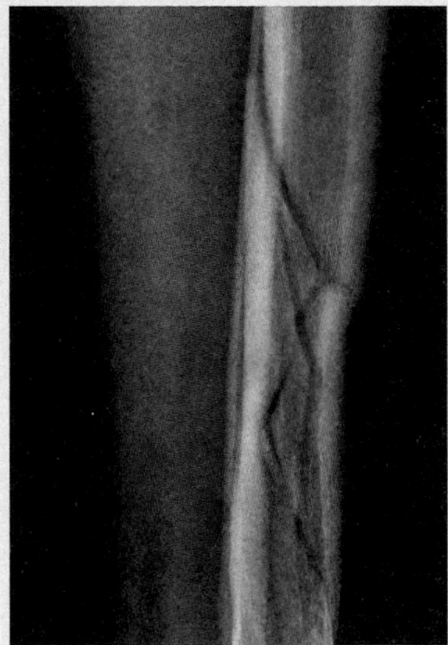

(e)

FIGURE 6.44
Stages (*a–d*) of the repair of a fracture. (*e*) An X ray of a healing fracture.

SUMMARY

6.1 Organization of the Skeletal System

1. The axial skeleton consists of the skull, auditory ossicles, hyoid bone, vertebral column, and rib cage.
2. The appendicular skeleton consists of the pectoral girdle, upper limbs, pelvic girdle, and lower limbs.

6.2 Functions of the Skeleton

1. The mechanical functions of bones include the support and protection of body tissues and organs. In addition, certain bones function as levers during movement.
2. The metabolic functions of bones include hemopoiesis and mineral storage.

6.3 Development of the Skeletal System

1. Bones are formed through endochondral ossification or through intramembranous ossification.
2. The six fontanels of the developing skull facilitate childbirth and permit rapid growth of the brain during infancy.

6.4 Bone Structure

1. Bones may be structurally classified as long, short, flat, or irregular.
2. The surface features of bones are classified as articulating surfaces, nonarticulating prominences, and depressions and openings.
3. A typical long bone has a diaphysis, or shaft, filled with marrow in the medullary cavity; epiphyses; epiphyseal plates for linear growth; and a covering of periosteum for growth in width and the attachment of ligaments and tendons.

6.5 Bone Tissue

1. Compact bone is the dense outer portion; spongy bone is the porous, vascular inner portion.
2. Osteoblasts are bone-forming cells, osteocytes are mature bone cells, and osteoclasts are bone-destroying cells.
3. In compact bone, the lamellae are the layers of inorganic matrix surrounding a central canal. Osteocytes are mature bone cells located within spaces between the lamellae, called lacunae.

6.6 Bone Growth

1. Bone growth is an orderly process determined by genetics, diet, and hormones.
2. In the adult, bone remodeling is a continual process involving osteoclasts in bone resorption and osteoblasts in forming new bone tissue.

6.7 Cranial Bones

1. The eight cranial bones include the frontal (1), parietal (2), temporal (2), occipital (1), sphenoid (1), and ethmoid (1).
2. The cranial bones enclose and protect the brain and provide surfaces for the attachment of muscles.

6.8 Facial Bones

1. The 14 facial bones include the nasal (2), maxilla (2), zygomatic (2), mandible (1), lacrimal (2), palatine (2), inferior nasal concha (2), and vomer (1).
2. The facial bones form the basic shape of the face, support the teeth, and provide for the attachment of the facial muscles.

6.9 Cavities and Sinuses of the Skull

1. The cranial cavity is divided into an anterior cranial fossa, a middle cranial fossa, and a posterior cranial fossa.
2. The oral cavity contains the teeth and tongue; the nasal cavity is divided by the nasal septum into two lateral halves; the orbital cavities contain the eyes; and the middle-ear cavities contain the auditory ossicles.
3. The paranasal sinuses drain into the nasal cavity. They decrease the weight of the skull and help to warm and moisten inhaled air.

6.10 Foramina of the Skull

1. Foramina are holes in bones for the passage of nerves or vessels.
2. The cranial nerves pass through foramina within cranial bones.

6.11 Vertebral Column

1. The vertebral column consists of 7 cervical, 12 thoracic, 5 lumbar, 5 fused sacral, and 3 to 5 fused coccygeal vertebrae.
2. The vertebral column provides a flexible support for the body, supports the head, and encloses and protects the spinal cord. It also provides attachment for the rib cage and the pelvic girdle.

6.12 Vertebral Regions

1. Cervical vertebrae have transverse foramina. The atlas is the first cervical vertebra, and the axis is the second cervical vertebra.
2. Thoracic vertebrae have facets for rib articulations. Lumbar vertebrae have heavy bodies and thick spinous processes. The sacrum consists of fused sacral vertebrae, and the coccyx consists of fused coccygeal vertebrae.

6.13 Rib Cage

1. The rib cage consists of a sternum, costal cartilages, and 12 paired ribs attached to the thoracic vertebrae.
2. The rib cage protects the heart and lungs and provides attachment sites for skeletal muscles.

6.14 Pectoral Girdle

1. The pectoral girdle consists of the paired scapulae and clavicles.
2. The structure of the pectoral girdle is adaptive for freedom of movement and extensive muscle attachment.

6.15 Upper Limb

1. The upper limb consists of the arm, forearm, and hand.
2. The arm (brachium) contains the humerus, which extends from the scapula to the elbow.
3. The forearm (antebrachium) contains the radius on the lateral side and the ulna on the medial side. The forearm extends from the elbow to the carpal region of the hand.
4. In the hand, there are 8 carpal bones, 5 metacarpal bones, and 14 phalanges.

6.16 Pelvic Girdle

1. The structure of the pelvic girdle and lower limb is adaptive for support and locomotion.
2. The pelvic girdle consists of the paired ossa coxae, joined at the symphysis pubis.
3. Each os coxae consists of an ilium, an ischium, and a pubis; in an adult they are fused.

6.17 Lower Limb

1. The lower limb consists of the thigh, leg, ankle, and foot.
2. The thigh contains the femur, which extends from the os coxae to the knee.
3. The patella (kneecap) is a sesamoid bone that protects the knee joint and strengthens the tendon of the anterior thigh muscle.
4. The leg contains the fibula on the lateral side and the tibia on the medial side. The leg extends from the knee to the tarsal region of the foot.
5. In the foot, there are 7 tarsal bones, 5 metacarpal bones, and 14 phalanges.

Review Activities

Objective Questions

1. Which of the following is *not* a bone of the appendicular skeleton? (*module 6.1*)
   a. the scapula
   b. a rib
   c. a metatarsal bone
   d. the patella

2. Which of the following statements is *false?* (*module 6.2*)
   a. Vitamin D is essential for proper bone development.
   b. Bones and teeth contain over 90% of the body's calcium.
   c. Red bone marrow is the site for hemopoiesis.
   d. Bones are rigid and hard because they are dead.

3. A fontanel is (*module 6.3*)
   a. a site for hemopoiesis.
   b. found only in a fetus.
   c. an opening into the skull.
   d. a membranous area on a forming skull.

4. Which of the following is *not* a long bone? (*module 6.4*)
   a. talus
   b. a phalanx
   c. a metatarsal bone
   d. fibula

5. Specialized bone cells that are responsible for bone resorption are (*module 6.5*)
   a. osteoblasts.
   b. osteocytes.
   c. osteons.
   d. osteoclasts.

6. The crista galli is a structural feature of which bone? (*module 6.7*)
   a. sphenoid bone
   b. ethmoid bone
   c. palatine bone
   d. temporal bone

7. Transverse foramina are located in (*module 6.12*)
   a. lumbar vertebrae.
   b. sacral vertebrae.
   c. thoracic vertebrae.
   d. cervical vertebrae.

8. The clavicle articulates with (*module 6.14*)
   a. the scapula and humerus.
   b. the humerus and manubrium.
   c. the manubrium and scapula.
   d. the manubrium, scapula, and humerus.

9. Which of the following bones has a conoid tubercle? (*module 6.14*)
   a. scapula
   b. humerus
   c. radius
   d. clavicle

10. The "elbow" of the ulna is formed by (*module 6.15*)
   a. the lateral epicondyle.
   b. the olecranon.
   c. the coronoid process.
   d. the styloid process.

11. Pelvimetry is a measurement of (*module 6.16*)
   a. the os coxae.
   b. the symphysis pubis.
   c. the pelvic brim.
   d. the lesser pelvis.

12. The obturator foramen is found within (*module 6.16*)
   a. the os coxae.
   b. the ilium.
   c. the sacrum.
   d. the skull.

Essay Questions

1. Define *osteon.* Sketch an osteon and label the osteocytes, lacunae, lamellae, central canal, and canaliculi. (*module 6.5*)
2. Describe how bones grow in length and in width. How are these processes similar? How do they differ? (*module 6.6*)
3. Describe the functions of the sella turcica, foramen magnum, petrous part of the temporal bone, perpendicular plate, and crista galli. (*module 6.7*)
4. Describe each of the long bones of the upper limb. (*module 6.15*)
5. Compare the pectoral and pelvic girdles as to structure, articulation with the axial skeleton, and function. (*modules 6.14 and 6.16*)
6. How do the male and female pelves differ structurally? (*module 6.16*)
7. List the processes of the bones of the lower limb that can be palpated. (*module 6.17*)

Labeling Exercise

Label the structures indicated on the figure to the right.

1. ______________________________
2. ______________________________
3. ______________________________
4. ______________________________
5. ______________________________
6. ______________________________

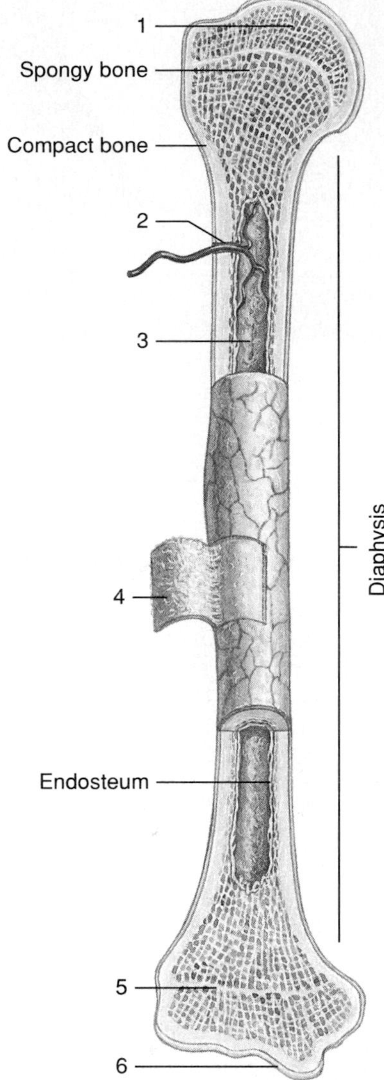

Critical Thinking Questions

1. Explain why a proper balance of vitamins, hormones, and minerals is essential in maintaining healthy bone tissue. Give examples of diseases or skeletal conditions that may occur if there is an imbalance of any of these three essential substances.
2. Upon examining an X ray of a 10-year-old's broken humerus, the physician expressed concern because the fracture was across an epiphyseal plate. Why the concern?
3. Many people think that the bones in our bodies are dead—understandable considering that we associate bones with graveyards, Halloween, and ghosts. Your kid sister is convinced of this. What information could you use to try to get her to change her mind?
4. You read in *National Geographic* that a team of archaeologists recently completed an examination of 18 skeletons from people buried under tons of volcanic ash 1,200 years ago. By analyzing the bones, the scientists were able to determine the sex, physical health (including a partial medical history), approximate age, and even the general profession of each of the 18 individuals. How could the examination of a preserved skeleton yield such a vast array of information?
5. Each body system is functionally associated with other body systems in the workings of the organism. Describe the functional relationship for each of the following: (a) red bone marrow and the blood of the circulatory system; (b) bone matrix and the storage of essential minerals; (c) periosteum and the tendinous attachment of muscles; (d) flat bones and the protection of vital body organs; and (e) red bone marrow and body immunity.

CHAPTER

7

"It is not the bones . . . that allow movement [of your body], but rather the articulations, or joints, between the bones."
page 155

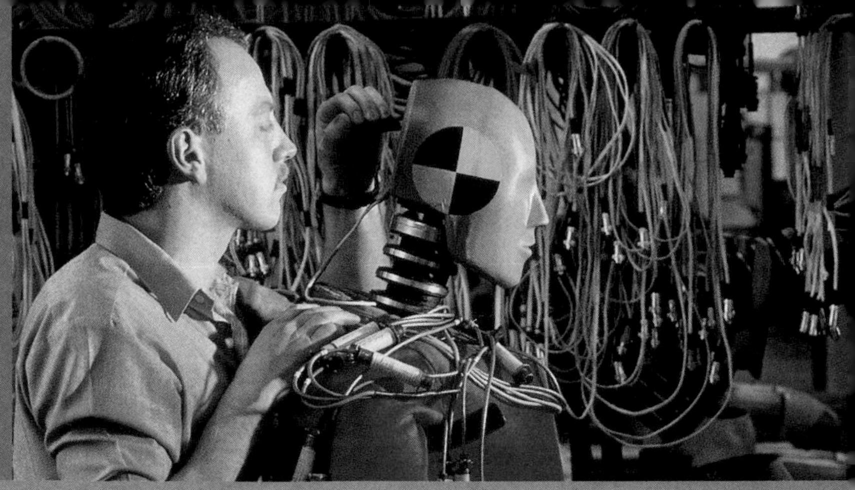

Articulations

Chapter Outline

Terms to Remember

articulation, 155
fibrous joint, 155
suture, 157
syndesmosis (*sin"des-mo'sis*), 157
gomphosis (*gom-fo'sis*), 158
cartilaginous (*kar"tĭ-laj'ĭ-nus*) joint, 158
symphysis (*sim'fĭ-sis*), 158
synchondrosis (*sin"kon-dro'sis*), 158
synovial (*sĭ-no've-al*) joint, 159
joint capsule, 159
synovial fluid, 159
meniscus (*mĕ-nis'kus*), 160
bursa, 160
flexion, 163
extension, 164
abduction, 164
adduction, 164
rotation, 164
circumduction, 165
protraction, 165
retraction, 165

Learning Objectives

After studying this chapter, you should be able to

- compare and contrast the three principal categories of joints, using specific joints as examples.
- describe the development of a synovial joint.
- discuss the three kinds of fibrous joints and indicate which one is especially important to a dentist.
- compare the two kinds of cartilaginous joints and indicate which is associated with bone growth.
- describe the structure of a synovial joint and discuss the importance of the joint capsule and synovial fluid.
- discuss the various kinds of synovial joints, noting the range of movement of each.
- discuss the various kinds of movements that are possible at synovial joints.

7.1 Classification of Joints

Articulations between the bones are structurally classified as fibrous joints, cartilaginous joints, and synovial joints.

One of the functions of the skeletal system is to permit movement of your body. It is not the bones themselves that allow movement, however, but rather the **articulations,** or **joints,** between the bones (fig. 7.1). Although the joints of the body are actually part of the skeletal system, this chapter is devoted entirely to them. Before embarking on a study of joints, you should be familiar with the structure and locations of the bones; thus, the previous chapter provides a foundation for this one. In the next chapter, you will learn how the muscles of the body cause the bones to move at the joints.

You probably associate the term *joint* with vigorous action, as in the movements of a gymnast or ballet dancer (fig. 7.2). However, joints also enable the ordinary activities of walking, writing, and speaking. And some joints, such as those of the skull, are so tightly fitted that they permit no movement at all.

Arthrology is the science concerned with the study of joints. If you were to take a course entitled **kinesiology,** you would study the functional relationship, or *biomechanics,* of the skeleton, joints, muscles, and innervation (nerve supply) as they work together to produce movement. As you study the joints in this chapter, use your own body to practice the movements being described. This will help you understand the adaptive advantage, as well as the limitations, of each type of movement.

The articulations of the body are grouped by their structure into three principal categories.

1. **Fibrous joints.** In fibrous joints, the bones are held together by fibrous connective tissue. These joints have no joint cavities.
2. **Cartilaginous joints.** In cartilaginous joints, the bones are held together by cartilage. These joints also lack joint cavities.
3. **Synovial joints.** Synovial joints have cartilage covering the articulating bones and ligaments that help to hold the joints together. These joints have fluid-filled joint cavities.

arthrology: Gk. *arthron,* joint; *logos,* study
kinesiology: Gk. *kinesis,* movement; *logos,* study

FIGURE 7.1
The articulations, or joints, between the bones of the skeleton permit a wide range of elegant movement in the human body.

FIGURE 7.2
A visual image of a dance sequence performed by a ballet dancer. Analyses of body movements are enhanced by computerized graphics.

7.2 Development of Synovial Joints

Synovial joints are formed by the end of the third month of fetal development. Movement at these joints occurs shortly thereafter, as the muscles of the fetus undergo contractions.

The appearance of *precartilaginous cells* at 6 weeks indicates developing synovial joints (fig. 7.3). As cartilage develops within a forming bone, a thin flattened sheet of cells forms around the cartilaginous model to become the perichondrium. These same cells are continuous across the gap between the adjacent developing bones. Surrounding the gap, the flattened cells become the **joint capsule.**

During the early part of the third month of development, the cells still remaining within the joint capsule begin migrating toward the epiphyses of the adjacent developing bones. The cleft formed by the outward migration eventually enlarges to become the **joint cavity.** Thin pads of hyaline cartilage develop on the surfaces of the epiphyses that are in contact with the joint cavity. These pads become the **articular cartilages** of the functional joint. As the joint continues to develop, a highly vascular **synovial membrane** forms on the inside of the joint capsule and begins secreting a watery *synovial fluid* into the joint cavity.

In certain developing synovial joints, the precartilaginous cells do not migrate away from the center of the joint cavity. Rather, they give rise to wedges of cartilage, called **menisci,** as in the knee joint, or the complete cartilage pads, called **articular discs,** as in the sternoclavicular joint.

The formation of most synovial joints is completed by the end of the third month. A short time later, fetal contractions cause movement at these joints. This first perceptible movement of the fetus in the uterus is called *quickening,* and it usually occurs around the sixteenth week of pregnancy. Fetal contractions are essential for the development of healthy joints. Joint movement nourishes the articular cartilage and prevents the fusion of connective tissues within the joint.

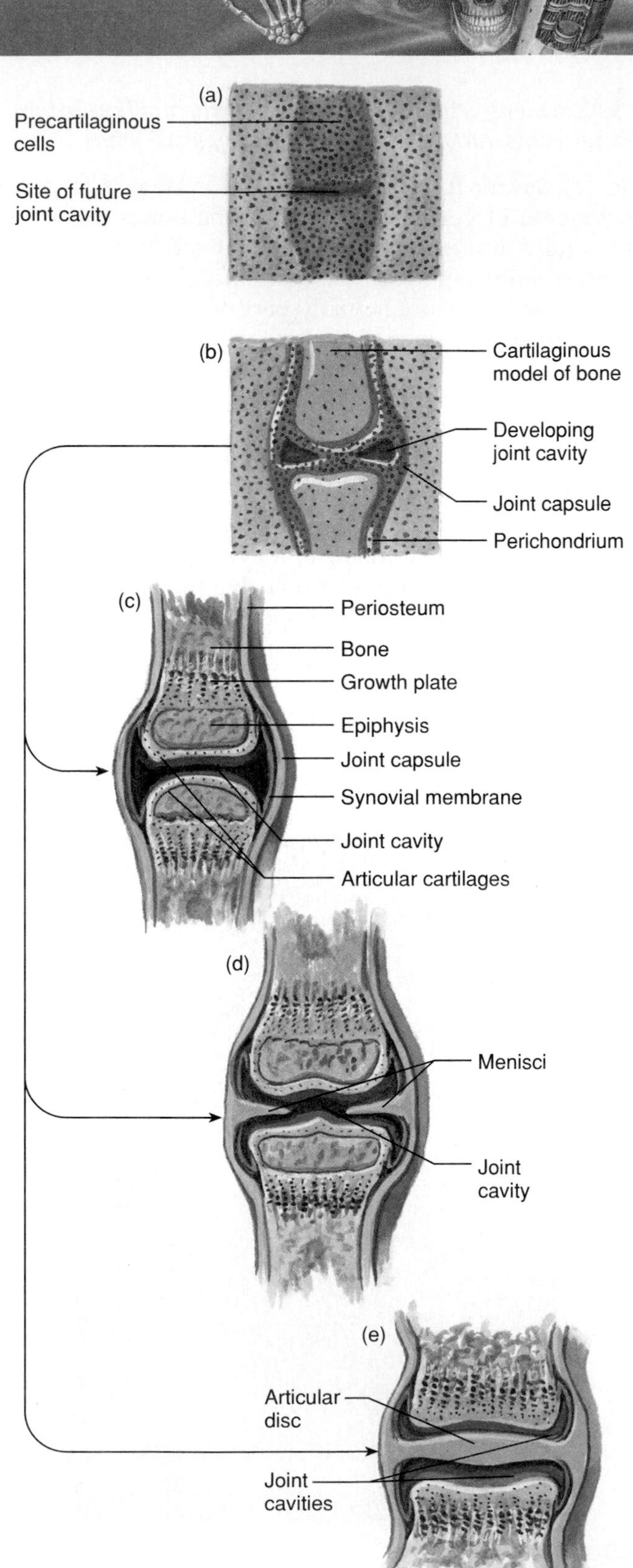

FIGURE 7.3
The development of synovial joints. (*a*) At 6 weeks, different densities of precartilaginous cells denote where the bones and joints will form. (*b*) At 9 weeks, a basic synovial model is present. At 12 weeks, the synovial joints are formed and have either (*c*) a free joint cavity (e.g., interphalangeal joint); (*d*) a cavity containing menisci (e.g., knee joint); or (*e*) a cavity with a complete articular disc (e.g., sternoclavicular joint).

7.3 Fibrous Joints and Cartilaginous Joints

The articulating bones in fibrous joints are bound by fibrous connective tissue, and the articulating bones in cartilaginous joints are bound by cartilage. Fibrous joints are of three types: sutures, syndesmoses, and gomphoses. Cartilaginous joints are of two types: symphyses and synchondroses.

Fibrous Joints

As the name suggests, the articulating bones of fibrous joints are tightly bound by fibrous connective tissue. Fibrous joints range from rigid and relatively immovable joints to those that are slightly movable. The three kinds of fibrous joints are *sutures, syndesmoses* (*sin″des-mo′sēz*), and *gomphoses* (*gom-fo′sēz*).

Sutures

Sutures are found only within the skull. They are characterized by a thin layer of dense irregular connective tissue that binds the articulating bones (fig. 7.4). Sutures form at about 18 months of age as bone replaces the pliable fontanels of an infant's skull (see fig. 6.5).

Sutures vary in the appearance of the articulating edge of bone. A **serrate suture** is characterized by jagged interlocking articulations. This is the most common type of suture, an example being the coronal suture between the frontal and parietal bones (see fig. 6.14). In a **lap suture,** the edge of one bone overlaps that of the articulating bone. The squamous suture (fig. 6.14) between the temporal and parietal bones is an example. In a **plane suture,** the edges of the articulating bones are fairly smooth. An example is the median palatine suture, where the two maxillae articulate to form the hard palate (see fig. 6.21).

Syndesmoses

In a syndesmotic joint, adjacent bones are held together by collagenous fibers or interosseous ligaments. A syndesmosis is characteristic of the distal ends of the tibia-fibula and the radius-ulna (fig. 7.5).

suture: L. *sutura,* sew
syndesmosis: Gk. *syndesmos,* binding together

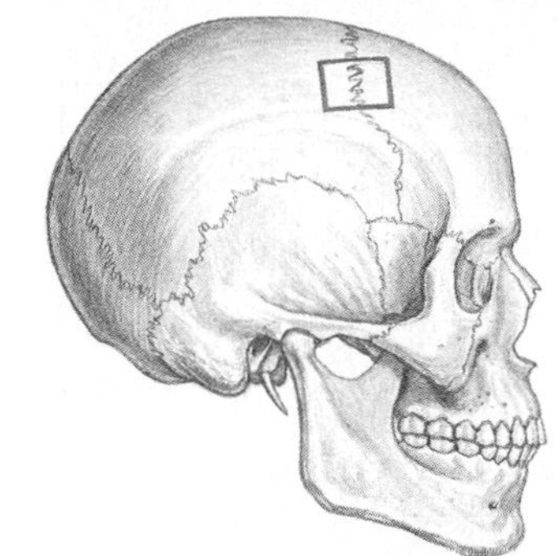

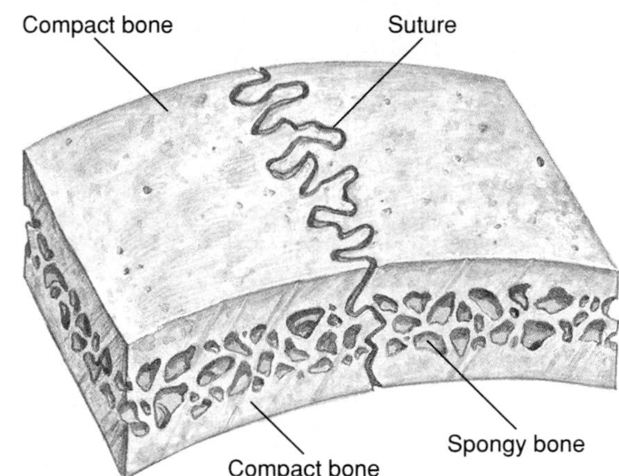

FIGURE 7.4
A section of the skull showing a suture.

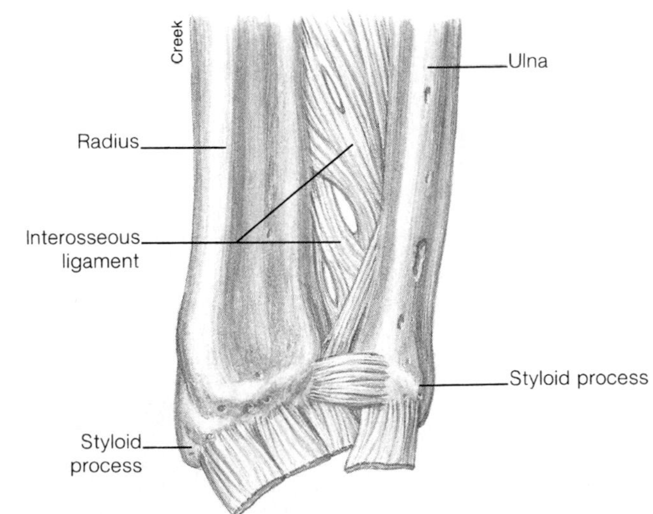

FIGURE 7.5
The side-to-side articulation of the ulna and radius forms a syndesmotic joint. An interosseous ligament tightly binds these bones and permits only slight movement between them.

Gomphoses

Gomphoses occur between the teeth and the supporting bones of the jaws (fig. 7.6). More specifically, a gomphosis is where the root of a tooth is attached to the periodontal ligament of the dental alveolus (tooth socket).

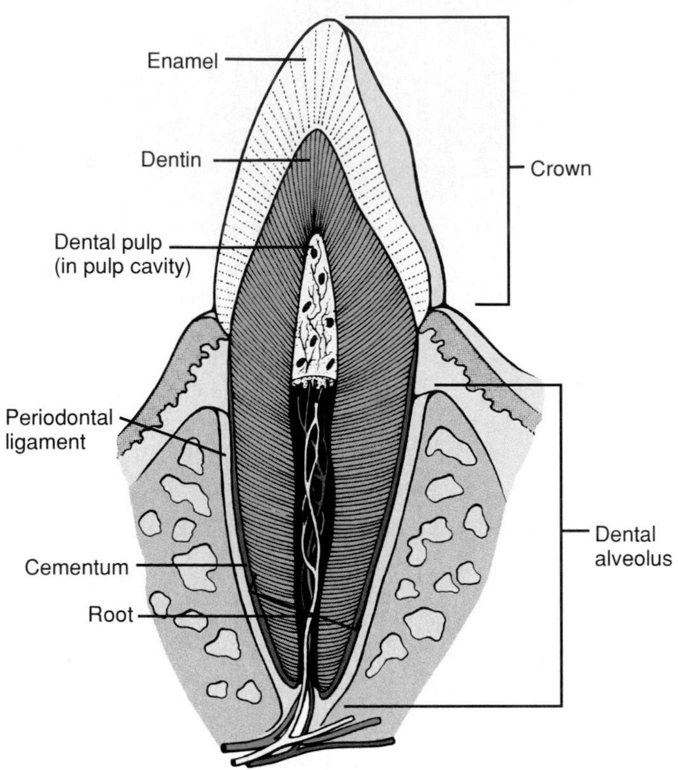

FIGURE 7.6
A gomphosis is a fibrous joint where a tooth is held in its socket.

Cartilaginous Joints

Cartilaginous (*kar"tĭ-laj'ĭ-nus*) joints allow limited motion in response to twisting or compression. The two types of cartilaginous joints are *symphyses* (*sim'fĭ-sēz*) and *synchondroses* (*sin"kon-dro'sēz*).

Symphyses

The adjoining bones of a symphysis joint are separated by a pad of fibrocartilage. This pad cushions the joint and allows slight movement. The symphysis pubis and the intervertebral joints (fig. 7.7) at the intervertebral discs are examples of symphyses. Although only limited motion is possible at each intervertebral joint, the combined movement of all the joints of the vertebral column results in extensive spinal action.

During pregnancy, hormones cause the pelvic joints to soften and become more flexible. This spreads the pelvic cavity and reduces trauma during childbirth by increasing the diameter of the pelvic outlet. Following the birth of a mother's first child, the diameter of her pelvis will be slightly wider than before.

Synchondroses

Synchondroses are cartilaginous joints that have hyaline cartilage between the bone segments. Some of these joints are temporary, forming the growth lines (epiphyseal

gomphosis: Gk. *gompho,* nail or bolt
symphysis: Gk. *symphysis,* growing together
synchondrosis: Gk. *syn,* together; *chondros,* cartilage

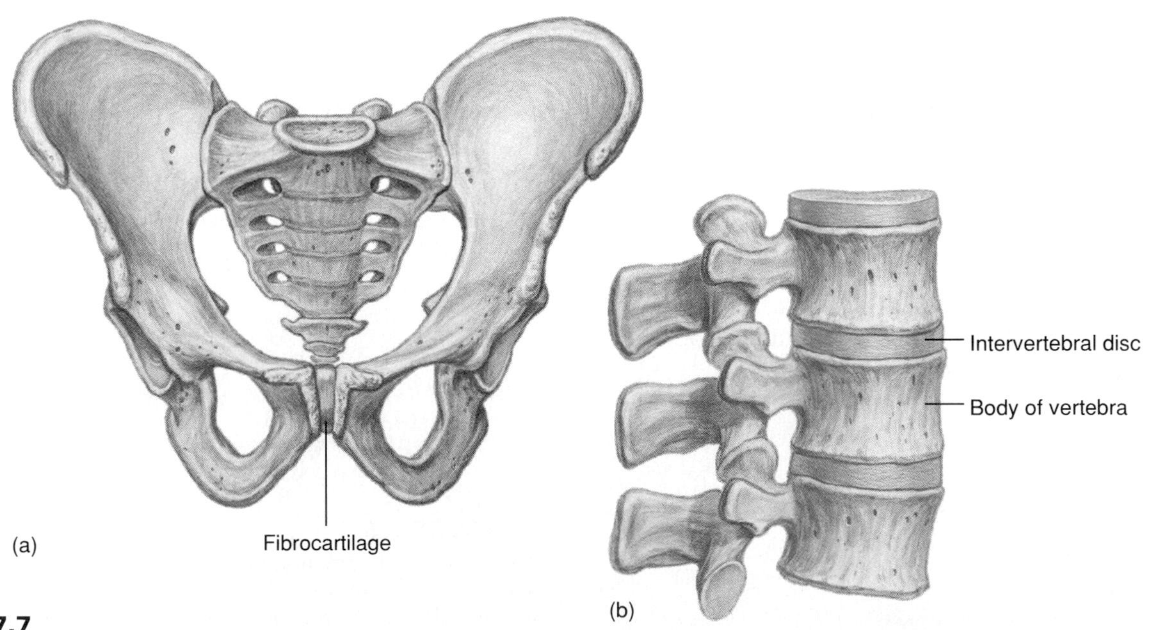

FIGURE 7.7
Examples of symphyses. (*a*) The symphysis pubis and (*b*) the intervertebral joints between vertebral bodies.

plates) between the diaphyses and epiphyses in the long bones of children (fig. 7.8). When growth is complete, these synchondrotic joints ossify and are no longer visible. The costal cartilages between the ends of the ribs and the costal cartilages that attach to the sternum are examples of synchondroses that do not ossify.

A fracture of a long bone in a child may be extremely serious if it involves the mitotically active epiphyseal plate of a synchondrotic joint. If such an injury is not treated, bone growth is likely to be retarded or arrested, resulting in limbs of unequal length.

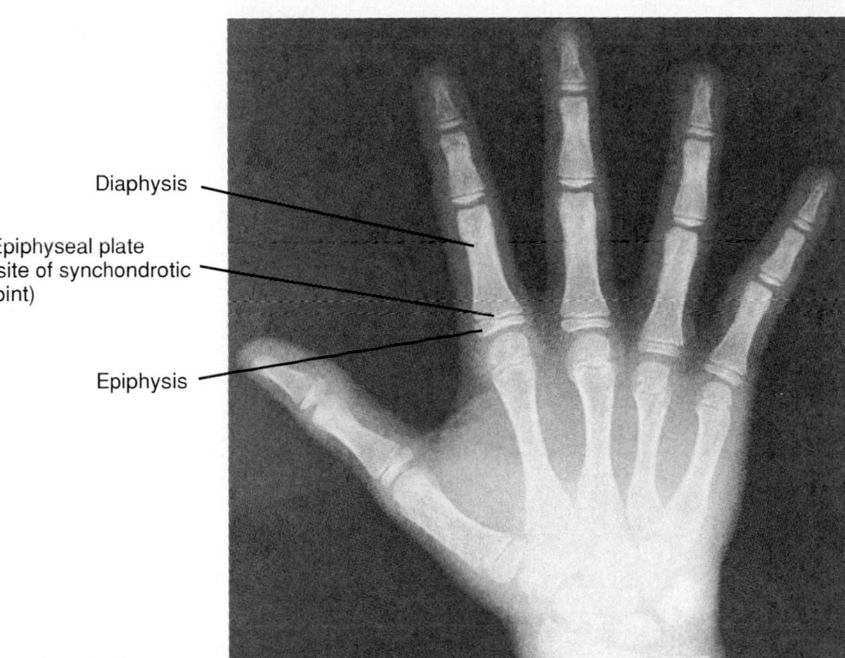

FIGURE 7.8
An X ray of the right hand of a 9-year-old child showing synchondrotic joints between the diaphyses and epiphyses of the long bones of the hand.

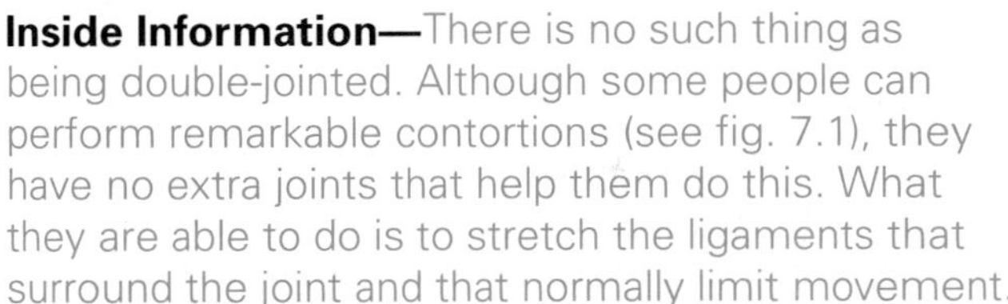

7.4 Synovial Joints

The freely movable synovial joints are enclosed by joint capsules containing synovial fluid. Types of synovial joints include gliding, hinge, pivot, condyloid, saddle, and ball-and-socket.

The most obvious type of articulation in the body is the freely movable **synovial** (*sĭ-no′ve-al*) **joint.** The function of synovial joints is to provide a wide range of precise, smooth movements, at the same time maintaining stability, strength, and, in certain aspects, rigidity in the body.

Synovial joints are the most complex and varied of the three major types of joints. A synovial joint's range of motion is determined by three factors:

1. the structure of the bones involved in the articulation (certain processes on bones, for example, may limit the range of motion preventing the overextension of a joint);
2. the strength and tautness of the associated ligaments, tendons, and joint capsule; and
3. the size, arrangement, and action of the muscles that span the joint.

Range of motion at synovial joints varies widely in different people, and is closely related to conditioning. Physical therapists often recommend range-of-motion (ROM) exercises to increase joint mobility, along with resistance exercises to strengthen the tissues that are part of the joint structure.

Inside Information—There is no such thing as being double-jointed. Although some people can perform remarkable contortions (see fig. 7.1), they have no extra joints that help them do this. What they are able to do is to stretch the ligaments that surround the joint and that normally limit movement.

Structure of a Synovial Joint

Synovial joints are enclosed by a fibroelastic **joint capsule,** which is filled with lubricating **synovial fluid** (fig. 7.9). The term *synovial* is derived from a Greek word meaning "egg white," which this fluid resembles. Synovial fluid is secreted by a thin **synovial membrane** that lines the inside of the capsule. Synovial fluid is similar to interstitial fluid (fluid surrounding cells of a tissue), and has a high concentration of hyaluronic acid, a lubricating substance. The bones that articulate in a synovial joint are capped with a smooth **articular cartilage,** which is only about 2 mm thick. Because articular cartilage lacks blood vessels, it has to be nourished by the synovial fluid.

Ligaments help to bind the bones together in a synovial joint and may be located within the joint cavity or on the outside of the capsule. Tough, fibrous, cartilaginous

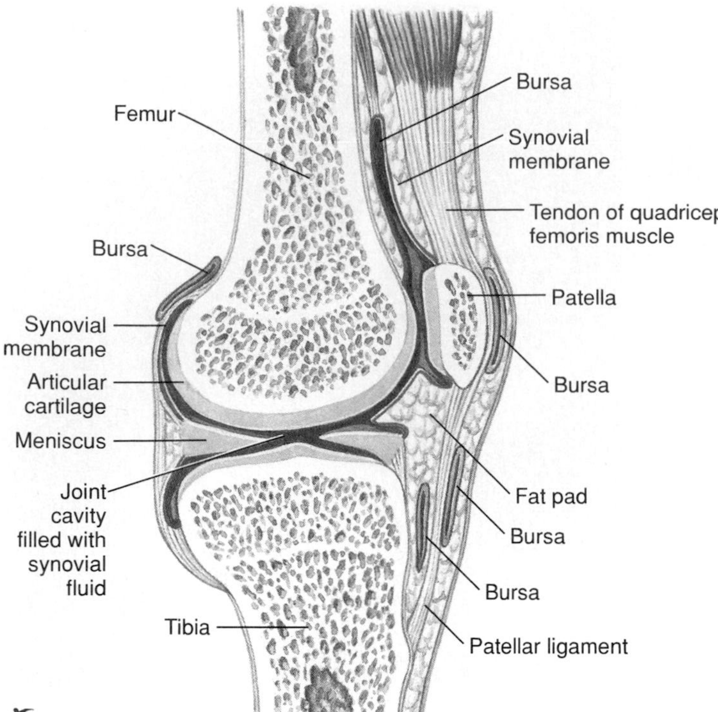

FIGURE 7.9
A synovial joint is represented by the knee joint, shown in a longitudinal section.

pads called **menisci** (*mĕ-nis'ki*—singular, *meniscus*) are located within the capsule of certain synovial joints (e.g., the knee joint). The menisci serve to cushion, as well as to guide, the articulating bones.

Articulating bones of synovial joints normally do not come in direct contact with one another. Articular cartilage caps the articular surface of each bone, and synovial fluid circulates through the joint during movement. Both minimize friction as well as cushion the articulating bones. Trauma or disease may render either of these two joint components nonfunctional, however. In this case, bony deposits will form, causing a type of arthritis.

Closely associated with some synovial joints are flattened, pouchlike sacs called **bursae** (singular, *bursa*), which are filled with synovial fluid (fig. 7.9). These closed sacs are commonly located between muscles or in sites where a tendon passes over a bone. They function to cushion muscles and assist movement of tendons or muscles over bony or ligamentous surfaces.

meniscus: Gk. *meniskos,* small moon
bursa: Gk. *byrsa,* bag or purse

Kinds of Synovial Joints

Synovial joints are classified into six main types according to their structure and the motion they permit. The six types are *gliding, hinge, pivot, condyloid, saddle,* and *ball-and-socket.*

Gliding

Gliding joints allow only side-to-side and back-and-forth movement. This is the simplest type of synovial joint movement. The articulating surfaces are nearly flat, or one may be slightly concave and the other slightly convex. The intercarpal and intertarsal joints are examples of gliding joints (fig. 7.10).

Hinge

The structure of a hinge joint permits bending in only one plane, much like the hinge of a door. In this type of articulation, the surface of one bone is concave, and the other convex (fig. 7.11). Hinge joints are the most common type of synovial joints. Examples include the knee, the elbow (humeroulnar articulation), and the joints between the phalanges.

Knee injuries are common among athletes. When their knees are wrenched or struck from the side during play, the ligaments and cartilage can be ripped and torn. The three C's—the anterior cruciate ligament, the collateral ligaments, and the cartilage—are the most likely sites of crippling injury. Joe Namath in football, Pete Maravich in basketball, and Bobby Orr in hockey are among the many star professionals whose playing days were cut short because of trauma to their knees.

Pivot

The movement in a pivot joint is limited to rotation about a central axis. In this type of articulation, the articular surface on one bone is conical or rounded and fits into a depression on another bone. Examples are the proximal articulation of the radius and ulna for rotation of the forearm, as in turning a doorknob, and the articulation between the atlas and axis that allows rotational movement of the head (fig. 7.12).

Condyloid

A condyloid (*kon'dĭ-loid*) articulation is structured so that an oval, convex surface of one bone fits into a concave depression on another bone. This permits movement in two directions (biaxial), as in up-and-down and side-to-side motions. The radiocarpal joint of the wrist and the metacarpophalangeal joints are examples of condyloid joints (fig. 7.13).

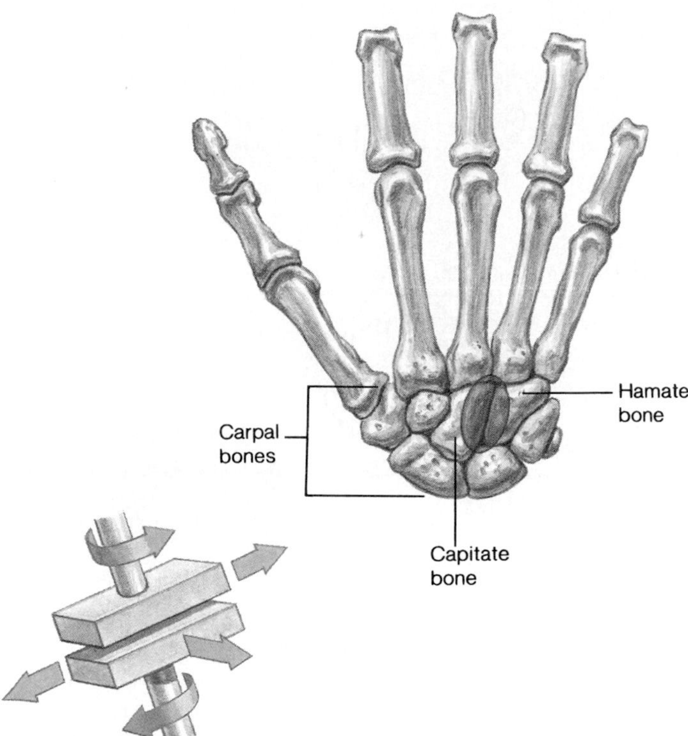

FIGURE 7.10
The intercarpal articulations in the wrist are examples of gliding joints in which the articulating surfaces of the adjacent bones are flattened or slightly curved. Note the diagrammatic representation showing the direction of possible movement.

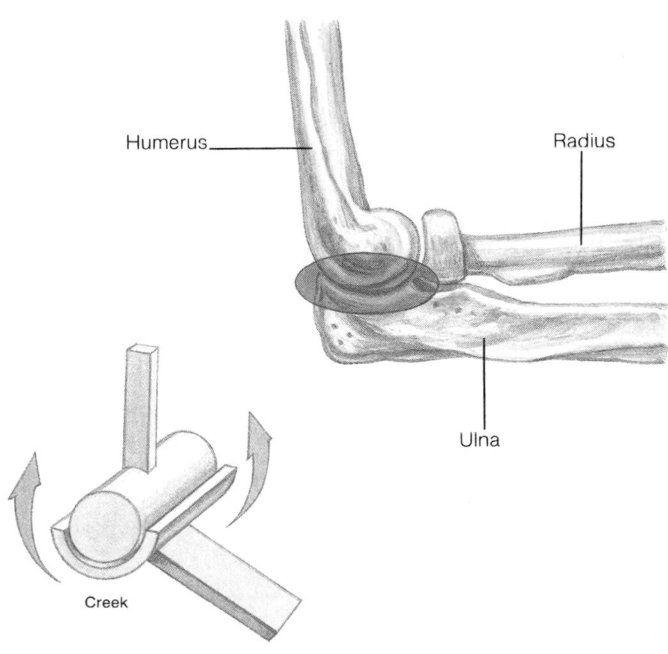

FIGURE 7.11
A hinge joint permits only a bending movement (flexion and extension). The hinge joint of the elbow involves the distal end of the humerus articulating with the proximal end of the ulna. Note the diagrammatic representation showing the direction of possible movement.

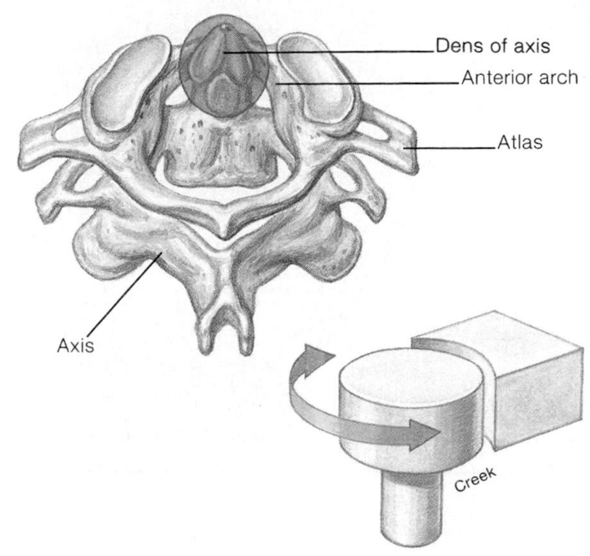

FIGURE 7.12
The atlas articulating with the axis forms a pivot joint that permits rotation. Note the diagrammatic representation showing the direction of possible movement. (Refer to fig. 7.11 and determine which articulating bones of the elbow region form a pivot joint.)

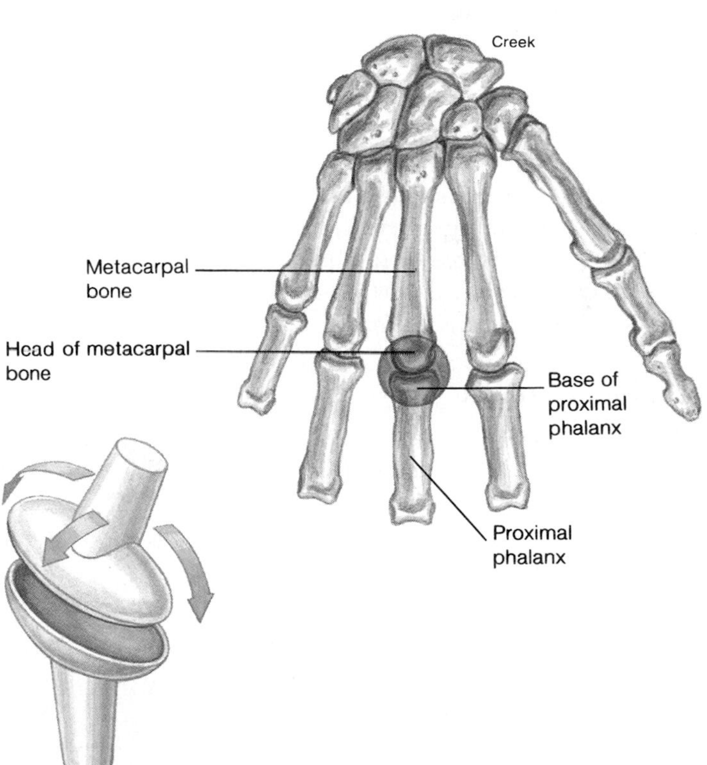

FIGURE 7.13
The metacarpophalangeal articulations of the hand are examples of condyloid joints in which the oval condyle of one bone articulates with the cavity of another. Note the diagrammatic representation showing the direction of possible movement.

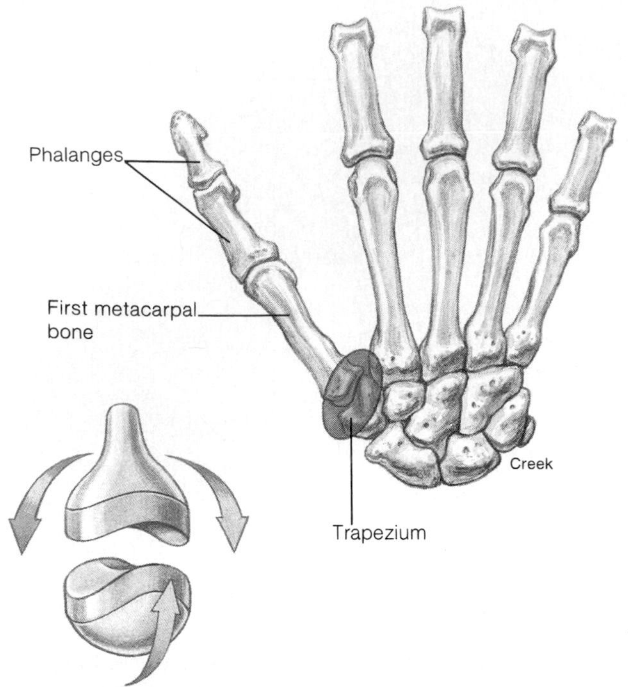

FIGURE 7.14
A saddle joint is formed as the trapezium articulates with the base of the first metacarpal bone. Note the diagrammatic representation showing the direction of possible movement.

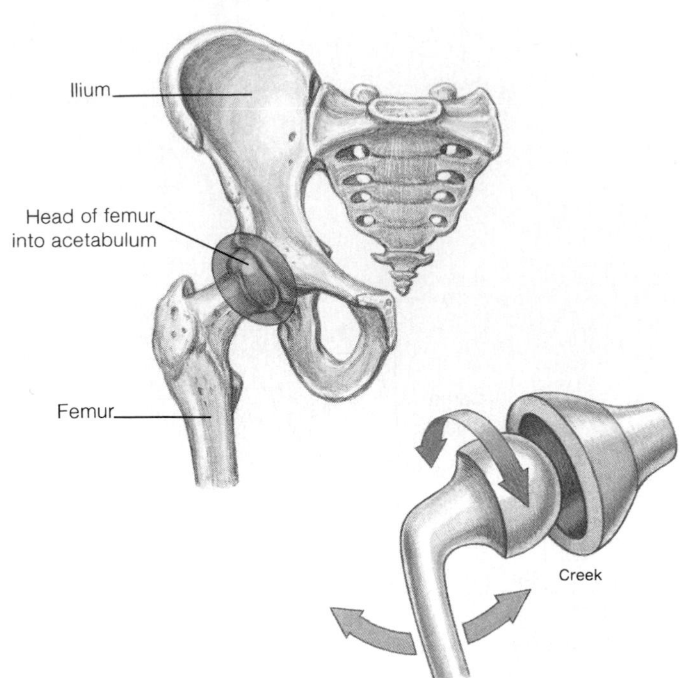

FIGURE 7.15
A ball-and-socket articulation illustrated by the hip joint. Note the diagrammatic representation showing the direction of possible movement.

Saddle

Each articular process of a saddle-shaped joint has a concave surface in one direction and a convex surface in another. This articulation is a modified condyloid joint that allows a wide range of movement. The only saddle joint in the body is located at the base of the thumb, where the trapezium of the carpus (wrist) articulates with the first metacarpal bone (fig. 7.14).

Ball-and-Socket

Ball-and-socket joints are formed by the articulation of a rounded convex surface with a cuplike cavity (fig. 7.15). This type of articulation provides the greatest range of movement of all the synovial joints. Examples are the hip and shoulder joints.

The shoulder joint is vulnerable to dislocations from sudden jerks on the arm, especially in children whose shoulder muscles are still relatively weak. Parents should be careful not to force a child to follow by yanking on the arm. Dislocation of the shoulder is extremely painful and may cause permanent damage or even muscle atrophy. The hip joint, by contrast, is not easily dislocated because of the depth of the joint socket (acetabulum) and the powerful muscles surrounding the joint.

A summary of the various types of joints is presented in table 7.1.

Table 7.1 Types of Articulations

Type	Structure	Movements	Example
Fibrous Joints	Articulating parts joined by fibrous connective tissue		
1. Suture	Edges of articulating bones frequently jagged; separated by thin layer of fibrous tissue	None	Sutures between bones of the skull
2. Syndesmoses	Articulating bones bound by interosseous ligament	Slightly movable	Joints between tibia-fibula and radius-ulna
3. Gomphoses	Teeth bound into dental alveoli of bone by periodontal ligament	None	Teeth secured into dental alveoli (sockets)
Cartilaginous Joints	Skeletal parts joined by fibrocartilage or hyaline cartilage		
1. Symphyses	Articulating bones separated by pad of fibrocartilage	Slightly movable	Intervertebral joints; symphysis pubis and sacroiliac joint
2. Synchondroses	Mitotically active hyaline cartilage between skeletal parts	None	Epiphyseal plates within long bones; costal cartilages of rib cage
Synovial Joints	Joint capsule containing synovial membrane and synovial fluid		
1. Gliding	Flattened or slightly curved articulating surfaces	Sliding	Intercarpal and intertarsal joints
2. Hinge	Concave surface of one bone articulates with convex surface of another	Bending motion in one plane	Knee; elbow; joints of phalanges
3. Pivot	Conical surface of one bone articulates with a depression of another	Rotation about a central axis	Atlantoaxial joint; proximal radioulnar joint
4. Condyloid	Oval condyle of one bone articulates with cavity of another	Biaxial movement	Radiocarpal joint; metacarpophalangeal joint
5. Saddle	Concave and convex surface on each articulating bone	Wide range of movements	Carpometacarpal joint of thumb
6. Ball-and-socket	Rounded convex surface of one bone articulates with cuplike socket of another	Movement in all planes and rotation	Shoulder and hip joints

7.5 Movements at Synovial Joints

Movements at synovial joints are broadly classified as angular and circular. These movements are produced by the contraction of muscles crossing the joints and attaching to or near the bones forming the articulations.

Angular Movements

Angular movements increase or decrease the joint angle produced by the articulating bones. The four types of angular movements are *flexion, extension, abduction,* and *adduction.* The descriptions of all joint movements are in reference to a person in anatomical position, as described in chapter 1 (see module 1.4).

flexion: L. *flectere,* to bend

Flexion

Flexion is a movement that decreases the joint angle on an anterior-posterior plane (fig. 7.16*a*). Examples are bending the elbow or knee. In the ankle joint, flexion occurs as the top surface of the foot is elevated. This movement is frequently called **dorsiflexion** (fig. 7.16*b*). Pressing the foot downward (as in rising on the toes) is called **plantar flexion.** The shoulder joint is flexed when the arm is brought forward, decreasing the joint angle.

Inside Information—Technically speaking, it is impossible to flex a muscle. A person contracts a muscle to flex a joint.

Extension

In extension, which is the reverse of flexion, the joint angle is increased (fig. 7.16*a*). Extension returns a body part to the anatomical position. In an extended joint, the angle between the articulating bones is 180°. An exception is the ankle joint, in which a 90° angle already exists between the foot and leg. Examples of extension are the straightening of the elbow or knee joints from flexion positions. **Hyperextension** occurs when a portion of the body is extended beyond the anatomical position so that the joint angle is greater than 180°. An example of hyperextension is bending the neck to tilt the head backward, as in looking at the sky.

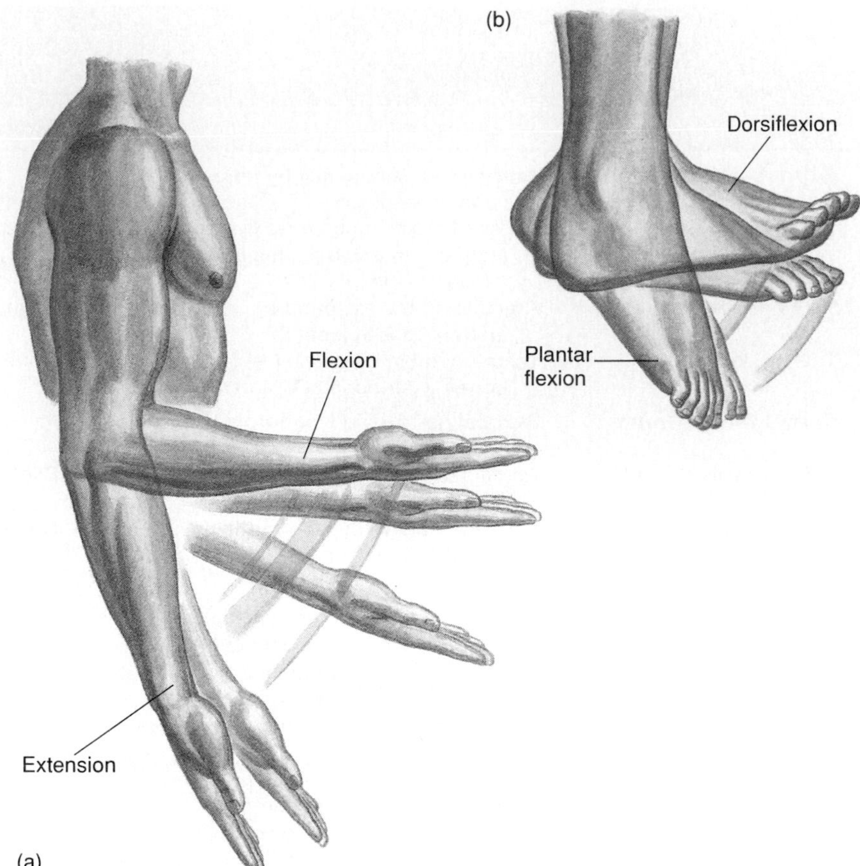

Abduction

Abduction is the movement of a body part away from the main axis of the body, or away from the midsagittal plane, in a lateral direction (fig. 7.16*c*). An example of abduction is moving the arms sideward, away from the body. Spreading the fingers apart is another example.

Adduction

Adduction, the opposite of abduction, is the movement of a body part toward the main axis of the body (fig. 7.16*c*). In the anatomical position, the arms and legs have been adducted toward ("*add*ed" to) the midplane of the body.

Circular Movements

Joints that permit circular movement are composed of a bone with a rounded or oval surface that articulates with a corresponding depression on another bone. The two types of circular movements are *rotation* and *circumduction.*

Rotation

Rotation is the movement of a bone around its own axis (see figs. 7.12 and 7.17*a*). Examples are turning the head from side to side and twisting at the waist.

Supination is a specialized rotation of the forearm that results in the palm being turned forward (anteriorly). In the supine position, the ulna and radius of the forearm are parallel and in anatomical position. **Pronation** is the opposite of supination. It is a rotational movement of the forearm that results in the palm being directed backward (posteriorly).

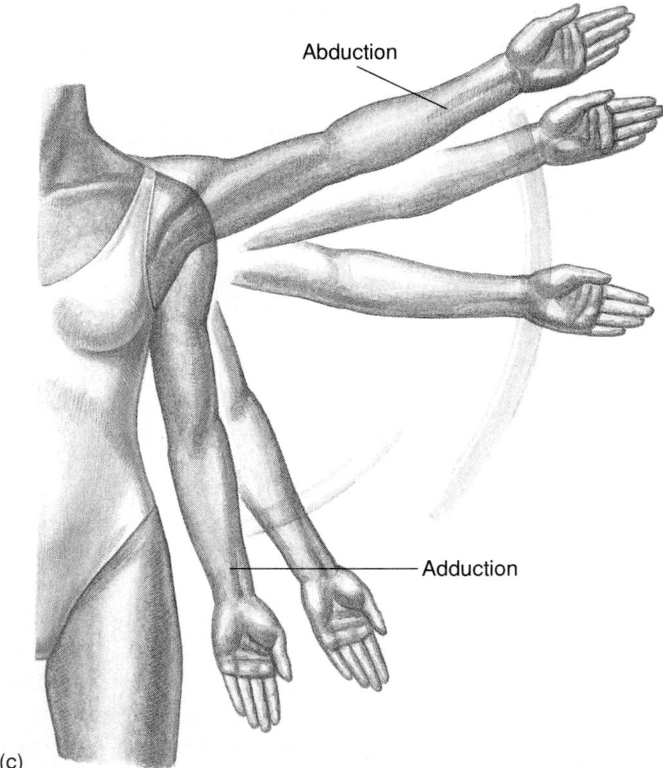

FIGURE 7.16
Angular movements within synovial joints include (*a*) flexion and extension, (*b*) dorsiflexion and plantar flexion, and (*c*) abduction and adduction.

extension: L. *ex,* out, away from; *tendere,* stretch
abduction: L. *abducere,* lead away
adduction: L. *adductus,* bring to
rotation: L. *rotare,* a wheel

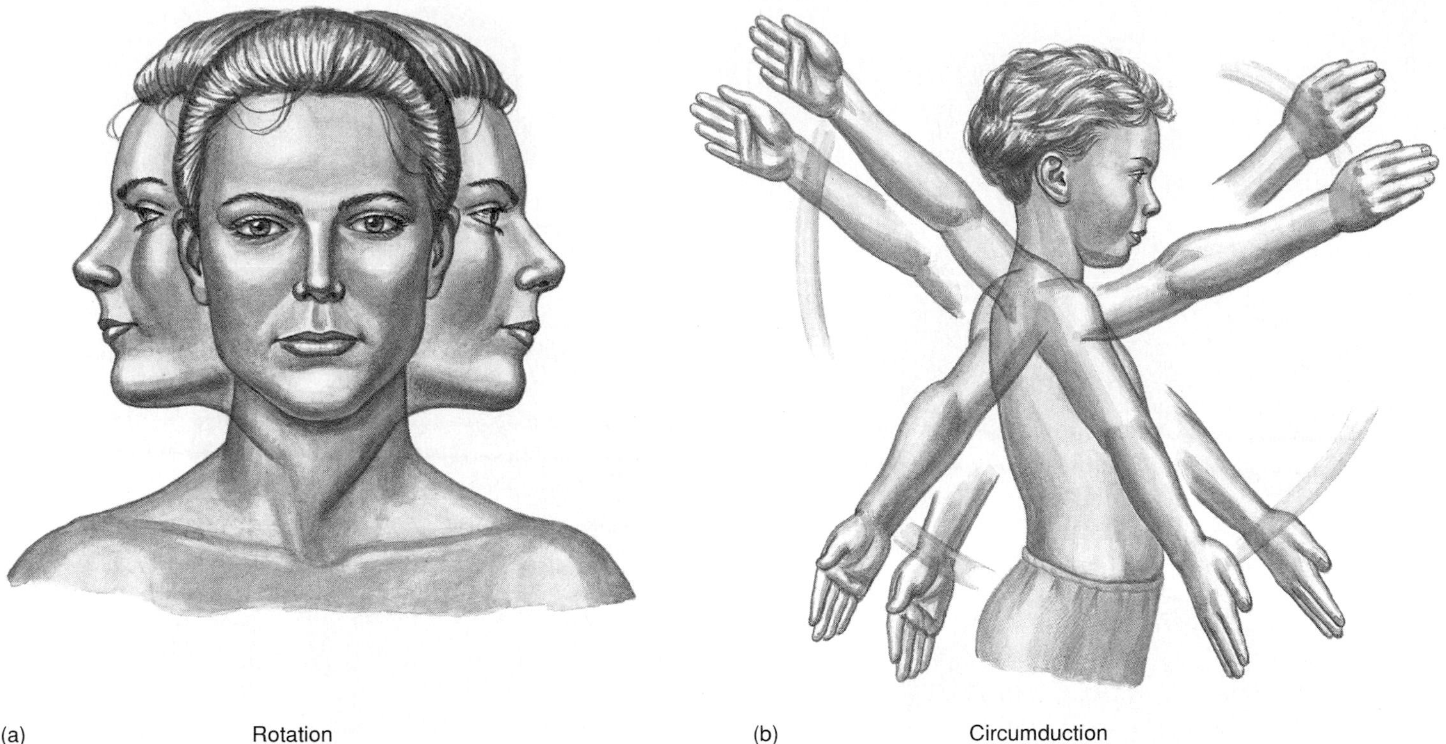

(a) Rotation (b) Circumduction

FIGURE 7.17
Circular movements within synovial joints include (*a*) rotation and (*b*) circumduction.

Applied to the foot, the term *pronation* describes a combination of eversion (see below) and abduction movements that result in a lowering of the longitudinal arch.

Circumduction

Circumduction is the circular movement of a body part so that a cone-shaped airspace is traced. The distal extremity performs the circular movement, and the proximal attachment at the joint serves as the pivot (fig. 7.17*b*). This type of movement is possible at the shoulder, wrist, trunk, hip, and ankle joints.

Special Movements

Because the terms used to describe most movements around an axis do not apply to all joints, other terms must be used. **Inversion** is the movement of the sole of the foot inward or medially (fig. 7.18*a*). **Eversion,** the opposite of inversion, is the movement of the sole outward or laterally.

The condition of the heels of your shoes can tell you whether you invert or evert your foot as you walk. If the heel is worn down on the outer side, you tend to invert your foot as you walk. If the heel is worn down on the inside, you tend to evert your foot.

Protraction is the movement of part of the body forward on a plane parallel to the ground, such as the thrusting out of the lower jaw (fig. 7.18*b*). **Retraction** is the opposite of protraction and is the pulling back of a protracted part of the body on a plane parallel to the ground.

Elevation is a movement that raises a body part. Examples include elevating the mandible to close the mouth and lifting the shoulders to shrug (fig. 7.18*c*). **Depression** is the opposite of elevation. Both the mandible and shoulders are depressed when moved downward.

Any of the synovial joints can be dislocated, some more easily than others. The temporomandibular joint between the temporal bone of the skull and the mandible is one that is dislocated often. This joint can be easily palpated by applying pressure to the area in front of your ear and opening and closing your mouth. It is most likely to dislocate when the mandible is depressed, as in yawning. Relocating the jaw is usually a simple task, however, and is accomplished by pressing downward on the lower molars while pushing the jaw backward.

Temporomandibular joint (TMJ) syndrome *is an ailment that afflicts an estimated 75 million Americans. The likely cause of TMJ syndrome is a malalignment of one or both temporomandibular joints. The symptoms of the condition vary from occasional pain to intense and continuous pain of the head, neck, shoulders, or back. Dizziness and ringing in the ears may also occur.*

Many of the movements permitted at synovial joints are visually summarized in figures 7.19 through 7.21.

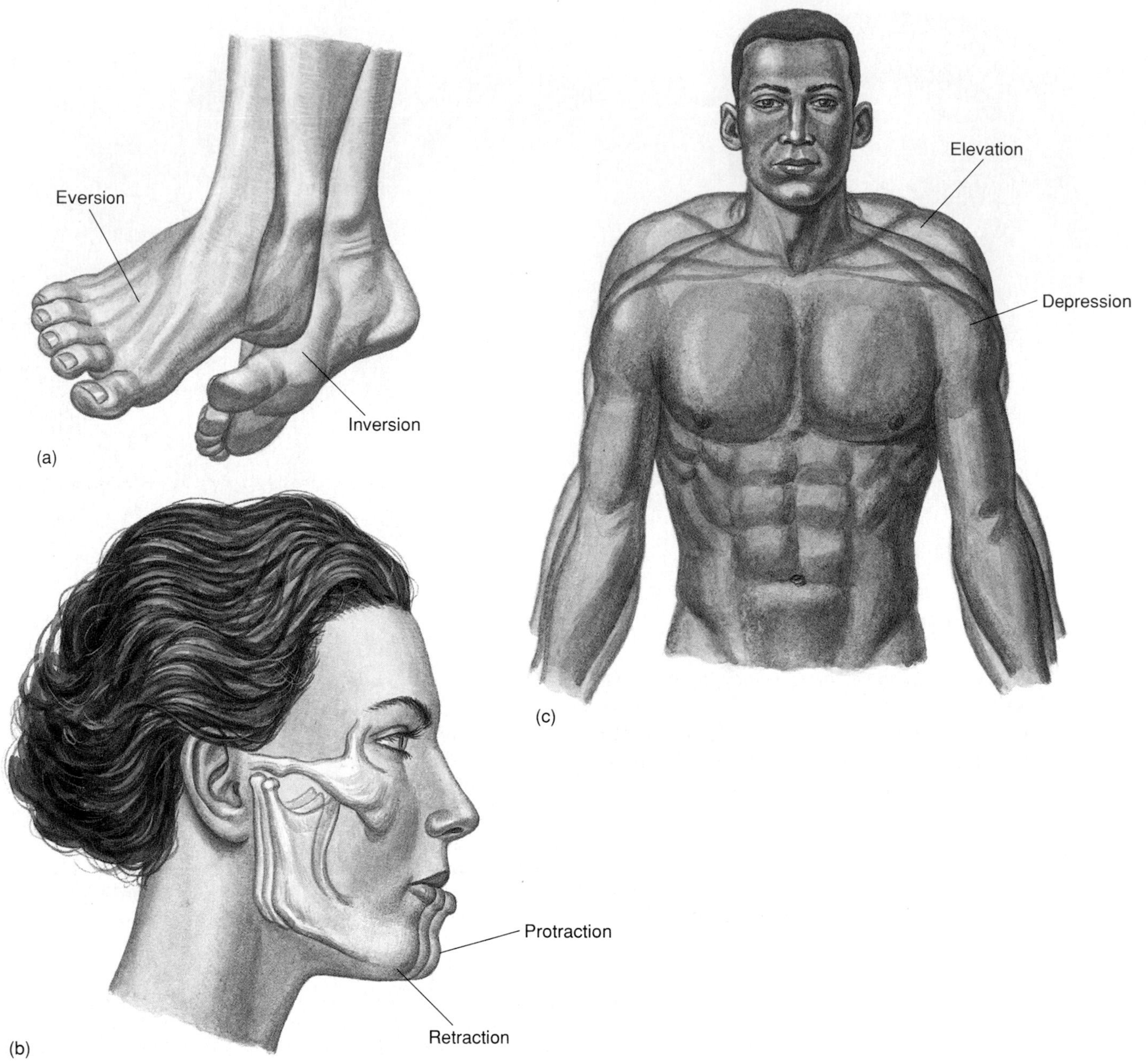

FIGURE 7.18

Special movements within synovial joints include (*a*) inversion and eversion, (*b*) protraction and retraction, and (*c*) elevation and depression.

FIGURE 7.19

A photographic summary of joint movements. (*a*) Adduction of shoulder, hip, and carpophalangeal joints; (*b*) abduction of shoulder, hip, and carpophalangeal joints; (*c*) rotation of vertebral column; (*d*) lateral flexion of vertebral column; (*e*) flexion of vertebral column; (*f*) hyperextension of vertebral column; (*g*) flexion of shoulder, hip, and knee joints of right side of body; extension of elbow and wrist joints; (*h*) hyperextension of shoulder and hip joints on right side of body; plantar flexion of right ankle joint.

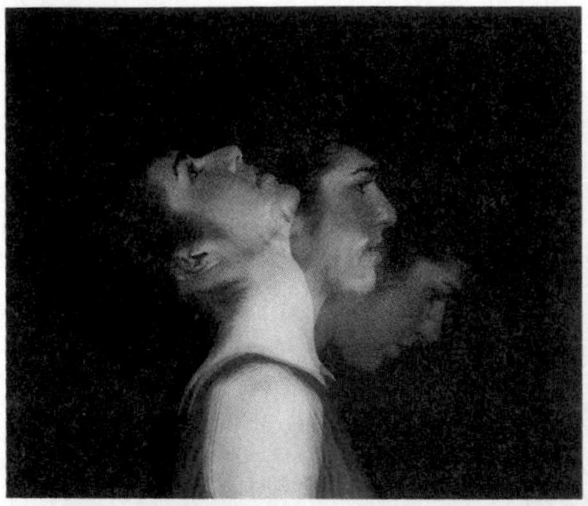

(a) 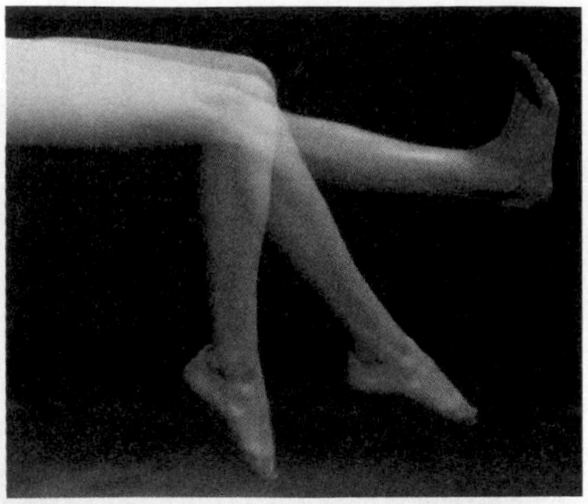 (b)  (c)

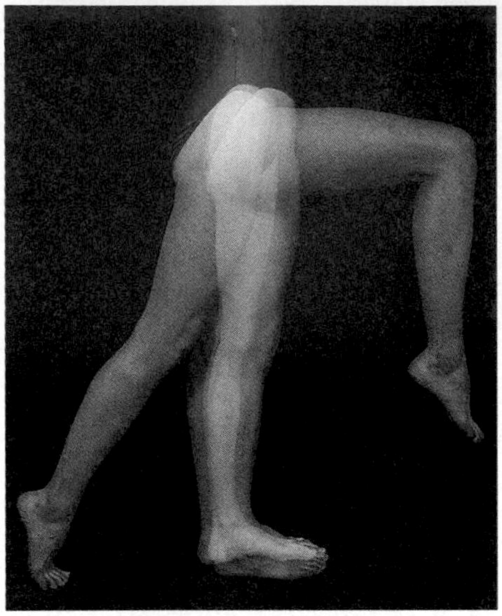

(d)  (e) 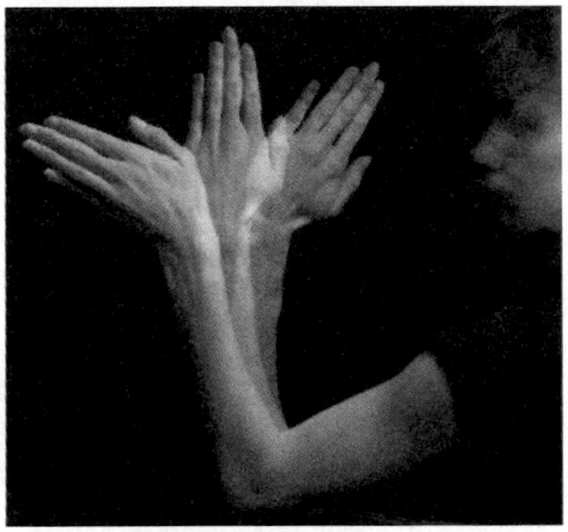 (f)

FIGURE 7.20

A visual summary of some angular movements at synovial joints. (*a*) Flexion, extension, and hyperextension in the cervical region; (*b*) flexion and extension at the knee joint, and plantar flexion and dorsiflexion at the ankle joint; (*c*) flexion and extension at the elbow joint, and flexion, extension, and hyperextension at the wrist joint; (*d*) flexion, extension, and hyperextension at the hip joint, and flexion and extension at the knee joint; (*e*) adduction and abduction of the arm and fingers; (*f*) posterior view of abduction and adduction of the hand at the wrist joint. Note that the range of abduction at the wrist joint is less extensive than the range of adduction as a result of the length of the styloid process of the radius.

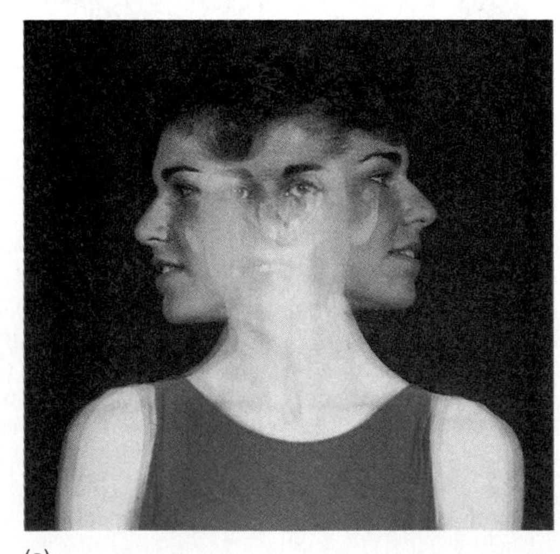

(a)

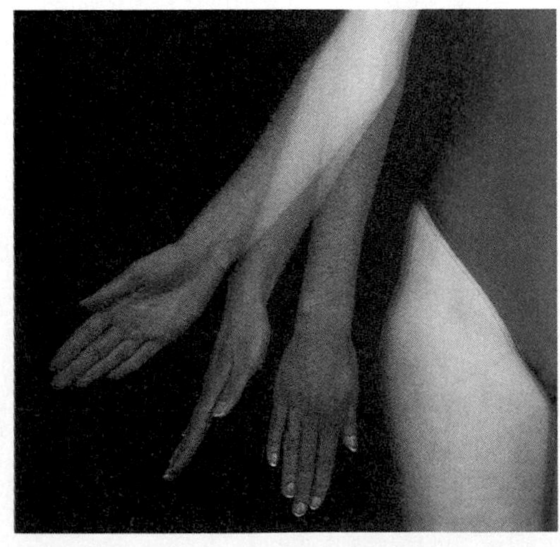

(b)

FIGURE 7.21

A visual summary of some rotational movements at synovial joints. (*a*) Rotation of the head at the cervical vertebrae, especially at the atlantoaxial joint, and (*b*) rotation of the forearm (antebrachium) at the proximal radioulnar joint.

Clinical Considerations

A synovial joint is a remarkable biologic system. Its self-lubricating action provides a shock-absorbing cushion between articulating bones and enables almost frictionless movement under tremendous loads and impacts. Under normal circumstances and in most people, the many joints of the body perform without problems throughout life. Joints are not indestructible, however, and are subject to various forms of trauma and disease.

Trauma to Joints

In a **strained joint,** unusual or excessive exertion stretches the tendons or muscles surrounding a joint. The damage is not serious. Strains are frequently caused by not "warming up" the muscles and not "stretching" the joints prior to exercise. A **sprain** is a tearing of the ligaments or tendons surrounding a joint. There are various grades of sprains, and the severity determines the treatment. Severe sprains damage articular cartilages and may require surgery. Sprains are usually accompanied by **synovitis,** an inflammation of the joint capsule.

Luxation, or **joint dislocation,** is derangement of the articulating bones that compose the joint. The shoulder and knee joints are the most vulnerable to dislocation. Without medical treatment, a dislocated joint will heal itself. This self-healing may be incomplete, however, leaving the person with a "trick knee," for example, that may unexpectedly give way. Professional treatment is recommended for luxations to properly realign the bones and avoid the possibility of impinging on nerves or blood vessels.

Bursitis is an inflammation of a bursa. Because of its closeness to the joint, bursitis may even affect the joint capsule. Bursitis may be caused by excessive stress on the bursa from overexertion, or it may be a local or widespread inflammatory process. As the bursa swells, the surrounding muscles become sore and stiff.

The most common cause of back pain is *strained muscles,* generally the result of overexertion. The second most frequent back ailment is a *herniated disc.* In this condition, the jellylike center of a disc squeezes out and pushes against a spinal nerve, causing excruciating pain. The third most frequent back problem involves dislocated articular facets between two vertebrae. This may be caused by sudden twisting of the vertebral column. Treatment of back ailments varies from bed rest to spinal manipulation to extensive surgery.

Curvature disorders are another problem of the vertebral column. **Kyphosis** (*ki-fo'sis*) (hunchback) is an exaggeration of the thoracic curve. **Lordosis** (swayback) is an excessive lumbar curve. **Scoliosis** (crookedness) is an abnormal lateral curvature of the vertebral column (fig. 7.22).

Inside Information—Quasimodo, the protagonist of Victor Hugo's novel *The Hunchback of Notre Dame* is the most famous fictional sufferer of kyphosis, the abnormally exaggerated backward curvature of the thoracic spine.

continued

kyphosis: Gk. *kyphos,* hunched
lordosis: Gk. *lordos,* curving forward
scoliosis: Gk. *skoliosis,* crookedness

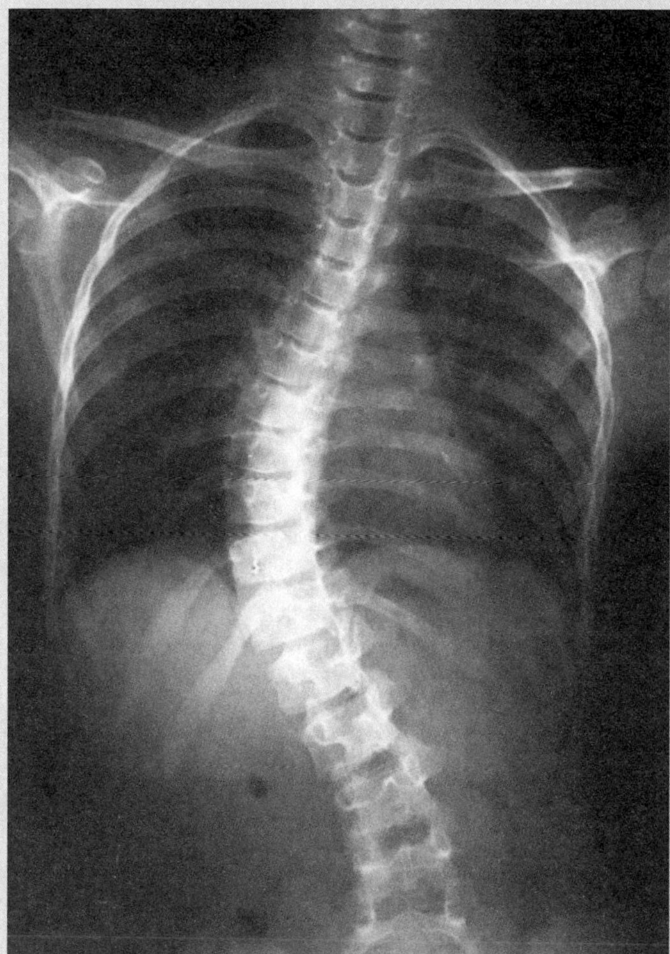

FIGURE 7.22
Scoliosis is a lateral curvature of the spine usually in the thoracic region. It may be congenital, acquired, or disease-related.

Clinical Considerations

Diseases of Joints

Arthritis is a generalized designation for over 50 different joint diseases, all of which have the symptoms of edema, inflammation, and pain. The causes are unknown, but certain types follow joint trauma or bacterial infection. Some types are genetic and others result from hormonal or metabolic disorders. The most common forms are *rheumatoid arthritis, osteoarthritis,* and *gouty arthritis.*

In **rheumatoid arthritis,** the synovial membrane thickens and becomes tender, and synovial fluid accumulates. This is generally followed by gradual deterioration of the articular cartilage, which eventually exposes bone tissue. When bone tissue is unprotected, joint ossification produces the crippling effect of this disease.

Osteoarthritis is a degenerative joint disease that results from aging and irritation of the joints. Although osteoarthritis is far more common than rheumatoid arthritis, it is usually less damaging. It is a progressive disease in which the articular cartilages gradually soften and disintegrate. Osteoarthritis most frequently affects the knee, hip, and intervertebral joints.

Gouty arthritis results from a metabolic disorder in which an abnormal amount of uric acid is retained in the blood and sodium urate crystals are deposited in the joints. It is as if "metabolic trash" was being disposed of in the wrong place. The salt crystals irritate the articular cartilage and synovial membrane, causing swelling, tissue deterioration, and pain.

Treatment of Joint Disorders

Arthroscopy (*ar-thros'kŏ-pe*) is a procedure widely used in diagnosing and, to a limited extent, treating joint disorders. Arthroscopic inspection involves making an incision through the skin and joint capsule through which the tubelike arthroscopic instrument is threaded (fig. 7.23). In arthroscopy of the knee, the articular cartilage,

rheumatoid: Gk. *rheuma,* a flowing
gout: L. *guta,* a drop (thought to be caused by "drops of viscous humors")

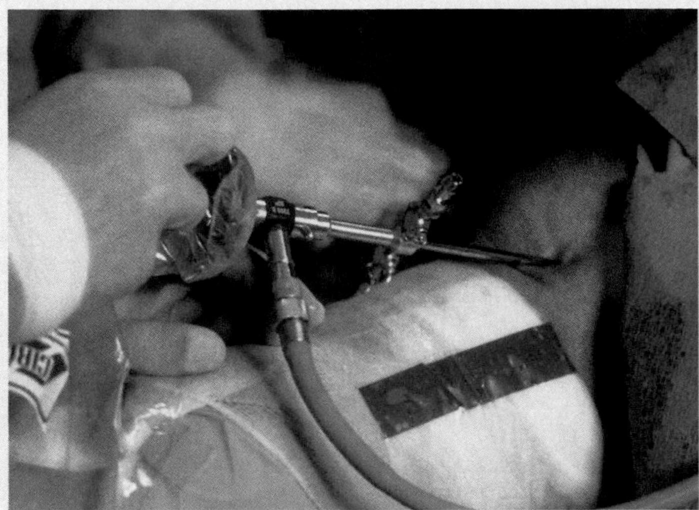

FIGURE 7.23
Arthroscopy. In this technique, a needlelike viewing arthroscope is threaded into the joint capsule through a tiny incision. The arthroscope has a fiberoptic light source that illuminates the interior of the joint; thus, the position of surgical instruments that may be inserted through other small incisions can be seen.

SUMMARY

7.1 Classification of Joints

1. Joints are formed as adjacent bones articulate. Arthrology is the science concerned with the study of joints, and kinesiology is the study of movements involving certain joints.
2. Structurally, joints are classified as fibrous, cartilaginous, and synovial types.

7.2 Development of Synovial Joints

1. Development of synovial joints is initiated during the sixth week as the perichondrium from adjacent developing bones spans the future joint cavity and becomes the joint capsule.
2. Synovial joints form as the epiphyses of adjacent bones develop and as the muscles that move the joints undergo contractions.

synovial membrane, menisci, and cruciate ligaments can be observed. Samples can be extracted, and pictures taken for further evaluation.

Remarkable advancements have been made in **joint prostheses** (*pros-the'sēz*) (fig. 7.24). These artificial articulations do not take the place of normal, healthy joints, but they are a viable option for reducing unbearable pain and increasing joint mobility in chronically disabled arthritis patients.

prosthesis: Gk. *pros,* in addition to; *thesis,* a setting down

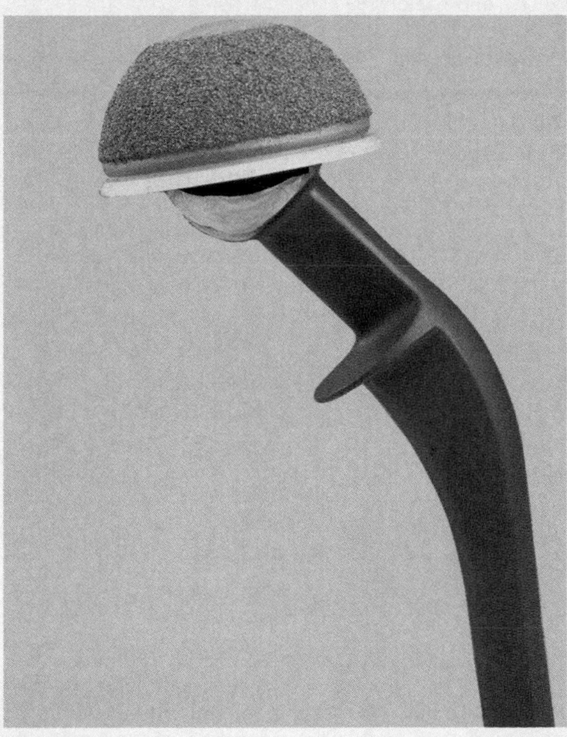

(a)

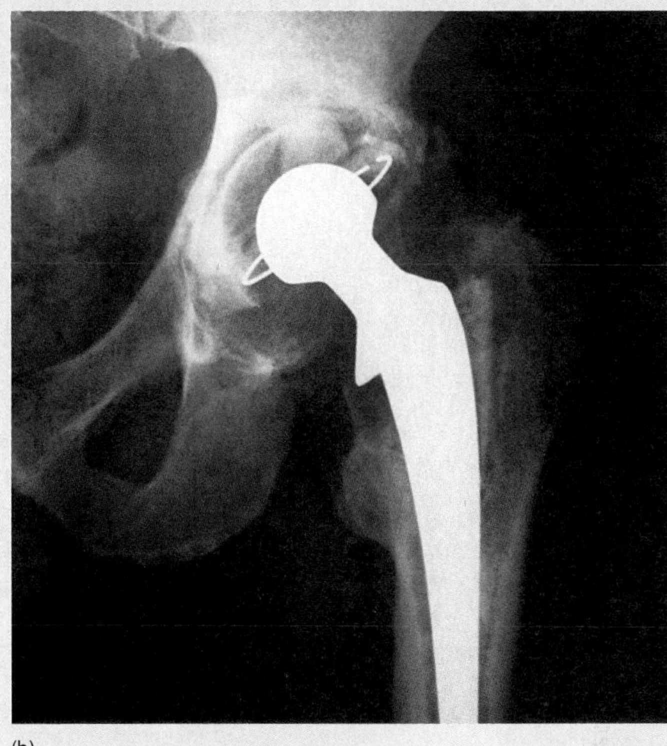

(b)

FIGURE 7.24

Joint prostheses (replacements) are available for all synovial joints. These "artificial joints" generally consist of a stainless steel portion (*a*) that articulates with a durable plastic portion. The X ray (*b*) shows a hip prosthesis in place.

7.3 Fibrous Joints and Cartilaginous Joints

1. Articulating bones in fibrous joints are tightly connected by fibrous connective tissue. Fibrous joints are either immovable or slightly movable. They are of three types: sutures, syndesmoses, and gomphoses.
2. Sutures are found only in the skull and are classified as serrate, lap, or plane.
3. Syndesmoses are found in the forearm and leg, where adjacent bones are held together by interosseous ligaments. Slight movement is permitted at syndesmoses.
4. Gomphoses are found only in the skull, where the teeth are bound into their sockets by periodontal ligaments.
5. Cartilaginous joints allow limited motion in response to twisting or compression. The two types are symphyses and synchondroses.
6. The symphysis pubis and the intervertebral disc joints are examples of symphyses.
7. Some synchondroses are temporary joints formed in the growth plates between the diaphyses and epiphyses in the long bones of children. Other synchondroses are permanent; for example, the costal cartilages of the rib cage.

7.4 Synovial Joints

1. The freely movable synovial joints are enclosed by joint capsules that contain synovial fluid. Types of synovial joints include gliding, hinge, pivot, condyloid, saddle, and ball-and-socket.

2. Synovial joints contain a joint cavity, synovial membrane, and articular cartilage. Some also contain articular discs, accessory ligaments, and associated bursae.
3. The movement of a synovial joint is determined by the structure of the articulating bones, the ligaments and tendons, and the muscles that act on the joint.

7.5 Movements at Synovial Joints

1. Movements at synovial joints are produced by the contraction of the skeletal muscles crossing the joints and attaching to or near the bones forming the articulations.
2. Angular movements increase or decrease the joint angle produced by the articulating bones. Flexion decreases the joint angle on an anterior-posterior plane; extension increases the same joint angle. Abduction is the movement of a body part away from the main axis of the body; adduction is the movement of a body part toward the main axis of the body.
3. Circular movements can occur only in joints that are composed of a bone with a rounded surface that articulates with a corresponding depression on another bone. Rotation is the movement of a bone around its own axis. Circumduction is a conelike movement of a body part.
4. Special joint movements include inversion and eversion, protraction and retraction, and elevation and depression.

Review Activities

Objective Questions

1. Which statement regarding joints is *false?* (*module 7.1*)
   a. Joints are the locations where two or more bones articulate.
   b. The definition of a joint implies that movement is permitted.
   c. Arthrology is the study of joints; kinesiology is the study of body movement.

2. Synovial fluid is secreted by (*module 7.2*)
   a. the synovial membrane.
   b. the joint capsule.
   c. the articular cartilage.
   d. a meniscus.

3. Synchondroses are a type of (*module 7.3*)
   a. fibrous joint.
   b. synovial joint.
   c. cartilaginous joint.

4. An interosseous ligament is characteristic of (*module 7.3*)
   a. a suture.
   b. a synchondrosis.
   c. a symphysis.
   d. a syndesmosis.

5. Which of the following joint type–function word pairs is *incorrect?* (*module 7.3*)
   a. synchondrosis/growth at the epiphyseal plate
   b. symphysis/movement at the intervertebral joint
   c. suture/strength and stability in the skull
   d. syndesmosis/movement of the jaw

6. Which of the following is *not* characteristic of all synovial joints? (*module 7.4*)
   a. articular cartilage
   b. synovial fluid
   c. a joint capsule
   d. a meniscus

7. The atlantoaxial and the proximal radioulnar synovial joints are specifically classified as (*module 7.4*)
   a. hinge joints.
   b. gliding joints.
   c. pivotal joints.
   d. condyloid joints.

8. Movement of a body part away from the main axis of the body, or away from the midsagittal plane, is the definition of (*module 7.5*)
   a. flexion.
   b. abduction.
   c. circumduction.
   d. adduction.

9. A thickening and tenderness of the synovial membrane and the accumulation of synovial fluid are signs of the development of (*Clinical Considerations*)
   a. arthroscopitis.
   b. gouty arthritis.
   c. osteoarthritis.
   d. rheumatoid arthritis.

Essay Questions

1. Why is the anatomical position so important in explaining the movements that are possible at joints? (*module 7.1*)
2. Describe the characteristics of fibrous joints and cartilaginous joints. (*module 7.3*)
3. Define *gomphosis* and explain why the gomphoses are of importance in the field of dentistry. (*module 7.3*)
4. What general feature is characteristic of all synchondroses? Give an example of a synchondrosis. (*module 7.3*)
5. What are the structural components of a synovial joint that determine the amount of movement permitted? (*module 7.4*)
6. What is synovial fluid? Where is it produced, and what are its functions? (*module 7.4*)
7. What is a bursa? Where are bursae found and what do they contain? (*module 7.4*)
8. Give an example of each of the six kinds of synovial joints and describe the range of movement possible at each. (*modules 7.4 and 7.5*)
9. What is meant by a sprained ankle? How does a sprain differ from a strain or a luxation? (*Clinical Considerations*)

Labeling Exercise

Label the structures indicated on the figure to the right.

1. ______________________________
2. ______________________________
3. ______________________________
4. ______________________________
5. ______________________________
6. ______________________________

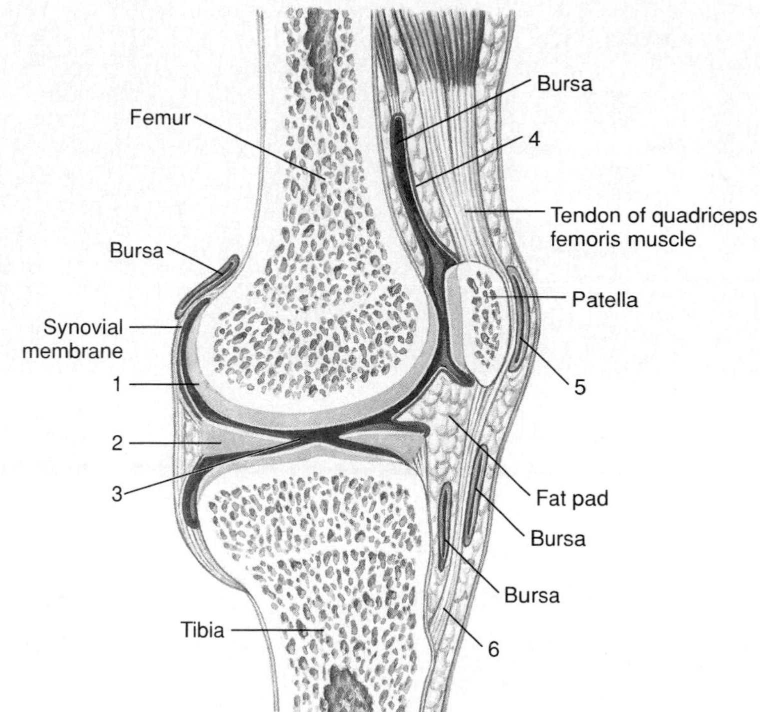

Critical Thinking Questions

1. Located in bursae and synovial joints, synovial fluid is contained at the locations where it is produced. It has been described as the perfect body lubricant. Discuss the value of synovial fluid and identify at least two clinical conditions involving this fluid.
2. If ball-and-socket joints permit the greatest range of movement, why aren't all the synovial joints of this type?
3. Identify four types of synovial joints located in the hand and state the types of movement permitted by each.
4. From the previous chapter on the skeletal system, you learned that the periosteum does not cover the articular cartilage. Review the functions of the periosteum and explain why it is not found in synovial joints.
5. In anticipation of what you will learn about muscles in the following chapter, explain why each movement has an opposite movement. For example, extension is the opposite movement of flexion, and abduction is the opposite of adduction.

"Being mentally 'psyched up' to accomplish an athletic feat involves voluntary activation of more motor units within the muscles." *page 181*

Muscular System

Chapter Outline

Terms to Remember

fascia (*fash' e-ă*), 176
sarcolemma (*sar"kŏ-lem' ă*), 178
myofibril, 178
myofilament, 178
actin, 178
myosin, 178
sarcomere (*sar' kŏ-mēr*), 179
twitch, 180
summation, 180
tetanus, 180
treppe (*trep' e*), 180
motor unit, 180
isotonic contraction, 181
isometric contraction, 181
tropomyosin (*tro"pŏ-mi' ŏ-sin*), 185
troponin (*tro' pŏ-nin*), 185
phosphocreatine (*fos"fo-kre' ă-tēn*), 186
oxygen debt, 187

Learning Objectives

After studying this chapter, you should be able to

- describe the arrangement of muscle fibers and connective tissue within a skeletal muscle.
- describe the banding pattern of a skeletal muscle fiber.
- explain the nature of a muscle twitch and describe how summation and tetanus are produced.
- distinguish between isometric and isotonic contractions.
- define *motor unit* and discuss the relationship between motor unit recruitment and dexterity.
- describe the sliding filament mechanism of contraction, noting how the bands in a muscle fiber change during contraction.
- discuss the role of calcium ions in muscle contraction and explain how electrical stimulation influences the availability of Ca^{++}.
- distinguish between fast-twitch and slow-twitch fibers and explain how muscles adapt to exercise training.
- locate the major muscles of the axial skeleton and give some examples of synergistic and antagonistic muscles.
- locate the major muscles of the appendicular skeleton and give some examples of synergistic and antagonistic muscles.

8.1 Introduction to the Muscular System

Skeletal muscles are composed of contractile fibers. When the fibers contract, they exert tension on the layers of associated connective tissues and tendons. Fascia binds muscles to each other and to other structures.

Over 600 muscles make up your muscular system and account for approximately 40% of your body weight. As you learned in chapter 1 (see module 1.3), an organ is a structure consisting of two or more tissue types that performs a specific function. In this context, each skeletal muscle of the body is an organ—it is composed of skeletal muscle tissue, connective tissue, and nervous tissue. And each muscle has a particular function, such as moving a finger or blinking an eyelid.

Muscles serve three general functions: (1) motion, (2) heat production, and (3) posture and body support.

1. **Motion.** The most obvious function performed by skeletal muscles is to move the body or parts of the body, as in walking, writing, chewing, and swallowing. The contraction of skeletal muscle is equally important in breathing and in moving internal body fluids. The stimulation of individual skeletal muscle fibers maintains a state of muscle contraction called *tonus,* which is important in the movement of blood and lymph.

   The involuntary contraction of smooth muscle tissue is essential for movement of materials through the body. Likewise, the involuntary contraction of cardiac muscle tissue continuously pumps blood throughout the body.
2. **Heat production.** Body temperature is held remarkably constant. Metabolism within the cells releases heat as an end product. Since muscles constitute approximately 40% of body weight and are in a continuous state of fiber activity, they are very important heat producers. The rate of heat production increases greatly as you exercise.
3. **Posture and body support.** The skeletal system provides a framework for your body, but your skeletal muscles maintain your posture and provide support around the flexible joints.

Muscles are usually described in groups according to their anatomical location and cooperative functions. The *muscles of the axial skeleton* include the facial muscles, neck muscles, and the anterior and posterior trunk muscles. The *muscles of the appendicular skeleton* include those that act on the pectoral and pelvic girdles and those that move the segments of the limbs. The principal superficial muscles are shown in figure 8.1.

Muscle Attachments

Skeletal muscles are attached to bone on each end by tough **tendons,** composed of dense fibrous connective tissue. A tendon binds a muscle to the periosteum of a bone. When a muscle contracts, it shortens, and this places tension on its tendons and attached bones. The muscle tension causes movement of the bones at a synovial joint (see chapter 7, module 7.4), where one of the attached bones generally moves more than the other. The more movable bony attachment of the muscle, known as the *insertion,* is pulled toward its less movable attachment, the *origin.*

Flattened, sheetlike tendons are called **aponeuroses** (*ap″ŏ-noo-ro′sēz*). An example is the galea aponeurotica, which is found on the top and sides of the skull (see fig. 8.18). In certain places in the body, tendons are enclosed by protective tendon sheaths that lubricate the tendons with synovial fluid. In the wrist and ankle, the entire group of tendons is covered by a thin but strong band of connective tissue called a **retinaculum** (*ret″i-nak′yoo-lum*) (see, for example, the extensor retinaculum in fig. 8.32*b*).

Inside Information—When the Greek warrior Achilles was an infant, his mother sought to make him invincible by dipping him in the River Styx. But since she held him by the heel as she immersed him, his heel never touched the water. Years later, Achilles was killed in battle when an adversary shot an arrow through his heel—his one vulnerable spot.

The tendo calcaneus (Achilles tendon) which attaches your calf muscles to your calcaneus (heel bone) gets its popular name from this mythological tale. Unlike many other tendons, it has no protective covering, and thus is vulnerable to inflammation and injury.

Associated Connective Tissue

Contracting muscle fibers would not be effective if they worked as isolated units. Each fiber is bound to adjacent fibers to form bundles, and the bundles in turn are bound to other bundles. With this arrangement, the contraction in one area of a muscle works in conjunction

aponeurosis: Gk. *aponeurosis,* change into a tendon
retinaculum: L. *retinere,* to hold back (retain)

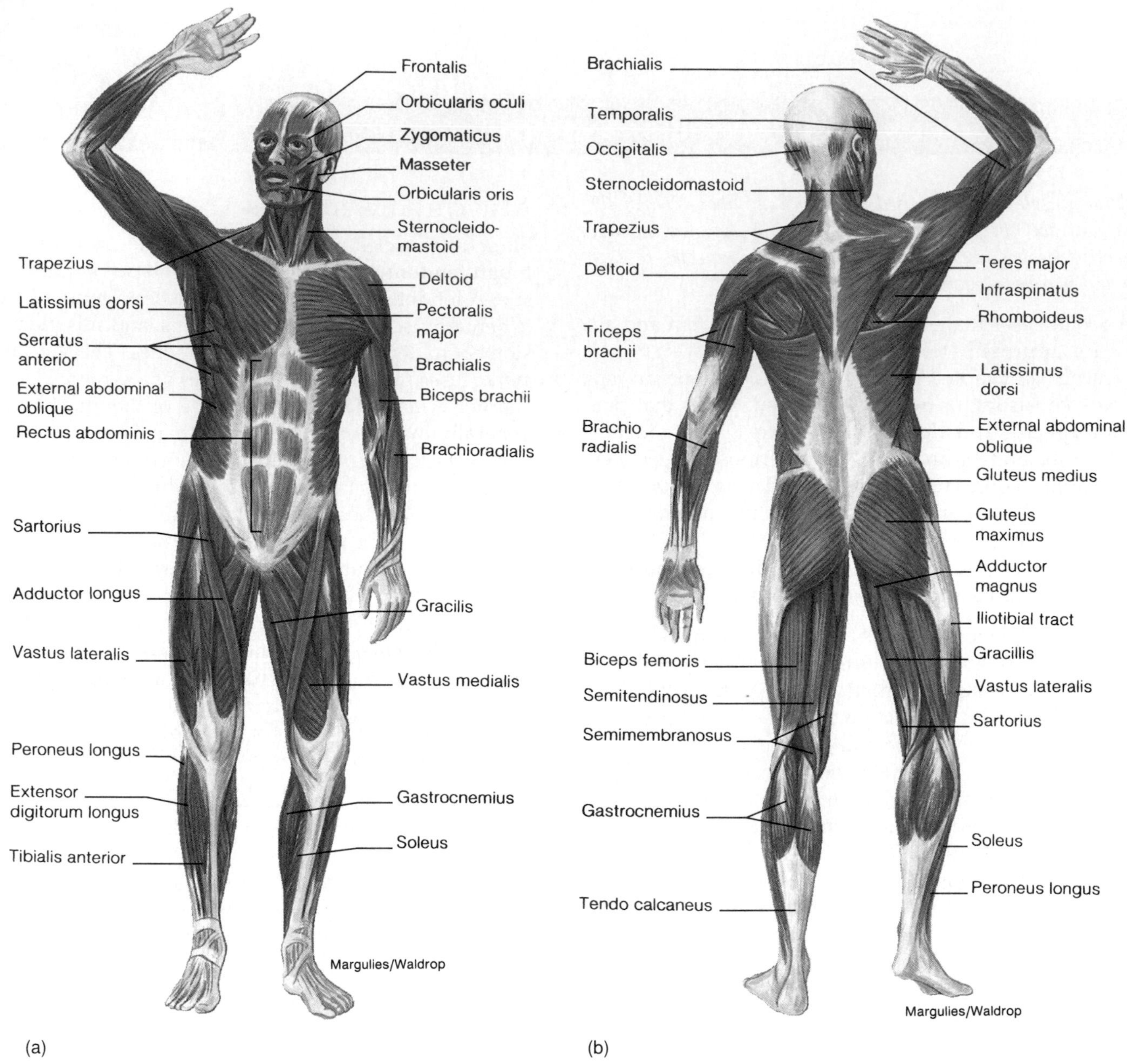

FIGURE 8.1
The principal skeletal muscles. (*a*) An anterior view and (*b*) a posterior view.

with contracting fibers elsewhere in the muscle. The binding substances within muscles are the associated connective tissues.

Connective tissue is structurally arranged within muscle to protect, strengthen, and bind muscle fibers into bundles and bind the bundles together (fig. 8.2). The individual fibers of skeletal muscles are surrounded by a fine sheath of connective tissue called **endomysium** (*en"do-mis'e-um*). The endomysium binds adjacent fibers together and supports capillaries and nerve endings serving the muscle. Another connective tissue, the **perimysium,** binds groups of fibers together into bundles called **fasciculi** (*fă-sik'yŭ-li*—singular, *fasciculus*). The perimysium supports blood vessels and nerve fibers serving the various fasciculi. The entire muscle is covered by the **epimysium,** which in turn is continuous with a tendon.

Fascia (*fash'e-ă*) is a fibrous connective tissue of varying thickness that covers muscle and attaches to the skin. *Superficial fascia* secures the skin to the underlying muscle.

endomysium: Gk. *endon,* within; *myos,* muscle
perimysium: Gk. *peri,* around; *myos,* muscle
fasciculus: L. *fascis,* bundle
epimysium: Gk. *epi,* upon; *myos,* muscle
fascia: L. *fascia,* a band or girdle

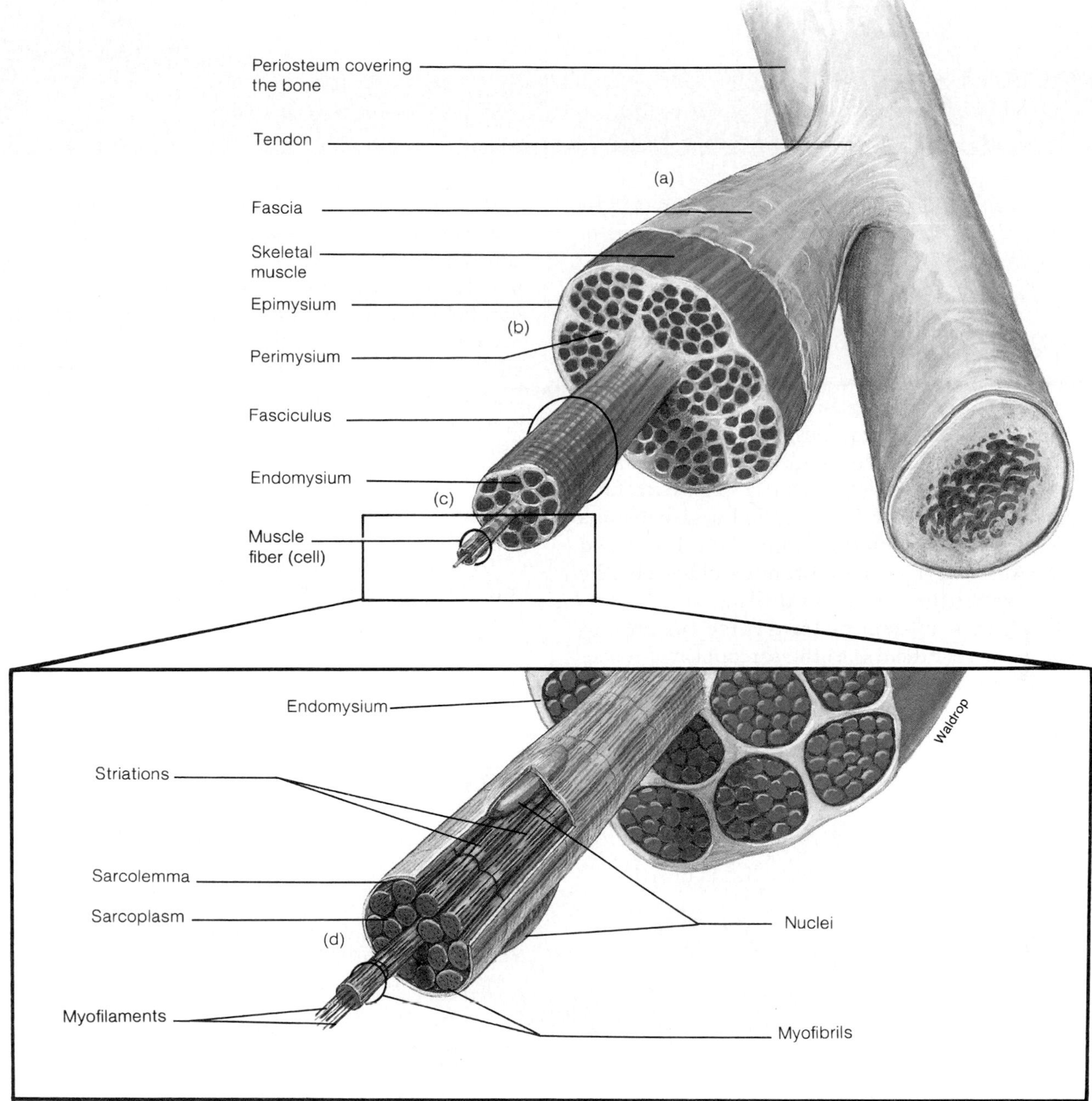

FIGURE 8.2

The relationship between muscle tissue and connective tissue. (*a*) The fascia and tendon attaches a muscle to the periosteum of a bone. (*b*) The epimysium surrounds the entire muscle, and the perimysium separates and binds the fasciculi (muscle bundles). (*c*) The endomysium surrounds and binds individual muscle fibers. (*d*) An individual muscle fiber is composed of myofibrils that contain myofilaments of actin and myosin.

The superficial fascia over the buttocks and abdominal wall is thick and laced with adipose tissue. By contrast, the superficial fascia under the skin on the back of the hand and on the facial region is thin. *Deep fascia* is an inward extension of the superficial fascia. It lacks adipose tissue and blends with the epimysium of muscle. Deep fascia surrounds adjacent muscles, compartmentalizing and binding them into functional groups.

Inside Information—The tenderness of meat is due in part to the amount of connective tissue present in a particular cut. A slice of meat from the ends of a muscle contains much more connective tissue than a cut through the belly of the muscle. Fibrous meat is difficult to chew and may present a social problem in trying to gracefully extract it from between the teeth.

8.2 Skeletal Muscle Fibers and Types of Muscle Contractions

Muscle fiber contraction in response to a motor impulse results from a sliding movement within the myofibrils in which the length of the sarcomeres is reduced.

Microscopic Structure of Muscle Fibers

Despite their unusual elongated shape, muscle cells have the same organelles as other cells: mitochondria, intracellular membranes, glycogen granules, and so forth. Unlike other cells in the body, however, skeletal muscle fibers are multinucleated and striated (striped) (fig. 8.3). Each fiber is surrounded by a cell membrane called a **sarcolemma** (*sar″kŏ-lem′ă*), and the cytoplasm within the cell is called **sarcoplasm.** A network of membranous channels, the **sarcoplasmic reticulum,** extends throughout the sarcoplasm (fig. 8.4). A system of **transverse tubules,** or **T tubules,** runs perpendicular to the sarcoplasmic reticulum and opens to the outside through the sarcolemma.

Also embedded in the sarcoplasm, and extending the entire length of the fiber, are numerous threadlike structures called **myofibrils** (fig. 8.5). These myofibrils are approximately one micrometer (1 μm) in diameter and extend in parallel from one end of the muscle fiber to the other. The myofibrils are so densely packed that other organelles—such as mitochondria and intracellular membranes—are restricted to the narrow cytoplasmic spaces that remain between adjacent myofibrils. Each myofibril is composed of even smaller protein filaments, or **myofilaments.** *Thin filaments* are about 6 nm in diameter and are composed of the protein **actin.** *Thick filaments* are about 16 nm in diameter and are composed of the protein **myosin.**

The characteristic dark and light striations of skeletal muscle myofibrils are due to the arrangement of these

actin: L. *actus,* motion, doing
myosin: L. *myosin,* within muscle

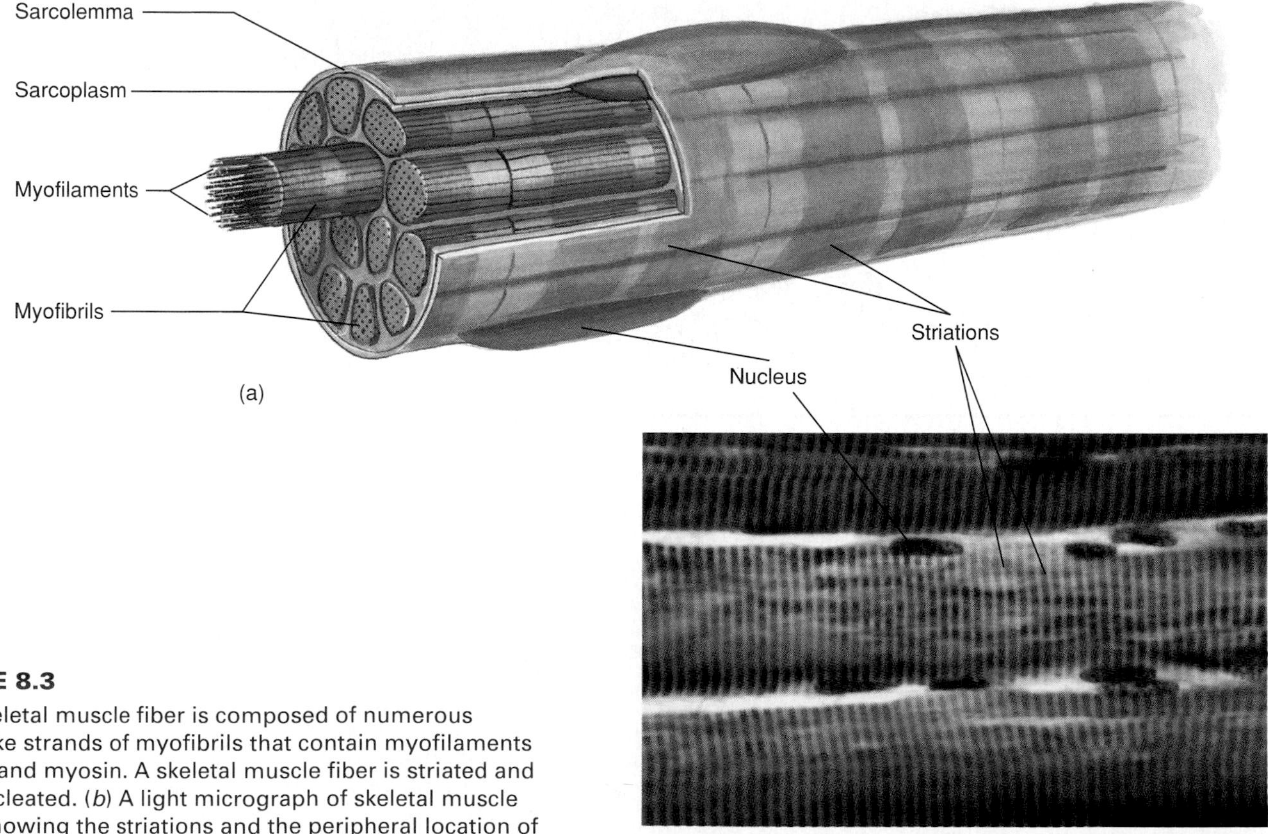

FIGURE 8.3
(*a*) A skeletal muscle fiber is composed of numerous threadlike strands of myofibrils that contain myofilaments of actin and myosin. A skeletal muscle fiber is striated and multinucleated. (*b*) A light micrograph of skeletal muscle fibers showing the striations and the peripheral location of the nuclei.

myofilaments. The dark bands are called **A bands,** and the light bands are called **I bands.** At high magnification, thin dark lines can be seen in the middle of the I bands. These are called **Z lines.** The arrangement of thick and thin filaments between a pair of Z lines forms a repeating pattern that serves as the basic subunit of skeletal muscle contraction. These subunits, from Z line to Z line, are known as **sarcomeres** (*sar'kŏ-mērz*) (fig. 8.5). A longitudinal section of a myofibril thus presents a side view of successive sarcomeres.

The I bands within a myofibril are the lighter areas that extend from the edge of one stack of thick myosin filaments to the edge of the next stack of thick filaments. They are light in appearance because they contain only thin filaments. The thin filaments, however, do not end at the edges of the I bands. Instead, each thin filament extends partway into the A bands on each side. Since thick and thin filaments overlap at the edges of each A band, the edges of the A band are darker in appearance than the central region. These central lighter regions of the A bands are called the *H zones* (for *helle,* a German word meaning "bright"). The central H zones thus contain only thick filaments that are not overlapped by thin filaments.

Inside Information—Skeletal muscle fibers are extremely long. Some may reach a length of 30 cm (12 in.) and extend the full length of a muscle.

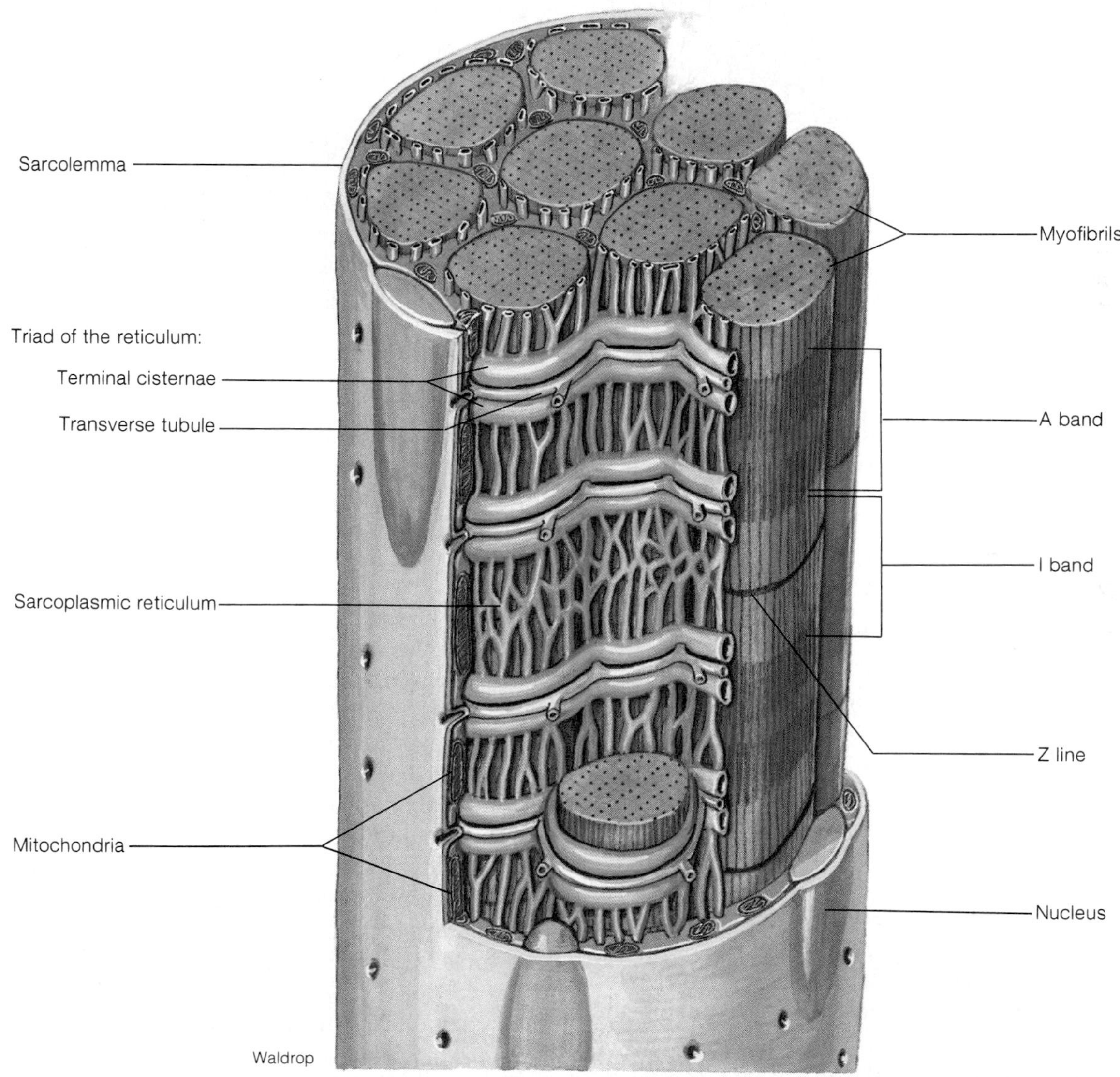

FIGURE 8.4
The structural relationship of the myofibrils of a muscle fiber to the sarcolemma, transverse tubules, and sarcoplasmic reticulum. (Note the position of the mitochondria.)

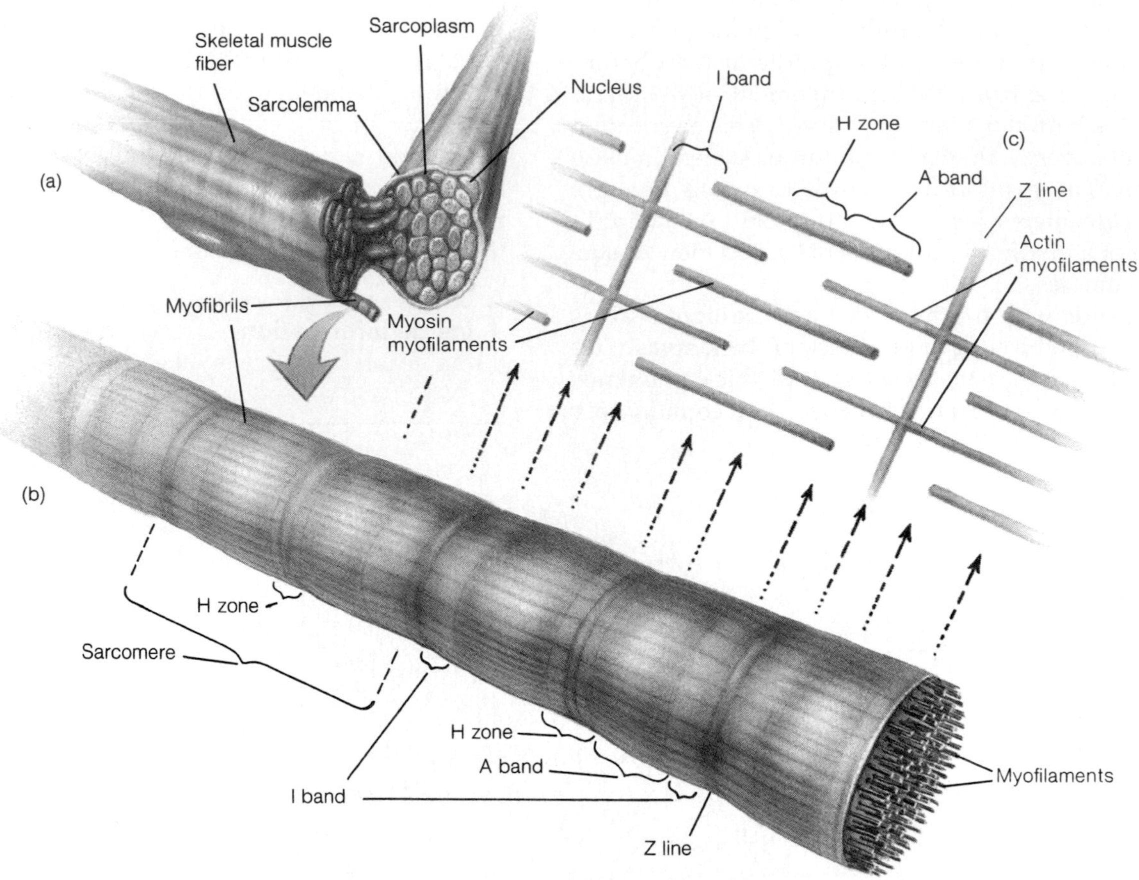

FIGURE 8.5
The structure of a myofibril. (*a*) The many myofibrils of a skeletal muscle fiber are arranged in compartments (*b*) called sarcomeres. (*c*) The characteristic striations of a sarcomere are due to the arrangement of thin and thick filaments, composed of actin and myosin, respectively.

Twitch, Summation, and Tetanus

When a muscle is stimulated with a single electric shock of a sufficient voltage, it quickly contracts and relaxes. This response is called a **twitch.** Increasing the stimulus voltage increases the strength of the twitch up to a maximum. The strength of a muscle contraction can thus be *graded,* or varied—an obvious requirement for the proper control of skeletal movements. If a second electric shock is delivered immediately after the first, it will produce a second twitch that may partially "ride piggyback" on the first. This phenomenon is called **summation.**

If the stimulator is set to deliver an increasing frequency of electric shocks automatically, the relaxation time between successive twitches will get shorter and shorter as the strength of contraction increases in amplitude. This effect is known as **incomplete tetanus.** Finally, at a particular "fusion frequency" of stimulation, there is no visible relaxation between successive twitches (fig. 8.6). Contraction is smooth and sustained, as it is during normal muscle contraction in vivo (within the body). This smooth, sustained contraction of normal activity is called **complete tetanus.** (This tetanus, a manner of muscle contraction, should not be confused with the disease of the same name, which is accompanied by a painful state of muscle contracture, or *tetany.*)

Stimulation of fibers within a muscle in vitro (outside the body) with an electric stimulator, or in vivo by nerve impulses through motor axons, usually results in *all-or-none contractions* of the individual fibers. In other words, when a threshold, or greater, stimulus is applied, each muscle fiber contracts maximally or not at all. Stronger muscle contractions are produced by the stimulation of greater numbers of muscle fibers. Skeletal muscles can thus produce **graded contractions** in which the strength is varied by variations in the number of fibers stimulated to contract.

Treppe

If the voltage of the electrical shocks delivered to an isolated muscle in vitro is gradually increased from zero, the strength of the muscle twitches will increase accordingly,

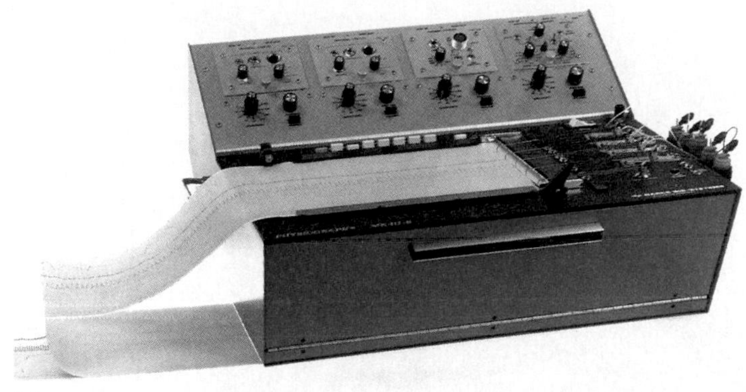

(a)

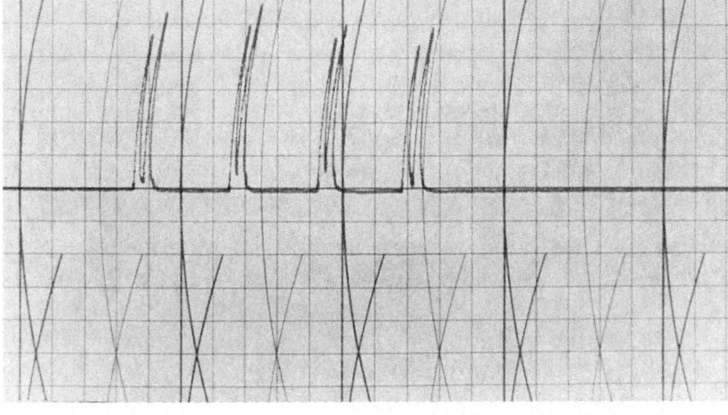

(b)

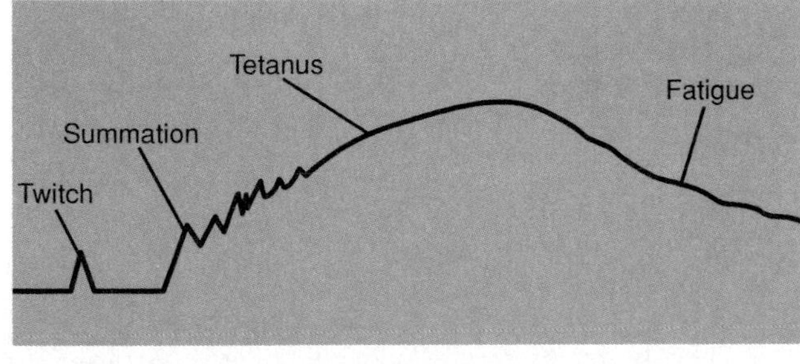

(c)

FIGURE 8.6

(*a*) A physiograph recorder. (*b*) A photograph and (*c*) an illustration of the behavior of an isolated leg muscle from a frog in response to electrical shocks.

up to a maximal value at which all of the muscle fibers are stimulated. This demonstrates the graded nature of the muscle contraction. If a series of electrical shocks at this one maximal voltage are given to a fresh muscle, each of the twitches evoked will be successively stronger, up to a somewhat higher maximum. This demonstrates **treppe** (*trep'e*), or the *staircase effect*. Treppe may represent a warmup effect, and is believed to be due to an increase in intracellular calcium ions (Ca^{++}) that are needed for muscle contraction (as discussed in module 8.4).

Innervation of Muscle Fibers

A **motor unit** consists of a single motor neuron and the group of muscle fibers supplied, or innervated, by the motor neuron (fig. 8.7). When a nerve impulse travels through a motor neuron, all of the muscle fibers served by the neuron contract simultaneously to their maximum. Most muscles have an *innervation ratio* of 1 motor neuron for each 100 to 150 muscle fibers. Precise, or fine, neural control over the strength of muscle contraction is achieved when many small motor units are involved. In the ocular muscles that position the eyes, for example, the innervation ratio (motor neuron:muscle fibers) of an average motor unit is 1 neuron per 10 muscle fibers. This affords a fine degree of control. In massive muscles that are responsible for powerful but crude contractions, such as those of the thigh, more than 1000 muscle fibers may be controlled by a single motor neuron.

Skeletal muscles are voluntary in that they can be consciously contracted. The magnitude of the task determines the number of motor units that are activated. Performing a light task, such as lifting a book, requires few motor units, whereas lifting a table requires many. Being mentally "psyched up" to accomplish an athletic feat involves voluntary activation of more motor units within the muscles. Seldom does a person utilize all the motor units within a muscle, but the secretion of epinephrine (*ep"ĭ-nef'rin*) from the adrenal gland does promote an increase in the force that can be produced when a given number of motor units are activated.

Isotonic and Isometric Contractions

In order for muscle fibers to shorten when they contract, they must generate a force that is greater than the forces that act to prevent movement of the muscle's insertion. Flexion of the elbow, for example, occurs against the force of gravity and the weight of the objects being lifted. The tension produced by the contraction of one muscle fiber is insufficient to overcome these opposing forces. However, the combined contractions of many muscle fibers may produce enough tension to overcome them and flex the elbow. In this case, the muscle and all its fibers shorten in length.

Contraction that results in visible muscle shortening is called **isotonic contraction** because the force of contraction remains relatively constant throughout the shortening process (fig. 8.8). If the opposing forces are too great or if the number of motor units activated is too few to shorten the muscle, however, an **isometric contraction** is produced, and movement does not occur.

isotonic: Gk. *isos*, equal; *tonos*, tension
isometric: Gk. *isos*, equal; *metron*, measure

FIGURE 8.7

A motor unit consists of a motor neuron coupled with the skeletal muscle fibers it serves. (Actually, most skeletal muscles have hundreds of motor units.)

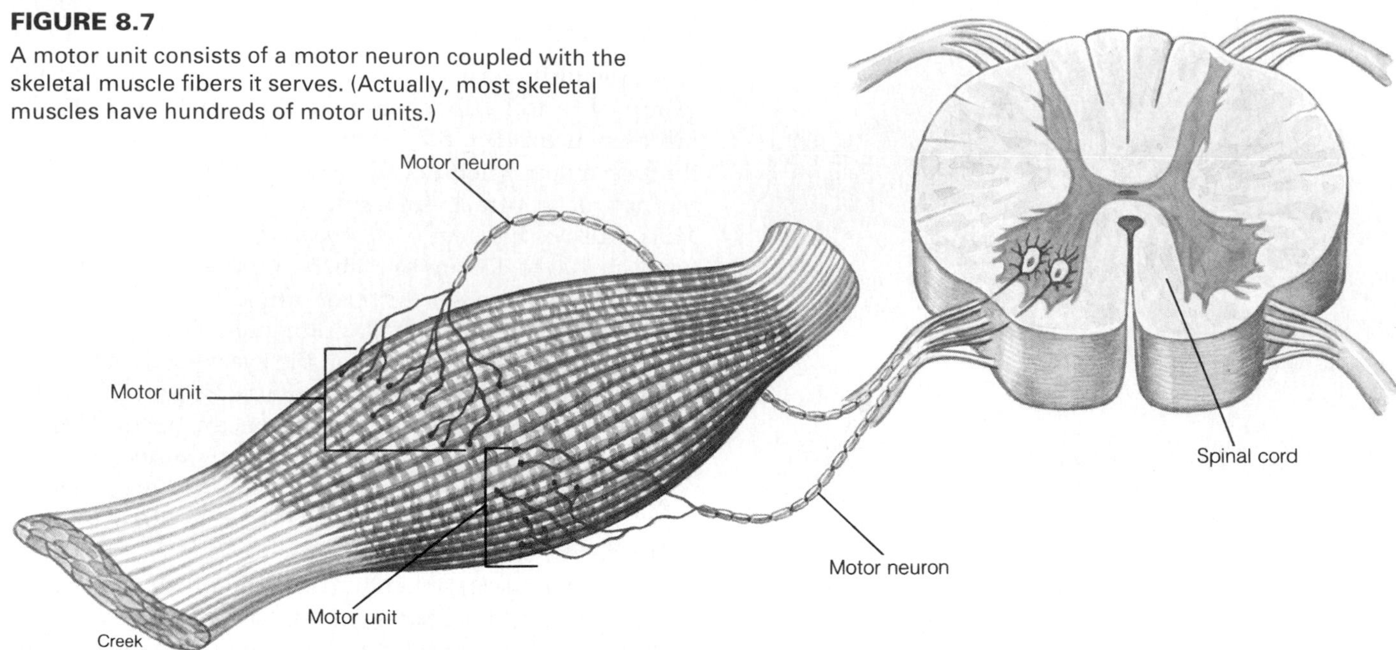

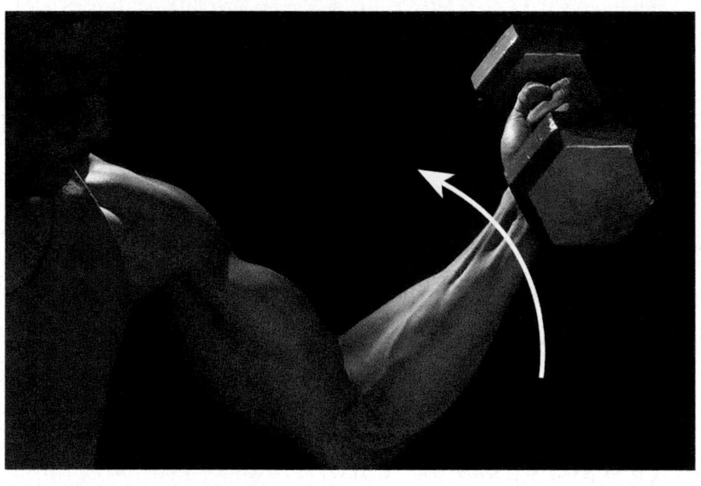

(a)

(b)

FIGURE 8.8

(*a*) Isotonic and (*b*) isometric contraction.

8.3 Sliding Filament Theory of Contraction

Movements of cross bridges that extend out from the thick to the thin myofilaments cause sliding of the myofilaments, and thus muscle tension and shortening. ATP is required for the cross-bridge cycle.

When a muscle contracts, it decreases in length as a result of the shortening of its individual fibers. Shortening of the muscle fibers, in turn, is produced by shortening of their myofibrils, which occurs as a result of the shortening of the sarcomeres. As the sarcomeres shorten in length, however, the A bands do *not* shorten, but instead move closer together. The I bands—which represent the distance between A bands of successive sarcomeres—decrease in length.

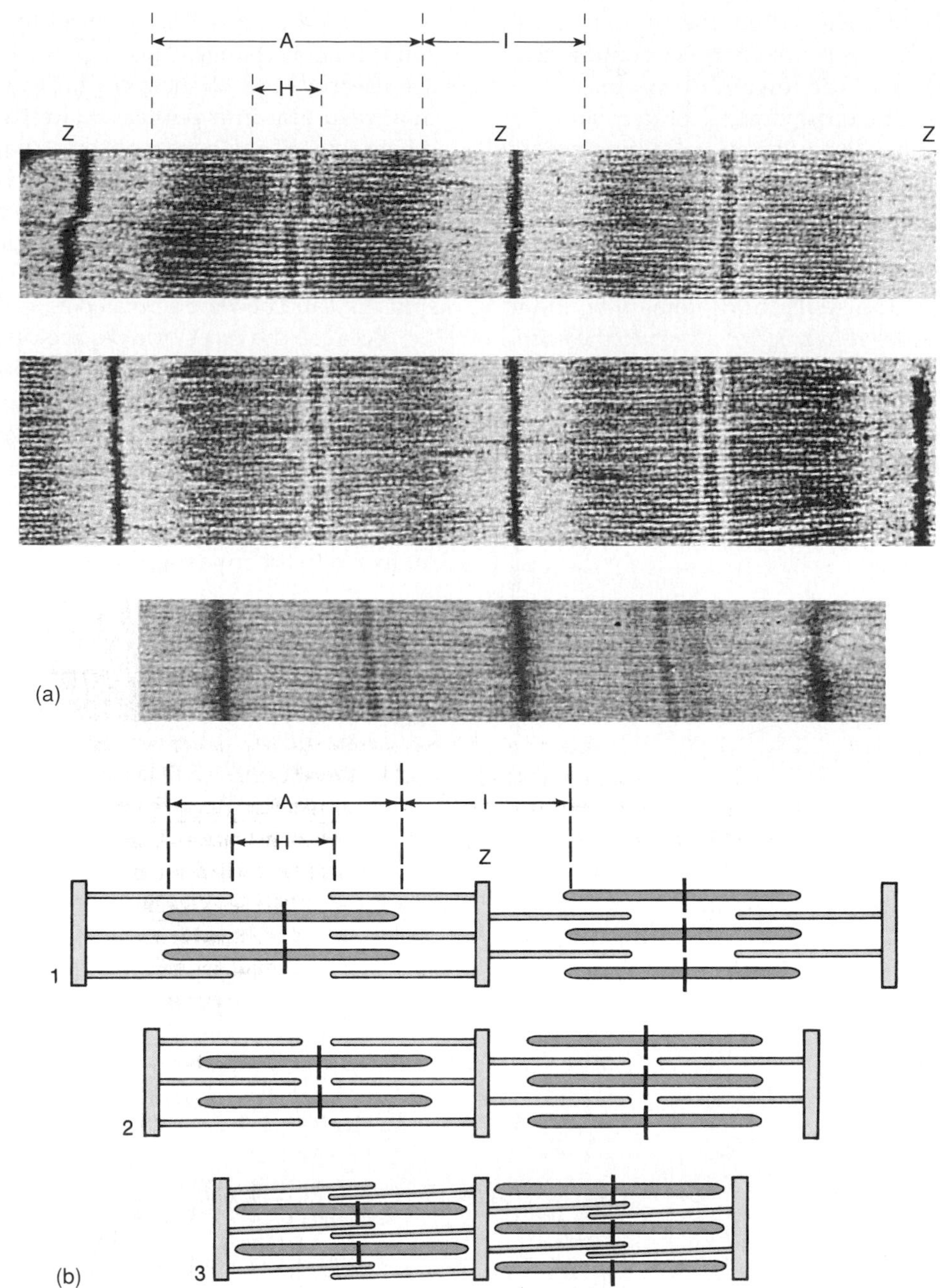

FIGURE 8.9
The sliding filament model of contraction. As the filaments slide, the Z lines are brought closer together. (1) Relaxed muscle, (2) partially contracted muscle, and (3) fully contracted muscle.

The thin actin filaments composing the I band, however, do not shorten. Close examination reveals that the length of the thick and thin filaments remains constant during muscle contraction. Shortening of the sarcomeres is produced not by shortening of the myofilaments, but rather by the *sliding* of thin filaments over and between the thick filaments. In the process of contraction, the thin filaments on either side of each A band slide deeper and deeper toward the center, producing increasing amounts of overlap with the thick filaments. The I bands (containing only thin filaments) and H bands (containing only thick filaments) thus get shorter during contraction (fig. 8.9).

Sliding of the myofilaments is produced by the action of numerous **cross bridges** that extend out from the myosin toward the actin. These cross bridges are part of the myosin proteins that extend from the axis of the thick filaments to form "arms" that terminate in globular "heads" (fig. 8.10). The orientation of cross bridges on one side of a sarcomere is opposite to that of the cross bridges on the other side. Therefore, when the myosin cross bridges attach to actin on each side of the sarcomere, they can pull the actin from each side toward the center.

Each globular head of a cross bridge contains an ATP-binding site closely associated with an actin-binding site (fig. 8.11). The globular heads function as **myosin ATPase** enzymes, splitting ATP into ADP and P_i. This reaction occurs before the cross bridges combine with actin, and indeed is required for activating the cross bridges so that they can attach to actin. The ADP and P_i remain bound to the myosin heads until the cross bridges attach to the actin.

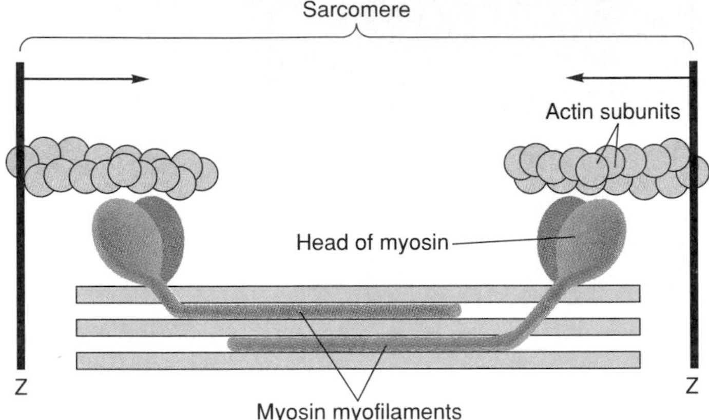

FIGURE 8.10
Myosin cross bridges are oriented in opposite directions on either side of a sarcomere.

When the cross bridges bind to actin, they undergo a conformation change. This has two effects: (1) ADP and P_i are released, and (2) the cross bridges change their orientation, resulting in a **power stroke** that pulls the thin filaments toward the center of the A bands (fig. 8.12). At the end of the power stroke, each cross bridge binds to a new ATP molecule. This bonding of the cross bridge to a new ATP causes the cross bridge to break its bond with actin and resume its resting orientation. The myosin ATPase will then split ATP and become activated as in the previous cycle.

Because the cross bridges are tiny, significant shortening of a muscle requires that the contraction cycles be repeated many times. The cross bridges must detach from the actin at the end of a power stroke, resume their resting orientation, and then reattach to the actin and repeat the cycle. The detachment of a cross bridge from actin at the end of a power stroke requires the bonding of a new ATP molecule to the myosin ATPase.

Rigor mortis, *the stiffening of the muscles following death, illustrates that the detachment of cross bridges is ATP driven. As ATP in the myofibrils of the skeletal muscles is depleted, the "rigor complexes" that form between myosin and actin cannot detach; thus, the muscles lock in the contracted position. With all the skeletal muscles involved, the body becomes as stiff as a board—a physical state that gradually dissipates over the next 2 to 3 days as the muscle proteins decompose.*

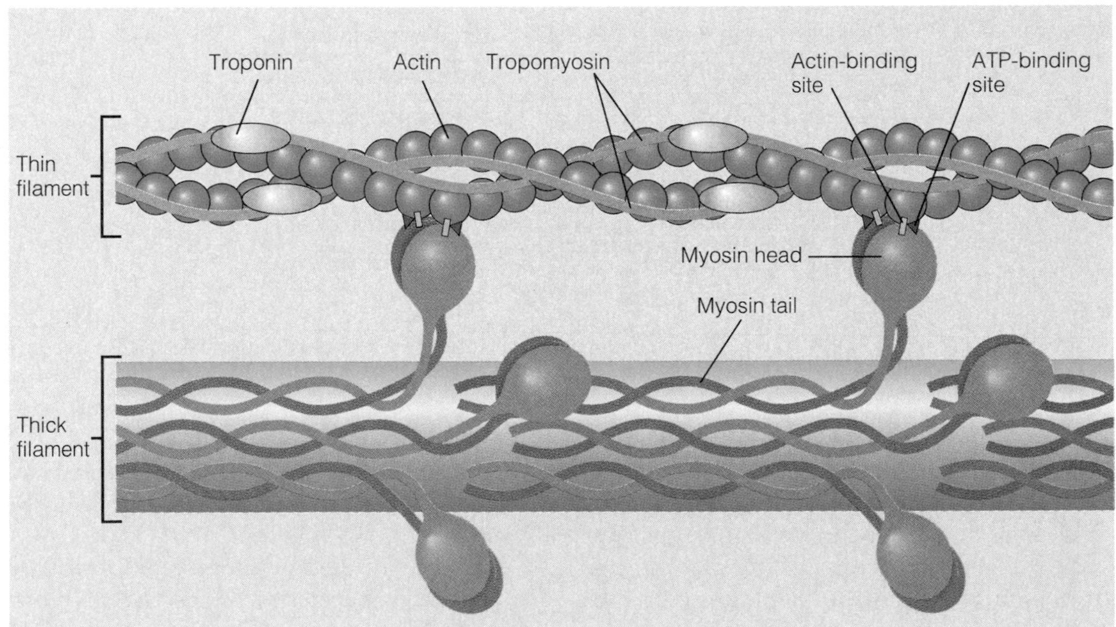

FIGURE 8.11
The structure of myosin showing its binding sites for ATP and for actin.

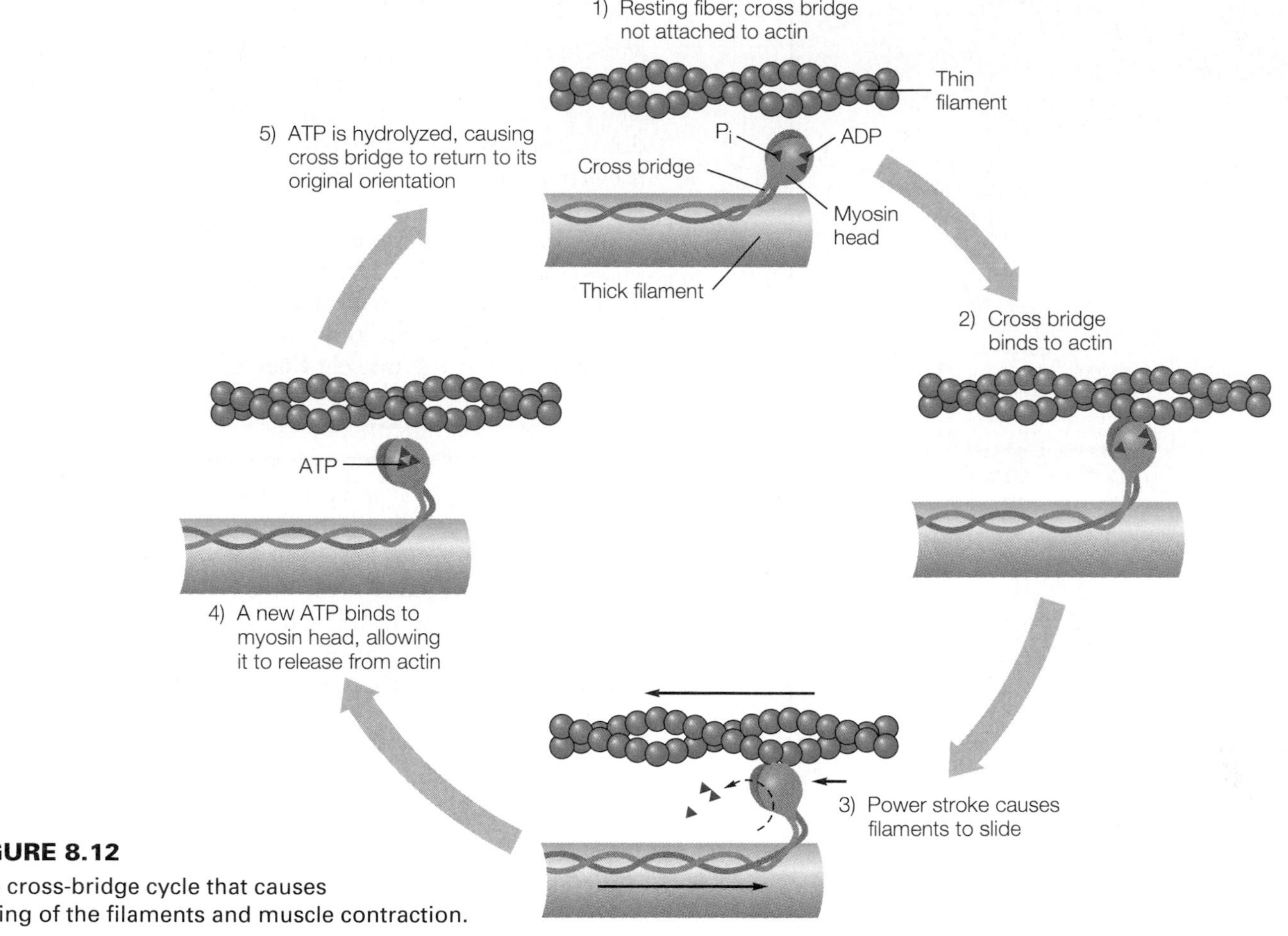
1) Resting fiber; cross bridge not attached to actin
Thin filament
Pi
ADP
Cross bridge
Myosin head
Thick filament
2) Cross bridge binds to actin
3) Power stroke causes filaments to slide
4) A new ATP binds to myosin head, allowing it to release from actin
ATP
5) ATP is hydrolyzed, causing cross bridge to return to its original orientation

FIGURE 8.12
The cross-bridge cycle that causes sliding of the filaments and muscle contraction.

8.4 Regulation of Contraction

Calcium ions, stored in the sarcoplasmic reticulum, serve as a control switch for muscle contraction. This function is aided by two proteins, troponin and tropomyosin.

The actin filament—or *F-actin*—is a polymer consisting of 300 to 400 globular subunits (*G-actin*), arranged in a double row and twisted to form a helix. A different type of protein, known as **tropomyosin** (*tro″pŏ′mi′ŏ-sin*), lies within the groove between the double row of G-actin (fig. 8.13). There are 40 to 60 tropomyosin molecules per thin filament, with each tropomyosin spanning a distance of approximately seven actin subunits. In a relaxed muscle, the position of the tropomyosin is such that it physically blocks the cross bridges from binding to specific attachment sites in the actin. Thus, in order for the myosin cross bridges to attach to actin, the tropomyosin must be moved.

Attached to the tropomyosin, rather than directly to the actin, is a third type of protein, called **troponin** (*trop′pŏ-nin*), within the thin filaments. Troponin and tropomyosin work together to regulate the attachment of cross bridges to actin, and thus serve as a switch for muscle contraction and relaxation.

Role of Ca^{++} in Muscle Contraction

In a relaxed muscle, the concentration of calcium ions (Ca^{++}) in the sarcoplasm is very low. When the muscle cell is stimulated to contract, certain mechanisms (to be discussed shortly) cause the concentration of Ca^{++} in the sarcoplasm to quickly rise. Some of this Ca^{++} attaches to a subunit of troponin, causing a conformation change that moves the troponin *and* its attached tropomyosin out of the way so that the cross bridges can attach to actin (fig. 8.13). Once the attachment sites on the actin are exposed, the cross bridges can bind to actin, undergo power strokes, and produce muscle contraction.

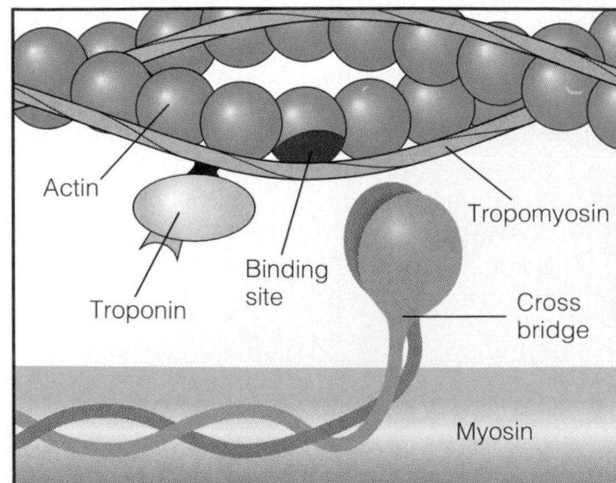

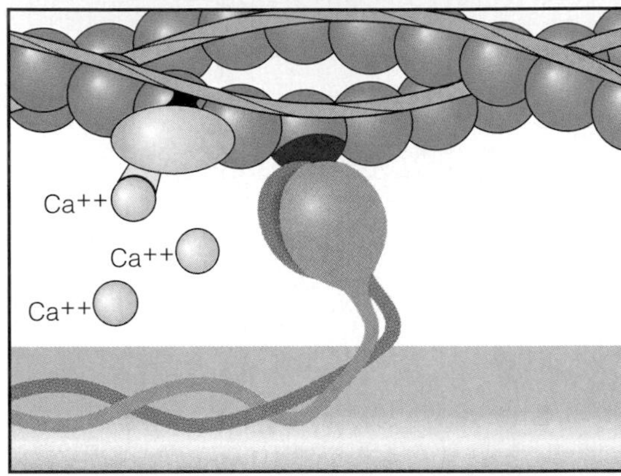

FIGURE 8.13

The attachment of Ca^{++} to troponin causes movement of the troponin-tropomyosin complex, which exposes binding sites on the actin. The myosin cross bridges can then attach to actin and undergo a power stroke.

Excitation-Contraction Coupling

Muscle relaxation is produced by the active transport of Ca^{++} out of the sarcoplasm into the sarcoplasmic reticulum (see fig. 8.4). The sarcoplasmic reticulum is a modified endoplasmic reticulum, consisting of interconnected sacs and tubes that surround each myofibril within the muscle cell.

Most of the Ca^{++} in a relaxed muscle fiber is stored within expanded portions of the sarcoplasmic reticulum known as *terminal cisternae* (*sis-ter'ne*). When a muscle fiber is stimulated to contract, the stored Ca^{++} is released from the sarcoplasmic reticulum so that it can attach to troponin. When the muscle fiber is no longer stimulated, the Ca^{++} from the sarcoplasm is actively transported into the sarcoplasmic reticulum.

The terminal cisternae of the sarcoplasmic reticulum are separated by only a very narrow gap from the *transverse tubules* (*T tubules*)—the narrow membranous "tunnels" formed from and continuous with the sarcolemma. The transverse tubules open to the extracellular environment through pores in the cell surface and are capable of conducting electrical impulses from the cell membrane.

Nerve impulses that arrive at the axon terminal of a motor neuron evoke the production of electrical impulses in the skeletal muscle fibers. These impulses are conducted along the sarcolemma and into the interior of the fiber along the transverse tubules. Impulses conducted by the transverse tubules cause the release of Ca^{++} from the sarcoplasmic reticulum. The released Ca^{++} binds to troponin, causing the displacement of tropomyosin and allowing the actin to bind to the myosin cross bridges. Muscle contraction is thus stimulated. In short, the release of Ca^{++} from the sarcoplasmic reticulum *couples electrical excitation to muscle contraction.*

8.5 Metabolism of Skeletal Muscles

Muscles obtain energy from the aerobic respiration of glucose and fatty acids, and from the anaerobic respiration of glucose. Not all muscle fibers are alike, and the various types respond differently to exercise.

Phosphocreatine and ATP

Skeletal muscles at rest obtain most of their energy from the aerobic respiration of fatty acids. Energy obtained by cell respiration is used to make ATP, which serves as the immediate source of energy for (1) the movement of cross bridges for muscle contraction and (2) the pumping of Ca^{++} into the sarcoplasmic reticulum for muscle relaxation. During exercise, muscle glycogen and blood glucose are also used as energy sources.

During sustained muscle activity, ATP may be utilized faster than it can be produced through cell respiration. At these times, the rapid renewal of ATP is extremely important. This is accomplished by combining ADP with phosphate derived from another high-energy phosphate compound called **phosphocreatine** (*fos"fo-kre'ă-tēn*), or **creatine phosphate** (fig. 8.14). During times of rest, the depleted reserve of phosphocreatine can be restored by the reverse reaction—phosphorylation of creatine with phosphate derived from ATP.

Maximal Oxygen Uptake and Oxygen Debt

Skeletal muscles respire anaerobically for the first 45 to 90 seconds of moderate-to-heavy exercise because the

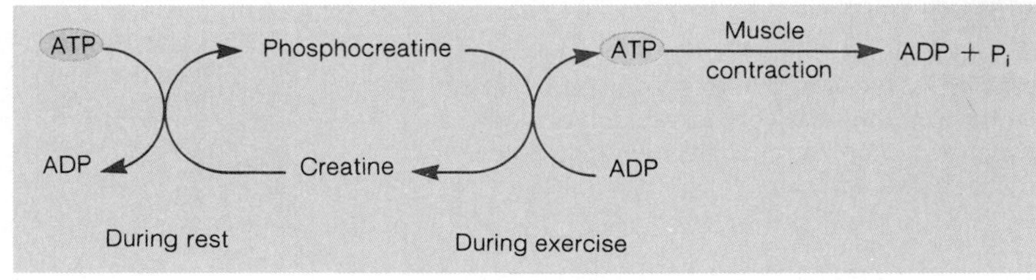

FIGURE 8.14
The production and utilization of phosphocreatine in muscles.

cardiopulmonary system requires this amount of time to sufficiently increase the oxygen supply to the exercising muscles. If exercise is moderate and the person is in good physical condition, aerobic respiration contributes the major portion of the skeletal muscle energy requirements following the first 2 minutes of exercise (fig. 8.15).

The maximum rate of oxygen consumption (by aerobic respiration) in the body is called the **maximal oxygen uptake.** The maximal oxygen uptake is determined primarily by a person's age, size, and sex. It is 15% to 20% higher for males than for females and highest at age 20 for both sexes. Some world-class athletes have maximal oxygen uptakes that are twice the average for their age and sex. This appears to be due largely to genetic factors, but training can increase the maximum oxygen uptake by about 20%.

When a person stops exercising, the rate of oxygen uptake does not immediately go back to pre-exercise levels; it returns slowly (the person continues to breathe heavily for some time afterward). This extra oxygen is used to repay the **oxygen debt** incurred during exercise. The oxygen debt includes oxygen that was withdrawn from hemoglobin in blood and myoglobin in muscle (see chapter 18, module 18.10); the extra oxygen required for metabolism by tissues warmed during exercise; and the oxygen needed for the metabolism of lactic acid produced during anaerobic respiration.

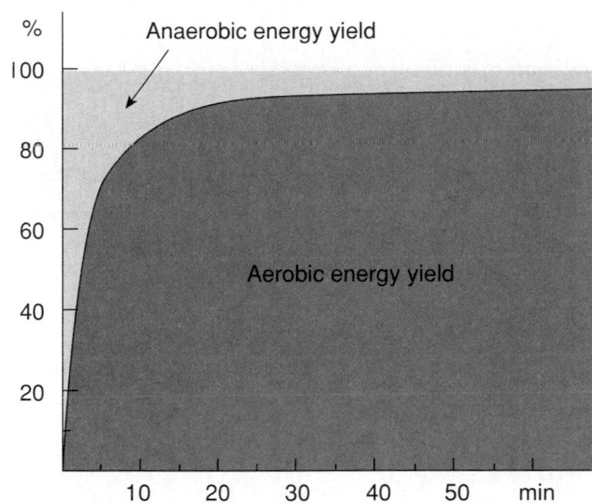

FIGURE 8.15
The relative contributions of anaerobic and aerobic respiration to the total energy in a well-trained person performing at maximal effort.

Slow- and Fast-Twitch Fibers

Skeletal muscle fibers can be divided on the basis of their contraction speed (time required to reach maximum tension) into **slow-twitch,** or **type I fibers,** and **fast-twitch,** or **type II fibers.** This distinction is associated with different myosin ATPase isoenzymes, designated "slow" and "fast," by which the two fiber types can be identified when they are appropriately stained (fig. 8.16). The ocular muscles that position the eyes, for example, have a high proportion of fast-twitch fibers and reach maximum tension in about 7.3 msec (milliseconds—thousandths of a second). The soleus muscle in the leg, by contrast, has a high proportion of slow-twitch fibers and requires about 100 msec to reach maximum tension (fig. 8.17).

FIGURE 8.16
Skeletal muscle (of a cat) stained to indicate the activity of myosin ATPase. Type II fibers exhibit higher ATPase activity than do type I fibers.

Since muscles like the soleus are *postural muscles,* they must be able to sustain a contraction for a long period of time without fatigue. This resistance to fatigue is aided by other characteristics of slow-twitch (type I) fibers that endow them with a high capacity for aerobic respiration. Slow-twitch fibers have a rich supply of capillaries, numerous mitochondria and aerobic respiratory enzymes, and a high concentration of *myoglobin.*

FIGURE 8.17
A comparison of the rates with which maximum tension is developed in three muscles. These include (*a*) the relatively fast-twitch ocular and (*b*) gastrocnemius muscles, and (*c*) the slow-twitch soleus muscle.

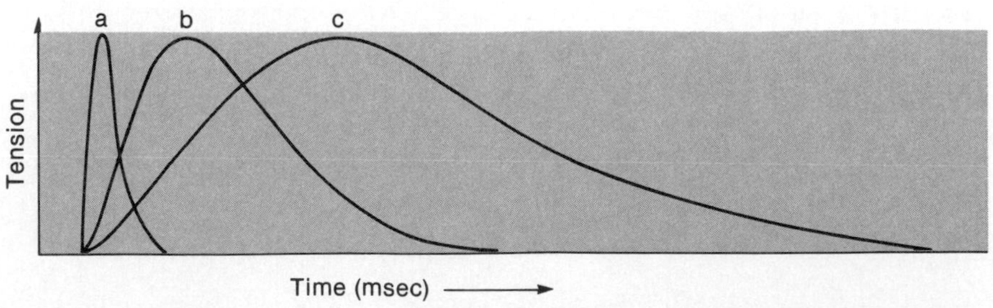

TABLE 8.1 Characteristics of Red, Intermediate, and White Muscle Fibers

	Red (Type I)	Intermediate (Type IIA)	White (Type IIB)
Diameter	Small	Intermediate	Large
Z-line thickness	Wide	Intermediate	Narrow
Glycogen content	Low	Intermediate	High
Resistance to fatigue	High	Intermediate	Low
Capillaries	Many	Many	Few
Myoglobin content	High	High	Low
Respiration type	Aerobic	Aerobic	Anaerobic
Twitch rate	Slow	Fast	Fast
Myosin ATPase content	Low	High	High

Myoglobin is a red pigment, similar to the hemoglobin in red blood cells, that improves the delivery of oxygen to the slow-twitch fibers. Because of their high myoglobin content, slow-twitch fibers are also called *red fibers.*

The thicker fast-twitch (type II) fibers have fewer capillaries and mitochondria, and less myoglobin; hence, these fibers are also called *white fibers.* Fast-twitch fibers are adapted to respire anaerobically by a large store of glycogen and a high concentration of glycolytic enzymes. In addition to the type I (slow-twitch) and type II (fast-twitch) fibers, human muscles may also contain an intermediate type. These intermediate fibers are fast-twitch but also have a high aerobic capability. They are sometimes called type IIA, to distinguish them from the anaerobically adapted fast-twitch fibers (which are then labeled IIB). The three fiber types are compared in table 8.1.

Muscle Fatigue

Muscle fatigue during a sustained maximal contraction, when all the motor units are being used and the rate of neural firing is maximal—as when lifting an extremely heavy weight—appears to be due to an accumulation of extracellular potassium ions (K^+). (Potassium ions leave both axons and muscle fibers during the production of electrical impulses.) This reduces the membrane potential of muscle fibers and interferes with their ability to produce impulses. Fatigue under these circumstances lasts only a short time, and maximal tension can again be produced after less than a minute's rest.

Fatigue during moderate exercise occurs as the slow-twitch fibers deplete their reserve glycogen and fast-twitch fibers are increasingly recruited. Since fast-twitch fibers obtain their energy through anaerobic respiration by converting glucose to lactic acid, this results in a rise in intracellular H^+ and a fall in pH. The decrease in muscle pH, in turn, inhibits the activity of key glycolytic enzymes so that the rate of ATP production is reduced. This may interfere with the ability of the sarcoplasmic reticulum to accumulate Ca^{++} by active transport, and thus interfere with the ability of the muscle to contract in response to electrical stimulation. The decreased cellular pH, in other words, produces fatigue by interfering with excitation-contraction coupling.

Adaptations to Exercise

When exercise is performed at low levels of effort, such that the oxygen consumption rate is less than 50% of its maximum, the energy for muscle contraction is obtained almost entirely from aerobic cell respiration. Anaerobic cell respiration, with its consequent production of lactic acid, contributes to the energy requirements as the exercise level rises and more than 60% of the maximal oxygen uptake is required. Highly trained endurance athletes, however, can continue to respire aerobically, with little lactic acid production, at up to 80% of their maximal oxygen uptake. These athletes thus produce less lactic acid at a given level of exercise than most of us; therefore, they are less subject to fatigue than the average person.

Since the fiber types are determined by their innervations, endurance training cannot change fast-twitch (type II) fibers to slow-twitch (type I) fibers. All fiber types, however, adapt to endurance training by an increase in myoglobin and aerobic respiratory enzymes. In fact, the maximal oxygen uptake can be increased by as much as 20% through this training. In addition to changes in aerobic capacity, fibers show an increase in their content of triglycerides, which serve as an alternate energy source and help to spare their stores of glycogen. A summary of the changes that occur as a result of endurance training is presented in table 8.2.

Table 8.2 The Effects of Endurance Training (Long-distance Running, Swimming, Bicycling, etc.) on Skeletal Muscles

1. Improved ability to obtain ATP from oxidative phosphorylation
2. Increased size and number of mitochondria
3. Less lactic acid produced per given amount of exercise
4. Increased myoglobin content
5. Increased intramuscular triglyceride content
6. Increased lipoprotein lipase (enzyme needed to utilize lipids from blood)
7. Increased proportion of energy derived from fat; less from carbohydrates
8. Lower rate of glycogen depletion during exercise
9. Improved efficiency in extracting oxygen from blood

8.6 Organization of Skeletal Muscles

Skeletal muscles are arranged in functional groups that are adaptive in causing particular movements. Within each muscle, the fibers are arranged in a specific pattern that provides specific functional capabilities.

Group Action of Muscles

Just as individual muscle fibers seldom contract independently, muscles generally do not contract separately but work as functional groups. Muscles that work together to promote a particular movement are called **synergists.** Muscles that oppose each other are known as **antagonists,** and are generally located on opposite sides of a limb. The two heads of the biceps brachii muscle, for example, contract together along with the brachialis muscle to *flex* the elbow joint. The triceps brachii muscle, the antagonist of the biceps brachii and brachialis muscles, *extends* the elbow as it is contracted.

The muscle that provides the major force in producing a desired action is called the **prime mover,** or **agonist.** The biceps brachii is the prime mover in flexing the elbow. When straightening (extending) the elbow, however, the triceps brachii serves as the prime mover; in this case, the biceps brachii is the antagonist.

In order to be effective at causing body movement, a muscle has to span a joint that permits movement as the muscle is contracted. As stated in module 8.1, the *origin* of a muscle is its point of attachment onto a bone that is stationary (not moved when the muscle contracts). The *insertion* is the opposite point of attachment onto the bone that is to be moved as the muscle contracts. In a group action, some muscles act as **fixators.** They stabilize the origin of the prime mover so that it can function more efficiently.

Muscle Architecture

Skeletal muscles may be classified on the basis of fiber arrangement as *parallel, convergent, pennate,* and *sphincteral* (circular) types (table 8.3). Each of these types provides certain capabilities for the muscles in which it is found.

Muscle-fiber architecture can be observed on a cadaver or other dissection specimen. If you have the opportunity to learn the muscles of the body from a cadaver, observe the fiber architecture of specific muscles and try to determine the advantages afforded to each muscle by its location and action.

synergist: Gk. *synergein,* cooperate
antagonist: Gk. *antagonistes,* struggle against

Naming of Muscles

One of your tasks as a student of anatomy and physiology is to learn the names of the principal muscles of the body. Although this task may seem overwhelming, keep in mind that most of the muscles are paired; that is, the right side is the mirror image of the left. To help you further, most muscles have names that are descriptive.

As you learn the muscles of the body, consider how each was named. Identify the muscle on the figure referenced in the text narrative and locate it on your own body as well. Use your body to act out its movement. Feel it contracting beneath your skin and note the movement that occurs at the joint. Learning the muscles in this way will simplify the task and make it more meaningful.

The following are some of the criteria by which the names of the muscles have been logically derived:

1. **Shape:** rhomboideus (like a rhomboid); trapezius (like a trapezoid); or denoting the number of heads of origin: triceps (three heads), biceps (two heads)
2. **Location:** pectoralis (in the chest, or pectus); intercostal (between ribs); brachii (upper arm)
3. **Attachment:** many facial muscles (zygomaticus, nasalis, temporalis); sternocleidomastoid (sternum, clavicle, and mastoid process of the skull)
4. **Size:** maximus (larger, largest); minimus (smaller, smallest); longus (long); brevis (short)
5. **Orientation of fibers:** rectus (straight); transverse (across); oblique (in a slanting or sloping direction)
6. **Relative position:** lateral, medial, abdominal, internal, and external
7. **Function:** adductor, flexor, extensor, pronator, and levator (lifter)

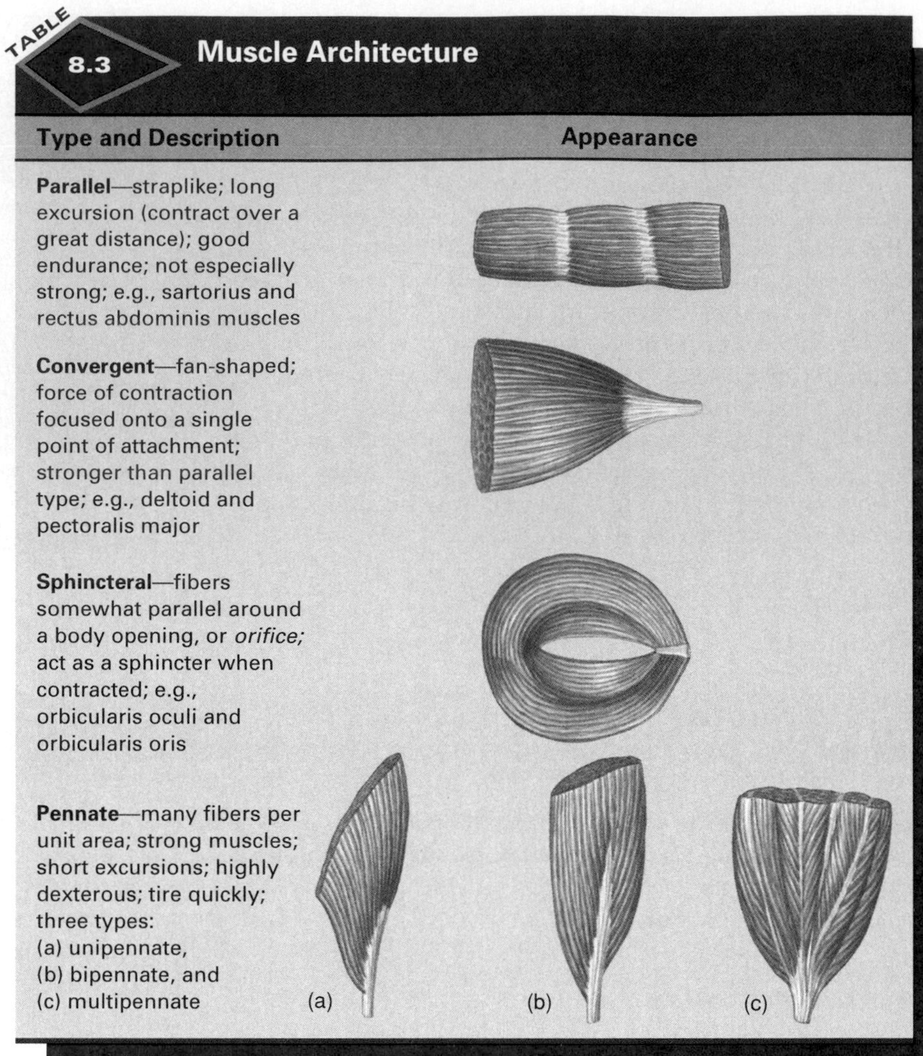

Table 8.3 Muscle Architecture

Type and Description	Appearance
Parallel—straplike; long excursion (contract over a great distance); good endurance; not especially strong; e.g., sartorius and rectus abdominis muscles	
Convergent—fan-shaped; force of contraction focused onto a single point of attachment; stronger than parallel type; e.g., deltoid and pectoralis major	
Sphincteral—fibers somewhat parallel around a body opening, or *orifice;* act as a sphincter when contracted; e.g., orbicularis oculi and orbicularis oris	
Pennate—many fibers per unit area; strong muscles; short excursions; highly dexterous; tire quickly; three types: (a) unipennate, (b) bipennate, and (c) multipennate	(a) (b) (c)

orifice: L. *orificium,* mouth; *facere,* to make
pennate: L. *pennatus,* feather

8.7 Muscles of the Head

Our well-developed facial muscles allow a wide range of facial expressions with which to communicate socially. We also have four pairs of mastication muscles for chewing, six ocular (eye) muscles of eye movement, and three pairs of tongue muscles for speaking and manipulating food.

Muscles of Facial Expression

The superficial muscles of facial expression are located on the scalp, face, and neck (fig. 8.18). Although highly variable in size and strength, these muscles all originate from the bones of the skull or flat, sheetlike tendons and insert onto the hypodermis of the skin (table 8.4). They are all

TABLE 8.4 Muscles of Facial Expression

Muscle	Origin	Insertion	Action
Epicranius	Galea aponeurotica and occipital bone	Skin of eyebrow and galea aponeurotica	Wrinkles forehead and moves scalp
Frontalis	Galea aponeurotica	Skin of eyebrow	Wrinkles forehead and elevates eyebrow
Occipitalis	Occipital bone and mastoid process	Galea aponeurotica	Moves scalp backward
Corrugator	Fascia above eyebrow	Root of nose	Draws eyebrow toward midline
Orbicularis oculi	Bones of medial orbit	Tissue of eyelid	Closes eye
Nasalis	Maxilla and nasal cartilage	Aponeurosis of nose	Dilates nostril
Orbicularis oris	Fascia surrounding lips	Mucosa of lips	Closes and purses lip
Levator labii superioris	Upper maxilla and zygomatic bone	Orbicularis oris and skin above lips	Elevates upper lip
Zygomaticus	Zygomatic bone	Superior corner of orbicularis oris	Elevates corner of mouth
Risorius	Fascia of cheek	Orbicularis oris at corner of mouth	Draws angle of mouth laterally
Depressor anguli oris	Mandible	Inferior corner of orbicularis oris	Depresses corner of mouth
Depressor labii inferioris	Mandible	Orbicularis oris and skin of lower lip	Depresses lower lip
Mentalis	Mandible (chin)	Orbicularis oris	Elevates and protrudes lower lip
Platysma	Fascia of neck and chest	Inferior border of mandible	Depresses lower lip
Buccinator	Maxilla and mandible	Orbicularis oris	Compresses cheek

corrugator: L. *corrugo,* a wrinkle
risorius: L. *risor,* a laughter
mentalis: L. *mentum,* chin
platysma: Gk. *platys,* broad
buccinator: L. *bucca,* cheek

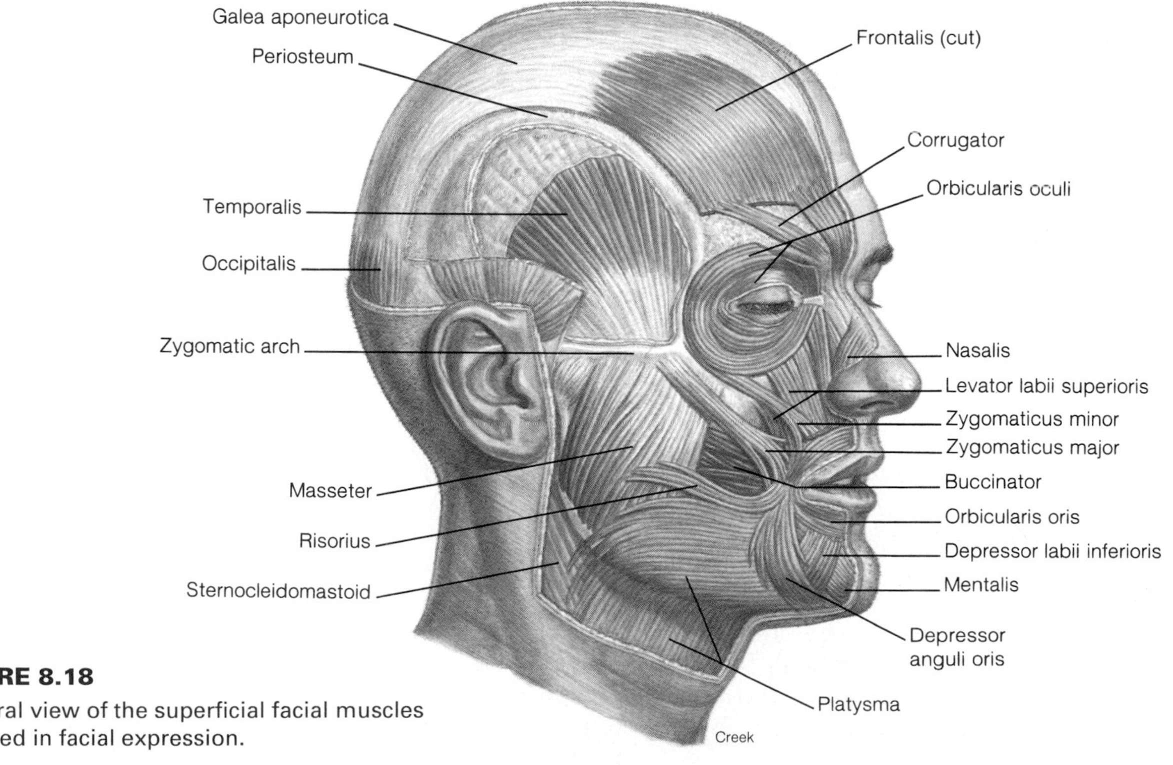

FIGURE 8.18
A lateral view of the superficial facial muscles involved in facial expression.

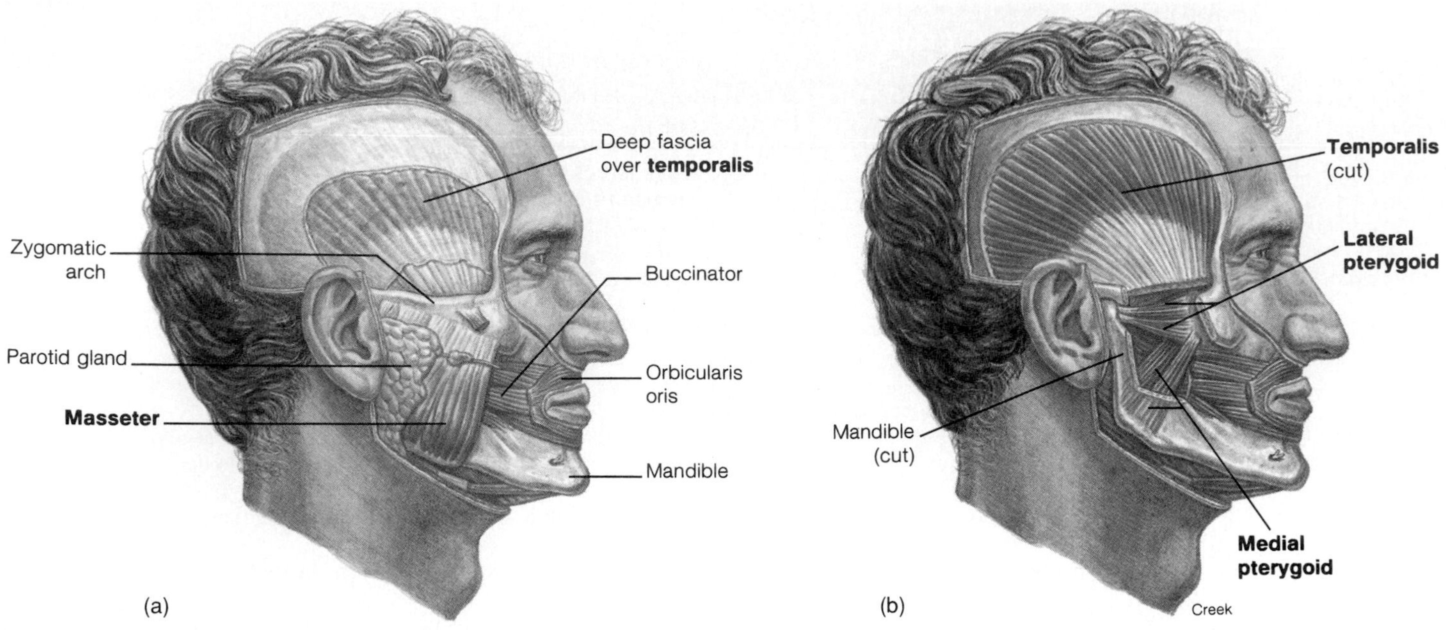

FIGURE 8.19

Muscles of mastication. (*a*) A superficial view and (*b*) a deep view showing the pterygoid muscles. (The muscles of mastication are labeled in boldface.)

TABLE 8.5 Muscles of Mastication

Muscle	Origin	Insertion	Action
Temporalis	Temporal fossa	Coronoid process of mandible	Elevates mandible
Masseter	Zygomatic arch	Lateral part of ramus of mandible	Elevates mandible
Medial pterygoid	Sphenoid bone	Medial aspect of mandible	Elevates mandible and moves mandible laterally
Lateral pterygoid	Sphenoid bone	Anterior side of mandibular condyle	Protracts mandible

masseter: Gk. *maseter,* chew
pterygoid: Gk. *pteron,* wing

innervated (served with neurons) by one of the two facial cranial nerves. The locations and points of attachments of most of the facial muscles are such that, when contracted, the muscles cause movements around the eyes, nostrils, or mouth.

The muscles of facial expression are of clinical concern for reasons involving the facial nerve (see fig. 10.21). Located right under the skin, the many branches of the facial cranial nerve are vulnerable to trauma. Facial lacerations and fractures of the skull frequently damage branches of this nerve. The extensive pattern of motor innervation becomes apparent in stroke victims and people with Bell's palsy. *The facial muscles on one side of the face are affected in these people, and that side of the face appears to sag.*

Muscles of Mastication

The large **temporalis** and **masseter** (*mă-se'ter*) muscles (fig. 8.19) are powerful elevators of the mandible in conjunction with the **medial pterygoid** (*ter'ĭ-goid*) muscle. The primary function of the medial and lateral pterygoid muscles is to provide grinding movements of the teeth. The **lateral pterygoid** also protracts the mandible (table 8.5).

Ocular Muscles

The movements of the eyeball are controlled by six extrinsic ocular (eye) muscles (fig. 8.20 and table 8.6). Five of these muscles arise from the margin of the optic foramen at the back of the orbital cavity and insert on the outer layer (sclera) of the eyeball. Four **rectus muscles** maneuver the eyeball in the direction indicated by their names (*superior, inferior, lateral,* and *medial*), and two

oblique muscles (*superior* and *inferior*) rotate the eyeball on its axis. The medial rectus on one side contracts with the medial rectus of the opposite eye when focusing on close objects. The superior oblique passes through a pulleylike cartilaginous loop, the *trochlea,* before attaching to the eyeball.

Another muscle, the **levator palpebrae superioris,** is located in the ocular region but is not attached to the eyeball. It extends into the upper eyelid and raises the eyelid when contracted.

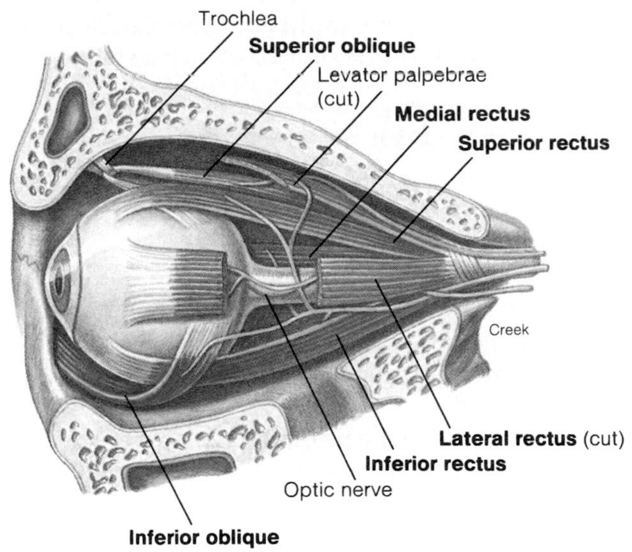

FIGURE 8.20
A lateral view of the extrinsic ocular muscles.

Muscles That Move the Tongue

The tongue is a highly specialized muscular organ that functions in speaking, manipulating food, cleansing the teeth, and swallowing. The *intrinsic tongue muscles* are located within the tongue and are responsible for its mobility and changes of shape. The *extrinsic tongue muscles* are those that originate on structures other than the tongue and insert onto it to cause gross tongue movement (fig. 8.21). The three paired extrinsic muscles are the **genioglossus, styloglossus,** and **hyoglossus.** When the anterior portion of the genioglossus is contracted, the tongue is depressed and thrust forward. If both genioglossus muscles are contracted together, the surface of the tongue becomes transversely concave. This muscle is used by infants in nursing; the tongue is positioned around the nipple with a concave groove channeled toward the pharynx.

TABLE 8.6 Ocular Muscles

Muscle	Cranial Nerve Innervation	Movement of Eyeball
Lateral rectus	Abducens	Lateral
Medial rectus	Oculomotor	Medial
Superior rectus	Oculomotor	Superior and medial
Inferior rectus	Oculomotor	Inferior and medial
Inferior oblique	Oculomotor	Superior and lateral
Superior oblique	Trochlear	Inferior and lateral

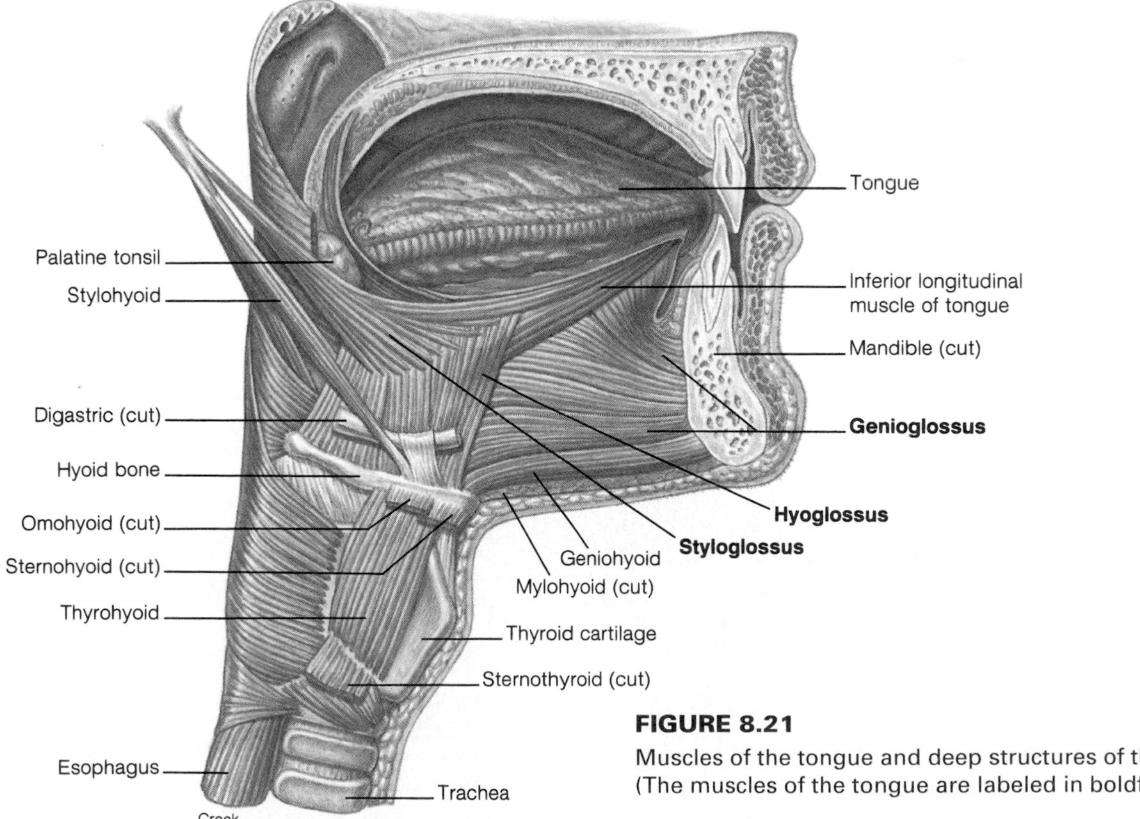

FIGURE 8.21
Muscles of the tongue and deep structures of the neck. (The muscles of the tongue are labeled in boldface.)

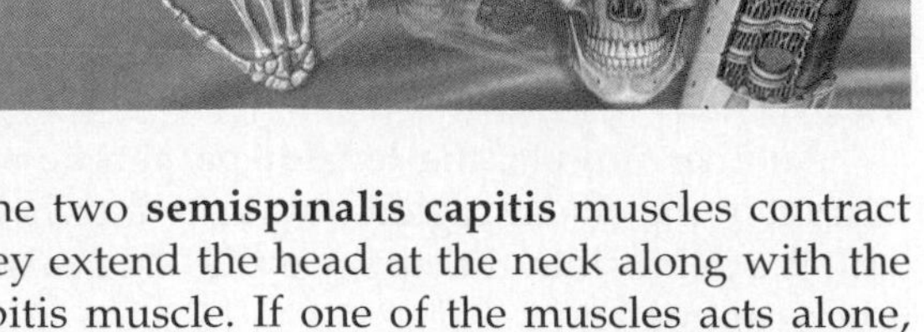

8.8 Muscles of the Neck

Muscles of the neck either support and move the head or are associated with structures within the neck region, such as the hyoid bone and larynx.

Posterior Neck Muscles

The posterior neck muscles include the *sternocleidomastoid* (originates anteriorly), *trapezius, splenius capitis, semispinalis capitis,* and *longissimus capitis* (figs. 8.22, 8.23 and table 8.7).

When contracted on one side, the **sternocleidomastoid** muscle turns the head sideways in a direction opposite the side on which the muscle is located. If both sternocleidomastoid muscles are contracted, the head is pulled forward and down. The sternocleidomastoid is covered by the platysma muscle (see fig. 8.18).

Although a portion of the **trapezius** muscle extends over the posterior neck region, it is primarily a superficial muscle of the back and will be discussed in module 8.12.

When the **splenius capitis** muscle contracts on one side, the head rotates and extends to one side. Contracted together, the splenius capitis muscles extend the head at the neck. Further contraction causes hyperextension of the neck and head.

When the two **semispinalis capitis** muscles contract together, they extend the head at the neck along with the splenius capitis muscle. If one of the muscles acts alone, the head is rotated to the side. Similarly, when contracted, the straplike **longissimus capitis** muscle extends the head at the neck, bends it to one side, or rotates it slightly.

Suprahyoid Muscles

This group of muscles located above the hyoid bone includes the *digastric, mylohyoid, geniohyoid,* and *stylohyoid* (figs. 8.21 and 8.23).

The **digastric** is a two-bellied muscle that can open the mouth or elevate the hyoid bone. As the **mylohyoid** muscle contracts, the floor of the mouth is elevated. It aids swallowing by forcing the food toward the back of the mouth.

The short, straplike **geniohyoid** muscle is deep to the mylohyoid. The geniohyoid extends from the medial surface of the mandible at the chin to the hyoid bone, which it elevates as it contracts. When the slender **stylohyoid** muscle contracts, it elevates and retracts the hyoid bone, thereby elongating the floor of the mouth during swallowing.

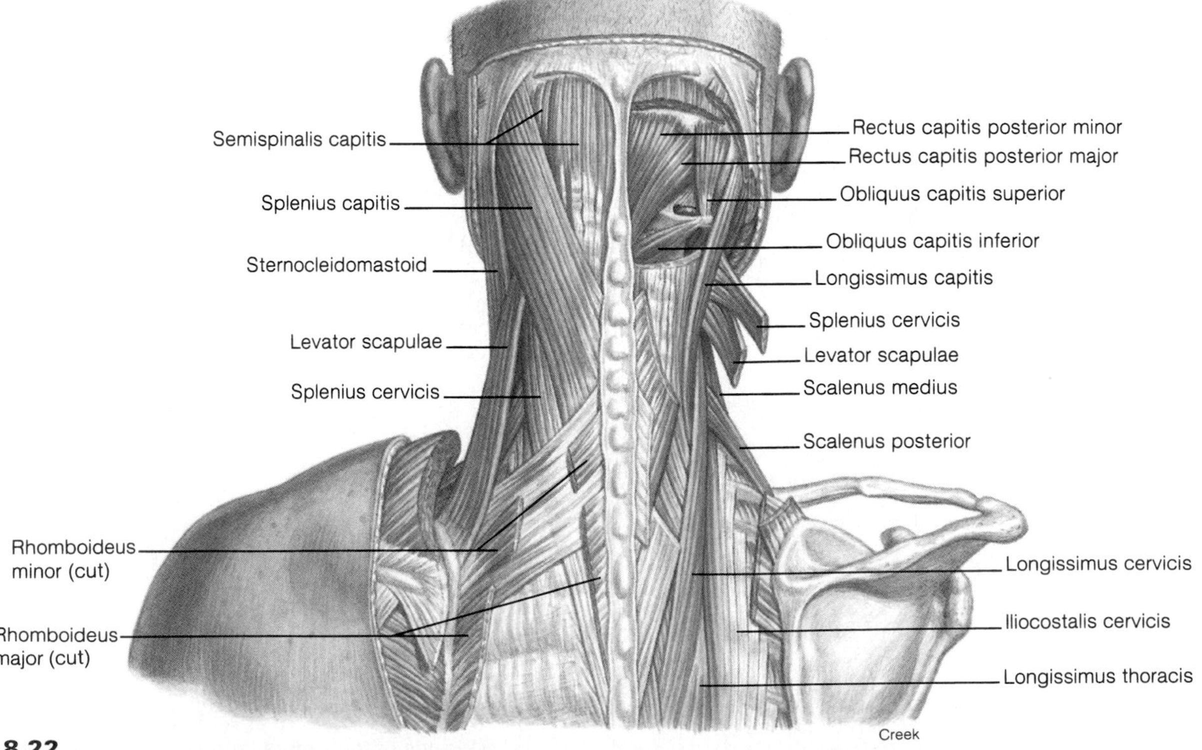

FIGURE 8.22
Deep muscles of the posterior neck and upper back regions.

TABLE 8.7 **Muscles of the Neck**

Muscle	Origin	Insertion	Action	Innervation
Sternocleidomastoid	Sternum and clavicle	Mastoid process of temporal bone	Turns head to side; flexes neck	Accessory n.
Digastric	Inferior border of mandible and mastoid process of temporal bone	Hyoid bone	Opens mouth; elevates hyoid bone	Trigeminal n. (ant. belly); facial n. (post. belly)
Mylohyoid	Inferior border of mandible	Body of hyoid bone and median raphe	Elevates hyoid bone and floor of mouth	Trigeminal n.
Geniohyoid	Medial surface of mandible at chin	Body of hyoid bone	Elevates hyoid bone	Hypoglossal n.
Stylohyoid	Styloid process of temporal bone	Body of hyoid bone	Elevates and retracts tongue	Facial n.
Sternohyoid	Manubrium	Body of hyoid bone	Depresses hyoid bone	Hypoglossal n.
Sternothyroid	Manubrium	Thyroid cartilage	Depresses thyroid cartilage	Hypoglossal n.
Thyrohyoid	Thyroid cartilage	Great cornu of hyoid bone	Depresses hyoid bone; elevates thyroid cartilage	Hypoglossal n.
Omohyoid	Superior border of scapula	Clavicle and body of hyoid bone	Depresses hyoid bone	Hypoglossal n.

digastric: L. *di,* two; Gk. *gaster,* belly
mylohyoid: Gk. *mylos,* akin to; *hyoeides,* pertaining to hyoid bone
omohyoid: Gk. *omos,* shoulder

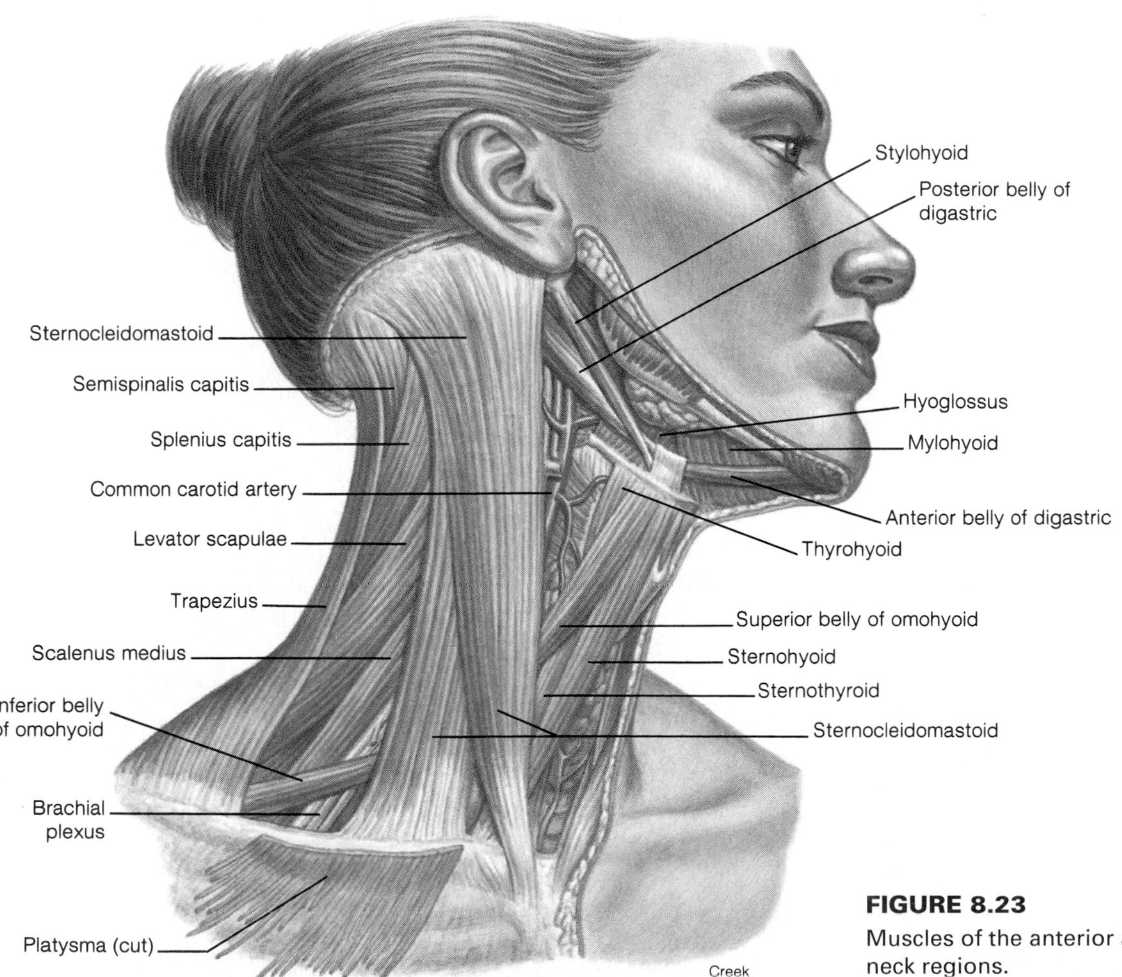

FIGURE 8.23
Muscles of the anterior and lateral neck regions.

Infrahyoid Muscles

This group of muscles located below the hyoid bone includes the *sternohyoid, sternothyroid, thyrohyoid,* and *omohyoid* (see fig. 8.21).

When contracted, the **sternohyoid** muscle depresses the hyoid bone. The **sternothyroid** muscle is immediately lateral to the sternohyoid, and when contracted it pulls the larynx downward. The **thyrohyoid** muscle elevates the larynx and lowers the hyoid. When the long, thin **omohyoid** muscle contracts, it depresses the hyoid.

The coordinated movements of the hyoid bone and the larynx are impressive. The hyoid bone does not articulate with any other bone, yet it has eight paired muscles attached to it. Two involve tongue movement, one lowers the jaw, one elevates the floor of the mouth, and four depress the hyoid bone or elevate the thyroid cartilage of the larynx.

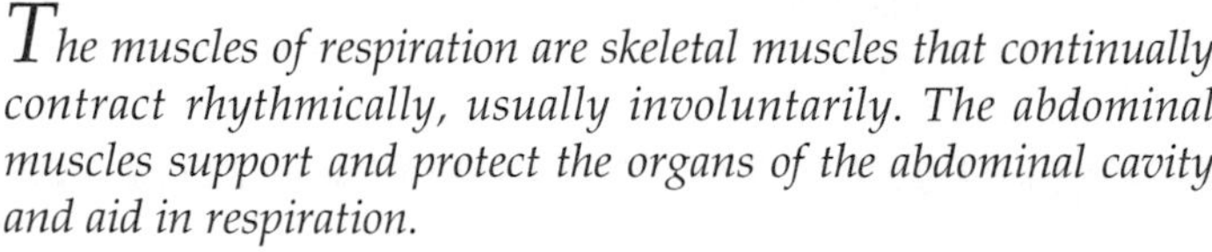

8.9 Muscles of Respiration and Muscles of the Abdominal Wall

The muscles of respiration are skeletal muscles that continually contract rhythmically, usually involuntarily. The abdominal muscles support and protect the organs of the abdominal cavity and aid in respiration.

Muscles of Respiration

During normal relaxed inspiration (inhalation), the contracting muscles are the **diaphragm, external intercostals,** and the interchondral portion of the **internal intercostals** (fig. 8.24). A downward contraction of the dome-shaped diaphragm causes a vertical increase in the thoracic dimension. A simultaneous contraction of the external intercostals and the interchondral portion of the internal intercostals produces an increase in the lateral dimension of the thorax. In addition, the **scalene** and **sternocleidomastoid** muscles of the neck may assist in inspiration by helping to elevate the ribs. The intercostal muscles are innervated by the intercostal nerves, and the diaphragm receives its stimuli through the phrenic nerves.

Expiration (exhalation) is primarily a passive process that occurs as the muscles of inspiration are relaxed and the rib cage recoils to its original position. During forced expiration, the interosseous portion of the **internal intercostals** contracts, causing the rib cage to be depressed. This portion of the internal intercostals lies under the external intercostals, and

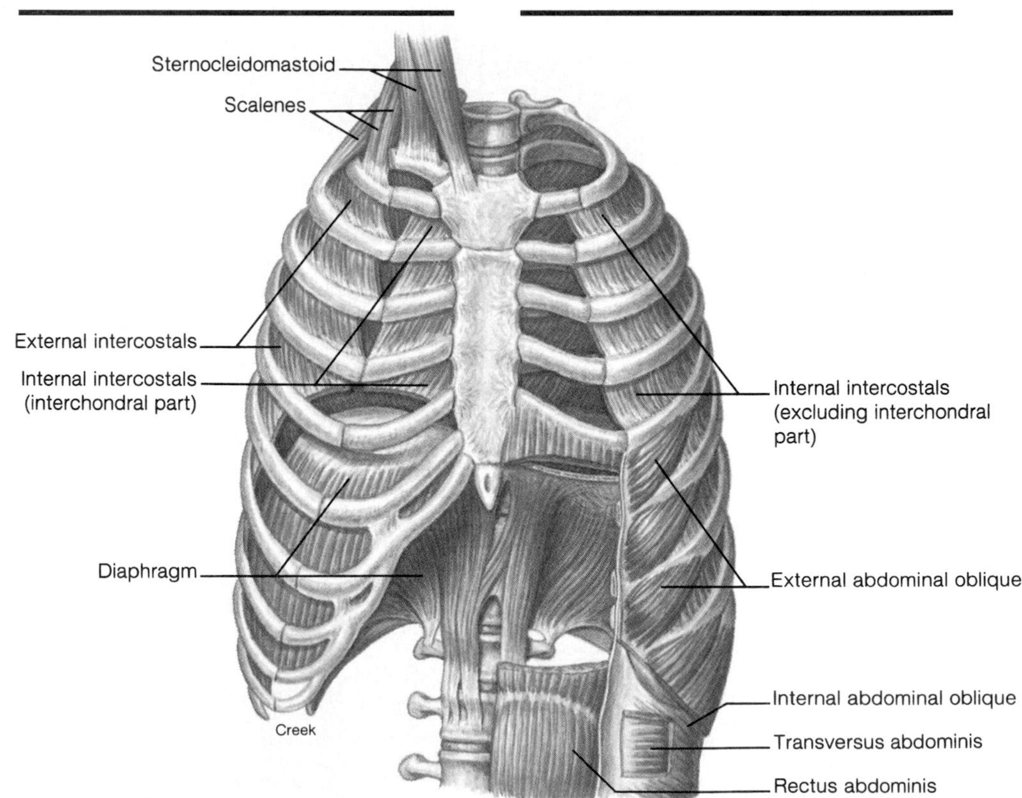

FIGURE 8.24
Muscles of respiration.

its fibers are directed downward and backward. The **abdominal muscles** may also contract during forced expiration, which increases pressure within the abdominal cavity and forces the diaphragm superiorly, squeezing additional air out of the lungs.

Muscles of the Abdominal Wall

The anterolateral abdominal wall is composed of four pairs of flat, sheetlike muscles: the *external abdominal oblique, internal abdominal oblique, transversus abdominis,* and *rectus abdominis* (figs. 8.24 and 8.25, and table 8.8). The movements of these muscles are varied and include rotating, flexing, and compressing the abdomen.

The **external abdominal oblique** is the strongest of the lateral abdominal muscles. Its fibers are directed inferiorly and medially. The **internal abdominal oblique** lies deep to the external abdominal oblique, and its fibers are directed at right angles to the external abdominal oblique. The **transversus abdominis** is the deepest muscle of the

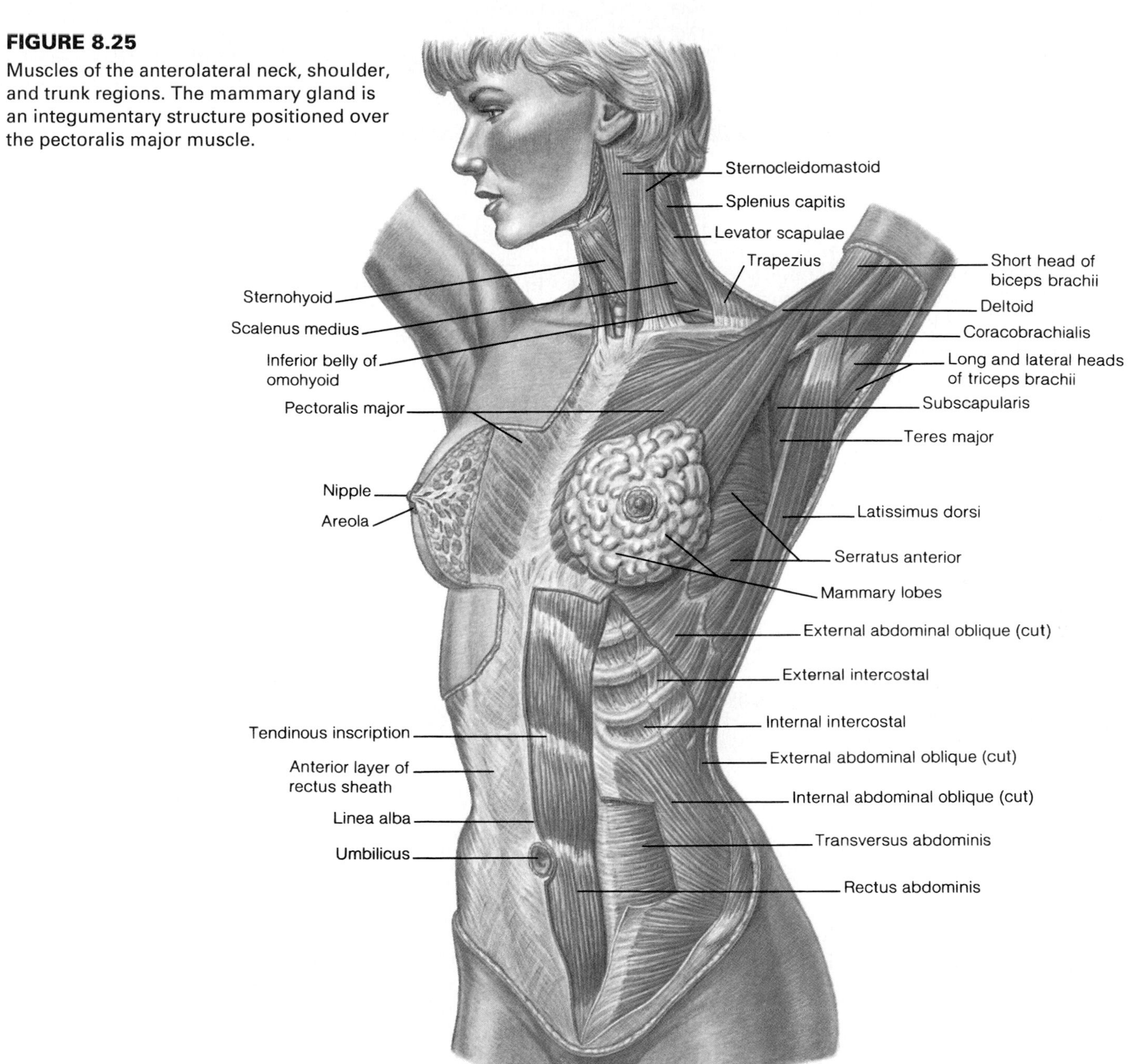

FIGURE 8.25
Muscles of the anterolateral neck, shoulder, and trunk regions. The mammary gland is an integumentary structure positioned over the pectoralis major muscle.

TABLE 8.8 **Muscles of the Abdominal Wall**

Muscle	Origin	Insertion	Action
External abdominal oblique	Lower eight ribs	Iliac crest and linea alba	Compresses abdomen; lateral rotation; draws thorax downward
Internal abdominal oblique	Iliac crest, inguinal ligament, and lumbodorsal fascia	Linea alba and costal cartilage of last three or four ribs	Compresses abdomen; lateral rotation; draws thorax downward
Transversus abdominis	Iliac crest, inguinal ligament, lumbar fascia, and costal cartilage of last six ribs	Xiphoid process, linea alba, and pubis	Compresses abdomen
Rectus abdominis	Pubic crest and symphysis pubis	Costal cartilage of fifth to seventh ribs and xiphoid process of sternum	Flexes vertebral column

rectus abdominis: L. *rectus,* straplike; *abdomino,* belly

three, and its fibers pass directly medially around the abdominal wall. The long, straplike **rectus abdominis** is entirely enclosed in a fibrous sheath formed from the aponeuroses of the other three abdominal muscles. The *linea alba* is a band of connective tissue on the midline of the abdomen that separates the two rectus abdominis muscles. Bands of connective tissue called *tendinous inscriptions* transect the rectus abdominis muscles at several points, causing the surface abdominal anatomy of a well-muscled person to appear segmented (see fig. 4, pg. 642).

Inside Information—Carefully examine a slice of bacon and you will see three distinct layers of muscle between the fat. These are the abdominal muscles from a pig, which is the source of bacon.

A surgical incision along the linea alba severs no muscles and few vessels or nerves, which is why it is quick to heal. It has been said that only a zipper would provide a more convenient entry to the abdominal cavity.

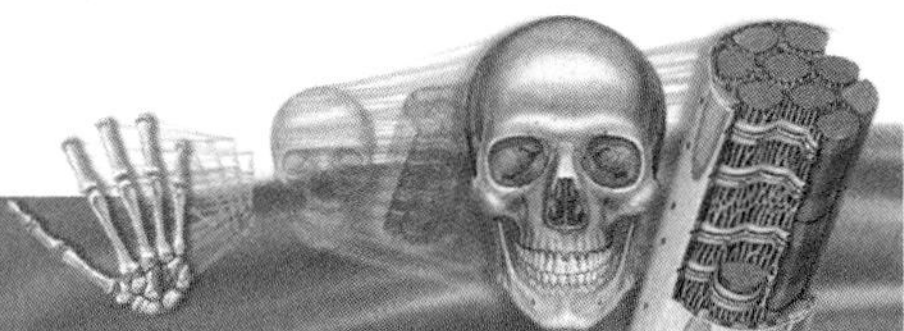

8.10 Muscles of the Pelvic Outlet

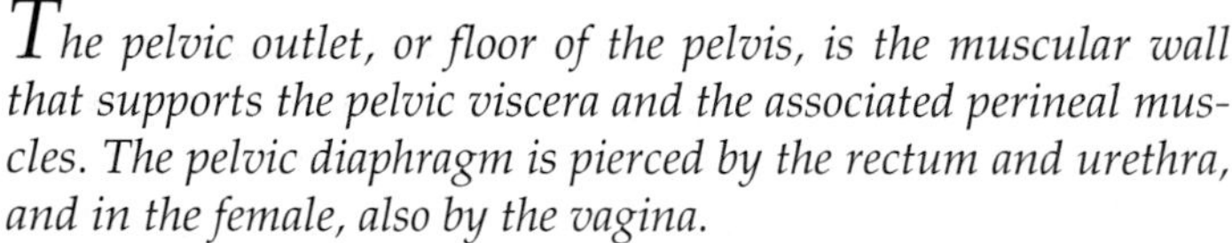

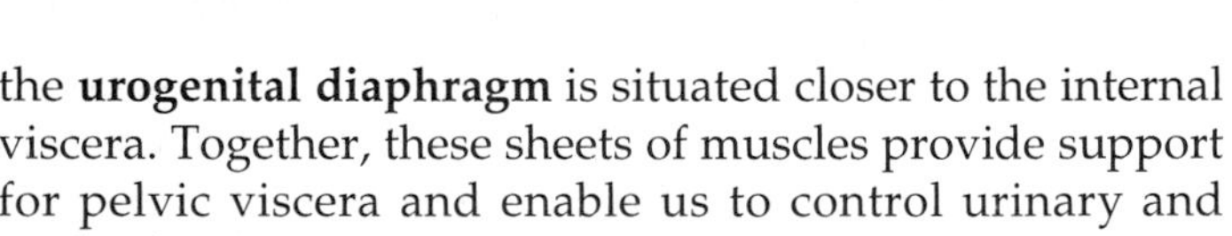

The pelvic outlet, or floor of the pelvis, is the muscular wall that supports the pelvic viscera and the associated perineal muscles. The pelvic diaphragm is pierced by the rectum and urethra, and in the female, also by the vagina.

Any sheet that separates cavities may be termed a diaphragm. The pelvic outlet—the entire muscular basin at the bottom of the pelvic cavity—contains two: the *pelvic diaphragm* and the *urogenital diaphragm.* The **pelvic diaphragm** is immediately deep to the external genitalia; the **urogenital diaphragm** is situated closer to the internal viscera. Together, these sheets of muscles provide support for pelvic viscera and enable us to control urinary and anal sphincters.

The pelvic diaphragm consists of the *levator ani* and the *coccygeus* muscles (table 8.9). The **levator ani** (fig. 8.26) is a thin sheet of muscle that helps to support the pelvic viscera and constrict the lower part of the rectum, pulling it forward and aiding defecation. The deeper, fan-shaped **coccygeus** aids the levator ani in its functions.

TABLE 8.9 **Muscles of the Pelvic Outlet**

Muscle	Origin	Insertion	Action
Levator ani	Spine of ischium and pubic bone	Coccyx	Supports pelvic viscera; aids in defecation
Coccygeus	Ischial spine	Sacrum and coccyx	Supports pelvic viscera; aids in defecation
Transversus perinei	Ischial tuberosity	Central tendon	Supports pelvic viscera
Bulbospongiosus	Central tendon	Males: base of penis; females: root of clitoris	Constricts urethral canal; constricts vagina
Ischiocavernosus	Ischial tuberosity	Males: pubic arch and crus of penis; females: pubic arch and crus of clitoris	Aids erection of penis or clitoris

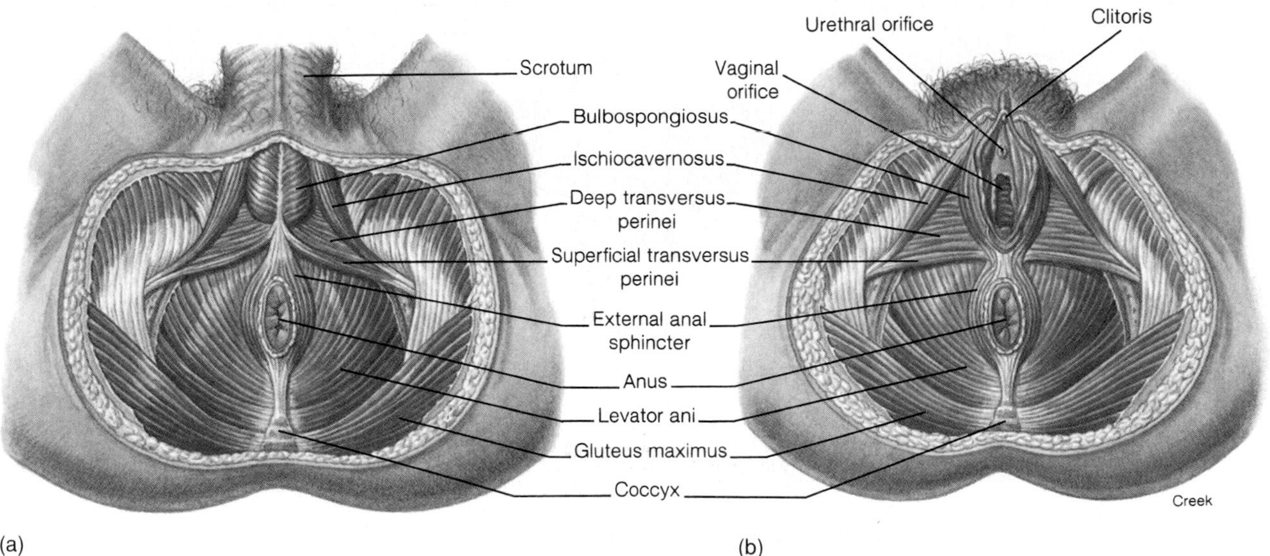

FIGURE 8.26
Muscles of the pelvic outlet. (*a*) Male and (*b*) female.

An episiotomy (ĕ-pe'ze-ot"ŏ-me) *is a surgical incision, for obstetrical purposes, of the vaginal orifice and a portion of the levator ani muscle of the perineum. Following a pudendal nerve block, an episiotomy may be done during childbirth to accommodate the head of an emerging fetus with minimal tearing of the tissues. After delivery, the cut is sutured.*

The *urogenital diaphragm* consists of two deep muscles—the **deep transversus perinei** and the **external anal sphincter** muscles. The external anal sphincter is a funnel-shaped constrictor muscle that surrounds the anal canal.

Inferior to the pelvic diaphragm are the *perineal muscles,* which provide the skeletal muscular support to the genitalia. They include the *bulbocavernosus, ischiocavernosus,* and the *superficial transversus perinei* muscles (fig. 8.26). The muscles of the pelvic diaphragm and the urogenital diaphragm are similar in the male and female, but the perineal muscles exhibit marked sex-based differences.

In males, the **bulbocavernosus** of one side unites with that of the opposite side to form a muscular constriction surrounding the base of the penis. When contracted, the two muscles constrict the urethral canal and assist in emptying the urethra. In females, these muscles are separated by the vaginal orifice, which they constrict as they contract. The **ischiocavernosus** inserts onto the pubic arch and crus of the penis in the male and onto the pubic arch and clitoris in the female. This muscle aids the erection of the penis in the male and the clitoris in the female.

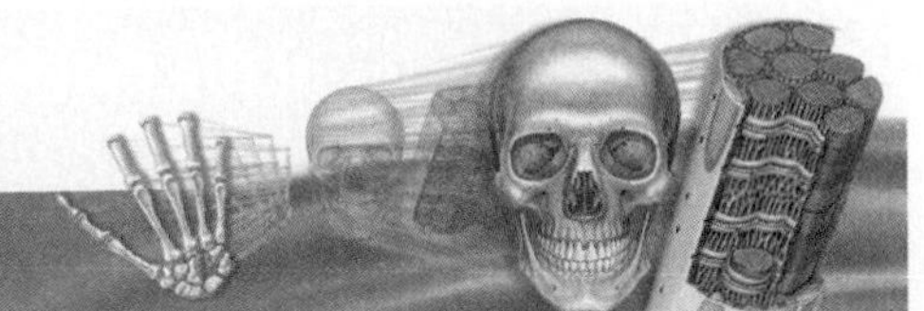

8.11 Muscles of the Vertebral Column

The complex muscles that move the vertebral column are strong because they have to provide support and movement in resistance to the effect of gravity.

The vertebral column can be flexed, extended, abducted, adducted, and rotated. The muscle that flexes the vertebral column, the rectus abdominis, has already been described as a paired straplike muscle of the anterior abdominal wall. The opposing extensor muscles (located on the posterior side of the vertebral column) must be stronger than the flexors because extension (such as lifting an object) is in opposition to gravity. The extensor muscles consist of a superficial group and a deep group. The principal muscles of the vertebral column are depicted in figure 8.27 and listed in table 8.10.

The **erector spinae** muscles constitute a massive superficial muscle group that extends from the sacrum to the skull. It actually consists of three groups of muscles:

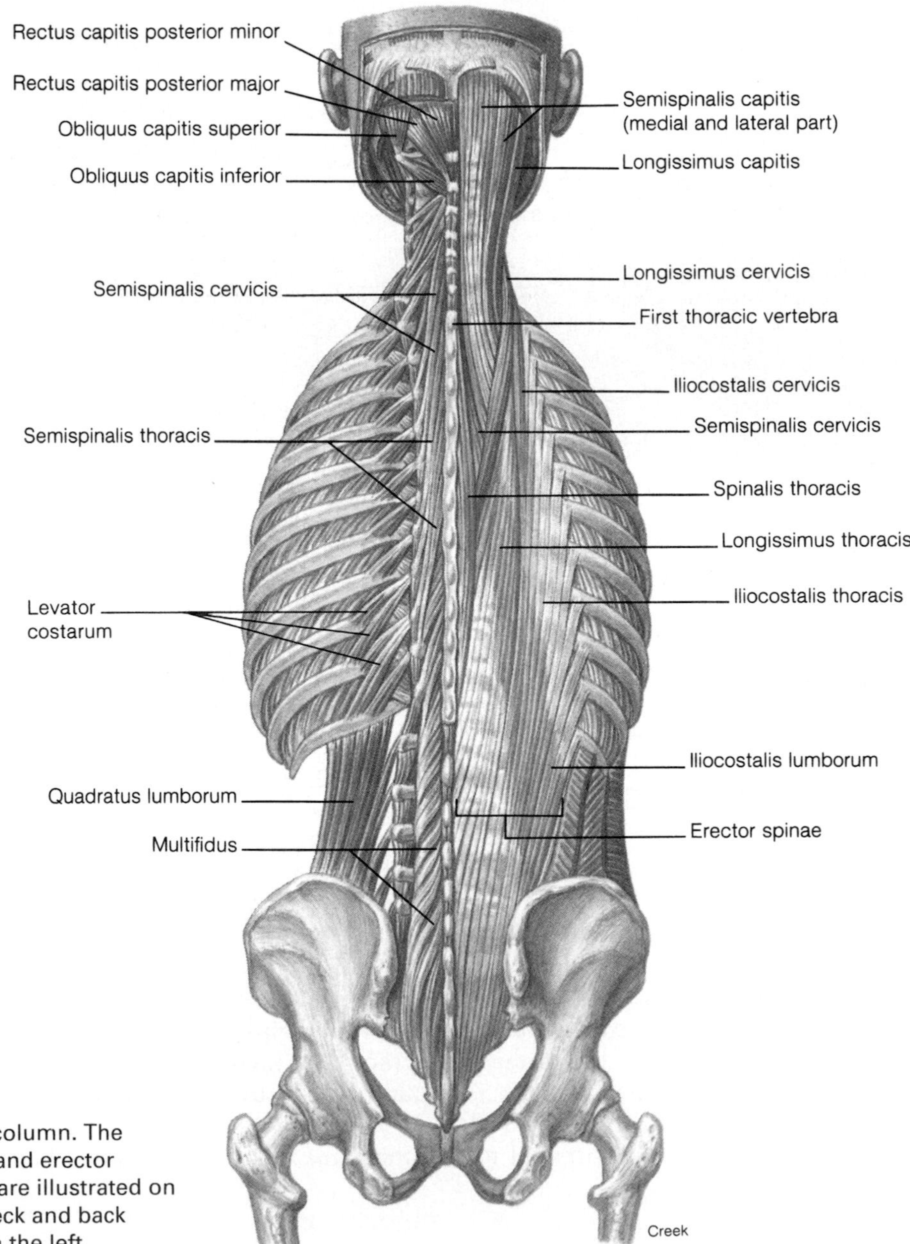

FIGURE 8.27

Muscles of the vertebral column. The superficial neck muscles and erector spinae group of muscles are illustrated on the right, and the deep neck and back muscles are illustrated on the left.

the **iliocostalis, longissimus,** and **spinalis** muscles. Each of these groups, in turn, consists of overlapping slips of muscle. The iliocostalis is the most lateral group, the longissimus is intermediate in position, and the spinalis, in medial position, is in contact with the spinous processes of the vertebrae. The erector spinae muscles laterally flex the vertebral column and are strong extensors.

The erector spinae muscles are frequently strained through improper lifting. A heavy object should not be lifted with the vertebral column flexed; instead, the thighs and knees should be flexed so that the pelvic and leg muscles can aid in the task.

Pregnancy may also put a strain on the erector spinae muscles. Pregnant women will try to counterbalance the effect of the protruding abdomen by hyperextending the vertebral column. This results in an exaggerated lumbar curvature, strained muscles, and a peculiar gait.

The deep **quadratus lumborum** muscle originates on the iliac crest and lower three lumbar vertebrae. It inserts on the transverse processes of the first four lumbar vertebrae and the inferior margin of the twelfth rib. When the right and left quadratus lumborum muscles contract together, the vertebral column in the lumbar region extends. Separate contraction causes lateral flexion of the spine.

TABLE 8.10 **Muscles of the Vertebral Column**

Muscle	Origin	Insertion	Action	Innervation
Quadratus lumborum	Iliac crest and lower three lumbar vertebrae	Twelfth rib and upper four lumbar vertebrae	Extends lumbar region; lateral flexion of vertebral column	Intercostal nerve T12 and lumbar nerves L2–L4
Erector spinae	Consists of three groups of muscles: iliocostalis, longissimus, and spinalis. The iliocostalis and longissimus are further subdivided into three groups on the basis of location along the vertebral column.			
Iliocostalis lumborum	Crest of ilium	Lower six ribs	Extends lumbar region	Posterior rami of lumbar nerves
Iliocostalis thoracis	Lower six ribs	Upper six ribs	Extends thoracic region	Posterior rami of thoracic nerves
Iliocostalis cervicis	Angles of third to sixth rib	Transverse processes of fourth to sixth cervical vertebrae	Extends cervical region	Posterior rami or cervical nerves
Longissimus thoracis	Transverse processes of lumbar vertebrae	Transverse processes of all the thoracic vertebrae and lower nine ribs	Extends thoracic region	Posterior rami of spinal nerves
Longissimus cervicis	Transverse processes of upper four or five thoracic vertebrae	Transverse processes of second to sixth cervical vertebrae	Extends cervical region; lateral flexion	Posterior rami of spinal nerves
Longissimus capitis	Transverse processes of upper five thoracic vertebrae and articular processes of lower three cervical vertebrae	Posterior margin of cranium and mastoid process of temporal bone	Extends head; acting separately, turns face toward that side	Posterior rami of middle and lower cervical nerves
Spinalis thoracis	Spinous processes of upper lumbar and lower thoracic vertebrae	Spinous processes of upper thoracic vertebrae	Extends vertebral column	Posterior rami of spinal nerves

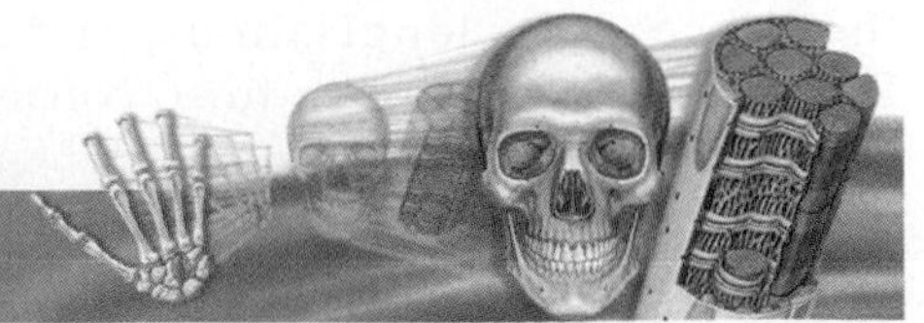

8.12 Muscles of the Shoulder and Upper Limb

The only attachment of the shoulder to the axial skeleton is at the sternoclavicular joint; therefore, strong muscles are necessary to stabilize this region. During brachial movement, the scapula has to be held fixed. Muscles that move the humerus originate on the scapula. The muscles of the upper limb vary from the large, powerful muscles of the brachium to the long, delicate muscles of the antebrachium (forearm).

Muscles That Act on the Shoulder

The muscles that act on the pectoral girdle originate on the axial skeleton and can be divided into anterior and posterior groups. The anterior group of muscles includes the **serratus anterior, pectoralis minor,** and **subclavius** muscles (fig. 8.28). The posterior group includes the **trapezius, levator scapulae,** and **rhomboideus** muscles (fig. 8.29).

Muscles That Act on the Humerus

Of the nine muscles that span the shoulder joint to insert on the humerus, only two—the **pectoralis major** and **latissimus dorsi** muscles—do not originate on the scapula. These two are designated as axial muscles, whereas the remaining seven are scapular muscles. The scapular muscles include the **deltoid, supraspinatus, infraspinatus, teres major, teres minor, subscapularis,** and **coracobrachialis** muscles. The actions of these muscles are summarized in table 8.11.

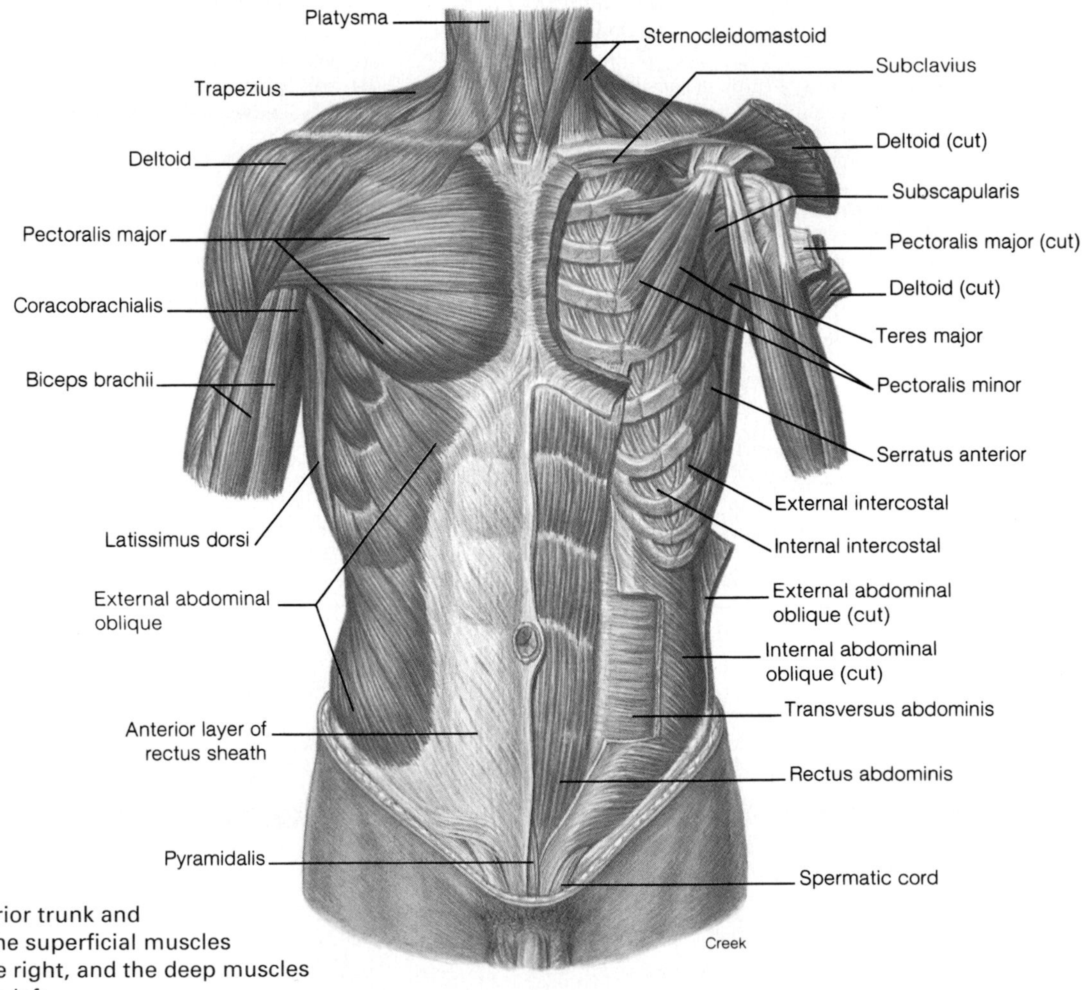

FIGURE 8.28
Muscles of the anterior trunk and shoulder regions. The superficial muscles are illustrated on the right, and the deep muscles are illustrated on the left.

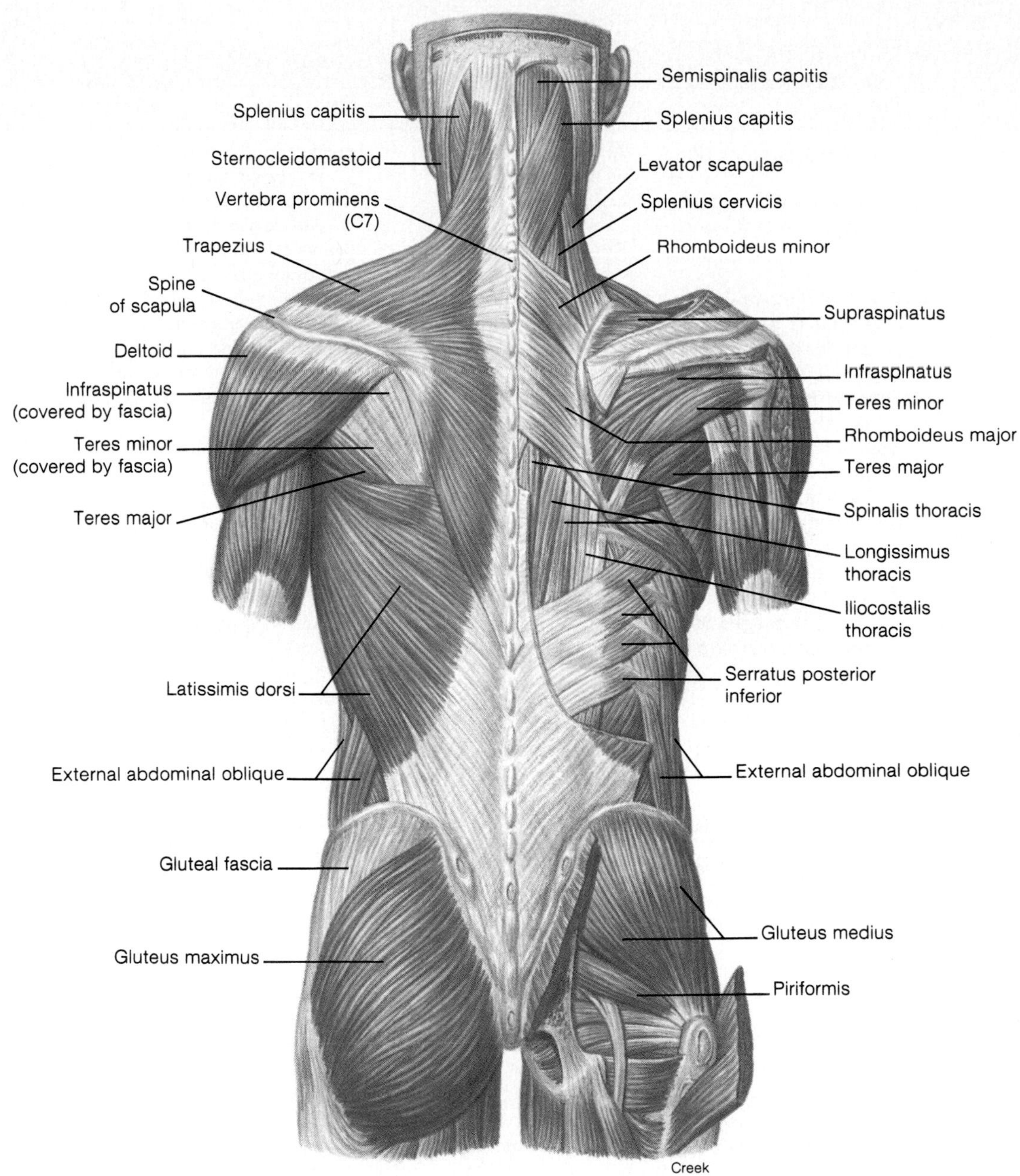

FIGURE 8.29
Muscles of the posterior neck, shoulder, trunk, and gluteal regions. The superficial muscles are illustrated on the left, and the deep muscles are illustrated on the right.

Muscles That Act on the Forearm

The powerful **biceps brachii,** which bulges when the elbow is flexed, is the most familiar muscle of the arm. It is positioned on the anterior surface of the humerus, yet has no attachments to the humerus (fig. 8.30). The biceps brachii has a dual origin. A medial tendinous head, the *short head,* arises from the coracoid process of the scapula, and the *long head* originates on the superior tuberosity of the glenoid cavity, passes through the shoulder joint, and descends in the intertubercular groove on the humerus. Both heads of the biceps brachii insert on the radial tuberosity. The **brachialis** is located on the distal anterior half of the humerus, deep to the biceps brachii. It contracts simultaneously with the biceps in flexing the forearm at the elbow joint.

The prominent **brachioradialis** is positioned along the lateral (radial) surface of the forearm. It, too, flexes the forearm at the elbow joint.

TABLE 8.11 Muscles That Act on the Pectoral Girdle and That Move the Humerus

Muscle	Origin	Insertion	Action	Innervation
Serratus anterior	Upper eight or nine ribs	Anterior vertebral border of scapula	Pulls scapula forward and downward	Long thoracic n.
Pectoralis minor	Sternal ends of third, fourth, and fifth ribs	Coracoid process of scapula	Pulls scapula forward and downward	Medial pectoral n.
Subclavius	First rib	Subclavian groove of clavicle	Draws clavicle downward	Spinal nerves C5, C6
Trapezius	Occipital bone and spines of seventh cervical and all thoracic vertebrae	Clavicle, spine of scapula, and acromion	Elevates, depresses, and adducts scapula; hyperextends neck; braces shoulder	Accessory nerve; spinal nerves C3, C4
Levator scapulae	First to fourth cervical vertebrae	Vertebral border of scapula	Elevates scapula	Dorsal scapular n.
Rhomboideus major	Spines of second to fifth thoracic vertebrae	Vertebral border of scapula	Elevates and adducts scapula	Dorsal scapular n.
Rhomboideus minor	Seventh cervical and first thoracic vertebrae	Vertebral border of scapula	Elevates and adducts scapula	Dorsal scapular n.
Pectoralis major	Clavicle, sternum, costal cartilages of second to sixth rib; rectus sheath	Crest of greater tubercle of humerus	Flexes, adducts, and rotates arm medially	Medial and lateral pectoral nn.
Latissimus dorsi	Spines of sacral, lumbar, and lower thoracic vertebrae; iliac crest and lower four ribs	Intertubercular groove of humerus	Extends, adducts, and rotates humerus medially; adducts arm	Thoracodorsal n.
Deltoid	Clavicle, acromion and spine of scapula	Deltoid tuberosity of humerus	Abducts arm; extends or flexes humerus	Axillary n.
Supraspinatus	Fossa—superior to spine of scapula	Greater tubercle of humerus	Abducts and laterally rotates humerus	Suprascapular n.
Infraspinatus	Fossa—inferior to spine of scapula	Greater tubercle of humerus	Rotates arm laterally	Suprascapular n.
Teres major	Inferior angle and lateral border of scapula	Crest of lesser tubercle of humerus	Extends humerus, or adducts and rotates arm medially	Lower subscapular n.
Teres minor	Axillary border of scapula	Greater tubercle and groove of humerus	Rotates arm laterally	Axillary n.
Subscapularis	Subscapular fossa	Lesser tubercle of humerus	Rotates arm medially	Subscapular n.
Coracobrachialis	Coracoid process of scapula	Shaft of humerus	Flexes and adducts shoulder joint	Musculocutaneous n.

---

serratus anterior: L. *serratus,* saw-shaped
trapezius: Gk. *trapezoeides,* trapezoid-shaped
rhomboideus major: Gk. *rhomboides,* rhomboid-shaped
pectoralis major: L. *pectus,* chest
latissimus dorsi: L. *latissimus,* widest
deltoid: Gk. *delta,* triangular
teres major: L. *teres,* rounded

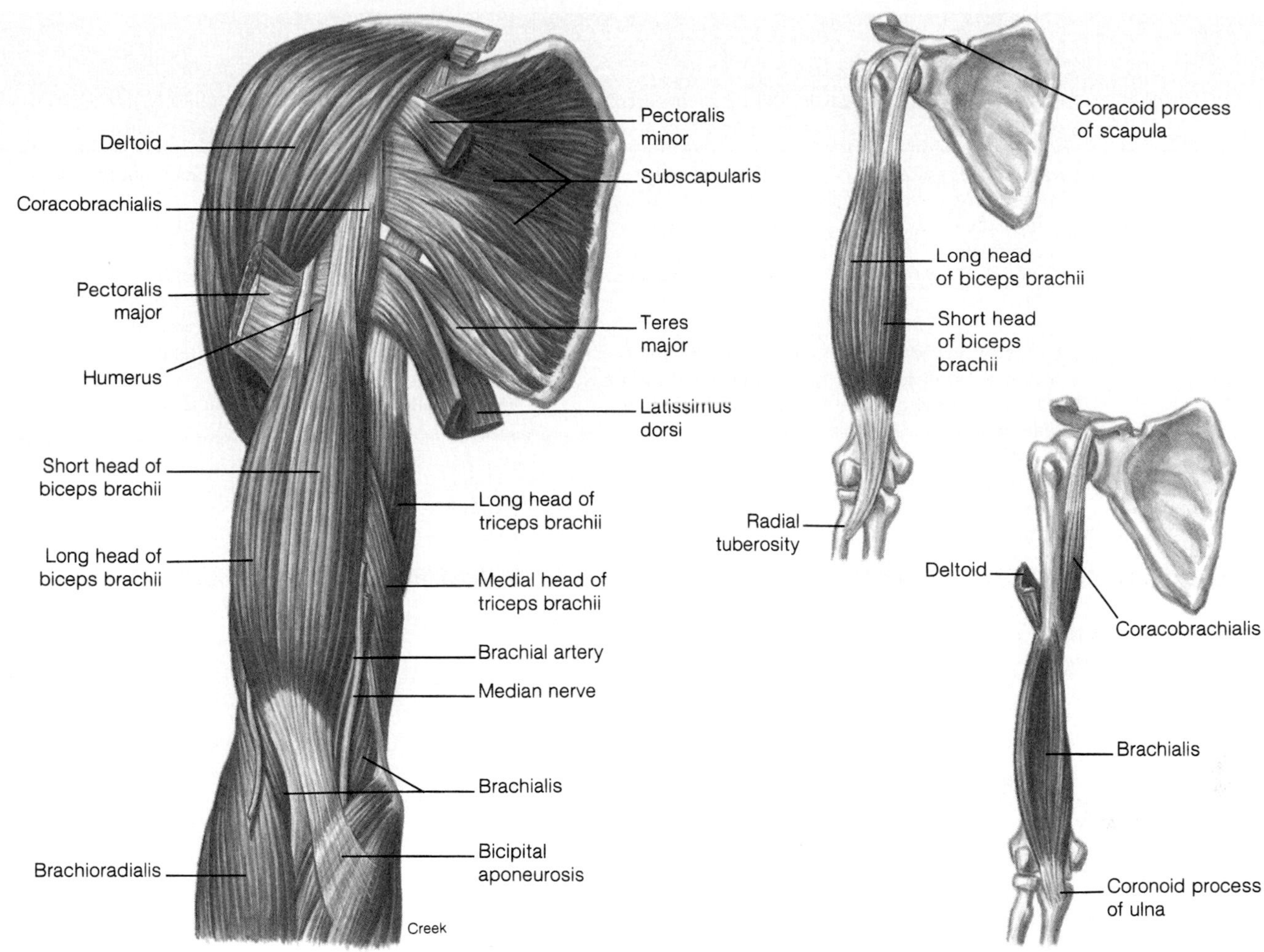

FIGURE 8.30
Muscles of the right shoulder and brachium.

The **triceps brachii,** located on the posterior surface of the brachium (fig. 8.31), extends the forearm at the elbow joint, in opposition to the action of the biceps brachii; thus, these two muscles are antagonists. The triceps brachii has three heads, or origins. Two of the three, the *lateral head* and *medial head,* arise from the humerus, whereas the *long head* arises from the infraglenoid tuberosity of the scapula. A common tendinous insertion attaches the triceps brachii to the olecranon of the ulna. The small **anconeus** muscle is a synergist of the triceps brachii in elbow extension.

A summary of the muscles that act on the forearm is presented in table 8.12.

Muscles of the Forearm That Act on the Wrist, Hand, and Fingers

Several of the muscles of the forearm act on two joints—the elbow and wrist (figs. 8.32 and 8.33). Others act on the joints of the wrist, hand, and fingers. Still others produce rotational movement at the radioulnar joint. Most of these muscles perform four primary actions on the hand and fingers: supination, pronation, flexion, and extension. Other actions of the hand include adduction and abduction. The muscles that act on the wrist, hand, and fingers are summarized in table 8.13.

Table 8.12 Muscles That Act on the Forearm

Muscle	Origin	Insertion	Action	Innervation
Biceps brachii	Coracoid process and tuberosity above glenoid cavity of scapula	Radial tuberosity	Flexes elbow joint; supinates forearm and hand at elbow joint	Musculocutaneous n.
Brachialis	Anterior shaft of humerus	Coronoid process of ulna	Flexes elbow joint	Musculocutaneous, median, and radial nn.
Brachioradialis	Lateral supracondylar ridge of humerus	Proximal to styloid process of radius	Flexes elbow joint	Radial n.
Triceps brachii	Tuberosity below glenoid cavity and lateral and medial surfaces of humerus	Olecranon of ulna	Extends elbow joint	Radial n.
Anconeus	Lateral epicondyle of humerus	Olecranon of ulna	Extends elbow joint	Radial n.

biceps brachii: L. *biceps,* two heads
triceps brachii: L. *triceps,* three heads
anconeus: Gk. *ancon,* elbow

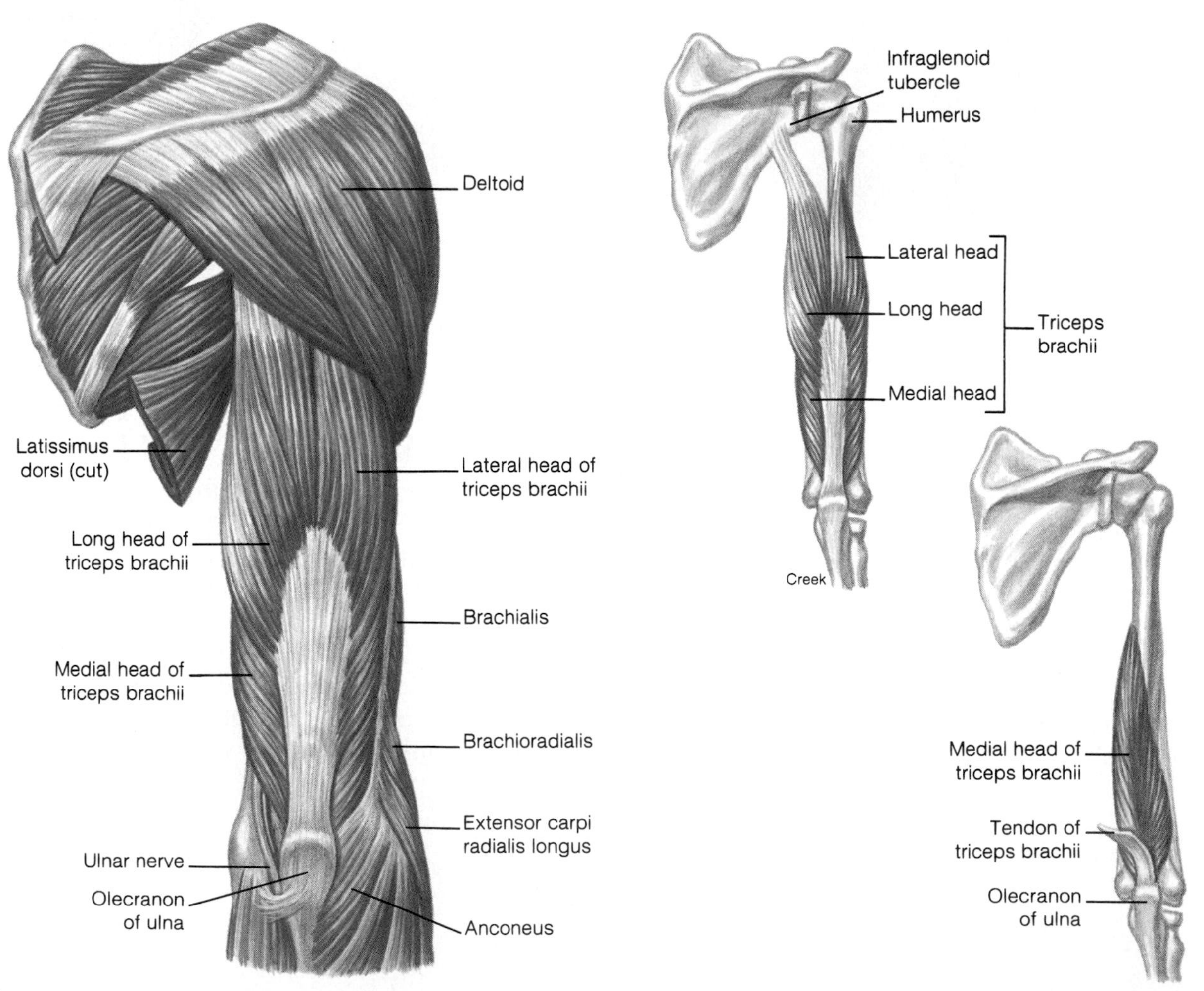

FIGURE 8.31
Muscles of the right posterior shoulder and brachium.

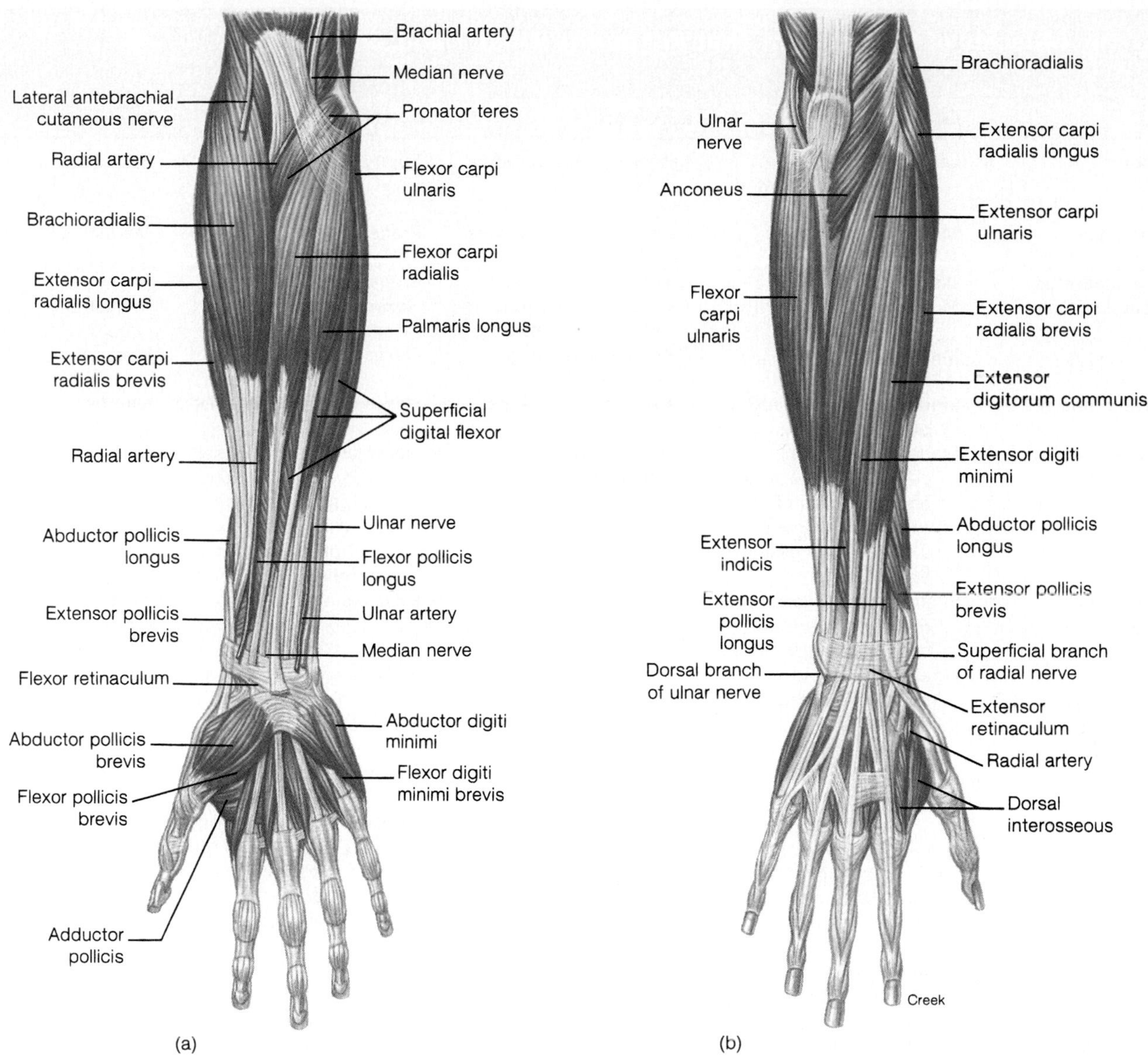
Brachial artery
Median nerve
Lateral antebrachial cutaneous nerve
Pronator teres
Radial artery
Flexor carpi ulnaris
Brachioradialis
Flexor carpi radialis
Extensor carpi radialis longus
Palmaris longus
Extensor carpi radialis brevis
Superficial digital flexor
Radial artery
Abductor pollicis longus
Ulnar nerve
Flexor pollicis longus
Extensor pollicis brevis
Ulnar artery
Median nerve
Flexor retinaculum
Abductor pollicis brevis
Abductor digiti minimi
Flexor pollicis brevis
Flexor digiti minimi brevis
Adductor pollicis
(a)
Brachioradialis
Ulnar nerve
Extensor carpi radialis longus
Anconeus
Extensor carpi ulnaris
Flexor carpi ulnaris
Extensor carpi radialis brevis
Extensor digitorum communis
Extensor digiti minimi
Extensor indicis
Abductor pollicis longus
Extensor pollicis longus
Extensor pollicis brevis
Dorsal branch of ulnar nerve
Superficial branch of radial nerve
Extensor retinaculum
Radial artery
Dorsal interosseous
Creek
(b)

FIGURE 8.32
Superficial muscles of the right forearm. (*a*) An anterior view and (*b*) a posterior view.

Table 8.13 Muscles of the Forearm That Act on the Wrist, Hand, and Fingers

Muscle	Origin	Insertion	Action	Innervation
Supinator	Lateral epicondyle of humerus and crest of ulna	Lateral surface of radius	Supinates hand	Radial n.
Pronator teres	Medial epicondyle of humerus	Lateral surface of radius	Pronates hand	Median n.
Pronator quadratus	Distal fourth of ulna	Distal fourth of radius	Pronates hand	Median n.
Flexor carpi radialis	Medial epicondyle of humerus	Base of second and third metacarpal bones	Flexes and abducts hand at wrist	Median n.
Palmaris longus	Medial epicondyle of humerus	Palmar aponeurosis	Flexes wrist	Median n.
Flexor carpi ulnaris	Medial epicondyle and olecranon	Carpal and metacarpal bones	Flexes and adducts wrist	Ulnar n.
Flexor digitorum superficialis	Medial epicondyle, coronoid process, and anterior border of radius	Middle phalanges of digits II–V	Flexes wrist and digits at carpophalangeal and interphalangeal joints	Median n.
Flexor digitorum profundus	Proximal two-thirds of ulna and interosseous ligament	Distal phalanges of digits II–V	Flexes wrist and digits at carpophalangeal and interphalangeal joints	Median and ulnar nn.
Flexor pollicis longus	Shaft of radius, interosseous membrane, and coronoid process of ulna	Distal phalanx of thumb	Flexes joints of thumb	Median n.
Extensor carpi radialis longus	Lateral supracondylar ridge of humerus	Second metacarpal bone	Extends and abducts wrist	Radial n.
Extensor carpi radialis brevis	Lateral epicondyle of humerus	Third metacarpal bone	Extends and abducts wrist	Radial n.
Extensor digitorum communis	Lateral epicondyle of humerus	Posterior surfaces of digits II–V	Extends wrist and phalanges at joints of carpophalangeal and interphalangeal joints	Radial n.
Extensor digiti minimi	Lateral epicondyle of humerus	Extensor aponeurosis of fifth digit	Extends joints of fifth digit and wrist	Radial n.
Extensor carpi ulnaris	Lateral epicondyle of humerus and olecranon	Base of fifth metacarpal bone	Extends and adducts wrist	Radial n.
Extensor pollicis longus	Middle shaft of ulna, lateral side	Base of distal phalanx of thumb	Extends joints of thumb; abducts joints of hand	Radial n.
Extensor pollicis brevis	Distal shaft of radius and interosseous ligament	Base of first phalanx of thumb	Extends joints of thumb; abducts joints of hand	Radial n.
Abductor pollicis longus	Distal radius and ulna and interosseous ligament	Base of first metacarpal bone	Abducts joints of thumb and joints of hand	Radial n.

supinator: L. *supin,* bend back
pronator teres: L. *pron,* bend forward
palmaris longus: L. *palma,* flat of hand
flexor digitorum profundus: L. *profundus,* deep
flexor pollicis longus: L. *pollex,* thumb

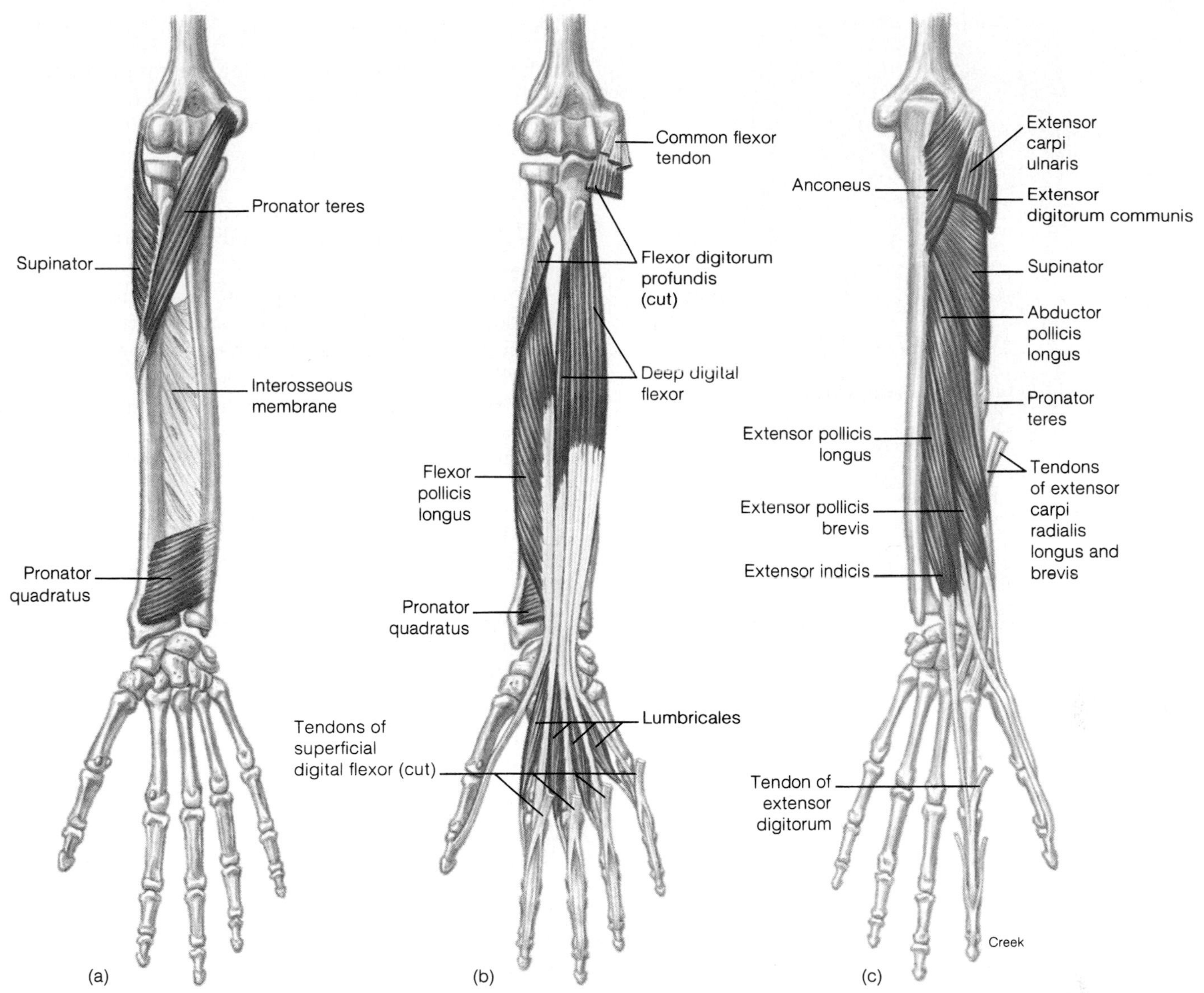

FIGURE 8.33
Deep muscles of the right forearm. (*a*) Rotators, (*b*) flexors, and (*c*) extensors.

8.13 Muscles of the Thigh and Lower Limb

The muscles that move the thigh originate from the pelvic girdle and insert on various places on the femur. These muscles stabilize a highly movable hip joint and provide support for the body in standing and walking. The muscles of the lower limb vary from the large, powerful muscles of the hip and thigh to the long, delicate muscles of the leg.

Muscles of the Hip That Act on the Thigh

The muscles that move the thigh are arranged in anterior, posterior, and medial groups. The *anterior muscles* (located in the fossa of the ilium) are the **iliacus** and **psoas** (*so'as*) **major** muscles (fig. 8.34); they are frequently referred to as a single muscle, the **iliopsoas.**

The *posterior muscles* that move the thigh include the **gluteus maximus, gluteus medius, gluteus minimus,** and **tensor fascia latae** muscles (fig. 8.35). A deep group of six lateral rotators of the thigh is positioned directly over the posterior aspect of the hip joint. They are, from superior to inferior, the **piriformis, superior gemellus, obturator internus, inferior gemellus, obturator externus,** and **quadratus femoris** muscles. The anterior and posterior group of muscles that move the thigh are summarized in table 8.14.

The *medial muscles* that move the thigh include the **gracilis, pectineus, adductor longus, adductor brevis,** and **adductor magnus** muscles (fig. 8.36). These muscles are summarized in table 8.15.

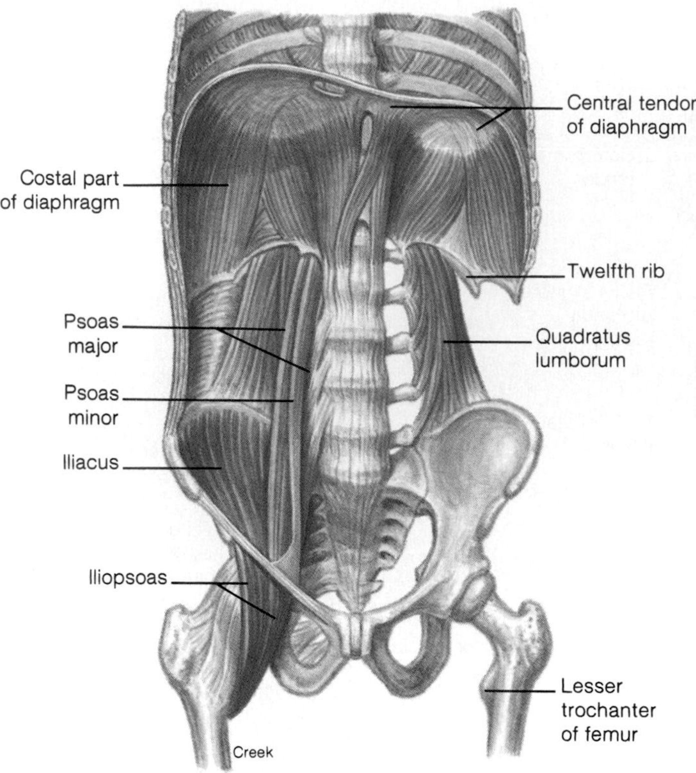

FIGURE 8.34
Anterior pelvic muscles that move the hip.

Muscles of the Thigh That Act on the Leg

The muscles that move the leg are divided according to function and position into anterior extensors and posterior flexors. The *anterior extensor muscles* that move the leg are the *sartorius* and *quadriceps femoris* (fig. 8.37 and table 8.16).

The **sartorius** obliquely crosses the anterior aspect of the thigh and can act on both the hip and knee joints. The **quadriceps femoris** is actually a composite of four distinct muscles that have separate origins but a common insertion on the patella via the *patellar tendon.* The patellar tendon is continuous over the patella and becomes the *patellar ligament* as it attaches to the head of the tibial tuberosity. The four muscles of the quadriceps femoris muscle are the **rectus femoris, vastus lateralis, vastus medialis,** and **vastus intermedius** muscles.

Inside Information—The sartorius is the longest muscle in your body. When it was first described, it was commonly referred to as the "tailor's muscle." This is because the sartorius muscles help to effect the cross-legged sitting position in which tailors are often depicted.

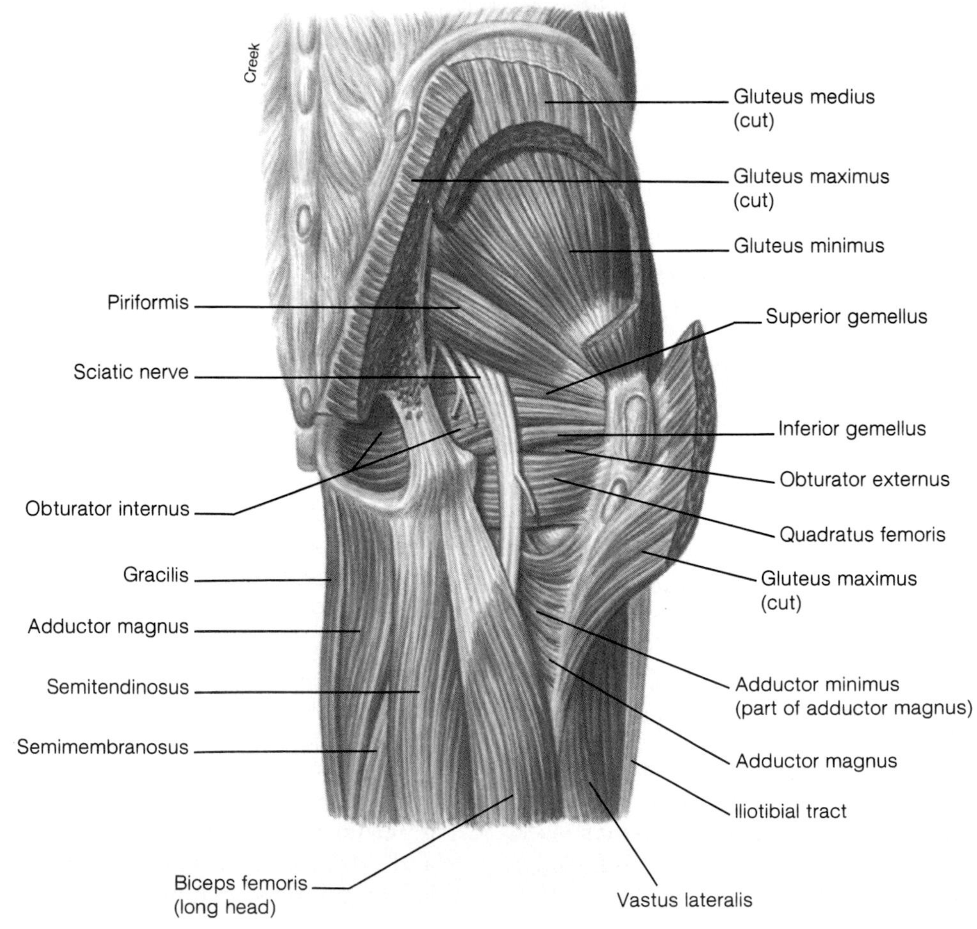

FIGURE 8.35
Deep gluteal muscles.

TABLE 8.14 Anterior and Posterior Muscles That Move the Thigh

Muscle	Origin	Insertion	Action	Innervation
Iliacus	Iliac fossa	Lesser trochanter of femur, along with psoas major	Flexes and rotates thigh laterally at hip joint; flexes joints of vertebral column	Femoral n.
Psoas major	Transverse processes of all lumbar vertebrae	Lesser trochanter, along with iliacus	Flexes and rotates thigh laterally at hip joint; flexes joints of vertebral column	Spinal nerves L2, L3
Gluteus maximus	Iliac crest, sacrum, coccyx, and aponeurosis of the lumbar region	Gluteal tuberosity and iliotibial tract	Extends and rotates thigh laterally at hip joint	Inferior gluteal n.
Gluteus medius	Lateral surface of ilium	Greater trochanter	Abducts and rotates thigh medially at hip joint	Superior gluteal n.
Gluteus minimus	Lateral surface of lower half of ilium	Greater trochanter	Abducts and rotates thigh medially at hip joint	Superior gluteal n.
Tensor fasciae latae	Anterior border of ilium and iliac crest	Iliotibial tract	Abducts thigh at hip joint	Superior gluteal n.

psoas major: Gk. *psoa,* loin
gluteus maximus: Gk. *gloutos,* rump

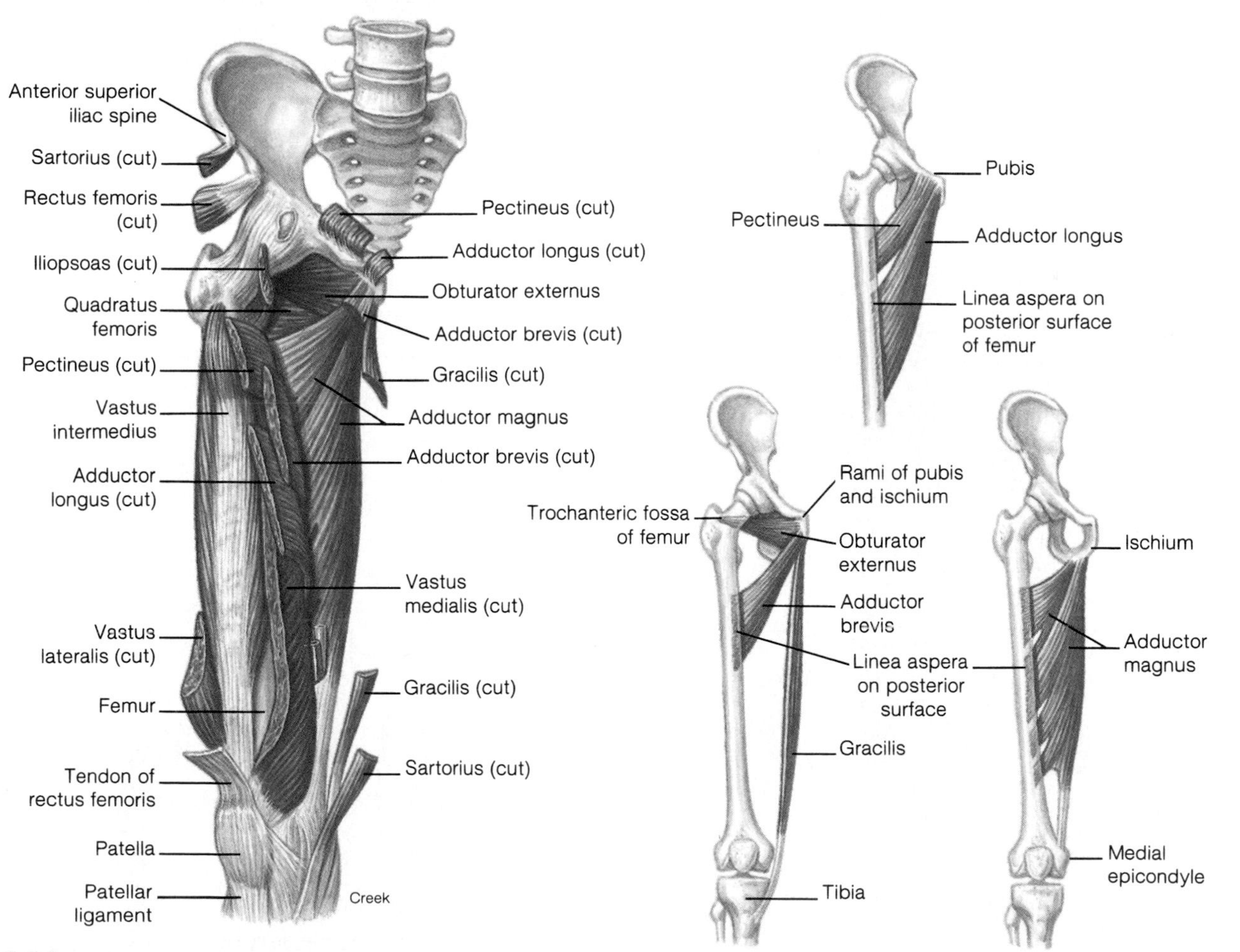

FIGURE 8.36
Adductor muscles of the right thigh.

TABLE 8.15 **Medial Muscles That Move the Thigh**

Muscle	Origin	Insertion	Action	Innervation
Gracilis	Inferior edge of symphysis pubis	Proximal medial surface of tibia	Adducts thigh at hip joint; flexes and rotates leg at knee joint	Obturator n.
Pectineus	Pectineal line of pubis	Distal to lesser trochanter of femur	Adducts and flexes thigh at hip joint	Femoral n.
Adductor longus	Pubis—below pubic crest	Linea aspera of femur	Adducts, flexes, and laterally rotates thigh at hip joint	Obturator n.
Adductor brevis	Inferior ramus of pubis	Linea aspera of femur	Adducts, flexes, and laterally rotates thigh at hip joint	Obturator n.
Adductor magnus	Inferior ramus of ischium and pubis	Linea aspera and medial epicondyle of femur	Adducts, flexes, and laterally rotates thigh at hip joint	Obturator n.

gracilis: Gk. *gracilis,* slender

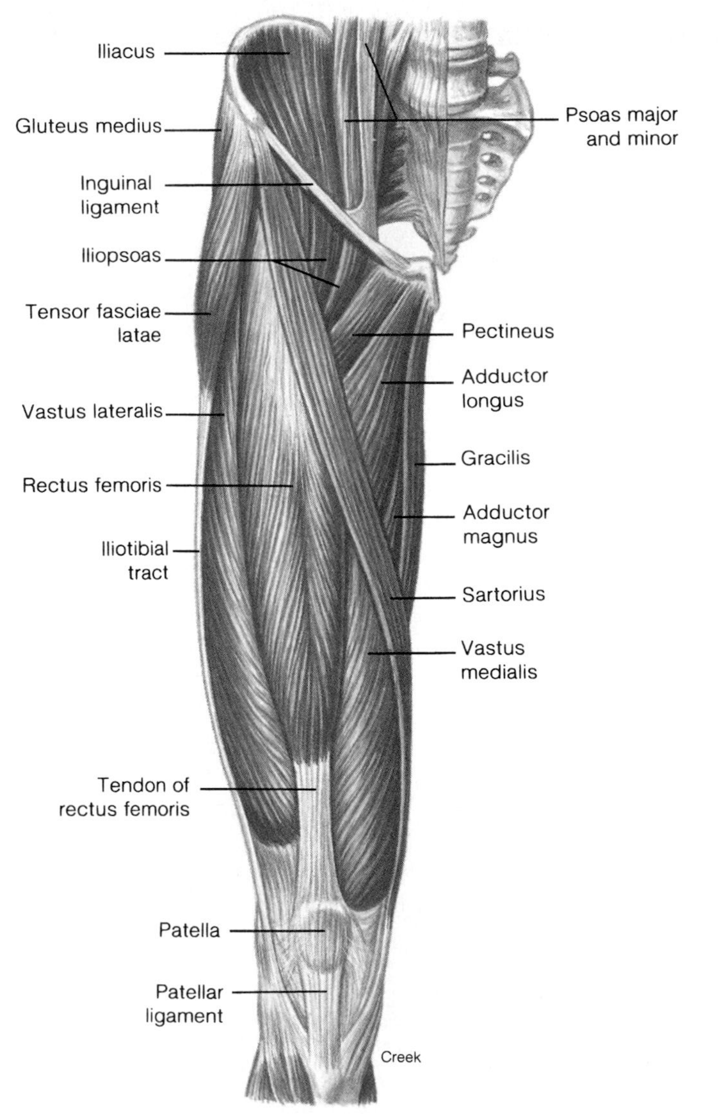

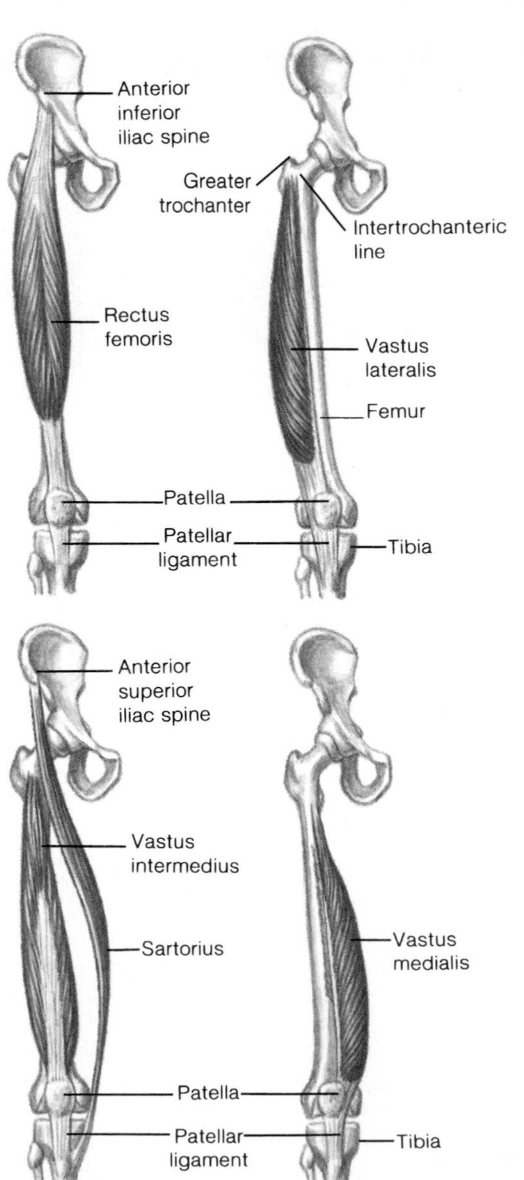

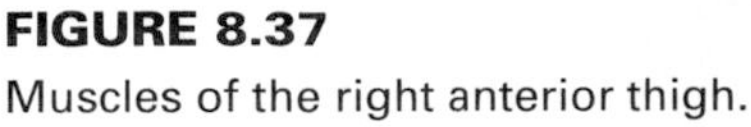
FIGURE 8.37
Muscles of the right anterior thigh.

TABLE 8.16 **Anterior Thigh Muscles That Move the Leg**

Muscle	Origin	Insertion	Action	Innervation
Sartorius	Anterior superior iliac spine	Medial surface of tibia	Flexes leg and thigh; abducts and rotates thigh laterally; rotates leg medially at hip joint	Femoral n.
Quadriceps femoris		Patella by common tendon, which continues as patellar ligament to tibial tuberosity	Extends leg at knee joint	Femoral n.
Rectus femoris	Anterior inferior iliac spine and lip of acetabulum			
Vastus lateralis	Greater trochanter and linea aspera of femur			
Vastus medialis	Medial surface and linea aspera of femur			
Vastus intermedius	Anterior and lateral surfaces of femur			

sartorius: L. *sartor*, a tailor (muscle used to cross legs in a tailor's position)

There are three *posterior flexor muscles* of the thigh, known as the **hamstrings** (fig. 8.38). These muscles are antagonistic to the quadriceps femoris muscles in flexing the knee. The three hamstring muscles are the **biceps femoris, semitendinosus,** and **semimembranosus** muscles (table 8.17).

Inside Information—The name *hamstrings* for the posterior flexors of the thigh derives from the old butchers' practice of using the tendons of these fleshy muscles to hang hams for smoke curing.

Injuries to the hamstrings are a common occurrence in some sports. An injury to one or more of these muscles usually occurs when sudden lateral or medial stress to the knee joint tears a muscle or tendon. Because of its structure and the stress placed upon it in competition, the knee joint is particularly susceptible to injury.

Muscles of the Leg That Act on the Ankle, Foot, and Toes

There are three groups of *crural muscles: anterior, lateral,* and *posterior.* The anteromedial aspect of the leg along the shaft of the tibia lacks muscle attachment.

The anterior crural muscles include the **tibialis anterior, extensor digitorum longus, extensor hallucis longus,** and **peroneus tertius** muscles (fig. 8.39). The lateral crural muscles are the **peroneus longus** and **peroneus brevis** muscles. The actions of these muscles are summarized in table 8.18.

The seven posterior crural muscles can be divided into a superficial and deep group (fig. 8.40). The superficial group is composed of the **gastrocnemius, soleus,** and **plantaris** muscles. The gastrocnemius and soleus insert onto the calcaneus via the common *tendo calcaneus* (*tendon of Achilles*), and are frequently referred to as a single muscle, the **triceps surae.** The four deep posterior crural muscles are the **popliteus, flexor hallucis longus, flexor digitorum longus,** and **tibialis posterior** muscles.

Various surface features of the body are shown in Appendix B, figures 1–7. As you review these figures, try to locate the muscles that are labeled on your own body.

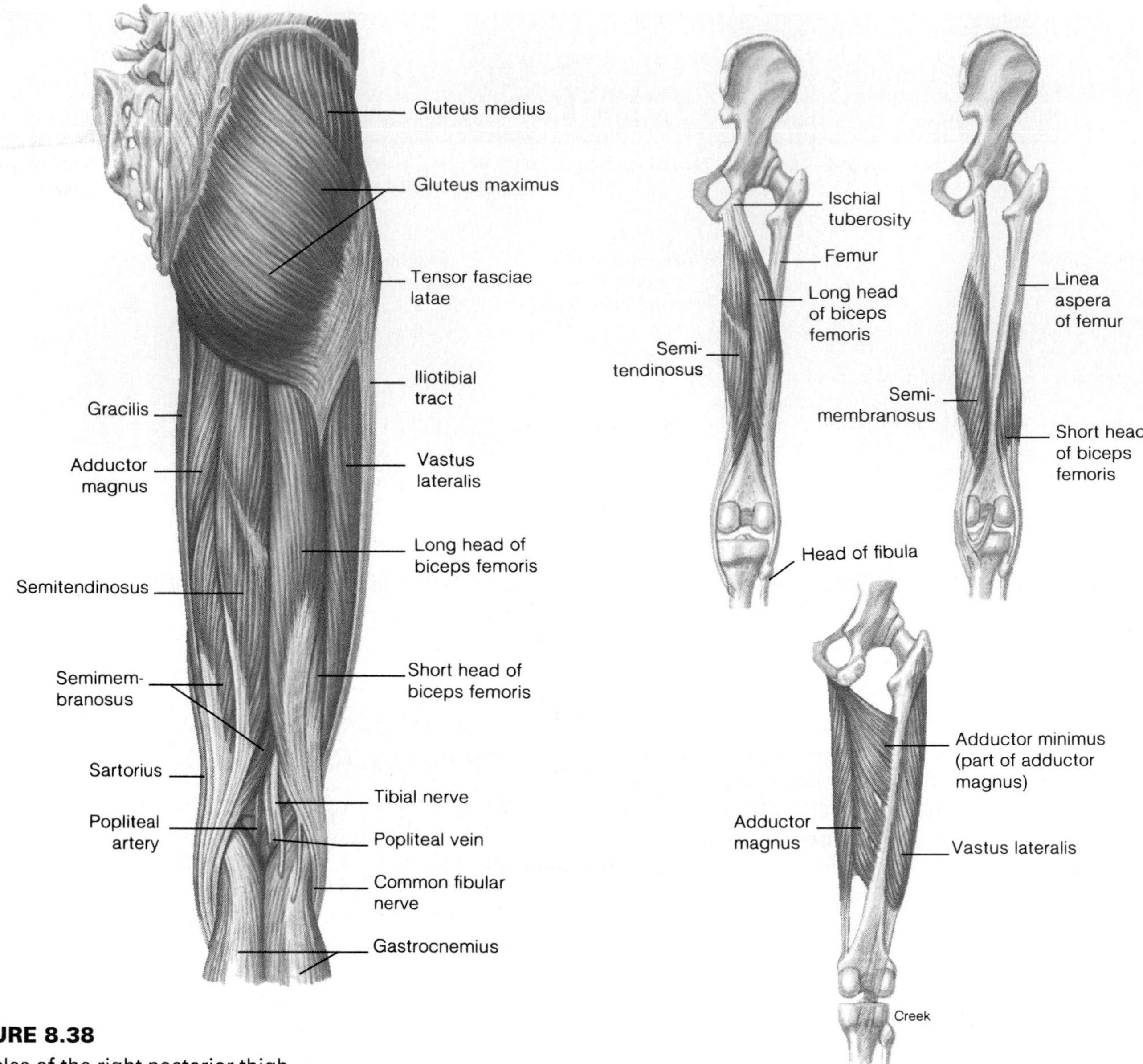

FIGURE 8.38
Muscles of the right posterior thigh.

TABLE 8.17 Posterior Thigh Muscles That Move the Leg

Muscle	Origin	Insertion	Action	Innervation
Biceps femoris	Long head—ischial tuberosity; short head—linea aspera of femur	Head of fibula and lateral epicondyle of tibia	Flexes leg at knee joint; extends and laterally rotates thigh at hip joint	Tibial n.
Semitendinosus	Ischial tuberosity	Proximal portion of medial surface of shaft of tibia	Flexes leg at knee joint; extends and medially rotates thigh at hip joint	Tibial n.
Semimembranosus	Ischial tuberosity	Medial epicondyle of tibia	Flexes leg at knee joint; extends and medially rotates thigh at hip joint	Tibial n.

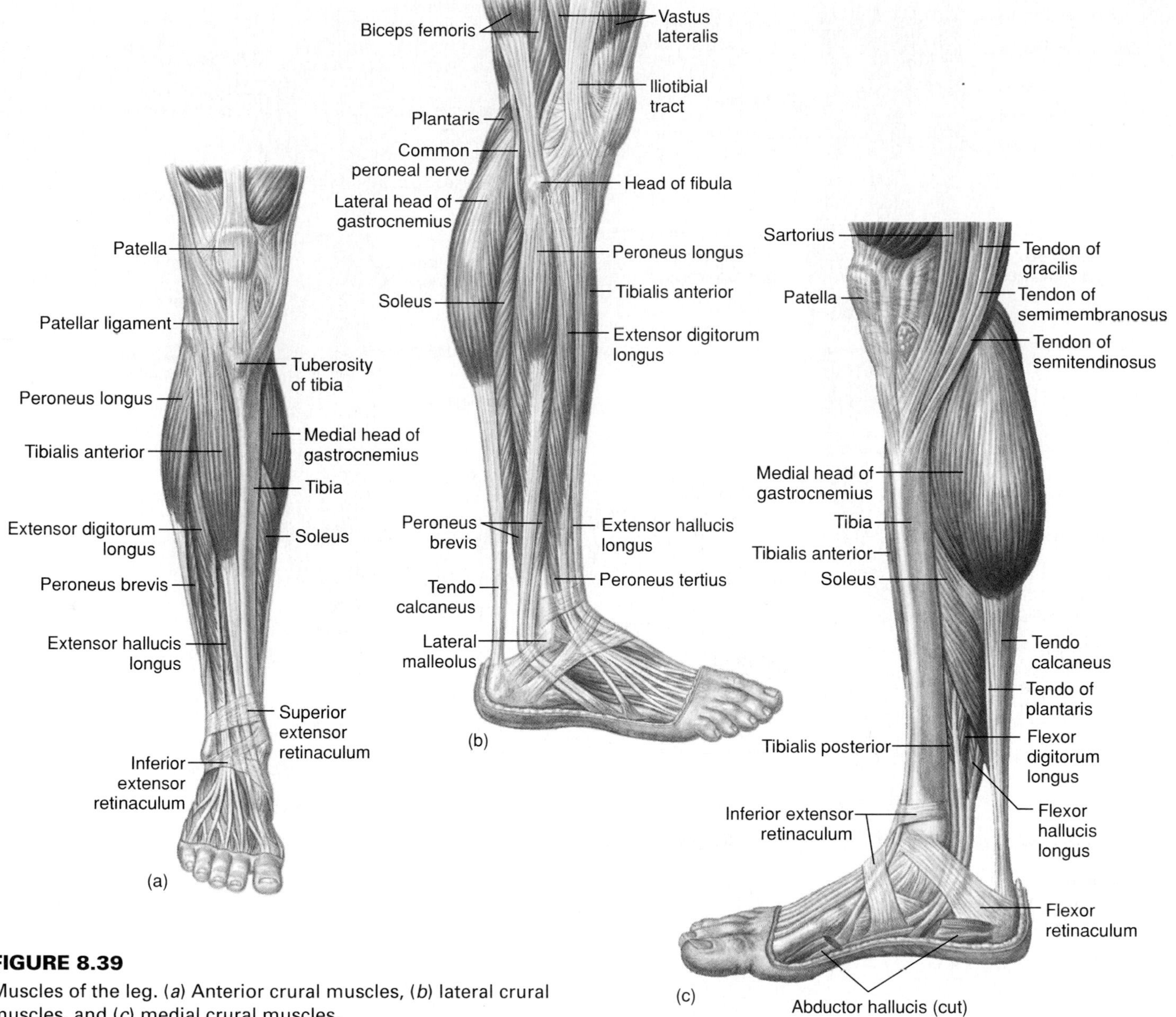

FIGURE 8.39

Muscles of the leg. (*a*) Anterior crural muscles, (*b*) lateral crural muscles, and (*c*) medial crural muscles.

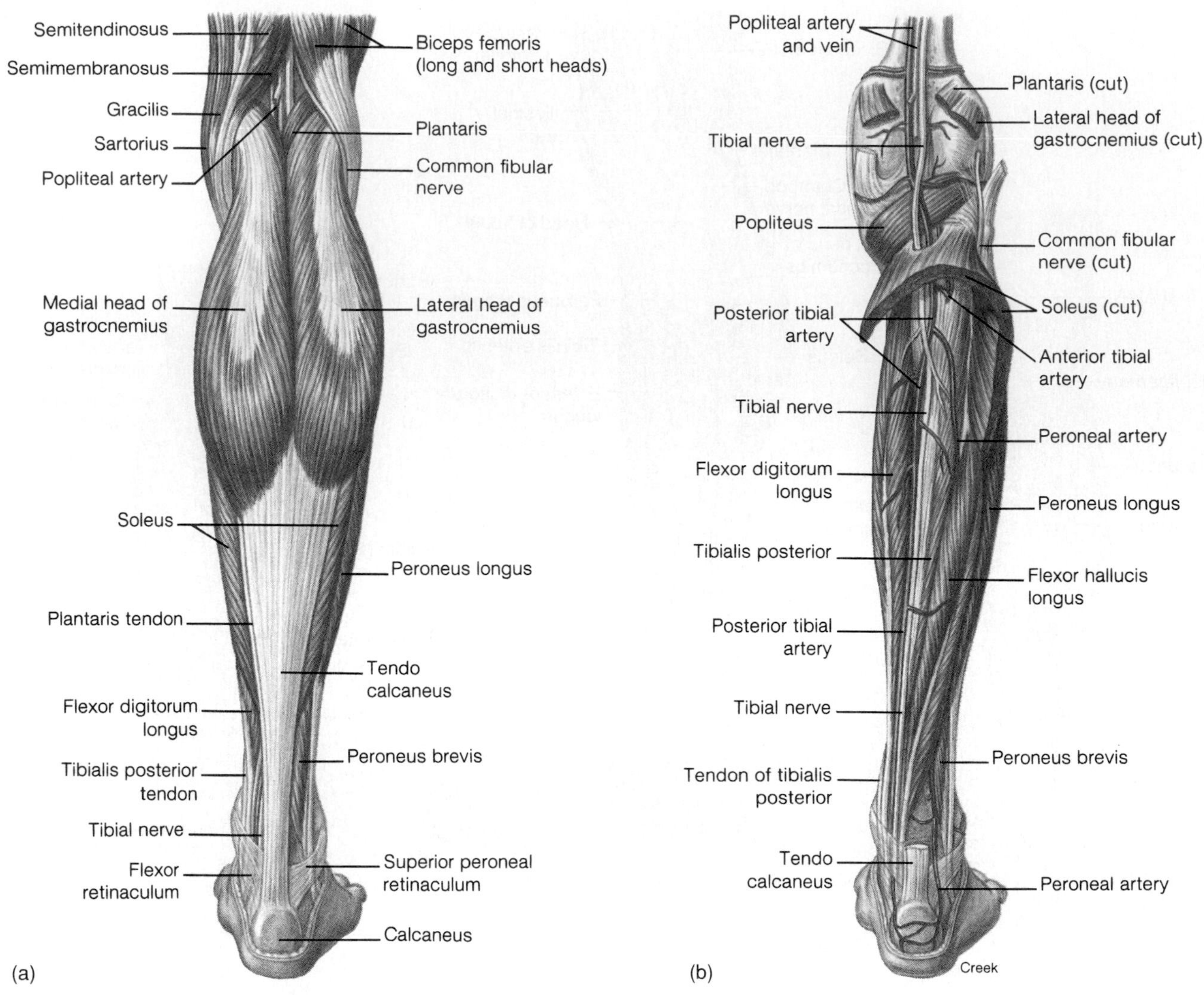

FIGURE 8.40
Muscles of the leg. (*a*) Superficial posterior crural muscles and (*b*) deep posterior crural muscles.

Table 8.18 Muscles of the Lower Leg That Move the Ankle, Foot, and Toes

Muscle	Origin	Insertion	Action	Innervation
Tibialis anterior	Lateral epicondyle and body of tibia	First metatarsal bone and first cuneiform	Dorsiflexes ankle, inverts foot at ankle	Deep fibular n.
Extensor digitorum longus	Lateral epicondyle of tibia and anterior surface of fibula	Extensor expansions of digits II–V	Extends digits II–V; dorsiflexes foot at ankle	Deep fibular n.
Extensor hallucis longus	Anterior surface of fibula and interosseous membrane	Distal phalanx of digit I	Extends joints of big toe; assists dorsiflexion of foot at ankle	Deep fibular n.
Peroneus tertius	Anterior surface of fibula and interosseous membrane	Dorsal surface of fifth metatarsal bone	Dorsiflexes and everts foot at ankle	Deep fibular n.
Peroneus longus	Lateral epicondyle of tibia and head and shaft of fibula	First cuneiform and metatarsal bone I	Plantar flexes and everts foot at ankle	Superficial fibular n.
Peroneus brevis	Lower aspect of fibula	Metatarsal bone V	Plantar flexes and everts foot at ankle	Superficial fibular n.
Gastrocnemius	Lateral and medial epicondyle of femur	Posterior surface of calcaneus	Plantar flexes foot at ankle; flexes knee joint	Tibial n.
Soleus	Posterior aspect of fibula and tibia	Calcaneus	Plantar flexes foot at ankle	Tibial n.
Plantaris	Lateral supracondylar ridge of femur	Calcaneus	Plantar flexes foot at ankle	Tibial n.
Popliteus	Lateral condyle of femur	Upper posterior aspect of tibia	Flexes and medially rotates leg at knee joint	Tibial n.
Flexor hallucis longus	Posterior aspect of fibula	Distal phalanx of big toe	Flexes joint of distal phalanx of big toe	Tibial n.
Flexor joints of digitorum longus	Posterior surface of tibia	Distal phalanges of digits II–V	Flexes joints of distal phalanges of digits II–V	Tibial n.
Tibialis posterior	Tibia and fibula and interosseous ligament	Navicular, cuneiform, cuboid, and metatarsal bones II–IV	Plantar flexes and inverts foot at ankle; supports arches	Tibial n.

extensor hallucis longus: L. *hallus,* great toe
peroneus tertius: Gk. *perone,* fibula; *tertius,* third
gastrocnemius: Gk. *gaster,* belly; *kneme,* leg
soleus: L. *soleus,* sole of foot
popliteus: L. *poples,* ham of the knee

SYNOPTIC SPOTLIGHT

Anything to Win: Anabolic Steroids

The merciless rigor of modern competitive sports, especially at the international level, the glory of victory, and the growing social and economical rewards of sporting success (in no way any longer related to reality) increasingly forces athletes to improve their performance by any means available.

—Statement by the Medical Commission of the International Olympic Committee, 1972

In their drive to excel, athletes have always searched for "something special" to give them a competitive advantage. Starting in the 1960s, weightlifters and bodybuilders began to think that they had found that magic something in the form of anabolic steroids. By the 1970s, these drugs already had become established as an integral feature of the high-technology scene of competitive sports—a fact attested to by the International Olympic Committee, which eventually included them on their list of banned doping substances. However, widespread use of anabolic steroids continues today, despite official bans and a lot of recent bad press.

Just what are these anabolic steroids? Do they provide a real physiological advantage for some of the athletes who use them? And if the answer is yes, do they present an ethical dilemma by violating the concept of fair play?

First synthesized in the 1950s, anabolic steroids are drugs that mimic the male sex hormone testosterone. By binding to specific receptor sites on muscle cells and various other tissues, testosterone greatly contributes to the male secondary sex characteristics and to the gender differences in muscle mass and strength that begin to develop at the onset of puberty. Testosterone's androgenic, or masculinizing, effects can be minimized by synthetically manipulating the chemical structure of the steroid so that its anabolic (protein-building) activity is emphasized for purposes of promoting muscular growth.

Anabolic steroids were first introduced by pharmaceutical companies for the treatment of patients who were deficient in natural androgens, or who had muscle-wasting diseases. They have a role in treating metastatic cancer, osteoporosis in postmenopausal women, and some types of anemia. Athletes who use them to increase muscle mass and strength often take a progressively increasing dose in both oral and injectable form (a method called "stacking") far in excess of the clinical therapeutic dose. In months prior to competition, they gradually decrease the dose to reduce chances of detection.

Although testimonials to the effectiveness of anabolic steroids abound, it is far from clear whether using them actually leads to improved athletic performance. For example, normal healthy men who take these drugs without training show no gains in muscle size or strength. Some of the weight gain that occurs has been shown to be due to retained body fluid rather than to performance-related bulk. In fact, studies have not shown conclusively that anabolic steroids increase lean muscle bulk and strength any more than simple weight-resistance training alone. To confound the problem, improvements in performance may occur not simply because of increased muscle mass but because the anabolic steroids act via the central nervous system to make athletes more aggressive and fatigue-resistant, and thus able to engage in traditional training regimens for longer periods of time.

Whether or not anabolic steroids exert a positive influence on performance, the side effects associated with heavy, prolonged use are well documented. These include hypertension, acne, edema, and damage to the liver, heart, and adrenal glands. Psychiatric symptoms can include hallucinations, paranoid delusions, and manic episodes. Users may also be predisposed to violent, antisocial behavior.

In men, anabolic steroids can cause infertility, impotence, atrophy of the testes, premature balding, and enlargement of the breasts (gynecomastia). Women can develop excessive body and facial hair, male-pattern balding, menstrual irregularities, and deepening of the voice. While not verified, it has been speculated that excessively rapid aging in women may be related to steroid use. In children and adolescents, steroids can produce premature closing of the epiphyses (bone plates), leading to stunted or retarded growth.

Particularly troublesome is the fact that use of anabolic steroids is no longer restricted to athletes. In a recent study, approximately 500,000 American high school students admitted to using or having used anabolic steroids. Improved athletic performance was the most common reason cited, but a large percentage of the students stated that they simply wanted to improve the appearance of their bodies. Regrettably, for some adolescents at least, the promise of merely looking like a powerful athlete is incentive enough to cause them to experiment with drugs that may endanger their health.

Clinical Considerations

Compared to the other systems of the body, the muscular system is extremely durable. If properly exercised, the muscles of the body can adequately serve a person for a lifetime. Muscles are capable of doing incredible amounts of work; through conditioning they can become even stronger.

Clinical considerations include the diagnosis of muscle conditions, functional conditions in muscles, and diseases of muscles.

Diagnosis of Muscle Conditions

The clinical symptoms of muscle diseases involve weakness, loss of muscle mass (atrophy), and pain. The most obvious diagnostic procedure is a clinical examination of the patient. Following this, it may be necessary to test the muscle function using **electromyography (EMG)** to measure conduction rates and motor unit activity within a muscle. Laboratory tests may include serum enzyme assays or muscle biopsies. A biopsy is perhaps the most definitive diagnostic tool. Progressive atrophy, polymyositis, and metabolic diseases of muscles can be determined through a biopsy.

Functional Conditions in Muscles

Muscles depend on systematic periodic contraction to maintain optimal health. Obviously, overuse or disease will cause a change in muscle tissue. The immediate effect of overexertion on muscle tissue is an accumulation of lactic acid, which results in fatigue and soreness. Excessive contraction of a muscle can also damage the fibers or associated connective tissue, resulting in a **strained muscle.**

When skeletal muscles are not contracted, either because the motor nerve supply is blocked or because the limb is immobilized (as when a broken bone is in a cast), the muscle fibers **atrophy,** or diminish in size. Atrophy is reversible if exercise is resumed, as after a healed fracture, but tissue death is inevitable if the nerves cannot be stimulated.

The fibers in healthy muscle tissue increase in size, or **hypertrophy,** if a muscle is systematically exercised. This increase in muscle size and strength is due not to an increase in the number of muscle cells but to the production of more myofibrils, accompanied by a strengthening of the associated connective tissue.

A **cramp** within a muscle is an involuntary, painful, and prolonged contraction. Cramps can occur while muscles are in use or at rest. The precise cause of cramps is unknown, but evidence indicates that they may be related to conditions within the muscle (e.g., calcium or oxygen deficiencies) or to stimulation of the motor neurons.

Torticollis, or **wryneck,** is an abnormal condition in which the head is inclined to one side as a result of the contraction of muscles on that side of the neck. This disorder may be either inborn or acquired.

Diseases of Muscles

Fibromyositis is an inflammation of both skeletal muscular tissue and the associated connective tissue. Its causes are not fully understood. Fibromyositis frequently occurs in the extensor muscles of the lumbar region of the vertebral column, where extensive aponeuroses exist. Fibromyositis of this region is called **lumbago,** or **rheumatism.**

Muscular dystrophy is a genetic disease characterized by a gradual atrophy and weakening of muscle tissue. There are several kinds of muscular dystrophy, none of whose etiology is completely understood. The most frequent type affects children and is sex-linked to the male child. As muscular dystrophy progresses, the muscle fibers atrophy and are replaced by adipose tissue. Most children who have muscular dystrophy die before the age of 20.

The disease **myasthenia gravis** is characterized by extreme muscle weakness and low endurance. It results from a defect in the transmission of impulses at the neuromuscular junction. Myasthenia gravis is believed to be an autoimmune disease, and it typically affects women between the ages of 20 and 40.

Poliomyelitis (polio) is actually a viral disease of the nervous system that causes muscle paralysis. The viruses are usually localized in the anterior (ventral) horn of the spinal cord, where they affect the motor nerve impulses to skeletal muscles.

Neoplasms (abnormal growths of new tissue) are rare in muscles, but when they do occur, they are usually malignant. **Rhabdomyosarcoma** (*rab″do-mi″ŏ-sar-ko′mă*) is a malignant tumor of skeletal muscle. It can arise in any skeletal muscle, and most often afflicts young children and elderly people.

SUMMARY

8.1 Introduction to the Muscular System

1. The primary functions of skeletal muscles are body motion, heat production, maintenance of posture, and body support.
2. A skeletal muscle attaches, by means of tendons, to its origin and insertion.
3. Individual muscle fibers are covered by endomysium. Muscle bundles, called fasciculi, are covered by perimysium. The entire muscle is covered by epimysium.

8.2 Skeletal Muscle Fibers and Types of Muscle Contractions

1. Skeletal muscle cells, called fibers, have dark A bands and light I bands.
2. Muscle fibers are innervated by motor neurons. All of the muscle fibers innervated by one motor neuron are part of a motor unit.
3. Each muscle fiber is composed of smaller subcellular units called myofibrils.
4. Each myofibril contains thick and thin filaments, the arrangement of which produces dark A bands and light I bands.
5. Muscle twitches can be made to summate, producing a smooth, sustained contraction called tetanus.
6. The warmup effect is called treppe.
7. Contractions that result in muscle shortening are isotonic; when no shortening occurs, the contractions are isometric.

8.3 Sliding Filament Theory of Contraction

1. Shortening of the sarcomeres is produced by sliding of thin actin filaments between the thick myosin filaments of the A bands.
2. Sliding of the filaments occurs as a result of the action of myosin cross bridges.
3. The cross-bridge heads function as myosin ATPase enzymes. Hydrolysis of ATP provides energy for the power strokes of the cross bridges.
4. After a power stroke, the bonding of a new ATP to a cross bridge allows the cross bridge to detach from actin.

8.4 Regulation of Contraction

1. Attachment of cross bridges to actin is inhibited by the protein tropomyosin when the muscle is at rest.
2. Another protein called troponin is attached to tropomyosin. When troponin binds Ca^{++}, the troponin-tropomyosin complex shifts its position so that the cross bridges can bind to actin and produce contraction.
3. In a relaxed muscle, the Ca^{++} is stored in the sarcoplasmic reticulum.
4. Electrical impulses, conducted by transverse tubules into the muscle fiber, stimulate the release of Ca^{++} from the sarcoplasmic reticulum.

8.5 Metabolism of Skeletal Muscles

1. In addition to being synthesized from glucose, ATP can be produced from ADP and phosphate derived from phosphocreatine.
2. Endurance training increases the maximal oxygen uptake.
3. The oxygen debt refers to the extra oxygen required by the body after exercise.
4. Slow-twitch muscle fibers are adapted for aerobic respiration; fast-twitch fibers are adapted for anaerobic respiration.
5. Muscle fatigue may occur as a result of production of lactic acid through anaerobic respiration.
6. Muscle fibers exhibit a variety of adaptations to endurance training.

8.6 Organization of Skeletal Muscles

1. Synergistic muscles contract together. Antagonistic muscles perform in opposition to a particular muscle group.
2. Muscles are named on the basis of shape, location, attachment, size, orientation of fibers, relative position, and function.
3. Muscles may be classified according to fiber arrangement as parallel, convergent, pennate, or sphincteral (circular).

8.7 Muscles of the Head

1. Muscles of facial expression are positioned superficially on the scalp, face, and neck. These muscles provide the complex facial expressions used in social communication.
2. Muscles of mastication include the temporalis, masseter, and pterygoids.
3. Movements of the eyeball are controlled by four rectus and two oblique muscles.
4. The paired genioglossus, styloglossus, and hyoglossus muscles move the tongue.

8.8 Muscles of the Neck

1. The posterior neck muscles include the sternocleidomastoid, trapezius, splenius capitis, semispinalis capitis, and longissimus capitis.
2. The suprahyoid muscles include the digastric, mylohyoid, geniohyoid, and stylohyoid.
3. The infrahyoid muscles include the sternohyoid, sternothyroid, thyrohyoid, and omohyoid.

8.9 Muscles of Respiration and Muscles of the Abdominal Wall

1. The principal muscles of inspiration are the diaphragm, external intercostals, and internal intercostals.
2. Expiration is primarily passive, but forceful expiration is achieved by contraction of the internal intercostals and the abdominal muscles.

8.10 Muscles of the Pelvic Outlet

1. The pelvic outlet consists of several muscles that form a pelvic diaphragm and urogenital diaphragm.

8.11 Muscles of the Vertebral Column

1. The erector spinae muscle group and the quadratus lumborum muscles support the vertebral column and permit movement in resistance to the effect of gravity.

8.12 Muscles of the Shoulder and Upper Limb

1. Three anterior muscles and three posterior muscles originate on the axial skeleton and act on the pectoral girdle.
2. Nine muscles span the shoulder joint and act on the humerus.
3. The biceps brachii, brachialis, brachioradialis, triceps brachii, and anconeus are the muscles that act on the forearm.
4. The muscles that cause wrist, hand, and most finger movements are positioned along the anterior and posterior aspects of the forearm.

8.13 Muscles of the Thigh and Lower Limb

1. The muscles that move the thigh are arranged in anterior, posterior, and medial groups. These muscles stabilize the hip joint and provide support in bipedal stance and locomotion.
2. The muscles that move the leg are divided into anterior extensors and posterior flexors. They originate on the pelvic girdle or thigh and are surrounded by fascial sheets, which are a continuation of the fascia lata and iliotibial tract.
3. The muscles that move the ankle, foot, and toes are arranged in anterior, lateral, and posterior groups.

Review Activities

Objective Questions

1. Skeletal muscles are bound to the hypodermis of the skin by (*module 8.1*)
   a. the endomysium.
   b. the perimysium.
   c. the fascia.
   d. the retinaculum.

2. Which of the following muscles have motor units with the highest innervation ratio? (*module 8.2*)
   a. leg muscles
   b. arm muscles
   c. muscles that move the fingers
   d. muscles of the trunk

3. When a skeletal muscle shortens during contraction, which of the following statements is *false?* (*module 8.3*)
   a. The A bands shorten.
   b. The H bands shorten.
   c. The I bands shorten.
   d. The sarcomeres shorten.

4. Which of the following statements about cross bridges is *false?* (*module 8.3*)
   a. They are composed of myosin.
   b. They bind to ATP after they detach from actin.
   c. They contain an ATPase.
   d. They split ATP before they attach to actin.

5. Electrical excitation of a muscle fiber *most directly* causes (*module 8.4*)
   a. movement of tropomyosin.
   b. attachment of the cross bridges to actin.
   c. release of Ca^{++} from the sarcoplasmic reticulum.
   d. splitting of ATP.

6. The energy for muscle contraction is *most directly* obtained from (*module 8.5*)
   a. phosphocreatine.
   b. ATP.
   c. anaerobic respiration.
   d. aerobic respiration.

7. Strong muscles that fatigue quickly and that have short excursions are classified as (*module 8.6*)
   a. parallel.
   b. convergent.
   c. pennate.
   d. circular.

8. The muscle that originates on the zygomatic arch and that elevates the mandible when contracted is (*module 8.7*)
   a. the masseter.
   b. the lateral pterygoid.
   c. the temporalis.
   d. the zygomaticus.

9. Which of the following statements concerning the abdominal wall is *false?* (*module 8.9*)
   a. Contraction of the muscles of the abdominal wall aids in forced inspiration (inhalation).
   b. The linea alba is a connective tissue band devoid of muscle.
   c. Tendinous inscriptions segment the rectus abdominis.
   d. Contraction of the rectus abdominis muscle flexes the vertebral column.

10. Which of the following muscles acts to constrict the urethral canal and the vagina? (*module 8.10*)
   a. transversus perinei
   b. bulbocavernosus
   c. coccygeus
   d. ischiocavernosus
   e. levator ani

11. Which of the following muscles is an antagonist to the longissimus thoracis? (*module 8.11*)
   a. spinalis thoracis
   b. quadratus lumborum
   c. longissimus cervicis
   d. rectus abdominis

12. Which of the following muscles is a synergist to the biceps brachii in flexing the elbow joint? (*module 8.12*)
   a. deltoid
   b. triceps
   c. brachialis
   d. anconeus

13. A flexor of the shoulder joint is (*module 8.12*)
   a. the pectoralis major.
   b. the supraspinatus.
   c. the trapezius.
   d. the latissimus dorsi.

14. Each of the following is a two-joint muscle (contracts over both the hip and knee joints) *except* (*module 8.13*)
   a. the gracilis.
   b. the sartorius.
   c. the rectus femoris.
   d. the semitendinosus.
   e. the vastus medialis.

Essay Questions

1. Compare and contrast the various binding connective tissues associated with skeletal muscles. *(module 8.1)*
2. Using the concept of motor units, explain how skeletal muscles produce graded and sustained contractions. *(module 8.2)*
3. Why don't cross bridges attach to the thin filaments when a muscle is relaxed? Trace the sequence of events that allows the cross bridges to attach to the thin filaments when a muscle is stimulated by a nerve impulse. *(module 8.3)*
4. Explain the role of Ca^{++} in muscle contraction. How is muscle relaxation produced? *(module 8.4)*
5. Explain the role of ATP in muscle contraction and in muscle relaxation. *(module 8.5)*
6. Describe the various types of muscle fiber architecture, and discuss the advantage of each. *(module 8.6)*
7. Refer to figure 8.18 and table 8.4 and determine which facial muscles could cause movements of the lips. *(module 8.7)*
8. Attempt to contract, one at a time, each of the neck muscles depicted in figure 8.23. *(module 8.8)*
9. Note the movement of the abdominal wall as the abdomen protrudes during a relaxed inspiration. With your hands on your rib cage, inspire deeply and feel the rib cage elevate. Identify the muscles involved in each of these movements. *(module 8.9)*
10. Which of the muscles of the pelvic outlet support the floor of the pelvic cavity? Which are associated with the genitalia? *(module 8.10)*
11. Hyperextend your wrist and note the tendons of the extensor muscles on the back of your hand. Refer to figure 8.32 and identify these tendons. *(module 8.12)*
12. Because of their size and strength, most of the thigh muscles can be observed and palpated as they are contracted. Identify these muscles on your own body and act out their movements. *(module 8.13)*

Labeling Exercise

Label the structures indicated to the right.

1. ______________________
2. ______________________
3. ______________________
4. ______________________
5. ______________________
6. ______________________

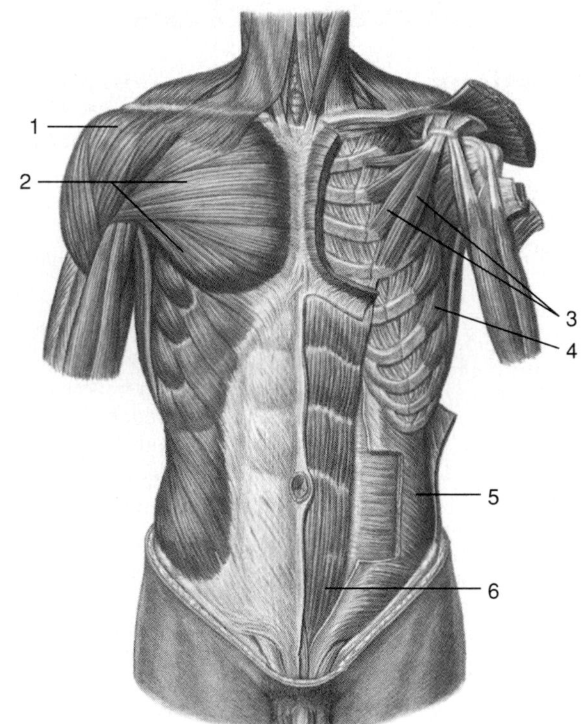

Critical Thinking Questions

1. Before any athletic event, well-coached athletes will perform calisthenics, as, for example, in the warm-up routine of basketball players before the opening buzzer. While this activity is certainly entertaining to watch, there is a physiological justification. What is it?
2. Different athletic activities require different muscle groups. How would you expect the body of a speed skater to be different from that of a kayak paddler? Which muscles would be highly developed in each of these athletes?
3. Both your soleus and gastrocnemius plantar flex your foot at the ankle, yet only the gastrocnemius can flex the knee. Explain why.
4. Locate and contract each of the following muscles on yourself: (a) pectoralis major, (b) deltoid, (c) triceps, (d) pronator teres, (e) rhomboideus major, (f) trapezius, (g) serratus anterior, and (h) latissimus dorsi. Based on function, describe exercises that would strengthen each of these muscles.
5. Some of the muscles in table 8.11 act as synergists to each other. Which ones?
6. You are having problems keeping one eye open, and your physician says that you have ocular ptosis. Which muscle do you think is malfunctioning?

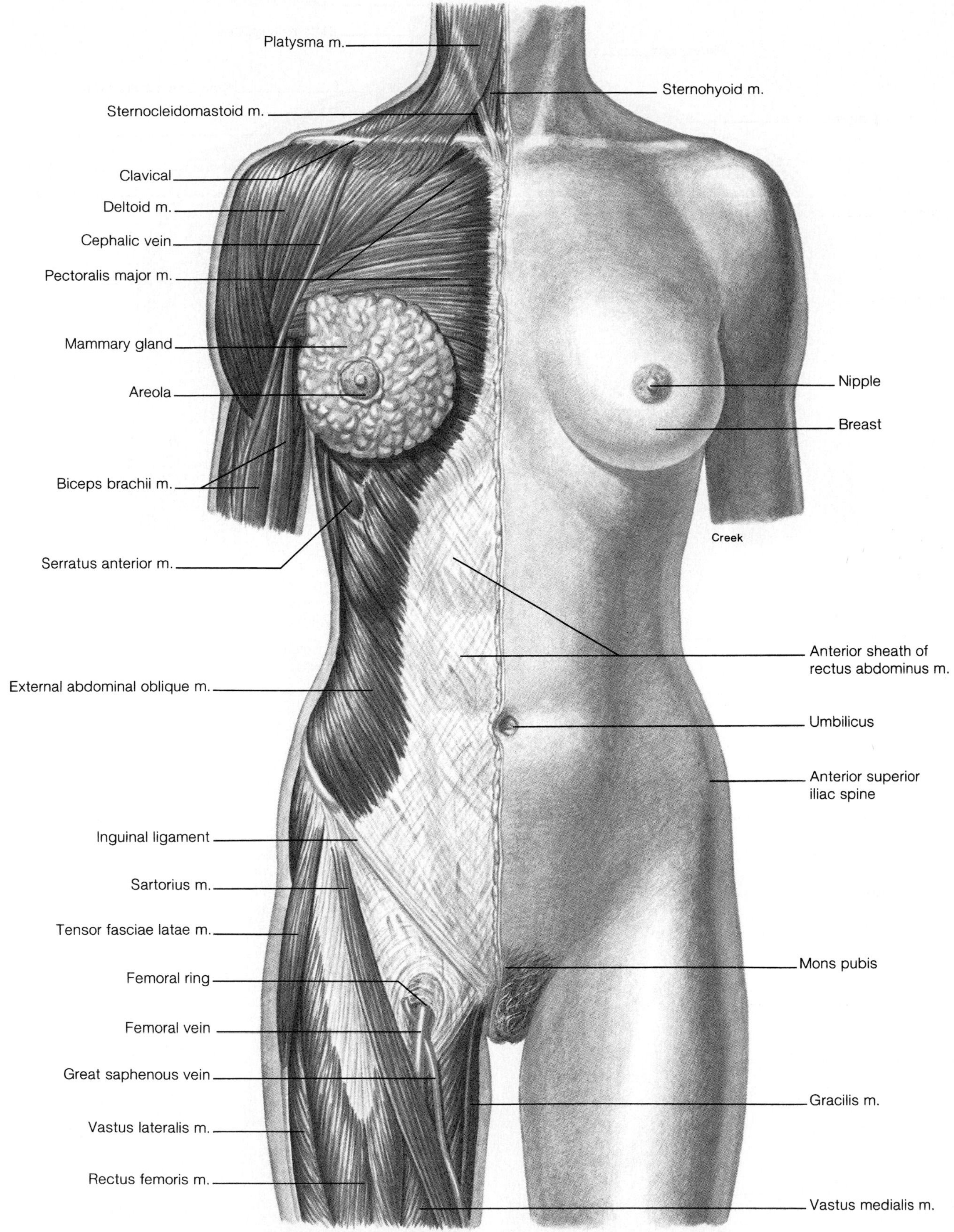

PLATE 1

An anterior view of the female trunk with the superficial muscles exposed on the right side. (*m.* stands for *muscle.*)

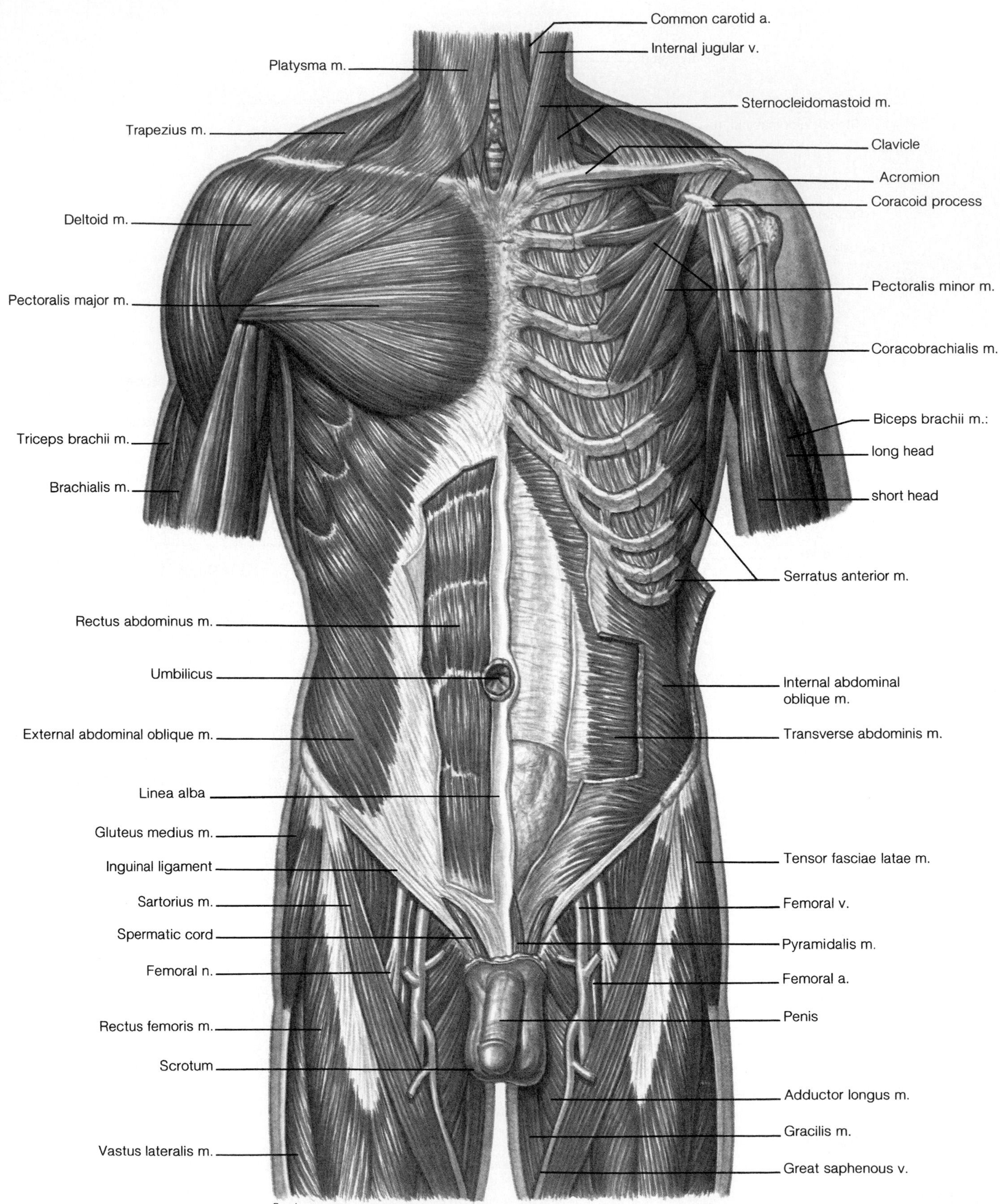

PLATE 2

An anterior view of the male trunk with the deeper muscle layers exposed. (*n.* stands for *nerve; a.* stands for *artery; v.* stands for *vein.*)

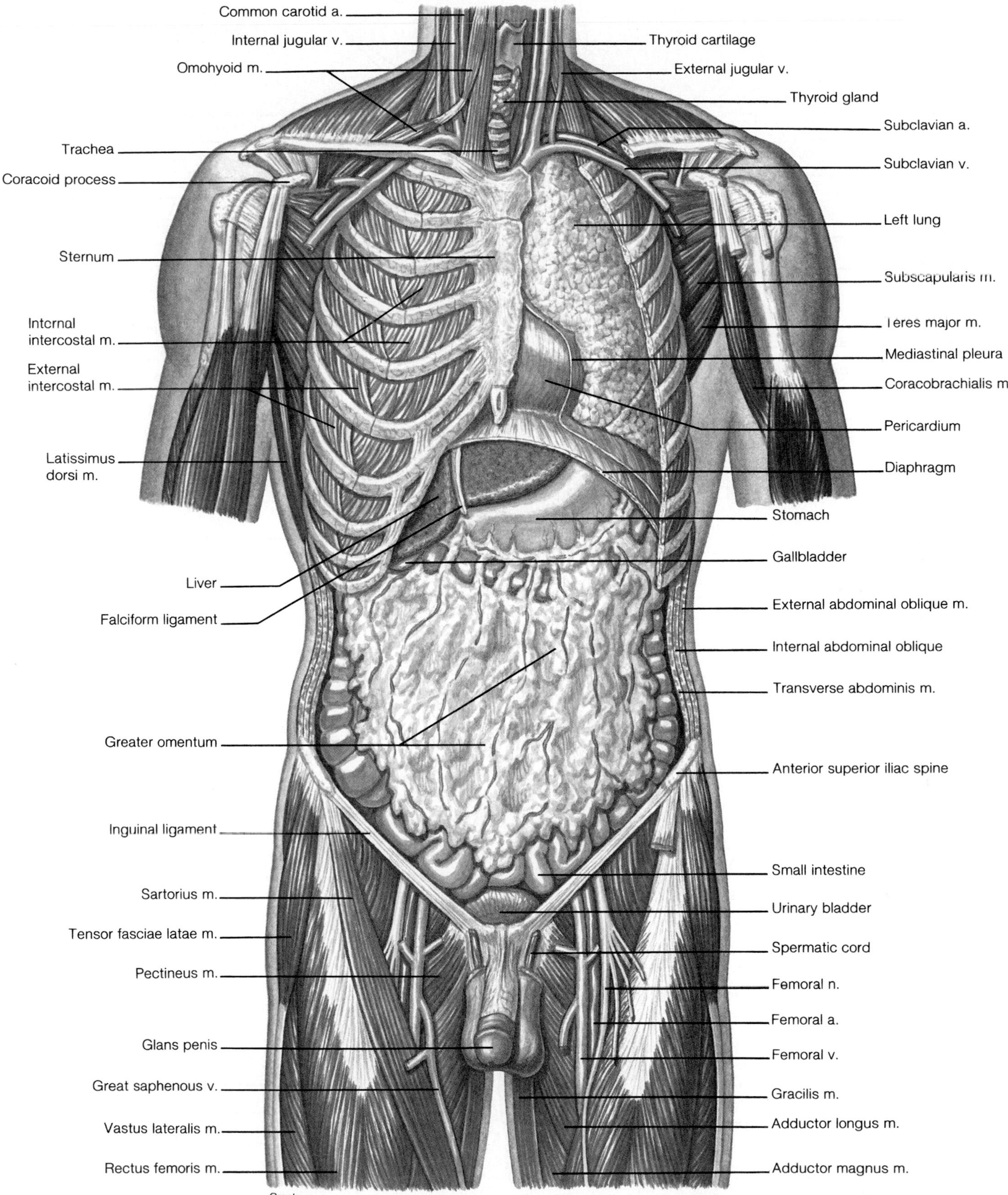

PLATE 3
An anterior view of the male trunk with the deep muscles removed and the abdominal viscera exposed.

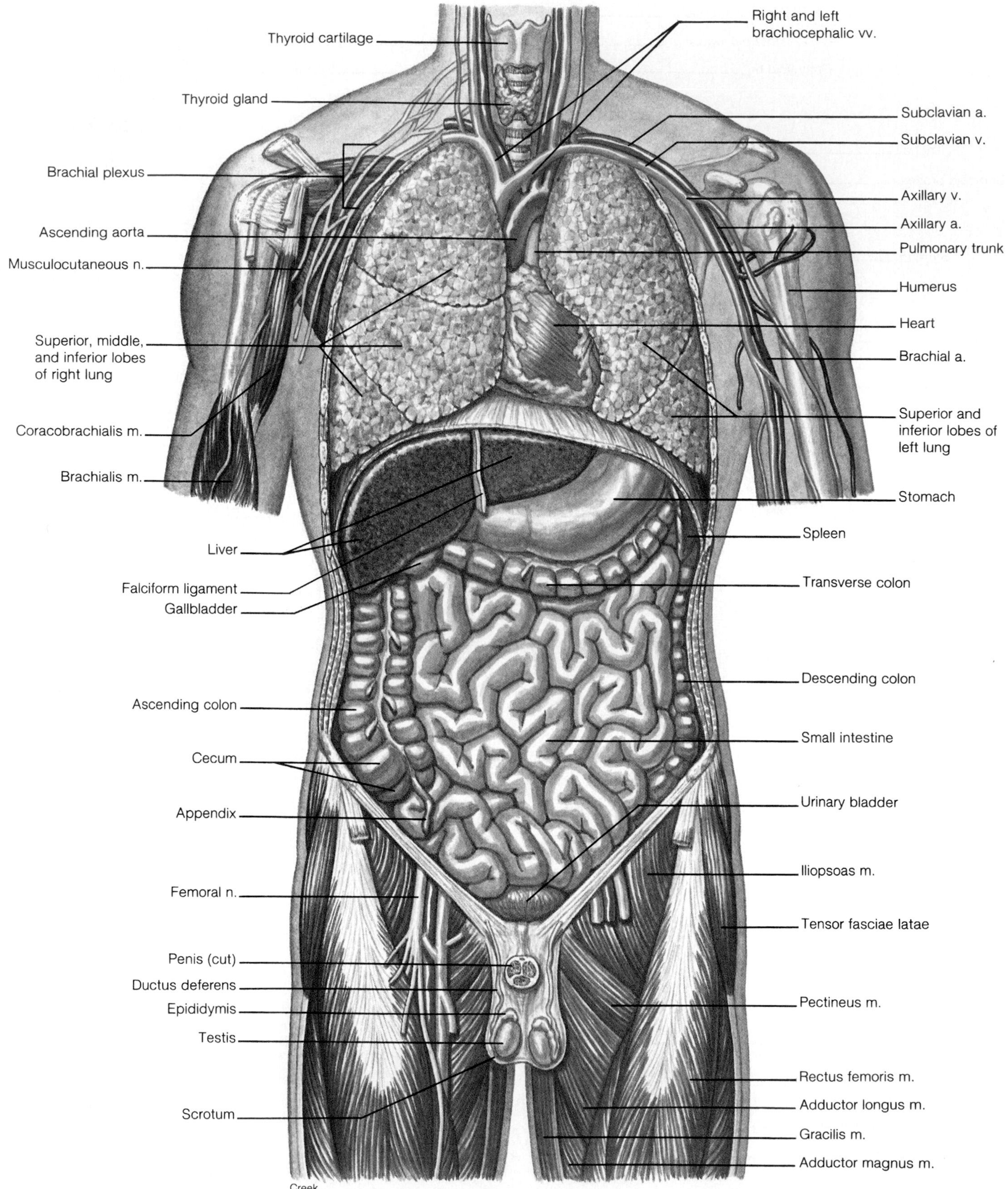

PLATE 4
An anterior view of the male trunk with the thoracic and abdominal viscera exposed.

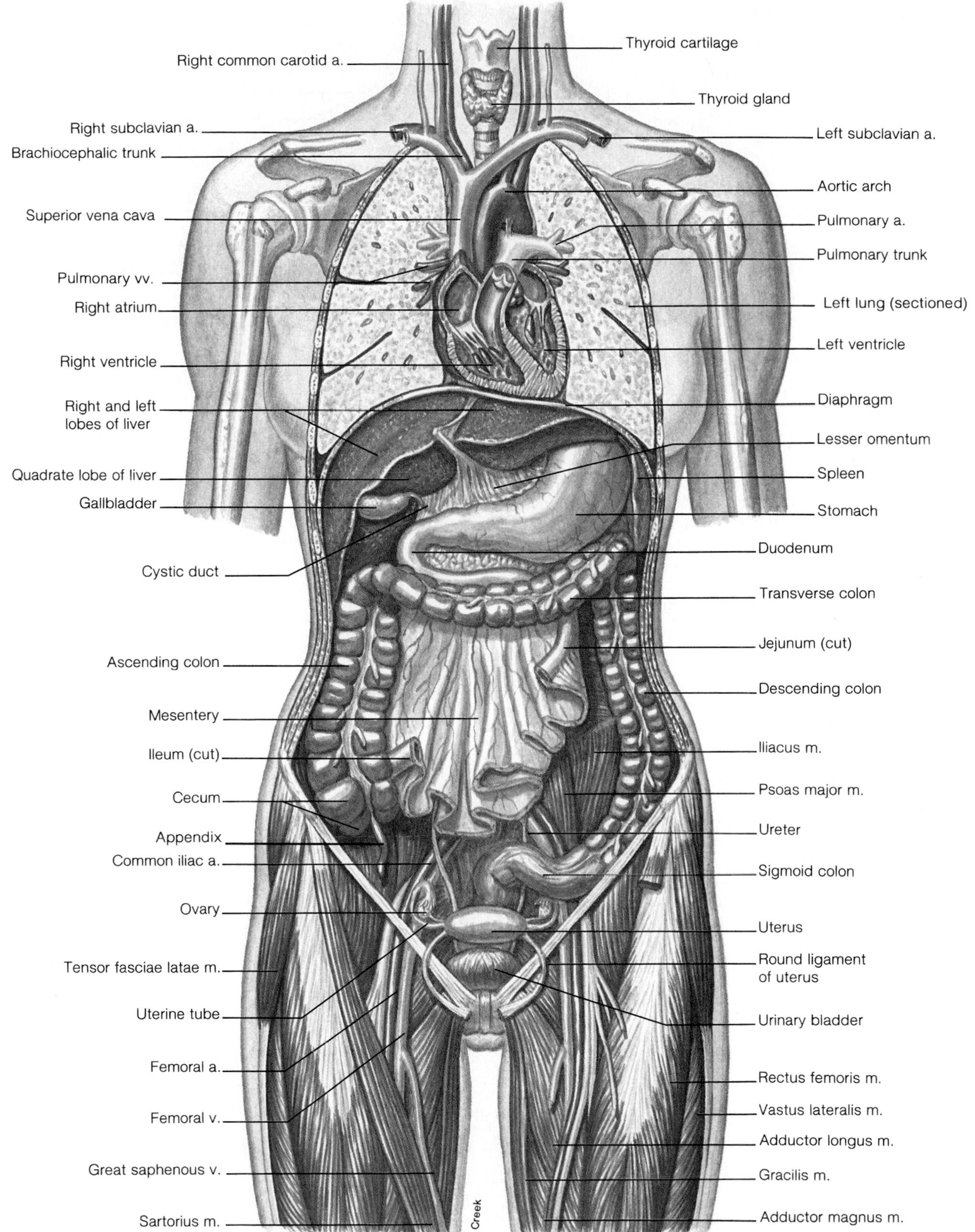

PLATE 5

An anterior view of the female trunk with the lungs, heart, and small intestine sectioned.

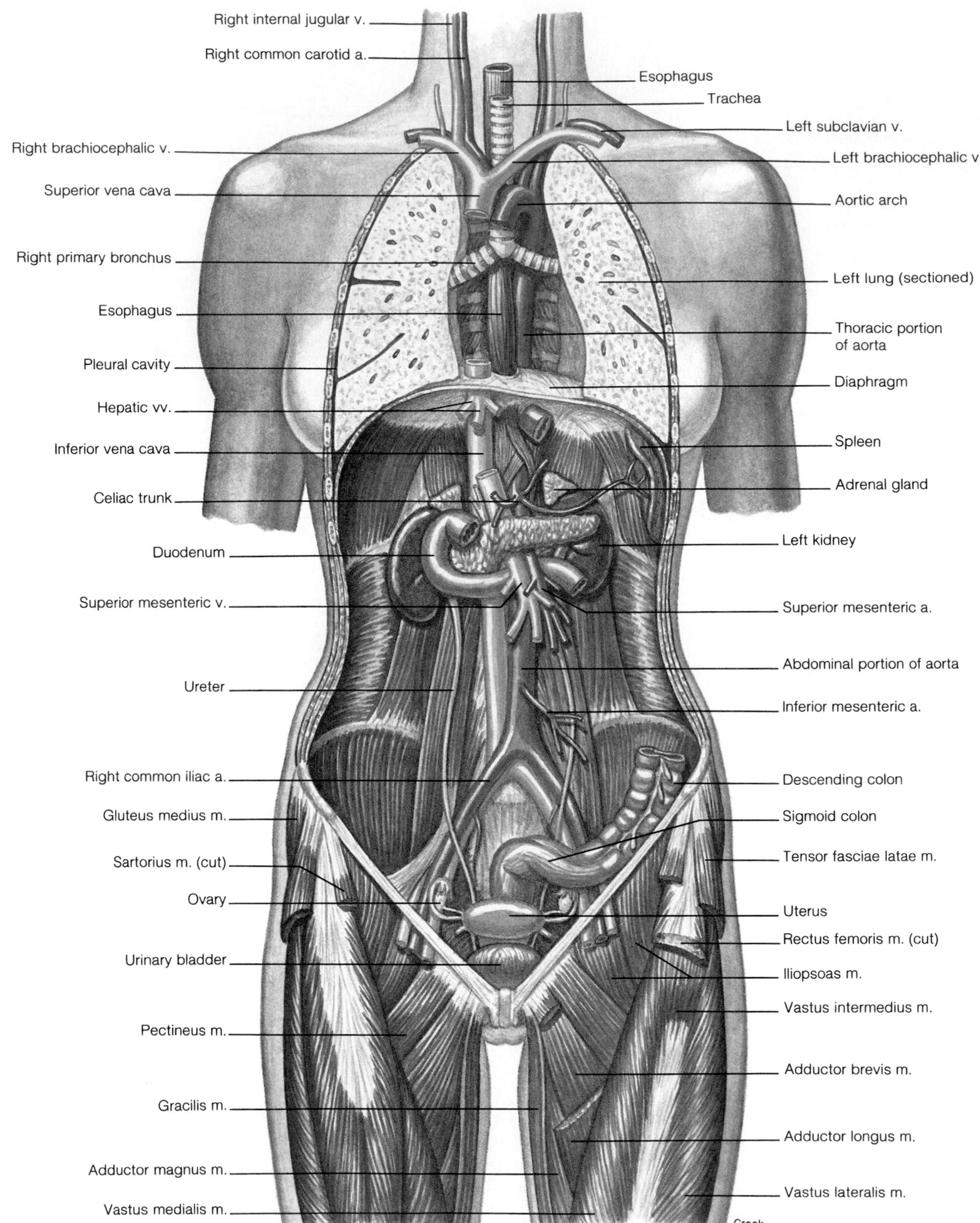
Right internal jugular v.
Right common carotid a.
Esophagus
Trachea
Left subclavian v.
Right brachiocephalic v.
Left brachiocephalic v.
Superior vena cava
Aortic arch
Right primary bronchus
Left lung (sectioned)
Esophagus
Thoracic portion of aorta
Pleural cavity
Diaphragm
Hepatic vv.
Inferior vena cava
Spleen
Celiac trunk
Adrenal gland
Left kidney
Duodenum
Superior mesenteric v.
Superior mesenteric a.
Abdominal portion of aorta
Ureter
Inferior mesenteric a.
Right common iliac a.
Descending colon
Gluteus medius m.
Sigmoid colon
Sartorius m. (cut)
Tensor fasciae latae m.
Ovary
Uterus
Rectus femoris m. (cut)
Urinary bladder
Iliopsoas m.
Vastus intermedius m.
Pectineus m.
Adductor brevis m.
Gracilis m.
Adductor longus m.
Adductor magnus m.
Vastus lateralis m.
Vastus medialis m.
Creek

PLATE 6
An anterior view of the female trunk.

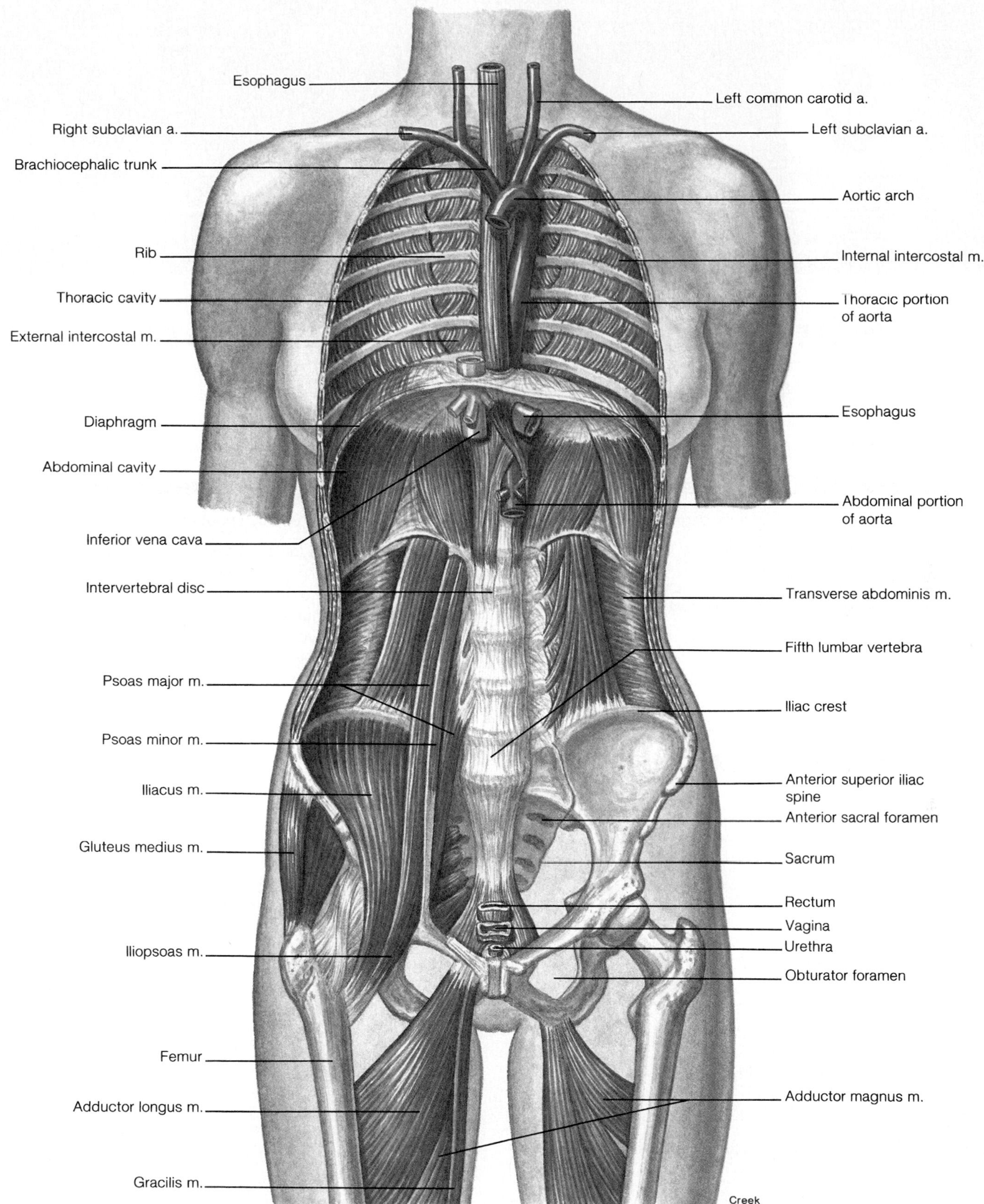

PLATE 7
An anterior view of the female trunk with the thoracic, abdominal, and pelvic visceral organs removed.

"Dendrites are thin, branched processes that extend from the cytoplasm of the [neuron] cell body." *page 232*

Functional Organization of Nervous Tissue

Chapter Outline

Terms to Remember

neuron (*noor'on*), 232
ganglion (*gang'gle-on*), 232
dendrite (*den'drīt*), 232
axon (*ak'son*), 232
nerve, 233
neuroglia (*noo-rog'le-ă*), 234
Schwann cell, 234
myelin (*mi'ĕ-lin*) sheath, 234
node of Ranvier (*rahn-ve-a'; ran'vēr*), 234
resting membrane potential, 236
depolarization, 236
repolarization, 236
refractory period, 237
saltatory (*sal'tă-to"re*) conduction, 239
synapse (*sin'aps*), 239
neurotransmitter, 239
acetylcholine (*ă-set"l-ko'lēn*), 239
acetylcholinesterase (*ă-set"l-ko"lĭ-nes'tĕ-rās*), 240

Learning Objectives

After studying this chapter, you should be able to

- describe the organization and basic functions of the nervous system.
- describe the parts of a neuron and define the terms *nerve, ganglion, tract,* and *nucleus.*
- identify the various types of neuroglial cells and explain how a myelin sheath is produced.
- explain how axons produce action potentials and describe the properties of action potentials.
- explain how unmyelinated and myelinated axons conduct action potentials.
- describe the steps of synaptic transmission when acetylcholine is used as a neurotransmitter.
- list some chemicals, other than acetylcholine, that function as neurotransmitters.

9.1 Introduction to the Nervous System

The central nervous system and the peripheral nervous system are structural components of the nervous system, whereas the autonomic nervous system is a functional component. Together they orient the body, coordinate body activities, permit the assimilation of experiences, and program instinctual behavior.

The immensely complex brain and its myriad of connecting pathways constitute the nervous system—the master controlling and communicating center of the body and the seat of all mental activity. The primary functions of the nervous system are threefold. First, it monitors changes (stimuli) within the body and in the external environment; second, it interprets the changes in a process called integration; and third, it effects responses by activating muscles or glands. These overlapping functions are critical for maintaining body homeostasis.

The nervous system is divided into the **central nervous system (CNS),** which includes the brain and spinal cord, and the **peripheral nervous system (PNS),** which includes the cranial nerves that arise from the brain and the spinal nerves that arise from the spinal cord (fig. 9.1 and table 9.1).

The **autonomic nervous system (ANS),** is a functional subdivision of the nervous system. The controlling centers of the ANS are within the brain and are considered part of the CNS; the peripheral portions of the ANS are subdivided into the **sympathetic** and **parasympathetic divisions.**

The central and peripheral nervous systems are considered in chapter 10, and the autonomic nervous system is the topic of chapter 11. In the following modules of this chapter, we will consider the workings of the cells of the nervous system—neurons and neuroglia.

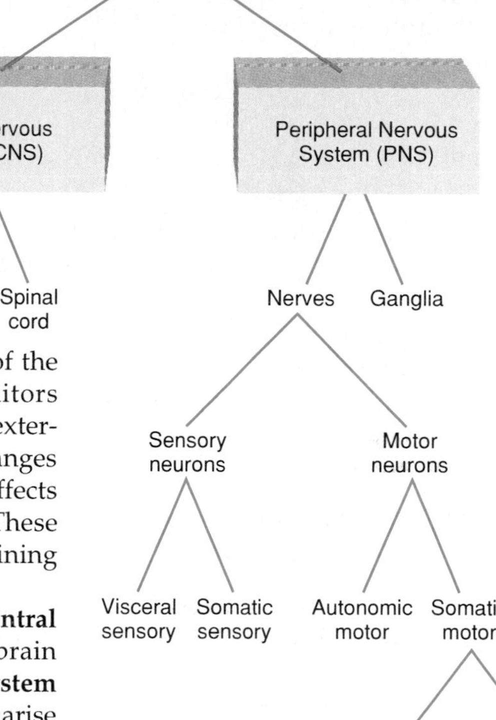

FIGURE 9.1 Subdivisions of the nervous system.

TABLE 9.1 Terminology Used to Describe the Nervous System

Term	Definition
Central nervous system (CNS)	Brain and spinal cord
Peripheral nervous system (PNS)	Nerves and ganglia
Sensory neuron (afferent neuron)	Neuron that transmits impulses from a sensory receptor to the CNS
Motor neuron (efferent neuron)	Neuron that transmits impulses from the CNS to an effector organ; for example, a muscle
Nerve	Cablelike collection of nerve fibers; may be "mixed" (contain both sensory and motor fibers)
Plexus	Network of interlaced nerves
Somatic motor nerve	Nerve that stimulates contraction of skeletal muscles
Autonomic motor nerve	Nerve that stimulates contraction (or inhibits contraction) of smooth muscle and cardiac muscle and that stimulates secretion of glands
Ganglion	Grouping of neuron cell bodies located outside the CNS
Nucleus	Grouping of neuron cell bodies within the CNS
Tract	Grouping of nerve fibers that interconnect regions of the CNS

9.2 Neurons and Nerves

Neuron processes contribute to nerves, and their cell bodies form ganglia and nuclei. The nervous system contains several different types of neurons.

Neurons are the basic structural and functional units of the nervous system (fig. 9.2). Although neurons vary widely in size and shape, they are not likely to be mistaken for any other type of cell. Typically, a neuron consists of a cell body, dendrites, and a single axon (figs. 9.2 and 9.3). Dendrites and axons can be referred to generically as *processes,* or extensions from the cell body.

The **cell body** is the enlarged portion of the neuron. It contains a well-defined nucleus and nucleolus, and serves as the "nutritional center" of the neuron, where macromolecules are produced. The cell body also contains granular *chromatophilic substances* (*Nissl bodies*), which are composed of a rough endoplasmic reticulum. The cell bodies within the CNS are frequently clustered into groups called **nuclei** (not to be confused with the nucleus of a cell). Cell bodies in the PNS usually occur in clusters called **ganglia** (*gang'gle-ă*) (table 9.1).

Dendrites (*den'drīts*) are thin, branched processes that extend from the cytoplasm of the cell body. Dendrites serve as a receptive area for stimuli and transmit electrical impulses to the cell body.

The **axon** (*ak'son*) is the process that conducts impulses away from the cell body. Axons vary in length from only a millimeter to as long as a meter or more (for axons that extend from the CNS to the foot). The origin of the axon near the cell body is called the *axon hillock,* and side branches that may extend from the axon are called *axon collaterals.* **Axon terminals** are the bulbular endings of an axon. Axons are frequently referred to as *nerve fibers.*

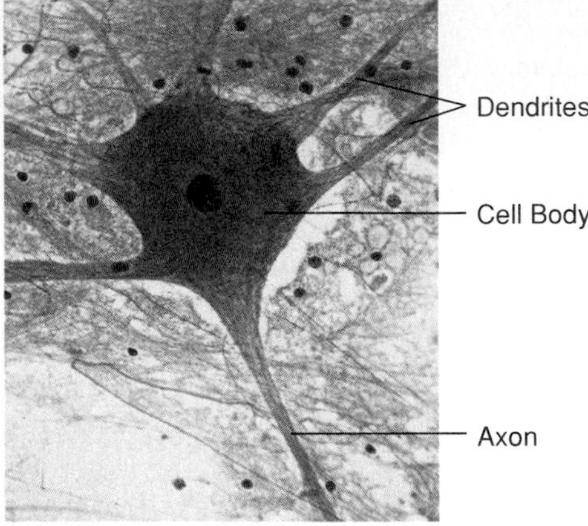

FIGURE 9.2
The neuron as seen in a photomicrograph of nerve tissue.

Inside Information—The axons of motor neurons that control the muscles of your great toe extend from your spinal cord to your foot, making these neurons the longest cells in the body.

Neurons may be classified according to their structure or function. The functional classification is based on the direction in which they conduct impulses. **Sensory** (*afferent*) **neurons** conduct impulses from sensory receptors into the CNS. **Motor** (*efferent*) **neurons** conduct impulses

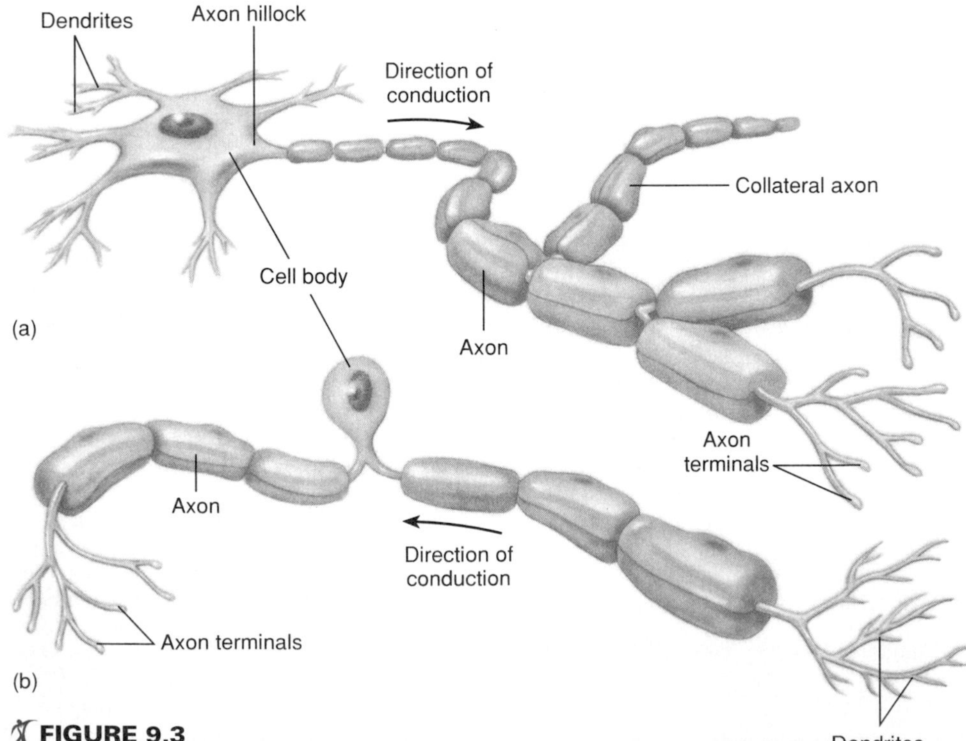

FIGURE 9.3
The structure of (*a*) a motor neuron and (*b*) a sensory neuron.

Nissl body: from Franz Nissl, German neuroanatomist, 1860–1919
ganglion: Gk. *ganglion,* swelling
dendrite: Gk. *dendron,* tree branch
axon: Gk. *axon,* axis

out of the CNS to effector organs (muscles and glands). **Association neurons,** or **interneurons,** are located entirely within the CNS and serve the associative, or integrative, functions of the nervous system (see module 10.11 and fig. 10.27). The term *innervation* refers to the nerve supply of a body part, which can be either sensory or motor.

The structural classification of neurons is based on the number of processes that extend from the cell body of the neuron (fig. 9.4). **Bipolar neurons** have two processes, one at either end; this type is found in the retina of the eye. **Multipolar neurons** (the most common type) have several dendrites and one axon extending from the cell body. Motor neurons are multipolar. **Pseudounipolar neurons** have a single short process that divides like a T to form a longer process. They are called pseudounipolar (*pseudo* = false) because they originate as bipolar neurons, but then their two processes converge and partially fuse during early embryonic development. Sensory neurons are pseudounipolar—one end of the process formed by the T receives sensory stimuli and produces nerve impulses; the other end of the T delivers these impulses to synapses within the brain or spinal cord.

A **nerve** is a bundle of nerve fibers outside the CNS. Most nerves are composed of both motor and sensory fibers, and are thus called *mixed nerves.* Some of the cranial nerves, however, contain only sensory fibers. These are the nerves that serve the special senses of smell, sight, hearing, and equilibrium.

Inside Information—Throughout ancient and medieval times, nerves were thought to be hollow structures like arteries and veins, and also like blood vessels, to carry fluid. The rarefied "fluid" of nerves, described by Galen and others, was referred to as "animal spirits."

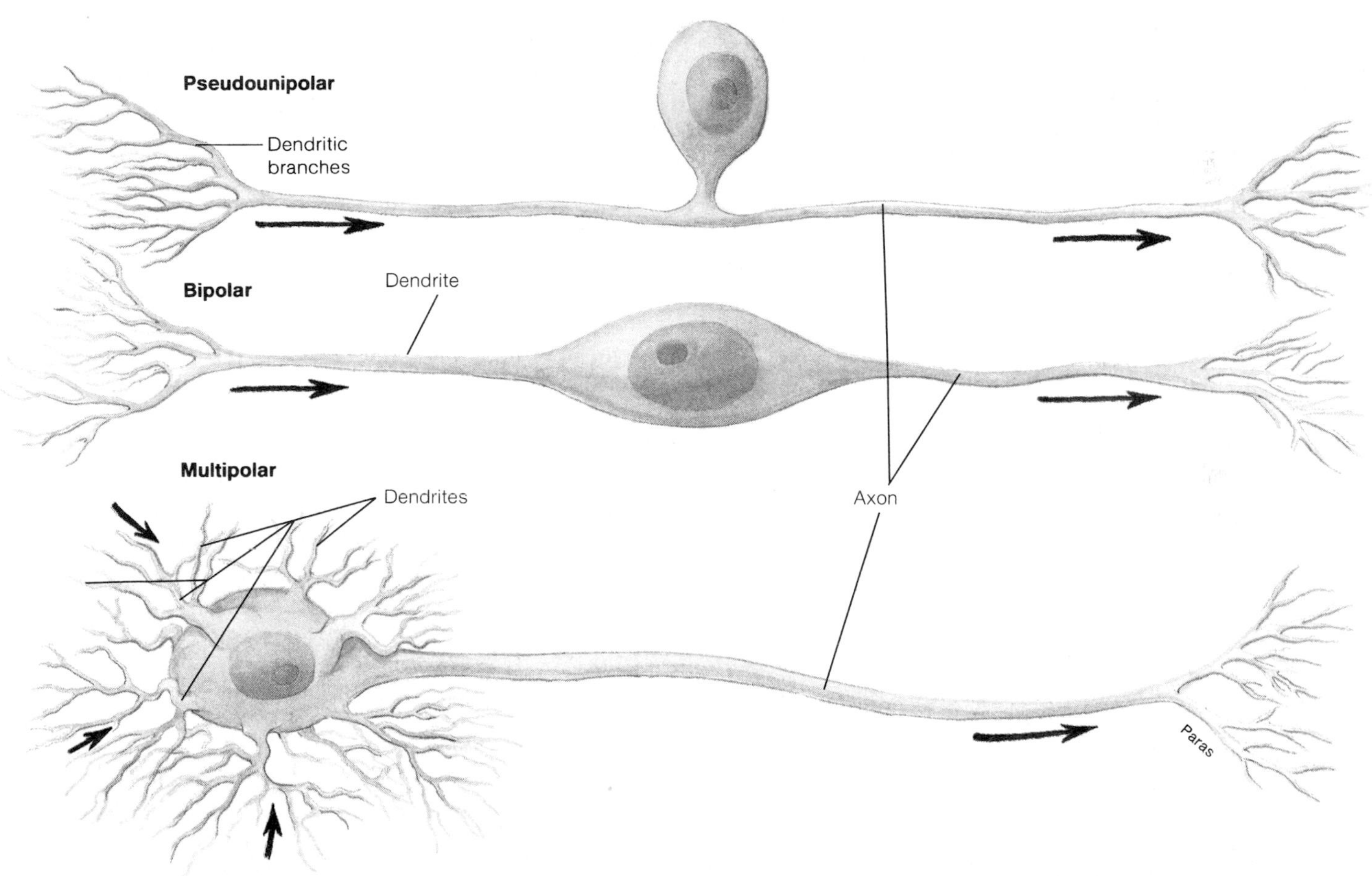

FIGURE 9.4
Three different types of neurons. (Arrows indicate the direction of nerve impulses.)

9.3 Neuroglia

Neuroglial cells do not produce impulses, but they support the functions of the nervous system in a variety of ways. One of the functions of neuroglia is the formation of a myelin sheath around axons.

Types of Neuroglia

Neuroglia (*noo-rog'le-ă*), or **glial** (*gle'al*) **cells,** are supportive cells of the nervous system that aid the function of neurons. There are six categories of neuroglial cells:

1. **Schwann cells** (*neurolemmocytes*), which form myelin sheaths around axons of the PNS;
2. **oligodendrocytes** (*ol" ĭ-go-den'dro-sīts*), which form myelin sheaths around axons of the CNS;
3. **microglia,** which migrate through the CNS and phagocytize foreign and degenerated material;
4. **astrocytes,** which help to regulate the passage of molecules from the blood to the brain;
5. **ependymal** (*ĕ-pen'dĭ-mal*) **cells,** which aid the movement of fluid within the CNS; and
6. **ganglionic cells** (*satellite cells*), which support neuron cell bodies within the ganglia of the PNS.

neuroglia: Gk. *neuron,* nerve; *glia,* glue
Schwann cell: Theodor Schwann, German histologist, 1810–82.
oligodendrocyte: Gk. *oligos,* few; L. *dens,* tooth; Gk. *kytos,* hollow (cell)
microglia: Gk. *mikros,* small; *glia,* glue
astrocyte: Gk. *aster,* star; *kytos,* hollow (cell)
nodes of Ranvier: from Louis Ranvier, French pathologist, 1835–1922

Myelin Sheath and the Sheath of Schwann

Some axons in the CNS and PNS are surrounded by a **myelin** (*mi'ĕ-lin*) **sheath,** which is composed of successive wrappings of the cell membrane of either Schwann cells (fig. 9.5) or oligodendrocytes. These are known as *myelinated axons.* Other axons do not have a myelin sheath, and are called *unmyelinated axons.* Axons that are less than 2 µm in diameter are usually unmyelinated; those that are larger are generally myelinated. Schwann cells form myelin in the PNS, whereas oligodendrocytes form myelin in the CNS.

In the process of myelin formation in the PNS, Schwann cells roll around the axon, much like a roll of electrician's tape is wrapped around a wire. Unlike electrician's tape, however, the wrappings are made in the same spot, so that each wrapping overlaps the others. The cytoplasm, meanwhile, becomes squeezed to the outer region of the Schwann cell, much as toothpaste is squeezed to the top of the tube as the bottom is rolled up (fig. 9.6). Each Schwann cell wraps only about 1 mm of axon, leaving gaps of exposed axon between the adjacent Schwann cells. These gaps in the myelin sheath are known as the **nodes of Ranvier** (*rahn-ve-a'; ran'vĕr*), also called *neurofibril nodes.* The successive wrappings of Schwann cell membrane provide insulation around the axon, leaving only the nodes of Ranvier exposed to produce nerve impulses.

The Schwann cells remain alive as their cytoplasm is squeezed to the outside of the myelin sheath. As a result, myelinated axons of the PNS, like their unmyelinated

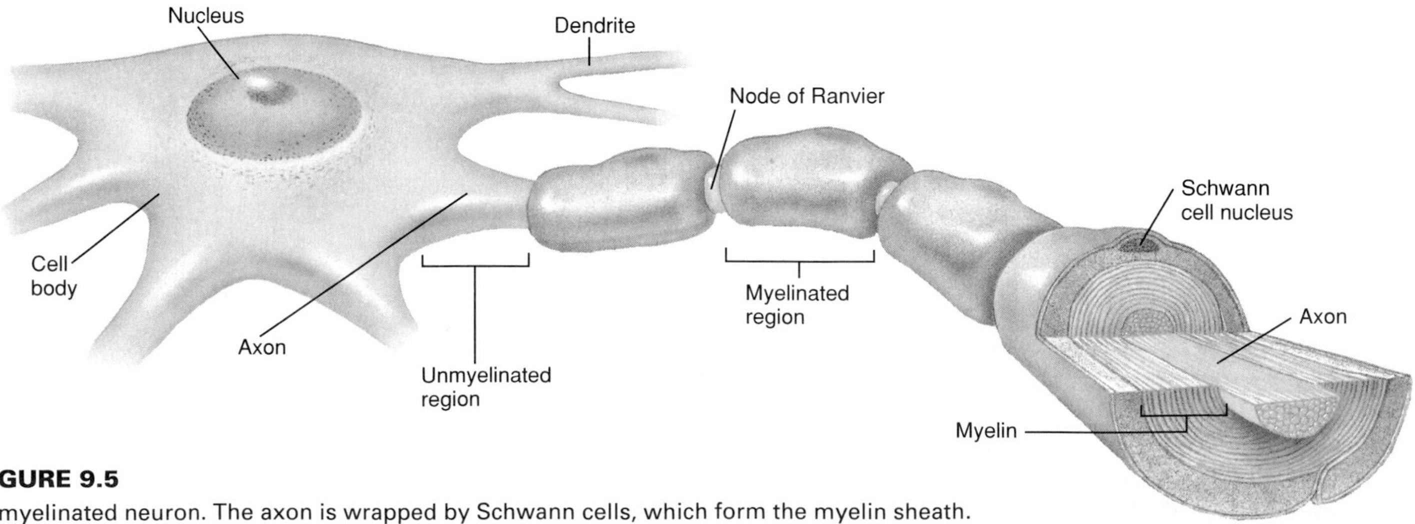

FIGURE 9.5
A myelinated neuron. The axon is wrapped by Schwann cells, which form the myelin sheath.

counterparts, are surrounded by a living sheath of Schwann cell cytoplasm that forms the **sheath of Schwann,** also known as a *neurilemmal sheath.*

The myelin sheaths of the CNS are formed by oligodendrocytes. Unlike a Schwann cell, which forms a myelin sheath around only one axon, each oligodendrocyte has extensions, like the tentacles of an octopus, that form myelin sheaths around several axons (fig. 9.7). The myelin sheaths around axons of the CNS give this tissue a white color, primarily because of the lipids present. Areas of the CNS that contain a high concentration of myelinated axons thus form the **white matter.** The **gray matter** of the CNS is composed of high concentrations of cell bodies and dendrites, which lack myelin sheaths. (The only dendrites that have myelin sheaths are those of the sensory neurons of the PNS.)

Inside Information—The reason newborns cannot see at first is because their optic nerves are not yet completely myelinated. And before they can walk with ease, they must wait about a year, until myelinization of the neurons innervating the muscles of their legs is complete.

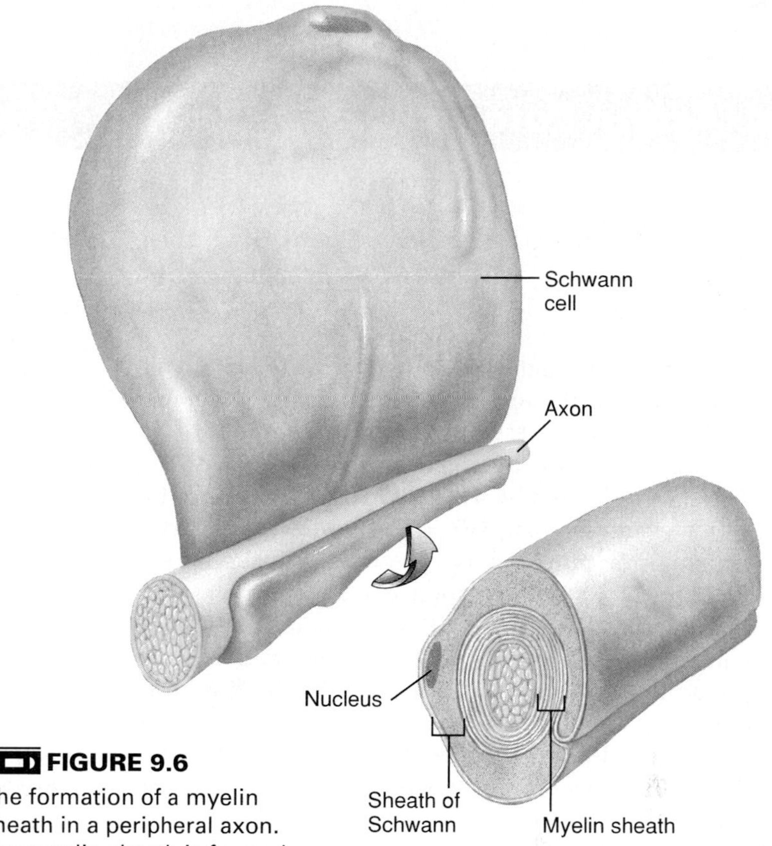

FIGURE 9.6
The formation of a myelin sheath in a peripheral axon. The myelin sheath is formed by successive wrappings of the Schwann cell membranes, leaving most of the Schwann cell cytoplasm outside the myelin. The sheath of Schwann is thus located outside the myelin sheath.

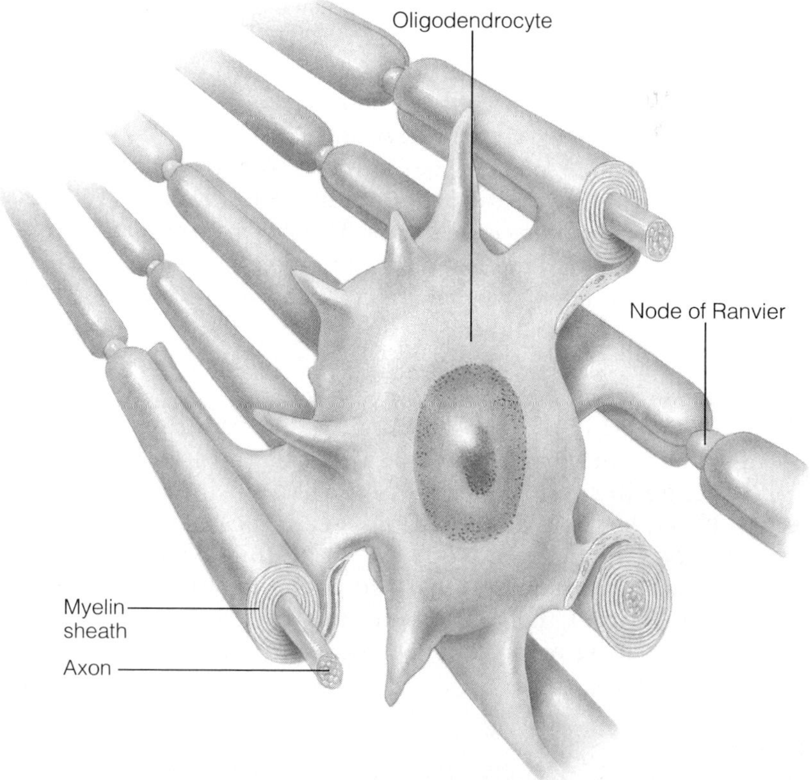

FIGURE 9.7
Formation of myelin sheaths in the central nervous system by an oligodendrocyte. One oligodendrocyte forms myelin around several axons.

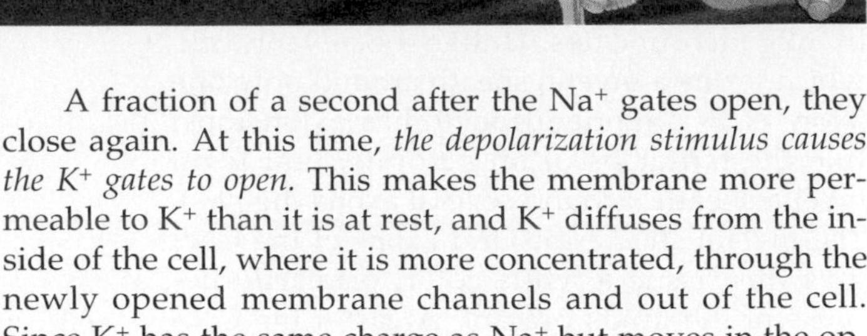

9.4 Action Potentials

In the axon membrane, ion channels for Na^+ and K^+ are guarded by gates that open in response to depolarization. Diffusion of Na^+ and K^+ produces action potentials, which are all-or-none events.

All cells in the body maintain a potential difference (voltage) across the cell membrane. This potential difference can be referred to as a **resting membrane potential.** The inside of the cell is negatively charged in comparison to the outside, as indicated by a negative sign in front of the voltage (for example, –65 mV). Although all cells have a resting membrane potential, only nerve and muscle cells can change this potential in response to a stimulus.

Changes in the potential difference across the membrane can be measured by the voltage developed between two electrodes—one placed inside the cell, the other placed outside the cell membrane at the region being recorded. The voltage between these two *recording electrodes* can be visualized by connecting the electrodes to an oscilloscope (fig. 9.8).

An upward deflection of the line indicates that the inside of the membrane has become less negative (or more positive) compared to the outside of the membrane. A downward deflection of the line, conversely, indicates that the inside of the cell has become more negative. If appropriate stimulation causes positive charges to flow into the cell, the line will deflect upward; this change is called **depolarization.** If the inside of the membrane becomes more negative, the line on the oscilloscope will deflect downward; this is called **hyperpolarization.**

The permeability of the membrane to Na^+, K^+, and other ions is regulated by parts of the ion channels through the cell membrane called **gates.** The gated channels of an axon are closed at the resting membrane potential of –65 mV, but they open when the membrane is depolarized to a certain threshold level. Since the opening and closing of these gates is regulated by the membrane voltage, the gates are said to be **voltage regulated.**

If a certain level of depolarization is achieved (from –65 mV to –55 mV for example), a sudden and very rapid change in the membrane potential will be observed. This is because, at a threshold level, *depolarization causes the Na^+ gates to open.* Now, the permeability properties of the membrane are changed and Na^+ diffuses from the outside of the cell, where it is more concentrated, through the newly opened membrane channels and into the cell.

A fraction of a second after the Na^+ gates open, they close again. At this time, *the depolarization stimulus causes the K^+ gates to open.* This makes the membrane more permeable to K^+ than it is at rest, and K^+ diffuses from the inside of the cell, where it is more concentrated, through the newly opened membrane channels and out of the cell. Since K^+ has the same charge as Na^+ but moves in the opposite direction, the outward diffusion of K^+ acts to reestablish the resting membrane potential. This process is called **repolarization** (fig. 9.9).

Figure 9.10*b* illustrates the movement of Na^+ and K^+ through the axon membrane in response to a depolarization stimulus. These changes in Na^+ and K^+ diffusion and the resulting changes in the membrane potential that they produce constitute an event called the **action potential,** or **nerve impulse,** as shown in figure 9.10*a.*

The amplitude of action potentials is **all or none,** rather than graded. When depolarization is below a threshold value, action potentials are not produced. When depolarization reaches threshold, a maximum potential

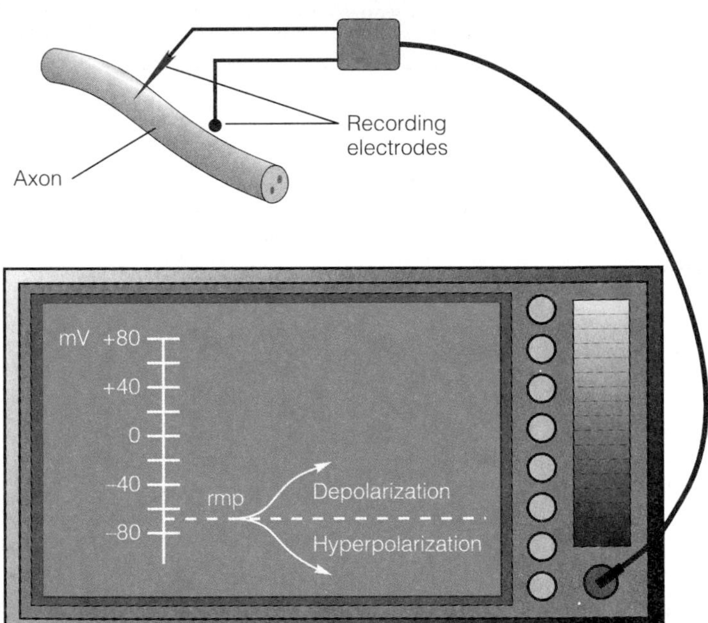

FIGURE 9.8
The difference in potential in millivolts (mV) between an intracellular and extracellular recording electrode is displayed on an oscilloscope screen. The resting membrane potential (rmp) of the axon may be reduced (depolarization) or increased (hyperpolarization).

change (the action potential, from −65 mV to +40 mV) is produced. A stronger stimulus does not produce a larger action potential. Instead, a stronger stimulus results in a *greater frequency* of action potentials.

As the stimulus strength is increased, the frequency of action potentials produced at that point will increase accordingly. As action potentials occur with increasing frequency, the time between successive action potentials becomes shorter. However, the interval between successive action potentials will never become so short that one action potential is produced before the preceding one has finished.

During the time that a patch of axon membrane is producing an action potential, it is incapable of responding to further stimulation. It is in what is known as the **refractory period.** It is because of this refractory period that action potentials cannot run together; they remain all-or-none events.

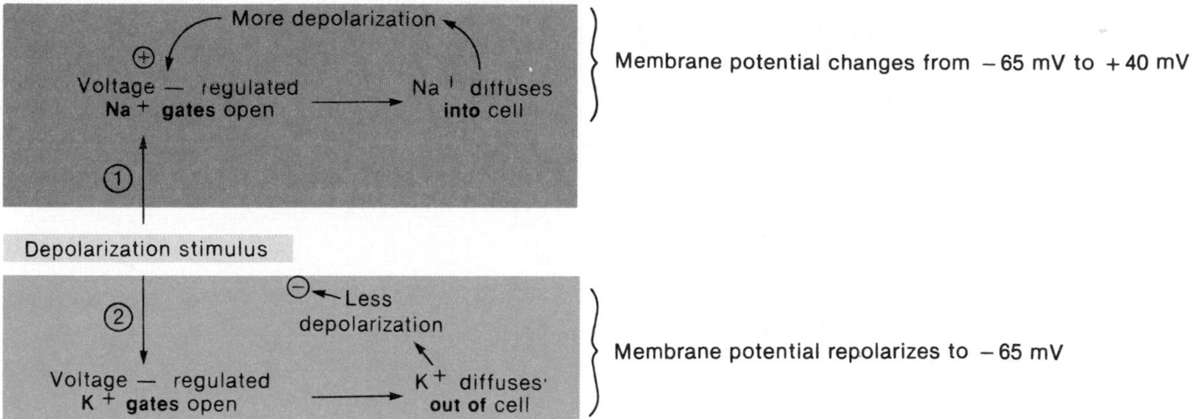

FIGURE 9.9

Depolarization of an axon has two effects: (1) Na^+ gates open and Na^+ diffuses into the cell, and (2) after a brief period, K^+ gates open and K^+ diffuses out of the cell. An inward diffusion of Na^+ causes further depolarization, which in turn causes further opening of Na^+ gates in a positive feedback (+) fashion. The opening of K^+ gates and outward diffusion of K^+ makes the inside of the cell more negative, and thus has a negative feedback effect (–) on the initial depolarization.

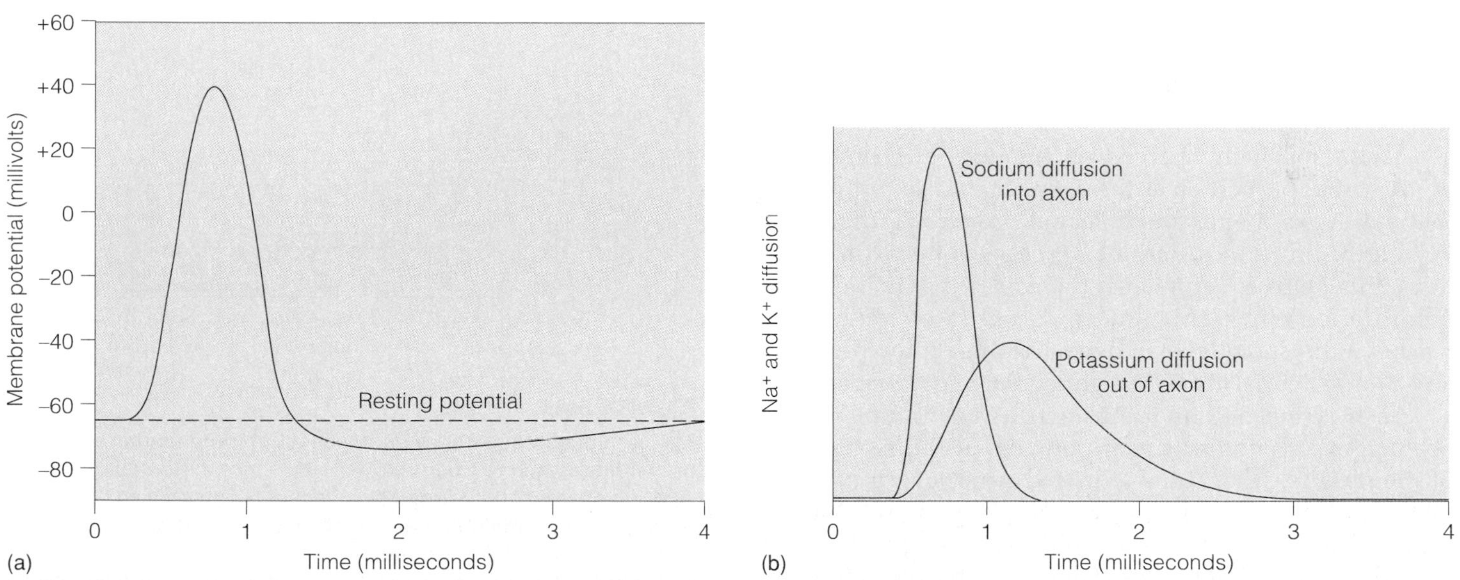

FIGURE 9.10

An action potential (*a*) is produced by an increase in sodium diffusion that is followed, with a short time delay, by an increase in potassium diffusion (*b*). This drives the membrane potential first toward the sodium equilibrium potential and then toward the potassium equilibrium potential.

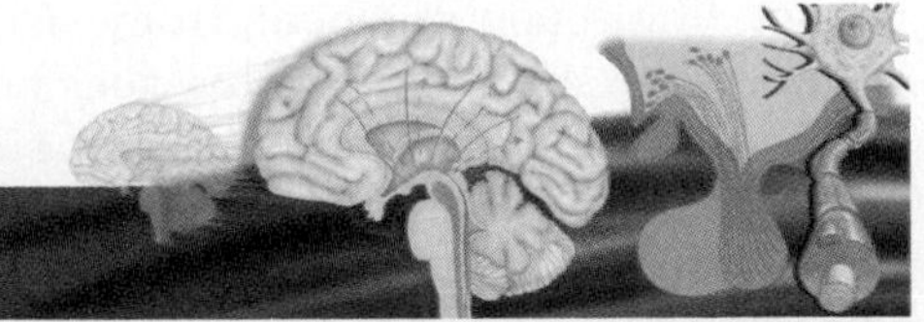

9.5 Conduction of Action Potentials

Each action potential depolarizes an adjacent region of the axon, causing it to produce its own action potential. Action potentials are thus regenerated along the axon.

If a pair of stimulating electrodes produces a depolarization that is too weak to reach threshold (about –55 mV), the change in membrane potential will be *localized* to within 1 to 2 mm of the point of stimulation. This is because the axon is a very poor conductor of charges.

The term *cable properties* is used to describe the ability of a neuron to transmit charges through its cytoplasm. These cable properties are quite poor because of the high internal resistance to the spread of charges and the leakage of many charges out of the axon through its membrane. If an axon had to conduct only through its cable properties, therefore, no axon could be more than a millimeter long.

Conduction in an Unmyelinated Axon

When stimulating electrodes artificially depolarize one point of an axon membrane to a threshold level, voltage-regulated gates open and an action potential is produced at that small region of axon membrane. For about the first millisecond of the action potential, when the membrane voltage changes from –65 mV to +30 mV, a current of Na^+ enters the cell by diffusion due to the opening of the Na^+ gates. Each action potential thus "injects" positive charges (Na^+) into the axon.

These positively charged sodium ions are conducted, by the cable properties of the axon, to an adjacent region that still has a membrane potential of –65 mV (fig. 9.11). Within the limits of the cable properties of the axon (1 to 2 mm), this helps to depolarize the adjacent region of axon membrane. When this adjacent region of membrane reaches a threshold level of depolarization, it too produces an action potential as its voltage-regulated gates open.

Each action potential thus acts as a stimulus for the production of another action potential at the next region of membrane. In an unmyelinated axon, every patch of membrane that contains Na^+ and K^+ gates can produce an action potential. Action potentials are thus produced at locations only a fraction of a micrometer apart all along the length of the axon.

Notice that action potentials are not really "conducted," although it is convenient to use that word. Each action potential is a separate, complete event that is repeated, or *regenerated,* along the axon's length. The action potential produced at the end of the axon is thus a completely new event that was produced in response to depolarization from the previous action potential. The last action potential has the same amplitude as the first. Action potentials are thus said to be **conducted without decrement** (without decreasing in amplitude).

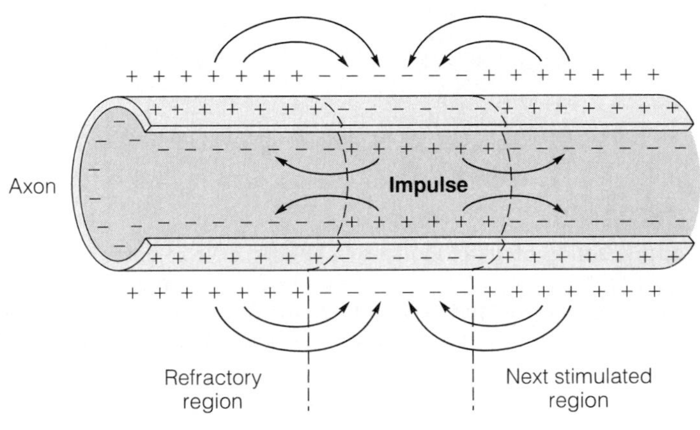

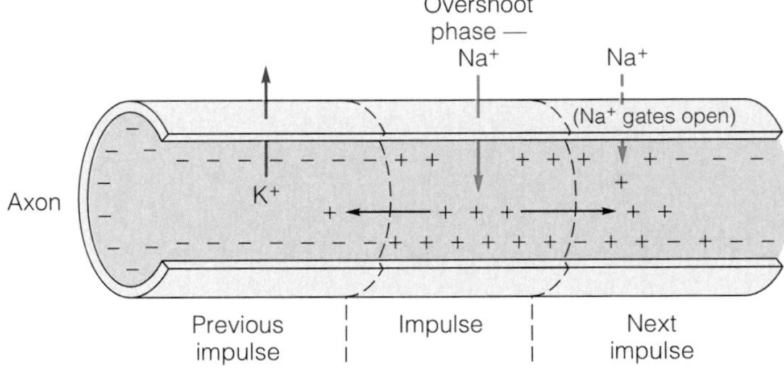

FIGURE 9.11

The conduction of a nerve impulse (action potential) in an unmyelinated nerve fiber (axon). Each action potential "injects" positive charges that spread to adjacent regions. The region that has just produced an action potential is refractory (is insensitive to further stimulation). The adjacent region is partially depolarized. As a result, its voltage-regulated Na^+ gates open, repeating the process.

Conduction in a Myelinated Axon

The myelin sheath provides insulation for the axon, preventing movements of Na^+ and K^+ through the membrane. If the myelin sheath were continuous, therefore, no action potentials could be produced. Fortunately, there are interruptions in the myelin (the nodes of Ranvier).

Since the cable properties of axons can conduct depolarizations only over a very short distance (1 to 2 mm), it follows that the nodes of Ranvier must be about this distance apart. In fact, studies have shown that Na^+ channels are highly concentrated at the nodes and almost absent between the nodes. Action potentials, therefore, occur only at the nodes of Ranvier (fig. 9.12), and they seem to "leap" from node to node—a process called **saltatory** (*sal'tă-to"re*) **conduction.** Because fewer action potentials need to be produced, saltatory conduction allows a faster rate of conduction than is possible in an unmyelinated fiber. Myelinated axons are thus the fastest conductors in the body, carrying impulses at speeds that may exceed 100 m/sec. These axons are involved in quick stretch reflexes in skeletal muscles. By contrast, the conduction rates of unmyelinated axons that mediate slow visceral responses do not exceed 10 m/sec.

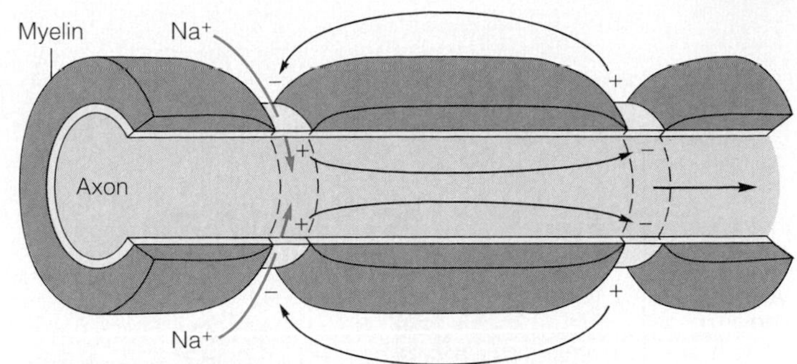

FIGURE 9.12
The conduction of the nerve impulse in a myelinated nerve fiber. Since the myelin sheath prevents inward Na^+ current, action potentials can be produced only at the interruptions in the myelin sheath, or nodes of Ranvier. This "leaping" of the action potential from node to node is known as saltatory conduction.

Inside Information—The velocity of a nerve impulse was first measured in 1852 by the German physicist, anatomist, and physiologist Hermann von Helmholtz. Perhaps von Helmholtz was responding to a challenge. Just 6 years earlier, his famous compatriot Johannes Müller had stated categorically that no person could ever succeed in accomplishing this feat.

9.6 Synaptic Transmission

Action potentials in the presynaptic neuron stimulate the release of a chemical neurotransmitter, which activates the postsynaptic cell.

A **synapse** (*sin'aps*) is the functional connection between a neuron and a second cell. In the CNS this other cell is also a neuron. In the PNS the other cell may be a neuron, or it may be an *effector cell,* located in either a muscle or a gland. Although the physiology of neuron-neuron synapses and neuron-muscle synapses is similar, the latter type are often called **myoneural,** or *neuromuscular,* **junctions.**

saltatory: L. *saltatio,* leap

Most synapses contain a tiny gap, called the **synaptic cleft,** between the axon terminal of the presynaptic neuron and the postsynaptic cell (fig. 9.13). When this gap was first observed, the idea arose that synaptic transmission might be chemical—that the axon terminal of the presynaptic neuron might release chemical **neurotransmitters** that stimulated the postsynaptic cell. Indeed this proved to be the case, and the neurotransmitter **acetylcholine** (*ă-set"l-ko'lēn*), or **ACh,** was the first to be discovered.

Neurotransmitter molecules within the axon terminals of presynaptic neurons are contained within many small, membrane-enclosed **synaptic vesicles.** In order for the neurotransmitter within these vesicles to be released into the synaptic cleft, the vesicle membrane must fuse with the axon membrane and release its contents. This process is known as *exocytosis.*

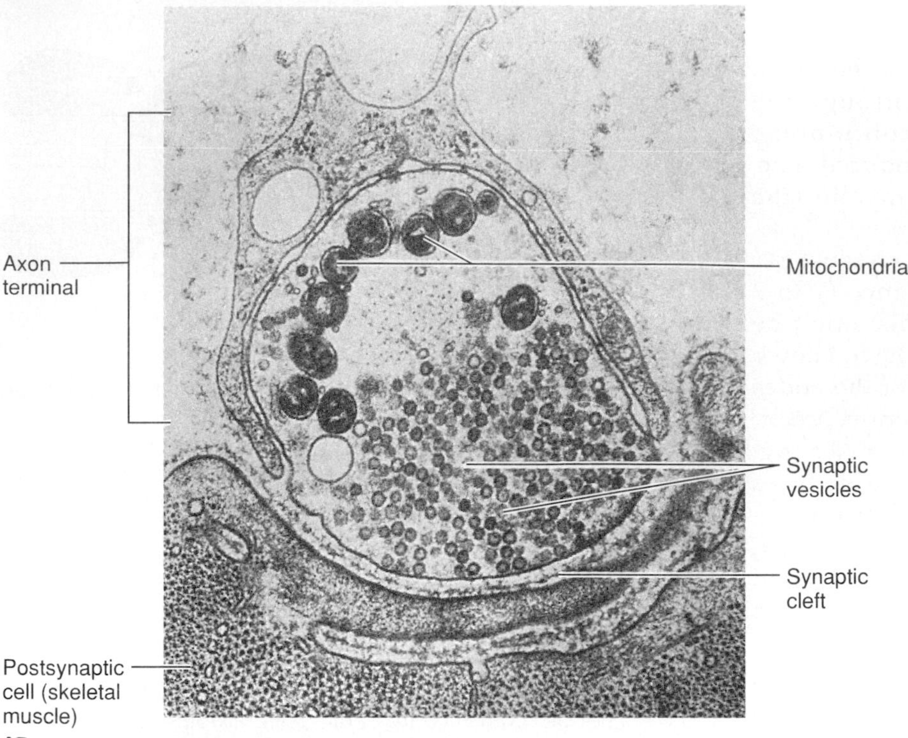

FIGURE 9.13
An electron micrograph of a chemical synapse showing synaptic vesicles at the axon terminal.

Chemically Regulated Gates

The synaptic vesicles each contain about 10,000 molecules of ACh. Once these molecules are released, they quickly diffuse across the narrow synaptic cleft to the membrane of the postsynaptic cell. Here they chemically bind to **receptor proteins** that are built into the postsynaptic membrane.

The bonding of ACh with its receptor protein causes the opening of ion channel gates for Na^+ and K^+ (fig. 9.14). These gates, located only in the postsynaptic membrane, are called **chemically regulated gates** because they open in response to bonding by a chemical (ACh).

Because of the characteristics of chemically regulated gates, neurotransmitters do not directly produce action potentials. Instead, opening of the chemically regulated gates causes depolarizations called **excitatory postsynaptic potentials (EPSPs).** When more ACh is released, the EPSPs are larger. EPSPs are thus graded instead of all or none, and they differ from action potentials in several other respects as well. The EPSP depolarization serves as the stimulus for the production of action potentials in the postsynaptic cells.

The effects of neurotransmitters are not always excitatory; they sometimes inhibit the postsynaptic cell. For example, stimulation of the vagus nerve to the heart causes the heart to slow its rate of beat. The axons of the vagus nerve innervating the heart release ACh, but the ACh produces a hyperpolarization at this synapse. That is, the inside of the cell becomes even more negative than at rest. This is caused by the opening of separate K^+ channels that allow K^+ to diffuse out of the postsynaptic cell. Hyperpolarizations produced by neurotransmitters are called **inhibitory postsynaptic potentials (IPSPs).** Similar inhibitory effects are also produced by many synapses in the central nervous system.

Acetylcholinesterase

The bond between ACh and its receptor protein exists for only a brief instant. The ACh-receptor complex quickly dissociates, but as long as free ACh is in the vicinity, it can be quickly re-formed. In order for activity in the postsynaptic cell to be controlled, ACh must be inactivated quickly after it is released. The inactivation of ACh is achieved by means of an enzyme called **acetylcholinesterase** (*ă-set″l-ko″lĭ-nes′tĕ-rās*) **(AChE),** which is present on the postsynaptic membrane or immediately outside the membrane, with its active site facing the synaptic cleft (fig. 9.15).

Nerve gas *exerts its odious effects by inhibiting AChE in skeletal muscles. Since ACh is not degraded, it can continue to combine with receptor proteins and can continue to stimulate the postsynaptic muscle cell, leading to spastic paralysis. Clinically, cholinesterase inhibitors are used to enhance the effects of ACh on muscle contraction when neuromuscular transmission is weak, as in the disease myasthenia gravis.*

Stimulation of Neurons

Within the CNS, the axon terminals of one neuron typically synapse with the dendrites or cell body of another. The dendrites and cell body thus serve as the receptive area of the neuron, and it is in these regions that receptor proteins for neurotransmitters and chemically regulated gates are located. The first voltage-regulated gates are located at the beginning of the axon, at the axon hillock. It is therefore here that action potentials are first produced.

In summary, EPSPs, conducted from the dendrites and cell body, serve as the normal stimuli for the production of action potentials in the axon hillock, and the action potentials at this point serve as the depolarization stimuli for the next region, and so on. This chain of events ends at the axon terminal, where the neurotransmitter is released. It is always the neurotransmitter that activates the postsynaptic cell.

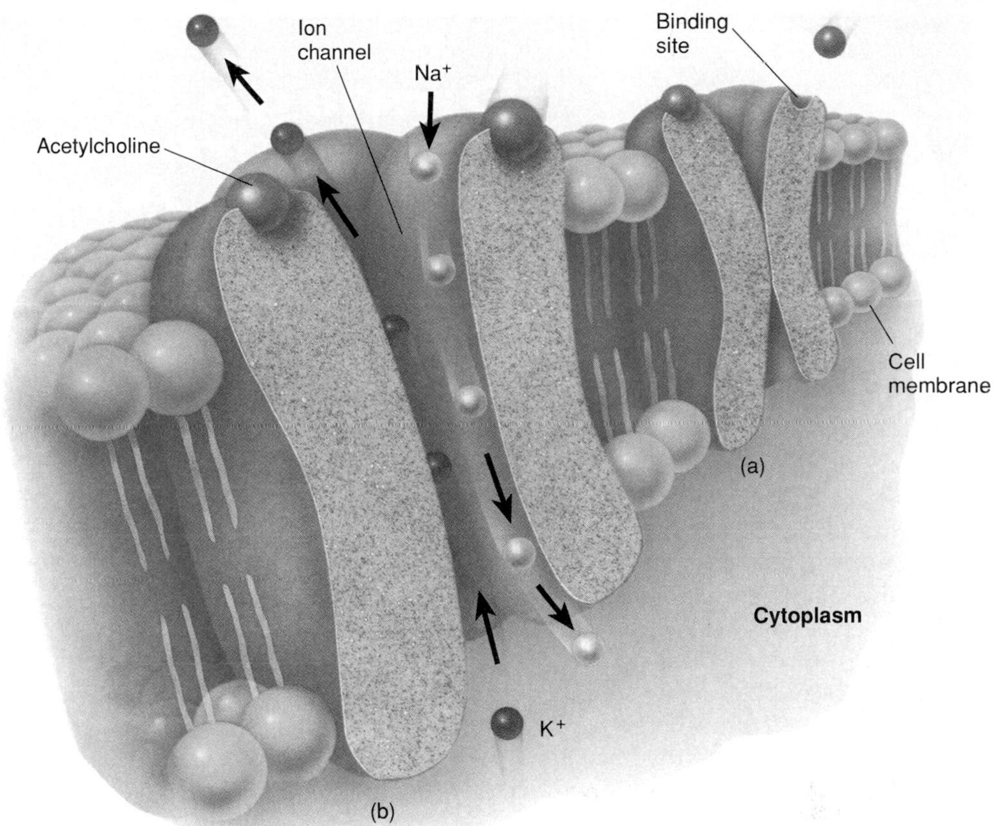

FIGURE 9.14
The binding of acetylcholine to receptor proteins causes the opening of chemically regulated gates in the postsynaptic membrane, as shown in (*a*). This results in the increased diffusion of Na^+ and K^+ through the membrane (*b*).

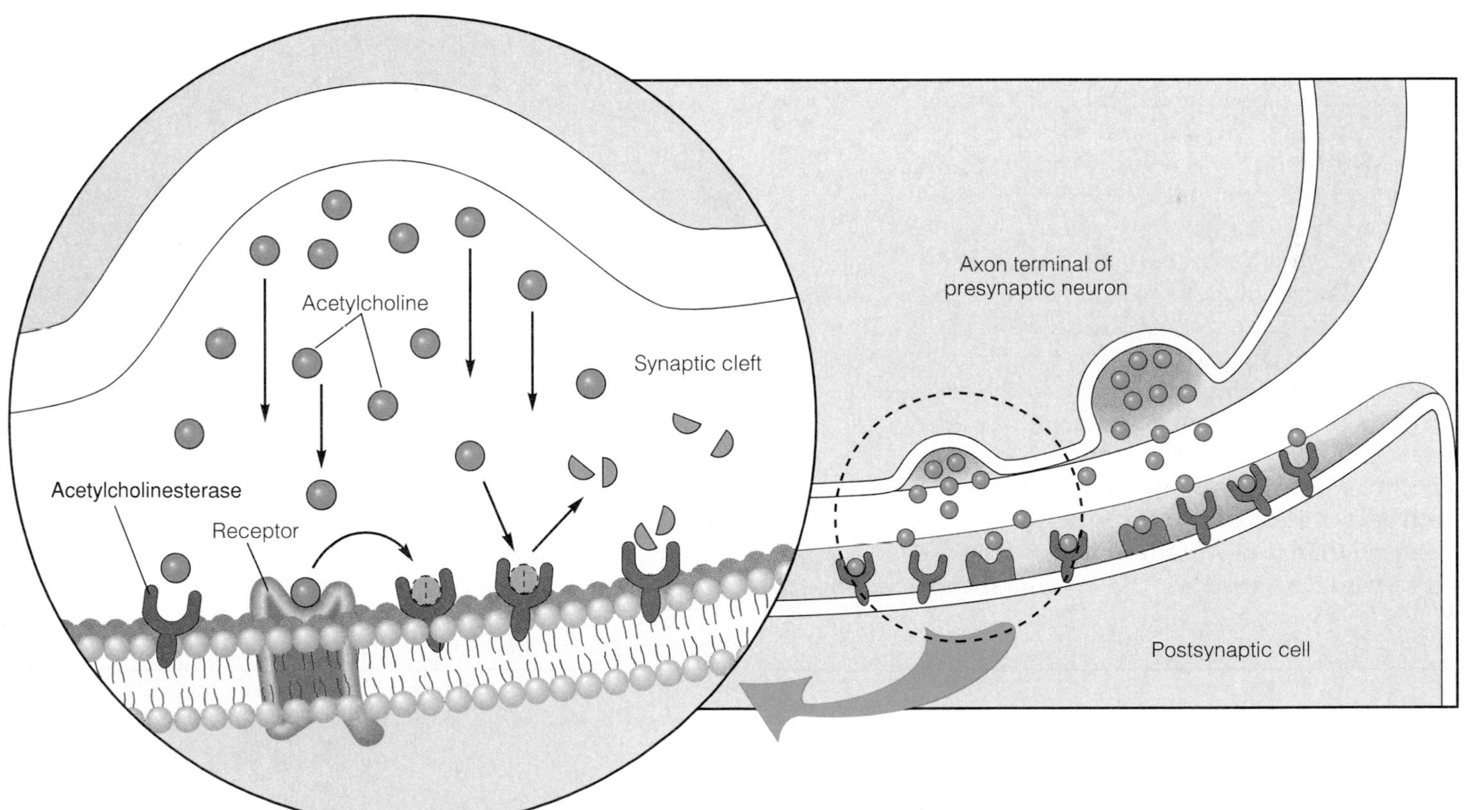

FIGURE 9.15
The release of acetylcholine (ACh) from the axon terminal of a presynaptic neuron. Acetylcholinesterase (AChE) in the postsynaptic membrane inactivates the ACh released into the synaptic cleft.

SYNOPTIC SPOTLIGHT

Damage Control: Regeneration of a Cut Axon

Although neurons cannot replicate (mitotically divide), in many cases they may be able to regenerate (repair themselves) after having been traumatized. In the case of a peripheral nerve, healing and regeneration are likely to occur if the two ends are aligned in close proximity. The distal portion of the axon that was severed from the cell body degenerates and is phagocytosed by Schwann cells and by macrophages that migrate into the trauma zone. Once the debris has been disposed of, the surviving Schwann cells, surrounded by the basement membrane, proliferate to form a **regeneration tube** (see the figure to the right), as the part of the axon that is connected to the cell body begins to grow and exhibit amoeboid movement. The Schwann cells of the regeneration tube are believed to secrete chemicals that attract the growing axon tip, and the tube itself helps to guide the regenerating axon to its proper destination. Even a major nerve fiber that has been severed can be surgically reconnected and the function of the nerve largely reestablished if the surgery is performed before tissue death has occurred.

Repair of neural damage in the central nervous system is much more problematic. In contrast to peripheral nerve fibers, severed axons of central nerve fibers never regenerate distances of more than 1 mm under normal circumstances. In the CNS, the myelin sheath is formed by oligodendrocytes. Unlike the Schwann cells in the PNS, these supporting cells do not produce growth-promoting chemicals, and they do not proliferate

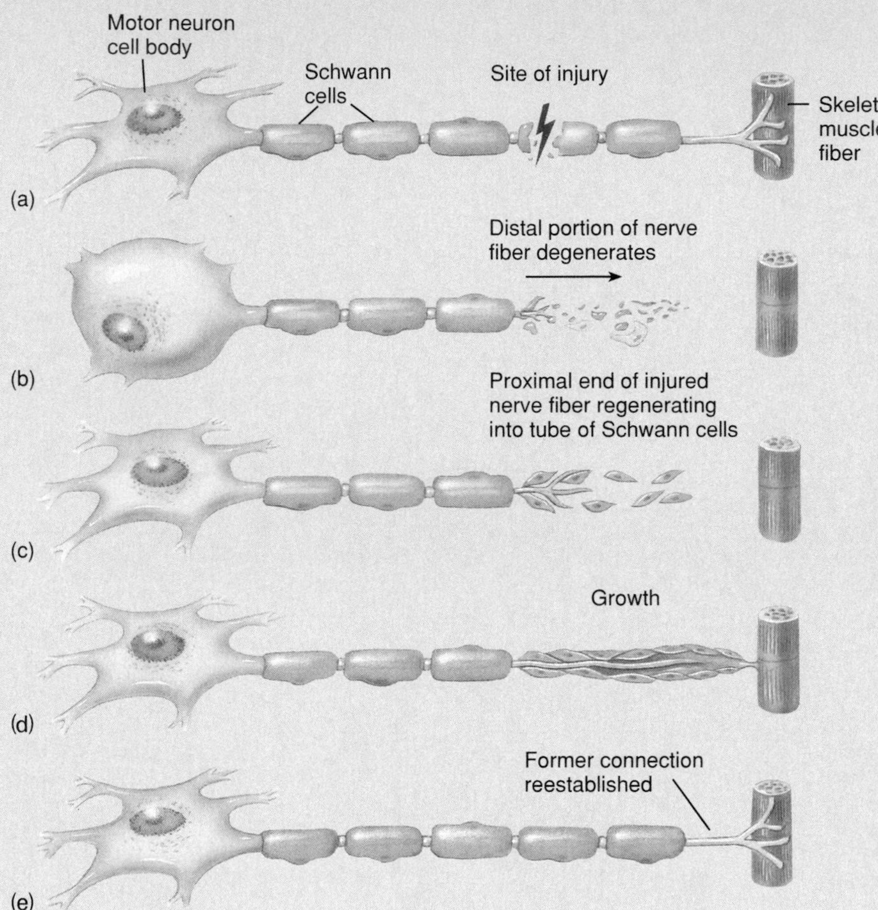

Axon regeneration in a peripheral nerve. (*a*) If a neuron is severed through a myelinated axon, the proximal portion may survive, but (*b*) the distal portion will degenerate through phagocytosis. The myelin sheath provides a pathway (*c* and *d*) for the regeneration of the axon, and (*e*) innervation is restored.

following an injury. Thus, when an axon is damaged in the CNS, no regeneration tube is formed to guide fiber regrowth. Further reducing chances of regeneration is the dense scar tissue that replaces the dead oligodendrocytes. This *glial scar,* produced by astrocytes to repair structural damage, effectively blocks the progress of sprouting axons.

Despite the fact that damage to the brain or spinal cord is generally considered irreversible, experiments in vitro suggest that even axons in the CNS can be stimulated to regenerate to an appreciable extent if an appropriate environment is provided. In a developing fetal brain, chemicals called **neurotrophins** promote nerve growth. Neurotrophins have also been shown to promote neuron regeneration in adult brains and spinal cords in some experimental animals. Additionally, certain chemicals, including *myelin-associated inhibitory proteins,* have been shown to inhibit axon regeneration. In a recent experiment, axon regeneration in the spinal cord of a rat was improved by simultaneously providing a neurotrophin and blocking the inhibitory proteins with antibodies.

In other promising research, scientists have found that Schwann cells transplanted to the CNS can promote axonal regeneration. Efforts are also underway to find a means of synthesizing the growth-promoting factor secreted by the Schwann cells. This research holds out hope for scores of accident and stroke victims whose nerve damage is currently irreparable.

Neurotransmitter Diversity

Acetylcholine was the first neurotransmitter to be discovered, and it is still probably the best understood. However, dozens of other chemicals have since been identified as neurotransmitters. Specific transmitter chemicals perform different functions, and disorders of neurotransmitter action are associated with specific neurological diseases.

Neurons in particular regions of the brain use **dopamine** as a neurotransmitter, and some of them are involved in the control of skeletal movements. When these neurons are diseased, a person may develop *Parkinson's disease.* Parkinson's disease, which is associated with symptoms of muscle tremors and rigidity, is a major cause of neurological disability in people over the age of 60. The patients are frequently treated with L-Dopa, the parent molecule of dopamine, which can enter the brain and be converted into dopamine.

People with schizophrenia *are often treated with drugs that block the formation of dopamine in the brain, suggesting that excessive dopamine as a neurotransmitter may be partly responsible for the schizophrenia. Indeed, people with Parkinson's disease who are given high dosages of L-Dopa may develop symptoms of schizophrenia. Recent findings that people with schizophrenia have increased amounts of dopamine receptor proteins in the forebrain are further evidence of a link between dopamine and schizophrenia.*

Norepinephrine is a neurotransmitter that is related to dopamine and to the hormone epinephrine (adrenalin). Norepinephrine is an important neurotransmitter in the brain and in the autonomic nervous system, as described in chapter 11 (see module 11.3). The amino acid **glutamic acid** is the major excitatory neurotransmitter in the brain, producing depolarizations (EPSPs). The amino acid **glycine,** and an amino acid derivative called **GABA** (for gamma-aminobutyric acid), are inhibitory neurotransmitters—they produce hyperpolarizations (IPSPs). GABA is, in fact, the most prevalent neurotransmitter in the brain. A deficiency of those neurons that use GABA is responsible for the uncontrolled movements seen in people with *Huntington's chorea.*

There are also a number of polypeptides that different neurons use as neurotransmitters. Among these are several molecules whose effects are similar to those of opium or morphine; hence, they are called *opioids.* The polypeptide **β-endorphin** is an opioid neurotransmitter that may block the transmission of pain. The opioid neurotransmitters may also promote a type of psychic reward system; for example, some people have speculated that they may be responsible for the "jogger's high."

Clinical Considerations

Diseases of the Myelin Sheath

Multiple sclerosis (*sklĕ-ro'sis*), or **MS,** is a disease that progressively destroys the myelin sheaths of neurons in multiple areas of the CNS. It begins slowly, usually in young adulthood, and is characterized by cycles in which symptoms worsen or become milder. Initially, lesions form on the myelin sheaths and soon develop into hardened *scleroses,* or scars (hence the name). Destruction of the myelin sheaths prohibits the normal conduction of impulses, resulting in a progressive loss of functions. Because myelin degeneration is so widely distributed, MS has a greater variety of symptoms than any other neurological disease. This characteristic, coupled with remissions, makes diagnosis of MS difficult, and not infrequently the patient may at first be considered neurotic.

As the disease progresses, the symptoms may include double vision (diplopia), spots in the visual field, blindness, tremor, numbness of the appendages, and locomotor difficulty. Eventually, the patient becomes bedridden, and death may occur anytime from 7 to 30 years after the first symptoms appear.

In **Tay–Sachs disease,** the myelin sheaths are destroyed by the excessive accumulation of one of the lipid components of the myelin. This occurs as a result of an enzyme deficit due to the inheritance of genes that are carried by the parents in a recessive state. Tay–Sachs disease, which is inherited primarily by individuals of Eastern European Jewish descent, appears before the infant is a year old. It causes blindness, loss of mental and motor ability, and ultimately death by the age of 3. Potential parents can tell if they are carriers for this condition by means of a blood test for the particular enzyme.

Problems of Neuromuscular Transmission

The muscle weakness characteristic of **myasthenia** (*mi"as-the'ne-ă*) **gravis** is due to the blockage of ACh receptors by antibodies secreted by the immune system. Paralysis in people who eat shellfish poisoned with *saxitoxin,* which is produced by unicellular organisms that cause the red tides, results from inhibition of the chemically regulated Na^+ gates at the neuromuscular junction. A similar inhibition and paralysis is produced by *tetrodotoxin,* a poison found in the puffer fish.

Alzheimer's Disease

Alzheimer's (*alts-hi-merz*) **disease,** also called *senile dementia-Alzheimer type* (*SDAT*), is characterized by progressive mental deterioration. The cause of Alzheimer's is not known, but there is evidence that it is associated with a loss of neurons that use acetylcholine as a neurotransmitter. These axons terminate in the hippocampus and cerebral cortex of the brain, which are areas concerned with memory storage.

This disease is associated with a deficiency of the enzyme responsible for producing acetylcholine from acetyl coenzyme A and choline. Treatment for Alzheimer's has been reported to be fairly effective with drugs that inhibit the activity of acetylcholinesterase; by inhibiting this enzyme, the breakdown of ACh in synapses is reduced, so that the action of ACh is improved.

Autopsies of people who have died of Alzheimer's disease reveal "neuritic plaques," which are composed of degenerating axons and deposits of amyloid protein. Similar plaques are seen in the brains of people with Down syndrome, a genetic disease caused by an extra chromosome number 21. Recently, scientists have discovered that the gene that codes for the amyloid protein in people with Alzheimer's disease is located in chromosome number 21, suggesting that Alzheimer's may be caused by genetic defects located on this chromosome.

---

multiple sclerosis: L. *multiplus,* many parts; Gk. *skleros,* hardened

Tay–Sachs disease: from Warren Tay, English physician, 1843–1927, and Bernard Sachs, American neurologist, 1858–1944

myasthenia: Gk. *myos,* muscle; *asthenia,* weakness

Alzheimer's disease: from Alois Alzheimer, German neurologist, 1864–1915

SUMMARY

9.1 Introduction to the Nervous System

1. The primary functions of the nervous system are to monitor, integrate, and respond to changes in the internal and external environments.
2. The nervous system is divided into the CNS, which includes the brain and spinal cord, and the PNS, which includes the nerves and ganglia.
3. The ANS has two principal subdivisions—the sympathetic and parasympathetic.

9.2 Neurons and Nerves

1. Neurons contain dendrites, a cell body, and an axon.
2. Neurons may be sensory, motor, or association neurons.
3. Neurons may be classified as pseudounipolar (sensory), bipolar, or multipolar.

9.3 Neuroglia

1. Axons in the PNS are surrounded by a sheath of Schwann. Some of them are also surrounded by a myelin sheath, which is composed of successive wrappings of the Schwann cell membrane.
2. Oligodendrocytes form myelin sheaths in the CNS.
3. Myelin sheaths form the white matter of the CNS. The gray matter contains the unmyelinated cell bodies and dendrites.

9.4 Action Potentials

1. The resting membrane potential of an axon is about –65 mV.
2. Depolarization causes voltage-regulated Na^+ gates to open, allowing Na^+ to diffuse into the neuron.
3. Depolarization next causes K^+ gates to open, allowing K^+ to diffuse out of the neuron.
4. The diffusion of Na^+ into the axon causes depolarization.
5. The diffusion of K^+ out of the axon causes repolarization, and completes the action potential.
6. Action potentials are all or none in amplitude; any depolarization stimulus above a threshold value produces an action potential of the same amplitude.
7. Stronger stimuli produce a greater frequency of action potentials but do not increase the amplitude of each action potential.

9.5 Conduction of Action Potentials

1. Each action potential serves as the depolarization stimulus for the production of an action potential at an adjacent region of the axon membrane. In an unmyelinated axon, this occurs only fractions of a micrometer apart.
2. Action potentials are produced only at the nodes of Ranvier in myelinated axons. The nerve impulse jumps from node to node, in what is called saltatory conduction.
3. Saltatory conduction of action potentials in a myelinated axon is faster than conduction in an unmyelinated axon.

9.6 Synaptic Transmission

1. Most synapses involve the release of a chemical neurotransmitter across a synaptic cleft.
2. Acetylcholine (ACh) was the first neurotransmitter to be discovered. It is stored in synaptic vesicles and released from the axon terminals by exocytosis in response to action potentials.
3. ACh opens chemically regulated gates in the postsynaptic membrane. This allows the diffusion of Na^+ into the cell to produce depolarization.
4. Depolarization produced in response to a chemical neurotransmitter is called an excitatory postsynaptic potential, or EPSP.
5. EPSPs serve as the depolarization stimulus for the production of action potentials.
6. ACh is inactivated by acetylcholinesterase (AChE) at the synapse.
7. Among the many other neurotransmitters that have been identified are dopamine, norepinephrine, and glutamic acid.

Review Activities

Objective Questions

1. A collection of neuron cell bodies located outside the CNS is called (*module 9.1*)
   a. a tract.
   b. a nerve.
   c. a nucleus.
   d. a ganglion.

2. Which of the following neuron types are pseudounipolar? (*module 9.2*)
   a. sensory neurons
   b. somatic motor neurons
   c. neurons in the retina
   d. autonomic motor neurons

3. The neuroglial cells that form myelin sheaths in the PNS are (*module 9.3*)
   a. oligodendrocytes.
   b. ganglionic cells.
   c. Schwann cells.
   d. astrocytes.
   e. microglia.

4. Depolarization of an axon is produced by (*module 9.4*)
   a. inward diffusion of Na^+.
   b. active extrusion of K^+.
   c. outward diffusion of K^+.
   d. inward active transport of Na^+.

5. Repolarization of an axon during an action potential is produced by (*module 9.4*)
   a. inward diffusion of Na^+.
   b. active extrusion of K^+.
   c. outward diffusion of K^+.
   d. inward active transport of Na^+.

6. As the strength of a depolarizing stimulus to an axon is increased, (*module 9.4*)
   a. the amplitude of action potentials increases.
   b. the duration of action potentials increases.

c. the speed with which action potentials are conducted increases.
d. the frequency with which action potentials are produced increases.

7. The conduction of action potentials in a myelinated nerve fiber is (*module 9.5*)
a. saltatory.
b. without decrement.
c. faster than in an unmyelinated fiber.
d. all of the above.

8. Which of the following is *not* a characteristic of action potentials? (*module 9.5*)
a. They are produced by voltage-regulated gates.
b. They are conducted without decrement.
c. Na^+ and K^+ gates open at the same time.
d. The membrane potential reverses polarity during depolarization.

9. Which of the following is *not* a characteristic of EPSPs? (*module 9.6*)
a. They are all or none in amplitude.
b. They decrease in amplitude with distance.
c. They are produced in dendrites and cell bodies.
d. They are graded in amplitude.

10. A drug that inactivates acetylcholinesterase (*module 9.6*)
a. inhibits the release of ACh from presynaptic endings.
b. inhibits the attachment of ACh to its receptor protein.
c. increases the ability of ACh to stimulate muscle contraction.
d. has all of the above effects.

Essay Questions

1. Discuss the organization and general functions of the nervous system. (*module 9.1*)
2. List the parts of a neuron and describe their functions. (*module 9.2*)
3. Explain the functional and structural classifications of neurons. (*module 9.2*)
4. Define *ganglia, nuclei,* and *nerve.* (*module 9.2*)
5. List the different types of neuroglial cells and describe their functions. (*module 9.3*)
6. Explain how a myelin sheath is formed in the PNS. How does this differ from myelin formation in the CNS? (*module 9.3*)
7. Define *depolarization* and *hyperpolarization.* Compare the characteristics of action potentials with those of synaptic potentials. (*modules 9.4 and 9.6*)
8. Explain how voltage-regulated gates produce an all-or-none action potential. (*module 9.4*)
9. Explain why conduction in a myelinated axon is faster than in an unmyelinated axon. (*module 9.5*)
10. Trace the course of events between the production of an EPSP and the generation of action potentials at the axon hillock. (*module 9.6*)

Labeling Exercise

Label the structures indicated on the figure to the right.

1. ______________________________
2. ______________________________
3. ______________________________
4. ______________________________
5. ______________________________
6. ______________________________

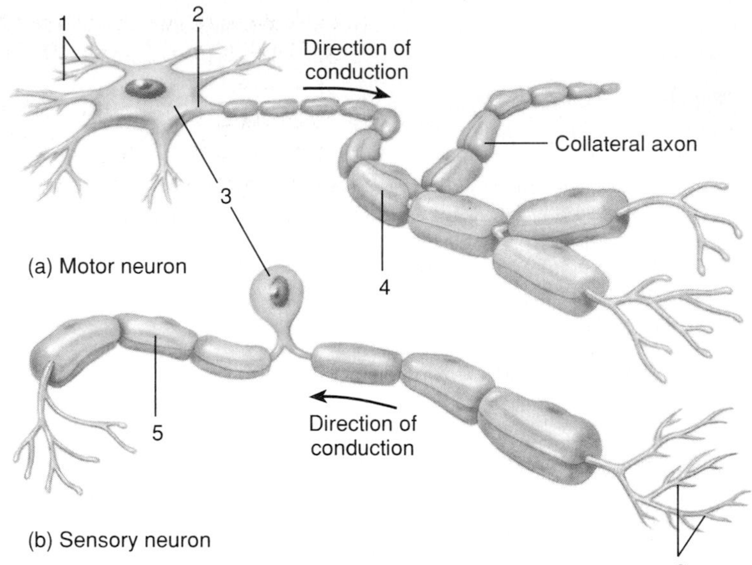

(a) Motor neuron

(b) Sensory neuron

Critical Thinking Questions

1. Why are pseudounipolar neurons not called "unipolar" neurons? What advantage might a sensory neuron gain by being pseudounipolar instead of bipolar or multipolar?
2. Can all peripheral neurons be said to have a sheath of Schwann—even those that do not have a myelin sheath? What advantage does a sheath of Schwann bestow that is lacking in the CNS?
3. Why is it important, in the production of an action potential, that the voltage-regulated Na^+ and K^+ gates open at different times? What might happen if they should both open at the same time?
4. Considering the physiological function of a myelin sheath, how would diseases of the myelin, such as multiple sclerosis or Tay–Sach's disease, affect nerve function?
5. People with the disease myasthenia gravis produce antibodies that block their ACh receptors. How do you suppose this affects their ability to contract muscles? How can such people be helped by drugs that block acetylcholinesterase?

CHAPTER 10

"[T]here is actually only one highly integrated nervous system that serves as the body's master control center and communications network." *page 248*

Nervous System

Chapter Outline

Terms to Remember

cerebrum (*ser'ĕ-brum*), 250
cerebral hemisphere, 250
gyrus, 252
sulcus (*sul'kus*), 252
electroencephalogram, 255
basal nuclei, 256
diencephalon (*di"en-sef'ă-lon*), 258
thalamus, 258
hypothalamus, 258
pituitary gland, 259
limbic system, 260
brain stem, 260
midbrain, 261
pons, 261
medulla oblongata (*mĕ-dul'ă ob"long-gă'tă*), 261
reticular formation, 262
cerebellum (*ser"ĕ-bel'um*), 262
meninges (*mĕ-nin'jēz*), 263
cerebrospinal fluid, 265
blood-brain barrier, 267
cranial nerve, 269
spinal nerve, 275
nerve plexus, 276
reflex arc, 277

Learning Objectives

After studying this chapter, you should be able to

- distinguish between the subdivisions of the nervous system.
- describe the structure of the cerebrum and list the functions of the cerebral lobes.
- discuss the clinical importance of the electroencephalogram and describe the various EEG patterns.
- describe the locations of the thalamus, hypothalamus, and pituitary gland and list the functions of these structures.
- locate the structures of the brain stem and describe their structure and functions.
- describe the locations of the limbic system and reticular formation and explain the role of each of these functional systems.
- locate the cerebellum and describe its structure and functions.
- explain how the meninges protect the CNS.
- describe the locations of ventricles of the brain and discuss the formation, function, and flow of cerebrospinal fluid.
- list the 12 pairs of cranial nerves and describe the body region and structures innervated by each.
- describe the structure of the spinal cord.
- describe the general distribution of a spinal nerve and discuss the location and function of each of the four nerve plexuses.
- define *reflex arc* and list the five components of this nerve pathway.

10.1 General Features of the Nervous System

Each of the subdivisions of the nervous system has distinguishing structural and functional features. Although we speak of the subdivisions as separate systems, there is actually only one highly integrated nervous system that serves as the body's master control center and communications network.

The nervous system's amazing abilities reside not in the individual neuron, but in the high level of organization of the billions of neurons that compose the system as a whole. Scientists are still striving to answer questions about the organization and functioning of this system; indeed, *neurology*—the study of the nervous system—has been referred to as the last frontier of anatomy and physiology.

As described in the opening module of the previous chapter (see fig. 9.1), the nervous system can be divided into the **central nervous system (CNS),** and the **peripheral nervous system (PNS).** The CNS consists of the *brain* and *spinal cord,* and the PNS consists of the nerves that arise from the brain—the *cranial nerves*—and from the spinal cord—the *spinal nerves* (fig. 10.1). All parts of the body are linked to the CNS through the peripheral nerves.

The **autonomic nervous system (ANS),** the focus of chapter 11, is a functional subdivision of the entire nervous system. The ANS comprises two subdivisions—the *sympathetic division,* which prepares the body for emergency situations, and the *parasympathetic division,* which restores and conserves body energy during periods of tranquility.

The nervous system is one of the first recognizable features of a human embryo. At about 3 weeks following conception, the **neural tube,** which is a hollow core of ectodermal cells along the back of the embryo, is formed. The front part of the neural tube develops into the brain, and the back part develops into the spinal cord. By the middle of the fourth week following conception, three distinct swellings are evident in the front part of the neural tube: the **prosencephalon** (*pros"en-sef'ă-lon*) (forebrain), the

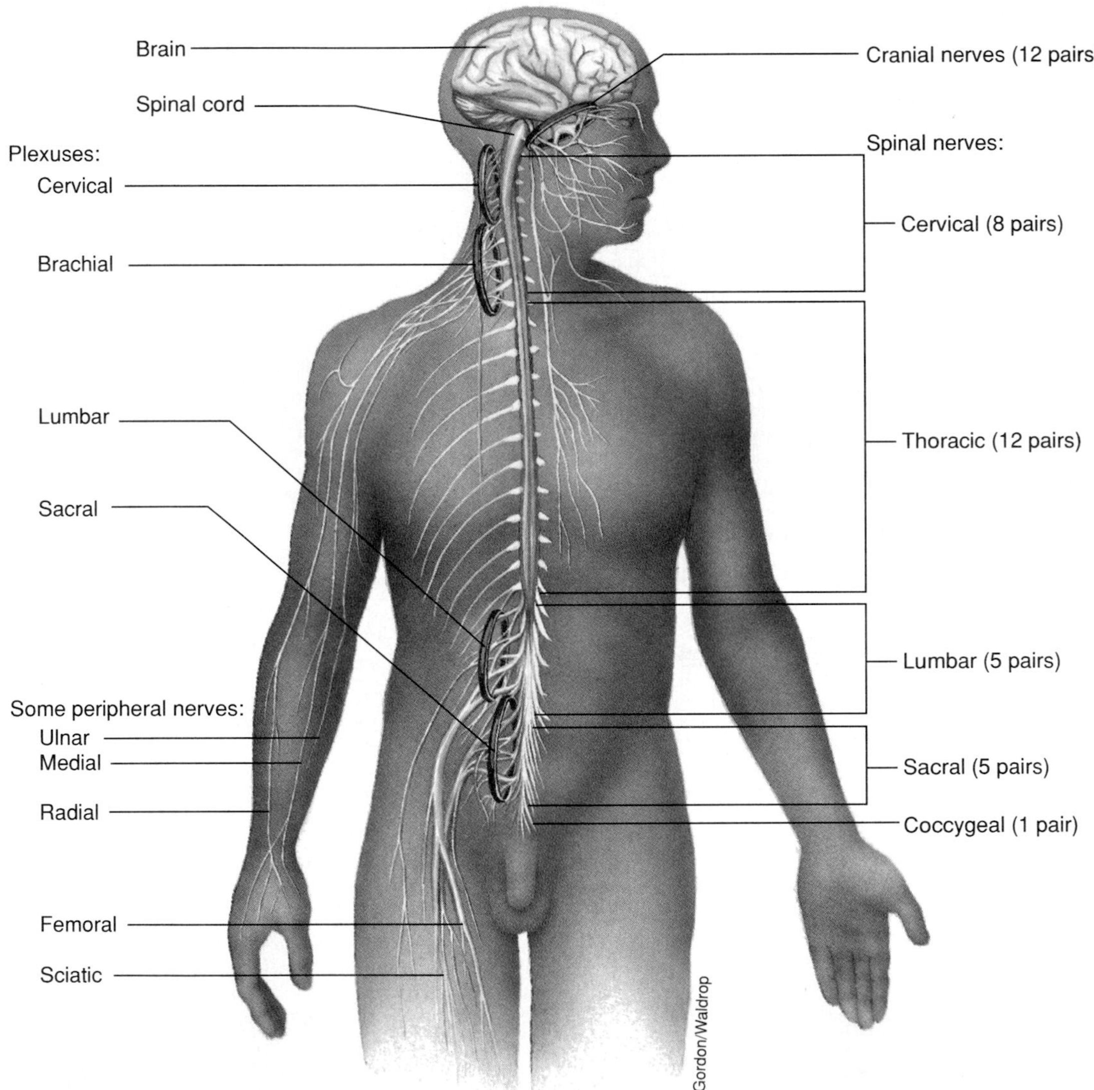

FIGURE 10.1
The nervous system. The central nervous system (CNS) consists of the brain and spinal cord. The peripheral nervous system (PNS) consists of cranial nerves and spinal nerves. Also part of the PNS are the plexuses and additional nerves that arise from the cranial and spinal nerves. The autonomic nervous system (ANS) is a functional division of the nervous system.

mesencephalon (midbrain), and the **rhombencephalon** (hindbrain) (fig. 10.2). Further development during the fifth week results in the formation of five regions: The **telencephalon** and the **diencephalon** (*di″en-sef′ă-lon*) form from the forebrain; the **mesencephalon** remains unchanged; and the **metencephalon** and **myelencephalon** form from the hindbrain. Specific structures develop in each of these regions, some of which are listed in table 10.1.

Inside Information—The brain is approximately one-fourth its final size at birth; the rest of the body is one-twentieth.

Characteristics of the Central Nervous System

The entire delicate CNS is protected by a bony encasement—the cranium surrounding the brain and the vertebral column surrounding the spinal cord. The **meninges** (*mĕ-nin′jēz*) form a protective membranous covering between the bone and the soft tissue of the CNS. The CNS is bathed in **cerebrospinal** (*ser″ĕ-bro-spi′nal*) **fluid** that circulates within the hollow spaces of the CNS.

The CNS is composed of **gray** and **white matter.** Gray matter consists of either nerve cell bodies and dendrites or of bundles of unmyelinated axons and neuroglia (see

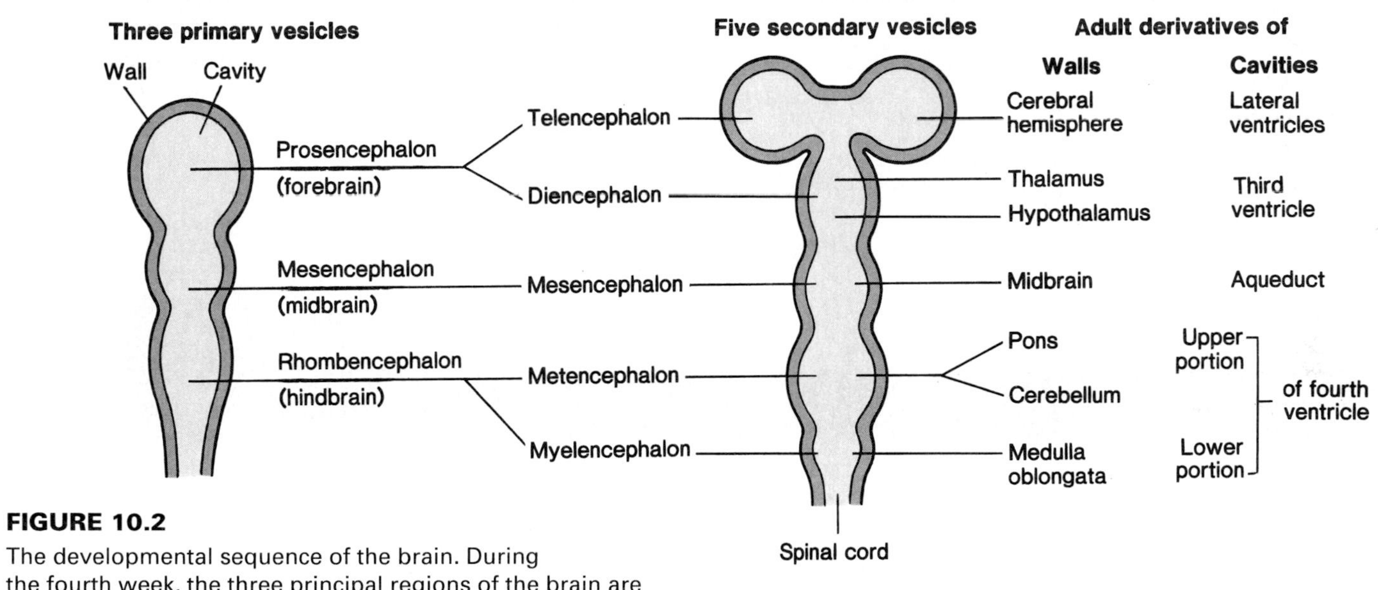

FIGURE 10.2
The developmental sequence of the brain. During the fourth week, the three principal regions of the brain are formed. During the fifth week, a five-regioned brain develops and specific structures begin to form.

TABLE 10.1 Derivation and Functions of the Major Brain Structures

	Region	Structure	Function
Prosencephalon (forebrain)	Telencephalon	Cerebrum	Control of most sensory and motor activities; reasoning, memory, intelligence, etc.; instinctual and limbic functions
	Diencephalon	Thalamus	Relay center; all impulses (except olfactory) going into cerebrum synapse here; some sensory interpretation; initial autonomic response to pain
		Hypothalamus	Regulation of food and water intake, body temperature, heartbeat, etc.; control of secretory activity in anterior pituitary; instinctual and limbic functions
		Pituitary gland	Regulation of other endocrine glands
Mesencephalon (midbrain)	Mesencephalon	Superior colliculi	Visual reflexes (hand-eye coordination)
		Inferior colliculi	Auditory reflexes
		Cerebral peduncles	Reflex coordination; contain many motor fibers
Rhombencephalon (hindbrain)	Metencephalon	Cerebellum	Balance and motor coordination
	Myelencephalon	Pons	Relay center; contains nuclei (pontine nuclei)
		Medulla oblongata	Relay center; contains many nuclei; visceral autonomic center (e.g., respiration, heart rate, vasoconstriction)

chapter 9, module 9.3). The gray matter of the brain exists as the outer convoluted *cortex layer* of the cerebrum and cerebellum. In addition, specialized gray matter clusters of cell bodies of neurons called *nuclei* are found deep within the white matter. White matter forms the tracts within the CNS and consists of clusters of myelinated axons and associated neuroglia.

The brain of an adult weighs nearly 1.5 kg (3–3.5 lb) and is composed of an estimated 100 billion neurons. Neurons communicate with one another by means of innumerable synapses between the axons and dendrites within the brain. Neurotransmission within the brain is regulated by numerous neurotransmitter chemicals (see module 10.7) found in specific brain regions and tracts.

Characteristics of the Peripheral Nervous System

The PNS is that portion of the nervous system outside the central nervous system (see fig. 10.1). Impulses to and from the brain and spinal cord are conveyed by the PNS. Sensory receptors within the sensory organs, neurons, nerves, ganglia, and plexuses (see chapter 9, table 9.1) are all part of the PNS, which serves virtually every part of the body. The sensory receptors are discussed in chapter 12.

The nerves of the PNS are classified as **cranial nerves** or **spinal nerves,** depending on whether they arise from the brain or the spinal cord. The terms *sensory nerve, motor nerve,* and *mixed nerve* relate to the direction in which the nerve impulses are being conducted. **Sensory nerves** consist of sensory (afferent) neurons that convey impulses toward the CNS. **Motor nerves** consist of motor (efferent) neurons that convey impulses away from the CNS. **Mixed nerves** are composed of both sensory and motor neurons, and therefore convey impulses in both directions. The reflexes considered in this chapter (see module 10.11) involve sensory and motor nerves of the PNS and specific portions of the CNS.

10.2 Cerebrum

The cerebrum, consisting of five paired lobes within two convoluted hemispheres, is concerned with higher brain functions, such as the perception of sensory impulses, the instigation of voluntary movement, the storage of memory, thought processes, and reasoning ability. The cerebrum is also concerned with instinctual and limbic (emotional) functions.

Structure of the Cerebrum

The cerebrum (*ser'ĕ-brum*), located in the region of the telencephalon, is the largest and most obvious portion of the brain (fig. 10.3). It accounts for about 80% of the mass of the brain and is responsible for the higher mental functions, including memory and reason. The cerebrum consists of the **right** and **left hemispheres,** which are incompletely separated by a **longitudinal cerebral fissure** (fig. 10.4). Portions of the two hemispheres are connected internally by the **corpus callosum** (*kă-lo'sum*), a large tract of white matter (see fig. 10.3*b*). Each cerebral hemisphere contains a central cavity called the **lateral ventricle,** which is filled with cerebrospinal fluid.

cerebrum: L. *cerebrum,* brain

The two cerebral hemispheres carry out different functions. In most people, the left hemisphere controls analytical and verbal skills, such as reading, writing, and mathematics. The right hemisphere is the source of spatial and artistic kinds of intelligence. The corpus callosum permits a sharing of learning and memory between the two hemispheres. Severing the corpus callosum is a radical treatment to control severe epileptic seizures. Although this surgery has proven successful, it results in the cerebral hemispheres functioning as separate structures, each with its own information, competing for control. A more recent and more effective technique of controlling epileptic seizures is a precise laser treatment of the corpus callosum.

The cerebrum consists of two layers. The surface layer, referred to as the **cerebral cortex,** is composed of gray matter that is 2 to 4 mm (0.08 to 0.16 in.) thick (fig. 10.5). Beneath the cerebral cortex is the thick **white matter** of the cerebrum, which constitutes the second layer. The cerebral cortex is characterized by numerous folds and grooves called *convolutions.* Convolutions form during early fetal development, when brain size increases rapidly and the

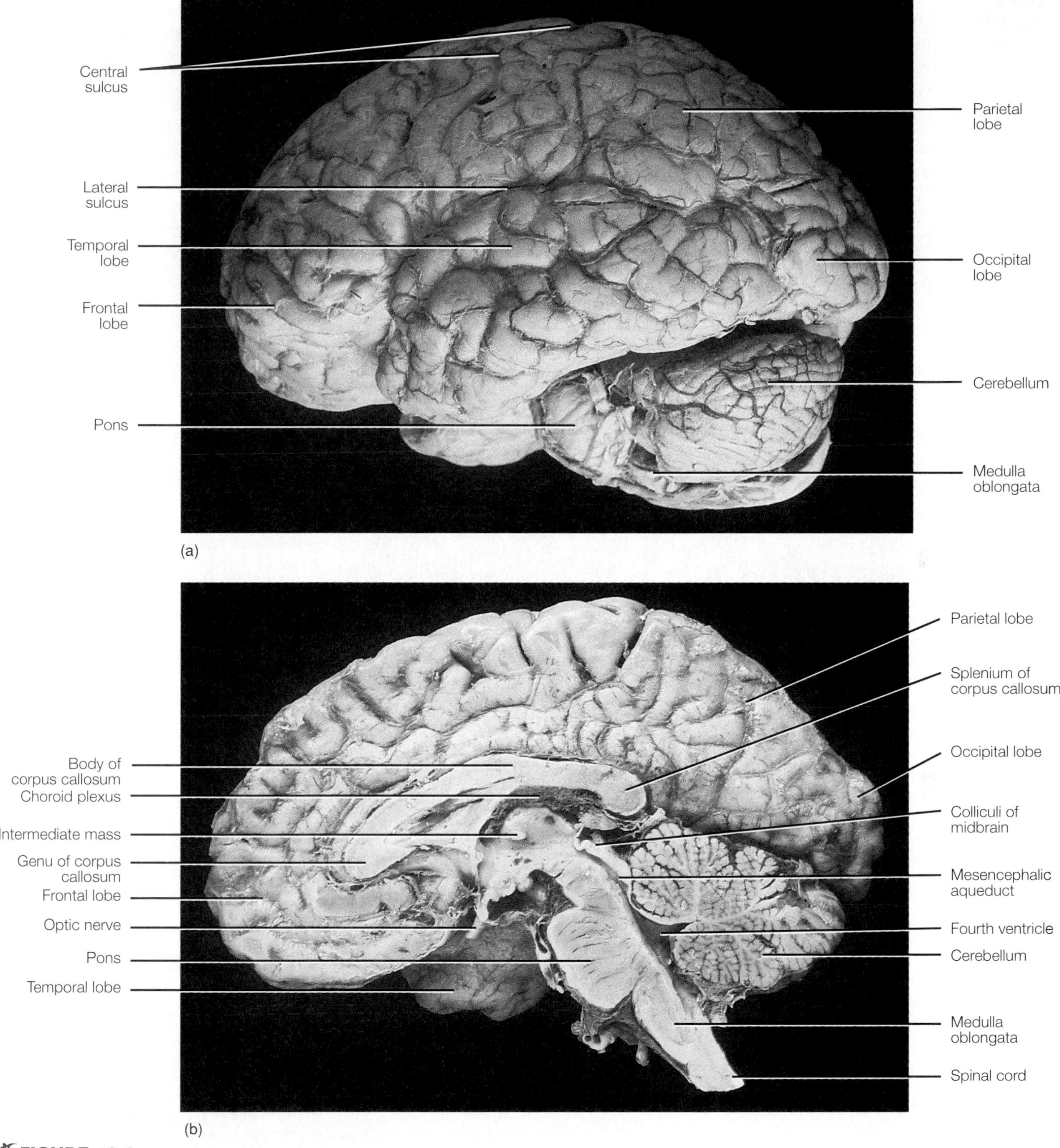

FIGURE 10.3

The brain. (*a*) A lateral view and (*b*) a sagittal view.

cerebral cortex enlarges out of proportion to the underlying white matter. The elevated folds of the convolutions are the **cerebral gyri** (singular, *gyrus*), and the depressed grooves are the **cerebral sulci** (*sul'si*—singular, *sulcus*) (see fig. 10.4). The convolutions effectively triple the area of the gray matter, which is composed of neuron cell bodies.

Inside Information—The brains of our hominid ancestors, the Neanderthals, were actually larger than our own—a fact that may have been related to their heavy, large bodies. Those portions of the brain most important for abstract thought, however, appear to be more highly developed and larger in modern humans.

Lobes of the Cerebrum

Each cerebral hemisphere is subdivided into lobes by especially deep sulci. Four of these lobes appear on the surface of the cerebrum and are named according to the overlying cranial bones (see fig. 10.4). The reasons for the separate cerebral lobes, as well as two cerebral hemispheres, have to do with specificity of function.

Frontal Lobe

The frontal lobe forms the anterior portion of each cerebral hemisphere (see fig. 10.4). A prominent deep furrow

gyrus: Gk. *gyros*, circle
sulcus: L. *sulcus*, a furrow or ditch

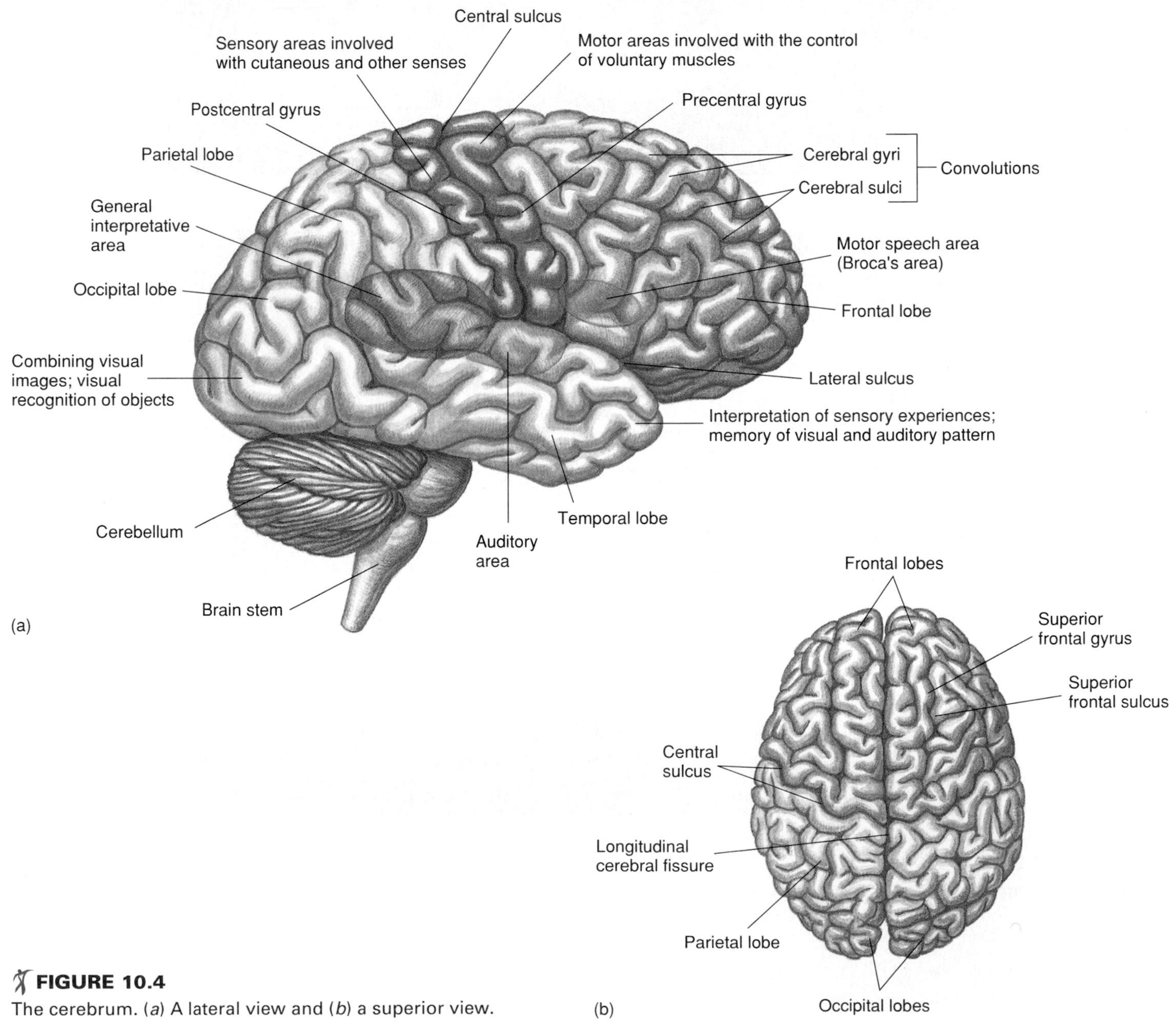

FIGURE 10.4
The cerebrum. (*a*) A lateral view and (*b*) a superior view.

called the **central sulcus** separates the frontal lobe from the parietal lobe. The central sulcus extends at right angles from the longitudinal fissure to the lateral sulcus. The **lateral sulcus** extends laterally from the inferior surface of the cerebrum to separate the frontal and temporal lobes. The **precentral gyrus,** an important motor area, is positioned immediately in front of the central sulcus. Frontal lobe functions include initiating voluntary motor impulses for the movement of skeletal muscles, analyzing sensory experiences, and providing responses relating to personality. The frontal lobes also mediate responses related to memory, emotions, reasoning, judgment, planning, and speaking. The **motor speech area** (Broca's area) is the principal speech center. It is located in the left inferior gyrus of the frontal lobe. Neural activity in the motor speech area causes selected stimulation of motor impulses in motor centers elsewhere in the frontal lobe, which in turn causes coordinated skeletal

Broca's area: From Pierre P. Broca, French neurologist, 1824–80

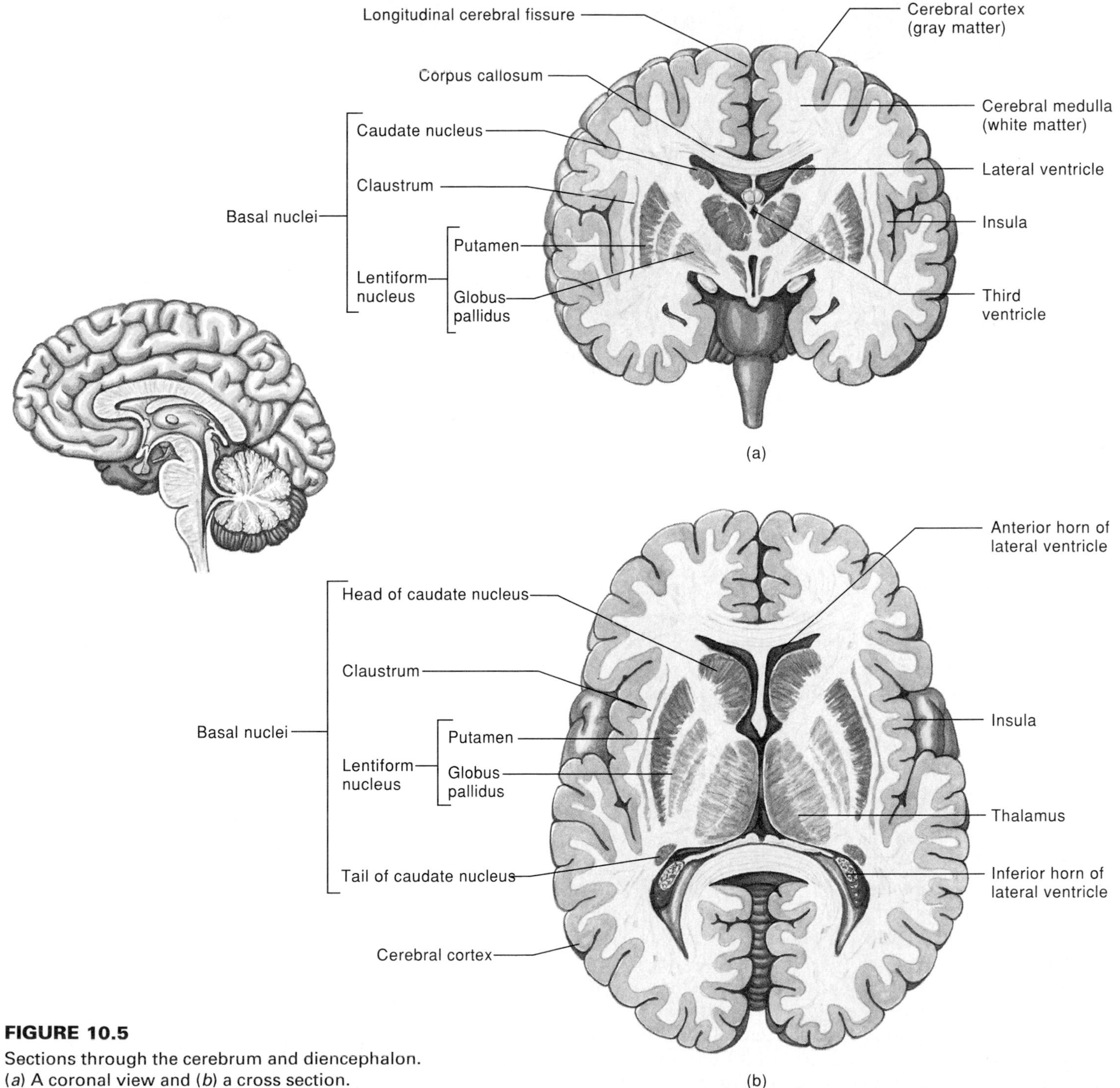

FIGURE 10.5
Sections through the cerebrum and diencephalon. (*a*) A coronal view and (*b*) a cross section.

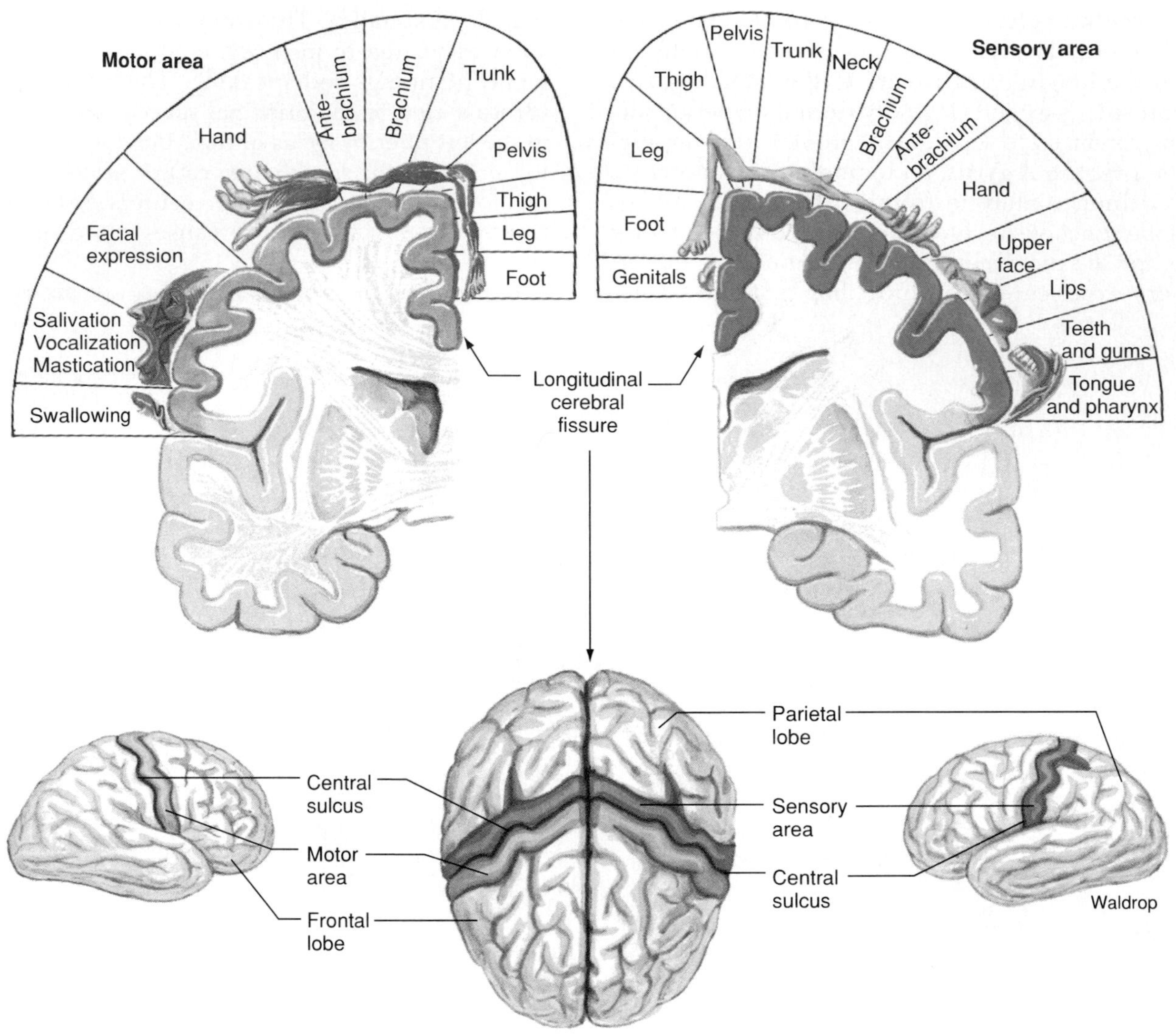

FIGURE 10.6
Motor and sensory areas of the cerebral cortex. Motor areas control skeletal muscles and sensory areas respond to somatesthetic stimuli.

muscle movement in the pharynx and larynx. At the same time, motor impulses are sent to the respiratory muscles to regulate air movement across the vocal cords. The combined muscular stimulation translates thought patterns into speech. Damage to the motor speech area results in absent or defective speech or language comprehension, a nerve defect known as *aphasia.*

Parietal Lobe

The parietal lobe is posterior to the central sulcus of the frontal lobe. An important sensory area called the **postcentral gyrus** (see fig. 10.4) is positioned immediately behind the central sulcus. The postcentral gyrus is designated as a *somatesthetic area* because it responds to stimuli from cutaneous and muscular receptors throughout the body.

The size of the portions of the precentral gyrus responsible for motor movement and the size of the portions of the postcentral gyrus that respond to sensory stimuli do not correspond to the size of the part of the body being served. Rather, they correspond to the number of motor units activated or to the density of the receptors (fig. 10.6). For example, because the hand has many motor units and sensory receptors, larger portions of the precentral and postcentral gyri serve it than serve the thorax, even though the thorax is much larger.

In addition to responding to somatesthetic stimuli, the parietal lobe functions in speech comprehension and in verbal articulation of thoughts and emotions. The parietal lobe also interprets the textures and shapes of objects as they are handled.

Temporal Lobe

The temporal lobe is located below the parietal lobe and the posterior portion of the frontal lobe. It is separated from both by the lateral sulcus (see fig. 10.4). The temporal lobe contains auditory centers that receive sensory

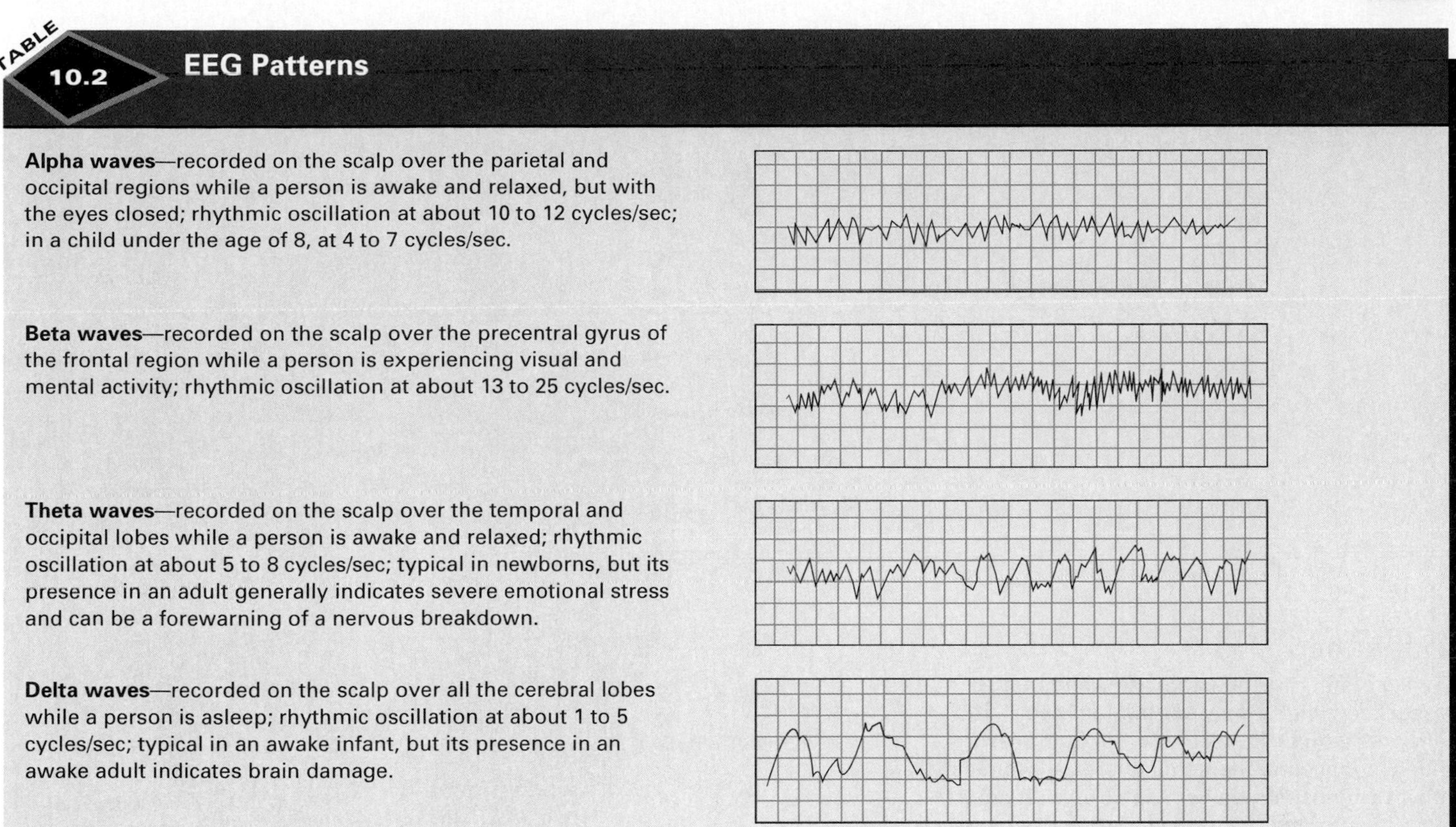

TABLE 10.2 EEG Patterns

Wave type	Description
Alpha waves	recorded on the scalp over the parietal and occipital regions while a person is awake and relaxed, but with the eyes closed; rhythmic oscillation at about 10 to 12 cycles/sec; in a child under the age of 8, at 4 to 7 cycles/sec.
Beta waves	recorded on the scalp over the precentral gyrus of the frontal region while a person is experiencing visual and mental activity; rhythmic oscillation at about 13 to 25 cycles/sec.
Theta waves	recorded on the scalp over the temporal and occipital lobes while a person is awake and relaxed; rhythmic oscillation at about 5 to 8 cycles/sec; typical in newborns, but its presence in an adult generally indicates severe emotional stress and can be a forewarning of a nervous breakdown.
Delta waves	recorded on the scalp over all the cerebral lobes while a person is asleep; rhythmic oscillation at about 1 to 5 cycles/sec; typical in an awake infant, but its presence in an awake adult indicates brain damage.

neurons from the cochlea of the ear. This lobe also interprets some sensory experiences and stores memories of both auditory and visual experiences.

Occipital Lobe

The occipital lobe forms the posterior portion of the cerebrum. As indicated in figure 10.4, it is not distinctly separated from the temporal and parietal lobes. The principal functions of the occipital lobe concern vision. The occipital lobe integrates eye movements by directing and focusing the eye. It is also responsible for visual association—correlating visual images with previous visual experiences and other sensory stimuli.

Insula

The insula is a deep portion of the cerebrum that cannot be viewed on the surface (see fig. 10.5). Sometimes called the fifth lobe, it is deep to the lateral sulcus and is covered by portions of the frontal, parietal, and temporal lobes. The insula integrates other cerebral activities and may have some function in memory.

insula: L. *insula,* island

Because of the size and position of the cerebrum, parts of it frequently suffer brain trauma. A concussion to the brain may cause a temporary or permanent impairment of cerebral functions. Much of what is known about cerebral function comes from observing body dysfunctions when specific regions of the cerebrum are traumatized.

Brain Waves

Neurons within the cerebral cortex continuously generate electrical activity. This activity can be recorded by electrodes attached to precise locations on the scalp, producing an **electroencephalogram (EEG).** An EEG pattern, commonly called *brain waves,* is the collective expression of millions of action potentials from neurons.

Brain waves are first emitted from a developing brain during early fetal development and continue throughout a person's life. The cessation of brain-wave patterns (a "flat EEG") may be a decisive factor in the legal determination of death.

Certain distinct EEG patterns signify healthy mental functioning. Deviations from these patterns are of clinical significance in diagnosing trauma, mental depression, hematomas, tumors, infections, and epileptic lesions. Normally, there are four kinds of EEG patterns, as described in table 10.2.

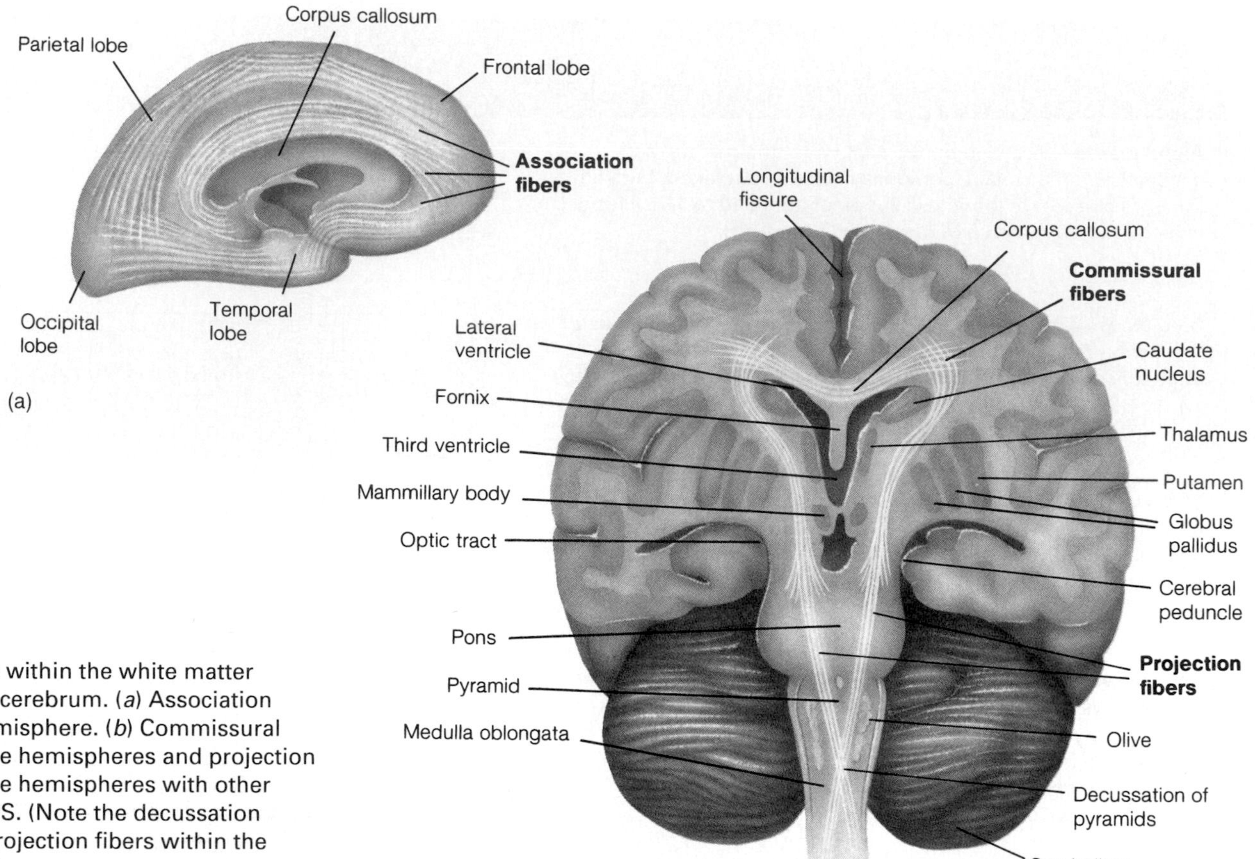

FIGURE 10.7

Types of fiber tracts within the white matter associated with the cerebrum. (*a*) Association fibers of a given hemisphere. (*b*) Commissural fibers connecting the hemispheres and projection fibers connecting the hemispheres with other structures of the CNS. (Note the decussation [crossing over] of projection fibers within the medulla oblongata.)

White Matter of the Cerebrum

The thick white matter of the cerebrum is deep to the cerebral cortex (see fig. 10.5) and consists of dendrites, myelinated axons, and associated neuroglia. These fibers form the billions of connections within the brain by which information is transmitted to the appropriate places in the form of electrical impulses. The three types of fiber tracts within the white matter are named according to location and the direction in which they conduct impulses (fig. 10.7).

1. **Association fibers** are confined to a given cerebral hemisphere and conduct impulses between neurons within that hemisphere.
2. **Commissural** (*kă-mĭ-shur'al*) **fibers** connect the neurons and gyri of one hemisphere with those of the other. The corpus callosum and anterior commissure (fig. 10.8) are composed of commissural fibers.
3. **Projection fibers** form the ascending and descending tracts that transmit impulses from the cerebrum to other parts of the brain and spinal cord, and from the spinal cord and other parts of the brain to the cerebrum.

lentiform: L. *lentis,* elongated
putamen: L. *putare,* to cut, prune
globus pallidus: L. *globus,* sphere; *pallidus,* pale

Basal Nuclei

The basal nuclei are specialized paired masses of gray matter located deep within the white matter of the cerebrum (fig. 10.9). The most prominent of the basal nuclei is the **corpus striatum,** so named because of its striped appearance. The corpus striatum is composed of several masses of nuclei. The **caudate nucleus** is the upper mass. A thick band of white matter lies between the caudate nucleus and the two masses underneath, collectively called the **lentiform nucleus.** The lentiform nucleus consists of a lateral portion, called the **putamen** (*pyoo-ta'men*), and a medial portion, called the **globus pallidus.** The **claustrum** is another portion of the basal nuclei. It is a thin layer of gray matter just deep to the cerebral cortex of the insula.

The basal nuclei are associated with other structures of the brain, particularly within the mesencephalon. The caudate nucleus and the putamen of the lentiform nucleus control unconscious contractions of certain skeletal muscles, such as those of the upper limbs involved in involuntary arm movements during walking. The globus pallidus regulates the muscle tone necessary for specific intentional body movements. Disease or trauma to the basal nuclei generally causes a variety of motor movement dysfunctions, including rigidity, tremor, and rapid and aimless movements.

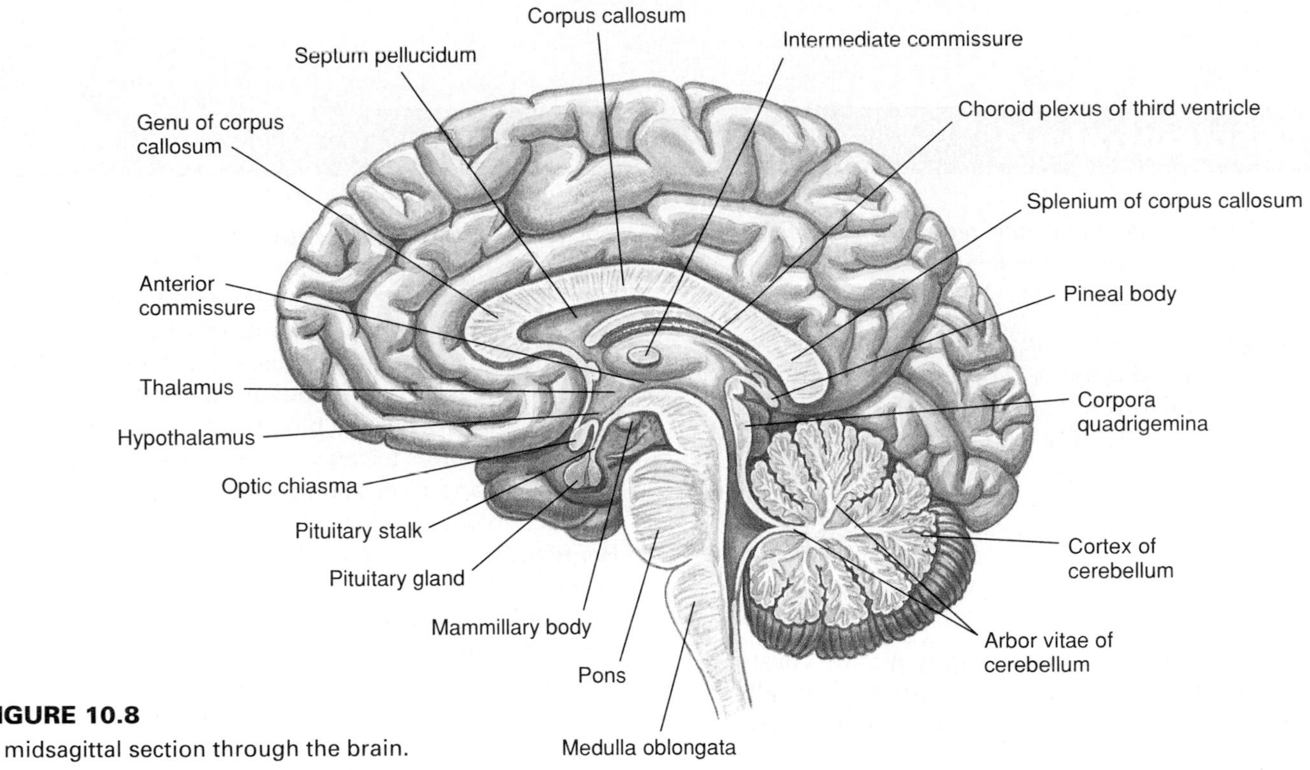

FIGURE 10.8
A midsagittal section through the brain.

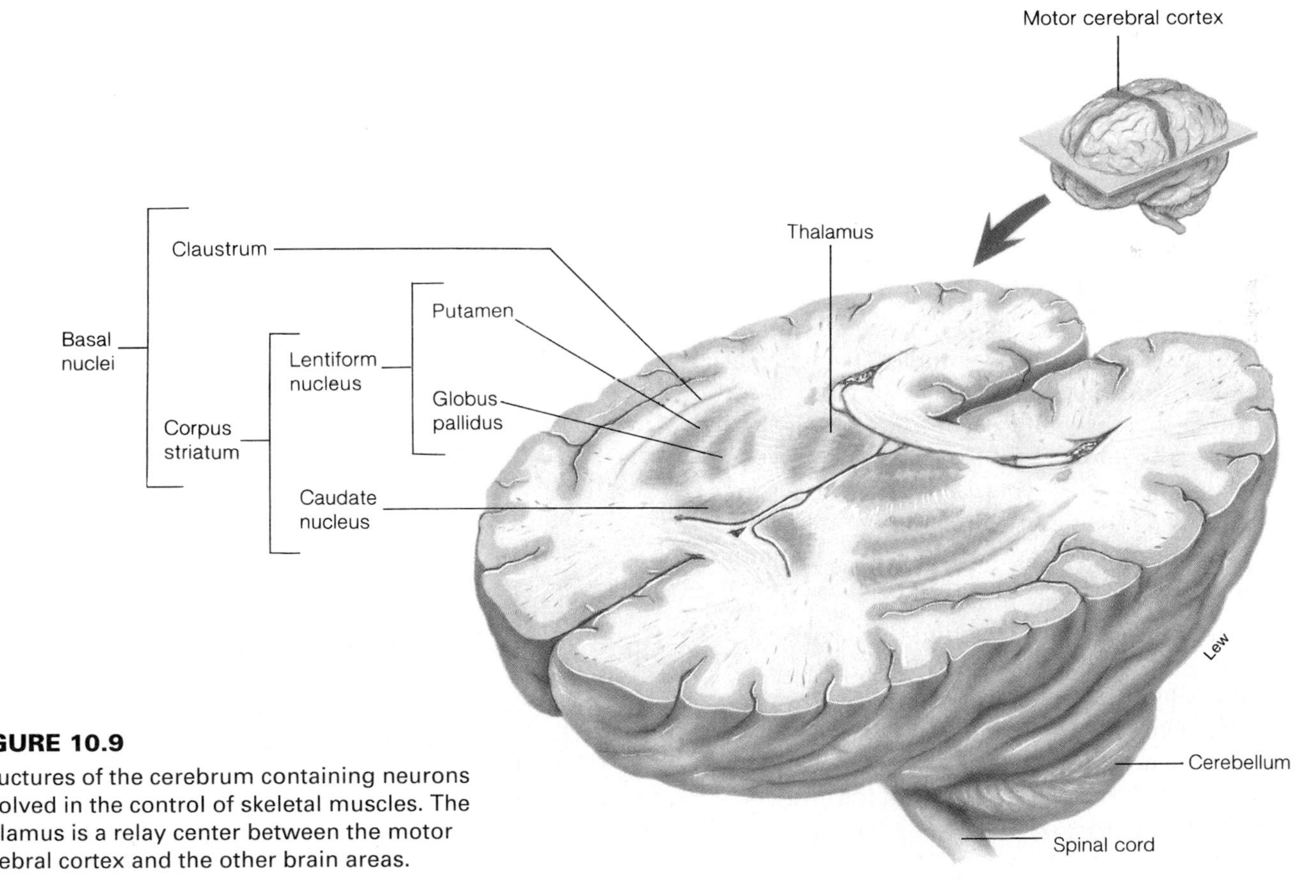

FIGURE 10.9
Structures of the cerebrum containing neurons involved in the control of skeletal muscles. The thalamus is a relay center between the motor cerebral cortex and the other brain areas.

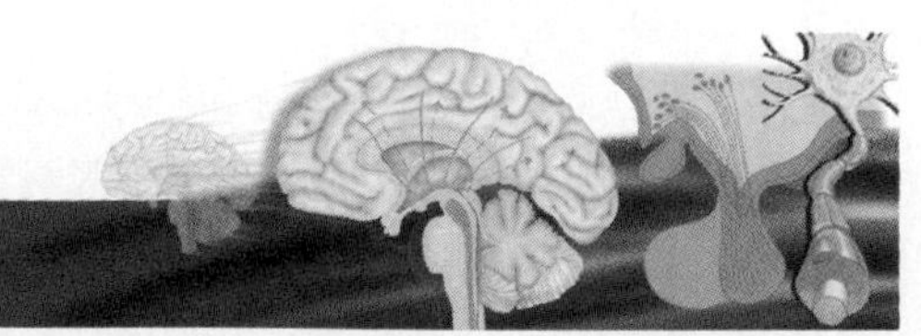

10.3 Diencephalon

The diencephalon is a major autonomic region of the brain that includes such vital structures as the thalamus, hypothalamus, epithalamus, and pituitary gland.

The **diencephalon** (*di"en-sef'ă-lon*) is the second subdivision of the forebrain and is almost completely surrounded by the cerebral hemispheres of the telencephalon. The third ventricle (see fig. 10.7) is a cavity located on the median plane within the diencephalon. The most important structures of the diencephalon are the *thalamus, hypothalamus, epithalamus,* and *pituitary gland.*

Thalamus

The thalamus is a large oval mass of gray matter, constituting nearly four-fifths of the diencephalon. It is actually a paired organ, with each portion positioned immediately below the lateral ventricle of its respective cerebral hemisphere (see figs. 10.7 through 10.9). The principal function of the thalamus is to act as a relay center for all sensory impulses, except smell, to the cerebral cortex. The thalamus also performs some sensory interpretation. The cerebral cortex discriminates pain and other tactile stimuli, but the thalamus responds to general sensory stimuli and provides crude awareness.

Hypothalamus

The hypothalamus is a small portion of the diencephalon located below the thalamus, where it forms the floor and part of the lateral walls of the third ventricle. The hypothalamus consists of several masses of nuclei interconnected with other parts of the nervous system (see figs. 10.8 and 10.11). Despite its small size, it performs numerous functions vital to overall body homeostasis. Most of these relate directly or indirectly to the regulation of autonomic activities of visceral organs. The principal autonomic and limbic (emotional) functions of the hypothalamus are as follows:

1. **Cardiovascular regulation.** Although the heartbeat is automatic, impulses from the hypothalamus cause autonomic acceleration or deceleration of the heart rate. Impulses from the posterior hypothalamus produce a rise in arterial blood pressure and an increase in the heart rate. Impulses from the anterior portion have the opposite effect.
2. **Body-temperature regulation.** Specialized nuclei within the hypothalamus are sensitive to changes in body temperature. If the blood flowing near these nuclei is above normal temperature, the hypothalamus initiates impulses that cause heat loss through sweating and vasodilation of cutaneous vessels. A below-normal blood temperature causes the hypothalamus to relay impulses that result in heat production and retention through shivering and contraction of cutaneous blood vessels.
3. **Regulation of water and electrolyte balance.** Specialized osmoreceptors in the hypothalamus monitor blood concentration. When there is a loss of water from blood, increasing the concentration of solutes, antidiuretic hormone (ADH) is produced by the hypothalamus and released from the posterior pituitary. At the same time, a thirst center within the hypothalamus produces feelings of thirst.
4. **Regulation of hunger and control of gastrointestinal activity.** The feeding center in the hypothalamus monitors blood glucose, fatty acid, and amino acid levels. Low levels of these substances in the blood are partially responsible for a feeling of hunger. When enough food has been eaten, the satiety center in the hypothalamus inhibits the feeding center. The hypothalamus also regulates glandular secretions and the peristaltic movements of the digestive tract.
5. **Regulation of sleeping and wakefulness.** The hypothalamus has both a sleep center and a wakefulness center that function with other parts of the brain to determine a person's level of conscious alertness.
6. **Sexual response.** Specialized sexual center nuclei in the hypothalamus respond to sexual stimulation of the tactile receptors within the genital organs. The experience of orgasm involves neural activity in the sexual center of the hypothalamus.
7. **Emotions.** A number of nuclei within the hypothalamus are associated with specific emotional responses, such as anger, fear, pain, and pleasure.

thalamus: L. *thalamus,* inner room

8. **Control of endocrine functions.** The hypothalamus produces neurosecretory chemicals that stimulate the anterior pituitary to release various hormones, which in turn regulate other endocrine glands (see chapter 13, module 13.3). The hypothalamus also produces the two hormones released by the posterior pituitary.

Epithalamus

The epithalamus is the posterior portion of the diencephalon that includes a thin roof over the third ventricle. The inside lining of the roof consists of a vascular **choroid plexus** where cerebrospinal fluid is produced (see module 10.6). A small cone-shaped mass, the **pineal gland** (also called the **pineal body**), extends outward from the posterior end of the epithalamus (see fig. 10.8). The pineal gland is thought to have a neuroendocrine function.

Pituitary Gland

The rounded, pea-shaped *pituitary gland,* or *hypophysis* (*hi-pof'ĭ-sis*), is positioned on the inferior aspect of the diencephalon. It is attached to the hypothalamus by the **infundibulum** (*in"fun-dib'yŭ-lum*) and is supported by the sella turcica of the sphenoid bone (fig. 10.10). The pituitary, which has an endocrine function (see chapter 13, module 13.2), is structurally and functionally divided into an **anterior pituitary,** also called the **adenohypophysis** (*ad"ĕ-no-hi-pof'ĭ-sis*), and a **posterior pituitary,** also called the **neurohypophysis.** The adenohypophysis is composed of glandular tissue, whereas the neurohypophysis is composed largely of neuroglia and nerve fibers.

pineal: L. *pinea,* pine cone
pituitary: L. *pituita,* phlegm (this gland was originally thought to secrete mucus into the nasal cavity)
infundibulum: L. *infundibulum,* funnel

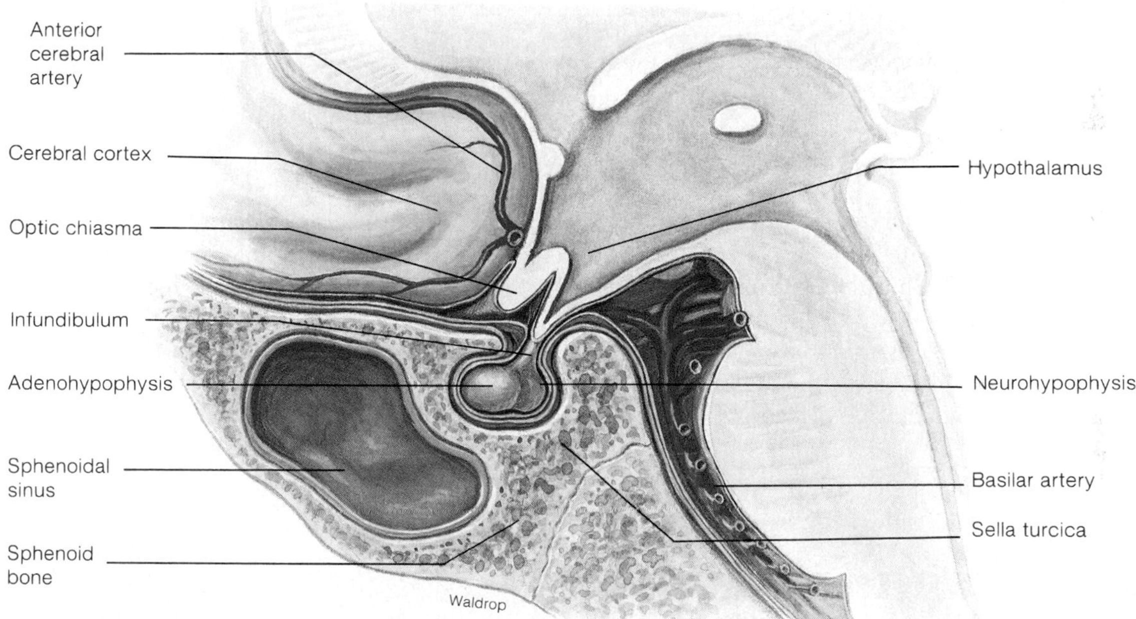

FIGURE 10.10
The pituitary gland is positioned within the sella turcica of the sphenoid bone.

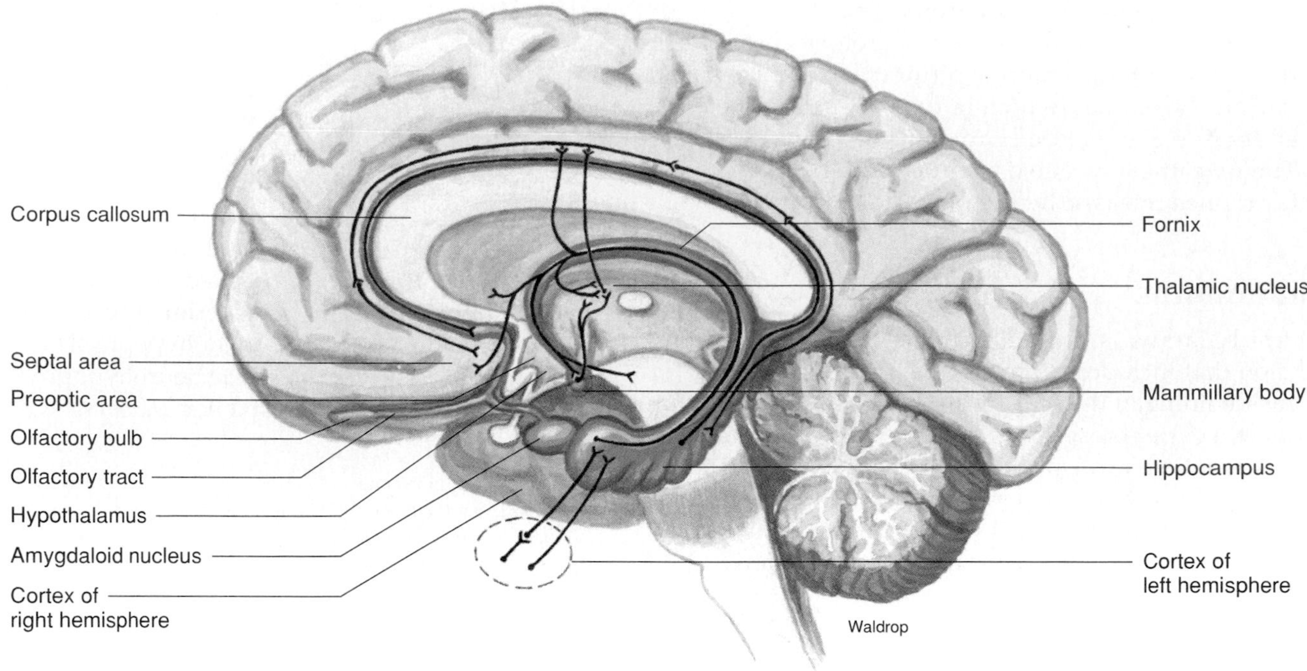

FIGURE 10.11
The limbic system and the pathways that interconnect the structures within this system. (Note that the left temporal lobe of the cerebral cortex has been removed.)

Limbic System

The limbic system consists of a group of fiber tracts and nuclei that form a ring (limbus) around the brain stem (fig. 10.11). It includes the cingulate gyrus of the cerebral cortex, the hypothalamus, the fornix (a fiber tract), the hippocampus, and the amygdaloid nucleus. These structures, which were derived early in the course of vertebrae evolution, were once called the *rhinencephalon* (*ri″nen-sef′ă-lon*), or "smell brain," because of their importance in the central processing of olfactory information.

In humans, these structures are centers involved in basic emotional drives—such as anger, fear, sex, and hunger—and in short-term memory. Complex circuits between the hypothalamus and other parts of the limbic system contribute visceral responses to emotions.

10.4 Brain Stem and Cerebellum

The brain stem consists of the midbrain, pons, and medulla oblongata. Each of these structures contains nuclei for autonomic functions of the body and their connecting tracts. Behind the brain stem is the cerebellum, whose principal function is the coordination of skeletal muscle contractions.

Brain Stem

The brain stem is the portion of the brain that attaches to the spinal cord. Its principal regions, from superior to inferior, are the *midbrain, pons,* and *medulla oblongata,* each of which is approximately an inch long.

Midbrain

The midbrain is the section of the brain stem between the diencephalon and the pons (fig. 10.12). Within the midbrain is the **mesencephalic aqueduct** (*cerebral aqueduct*) (see module 10.6), which connects the third and fourth ventricles. The midbrain also contains the corpora quadrigemina and cerebral peduncles.

The **corpora quadrigemina** (*kwad"rĭ-jem'ĭ-nă*) are the four rounded elevations on the posterior portion of the midbrain (fig. 10.12). The **superior colliculi** are the two upper elevations, which are concerned with visual reflexes. The **inferior colliculi** are the two inferior elevations, which are concerned with auditory reflexes. The **cerebral peduncles** (*pĕ-dung'k'lz*) are a pair of cylindrical structures composed of ascending and descending projection fiber tracts that support the cerebrum and connect it to other regions of the brain and to the spinal cord.

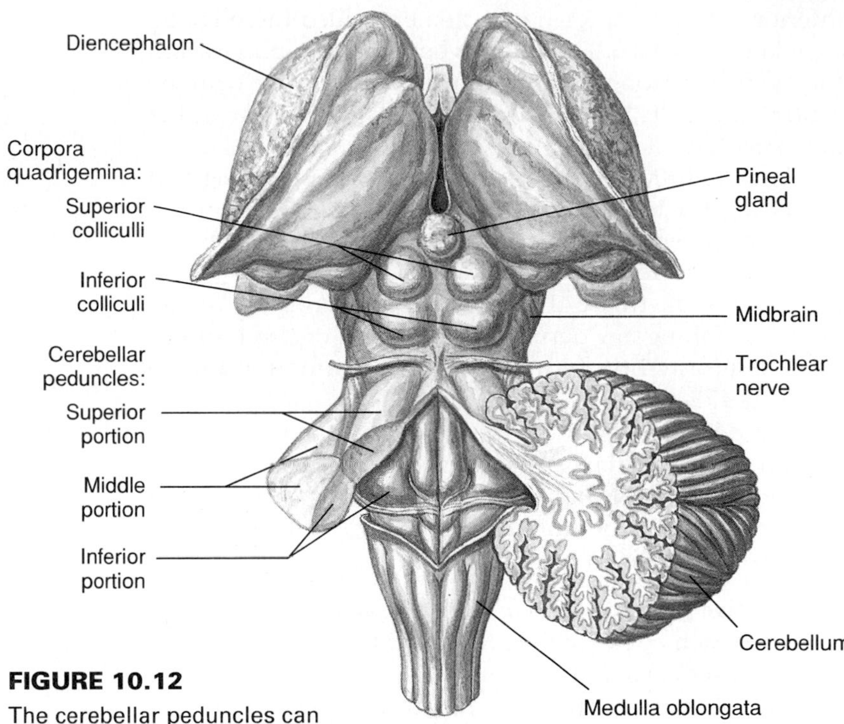

FIGURE 10.12
The cerebellar peduncles can be seen when the cerebellar hemisphere has been removed from its attachment to the brain stem.

Pons

The pons is seen as a rounded bulge on the underside of the brain, between the midbrain and the medulla oblongata (figs. 10.7 and 10.13). Scattered throughout the pons are several nuclei associated with specific cranial nerves. Other nuclei of the pons function with nuclei of the medulla oblongata to regulate the rate and depth of breathing. The two respiratory centers of the pons are called the **apneustic** and the **pneumotaxic areas** (fig. 10.13).

Medulla Oblongata

The medulla oblongata (*mĕ-dul'ă ob"long-gă'tă*) is a bulbous structure within the myelencephalon of the brain stem. It is continuous with the pons anteriorly, ending at the spinal cord at the level of the foramen magnum (figs. 10.7 and 10.13). Externally, the medulla oblongata resembles the spinal cord, except for the two elevated triangular structures called **pyramids** on the

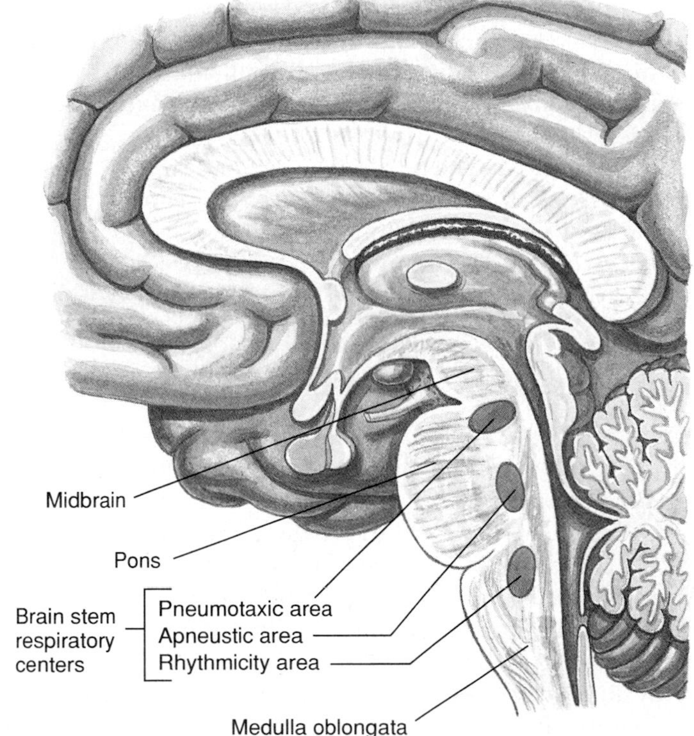

FIGURE 10.13
Nuclei within the pons and medulla oblongata that constitute the respiratory center.

corpora quadrigemina: L. *corpus*, body; *quadri*, four; *geminus*, twin
colliculus: L. *colliculus*, small mound
pons: L. *pons*, bridge
medulla: L. *medulla*, marrow

inferior side, and an oval enlargement called the **olive** on each lateral surface. The **fourth ventricle,** the space within the medulla oblongata, is continuous posteriorly with the central canal of the spinal cord, and anteriorly with the mesencephalic aqueduct (see fig. 10.19).

The medulla oblongata contains the descending and ascending tracts through which impulses are conducted between the spinal cord and various parts of the brain. Most of the fibers within these tracts cross over to the opposite side through the pyramidal region of the medulla oblongata, permitting one side of the brain to receive information from, and send information to, the opposite side of the body (see fig. 10.7).

The gray matter of the medulla oblongata consists of several nuclei for the cranial nerves and sensory relay. In addition, the medulla oblongata contains three nuclei that function as autonomic centers for controlling vital visceral functions.

1. **Cardiac center.** Both *inhibitory* and *acceleratory fibers* arise from nuclei of the cardiac center. Inhibitory impulses constantly travel through the vagus (tenth cranial) nerves to slow the heartbeat. Accelerator impulses travel through the spinal cord and eventually innervate the heart through fibers within the upper five thoracic spinal nerves.
2. **Vasomotor center.** Nuclei of the vasomotor center send impulses via the spinal cord and spinal nerves to the smooth muscles of arteriole walls, causing them to constrict or dilate, and thus regulate blood pressure and blood flow.
3. **Respiratory center.** The respiratory center of the medulla oblongata controls the rate and depth of breathing and functions in conjunction with the respiratory nuclei of the pons (fig. 10.13) to produce rhythmic breathing.

Reticular Formation

Throughout the brain stem is a complex network of nuclei and nerve fibers called the reticular formation. The reticular formation and its connections function as the *reticular activating system* (*RAS*) in arousing the cerebrum. Portions of the reticular formation are located in the spinal cord, pons, midbrain, and parts of the hypothalamus and thalamus (fig. 10.14). Nuclei within the reticular formation generate a continuous flow of impulses unless they are inhibited by other parts of the brain. The principal functions of the RAS are to keep the cerebrum in a state of alert consciousness and to selectively monitor the sensory sensations being sent to the cerebrum. The RAS also helps the cerebellum activate selective motor units to maintain muscle tonus and produce smooth, coordinated contractions of skeletal muscles.

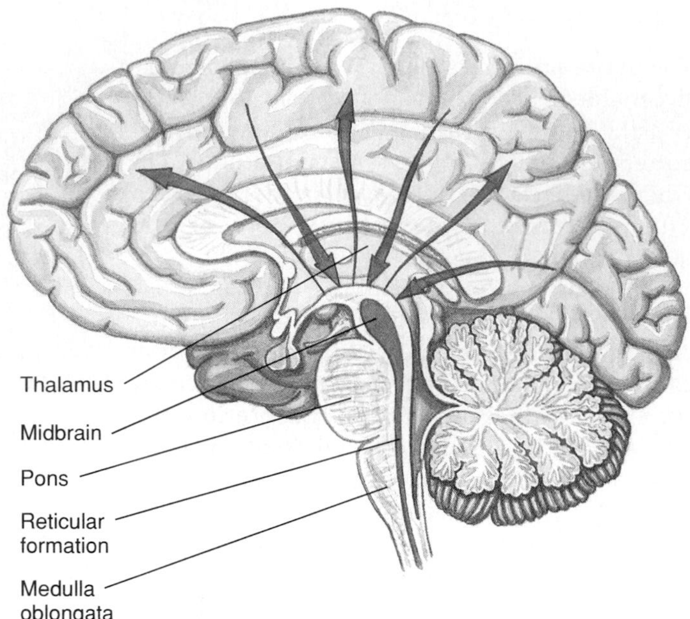

FIGURE 10.14
The reticular activating system. The arrows indicate the direction of impulses along nerve pathways that connect with the RAS.

The RAS is sensitive to changes in and trauma to the brain. The sleep response is thought to occur because of a decrease in activity within the RAS, perhaps due to the secretion of specific neurotransmitters. A blow to the head or certain drugs and diseases may damage the RAS, causing unconsciousness. A coma *is a state of unconsciousness and inactivity of the RAS that even the most powerful external stimuli cannot disturb.*

Cerebellum

The cerebellum (*ser″ĕ-bel′um*) is the second largest structure of the brain. It is located in the metencephalon and occupies the inferior and posterior aspect of the cranial cavity. The cerebellum consists of two **hemispheres** and a central constricted area called the **vermis** (fig. 10.15). **Cerebellar peduncles** support the cerebellum and provide it with tracts for communicating with the rest of the brain (see fig. 10.12).

Like the cerebrum, the cerebellum has a thin outer layer of gray matter, the **cerebellar cortex,** and a thick deeper layer of white matter. The cerebellum is convoluted into a series of slender parallel gyri. The tracts of white matter within the cerebellum have a distinctive

cerebellum: L. *cerebellum,* diminutive of *cerebrum,* brain
vermis: L. *vermis,* worm
peduncle: L. *peduncle,* diminutive of *pes,* foot

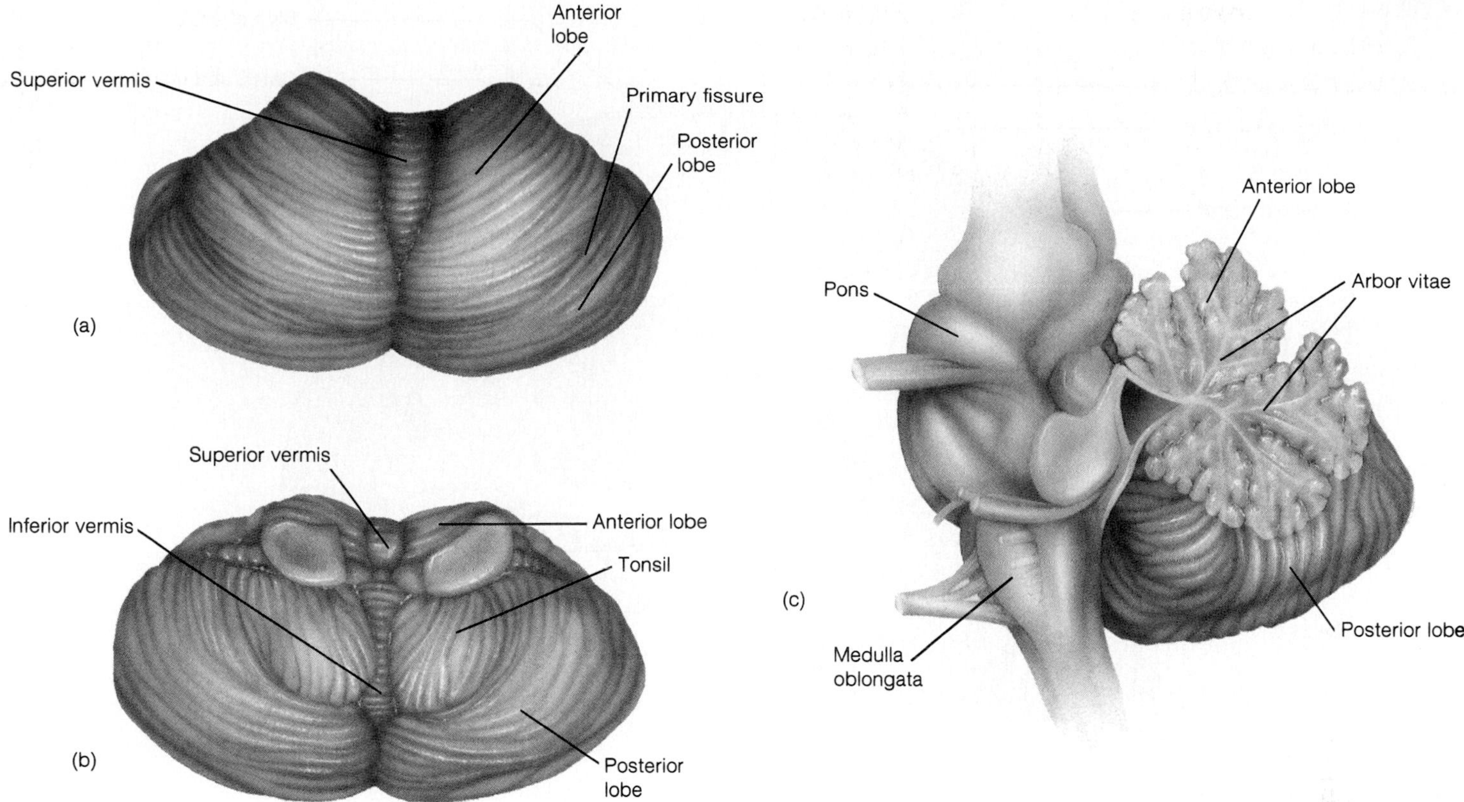

FIGURE 10.15
The structure of the cerebellum. (*a*) A superior view, (*b*) an inferior view, and (*c*) a sagittal view.

branching pattern called the **arbor vitae** that can be seen in a sagittal view (fig. 10.15*c*).

The principal function of the cerebellum is coordinating skeletal muscle contractions by recruiting precise motor units within the muscles. Impulses for voluntary muscular movement originate in the cerebral cortex and are coordinated by the cerebellum. The cerebellum constantly initiates impulses to selective motor units for maintaining posture and muscle tone. The cerebellum also adjusts to incoming impulses from proprioceptors within muscles, tendons, joints, and special sense organs. A *proprioceptor* is a sensory nerve ending that is sensitive to changes in the tension of a muscle or tendon (see chapter 12, module 12.1).

arbor vitae: L. *arbor,* tree; *vitae,* life

Trauma to the cerebellum frequently causes an impairment of skeletal muscle function. Movements become jerky and uncoordinated in a condition known as ataxia. Nystagmus, *constant motion of the eyes, may also occur. In addition, a loss of equilibrium results in a disturbance of gait. Alcohol intoxication causes similar uncoordinated body movements.*

10.5 Meninges

The CNS is covered by protective meninges; namely, the dura mater, the arachnoid mater, and the pia mater.

The central nervous system is protected by three connective tissue membranous coverings called the **meninges** (singular, *meninx*) (figs. 10.16 and 10.17). Individually, from the outside in, they are known as the dura mater, the arachnoid mater, and the pia mater.

meninges: L. plural form of *meninx,* membrane

The **dura mater,** as its name suggests, is composed primarily of tough, fibrous connective tissue. The **cranial dura**

dura mater: L. *dura,* hard; *mater,* mother

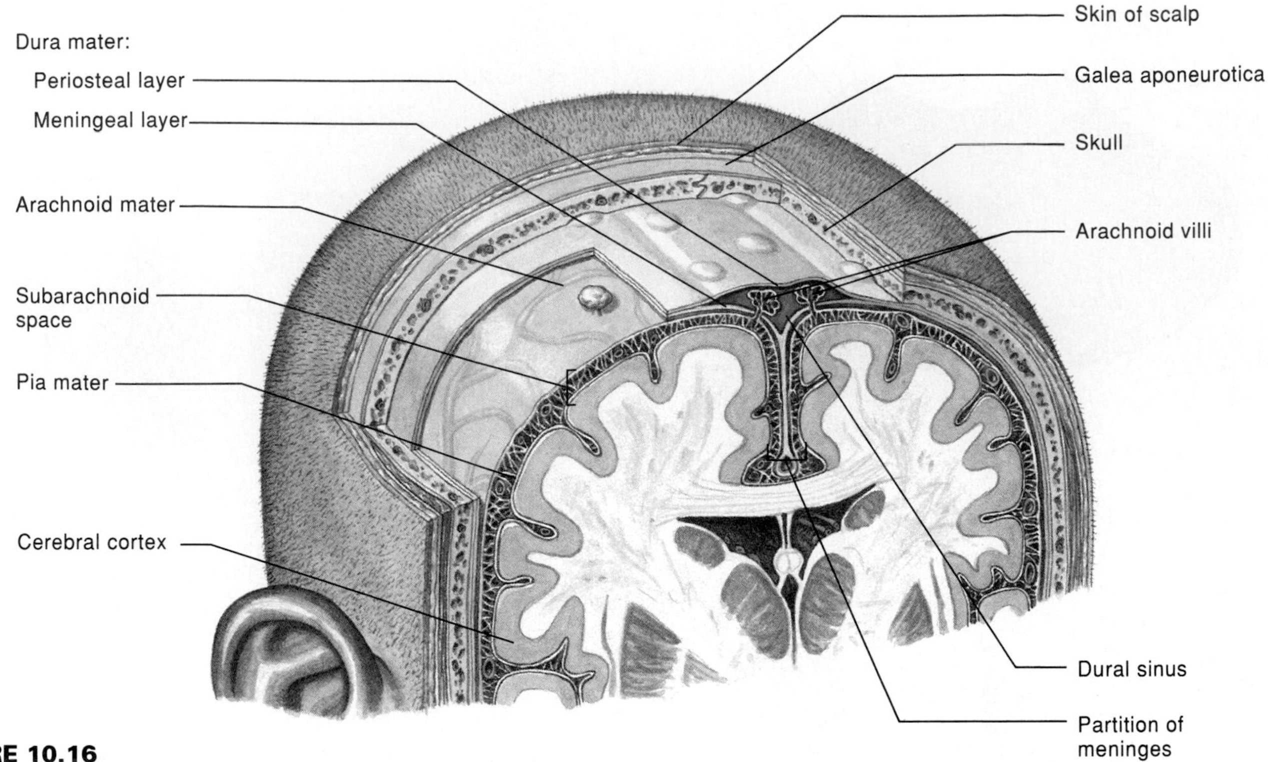

FIGURE 10.16
Meninges and associated structures surrounding the brain.

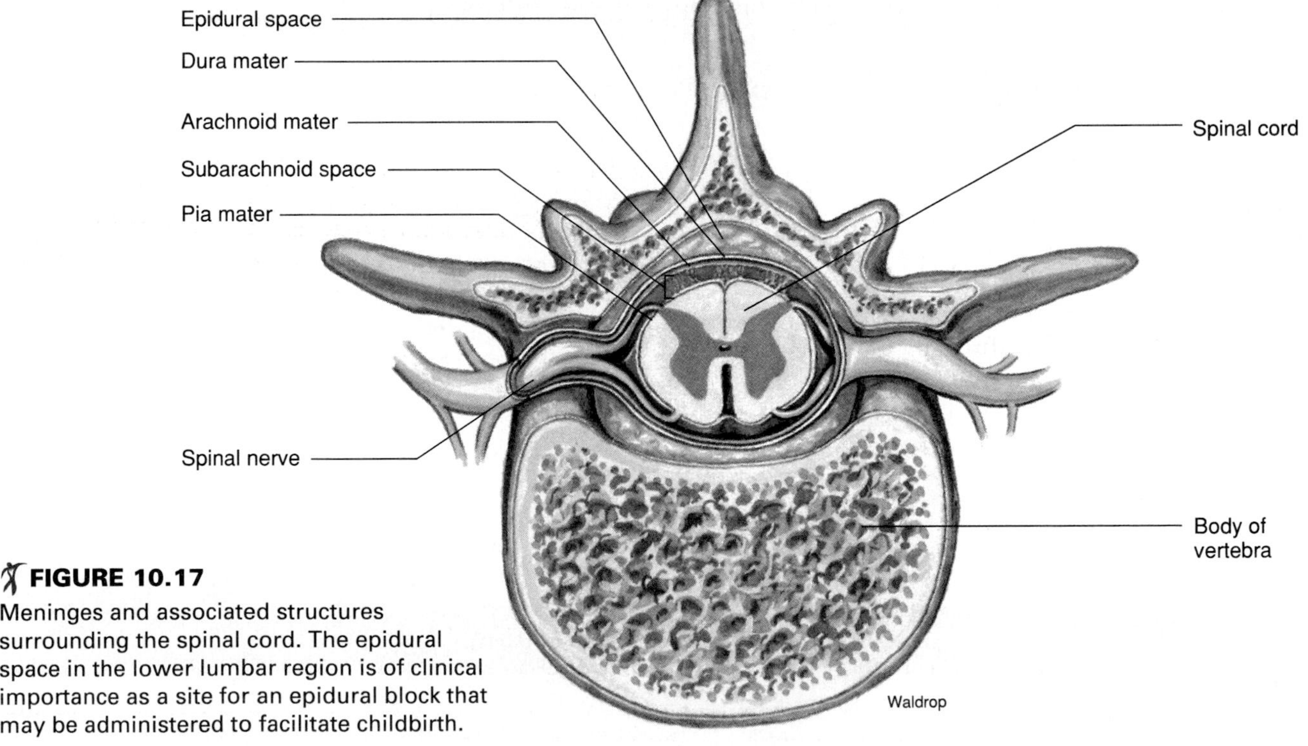

FIGURE 10.17
Meninges and associated structures surrounding the spinal cord. The epidural space in the lower lumbar region is of clinical importance as a site for an epidural block that may be administered to facilitate childbirth.

mater is a double-layered structure surrounding the brain. The outer layer adheres lightly to the inner surface of the cranium, where it constitutes the periosteum. The thin inner layer follows the general contour of the brain. In certain regions, the layers are separated, enclosing **dural sinuses** (fig. 10.16) that collect venous blood and drain it to the internal jugular veins of the neck.

The **spinal dura mater** is not double-layered. It forms a tough, tubular **dural sheath** that surrounds the spinal cord. The **epidural space** is the area between the spinal dura mater and the vertebrae forming the vertebral canal. It is highly vascular and contains loose and adipose connective tissues that form a protective pad around the spinal cord.

The **arachnoid mater,** or simply the **arachnoid,** is the middle of the three meninges. This delicate, netlike membrane spreads over the CNS but generally does not extend into the sulci or fissures of the brain. The **subarachnoid space,** located between the arachnoid and the deepest meninx, the pia mater, contains cerebrospinal fluid (see module 10.6). The subarachnoid space is maintained by delicate, weblike strands that connect the arachnoid and pia mater (fig. 10.16).

The thin **pia mater,** which is attached to the surfaces of the brain and spinal cord, is composed of loose connective tissue. It is highly vascular and functions to support the vessels that nourish the underlying cells of the brain and spinal cord. The pia mater is specialized over the ventricles, where it contributes to the formation of the choroid plexuses along with the arachnoid mater.

Inside Information—Use of the word *mater* (meaning "mother") in the names of the meninges dates back to a theory of the medieval Arabs, who believed that out of these membranes all the other body membranes were formed.

Meningitis *is an inflammation of the meninges, usually caused by bacteria or viruses. The arachnoid and the pia mater are the two meninges most frequently affected. Meningitis is accompanied by high fever and severe headache. Complications may cause sensory impairment, paralysis, or mental retardation. Untreated meningitis generally results in coma and death.*

arachnoid: L. *arachnoides,* like a cobweb
pia mater: L. *pia,* soft or tender; *mater,* mother

10.6 Ventricles and Cerebrospinal Fluid

The ventricles, central canal, and subarachnoid space contain cerebrospinal fluid. This fluid is formed by the active transport of substances from blood plasma in the choroid plexuses.

Cerebrospinal fluid (CSF) is a clear, lymphlike fluid that serves as a protective cushion around and within the CNS. The fluid also buoys the brain. CSF circulates through the various **ventricles** of the brain, the **central canal** of the spinal cord, and the **subarachnoid space** around the entire CNS. The CSF returns to the circulatory system by draining through the walls of the **arachnoid villi,** which are venous capillaries.

Ventricles of the Brain

The ventricles of the brain are connected to one another and to the central canal of the spinal cord (fig. 10.18). Each of the two **lateral ventricles** (first and second ventricles) is located in one of the hemispheres of the cerebrum, inferior to the corpus callosum. The **third ventricle** is located in the diencephalon, between the thalami. Each lateral ventricle is connected to the third ventricle by a narrow oval opening called the **interventricular foramen.** The **fourth ventricle** is located in the brain stem and cerebellum. The **mesencephalic** (cerebral) **aqueduct** passes through the midbrain to link the third and fourth ventricles. The fourth ventricle also communicates posteriorly with the central canal. CSF exits from the fourth ventricle into the subarachnoid space (fig. 10.19).

Cerebrospinal Fluid

Cerebrospinal fluid buoys the CNS and protects it from mechanical injury. The brain weighs about 1500 grams, but suspended in CSF its buoyed weight is about 50 grams. This means that the brain has a near neutral buoyancy; at a true neutral buoyancy, an object does not float or sink but is suspended in its fluid environment.

In addition to buoying the CNS, CSF reduces the damaging effect of an impact to the head by spreading the force over a larger area. It also helps to remove metabolic wastes from nervous tissue. Since the CNS lacks lymphatic circulation, the CSF moves cellular wastes into the venous return at its places of drainage.

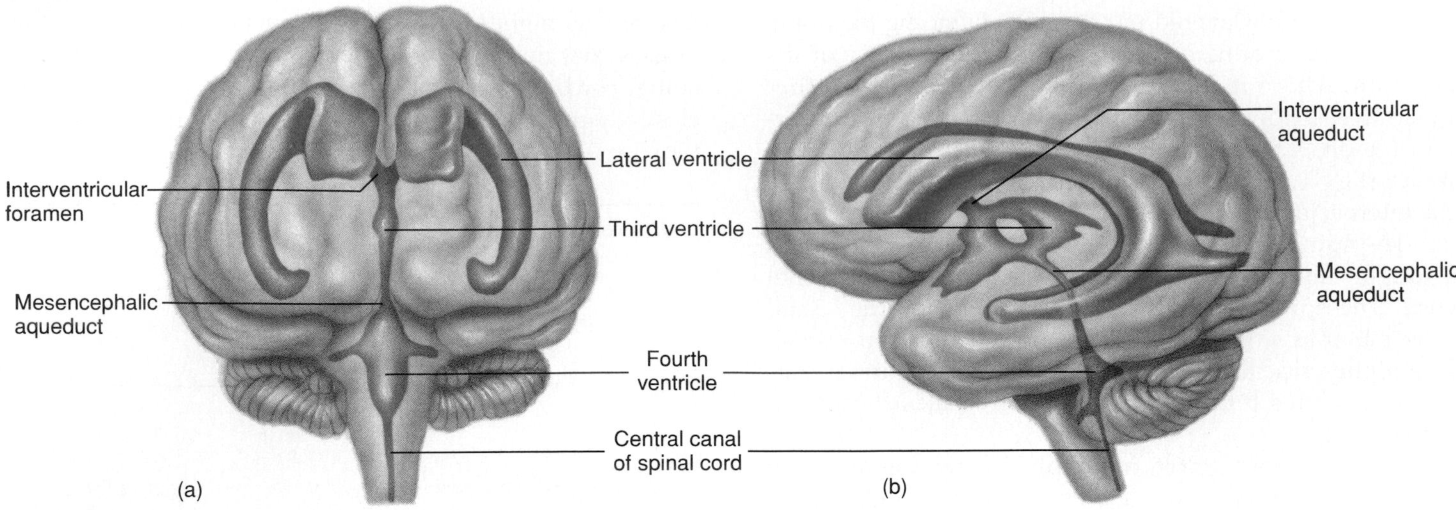

FIGURE 10.18
The ventricles of the brain. (*a*) An anterior view and (*b*) a lateral view.

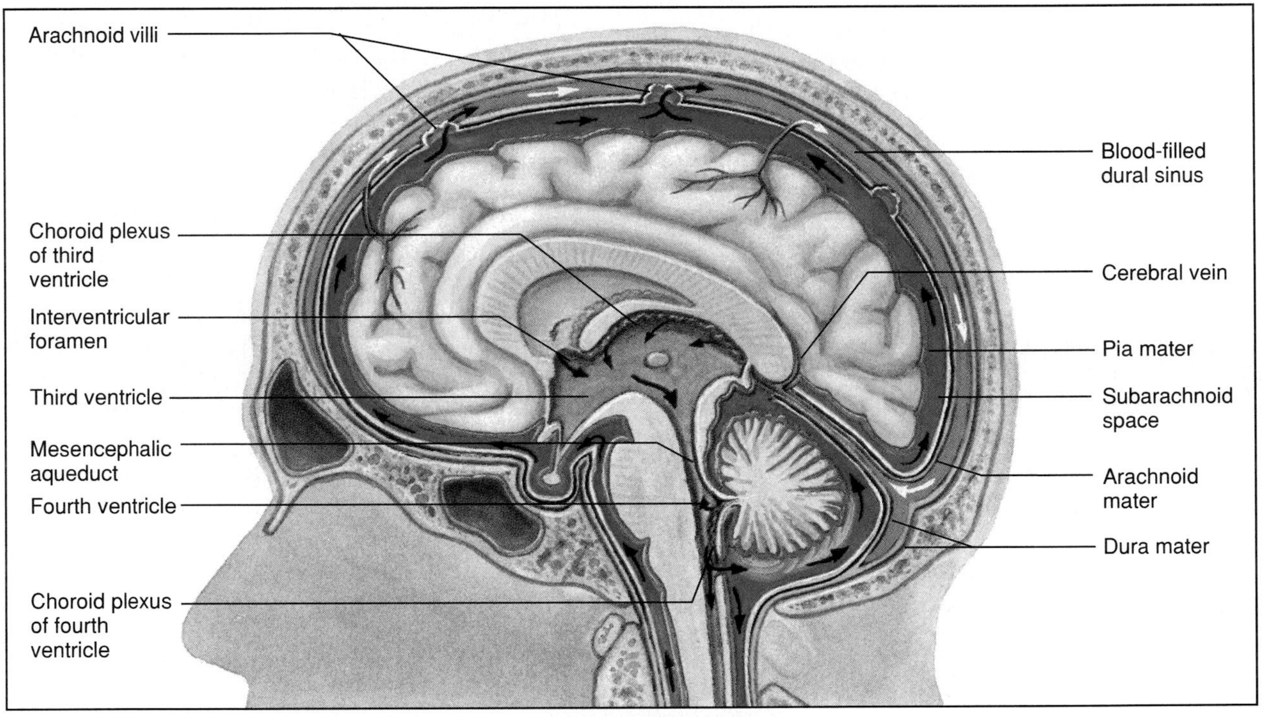

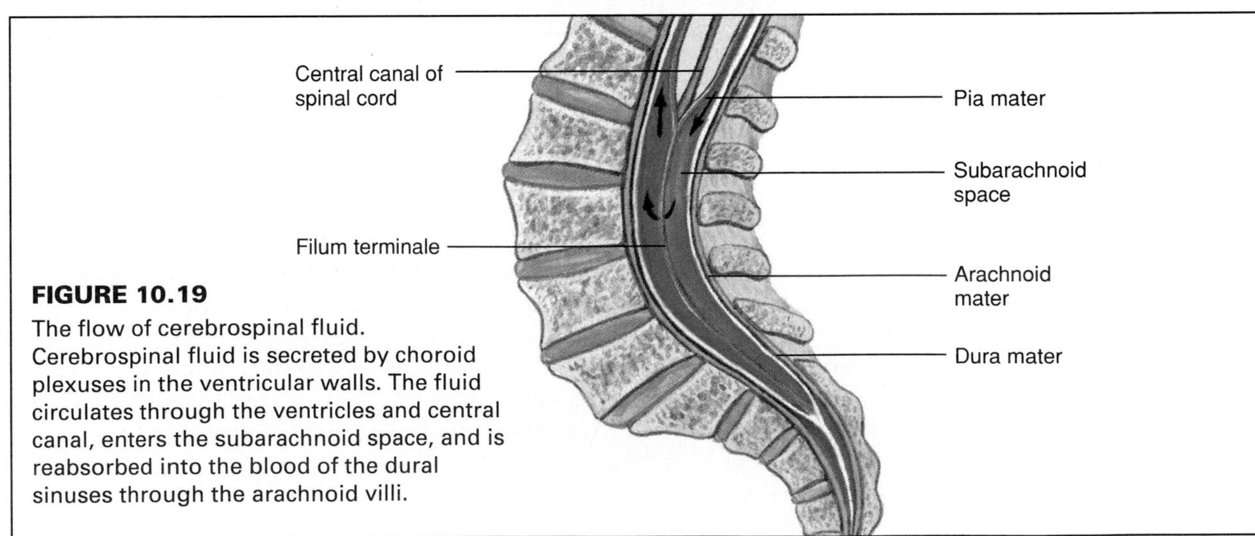

FIGURE 10.19
The flow of cerebrospinal fluid. Cerebrospinal fluid is secreted by choroid plexuses in the ventricular walls. The fluid circulates through the ventricles and central canal, enters the subarachnoid space, and is reabsorbed into the blood of the dural sinuses through the arachnoid villi.

The clear, watery CSF is continuously produced from materials within the blood by masses of specialized capillaries called **choroid plexuses.** Various components of the plexuses form a *blood–cerebrospinal fluid barrier* that prohibits harmful substances from entering the CSF. CSF is circulated by movement of the ciliated ependymal cells that cover the choroid plexuses and line the central canal.

Up to 800 ml of CSF are produced daily, although only 140–200 ml are actually bathing the CNS at any given moment. CSF is similar in composition to blood plasma from which it is formed, containing proteins, glucose, urea, and white blood cells. Comparing electrolytes, CSF contains more sodium, chloride, magnesium, and hydrogen, and fewer calcium and potassium ions than blood plasma.

Internal hydrocephalus *is a condition in which CSF builds up within the ventricles of the brain. It is more common in infants, whose cranial sutures have not yet ossified, than in older individuals. If the pressure is excessive, the condition may have to be treated surgically.*

External hydrocephalus, *an accumulation of fluid within the subarachnoid space, usually results from an obstruction of drainage at the arachnoid villi. The external pressure compresses neural tissue and is likely to cause brain damage.*

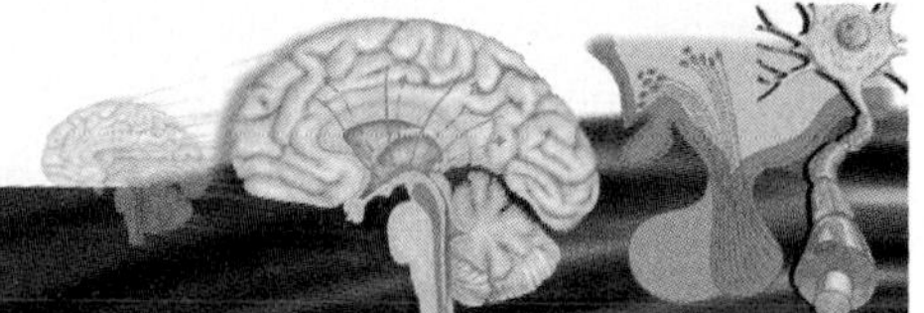

10.7 Brain Metabolism

Nervous tissue has the highest metabolic rate in the body. The extremely sensitive brain requires a continuous supply of oxygen and glucose and is absolutely dependent on a stable environment.

The brain accounts for only 2% of a person's body weight, and yet it receives approximately 20% of the total resting cardiac output. This amounts to a flow of about 750 ml of blood per minute, which is maintained at relatively constant levels. This continuous flow is so crucial that a failure of cerebral circulation for as short an interval as 10 seconds causes unconsciousness.

The brain is composed of perhaps the most sensitive tissue of the body. Due to its high metabolic rate, not only does the brain require continuous oxygen, but it also requires a continuous nutrient supply and rapid removal of wastes. The brain is also very sensitive to certain toxins and drugs. The cerebrospinal fluid serves the metabolic needs of the brain by helping to provide nutrients and remove wastes.

The brain of a newborn is especially sensitive to oxygen deprivation or to excessive oxygen. If complications arise during childbirth and the oxygen supply from the mother's blood to the baby is interrupted while it is still in the birth canal, the infant may be stillborn or suffer brain damage that can result in cerebral palsy, epilepsy, paralysis, or mental retardation.

Before the dangers of oxygen toxicity were realized, premature infants were sometimes given oxygen treatments that resulted in damage to the retina and blindness.

To a large extent, homeostasis of the brain is maintained by the **blood-brain barrier (BBB).** The BBB is a structural arrangement of capillaries, surrounding connective tissue, and specialized neuroglial cells called astrocytes (see chapter 9, module 9.3) that selectively determine which substances can move from the plasma of the blood to the extracellular fluid of the brain. Certain substances, such as water, oxygen, carbon dioxide, glucose and lipid-soluble compounds (alcohol, for example), pass readily through the BBB. Certain inorganic ions (Ca^{++} and K^{+}) pass more slowly, so that the concentrations of these ions in the brain differ from those in the blood plasma. Other substances, such as proteins, lipids, creatine, urea, inulin, certain toxins, and most antibiotics, are restricted in passage. The BBB is an important factor to consider when planning drug therapy for neurological disorders.

The BBB presents difficulties in the chemotherapy of brain diseases because drugs that could enter other organs may not be able to enter the brain. Parkinson's disease, *for example, is caused by a lack of dopamine, which normally is produced by certain neurons in the brain. Levodopa (L-Dopa), a precursor molecule to dopamine, is administered to Parkinson's patients (rather than dopamine itself) because L-Dopa can cross the BBB, whereas dopamine cannot. Within the brain, L-Dopa is converted to dopamine which reduces the symptoms of Parkinson's disease.*

A variety of chemicals in the CNS function as **neurotransmitters** (see chapter 9, module 9.6). Numerous neurotransmitters have been identified within the brain (table 10.3), yet most of their functions are little understood. It

TABLE 10.3 Selected Neurotransmitters in the Central Nervous System

Neurotransmitter	Location and Function
Acetylcholine (ACh)	Abundant throughout the cerebral cortex and in portions of the ANS; stimulates synaptic transmission and is thought to enhance learning
Substance P	Abundant in spinal cord nerve tracts, thalamus, and parietal lobes of cerebrum; stimulates perception of pain
Enkephalins (several kinds)	Abundant in spinal cord nerve tracts, brain stem, hypothalamus, thalamus, limbic system, and parietal lobes of cerebrum; reduce sensitivity to pain by suppressing substance P
Endorphins (several kinds)	Abundant in spinal cord nerve tracts, brain stem, pituitary gland, hypothalamus, thalamus, and parietal lobes of cerebrum; reduce sensitivity to pain by suppressing substance P; also thought to influence memory and learning, sex drive, and body temperature
Norepinephrine	Abundant in spinal cord, midbrain, hypothalamus, limbic system, and portions of cerebral cortex; stimulates synaptic transmission and is thought to enhance awareness and emotions, as well as influence dreaming during sleep
Dopamine	Abundant in midbrain, hypothalamus, and limbic system; involved in emotional responses and involuntary movements of skeletal muscles
Serotonin	Abundant in spinal cord, brain stem, cerebellum, hypothalamus, and limbic system; inhibits conscious awareness; thought to be involved in inducing sleep and regulating temperature
Gamma-aminobutyric acid (GABA)	Abundant in cerebellum, midbrain, hypothalamus, thalamus, and occipital lobes of cerebrum; inhibits synaptic transmission

does seem, however, that each neurotransmitter is concentrated in specific locations and serves a specific function, and that a neurochemical balance is necessary for normal brain function.

Measurable increases in regional blood flow within the brain and in glucose and oxygen metabolism accompany mental functions, including perception and emotion. These metabolic changes can be assessed through the use of *positron emission tomography* (*PET*). The technique of a PET scan (fig. 10.20) is based on injecting radioactive tracer molecules labeled with carbon-11, fluorine-18, and oxygen-15 into the bloodstream and photographing the gamma rays that are emitted from the patient's brain through the skull. PET scans are of value in studying neurotransmitters and neuroreceptors, as well as the basic metabolism of the brain.

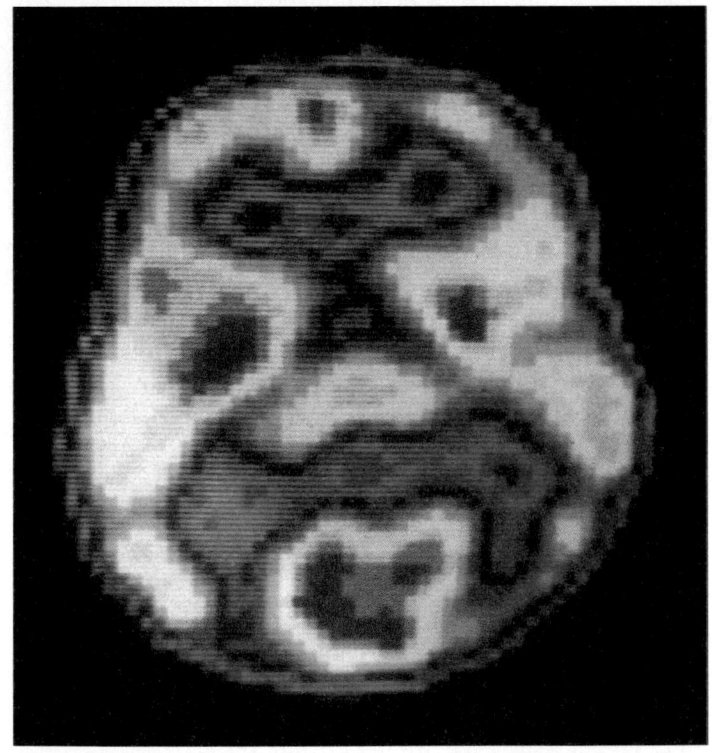

FIGURE 10.20

A positron emission tomographic (PET) scan (in cross section) of an unmedicated patient with schizophrenia. The red areas indicate high glucose use. The scan shows highest glucose uptake in the brain's visual centers.

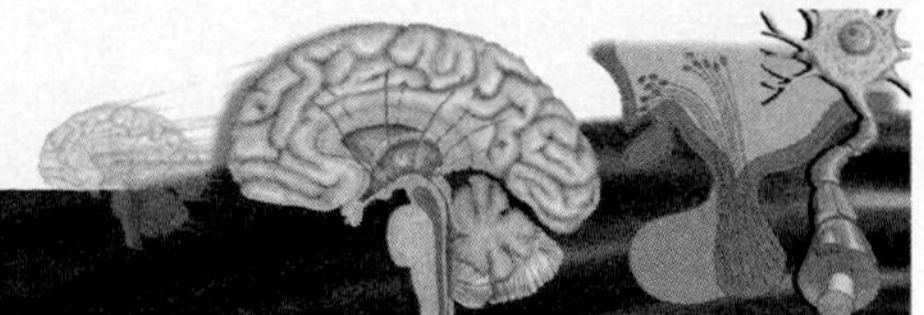

10.8 Cranial Nerves

Twelve pairs of cranial nerves emerge from the inferior surface of the brain and pass through the foramina of the skull to innervate structures in the head, neck, and visceral organs of the trunk.

Of the 12 pairs of **cranial nerves,** 2 pairs arise from the forebrain and 10 pairs arise from the brain stem (fig. 10.21). The cranial nerves are designated by Roman numerals and names. The Roman numerals refer to the order in which the nerves are positioned from the front of the brain to the back. The names indicate the structures innervated or the principal functions of the nerves. Although most cranial nerves are mixed, three cranial nerves are associated with the special senses and consist of sensory neurons only. These are the olfactory, optic, and vestibulocochlear nerves. The cell bodies of these nerves are located in ganglia outside the brain. The cranial nerves are summarized in table 10.4.

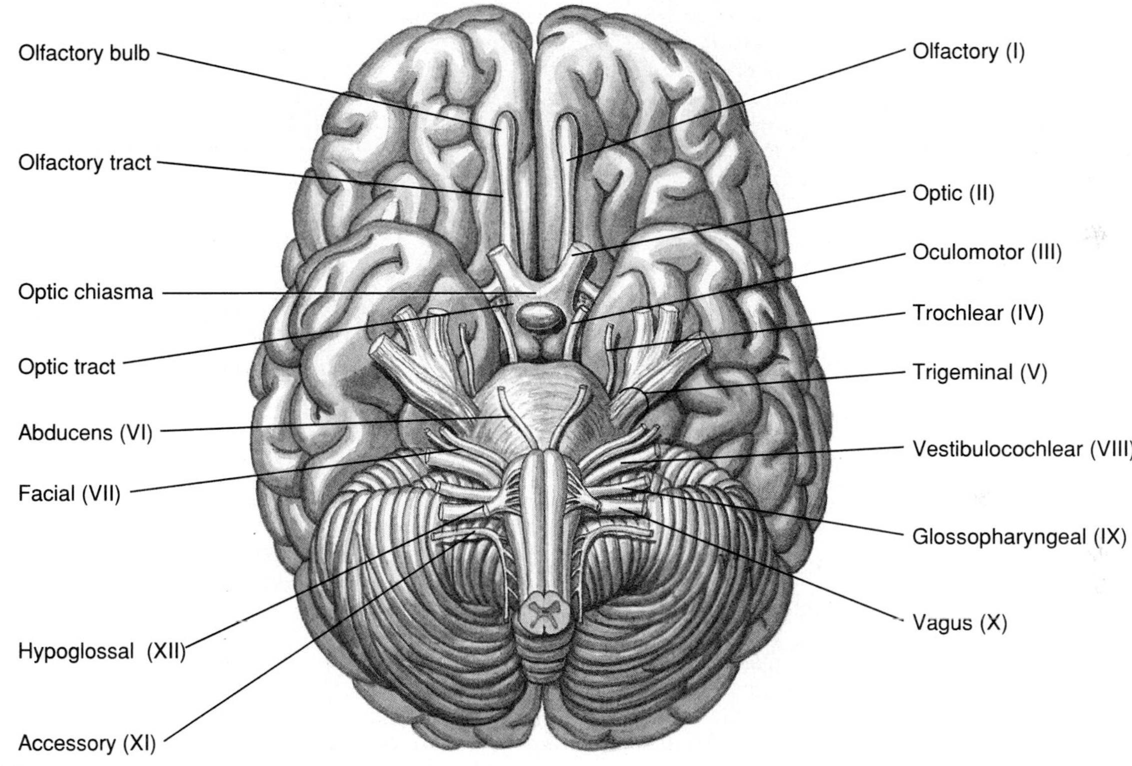

FIGURE 10.21
The cranial nerves.

Table 10.4 Cranial Nerves

Nerve	Description	Function
Olfactory (I)	Arises in nasal mucosa; terminates in olfactory areas of cerebral cortex	Sensory only—smell
Optic (II)	Arises in retina of eye; terminates in visual areas of cerebral cortex	Sensory only—sight
Oculomotor (III)	*Motor part:* Arises in midbrain and serves upper eyelid and four extrinsic eye muscles (superior oblique and superior, medial, and inferior recti muscles); parasympathetic innervation to ciliary body and iris of eye *Sensory part:* Arises from proprioceptors in extrinsic eye muscles; terminates in midbrain	Mixed nerve (primarily motor) *Motor part:* Moves eyelid and eyeball; alters shape of lens for near vision; constricts pupil *Sensory part:* Proprioception from muscles innervated with motor fibers
Trochlear (IV)	*Motor part:* Arises in midbrain and serves superior oblique extrinsic eye muscle *Sensory part:* Arises from proprioceptors in superior oblique muscle; terminates in midbrain	Mixed nerve (primarily motor) *Motor part:* Moves eyeball *Sensory part:* Proprioception from muscle innervated with motor fibers
Trigeminal (V) Ophthalmic nerve Maxillary nerve Mandibular nerve	*Motor part:* Arises in pons and serves muscles of mastication (through mandibular nerve) *Sensory parts:* Arise from facial and oral regions; terminate in pons via trigeminal ganglion	Mixed nerve *Motor part:* Mandibular nerve conveys motor impulses to muscles of mastication *Sensory parts:* Ophthalmic nerve conveys sensory impulses from cornea and skin of nose, forehead, and scalp. Maxillary nerve conveys sensory impulses from nasal mucosa, upper teeth and gums, palate, upper lip, and skin of cheek. Mandibular nerve conveys sensory impulses from temporal region, tongue, lower teeth and gums, and skin of chin and lower jaw.
Abducens (VI)	*Motor part:* Arises in pons and serves lateral rectus extrinsic eye muscle *Sensory part:* Arises from proprioceptors in lateral rectus muscle; terminates in pons	Mixed nerve *Motor part:* Moves eyeball *Sensory part:* Proprioception from muscle innervated with motor fibers
Facial (VII)	*Motor part:* Arises in pons and serves muscles of facial expression; parasympathetic fibers of ANS serve tear and salivary (lingual and submandibular) glands *Sensory part:* Arises from taste buds on anterior two-thirds of tongue; terminates in gustatory areas of cerebral cortex; proprioceptors in muscles of facial expression	Mixed nerve *Motor part:* Moves muscles of facial expression and causes secretion of tears and saliva *Sensory part:* Senses taste (sweet) from tip of tongue

olfactory: L. *olfacere,* smell out
optic: L. *optica,* see
trochlear: Gk. *trochos,* a wheel
trigeminal: L. *trigeminus,* three born together
ophthalmic: L. *ophthalmia,* region of the eye

Nerve	Description	Function
Vestibulocochlear (VIII) Vestibular nerve Cochlear nerve	Vestibular nerve arises in vestibular organs (semicircular canals, saccule, and utricle) of inner ear; terminates in thalamus Cochlear nerve arises in cochlea of inner ear; terminates in thalamus	Sensory only Vestibular nerve conveys sensory impulses associated with equilibrium Cochlear nerve conveys sensory impulses associated with hearing
Glossopharyngeal (IX)	*Motor part:* Arises in medulla oblongata and serves pharyngeal muscles used in swallowing; parasympathetic fibers of ANS serve salivary (parotid) gland *Sensory part:* Arises from taste buds on posterior one-third of tongue, carotid sinus, and proprioceptors in pharyngeal muscles; terminates in thalamus	Mixed nerve *Motor part:* Causes secretion of saliva *Sensory part:* Senses taste (sour, salty, and bitter) from sides and back of tongue; aids regulation of blood pressure
Vagus (X)	*Motor part:* Arises in medulla oblongata and serves pharyngeal and laryngeal muscles; parasympathetic fibers of ANS innervate involuntary muscles and glands of the gastrointestinal tract *Sensory part:* Arises from organs served by the motor part; terminates in medulla oblongata and pons	Mixed nerve *Motor part:* Causes contraction of pharyngeal muscles (swallowing) and laryngeal muscles (phonation); visceral muscle movement *Sensory part:* Senses certain tastes; senses pressure on tympanic membrane; senses visceral discomfort
Accessory (XI)	*Motor parts:* One part arises in medulla oblongata and serves pharyngeal and laryngeal muscles; another part arises in cervical spinal cord and supplies sternocleidomastoid and trapezius muscles *Sensory parts:* Arise from organs served by the motor parts; terminate in brain stem	Mixed nerve (primarily motor) *Motor part:* Causes contraction of pharyngeal muscles in swallowing and speech; causes contraction of muscles that move head, neck, and shoulders *Sensory part:* Proprioception from muscles innervated with motor fibers
Hypoglossal (XII)	*Motor part:* Arises in medulla oblongata and serves muscles of the tongue *Sensory part:* Arises from muscles of the tongue; terminates in medulla oblongata	Mixed nerve (primarily motor) *Motor part:* Causes movement of tongue during speech and swallowing *Sensory part:* Proprioception from muscles innervated with motor fibers

Note: A mnemonic device that may help you remember the cranial nerves in the order that they emerge from the brain is "On old Olympus's towering top, a fat vicious goat vandalized a hat." The initial letters of the words in the saying correspond to the initial letters of the cranial nerves.

vestibulocochlear: L. *vestibulum,* chamber; *cochlea,* snail shell
glossopharyngeal: L. *glossa,* tongue; Gk. *pharynx,* throat
vagus: L. *vagus,* wandering
hypoglossal: Gk. *hypo,* under; L. *glossa,* tongue

10.9 Spinal Cord

The spinal cord consists of centrally located gray matter, involved in reflexes, and peripherally located ascending and descending tracts of white matter that conduct impulses to and from the brain.

The **spinal cord** is the portion of the CNS that extends through the vertebral canal of the vertebral column (fig. 10.22). It is continuous with the brain through the foramen magnum of the skull. The spinal cord has two principal functions.

1. **Impulse conduction.** It provides a means of neural transport to and from the brain through tracts of white matter. Ascending tracts conduct impulses from the sensory receptors of the body to the brain. Descending tracts conduct motor impulses from the brain to the muscles and glands.
2. **Reflex integration.** It serves as a center for spinal reflexes (see module 10.11). Specific nerve pathways enable movements that are not initiated voluntarily by the brain.

Structure of the Spinal Cord

The spinal cord extends inferiorly from the position of the foramen magnum of the skull to the level of the first lumbar vertebra (L1). Two prominent enlargements can be seen in a posterior view (fig. 10.22). The cervical enlargement is located between the third cervical vertebra (C3) and the second thoracic vertebra (T2). Nerves emerging from this region serve the upper limbs. The lumbar enlargement lies between the ninth and twelfth thoracic vertebrae. Nerves from the lumbar enlargement supply the lower limbs.

The tapering terminal portion of the spinal cord is called the **conus medullaris.** The **filum terminale** is a fibrous strand of connective tissue that extends inferiorly from the conus medullaris at the level of L1 to the coccyx. Nerve roots radiate inferiorly from the conus medullaris through the vertebral canal. Collectively they are referred to as the **cauda equina** because they resemble a horse's tail.

filum terminale: L. *filum*, filament; *terminus*, end
cauda equina: L. *cauda*, tail; *equus*, horse

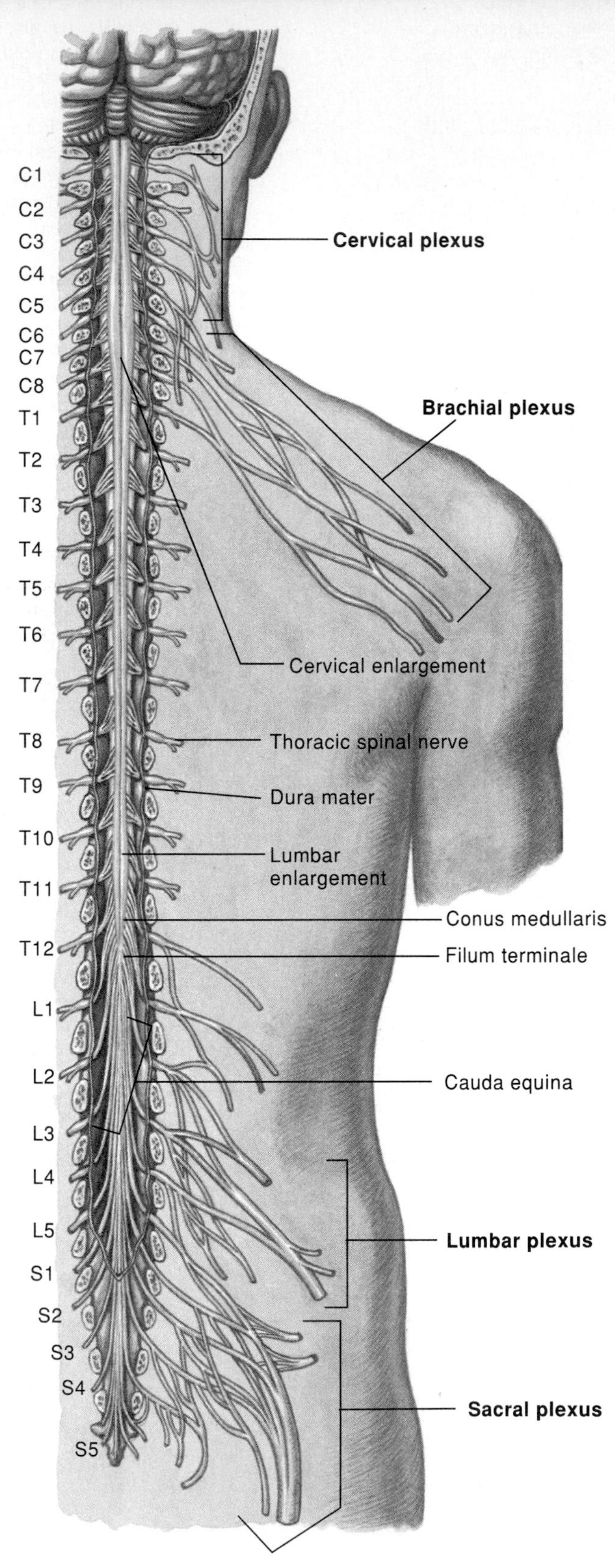

FIGURE 10.22
The spinal cord and plexuses. (The plexuses are indicated in boldface.)

Two grooves, an **anterior median fissure** and a **posterior median sulcus,** extend the length of the spinal cord and partially divide the cord into right and left portions (fig. 10.23). The gray matter of the spinal cord is centrally located and surrounded by white matter. For the most part, it is composed of nerve cell bodies, neuroglia, and unmyelinated association neurons (interneurons). The white matter primarily consists of bundles, or tracts, of myelinated fibers of sensory and motor neurons.

The core of gray matter roughly resembles the letter H (fig. 10.23). Projections of the gray matter within the spinal cord are called *horns* and are named according to the direction in which they project. The paired **posterior horns** extend posteriorly, and the paired **anterior horns** project anteriorly. A pair of short **lateral horns** extend to the sides and are located between the other two pairs. Lateral horns are prominent only in the thoracic and upper lumbar regions. The transverse bar of gray matter that connects the paired horns across the center of the spinal cord is called the **gray commissure.** Within the gray commissure is the **central canal.** It is continuous with the ventricles of the brain and is filled with cerebrospinal fluid.

Spinal Cord Tracts

The tracts of white matter within the spinal cord are designated as either ascending or descending tracts, based on the direction in which they transmit nerve impulses. **Ascending tracts** conduct sensory impulses from body parts to the brain. **Descending tracts** conduct motor impulses from the brain to muscles and glands.

The nerve fibers within the tracts are generally myelinated and have specific sites of origin and termination. The fibers of the tracts either remain on the same side of the brain and spinal cord or cross over within the medulla oblongata or spinal cord. The crossing-over of nerve tracts is referred to as *decussation* (*de″kus-a′shun*). Figure 10.24 illustrates an ascending tract that decussates within the medulla oblongata, and figure 10.25 illustrates a descending tract that decussates within the medulla oblongata.

The composite names of most spinal cord tracts reflect their origin and termination. The names of the ascending tracts usually begin with *spino-,* indicating their origin in the spinal cord. For example, a **spinothalamic tract** begins at a synapse in the spinal cord and transmits sensory impulses associated with pain to the thalamus. The descending tracts are grouped according to place of origin as either *corticospinal* or *extrapyramidal.* **Corticospinal tracts** descend from the cerebral cortex to the lower motor neurons. Most (about 85%) of the corticospinal fibers decussate in the pyramids of the medulla oblongata; the remaining 15% do not cross from one side to the other. Because of this crossing pattern, the right cerebral hemisphere primarily controls the musculature on the left side of the body, whereas the left cerebral hemisphere controls the right musculature.

Extrapyramidal tracts originate in medulla oblongata of the brain stem region. These tracts function in association with nuclei within the brain that maintain balance and posture.

decussation: L. *decussare,* to form an X intersection

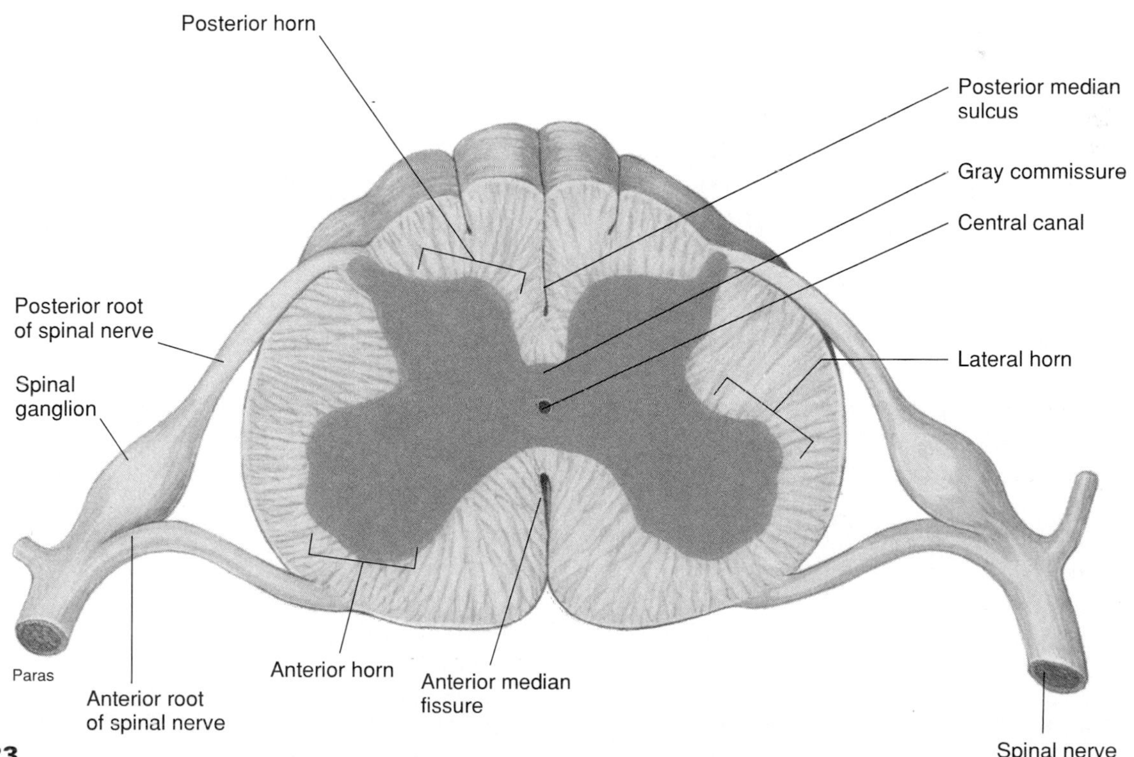

FIGURE 10.23
The spinal cord in cross section.

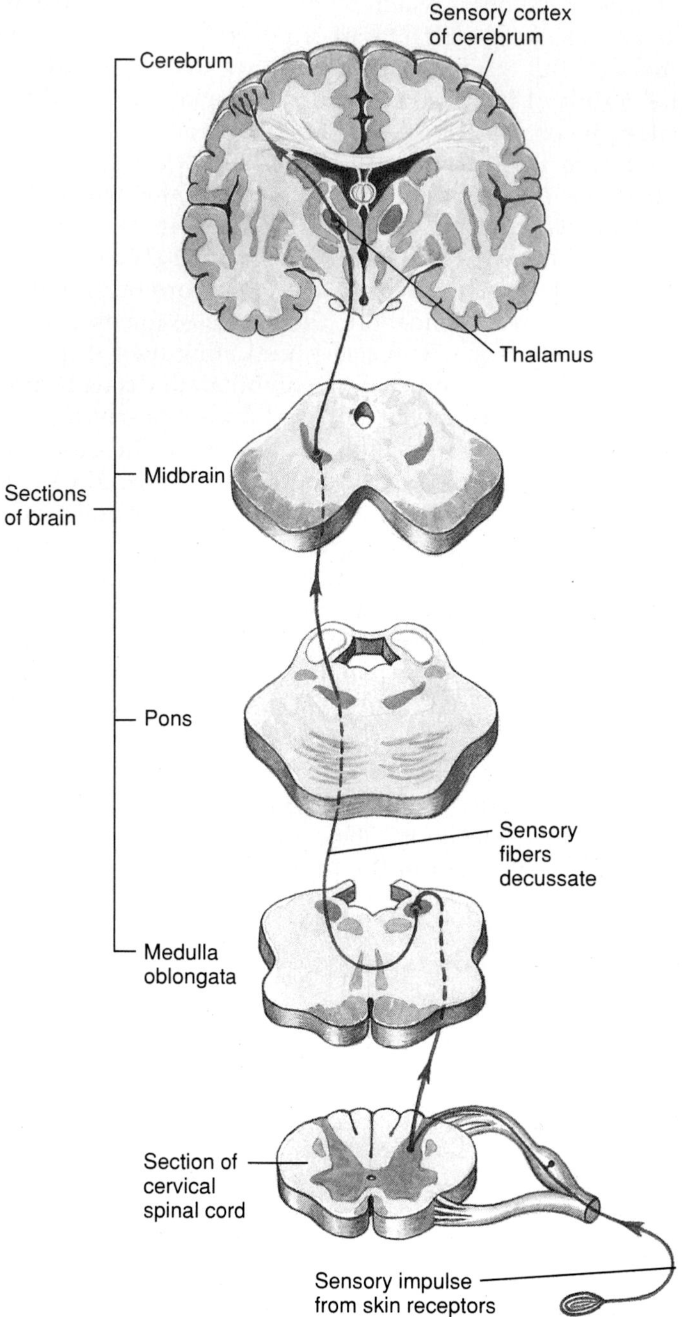

FIGURE 10.24

An ascending tract composed of sensory fibers that decussate (cross over) in the medulla oblongata of the brain stem.

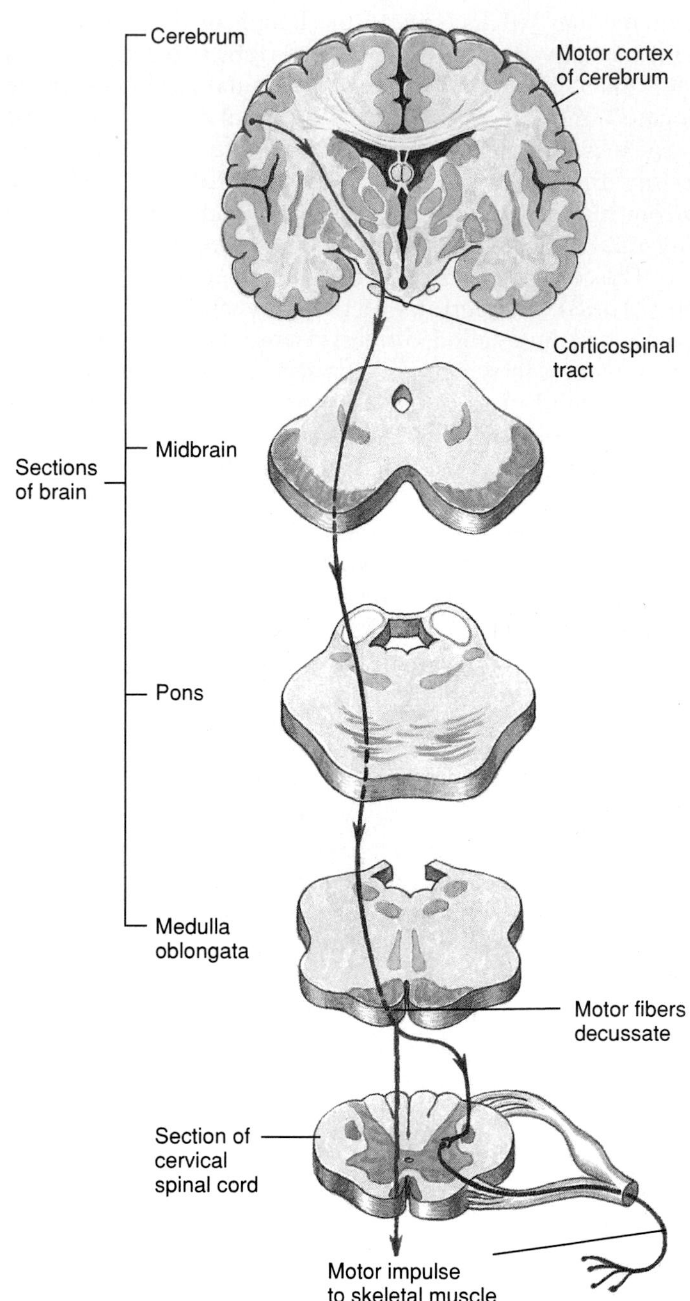

FIGURE 10.25

A descending corticospinal tract composed of motor fibers that decussate in the medulla oblongata of the brain stem.

10.10 Spinal Nerves and Nerve Plexuses

Each of the 31 pairs of spinal nerves is formed by the union of a posterior and an anterior spinal root that emerges from the spinal cord through an intervertebral foramen. In the cervical, brachial, lumbar, and sacral regions, the spinal nerves come together and then split again as networks of nerves called plexuses.

Spinal Nerves

The 31 pairs of spinal nerves (see fig. 10.22) are grouped as follows: 8 cervical, 12 thoracic, 5 lumbar, 5 sacral, and 1 coccygeal. A spinal nerve is a mixed nerve, composed of both sensory and motor fibers. It is attached to the spinal cord by a **posterior (dorsal) root,** composed of sensory fibers, and an **anterior (ventral) root,** composed of motor fibers (fig. 10.26). The posterior root contains an enlargement called the **spinal (sensory) ganglion,** where the cell bodies of sensory neurons are located. The axons of sensory neurons convey sensory impulses through the posterior root and into the spinal cord, where synapses occur with dendrites of other neurons. The anterior root consists

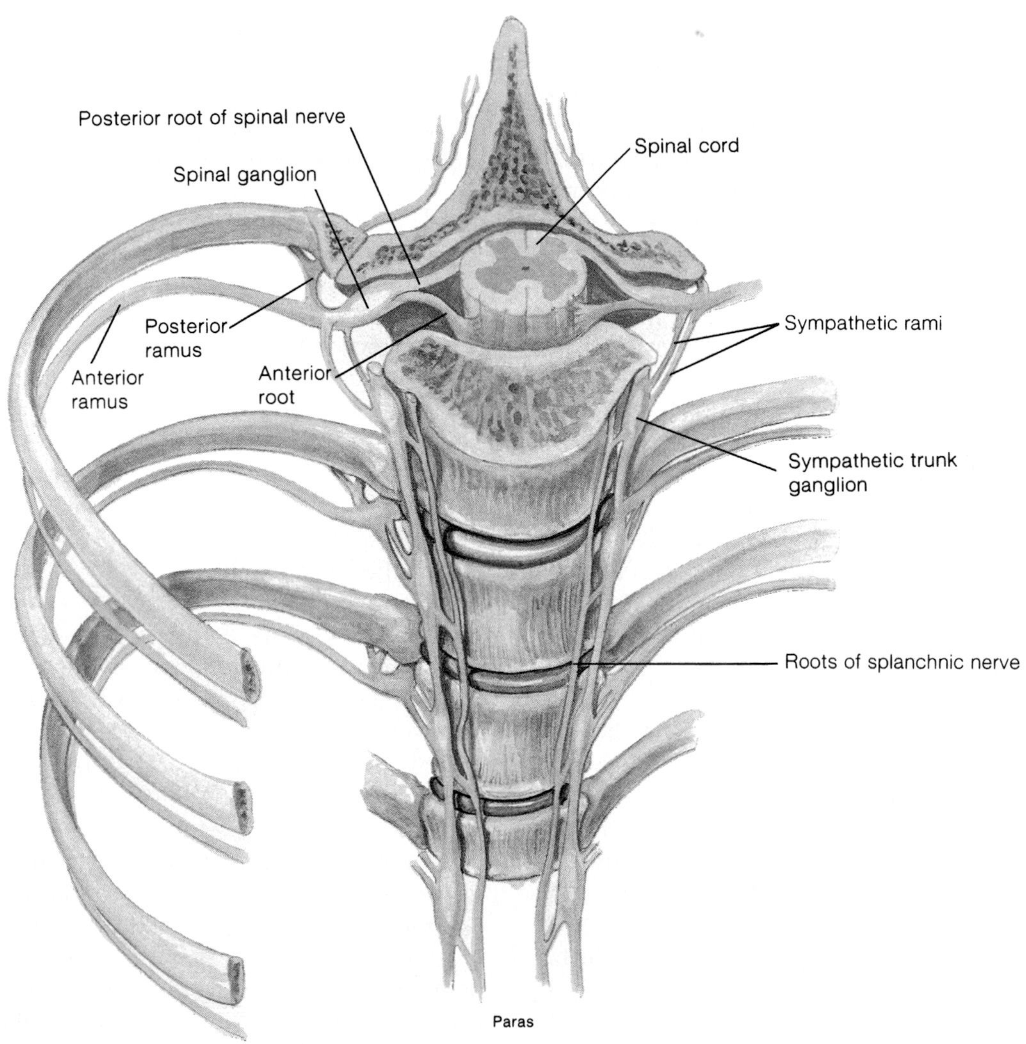

FIGURE 10.26
A section of the spinal cord and thoracic spinal nerves.

of axons of motor neurons that convey motor impulses away from the CNS. A spinal nerve is formed as the fibers from the posterior and anterior roots converge and emerge through an intervertebral foramen.

The disease herpes zoster, *also known as* shingles, *is a viral infection of the spinal ganglia. Herpes zoster causes painful clusters of fluid-filled vesicles in the skin that follow the route of the affected peripheral sensory neurons. The disease develops in adults who were first exposed to the virus as children, and is usually self-limited. Treatment may involve large doses of the antiviral drug acyclovir (Zorivax).*

A spinal nerve divides immediately after it emerges through the intervertebral foramen. The **posterior ramus** is the branch that innervates the muscles, joints, and skin of the back along the vertebral column (fig. 10.26). The **anterior ramus** is the branch that innervates the muscles and skin on the lateral and anterior side of the trunk. Combinations of the anterior rami innervate the limbs.

The **rami communicantes** are two branches from each spinal nerve that connect to a **sympathetic trunk ganglion,** which is part of the autonomic nervous system. The rami communicantes are composed of a **gray ramus,** containing unmyelinated fibers, and a **white ramus,** containing myelinated fibers. This arrangement is described in more detail in chapter 11 (see module 11.2).

Nerve Plexuses

Except in thoracic nerves T2–T12, the anterior rami of the spinal nerves come together and then split again as networks of nerves referred to as **nerve plexuses.** There are four paired plexuses of spinal nerves: the cervical, the brachial, the lumbar, and the sacral (see fig. 10.22). Nerves that emerge from the plexuses are named according to the structures they innervate or the general course they take.

The **cervical plexus** is located deep in the neck, lateral to the first five cervical vertebrae. Nerves of the cervical plexus innervate the skin and muscles of the neck and portions of the head and shoulders. The *phrenic nerve* that innervates the diaphragm arises from the cervical plexus. This means that for each inhalation during breathing, nerve impulses originating in the respiratory centers within the brain (the pons and medulla oblongata) are transmitted through the upper spinal cord, through specific parts of the paired cervical plexuses, to the paired phrenic nerves that innervate the diaphragm.

The **brachial plexus** is located in the axillary region, lateral to the last three cervical vertebrae and the first thoracic vertebra. Nerves from the brachial plexus innervate the skin, joints, and muscles of the shoulder and upper limb. Five major nerves arise from the brachial plexus, as described in table 10.5.

Trauma to the brachial plexus sometimes occurs, especially if the clavicle, upper ribs, or lower cervical vertebrae are seriously fractured. Occasionally, the brachial plexus of a newborn is severely strained during a difficult delivery, when the baby is pulled through the birth canal. In such cases, the arm of the injured side is paralyzed and eventually withers as the muscles atrophy in relation to the extent of the injury.

Inside Information—Hitting the "funny bone" at the back of the elbow is actually trauma to the ulnar nerve. The tingling sensation in the hand indicates the innervation pattern of this nerve.

The **lumbar plexus** is positioned to the side of the first four lumbar vertebrae. The nerves that arise from the lumbar plexus innervate structures of the lower abdomen and anterior and medial portions of the lower limb. The *femoral* and the *obturator nerves* are two of several that arise from the lumbar plexus. The femoral nerve innervates the skin of the anterior and medial aspect of the thigh and medial aspect of the leg and foot, and the anterior muscles of the thigh and extensor muscles of the leg. The obturator nerve innervates the skin of the medial aspect of the thigh and the adductor muscles of the lower limb.

The **sacral plexus** is positioned immediately below the lumbar plexus. Because some nerves of the sacral plexus also contain fibers from nerves of the lumbar plexus, these two plexuses are frequently described collectively as the **lumbosacral plexus.** The *sciatic (si-at'ik) nerve* is the largest nerve arising from the sacral plexus. It is also the largest nerve in the body. It extends from beneath the gluteal muscles down the posterior aspect of the thigh. The sciatic nerve is actually composed of two nerves—the *tibial* and *common fibular nerves,* wrapped in a common connective tissue sheath. The tibial nerve innervates the skin of the posterior surface of the leg and sole of the foot, and the posterior leg muscles. The common fibular nerve innervates the skin on the anterior surface of the leg and front of the foot, and the lateral and anterior leg muscles.

TABLE 10.5 **Major Nerves of the Brachial Plexus and a Schematic of Their Origins**

Nerve	Innervation	Schematic
Axillary	Skin of shoulder; shoulder joint; deltoid and teres minor muscles	
Radial	Skin of posteriolateral surface of arm, forearm, and hand; posterior muscles of arm and forearm	
Musculocutaneous	Skin of lateral surface of forearm; anterior muscles of arm	
Ulnar	Skin of medial third of hand; flexor muscles of anterior forearm, medial palm, and hand	
Median	Skin of lateral two-thirds of hand; flexor muscles of anterior forearm, lateral palm, and hand	

Upper trunk
Lower trunk
Lateral cord
Musculocutaneous nerve
C5
C6
C7
C8
T1
Medial cord
Posterior cord
Radial nerve
Median nerve
Ulnar nerve
Axillary nerve

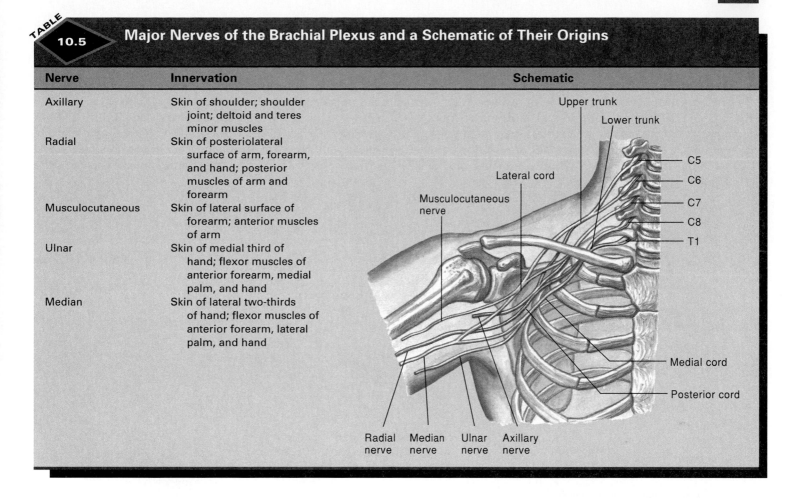

Because of its position, the sciatic nerve is of tremendous clinical importance. A posterior dislocation of the hip joint will generally injure the sciatic nerve. A herniated disc or pressure from the uterus during pregnancy may damage the nerve roots, resulting in a condition called sciatica. *Sciatica is characterized by a sharp pain in the gluteal region and by pain that extends down the posterior side of the thigh. An improperly administered injection into the buttock may injure the sciatic nerve itself.*

Even a temporary compression of the sciatic nerve as a person sits on a hard surface for a period of time may result in the perception of tingling in the limb as the person stands up. The limb is said to have "gone to sleep."

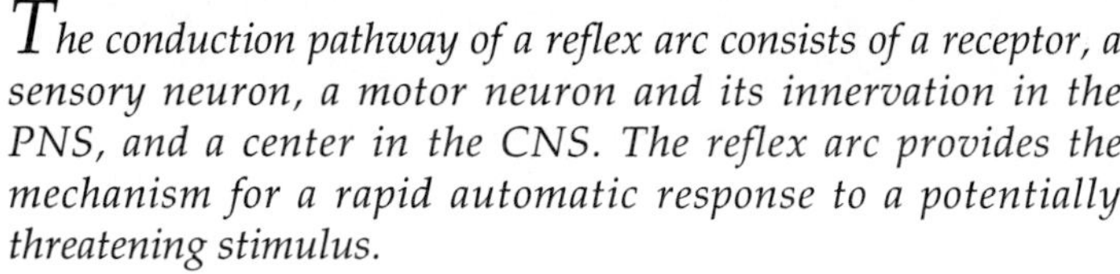

10.11 Reflex Arcs

The conduction pathway of a reflex arc consists of a receptor, a sensory neuron, a motor neuron and its innervation in the PNS, and a center in the CNS. The reflex arc provides the mechanism for a rapid automatic response to a potentially threatening stimulus.

Specific **nerve pathways** provide routes by which impulses travel through the nervous system. Frequently, a nerve pathway begins with impulses being conducted to the CNS through sensory receptors and sensory neurons of the PNS. Once within the CNS, impulses may immediately travel back through motor portions of the PNS to activate specific skeletal muscles, glands, or smooth muscles. Impulses also may be sent simultaneously to other parts of the CNS through ascending tracts within the spinal cord.

The simplest type of nerve pathway is a **reflex arc** (fig. 10.27). A reflex arc implies an automatic, unconscious, protective response to a situation in an attempt to maintain body homeostasis. Impulses are conducted over a short route from sensory to motor neurons, and only two or three neurons are involved. The five components of a reflex arc are the receptor, sensory neuron, center, motor neuron, and effector. The **receptor** includes the dendrite of a sensory neuron and the place where the electrical impulse is initiated. The **sensory neuron** relays the impulse through the posterior root to the CNS. The **center** is located within the CNS and usually involves an *association neuron* (*interneuron*). It is here that the arc is made and other impulses are sent through synapses to other parts of the body. The **motor neuron** conducts the

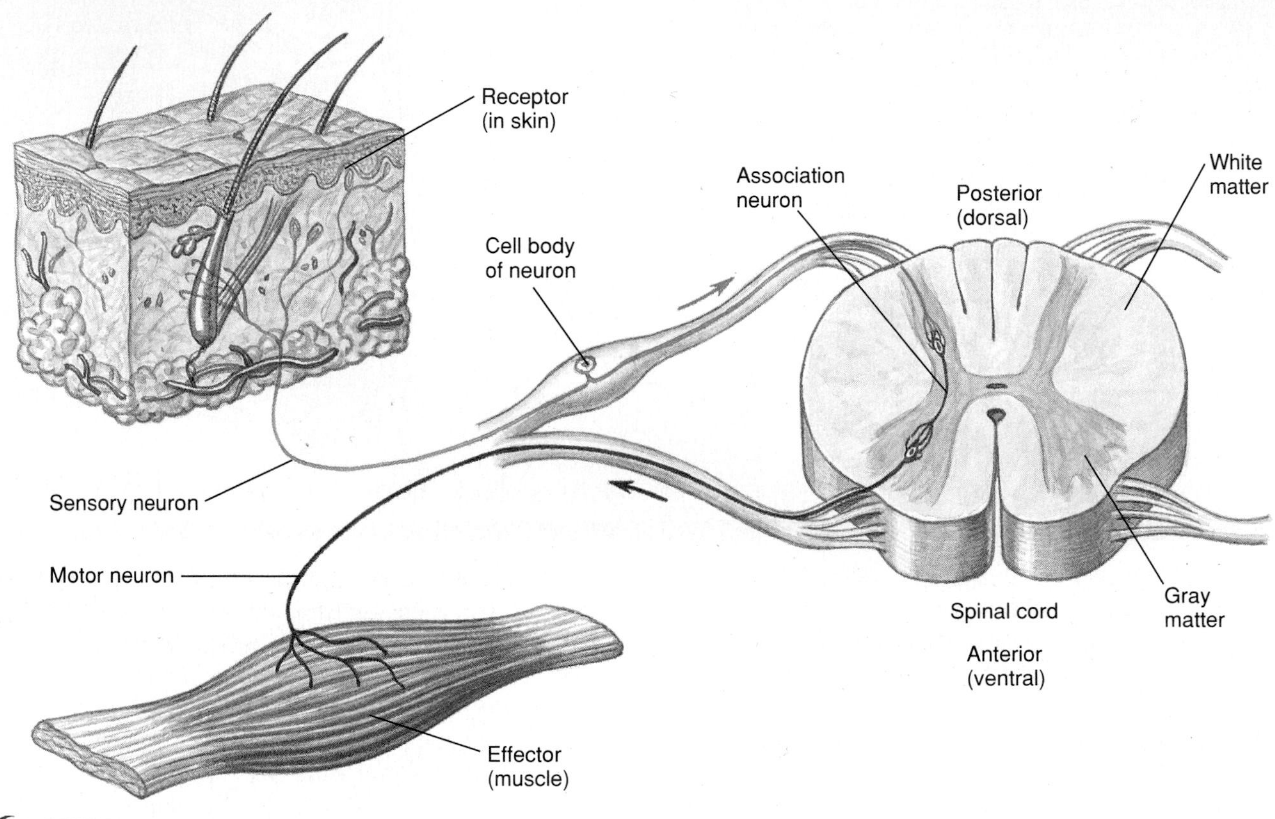

FIGURE 10.27
The reflex arc.

impulse to an **effector organ** (generally a skeletal muscle). The response of the effector is called a *reflex action* or, simply, a *reflex* (fig. 10.28).

The protective value of reflexes is obvious. For example, you quickly pull your hand away when you touch something hot. The reflexive action has occurred even before you feel the pain. The center for reflexes can also be within the brain. A sudden loud sound, for example, will cause you to reflexively move your head away from the source of the sound. Also, you close your eyelids reflexively in response to an object suddenly moving toward your eyes.

Part of a routine physical examination involves testing a person's reflexes. The condition of the nervous system, particularly the functioning of the synapses, may be determined by examining reflexes. In case of injury to some portion of the nervous system, testing certain reflexes may indicate the location and extent of the injury. Also, an anesthesiologist may try to initiate a reflex to ascertain the effect of an anesthetic.

anesthesia: Gk. *an,* without; *aisthesis,* sensation

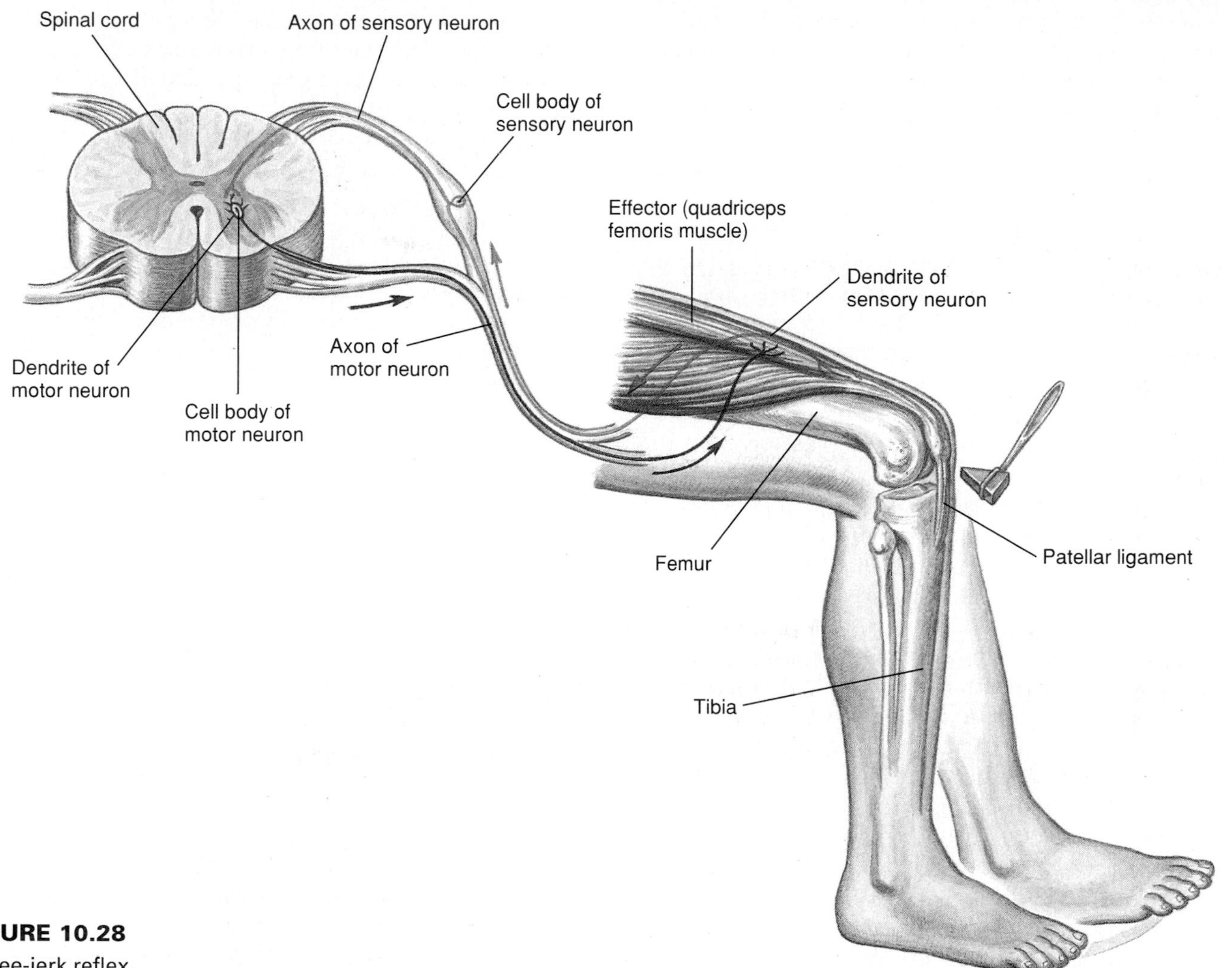

FIGURE 10.28
The knee-jerk reflex.

SYNOPTIC SPOTLIGHT

Male and Female Brains: Are Women Really from Venus and Men from Mars?

Men Are from Mars, Women Are from Venus, the catchy title of the 1992 bestseller by John Gray, succinctly expresses an ancient dilemma. What—if anything—do men's and women's brains do differently? Ignoring the obvious differences in reproductive organs and body contours, there are certainly differences in the way men and women behave. Neurophysiologists have been seeking explanations for these behavior differences for decades. Some of the key questions involved include: Does the anatomy of a male brain differ from that of a female brain? How? Do structural differences mean that the brain operates differently in men and women? Are there differences in intellectual ability? Tantalizing preliminary evidence is suggesting answers to these questions, but as yet there are no final resolutions.

The general statement that men and women respond and behave differently under the same circumstances is true; demonstrable differences in behavior and in general brain operations have been carefully documented and described. For example, from the crib, male babies tend to be more aggressive and females more passive—tendencies that persist through life. As children, girls learn to read earlier than their male counterparts and are better at tasks involving the alphabet. As teens, boys tend to do better in mathematics and girls tend to excel at verbal tasks. As adults, women continue to exhibit superior verbal fluency and to better deduce the meaning of unfamiliar words from context cues than men. In spatial operations, men have the edge in such skills as negotiating a maze, reading a map, and quickly discriminating between right and left. Men also perform better than women when asked to visualize an object and to imagine rotating it. On the other hand, women tend to perform better than men when asked to look at objects of different shapes, sizes, and colors, and then to group them in some order.

While certain generalities about behavioral differences between men and women appear to be well founded, it is important to remember that evidence for these findings is based on statistical averages. Not *all* boys are better at math, for example, than *all* girls; some girls outperform most boys. Likewise, some boys outperform most girls in the verbal domain. However, even if the findings cannot be applied to individual cases, most scientists agree that there are demonstrable differences in behavior between the sexes. It should not be surprising, then, that there are also distinct anatomical differences in the brains of men and women.

The fact that male and female brains do differ structurally in a number of ways was first demonstrated in the 1970s. For example, a specific region in the hypothalamus—called the *sexually dimorphic nucleus*—was found to have a distinctive synaptic pattern in each sex. In a series of experiments on monkeys, males castrated shortly after birth developed the female pattern, while females given injections of testosterone developed the male pattern. Based on these and later observations, scientists concluded that fetal testosterone is responsible not only for masculinizing the developing genitals, but the brain as well.

It has also been reported that women have more neurons in their temporal lobes than men. This may explain why the left temporal lobe in women is longer. In addition, the corpus callosum tends to be wider in women, particularly in the posterior regions, than in men. Brain scans reveal better connections between the hemispheres in women, which is consistent with the structural differences in the corpus callosum, and which supports the hypothesis that the female brain is less lateralized. Finally, clusters of neurons in the spinal cord that serve the external genitals are larger in men than in women. As with the sexually dimorphic nucleus of the hypothalamus, testosterone exposure before birth produces the male pattern of these neurons.

While the existence of structural differences in the male and female human brain is clear, it is less clear what they signify. To what extent do they account for differences in abilities and characteristics? And how big a role do environmental factors play in determining maleness or femaleness? It is possible that there will never be universally applicable answers to these questions. But for now it is safe to say that men and women *are* different in both their anatomy and their behavior.

Clinical Considerations

If a physician suspects abnormalities involving the nervous system, a number of tests and procedures can be performed. In a **lumbar puncture,** a fine needle is inserted between the third and fourth lumbar vertebrae, and a sample of cerebrospinal fluid is withdrawn from the subarachnoid space. Samples of CSF may be examined for blood, indicating hemorrhaging, or for abnormally high white blood cell counts, indicating infection. Excessive pressure of the CSF may indicate certain diseases or trauma.

The condition of the arteries of the brain can be determined through a **cerebral angiogram.** In this technique, a radiopaque substance is injected into the common carotid arteries and allowed to disperse through the cerebral vessels. Bulging vessels (aneurysms) and vascular constrictions or displacements by tumors may then be revealed on X rays.

The development of the **CT scanner,** or **computerized axial tomographic scanner,** revolutionized the diagnosis of brain disorders. More recently, the use of **MRI,** or **magnetic resonance imaging,** has immensely improved neurological diagnostic procedures. In MRI, a sharply detailed image of a patient's brain or spinal cord is projected onto a television screen, enabling quick and accurate diagnosis of tumors, aneurysms, blood clots, herniated discs, and hemorrhage.

Injuries

Although the brain and spinal cord are well protected within a bony encasement, they are sensitive organs, highly susceptible to injury. A **concussion** is a violent jarring or shaking injury to the brain caused by a blow to the head. Bones of the skull may or may not be fractured. A concussion usually results in a brief period of unconsciousness, followed by mild **delirium** in which the patient is in a state of confusion. **Amnesia** is a more intense disorientation in which the patient suffers various degrees of memory loss.

A person who survives a severe head injury may be *comatose* for a short or an extended period of time. A **coma** is a state of unconsciousness from which the patient cannot be aroused, even by the most intense external stimuli. The area of the brain most likely to cause a coma from trauma is the reticular activating system. Although a head injury is the most common cause of coma, chemical imbalances associated with certain diseases (e.g., diabetes) or the ingestion of drugs or poisons may also be responsible.

The flexibility of the vertebral column is essential for body movement, but this flexibility also leaves the spinal cord and spinal nerves somewhat vulnerable to trauma. Falls or severe blows to the back are a common cause of injury. A skeletal injury, such as a fracture, dislocation, or compression of the vertebrae, usually traumatizes nervous tissue as well. The consequences of the trauma depend on the location and severity of the injury and the medical treatment the patient receives. If nerve fibers of the spinal cord are severed, motor or sensory functions will be permanently lost.

Paralysis is a permanent loss of motor control, usually resulting from disease or a lesion of the spinal cord or specific nerves. Paralysis of both lower limbs is called **paraplegia.** Paralysis of both the upper and lower limb on the same side is called **hemiplegia,** and paralysis of all four limbs is **quadriplegia.**

Whiplash is a sudden hyperextension and flexion of the cervical vertebrae, as may occur, for example, in a rear-end automobile collision. Recovery from a minor whiplash (muscle and ligament strains) is generally complete but slow. Severe whiplash (spinal cord compression) may cause permanent paralysis to the structures below the level of injury.

Disorders and Diseases of the Nervous System

Headaches are the most common ailment of the CNS. Most headaches are due to dilated blood vessels in the meninges of the brain. Headaches are generally not symptomatic of brain disorders; rather, they tend to be associated with physiological stress, eyestrain, or fatigue. Persistent and intense headaches may indicate a more serious problem, such as a brain tumor.

A **migraine** is a specific type of headache commonly preceded or accompanied by visual impairments and gastrointestinal unrest. It is not known why only 5%–10% of the population periodically suffer from migraines or why they are more common in women. Fatigue, allergy, or emotional stress tends to trigger migraines.

Fainting is a brief loss of consciousness that may result from a rapid pooling of blood in the lower limbs. It may occur when a person rapidly arises from a reclined position, or receives a blow to the head, or experiences an intense psychologic stimulus, as in viewing a cadaver for the first time. Fainting is of more concern when it is symptomatic of a particular disease.

continued

amnesia: L. *amnesia,* forgetfulness
comatose: Gk. *koma,* deep sleep
paralysis: Gk. *paralysis,* loosening
paraplegia: Gk. *para,* beside; *plessein,* to strike

Clinical Considerations

Dyslexia is a defect in the language center within the brain. In dyslexia, people reverse the order of letters in syllables, of syllables in words, and of words in sentences. The sentence: "The man saw a red dog," for example, might be read by the dyslexic as "A red god was the man." Dyslexia is believed to result from the failure of one cerebral hemisphere to respond to written language, perhaps due to structural defects. It can usually be overcome by intense remedial instruction in reading and writing.

The nervous system is vulnerable to a variety of organisms and viruses that may cause abscesses or infections. **Meningitis** is an infection of the meninges. It may be confined to the spinal cord, in which case it is referred to as *spinal meningitis,* or it may involve the brain and associated meninges, in which case it is known as *encephalitis.* When both the brain and spinal cord are involved, the correct term is *encephalomyelitis* (*en-sef"ă-lo-mi"ĕ-li'tis*). Viral meningitis is more serious than bacterial meningitis. Nearly 20% of viral encephalitides are fatal. The infectious organisms that cause meningitis probably enter the body through respiratory passageways.

Epilepsy is a relatively common brain disorder with a strong hereditary basis, but it also can be caused by head injuries, tumors, and childhood infectious diseases. A person with epilepsy may periodically suffer from an epileptic seizure, which has various symptoms depending on the type of epilepsy.

Two common kinds of epilepsy are petit mal (*pet'e-mal'*) and grand mal. **Petit mal** occurs almost exclusively in children between the ages of 3 and 12. A child experiencing a petit mal seizure loses contact with reality for 5 to 30 seconds but does not lose consciousness or have convulsions. There may be slight uncontrollable facial gestures or eye movements, however, and the child will stare, as if in a daydream. During a petit mal seizure, the thalamus and hypothalamus produce an extremely depressed EEG pattern of 3 waves per second. Children with petit mal usually outgrow the condition by age 10 and generally require no medication.

Grand mal is a more serious form of epilepsy characterized by periodic convulsive seizures that generally render a person unconscious. Grand mal epileptic seizures are accompanied by rapid EEGs of 25 to 30 waves per second. This sudden increase from the normal of about 10 waves per second may cause an extensive stimulation of motor units and, therefore, uncontrollable skeletal muscle activity. During a grand mal seizure, a person loses consciousness, convulses, and may lose urinary bladder and bowel control. The unconsciousness and convulsions usually last a few minutes, after which the muscles relax and the person awakes but remains disoriented for a short time.

dyslexia: Gk. *dys,* bad; *lexis,* speech
epilepsy: Gk. *epi,* upon; *lepsis,* seize
petit mal: L. *pitinnus,* small child; *malus,* bad

SUMMARY

10.1 General Features of the Nervous System

1. The central nervous system (CNS) consists of the brain and spinal cord and contains gray and white matter. It is covered with meninges and is bathed in cerebrospinal fluid.
2. The prosencephalon, mesencephalon, and rhombencephalon develop from the embryonic neural tube. The telencephalon and diencephalon develop from the forebrain (prosencephalon). The metencephalon and myelencephalon develop from the hindbrain (rhombencephalon).
3. The peripheral nervous system (PNS) is that portion of the nervous system outside the CNS. The PNS includes cranial nerves, arising from the brain, and spinal nerves, arising from the spinal cord.
4. The autonomic nervous system (ANS) comprises two major subdivisions—the sympathetic and parasympathetic divisions.

Epilepsy almost never affects intelligence. It can be effectively treated with drugs in about 85% of the cases.

Inside Information—It was Hippocrates, or possibly one of his disciples, who first maintained that epilepsy was a disease like any other—due to some organic failure and potentially curable without resorting to magic. But even down to modern times, the belief has persisted that grand mal seizures, with their violent muscle spasms, are a consequence of demonic possession ("seizure") of the body.

Cerebral palsy is a motor nerve disorder characterized by paresis (partial paralysis) and lack of muscular coordination. It is caused by damage to the motor areas of the brain during prenatal development, birth, or infancy. During neural development within an embryo, radiation or bacterial toxins (such as from German measles) transferred through the placenta of the mother may cause cerebral palsy. Oxygen deprivation due to complications at birth and hydrocephalus in a newborn also may cause cerebral palsy. The three areas of the brain most severely affected by this disease are the cerebral cortex, the basal nuclei, and the cerebellum. The type of cerebral palsy is determined by the particular region of the brain that is affected.

Some degree of mental retardation occurs in about 65% of cerebral palsy victims. Partial blindness, deafness, and speech problems frequently accompany this disease. Cerebral palsy is nonprogressive; that is, the impairments do not worsen as a person ages. However, neither are there physical improvements.

Poliomyelitis, or *infantile paralysis,* is primarily a childhood disease caused by a virus that destroys nerve cell bodies within the anterior horn of the spinal cord. This degenerative disease is characterized by fever, severe headache, stiffness and pain, and the loss of certain reflexes. Muscle paralysis sets in quickly, and eventually the muscles may atrophy. Death can result if the virus invades the vasomotor and respiratory nuclei within the medulla oblongata or anterior horn cells controlling respiratory muscles. Poliomyelitis has been effectively controlled with immunization.

Cerebrovascular accident (CVA) is the most common disease of the nervous system. It is the third most frequent cause of death in the United States, and perhaps the number one cause of disability. The term **stroke** is frequently used as a synonym for CVA, but actually a stroke refers to the sudden and dramatic appearance of a neurological defect. *Cerebral thrombosis,* in which a thrombus, or clot, forms in an artery of the brain, is the most common cause of CVA.

Patients who recover from CVA frequently suffer partial paralysis and mental disorders, such as loss of language skills. The dysfunction depends upon the severity of the CVA and the regions of the brain that were injured. Patients surviving a CVA can often be rehabilitated, but approximately two-thirds die within 3 years of the initial damage.

10.2 Cerebrum

1. The cerebrum, consisting of two convoluted hemispheres, is concerned with higher brain functions, such as the perception of sensory impulses, the instigation of voluntary movement, the storage of memory, thought processes, and reasoning ability.
2. The cerebral cortex of the cerebral hemispheres is convoluted with gyri and sulci.
3. Each cerebral hemisphere is subdivided into frontal, parietal, temporal, and occipital lobes. The insula, sometimes called the fifth lobe, lies deep within the cerebrum; it cannot be seen in an external view.
4. Brain waves, generated by the cerebral cortex, are recorded as an electroencephalogram and may provide valuable diagnostic information.
5. The white matter of the cerebrum consists of association, commissural, and projection fibers.
6. Basal nuclei are specialized masses of gray matter located within the white matter of the cerebrum.

10.3 Diencephalon

1. The diencephalon is a major autonomic region of the brain.
2. The thalamus is an ovoid mass of gray matter that functions as a relay center for sensory impulses and responds to pain.
3. The hypothalamus is an aggregation of specialized nuclei that regulate many visceral activities. It also serves emotional and instinctual functions.
4. The epithalamus contains the pineal gland and the vascular choroid plexus over the roof of the third ventricle.
5. The pituitary gland is divided into an anterior pituitary, composed of glandular tissue, and a posterior pituitary, composed of neural tissue. Both have endocrine functions.
6. The limbic system, composed of several fiber tracts and nuclei, is involved in basic emotional drives and short-term memory.

10.4 Brain Stem and Cerebellum

1. The brain stem consists of the midbrain, pons, and medulla oblongata.
2. Within the midbrain, the corpora quadrigemina are concerned with visual and auditory reflexes. The cerebral peduncles of the midbrain contain ascending and descending fiber tracts that connect the cerebrum to other regions of the brain and to the spinal cord.
3. The pons contains nuclei for certain cranial nerves and for the regulation of respiration.
4. The medulla oblongata is composed of the ascending and descending tracts of the spinal cord. It contains nuclei for several autonomic functions.
5. Extending throughout the brain stem is a system of neurons called the reticular formation. The reticular activating system of the reticular formation arouses the cerebrum.
6. The cerebellum consists of two hemispheres connected by the vermis. It is composed of a series of white matter tracts called the arbor vitae, which are surrounded by a thin convoluted cortex of gray matter. The cerebellum is concerned with coordinated contractions of skeletal muscles.

10.5 Meninges

1. The strong outer dura mater is double-layered where it surrounds the brain. Surrounding the spinal cord, it is single-layered.
2. The arachnoid mater is a netlike menix surrounding the subarachnoid space, which contains cerebrospinal fluid.
3. The thin pia mater adheres to the contour of the CNS.

10.6 Ventricles and Cerebrospinal Fluid

1. The lateral (first and second), third, and fourth ventricles are interconnected chambers within the brain that are continuous with the central canal of the spinal cord.
2. The ventricles are filled with cerebrospinal fluid, which also flows throughout the subarachnoid space.
3. Cerebrospinal fluid is continuously secreted by the choroid plexuses and is absorbed into the blood at the arachnoid villi.

10.7 Brain Metabolism

1. The brain accounts for only about 2% of a person's body weight and yet requires 20% of the cardiac output.
2. Neurotransmitter chemicals play a major role in the functioning of the brain.
3. The blood-brain barrier determines which substances within blood plasma can enter the extracellular fluid of the brain.

10.8 Cranial Nerves

1. Twelve pairs of cranial nerves emerge from the inferior surface of the brain to innervate structures in the head, neck, and visceral organs of the trunk.
2. The olfactory, optic, and vestibulocochlear cranial nerves are sensory only; the trigeminal, glossopharyngeal, and vagus are mixed; and the others are primarily motor, with a few proprioceptive/sensory fibers.

10.9 Spinal Cord

1. The spinal cord has a cervical enlargement, a lumbar enlargement, and two longitudinal grooves that partially divide it into right and left halves.
2. The conus medullaris is the terminal portion of the spinal cord, and the cauda equina are nerve roots that radiate inferiorly from that point.
3. Ascending and descending spinal cord tracts are contained in the white matter of the spinal cord. Many of the fibers contained in these tracts decussate (cross over) in the spinal cord or in the medulla oblongata.

10.10 Spinal Nerves and Nerve Plexuses

1. Each of the 31 pairs of spinal nerves is formed by the union of an anterior (ventral) and posterior (dorsal) spinal root that emerges from the spinal cord through an intervertebral foramen to innervate a particular body region.
2. Each spinal nerve is a mixed nerve consisting of a posterior root of sensory fibers and an anterior root of motor fibers.
3. Except in the thoracic nerves T2–T12, the anterior rami of the spinal nerves come together and then split again as networks of nerves called plexuses. There are four plexuses of spinal nerves—the cervical, the brachial, the lumbar, and the sacral—the last two of which are often referred to as the lumbosacral plexus.

10.11 Reflex Arcs

1. The conduction pathway of a reflex arc consists of a receptor, a sensory neuron, a motor neuron and its innervation in the PNS, and a center (usually an association neuron) in the CNS. The reflex arc enables a rapid automatic response to a potentially threatening stimulus.
2. A reflex arc is the simplest type of nerve pathway.

Review Activities

Objective Questions

1. The central nervous system consists of the brain and (*module 10.1*)
   a. the cranial nerves.
   b. the spinal nerves.
   c. the spinal cord.
   d. the ganglia.

2. The principal connection between the cerebral hemispheres is (*module 10.2*)
   a. the corpus callosum.
   b. the pons.
   c. the intermediate mass.
   d. the precentral gyrus.

3. Which of the following statements about the hypothalamus is *false?* (*module 10.3*)
   a. It regulates body temperature.
   b. It causes autonomic acceleration or deceleration of the heart rate.
   c. It controls the respiratory rate.
   d. It monitors the concentration of the blood and produces antidiuretic hormone (ADH).

4. Respiratory nuclei are found in both (*module 10.4*)
   a. the pons and the cerebellum.
   b. the cerebellum and the medulla oblongata.
   c. the pons and the medulla oblongata.
   d. the cerebrum and the cerebellum.

5. The outer meninx in contact with the cranium is (*module 10.5*)
   a. the pia mater.
   b. the arachnoid mater.
   c. the choroid mater.
   d. the dura mater.

6. The fourth ventricle is located within (*module 10.6*)
   a. the cerebrum.
   b. the diencephalon.
   c. the spinal cord.
   d. the brain stem.

7. What percentage of the cardiac output consistently flows to the brain? (*module 10.7*)
   a. 2%
   b. 10%
   c. 20%
   d. 50%

8. Which paired cranial nerve is being tested as you visually follow a physician's finger that is brought close to your nose, causing you to become cross-eyed? (*module 10.8*)
   a. abducens
   b. oculomotor
   c. trochlear
   d. optic

9. Which of the following structures is (are) *not* associated with the spinal cord? (*module 10.9*)
   a. central canal
   b. choroid plexus
   c. association neurons
   d. anterior horn
   e. corticospinal tracts

10. Which of the following is *not* a spinal nerve plexus? (*module 10.10*)
   a. brachial plexus
   b. cervical plexus
   c. lumbar plexus
   d. thoracic plexus
   e. sacral plexus

11. A reflex arc (*module 10.11*)
   a. always involves the spinal cord.
   b. occurs only in the appendages.
   c. allows a quick decision for voluntary movement.
   d. follows a precise neurological pathway between the receptor site and the effector organ.

Essay Questions

1. List the five regions of the brain and indicate at least one major structure found in each. (*module 10.1*)
2. Diagram a lateral view of the cerebrum and label the four superficial lobes and the fissures that separate them. List the general functions of each of the paired cerebral lobes. (*module 10.2*)
3. Describe the location of the pituitary gland relative to the rest of the diencephalon and the rest of the brain. Discuss the functional relationship between the hypothalamus and the pituitary gland. (*module 10.3*)
4. Describe the location of the reticular formation. State two functions of the RAS. (*module 10.4*)
5. Describe the location of the dural sinuses and the epidural space. (*module 10.5*)
6. Explain the functions of cerebrospinal fluid. Where is CSF produced and where does it drain? (*module 10.6*)
7. Which cranial nerves are involved in tasting, chewing and manipulating food, and in swallowing? (*module 10.8*)
8. Diagram a cross section of the spinal cord and label the structures of the gray matter and the white matter. Describe the location of the spinal cord. (*module 10.9*)
9. Describe a spinal nerve. Explain the statement that all spinal nerves are mixed nerves. (*module 10.10*)
10. What is a reflex arc? Explain how reflexes are important in maintaining body homeostasis. (*module 10.11*)

Labeling Exercise

Label the structures indicated on the figure to the right.

1. ____________________
2. ____________________
3. ____________________
4. ____________________
5. ____________________
6. ____________________

Central sulcus
3
Postcentral gyrus
2
Cerebral gyri
Cerebral sulci
Motor speech area (Broca's area)
Occipital lobe
4
5
6
1
Brain stem

Critical Thinking Questions

1. List the body systems and describe the functions over which the hypothalamus has some control.
2. In a patient's case study, any observed or suspected structural/functional abnormalities of body structures are reported. Prepare a brief case study of a patient who has suffered severe trauma to the medulla oblongata from a blow to the back of the skull.
3. After being revived from a knockout punch during the world featherweight championship fight, the ringside physician determined that the boxer had sustained oculomotor and facial nerve damage when his right zygomatic bone was shattered from a hard left hook. What symptoms might the boxer have displayed that would have caused the physician to come to this conclusion?
4. Trauma to the spinal cord is always serious. The higher it occurs in the spine, the more serious it is. Why should this be so? Explain why a reflex arc below the level of the trauma may continue to function for a short while, regardless of the extent of the trauma. What may be indicated if a patient with a spinal cord injury can move his lower limbs but has no sensory perception?
5. A 63-year-old truck driver made an appointment with the company doctor because of pain and numbness in his left leg. Following a routine physical exam, the physician scheduled the man for a magnetic resonance imaging of his lumbar region. The patient's MR scan (shown to the right) indicates that he has a herniated disc (arrow) in the lumbar region. Explain how such a condition could develop and account for the man's symptoms. Discuss the possible treatment of this condition. (You may have to refer to medical textbooks in your library to find the answer.)

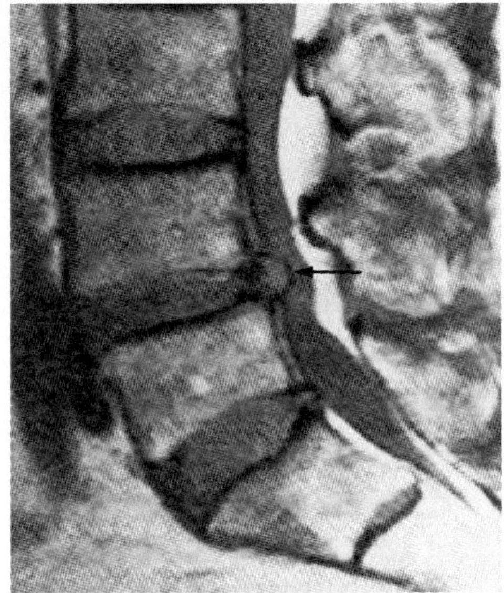

CHAPTER
11
"[W]e have the remarkable ability to consciously influence autonomic activity using techniques such as biofeedback and meditation." page 296

Autonomic Nervous System

Chapter Outline

Terms to Remember

preganglionic neuron, 289
postganglionic neuron, 289
sympathetic division, 289
parasympathetic division, 289
paravertebral ganglia, 290
collateral ganglia, 290
sympathoadrenal system, 292
adrenergic (*ad"rĕ-ner'jik*) receptors, 293
terminal ganglia, 294
cholinergic (*ko"lĭ ner'jik*) receptors, 297

Learning Objectives

After studying this chapter, you should be able to

- define the autonomic nervous system as a functional component of the nervous system.
- compare autonomic and somatic motor neurons with regard to location and function.
- describe the structure of the sympathetic division of the autonomic nervous system and the relationship between the sympathetic division and the adrenal medulla.
- describe the functions of the sympathetic division and explain why the phrase "fight or flight" applies to this division.
- describe how autonomic nerve fibers are classified as either cholinergic or adrenergic.
- describe the structure of the parasympathetic division of the autonomic nervous system.
- describe the functions of the parasympathetic division and compare the effects of sympathetic and parasympathetic stimulation on different body organs.

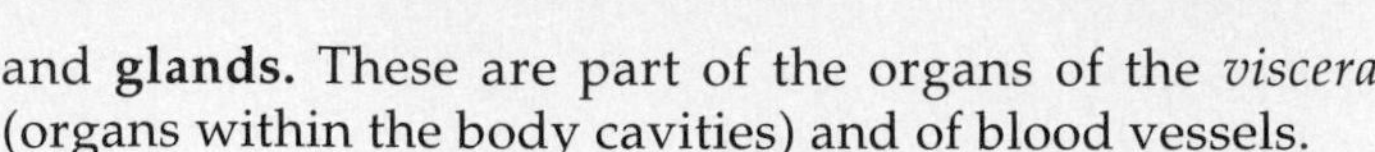

11.1 Autonomic Neurons

Autonomic neurons are of two types: preganglionic and postganglionic. The postganglionic neurons regulate the function of the involuntary effector organs.

The autonomic portion of the nervous system is concerned with maintaining homeostasis within the body by fine-tuning visceral activity in response to changing conditions, and for the most part without our conscious awareness. Although the **autonomic nervous system (ANS)** is composed of specific parts of the central nervous system and the peripheral nervous system, it has a certain amount of functional independence.

Autonomic motor nerves innervate organs whose functions are not usually under voluntary control. The effectors that respond to autonomic regulation include **cardiac muscle** (the heart), **smooth** (visceral) **muscles,** and **glands.** These are part of the organs of the *viscera* (organs within the body cavities) and of blood vessels.

As discussed in chapter 10 (see module 10.1), neurons of the peripheral nervous system (PNS) that conduct impulses away from the central nervous system (CNS) are known as *motor,* or *efferent, neurons.* There are two major categories of motor neurons—somatic and autonomic. Somatic motor neurons have their cell bodies within the CNS and have axons that extend to skeletal muscles (fig. 11.1, *left*). These axons synapse with skeletal muscle cells and stimulate these cells to contract.

Unlike somatic motor neurons, which conduct impulses along a single axon from the spinal cord to the neuromuscular junction, autonomic motor control involves two neurons in the motor pathway from the CNS. The first of these neurons has its cell body in the gray matter

Somatic motor reflex
Association neuron
Posterior root ganglion
Somatic motor neuron
Sensory neuron
Paras

Autonomic motor reflex
Association neuron
Posterior root ganglion
Preganglionic neuron
Autonomic ganglion
Postganglionic neuron
Sensory neuron
Viscera

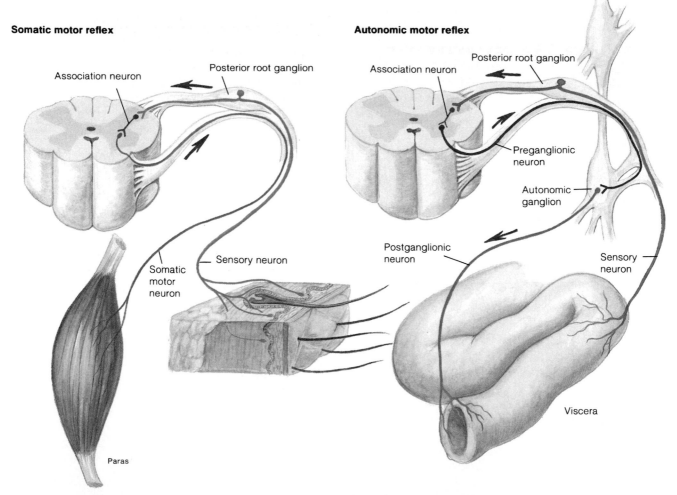

FIGURE 11.1
A comparison between a somatic motor reflex and an autonomic motor reflex.

of the brain or spinal cord. The axon of this neuron does not directly innervate the effector organ; rather, it synapses with a second neuron within an *autonomic ganglion* (a ganglion is a collection of cell bodies outside the CNS). The first neuron is thus called a **preganglionic neuron.** The second neuron in this pathway is called a **postganglionic neuron.** Its axon extends from the autonomic ganglion and synapses with the cells of an effector organ (fig. 11.1, *right*).

The autonomic nervous system is divided into the **sympathetic** and the **parasympathetic divisions.** These two parts of the autonomic nervous system differ in their structure and function. In figure 11.2, the sympathetic division is shown in red and the parasympathetic division in blue. For both divisions, the solid lines indicate preganglionic neurons (originating from the CNS), and the dashed lines represent postganglionic neurons (originating from autonomic ganglia). Notice that the preganglionic parasympathetic neurons (fig. 11.2, solid blue lines) originate from the brain and sacral regions of the spinal cord. For this reason, the parasympathetic division is also called the *craniosacral division.* The preganglionic sympathetic neurons (fig. 11.2, solid red lines) originate from the thoracic and lumbar regions of the spinal cord; hence, the sympathetic division is also called the *thoracolumbar division.* Examination of figure 11.2 reveals that it is always the postganglionic neurons (fig. 11.2, dashed lines, both blue and red) that actually synapse with the muscle and gland cells.

ganglion: Gk. *ganglion,* a swelling

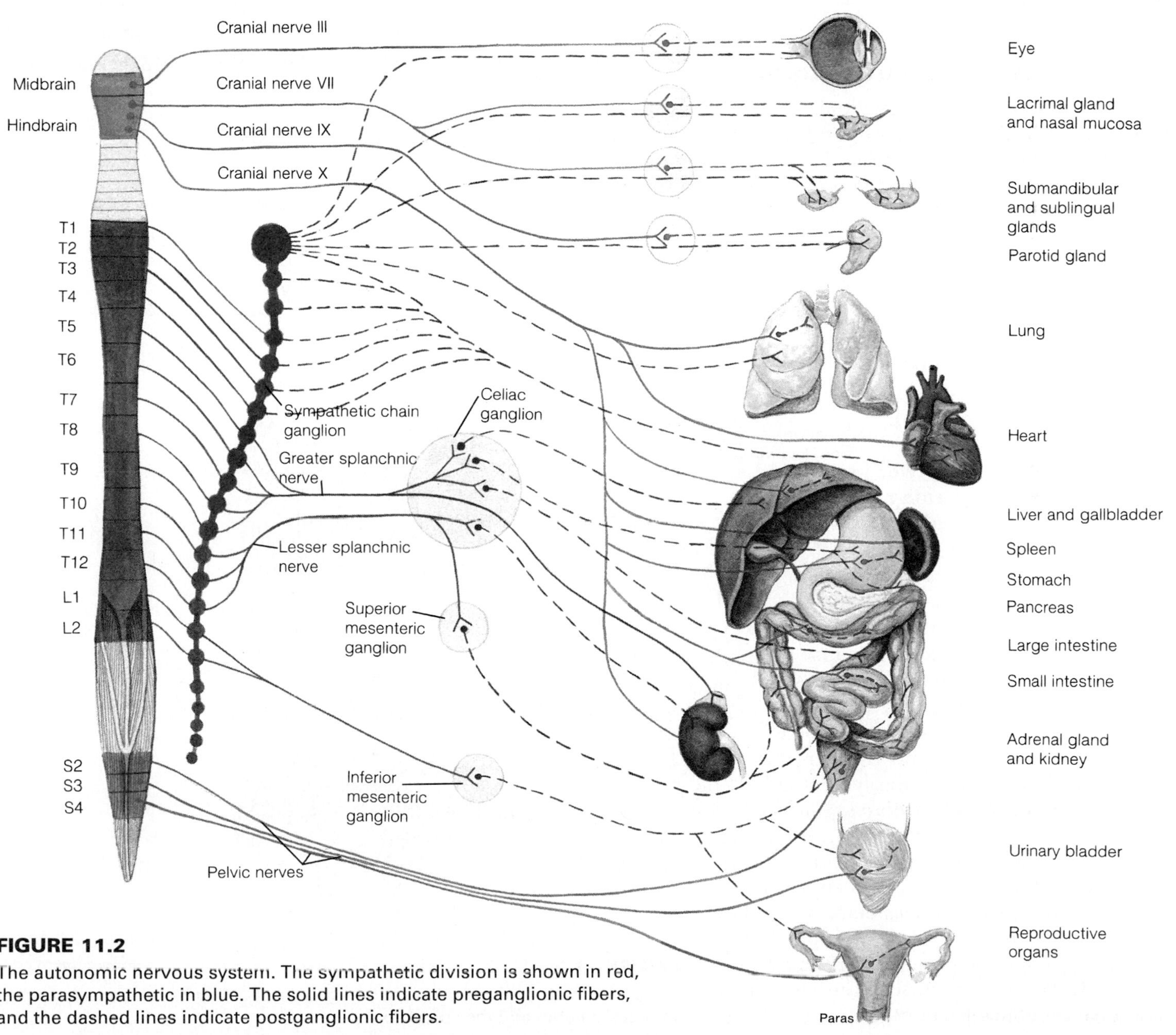

FIGURE 11.2
The autonomic nervous system. The sympathetic division is shown in red, the parasympathetic in blue. The solid lines indicate preganglionic fibers, and the dashed lines indicate postganglionic fibers.

The release of the neurotransmitter acetylcholine (ACh) from somatic motor neurons always stimulates the effector organ (skeletal muscles). By contrast, some autonomic axons release transmitters that inhibit the activity of their effectors (see chapter 9, module 9.6). An increase in the activity of the vagus nerve that supplies inhibitory fibers to the heart, for example, will slow the heart rate. This effect is antagonistic to that of the sympathetic nerve activity, which stimulates an increase in cardiac rate.

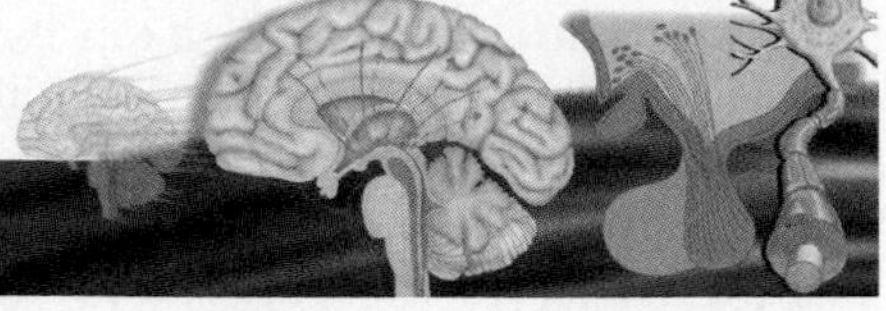

11.2 Structure of the Sympathetic Division

Most preganglionic neurons of the sympathetic division synapse in a chain of ganglia just outside the spinal cord, or in collateral ganglia within the abdominal cavity. Some preganglionic sympathetic axons synapse within the adrenal medulla, thus stimulating this gland.

Sympathetic Chain of Ganglia

The cell bodies of the preganglionic sympathetic neurons are located in the spinal cord, and their axons exit the spinal cord from the first thoracic (T1) to the second lumbar (L2) level. These axons leave the spinal cord in the anterior (ventral) roots of spinal nerves, and so accompany the somatic motor axons as they join with sensory neurons to form the spinal nerves. The sympathetic axons, however, soon separate from the somatic motor fibers and take a "side road" to the sympathetic ganglia (fig. 11.3). These ganglia, called **paravertebral ganglia,** lie in two vertical rows—one along each side of the spinal cord. Ganglia within each row are interconnected, forming a *sympathetic chain of ganglia* containing the cell bodies of the postganglionic sympathetic neurons. Most of the preganglionic axons synapse with the postganglionic neurons within this chain of ganglia.

Since the preganglionic sympathetic fibers are myelinated and thus appear white, the "side roads" to the sympathetic ganglia are called *white rami communicantes* (singular, *ramus communicans*). Since the fibers of the postganglionic sympathetic neurons are unmyelinated and thus appear gray, they form the *gray rami communicantes* as they go back to rejoin the fibers of the spinal nerve (fig. 11.4). It is these postganglionic axons that innervate the smooth muscle, cardiac muscle, and glands of the effector organs.

Collateral Ganglia

Many preganglionic fibers that exit the spinal cord in the abdomen pass through the sympathetic chain of ganglia without synapsing. Beyond the sympathetic chain, these preganglionic fibers form *splanchnic (splangk'nik) nerves.* Preganglionic fibers in the splanchnic nerves synapse in **collateral ganglia** (fig. 11.5).

ramus: L. *ramus,* a branch
splanchnic: Gk. *splanchono-,* relating to viscera

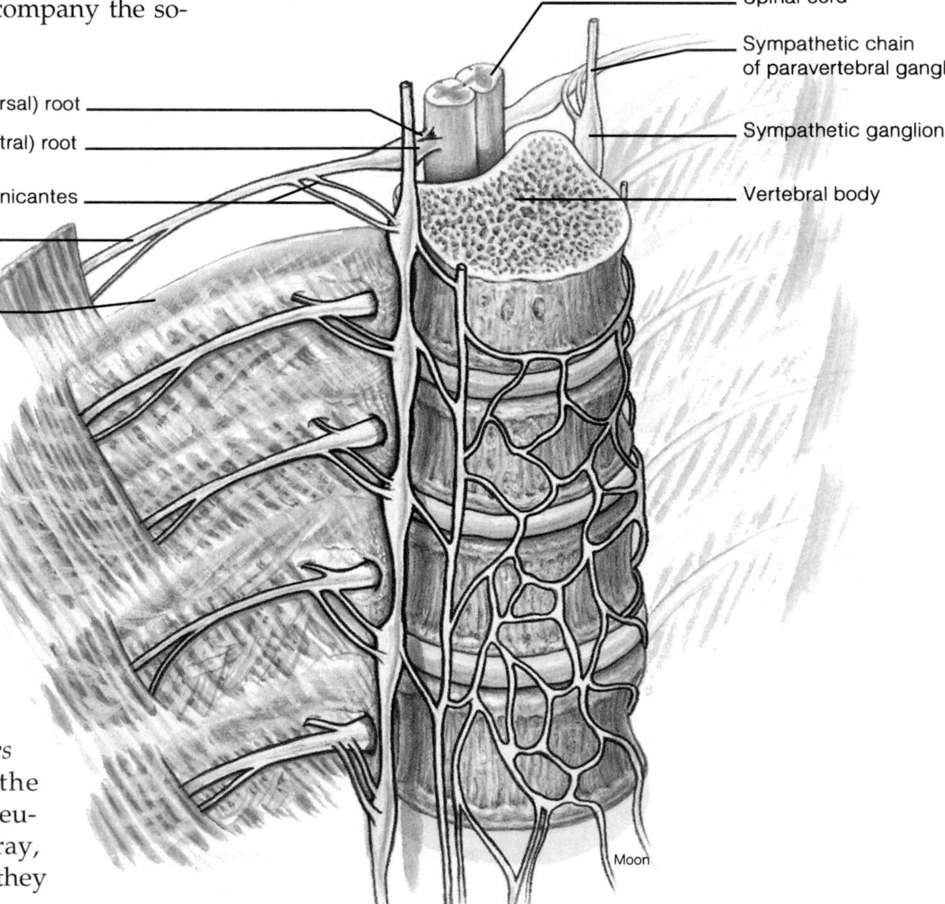

FIGURE 11.3
The sympathetic chain of paravertebral ganglia showing its relationship to the vertebral column and the spinal cord.

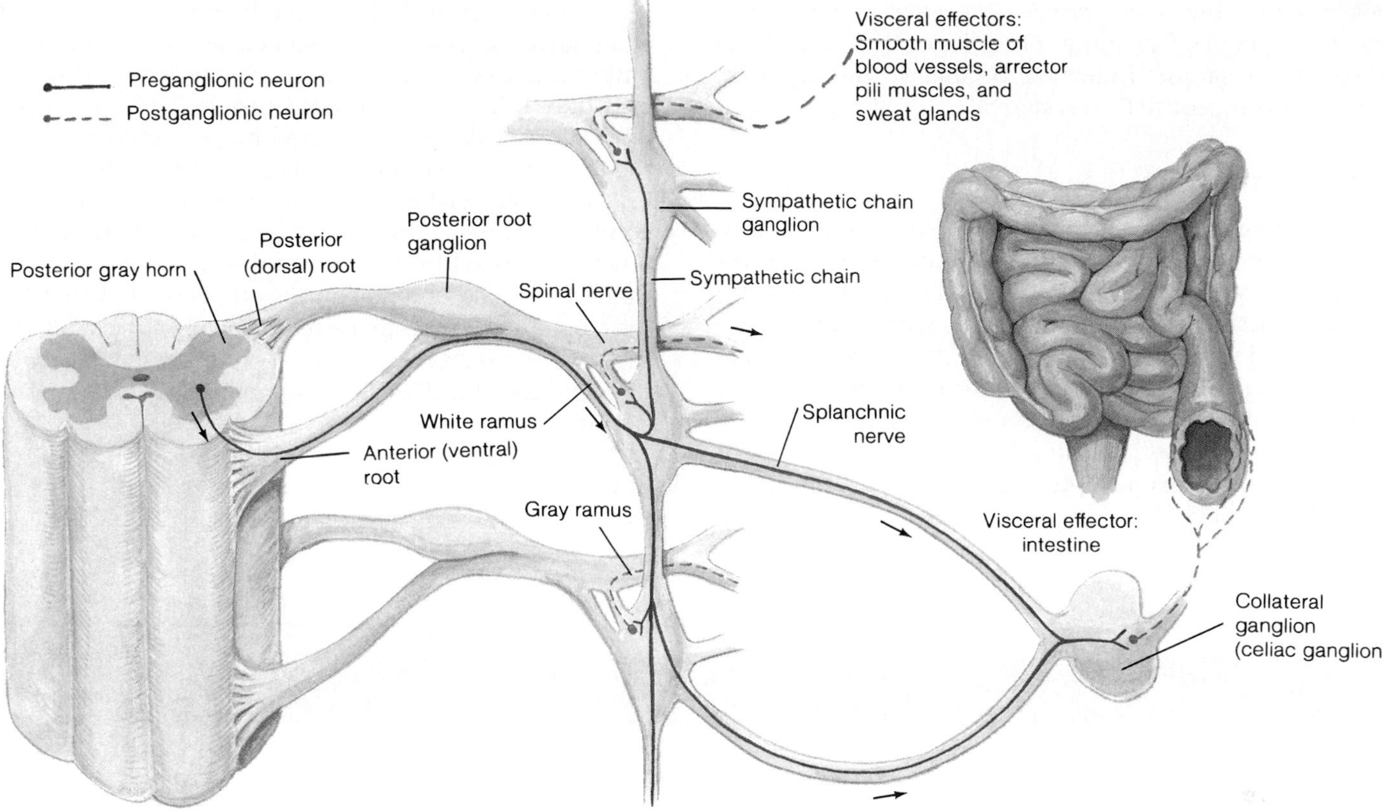

FIGURE 11.4

Sympathetic chain ganglia, the sympathetic chain, and rami communicantes of the sympathetic division of the ANS. (Solid lines = preganglionic fibers; dashed lines = postganglionic fibers.)

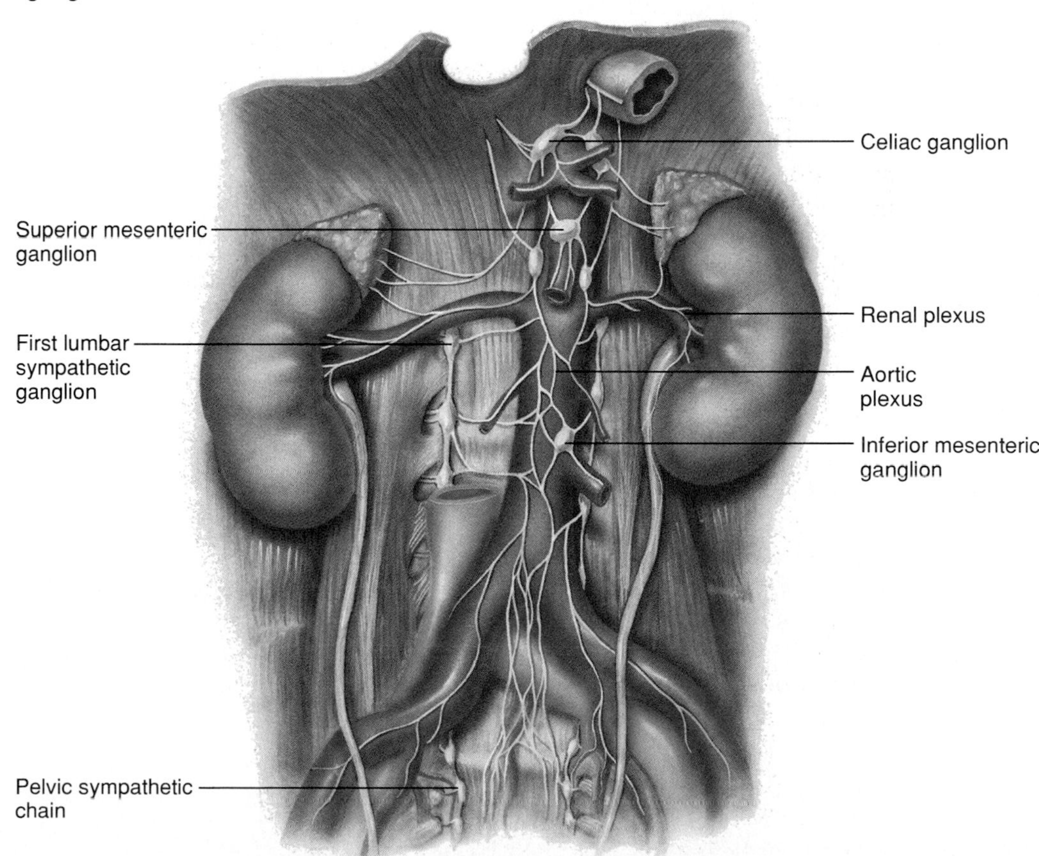

FIGURE 11.5

The collateral sympathetic ganglia: the celiac and the superior and inferior mesenteric ganglia.

These include the *celiac, superior mesenteric,* and *inferior mesenteric ganglia.* Postganglionic fibers that arise from the collateral ganglia innervate organs of the digestive, urinary, and reproductive systems.

Adrenal Glands

The adrenal glands are located above each kidney. Each adrenal is composed of two parts: an outer **adrenal cortex** and an inner **adrenal medulla.** These two parts are actually two functionally different glands, with different embryonic origins, different hormones, and different regulatory mechanisms. The adrenal cortex secretes steroid hormones; the adrenal medulla secretes the hormone **epinephrine** (*ep″ĭ-nef′rin*) (also called *adrenaline*) and, to a lesser degree, **norepinephrine,** when it is stimulated by the sympathetic division of the ANS.

The adrenal medulla can be thought of as a modified sympathetic ganglion; its cells are derived from the same embryonic tissue that forms postganglionic sympathetic neurons. Like a sympathetic ganglion, the cells of the adrenal medulla are innervated by preganglionic sympathetic fibers. Unlike postganglionic neurons, however, the cells of the adrenal medulla do not have axons and do not conduct impulses. Instead, they secrete their hormones when stimulated by the preganglionic sympathetic axons. Epinephrine and norepinephrine are therefore secreted into the blood by the adrenal medulla whenever the sympathetic division of the ANS is active. For this reason, the two are often grouped together as the **sympathoadrenal system.**

adrenal: L. *ad,* to; *renes,* kidney
cortex: L. *cortex,* bark
medulla: L. *medulla,* marrow

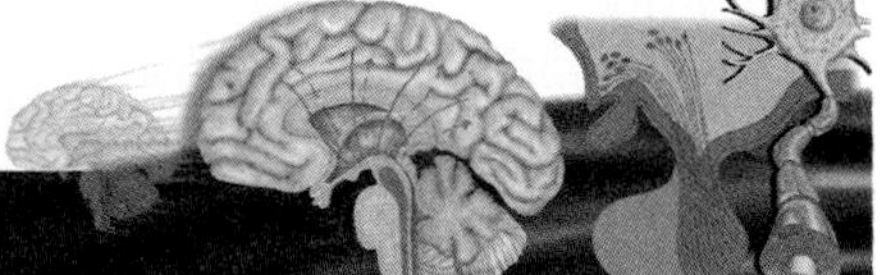

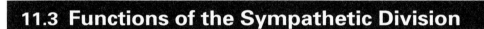

11.3 Functions of the Sympathetic Division

The sympathetic nerves that innervate visceral organs release norepinephrine, and this neurotransmitter, together with epinephrine secreted by the adrenal gland, prepares the body for fight or flight.

Acetylcholine (ACh) is the neurotransmitter of all preganglionic fibers (both sympathetic and parasympathetic), as illustrated in figure 11.6. Transmission at these synapses is thus said to be **cholinergic** (*ko″lĭ-ner′jik*). The neurotransmitter released by most postganglionic sympathetic nerve fibers is **norepinephrine** (*noradrenaline*). Transmission at these synapses is thus said to be **adrenergic** (*ad″rĕ-ner′jik*). Since it is the postganglionic neurons that innervate the visceral organs, the actions of the sympathetic division on the functions of the body are adrenergic effects.

Adrenergic stimulation—by epinephrine secreted into the blood by the adrenal medulla (fig. 11.6), and by norepinephrine released from sympathetic nerve endings—has different effects in different organs. Epinephrine and norepinephrine stimulate the heart to contract more rapidly and forcefully, and some blood vessels (those in the viscera and skin) to constrict, for example, but they cause dilation of other blood vessels (those in the skeletal muscles and heart), as well as dilation of the bronchioles (narrow airways) of the lungs. These different effects can be understood in light of the central "theme" of the sympathetic division: **fight or flight.**

The sympathoadrenal system prepares the body for physical exertion. This explains the increased pumping activity of the heart and the diversion of blood from the viscera and skin to the skeletal muscles and heart. In addition, stimulation through sympathetic nerves reduces the activity of the digestive system, so that blood can be shunted to the exercising muscle, and it stimulates the secretion of sweat glands. The "cold sweat" produced when you find yourself in a threatening situation results from blood vessel constriction and sweat gland secretion in the skin when the sympathetic division is activated.

Inside Information—Walter B. Cannon, a famous American physiologist of the early twentieth century, first used the phrase "fight or flight" to describe the series of biochemical changes in the body that constitutes an arousal reaction. It was also Cannon who coined the term *homeostasis* for the constancy of internal conditions that the body must actively maintain.

The different effects of sympathetic activity are associated with different **adrenergic receptor proteins** for epinephrine and norepinephrine. There are two major classes of these receptor proteins, designated **alpha-** (α)

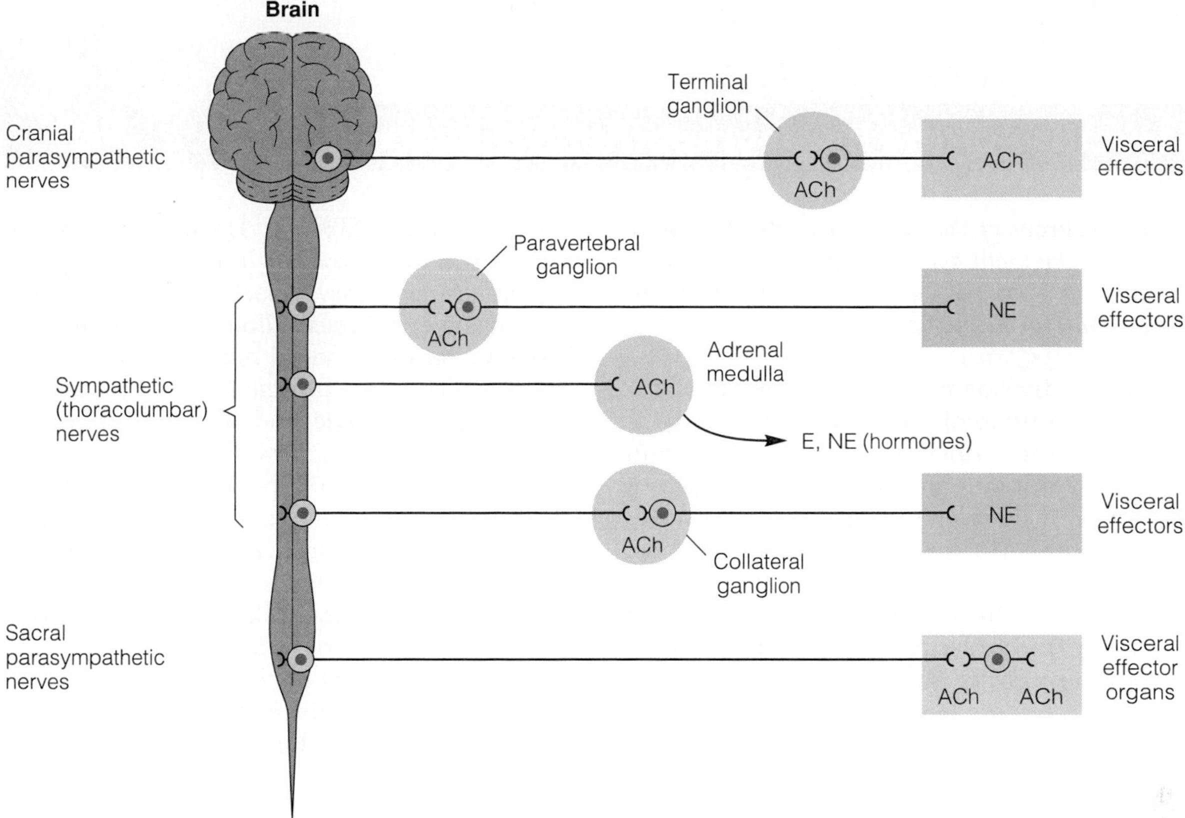

FIGURE 11.6

Neurotransmitters of the autonomic motor system (ACh = acetylcholine; NE = norepinephrine; E = epinephrine). Those nerves that release ACh are called cholinergic; those nerves that release NE are called adrenergic. The adrenal medulla secretes both epinephrine (85%) and norepinephrine (15%) as hormones into the blood.

and **beta- (β) adrenergic receptors.** Blood vessels in the skin, for example, contain alpha-adrenergic receptors, and their stimulation by norepinephrine causes constriction of the blood vessels. The binding of norepinephrine and epinephrine to beta-adrenergic receptors in the heart stimulates its pumping activity, and the binding of these hormones to beta-adrenergic receptors in the bronchioles promotes dilation of these airways.

There are two subtypes of each category of adrenergic receptor. These are designated by subscripts: α_1 and α_2; β_1 and β_2. Scientists have developed compounds that selectively bind to one type or the other of adrenergic receptor. If the adrenergic effect is promoted when the drug binds to the receptor, the drug is said to be an *agonist* of that effect. If the drug blocks the adrenergic effect, it is called an *antagonist.* Pharmaceutical drugs that selectively promote or block sympathoadrenal effects are in wide clinical use.

Many people with hypertension used to be treated with a beta-blocking drug known as propranolol. *This drug blocks β_1 receptors, which are located in the heart, and thereby lowers the cardiac rate and blood pressure. However, propranolol also blocks β_2 receptors, which are located in the bronchioles of the lungs. This reduces the bronchodilation effect of epinephrine, producing bronchoconstriction and asthma in susceptible people. A more specific β_1 antagonist,* atenolol, *is now used instead of propranolol to slow the cardiac rate and lower blood pressure without the adverse effect of bronchoconstriction.*

At one time, people with asthma inhaled an epinephrine spray that stimulated β_1 receptors in the heart as well as β_2 receptors in the airways. Now, drugs such as terbutaline *that selectively function as β_2 agonists are more commonly used to treat asthma.*

11.4 Structure of the Parasympathetic Division

The preganglionic neurons of the parasympathetic division of the ANS originate in the brain and sacral region of the spinal cord. They synapse in terminal ganglia located next to, or within, the organs that they innervate.

The parasympathetic division is also known as the *craniosacral division* of the autonomic nervous system. This is because its preganglionic fibers originate in the brain (specifically, in the midbrain and medulla oblongata of the brain stem) and in the second through fourth sacral levels of the spinal cord. Unlike sympathetic fibers, most parasympathetic fibers do not travel within spinal nerves. As a result, cutaneous effectors (blood vessels, sweat glands, and arrector pili muscles) and blood vessels in skeletal muscles receive sympathetic but not parasympathetic innervation.

The preganglionic fibers synapse in ganglia that are located next to—or actually within—the organs they innervate. These parasympathetic ganglia are generally called **terminal ganglia,** although a few parasympathetic ganglia in the head are also known by more specific names (table 11.1).

Four of the 12 pairs of cranial nerves contain preganglionic parasympathetic fibers. These are the oculomotor (III), facial (VII), glossopharyngeal (IX), and vagus (X) nerves. Parasympathetic fibers within the first three of these cranial nerves synapse in ganglia located in the head; fibers in the vagus nerve synapse in terminal ganglia located in many regions of the body.

The oculomotor nerve contains somatic motor and parasympathetic fibers that originate in the oculomotor nuclei of the midbrain. These parasympathetic fibers synapse in the *ciliary ganglion,* whose postganglionic fibers innervate the ciliary muscle and constrictor fibers in the iris of the eye. Preganglionic fibers that originate in the pons travel in the facial nerve to the *pterygopalatine ganglion,* which sends postganglionic fibers to the nasal mucosa, pharynx, palate, and lacrimal glands. Another group of fibers in the facial nerve terminate in the *submandibular ganglion,* which sends postganglionic fibers to the submandibular and sublingual glands. Preganglionic fibers of the glossopharyngeal nerve synapse in the *otic ganglion,* which sends postganglionic fibers to the parotid gland.

Nuclei in the medulla oblongata contribute preganglionic fibers to the very long vagus (tenth cranial) nerves. These preganglionic fibers travel through the neck to the thoracic cavity, and through the esophageal opening in the diaphragm to the abdominal cavity (fig. 11.7). In each region, some of these preganglionic fibers leave the main trunks of the vagus nerves and enter the visceral organs. Within these organs are terminal ganglia. It is here that the preganglionic axons synapse with the postganglionic neurons. The preganglionic vagus fibers are thus quite long, providing parasympathetic innervation to the heart, lungs, esophagus, stomach, pancreas, liver, small intestine, and the upper half of the large intestine. The postganglionic axons arising from the terminal ganglia, by contrast, are short, but it is these fibers that actually synapse with the smooth muscle, cardiac muscle, and glands.

vagus: L. *vagus,* wandering

Table 11.1 The Parasympathetic (Craniosacral) Division

Nerve	Origin of Preganglionic Fibers	Location of Terminal Ganglia	Effector Organs
Oculomotor (third cranial) nerve	Midbrain (cranial)	Ciliary ganglion	Eye (smooth muscle in iris and ciliary body)
Facial (seventh cranial) nerve	Medulla oblongata (cranial)	Pterygopalatine and submandibular ganglia	Lacrimal, mucous, and salivary glands
Glossopharyngeal (ninth cranial) nerve	Medulla oblongata (cranial)	Otic ganglion	Parotid gland
Vagus (tenth cranial) nerve	Medulla oblongata (cranial)	Terminal ganglia in or near organ	Heart, lungs, gastrointestinal tract, liver, pancreas
Pelvic spinal nerves	S2–S4 (sacral)	Terminal ganglia near organs	Lower half of large intestine, rectum, urinary bladder, and reproductive organs

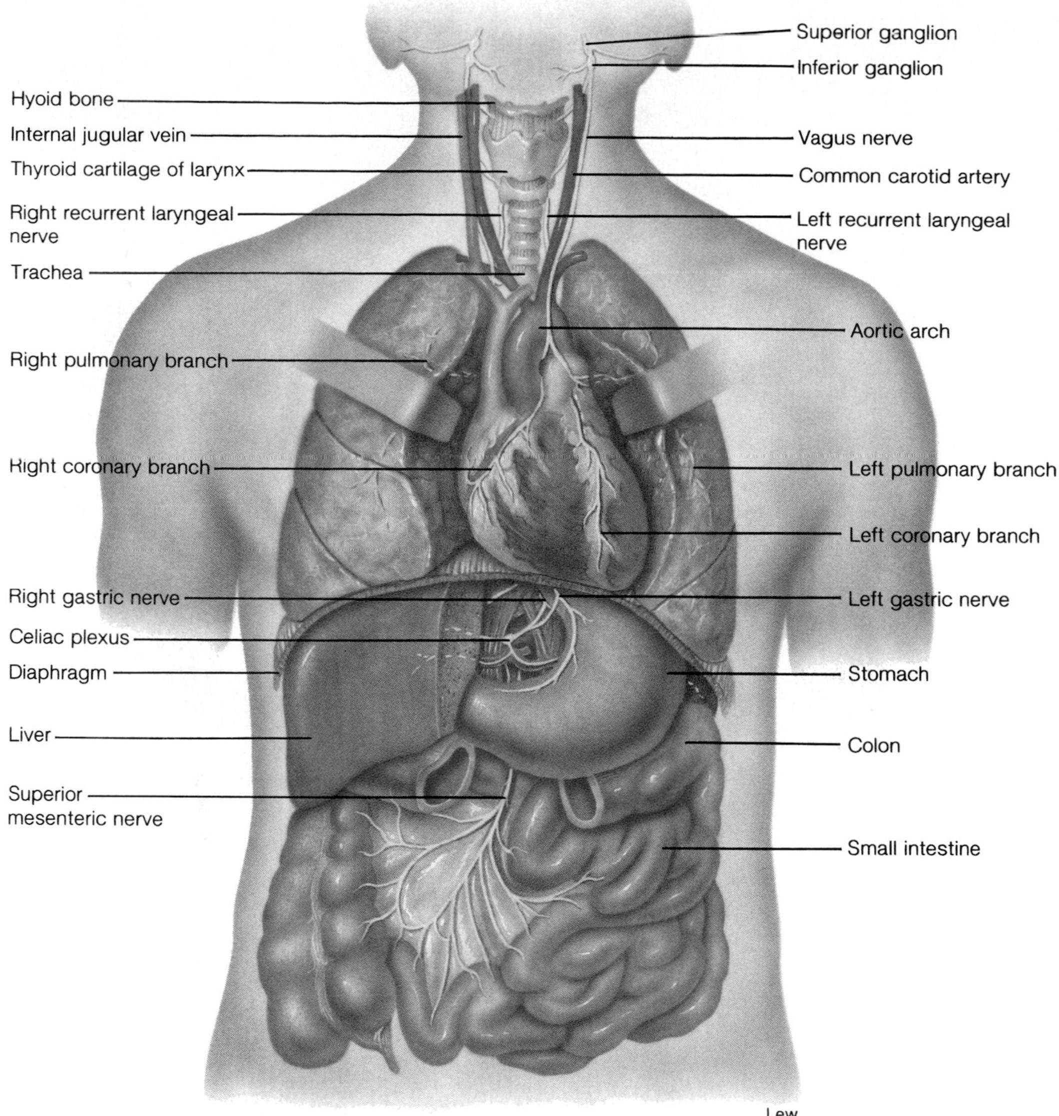

FIGURE 11.7
The vagus nerve and its branches provide parasympathetic innervation to internal organs in many regions of the body. All nerves shown are derived from the vagus nerve.

Preganglionic fibers from the sacral levels of the spinal cord provide parasympathetic innervation to the lower half of the large intestine, rectum, and to the urinary and reproductive systems. These fibers, like those of the vagus, synapse with terminal ganglia located within the effector organs.

11.5 Functions of the Parasympathetic Division

Parasympathetic stimulation slows the heart rate and promotes the activities of the digestive system. These effects are largely produced through the action of ACh in the effector organs.

In general, the effects of parasympathetic stimulation are opposite to those of sympathetic stimulation (table 11.2). Indeed, the parasympathetic division is sometimes called the "resting and digesting" division. The parasympathetic axons to the heart, for example, slow its rate of beat. This is antagonistic to the effects of the sympathetic axons to the heart, which increase the heart rate. Similarly, parasympathetic nerves stimulate constriction of the

TABLE 11.2 **Effects of Autonomic Nerve Stimulation on Various Visceral Effector Organs**

Effector Organ	Sympathetic Effect	Parasympathetic Effect
Eye		
Iris (pupillary dilator muscle)	Dilation of pupil	—
Iris (pupillary sphincter muscle)	—	Constriction of pupil
Ciliary muscle	Relaxation (for far vision)	Contraction (for near vision)
Glands		
Lacrimal (tear)	—	Stimulation of secretion
Sweat	Stimulation of secretion	—
Salivary	Decreased secretion; saliva becomes thick	Increased secretion; saliva becomes thin
Stomach	—	Stimulation of secretion
Intestine	—	Stimulation of secretion
Adrenal medulla	Stimulation of hormone secretion	—
Heart		
Rate	Increased	Decreased
Conduction	Increased rate	Decreased rate
Strength	Increased	—
Blood vessels	Mostly constriction; affects all organs	Dilation in a few organs (e.g., penis)
Lungs		
Bronchioles (tubes)	Dilation	Constriction
Mucous glands	Inhibition of secretion	Stimulation of secretion
Gastrointestinal tract		
Motility	Inhibition of movement	Stimulation of movement
Sphincters	Closing stimulated	Closing inhibited
Liver	Stimulation of glycogen hydrolysis	—
Adipocytes (fat cells)	Stimulation of fat hydrolysis	—
Pancreas	Inhibition of exocrine secretions	Stimulation of exocrine secretions
Spleen	Stimulation of contraction	—
Urinary bladder	Muscle tone aided	Stimulation of contraction
Arrector pili muscles	Stimulation of hair erection, causing goosebumps	—
Uterus	If pregnant, contraction; if not pregnant, relaxation	—
Penis	Ejaculation	Erection (due to vasodilation)

bronchioles of the lungs in contrast to the effects of sympathetic nerves, which cause bronchodilation. As described in chapter 12 (see fig. 12.12), parasympathetic nerves in the iris (colored portion) of the eye stimulate a circular layer of smooth muscle to contract, making the pupil smaller, whereas sympathetic nerves act on a different muscle layer, causing the pupil to enlarge.

Parasympathetic nerves stimulate secretions of mucous glands and digestive glands. They also stimulate contractions of the smooth muscle layers of the gastrointestinal tract, promoting the movement of food through this tract. In addition, parasympathetic nerves promote dilation of blood vessels in the digestive system. Thus, in contrast to the effects of sympathetic nerve activity, parasympathetic nerve activity promotes digestive functions.

Inside Information—Traditionally, the distinction between the somatic system and the autonomic nervous system was drawn on the basis that the former is under conscious control whereas the latter is not. Recently, however, it has been discovered that we have the remarkable ability to consciously influence autonomic activity using techniques such as biofeedback and meditation. This "discovery" comes as old news to Indian yogis, who have been exploiting this ability for generations.

The effects of parasympathetic stimulation on the visceral organs are generally produced by the binding of ACh to its receptor protein, and thus are classified as *cholinergic effects.* There are two subtypes of cholinergic receptors. Instead of being designated alpha and beta, like the adrenergic receptors, the cholinergic receptor subtypes are named according to the poisons that stimulate these receptors. **Nicotinic** (*nik″ŏ-tin′ik*) **receptors** are ACh receptors that are also stimulated by the drug nicotine (from tobacco plants). **Muscarinic** (*mus″kă-rin′ik*) **receptors** are ACh receptors that are also stimulated by muscarine, a poison derived from certain mushrooms. The ACh released by preganglionic axons (and by somatic motor axons) stimulates nicotinic receptors. The ACh released by postganglionic parasympathetic axons stimulates muscarinic receptors. Parasympathetic nerve effects, therefore, involve the binding of ACh to muscarinic ACh receptors in the cells of the target organs.

The different types of ACh receptors can also be blocked by different drugs. *Curare,* which blocks nicotinic receptors, is used clinically to promote relaxation of skeletal muscles. *Atropine,* also called *belladonna,* blocks the muscarinic receptors, and so inhibits the effects of parasympathetic nerves.

Atropine is derived from the deadly nightshade plant (Atropa belladonna). Extracts of this plant were used by women during the Middle Ages to dilate their pupils (atropine inhibits parasympathetic stimulation of the iris), which was thought to enhance their beauty (in Italian, bella *= beautiful;* donna *= woman). Atropine is used clinically today to dilate pupils during eye examinations, to dry mucous membranes of the respiratory tract prior to general anesthesia, to inhibit excessive gastric acid secretion, and to reduce spasmodic contractions of the lower gastrointestinal tract.*

Summary

11.1 Autonomic Neurons

1. Preganglionic neurons of the autonomic nervous system originate in the CNS and synapse with postganglionic neurons, which are located in autonomic ganglia.
2. It is always the postganglionic neuron that stimulates the involuntary effectors—smooth muscle, cardiac muscle, and glands.
3. The autonomic nervous system is subdivided into sympathetic and parasympathetic divisions.

11.2 Structure of the Sympathetic Division

1. Preganglionic axons leave the thoracic and lumbar regions of the spinal cord in the anterior (ventral) roots of spinal nerves.
2. Postganglionic neurons are located in a double chain of paravertebral ganglia, outside of the spinal cord, and in collateral ganglia within the abdominal cavity.
3. Some preganglionic axons enter the adrenal gland and stimulate the adrenal medulla to secrete the hormone epinephrine, and to a lesser degree, norepinephrine.

11.3 Functions of the Sympathetic Division

1. The sympathetic nerves and the adrenal medulla—functioning together as the sympathoadrenal system—prepare the body for "fight or flight."
2. The sympathoadrenal system stimulates the pumping activity of the heart and the diversion of blood from the viscera and skin toward the skeletal muscles and heart.
3. Sympathetic nerve effects are produced by the action of the neurotransmitter norepinephrine, which binds to adrenergic receptor proteins in the target cells.
4. There are two subtypes of adrenergic receptors, alpha and beta, that are active in different visceral organs.

11.4 Structure of the Parasympathetic Division

1. Preganglionic neurons leave the midbrain, medulla oblongata, and sacral region of the spinal cord and synapse with terminal ganglia throughout the body.
2. The terminal ganglia in the head have individual names and are located next to the innervated organs.
3. Most terminal ganglia in the thoracic and abdominal cavities are located within the visceral organs and receive preganglionic axons from the vagus nerve.

11.5 Functions of the Parasympathetic Division

1. In general, parasympathetic nerve stimulation produces responses that are opposite to those produced by sympathetic nerves.
2. Parasympathetic nerve stimulation increases the activity of the digestive system, stimulates mucous gland secretion, and promotes dilation of blood vessels in the gastrointestinal tract.
3. Parasympathetic nerve effects are produced by the release of ACh as a neurotransmitter.

Review Activities

Objective Questions

1. Which of the following statements about autonomic neurons is *true?* (*module 11.1*)
   a. They are motor neurons.
   b. Preganglionic neurons have cell bodies in the CNS.
   c. Postganglionic neurons innervate smooth muscles, cardiac muscle, and glands.
   d. All of the above are true.

2. Which of the following statements about the superior mesenteric ganglion is *true?* (*module 11.2*)
   a. It is a parasympathetic ganglion.
   b. It is a paravertebral sympathetic ganglion.
   c. It is located in the head.
   d. It contains the cell bodies of postganglionic sympathetic neurons.

3. The adrenal medulla is stimulated by (*module 11.2*)
   a. preganglionic sympathetic axons.
   b. postganglionic sympathetic axons.
   c. preganglionic parasympathetic axons.
   d. postganglionic parasympathetic axons.

4. Which of the following fibers release norepinephrine? (*module 11.3*)
   a. preganglionic sympathetic axons
   b. postganglionic sympathetic axons
   c. preganglionic parasympathetic axons
   d. postganglionic parasympathetic axons

5. Which of the following does *not* occur as a result of sympathetic nerve stimulation? (*module 11.3*)
   a. constriction of blood vessels in the skin
   b. dilation of bronchioles
   c. dilation of blood vessels in the gastrointestinal tract
   d. an increase in the heart rate

6. Propranolol is a beta-adrenergic receptor blocker. It could therefore be used to (*module 11.3*)
   a. dilate blood vessels.
   b. slow the heart rate.
   c. increase blood pressure.
   d. increase salivary gland secretion.

7. Parasympathetic ganglia are located (*module 11.4*)
   a. in a chain parallel to the spinal cord.
   b. in the posterior (dorsal) roots of spinal nerves.
   c. next to or within the organs innervated.
   d. in the brain.

8. Postganglionic axons from parasympathetic ganglia (*module 11.5*)
   a. release ACh.
   b. stimulate muscarinic receptors.
   c. generally antagonize the effects of sympathetic nerve stimulation.
   d. do all of the above.

9. Atropine blocks parasympathetic nerve effects. It could therefore be used to do all of the following *except* (*module 11.5*)
   a. dilate the pupils.
   b. decrease mucus secretion.
   c. decrease gastric acid secretion.
   d. decrease the heart rate.

Essay Questions

1. Compare the distribution of ganglia in the sympathetic and parasympathetic divisions. (*module 11.1*)
2. Describe the relationship between the sympathetic division and the adrenal medulla. (*module 11.2*)
3. Describe the different types of adrenergic receptors and explain how their differences are clinically exploited. (*module 11.3*)
4. Identify the origin of the vagus nerve and the location of the terminal ganglia that it innervates. (*module 11.4*)
5. Compare the effects of adrenergic and cholinergic stimulation on the cardiovascular and digestive systems. (*module 11.5*)
6. Compare nicotinic and muscarinic ACh receptors in terms of their location and how they are stimulated. (*module 11.5*)

Labeling Exercise

Label the structures indicated on the figure to the right.

1. ______________________
2. ______________________
3. ______________________
4. ______________________
5. ______________________

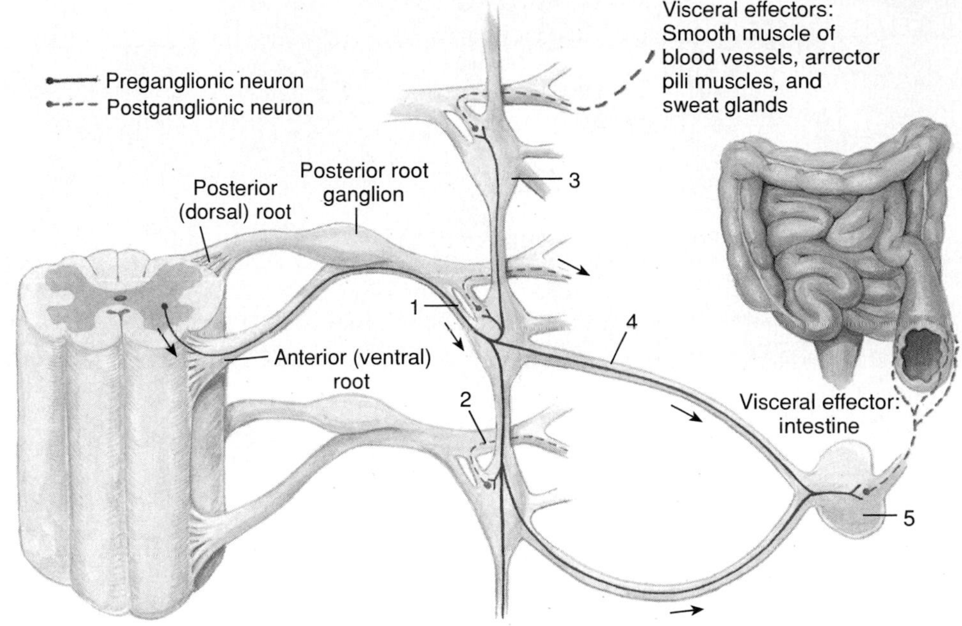

Critical Thinking Questions

1. Could parasympathetic ganglia form a chain of ganglia, analogous to the sympathetic chain? Why or why not?
2. If the adrenal medulla of each adrenal gland is removed from a laboratory rat, the rat will still live. Why? (Hint: Is there a "backup" for the adrenal medulla?)
3. Suppose a man has fainted and you lift his wrist to feel for a pulse. How would his skin feel? How would you characterize his pulse? What specific roles would autonomic nerves have in these effects?
4. When you look up the ingredients in a cold medicine, you see that one is listed as an "alpha (α_1) agonist," given to promote vasoconstriction in the nasal mucosa. Another medication, given to treat high blood pressure, contains a "beta-adrenergic blocker (antagonist)." Would these drugs be likely to promote an asthma attack in a susceptible person? Explain.
5. Why would someone be given a prescription for atropine if they had gastritis? Why would the person's mouth feel dry after taking this drug?

CHAPTER

12

"Although we usually speak of five senses, in reality we possess many more." *page 301*

The Senses

Chapter Outline

Terms to Remember

exteroceptor (*ek"stĕ-ro-sep'tor*), 301
visceroceptor (*vis"er-ō-sep'tor*), 302
proprioceptor (*pro"pre-o-sep'tor*), 302
olfaction, 305
gustation, 305
conjunctiva (*kon"jungk-ti'vă*), 308
sclera (*skler'ă*), 310
cornea (*kor'ne-ă*), 310
choroid (*kor'oid*), 310
ciliary (*sil'e-er"e*) body, 310
lens, 310
iris, 312
retina (*ret'ĭ-nă*), 312
tympanic membrane, 319
auditory ossicles, 319
perilymph, 320
endolymph, 320
semicircular canals, 321
cochlea (*kok'le-ă*), 321
spiral organ, 322

Learning Objectives

After studying this chapter, you should be able to

- compare and contrast somatic, visceral, and special senses.
- distinguish between exteroceptors, visceroceptors, and proprioceptors, and give examples of each.
- explain the significance of pain and give examples of referred pain and phantom pain.
- describe the structures involved in the sense of smell and trace the sensory pathway of olfaction.
- describe the structures involved in the sense of taste and identify the cranial nerves that serve the tongue with sensory fibers.
- describe the accessory structures of the eye and the structure of the eyeball.
- trace the path of light rays through the eye and explain how these rays are focused on distant and near objects.
- describe the structures of the ear that relate to hearing, including their locations and functions.
- trace the path of sound waves through the ear and explain how these waves are transmitted and converted to nerve impulses.
- explain the mechanisms by which balance is maintained.

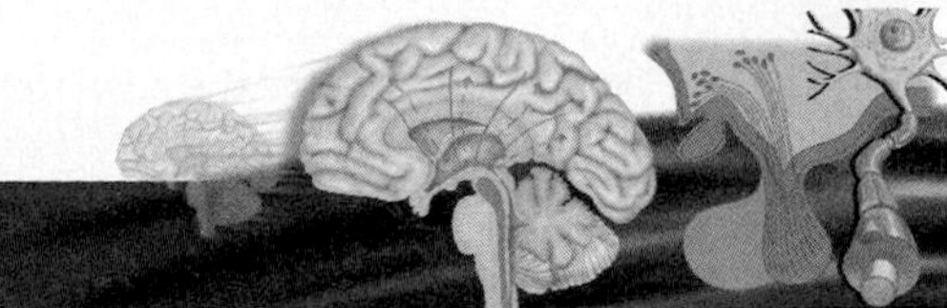

12.1 Overview of the Senses

Sensory organs are specialized extensions of the nervous system. They contain sensory neurons adapted to respond to specific stimuli and conduct messages to the brain. The senses are classified as general or special based on the structure of the receptors; they are also classified as somatic or visceral based on the location of the receptors.

The sense organs are actually extensions of the nervous system that respond to changes in the internal and external environment and transmit nerve impulses to the brain. They have been described as windows for the brain because it is through them that we achieve awareness of our environment. If we could not "sense" our environment and respond appropriately to maintain homeostasis, we could not survive very well on our own. The conscious interpretation of sensations in the brain is referred to as *perception.* Perceptions are the creations of our brain; in other words, we see, hear, and feel with our brain.

Only impulses reaching the cerebral cortex of the brain are consciously interpreted. If impulses end in the spinal cord or brain stem, they initiate a reflexive motor response rather than a conscious perception. Impulses reaching the cerebral cortex travel through nerve fibers composed of sensory, or ascending, tracts. Clusters of neuron cell bodies, called *nuclei,* are synaptic sites along sensory tracts within the CNS. The nuclei that sensory impulses pass through before reaching the cerebral cortex are located in the spinal cord and brain stem.

Classification of Receptors

Structurally, the sensory receptor can be the dendrites of sensory neurons, which are either free (such as those in the skin that respond to pain and temperature) or encapsuled within nonneural structures (table 12.1). Other receptors form from epithelial cells that synapse with sensory dendrites. These include taste buds on the tongue, photoreceptors in the eyes, and hair cells in the inner ears.

> *The highly sensitive fingertips are used in reading braille.* Braille *symbols consist of dots that are raised 1 mm from the surface of the page and separated from each other by 2.5 mm. Experienced braille readers can scan words at about the same speed that a sighted person can read aloud—a rate of about 100 words per minute.*

Although we usually speak of five senses, in reality we possess many more. The senses of the body can be classified as *general* or *special* according to the degree of complexity of their receptors and sensory pathways. **General senses** are widespread through the body and are structurally simple. Examples are touch, pressure, temperature, and pain. **Special senses** are localized in complex receptor organs and have extensive neural pathways (tracts) in the brain. Among the special senses are taste, smell, sight, hearing, and balance.

The senses can also be classified as *somatic* or *visceral* according to the location of the receptors. **Somatic senses** are those in which the receptors are localized within the body wall. These include the cutaneous (skin) receptors and those within muscles, tendons, and joints. **Visceral senses** are those in which the receptors are located within visceral organs. Both classification schemes may be used in describing some senses; for example, hearing (a special somatic sense) or pain from the gastrointestinal tract (a general visceral sense).

Senses are also classified according to the location of the receptors and the types of stimuli to which they respond. There are three basic kinds of receptors: *exteroceptors (ek"stĕ-ro-sep'torz), visceroceptors (vis"er-ŏ-sep'torz),* and *proprioceptors (pro"pre-o-sep'torz).*

Exteroceptors

Exteroceptors are specialized sensory nerve cells that are located near the surface of the body, where they respond to stimuli from the external environment. They include the following:

1. **cones and rods** in the retina of the eye—*photoreceptors;*
2. **hair cells** in the spiral organ (organ of Corti) within the inner ear—*mechanoreceptors;*
3. **olfactory receptors** in the nasal epithelium of the nasal cavity—*chemoreceptors;*
4. **taste receptors** on the tongue—*chemoreceptors;* and
5. **skin receptors** within the dermis—*tactile receptors* for touch; *mechanoreceptors,* for pressure; *thermoreceptors* for temperature; and *nociceptors (no"sĭ-sep'torz)* for pain. Pain receptors are located throughout the body, but only those located within the skin are classified as exteroceptors.

---

braille: from Louis Braille, French teacher of the blind, 1809–52
somatic: Gk. *somatikos,* body
visceral: L. *viscera,* body organs
nociceptor: L. *nocco,* to injure; *ceptus,* taken

Table 12.1 Cutaneous Receptors

Type	Location	Function
Corpuscles of touch (Meissner's corpuscles) (mechanoreceptors)	Papillae of dermis; numerous in hairless portions of body (eyelids, fingertips, lips, nipples, external genitalia)	Detect light motion against surface of skin
Free nerve endings (thermoreceptors; pain receptors)	Lower layers of epidermis	Detect changes in temperature; detect tissue damage
Root hair plexuses (tactile receptors)	Around hair follicles	Detect movement of hair
Lamellated (pacinian) corpuscles (mechanoreceptors)	Hypodermis; synovial membranes; perimysium; within certain visceral organs	Detect deep pressure and high-frequency vibration
Organs of Ruffini (mechanoreceptors)	Lower layers of dermis	Detect deep pressure and stretch
Bulbs of Krause (mechanoreceptors)	Dermis	Detect light pressure and low-frequency vibration

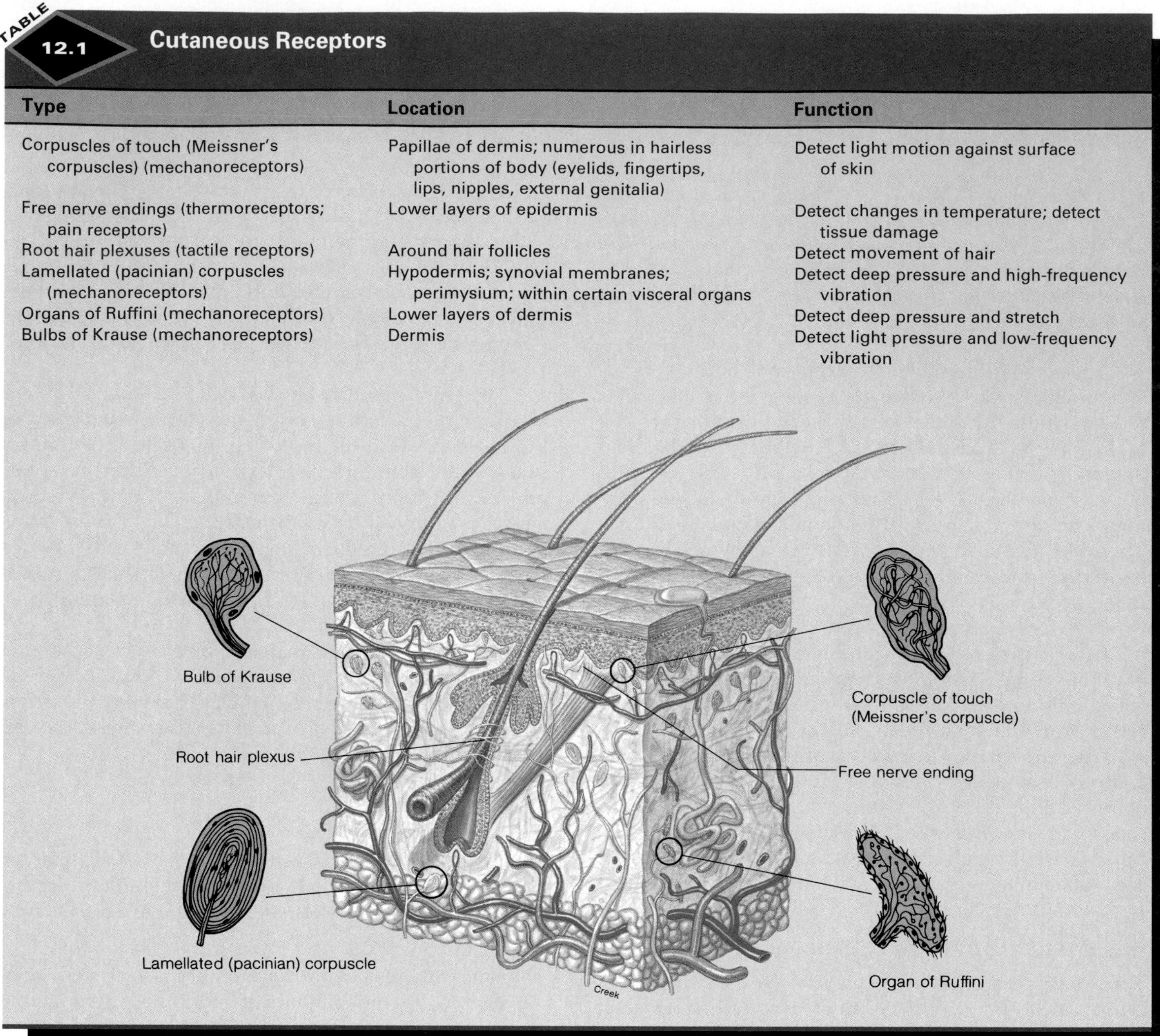

corpuscle: L. *corpusculum,* diminutive of *corpus,* body
Meissner's corpuscle: from George Meissner, German histologist, 1829–1905
pacinian corpuscle: from Filippo Pacini, Italian anatomist, 1812–83

organ of Ruffini: from Angelo Ruffini, Italian anatomist, 1864–1929
bulb of Krause: from Wilhelm J. F. Krause, German anatomist, 1833–1910

Visceroceptors

As the name implies visceroceptors are sensory nerve cells that produce sensations arising from the viscera, such as internal pain, hunger, thirst, fatigue, or nausea. Specialized visceroceptors located within the circulatory system are sensitive to changes in blood pressure; these are called *baroreceptors.*

Proprioceptors

Proprioceptors are sensory nerve cells that relay information about body position, equilibrium, and movement. They are located in the inner ear, joints, tendons, and muscles.

proprioceptor: L. *proprius,* one's own; *ceptus,* taken

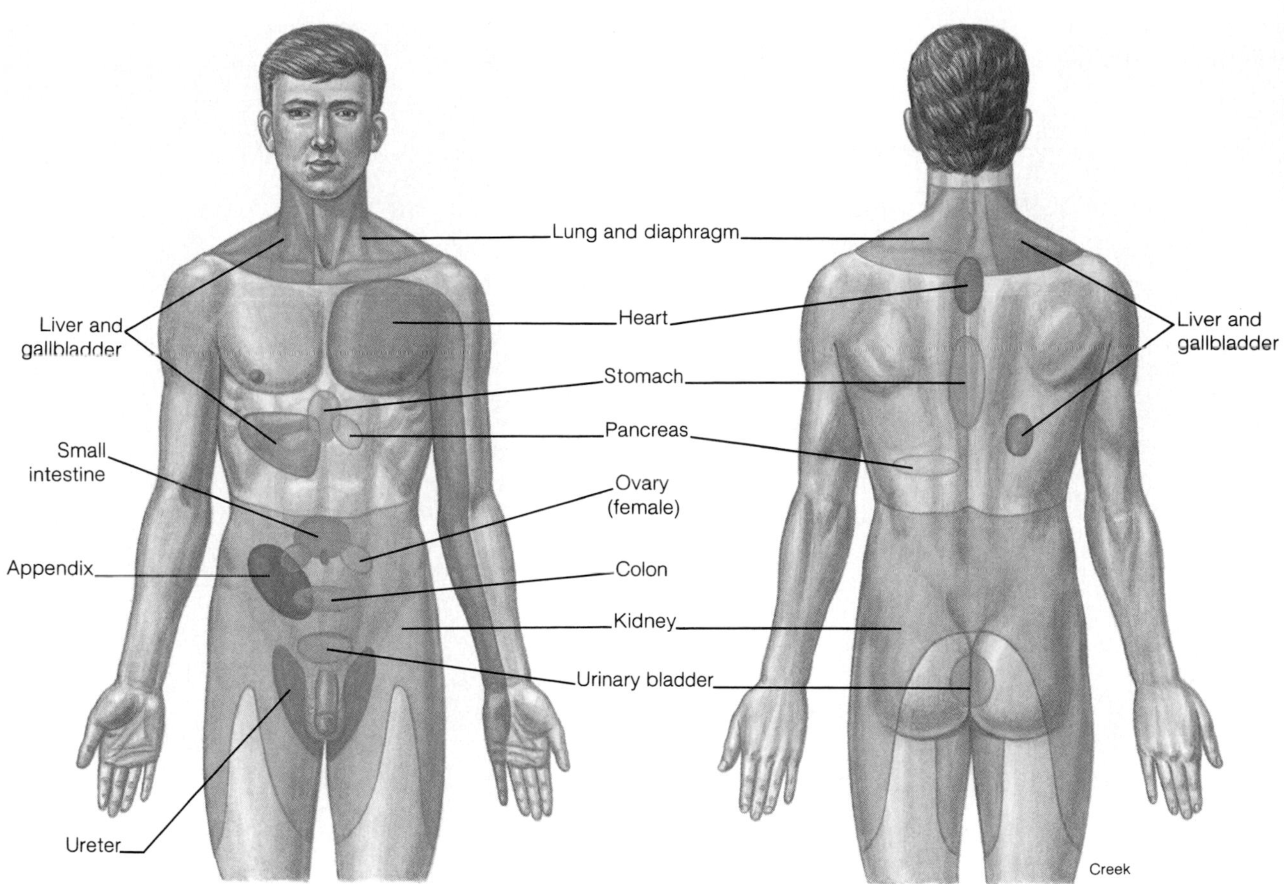

FIGURE 12.1
Sites of referred pain are areas of the body distant from the organs in which the pain actually arises.

Pain Receptors

Pain receptors respond to damage to tissues and are activated by all types of stimuli. The principal receptors for pain are the free nerve endings, several million of which are distributed throughout the skin and internal tissues. Pain receptors are sparse in most visceral organs and absent entirely within the nervous tissue of the brain. Although the free nerve endings are specialized to respond to tissue damage, all of the cutaneous receptors will relay impulses that are interpreted as pain if stimulated excessively.

The protective value of pain receptors is that they alert our consciousness when something is irritating or damaging the body. A person in pain is, of course, motivated to take some action to reduce it. The few unfortunate people born with a genetic disorder that makes them insensitive to pain usually die young from tissue deterioration or infections resulting from their wounds.

The sensation of pain can be clinically classified as **somatic pain** or **visceral pain.** Stimulation of the cutaneous pain receptors results in the perception of superficial somatic pain. Deep somatic pain comes from stimulation of receptors in skeletal muscles, joints, and tendons.

Stimulation of the receptors within the viscera causes the perception of visceral pain. Through precise neural pathways, the brain is able to perceive the area of stimulation and project the pain sensation back to that area. The sensation of pain from certain visceral organs, however, may not be perceived as arising from these organs but from other somatic locations. This phenomenon is known as **referred pain** (fig. 12.1). The sensation of referred pain is consistent from one person to another and is clinically important in diagnosing organ dysfunctions. The pain of a heart attack, for example, may be perceived subcutaneously over the heart and down the medial side of the left arm. Ulcers of the stomach may cause pain that is perceived as coming from the upper central region of the trunk. The particular nerve pathways established during embryonic development provide an explanation for referred pain.

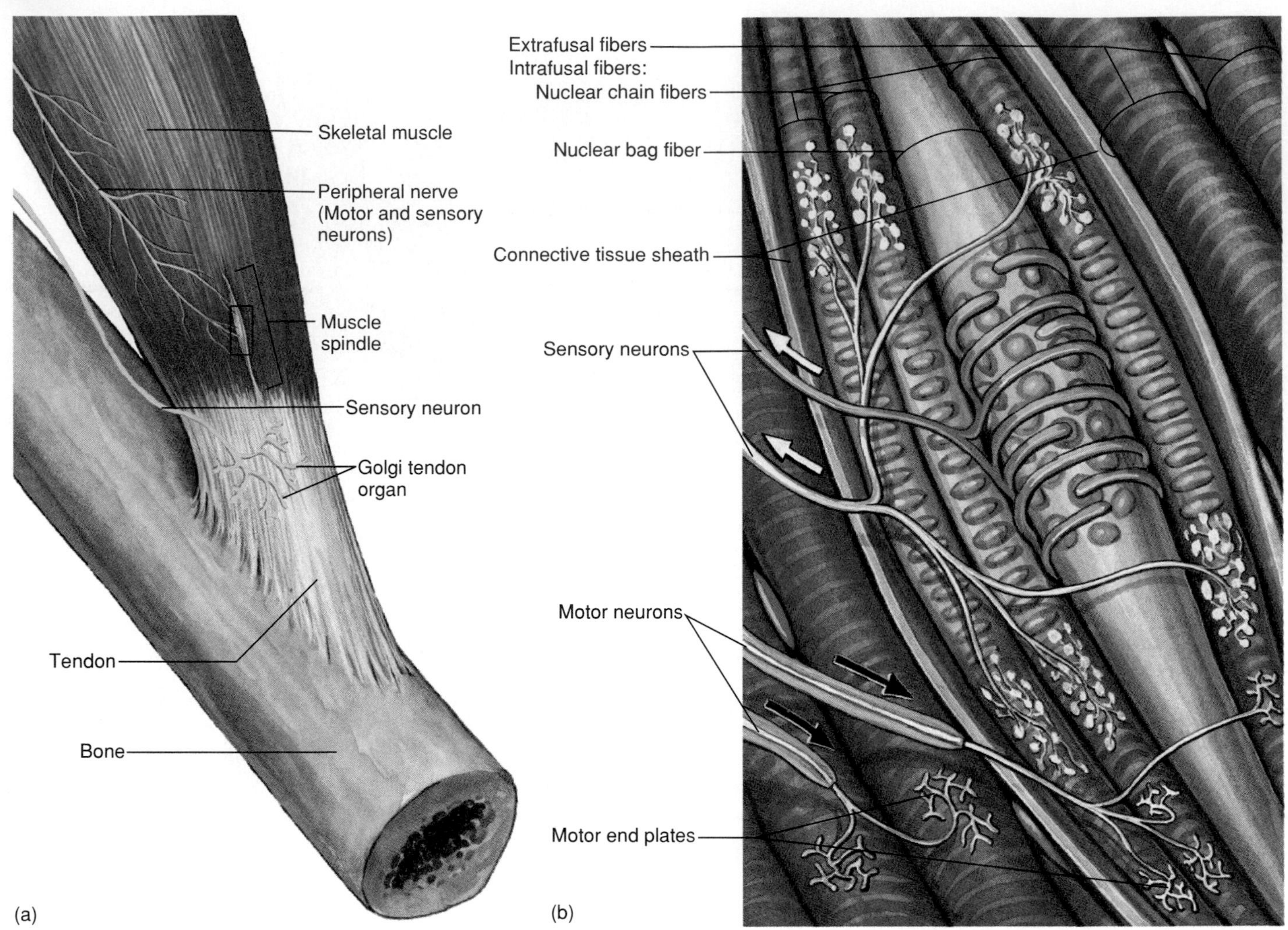

FIGURE 12.2

The structure of a Golgi tendon organ and muscle spindles within a skeletal muscle. (*a*) The location of most muscle spindles. (*b*) A magnification of the structure and innervation of muscle spindles.

Phantom pain is frequently experienced by amputees, who continue to feel pain from the body part that was amputated as if it were still there. After amputation, the severed sensory neurons heal and function in the remaining portion of the body limb. Although it is not known why impulses that are interpreted as pain are periodically sent through these neurons, the sensations evoked in the brain are projected to the amputated region, resulting in the feeling that the missing part is painful.

Proprioceptors

Proprioceptors advise the brain of our own movements (*proprius* means "one's own") by responding to changes in stretch and tension. Proprioceptor information is used to adjust the strength and timing of muscle contraction to produce coordinated movements.

The various types of proprioceptors are located in and around synovial joints, in skeletal muscle, between tendons and muscles, and in the inner ear.

1. **Joint kinesthetic receptors** are located in the connective tissue capsule in synovial joints, where they are stimulated by changes in body position as the joints are moved.
2. **Muscle spindles** are located in skeletal muscle, particularly in the muscles of the limbs. They consist of the endings of sensory neurons spiraled around specialized individual muscle fibers (fig. 12.2). Muscle spindles are stimulated by an increase in muscle tension due to the lengthening or stretching of the individual spindles, and thus provide information about the length of the muscle and the speed of muscle contraction.

3. **Neurotendinous receptors** (*Golgi tendon organs*) are located where a muscle attaches to a tendon (fig. 12.2). They are stimulated by the tension produced in a tendon when the attached muscle is either stretched or contracted.
4. **Sensory hair cells** of the inner ear are located in a tubular structure called the *membranous labyrinth.* Their function in equilibrium is discussed in module 12.6.

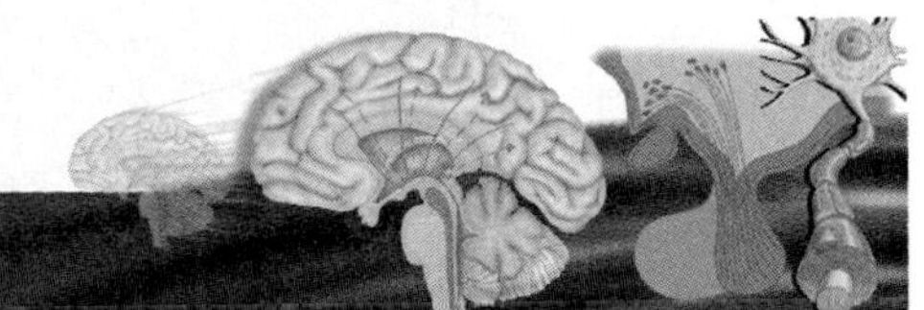

12.2 Senses of Smell and Taste

Olfactory (smell) and gustatory (taste) receptors are specialized dendritic endings of sensory neurons that respond to chemical stimuli. They transmit sensations through specific cranial nerves and neurological pathways directly to the cerebral cortex for perception.

Sense of Smell: Olfaction

Because we do not rely on smell for communicating or for finding food, the olfactory sense is probably the least important of our senses. It is more important in detecting the presence of an odor than its intensity. Smell functions closely with taste in that the receptors for both are *chemoreceptors,* which require dissolved substances for stimuli.

Olfactory receptor cells are located in the nasal epithelium within the roof of the nasal cavity on both sides of the nasal septum (fig. 12.3). Olfactory cells are moistened by the surrounding glandular goblet cells. The cell bodies of the bipolar olfactory cells lie between the supporting columnar cells. The free end of each olfactory cell contains several dendritic endings, called **olfactory hairs,** which lie in a thin mucous film on the epithelial surface. These hairs respond to airborne molecules that enter the nasal cavity.

Inside Information—Compared to other mammals, we do not have a keen sense of smell. For example, for a person to detect the odor of acetic acid, there has to be a concentration of at least 50 million molecules per cubic centimeter of air; for a dog, it is only half a million. Even so, trained fragrance testers can discern thousands of different scents, and can remember them from one testing session to the next.

The unmyelinated axons of the olfactory cells unite to form the **olfactory nerves,** which traverse the foramina of the cribriform plate and terminate in the paired masses of gray and white matter called the **olfactory bulbs.** The olfactory bulbs lie to the sides of the crista galli of the ethmoid bone (see fig. 6.22), beneath the frontal lobes of the cerebrum. Within the olfactory bulbs, the neurons of the olfactory nerves synapse with dendrites of the olfactory tract. Sensory impulses are conveyed along the olfactory tract and into the olfactory portion of the cerebral cortex, where they are perceived as odor.

The olfactory pathway is closely linked to the limbic system of the brain, which plays an important role in emotion and memory. Perhaps this explains why a particular smell sensation, more powerfully than other sensations, can evoke emotionally charged memories.

Sense of Taste: Gustation

The taste receptors are located in specialized sensory organs called **taste buds,** which are numerous on the surface of the tongue (fig. 12.4). The cylindrical taste bud is composed of many **receptor cells** (fig. 12.4*c*). The exposed surface of the receptor cells support **gustatory hairs** that project through an opening in the taste bud called the

gustatory: L. *gustare,* to taste

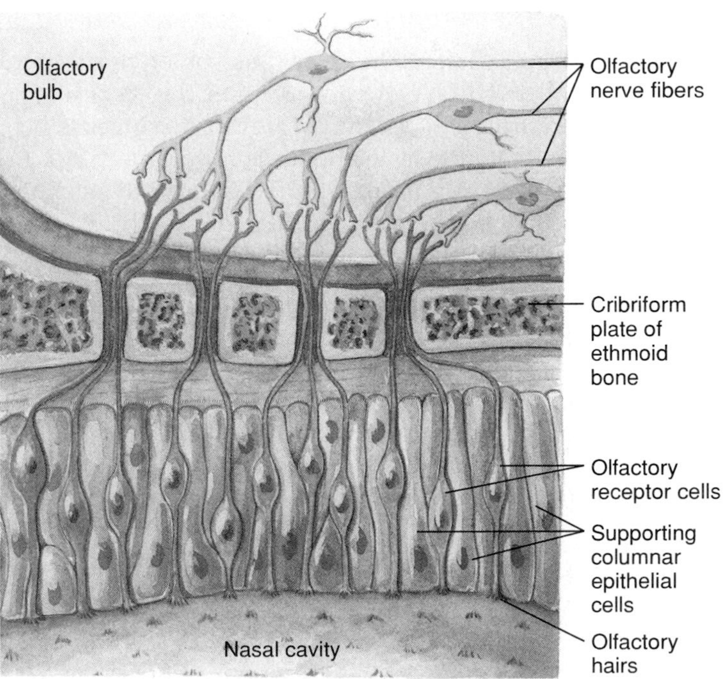

FIGURE 12.3
The olfactory receptor area at the roof of the nasal cavity.

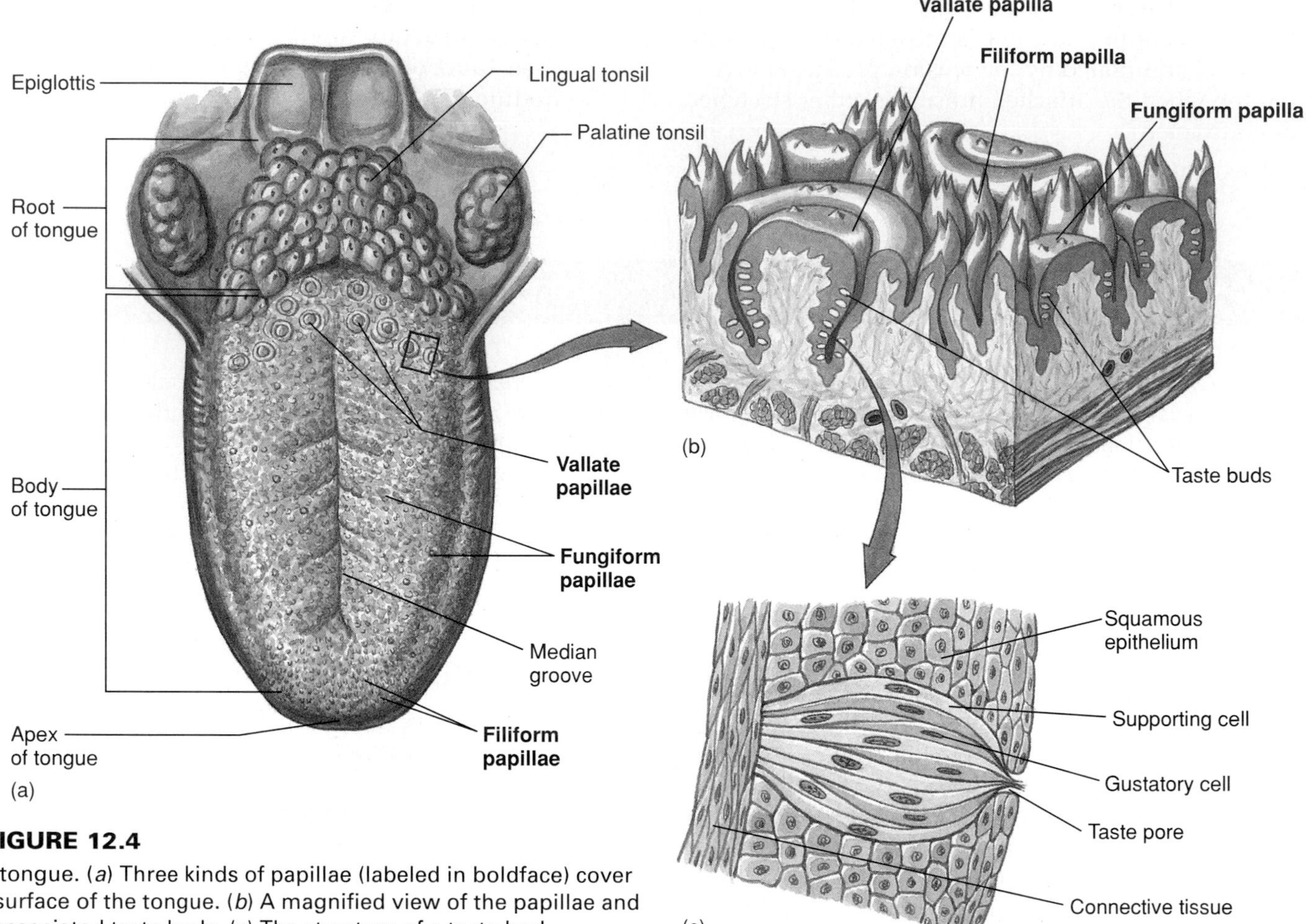

FIGURE 12.4
The tongue. (*a*) Three kinds of papillae (labeled in boldface) cover the surface of the tongue. (*b*) A magnified view of the papillae and the associated taste buds. (*c*) The structure of a taste bud.

taste pore. The gustatory hairs are the sensitive portion of the receptor cells. Saliva provides the necessary moistened environment for molecules of dissolved food particles to activate the taste bud receptor cells.

Most taste buds are found in projections of the tongue mucosa called *papillae.* These papillae are of three basic types (fig. 12.4*a,b*):

1. **vallate papillae**—the largest but least numerous papillae, which are arranged in an inverted V-shape pattern on the back of the tongue;
2. **fungiform papillae**—knoblike papillae, present on the tip and sides of the tongue; and
3. **filiform papillae**—short, threadlike papillae, located on the anterior two-thirds of the tongue.

Taste buds are found only in the vallate and fungiform papillae. The filiform papillae, although the most numerous of the human tongue papillae, are not involved in the perception of taste. Their outer cell layers are continuously converted into scalelike projections, which give the tongue's surface its somewhat abrasive feel.

papilla: L. *papillae,* nipple

There are only four basic tastes, which are sensed most acutely on particular parts of the tongue. These are *sweet* (tip of tongue), *sour* (sides of tongue), *bitter* (back of tongue), and *salty* (over most of the tongue, especially on the sides). The wide variety of tastes we perceive presumably results from combining the four basic types. However, because the sense of taste is intimately associated with the sense of smell, most foods taste bland to someone suffering from a cold. In fact, when your nose is stuffed, you lose about 80% of your ability to detect taste.

Sour taste is produced by hydrogen ions (H^+); all acids therefore taste sour. Most organic molecules, particularly sugars, taste sweet to varying degrees. Only table salt (NaCl) has a pure salty taste. Other salts, such as KCl (commonly used in place of NaCl by people with hypertension), taste salty but have bitter overtones. Bitter taste is evoked by quinine and seemingly unrelated molecules.

The sensory pathway that relays taste sensations to the brain involves mainly two cranial nerves (fig. 12.5). Taste buds on the posterior third of the tongue have a sensory pathway through the *glossopharyngeal nerve,* whereas the anterior two-thirds of the tongue is served by the *chorda tympani branch of the facial nerve.*

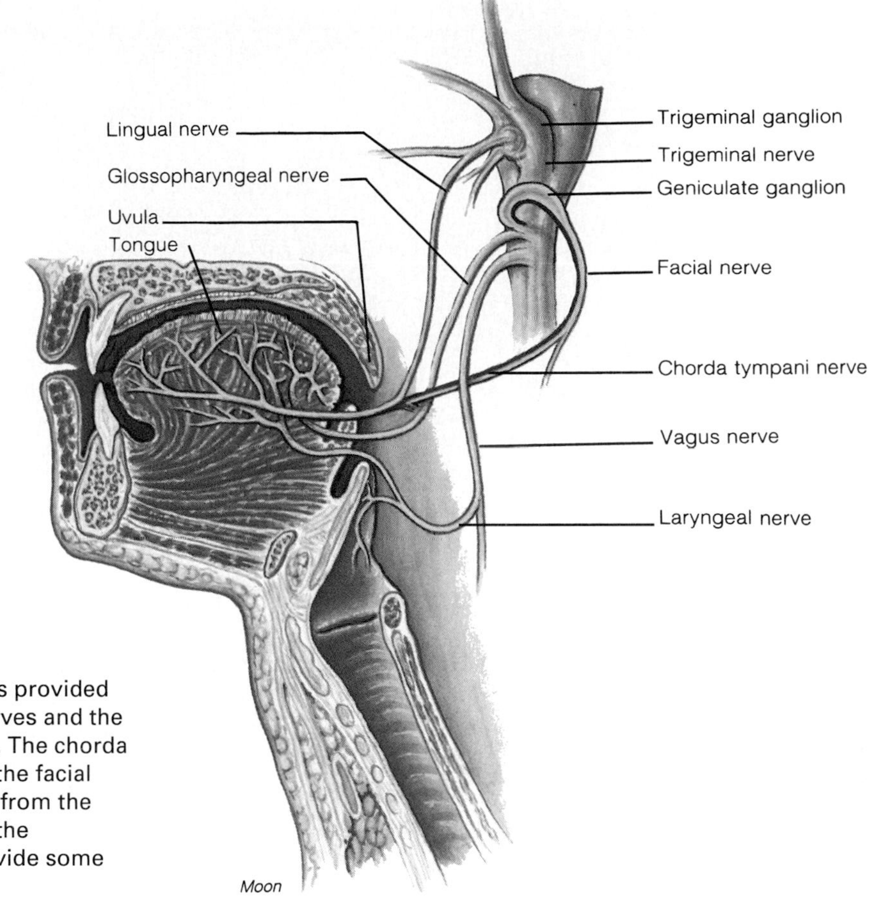

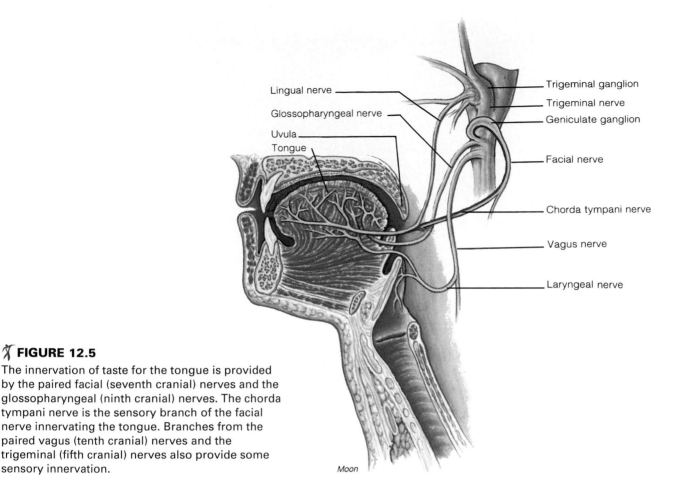

FIGURE 12.5
The innervation of taste for the tongue is provided by the paired facial (seventh cranial) nerves and the glossopharyngeal (ninth cranial) nerves. The chorda tympani nerve is the sensory branch of the facial nerve innervating the tongue. Branches from the paired vagus (tenth cranial) nerves and the trigeminal (fifth cranial) nerves also provide some sensory innervation.

12.3 Sense of Sight: Accessory Structures of the Eye

Accessory structures of the eye either protect the eyeball or provide eye movement. Protective structures include the orbit, eyebrow, eyelids, eyelashes, conjunctiva, and lacrimal apparatus. Eyeball movements are made possible by extrinsic eye muscles that extend from the orbit to the sclera of the eye.

Protective Structures of the Eye

Each eyeball is positioned in a protective bony depression in the skull called the **orbit** (see fig. 6.16). The **eyebrows** (fig. 12.6) are integumentary structures positioned on the skin along the superior orbital ridges. Eyebrows effectively shade the eyes from the sun and prevent perspiration or falling particles from getting into the eyes. Underneath the skin of each eyebrow is part of the orbicularis oculi muscle and part of the corrugator muscle (see fig. 8.18). Contraction of either of these muscles causes the eyebrow to move, often reflexively, to protect the eye.

Eyelids (*palpebrae*) develop as reinforced folds of skin with attached skeletal muscle that enables them to move. Contraction of the orbicularis oculi muscle closes the eyelids over the eye, and contraction of the levator palpebrae

palpebra: L. *palpebra,* eyelid (related to *palpare,* to pat gently)

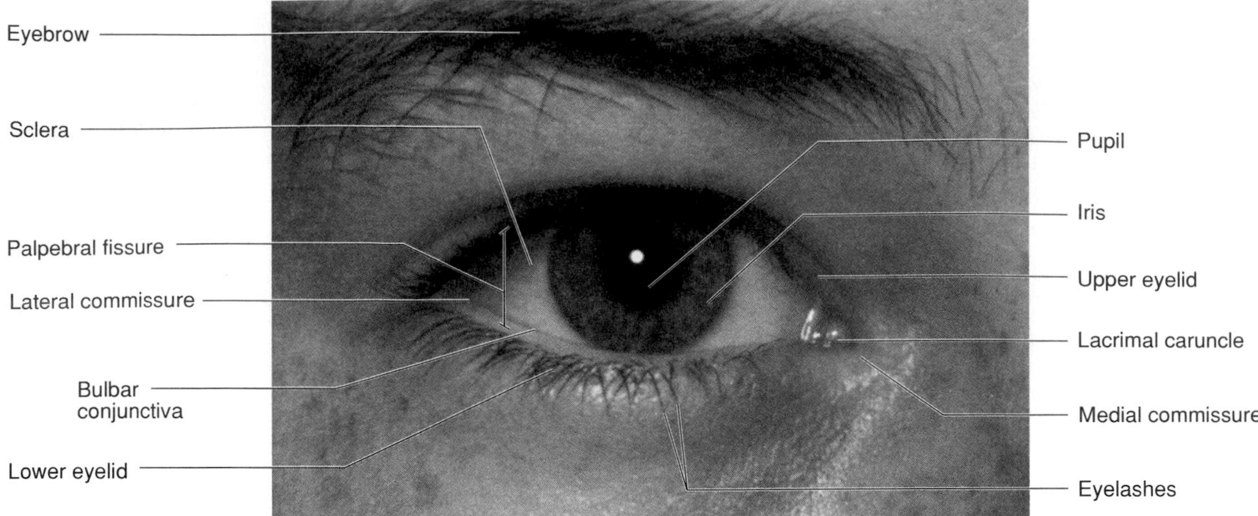

FIGURE 12.6
The surface anatomy of the eye.

superioris muscle elevates the upper eyelid to expose the eye. To avoid a blurred image, the eyelid will generally blink when the eyeball moves to a new position of fixation.

Inside Information—On the average, you blink once every 7 seconds. Blinking keeps your eyes from drying out by moving fluid over the surface of the eyeballs. Reflexively blinking as a moving object approaches your eye is obviously of great protective value.

The **palpebral fissure** (fig. 12.6) is the space between the upper and lower eyelids. The shape of the palpebral fissure is elliptical when the eyes are open. The **commissures** (*canthi*) of the eye are the medial and lateral corners, where the eyelids come together. The medial commissure is broader than the lateral commissure and is characterized by a small, reddish fleshy elevation called the **lacrimal caruncle** (*kar'ung-kul*) that contains sebaceous and sweat glands. The caruncle produces the whitish secretion, commonly called "sleep dust," that sometimes collects during sleep.

In people of Asian descent, a fold of skin of the upper eyelid, called the *epicanthic fold,* may normally cover part of the medial commissure. An epicanthic fold may also be present in some infants with Down syndrome.

Each eyelid supports a row of numerous **eyelashes,** which protect the eye from airborne particles. The shaft of each eyelash is surrounded by a root hair plexus that makes the hair sensitive enough to cause a reflexive closure of the lids.

In addition to the layers of the skin and the underlying connective tissue and muscle, each eyelid contains a tarsal plate, tarsal glands, and conjunctiva. The fibrous **tarsal plates** maintain the shape of the eyelids (fig. 12.7). Specialized sebaceous glands, called **tarsal glands,** are embedded within the tarsal plates along the exposed inner surfaces of the eyelids. The oily secretions of the tarsal glands help to prevent the eyelids from sticking to each other.

The **conjunctiva** (*kon"jungk-ti'vă*) is a thin mucus-secreting epithelial membrane that lines the interior surface of each eyelid and exposed anterior surface of the eyeball (fig. 12.7). The *palpebral conjunctiva* is the portion that covers the tarsal plates of the eyelids. Where the conjunctiva reflects (folds back) onto the surface of the eyeball, it is known as the *bulbar conjunctiva.* This portion is transparent and especially thin where it covers the cornea. Because the conjunctiva is continuous from the eyelids to the anterior surface of the eyeball, a space called the **conjunctival sac** is present when the eyelids are closed. The conjunctival sac protects the eyeball by preventing objects (including a contact lens) from passing beyond the confines of the sac. The conjunctiva heals rapidly if scratched.

The **lacrimal apparatus** consists of the **lacrimal gland,** which secretes the lacrimal fluid (tears), and a series of ducts that drain the tears into the nasal cavity (fig. 12.8). *Lacrimal fluid* is a lubricating mucous secretion that contains a bactericidal enzyme called *lysozyme.* This enzyme reduces the likelihood of infections.

Humans are the only animals known to weep in response to emotional stress. While crying, the volume of lacrimal secretion is so great that the tears may spill over the edges of the eyelids and the nasal cavity fills with fluid. The crying response results from stimulation of the lacrimal glands by parasympathetic motor neurons.

commissure: L. *commissure,* a joining
caruncle: L. *caruncula,* diminutive of *caro,* flesh
tarsal: Gk. *tarsos,* flat basket

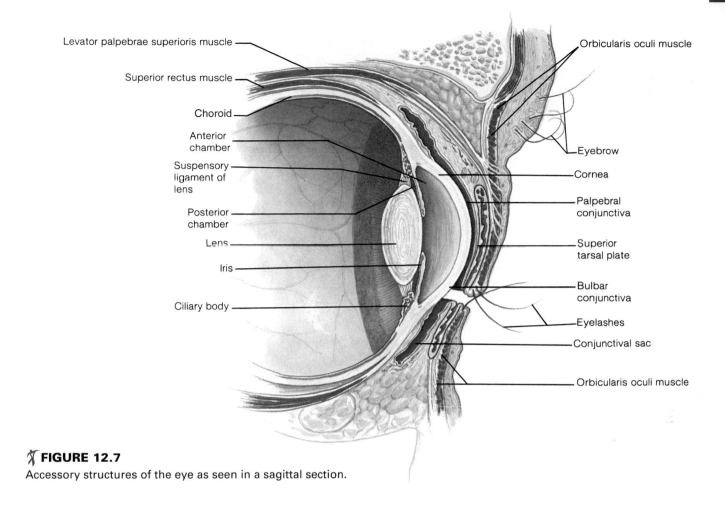

FIGURE 12.7
Accessory structures of the eye as seen in a sagittal section.

Extrinsic Eye Muscles

The six extrinsic eye muscles control the movements of the eyeball (see fig. 8.20). Each extrinsic eye muscle originates from the bony orbit and inserts by a tendinous attachment to the tough outer tunic of the eyeball. Four *recti muscles* (singular, *rectus*) maneuver the eyeball in the direction indicated by their names (*superior, inferior, lateral,* and *medial*); the two *oblique muscles* (*superior* and *inferior*) rotate the eyeball. Although stimulation of each muscle causes a precise movement of the eyeball, most of the movements involve the combined contraction of usually two muscles.

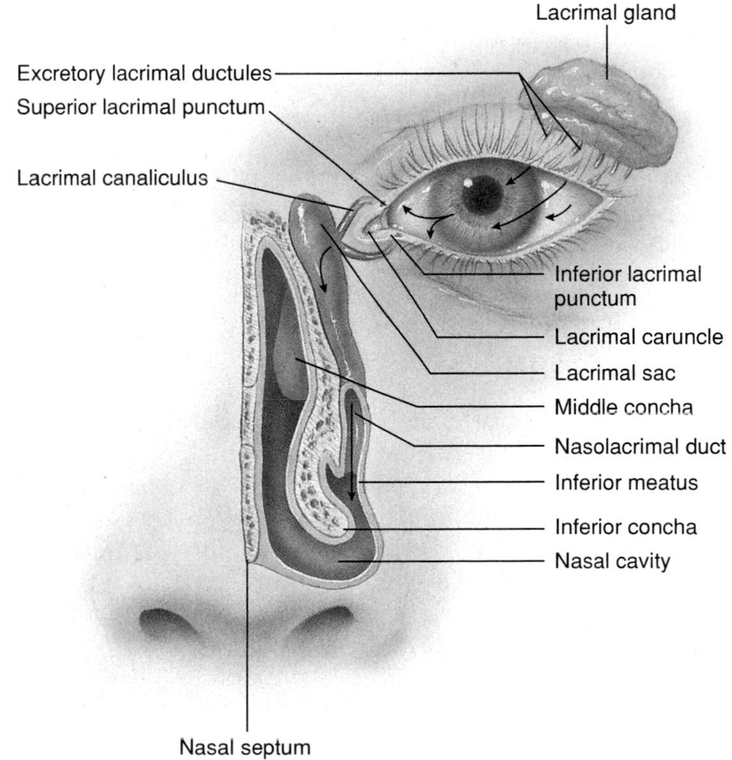

FIGURE 12.8
The lacrimal apparatus consists of the lacrimal gland and the drainage pathway of lacrimal fluid into the nasal cavity. The lacrimal gland produces lacrimal fluid (tears), which moistens and cleanses the conjunctiva that lines the underside of the eyelids and covers the anterior surface of the eyeball.

12.4 Sense of Sight: Structure of the Eye

The eye of an adult is spherical and is approximately 25 mm (1 in.) in diameter. The wall of the eye consists of three basic layers: the fibrous tunic, the vascular tunic, and the internal tunic, or retina. Rods and cones in the retina are sensitive to light.

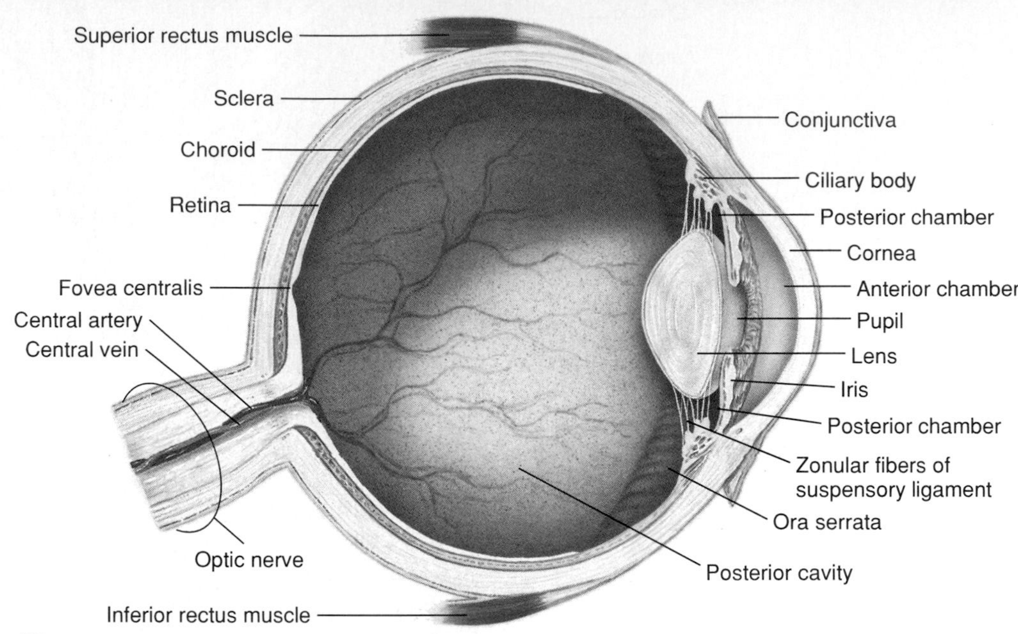

FIGURE 12.9
The internal anatomy of the eyeball.

Fibrous Tunic

The fibrous tunic is the outer layer of the eye. It is divided into two regions: the posterior five-sixths is the opaque *sclera,* and the anterior one-sixth is the transparent *cornea* (fig. 12.9 and table 12.2).

The toughened **sclera** (*skler'ă*) is the white of the eye. It is composed of tightly bound connective tissue, which gives shape to the eye and protects its inner structures. The large *optic nerve* exits through the sclera at the back of the eye. The transparent **cornea** (*kor'ne-ă*) is convex to refract (bend in a converging pattern) incoming light rays.

A cornea that does not transmit or refract light effectively can be replaced. In a surgical procedure called a corneal transplant (keratoplasty), *the abnormally shaped, scarred, or diseased cornea is removed, and a donor cornea is sutured into place.*

Vascular Tunic

The vascular tunic, or **uvea** (*yoo've-ă*) of the eye consists of the *choroid,* the *ciliary body,* and the *iris* (fig. 12.9).

The **choroid** (*kor'oid*) is a thin, highly vascular layer that lines most of the internal surface of the sclera. The **ciliary** (*sil'e-er"e*) **body** is the thickened anterior portion of the vascular tunic that forms an internal muscular ring toward the front of the eyeball. Bands of smooth muscle fibers, collectively called the *ciliary muscles,* are found within the ciliary body. Numerous extensions of the ciliary body called ciliary processes attach to the *zonular fibers,* which in turn attach to the *lens capsule* (see fig. 12.14). Collectively, the zonular fibers constitute the **suspensory ligament.** The transparent **lens** consists of tight layers of proteins arranged like the layers of an onion. The thin, clear **lens capsule** encloses the lens and provides attachment for the suspensory ligament.

The shape of the lens determines the degree to which light rays that pass through will be refracted, or bent. Constant tension of the suspensory ligament, when the ciliary muscles are relaxed, flattens the lens somewhat (fig. 12.10). Contraction of the ciliary muscles relaxes the suspensory ligament and makes the lens more spherical. The constant tension within the lens capsule causes the surface of the lens to become more convex when the suspensory ligament is not taut. A flattened lens permits viewing a distant object, whereas a rounded lens permits viewing a close object.

sclera: Gk. *skleros,* hard
optic: L. *optica,* see
cornea: L. *cornu,* horn
choroid: Gk. *chorion,* membrane
zonular: L. *zona,* a girdle

TABLE 12.2 **Location and Functions of Structures of the Eyeball**

Tunic and Structure	Location	Composition	Function
Fibrous tunic	Outer layer of eyeball	Avascular connective tissue	Gives shape to eyeball
Sclera	Posterior outer layer; white of the eye	Tightly bound elastic and collagen fibers	Supports and protects eyeball
Cornea	Anterior surface of eyeball	Tightly packed dense connective tissue—transparent and convex	Transmits and refracts light
Vascular tunic (uvea)	Middle layer of eyeball	Highly vascular pigmented tissue	Supplies blood; prevents reflection
Choroid	Middle layer in posterior portion of eyeball	Vascular layer	Supplies blood to eyeball
Ciliary body	Anterior portion of vascular tunic	Smooth muscle fibers and glandular epithelium	Supports lens through suspensory ligament and determines its thickness; secretes aqueous humor
Iris	Anterior portion of vascular tunic, continuous with ciliary body	Pigment cells and smooth muscle fibers	Regulates diameter of pupil, and hence the amount of light entering the vitreous chamber
Internal tunic	Inner layer of eyeball	Tightly packed photoreceptors, neurons, blood vessels, and connective tissue	Provides location and support for rods and cones
Retina	Principal portion of internal tunica	Photoreceptor neurons (rods and cones), bipolar neurons, and ganglion neurons	Photoreception; transmits impulses
Lens (not part of any tunic)	Between posterior and vitreous chambers; supported by suspensory ligament of ciliary body	Tightly arranged protein fibers; transparent	Refracts light and focuses onto fovea centralis

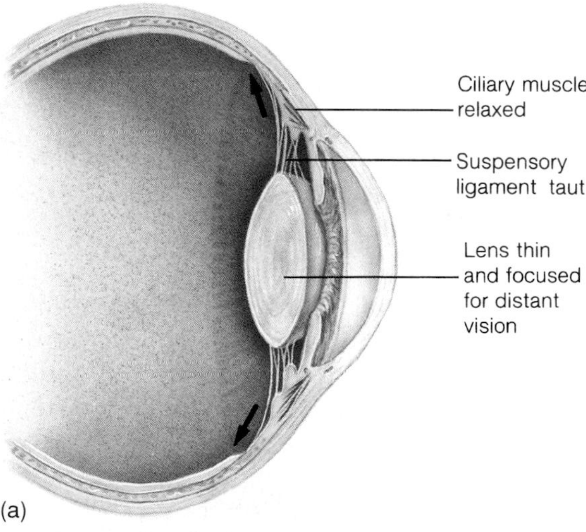

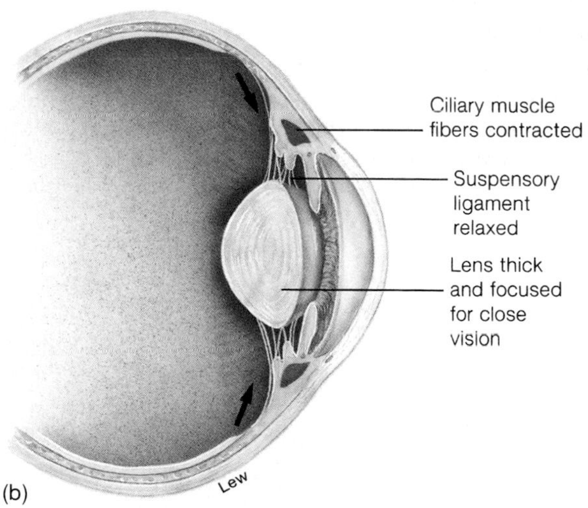

FIGURE 12.10
Changes in the shape of the lens to bring light rays into sharp focus on the retina. (*a*) The lens is flattened for distant vision when the ciliary muscle fibers are relaxed and the suspensory ligament is taut. (*b*) The lens is more spherical for close-up vision when the ciliary muscle fibers are contracted and the suspensory ligament is relaxed.

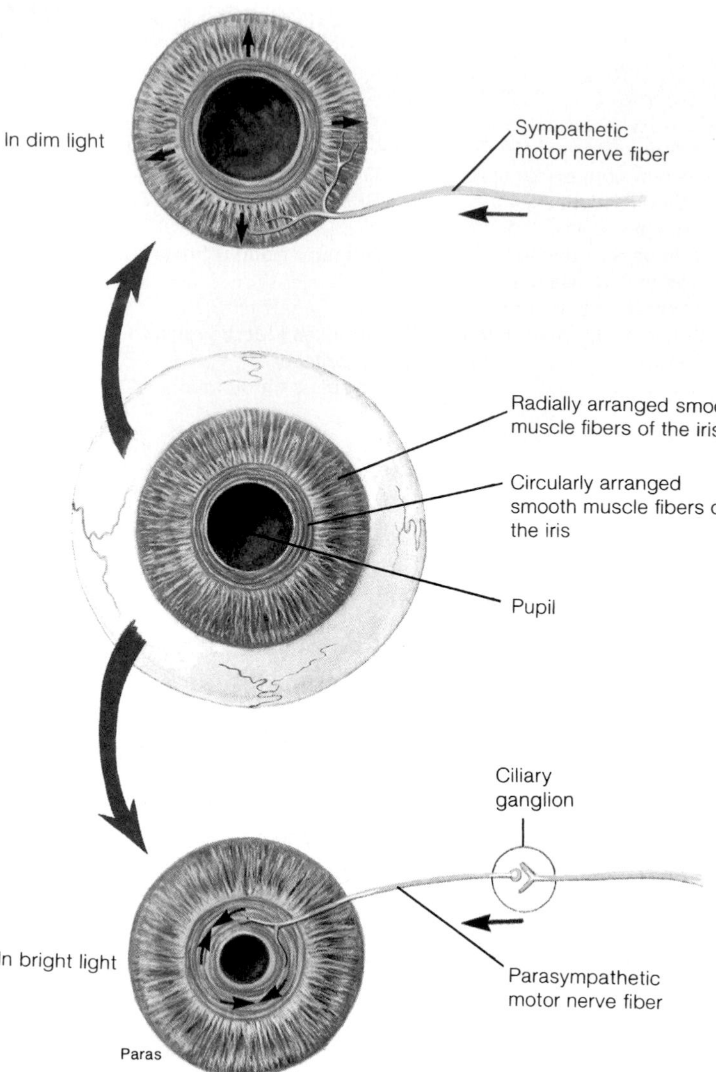

FIGURE 12.11
Dilation and constriction of the pupil. In dim light, the radially arranged smooth muscle fibers are stimulated to contract by sympathetic stimulation, dilating the pupil. In bright light, the circularly arranged smooth muscle fibers are stimulated to contract by parasympathetic stimulation, constricting the pupil.

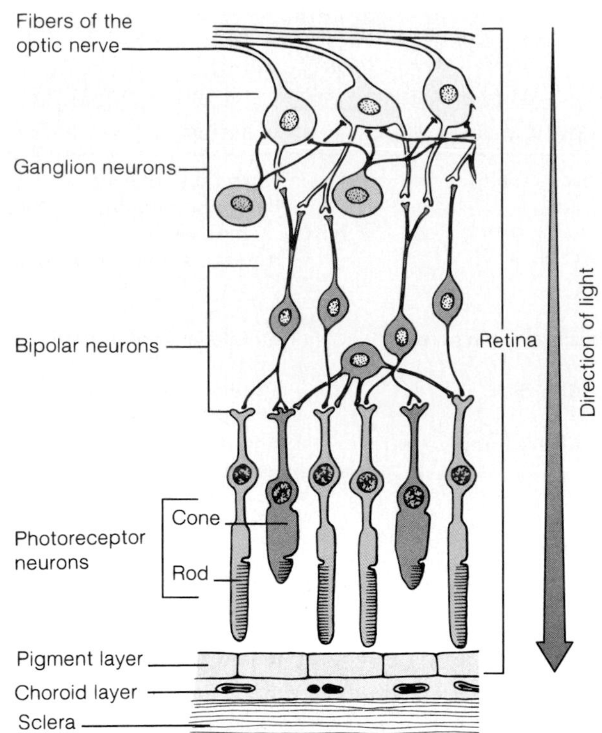

FIGURE 12.12
The layers of the retina. The retina is inverted, so that light must pass through various layers of nerve cells before reaching the rods and cones.

Inside Information—The amount of dark pigment, melanin, in the iris is what determines its color. In newborns, melanin is concentrated in the folds of the iris, so that all newborn babies have blue eyes. After a few months, the melanin moves to the surface of the iris and gives the baby his or her permanent eye color, ranging from steel blue to dark brown.

The **iris** is the anterior portion of the vascular tunic, and is continuous with the choroid. The iris is viewed from the outside as the colored portion of the eyeball (see figs. 12.6 and 12.9). It consists of smooth muscle fibers arranged in a circular and a radial pattern. Contraction of the smooth muscle fibers changes the diameter of the **pupil,** an opening in the center of the iris. Contraction of the radially arranged fibers of the iris, stimulated by dim light, dilates the pupil and increases the amount of light entering the eyeball (fig. 12.11). Contraction of the circularly arranged fibers, in response to bright light, constricts the pupil and permits less light to enter.

Internal Tunic (Retina)

The **retina** (*ret'ĭ-nă*) covers the choroid as the innermost layer of the eye. It consists of an outer *pigmented layer,* in contact with the choroid, and an inner *nervous layer,* or *visual portion.* The thick nervous layer of the retina terminates in a jagged margin near the ciliary body called the **ora serrata.** The thin pigmented layer extends over the back of the ciliary body and iris.

The nervous layer of the retina is composed of three principal layers of neurons. Listing them in the order in which they conduct impulses, they are **rod** and **cone cells, bipolar neurons,** and **ganglion neurons** (fig. 12.12). In terms of the passage of light, however, the order is

iris: Gk. *irid,* rainbow
ora serrata: L. *ora,* margin; *serra,* saw

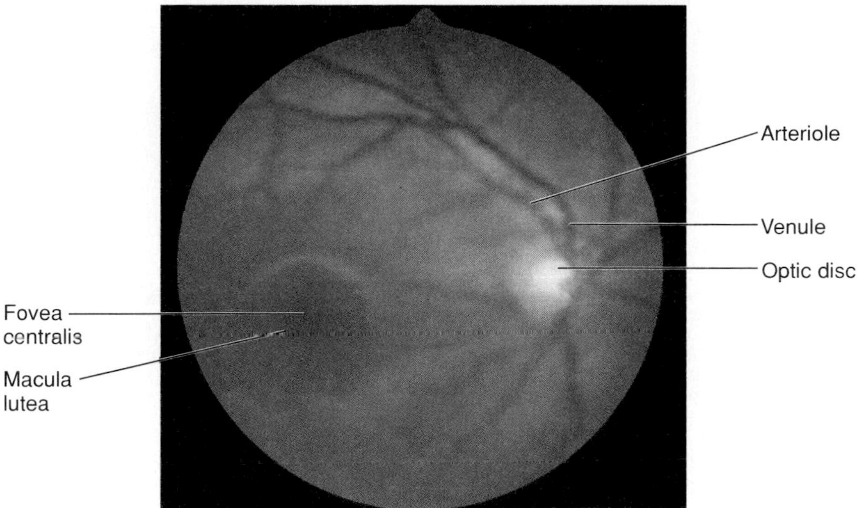

FIGURE 12.13
A view of the retina as seen with an ophthalmoscope. Optic nerve fibers leave the eyeball at the optic disc to form the optic nerve.

reversed. Light must first pass through the layer of ganglion cells and then the bipolar cells before reaching and stimulating the rods and cones.

Rods and cones are photoreceptors. Rods number over 100 million per eye and are longer and thinner than cones. Rods are positioned in the peripheral parts of the retina, where they respond to dim light to provide black-and-white vision. They also respond to form and movement but provide poor visual acuity (sharp focus). Cones, which number about 7 million per eye, are less sensitive than rods to light, but the cones provide color vision and greater visual acuity.

Both rods and cones perceive light using light-sensitive molecules of pigment. These pigments dissociate in response to absorbed light energy, and their decomposition activates an enzyme that triggers a series of reactions leading to the generation of an action potential. In rods, the pigment is called **rhodopsin.** Rhodopsin is formed from a light-absorbing molecule called *retinene* (a derivative of vitamin A) and a protein called *opsin.* Soon after rhodopsin dissociates, it is resynthesized in preparation for the next light ray. During the period of resynthesis, sensitivity to light is low, which accounts for the fact that you are unable to see for a few seconds as you enter a darkened room.

Cone cells contain pigments similar to rhodopsin, but the protein component of the pigment differs. Actually, there are three types of cones, each with a different protein component in the pigment. Each cone type is primarily sensitive to one of three colors—red, green, or blue. As with the rods, dissociation of the pigment leads to the production of an action potential in the ganglion cells.

Cones are concentrated in a depression near the center of the retina called the **fovea centralis,** which is the area of keenest vision (fig. 12.13). Surrounding the fovea centralis is the yellowish **macula lutea,** which also has an abundance of cones. There are no photoreceptors where the optic nerve is attached to the eye. This area is a *blind spot* and is referred to as the **optic disc.** Normally, a person is unaware of the blind spot because the eyes continually move about, an object is viewed from a different angle with each eye, and an object whose image falls on the blind spot of one retina will fall on receptors of the other retina.

Both the choroid and the retina are richly supplied with blood. Two *ciliary arteries* pierce the sclera at the posterior aspect of the eyeball (see figs. 12.9 and 12.13) and cross the choroid to the ciliary body and base of the iris. The *central vein* drains blood from the eyeball through the optic disc.

An examination of the internal eyeball with an ophthalmoscope is frequently part of a physical examination (fig. 12.13). An abnormal appearance of the capillary vessels on the retina may be an indication of certain diseases, such as atherosclerosis *or* diabetes. *As described in the Synoptic Spotlight (see page 327), an ophthalmoscopic examination can be particularly revealing with respect to a person's general health.*

Chambers of the Eye

The interior of the eye is separated by the lens into two main cavities—the **anterior cavity** and the **posterior cavity.** The anterior cavity is subdivided by the iris into an **anterior chamber** and a **posterior chamber** (fig. 12.14). The anterior chamber is located between the cornea and the iris. The posterior chamber is located between

rhodopsin: Gk. *rhodon,* rose; *ops,* eye
fovea: L. *fovea,* small pit
macula lutea: L. *macula,* spot; *luteus,* yellow

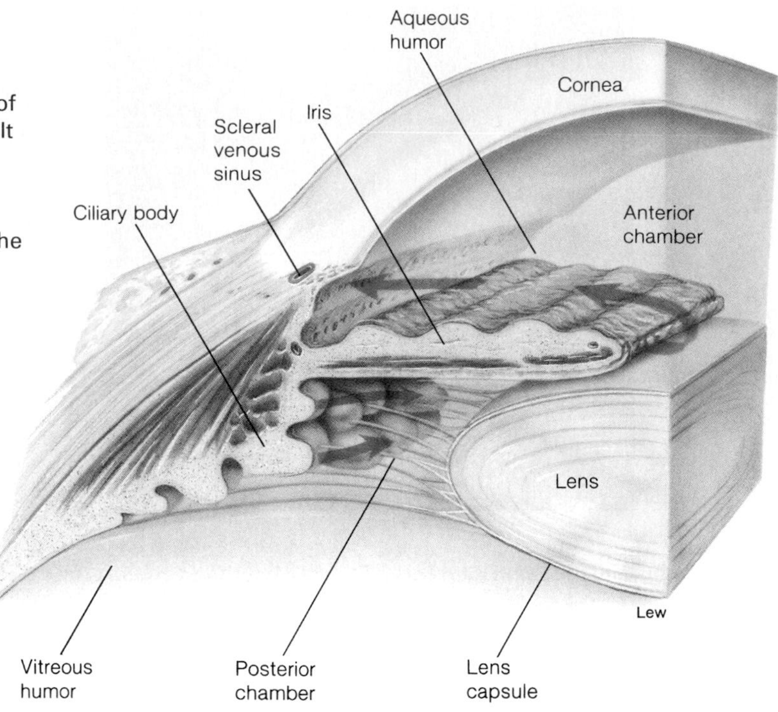

FIGURE 12.14
Aqueous humor maintains the intraocular pressure within the anterior and posterior chambers of the anterior cavity of the eyeball. It is secreted into the posterior chamber, flows through the pupil into the anterior chamber, and drains from the eyeball through the scleral venous sinus.

the iris and the lens. The anterior and posterior chambers connect through the pupil and are filled with a watery fluid called the *aqueous humor.* A constant daily production of 5 to 6 ml of aqueous humor maintains an internal eye pressure (intraocular pressure) of about 24 mm of mercury within the anterior and posterior chambers. Aqueous humor drains from the eyeball through the **scleral venous sinus** (*canal of Schlemm*) into the blood.

Behind the lens and surrounded by the retina is the large posterior cavity, which is filled with a transparent jellylike *vitreous humor.* Vitreous humor also contributes to the internal eye pressure to maintain the shape of the eyeball and to hold the retina against the choroid. Unlike aqueous humor, vitreous humor is not continuously produced, but is formed prenatally.

Puncture wounds to the eye are especially dangerous because they may cause blindness. Protective goggles or shields should be used in hazardous occupations, certain sports, and in chemistry and microbiology labs. If you should sustain a puncture wound, do not remove the object *if it is still impaling the eyeball. Removal may allow the humors to drain, causing loss of internal eye pressure, a detached retina, and possibly blindness.*

12.5 Physiology of Vision

The visual process includes the transmission and refraction of light rays, accommodation of the lens, constriction of the pupil, and convergence of the eyes. Neural pathways from the retina to specific areas of the brain assist coordination and permit visual perception.

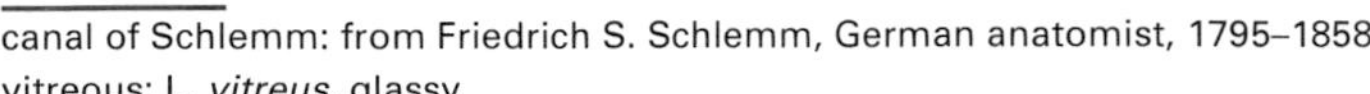

canal of Schlemm: from Friedrich S. Schlemm, German anatomist, 1795–1858
vitreous: L. *vitreus,* glassy

Processing Light within the Eyeball

The formation of an image on the retina requires five basic processes:

1. **transmission of light rays** through transparent media of the eye;
2. **refraction of light rays** through media of different densities;
3. **accommodation of the lens** to focus the light rays;

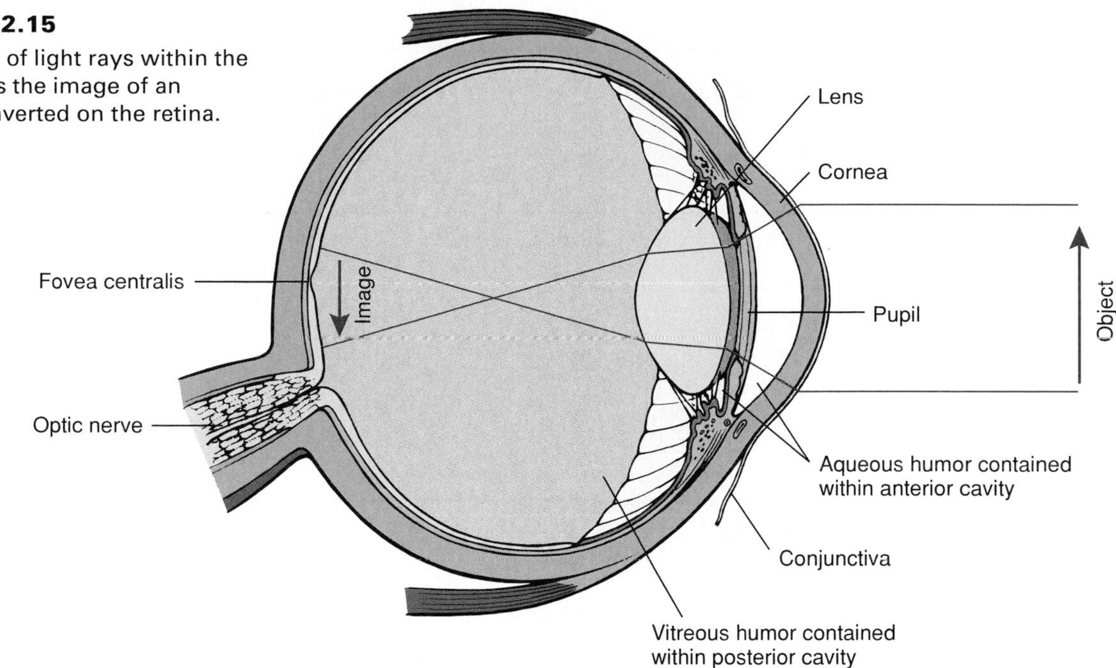

FIGURE 12.15
The refraction of light rays within the eyeball causes the image of an object to be inverted on the retina.

4. **constriction of the pupil** by the iris to regulate the amount of light entering the posterior cavity; and
5. **convergence of the eyeballs,** so that visual acuity is maintained.

Light rays entering the eye must pass through five transparent media before they stimulate the photoreceptors. In sequence, they are the conjunctiva, cornea, aqueous humor, lens, and vitreous humor (fig. 12.15).

Refraction is the bending of light rays. The convex cornea is the principal refractive medium; the fluids within the eye produce only minimal refraction. Because only the lens can be altered in shape, it is particularly important for refining and altering refraction to achieve a precise visual image. The refraction of light rays is so extensive that the visual image is formed upside down on the retina. Nerve impulses of the image in this position are relayed to the visual cortex of the occipital lobe, where the inverted image is interpreted as right side up.

Accommodation is the automatic adjustment of the curvature of the lens by contraction of ciliary muscles to bring light rays into sharp focus on the retina. This muscle contraction causes the suspensory ligament to relax and the lens to become thicker (see fig. 12.10). A thicker, more convex lens causes the greater refraction of light required for viewing close objects. There is a limit at how round a lens may become, however. This is why focusing is blurred if an object is held closer than about 6 inches from the eyes.

Constriction of the pupil, which is caused by contraction of the circular muscle fibers of the iris (see fig. 12.11), is important for two reasons. First, a reflexive pupil constriction reduces the amount of light that enters the posterior cavity, thereby protecting the retina from sudden or intense bright light. More importantly, a reduced pupil diameter prevents light rays from entering the posterior cavity through the periphery of the lens, which would cause blurred vision. Autonomic constriction of the pupil and accommodation of the lens occur simultaneously.

Convergence refers to the movement of the eyeballs toward the nose when viewing close objects. In fact, focusing on an object close to the tip of the nose causes a person to appear cross-eyed. The eyes must converge when viewing close objects because only then can the light rays focus on the same portions in both retinas.

Amblyopia exanopsia, *commonly called "lazy eye," is a condition of ocular muscle weakness. This causes a deviation of one eye, so that there is not a concurrent convergence of both eyeballs. With this condition, two images are received by the optic cortex—one of which is suppressed to avoid* diplopia (double vision), *or images of unequal clarity.*

Visual Spectrum

The eyes convert the energy of the *electromagnetic spectrum* (fig. 12.16) into nerve impulses. Only a limited part of this spectrum can excite the photoreceptors. Electromagnetic

amblyopia: Gk. *amblys,* dull; *ops,* vision

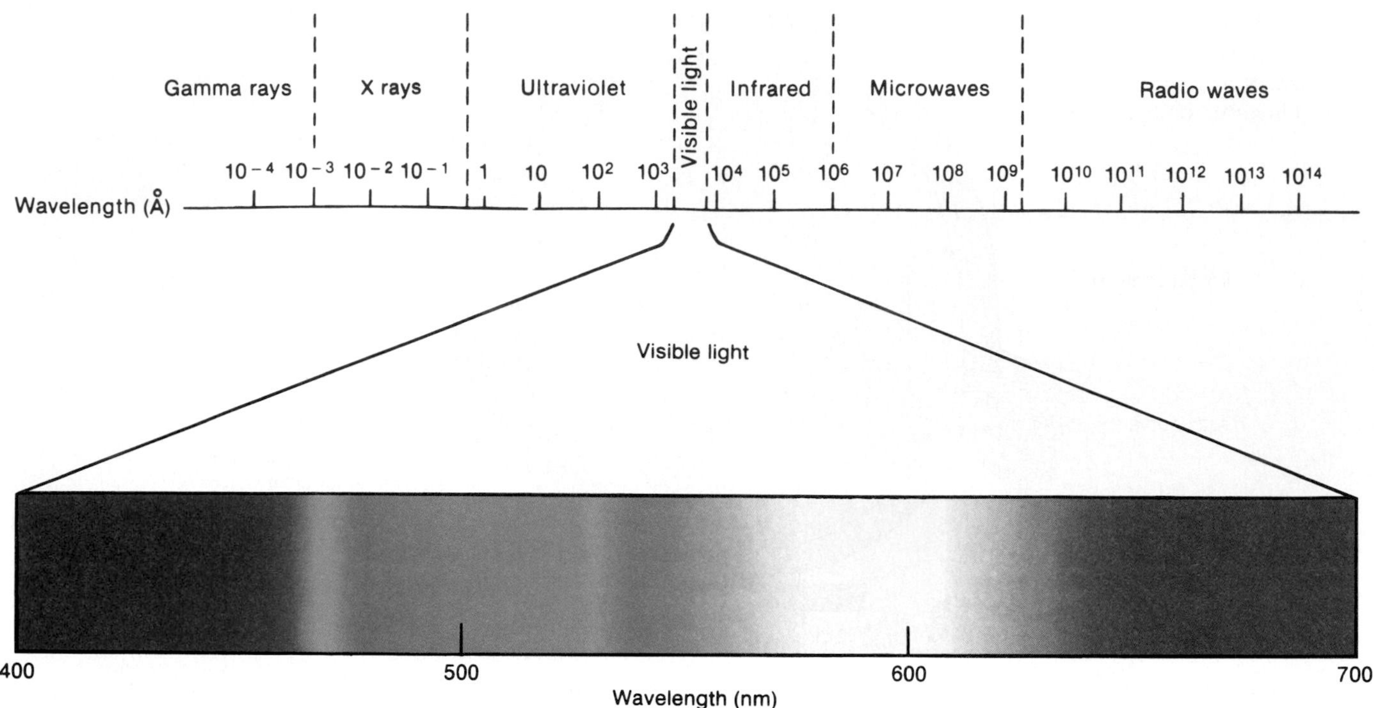

FIGURE 12.16

The electromagnetic spectrum (*top*) is shown in Angstrom units (1Å = 10^{-10} meter). The visible spectrum (*bottom*) comprises only a small range of this spectrum and is shown in nanometer units (1 nm = 10^{-9} meter).

energy with wavelengths between 400 and 700 nanometers (nm) constitute *visible light.* Light of longer wavelengths, which are in the infrared regions of the spectrum, does not have enough energy to excite photoreceptors but is perceived as heat. Ultraviolet light, which has shorter wavelengths and more energy than visible light, is filtered out by the yellow color of the eye's lens. Certain insects, such as honeybees, and people who have had their lenses removed, can see light in the ultraviolet range.

If one of the types of cones is deficient or missing from the nervous layer of the retina, as occurs in about 5% of Americans, the person will be unable to distinguish certain colors from others, a condition called colorblindness. *This condition is inherited, and it affects males much more frequently than females.*

Inside Information—A driver with red-green colorblindness—the most common type—is able to distinguish traffic signals without difficulty because yellow has been added to the red light, and blue has been added to the green.

Neural Pathways of Vision and Processing of Visual Information

Rods and cones are the functional units of sight in that they respond to light rays and produce nerve impulses. Impulses from the rods and cones pass through bipolar neurons to ganglion neurons (see fig. 12.12). The two **optic nerves** (one from each eye) converge at the **optic chiasma** (*ki-as'ma*) (fig. 12.17). All the fibers arising from the medial (nasal) half of each retina cross to the opposite side. Those fibers of the optic nerve that arise from the lateral (temporal)

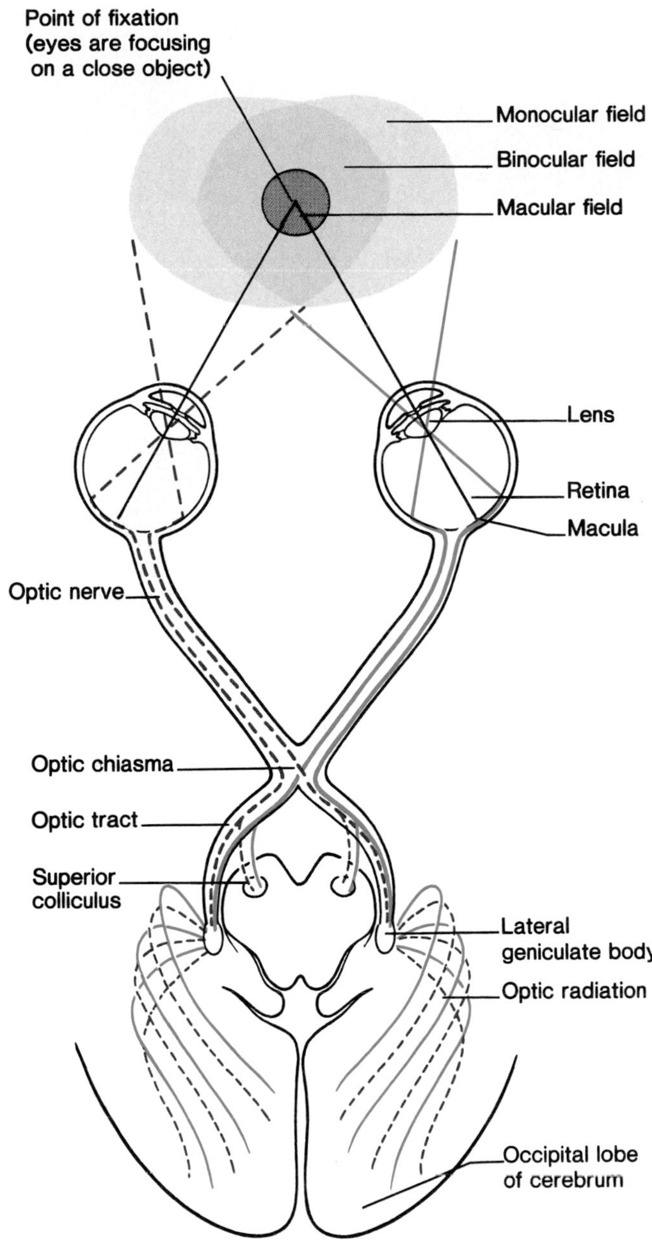

FIGURE 12.17

Visual fields of the eyes and neural pathways for vision. An overlapping of the visual field of each eye provides binocular vision, which is the ability to perceive depth.

half of the retina do not cross, however. The **optic tract** is a continuation of optic nerve fibers from the optic chiasma. It is composed of fibers arising from the retinas of both eyes.

As the optic tracts enter the brain, some of the fibers in the tracts terminate in the **superior colliculi** (*ko-lik'yoo-li*) of the midbrain (see chapter 10, module 10.4). These fibers and the motor pathways they activate constitute the *tectal system,* which is responsible for body-eye coordination. The tectal system is also involved in the control of the intrinsic eye muscles—the iris and the muscles of the ciliary body. Shining a light into one eye stimulates the *pupillary reflex,* causing both pupils to constrict. This is caused by parasympathetic stimulation through the superior colliculi. Postganglionic neurons in the ciliary ganglia behind the eyes, in turn, stimulate constrictor fibers in the iris. Contraction of the ciliary body during accommodation also involves the superior colliculi.

Approximately 70% to 80% of the fibers in the optic tract pass to the *lateral geniculate* (*jĕ-nik'yŭ-lit*) *body of the thalamus.* Here, the fibers synapse with neurons whose axons constitute a pathway called the *optic radiation.* The optic radiation transmits impulses to the *striate cortex area of the occipital cerebral lobe.* This entire arrangement of visual fibers, known as the *geniculostriate system,* is responsible for perception of the visual field.

The nerve fibers that cross at the optic chiasma arise from the retinas in the medial portions of both eyes. The rods and cones of these fibers are stimulated by light entering the eye from the periphery. If the optic chiasma were cut longitudinally, peripheral vision would be lost and leave only "tunnel vision." If an optic tract were cut, both eyes would be partially blind—the lateral field of vision lost for one eye, and the medial field of vision lost for the other.

For visual information to have meaning, it must be associated with past experience and coupled with information from other senses. Some of this higher processing occurs in the inferior temporal lobes of the cerebral cortex. Experimental removal of these areas from monkeys impairs their ability to remember visual tasks that they had previously learned and hinders their ability to associate visual images with the significance of the objects viewed. Monkeys with their inferior temporal lobes removed, for example, will fearlessly handle a snake.

12.6 Senses of Hearing and Equilibrium: Structure of the Ear

Structures of the outer, middle, and inner ear are involved in the sense of hearing. The inner ear also contains structures that provide a sense of balance and equilibrium.

The ear is the organ of hearing and equilibrium. It contains receptors that respond to movements of the head and receptors that convert sound waves into nerve impulses. Impulses from both receptor types are transmitted through the vestibulocochlear (eighth cranial) nerve to the brain for interpretation. The ear consists of three principal regions: the *outer ear,* the *middle ear,* and the *inner ear.*

Outer Ear

The outer ear consists of the **auricle** and the **external auditory canal.** The external auditory canal is the fleshy tunnel within the bony passageway called the *external auditory* (or *acoustic*) *meatus.* The auricle is the visible fleshy appendage attached to the side of the head. It consists of a cartilaginous framework of elastic connective tissue covered with skin. The rim of the auricle is the **helix,** and the inferior fleshy portion is the **earlobe** (fig. 12.18).

The external auditory canal is a slightly S-shaped canal about 2.5 cm (1 in.) long, extending slightly upward from the auricle to the tympanic membrane (fig. 12.19). Specialized wax-secreting glands, called **ceruminous** (*sĕ-roo'mĭ-nus*) **glands,** are located in the skin, deep within

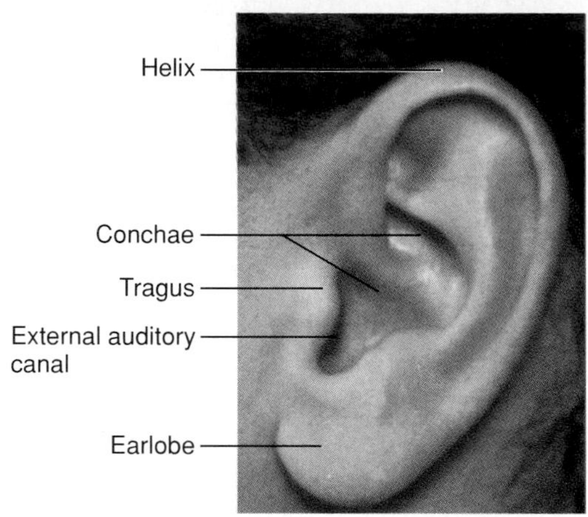

FIGURE 12.18
The surface anatomy of the auricle of the ear.

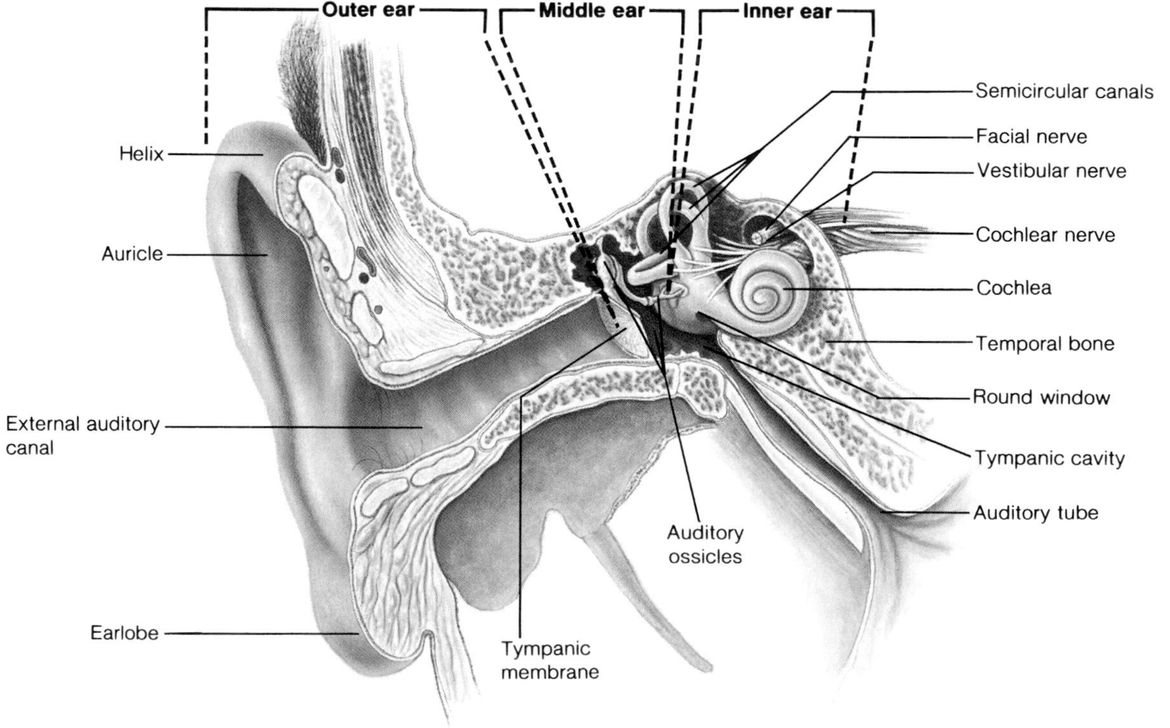

FIGURE 12.19
The ear. (Note the outer, middle, and inner regions indicated by dashed lines.)

the canal. *Cerumen* (ear wax) secreted from these glands keeps the tympanic membrane soft and waterproof. The bitter cerumen is probably an insect repellent as well.

The **tympanic membrane** (eardrum) is a thin, semitransparent partition of fibrous connective tissue that forms an airtight seal between the external auditory canal and the middle ear. It moves like a small drum in response to bombardment of sound waves.

Inspecting the tympanic membrane with an otoscope *during a physical examination provides clinical information about the condition of the middle ear. The color of the membrane, its curvature, the presence of lesions, and the position of the malleus of the middle ear are features of particular importance. If ruptured, the tympanic membrane can generally regenerate, healing itself within days.*

Middle Ear

The middle ear is an air-filled chamber called the **tympanic cavity** (fig. 12.20) in the petrous part of the temporal bone (see fig. 6.22). A bony partition containing the **vestibular window** (*oval window*) and **cochlear window** (*round window*) separates the middle ear from the inner ear. The tympanic cavity is connected to the *mastoidal air cells* within the mastoid process of the temporal bone. The **auditory** (eustachian) **tube** connects the tympanic cavity anteriorly with the nasopharynx and serves to equalize air pressure on both sides of the tympanic membrane.

Three *auditory ossicles* (see fig. 6.19) cross the tympanic cavity from the tympanic membrane to the vestibular window (fig. 12.20). These tiny bones (the smallest in the body), from outer to inner, are the **malleus** (hammer), **incus** (anvil), and **stapes** (stirrup). As the auditory ossicles transmit vibrations from the tympanic membrane, they act as a lever system to increase the force of vibration on the vestibular window. In addition, the force of vibration is intensified as it is transmitted from the relatively large surface of the tympanic membrane to the smaller surface area of the vestibular window. The combined effect increases the force of the vibrations about 20 times.

otoscope: Gk. *otikos,* ear; *skopein,* to examine
eustachian tube: from Bartolommeo E. Eustachio, Italian anatomist, 1520–74
malleus: L. *malleus,* hammer
incus: L. *incus,* anvil
stapes: L. *stapes,* stirrup

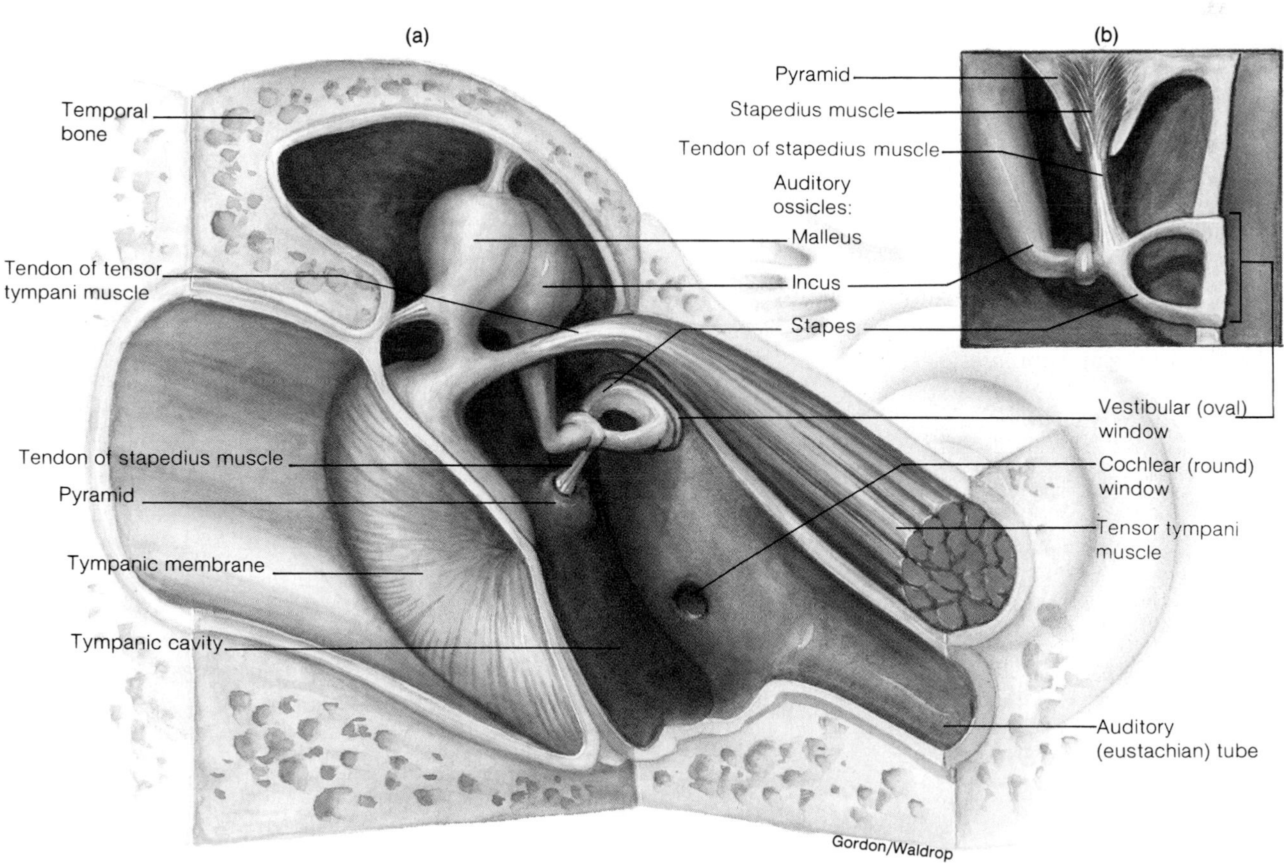

FIGURE 12.20
(*a*) The auditory ossicles and associated structures within the tympanic cavity. (*b*) The stapedius muscle arises from a bony protrusion called the pyramid.

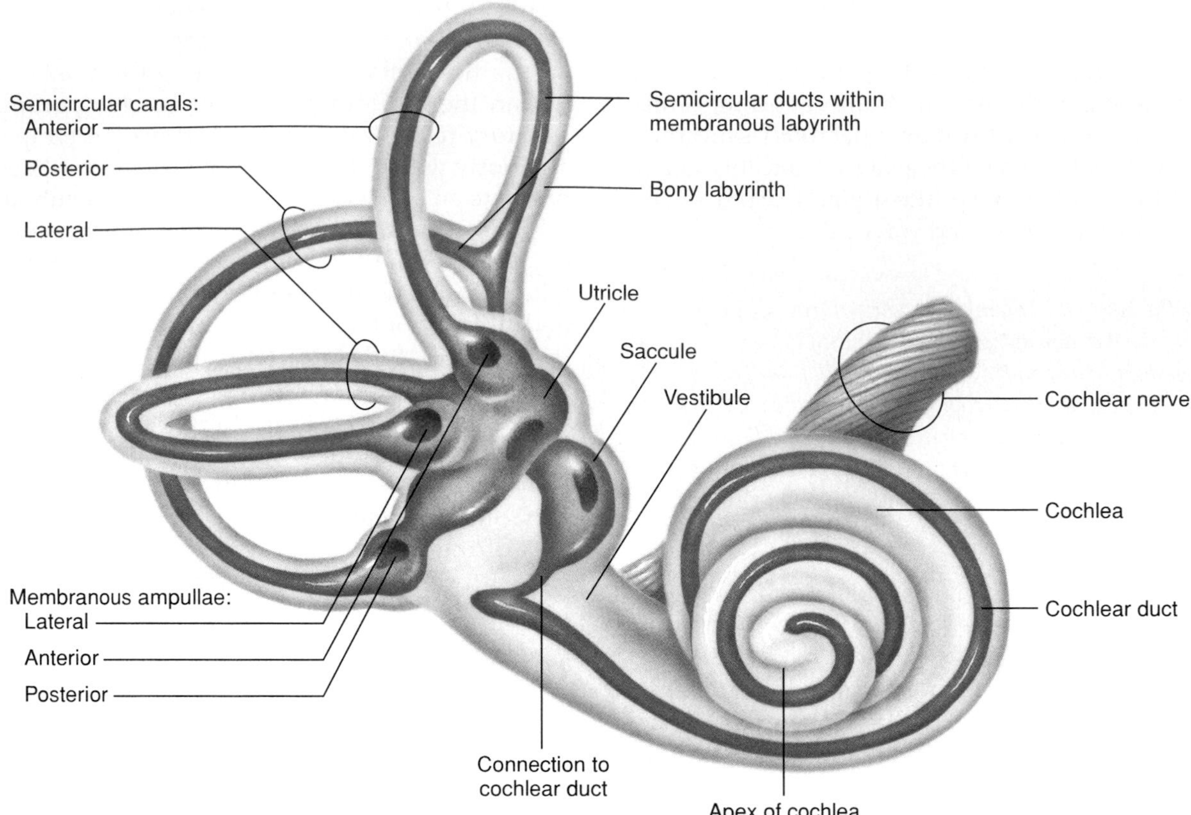

FIGURE 12.21
The labyrinths of the inner ear. The membranous labyrinth (darker color) is contained within the bony labyrinth.

Two small skeletal muscles, the **tensor tympani** muscle and the **stapedius** muscle, attach to the malleus and stapes, respectively, and contract reflexively to protect the inner ear against loud noises. When contracted, the tensor tympani muscle pulls the malleus inward, and the stapedius muscle pulls the stapes outward. This combined action reduces the force of vibration of the auditory ossicles.

An equalization of air pressure on the two sides of the tympanic membrane is important in hearing. When atmospheric pressure is reduced, as occurs when traveling to higher altitudes, the tympanic membrane bulges outward in response to the greater air pressure within the tympanic cavity. The bulging is painful and may impair hearing by reducing flexibility. The auditory tube, which is collapsed most of the time in adults, opens during swallowing or yawning and allows the air pressure on the two sides of the tympanic membrane to equalize.

Inner Ear

The entire structure of the *inner ear* is referred to as the **labyrinth** (*lab'ĭ-rinth*). The labyrinth is divided into two parts: an outer *bony labyrinth* and a *membranous labyrinth,* which lies within the bony labyrinth (fig. 12.21). The space between the bony labyrinth and the membranous labyrinth is filled with a fluid called *perilymph,* which is secreted by cells lining the bony canals. Inside the tubular chambers of the membranous labyrinth is yet another fluid called *endolymph.* These two fluids provide a liquid-conducting medium for the vibrations involved in hearing and the maintenance of equilibrium.

The bony labyrinth is structurally and functionally divided into three areas: the *vestibule, semicircular canals,* and *cochlea.* The functional organs for hearing and equilibrium are located in these areas.

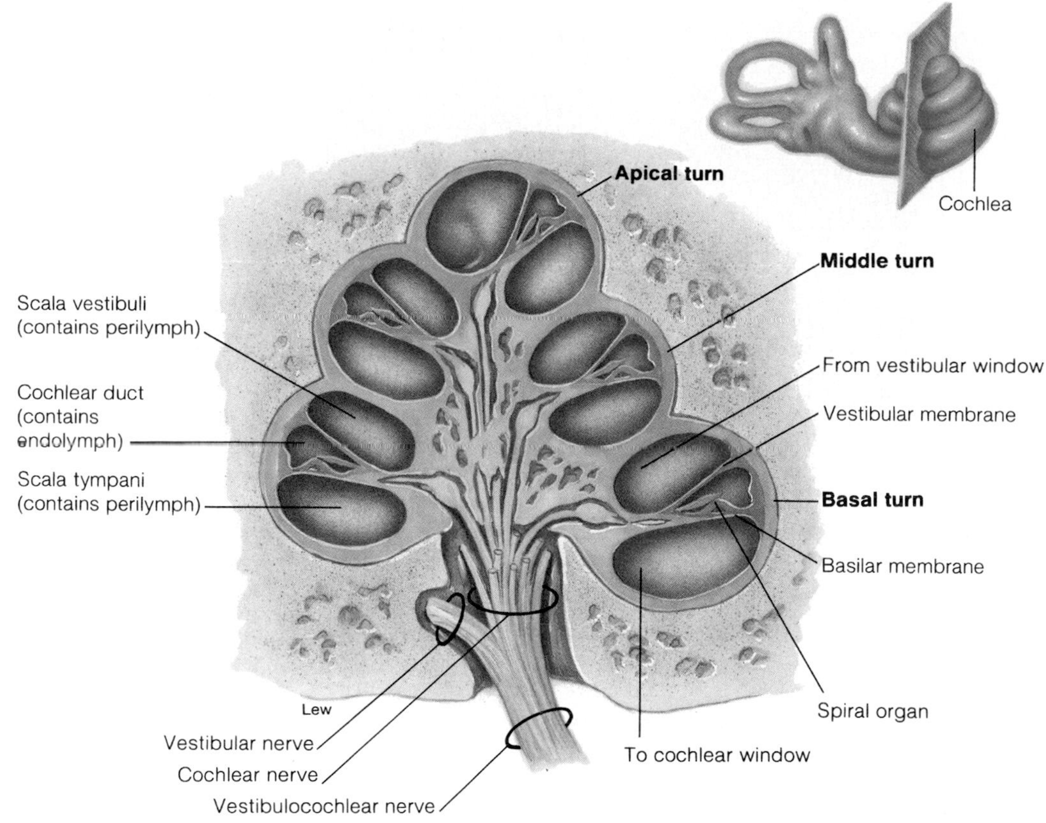

FIGURE 12.22
A cross section of the cochlea showing its three turns (indicated in boldface type) and its three compartments—scala vestibuli, cochlear duct, and scala tympani.

Vestibule

The vestibule is the central portion of the bony labyrinth. It contains the *vestibular* (oval) *window,* into which the stapes fits, and the *cochlear* (round) *window* on the opposite end (see fig. 12.20).

The membranous labyrinth within the vestibule consists of two connected sacs called the **utricle** (*yoo'trĭ-kul*) and the **saccule** (*sak'yool*). The utricle is larger than the saccule and lies in the upper back portion of the vestibule. Both the utricle and saccule contain receptors that are sensitive to gravity and linear movement of the head.

Semicircular Canals

Posterior to the vestibule are the three bony semicircular canals, placed at nearly right angles to each other. The thinner semicircular ducts form the membranous labyrinth within the semicircular canals. Each of the three semicircular ducts has a membranous *ampulla* at one end and connects with the upper back part of the utricle. Receptors within the semicircular ducts are sensitive to angular acceleration and deceleration of the head, as in rotational movement.

Cochlea

The snail-shaped cochlea (*kok'le-ă*) is coiled two and a half times around a central core of bone (fig. 12.22). There are three chambers in the cochlea. The upper chamber, the **scala** (*ska'lă*) **vestibuli,** begins at the vestibular window and extends to the apex (end) of the coiled cochlea. The lower chamber, the **scala tympani,** begins at the apex and terminates at the

cochlea: L. *cochlea,* snail shell
scala: Gk. *scala,* staircase

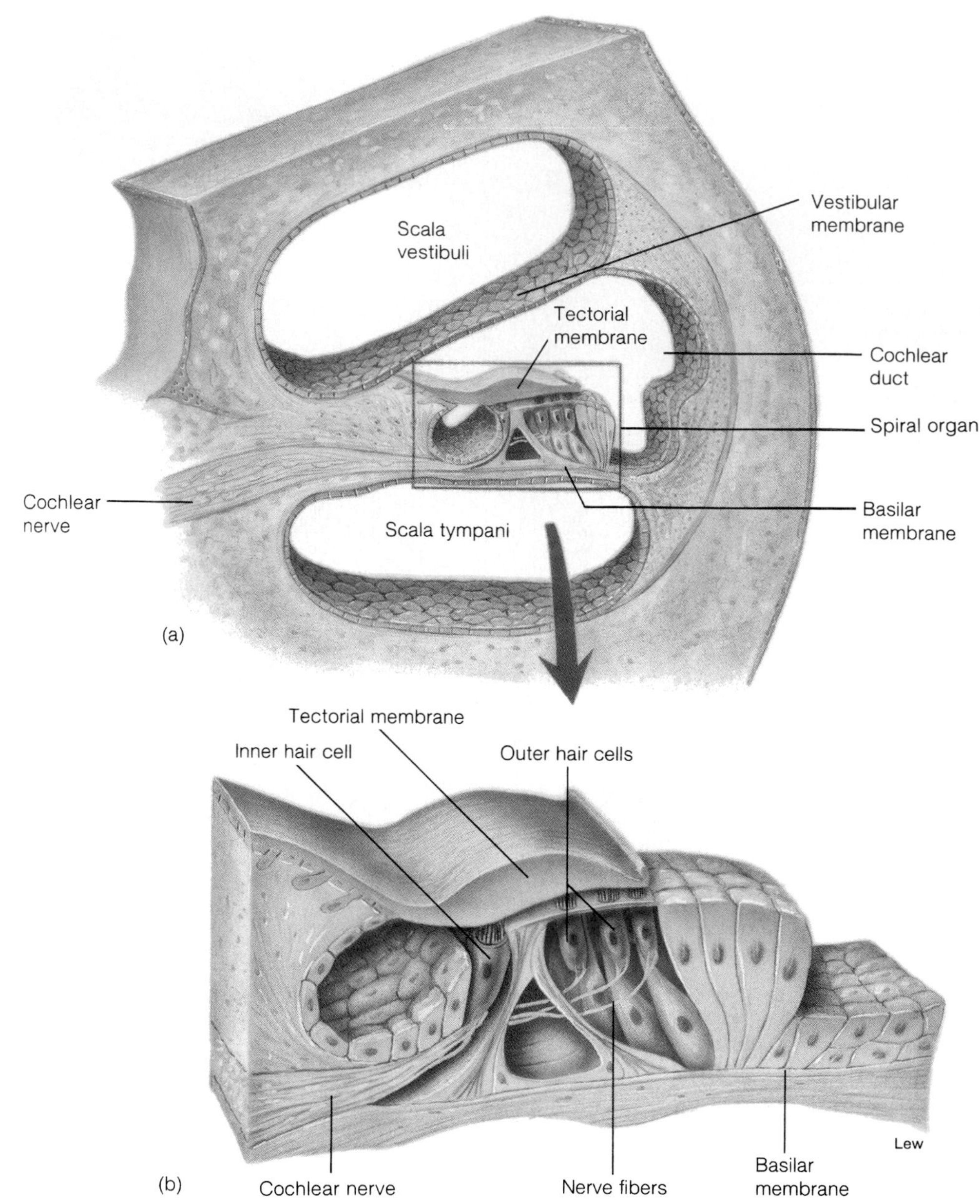

FIGURE 12.23
A section of the cochlea showing (*a*) the spiral organ within the cochlear duct. (*b*) The spiral organ in greater detail.

cochlear window. Both the scala vestibuli and the scala tympani are filled with perilymph. They are completely separated, except at the narrow apex of the cochlea, called the **helicotrema** (*hel″ĭ-kŏ-tre′mă*), where they are continuous (see fig. 12.24). Between the scala vestibuli and the scala tympani is the **cochlear duct,** the triangular middle chamber of the cochlea. The roof of the cochlear duct is called the **vestibular membrane,** and the floor is called the **basilar membrane.** The cochlear duct, which is filled with endolymph, ends at the helicotrema.

Within the cochlear duct is a specialized structure called the **spiral organ** (organ of Corti). The sound receptors that transform mechanical vibrations into nerve impulses are located along the basilar membrane of this structure, making it the functional unit of hearing. The epithelium of the spiral organ consists of supporting cells and hair cells (fig. 12.23). The bases of the hair cells are anchored in the basilar membrane, and their tips are embedded in the **tectorial membrane,** which forms a gelatinous covering over them.

helicotrema: Gk. *helix,* a spiral; *trema,* a hole
organ of Corti: from Alfonso Corti, Italian anatomist, 1822–88

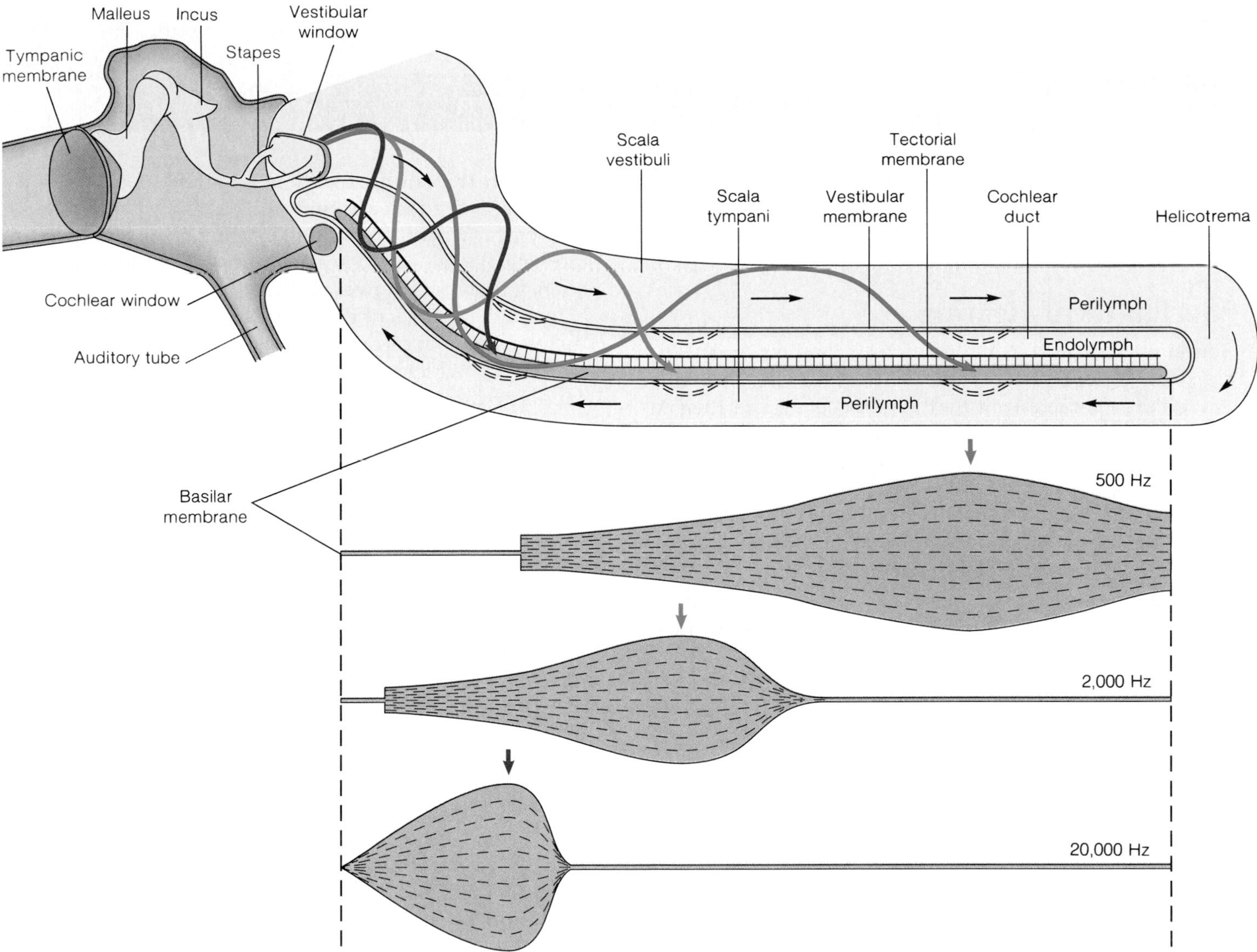

FIGURE 12.24

Sounds of low frequency cause pressure waves of perilymph to pass through the helicotrema. Sounds of higher frequency cause pressure waves to "shortcut" through the cochlear duct. This causes displacement of the basilar membrane and stimulation of the hair cells against the tectorial membrane, which is central to the transduction of sound waves into nerve impulses. (The frequency of sound waves is measured in hertz [Hz].)

Based on the structures involved, there are two types of deafness. Conduction deafness *is caused by an interference with the sound waves through the outer or middle ear. Conduction problems include impacted cerumen (wax), a ruptured tympanic membrane, a severe middle-ear infection, and adhesions (tissue growths) of one or more auditory ossicles* (otosclerosis). *Medical treatment usually improves the hearing loss from conduction deafness.*

Perceptive deafness *results from disorders that affect the inner ear, the cochlear nerve or nerve pathway, or the auditory centers within the brain. Perceptive impairment ranges in severity from the inability to hear certain frequencies to deafness. Such deafness may be caused by a number of factors, including diseases, trauma, and genetic or developmental problems. Elderly people frequently experience some perceptive deafness. The ability to perceive high-frequency sounds is generally lost first. Hearing aids may help patients with perceptive deafness. This type of deafness is permanent, however, because it involves destruction of sensory structures that cannot regenerate.*

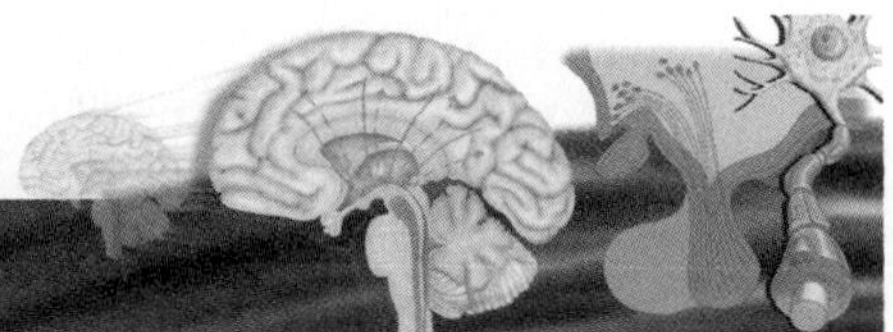

12.7 Physiology of Hearing and Equilibrium

The spiral organ is the functional unit of hearing. The vestibular organs assist in maintaining equilibrium as the body changes positions.

Mechanics of Hearing

Sound waves travel in all directions from their source, like ripples in a pond after a stone is dropped. These waves of energy are characterized by their frequency and their intensity. The *frequency,* or number of waves that pass a given point in a given time, is measured in *hertz* (*Hz*). The *pitch* of a sound is directly related to its frequency—the higher the frequency of a sound, the higher its pitch. For example, striking the high C on a piano produces a high frequency of sound that has a high pitch.

The *intensity,* or loudness of a sound, is directly related to the amplitude of the sound waves. Sound intensity is measured in units known as *decibels* (*dB*). A sound that is barely audible—at the threshold of hearing—has an intensity of 0 dB. Every 10 decibels indicates a tenfold increase in sound intensity: a sound is 10 times higher than threshold at 10 dB, 100 times higher at 20 dB, a million times higher at 60 dB, and 10 billion times higher at 100 dB. The healthy human ear can detect very small differences in sound intensity—from 0.1 to 0.5 dB.

Inside Information—A snore can be as loud as 70 dB, as compared with 105 dB for a power mower. Frequent or prolonged exposure to sounds with intensities over 90 dB (including amplified rock music) can result in hearing loss.

Sound waves funneled through the external auditory canal produce extremely small vibrations of the tympanic membrane. Movements of the tympanic membrane during ordinary speech (with an average intensity of 60 dB) are estimated to be equal to the diameter of a molecule of hydrogen.

Sound waves passing through the solid medium of the auditory ossicles are amplified about 20 times as they reach the footplate of the stapes, which is seated within the vestibular window. As the vestibular window is displaced, pressure waves travel through the fluid medium of the scala vestibuli (figs. 12.22 and 12.24) and pass around the helicotrema to the scala tympani. Movements of perilymph within the scala tympani, in turn, displace the cochlear window into the tympanic cavity.

When the sound frequency (pitch) is sufficiently low, there is adequate time for the pressure waves of perilymph within the scala vestibuli to travel through the helicotrema to the scala tympani. As the sound frequency increases, however, these pressure waves do not have time to travel all the way to the apex of the cochlea. Instead, they are transmitted through the vestibular membrane, which separates the scala vestibuli from the cochlear duct, and through the **basilar membrane,** which separates the cochlear duct from the scala tympani, to the perilymph of the scala tympani. The distance that the pressure waves travel, therefore, decreases as the sound frequency increases.

Sounds of low pitch (with frequencies below 50 Hz) cause movements of the entire length of the basilar membrane—from the base to the apex. Higher sound frequencies result in maximum displacement of the basilar membrane closer to its base, as illustrated in fig. 12.24.

Displacement of the basilar membrane and hair cells by movements of perilymph causes the hair cell microvilli that are embedded in the tectorial membrane to bend. This stimulation excites the sensory cells, which causes the release of an unknown neurotransmitter that excites sensory endings of the cochlear nerve.

Cochlear sensory neurons in the *vestibulocochlear* (eighth cranial) *nerve* synapse with neurons in the medulla oblongata, which project to the inferior colliculi of the midbrain. Neurons in this area in turn project to the thalamus, which sends axons to the auditory cortex of the temporal lobe, where the auditory sensations (nerve impulses) are perceived as sound.

Ménière's disease *afflicts the inner ear and may cause hearing loss as well as equilibrium disturbance. The causes of Ménière's disease are not completely understood, but they are thought to be related to a dysfunction of the autonomic nervous system that causes a vasoconstriction within the inner ear. The disease is characterized by recurrent periods of* vertigo *(dizziness and a sensation of rotation),* tinnitus *(ringing in the ears), and progressive deafness in the affected ear. Ménière's disease is chronic and affects both sexes equally. It is most common in elderly persons.*

Ménière's disease: from Prosper Ménière, French physician, 1799–1862

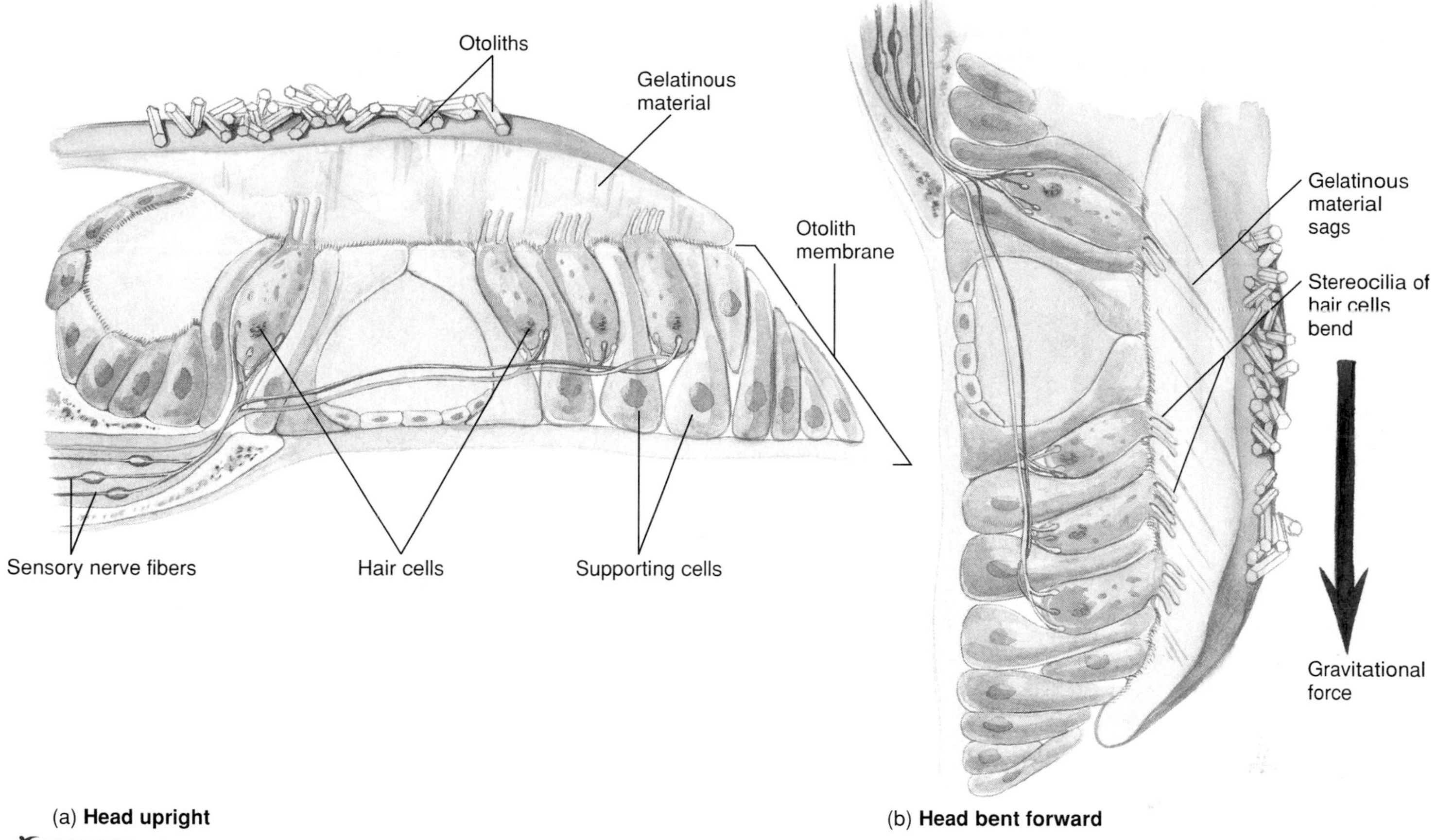

FIGURE 12.25

The otolith organ. (*a*) When the head is upright, the weight of the otoliths applies direct pressure to the sensitive stereocilia of the hair cells. (*b*) As the head is tilted forward, the stereocilia bend in response to gravity and stimulate the sensory nerve fibers.

Mechanics of Equilibrium

Maintaining our balance is a complex process that depends on continuous input from sensory neurons in the *vestibular organs* of both inner ears. Although the vestibular organs are the principal source of sensory information for equilibrium, the photoreceptors of the eyes, tactile receptors within the skin, and proprioceptors of tendons, muscles, and joints also provide sensory information that is needed to maintain equilibrium.

The vestibular organs provide the CNS with two kinds of receptor information. One kind is provided by receptors within the *saccule* and *utricle,* which are sensitive to gravity and to linear acceleration and deceleration of the head, as occur when riding in a car. The other is provided by receptors within the *semicircular ducts,* which are sensitive to rotational movements, as occur when turning the head, spinning, or tumbling.

Saccule and Utricle

Receptor hair cells of the saccule and utricle are located in a small thickened area of the walls of these structures called the **otolith organ** (fig. 12.25). Stereocilia of the hair cells project into a gelatinous material that supports microscopic crystals of calcium carbonate, called **otoliths** (*statoconia*). The otoliths increase the weight of the gelatinous mass, which results in a higher inertia (resistance to change in movement).

When a person is upright, the stereocilia of the utricle project vertically into the otolith membrane, whereas those of the saccule project horizontally. During forward acceleration, the otolith membrane lags behind the hair cells, so the hairs of the utricle are bent backward. This is similar to the backward thrust of the body when a car accelerates rapidly forward. The inertia of the otolith membrane similarly causes the hairs of the saccule to be pushed upward when a person jumps from a raised platform. Thus, because of the orientation of their hair cell processes, the utricle is more sensitive to horizontal acceleration, and the saccule is more sensitive to vertical acceleration. The changed pattern of action potentials in sensory nerve fibers that results from stimulation of the hair cells allows us to maintain our equilibrium with respect to gravity during linear acceleration.

otolith: Gk. *otos,* ear; *lithos,* stone

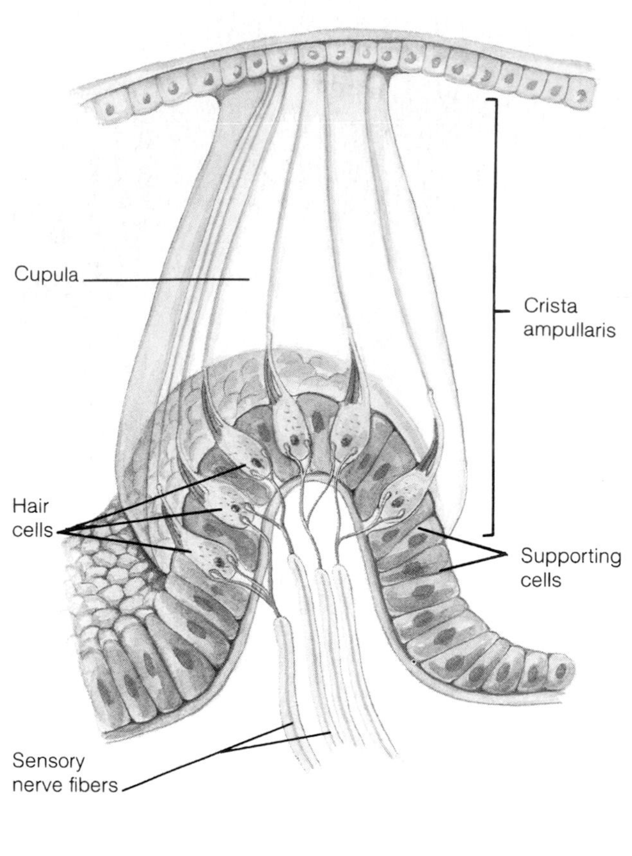

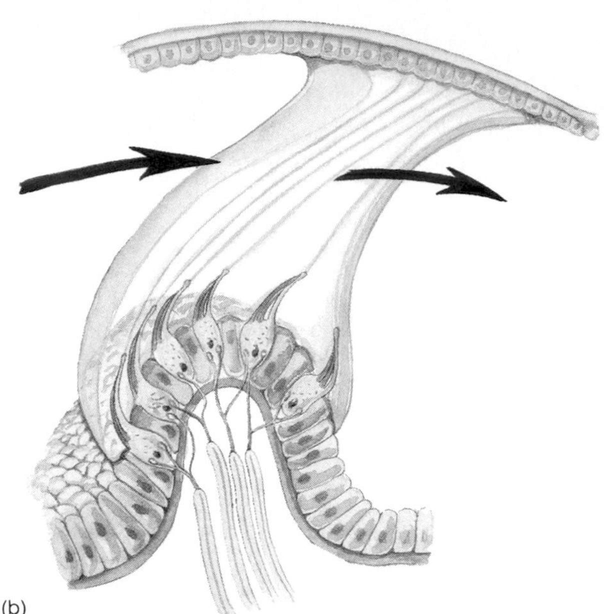

FIGURE 12.26
(*a*) The cupula and hair cells within the semicircular ducts. (*b*) Movement of the endolymph during rotation causes the cupula to displace and stimulate the hair cells.

Semicircular Canals

Receptor neurons of the semicircular canals are located in the **ampulla** at the base of each semicircular duct. The **crista ampullaris** (*am"poo-lar'is*) is an elevated area of the ampulla that contains numerous hair cells and supporting cells (fig. 12.26). Like the saccule and utricle, the hair cells have extensions that project into a dome-shaped gelatinous mass called the **cupula** (*kyoop'yoo-lă*). When the hair cells within the cupula are bent by rapid displacement of the fluid within the semicircular ducts, as in spinning around, the sensory receptors are stimulated.

Neural Pathways

Stimulation of the hair cells in the vestibular organs activates the sensory neurons of the *vestibulocochlear* (eighth cranial) *nerve.* These fibers transmit impulses to the cerebellum and to the vestibular nuclei of the medulla oblongata. The vestibular nuclei, in turn, send fibers to the oculomotor center of the brain stem and to the spinal cord. Neurons in the oculomotor center control eye movements, and neurons in the spinal cord stimulate movements of the head, neck, and limbs. Movements of the eyes and body produced by these pathways serve to maintain balance and track the visual field during rotation.

The dizziness and nausea that some people experience when they spin rapidly is explained by the activity occurring within the vestibular organs. When a person first begins to spin, the inertia of the endolymph within the semicircular ducts causes the cupula to bend in the opposite direction. As the spin continues, however, the endolymph and the cupula will eventually be moving in the same direction and at the same speed. If movement is suddenly stopped, the greater inertia of the endolymph causes it to continue moving in the direction of spin and to bend the cupula in that direction.

Bending of the cupula after movement has stopped affects control of the eyes and body. The eyes tend to slowly drift in the direction of the previous spin, and then are rapidly jerked back to the midline position, producing involuntary movements called postrotary vestibular nystagmus (nĭ-stag'mus). *People experiencing this effect may feel that they are spinning, or that the room is. The loss of equilibrium that results is called* vertigo. *If the vertigo is sufficiently severe, or if the person is particularly susceptible, the autonomic nervous system may become involved. This can produce dizziness, pallor, sweating, and nausea.*

cupula: L. *cupula,* cup-shaped

SYNOPTIC SPOTLIGHT

More Than Meets the Eye

After finally completing a residency at a respectable big-city hospital, you are offered a job with the Docs-R-Us health maintenance organization. It's not exactly the position of your dreams, but you feel a sense of urgency to start paying off your medical school debts. Although seeing a patient every 5 minutes was never your idea of a utopian medical practice, you agree to the stipulations in the contract and sign on the dotted line.

The bold corporate vision of Docs-R-Us is to do for medicine what fast food has done for the restaurant business. Consistent with that vision is the clause in your contract requiring you to limit your physical examination to inspection of only one organ. You are asked to declare which of all the organs in the body you will examine in all your patients. After pondering this for a moment—which organ will provide the astute examiner the most information about the health of the entire body—you select the eye.

Your first patient is a man in his mid-50s, dressed in a business suit, and looking like he's obviously enjoyed a few too many cheeseburgers. He apologizes for being late and mumbles something about how rush-hour traffic always gets his blood boiling.

"Doc, I've been having these awful chest pains. Can you fix me?"

Almost reflexively, you start to reach for your stethoscope. But remembering your contract, you suppress the urge. Instead, you reach into your desk and pull out an ophthalmoscope.

The ophthalmoscope is an instrument that consists of a mirror that reflects light into the eye, a viewing aperture, and a series of lenses that magnify the interior of the eyeball. Adjustment of the lens refraction allows the examiner to visualize different parts of the eye. Focusing closely allows inspection of anterior structures, such as the conjunctiva, cornea, iris, and anterior chamber. Focusing deeper and peering through the pupil, the posterior cavity, retina, and fundus can be illuminated.

As you scan the patient's chart, you notice that no blood pressure measurement has been recorded. More cutbacks, obviously. No matter, you think to yourself as you shine the light into the man's eye. You recall from medical school that hypertension (high blood pressure) leaves clues on the retina that a skilled observer should be able to pick up. Sure enough, the retinal arterioles are significantly narrowed. Some of them even resemble copper wire, with thickened walls and a reddish-brown discoloration. These findings are characteristic of long-standing hypertension.

As you explore the retina further, you notice several tiny red spots and recognize them as aneurysms associated with diabetes mellitus. These small-vessel malformations occur in 60% of patients who have had diabetes for 10 years or more. Diabetic patients also develop other retinal changes, as well as cataracts and retinal detachments—but these are notably absent in this patient.

Just before you complete the retinal exam, you make one final discovery. You see a couple of yellow spots known as Hollenhorst plaques. These are caused by particles of cholesterol, usually originating in the carotid artery, that have broken free and come to rest in the retinal arterioles. They are frequently a sign of impending serious cardiovascular disease.

"So whaddya think, Doc? Probably just gas?"

Of the known risk factors for heart attack, this man has at least three—diabetes, hypertension, and atherosclerosis. Thank goodness he doesn't smoke. All this you determined without vital sign measurements, a cardiac exam, or laboratory studies. The administrators would be so proud of you.

For a moment you reflect on what an excellent choice you made in selecting the eye as the subject of your exam. Through careful inspection of the eye, a huge array of systemic disorders can be diagnosed. A yellowish (jaundiced) sclera is an indication of liver and/or gallbladder problems. An abnormally dilated pupil is an indication of shock or of various nervous system disorders. Increased pressure in the brain from infection, stroke, or trauma leads to blurring of the optic disc. Sickle cell anemia produces convoluted retinal vessels. High copper levels in the body (Wilson's disease) is often diagnosed by a brown band around the iris. Small-vessel bleeding in the retina reflects certain protein-producing malignancies. Clouding of the cornea can be the first clue to some inherited disorders of sugar metabolism. As the English physician Peter Latham advised his medical students more than a century ago, "In the eye you will see all diseases in miniature, and you will see them as through a glass."

"I'm sorry, sir," you tell your patient, "I'm going to have to admit you to the hospital for a workup for possible heart disease. Please get your things together and follow the nurse."

"It's a little sudden, but I guess so. Thanks Doc." He pauses as he heads for the door. "Say, Doc, before I head up to the hospital room you think I could step out for a smoke?"

A moment after he leaves, the receptionist steps in to inform you that your next two patients have canceled. "Good," you say, "that will give me at least 10 minutes to scan the paper for another job."

Clinical Considerations

Functional Problems, Infections, and Diseases of the Eye

Few people have perfect vision. Slight variations in the shape of the eyeball or curvature of the cornea or lens cause an image to be focused in front of or behind the retina. Most variations are slight, however, and the error of refraction goes unnoticed. Visual problems that are not corrected may cause blurred vision, fatigue, chronic headaches, and depression.

Myopia (nearsightedness) is an elongation of the eyeball. As a result, light rays focus at a point in the vitreous humor in front of the retina (fig. 12.27). Only light rays from close objects can be focused clearly on the retina; distant objects appear blurred, hence the common term nearsightedness. **Hyperopia** (farsightedness) is a condition in which the eyeball is too short, which causes light rays to be brought to a focal point behind the retina. **Presbyopia** is a condition in which the lens tends to lose its elasticity and ability to accommodate. It is relatively common in people over 50 years of age. In order to read print on a page, a person with presbyopia must hold the page farther from the eyes than the normal reading distance. **Astigmatism** is a condition in which an irregular curvature of the cornea or lens distorts the refraction of light rays. If a person with astigmatism views a circle, the image will not appear clear in all 360 degrees; the part of the circle that appears blurred can be used to map the astigmatism.

As an alternative to a concave lens, a relatively new surgical procedure called **radial keratotomy** (fig. 12.28) is sometimes used to treat mild to moderate myopia. In this technique, 8 to 16 microscopic slashes, like the spokes of a wheel, are made in the cornea from the center to the edge. The pressure inside the eyeball bulges the weakened cornea and flattens its center, changing the focal length of the eyeball. Even newer is a laser surgery called *photorefractive keratectomy*. In this 1- to 2-minute operation, the cornea is flattened by vaporizing microscopic slivers from its surface.

A **cataract** is a clouding of the lens that leads to a gradual blurring of vision and the eventual loss of sight (fig. 12.29). A cataract is not a growth within or upon the eye, but rather a chemical change in the protein of the lens. It is caused by injury, poisons, infections, or age degeneration. Recent evidence indicates that even excessive UV light may cause cataracts. A cataract can be removed surgically, and vision restored, by implanting a tiny intraocular lens that either clips to the iris or is secured into the vacant lens capsule.

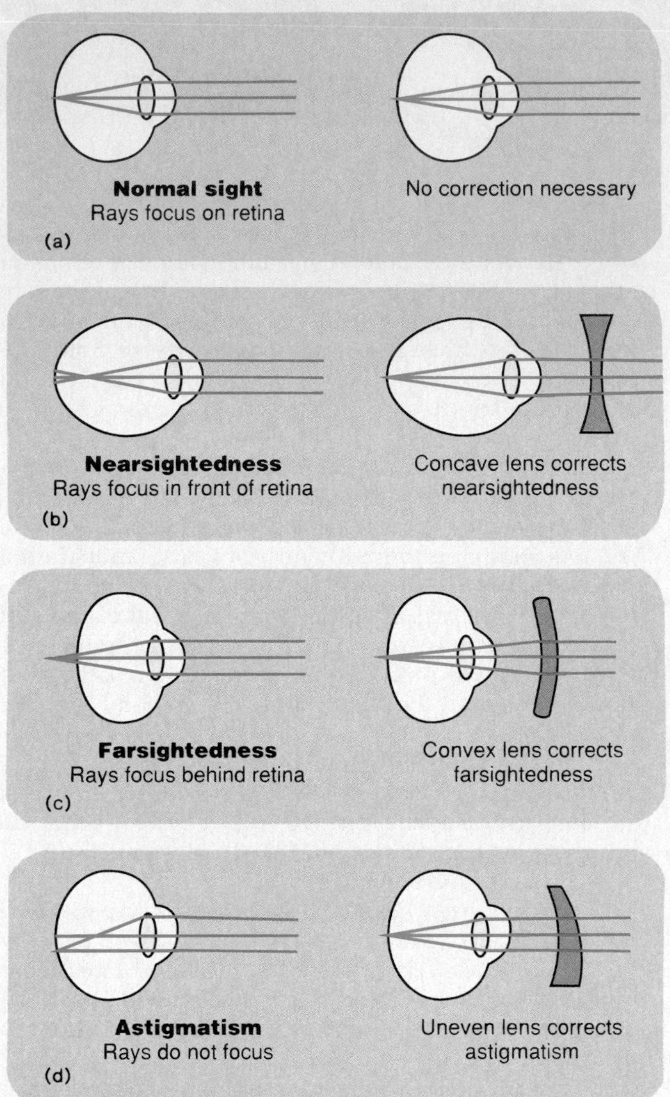

FIGURE 12.27
In a normal (emmetropic) eye (*a*), parallel rays of light are brought to a focus on the retina by refraction in the cornea and lens. If the eye is too long, as in myopia (*b*), the focus is in front of the retina. This can be corrected by a concave lens. If the eye is too short, as in hyperopia (*c*), the focus is behind the retina. This is corrected by a convex lens. In astigmatism (*d*), light refraction is uneven because of irregularities in the shape of the cornea or lens.

myopia: Gk. *myein,* to shut; *ops,* eye
hyperopia: Gk. *hyper,* over; *ops,* eye
presbyopia: Gk. *presbys,* old man; *ops,* eye
astigmatism: Gk. *a,* without; *stigma,* point
cataract: Gk. *katarrhegnynai,* to break down

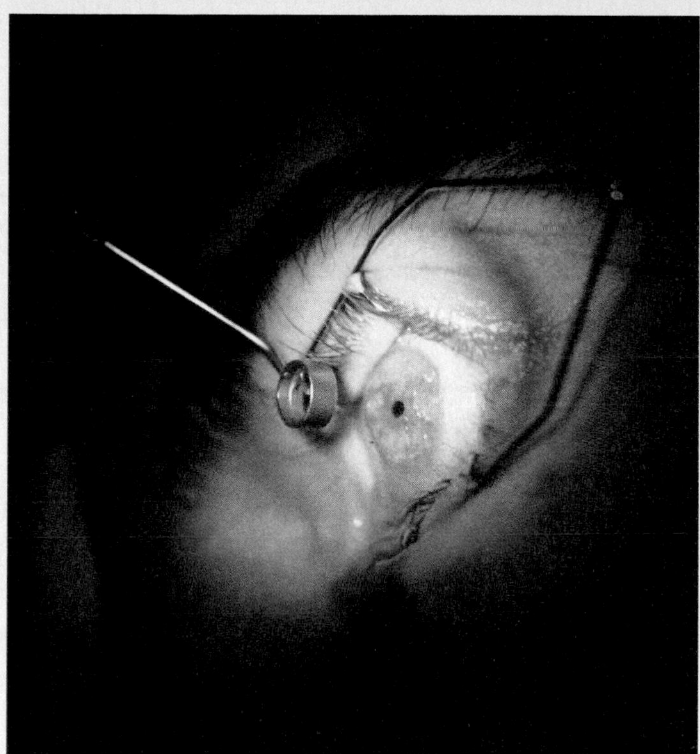

FIGURE 12.28
Radial keratotomy is a relatively new surgical procedure that actually changes the shape of the eyeball.

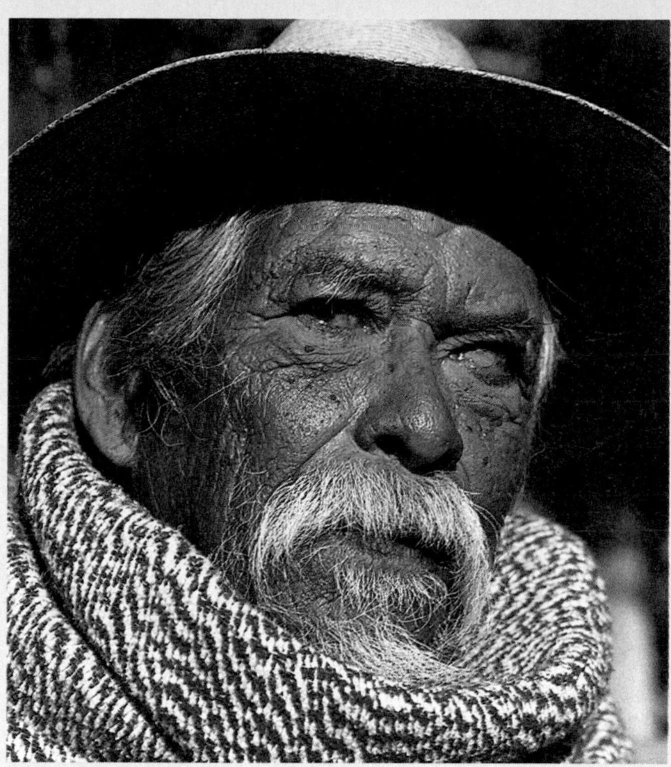

FIGURE 12.29
An untreated cataract causes blindness in the afflicted eye because light rays cannot penetrate a cloudy lens. The pupil of the left eye of the person shown above appears white due to reflected light rays from the cataracted lens.

Retinal detachment is a separation of the nervous or visual layer of the retina from the underlying pigment layer. It generally begins as a minute tear in the retina that gradually extends as vitreous humor accumulates between the layers. Retinal detachment may result from hemorrhage (bleeding), a tumor, degeneration, or trauma from a violent blow to the eye. A detached retina may be repaired by using laser beams or intense heat to destroy the tissue beneath the tear and rejoin the layers.

Glaucoma is an abnormal condition of high pressure within an eye. It is a common cause of blindness. In glaucoma, aqueous humor does not drain through the scleral venous sinus as quickly as it is produced. This causes compression of the blood vessels in the eye, and compression of the optic nerve. Retinal cells die and the optic nerve may atrophy, producing blindness.

Conjunctivitis (inflammation of the conjunctiva) may result from sensitivity to light, allergens, or an infection caused by viruses or bacteria. Bacterial conjunctivitis is commonly called "pinkeye."

A **chalazion** (*kă-la'ze-on*) is a tumor or cyst on the eyelid that results from infection of the tarsal glands and a subsequent blockage of the ducts of these glands.

A **stye** (*hordeola*) is a relatively common mild infection of the hair follicle of an eyelash or the associated sebaceous gland. A stye may spread readily from one eyelash to another if untreated. Poor hygiene and excessive use of cosmetics may contribute to the development of styes.

Inside Information—Benjamin Franklin, impatient with having to switch his spectacles constantly for near and far vision, created the first pair of bifocals in 1784. In Franklin's time, the two lenses were joined together in a metallic frame. One-part bifocal lenses did not become available until 1910.

continued

glaucoma: Gk. *glaukos,* gray
chalazion: Gk. *chalazion,* hail; a small tubercle

Clinical Considerations

Functional Problems, Infections, and Diseases of the Ear

External otitis is a general term for infections of the outer ear. The causes of external otitis range from dermatitis to fungal and bacterial infections.

Acute purulent otitis media is a middle-ear infection. Pathogens of this disease usually enter through the auditory tube, often following a cold or tonsillitis. As a middle-ear infection progresses to the inflammatory stage, the auditory tube closes and drainage is blocked. An intense earache is a common symptom of a middle-ear infection.

Repeated middle-ear infections, particularly in children, usually call for an incision of the tympanic membrane called a **myringotomy** (*mir″ing-got′ŏ-me*). This is followed by the implantation of a tiny tube within the tympanic membrane (fig. 12.30). The tube, which is eventually sloughed out of the ear, permits the infection to heal and helps prohibit further infections by keeping the auditory tube open.

Perforation of the tympanic membrane may occur as the result of infections or trauma. The membrane might be ruptured, for example, by a sudden, intense noise. Spontaneous perforation of the membrane usually heals rapidly, but scar tissue may form and lessen the sensitivity to sound vibrations.

Otosclerosis is a progressive deterioration of the normal bone in the bony labyrinth of the inner ear and its replacement with vascular spongy bone. Eventually, the condition results in deafness, although an operation to replace the stapes in the middle ear usually restores hearing.

Inside Information—By the time Ludwig van Beethoven had completed his Ninth Symphony in 1824, he had lost almost all of his hearing. Most physicians believe that Beethoven suffered from otosclerosis, a condition that could not be treated in his time. Today it is often possible to correct otosclerosis with hearing aids or with surgery to replace the stapes.

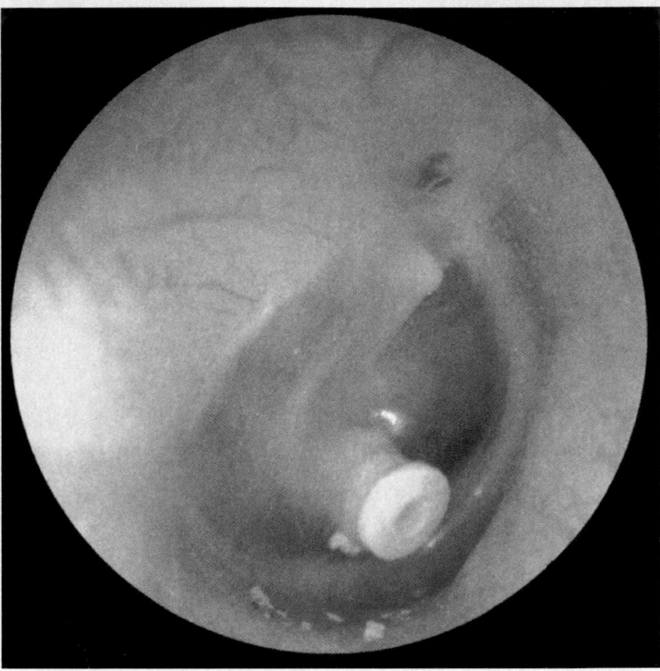

FIGURE 12.30
An implanted ventilation tube in the tympanic membrane following a myringotomy.

SUMMARY

12.1 Overview of the Senses

1. Sensory organs are specialized extensions of the nervous system that respond to specific stimuli and conduct nerve impulses.
2. A stimulus to a receptor that conducts an impulse to the brain is necessary for perception.
3. The senses are classified according to the structure or location of the receptors, or on the basis of the stimuli to which the receptors respond.

12.2 Senses of Smell and Taste

1. Olfactory receptors of the olfactory nerve respond to chemical stimuli and transmit the sensation of olfaction to the cerebral cortex.
2. Taste receptors in taste buds respond to chemical stimuli and transmit the sensation of gustation to the cerebral cortex. The kinds of taste sensation are sweet, salty, sour, and bitter.

12.3 Sense of Sight: Accessory Structures of the Eye

1. Protective structures of the eye include the eyebrow, eyelids, eyelashes, glands, and conjunctiva.
2. Six extrinsic ocular muscles control the movements of the eyeball.

12.4 Sense of Sight: Structure of the Eye

1. The eyeball consists of the fibrous tunic, which is divided into the sclera and cornea; the vascular tunic, which consists of the choroid, the ciliary body, and the iris; and the internal tunic, or retina, which consists of an outer pigmented layer and an inner nervous layer.
2. The anterior cavity (subdivided into the anterior and posterior chambers) contains aqueous humor; the posterior cavity contains vitreous humor.

12.5 Physiology of Vision

1. The visual process includes the transmission and refraction of light rays, accommodation of the lens, constriction of the pupil, and convergence of the eyeballs.
2. Neural pathways from the retina to the superior colliculus of the midbrain help to regulate eye and body movements.

12.6 Senses of Hearing and Equilibrium: Structure of the Ear

1. The outer ear consists of the auricle and the external auditory canal.
2. The middle ear (tympanic cavity) contains the auditory ossicles and the auditory muscles. This chamber is bounded by the tympanic membrane and the vestibular (oval) and cochlear (round) windows.
3. The tympanic cavity connects to the pharynx through the auditory tube.
4. The inner ear contains the spiral organ for hearing. It also contains the semicircular canals, saccule, and utricle (located in the vestibule) for maintaining balance and equilibrium.

12.7 Physiology of Hearing and Equilibrium

1. Sound waves are funneled through the external auditory canal, causing the tympanic membrane to vibrate. As vibrations are passed through the auditory ossicles to the vestibular window, they are amplified. Movement of the vestibular window causes displacement of perilymph and movement of the basilar membrane of the spiral organ. Action potentials are generated through the cochlear nerve as the hair cells along the basilar membrane contact the tectorial membrane.
2. Otoliths within the vestibule of the inner ear respond to the effect of gravity in maintaining the body in a position of static equilibrium.
3. Each semicircular canal within the inner ear contains a crista ampullaris, which has hair cells embedded in a gelatinous cupula. Displacement of endolymph during a spinning movement stimulates the hair cells of the semicircular canals resulting in the transmission of action potentials through the vestibular nerve. It is through these sensations that kinetic equilibrium is monitored.

Review Activities

Objective Questions

1. Conscious interpretation of changes in the internal and external environment is called (*module 12.1*)
a. responsiveness.
b. perception.
c. sensation.
d. accommodation.

2. Specialized visceroceptors that are sensitive to changes in blood pressure are called (*module 12.1*)
a. baroreceptors.
b. exteroceptors.
c. proprioceptors.
d. nociceptors.

3. The sensation of visceral pain perceived as arising from another somatic location is known as (*module 12.1*)
a. related pain.
b. phantom pain.
c. referred pain.
d. parietal pain.

4. Large papillae arranged in an inverted V-pattern at the back of the tongue are called (*module 12.2*)
a. fungiform papillae.
b. ciliate papillae.
c. vallate papillae.
d. filiform papillae.

5. Contraction of the orbicularis oculi muscle (*module 12.3*)
a. elevates the upper eyelid.
b. moves the eyeball.
c. closes the eyelids.
d. constricts the pupil.

6. The thin, highly vascular layer that underlies most of the sclera is (*module 12.4*)
a. the conjunctiva.
b. the cornea.
c. the retina.
d. the choroid.

7. Aqueous humor drains from the anterior and posterior chambers of the eye through (*module 12.4*)
a. the lacrimal duct.
b. the fovea centralis.
c. the scleral venous sinus.
d. the optic disc.

8. Pupils dilate in response to the contraction of (*module 12.4*)
a. the ciliary muscles.
b. the circular iris muscles.
c. the radial iris muscles.
d. the orbicularis oculi muscles.

9. Automatic adjustment of the curvature of the lens by ciliary muscle contraction is called (*module 12.5*)
a. convergence.
b. accommodation.
c. focusing.
d. constriction.

10. The middle ear is separated from the inner ear by (*module 12.6*)
   a. the cochlear window.
   b. the tympanic membrane.
   c. the vestibular window.
   d. both a and c.

11. Stimulation of hair cells in the semicircular canals results from the movement of (*module 12.7*)
   a. endolymph.
   b. aqueous humor.
   c. perilymph.
   d. otoliths.

Essay Questions

1. List the senses of the body and differentiate between the special and somatic senses. How are these two classes of senses similar? (*module 12.1*)
2. List the functions of proprioceptors and differentiate between the various types. What role do proprioceptors play in the kinesthetic sense? (*module 12.1*)
3. Compare and contrast olfaction and gustation. Identify the cranial nerves that serve both of these senses. (*module 12.2*)
4. Describe the accessory structures of the eye and list their functions. (*module 12.3*)
5. Describe the arrangement of rods and cones in the periphery of the eye and in the fovea centralis. What are the effects of light on these photoreceptors? (*module 12.4*)
6. Explain the process by which light rays are focused onto the fovea centralis. (*module 12.5*)
7. Diagram the ear and label the structures of the outer, middle, and inner ear. (*module 12.6*)
8. Trace a sound wave through the structures of the outer, middle, and inner ear. (*module 12.7*)
9. What are the vestibular organs? How do they function in maintaining equilibrium and balance? (*module 12.7*)

Labeling Exercise

Label the structures indicated on the figure to the right.

1. ______________________________
2. ______________________________
3. ______________________________
4. ______________________________
5. ______________________________
6. ______________________________

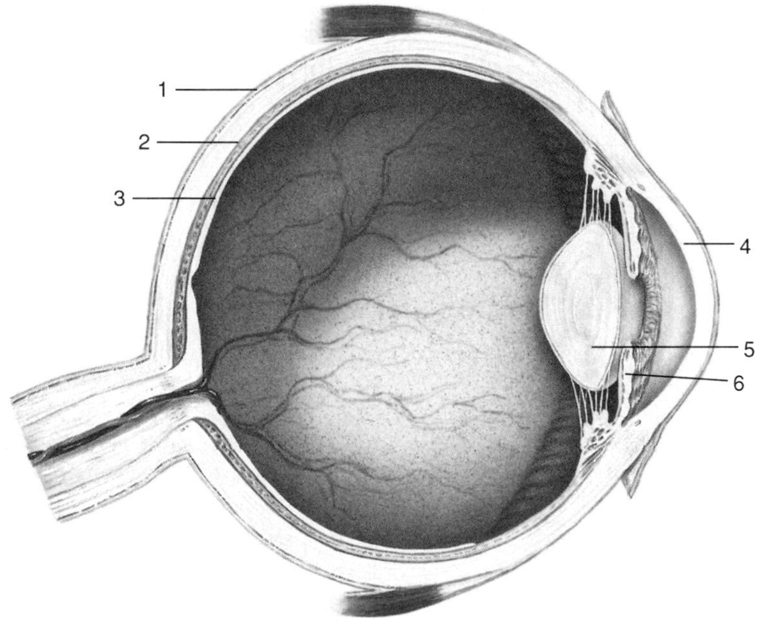

Critical Thinking Questions

1. When excessively stimulated, any sensory receptor will convey sensations that are interpreted as pain. For example, a bright flash of light or a loud sound is painful. Explain the adaptive value of pain. Where in the brain does pain perception occur? (Hint: See chapter 10, module 10.2).
2. A displaced contact lens from the surface of the cornea is irritating, but it is not of clinical concern because it will stay in contact with the bulbar conjunctiva. Considering the anatomy of the eye, why can't a contact lens be displaced into the orbit?
3. Located in the dense petrous part of the temporal bone, the structures of the inner ear are well protected. Discuss other features of the ear that protect the sense of hearing.
4. Define *myringotomy* and explain why this procedure is usually successful in treating children who suffer recurring middle-ear infections. (To help you answer this question, consider the advantage of punching two holes in an oil can—one across from the other—as opposed to punching a single hole.)
5. A child is referred to an ENT (ear, nose, and throat) specialist because of a ruptured tympanic membrane sustained during an explosion. In testing for hearing loss, the physician has the child respond to the sound of a vibrating tuning fork. Interestingly, the sound is most clearly heard when the vibrating tuning fork is placed on the mastoid process of the temporal bone. What is the explanation for this?

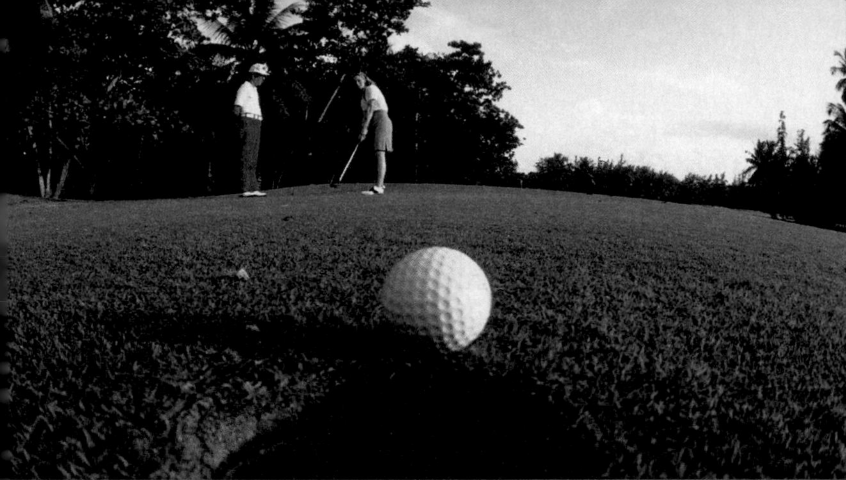

CHAPTER

13

"Endocrine glands secrete chemical hormones into the blood . . . to be carried to their target organs." *page 334*

Endocrine System

Chapter Outline

Terms to Remember

endocrine (*en' dŏ-krin*) gland, 334
hormone, 334
catecholamine (*kat"ĕ-kol' ă-mēn*), 335
polypeptide, 335
steroid, 335
pituitary (*pĭ-too' ĭ-ter-e*) gland, 336
posterior pituitary (neurohypophysis) (*noor"o-hi-pof' ĭ-sis*), 336
anterior pituitary (adenohypophysis) (*ad"n-o-hi-pof' ĭ-sis*), 336
trophic hormone, 337
hypothalamo-hypophyseal (*hi"pō-fiz' e-al*) tract, 339
adrenal (*ă-dre' nal*) gland, 341
adrenal cortex, 341
adrenal medulla, 341
thyroid gland, 343
parathyroid gland, 344
pancreatic islet, 345
pineal (*pin' e-al*) gland, 345
thymus, 347
gonad, 347
placenta (*plă-sen' tă*), 347
cyclic AMP, 349

Learning Objectives

After studying this chapter, you should be able to

- list the chemical categories of hormones and give examples of hormones in each of these categories.
- compare the actions of hormones with those of neurotransmitters.
- describe the structure of the pituitary gland and list the pituitary hormones.
- explain how the hypothalamus controls the posterior pituitary and the anterior pituitary, and describe negative feedback control of the anterior pituitary.
- describe the structure and function of the adrenal medulla and adrenal cortex.
- describe the structure and function of the thyroid and parathyroid glands.
- list the hormonal secretions of the pancreatic islets, pineal gland, thymus, stomach, small intestine, and placenta.
- explain how steroid hormones and thyroxine influence their target cells.
- explain how epinephrine and protein hormones influence their target cells.

13.1 Introduction to the Endocrine System

Working with the nervous system, the endocrine system regulates body activities. Endocrine glands secrete chemical hormones into the blood or interstitial tissue to be carried to their target organs.

The **endocrine system** consists of ductless **endocrine** (*en'dŏ-krin*) **glands** that secrete biologically active chemicals called **hormones** into the blood or surrounding interstitial fluid. Hormones affect the metabolism of their *target organs* and, by this means, help to regulate body metabolism, body growth, and reproduction.

The endocrine system functions closely with the nervous system in regulating and integrating body processes. The activity of most body organs is influenced by both the nervous and endocrine systems. For example, the biceps brachii contracts in response to motor nerve impulses to the muscle. The secretion of the hormone epinephrine (adrenaline) helps to make the contraction more forceful and to sustain it. In general, the action of nerve impulses is fast and the effects are of short duration, whereas the action of hormones is relatively slow and the effects are prolonged.

Inside Information—The endocrine system is unique in that its glands are widely scattered throughout the body, with no anatomical continuity. By contrast, the organs of the other body systems are physically connected in some fashion.

Many endocrine glands are discrete organs (fig. 13.1) whose primary functions are the production and secretion of hormones. Such "typical" endocrine glands include the pituitary, thyroid, and adrenal glands. The concept of the endocrine system, however, must be extended beyond the organs commonly described as endocrine glands to

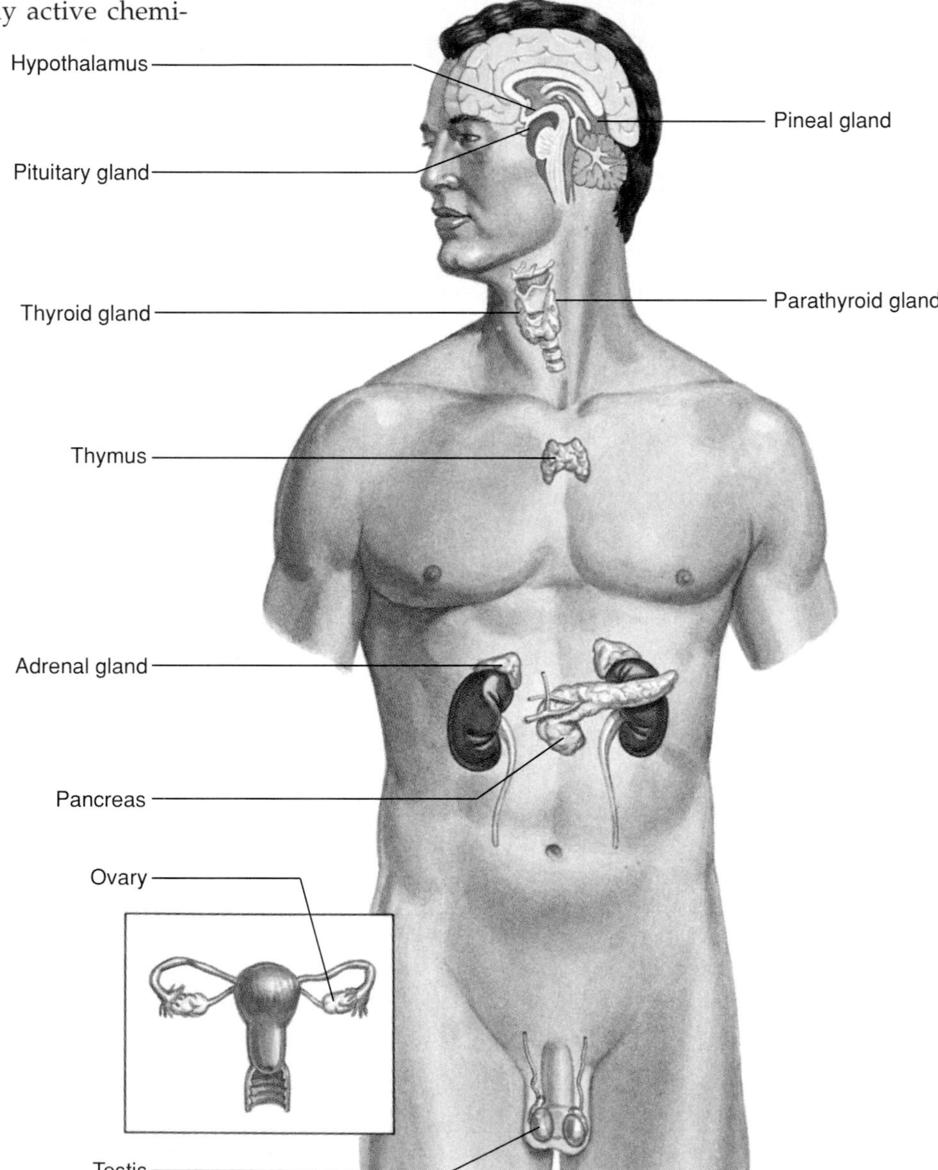

FIGURE 13.1
The location of the major endocrine organs of the human body.

endocrine: Gk. *endon,* within; *krinein,* to separate
hormone: Gk. *hormon,* to set in motion

include all of the organs that secrete hormones. In recent years, it has been discovered that organs such as the skin, kidneys, heart, liver, and brain secrete hormones in addition to performing other functions in the body.

Chemical Classification of Hormones

Hormones secreted by different endocrine glands vary in their chemical structure. All hormones, however, can be grouped into three general chemical categories:

1. **catecholamines** (*kat"ĕ-kol'ă mēnz*), including epinephrine and norepinephrine;
2. **polypeptides** and **glycoproteins,** including shorter chain polypeptides such as antidiuretic hormone and insulin, and large glycoproteins such as thyroid-stimulating hormone; and
3. **lipophilic hormones** (those that are lipid soluble). This category includes **steroids,** such as cortisol and estrogens, and the peptide hormone thyroxine.

Steroid hormones, which are derived from cholesterol (fig. 13.2), are lipids and thus are not water soluble. The gonads—testes and ovaries—secrete *sex steroids;* the adrenal cortex secretes *corticosteroids,* including cortisol and aldosterone among others.

The major thyroid hormones are composed of two derivatives of the amino acid tyrosine bonded together. These are the only hormones that contain iodine (fig. 13.3). When the hormone contains four iodine atoms, it is called *tetraiodothyronine* (*tet"ră-i"ŏ-dŏ-thi'ro-nēn*) (T_4), or *thyroxine.* When it contains three atoms of iodine, it is called *triiodothyronine* (*tri"i-ŏ"dŏ-thi'ro-nēn*) (T_3). Although these hormones are peptides and not steroids, they are similar to steroids in that they are relatively small, nonpolar molecules. Because of their chemical similarities, steroid and thyroid hormones are active when taken orally (as a pill); sex steroids are administered as contraceptive pills, and thyroid hormone pills are taken by people whose thyroid is deficient (who are hypothyroid). Other types of hormones cannot be taken orally because they would be digested into inactive fragments before being absorbed into the blood.

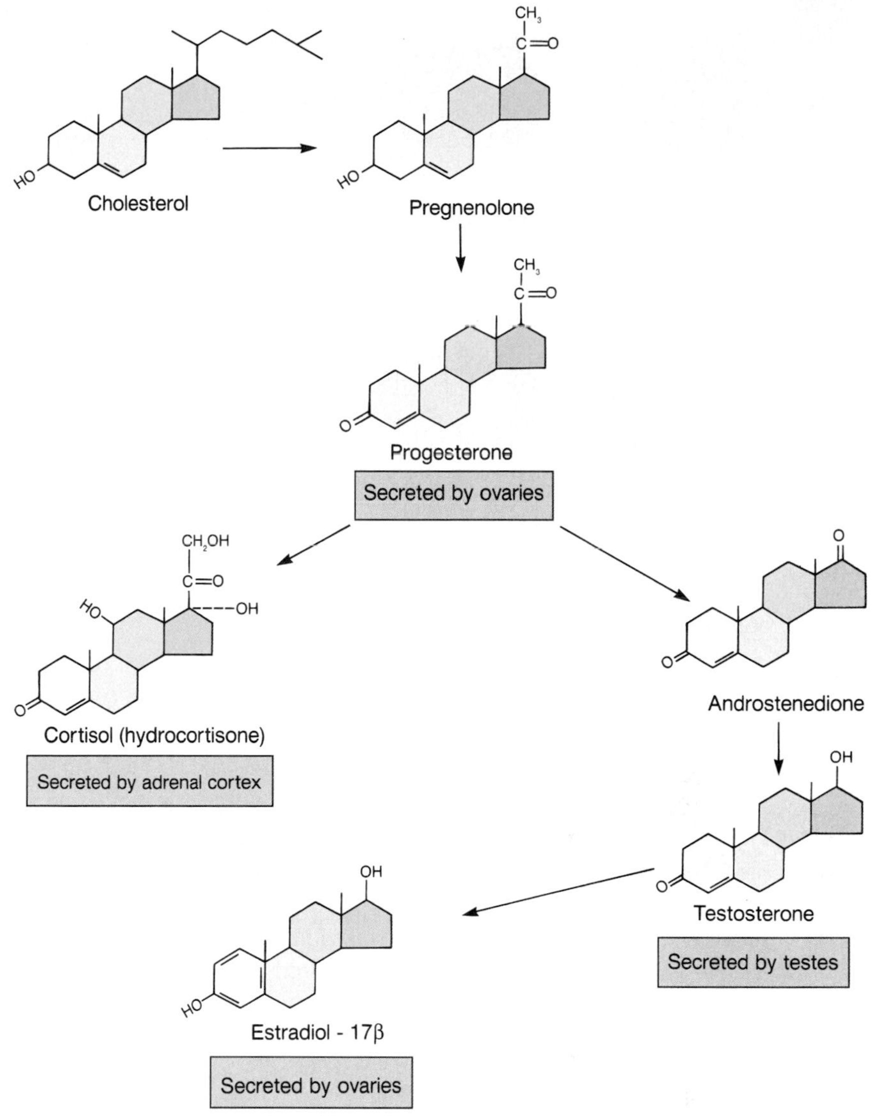

FIGURE 13.2
Simplified biosynthetic pathways for steroid hormones. Notice that progesterone (a hormone secreted by the ovaries) is a common precursor in the formation of all other steroid hormones, and that testosterone (the major androgen secreted by the testes) is a precursor in the formation of estradiol-17β, the major estrogen secreted by the ovaries.

Common Aspects of Neural and Endocrine Regulation

The fact that endocrine regulation is chemical might lead you to believe that it is fundamentally different from neural control systems that depend on the electrical properties of cells. This assumption is incorrect. As explained in chapter 9 (see module 9.6), nerve impulses through axons stimulate the cells they innervate through the

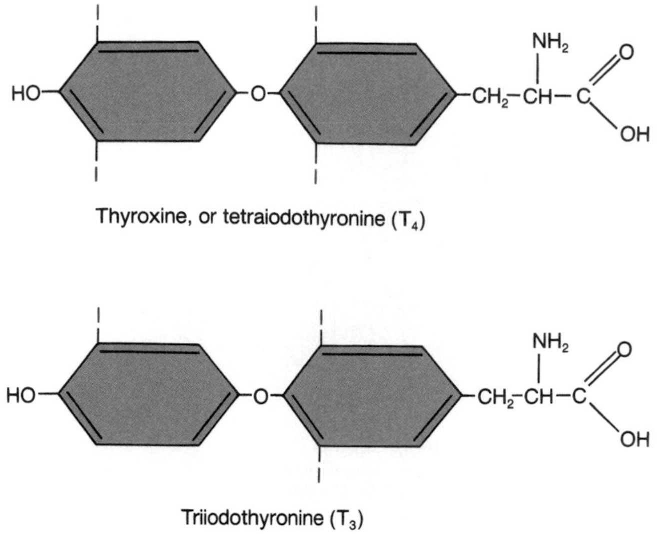

Thyroxine, or tetraiodothyronine (T_4)

Triiodothyronine (T_3)

FIGURE 13.3
The thyroid hormones, thyroxine (T_4) and triiodothyronine (T_3), are secreted in a ratio of 9 to 1.

release of a chemical neurotransmitter. Neurotransmitters do not travel in the blood as do hormones; instead, they diffuse only a very short distance across a synapse. In other respects, however, the actions of neurotransmitters and hormones are very similar.

Regardless of whether a particular chemical is acting as a neurotransmitter or as a hormone, the following conditions must apply in order for it to function in physiological regulation.

1. The target cells must have specific *receptor proteins* that combine with the neurotransmitter or hormone molecule.
2. The combination of the regulator molecule with its receptor proteins must cause a specific sequence of changes in the target cells.
3. There must be a mechanism to quickly turn off the action of the regulator molecule. This mechanism, which involves rapid removal and/or chemical inactivation of regulator molecules, is essential because without an "off-switch" physiological control would be impossible.

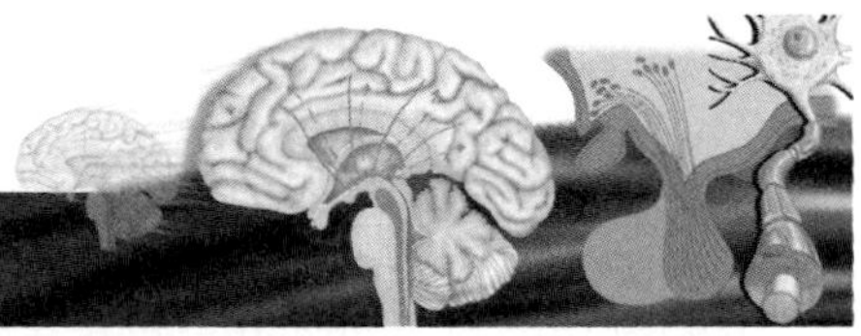

13.2 Pituitary Gland

The pituitary gland is structurally and functionally divided into two parts. The anterior pituitary produces and secretes a variety of hormones, whereas the posterior pituitary stores and releases two hormones that are produced within neuron cell bodies in the hypothalamus.

The **pituitary** (*pĭ-too'ĭ-ter-e*) **gland,** or **hypophysis** (*hi-pof'ĭ-sis*), is located on the inferior aspect of the brain in the region of the diencephalon (see chapter 10, module 10.3). It is a rounded, pea-shaped gland, about 1.3 cm (0.5 in.) in diameter, and is secured in a depression in the sphenoid bone called the sella turcica (see fig. 6.22). The pituitary gland is attached to the hypothalamus by a stalklike structure called the **pituitary stalk** (fig. 13.4).

The pituitary gland is structurally and functionally divided into a posterior lobe, called the **neurohypophysis** (*noor"o-hi-pof'ĭ-sis*), and an anterior lobe, or **adenohypophysis** (*ad"n-o-hi-pof'ĭ-sis*). Embryologically, the adenohypophysis develops from a pouchlike extension of the oral epithelium (fig. 13.5). The neurohypophysis, by contrast,

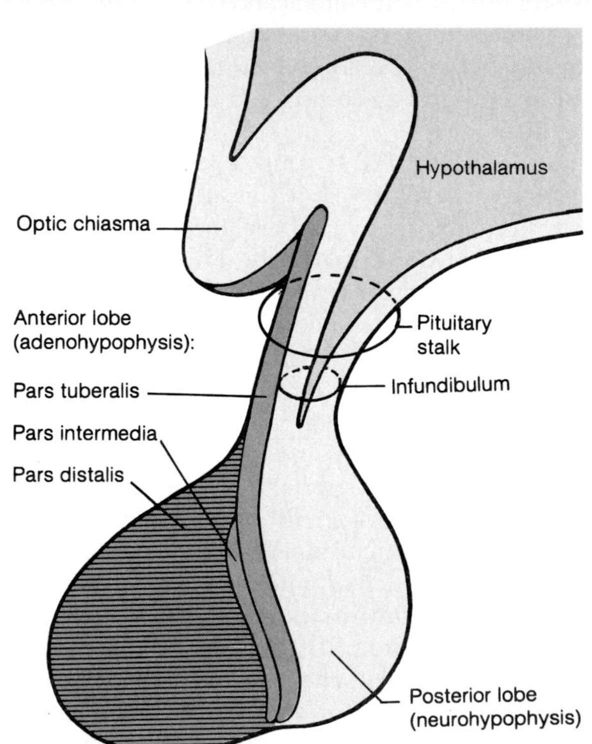

FIGURE 13.4
The structure of the pituitary gland.

pituitary: L. *pituita,* phlegm (the pituitary gland was originally thought to secrete mucus into the nasal cavity)
adenohypophysis: Gk. *adeno,* gland; *hypo,* under; *physis,* a growing

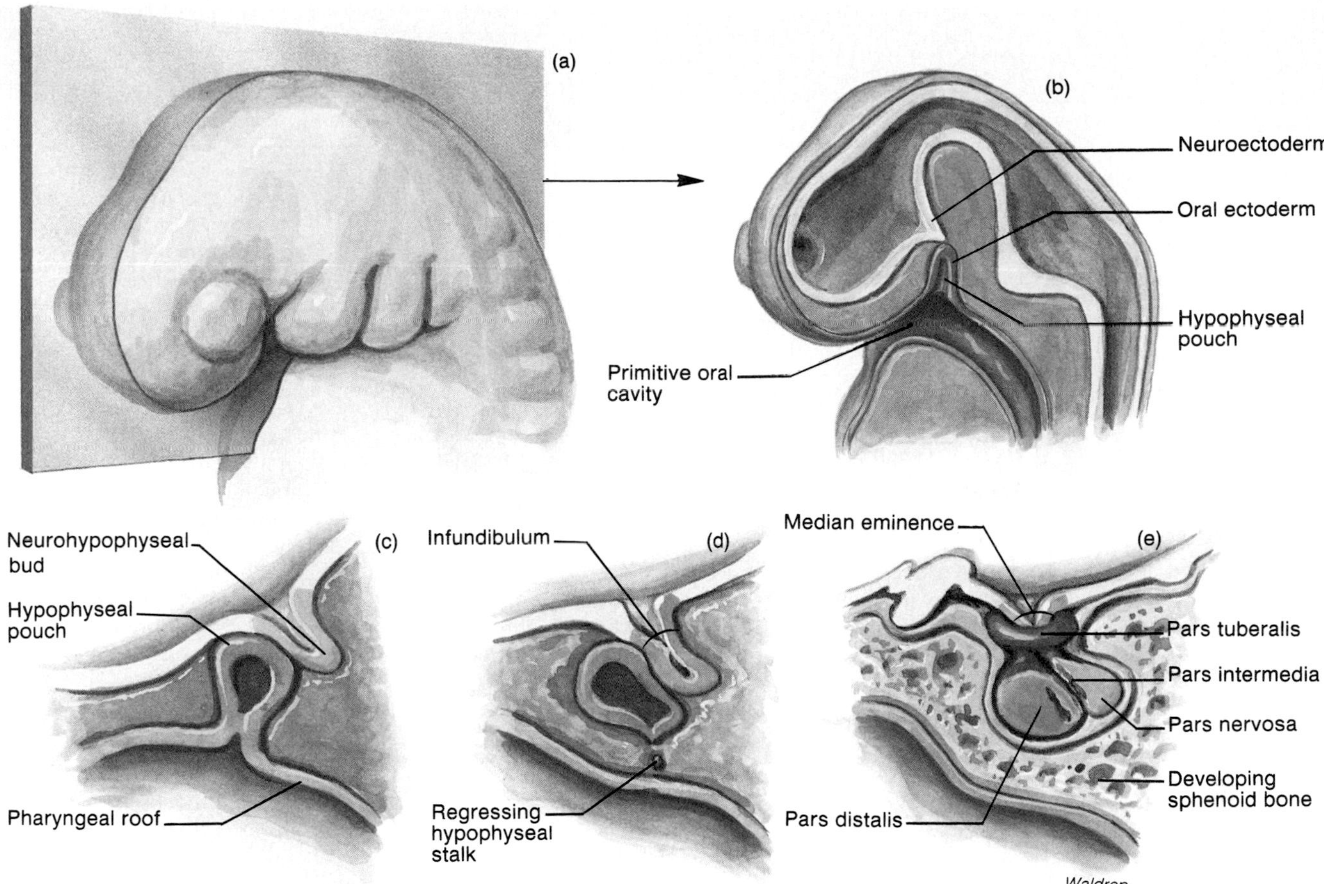

FIGURE 13.5

The development of the pituitary gland. (*a*) The head end of the embryo at 4 weeks showing the position of a midsagittal cut seen in the developmental sequence *b–e.* The pituitary gland arises from a specific portion of the neuroectoderm, called the neurohypophyseal bud, which evaginates downward during the 4th and 5th weeks respectively in (*b*) and (*c*), and from a specific portion of the oral ectoderm, called the hypophyseal (Rathke's) pouch, which evaginates upward from a specific portion of the primitive oral cavity. At 8 weeks (*d*), the hypophyseal pouch is no longer connected to the pharyngeal roof of the oral cavity. During the fetal stage (*e*), the development of the pituitary gland is completed.

develops embryologically from an inferior extension of nervous tissue from the diencephalon of the brain.

It follows from its embryonic origin that the neurohypophysis is composed of neural tissue. It consists of the *pars nervosa,* also known as the **posterior pituitary,** which stores and releases hormones, and the *infundibulum* (*in″fun-dib′yŭ-lum*), the portion of the pituitary stalk that connects the posterior pituitary to the hypothalamus. Nerve fibers extend through the infundibulum along with small neuroglia-like cells called *pituicytes.*

Since the adenohypophysis is derived from epithelial tissue, it is the truly glandular portion of the pituitary. The adenohypophysis consists of three parts.

1. The *pars distalis* is the rounded portion, and is the major endocrine part of the gland. The pars distalis is also known as the **anterior pituitary.**
2. The *pars tuberalis* is the thin extension, in contact with the infundibulum.
3. The *pars intermedia* is located between the anterior and posterior parts of the pituitary.

The hormones secreted by the anterior pituitary are called **trophic hormones.** The term *trophic* means "food." Although the anterior pituitary hormones are not actually food for their target organs, this term is used because high amounts of these hormones stimulate hypertrophy in their target organs, whereas insufficient amounts cause their target organs to atrophy. In designating the hormones of the anterior pituitary, therefore, "trophic" (conventionally shortened to *tropic,* which has a different meaning—"attracted to") is incorporated into the names. The hormones of the pars distalis are **ACTH, TSH, FSH, LH, GH,** and **prolactin.** The full names and characteristics of these hormones are given in table 13.1.

Infundibulum: L. *infundibulum,* a funnel

Table 13.1 Anterior Pituitary Hormones

Hormone	Target Tissue	Stimulated by Hormone	Regulation of Secretion
ACTH (adrenocorticotropic hormone)	Adrenal cortex	Secretion of glucocorticoids	Stimulated by CRH (corticotropin-releasing hormone); inhibited by glucocorticoids
TSH (thyroid-stimulating hormone)	Thyroid gland	Secretion of thyroid hormones	Stimulated by TRH (thyrotropin-releasing hormone); inhibited by thyroid hormones
GH (growth hormone)	Most tissue	Protein synthesis and growth; lipolysis and increased blood glucose	Inhibited by somatostatin; stimulated by growth hormone–releasing hormone
FSH (follicle-stimulating hormone) and LH (luteinizing hormone)	Gonads	Gamete production and sex hormone secretion	Stimulated by GnRH (gonadotropin-releasing hormone); inhibited by sex steroids
Prolactin	Mammary glands and other accessory sex organs	Milk production, controversial actions in other organs	Inhibited by PIH (prolactin-inhibiting hormone)
LH (luteinizing hormone)	Gonads	Sex hormone secretion; ovulation and corpus luteum formation	Stimulated by GnRH

The pars intermedia of the adenohypophysis produces different forms of **melanocyte-stimulating hormone (MSH),** which in lower vertebrates cause a darkening of the skin as a means of camouflage. This reaction involves a cell type that is absent in humans. Although continued administration of MSH to humans over a period of several days will cause the skin to darken, MSH does not appear to have this effect in normal physiology. Interestingly, ACTH, secreted from the pars distalis, contains the MSH amino acid sequence as part of its structure, and abnormally high levels of ACTH will cause a darkening of the skin (see the "Clinical Considerations" section on Addison's disease).

The **posterior pituitary** releases only two hormones, both of which are produced in neuron cell bodies within the hypothalamus. These hormones, which are merely stored in the posterior pituitary, are *antidiuretic* (*an"te-di"yŭ-ret'ik*) *hormone* and *oxytocin.*

Antidiuretic hormone (ADH), also called **vasopressin,** promotes the retention of water by the kidneys, so that less water is excreted in the urine and more water is retained in the blood. This hormone also causes vasoconstriction in experimental animals, but the significance of this effect in humans is controversial.

Diabetes insipidus *results from a marked decrease in ADH secretion caused by trauma or disease to the hypothalamus or neurohypophysis. The symptoms of this disease are polyuria (voiding of excessive dilute urine), concentrated body fluids with dehydration, and a particularly heightened sensation of thirst.*

Oxytocin in females stimulates contractions of the uterus during labor and contractions of the mammary gland alveoli and ducts, producing the milk-ejection reflex during lactation. In males, a rise in oxytocin at the time of ejaculation has been measured, but the physiological significance of this hormone in males remains to be determined.

Injections of oxytocin may be given to a woman to induce labor if the pregnancy is prolonged or if the fetal membranes have ruptured and there is a danger of infection. Oxytocin administration after parturition causes the uterus to regress in size and squeezes the blood vessels, thus minimizing the danger of hemorrhage.

13.3 Regulation of Pituitary Gland Secretion

Both parts of the pituitary gland are regulated by the hypothalamus. Hypothalamic regulation is neural for the posterior pituitary and hormonal for the anterior pituitary.

Hypothalamic Control of the Posterior Pituitary

The two hormones released by the posterior pituitary (ADH and oxytocin) are actually produced in neuron cell bodies of the *supraoptic nuclei* and *paraventricular nuclei* of the hypothalamus. The hormones are transported along axons of the **hypothalamo-hypophyseal tract** (fig. 13.6) to the posterior pituitary, which stores and later releases these hormones. The posterior pituitary is thus more a storage organ than a true gland.

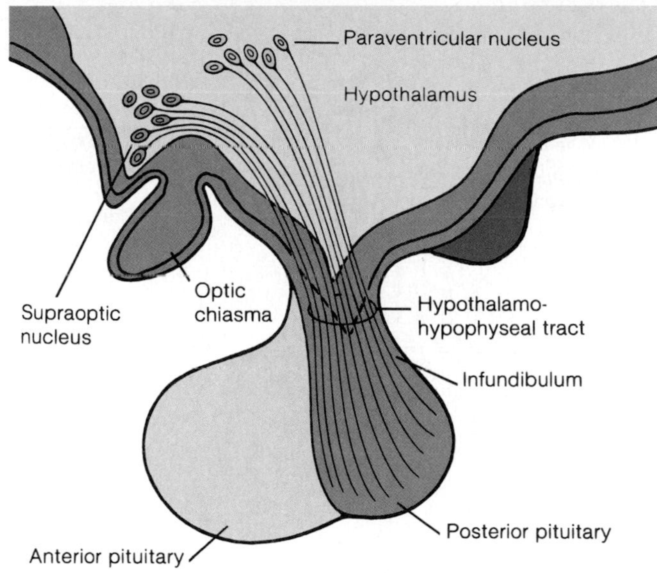

FIGURE 13.6
The posterior pituitary, or neurohypophysis, stores and secretes oxytocin and antidiuretic hormone produced in neuron cell bodies within the supraoptic and paraventricular nuclei of the hypothalamus. These hormones are transported to the posterior pituitary by nerve fibers of the hypothalamo-hypophyseal tract.

Hypothalamic Control of the Anterior Pituitary

Since axons do not enter the anterior pituitary, hypothalamic control of the anterior pituitary is achieved through hormonal rather than neural regulation. Neurons in the hypothalamus produce *releasing hormones* and *inhibiting hormones,* which are transported to axon terminals in the basal portion of the hypothalamus. This region, known as the *median eminence,* contains blood capillaries that are drained by venules in the stalk of the pituitary gland.

The venules that drain the median eminence deliver blood to a second capillary bed in the anterior pituitary. Since this second capillary bed receives venous blood from the first (is located "downstream" from the first), there is a vascular link between the median eminence and anterior pituitary. This arrangement of blood vessels (two capillary beds in series) is rare, and is described as a *portal system* (see chapter 15, module 15.6, for a discussion of the hepatic portal system). The vascular link between the hypothalamus and the anterior pituitary is thus called the **hypothalamo-hypophyseal portal system.**

Polypeptide hormones are secreted into the hypothalamo-hypophyseal portal system by neurons of the hypothalamus. These hormones regulate the secretions of the anterior pituitary (fig. 13.7). Thyrotropin-releasing hormone **(TRH)** stimulates the secretion of TSH, and corticotropin-releasing hormone **(CRH)** stimulates the secretion of ACTH. A single releasing hormone, gonadotropin-releasing hormone, or **GnRH,** stimulates the secretion of both gonadotropic hormones (FSH and LH) from the anterior pituitary. The secretion of prolactin and of growth hormone from the anterior pituitary is primarily regulated by hypothalamic inhibitory hormones, known as **PIH** (prolactin-inhibiting hormone) and **somatostatin,** respectively. A separate growth hormone–releasing hormone **(GHRH)** has also been discovered.

Feedback Control of the Anterior Pituitary

Anterior pituitary secretion of ACTH, TSH, and the gonadotropins (FSH and LH) is controlled by **negative feedback inhibition** from the target gland hormones. Secretion of ACTH is inhibited by a rise in corticosteroid secretion,

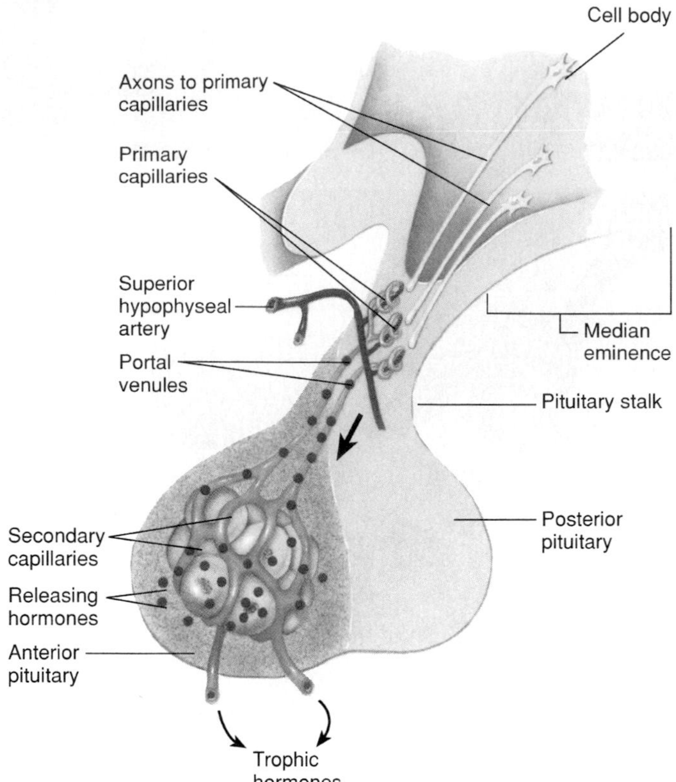

FIGURE 13.7

Neurons in the hypothalamus secrete releasing hormones (shown as dots) into the blood vessels of the hypothalamo-hypophyseal portal system. These releasing hormones stimulate the anterior pituitary to secrete its hormones into the general circulation.

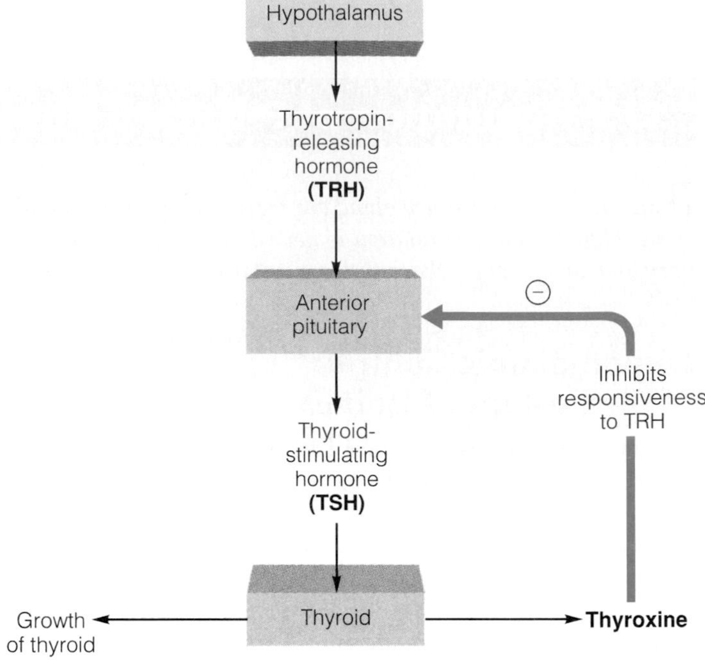

FIGURE 13.8

The secretion of thyroxine from the thyroid is stimulated by the thyroid-stimulating hormone (TSH) from the anterior pituitary. The secretion of TSH is stimulated by the thyrotropin-releasing hormone (TRH), which is secreted from the hypothalamus into the hypothalamo-hypophyseal portal system. This stimulation is balanced by the negative feedback inhibition of thyroxine, which decreases the responsiveness of the anterior pituitary to stimulation by TRH.

for example, and TSH is inhibited by a rise in the secretion of thyroxine from the thyroid. These negative feedback relationships are easily demonstrated by removal of the target glands. Castration (surgical removal of the gonads), for example, produces a rise in the secretion of FSH and LH. These effects demonstrate that, under normal conditions, the target glands exert an inhibitory effect on the anterior pituitary. This inhibitory effect can occur at either of two levels:

1. the target gland hormones can act on the hypothalamus and inhibit the secretion of releasing hormones, or
2. the target gland hormones can act on the anterior pituitary and inhibit its response to the releasing hormones.

Thyroxine, for example, appears to inhibit the response of the anterior pituitary to TRH and thus acts to reduce TSH secretion (fig. 13.8). Sex steroids, by contrast, reduce the secretion of gonadotropins by inhibiting both GnRH secretion and the ability of the anterior pituitary to respond to stimulation by GnRH (fig. 13.9).

In addition to negative feedback control of the anterior pituitary, there is an example of a hormone from a target

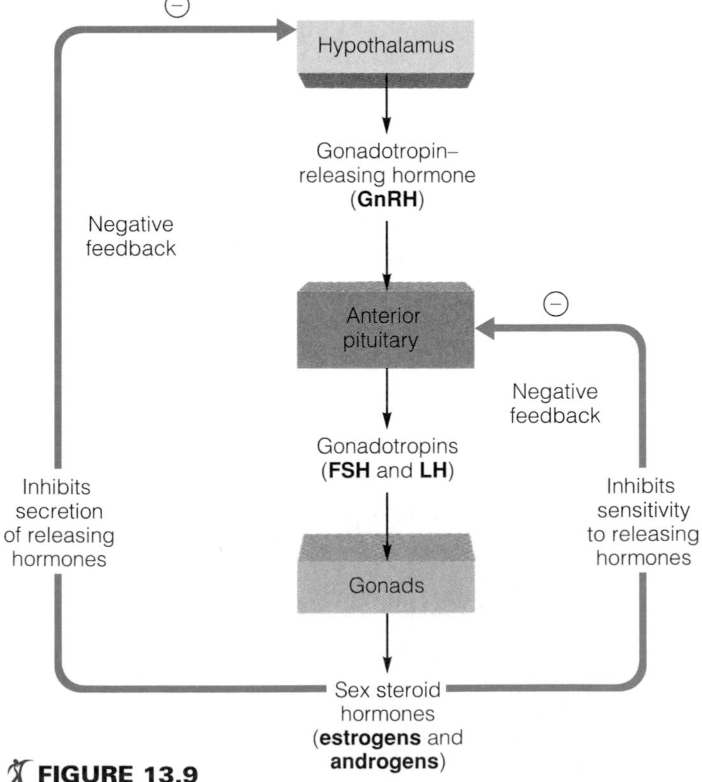

FIGURE 13.9

Negative feedback control of gonadotropin secretion.

organ whose action actually stimulates the secretion of an anterior pituitary hormone. Toward the middle of the menstrual cycle, the rising secretion of estradiol from the ovaries stimulates the anterior pituitary to secrete a "surge" of LH, which results in ovulation. This effect is commonly referred to as a *positive feedback* to distinguish it from the more usual negative feedback inhibition of target gland hormones on anterior pituitary secretion. Interestingly, higher levels of estradiol at a later stage of the menstrual cycle exert the opposite effect—negative feedback inhibition—on LH secretion. The control of gonadotropin secretion is discussed in more detail in chapter 23 (see module 23.4).

13.4 Adrenal Glands

The adrenal cortex is stimulated by ACTH and secretes corticosteroid hormones. The adrenal medulla is stimulated by sympathetic nerves and secretes epinephrine.

Structure of the Adrenal Glands

The adrenal (*ă-dre'nal*) glands (also called *suprarenal glands*) are paired organs that cap the superior borders of the kidneys (fig. 13.10*a*). The adrenal glands, along with the kidneys, are embedded against the muscles of the back in a protective pad of fat. Each adrenal gland consists of an outer adrenal cortex and inner adrenal medulla.

The **adrenal cortex** is subdivided into three zones: an outer *zona glomerulosa* (*glo-mer"yoo-lo'să*), an intermediate *zona fasciculata* (*fă-sik"yoo-lă'tă*), and an inner *zona reticularis* (fig. 13.10*b*). The **adrenal medulla** is composed of tightly packed clusters of *chromaffin* (*kro-maf'in*) *cells,* which are innervated by preganglionic sympathetic fibers.

The adrenal cortex is derived from embryonic mesoderm, whereas the adrenal medulla is derived from the neural crest (the same embryonic tissue that forms the peripheral nervous system). Like the anterior and posterior lobes of the pituitary, therefore, the adrenal cortex and adrenal medulla are really two different endocrine tissues. Although they are located in the same organ, they have different embryonic origins, secrete different hormones, and are regulated by different control systems.

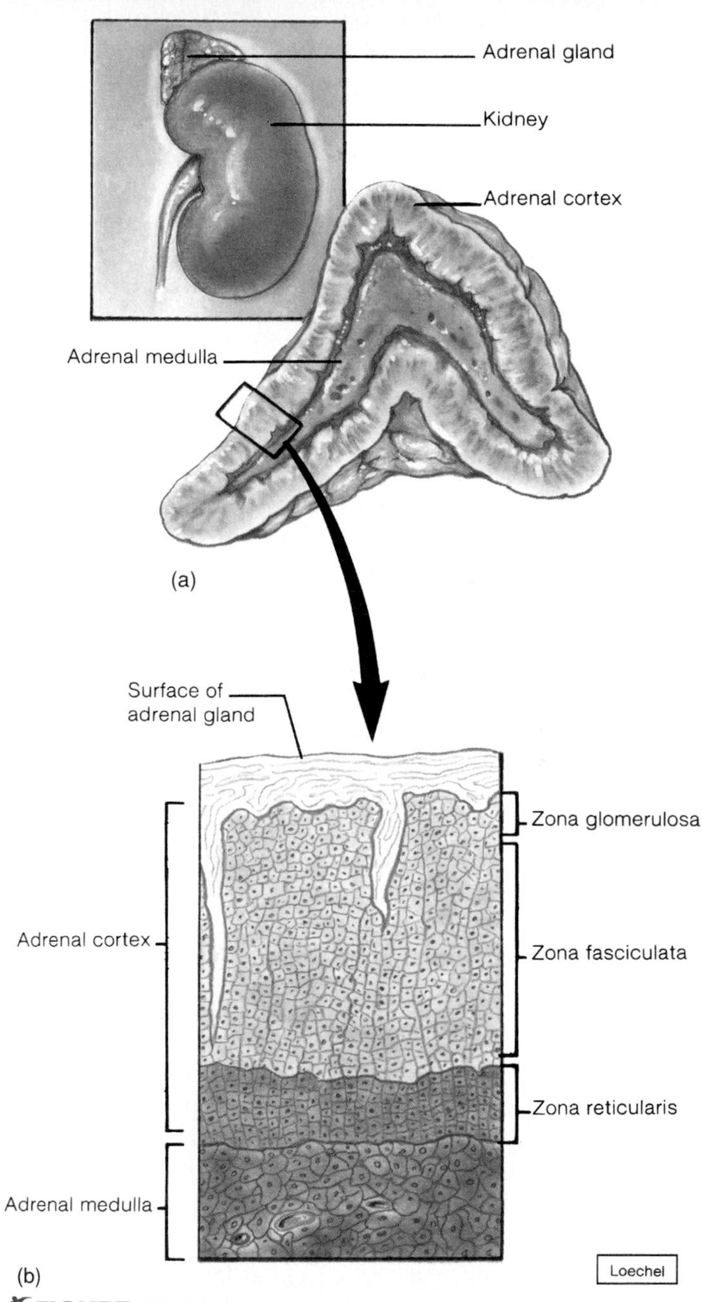

FIGURE 13.10
The structure of the adrenal gland. (*a*) The gross structure and (*b*) the histological structure showing the three zones of the adrenal cortex.

Functions of the Adrenal Cortex

The adrenal cortex secretes steroid hormones called **corticosteroids** (*kor"tĭ-ko-ster'oidz*), or **corticoids,** for short.

Corticosteroids are grouped into three functional categories:

1. **mineralocorticoids,** which regulate Na^+ and K^+ balance (by acting on the kidneys);
2. **glucocorticoids,** which regulate the metabolism of glucose and other organic molecules; and
3. **sex steroids,** which are weak androgens (and lesser amounts of estrogens) that supplement the sex steroids secreted by the gonads.

adrenal: L. *ad,* to; *renes,* kidney

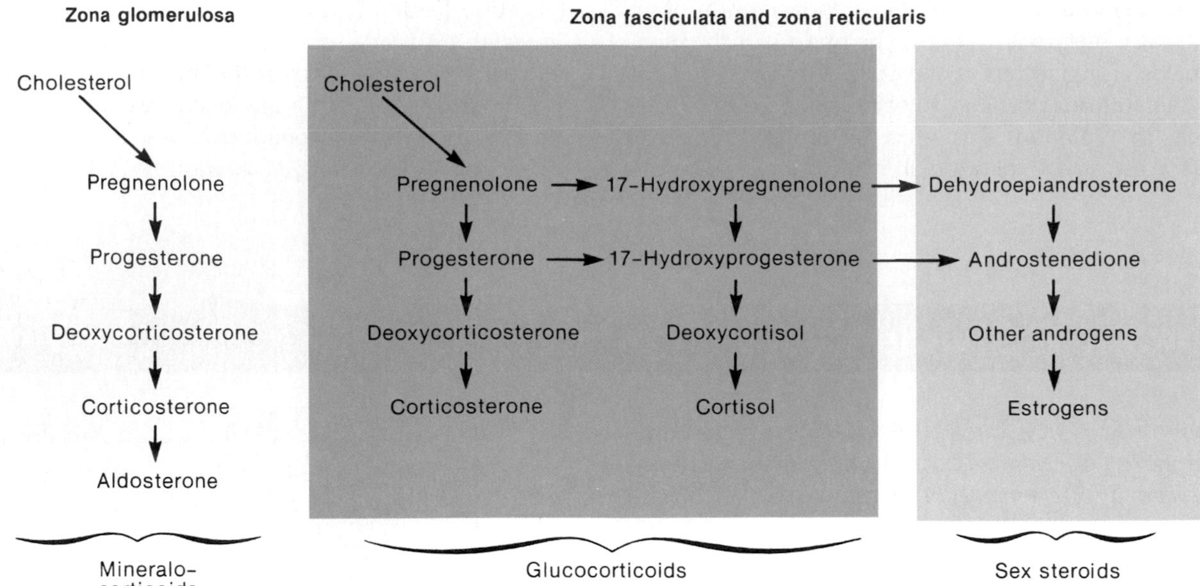

FIGURE 13.11
Simplified pathways for the synthesis of steroid hormones in the adrenal cortex. The adrenal cortex produces steroids that regulate Na^+ and K^+ balance (mineralocorticoids), steroids that regulate glucose balance (glucocorticoids), and small amounts of sex steroid hormones.

Aldosterone, the most potent of the mineralocorticoids, is produced in the zona glomerulosa (fig. 13.11). The predominant glucocorticoid is *cortisol* (*hydrocortisone*), which is secreted by the zona fasciculata and possibly by the zona reticularis as well. The secretion of cortisol is stimulated by ACTH (fig. 13.12). The secretion of aldosterone is controlled by other mechanisms related to blood volume and electrolyte balance (see chapter 19, module 19.7).

Secretion of ACTH from the anterior pituitary, and consequently of cortisol from the adrenal cortex, occurs when a person is under stress. The discovery that stress activates the pituitary-adrenal axis was made by a Canadian physiologist, Hans Selye. Further study led Selye to conclude that activation of the adrenal glands is part of "a nonspecific response of the body to readjust itself following any demand made upon it." Selye termed this nonspecific response the **general adaptation syndrome (GAS).**

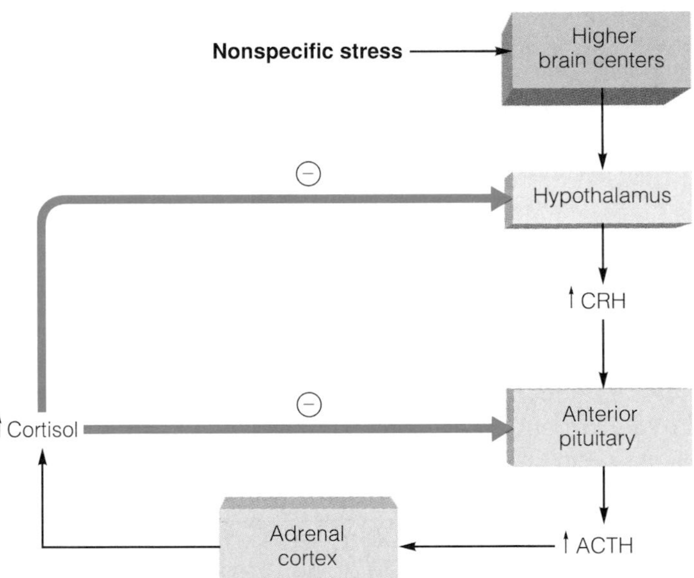

FIGURE 13.12
The activation of the pituitary-adrenal axis by nonspecific stress. Negative feedback control of the adrenal cortex is also shown.

Functions of the Adrenal Medulla

The chromaffin cells of the adrenal medulla secrete epinephrine and norepinephrine in an approximate ratio of 4 to 1, respectively. The effects of these hormones are similar to the effects caused by stimulation of the sympathetic nervous system.

The adrenal medulla is innervated by sympathetic nerve fibers, which are activated when the sympathetic division of the autonomic nervous system is stimulated. Many stressors, therefore, activate the adrenal medulla as well as the adrenal cortex. Activation of the adrenal medulla, together with the sympathetic division of the autonomic nervous system, prepares the body for greater physical performance. This is the *fight-or-flight response,* as discussed in chapter 11 (see module 11.3).

Glucocorticoids, such as hydrocortisone, can inhibit the immune system. Indeed, these steroids are often used to treat various inflammatory diseases and to suppress the immune rejection of a transplanted organ. It seems reasonable, therefore, that the elevated glucocorticoid secretion that can accompany stress may inhibit the ability of the immune system to protect against disease. In fact, there is evidence that prolonged stress results in an increased incidence of cancer and other diseases.

13.5 Thyroid and Parathyroid Glands

The thyroid gland's principal hormone, thyroxine, helps to regulate the metabolic rate and is needed for proper growth and development. Parathyroid hormone helps to regulate calcium balance in the blood.

Structure of the Thyroid Gland

The thyroid gland is located just below the larynx, in front of the trachea (fig. 13.13). It consists of two lobes connected by a mass of tissue called the *isthmus.* The thyroid is the largest of the pure endocrine glands, with a weight of approximately 25 g. Because of its rich blood supply, it appears redder than the tissues surrounding it.

On a microscopic level, the thyroid gland consists of many spherical hollow sacs called **thyroid follicles** (fig. 13.14). These follicles are lined with a simple cuboidal epithelium composed of *follicular cells.* The follicular cells synthesize the principal thyroid hormone *thyroxine.* Stored inside the follicles is a protein-rich fluid called *colloid.* Between the follicles are epithelial cells called *parafollicular cells,* which produce a hormone called *calcitonin* (or *thyrocalcitonin*).

thyroid: Gk. *thyros,* oblong shield

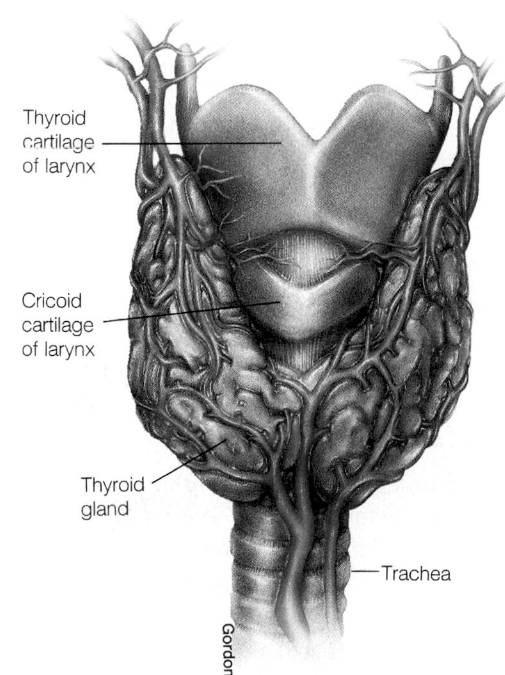

FIGURE 13.13
An anterior view of the thyroid gland showing its relationship to the larynx and trachea.

Thyroid Hormones

Iodine obtained in the diet is essential for the synthesis of thyroxine. Seafood is a source of iodine, and commercial salt generally contains iodine as an additive. Absorbed iodine is transported through the blood to the thyroid

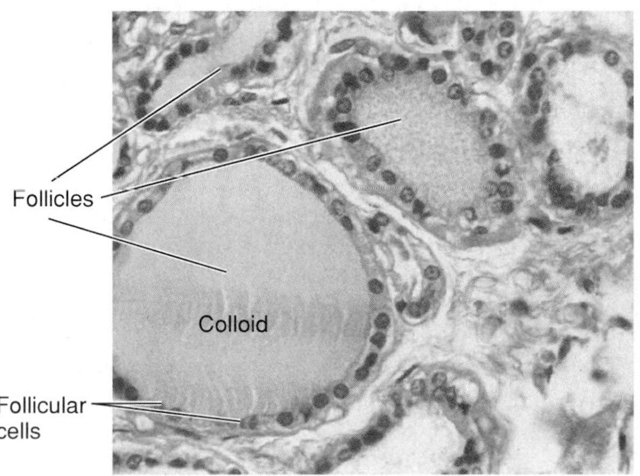

FIGURE 13.14
The histology of the thyroid gland showing numerous thyroid follicles. Each follicle consists of follicular cells surrounding the fluid known as colloid.

gland, where an active transport mechanism called an iodine pump moves the iodides (anions of iodine) into the follicle cells. The iodides are then converted (oxidized) to iodine and secreted into the colloid. Once in the colloid, the iodine attaches to specific amino acids (tyrosines) within the polypeptide chain of protein called **thyroglobulin.** Attachment of one iodine to a tyrosine results in *monoiodotyrosine (MIT)*; attachment of two iodines results in *diiodotyrosine (DIT)*. Within the colloid, enzymes link DIT and MIT together. Two linked DITs produce **tetraiodothyronine—T_4,** or **thyroxine.** A coupling of MIT and DIT produces **triiodothyronine** or **T_3**. At this point, T_4 and T_3 are still attached to thyroglobulin. Upon stimulation by TSH, the cells of the follicle take up a small volume of colloid by pinocytosis, hydrolyze the T_3 and T_4 from the thyroglobulin, and secrete the free hormones into the blood.

Through the activation of genes, thyroid hormones stimulate protein synthesis, promote maturation of the nervous system, and increase the rate of energy utilization by the body. The development of the central nervous system is particularly dependent on thyroid hormones, and a deficiency of these hormones during early childhood can cause serious mental retardation. As discussed in chapter 21 (see module 21.7), the *basal metabolic rate (BMR)*—which is the minimum rate of caloric expenditure by the body—is determined to a large degree by the level of thyroid hormones in the blood.

The stimulatory effect of TSH on the thyroid is dramatically revealed in people who develop an iodine-deficiency (endemic) goiter. *In the absence of sufficient dietary iodine, the thyroid cannot produce adequate amounts of T_4 and T_3. The resulting lack of negative feedback inhibition causes abnormally high levels of TSH secretion. High TSH cannot stimulate increased thyroxine secretion under these circumstances, but it does stimulate abnormal growth of the thyroid gland (fig. 13.15). Such abnormal growth is known as a* goiter.

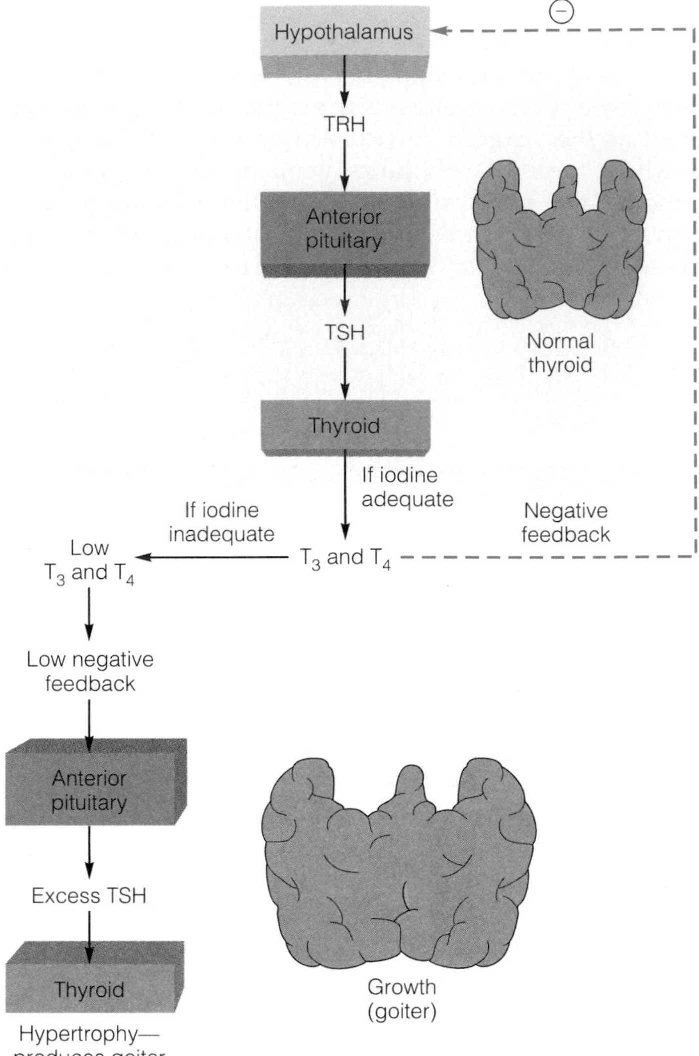

FIGURE 13.15
Lack of adequate iodine in the diet interferes with the negative feedback control of TSH secretion, resulting in the formation of an endemic goiter.

Inside Information—In certain parts of the world, including central Europe and the Great Lakes region of North America, iodine is essentially absent from the soil. Before iodine was added to table salt, the Great Lakes region was sometimes referred to as the "goiter belt."

Calcitonin, released by the parafollicular cells of the thyroid, works in concert with parathyroid hormone (discussed below) to regulate calcium levels in the blood. Calcitonin inhibits the breakdown of bone tissue and stimulates the excretion of calcium by the kidneys. Both actions result in the lowering of blood calcium levels.

Parathyroid Glands

The small, flattened parathyroid glands are embedded in the posterior surfaces of the lateral lobes of the thyroid gland (fig. 13.16). There are usually four parathyroid glands: a *superior pair* and an *inferior pair*. Each

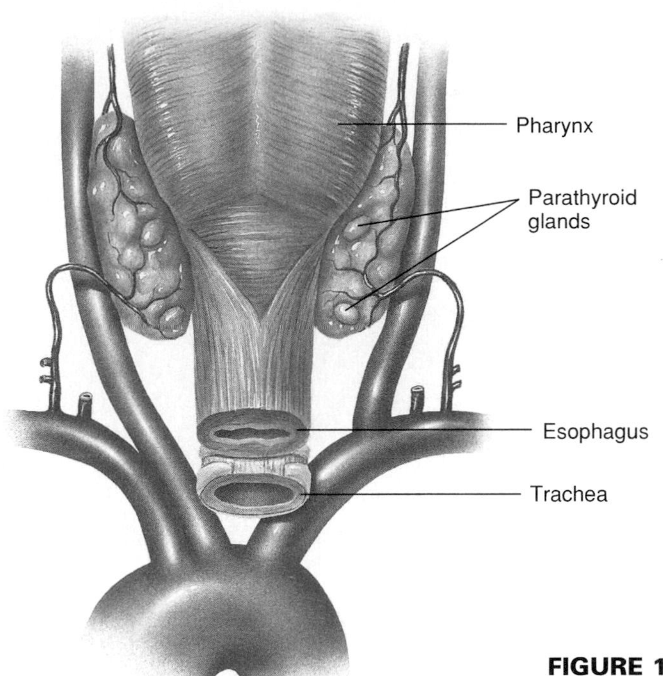

FIGURE 13.16
A posterior view of the parathyroid glands.

parathyroid gland is a small yellowish-brown body 3 to 8 mm (0.1 to 0.3 in.) long, 2 to 5 mm (0.07 to 0.2 in.) wide, and about 1.5 mm (0.05 in.) deep.

The single hormone secreted by the parathyroid glands is called **parathyroid hormone (PTH).** This hormone promotes a rise in blood calcium levels by acting on the small intestine, bones, and kidneys; thus, it opposes the effects of calcitonin, released by the thyroid gland. PTH activates vitamin D, which increases the amount of calcium absorbed from the small intestine into the bloodstream. It also increases the breakdown of bone tissue and decreases the rate at which calcium is lost in the urine. These actions help to raise the blood calcium and thus to defend the body against a fall in blood calcium concentrations.

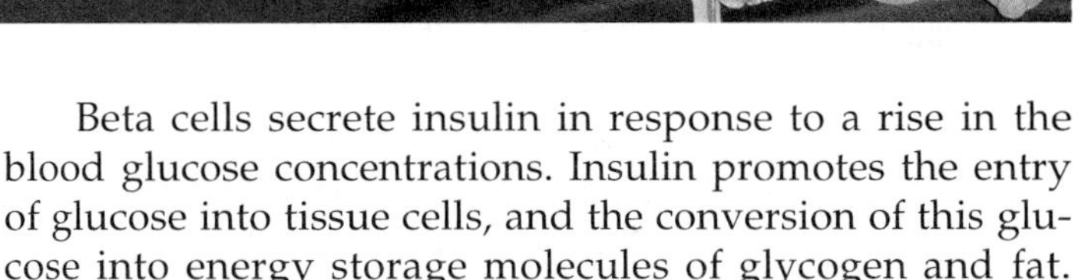

13.6 Pancreas and Other Endocrine Glands

The pancreatic islets secrete two hormones, insulin and glucagon, which are critically involved in the regulation of blood glucose levels in the body. Numerous other glands secrete hormones that regulate metabolism, digestion, immune function, and reproduction.

Pancreatic Islets (Islets of Langerhans)

The pancreas is both an endocrine and an exocrine gland. The gross structure of this gland and its exocrine functions in digestion are described in chapter 20 (see module 20.9). The endocrine portion of the pancreas consists of scattered clusters of cells called the **pancreatic islets,** or *islets of Langerhans* (fig. 13.17). On a microscopic level, the most conspicuous cells in the islets are the *alpha* and *beta cells* (fig. 13.18). The alpha cells secrete the hormone **glucagon,** and the beta cells secrete **insulin.**

Alpha cells secrete glucagon in response to a fall in the blood glucose concentrations. Glucagon stimulates the liver to convert glycogen to glucose, which causes the blood glucose level to rise. Glucagon also stimulates the hydrolysis of stored fat and the consequent release of free fatty acids into the blood.

Beta cells secrete insulin in response to a rise in the blood glucose concentrations. Insulin promotes the entry of glucose into tissue cells, and the conversion of this glucose into energy storage molecules of glycogen and fat. Insulin also aids the entry of amino acids into the cells, and the production of cellular protein.

The actions of insulin and glucagon are antagonistic. After a meal, insulin secretion is increased and glucagon secretion is decreased; fasting, by contrast, causes a rise in glucagon and a decline in insulin secretion. The metabolic effects of insulin and glucagon are discussed in more detail in chapter 21 (see module 21.8).

Pineal Gland

The small, cone-shaped pineal (*pin'e-al*) gland, also called the **pineal body,** is located in the roof of the third ventricle, where it is encapsulated by the meninges covering the

islets of Langerhans: from Paul Langerhans, German anatomist, 1847–88
insulin: L. *insula,* island
pineal: L. *pinea,* pine cone

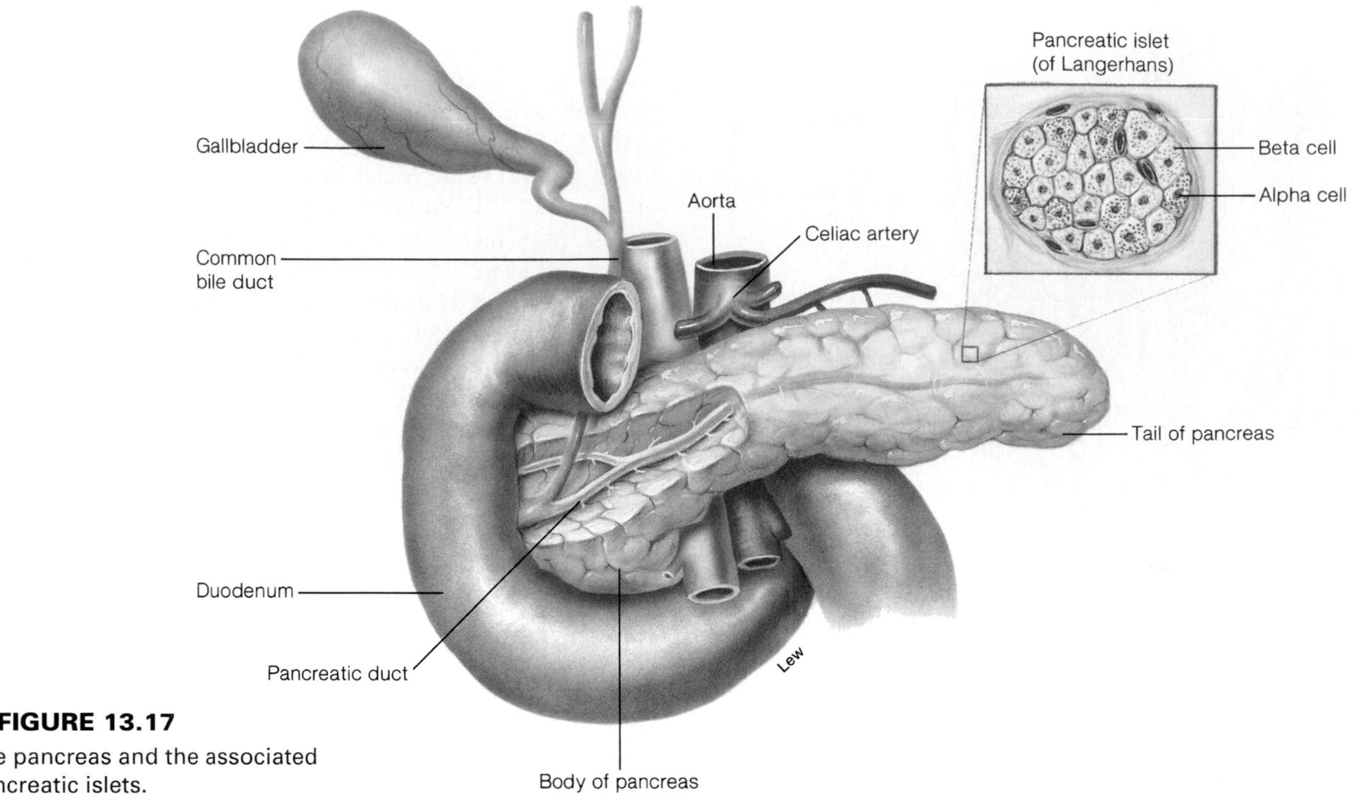

FIGURE 13.17
The pancreas and the associated pancreatic islets.

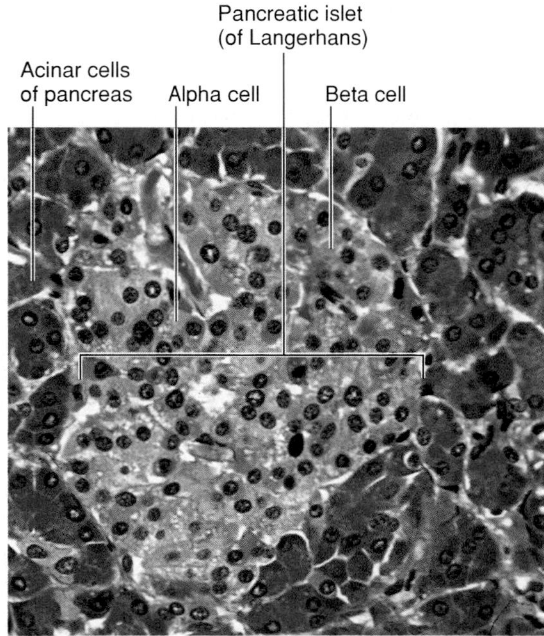

FIGURE 13.18
The histology of the pancreatic islets.

brain (see fig. 10.8). The gland begins to regress in size at about age 7, and in the adult it appears as a thickened strand of fibrous tissue.

The principal hormone of the pineal gland is **melatonin.** Production and secretion of this hormone is stimulated by activity of the *suprachiasmatic nucleus* (*SCN*) in the hypothalamus of the brain via activation of sympathetic neurons to the pineal gland. Activity of the SCN, and thus secretion of melatonin, is highest at night. During the day, neural pathways from the retina of the eyes depress the activity of the SCN, reducing sympathetic stimulation of the pineal gland, and thus decreasing melatonin secretion.

Since cycles of light and dark directly influence the levels of melatonin secretion, other circadian *(daily)* rhythms *of the body might be the result of the ebb and flow of melatonin. Derangements of such cycles—in* jet-lag, *for example—are tentatively ascribed to the effects of melatonin. Also, phototherapy used in the treatment of* seasonal affective disorder (SAD), *or "winter depression," might be effective because such treatment inhibits melatonin secretion.*

Inside Information—People who work the "graveyard shift" exhibit reversed melatonin secretion patterns. At night (when they are exposed to light), low levels of hormone are secreted. High levels are secreted during their daytime sleeping hours (when the lights are out).

It has long been suspected that melatonin inhibits the pituitary-gonad axis. Indeed, a decrease in melatonin secretion in many lower vertebrates is responsible for maturation of the gonads during the reproductive season. Excessive melatonin secretion in humans is associated with a delay in the onset of puberty; however, the role of melatonin in sexual maturation is still highly controversial.

Thymus

The thymus is a bilobed organ positioned in the upper mediastinum, in front of the aorta and behind the manubrium of the sternum (fig. 13.19). Although the size of the thymus varies considerably, it is relatively large in newborns and children, and sharply regresses in size after puberty. Besides decreasing in size, the thymus of adults becomes infiltrated with strands of fibrous and fatty connective tissue.

The thymus serves as the site of production of **T cells** (thymus-dependent cells), which are the lymphocytes involved in cell-mediated immunity (see chapter 17, module 17.7). In addition to providing T cells, the thymus secretes a number of hormones that are believed to stimulate T cells after they leave the thymus.

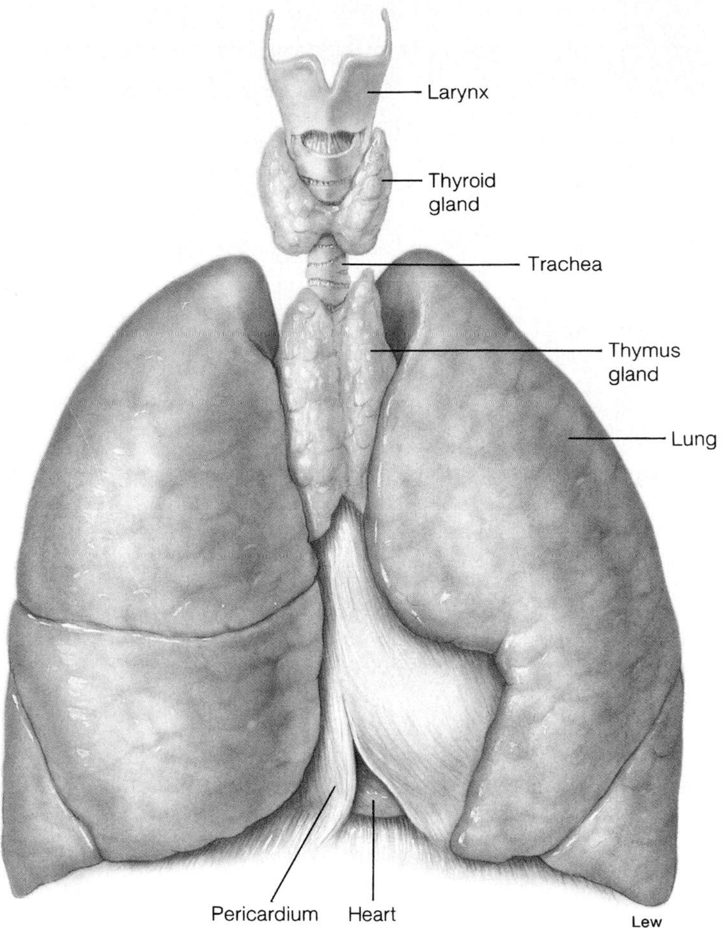

FIGURE 13.19
The thymus is a bilobed organ within the mediastinum of the thorax.

Stomach and Small Intestine

The stomach and small intestine secrete a number of hormones that act on the gastrointestinal tract (GI tract) itself and also on the pancreas and gallbladder (see chapter 20, module 20.11). The effects of these hormones, in conjunction with regulation by the autonomic nervous system, coordinate the activities of different regions of the GI tract and the secretions of pancreatic juice and bile.

Gonads and Placenta

The gonads **(testes** and **ovaries)** secrete *sex steroids.* These include male sex hormones, or *androgens,* and female sex hormones—*estrogens* and *progestogens.* The principal hormones in each of these categories are *testosterone, estradiol-17β,* and *progesterone,* respectively.

The placenta (*plă-cen'tă*) is the organ responsible for nutrient and waste exchange between the fetus and mother. The placenta is also an endocrine gland; it secretes large amounts of estrogens and progesterone, as well as a number of polypeptide and protein hormones that resemble some hormones secreted by the anterior pituitary. The physiology of the placenta and other aspects of reproductive endocrinology are covered in chapters 22 through 24.

13.7 Mechanism of Steroid and Thyroxine Action

Steroid and thyroid hormones are small, nonpolar molecules. They can easily enter their target cells, where they combine with receptor proteins and activate specific genes.

Although each hormone exerts its own characteristic effects on specific target cells, hormones that are in the same chemical category have similar mechanisms of action (table 13.2). These similarities involve the location of cellular receptor proteins and the events that occur in the target cells after the hormone has combined with its receptor protein.

Steroid and thyroid hormones are small molecules. Both groups are also nonpolar (lipophilic), and thus are not very water soluble. Unlike other hormones, therefore, steroids and thyroid hormones (primarily thyroxine) do not travel dissolved in the aqueous portion of the plasma. Rather,

thymus: Gk. *thymos,* a warty outgrowth
placenta: L. *placenta,* flat cake

TABLE 13.2 **Functional Categories of Hormones, Based on the Location of Their Receptor Proteins and Mechanisms of Action**

Hormone Type	Secreted by	Location of Receptors	Effects of Hormone-Receptor Interaction
Catecholamines	Adrenal medulla	Outer surface of cell membrane	Stimulates production of intracellular "second messenger," which activates previously inactive enzymes
Polypeptides and glycoproteins	All glands except adrenal cortex, gonads, and thyroid	Outer surface of cell membrane	Stimulates production of intracellular "second messenger," which activates previously inactive enzymes
Steroids	Adrenal cortex, testes, ovaries	Cytoplasm of target cells	Stimulates translocation of hormone-receptor complex to nucleus and activation of specific genes
Thyroxine (T_4)	Thyroid	Nucleus of target cells	After conversion to triiodothyronine (T_3), activates specific genes

they are transported to their target cells attached to plasma carrier proteins. These hormones then dissociate from the carrier proteins in the blood and easily pass through the lipid component of the target cell's membrane.

Steroid Hormones

Once through the cell membrane, some steroid hormones attach to *cytoplasmic receptor proteins* in the target cells. The steroid hormone–receptor protein complex then *translocates* to the nucleus and attaches by means of the receptor proteins to the chromatin. The sites of attachment in the chromatin, termed *acceptor sites,* are specific for the target tissue. It has been suggested that one part of the receptor binds to a chromosomal protein, while a different part binds to DNA.

The attachment of the receptor protein–steroid complex to the acceptor site "turns on" genes. Specific genes become activated and produce nuclear RNA, which is then processed into messenger RNA (mRNA). This new mRNA enters ribosomes and codes for the production of new proteins. Since some of these newly synthesized proteins may be enzymes, the metabolism of the target cell is changed in a specific manner (fig. 13.20). Steroid hormones, in short, affect their target cells by stimulating genetic transcription (RNA synthesis), which is followed by genetic translation (protein synthesis).

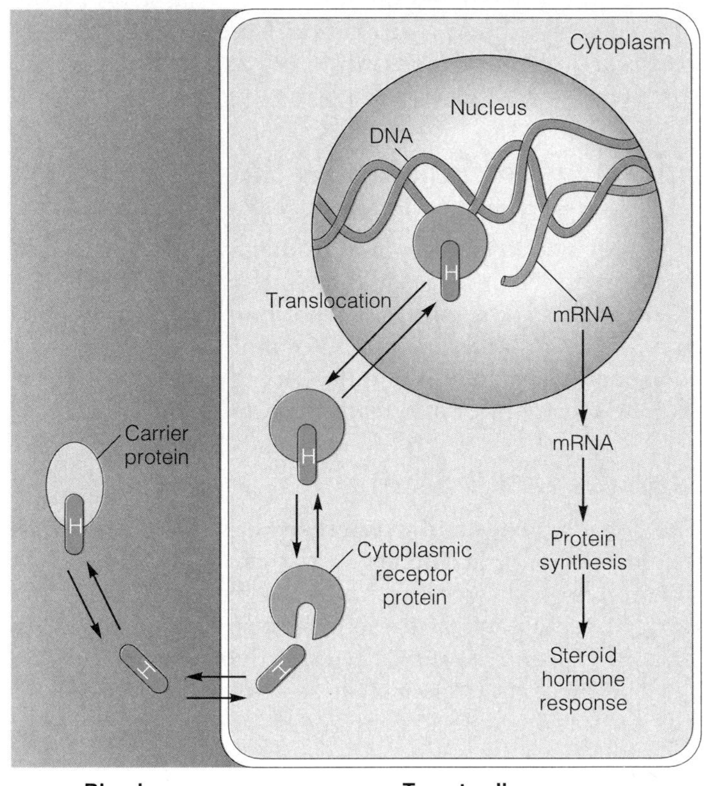

FIGURE 13.20
The mechanism of the action of a steroid hormone (H) on the target cells.

Thyroxine

As described in module 13.5, the major hormone secreted by the thyroid gland is thyroxine, or tetraiodothyronine (T_4). Like steroid hormones, thyroxine travels in the blood attached to carrier proteins—primarily, *thyroxine-binding globulin* (*TBG*). The thyroid also secretes a small amount of triiodothyronine, or T_3. The carrier proteins have a higher affinity (bond strength) for T_4 than for T_3, however, and as a result, the amount of unbound (or "free") T_3 is about 10 times greater than the amount of free T_4 in the blood plasma.

Only the free thyroxine and T_3 can enter target cells; the protein-bound thyroxine serves as a reservoir of this hormone in the blood (this is why it takes a couple of weeks after surgical removal of the thyroid for the symptoms of hypothyroidism to develop). Once the free thyroxine passes into the target cell cytoplasm, it is enzymatically changed into T_3. This is a critical step, since it is only the T_3 that is active within the target cells. Thyroxine is an inactive precursor of the active hormone. We can call such inactive precursors of hormones **prehormones.**

Receptor proteins for T_3 are already in the nucleus attached to chromatin, but these receptors cannot activate genes until the receptors bind to T_3 that enters the nucleus from the cytoplasm. The attachment of T_3 to the chromatin-bound receptor proteins activates genes and results in the production of new mRNA and new proteins. This sequence of events is summarized in figure 13.21.

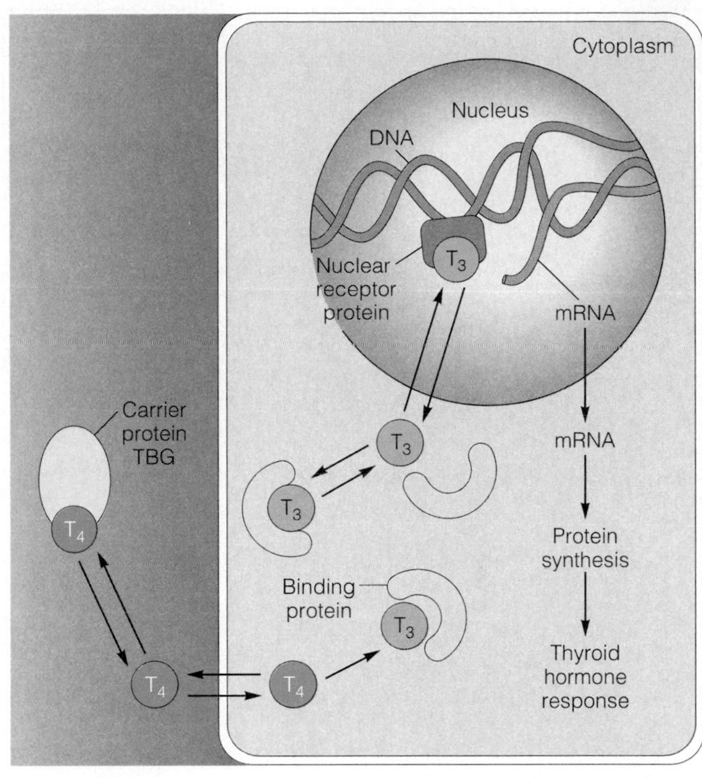

FIGURE 13.21
The mechanism of action of T_3 (triiodothyronine) on the target cells.

13.8 Mechanism of Catecholamine and Polypeptide Action

In order to stimulate a target cell, catecholamine and polypeptide hormones require the action of an intracellular second messenger. For many such hormones, the second messenger is cyclic AMP.

Catecholamine and polypeptide hormones cannot pass through the lipid barrier of the target cell membrane. Although some of these hormones enter the cell by pinocytosis, most of their effects are believed to result from their interaction with receptor proteins in the outer surface of the target cell membrane. Since they do not have to enter the target cells to exert their influence, other molecules must mediate the actions of these hormones within the target cells. If you think of hormones as "messengers" from the endocrine glands, the intracellular mediators of the hormone's action can be called **second messengers.**

Cyclic adenosine monophosphate (abbreviated **cAMP**) was the first "second messenger" to be discovered and is the best understood. The effects of epinephrine and norepinephrine induced by binding to beta-adrenergic receptors (see chapter 11, module 11.3) are due to cAMP production within the target cells. It has also been demonstrated that the effects of many (but not all) polypeptide hormones are also mediated by cAMP.

The bonding of these hormones with their membrane receptor proteins activates an enzyme called **adenylate cyclase** (*a-den'l-it si'klās*). This enzyme is built into the cell membrane and, when activated, it catalyzes the following reaction:

$$\text{ATP} \xrightarrow{\text{adenylate cyclase}} \text{cAMP} + \text{PP}_i$$

Adenosine triphosphate (ATP) is thus converted into cAMP and two inorganic phosphates (*pyrophosphate,* abbreviated PP_i). As a result of the interaction of the hormone with its receptor and the activation of adenylate

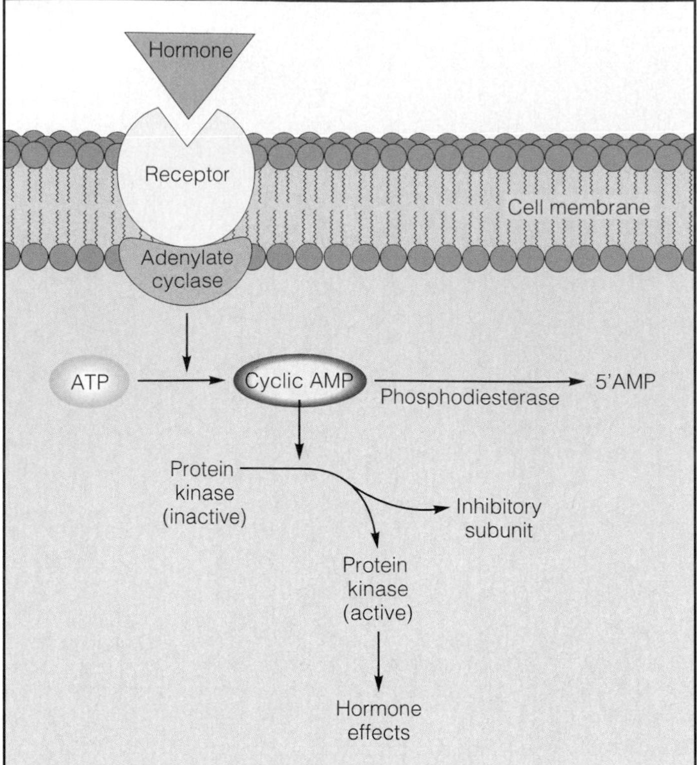

FIGURE 13.22
Cyclic AMP (cAMP) as a second messenger in the action of catecholamine and polypeptide hormones.

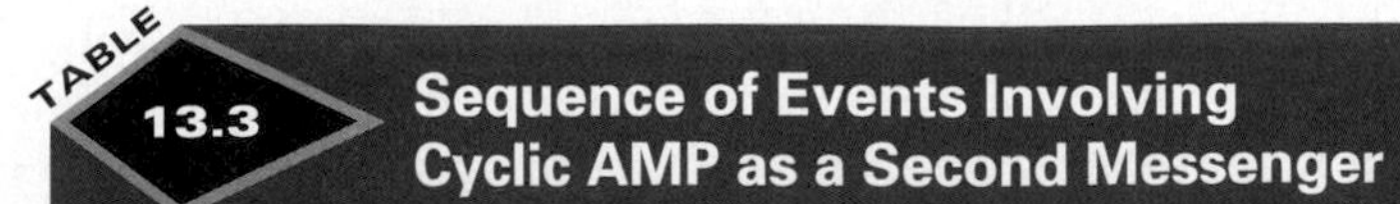

TABLE 13.3 Sequence of Events Involving Cyclic AMP as a Second Messenger

1. The hormones combine with their receptors on the outer surface of target cell membranes.
2. Hormone-receptor interaction stimulates activation of adenylate cyclase on the cytoplasmic side of the membranes.
3. Activated adenylate cyclase catalyzes the conversion of ATP to cyclic AMP (cAMP) within the cytoplasm.
4. Cyclic AMP activates protein kinase enzymes that were already present in the cytoplasm in an inactive state.
5. Activated cAMP-dependent protein kinase transfers phosphate groups (phosphorylates) to other enzymes in the cytoplasm.
6. The activity of specific enzymes is either increased or inhibited by phosphorylation.
7. Altered enzyme activity mediates the target cell's response to the hormone.

cyclase, therefore, the intracellular concentration of cAMP is increased. Cyclic AMP activates a previously inactive enzyme in the cytoplasm called **protein kinase.** The inactive form of this enzyme consists of two subunits: a catalytic subunit and an inhibitory subunit. The enzyme is produced in an inactive form and becomes active only when cAMP attaches to the inhibitory subunit. Bonding of cAMP with the inhibitory subunit causes this subunit to dissociate from the catalytic subunit, which then becomes active (fig. 13.22). In summary, the hormone—acting through an increase in cAMP production—increases the activity of protein kinase within its target cells.

Active protein kinase catalyzes the attachment of phosphate groups to different proteins in the target cells. As a result, some enzymes are activated and others are inactivated. Thus, cyclic AMP, acting through protein kinase, modulates the activity of enzymes that are already present in the target cell. This alters the metabolism of the target tissue in a manner characteristic of the actions of that specific hormone (table 13.3).

Like all biologically active molecules, cAMP must be rapidly inactivated for it to function effectively as a second messenger in hormone action. This function is served by an enzyme called **phosphodiesterase** (*fos"fo-di-es'tĕ-rās*), which hydrolyzes cAMP into inactive fragments. As a result of the action of phosphodiesterase, the stimulatory effect of a hormone that uses cAMP as a second messenger depends upon the continuous generation of new cAMP molecules, and thus depends upon the level of secretion of the hormone.

Drugs that inhibit the activity of phosphodiesterase prevent the breakdown of cAMP and result in increased concentrations of cAMP within the target cells. The drug theophylline *and its derivatives, for example, are used clinically to raise cAMP levels within bronchiolar smooth muscle. This duplicates and enhances the effect of epinephrine on the bronchioles (producing dilation) in people who suffer from asthma.* Caffeine, *a compound related to theophylline, is also a phosphodiesterase inhibitor, and thus produces its effects by raising the cAMP concentrations within tissue cells.*

In addition to cAMP, other substances can serve as second messengers in the actions of different hormones in different target tissues. The requirement for different second-messenger systems is obvious when we consider the case of antagonistic hormones. For example, epinephrine stimulates the breakdown of glycogen in liver cells, whereas insulin promotes the synthesis of glycogen. Since epinephrine's action is mediated by cAMP, insulin (a polypeptide hormone) must clearly use a different second-messenger system. Although a variety of second-messenger systems have been discovered, scientists still do not completely understand insulin's mechanism of action.

Clinical Considerations

Disorders of the Pituitary Gland

Inadequate growth hormone secretion during childhood causes **pituitary dwarfism.** Bone growth is delayed and the epiphyseal plates ossify before normal height is reached. Hyposecretion of growth hormone in an adult produces a rare condition called **pituitary cachexia** (*kă-kak'se-ă*). One of the symptoms of this disease is premature aging caused by tissue atrophy. Oversecretion of growth hormone during childhood, by contrast, causes **gigantism,** an abnormal increase in the length of long bones. Some individuals with this condition reach a height of 8 feet or more. Excessive growth hormone secretion in an adult does not cause further growth in the length of bones because the epiphyseal plates have already ossified. Hypersecretion of growth hormone in an adult causes **acromegaly** (*ak"ro-meg'ă-le*), in which the person's appearance gradually changes as a result of thickening of bones and growth of soft tissues, particularly those of the face, hands, and feet.

Inside Information—Heights that are well above or below average are not necessarily abnormal. Members of the Sudanese Dinka tribe often grow to be more than 7 1/2 feet tall—a trait that is normally inherited. The Pygmies of equatorial Africa, whose height is usually less than 5 feet, also are not abnormal in stature.

Disorders of the Adrenal Glands

Addison's disease is caused by inadequate secretion of both glucocorticoids and mineralocorticoids, which results in hypoglycemia, sodium and potassium imbalance, dehydration, hypotension, rapid weight loss, and generalized weakness. A person with this condition who is not treated with corticosteroids will die within a few days because of the severe electrolyte imbalance and dehydration. Another symptom of this disease is darkening of the skin. This is caused by excessive secretion of ACTH and possibly MSH (because MSH is derived from the same parent molecule as ACTH) as a result of lack of negative feedback inhibition of the pituitary by corticosteroids.

Inside Information—Former president John F. Kennedy took great pains to conceal the fact that he had Addison's disease. Repeatedly, the Kennedy press team issued reports that the president's bronzed skin was the result of malaria contracted during his war service.

Hypersecretion of corticosteroids results in **Cushing's syndrome.** This condition is generally caused by a tumor of the adrenal cortex or by oversecretion of ACTH from the adenohypophysis. Cushing's syndrome is characterized by changes in carbohydrate and protein metabolism, hyperglycemia, hypertension, and muscular weakness. Metabolic problems give the body a puffy appearance and can cause structural changes characterized as "buffalo hump" and "moon face." Similar effects are seen in people who receive prolonged treatment with corticosteroids, which are administered to reduce inflammation and inhibit the immune response.

Disorders of the Thyroid

The infantile form of **hypothyroidism** is known as **cretinism** (*kre'tĭ-nizm*). The clinical symptoms of cretinism include stunted growth, thickened facial features, abnormal bone development, mental retardation, low body temperature, and general lethargy. If cretinism is diagnosed early, it may be successfully treated by administering thyroxine.

Hypothyroidism in an adult causes **myxedema** (*mik"sĕ-de'mă*). This disorder affects body fluids, causing edema and increasing blood volume, hence increasing blood pressure. Symptoms of myxedema include a low metabolic rate, lethargy, and a tendency to gain weight. This condition is treated with thyroxine or with triiodothyronine, which are taken orally (as pills).

continued

cachexia: Gk. *kakao,* bad; *hexis,* habit
acromegaly: Gk. *akron,* extremity; *megas,* large
Addison's disease: from Thomas Addison, English physician, 1793–1860
Cushing's syndrome: from Harvey Cushing, American physician, 1869–1939
myxedema: Gk. *myxa,* mucus; *oidema,* swelling

Clinical Considerations

A *goiter* is an abnormal growth of the thyroid gland. When this condition results from inadequate dietary intake of iodine, the condition is termed **endemic goiter** (fig. 13.23). In this case, growth of the thyroid is due to excessive TSH secretion, with results from low levels of thyroxine secretion. Endemic goiter is thus associated with hypothyroidism.

Graves' disease, also called *toxic goiter,* involves growth of the thyroid associated with hypersecretion of thyroxine. This hyperthyroidism is produced by antibodies that act like TSH and stimulate the thyroid; it is an autoimmune disease. As a consequence of high levels of thyroxine, the metabolic rate and heart rate increase, the person loses weight, and the autonomic nervous system induces excessive sweating. In about half of the cases, *exophthalmos* (*ek"sof-thal'mos*), or bulging of the eyes, also develops (fig. 13.24) because of edema in the tissues of the eye sockets and swelling of the extrinsic eye muscles.

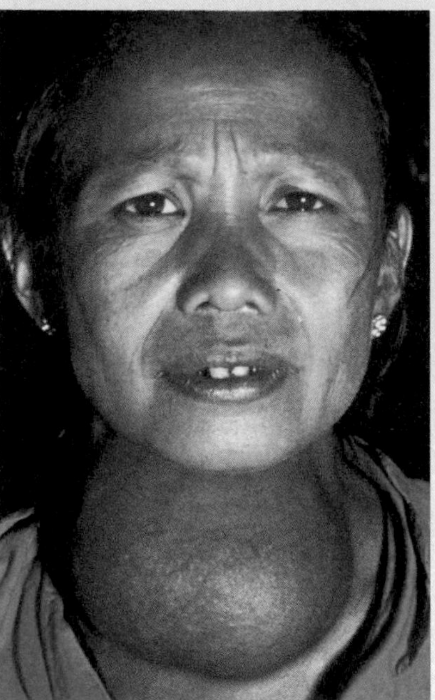

FIGURE 13.23
A simple, or endemic, goiter is caused by insufficient iodine in the diet.

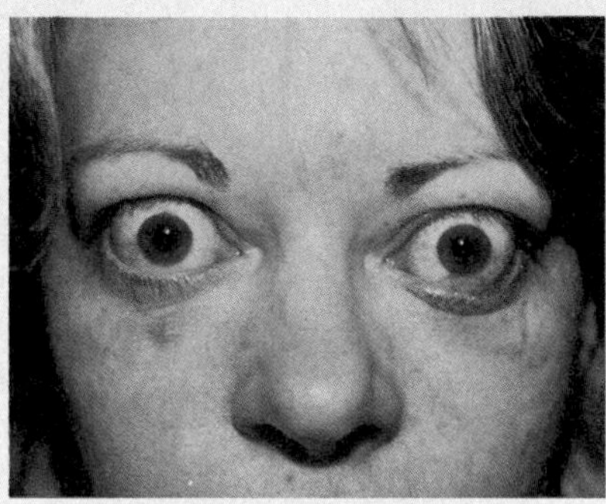

FIGURE 13.24
Hyperthyroidism is characterized by an increased metabolic rate, weight loss, muscular weakness, and nervousness. The eyes may also protrude.

Inside Information—Roughly 1% of the population has Graves' disease, including former president George Bush and former first lady Barbara Bush. The development of this disease in both husband and wife represents a medically fantastic coincidence.

Disorders of the Parathyroid Glands

Surgical removal of the parathyroid glands sometimes unintentionally occurs when the thyroid is removed because of a tumor or the presence of Graves' disease. The resulting fall in parathyroid hormone (PTH) causes a decrease in blood calcium levels, which can lead to severe muscle tetany. *Hyperparathyroidism* is usually caused by a tumor that secretes excessive amounts of PTH. This stimulates demineralization of bone, which makes the bones soft and raises the blood levels of calcium and phosphate. As a result of these changes, bones are subject to deformity and fracture, and stones composed of calcium phosphate are likely to develop in the urinary tract.

Disorders of the Pancreatic Islets

Diabetes mellitus is characterized by fasting hyperglycemia and the presence of glucose in the urine. There are two forms of this disease. **Type I,** or **insulin-dependent diabetes mellitus,** is caused by destruction of the beta cells and the resulting lack of insulin secretion. **Type II,** or **non-insulin-dependent diabetes mellitus** (which is the more common form), is caused by decreased tissue sensitivity to the effects of insulin, so that larger than normal amounts are required to produce a normal effect. Both types of diabetes mellitus are also associated with abnormally high levels of glucagon secretion. Diabetes mellitus is discussed in more detail in the "Clinical Considerations" section of chapter 21.

People who have a genetic predisposition for type II diabetes mellitus often first develop *reactive hypoglycemia.* In this condition, the rise in blood glucose that follows the ingestion of carbohydrates stimulates excessive secretion of insulin, which in turn causes the blood glucose levels to fall below the normal range. This can result in weakness, changes in personality, and mental disorientation.

Graves' disease: from Robert Graves, Irish physician, 1796–1853
exophthalmos: Gk. *ex,* out; *ophthalmos,* eyeball
diabetes: Gk. *diabetes,* to pass through a siphon

SUMMARY

13.1 Introduction to the Endocrine System

1. Endocrine glands secrete biologically active molecules called hormones into the blood for distribution to target tissues.
2. Hormones may be catecholamines, polypeptides, glycoproteins, or steroids.

13.2 Pituitary Gland

1. The pituitary, or hypophysis, consists of an anterior lobe, or adenohypophysis, and a posterior lobe, or neurohypophysis.
2. The adenohypophysis is derived from embryonic epithelium, whereas the neurohypophysis is derived from a downgrowth of the diencephalon.
3. The anterior pituitary secretes a number of trophic hormones that regulate other endocrine glands.
4. The posterior pituitary releases ADH and oxytocin, which are produced in the hypothalamus.

13.3 Regulation of Pituitary Gland Secretion

1. The hypothalamus regulates the posterior pituitary by way of a tract of nerve fibers.
2. The hypothalamus regulates the anterior pituitary by way of releasing and inhibiting hormones, which are secreted into the hypothalamo-hypophyseal portal system.
3. Secretion of anterior pituitary hormones is also regulated by negative feedback.

13.4 Adrenal Glands

1. Each adrenal gland consists of an outer adrenal cortex and an inner adrenal medulla.
2. The cortex, derived from mesoderm, secretes corticosteroid hormones.
3. Secretion of glucocorticoids (e.g., cortisol) by the adrenal cortex is stimulated by ACTH from the anterior pituitary.
4. The adrenal medulla, derived from the neural crest, secretes catecholamine hormones under stimulation from preganglionic sympathetic nerves during the fight-or-flight response.

13.5 Thyroid and Parathyroid Glands

1. The thyroid gland is composed of hollow follicles.
2. The thyroid hormones are tetraiodothyronine (T_4, or thyroxine) and triiodothyronine (T_3).
3. The parafollicular cells of the thyroid produce calcitonin.
4. The parathyroid glands secrete parathyroid hormone (PTH), which, in concert with calcitonin, helps to regulate Ca^{++} balance.

13.6 Pancreas and Other Endocrine Glands

1. The endocrine structures are the pancreatic islets, which secrete insulin and glucagon.
2. The pineal gland, located in the brain, secretes the hormone melatonin.
3. The thymus, in the upper mediastinum, secretes hormones that regulate the immune system.
4. The stomach and small intestine secrete a number of hormones that act on the digestive tract itself and that help to regulate pancreatic juice and bile secretions.
5. The gonads secrete sex steroid hormones, and the placenta secretes a variety of hormones, including large amounts of estrogens and progesterone.

13.7 Mechanism of Steroid and Thyroxine Action

1. Steroid hormones combine with cytoplasmic receptors, translocate to the nucleus, and activate specific genes.
2. Thyroid hormones attach to chromatin-bound receptors in the nucleus. This attachment activates specific genes, thereby stimulating RNA and protein synthesis.

13.8 Mechanism of Catecholamine and Polypeptide Action

1. The beta-adrenergic effects of epinephrine and the actions of many polypeptide hormones are mediated by the cyclic AMP (cAMP) as a second messenger.
2. Combination of the hormone with a membrane receptor stimulates cAMP production within the target cell.
3. Cyclic AMP activates protein kinase, which, in turn, activates other enzymes.

Review Activities

Objective Questions

1. Steroid hormones are secreted by (*module 13.1*)
  a. the adrenal cortex.
  b. the gonads.
  c. the thyroid.
  d. both a and b.
  e. both b and c.

2. Which of the following statements about the pituitary gland is *false?* (*module 13.2*)
  a. The pars distalis is also known as the anterior pituitary.
  b. Growth hormone and prolactin are secreted by the pars nervosa.
  c. The adenohypophysis is derived from oral epithelium.
  d. The neurohypophysis is an extension of the brain.

3. Which of the following statements about hypothalamic releasing hormones is *true?* (*module 13.3*)
  a. They are secreted into capillaries in the median eminence.
  b. They are transported by portal veins to the anterior pituitary.
  c. They stimulate the secretion of specific hormones from the anterior pituitary.
  d. All of the above are true.

4. Which of the following statements about the adrenal cortex is *true?* (*module 13.4*)
  a. It is not innervated by nerve fibers.
  b. It secretes some androgens.
  c. The zona granulosa secretes aldosterone.
  d. The zona fasciculata is stimulated by ACTH.
  e. All of the above are true.

5. Which of the following stimulates the adrenal medulla to secrete epinephrine? (*module 13.4*)
  a. TSH
  b. ACTH
  c. CRH
  d. sympathetic neurons

6. The hormone primarily responsible for setting the basal metabolic rate and for promoting the maturation of the brain is (*module 13.5*)
  a. cortisol.
  b. ACTH.
  c. TSH.
  d. thyroxine.

7. The secretion of which of the following hormones would be *increased* in a person with endemic goiter? (*module 13.5*)
  a. TSH
  b. thyroxine
  c. triiodothyronine
  d. all of the above

8. Which of the following statements about the hormone insulin is *true?* (*module 13.6*)
  a. It is secreted by alpha cells in the pancreatic islets.
  b. It is secreted in response to a rise in blood glucose.
  c. It stimulates the production of glycogen and fat.
  d. Both a and b are true.
  e. Both b and c are true.

9. Steroid hormones (*module 13.7*)
  a. bind to receptor proteins located on the cell membrane.
  b. cause the production of cAMP.
  c. activate protein kinase.
  d. stimulate the production of mRNA.

10. Which of the following hormones can use cAMP as a second messenger? (*module 13.8*)
  a. testosterone
  b. cortisol
  c. thyroxine
  d. epinephrine

Essay Questions

1. Define *endocrine gland* and *hormone.* Describe the different chemical classes of hormones and list specific hormones within each class. (*module 13.1*)
2. Compare hormones and neurotransmitters. (*module 13.1*)
3. List the hormones of the anterior pituitary and state their target organs. (*module 13.2*)
4. Explain how the hypothalamus regulates the secretions of the anterior pituitary. (*module 13.3*)
5. Describe the location, structure, and embryonic development of the adrenal glands. (*module 13.4*)
6. List the different classes of corticosteroids and give an example of a hormone within each class. (*module 13.4*)
7. Explain why a person with iodine deficiency is hypothyroid and has a goiter. (*module 13.5*)
8. Describe the structure of the pancreatic islets. List the pancreatic hormones and describe their actions. (*module 13.6*)
9. Trace the course of events by which a steroid hormone stimulates its target cells to produce new enzymes. (*module 13.7*)
10. Explain why second messengers are needed in the action of many hormones. Give examples of hormones that require the use of second messengers. (*module 13.8*)
11. Explain how cAMP functions as a second messenger. (*module 13.8*)

Labeling Exercise

Label the structures indicated on the figure to the right.

1. ______________________________
2. ______________________________
3. ______________________________
4. ______________________________
5. ______________________________

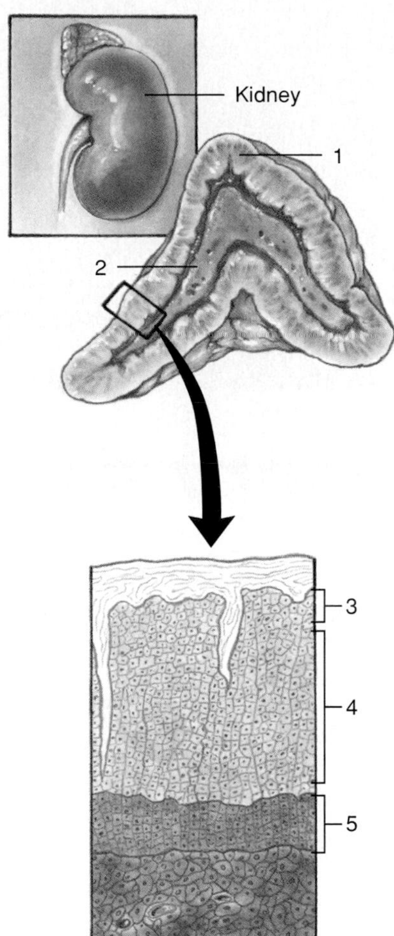

Critical Thinking Questions

1. Why do catecholamines require second messengers whereas the lipophilic hormones do not? Relate this difference to the location of their cellular receptor proteins.
2. The anterior pituitary has often been called a "master gland." In what sense might this description be correct? In what sense is it misleading and incorrect?
3. Are the anterior pituitary and posterior pituitary just two regions of one gland, or are they two different glands? Relate your answer to the hormones secreted, the regulation of secretion, and the embryonic origin of the anterior pituitary and posterior pituitary.
4. What blood test could you perform to see if you were under stress? Explain your answer.
5. Suppose a person with a normal thyroid gland took thyroid pills. Would he or she lose weight? Why or why not? What endocrine changes would be produced by those pills? If, after a few months, the person stopped taking the pills, what would happen to his or her body weight? Explain.

CHAPTER

14

"It is estimated that about 2.5 million erythrocytes are produced every second in order to replace the number that are continuously being destroyed by the spleen and liver." page 362

Circulatory System: Blood

Chapter Outline

Terms to Remember

heart, 357
artery, 357
vein, 357
capillary, 357
lymph vessel, 357
formed element, 357
blood plasma, 357
albumin (*al-byoo' min*), 358
globulin (*glob' yoo-lin*), 358
fibrinogen (*fi-brin' ō-jen*), 358
serum, 358
erythrocyte (*ĕ-rith' rŏ-sīt*), 359
leukocyte (*loo' kō-sīt*), 359
platelet (thrombocyte) (*throm' bŏ-sīt*), 360
myeloid tissue, 361
lymphoid tissue, 361
erythropoietin (*ĕ-rith"ro-poi-e' tin*), 362
ABO system, 362
agglutinate (*ă-gloot' n-āt*), 364
Rh factor, 364

Learning Objectives

After studying this chapter, you should be able to

- describe the organization of the circulatory system and state the general functions of its major components.
- identify the components of blood plasma and describe their functions.
- describe the physical characteristics and functions of erythrocytes, leukocytes, and platelets (thrombocytes).
- explain how erythrocytes and leukocytes are formed and state where these processes occur.
- describe the ABO and Rh blood groups and explain how blood typing is performed.
- explain how blood clots are formed and dissolved.

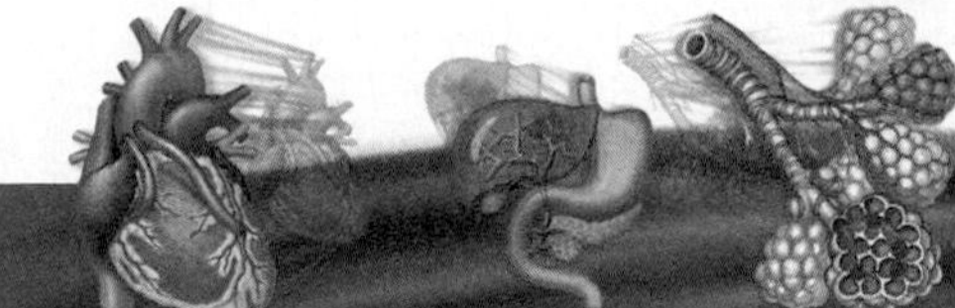

14.1 Introduction to the Circulatory System

Blood consists of fluid plasma and formed elements that circulate within a closed system of vessels as a result of the pumping activity of the heart.

Blood serves to transport nutrients to the trillions of cells of the body and carry away their secretions and waste products. The blood, however, can also serve as a delivery system for disease-causing viruses, bacteria, and their toxins. To guard against this, the circulatory system has protective mechanisms: the white blood cells and lymphatic system. In order to perform its various functions in maintaining homeostasis, the circulatory system works together with the respiratory, urinary, digestive, endocrine, and integumentary systems.

Major Components of the Circulatory System

The circulatory system is frequently divided into the **cardiovascular system,** which consists of the heart and blood vessels, and the **lymphatic system,** which consists of lymph vessels, lymph nodes, and lymphoid organs (spleen, thymus, and tonsils).

The **heart** is a four-chambered double pump. Its pumping action creates the pressure head needed to push blood in the vessels to the lungs and body cells. At rest, the heart of an adult pumps about 5 liters of blood per minute. It takes only about a minute for the blood to be circulated to the most distal part of the body and back to the heart.

Blood vessels form a tubular network that permits blood to flow from the heart to all the living cells of the body and then to return, completing the circuit. *Arteries* carry blood away from the heart, whereas *veins* return blood to the heart. Arteries and veins are continuous with each other through a series of connecting smaller blood vessels.

Inside Information—If laid end to end, it is estimated that your blood vessels would encircle the globe twice over—more than 60,000 miles of tubing!

Arteries branch extensively to form a "tree" of progressively smaller vessels. Those that are microscopic in diameter are called *arterioles.* Blood passes from the arterial to the venous system in *capillaries*—the thinnest and most numerous of the blood vessels. All exchanges of fluid, nutrients, and wastes between the blood and tissues occur across the walls of capillaries. Blood flows through capillaries and into microscopic veins called *venules.* The venules deliver blood to progressively larger veins that eventually return the blood to the heart.

As blood plasma passes through capillaries, the hydrostatic pressure of the blood forces some of the fluid out of the capillary walls. Fluid derived from blood plasma that passes out of capillary walls into the surrounding tissues is called *interstitial (in"ter-stish'al) fluid,* or *tissue fluid.* Some of this fluid returns directly to capillaries, and some enters **lymph vessels** located in the connective tissues surrounding the blood vessels. Fluid in lymph vessels is called *lymph (limf).* The lymph vessels converge into larger ducts that return the lymph to the blood of the circulatory system. **Lymph nodes,** distributed along the lymph vessels, cleanse the lymph prior to its return to the blood. Chapter 17 is devoted to the lymphatic system and immunity.

Composition of Blood

The total blood volume in the average adult is about 5 liters, constituting 8% of the total body weight. Blood leaving the heart is referred to as *arterial blood.* Arterial blood, with the exception of that going to the lungs, is bright red because of the high concentration of oxyhemoglobin (formed by the combination of oxygen and hemoglobin) in the red blood cells. *Venous blood* is blood returning to the heart. Except for the venous blood returning from the lungs, it is oxygen-poor, and therefore a darker red than the oxygen-rich arterial blood.

Inside Information—A "unit" (half a liter) of blood is drained when you donate blood. This represents approximately one-tenth of your total blood volume.

Blood is composed of a cellular portion, called **formed elements,** and a fluid portion, called **blood plasma.** When a blood sample is centrifuged, the heavier formed elements become packed into the bottom of the tube, leaving blood plasma at the top (fig. 14.1). The formed elements constitute approximately 45% of the total blood volume, a percentage known as the *hematocrit (HCT).* The blood plasma accounts for the remaining 55%. The hematocrit closely approximates the percentage of red blood cells per given volume of blood and is an important indicator of the oxygen-carrying capacity of blood.

hematocrit: Gk. *haima,* blood; *krino,* to separate

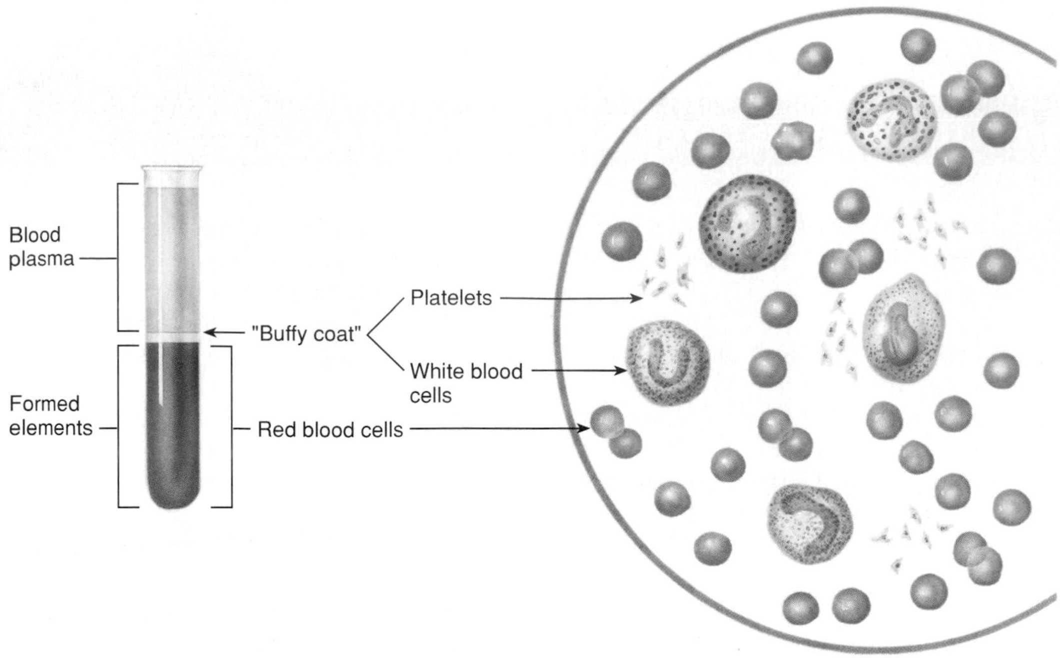

FIGURE 14.1

The formed elements (blood cells) become packed at the bottom of the test tube when blood is centrifuged, leaving the fluid blood plasma at the top of the tube. Red blood cells are the most abundant of the formed elements. White blood cells and platelets form only a thin, light-colored "buffy coat" at the interface between the packed red blood cells and the blood plasma.

Blood Plasma

Blood plasma is a straw-colored liquid consisting of water and many dissolved solutes. Its essential importance is that the plasma, together with fluids derived from it (interstitial fluid and cerebrospinal fluid), serve as the extracellular environment of all of the cells in the body. Because of this essential role, the volume, composition, and concentration of the blood plasma are finely regulated.

The total volume and concentration of plasma are important in relation to the regulation of blood pressure, thirst, kidney function, and other aspects of physiology. Sodium ion (Na^+) is the major solute in the plasma, and its concentration determines (by osmosis) the amount of plasma water, and hence the blood volume. Eating food rich in salt (NaCl), for example, will make you thirsty, and, when you drink, will result in an increased plasma volume. The blood plasma contains other ions as well. Plasma bicarbonate (HCO_3^-), for example, is maintained within a normal range of concentrations and is needed to maintain acid-base balance. Plasma potassium (K^+), as another example, must be kept within a very narrow range of concentrations or the heart will stop beating. Normal concentrations of plasma glucose, proteins, cholesterol, and other organic molecules are also required for health.

Plasma proteins constitute 7%–9% of the blood plasma. The three types of proteins are *albumins, globulins,* and *fibrinogen.* **Albumins** (*al-byoo'minz*) account for most (60%–80%) of the plasma proteins and are the smallest in size. Produced by the liver, they create the osmotic pressure that draws water from the surrounding interstitial fluid into the capillaries. This action is needed to maintain blood volume and pressure. **Globulins** (*glob'yoo-linz*) are grouped into three subtypes: **alpha globulins, beta globulins,** and **gamma globulins.** The alpha and beta globulins are produced by the liver and function to transport lipids and fat-soluble vitamins in the blood. Gamma globulins are antibodies produced by lymphocytes (one of the formed elements found in blood and lymphoid tissues) and function in immunity. **Fibrinogen** (*fi-brin'ŏ-jen*), which constitutes only about 4% of the total plasma proteins, is an important clotting factor produced by the liver. During the process of clot formation (described later in the chapter—module 14.5), fibrinogen is converted into insoluble threads of *fibrin.* The fluid from clotted blood, which is called *serum,* thus does not contain fibrinogen but is otherwise identical to plasma.

albumin: L. *albumen,* white
globulin: L. *globulus,* small globe
serum: L. *serum,* liquid

14.2 Formed Elements of Blood

The formed elements of blood include erythrocytes, leukocytes, and platelets (thrombocytes). These function in oxygen transport, immunity, and blood clotting, respectively.

Erythrocytes

Erythrocytes (*ĕ-rith'rŏ-sīts*), or **red blood cells (RBCs),** are biconcave discs—flattened, with a depressed center—2.2 μm thick. Their unique shape relates to their function of transporting oxygen; it provides an increased surface area through which gas can diffuse (fig. 14.2). The oxygen-transport function of erythrocytes is provided by their content of *hemoglobin,* which binds oxygen and gives blood its red color. Oxygen transport is described in more detail in chapter 18 (see module 18.10). Erythrocytes lack nuclei and have few organelles. Because of these deficiencies, their circulating life span is only about 120 days, after which they are destroyed by phagocytic cells in the liver, spleen, and bone marrow.

Anemia *refers to any condition in which there is an abnormally low hemoglobin concentration and/or red blood cell count. The most common type is* iron-deficiency anemia, *which is due to a deficient intake or absorption of iron, or to excessive iron loss. Iron is an essential component of the hemoglobin molecule. In* pernicious anemia, *the production of red blood cells is insufficient because of lack of vitamin B_{12}, which results from an inability of the stomach to produce intrinsic factor. In the absence of intrinsic factor, vitamin B_{12} cannot be absorbed.* Aplastic anemia *is anemia due to destruction of the bone marrow, which may be caused by chemicals (including benzene and arsenic) or by X rays.*

Leukocytes

Leukocytes (*loo'kŏ-sīts*), or **white blood cells (WBCs),** are larger than erythrocytes and are different in other ways as well. Leukocytes contain nuclei and mitochondria and can move in an ameboid fashion (erythrocytes are not able to move independently). Because of their ameboid ability, leukocytes can squeeze through pores in capillary walls and move to a site of infection, whereas erythrocytes usually remain confined within blood vessels. The movement of leukocytes through capillary walls is called *diapedesis* (see fig. 17.5).

Leukocytes are classified according to their stained appearance. Those leukocytes that have large granules in their cytoplasm are called **granular leukocytes,** or **granulocytes.** Those that contain very small granules that cannot be seen easily with the light microscope are called **agranular leukocytes,** or **agranulocytes.** The granular leukocytes include *eosinophils* (*e"ŏ-sin'ŏ-filz*),

diapedesis: Gk. *dia,* through; *pedester,* on foot

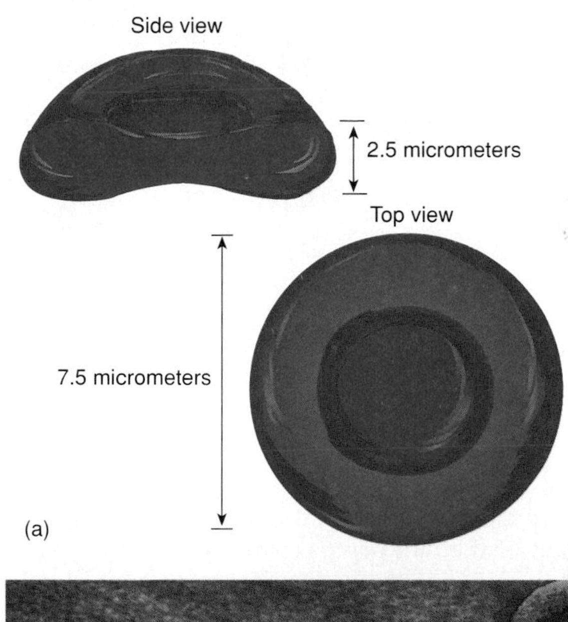

FIGURE 14.2
Erythrocytes. (*a*) A diagram and (*b*) a scanning electron micrograph.

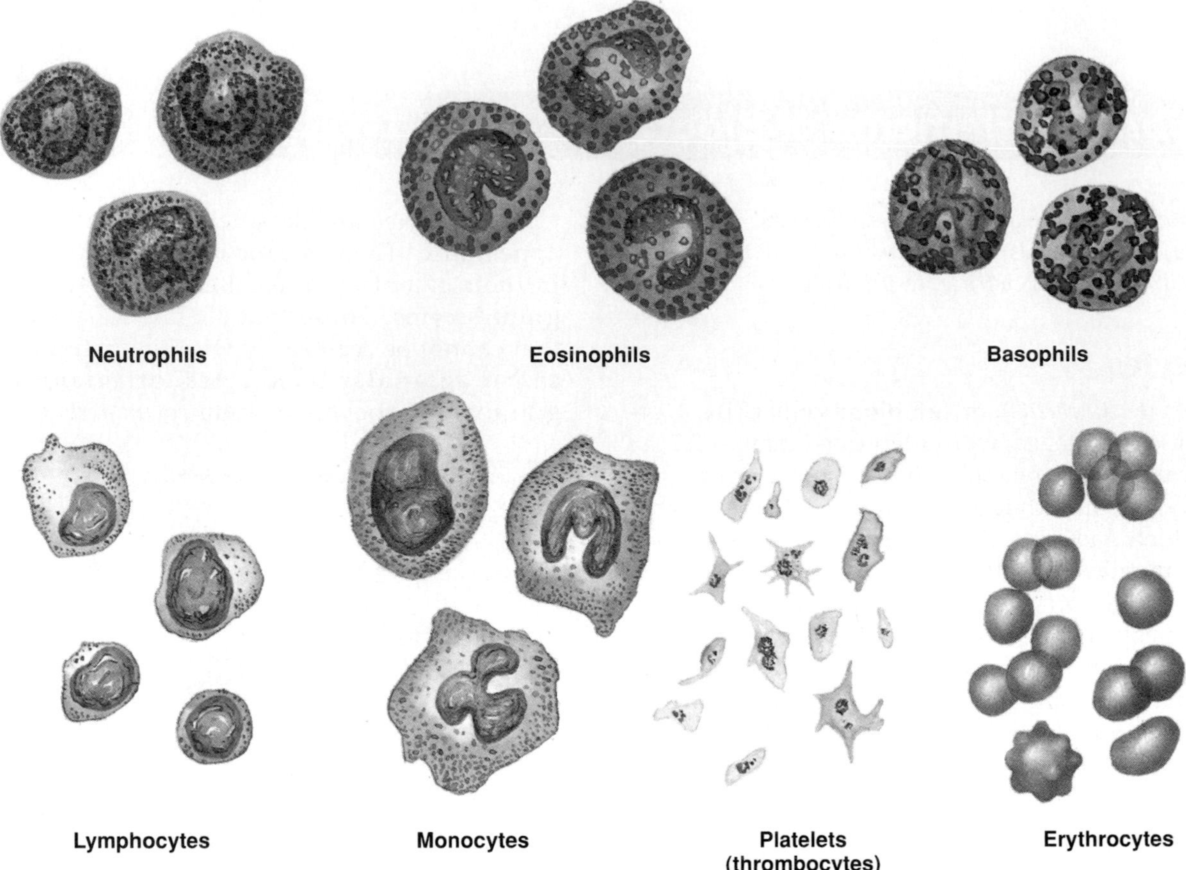

FIGURE 14.3
The formed elements of blood.

basophils, and *neutrophils,* classified on the basis of their staining properties (fig. 14.3). Neutrophils are the most abundant of the leukocytes (table 14.1), and their oddly shaped nuclei give them the alternate name of *polymorphonuclear* (*pol"e-mor"fo-noo'kle-ar*) (*PMN*) *neutrophils.*

Agranular leukocytes are of two types: *lymphocytes* and *monocytes.* Lymphocytes are usually the second most numerous type of leukocyte in the blood. They are small cells with round nuclei and little cytoplasm. It is the lymphocytes that confer specific immunity, including the production of antibodies. *Monocytes,* by contrast, are the largest of the leukocytes and generally have kidney- or horseshoe-shaped nuclei. Monocytes are phagocytic cells that participate with lymphocytes in the production of an immune response. The immune functions of the different white blood cells are described in detail in chapter 17.

Platelets (Thrombocytes)

Platelets, sometimes called *thrombocytes* (*throm'bŏ-sīts*), are the smallest of the formed elements. Actually, they are fragments of large bone marrow cells called *megakaryocytes,* which is why the term *formed elements* may be used rather than *blood cells* to describe erythrocytes, leukocytes, and platelets. The fragments that enter the circulation as platelets lack nuclei but, like leukocytes, are capable of ameboid movement. Platelets survive about 5 to 9 days before being destroyed by the spleen and liver. They function in blood clotting, as will be described in module 14.5.

thrombocytes: Gk. *thrombos,* clot; *kytos,* hollow (cell)

Table 14.1 Formed Elements of Blood

Component	Description	Number Present	Function
Erythrocyte (red blood cell)	Biconcave disc without nucleus; contains hemoglobin; survives 100 to 120 days	4,000,000 to 6,000,000/mm^3	Transports oxygen and carbon dioxide
Leukocytes (white blood cells)		5,000 to 10,000/mm^3	Aid in defense against infection by microorganisms
Granulocytes	About twice the size of red blood cells; large cytoplasmic granules; survive 12 hours to 3 days		
1. Neutrophil	Nucleus with 2 to 5 lobes; cytoplasmic granules stain slightly pink	54% to 62% of white cells present	Phagocytic
2. Eosinophil	Nucleus bilobed; cytoplasmic granules stain red in eosin stain	1% to 3% of white cells present	Helps to detoxify foreign substances; secretes enzymes that break down clots
3. Basophil	Nucleus lobed; cytoplasmic granules stain blue in hematoxylin stain	Less than 1% of white cells present	Releases anticoagulant heparin
Agranulocytes	Cytoplasmic granules not visible; survive 100 to 300 days (some much longer)		
1. Monocyte	2 to 3 times larger than red blood cell; nuclear shape varies from round to lobed	3% to 9% of white cells present	Phagocytic
2. Lymphocyte	Only slightly larger than red blood cell; nucleus nearly fills cell	25% to 33% of white cells present	Provides specific immune response (including antibodies)
Platelet (thrombocyte)	Cytoplasmic fragment; survives 5 to 9 days	25,000 to 450,000/mm^3	Enables clotting

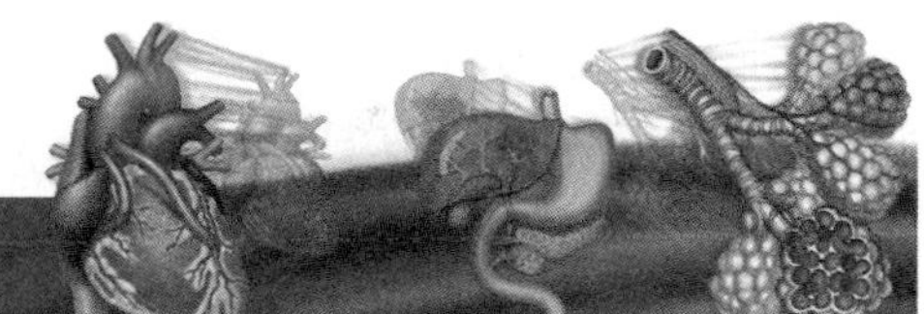

14.3 Hemopoiesis

The different blood cells are derived from a common bone marrow ancestor, the stem cell. Regulatory molecules determine the development of the blood cells along particular lines.

Blood cells are constantly formed through a process called **hemopoiesis** (*hem"ŏ-poi-e'sis*), also called **hematopoiesis.** The term *erythropoiesis* refers to the formation of erythrocytes, and *leukopoiesis* to the formation of leukocytes. These processes occur in two classes of tissues. **Myeloid tissue** is the red bone marrow of the long bones, ribs, sternum, bodies of vertebrae, and portions of the skull. **Lymphoid tissue** includes the lymph nodes, tonsils, spleen, and thymus. All of the different types of blood cells are produced in the bone marrow; lymphocytes, however, are also produced in the lymphoid tissue.

The major purpose of a bone marrow transplant is to provide competent hemopoietic stem cells to the recipient. If the bone marrow is returned to the same person, the procedure is called an autotransplant. *If the donor and recipient are different people, it is termed an* allogenic transplant.

hemopoiesis: Gk. *haima,* blood; *poiesis,* production

Erythropoiesis is an extremely active process. It is estimated that about 2.5 million erythrocytes are produced every second in order to replace the number that are continuously being destroyed by the spleen and liver. During the destruction of erythrocytes, iron is salvaged and returned to the red bone marrow, where it is used again in the formation of erythrocytes. As mentioned previously, the life span of an erythrocyte is approximately 120 days. Most agranular leukocytes remain functional for 100 to 300 days under normal conditions. The life span of granular leukocytes, by contrast, is extremely short—a mere 12 hours to 3 days.

Inside Information—At any given time, a staggering 25 trillion (25×10^{13}) erythrocytes are contained within the blood of an average adult—enough, if spread out, to cover four tennis courts.

Hemopoiesis begins the same way in both myeloid and lymphoid tissue (fig. 14.4). A population of undifferentiated (unspecialized) cells gradually differentiates (specializes) to become "stem cells," which give rise to the blood cells. These stem cells can duplicate themselves by mitosis, thus ensuring that the parent population never becomes depleted. As the cells become differentiated, they develop membrane receptors that respond to specific hormones, pushing them toward further development along particular lines. The earliest stem cells that can be distinguished under a microscope are the *erythroblasts* (which become erythrocytes), *myeloblasts* (which become granular leukocytes), *lymphoblasts* (which become lymphocytes), *monoblasts* (which become monocytes), and the *megakaryoblasts* (which become the cells that produce platelets).

The production of different subtypes of lymphocytes is stimulated by chemicals called **lymphokines** (discussed in connection with the immune system in chapter 17—module 17.7). The production of red blood cells is stimulated by a hormone called **erythropoietin** (*ĕ-rith″ro-poi-e′tin*), which is secreted by the kidneys. Erythropoietin stimulates the production of erythrocytes in the bone marrow, and is secreted whenever the delivery of oxygen to the kidneys is lower than normal. In this circumstance—which can occur, for example, when a person lives at high altitude—the increased production of red blood cells allows the blood to carry a higher concentration of oxygen to the tissues.

Considering that the kidneys produce erythropoietin, and that erythropoietin is needed to stimulate red blood cell production, it is not surprising that patients with kidney disease may also suffer from anemia. It has recently been shown that this anemia can be effectively treated by giving these patients human erythropoietin. This exciting therapeutic breakthrough was made possible by the commercial production of erythropoietin using genetic engineering techniques.

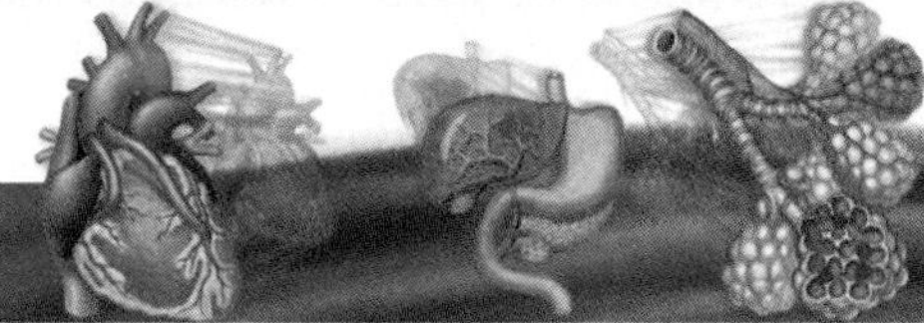

14.4 Blood Typing

The surfaces of erythrocytes contain genetically determined glycolipids called antigens. These antigens have been categorized into blood groups, the two most important of which are the ABO and Rh blood groups.

All cells in the body, including erythrocytes, contain certain molecules called antigens on their surfaces. One person's antigens may be recognized as foreign if transfused into another person, and trigger an immune response. As part of an immune response, particular lymphocytes secrete a class of proteins called *antibodies,* which bind in a specific fashion to antigens. In the case of an incompatible blood transfusion, the interaction between antigens and antibodies can have fatal consequences.

ABO System

At least 20 different blood groups have been identified in red blood cells, but the major group is known as the **ABO system.** In terms of the antigens present on the red blood cell surface, a person may be **type A** (with only A antigens), **type B** (with only B antigens), **type AB** (with both A and B antigens), or **type O** (with neither A nor B antigens). It should be emphasized that the blood type denotes the class of antigens (chemically, a type of glycolipid) found on the red blood cell surface.

Inside Information—The most common blood type is O, accounting for about 46% of the world's population. In certain places, however, other blood types predominate. In Norway, for example, most people have type A blood.

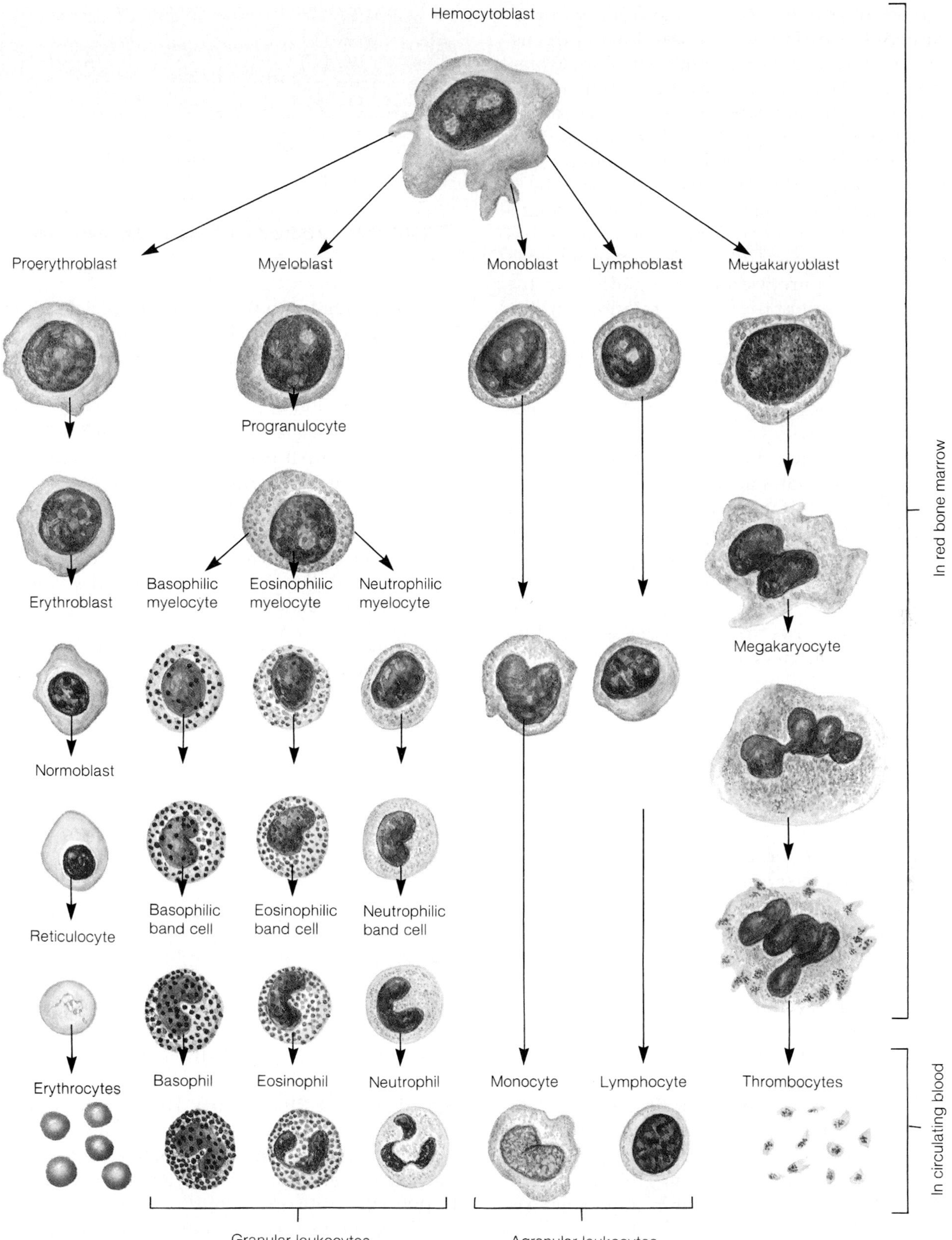

FIGURE 14.4
The process of hemopoiesis. Formed elements begin as hemocytoblasts and differentiate into the various kinds of blood cells, depending on the needs of the body.

Each person inherits two genes (one from each parent) that control the production of the ABO antigens. The genes for A or B antigens are dominant, since O simply means the absence of A and B. A person who is type A, therefore, may have inherited an A gene from each parent (may have the genotype AA), or an A gene from one parent and an O gene from the other parent (and thus have the genotype AO). Likewise, a person who is type B may have the genotype BB or BO. It follows that a type O person inherited an O gene from each parent (has the genotype OO), whereas a type AB person inherited the A gene from one parent and the B gene from the other (there is no dominant-recessive relationship between A and B).

The immune system is tolerant to (does not attack) its own red blood cell antigens. A person who is type A, for example, does not produce anti-A antibodies. Surprisingly, however, people with type A blood *do* make antibodies against the B antigen, and conversely, people with blood type B make antibodies against the A antigen. This is believed to result from the presence in the plasma of preformed antibodies, made in response to some common bacteria in the digestive tract. These antibodies act against antigens not present on a person's own red blood cells. A person who is type A, therefore, acquires antibodies that can react with B antigens by exposure to the bacteria, but, because of tolerance mechanisms, does not develop antibodies that can react with A antigens.

People who are type AB develop tolerance to both these antigens, and thus do not produce either anti-A or anti-B antibodies. Those who are type O, by contrast, do not develop tolerance to either antigen; therefore, they have both anti-A and anti-B antibodies in their blood plasma (table 14.2).

Before transfusions are performed, a *major cross match* is made by mixing serum from the recipient with blood cells from the donor. If the types do not match—if the donor is type A, for example, and the recipient is type B—the recipient's antibodies attach to the donor's red blood cells and form bridges that cause the cells to clump together, or **agglutinate** (fig. 14.5). Because of this agglutination (*ă-gloot″n-a′shun*) reaction, the A and B antigens are sometimes called *agglutinogens,* and the antibodies against them are called *agglutinins.* Transfusion errors that result in such agglutination in the blood can produce a blockage of small blood vessels throughout the body. The clumped red blood cells then begin to rupture (hemolyze), releasing their hemoglobin into the bloodstream. This liberated hemoglobin may cause severe kidney damage.

In emergencies, type O blood has been given to people who are type A, B, AB, or O. Since type O red blood cells lack both A and B antigens, the recipient's antibodies cannot agglutinate the donor's red blood cells. Type O is therefore a *universal donor,* but only as long as the plasma is removed and only the cells given in a transfusion. This is because the blood plasma from a type O person would agglutinate type A, type B, or type AB red blood cells. Likewise, type AB people are *universal recipients* because they lack anti-A and anti-B antibodies and thus cannot agglutinate donor red blood cells. (Donor blood plasma, however, could agglutinate recipient red blood cells if the transfusion volume were too large.) Because of the dangers involved, the universal donor and recipient concept in blood transfusions is strongly discouraged.

TABLE 14.2 The ABO System of Antigens on Red Blood Cells

Antigen on RBCs	Antibody in Plasma
A	Anti-B
B	Anti-A
O	Anti-A and anti-B
AB	Neither anti-A nor anti-B

Inside Information—When Europeans first experimented with blood transfusions in the seventeenth century, so many died that the procedure was outlawed in England, France, and Italy. No one knew that shared blood must match until 1901, when the human blood groups were discovered by the Austrian-born American immunologist Karl Landsteiner. For this discovery, Landsteiner was awarded the 1930 Nobel Prize for physiology/medicine.

Rh Factor

Another group of antigens found in most red blood cells is the *Rh factor* (Rh stands for rhesus monkey, in which these antigens were first discovered). People who have these antigens are said to be **Rh positive,** whereas those who do not are **Rh negative.** There are fewer Rh negative people (about 15% of the U.S. population) because this condition is recessive to Rh positive. The Rh factor is of particular significance when Rh negative mothers give birth to Rh positive babies.

Since the fetal and maternal blood are normally kept separate in the uterus, the Rh negative mother is not usually exposed to the Rh antigen of the fetus during the

FIGURE 14.5
The agglutination (clumping) of red blood cells occurs when cells with A-type antigens are mixed with anti-A antibodies and when cells with B-type antigens are mixed with anti-B antibodies. No agglutination would occur with type O blood (not shown).

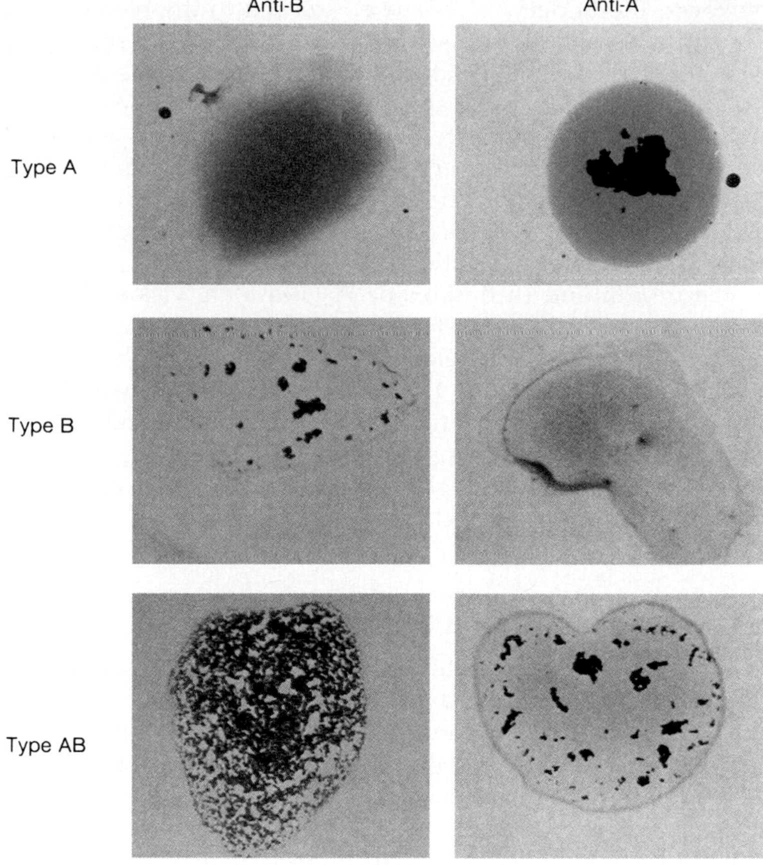

pregnancy. At the time of birth, however, a variable degree of exposure may occur, and the mother's immune system may become sensitized and produce antibodies against the Rh antigen. This does not always occur because the exposure may be minimal and because Rh negative women vary in their sensitivity to the Rh factor. However, if the woman does produce antibodies against the Rh factor, these antibodies can cross the placenta in subsequent pregnancies and cause hemolysis (destruction) of the Rh positive red blood cells of the fetus. The baby is therefore born anemic, with a condition called *erythroblastosis fetalis,* or *hemolytic disease of the newborn.*

Erythroblastosis fetalis can be prevented by injecting the Rh negative mother with antibodies against the Rh factor (a trade name for this is preparation RhoGAM—the GAM is short for gamma globulin, the class of plasma proteins in which antibodies are found) within 72 hours after the birth of each Rh positive baby. The injected antibodies inactivate the Rh antigens, and thus prevent the mother from becoming actively immunized to them.

14.5 Blood Clotting

When a blood vessel is cut, blood platelets stick to exposed collagen and to each other to form a platelet plug. This mass of platelets is reinforced by threads of fibrin, formed by activation of clotting factors in the blood plasma.

Functions of Platelets

In the absence of vessel damage, platelets (thrombocytes) are repelled from each other and from the endothelial lining of vessels. When the endothelium of vessels is damaged, however, the underlying tissue is exposed to the blood. Platelets are then able to stick to exposed collagen proteins that have become coated with a protein secreted by the endothelial cells. Once attached to the collagen, the secretory granules of the platelets release their products and begin to break down, or *degranulate.* These products include *ADP* (adenosine diphosphate), *serotonin,* and a prostaglandin called *thromboxane A_2.* This event is known as the **platelet release reaction.**

Serotonin and thromboxane A_2 stimulate vasoconstriction, which helps to decrease blood flow to the injured vessel. Also, the release of ADP and thromboxane A_2 from platelets that are stuck to exposed collagen makes other platelets in the vicinity "sticky," so that they adhere to those already stuck to the collagen. This produces a **platelet plug** in the damaged vessel, which is strengthened by the activation of plasma clotting factors.

Clotting Factors: Formation of Fibrin

The platelet plug is strengthened by a meshwork of insoluble protein fibers known as **fibrin.** Blood clots therefore contain platelets, fibrin, and usually trapped red blood cells that give the clot a red color. Finally, contraction of the platelets in the process of *clot retraction* forms a more compact and effective plug. Fluid squeezed from the clot as it retracts is called **serum.** As mentioned in module 14.1, serum is blood plasma without fibrinogen (the soluble precursor of fibrin).

There are two pathways that result in the conversion of fibrinogen into fibrin. Blood left in a test tube will clot without the addition of any external chemicals; the pathway that produces this clot is thus called the **intrinsic pathway.** The intrinsic pathway also produces clots in damaged blood vessels when collagen is exposed to blood plasma. Damaged tissues, however, release a chemical that initiates a "shortcut" to the formation of fibrin. Since this chemical is not found within the blood, the shorter pathway is called the **extrinsic pathway.** The extrinsic and intrinsic pathways are diagrammed in figure 14.6. These pathways involve a cascade of sequential activations, and the two pathways eventually merge to form a final common pathway to produce fibrin polymers.

Dissolution of Clots

Within a few days after a clot has formed, it is usually dissolved. As the damaged blood vessel wall is repaired, activated factor XII promotes the conversion of a precursor molecule into the active form called **kallikrein** (*ka″lĭ-kre′in*). Kallikrein catalyzes the conversion of *plasminogen* into the active molecule called **plasmin.** Plasmin is an enzyme that digests fibrin into "split products," thus promoting dissolution of the clot.

A number of plasminogen activators are used clinically to promote dissolution of clots. A recent development is the commercial availability of one of the endogenous compounds, called tissue plasminogen activator (TPA), *which is produced using genetically engineered bacteria.* Streptokinase *is a potent and more widely used activator of plasminogen, but is classified as exogenous because it is a normal bacterial product. Streptokinase and TPA may be injected into the general circulation or injected specifically into a coronary vessel that has become occluded by a thrombus (blood clot).*

Blood can be prevented from clotting in a test tube by the addition of *anticoagulants.* The various anticoagulants have different mechanisms of action, as described in table 14.3. People may acquire clotting disorders through the administration of anticoagulants or because of a vitamin K deficiency, or they may inherit a particular clotting disorder. Some of the more common of these disorders are listed in table 14.3.

Inside Information—The "Grandmother of Europe," Queen Victoria of England (1819–1901), was a carrier of hemophilia; unwittingly, she contaminated the inheritance of much of the royalty of Europe. The British royal family was spared because Victoria's son King Edward VII did not inherit the defective factor IX.

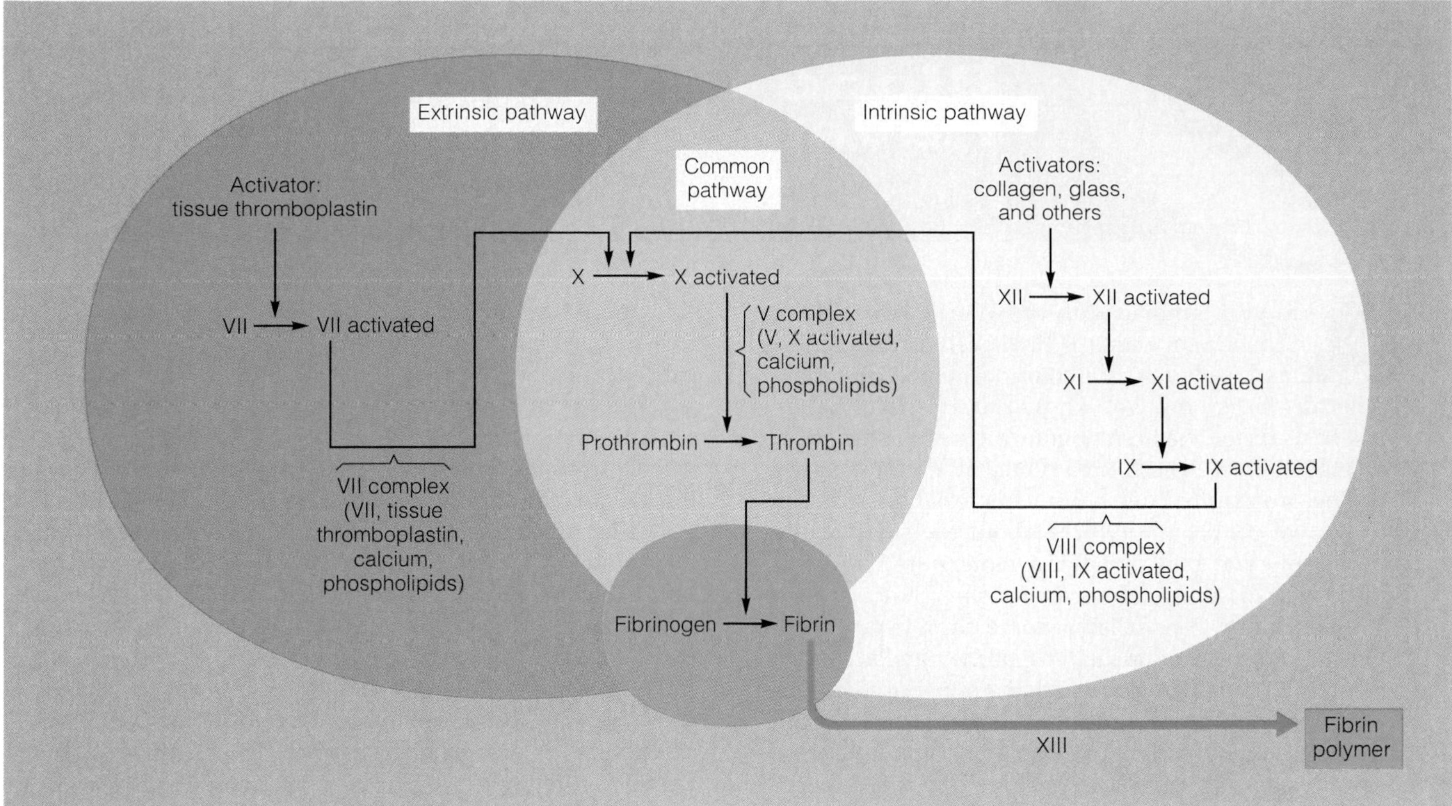

FIGURE 14.6
The extrinsic and intrinsic clotting pathways that lead to the formation of insoluble threads of fibrin polymers.

TABLE 14.3 Some Acquired and Inherited Clotting Disorders and a Listing of Anticoagulant Drugs

Category	Cause of Disorder	Comments
Acquired clotting disorders	Vitamin K deficiency	Inadequate formation of prothrombin and other clotting factors in the liver
Inherited clotting disorders	Hemophilia A (defective factor VIII_{AHF})	Recessive trait carried on X chromosome; results in delayed formation of fibrin
	von Willebrand's disease (defective factor VIII_{VWF})	Dominant trait carried on autosomal chromosome; impaired ability of platelets to adhere to collagen in subendothelial connective tissue
	Hemophilia B (defective factor IX), also called Christmas disease	Recessive trait carried on X chromosome; results in delayed formation of fibrin
	Anticoagulants	
Aspirin	Inhibits prostaglandin production, resulting in defective platelet release reaction	
Coumarin	Competes with the action of vitamin K	
Heparin	Inhibits activity of thrombin	
Citrate	Combines with Ca^{++}, and thus inhibits the activity of many clotting factors	

SYNOPTIC SPOTLIGHT

Blood Centers—Havens of Safety?

Are you familiar with the World War I slogan "Uncle Sam wants YOU?" Well, Uncle Sam still may want you, but your local blood center may not. Because it is now well known that certain diseases can be transmitted via homologous blood transfusions (in which blood is received from a donor other than oneself), screening procedures have become increasingly stringent. Even so, we continue to worry about the safety of our blood supply and how our blood centers are monitored. Such concern is reasonable considering that any of us may need a blood transfusion at some point in our lives.

Blood centers are unusual in that they are largely private enterprises that are governed and supervised by public agencies. No longer is the Red Cross the only agency that handles blood nationally. In many communities, the local Red Cross group handles *no* blood, instead focusing its energies on disaster relief and CPR and first aid training. (As the Red Cross is currently structured, the local community can elect to manage the blood "business," and many communities have decided to let agencies other than the Red Cross handle their blood.)

Whatever system has been established by the local community, the agency that accepts human blood donations is closely monitored by the Food and Drug Administration, and is subject to its regulations. Whenever violations persist past the time allowed for correction, the FDA can suspend or cancel the center's license and literally shut it down. In addition to governmental supervision, most blood centers are members of one or more professional organizations. These organizations, which include the American Association of Blood Banks and the Council of Community Blood Centers, all set professional accrediting standards for safety procedures in the screening and handling of blood. Hence, there is no such thing as a locally autonomous blood center.

Why would a center reject the offer of blood? Simply because the only acceptable blood donors are healthy individuals—people who have no history of diseases that could be transmitted through the blood. Perhaps the two most feared diseases that can be passed on in this manner are AIDS and hepatitis virus, both of which are incurable. Prescreening questionnaires are designed to eliminate individuals who are likely to be at risk for these diseases. Other health factors are also considered in potential donors; for example, individuals weighing less than 100 pounds are not permitted to donate blood.

Volunteer donors are *not* at risk for contamination. The needles and containers that hold the blood are discarded after being used once. The person drawing the single pint of blood that is being donated is either a trained medical technician or nurse. These health professionals have been educated as to how to avoid practices that might lead to contamination.

After blood has been collected, it is tested extensively to provide the potential recipient the maximum protection against acquiring disease. Of particular concern are the various tests for different forms of hepatitis and HIV (the virus that causes AIDS). Regarding hepatitis, at least three tests are performed to check for antibodies or antigens that would indicate the presence of hepatitis B or C. As for HIV, all blood centers routinely check the blood for the HIV-1 antibody, which, if present, would indicate that the blood was contaminated. As a further safeguard, testing for the HIV-2 antibody is expected to be mandated shortly. The FDA also is currently recommending the use of yet another blood test (now in the last stage of clinical trials) for HIV antibodies. This test can detect the virus 6 days earlier than other tests. It is hoped that it will prevent transmission of a few more cases of AIDS annually. In addition to tests for hepatitis and HIV, blood is routinely checked for syphilis and leukemia.

Statistics indicate that homologous blood transfusion is safer than it has ever been. The improved testing procedures in current use have minimized the likelihood of getting hepatitis from a transfusion to less than 3%. As recently as the 1960s, up to 25% of patients who received banked blood also contracted hepatitis. Once the medical community became aware that HIV could be transmitted through blood transfusions, industry-wide safeguards were designed and applied, so that now the risk of getting AIDS from a transfusion is less than 1%.

Despite public awareness that the risk of transfusion-acquired hepatitis or HIV infection is extremely low, any risk at all is considered significant by most people. In fact, increasing numbers of patients with upcoming elective surgery are opting for an *autologous preoperative transfusion;* that is, they are storing their own blood prior to surgery. Given that a "zero-risk" blood supply is generally considered impossible by medical researchers, and because hepatitis and AIDS pose such formidable threats, it is expected that alternatives to homologous blood transfusions will continue to be sought.

Summary

14.1 Introduction to the Circulatory System

1. The cardiovascular system includes the heart and blood vessels.
2. Arteries carry blood away from the heart; veins carry blood to the heart.
3. Blood consists of blood plasma and formed elements.
4. Blood plasma consists of water with Na^+ and other dissolved solutes. The plasma proteins include albumin and alpha, beta, and gamma globulin.

14.2 Formed Elements of Blood

1. Erythrocytes, or red blood cells, are biconcave discs that lack nuclei and mitochondria.
2. Erythrocytes contain hemoglobin, which functions to transport oxygen to the body cells.
3. Leukocytes, or white blood cells, are divided into two major classes: granular and agranular.
4. The granular leukocytes include neutrophils, eosinophils, and basophils. The agranular leukocytes include lymphocytes and monocytes.
5. Platelets (thrombocytes) are cytoplasmic fragments of large bone marrow cells called megakaryocytes. Platelets function in blood clotting.

14.3 Hemopoiesis

1. Myeloid tissue is the red bone marrow that produces all of the formed elements of blood.
2. Lymphoid tissue, which includes the lymph nodes, tonsils, spleen, and thymus, produces lymphocytes.
3. Hemopoiesis, or blood cell formation, begins with a common stem cell that differentiates along different lines.
4. Formation of erythrocytes is stimulated by erythropoietin, secreted by the kidneys.
5. Differentiation and proliferation of different types of leukocytes are stimulated by lymphokines.

14.4 Blood Typing

1. People with blood type A have type A antigens on the surface of their red blood cells and antibodies against blood type B in their blood plasma. People with blood type B have type B antigens and antibodies against blood type A.
2. People with blood type AB lack antibodies against the A and B blood types; they are universal recipients; people with type O blood are universal donors because their red blood cells lack A and B antigens.
3. If blood of different types is mixed, an agglutination reaction will occur.
4. People who are Rh positive have the Rh antigen. This condition is dominant to Rh negative.

14.5 Blood Clotting

1. Platelets adhere to exposed collagen and undergo a release reaction that makes them stick to other platelets. In this way, they form a platelet plug.
2. The platelet plug is reinforced by a meshwork of insoluble fibrin polymers, created by sequential activation of a number of clotting factors.
3. Conversion of fibrinogen into fibrin occurs in response to thrombin, which is produced from prothrombin in either a longer intrinsic pathway or a shorter extrinsic pathway.
4. The clot is eventually dissolved by the action of plasmin, which digests fibrin.

Review Activities

Objective Questions

1. Which of the following is the correct sequence of blood flow? *(module 14.1)*
   a. arteries-capillaries-venules-arterioles-veins
   b. veins-venules-capillaries-arterioles-arteries
   c. capillaries-arteries-arterioles-venules-veins
   d. arteries-arterioles-capillaries-venules-veins

2. Which of the following is an agranular leukocyte? *(module 14.2)*
   a. eosinophil
   b. neutrophil
   c. basophil
   d. monocyte

3. Hemoglobin is normally found in *(module 14.2)*
   a. erythrocytes.
   b. blood plasma.
   c. granular leukocytes.
   d. agranular leukocytes.
   e. all of the above.

4. Which of the following stem cells give rise to granular leukocytes? *(module 14.3)*
   a. myeloblasts
   b. erythroblasts
   c. lymphoblasts
   d. megakaryoblasts

5. Production of which of the following blood cells is stimulated by a hormone secreted by the kidneys? *(module 14.3)*
   a. lymphocytes
   b. monocytes
   c. erythrocytes
   d. neutrophils
   e. platelets

6. Which of the following is the safest to use in the transfusion of a person with type B blood? *(module 14.4)*
   a. type O blood
   b. type A blood
   c. type B blood
   d. type AB blood

7. Antibodies against both type A and type B antigens are found in the blood plasma of a person who is *(module 14.4)*
   a. type A.
   b. type B.
   c. type AB.
   d. type O.
   e. any of the above.

8. The activation of factor X is (*module 14.5*)

a. part of the intrinsic pathway only.
b. part of the extrinsic pathway only.
c. part of both the intrinsic and extrinsic pathways.
d. not part of either the intrinsic or the extrinsic pathway.

9. Which of the following statements about platelets is *true?* (*module 14.5*)

a. They form a plug by sticking to each other.
b. They release chemicals that stimulate vasoconstriction.
c. They provide phospholipids needed for the intrinsic pathway.
d. All of the above are true.

10. Which of the following statements about plasmin is *true?* (*module 14.5*)

a. It is involved in the intrinsic clotting system.
b. It is involved in the extrinsic clotting system.
c. It functions in fibrinolysis.
d. It promotes the formation of emboli.

Essay Questions

1. Explain how the circulatory system is organized and describe its major components. (*module 14.1*)
2. Describe the structure and function of erythrocytes. (*module 14.2*)
3. Describe the appearance and function of the various types of leukocytes. (*module 14.2*)
4. Define *myeloid* and *lymphoid tissue.* (*module 14.3*)
5. Explain the significance of erythropoietin and describe the stages of red blood cell production. (*module 14.3*)
6. Describe the origin, structure, and function of platelets (thrombocytes). (*module 14.3*)
7. Explain what is meant by blood types and describe how they are inherited. (*module 14.4*)
8. Explain how hemolytic disease of the newborn is produced. (*module 14.4*)
9. Explain why agglutination occurs when different blood types are mixed. (*module 14.5*)

Labeling Exercise

Label the blood cells shown in the figure to the right.

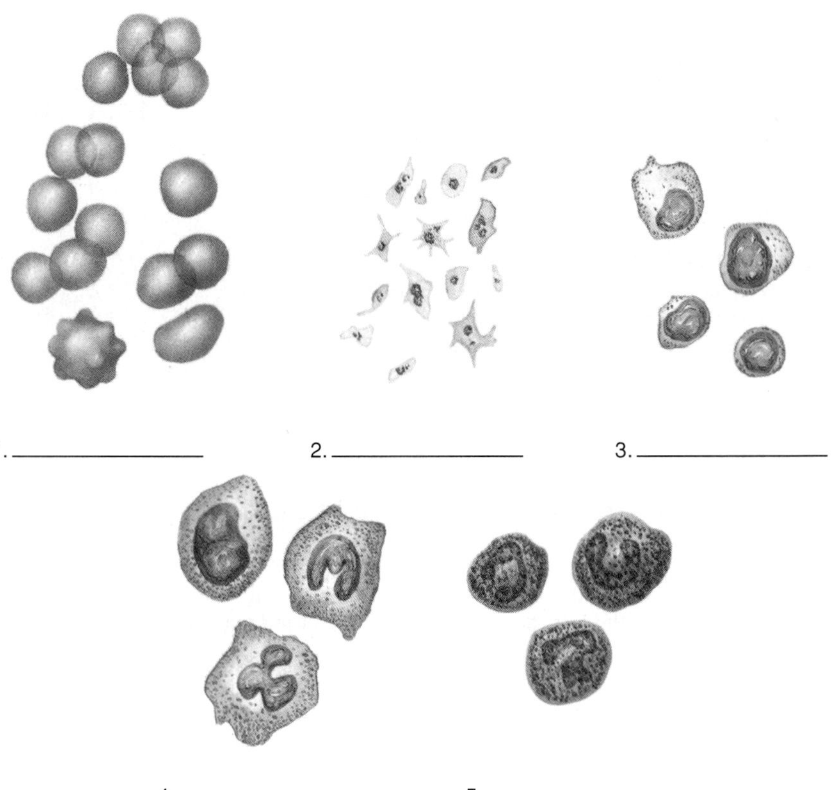

Critical Thinking Questions

1. You have AB blood. Can you determine the blood types of your parents? Could your parents have had a child with type O blood?
2. You have just had a baby with erythroblastosis fetalis. This was your first baby and you were treated with RhoGAM after this delivery. Should you consider having another child? What should you take into account before making such a decision? Also, should you now stop donating blood at your local center? Why or why not?
3. Is there a physiological advantage to forming the platelet plug prior to completion of clot formation? Explain your response.
4. At a recent medical examination, you weighted in at 120 pounds, or 54 kg. Your physician commented that you have about 4,500 ml of blood. Is this cause for concern? Why or why not?
5. Aspirin reduces the ability of blood platelets to stick together. Why are people told to avoid taking aspirin before donating blood? Considering the circulating life span of platelets, how long would you have to wait after taking aspirin before its effects on the platelets were eliminated.

CHAPTER 15

"The [four-chambered] structure of the heart enables it to serve as a transport system pump that keeps blood continuously circulating through the blood vessels of the body." *page 372*

Circulatory System: Heart and Blood Vessels

Chapter Outline

Terms to Remember

pericardium (*per"ĭ-kar' de-um*), 372
myocardium (*mi"ō-kar' de-um*), 372
endocardium (*en"do-kar' de-um*), 373
atrium (*a' tre-um*), 373
ventricle, 373
atrioventricular valves, 373
semilunar valves, 373
systole (*sis' tŏ-le*), 377
diastole (*di-as' tŏ-le*), 377
electrocardiogram, 379
artery, 381
capillary, 381
vein, 381
aorta (*a-or' tă*), 383
superior vena cava (*ve' nă ka' vă*), 389
inferior vena cava, 389
hepatic portal system, 393
placenta (*plă-sen' tă*), 395
umbilical vein, 395
foramen ovale (*o-val' e*), 396
umbilical artery, 396

Learning Objectives

After studying this chapter, you should be able to

- describe the location of the heart in relation to the other thoracic organs.
- define *pericardium* and explain the importance of the pericardial fluid.
- describe the structure and functions of the three layers of the heart wall.
- identify the chambers, valves, and associated vessels of the heart.
- trace the flow of blood through the heart and distinguish between the pulmonary and systemic circulations.
- describe the locations of the conduction tissues of the heart and trace the pathway of impulse conduction.
- describe the ECG deflections and explain the diagnostic value of an ECG.
- describe the structure and functions of arteries, capillaries, and veins.
- describe the arterial pathways to the principal body organs and regions.
- describe the venous drainage from the principal body organs and regions.
- describe the fetal circulation.

15.1 Structure of the Heart

The structure of the heart enables it to serve as a transport system pump that keeps blood continuously circulating through the blood vessels of the body.

General Description

The hollow, four-chambered, muscular heart is roughly the size of a clenched fist. It averages 255 grams in adult females and 310 grams in adult males. The heart is located within the thoracic cavity between the lungs, in the region known as the *mediastinum.* Roughly triangular, the cone-shaped *apex* of the heart is directed downward to the left and rests on the diaphragm. The *base* of the heart—the broad upper part where the large vessels attach—is directed toward the right. Thus, the heart assumes an oblique position in the mediastinum, tipping slightly to the left.

The heart is enclosed and protected by a loose-fitting serous sac called the **pericardium** (*per"ĭ-kar'de-um*) (fig. 15.1). The pericardium separates the heart from the other thoracic organs. It forms the wall of the *pericardial cavity,* which contains a watery *pericardial fluid.* The pericardium is actually composed of an outer *fibrous pericardium* and an inner *serous pericardium* (table 15.1). It is the serous pericardium that produces the lubricating pericardial fluid, which allows the heart to beat in kind of a frictionless bath.

Inside Information—The heart contracts an estimated 42 million times a year, pumping approximately 700,000 gallons of blood.

Heart Wall

The wall of the heart is composed of three layers, as shown in table 15.1. The **epicardium** (*ep"ĭ-kar'de-um*), the visceral layer of the serous pericardium, covers the heart surface; it forms the protective outer sheath of the heart. The **myocardium** (*mi"ŏ-kar'de-um*) is the thick middle layer of the heart, composed of cardiac muscle tissue. It is

pericardium: Gk. *peri,* around; *kardia,* heart

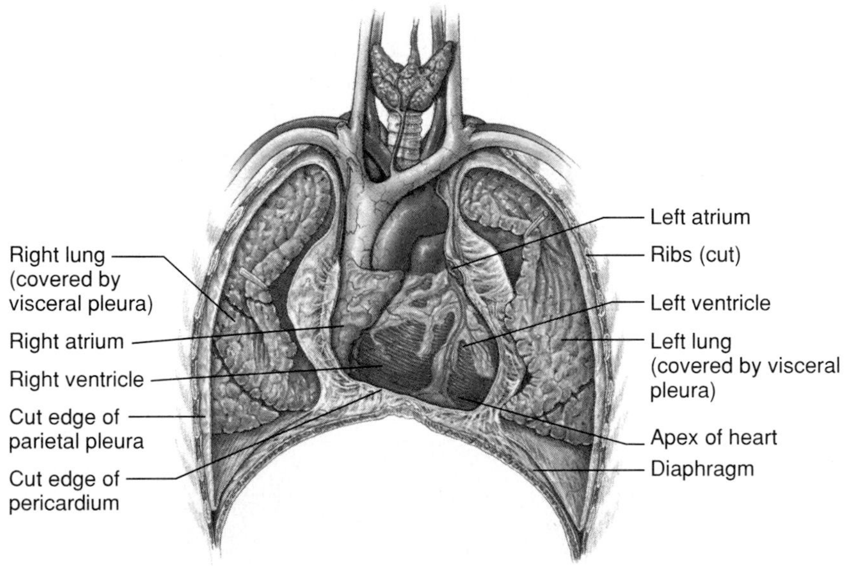

FIGURE 15.1
The position of the heart and associated serous membranes within the thoracic cavity.

TABLE 15.1 **Layers of the Heart Wall**

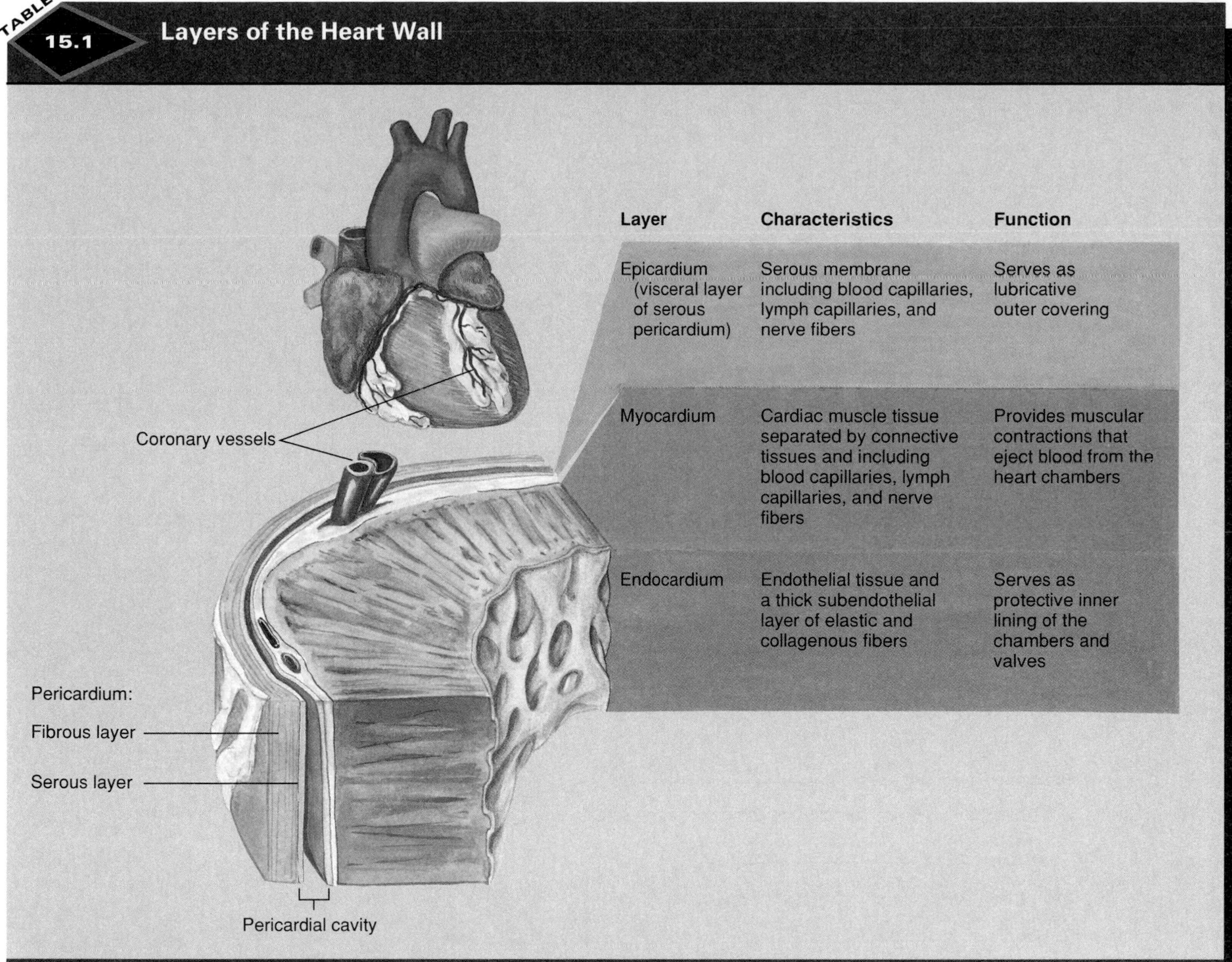

Layer	Characteristics	Function
Epicardium (visceral layer of serous pericardium)	Serous membrane including blood capillaries, lymph capillaries, and nerve fibers	Serves as lubricative outer covering
Myocardium	Cardiac muscle tissue separated by connective tissues and including blood capillaries, lymph capillaries, and nerve fibers	Provides muscular contractions that eject blood from the heart chambers
Endocardium	Endothelial tissue and a thick subendothelial layer of elastic and collagenous fibers	Serves as protective inner lining of the chambers and valves

responsible for the pumping action of the heart. The **endocardium** (*en"do-kar'de-um*) is the inner layer that is continuous with the endothelium of blood vessels.

Chambers and Valves

The heart is a double pump, consisting of upper **right** and **left atria** (singular, *atrium* [*a'tre-um*]) and lower **right** and **left ventricles.** The atria contract and empty simultaneously into the ventricles (fig. 15.2), which also contract in unison. The atria are separated by the thin *interatrial septum.* The ventricles are separated by the thick, muscular *interventricular septum.* **Atrioventricular valves** (*AV valves*) lie between the atria and ventricles, and **semilunar valves** are located at the bases of the two large vessels leaving the heart. Heart valves (table 15.2) maintain one-way flow of blood.

The following discussion follows the sequence in which blood flows through the atria, ventricles, and valves. It is important to keep in mind that the right side of the heart (right atrium and right ventricle) receives deoxygenated blood (blood low in oxygen) and pumps it to the lungs. The left side of the heart (left atrium and left ventricle) receives oxygenated blood (blood high in oxygen) from the lungs and pumps it to the body.

atrium: L. *atrium,* chamber
ventricle: L. *ventriculus,* diminutive of *venter,* belly

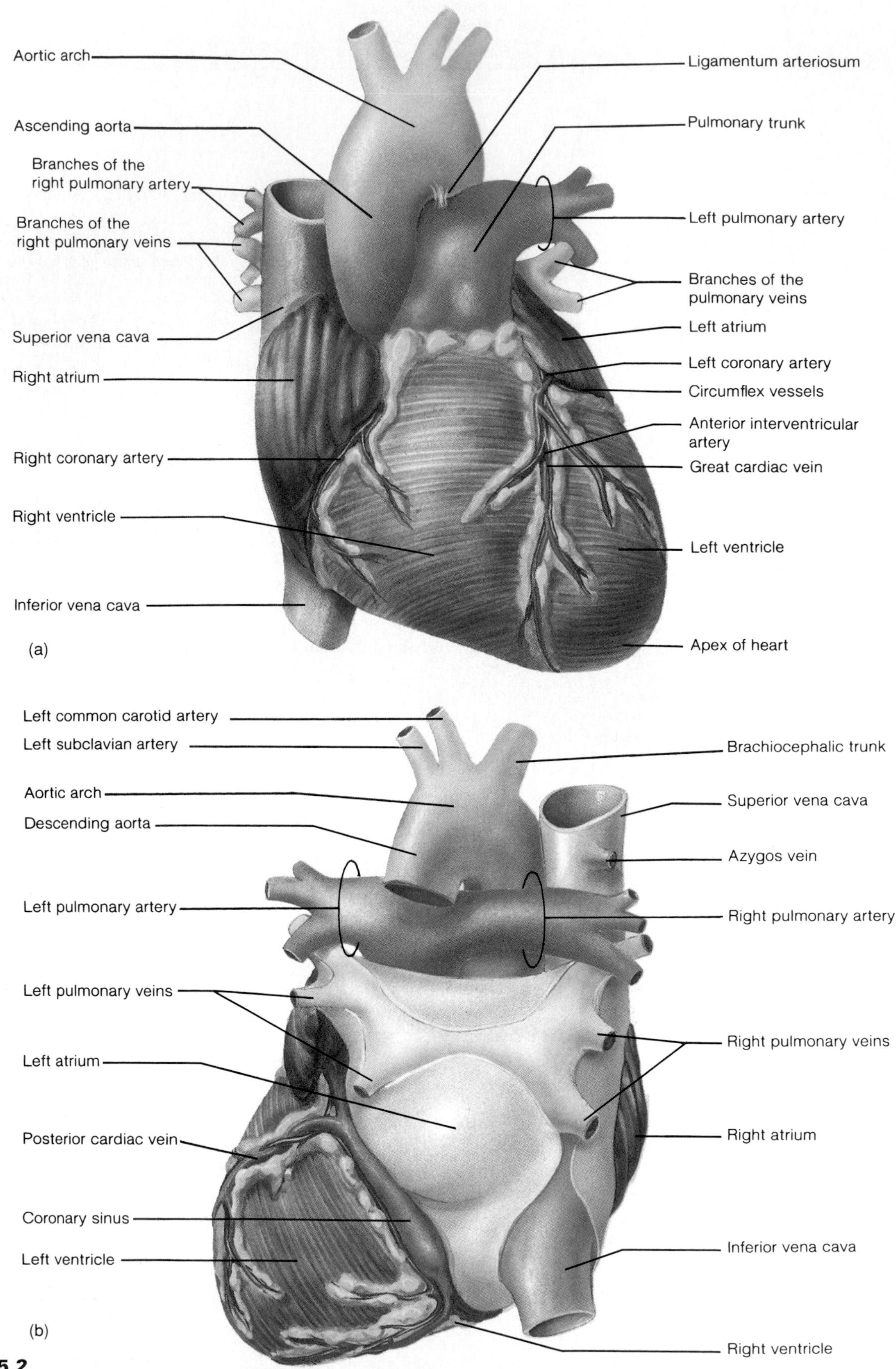

FIGURE 15.2
The structure of the heart. (*a*) An anterior view, (*b*) a posterior view, and (*c*) an internal view.

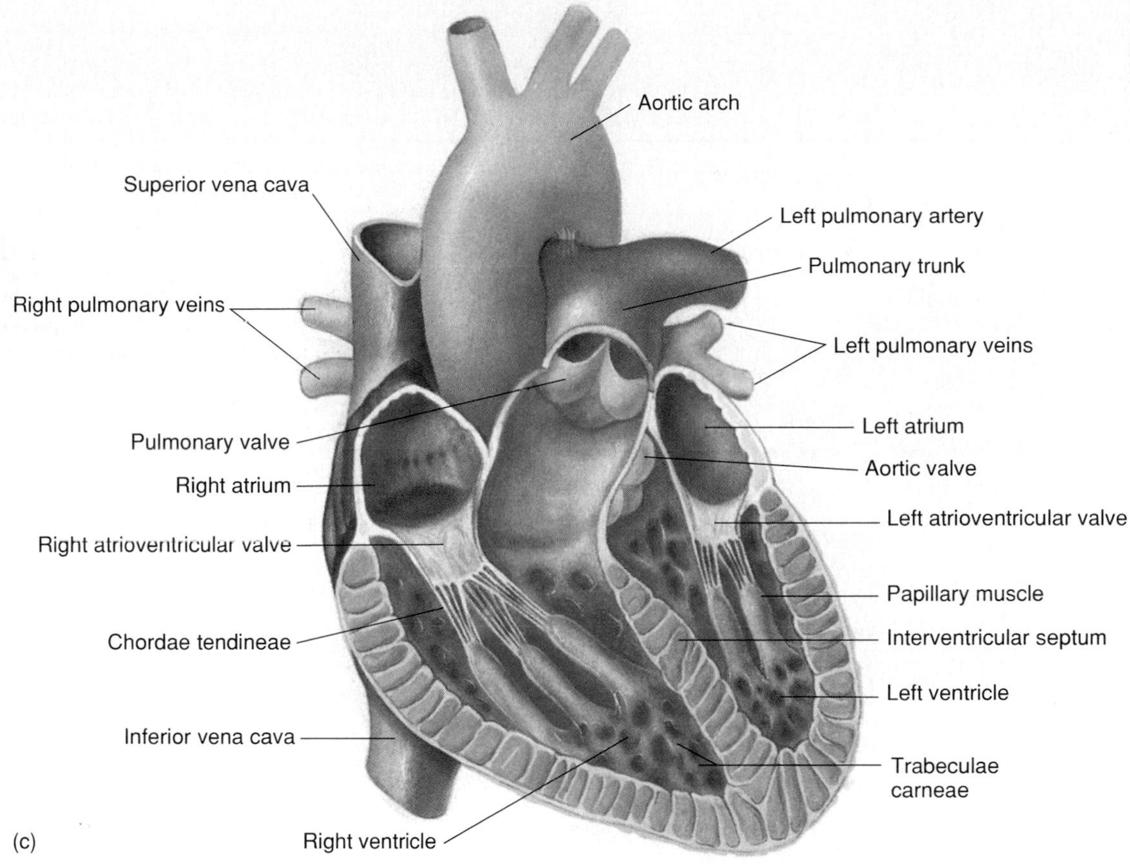

FIGURE 15.2
Continued.

The **right atrium** receives venous blood from the *superior vena cava,* which drains the upper portion of the body, and from the *inferior vena cava,* which drains the lower portion (fig. 15.2). The **coronary sinus** is an additional opening into the right atrium that receives venous blood from the myocardium of the heart itself.

Blood from the right atrium passes through the **tricuspid valve** to fill the **right ventricle.** The tricuspid valve is an AV valve that gets its name from its three valve *cusps.* The cusps of the AV valves are held in position by strong tendinous cords called **chordae tendineae.** The chordae tendineae are secured to the ventricular walls by cone-shaped **papillary muscles.** These structures prevent the AV valves from everting, like an umbrella in a strong wind, when the ventricles contract.

Ventricular contraction causes the tricuspid valve to close and the blood to leave the right ventricle through the *pulmonary trunk* and to enter the lungs through the *right* and *left pulmonary arteries.* The **pulmonary semilunar valve** lies at the base of the pulmonary trunk, where it prevents the backflow of ejected blood into the right ventricle.

After gas exchange has occurred within the capillaries of the lungs, oxygenated blood is transported to the **left atrium** through four *pulmonary veins,* two from each lung. The **left ventricle** receives blood from the left atrium. These two chambers are separated by the **bicuspid valve** (*mitral valve*). When the left ventricle is relaxed, this AV valve is open, allowing blood to flow from the atrium to the ventricle; when the left ventricle contracts, the bicuspid valve closes, preventing the backflow of blood into the atrium. The **aortic semilunar valve** is located at the base of the ascending aorta as it leaves the left ventricle. The aortic semilunar valve opens when the left ventricle contracts. When the left ventricle relaxes, it closes as a result of the arterial pressure of the blood. Thus, this valve prevents the backflow of blood into the relaxed left ventricle.

On the inner surface of both ventricular walls are distinct ridges called **trabeculae carneae** (fig. 15.2*c*) that reinforce the endocardium of the ventricles.

The *pulmonary circulation* consists of blood vessels that transport blood from the right ventricle to the lungs for gas exchange, and then to the left atrium of the heart. The *systemic circulation* involves all the vessels of the body that

---

vena cava: L. *vena,* vein; *cava,* empty
chordae tendineae: L. *chorda,* string; *tendere,* to stretch
bicuspid: L. *bi,* two; *cuspis,* tooth point or spike
mitral: L. *mitra,* like a bishop's mitre
trabeculae carneae: L. *trabecula,* small beams; *carneus,* flesh

TABLE 15.2 Valves of the Heart

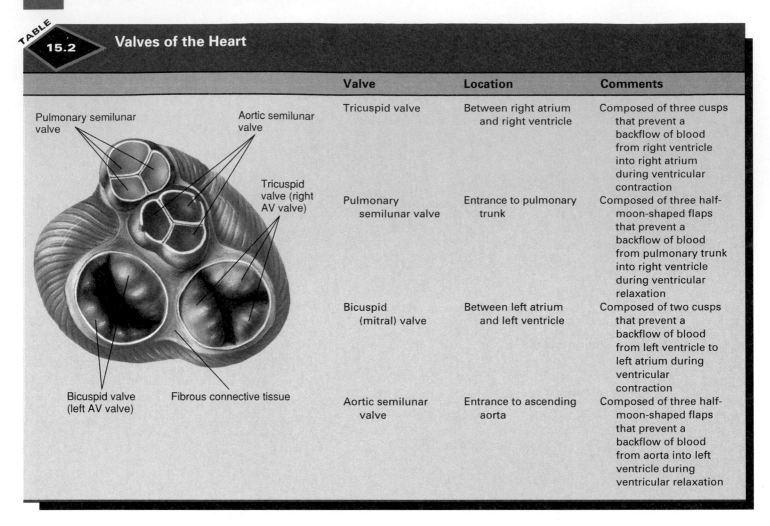

Valve	Location	Comments
Tricuspid valve	Between right atrium and right ventricle	Composed of three cusps that prevent a backflow of blood from right ventricle into right atrium during ventricular contraction
Pulmonary semilunar valve	Entrance to pulmonary trunk	Composed of three half-moon-shaped flaps that prevent a backflow of blood from pulmonary trunk into right ventricle during ventricular relaxation
Bicuspid (mitral) valve	Between left atrium and left ventricle	Composed of two cusps that prevent a backflow of blood from left ventricle to left atrium during ventricular contraction
Aortic semilunar valve	Entrance to ascending aorta	Composed of three half-moon-shaped flaps that prevent a backflow of blood from aorta into left ventricle during ventricular relaxation

are not part of the pulmonary circulation. It functions to transport blood to and from all of the cells and tissues of the body, except to that part of the lung served by the pulmonary vessels.

Coronary Circulation

The wall of the heart has its own supply of blood vessels to meet its vital needs. The flow of blood through the numerous vessels in the myocardium is called the *coronary circulation.* The main coronary vessels are the *left* and *right coronary arteries* that arise from the aorta, immediately beyond the aortic semilunar valve (see fig. 15.2*a*). Each coronary artery branches and subbranches to deliver oxygen and nutrients to the cells of the heart wall and to collect carbon dioxide and other wastes. Most of the deoxygenated blood leaves the heart wall via the *coronary sinus* on the posterior surface of the heart. As previously mentioned, the coronary sinus empties into the right atrium.

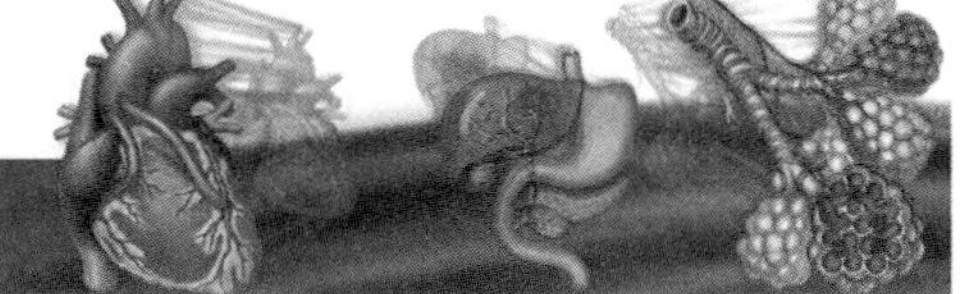

15.2 Conduction System and the Cardiac Cycle

Contraction of the atria and ventricles and movement of blood through the heart during the cardiac cycle is produced by intrinsic depolarization through the nodal tissues of the heart.

Cardiac muscle has an *intrinsic rhythmicity*—that is, the basic heartbeat originates in and is conducted through the heart without external stimulation. The **conduction system** consists of specialized strands of cardiac muscle tissue, called *nodal tissues,* that coordinate cardiac contraction. The *cardiac cycle,* to be discussed shortly, depends on the ability of the nodal tissues to initiate and distribute electrical impulses. The components of the conduction system are the **sinoatrial node** (*SA node*), **atrioventricular node** (*AV node*), **atrioventricular bundle** (*bundle of His*), and the **conduction myofibers** (*Purkinje fibers*). The locations of these components are indicated in figure 15.3.

The SA node, or *pacemaker,* initiates the cardiac cycle by producing an electrical impulse (fig. 15.4) that spreads over both atria, causing them to contract simultaneously and force blood into the ventricles. The impulse then passes to the AV node and continues through the atrioventricular bundle. The atrioventricular bundle divides into right and left bundle branches, which are continuous with the conduction myofibers within the heart wall. Stimulation of these fibers causes the ventricles to contract.

Inside Information—Even if all nerve connections to the heart are severed, the heart will continue to beat. And even when separated from the heart and placed in a saline solution, individual cardiac muscle cells can be seen randomly contracting when viewed under the microscope.

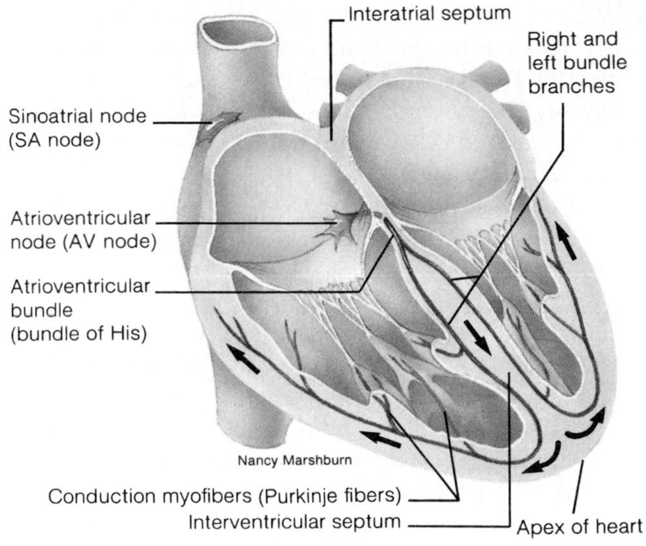

Interatrial septum
Right and left bundle branches
Sinoatrial node (SA node)
Atrioventricular node (AV node)
Atrioventricular bundle (bundle of His)
Nancy Marshburn
Conduction myofibers (Purkinje fibers)
Interventricular septum
Apex of heart

FIGURE 15.3

The conduction system of the heart.

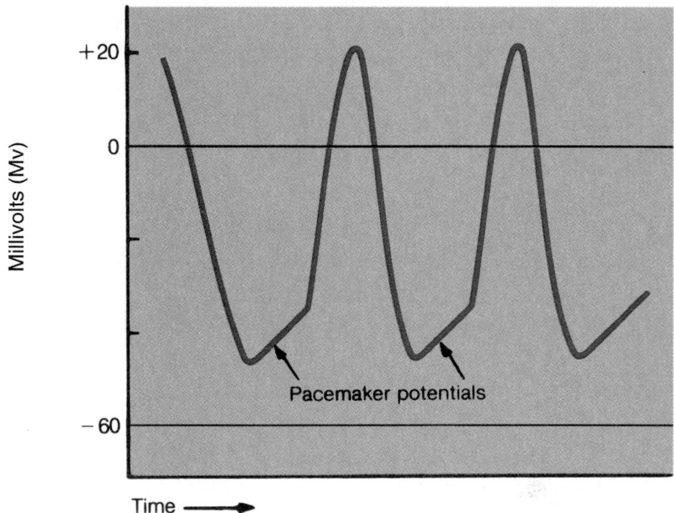

FIGURE 15.4

Pacemaker potentials and action potentials in the SA node.

The **cardiac cycle** refers to the repeating pattern of contraction and relaxation of the heart. The contraction phase is called **systole** (*sis'tŏ-le*), and the relaxation phase is called **diastole** (*di-as'tŏ-le*). When these terms are used without reference to specific chambers, they refer to the ventricles. It should be noted, however, that the atria also contract and relax. Atrial contraction occurs toward the end of diastole, when the ventricles are relaxed; when the ventricles contract during systole, the atria are relaxed.

The heart thus has a two-step pumping action. The right and left atria contract almost simultaneously, followed by contraction of the right and left ventricles 0.1 to 0.2 second later. During the time when both the atria and ventricles are relaxed, the venous return of blood fills the atria. The buildup of pressure that results causes the AV valves to open and blood to flow from the atria to the ventricles. It is estimated that the ventricles are about 80% filled with blood even before the atria contract. Contraction of the atria adds the final 20%. The total amount of blood in the ventricles at the end of diastole is called the *end-diastolic volume* (*EDV*).

Contraction of each ventricle ejects an amount of blood known as the *stroke volume* (*SV*), which is about two-thirds of the blood that the ventricles contained prior to contraction (the end-diastolic volume). This leaves one-third of the initial amount left in the ventricles as the *end-systolic volume* (*ESV*). The ventricles then fill with blood during the next cycle. At an average cardiac rate of 75 beats per minute, each cycle lasts 0.8 second; 0.5 second is spent in diastole, and systole takes 0.3 second.

When the heart is in diastole, pressure in the systemic arteries averages about 80 mmHg. The following events in the cardiac cycle then occur.

1. The ventricles begin their contraction and the intraventricular pressure rises, causing the AV valves to snap shut. This produces a characteristic "lub" sound. At this time, the ventricles are neither being filled with blood (the AV valves are closed) nor ejecting blood (the intraventricular pressure has not risen sufficiently to open the semilunar valves). This is the phase of *isovolumetric contraction.*

bundle of His: from Wilhelm His Jr., Swiss physician, 1863–1934

Purkinje fibers: from Johannes E. von Purkinje, Bohemian anatomist, 1787–1869

systole: Gk. *systole,* contraction

diastole: Gk. *diastole,* dilation or expansion

2. When the pressure in the left ventricle becomes greater than the pressure in the aorta, the aortic semilunar valve opens and the phase of *ejection* begins. The pressure in the left ventricle and aorta rises to about 120 mmHg (fig. 15.5) as the ventricular volume decreases.
3. As the pressure in the left ventricle falls below that of the aorta, the semilunar valves snap shut, which produces a "dub" sound. The pressure in the aorta falls to 80 mmHg, while the pressure in the left ventricle falls to 0 mmHg.
4. During *isovolumetric relaxation,* the AV and semilunar valves are closed and the ventricles are expanding.
5. When the pressure in the ventricles falls below the pressure in the atria, the AV valves open and a phase of *rapid filling* of the ventricles occurs.
6. *Atrial systole* empties the remaining blood into the ventricles immediately prior to the next phase of isovolumetric contraction.

The heart valves are positioned deep to the sternum, which tends to muffle and diffuse the valvular sounds. For this reason, a physician will listen with a stethoscope for the heart sounds at locations designated as *valvular auscultatory areas.* These areas are named according to the valve that can be detected (fig. 15.6).

Heart sounds are important because they provide information about the condition of the heart valves and other aspects of the heart. Abnormal sounds, referred to as heart murmurs, *are caused by valvular leakage or turbulence of the blood as it passes through the heart.*

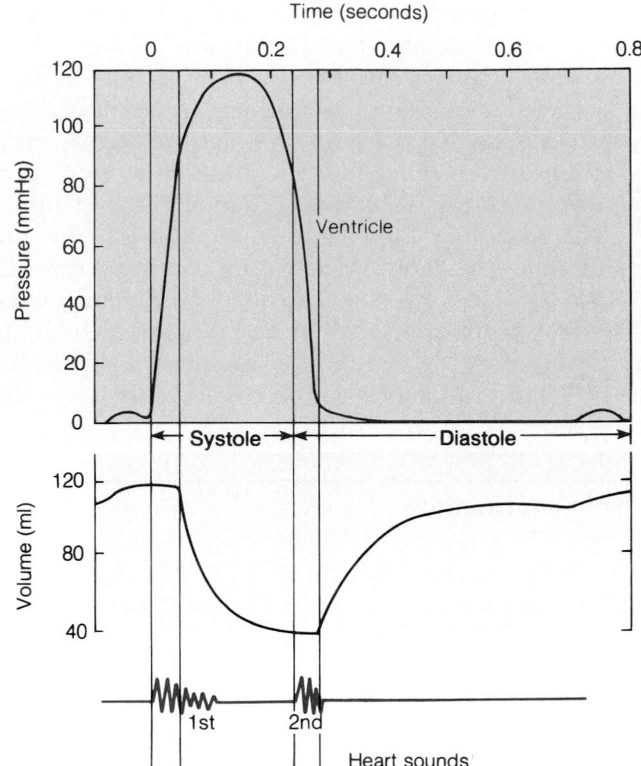

FIGURE 15.5
The relationship between the heart sounds and the left intraventricular pressure and volume. Closing of the AV valves occurs during the early part of contraction, when the intraventricular pressure rises prior to ejection of blood. Closing of the semilunar valves occurs at the beginning of ventricular relaxation, just prior to filling. The first and second sounds thus appear during the stages of isovolumetric contraction and isovolumetric relaxation (*iso* = same).

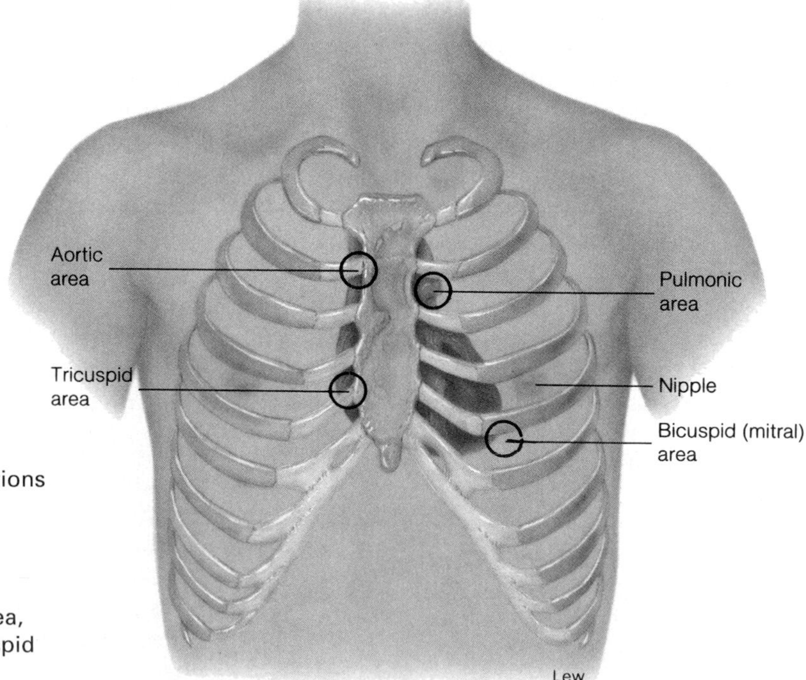

FIGURE 15.6
The valvular auscultatory areas are the routine stethoscope positions for listening to the heart sounds. The aortic area is for detection of the aortic semilunar valve; the pulmonic area, the pulmonary semilunar valve; the tricuspid area, the tricuspid valve; and the bicuspid area, the bicuspid valve.

15.3 Electrical Activity of the Heart

The electrical impulses that pass through the nodal tissue of the conduction system during the cardiac cycle can be recorded using an electrocardiograph to produce an electrocardiogram.

Pacemaker Potential

The cells of the SA node do not maintain a resting membrane potential in the same manner as resting neurons. Instead, during diastole, there is a spontaneous depolarization of the SA node called the **pacemaker potential.** The membrane potential begins at about –60 mV and gradually depolarizes to –40 mV, which is the threshold for producing an action potential in these cells. This spontaneous depolarization is produced by the diffusion of calcium ions (Ca^{++}) through openings in the membrane called *slow calcium channels.* At the threshold level of depolarization, other channels—called *fast calcium channels*—open, and Ca^{++} rapidly diffuses into the cells. Repolarization is produced by the opening of potassium (K^{+}) gates and outward diffusion of K^{+}. Once repolarization to –60 mV has been achieved, a new pacemaker potential begins that again culminates at the end of diastole, with a new action potential.

Although all cardiac muscle cells are capable of producing action potentials spontaneously, they are produced most rapidly in the SA node, which has more slow calcium channels than other regions of the heart. It is the characteristic rhythm of the SA node, called the *sinus rhythm,* that sets the pace for the heart as a whole, coordinating the contractions of the individual cardiac muscle cells. Under unusual circumstances, other regions of the heart, called *ecoptic pacemakers,* or *ecoptic foci,* can function as pacemakers, but the rhythms they set will be slower than that normally set by the SA node.

Electrocardiogram

A pair of surface electrodes placed on the skin over the heart will record a repeating pattern of electrical changes. As action potentials spread through the conduction system from the atria to the ventricles, the voltage measured between these two electrodes will vary. Recordings of these changes provide a "picture" of the electrical activity of the heart. Changing the position of the electrodes provides different perspectives, enabling an observer to gain a more comprehensive picture of the electrical events.

The body is a good conductor of electricity because tissue fluids contain a high concentration of ions that move (creating a current) in response to potential differences. Potential differences generated by the heart are thus conducted to the body surface, where they can be recorded by electrodes on the skin. The recording is called an **electrocardiogram (ECG or EKG);** the recording device is called an **electrocardiograph.**

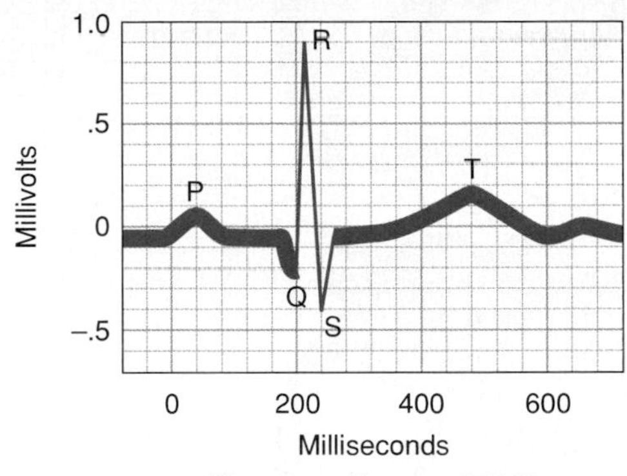

FIGURE 15.7
The electrocardiogram indicates the conduction of electrical impulses through the heart (see fig. 15.3) and measures and records both the intensity of this electrical activity (in millivolts) and the time intervals involved.

An ECG of a normal cardiac cycle is illustrated in figure 15.7. The *wave deflections,* designated P, QRS, and T, are produced as specific events of the cardiac cycle occur. Any heart disease that disturbs the electrical activity will produce characteristic changes in one or more of these waves, so understanding the normal wave-deflection patterns is clinically important.

As shown in figure 15.8, the principal aspects of an ECG are as follows:

P wave. Stimulation of the SA node produces the P wave, a small upward deflection that accompanies depolarization of the atrial fibers. The actual contraction of the atria follows the P wave by a fraction of a second. The ventricles of the heart are in diastole during the expression of the P wave. A missing or abnormal P wave may suggest a dysfunction of the SA node.

P-R interval. On the recording, this is the period of time from the start of the P wave to the beginning of the QRS wave. This interval indicates the amount of time required for the SA depolarization to reach the ventricles. A prolonged P-R interval suggests a conduction problem from the AV node.

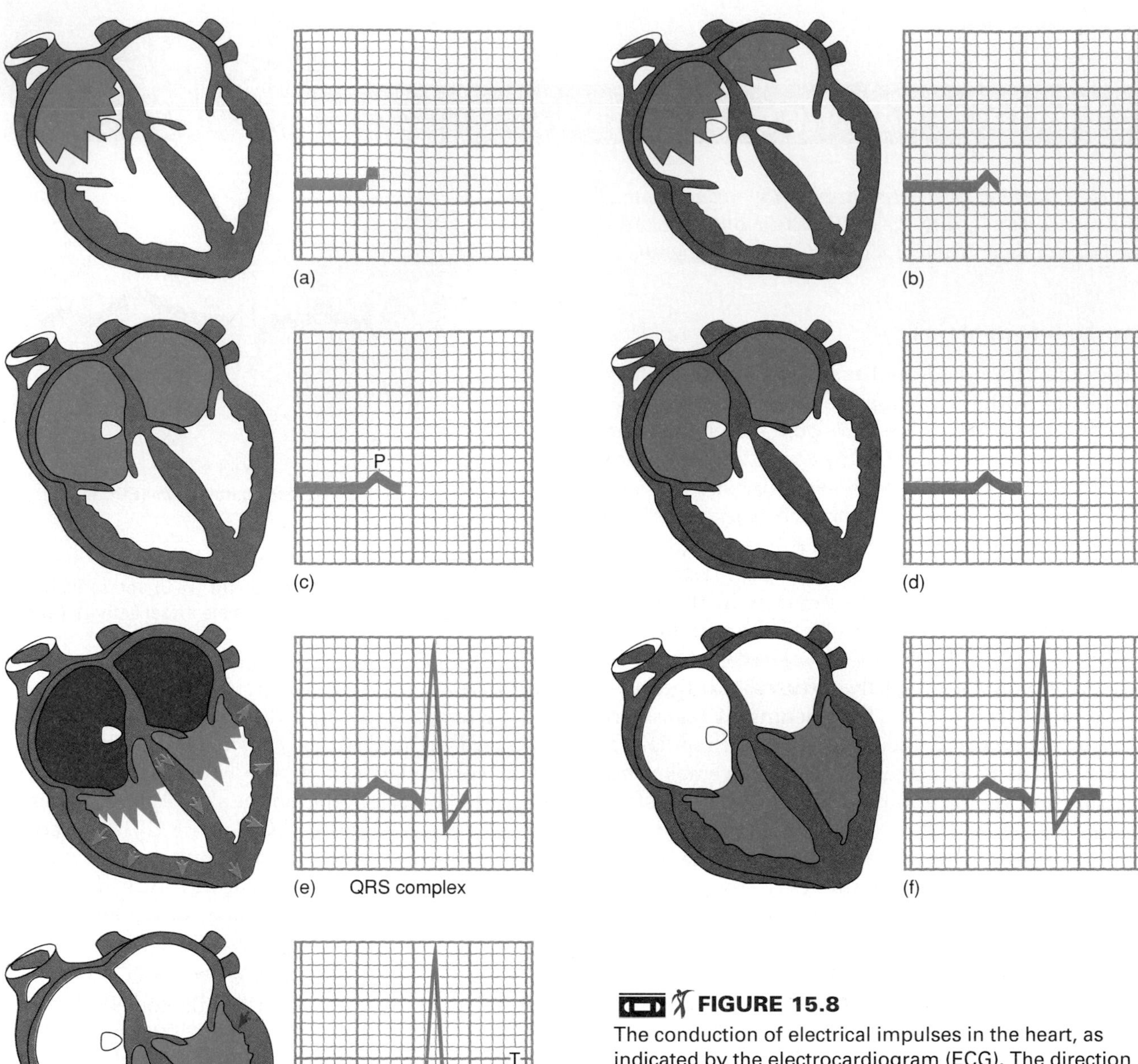

FIGURE 15.8
The conduction of electrical impulses in the heart, as indicated by the electrocardiogram (ECG). The direction of the arrows in (*e*) indicates that depolarization of the ventricles occurs from the inside (endocardium) out (to the epicardium), whereas the arrows in (*g*) indicate that repolarization of the ventricles occurs in the opposite direction.

QRS wave. The QRS wave begins as a short downward deflection, continues as a sharp upward spike, and ends as a downward deflection. The QRS wave indicates the depolarization of the ventricles. During this interval, the ventricles are in systole and blood is being ejected from the heart. It is also during this interval that the atria repolarize, but this event is hidden by the greater depolarization occurring in the ventricles. An abnormal QRS wave generally indicates heart problems of the ventricles. An enlarged R spike, for example, generally indicates enlarged ventricles.

S-T interval. The time duration known as the S-T interval represents the period between the completion of ventricular depolarization and initiation of repolarization. The S-T interval is depressed when the heart receives insufficient oxygen; in acute myocardial infarction, it is elevated.

T wave. The T wave is produced by ventricular repolarization. An arteriosclerotic heart will produce altered T waves, as will various other heart diseases.

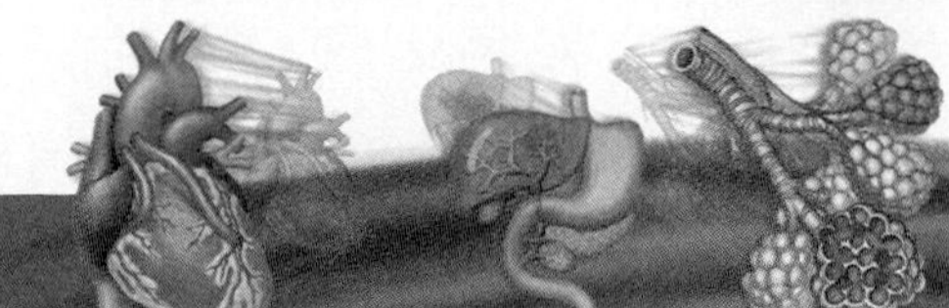

15.4 Blood Vessels

The structure of arteries and veins allows them to transport blood from the heart to the capillaries and back to the heart. The structure of capillaries permits the exchange of small amounts of blood plasma and dissolved molecules between the blood and surrounding tissues.

Blood vessels form a closed tubular network that permits blood to flow from the heart to all the living cells of the body and then back to the heart. Blood leaving the heart passes through vessels of progressively smaller diameters referred to as **arteries, arterioles,** and **capillaries.** Blood returning to the heart from the capillaries passes through vessels of progressively larger diameters called **venules** and **veins.**

Before the invention of the microscope, scientists could not understand how blood got from arteries into veins (see *Module 1.1*). The capillaries, too small to see with the unaided eye, had not yet been discovered. Capillaries are the smallest of blood vessels; they are the *functional units* of the circulatory system in which exchanges of gases and nutrients between the blood and the tissues occur.

The walls of arteries and veins consist of three layers, or *tunics,* as shown in figure 15.9.

1. The **tunica externa,** the outermost layer, is composed of loose connective tissue.
2. The **tunica media,** the middle layer, is composed of smooth muscle. The tunica media of arteries has variable amounts of elastic fibers.
3. The **tunica intima,** the innermost layer, is composed of simple squamous epithelium and connective tissue. The layer of simple squamous epithelium is sometimes referred to as the *endothelium.* This layer lines the inner wall of all blood vessels. Capillaries consist of endothelium only.

Although arteries and veins have the same basic structure, there are some important differences between the two types of vessels. Arteries have more muscle than do comparably sized veins. As a result, arteries appear round in cross section, whereas veins are usually partially collapsed. In addition, many veins have valves, which are absent in arteries.

tunic: L. *tunica,* covering or coat

Arteries

Large arteries, such as the aorta, expand when the pressure of the blood rises as a result of systole; they recoil, like a stretched rubber band, when the blood pressure falls during diastole. This elastic recoil helps to produce a smoother flow of blood through the smaller arteries and still smaller arterioles.

Small arteries and arterioles are less elastic than the larger arteries and have a thicker layer of smooth muscle in proportion to their diameters. Unlike the larger arteries, therefore, the smaller arteries and arterioles retain a relatively constant diameter as the pressure of the blood rises and falls during the heart's pumping activity.

Capillaries

Of all the components of the cardiovascular system, capillaries have the simplest structure. They are little more than tubes—just a single cell layer thick and approximately 1 mm long. The entire body is permeated with a fine mesh of these capillaries; in fact, their combined length is on the order of 60,000 miles. So extensive is their branching, that no cell in the body is more than a fraction of a millimeter away from any capillary; moreover, these tiny vessels provide a total surface area of 1,000 square miles for exchanges between blood and interstitial fluid.

Despite their large number, capillaries contain only about 250 ml of blood at any time, out of a total blood volume of about 4,000 ml (most is contained within veins). The amount of blood flowing through a capillary bed is determined in part by the action of the *precapillary sphincter muscles.* Blood flow to an organ is regulated by contraction of these smooth muscle sphincters and by the degree of resistance to blood flow provided by the small arteries and arterioles.

Veins

Veins are vessels that carry blood from capillaries back to the heart. The blood is delivered from microscopic vessels called *venules* into progressively larger vessels that empty into the large veins. The average pressure in the veins is only 2 mmHg, compared to an average arterial pressure of about 100 mmHg. The low venous pressure is insufficient to return blood to the heart, particularly from the lower limbs. Veins, however, pass between skeletal muscle groups that provide a massaging action as they contract.

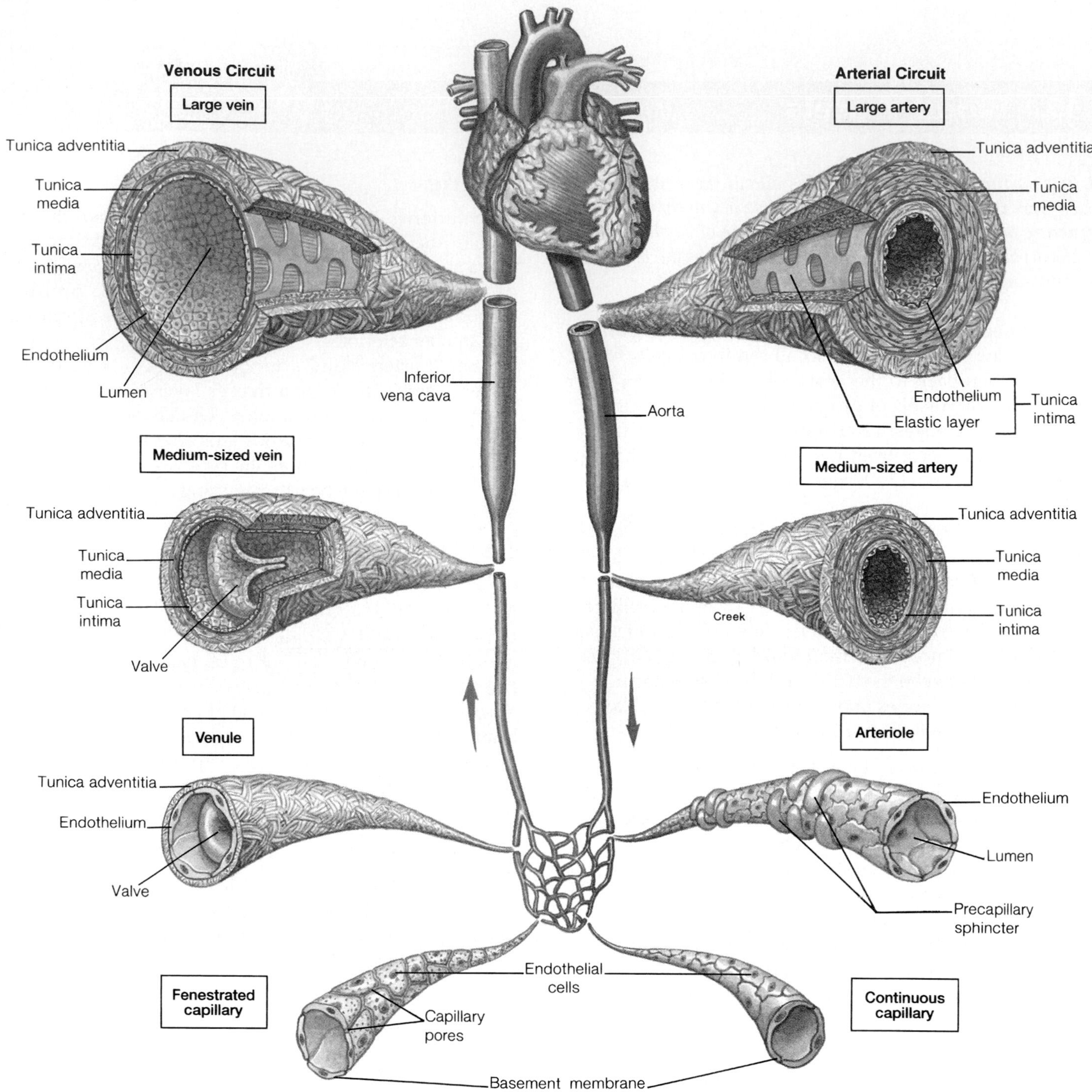

FIGURE 15.9
Relative thickness and composition of the tunics in comparable arteries and veins.

These *skeletal muscle pumps* intermittently increase the venous pressure in the region of vein squeezed by the surrounding muscles. When the pumps are less active, as when a person stands still or is bedridden, blood accumulates in the veins and causes them to bulge. When a person is more active, blood returns to the heart at a faster rate, and less is left in the venous system. A one-way flow of blood to the heart is maintained by the presence of venous valves, which close in response to the increased pressure.

Inside Information—In medieval times, it was thought that diseases caused blood to stagnate in certain parts of the body, and that letting it out revitalized the patient. Bloodletting was accomplished by barber-surgeons, who would cut open a vein at the site of the disorder. In addition, leeches were frequently applied to a patient's skin as another means of draining accumulated blood.

15.5 Principal Arteries of the Body

The aorta ascends from the left ventricle, arches to the left, and descends through the trunk. Branches of the aorta carry oxygenated blood to all the cells of the body.

Aortic Arch

The major systemic artery is the **aorta** (*a-or'tă*), from which all the principal arteries of the body arise (fig. 15.10). The systemic vessel that ascends from the left ventricle of the heart is called the **ascending aorta.** From it arises the **right** and **left coronary arteries,** which serve the myocardium of the heart. The aorta arches to the left and posteriorly over the pulmonary arteries as the **aortic arch.** Three vessels arise from the aortic arch: the *brachiocephalic trunk,* the *left common carotid artery,* and the *left subclavian artery,* as shown in figure 15.11.

The **brachiocephalic trunk** is the first vessel to branch from the aortic arch. As its name suggests, it supplies blood to the structures of the shoulder, arm, and head on the right side of the body. The brachiocephalic trunk splits at the junction of the sternum and the right clavicle into the **right common carotid artery,** which extends to the right side of the neck and head, and the **right subclavian artery,** which carries blood to the right shoulder and upper limb.

The remaining two branches from the aortic arch are the *left common carotid* and the *left subclavian arteries.* The **left common carotid artery** transports blood to the left side of the neck and head, and the **left subclavian artery** supplies the left shoulder and upper limb.

Arteries of the Neck and Head

The common carotid arteries course upward in the neck along both sides of the trachea (fig. 15.12). Each common carotid artery branches into the **internal** and **external carotid arteries** slightly below the angle of the mandible. At the base of the internal carotid artery is a slight dilation called the **carotid sinus.** The carotid sinus contains *baroreceptors,* which monitor blood pressure. Surrounding the carotid sinus are **carotid bodies,** which contain *chemoreceptors* that respond to chemical changes in the blood.

The brain is supplied with arterial blood that arrives through four vessels that eventually unite on the inferior surface of the brain in the area surrounding the pituitary gland (fig. 15.13). The four vessels are the paired *vertebral arteries* and the paired *internal carotid arteries.* The value of four separate vessels coming together at one location is that if one becomes blocked, the three alternate routes may still provide an adequate blood supply to the brain.

The **vertebral arteries** arise from the subclavian arteries at the base of the neck (fig. 15.13). They pass superiorly through the transverse foramina of the cervical vertebrae and enter the skull through the foramen magnum. Within the cranium, the two vertebral arteries unite to form the **basilar artery** at the level of the pons. The basilar artery ascends along the inferior surface of the brain stem and terminates by forming two **posterior cerebral arteries** that supply the posterior portion of the

carotid: Gk. *karotikos,* stupefying (stupor can be induced by finger pressure in the region of the carotid sinus)

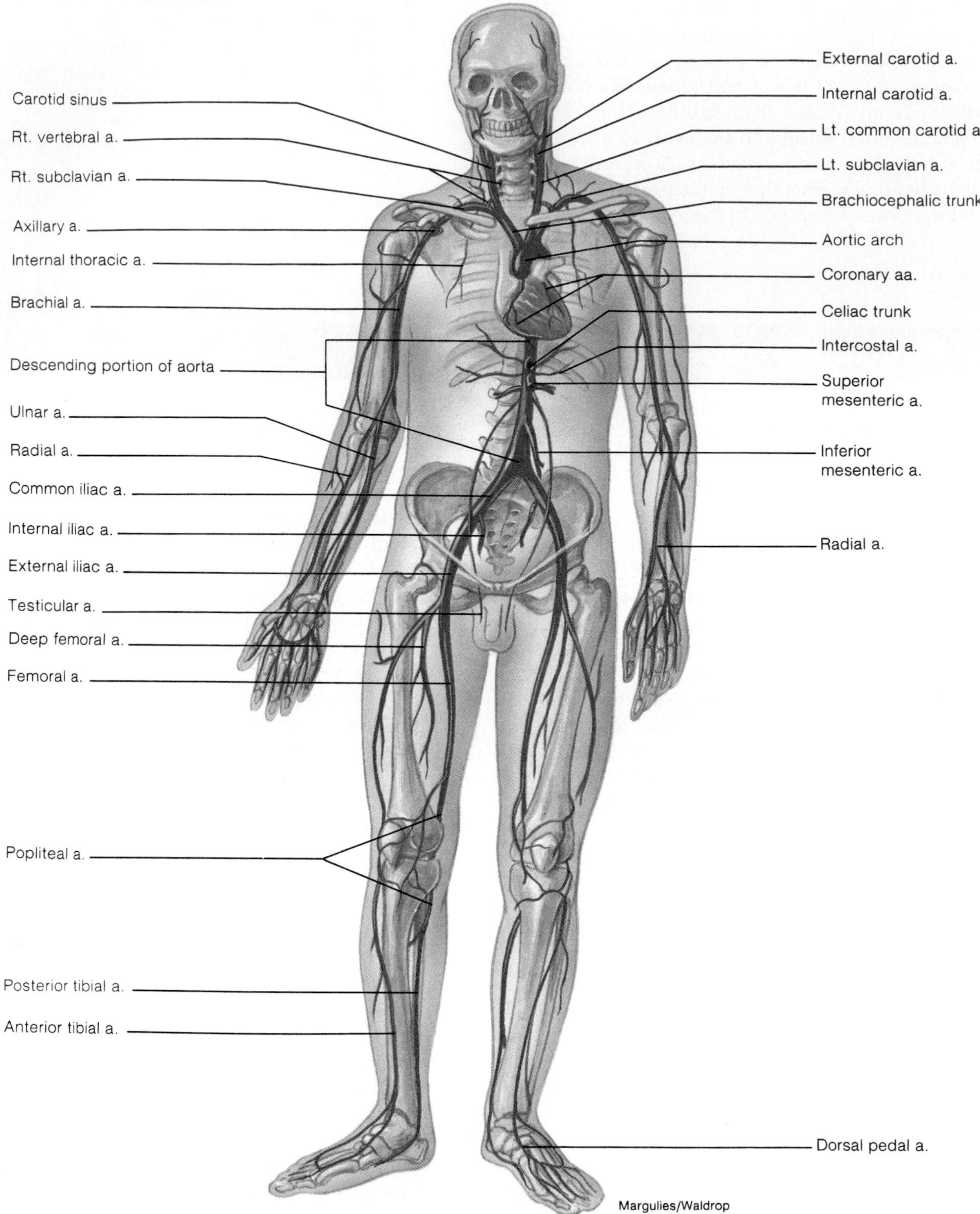

FIGURE 15.10
Principal arteries of the body (a. = artery; aa. = arteries).

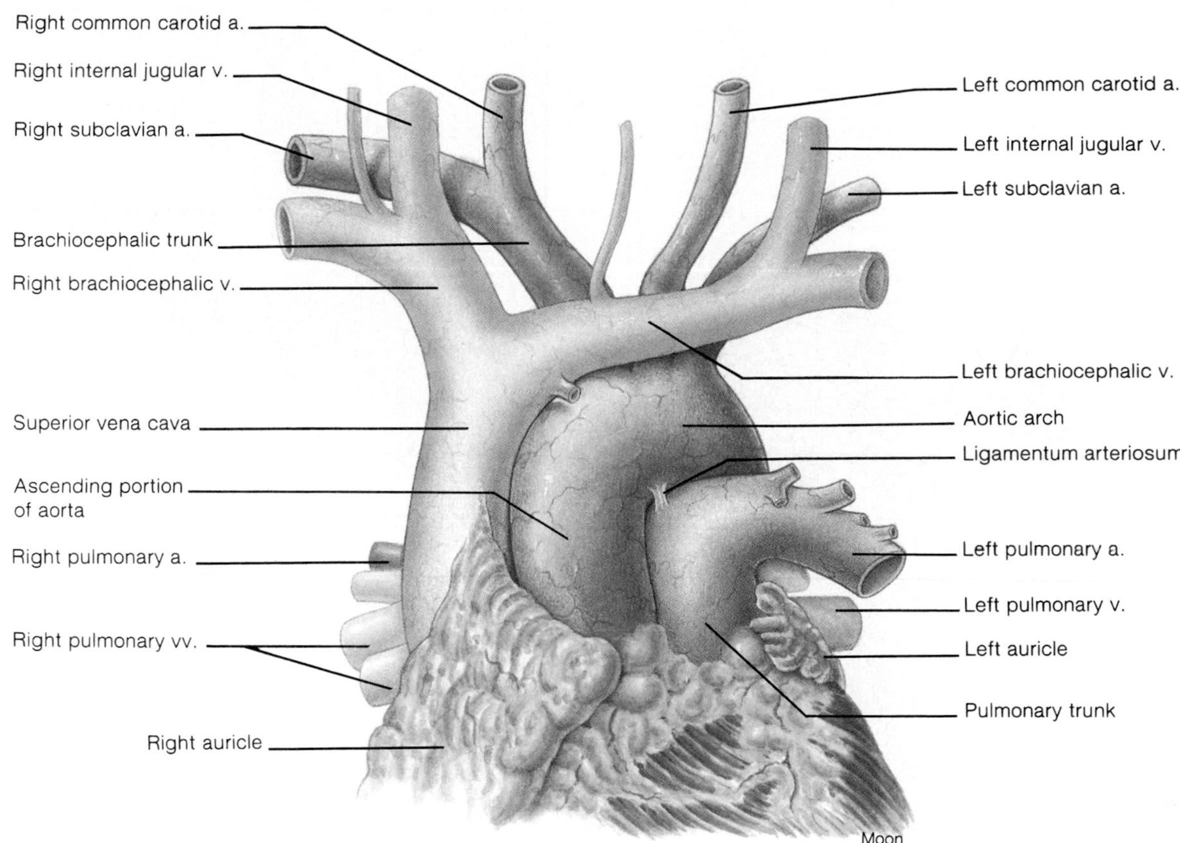

FIGURE 15.11
The structural relationship between the major arteries and veins to and from the heart (v. = vein; vv. = veins).

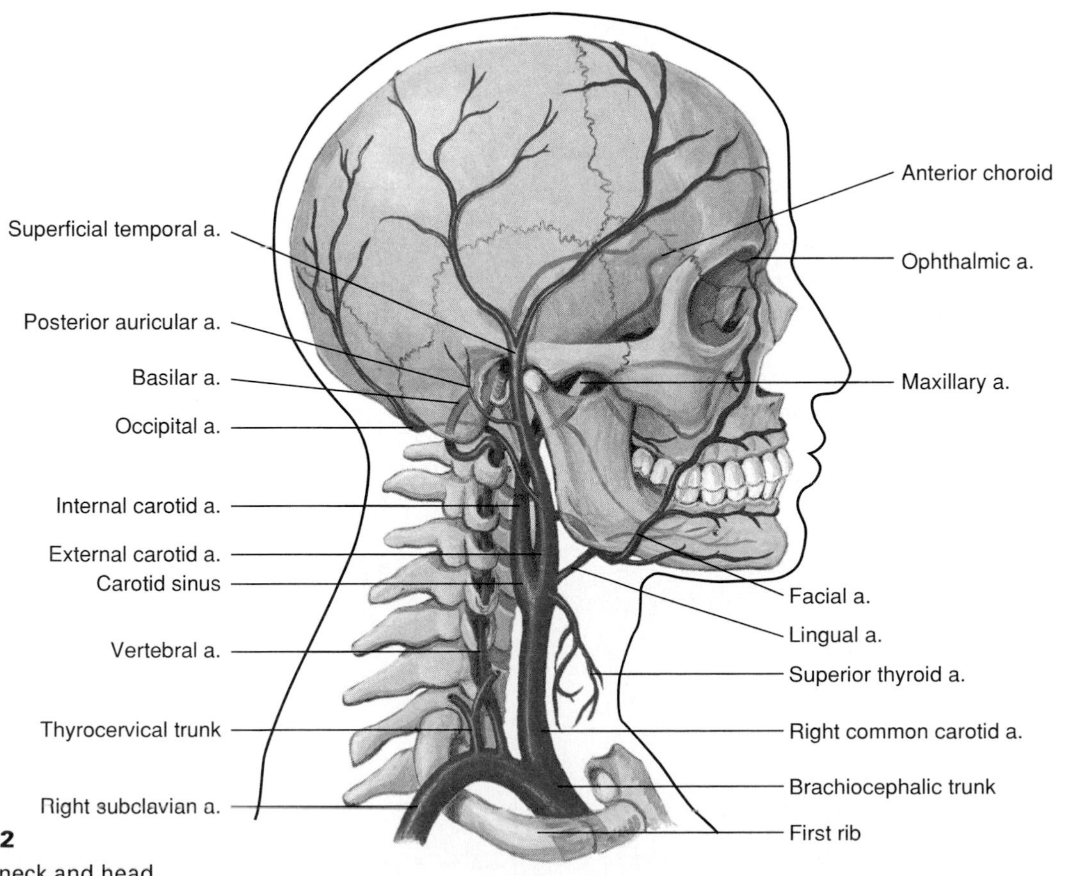

FIGURE 15.12
Arteries of the neck and head.

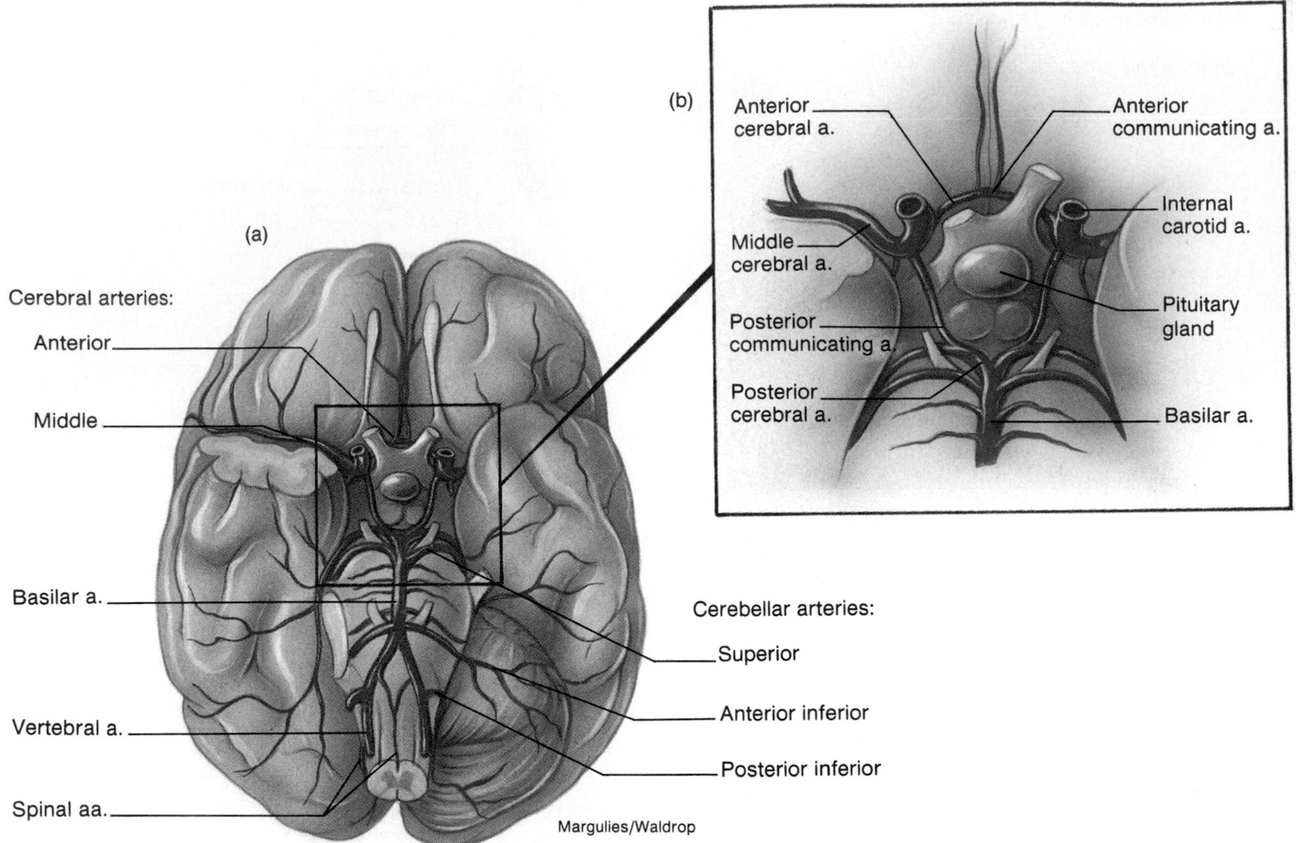

FIGURE 15.13

Arteries that supply blood to the brain. (*a*) An inferior view of the brain and (*b*) a close-up view of the region of the pituitary gland. The cerebroarterial circle (circle of Willis) consists of the arteries that encircle the pituitary gland.

cerebrum. The **posterior communicating arteries** are branches that arise from the posterior cerebral arteries and from part of the **cerebral arterial circle** (circle of Willis) surrounding the pituitary gland.

Each **internal carotid artery** branches from the common carotid artery and ascends in the neck to the base of the skull, where it enters the carotid canal of the temporal bone. Once it is on the inferior surface of the brain, several branches arise from the internal carotid artery. Three of the more important ones are the *ophthalmic artery,* which supplies the eye and associated structures, and the *anterior* and *middle cerebral arteries,* which provide blood to the cerebrum.

circle of Willis: from Thomas Willis, English physician, 1621–75

The paired **external carotid arteries** give off several branches as they extend upward along the side of the neck and head. These branches, which are named according to the areas or structures they serve, include the *superior thyroid, lingual, facial, occipital,* and *posterior auricular arteries.* Each external carotid artery terminates at a level near the mandibular condyle by dividing into **maxillary** and **superficial temporal arteries.**

Arteries of the Shoulder and Upper Limb

Each subclavian artery passes laterally deep to the clavicle, carrying blood toward the arm (figs. 15.10 and 15.14). From each subclavian arises a **vertebral artery** (already described); a short **thyrocervical trunk** (see fig. 15.12) that

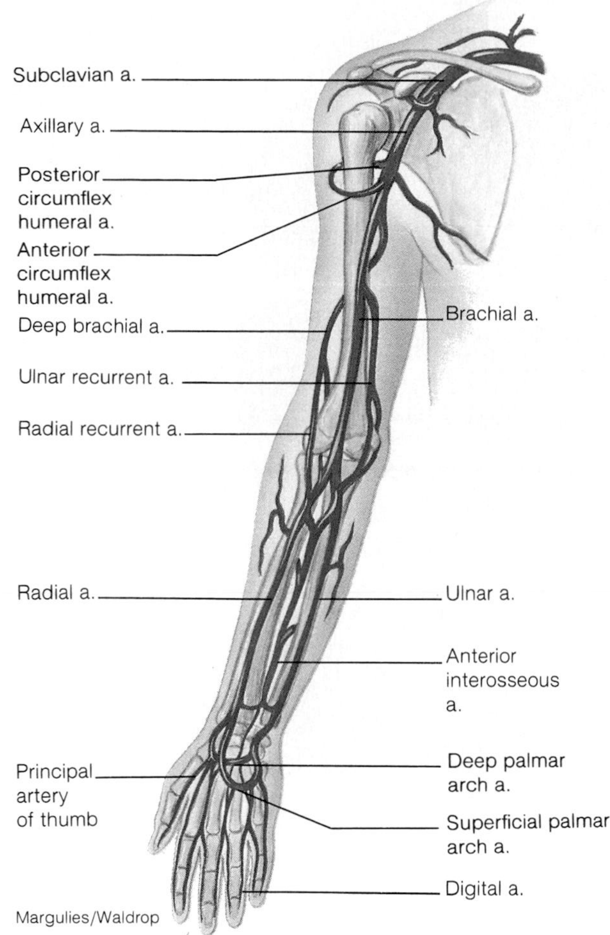

FIGURE 15.14
Arteries of the right shoulder and upper limb, shown in anterior view.

serves the thyroid gland, trachea, and larynx; and an **internal thoracic artery** that serves the thoracic wall, thymus, and pericardium.

The subclavian artery becomes the **axillary artery** as it passes into the axillary region. The **brachial artery** is the continuation of the axillary artery through the brachial region (fig. 15.14). The brachial artery is a major pressure point and the most common site for measuring blood pressure.

Just proximal to the cubital fossa, the brachial artery branches into the **radial** and **ulnar arteries,** which supply blood to the forearm and a portion of the hand and fingers. The radial artery is important as a site for recording the pulse near the wrist. At the wrist, the ulnar and radial arteries come together to form the **palmar arches** in the hand, from which **digital arteries** extend into the fingers.

Branches of the Aorta

The *thoracic aorta* is a continuation of the aortic arch as it descends through the thoracic cavity to the diaphragm. This large vessel gives off branches to the organs and muscles of the thoracic region. These branches include the **anterior intercostal arteries,** which supply blood to the intercostal muscles, and the **phrenic arteries,** which supply the diaphragm.

The *abdominal aorta* is the segment of the aorta between the diaphragm and the level of the fourth lumbar vertebra. Among the major arteries that arise from the abdominal aorta is the **celiac trunk.** This very large unpaired vessel arises anteriorly, just below the diaphragm. It divides immediately into three arteries: the **splenic artery,** going to the spleen; the **left gastric artery,** going to the stomach; and the **hepatic artery,** going to the liver (fig. 15.15). The **superior mesenteric artery** is another unpaired vessel that arises anteriorly, just below the celiac trunk. It supplies blood to the small intestine, the cecum, the appendix, the ascending colon, and the transverse colon.

The next major vessels to arise from the abdominal aorta are the paired **renal arteries,** which carry blood to the kidneys. The **gonadal arteries** are small paired vessels that arise from the abdominal aorta, just below the renal arteries. These are known as the **testicular arteries** in the male and the **ovarian arteries** in the female.

The **inferior mesenteric artery** is the last major branch of the abdominal aorta. It is an unpaired anterior vessel that arises just before the branching of the aorta to form the common iliac arteries. The inferior mesenteric artery supplies blood to the descending colon and rectum.

Arteries of the Pelvis and Lower Limb

The abdominal aorta terminates in the lower pelvic area as it splits into the **right** and **left common iliac arteries.** These vessels pass downward approximately 5 cm on their respective sides and terminate as they divide into the *internal* and *external iliac arteries.*

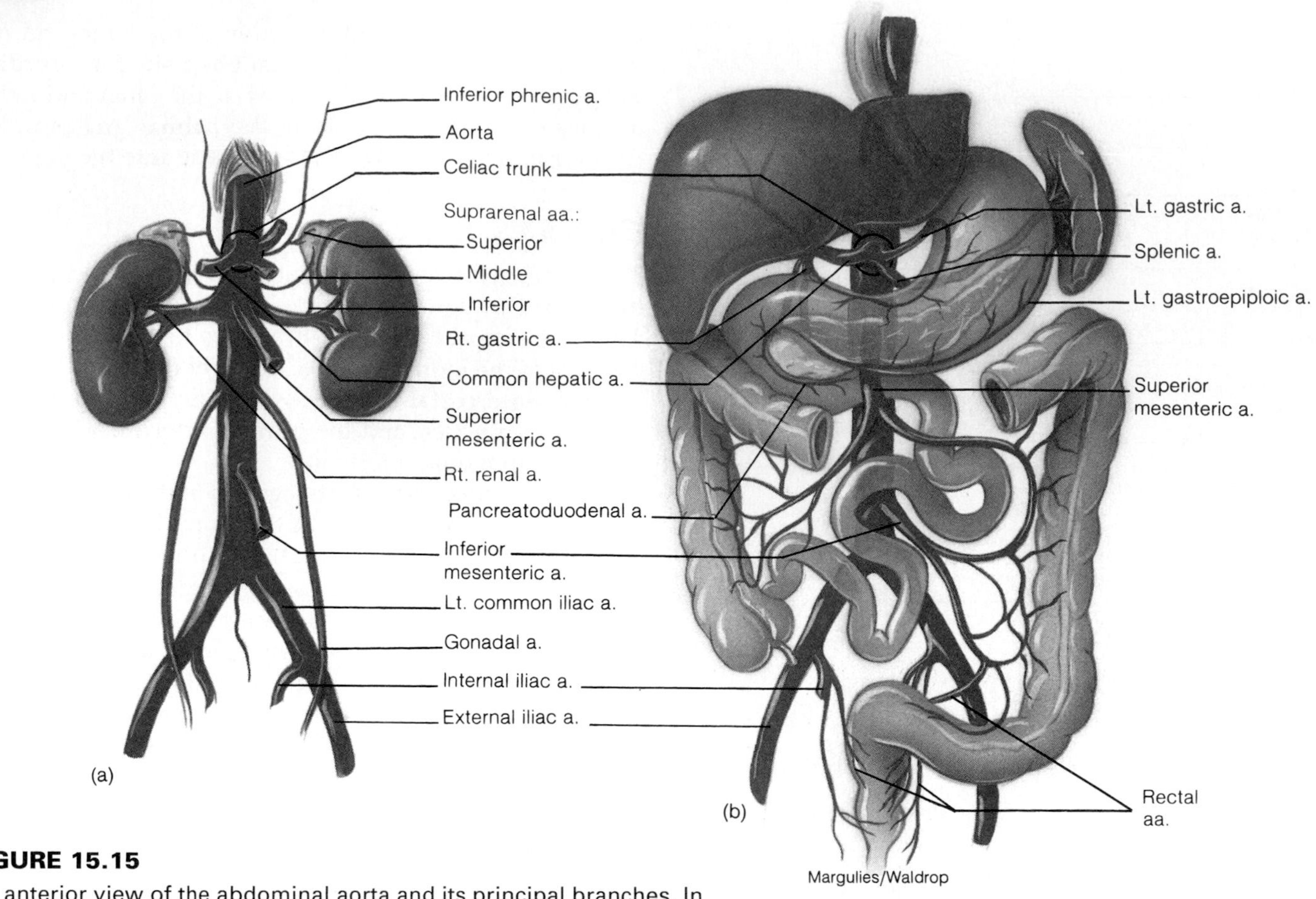

FIGURE 15.15
An anterior view of the abdominal aorta and its principal branches. In (*a*) the abdominal viscera have been removed, and in (*b*) they are intact.

The **internal iliac artery** has extensive branches to supply arterial blood to the gluteal muscles and the organs of the pelvic region (fig. 15.16). The internal visceral organs of the pelvis are served by the **middle rectal artery** and the **vesicular arteries** to the urinary bladder. In addition, **uterine** and **vaginal arteries** branch from the internal iliacs to serve the reproductive organs in females. The muscles of the buttock are served by the **superior** and **inferior gluteal arteries.** Some of the upper medial thigh muscles are supplied with blood from the **obturator artery.** The **internal pudendal artery** of the internal iliac serves the external genitalia of males and females, delivering blood during sexual stimulation.

Inside Information—Erection of the penis in the male and the corresponding swelling of erectile tissues in the female are vascular events controlled by the autonomic nervous system.

The **external iliac artery** passes out of the pelvic cavity beneath the inguinal ligament and becomes the **femoral artery.** The femoral artery passes through an area called the *femoral triangle* on the upper medial portion of the thigh, at which point it is close to the surface. Here, it can be palpated and serves as an important pressure point. Several vessels arise from the femoral to serve the thigh region. The largest of these, the **deep femoral artery,** passes posteriorly to serve the hamstring muscles. The **lateral** and **medial femoral circumflex arteries** encircle the proximal end of the femur and serve muscles in this region. The femoral artery becomes the **popliteal artery** as it passes across the posterior aspect of the knee (fig. 15.16).

The popliteal artery supplies small branches to the knee joint and then divides into an **anterior tibial artery** and **posterior tibial artery.** These vessels traverse the anterior and posterior aspects of the leg, respectively, providing blood to the muscles of these regions and to the foot.

At the ankle, the anterior tibial artery becomes the **dorsal pedal artery,** which serves the ankle and top of the foot. It then contributes to the formation of the **plantar arch** of the foot.

The posterior tibial artery sends a large **peroneal artery** to serve the peroneal muscles of the leg. At the ankle, the posterior tibial splits into the **lateral** and **medial plantar arteries** that supply the muscles and structures on the sole of the foot. The lateral plantar artery joins the dorsal pedal artery to form the **plantar arch.** This arterial arrangement is similar to that in the hand. **Digital arteries** arise from the plantar arch to supply the toes with blood.

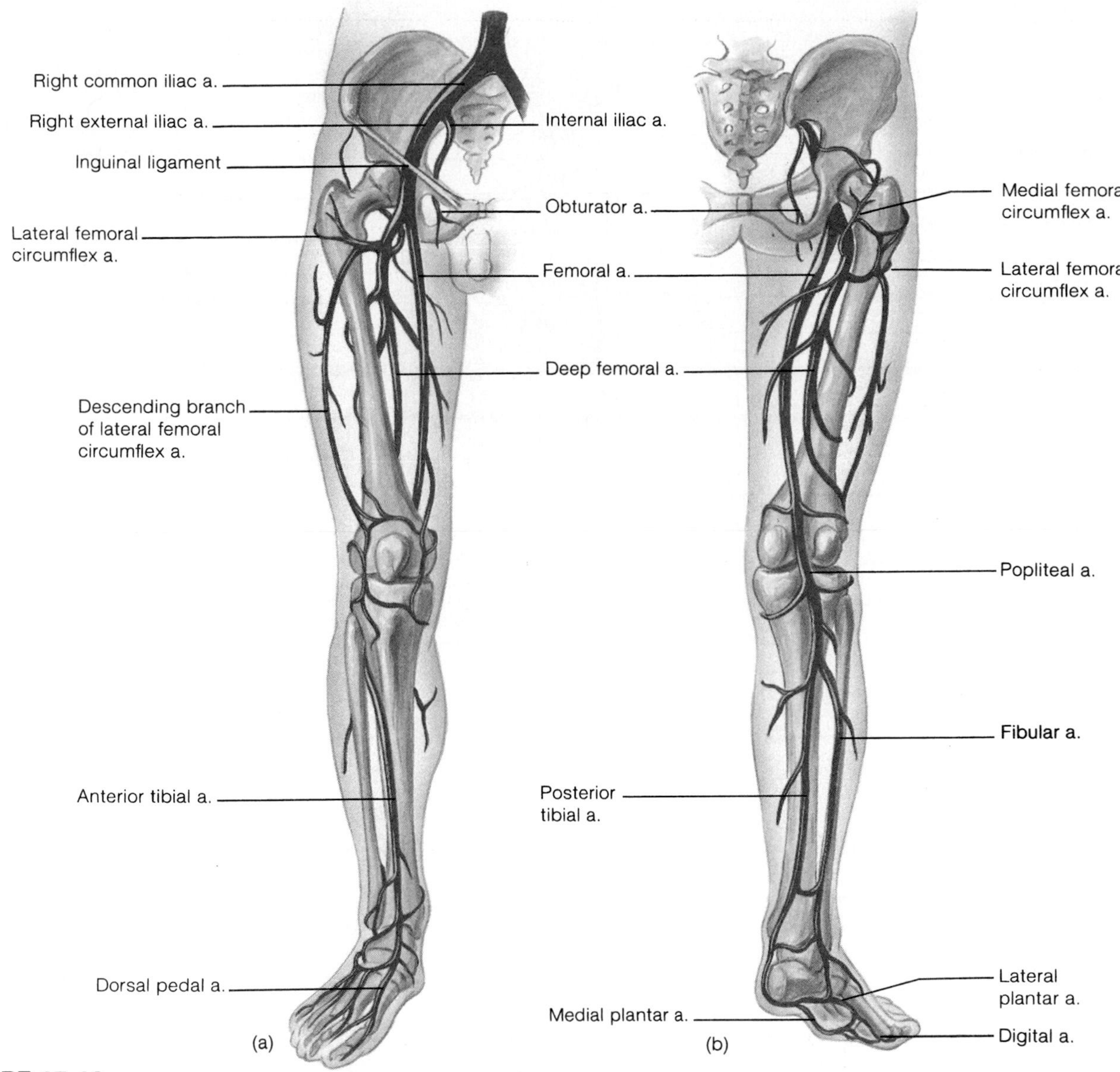

FIGURE 15.16
Arteries of the right hip and lower limb. (*a*) An anterior view and (*b*) a posterior view.

15.6 Principal Veins of the Body

After systemic blood has passed through the tissues and its oxygen is depleted, it returns through veins of progressively larger diameters to the right atrium of the heart.

In the venous portion of the circulation, blood flows from the smaller vessels into larger ones, so that a vein receives blood from smaller tributaries instead of giving off branches as an artery does. The veins from all parts of the body converge into two major vessels that empty into the right atrium: the **superior vena cava** (*ve'nă ka'vă*) and **inferior vena cava** (fig. 15.17). Veins are more numerous than arteries and are both superficial and deep. Superficial veins generally can be seen just beneath the skin and are clinically important in drawing blood and giving injections. Deep veins are close to the principal arteries and are usually similarly named. As with the arteries, veins are named according to the region in which they are found or the organ that they serve. (Note that when a vein serves an organ, it drains blood *away* from it.)

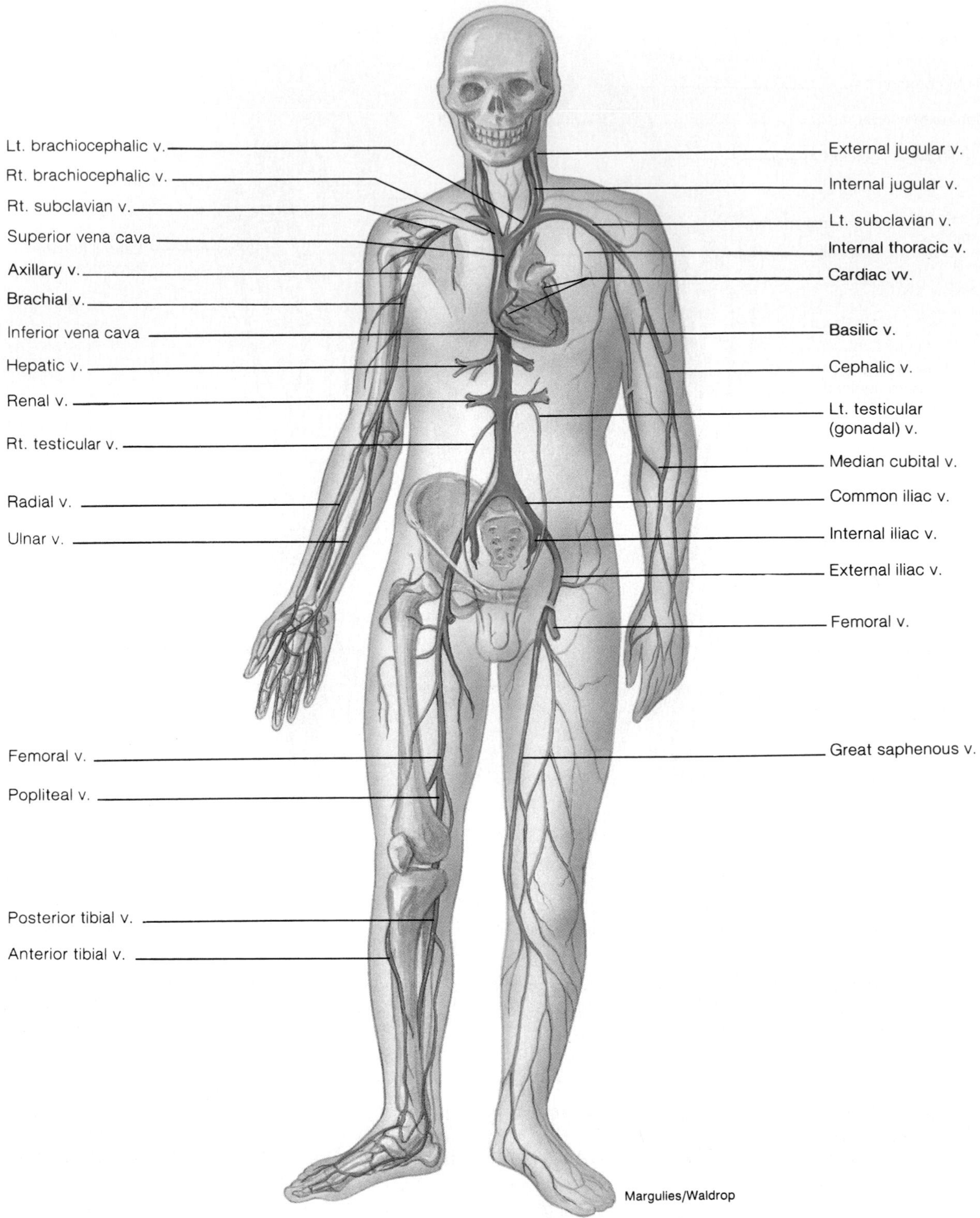

FIGURE 15.17
Principal veins of the body. Superficial veins are depicted in the left limbs, and deep veins in the right limbs.

Veins of the Head and Neck

Blood from the scalp, portions of the face, and the superficial neck regions is drained by the **external jugular veins** (fig. 15.18) that empty into the **right** and **left subclavian veins,** located just behind the clavicles. The paired **internal jugular veins** drain blood from the brain, meninges, and deep regions of the face and neck. Each internal jugular vein passes inferiorly down the neck adjacent to the common carotid artery and the vagus nerve. All three of these structures are positioned behind the sternocleidomastoid muscle and are surrounded by a protective membrane called the *carotid sheath.* The internal jugular empties into the **subclavian vein,** and the union of these two vessels forms the large **brachiocephalic vein** on each side. The two brachiocephalic veins then merge to form the superior vena cava, which empties into the right atrium of the heart (see fig. 15.11).

Veins of the Upper Limb

The upper limb has both deep and superficial venous drainage (fig. 15.19). The deep veins accompany the arteries of the same region and are similarly named. Both the **radial vein** on the lateral side of the forearm and the **ulnar vein** on the medial side drain blood from the **palmar arch** of the hand. The radial and ulnar veins join in the cubital fossa to form the **brachial vein,** which continues up the medial side of the brachium.

The main superficial vessels of the upper limb are the **basilic vein** and the **cephalic vein.** The basilic vein passes on the ulnar side of the forearm and the medial side of the arm. Near the head of the humerus, the basilic vein merges with the brachial vein to form the **axillary vein.**

The cephalic vein drains the superficial portion of the hand and forearm on the radial side, and then continues up the lateral side of the arm. In the shoulder region, the cephalic vein pierces the fascia and joins the axillary vein. The axillary vein then passes the first rib to form the subclavian vein, which unites with the external jugular to form the brachiocephalic vein of that side.

Superficially, in the cubital fossa of the elbow, the **median cubital vein** ascends from the cephalic vein on the lateral side to connect with the basilic vein on the medial side. The median cubital vein is a frequent site for venipuncture in order to remove a sample of blood or to add fluids to the blood.

Veins of the Thorax

The superior vena cava, formed by the union of the two brachiocephalic veins, drains venous blood from the head, neck, and upper limbs directly into the right atrium of the heart. These large vessels lack the valves that are characteristic of most other veins in the body.

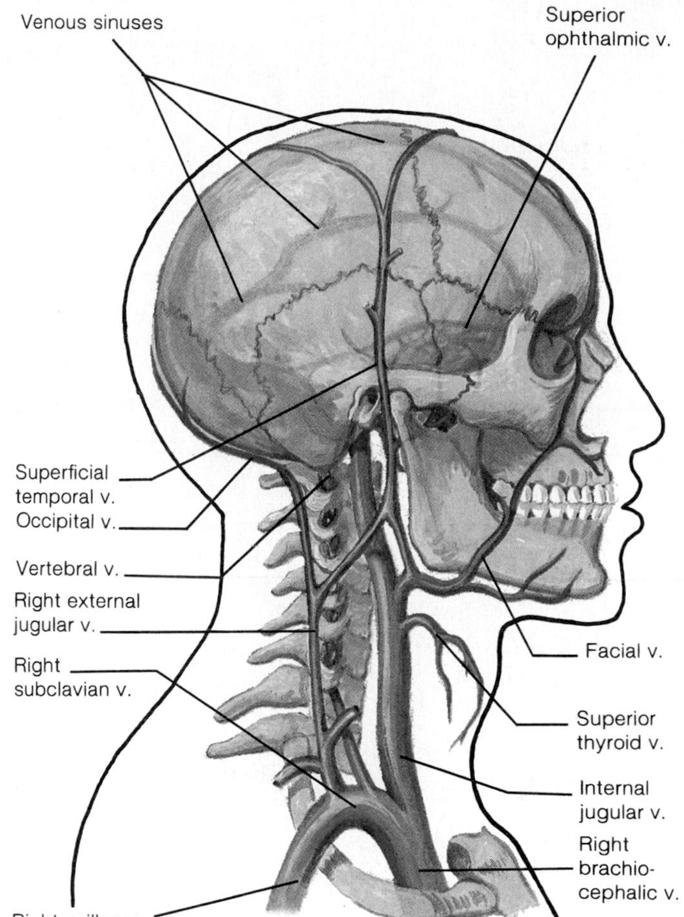

FIGURE 15.18
Veins that drain the head and neck.

In addition to receiving blood from the brachiocephalic veins, the superior vena cava collects blood from the *azygos (az'ĭ-gos) system of veins* arising from the posterior thoracic wall (fig. 15.20). The **azygos vein** extends superiorly along the posterior abdominal and thoracic walls on the right side of the vertebral column.

Veins of the Lower Limb

The lower limbs, like the upper limbs, have both a deep and a superficial group of veins (fig. 15.21). The deep veins accompany corresponding arteries and have more valves than do the superficial veins. The deep veins will be described first.

jugular: L. *jugulum,* throat or neck
azygos: Gk. *a,* without; *zygon,* yoke

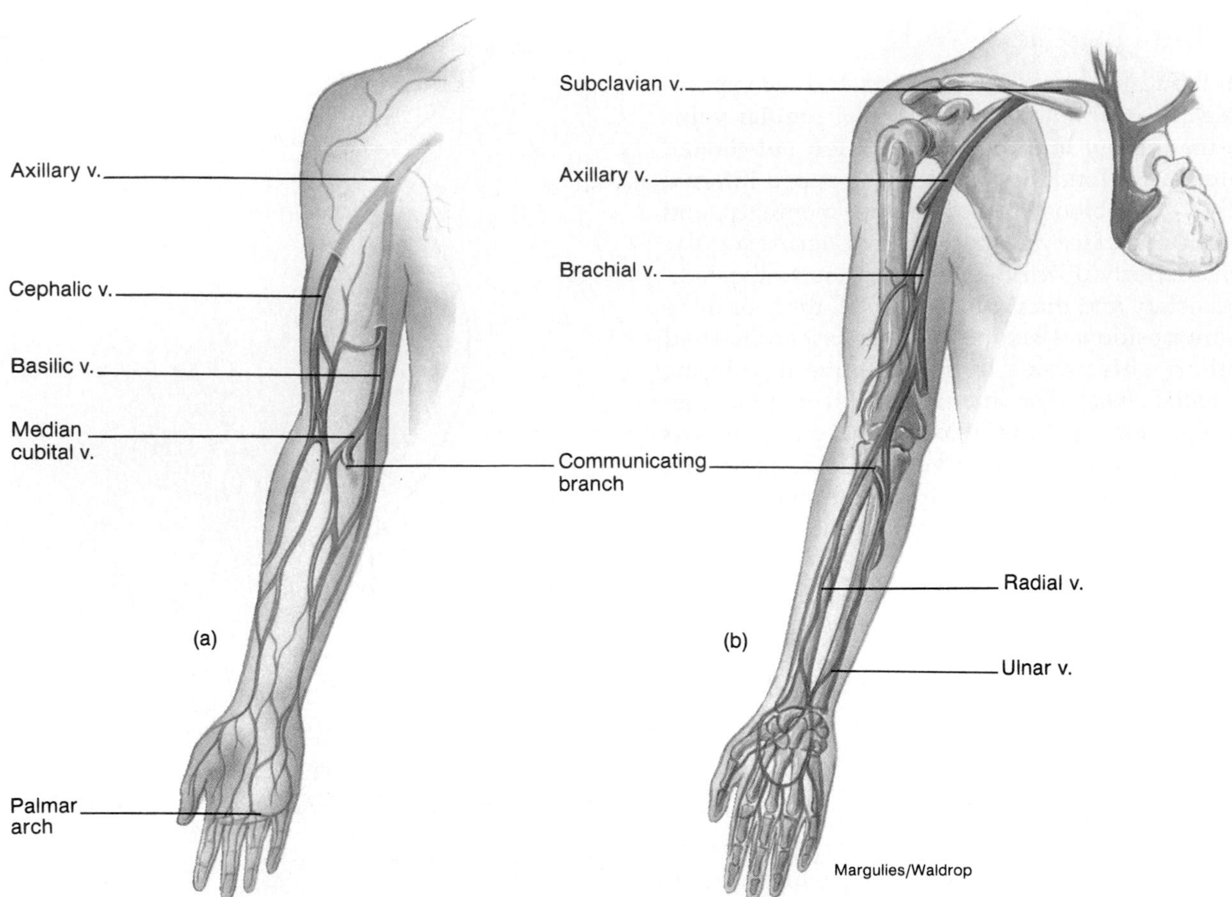

FIGURE 15.19
An anterior view of the veins that drain the upper right shoulder and limb. (*a*) Superficial veins and (*b*) deep veins.

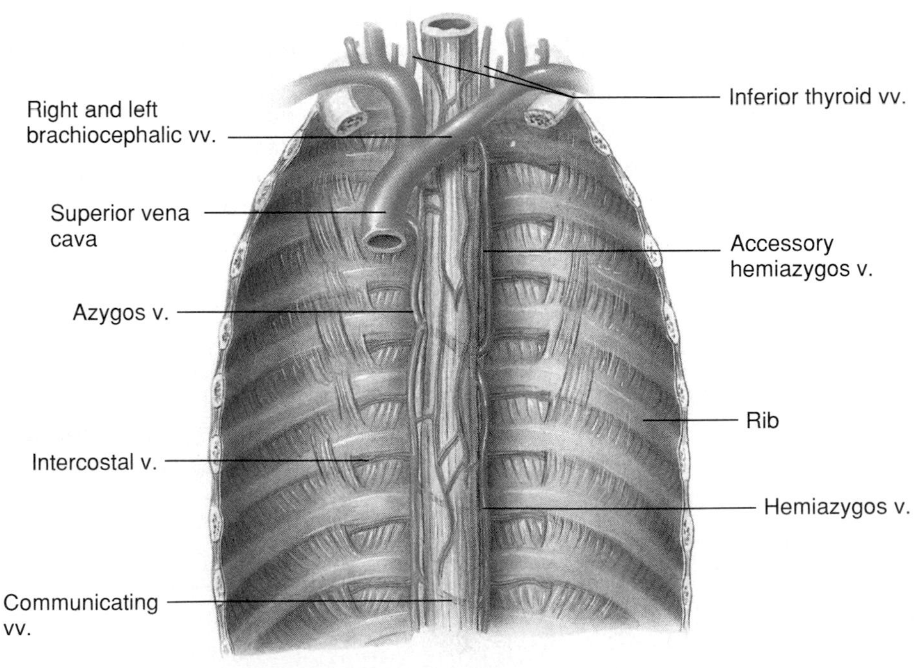

FIGURE 15.20
Veins of the thoracic region. (The lungs and heart have been removed.)

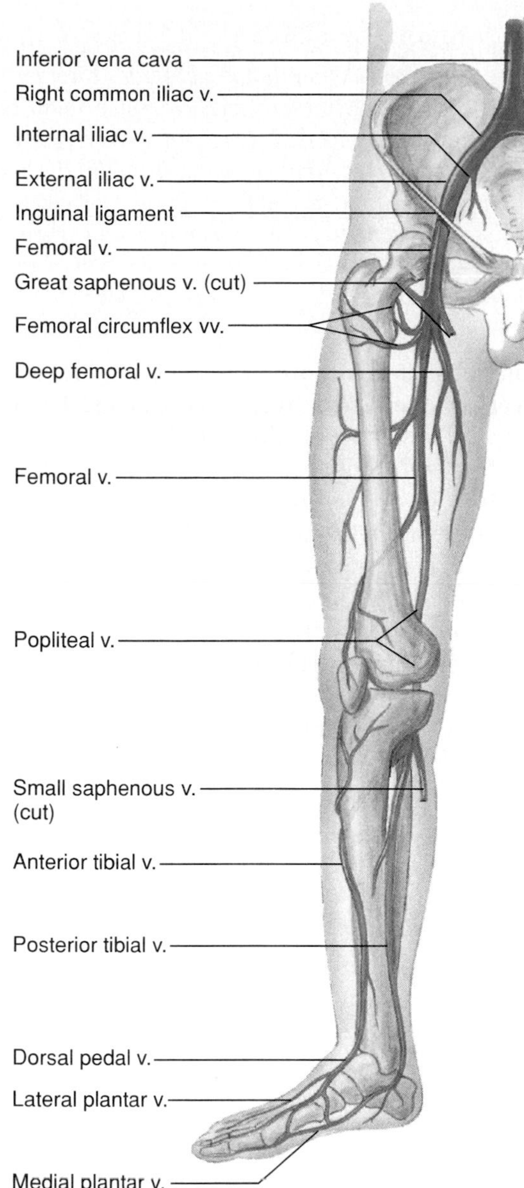

FIGURE 15.21
Veins of the right hip and lower limb.

The **posterior** and **anterior tibial veins** originate in the foot and course upward behind and in front of the tibia to the back of the knee, where they merge to form the **popliteal vein.** The popliteal vein receives blood from the knee region. Just above the knee, this vessel becomes the **femoral vein.** The femoral vein in turn continues up the thigh and receives blood from the **deep femoral vein** near the groin. Just above this point, the femoral vein receives blood from the **great saphenous** (*să-fe'nus*) **vein** and then becomes the **external iliac vein** as it passes under the *inguinal ligament.* The external iliac curves upward to the level of the sacroiliac joint, where it merges with the **internal iliac vein.** At the level of the fifth lumbar vertebra, the right and left common iliac veins unite to form the large inferior vena cava.

The superficial veins of the lower extremity are the **small** and **great saphenous veins.** The small saphenous vein arises from the lateral side of the foot, travels posteriorly along the surface of the calf of the leg, and descends deep to enter the popliteal vein behind the knee. The great saphenous vein originates at the arch of the foot and ascends superiorly along the medial aspect of the leg and thigh before emptying into the femoral vein.

saphenous: L. *saphena,* the hidden one

Inside Information—Extending from the arch of the foot and ascending to the medial aspect of the thigh, the great saphenous vein is the longest vessel of the body.

Veins of the Abdominal Region

The **inferior vena cava**—the largest vessel in the body in diameter—is formed by the union of the two common iliac veins that drain the lower extremities. It parallels the abdominal aorta on the right side as it ascends through the abdominal cavity to penetrate the diaphragm and enter the right atrium. Several veins empty into the inferior vena cava (see fig. 15.17); they correspond in name and position to arteries previously described.

Four paired **lumbar veins** (not shown) drain the posterior abdominal wall and spine. The **renal veins** drain blood from the kidneys and ureters into the inferior vena cava. The **right testicular vein** in males (or the **right ovarian vein** in females) drains the corresponding gonads, and the **right suprarenal vein** drains the right adrenal gland. These veins empty into the inferior vena cava. The **left testicular vein** (or **left ovarian vein**) and the **left suprarenal vein,** by contrast, empty into the left renal vein. The **inferior phrenic veins** receive blood from the inferior side of the diaphragm and empty into the inferior vena cava. **Right** and **left hepatic veins** originate from the capillary sinusoids of the liver and empty into the inferior vena cava immediately below the diaphragm.

Note that the inferior vena cava does not receive blood directly from the GI tract, pancreas, or spleen. Instead, the venous blood from these organs first passes through capillaries in the liver.

Hepatic Portal System

A *portal system* is a pattern of circulation in which the vessels that drain one group of capillaries deliver blood to a second group of capillaries, which in turn are drained by more usual systemic veins that carry blood to the right atrium of the heart. There are thus two capillary beds in series. The hepatic portal system is composed of veins that drain blood from capillaries in the intestines, pancreas, spleen, stomach, and gallbladder into capillaries in the liver (called *sinusoids*) and of the **right** and **left hepatic veins** that drain the liver and empty into the inferior vena cava (fig. 15.22). As a consequence of the hepatic portal system, the absorbed products of digestion must first pass through hepatic sinusoids in the liver before entering the general circulation.

The hepatic portal vein is the large vessel that receives blood from the digestive organs. It is formed by a union of the **superior mesenteric vein,** which drains nutrient-rich blood from the small intestine, and the **splenic vein.** The splenic vein drains the spleen but is enlarged because of a convergence of the following three tributaries: (1) the **inferior mesenteric vein** from the large intestine, (2) the **pancreatic vein** from the pancreas, and (3) the **left gastroepiploic vein** from the stomach. The right gastroepiploic vein, also from the stomach, drains directly into the superior mesenteric vein.

Three additional veins empty into the portal vein. The **right** and **left gastric veins** drain the lesser curvature of the stomach, and the **cystic vein** drains blood from the gallbladder.

One of the functions of the liver is to detoxify harmful substances, such as alcohol, that are absorbed into the blood from the small intestine. Since excessive quantities of alcohol cannot be processed in a single pass through the liver, a person becomes intoxicated. Eventually, the liver is able to process the alcohol as the circulating blood is repeatedly exposed to the liver sinusoids via the hepatic artery. Long-term alcohol abuse may result in cirrhosis *of the liver, as normal liver tissue is destroyed.*

In summary, it is important to note that the sinusoids of the liver receive blood from two sources. The hepatic artery supplies oxygen-rich blood to the liver, whereas the hepatic portal vein transports nutrient-rich blood from the small intestine for processing. These two blood sources become mixed in the liver sinusoids. Liver cells exposed to this blood obtain their nourishment from it and are uniquely suited (because of their anatomical position and enzymatic ability) to modify the chemical nature of venous blood that enters the general circulation from the GI tract.

gastroepiploic: Gk. *gastros,* stomach; *epiplein,* to float on (referring to greater omentum)

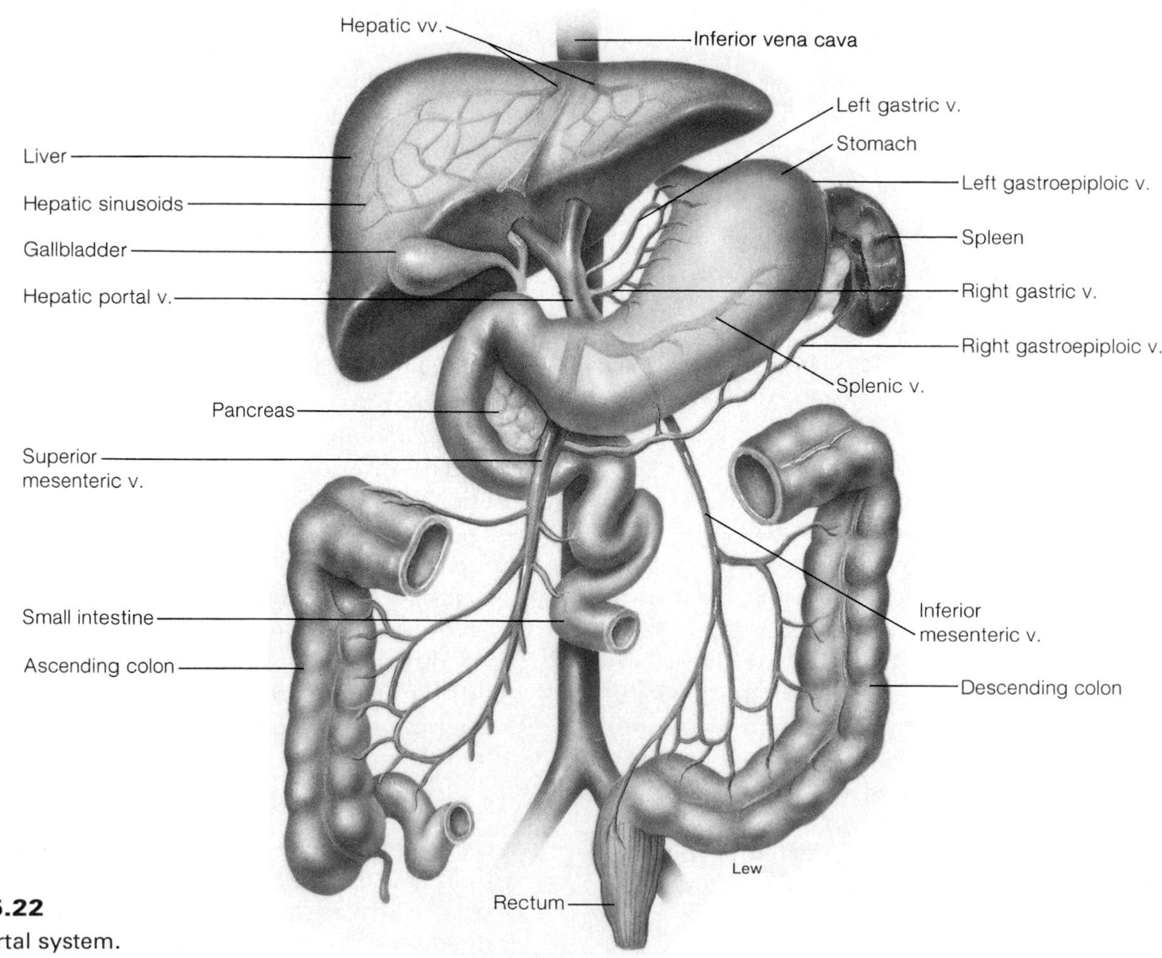

FIGURE 15.22
The hepatic portal system.

15.7 Fetal Circulation

All of the respiratory, excretory, and nutritional needs of the fetus are provided for by diffusion across the placenta instead of by the fetal lungs, kidneys, and gastrointestinal tract. Fetal circulation is adaptive to these conditions.

The circulation of blood through a fetus is by necessity different from blood circulation in a newborn (fig. 15.23). Respiration, the procurement of nutrients, and the elimination of metabolic wastes occur through the maternal blood instead of through the organs of the fetus. The capillary exchange between the maternal and fetal circulation occurs within the **placenta** (*plă-sen'tă*). This remarkable structure, which includes maternal and fetal capillary beds, is discharged following delivery as the afterbirth.

The **umbilical cord** is the connection between the placenta and the fetal umbilicus. It includes one **umbilical vein** and two **umbilical arteries,** surrounded by a gelatinous substance. Oxygenated and nutrient-rich blood flows through the umbilical vein toward the inferior surface of the fetal liver. At this point, the umbilical vein splits into two branches. One branch merges with the hepatic portal vein, and the other branch, called the **ductus venosus,** enters the

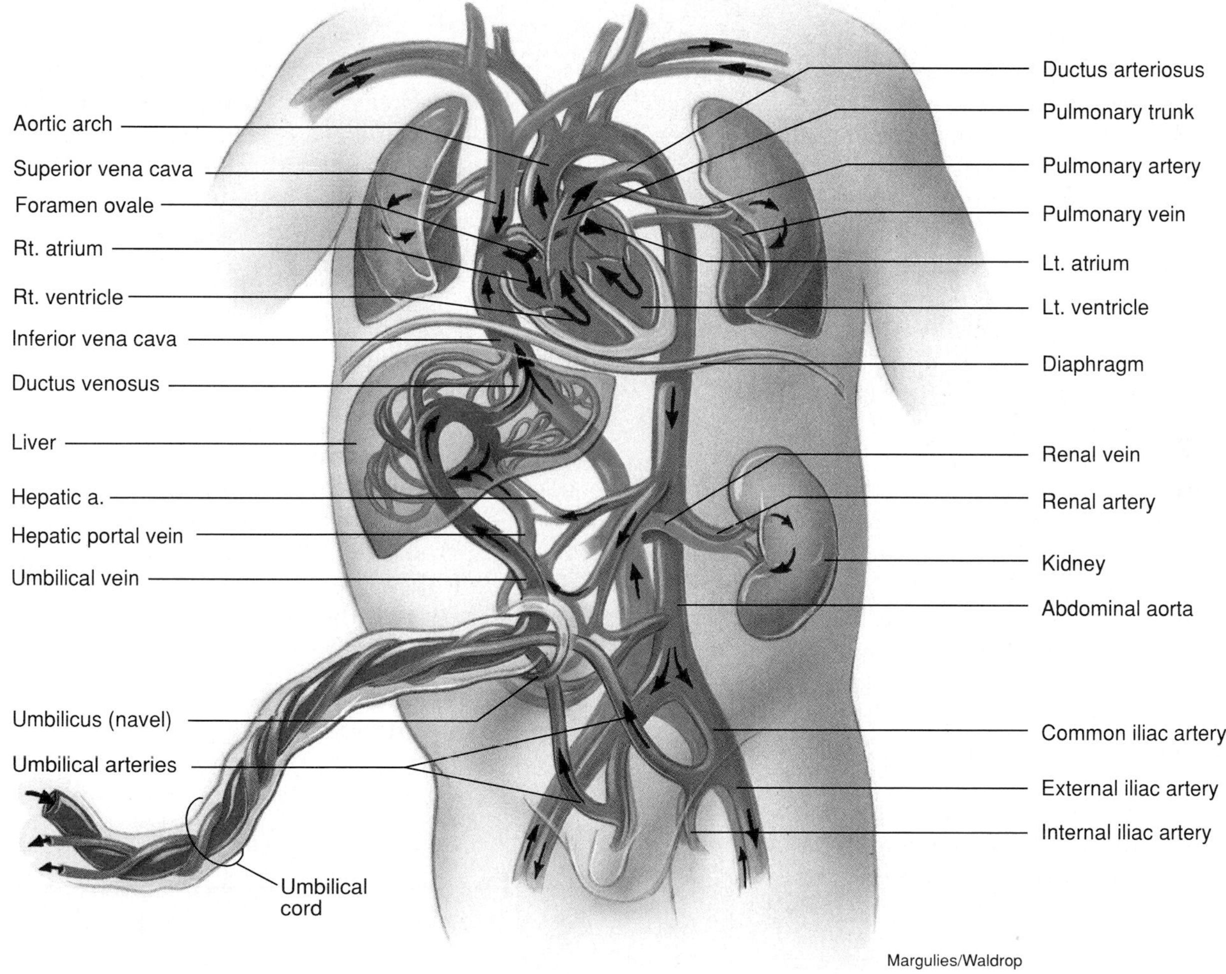

FIGURE 15.23
Fetal circulation. (Arrows indicate the direction of blood flow.)

TABLE 15.3 **Cardiovascular Structures of the Fetus and Changes in the Newborn**

Structure	Description	Function	Newborn Transformation
Umbilical vein	Connects the placenta to the liver; forms a major portion of the umbilical cord	Transports nutrient-rich oxygenated blood from the placenta to the fetus	Forms the round ligament of the liver
Ductus venosus	Venous shunt within the liver to connect with the inferior vena cava	Transports oxygenated blood directly into the inferior vena cava	Forms the ligamentum venosum, a fibrous cord in the liver
Foramen ovale	Opening between the right and left atria	Acts as a shunt to bypass the pulmonary circulation	Closes at birth and becomes the fossa ovalis, a depression in the interatrial septum
Ductus arteriosus	Connects the pulmonary trunk and aortic arch	Acts as a shunt to bypass the pulmonary circulation	Closes shortly after birth, atrophies, and becomes the ligamentum arteriosum
Umbilical arteries	Arise from the internal iliac arteries and enter the umbilical cord	Transport blood from the fetus to the placenta	Atrophies to become the lateral umbilical ligaments

inferior vena cava. Thus, oxygenated blood is mixed with venous blood returning from the lower limbs of the fetus before it enters the heart. The umbilical vein is the only vessel of the fetus that carries fully oxygenated blood.

The inferior vena cava empties into the right atrium of the fetal heart. Most of the blood passes from the right atrium into the left atrium through the **foramen ovale** (*o-val'e*), an opening between the two atria. Here, it mixes with a small quantity of blood returning through the pulmonary circulation. The blood then passes into the left ventricle, from which it is pumped into the aorta and through the body of the fetus. A small amount of blood entering the right atrium passes into the right ventricle and out of the heart via the pulmonary trunk. Since the lungs of the fetus are not functional, only a small portion of the blood continues through the pulmonary circulation (the resistance to blood flow is very high in the collapsed fetal lungs). Most of the blood in the pulmonary trunk passes through the **ductus arteriosus** into the aortic arch, where it mixes with blood coming from the left ventricle. Blood is eventually returned to the placenta by the two **umbilical arteries** that arise from the internal iliac arteries.

Note that, in the fetus, oxygen-rich blood is transported by the inferior vena cava to the heart and via the foramen ovale and ductus arteriosus to the systemic circulation.

Important changes occur in the circulatory system at birth. The foramen ovale, ductus arteriosus, ductus venosus, and the umbilical vessels are no longer necessary. The foramen ovale abruptly closes with the first breath of air because the reduced pressure in the right side of the heart causes a flap to cover the opening. This reduction in pressure when the lungs fill with air occurs because the vascular resistance to blood flow in the pulmonary circulation falls far below that of the systemic circulation. The pressure in the inferior vena cava and right atrium falls as a result of the loss of the placental circulation.

The constriction of the ductus arteriosus occurs gradually over a period of about 6 weeks after birth as the vascular smooth muscle fibers constrict in response to the higher oxygen concentration in the postnatal blood. The remaining structure of the ductus arteriosus gradually atrophies and becomes nonfunctional as a blood vessel. Transformation of the unique fetal circulatory system is summarized in table 15.3.

Inside Information—Although atherosclerosis is the bane of prosperous modern societies, it is not a uniquely modern phenomenon. Even in ancient Egyptian mummies, signs of this vascular disorder have been found.

SYNOPTIC SPOTLIGHT

Maintaining a Healthy Heart: Does Cholesterol Really Matter?

It's almost lunchtime, and you're yearning for a juicy hamburger, topped with cheese and bacon. You know your cholesterol is high, and you've been warned to watch your diet, but the temptation to indulge is strong. It occurs to you that now might be the time to find out a bit more about cholesterol and the role it plays in your body chemistry. Then, perhaps, it might be easier to avoid the foods that are bad for you.

Cholesterol is an odorless, waxy substance that is part of all animal cells, including those of humans. It is both manufactured by your body and introduced through your diet in such foods as egg yolks, dairy products, meats, fish, and poultry. In recent years it has come under heavy attack as a major cause of heart disease, yet it is absolutely essential to your health. For example, cholesterol is a key substance in each cell membrane and a component of many of the hormones produced in the body. It is necessary for brain and nerve development in infants, and it is also the starting material from which the liver produces bile acids, which are required for the digestion of fats. Unfortunately, however, not all of the cholesterol in the bloodstream is processed in beneficial ways. When there is too much of it, it may eventually be deposited in the walls of arteries and contribute to the formation of plaque. As a result, arteries become narrower and narrower, much like what happens when a water pipe becomes clogged with deposits of sludge.

Since cholesterol is a lipid, it is not soluble in water (or blood); in order to travel through the bloodstream, it must be wrapped in protein. The new molecules that are produced by this process are called **lipoproteins.** In general, the higher the percentage of lipid in the lipoprotein, the lower its density; the higher the percentage of protein, the greater its density. Thus, the lipoproteins of the blood include the **VLDLs** (very low-density lipoproteins), **LDLs** (low-density lipoproteins), and **HDLs** (high-density lipoproteins).

VLDLs carry some cholesterol but mainly triglycerides, which are produced in the liver from excess calories eaten. When the VLDLs travel through the bloodstream, the triglycerides are removed to be used as energy or stored as fat. During this process, the VLDLs are gradually converted to LDLs, which carry most of the blood cholesterol. It is the LDLs that appear to be responsible for depositing unused cholesterol in the artery walls; hence they are sometimes referred to as the "bad guys."

The third group of lipoproteins, the HDLs, transport cholesterol from the cells back to the liver to be converted into bile, much of which is eventually excreted in the feces. Since HDLs may also carry cholesterol away from cells in the artery walls, they may reduce plaque buildup, which is why they are sometimes called the "good guys."

The relationship between your total plasma cholesterol and your HDLs is one indicator of your risk for cardiovascular disease. This relationship is referred to as your *ratio.* The average American male has a total cholesterol (TC) to HDL ratio of 4.5. In American women, the average ratio is 4.0. This accounts in part for the fact that, in this country at least, women have a lower rate of heart disease than do men. According to a number of large studies, total cholesterol levels below 190 mg per 100 ml of blood (190 mg/dl), with a TC/HDL ratio below 4, seem to indicate a relatively low risk for cardiovascular disease. But these numbers are only one aspect of your risk. Other factors that must be considered are personal medical history, a family history of heart disease, and various lifestyle factors such as cigarette smoking.

The relative amounts of saturated and unsaturated fats in your diet also have an important effect on cholesterol levels. In general, food products from land animals contain saturated fats, whereas those from aquatic animals contain unsaturated fats. Saturated fats stimulate the synthesis of cholesterol by the liver, while inhibiting its excretion from the body; hence, saturated fats raise plasma cholesterol levels. By contrast, unsaturated fats enhance the excretion of cholesterol and its catabolism to bile salts, thereby helping to reduce plasma cholesterol levels.

Besides dietary adjustments, endurance, or aerobic, exercise is an essential aspect of any program to lower cholesterol. For most people aerobic exercise will raise levels of "good" HDL-cholesterol and tend to lower triglycerides. In addition, regular aerobic exercise will help to reduce body fat, and lower body fat is known to be associated with lower LDL-cholesterol. In short, diet and exercise go hand in hand in determining cholesterol levels and your probable coronary risk.

It is human nature to be concerned about your physical well-being and to try to do what you can to avoid potential health hazards. Now that you've made an effort to better understand what's going on with the cholesterol in your bloodstream and how this relates to your cardiovascular health, you may be more inclined to skip the burger for lunch and choose a salad instead.

Clinical Considerations

Heart Diseases

Heart diseases can be classified as either *congenital* or *acquired.* Congenital heart problems result from abnormalities in embryonic development and may be attributed to heredity, nutritional problems (poor diet) of the pregnant mother, or exposure to viral infections (e.g., rubella) while in the uterus. Congenital heart diseases occur in approximately 3 of every 100 births and account for about 50% of early childhood deaths. Many serious congenital heart defects can be corrected surgically, however, and others are not of a serious nature.

Heart murmurs may be either congenital or acquired. Generally, they are of no clinical significance. Nearly 10% of all people have heart murmurs, ranging from slight to severe. The most common type of congenital heart problem is a *septal defect* (fig. 15.24). An **arterial septal defect,** or **patent foramen ovale,** results from a failure of the fetal foramen ovale to close at the time of birth. A **ventricular septal defect** is caused by the abnormal development of the interventricular septum. This condition may interfere with closure of the AV valves and may be indicated by *cyanosis* (bluish coloration) and abnormal heart sounds. **Pulmonary stenosis** is a narrowing of the opening into the pulmonary trunk from the right ventricle. It may lead to a *pulmonary embolism* (clot), and is usually recognized by extreme lung congestion.

The **tetralogy of Fallot** is a combination of four defects within the heart of a newborn (fig. 15.25): (1) a ventricular septal defect, (2) an overriding aorta, (3) pulmonary stenosis (narrowing), and (4) right ventricular hypertrophy (enlargement). The pulmonary stenosis obstructs blood flow to the lungs and causes hypertrophy of the right ventricle. In an overriding aorta, the ascending portion arises midway between the right and left ventricles instead of from the left ventricle only. Open-heart surgery is necessary to correct tetralogy of Fallot, and the overall mortality rate is about 5%.

A tissue is said to be *ischemic* (*ĭ-ske'mik*) when it receives an inadequate supply of oxygen because of reduced blood flow. The most common cause of **myocardial**

tetralogy of Fallot: from Étienne-Louis A. Fallot, French physician, 1850–1911

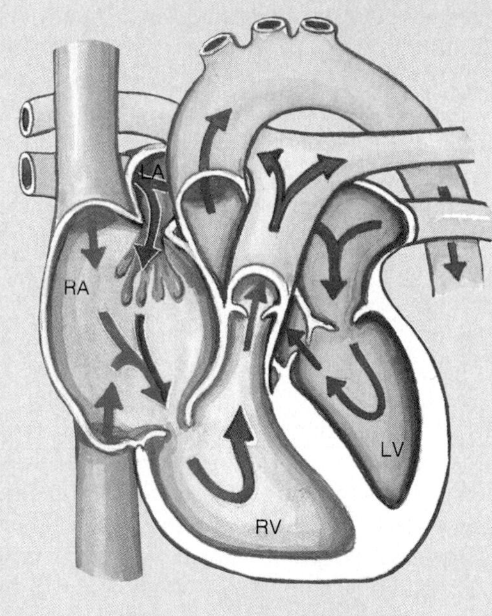

Septal defect in atria

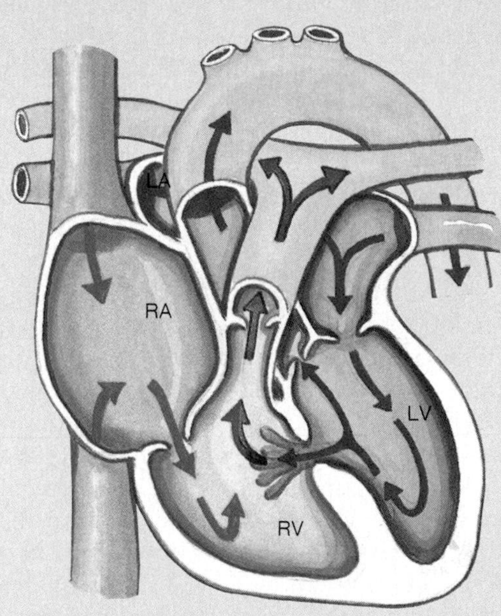

Septal defect in ventricles

FIGURE 15.24
Abnormal patterns of blood flow due to septal defects.

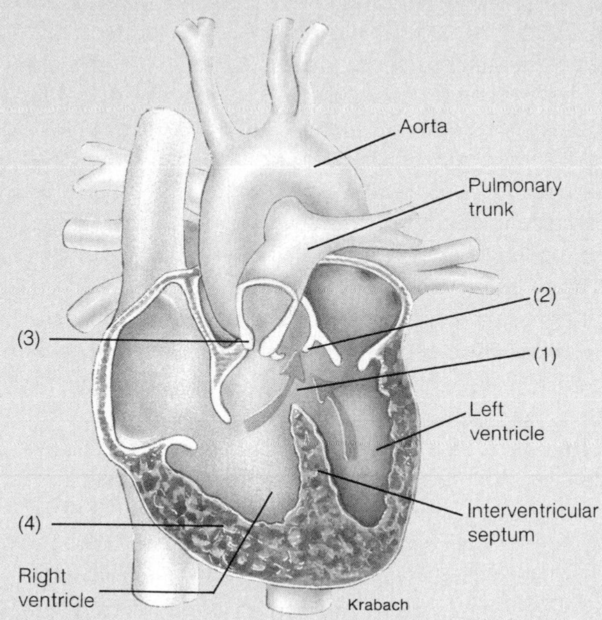

FIGURE 15.25

The tetralogy of Fallot. The four defects of this anomaly are (1) a ventricular septal defect, (2) an overriding aorta, (3) pulmonary stenosis (constriction), and (4) right ventricular hypertrophy (enlargement).

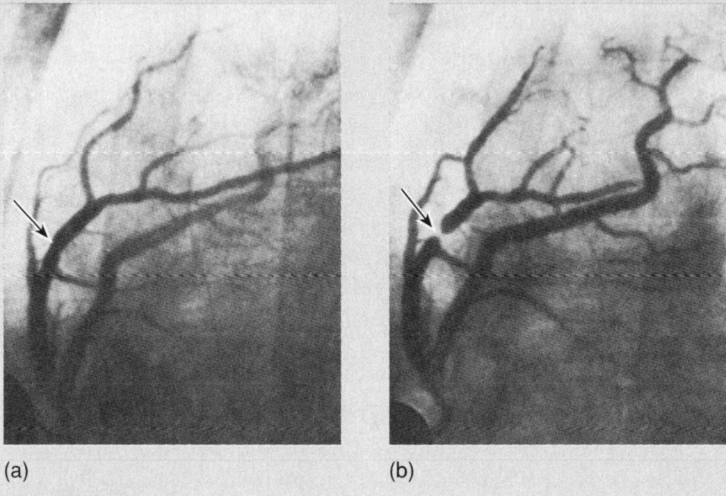

(a) (b)

FIGURE 15.26

An arteriogram of the left coronary artery in a patient *(a)* when the ECG was normal and (*b*) when the ECG showed evidence of myocardial ischemia. Notice that a coronary artery spasm—see arrow in (*b*)—appears to accompany the ischemia.

Donald S. Bain from Hurst et al: The Heart, *5/e, Fig. 47.4, pg. 1165. Reproduced with permission of the McGraw-Hill Companies.*

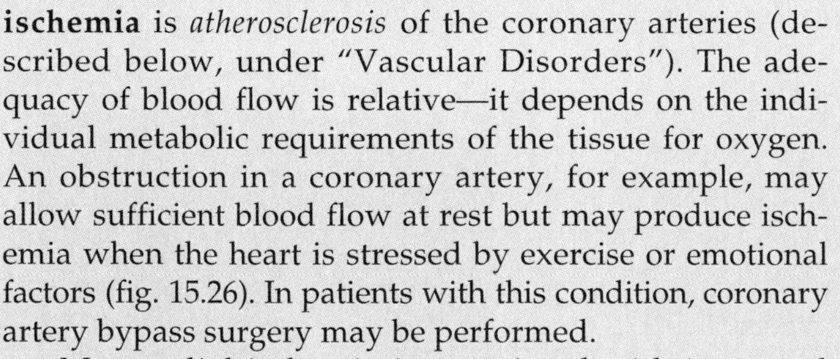

ischemia is *atherosclerosis* of the coronary arteries (described below, under "Vascular Disorders"). The adequacy of blood flow is relative—it depends on the individual metabolic requirements of the tissue for oxygen. An obstruction in a coronary artery, for example, may allow sufficient blood flow at rest but may produce ischemia when the heart is stressed by exercise or emotional factors (fig. 15.26). In patients with this condition, coronary artery bypass surgery may be performed.

Myocardial ischemia is associated with increased concentrations of blood lactic acid produced by anaerobic respiration of the ischemic tissue. This condition often causes substernal pain, which may also be referred to the left shoulder and arm, as well as to other areas (see fig. 12.1). This referred pain is called **angina pectoris.** People with angina frequently take nitroglycerin or related drugs that help to relieve the ischemia and pain. These drugs are effective because they stimulate vasodilation, which improves circulation to the heart and decreases the work that the heart must perform to eject blood into the arteries.

Myocardial cells are adapted to respire aerobically and cannot be deprived of oxygen for even a short time. If, after a few minutes, ischemia and anaerobic respiration continue, *necrosis* (cellular death) may occur in the areas most deprived of oxygen. A sudden, irreversible injury of this kind is called a **myocardial infarction,** or **MI.** The cardiac cells that die during an MI are replaced by scar tissue that cannot contract; hence, the heart muscle loses at least some of its strength. The lay term "heart attack" usually refers to a myocardial infarction.

Inside Information—Cardiovascular disease is responsible for nearly half of all deaths in American adults.

Clinical Considerations

Vascular Disorders

Hypertension, or high blood pressure, is the most common type of vascular disorder. In hypertension, the resting systolic blood pressure exceeds 140 mmHg. An estimated 22 million adult Americans are afflicted by hypertension. About 15% of the cases are the result of other body problems, such as kidney diseases, adrenal hypersecretion, or arteriosclerosis. If hypertension cannot be controlled by diet, exercise, or drugs that reduce the blood pressure, it can damage vital body organs, such as the heart and kidneys.

Arteriosclerosis, or hardening of the arteries, is a generalized degenerative vascular disorder that results in loss of elasticity and thickening of the arteries. **Atherosclerosis** is a specific type of arteriosclerosis in which a mass of plaque material called an *atheroma* forms on the tunica intima, narrowing the lumina of the arteries and prohibiting the normal flow of blood (fig. 15.27). As a further complication, an atheroma often creates a rough surface that can initiate the formation of a blood clot called a *thrombus.* An *embolism,* or *embolus,* is a detached thrombus that travels through the bloodstream and lodges so as to obstruct or occlude a blood vessel. An embolism lodged in a coronary artery is called a **coronary thrombus;** in a vessel of the lung it is a **pulmonary thrombus;** and in the brain it is a **cerebral thrombus,** which could cause a *stroke.*

The causes of atherosclerosis are not well understood, but the disease seems to be associated with improper diet, (see the preceding Synoptic Spotlight). Smoking, hypertension, obesity, lack of exercise, and heredity also appear to be causative factors.

Aneurysms, coarctation, and varicose veins are all types of vascular disfigurations. An **aneurysm** (*an'yŭ-riz-em*) is an expansion or bulging of the heart, aorta, or any other artery. Aneurysms are caused by weakening of the tunics and may rupture or lead to embolisms. **Coarctation** is a constriction of a segment of a vessel, usually the aorta, and is frequently caused by the tightening of a remnant of the ductus arteriosus around the vessel. **Varicose veins** are weakened veins that become stretched and swollen. They are most common in the legs because the force of gravity tends to weaken the valves and overload the veins. Varicose veins can also occur in the rectum, in which case they are called **hemorrhoids. Phlebitis** (*flĕ-bi'tis*) is inflammation of a vein. It may develop as a result of trauma or as an aftermath of surgery. Frequently, however, it appears for no apparent reason. Phlebitis interferes with normal venous blood flow.

arteriosclerosis: Gk. *arterio,* artery; *skleros,* hard
atheroma: Gk. *athere,* mush; *oma,* tumor
thrombus: Gk. *thrombos,* a clot
embolism: Gk. *embolos,* a plug

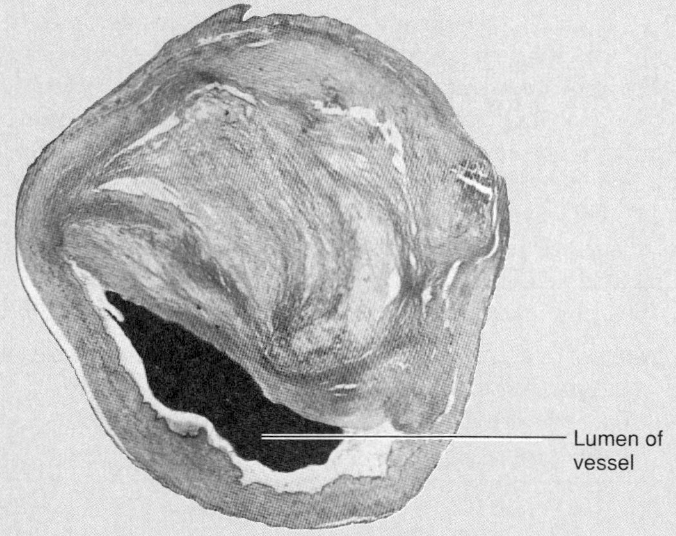

(a)

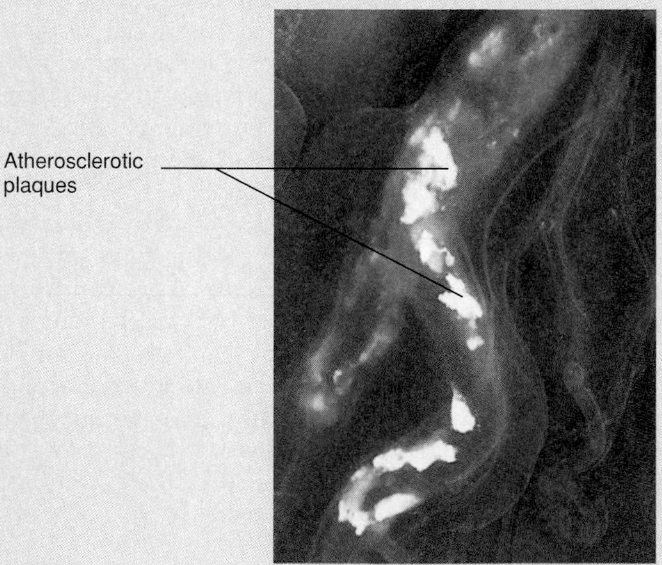

(b)

FIGURE 15.27
Atherosclerosis. (*a*) The lumen of a coronary artery is almost completely blocked by an atheroma. (*b*) A close-up view of the cleared left anterior descending coronary artery containing calcified atherosclerotic plaques. The heart is of an 85-year-old female.

SUMMARY

15.1 Structure of the Heart

1. The heart is located within the thoracic cavity, in the region of the mediastinum.
2. The pericardium is a loose-fitting protective sac surrounding the heart. The inner serous pericardium produces pericardial fluid that lubricates the heart as it beats.
3. The epicardium forms the protective outer sheath of the heart. The myocardium is the thick middle layer of the heart, composed of cardiac muscle tissue. The endocardium is the inner layer that is continuous with the endothelium of blood vessels.
4. The right and left sides of the heart pump blood through the pulmonary and systemic circulations, respectively.
5. The right ventricle pumps blood to the lungs. This blood then returns to the left atrium.
6. The left ventricle pumps blood into the aorta and systemic arteries. This blood then returns to the right atrium.
7. The heart contains right and left atrioventricular valves (the tricuspid and bicuspid valves, respectively); a pulmonary semilunar valve; and an aortic semilunar valve.
8. The myocardium of the heart is supplied with blood through the coronary circulation. Right and left coronary arteries branch from the aorta to serve the muscular wall of the heart, and the coronary sinus collects and empties the blood into the right atrium.

15.2 Conduction System and the Cardiac Cycle

1. The heart is a two-step pump; first the atria contract, and then the ventricles contract.
2. During diastole (ventricular relaxation), first the atria and then the ventricles fill with blood.
3. The ventricles are about 80% filled before the atria contract and add the final 20% to the end-diastolic volume.
4. When the ventricles contract at systole, the pressure within them first rises sufficiently to close the AV valves and then rises sufficiently to open the semilunar valves, allowing blood to be ejected out of the heart.

15.3 Electrical Activity of the Heart

1. Contraction of the atria and ventricles is produced by action potentials that originate in the sinoatrial (SA) node.
2. These electrical waves spread over the atria and then to the atrioventricular (AV) node.
3. From the AV node, the impulses are conducted by the atrioventricular bundle and conduction myofibers into the ventricular walls.
4. A recording of the pattern of electrical conduction is called an electrocardiogram (ECG).
5. The P wave is caused by depolarization of the atria; the QRS wave is caused by depolarization of the ventricles; the T wave is produced by repolarization of the ventricles.

15.4 Blood Vessels

1. Arteries and veins have a tunica externa, tunica media, and tunica intima.
2. Arteries have thicker muscle layers in proportion to their diameters than do veins because arteries must withstand a higher blood pressure.
3. Thinner walled veins have venous valves that direct blood to the heart when the veins are compressed by the skeletal muscles.
4. Capillaries are composed of endothelial cells only. They are the basic functional units of the circulatory system.

15.5 Principal Arteries of the Body

1. Three arteries arise from the aortic arch: the brachiocephalic trunk, the left common carotid artery, and the left subclavian artery. The brachiocephalic trunk divides into the right common carotid artery and the right subclavian artery.
2. The head and neck receive an arterial supply from branches of the internal and external carotid arteries and the vertebral arteries.
   a. The brain receives blood from the paired internal carotid arteries and the paired vertebral arteries, which form the cerebral arterial circle (circle of Willis) around the pituitary gland.
   b. The external carotid artery gives off many branches that supply the head and neck.
3. The shoulder and upper limb is served by the subclavian artery and its branches.
   a. The subclavian artery becomes first the axillary artery and then the brachial artery as it enters the arm.
   b. The brachial artery branches to form the radial and ulnar arteries, which supply blood to the forearm and a portion of the hand and fingers.
4. The abdominal aorta gives off the following branches: the inferior phrenic, celiac trunk, superior mesenteric, renal, suprarenal, testicular (or ovarian), and interior mesenteric arteries.
5. The common iliac arteries divide into the internal and external iliac arteries, which supply branches to the pelvis and lower extremities.

15.6 Principal Veins of the Body

1. Blood from the head and neck is drained by the external and internal jugular veins. Blood from the brain is drained by the internal jugular veins.
2. The upper limb and shoulder region is drained by superficial and deep veins.
3. In the thorax, the superior vena cava is formed by the union of the two brachiocephalic veins and also collects blood from the azygos system of veins.

4. The lower limb is drained by both superficial and deep veins. At the level of the fifth lumbar vertebra, the right and left common iliac veins unite to form the inferior vena cava.
5. Blood from capillaries in the GI tract is drained via the hepatic portal vein to the liver. This venous blood then passes through the hepatic sinusoids and is drained from the liver in the hepatic veins.
6. A portal system is a pattern of circulation in which blood flows from one capillary bed directly toward another, rather than toward the heart.

15.7 Fetal Circulation

1. Structural adaptations in the fetal circulatory system reflect the fact that oxygen and nutrients are obtained from the placenta rather than from the fetal lungs and GI tract.
2. Fully oxygenated blood is carried only in the umbilical vein, which drains the placenta. This blood is carried via the ductus venosus to the inferior vena cava of the fetus.
3. Partially oxygenated blood passes from the right to the left atrium via the foramen ovale, and from the pulmonary trunk to the aorta via the ductus arteriosus.

Review Activities

Objective Questions

1. Which of the following is *not* found within the ventricles of the heart? (*module 15.1*)
   a. chordae tendinae
   b. papillary muscles
   c. trabeculae carnae
   d. coronary sinus
   e. endocardium

2. All arteries in the body contain oxygen-rich blood with the exception of (*module 15.1*)
   a. the aorta.
   b. the pulmonary arteries.
   c. the renal arteries.
   d. the coronary arteries.

3. Venous blood from the coronary circulation directly enters (*module 15.1*)
   a. the inferior vena cava.
   b. the superior vena cava.
   c. the right atrium.
   d. the left atrium.

4. The "lub," or first heart sound, is produced by the closing of (*module 15.2*)
   a. the aortic semilunar valve.
   b. the pulmonary semilunar valve.
   c. the right tricuspid (atrioventricular) valve.
   d. the left bicuspid (atrioventricular) valve, or mitral valve.
   e. both atrioventricular valves.

5. The "lub" heart sound is produced at (*module 15.2*)
   a. the beginning of systole.
   b. the end of systole.
   c. the beginning of diastole.
   d. the end of diastole.

6. The QRS wave of an ECG is produced by (*module 15.3*)
   a. depolarization of the atria.
   b. repolarization of the atria.
   c. depolarization of the ventricles.
   d. repolarization of the ventricles.

7. The functional units of the circulatory system are (*module 15.4*)
   a. arteries.
   b. capillaries.
   c. erythrocytes (red blood cells).
   d. leukocytes (white blood cells).

8. Which of the following does/do *not* arise from the aortic arch? (*module 15.5*)
   a. brachiocephalic trunk
   b. coronary arteries
   c. left common carotid artery
   d. left subclavian artery

9. Which of the following does *not* supply blood to the brain? (*module 15.5*)
   a. external carotid artery
   b. internal carotid artery
   c. vertebral artery
   d. basilar artery

10. The hepatic portal system does *not* receive blood from (*module 15.6*)
   a. the gallbladder.
   b. the spleen.
   c. the stomach.
   d. the large intestine.
   e. the kidneys.

Essay Questions

1. Describe the location and structure of the heart. (*module 15.1*)
2. Trace the flow of blood through the heart and distinguish between the pulmonary and systemic circulations. (*module 15.1*)
3. Define *nodal tissues, conduction system, systole,* and *diastole.* (*module 15.2*)
4. Describe the cardiac cycle. What causes the lub and dub sounds of heart activity? (*module 15.2*)
5. Define *pacemaker potential.* (*module 15.3*)
6. Diagram and label an ECG. Indicate the electrical events in the heart that produce the wave deflections. (*module 15.3*)
7. Describe the basic structural pattern of arteries and veins. Explain how the structural differences in these vessels relate to differences in blood pressure. (*module 15.4*)
8. Describe the structure of capillaries and explain why these vessels are considered the basic functional units of the circulatory system. (*module 15.4*)
9. Describe the arterial pathway from the subclavian artery to the digital arteries. (*module 15.5*)
10. List the arteries that supply blood to the lower abdominal wall, the external genitalia, the hamstring muscles, the knee joint, and the foot. (*module 15.5*)
11. Describe the hepatic portal system and comment on the functional significance of this system. (*module 15.6*)
12. Describe the path of blood from the fetal heart through the placenta and back to the fetal heart. (*module 15.7*)
13. Identify five fetal circulatory structures that cease to function in a newborn. (*module 15.7*)

Labeling Exercise

Label the structures indicated on the figure to the right.

1. ______________________
2. ______________________
3. ______________________
4. ______________________
5. ______________________
6. ______________________

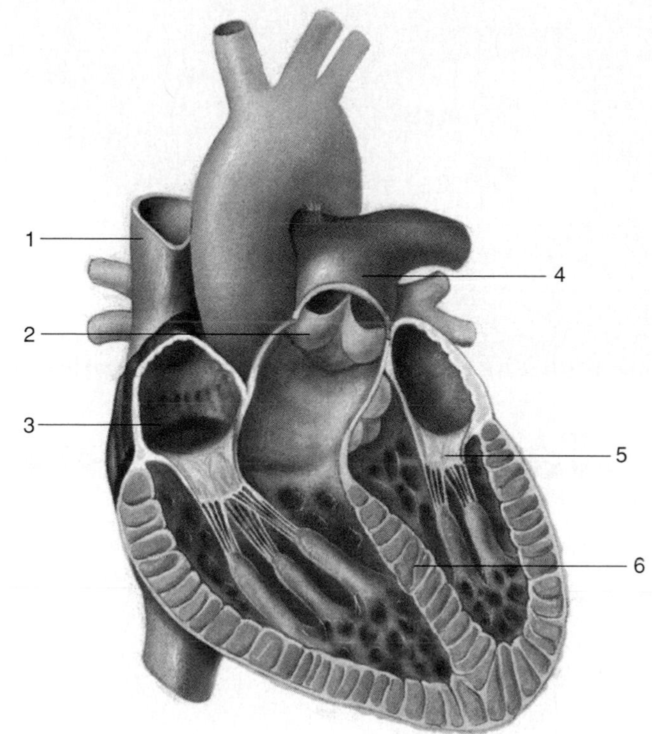

Critical Thinking Questions

1. A hospitalized 45-year-old man developed a thrombus (blood clot) in his lower thigh following severe trauma to his knee. The patient's physician explained that although the clot was near the great saphenous vein, the main concern was the occurrence of a pulmonary embolism. Explain the physician's reasoning and list, in sequence, the vessels the clot would have to pass through to become a pulmonary embolism.
2. An arterial pressure point is where the blood flow to a body region can be diminished by compressing an artery between the skin and a bony structure. Describe the location of the clinically important arterial pressure points and explain how this information could be put to use in giving a victim first-aid treatment. What precautions should be taken in applying pressure at an arterial pressure point?
3. Your sister's baby girl was diagnosed as having a congenital patent foramen ovale in her heart. Describe what this means and explain why the baby is cyanotic.
4. Examine the ECG profile in figure 15.7 and notice the time delay between the P and the QRS deflections. Also notice that the QRS deflection is much higher than the P deflection. Explain the significance of these two observations.
5. In our concern about coronary artery disease and heart attacks, most of our attention is focused on the ventricles rather than on the atria. Can you give a sound physiological explanation for this?

CHAPTER

16

"Principles that govern fluids and their movements influence cardiac activity and stroke volume." *page 406*

Cardiac Output and Blood Flow

Chapter Outline

Terms to Remember

stroke volume, 405
Frank–Starling law of the heart, 407
compliance, 408
interstitial (*in"ter-stish'al*) fluid, 409
oncotic (*on-kot'ik*) pressure, 410
Starling forces, 410
edema (*ĕ-de'mă*), 411
antidiuretic (*an"te-di"yŭ-ret'ik*) hormone (ADH), 412
aldosterone (*al-dos'ter-ōn*), 413
angiotensin II, 413
autoregulation, 414
hyperemia (*hi"per-e'me-ă*), 414
baroreceptor (*bar"o-re-sep'tor*), 416
Korotkoff (*kŏ-rot'kof*) sounds, 419
hypertension, 420

Learning Objectives

After studying this chapter, you should be able to

- describe how the cardiac rate is regulated.
- explain the Frank–Starling law of the heart and describe how contractile strength is influenced by epinephrine.
- discuss the factors that influence the return of venous blood to the heart.
- explain how interstitial fluid is produced and how it is returned to the vascular system.
- describe the relationship between blood volume and urine volume, and explain how hormones help to regulate the kidney's control over the blood volume.
- explain how blood can be diverted from one organ to another and how blood flow is regulated extrinsically and by intrinsic control mechanisms.
- identify the factors that influence blood pressure and explain the baroreceptor reflex mechanism.
- explain how blood pressure is measured.

16.1 Cardiac Output and Cardiac Rate

The cardiac output is a measure of the pumping ability of the heart and is affected by mechanisms that regulate the cardiac rate and stroke volume.

Only if the blood is kept circulating can it be effective in meeting the needs of the individual cells of the body. The heart is the driving force of circulation. Starting during embryonic development, it works ceaselessly to pump the blood through the tubular network of blood vessels. The pumping activity of the heart, the blood in the vessels, and the routing of the blood to the "neediest" organs are interrelated, with each factor influencing the others.

As we learned in the previous chapter (see module 15.2), the heart fills with blood during its relaxation phase, or **diastole,** and ejects the blood during its contraction phase, or **systole.** The amount of blood ejected by the heart is measured as the **cardiac output,** which is equal to the volume of blood pumped per minute by each ventricle. The average resting **cardiac rate** in an adult is 70 beats per minute; the average **stroke volume** (volume of blood pumped per beat by each ventricle) is 70 to 80 ml per beat. To calculate cardiac output, these two variables are multiplied, yielding an average cardiac output of about 5.5 L per minute:

$$\underset{(\text{ml/min})}{\text{cardiac output}} = \underset{(\text{ml/beat})}{\text{cardiac rate}} \times \underset{(\text{beats/min})}{\text{stroke volume}}$$

Cardiac output is linked to our blood volume. The **total blood volume** in a human body is also about 5.5 L. This means that each ventricle pumps the equivalent of the total blood volume each minute under resting conditions. Put another way, it takes about a minute for a drop of blood to flow through the systemic and pulmonary circuits. An increase in cardiac output, as occurs during exercise, must thus be accompanied by an increased rate of blood flow through the body's vessels. This is accomplished through regulation of the cardiac rate and stroke volume.

Inside Information—Generally, the smaller the organism, the faster the heartbeat. Women's hearts average 6 to 8 more beats per minute than men's hearts, and newborns may have a heartbeat as fast as 130 times per minute.

In the complete absence of neural influences, the heart will continue to beat according to the rhythm set by the SA node. This automatic rhythm is produced by a spontaneous depolarization of the SA node cells, so that they automatically produce action potentials. As described in chapter 15 (see module 15.2), these action potentials are conducted through gap junctions in the atria and through the conducting tissues into the myocardium of the ventricles.

Although nerve activity is not required for the production of action potentials, sympathetic stimulation and parasympathetic stimulation (through the vagus nerves) to the heart continuously modify the rate of the spontaneous depolarization of the SA node. Norepinephrine, released primarily by sympathetic nerve endings, and epinephrine, secreted by the adrenal medulla, increase the spontaneous firing rate of the SA node. Acetylcholine, released from parasympathetic endings, inhibits the SA node and thus decreases the rate of its spontaneous firing (fig. 16.1). The actual pace set by the SA node at any time represents the net effect of these antagonistic influences.

Autonomic innervation of the SA node is the major means by which cardiac rate is regulated. Autonomic stimulation does, however, affect cardiac rate by other mechanisms to a lesser degree. Sympathetic activity in the musculature of the atria and ventricles increases the strength of contraction and has several other effects (table 16.1).

During exercise, the cardiac rate increases as a result of decreased vagus nerve inhibition of the SA node, leading to the production of more frequent action potentials. These increases in cardiac rate can be supplemented by increased sympathetic nerve stimulation. The opposite relationship can be seen in well-conditioned hearts. The resting bradycardia (slow heart rate) of endurance-trained athletes is due largely to high vagus nerve activity.

The activity of the autonomic innervation of the heart is coordinated by **cardiac control centers** in the *medulla oblongata* of the brain stem. These cardioaccelaratory and cardioinhibitory centers, in turn, are affected by higher brain areas and by sensory feedback from pressure receptors, or *baroreceptors,* in the aorta and carotid arteries. In this way, a fall in blood pressure can produce a reflex increase in the cardiac rate. The *baroreceptor reflex* is discussed in more detail in connection with blood pressure regulation later in this chapter (see module 16.7).

Inside Information—An athlete's heartbeat is often slower than the average rate because endurance training has allowed it to eject more blood with each beat. A highly trained athlete can achieve a cardiac output as high as 35–40 l/min, whereas a "couch potato" may be lucky to attain a cardiac output of 20 l/min during very heavy exercise. The "couch potato," as a result, will fatigue much more quickly and be left panting in the dust.

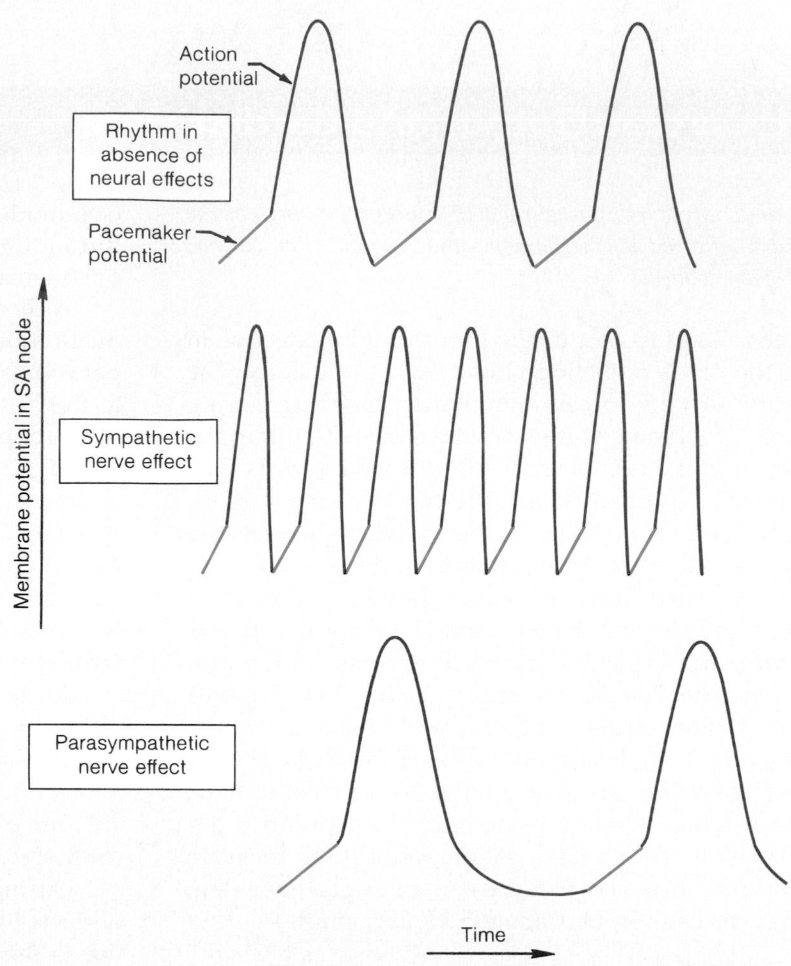

FIGURE 16.1
The rhythm set by the pacemaker potentials in the SA node. Sympathetic nerve activity decreases the rate of spontaneous depolarization, thus influencing the rate at which action potentials are produced.

TABLE 16.1 Effects of Autonomic Nerve Activity on the Heart

Region Affected	Effects of Sympathetic Activity	Effects of Parasympathetic Activity
SA node	Increased rate of diastolic depolarization; increased cardiac rate	Decreased rate of diastolic depolarization; decreased cardiac rate
AV node	Increased conduction rate	Decreased conduction rate
Atrial muscle	Increased strength of contraction	Decreased strength of contraction
Ventricular muscle	Increased strength of contraction	No significant effect

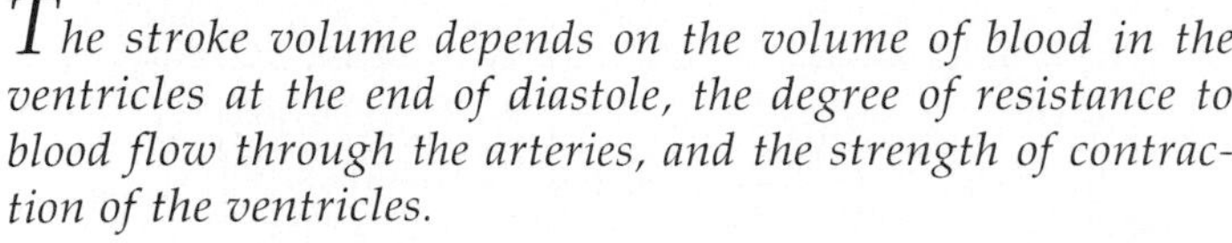

16.2 Regulation of Stroke Volume

The stroke volume depends on the volume of blood in the ventricles at the end of diastole, the degree of resistance to blood flow through the arteries, and the strength of contraction of the ventricles.

Principles that govern fluids and their movements influence cardiac activity and stroke volume. Before looking at cardiac activity, a few reminders are needed. First, remember that anywhere fluids are present, they exert pressure against their container. Human body fluids are mostly water, so this pressure is referred to as a **hydrostatic pressure.** When referring to blood, that same hydrostatic pressure is simplified to **blood pressure.** Second, as you know from experience, fluids always move from an area of high pressure to an area of low pressure. Lastly, when liquids are moving through tubes, the friction of the molecules against the wall of the tube opposes, or resists, the flow. This opposition is called **frictional resistance,** or simply **resistance.**

As discussed in the previous module, **stroke volume** is the volume of blood pumped per beat by each ventricle. The stroke volume is regulated by three variables:

1. **end-diastolic volume (EDV),** which is the volume of blood in the ventricles at the end of diastole;
2. **total peripheral resistance,** which is the frictional resistance, or impedance, to blood flow in the arteries; and
3. **contractility,** or strength, of ventricular contraction.

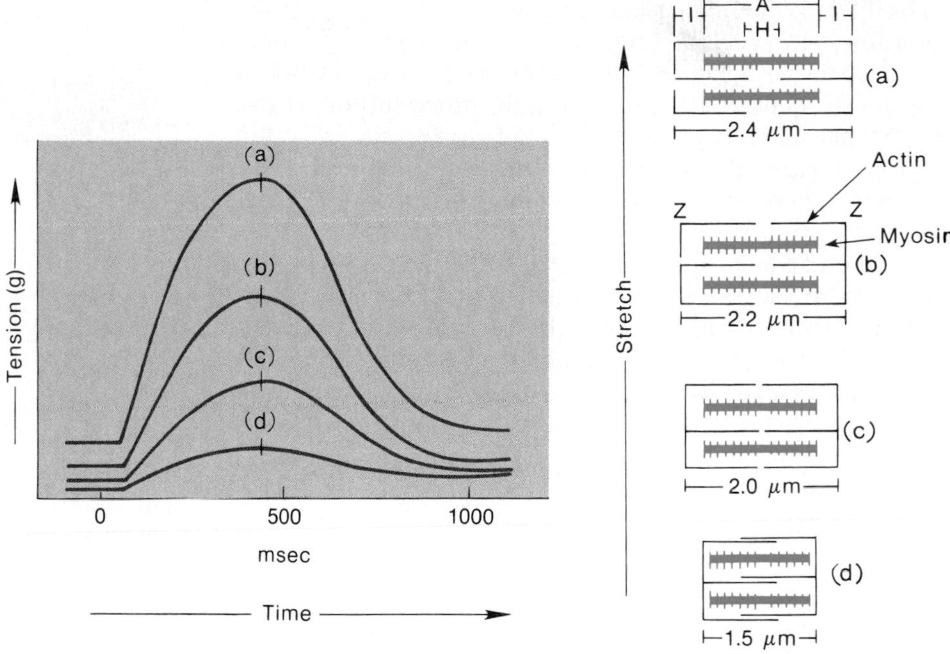

FIGURE 16.2
The Frank–Starling law of the heart. When the heart muscle is subjected to increased amounts of stretch, it contracts more strongly. As a result of the increased contraction strength (shown as tension), the time required to reach maximum contraction is the same, regardless of the degree of stretch.

The stroke volume is directly proportional to the EDV; an increase in EDV results in an increase in stroke volume. The stroke volume is also directly proportional to contractility. Thus, when the ventricles contract more forcefully, they empty more completely and therefore pump more blood. The reverse is also true.

In order to eject blood, the pressure generated in a ventricle when it contracts must be greater than the pressure in the arteries. As blood begins to be ejected from the ventricle, the added volume of blood in the arteries causes a rise in pressure against the "bottleneck" presented by the peripheral resistance. Ejection of blood stops shortly after the aortic pressure becomes equal to the intraventricular pressure.

Frank–Starling Law of the Heart

The proportion of the end-diastolic volume that is ejected depends upon the strength of ventricular contraction. Normally, contractile strength is sufficient to eject 70 to 80 ml of blood out of a total end-diastolic volume of 110 to 130 ml. The *ejection fraction* is thus about 60%. More blood is pumped per beat as the EDV increases, and thus the ejection fraction remains relatively constant over a range of end-diastolic volumes. In order for this to be true, the strength of ventricular contraction must increase as the end-diastolic volume increases.

Experiments demonstrate that the strength of ventricular contraction varies directly with the end-diastolic volume. Even in experiments where the heart is removed from the body (and is thus not subject to neural or hormonal regulation), and where the heart is filled with blood flowing from a reservoir, an increase in EDV within the physiological range results in increased contraction strength and, therefore, in increased stroke volume. This relationship between EDV, contractile strength, and stroke volume is thus a built-in, or *intrinsic,* property of heart muscle, and is known as the **Frank–Starling law of the heart.**

The intrinsic control of contractile strength and stroke volume is due to variations in the degree to which the myocardium is stretched. As the EDV rises, the myocardium is increasingly stretched and the sarcomeres of cardiac muscle become longer (fig. 16.2). Within physiological limits, the longer the sarcomeres become, the more powerfully they are able to contract.

Frank–Starling law of the heart: from Otto Frank, German physiologist, 1865–1944, and Ernest Henry Starling, English physiologist, 1866–1927

Extrinsic Control of Contractility

The EDV is the major intrinsic factor influencing stroke volume. However, SV can also be enhanced by *extrinsic factors* (nerve impulses and hormones). **Contractility** is an increase in contractile strength, independent of the EDV, that is brought about by changes in activity of the sympathoadrenal system. Both norepinephrine from sympathetic nerve endings and epinephrine from the adrenal medulla produce an increase in contractile strength.

The cardiac output is affected in two ways by the activity of the sympathoadrenal system: the heart contracts more strongly and it beats more rapidly (fig. 16.3). Both of these changes lead to an increase in the cardiac output. Stimulation of the parasympathetic nerves decreases the cardiac rate, but does not directly affect the contractile strength of the ventricles. Stimulation by parasympathetic nerves, then, reduces the cardiac output.

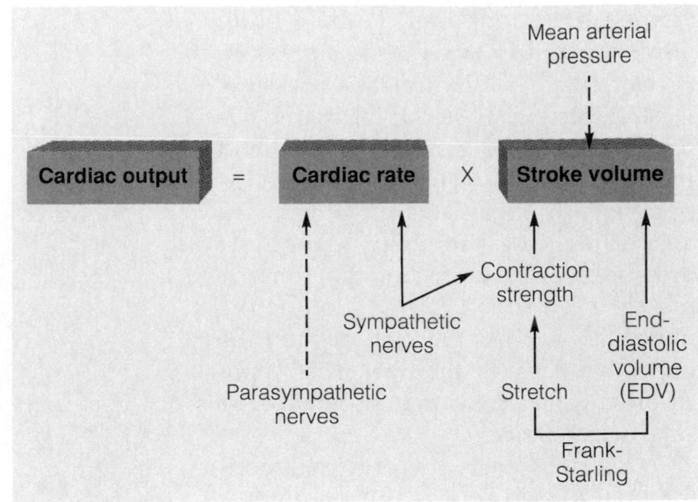

FIGURE 16.3
The regulation of cardiac output. Factors that stimulate cardiac output are shown as solid arrows; factors that inhibit cardiac output are shown as dashed arrows.

16.3 Venous Return

The flow of blood through veins to the heart depends on the venous pressure and the blood volume.

The end-diastolic volume—and hence the stroke volume and cardiac output—is controlled by factors that affect the return of blood to the heart. Since blood returns to the heart in veins, this flow can be called the **venous return.** The rate at which the atria and ventricles are filled with venous blood depends on the *total blood volume* and the *venous pressure* (pressure in the veins). It is the venous pressure that serves as the driving force for the return of blood to the heart.

Venous Pressure

Because the walls of veins are thinner and less muscular than those of arteries, the veins have a higher **compliance,** or ability to expand under pressure. This means that a given amount of pressure will cause more distension (expansion) in the veins than in arteries, so that the veins can hold more blood. Approximately two-thirds of the total blood volume is located in the veins (fig. 16.4). Veins are therefore called *capacitance vessels,* by analogy with electronic devices called capacitors that store electrical charges.

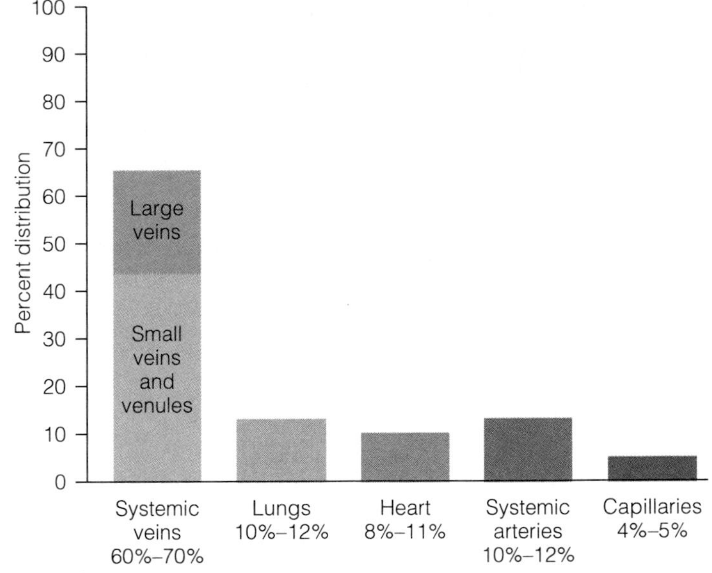

FIGURE 16.4
The distribution of blood within the body at rest.

Although veins contain almost 70% of the total blood volume, the mean venous pressure is only 2 mmHg, compared to a mean arterial pressure of 90 to 100 mmHg. The lower venous pressure is due in part to a pressure drop between arteries and capillaries and in part to the high venous compliance.

The venous pressure is highest in the tiniest veins—the venules—(10 mmHg) and lowest at the junction of the venae cavae with the right atrium (0 mmHg). This pressure difference is needed to drive the blood back to the heart. The major factors that influence the venous pressure and venous return are

1. *sympathetic nerve stimulation,* which causes smooth muscle contraction in the walls of veins, thus reducing their compliance and increasing the venous pressure;
2. *skeletal muscle pumps,* which squeeze veins during muscle contraction, thus increasing the venous pressure; and
3. *a pressure difference between the thoracic and abdominal cavities,* which promotes the flow of blood from abdominal to thoracic veins and back toward the heart.

When skeletal muscles contract, they function as "pumps" by virtue of their squeezing action on veins (described in chapter 15, module 15.4). Contraction of the diaphragm during inhalation also improves venous return. As the diaphragm contracts, it lowers, thus increasing the thoracic volume and decreasing the abdominal volume. This creates a partial vacuum in the thoracic cavity and a higher pressure in the abdominal cavity—conditions that favor blood flow from abdominal to thoracic veins (fig. 16.5).

Blood and Interstitial Fluid

Blood volume represents one part, or compartment, of the total body water. Approximately two-thirds of the total body water is contained within cells—in the *intracellular compartment.* The remaining one-third is contained in the *extracellular compartment.* This extracellular fluid is normally distributed so that about 80% is contained in the tissues as **interstitial** (*in"ter-stish'al*) **fluid** (tissue fluid), with the blood plasma accounting for the remaining 20% (fig. 16.6).

The distribution of water between the intracellular and extracellular fluid compartments is determined by a delicate balancing act of opposing forces acting at the capillaries. Blood pressure, for example, promotes the formation of interstitial fluid from blood plasma, whereas

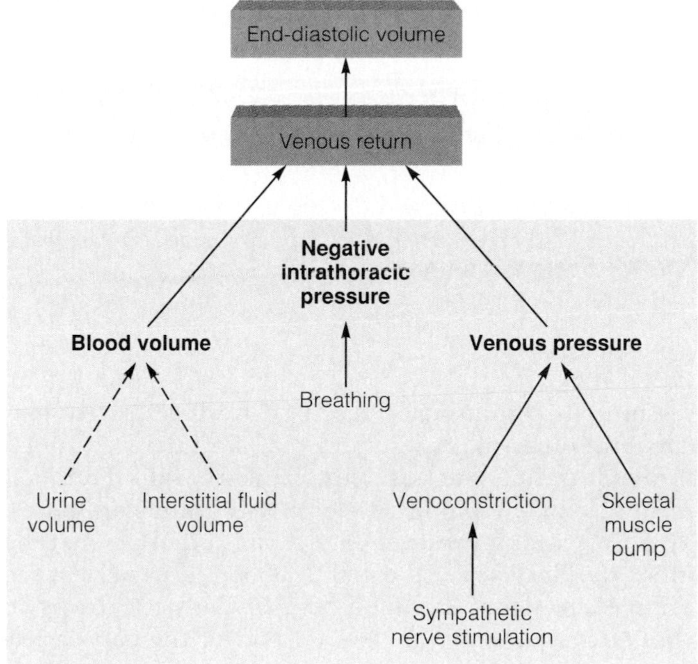

FIGURE 16.5
Variables that affect venous return, and thus end-diastolic volume. Direct relationships are indicated by solid arrows; inverse relationships are shown with dashed arrows.

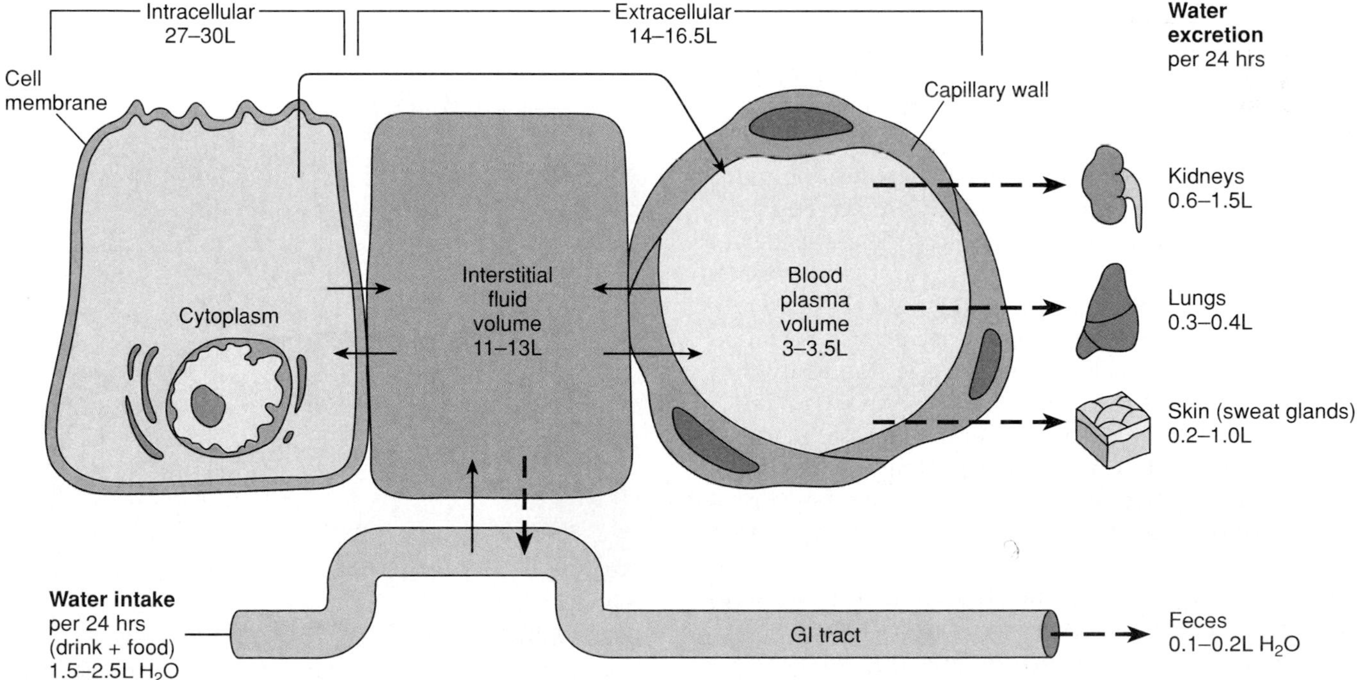

FIGURE 16.6
The daily intake and excretion of body water and its distribution between different intracellular and extracellular compartments.

osmotic forces draw water from the tissues into the circulatory system. The total volume of intracellular and extracellular fluid is normally maintained constant by a balance between water loss and water gain. Mechanisms that affect drinking, urine volume, and the distribution of water between blood plasma and the interstitial fluid thus help to regulate blood volume and, by this means, help to regulate cardiac output and blood flow.

16.4 Exchange of Fluid across Capillaries

Fluid is exchanged between the vascular and interstitial compartments, and excess interstitial fluid is drained by lymphatic vessels.

Starling Forces

Tissue cells continuously receive a fresh supply of glucose and other plasma solutes that are filtered through tiny endothelial channels in the capillary walls. Filtration results from the blood pressure within the capillaries. This hydrostatic pressure, which is exerted against the inner capillary wall, is equal to about 37 mmHg at the arteriolar end of systemic capillaries and drops to about 17 mmHg at the venular end of the capillaries. The **net filtration pressure** is equal to the hydrostatic pressure of the blood in the capillaries minus the hydrostatic pressure of the interstitial fluid outside the capillaries, which opposes filtration. If, as an extreme example, these two values were equal, no filtration would occur. The magnitude of the tissue hydrostatic pressure varies from organ to organ. A more realistic value of the hydrostatic pressure in the interstitial fluid is 1 mmHg, so that the net filtration pressure is 37 – 1 = 36 mmHg at the arteriolar end of the capillary and 17 – 1 = 16 mmHg at the venular end.

Glucose and other small organic molecules, inorganic salts, and ions are filtered along with water through the capillary channels. The concentrations of these substances in interstitial fluid are thus the same as in blood plasma. The protein concentration of interstitial fluid (2 g/100 ml), however, is less than the protein concentration of blood plasma (6–8 g/100 ml). This difference is due to the fact that the filtration of the large protein molecules is restricted by the small size of the capillary pores. The osmotic pressure exerted by plasma proteins is called the **oncotic** (*on-kot'ik*) **pressure.** The oncotic pressure promotes the osmotic movement of water into the capillaries, and is estimated to be 25 mmHg.

The opposing forces of net filtration pressure and oncotic pressure that affect the distribution of fluid across the capillary are known as **Starling forces.** As a result of Starling forces, fluid leaves the capillaries at the arteriolar end and returns to the capillaries at the venular end (fig. 16.7).

Blood plasma and interstitial fluid are thus continuously exchanged. The return of fluid to the circulatory system at the venular ends of the capillaries, however, does not exactly equal the amount filtered at the arteriolar ends. According to some estimates, approximately 85% of the capillary filtrate is returned directly to the capillaries; the remaining 15% (amounting to at least 2 L per day) is returned to the circulatory system by way of the lymphatic system.

Lymphatic Drainage

Excessive accumulation of interstitial fluid and filtered proteins is normally prevented by drainage of interstitial fluid into highly permeable, blind-ended **lymphatic capillaries** (fig. 16.8). Interstitial fluid that enters these lymphatic capillaries is known as *lymph.* The lymphatic capillaries unite to form larger and larger vessels called

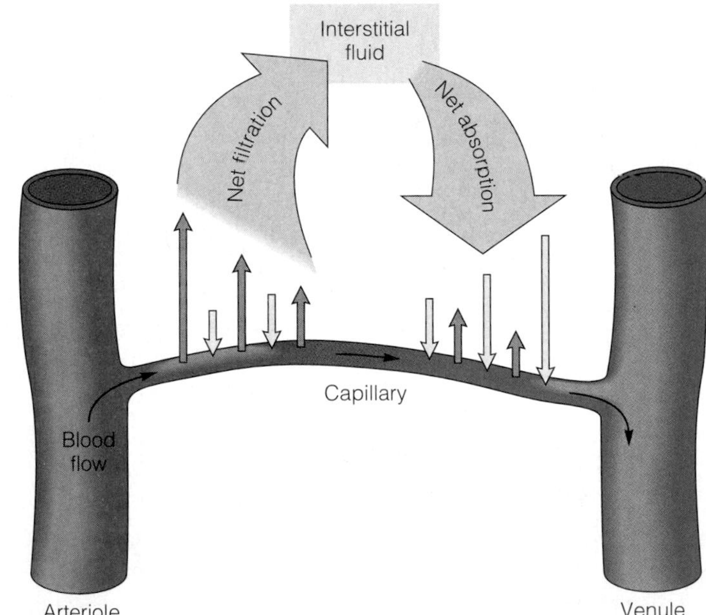

FIGURE 16.7
Interstitial fluid is formed by filtration (orange arrows) as a result of blood pressure at the arteriolar ends of capillaries and is returned to venular ends of capillaries by the oncotic pressure of blood plasma proteins.

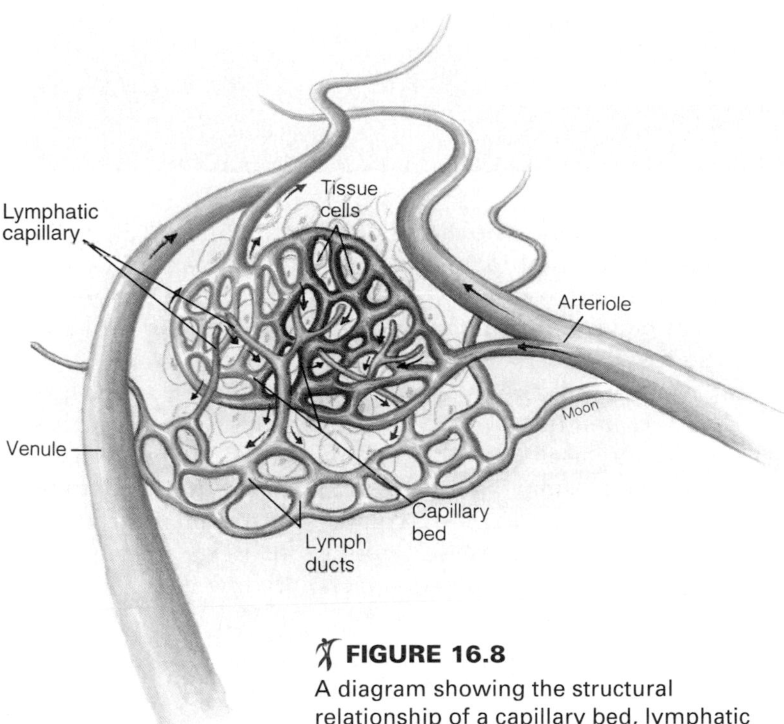

FIGURE 16.8
A diagram showing the structural relationship of a capillary bed, lymphatic capillaries, and lymph ducts.

lymphatic vessels. Lymph is transported by the lymphatic vessels into two large ducts that drain the lymph into the *right* and *left subclavian veins* (see fig. 17.1). In this way, the fluid is ultimately returned to the circulatory system from which it was originally derived.

An excessive accumulation of interstitial fluid is known as **edema** (*ĕ-de'mă*). The possible causes of edema are summarized in table 16.2.

In the tropical disease filariasis *(fil'a-ri"a-sis), mosquitoes transmit a parasitic roundworm to humans. The larvae of these parasites invade lymphatic vessels, become adults, and form masses of worms that block lymphatic drainage. The edema that results can be so severe as to produce an elephant-like appearance, with thickening and cracking of the skin. Because of these symptoms, this disease is commonly called* elephantiasis *(fig. 16.9).*

FIGURE 16.9
Parasitic larvae that block lymphatic drainage produce tissue edema, resulting in elephantiasis.

TABLE 16.2 Causes and Effects of Edema

Cause	Effect
Increased blood pressure or venous obstruction	Increases capillary filtration pressure so that more tissue fluid is formed at the arteriolar ends of capillaries.
Increased tissue protein concentration	Decreases osmosis of water into the venular ends of capillaries. Usually a localized tissue edema due to leakage of plasma proteins through capillaries during inflammation and allergic reactions. Myxedema due to hypothyroidism is also included in this category.
Decreased plasma protein concentration	Decreases osmosis of water into the venular ends of capillaries. May be caused by liver disease (which can be associated with insufficient plasma protein production), kidney disease (due to leakage of plasma protein into urine), or protein malnutrition.
Obstruction of lymphatic vessels	Infections by filaria roundworms transmitted by certain species of mosquito block lymphatic drainage, causing edema and tremendous swelling of the affected areas.

16.5 Renal Regulation of Blood Volume

The kidneys produce urine as a blood filtrate. Antidiuretic hormone promotes renal (kidney) retention of water, and aldosterone promotes renal retention of both salt and water.

Since the urine is derived from blood, the blood volume is regulated primarily through regulation of urine production. The formation of urine by the kidneys begins in the same manner as the formation of interstitial fluid—by filtration of blood plasma through capillary pores. These renal capillaries are known as *glomeruli,* and the filtrate they produce enters a system of tubules that transports and modifies the filtrate. The kidneys produce about 180 L of blood filtrate per day, but since the volume of blood in the body is only 5.5 L, it is clear that most of this filtrate must be returned to the circulatory system and recycled. The volume of urine excreted is only 1 to 2 L per day; 98% to 99% of the amount filtered is returned, or *reabsorbed,* back into the circulatory system.

The volume of urine excreted, and thus the blood volume, can be varied by changes in the reabsorption of filtrate. If 99% of the filtrate is reabsorbed, for example, 1% must be excreted. Decreasing the reabsorption by only 1%—to 98%—would double the volume of urine excreted. The percentage of reabsorption by the kidneys—and thus the urine volume and blood volume—is adjusted according to the needs of the body by the action of specific hormones that act on the kidneys. Through their effects on the kidneys, and the resulting changes in blood volume, these hormones serve important functions in the regulation of the circulatory system.

Regulation by Antidiuretic Hormone (ADH)

One of the major hormones involved in the regulation of blood volume is **antidiuretic** (*an"te-di"yŭ-ret"ik*) **hormone (ADH),** also known as *vasopressin.* The release of ADH from the posterior pituitary occurs when neurons called **osmoreceptors** in the hypothalamus detect an increase in plasma concentration.

The plasma concentration can be increased either by dehydration or by excessive salt intake. Stimulation of osmoreceptors produces sensations of thirst, leading to increased water intake, and to increased release of ADH from the posterior pituitary. Through mechanisms that will be discussed in chapter 19 (see module 19.5), ADH acts on the renal tubules to promote the reabsorption of water from the filtrate into the blood. A smaller volume of urine is thus excreted as a result of the action of ADH (fig. 16.10).

A person who is dehydrated, or who consumes excessive amounts of salt, thus drinks more and urinates less. This raises the blood volume and, in the process, dilutes the plasma to lower its previously elevated concentration. The rise in blood volume that results from these mechanisms is extremely important in stabilizing the condition of a dehydrated person with low blood volume and pressure.

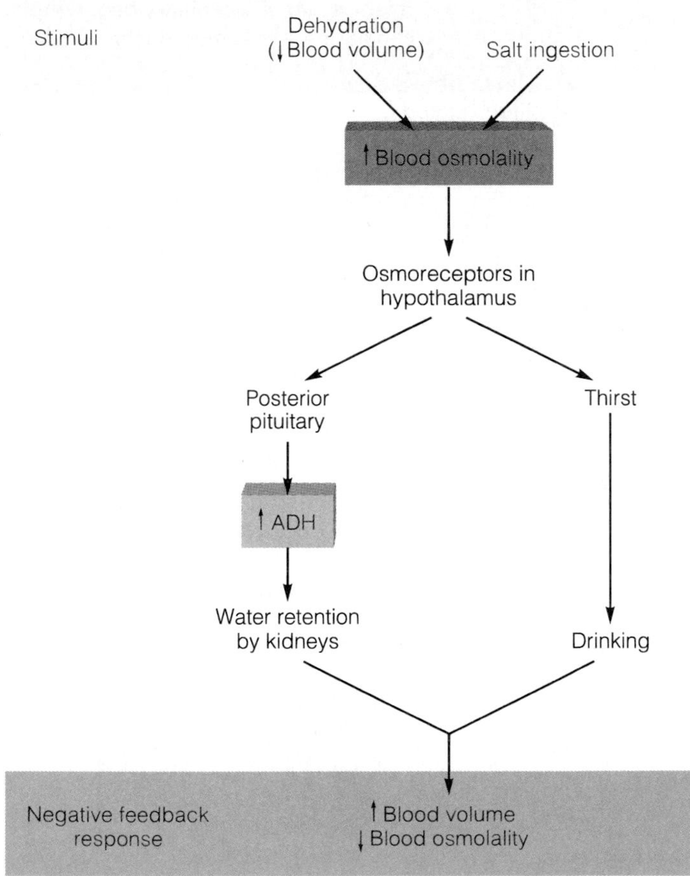

FIGURE 16.10
The negative feedback control of blood volume and blood osmolality.

Regulation by Aldosterone

From the preceding discussion, it is clear that a certain amount of dietary salt is required to maintain blood volume and pressure. Since Na^+ and Cl^- are easily filtered in the kidneys, a mechanism must exist to promote reabsorption and retention of salt when the dietary salt intake is too low. **Aldosterone** (*al-dos'ter-ōn*), a steroid hormone secreted by the adrenal cortex, stimulates the reabsorption of salt by the kidneys. Retention of salt indirectly promotes retention of water, and thereby helps to stabilize the blood volume.

Inside Information—Throughout most of human history, salt has been in short supply, and was therefore highly valued. Roman soldiers were paid partly in salt, a practice from which the word *salary* derives. Salt was also used to purchase slaves—hence the phrase "worth his salt."

Today, ironically, an excessive intake of salt has become a health hazard, particularly for people with hypertension (see "Clinical Considerations," p. 420). The daily intake of salt in many industrialized countries is 10 grams, or three times as much as is needed by most people.

The secretion of aldosterone is stimulated by salt deprivation, when the blood volume and pressure are reduced. When the blood flow and pressure are reduced in the renal artery, a group of cells in the kidneys called the **juxtaglomerular apparatus** secretes the enzyme **renin** into the blood. This enzyme cleaves a ten-amino-acid polypeptide called *angiotensin I* from a plasma protein called *angiotensinogen*. As angiotensin I passes through the capillaries of the lungs and other organs, an *angiotensin-converting enzyme* removes two amino acids. This leaves an eight-amino-acid polypeptide called **angiotensin II** (fig. 16.11). Conditions of salt deprivation, low blood volume, and low blood pressure, in summary, cause increased production of angiotensin II in the blood.

Angiotensin II promotes a rise in blood pressure in two ways. First, it directly stimulates vasoconstriction of arterioles, raising the peripheral resistance; second, it indirectly stimulates an increase in blood volume by acting on the adrenal cortex to promote aldosterone secretion. The relationship between angiotensin II and aldosterone is sometimes described as the *renin-angiotensin-aldosterone system*.

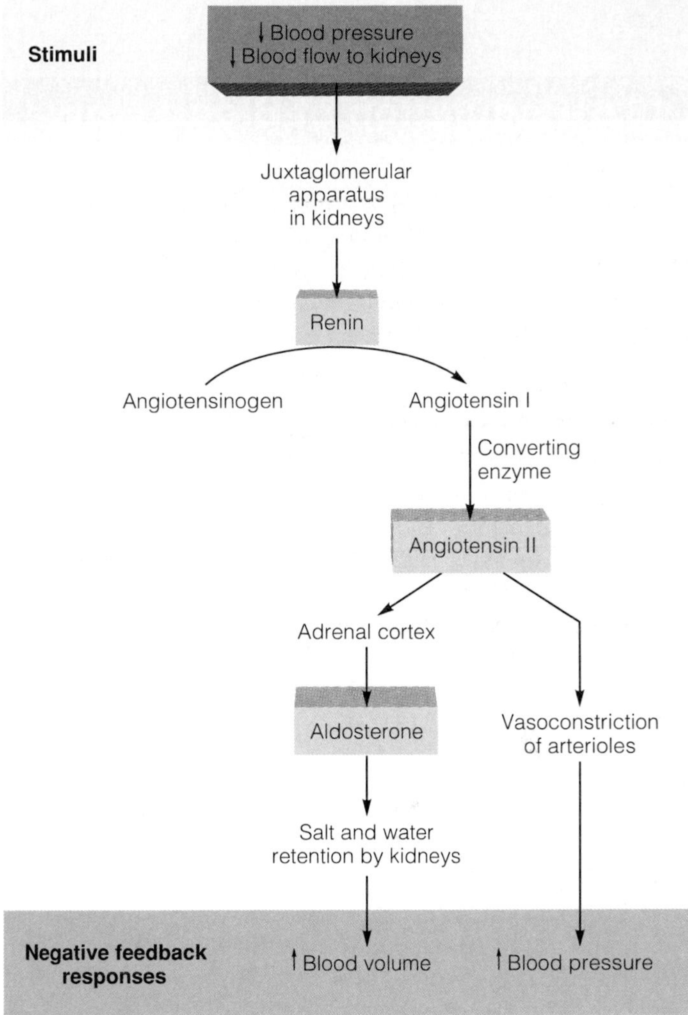

FIGURE 16.11
The negative feedback control of blood volume and pressure by the renin-angiotensin-aldosterone system.

One of the newer classes of drugs that can be used to treat hypertension (high blood pressure) are the angiotensin-converting enzyme, *or* ACE, inhibitors. *These drugs (such as* captopril*) block the formation of angiotensin II, and thus promote vasodilation, which decreases the total peripheral resistance. Because the workload of the heart is thereby reduced, these drugs are also useful in treating left ventricular hypertrophy and congestive heart failure.*

16.6 Regulation of Blood Flow

Blood flow through the arteries and arterioles of an organ can be controlled by constriction and dilation of the blood vessels. These changes in vessel diameter are regulated by sympathetic stimulation and by local conditions within the blood vessels and the organ.

Extrinsic Regulation of Blood Flow

There are two means of controlling and shifting blood flow—one from outside the system, the other from within. The term *extrinsic regulation* refers to control by the autonomic nervous system and endocrine system (table 16.3). This type of control differs from *intrinsic regulation,* in which regulatory mechanisms are built in, or intrinsic to, the cardiovascular system.

Activation of the sympathoadrenal system produces an increase in the cardiac output, and thus an increase in the total blood flow through the body. The binding of norepinephrine to alpha-adrenergic receptors (chapter 11, module 11.3) in the blood vessels of the skin and visceral organs leads to constriction of these blood vessels (vasoconstriction), and thus to a reduction in blood flow through the skin and viscera. When the sympathoadrenal system is activated in the fight-or-flight response, blood is diverted, or *shunted,* from the skin, GI tract, and kidneys.

Arterioles in skeletal muscles and the heart, by contrast, dilate during the fight-or-flight response. This vasodilation is partly produced by epinephrine secreted from the adrenal medulla, which stimulates beta-adrenergic receptors in the arterioles of those organs. During the fight-or-flight response, therefore, the viscera and skin receive a reduced blood flow due to the alpha-adrenergic effects of vasoconstriction in these organs, and the skeletal muscles receive an increased blood flow. This diversion of blood to the skeletal muscles during emergency conditions may give their response to the emergency an "extra edge."

Intrinsic Regulation of Blood Flow

Intrinsic control mechanisms are classified as either *myogenic* or *metabolic.* Some organs, the brain and kidneys in particular, use these intrinsic mechanisms to maintain relatively constant flow rates in the face of wide fluctuations in blood pressure. This ability is termed **autoregulation.**

Myogenic regulation can be seen in the cerebral arteries supplying blood to the brain. Changes in systemic arterial pressure are compensated for in the brain (and in some other organs) by the appropriate responses of vascular smooth muscle. A decrease in arterial pressure causes cerebral vessels to dilate; high blood pressure, by contrast, causes cerebral vessels to constrict. This myogenic regulation helps to maintain adequate cerebral blood flow yet protects against the damaging effects that uncompensated high blood pressure would have on vessels within the cranium.

Metabolic regulation occurs in the vessels of skeletal and cardiac muscles. The localized chemical conditions that promote vasodilation in these organs include

1. *decreased oxygen concentrations* that result from increased metabolic rate;
2. *increased carbon dioxide concentrations;*
3. *decreased tissue pH* (due to CO_2, lactic acid, and other metabolic products); and
4. *release of adenosine or* K^+ from the tissue cells.

The vasodilation that occurs in response to tissue metabolism can be demonstrated by constricting the blood supply to an area for a short time and then removing the constriction. The constriction allows metabolic products to accumulate; when the constriction is removed, the metabolic products that have accumulated cause vasodilation. The tissue thus appears red and feels warm. This response is called **reactive hyperemia** (*hi″per-e′me-ă*). A similar increase in blood flow occurs in skeletal muscles and the heart as a result of increased metabolism. This is called **active hyperemia.**

Intrinsic control mechanisms are summarized in table 16.4.

TABLE 16.3 **Extrinsic Control of Vascular Resistance and Blood Flow**

Extrinsic Agent	Effect	Comments
Sympathetic nerves		
Alpha-adrenergic	Vasoconstriction	Vasoconstriction is the dominant effect of sympathetic nerve stimulation on the vascular system, and it occurs throughout the body.
Beta-adrenergic	Vasodilation	There is some activity in arterioles in skeletal muscles and in coronary vessels, but effects are masked by dominant alpha-receptor-mediated constriction.
Cholinergic	Vasodilation	Effects are localized to arterioles in skeletal muscles and are produced only during a defense (fight-or-flight) reaction.
Parasympathetic nerves	Vasodilation	Effects are restricted primarily to the gastrointestinal tract, external genitalia, and salivary glands and have little effect on total peripheral resistance.
Angiotensin II	Vasoconstriction	A powerful vasoconstrictor produced as a result of secretion of renin from the kidneys, it may help to maintain adequate filtration pressure in the kidneys when systemic blood flow and pressure are reduced.
ADH (vasopressin)	Vasoconstriction	Although the effects of this hormone on vascular resistance and blood pressure in anesthetized animals are well documented, the importance of these effects in conscious humans is controversial.
Histamine	Vasodilation	Histamine promotes localized vasodilation during inflammation and allergic reactions.
Bradykinins	Vasodilation	Bradykinins are polypeptides secreted by endothelium and sweat glands that promote local vasodilation.
Prostaglandins	Vasodilation or vasoconstriction	Prostaglandins are cyclic fatty acids that can be produced by most tissues, including blood vessel walls. Prostaglandin I_2 is a vasodilator, whereas thromboxane A_2 is a vasoconstrictor. The physiological significance of these effects is presently controversial.

TABLE 16.4 **Intrinsic Control of Vascular Resistance and Blood Flow**

Category	Agent (↑ = increase; ↓ = decrease)	Mechanisms	Comments
Myogenic	↑ Blood pressure	Stretching of the arterial wall as the blood pressure rises directly stimulates increased smooth muscle tone (vasoconstriction).	It helps to maintain relatively constant rates of blood flow and pressure within an organ despite changes in systematic arterial pressure (autoregulation).
Metabolic	↓ Oxygen ↑ Carbon dioxide ↓ pH ↑ Adenosine ↑ K	Local changes in gas and metabolite concentrations act directly on vascular smooth muscle walls to produce vasodilation in the systemic circulation. The importance of different agents varies in different organs.	It aids in autoregulation of blood flow and also helps to shunt increased amounts of blood to organs with higher metabolic rates (active hyperemia).

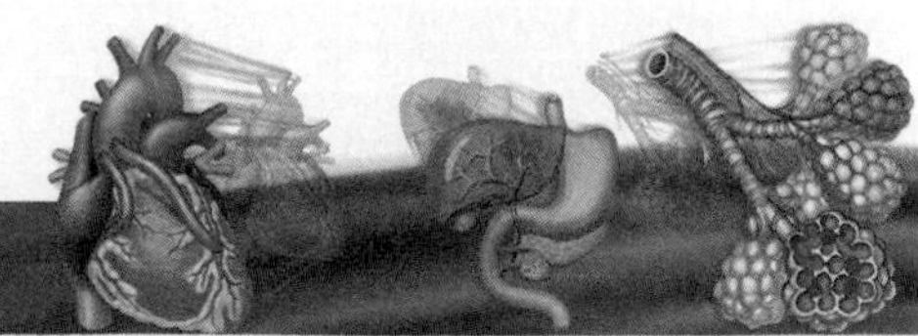

16.7 Regulation of Blood Pressure

Blood pressure in the arterial system is regulated by changes in the cardiac rate, blood volume, and total peripheral resistance that are coordinated by centers within the medulla oblongata.

Factors Affecting Blood Pressure

Resistance to flow in the arterial system is greatest in the arterioles because these vessels have the smallest diameters. Blood flow rate and pressure are thus reduced in the capillaries, which are located downstream of the high resistance imposed by the arterioles. The blood pressure and flow rate within the capillaries are further reduced by the fact that their total cross-sectional area is much greater, due to their larger number, than the cross-sectional areas of arteries and arterioles (fig. 16.12).

Variations in the diameter of arterioles due to vasoconstriction and vasodilation thus simultaneously affect both blood flow through capillaries and the *arterial blood pressure* "upstream" from the capillaries. The most important variables affecting blood pressure are the *cardiac rate, stroke volume* (determined primarily by the *blood volume*), and *total peripheral resistance.* An increase in any of these, if not offset by a decrease in another variable, will result in an increased blood pressure.

Blood pressure can thus be regulated by the kidneys, which control blood volume, and by the sympathoadrenal system. Increased activity of the sympathoadrenal system can raise blood pressure by stimulating vasoconstriction of arterioles (raising total peripheral resistance) and by promoting an increased cardiac output.

Baroreceptor Reflex

In order for blood pressure to be maintained within limits, specialized receptors for pressure are needed. **Baroreceptors** (*bar"o-re-sep'torz*) are stretch receptors located in the *aortic arch* and in the *carotid sinuses.* An increase in pressure causes the walls of these arterial regions to stretch and stimulate the activity of sensory nerve endings. A fall in pressure below the normal range, by contrast, reduces the frequency of action potentials produced by these sensory nerve fibers.

Sensory nerve activity from the baroreceptors ascends via the vagus and glossopharyngeal nerves to the medulla oblongata, which directs the autonomic nervous system to

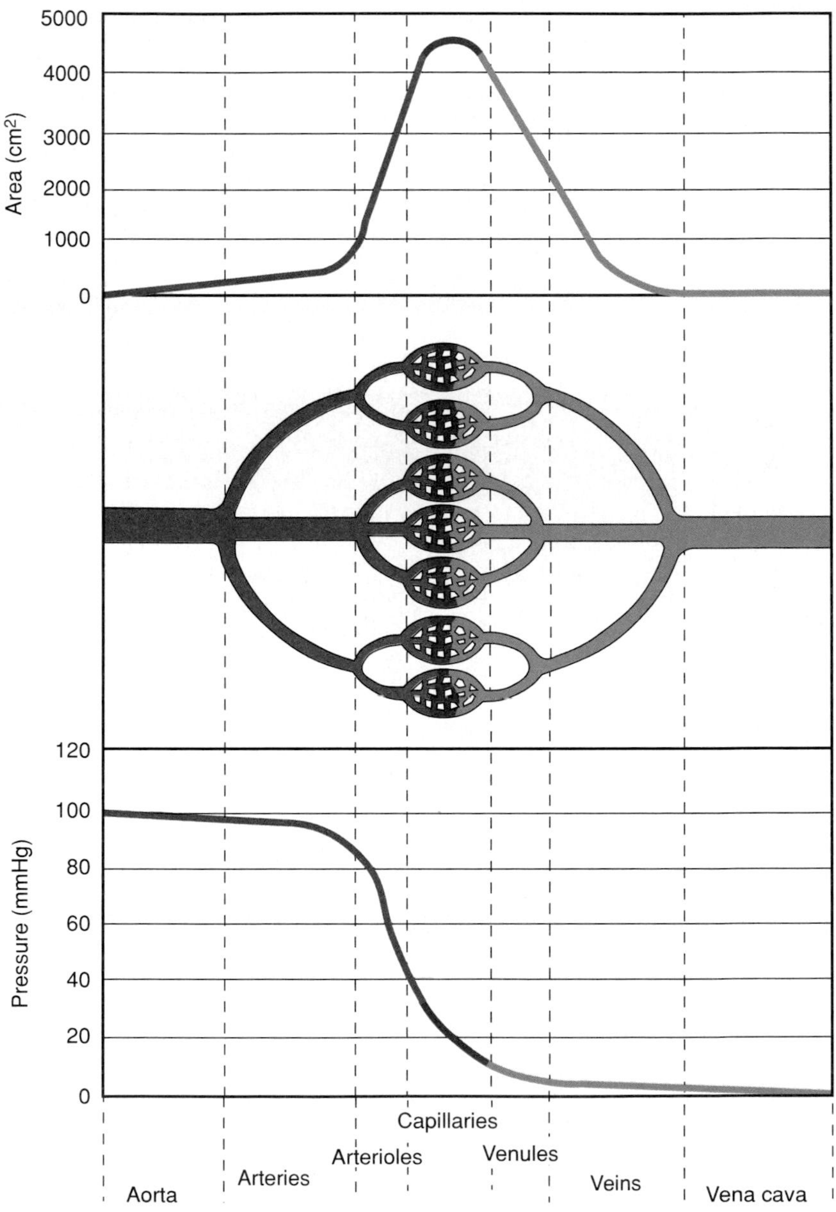

FIGURE 16.12
As blood passes from the aorta to the smaller arteries, arterioles, and capillaries, the cross-sectional area increases as the pressure decreases.

baroreceptor: Gk. *baros,* pressure; L. *receiver,* to receive

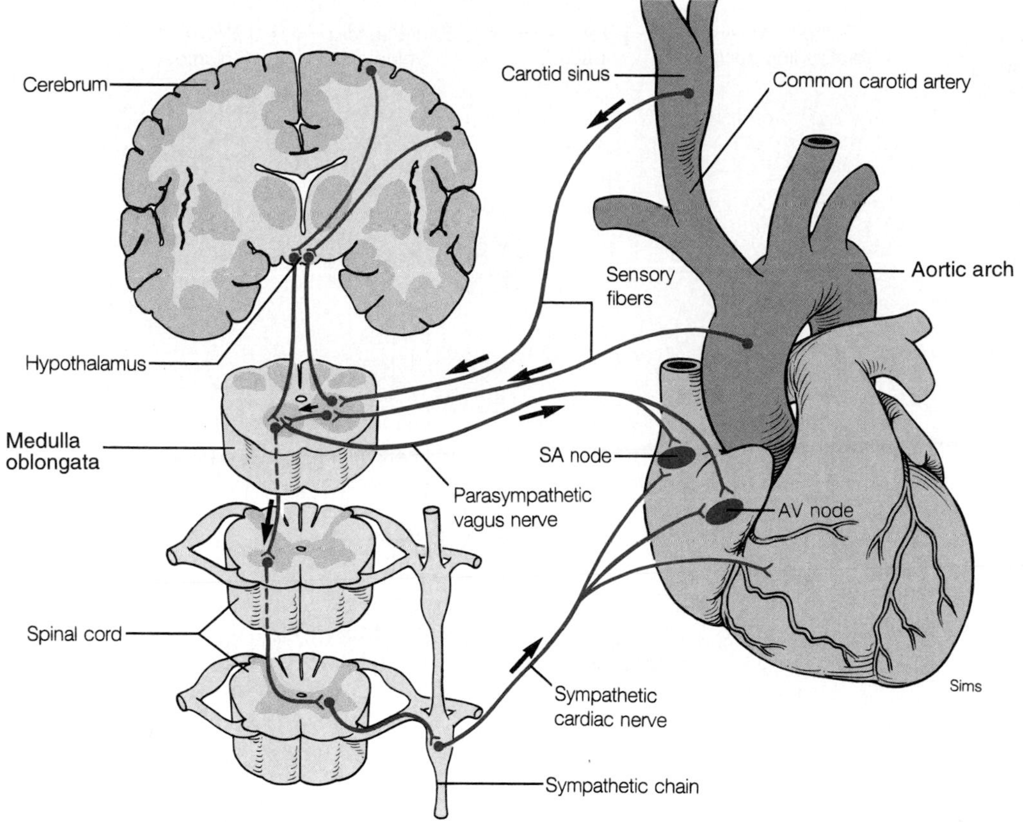

FIGURE 16.13

The baroreceptor reflex. Sensory stimuli from baroreceptors in the carotid sinus and the aortic arch, acting via control centers in the medulla oblongata, affect the activity of sympathetic and parasympathetic nerve fibers in the heart.

respond appropriately. **Vasomotor control centers** in the medulla oblongata control vasoconstriction/vasodilation, and hence help to regulate total peripheral resistance. **Cardiac control centers** in the medulla oblongata regulate the cardiac rate (fig. 16.13).

When a person goes from a reclining to a standing position, there is a shift of 500 to 700 ml of blood from the veins of the thoracic cavity to veins in the lower limbs, which expand to contain the extra volume of blood. This pooling of blood reduces the venous return and cardiac output. The resulting fall in blood pressure is almost immediately compensated for by the baroreceptor reflex. A decrease in baroreceptor sensory information traveling in the ninth and tenth cranial nerves to the medulla oblongata inhibits parasympathetic activity and stimulates sympathetic activity. This results in increased cardiac rate and vasoconstriction—responses that help to maintain an adequate blood pressure upon standing (fig. 16.14).

Because the baroreceptor reflex may require a few seconds to be fully effective, many people feel dizzy and disoriented if they stand up too quickly. If baroreceptor sensitivity is abnormally reduced, perhaps by atherosclerosis, an uncompensated fall in pressure may occur upon standing. This condition—called postural, *or* orthostatic, hypotension *(hypotension = low blood pressure)—can cause a person to feel extremely dizzy or even faint because of inadequate perfusion of the brain.*

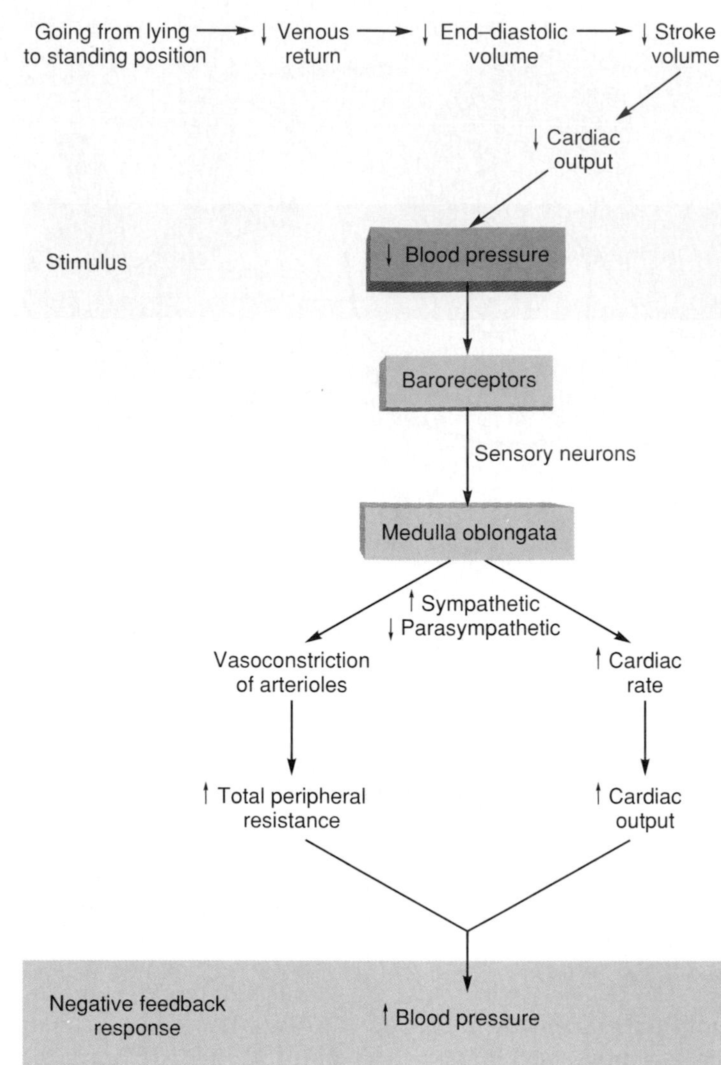

FIGURE 16.14
The negative feedback control of blood pressure by the baroreceptor reflex. This reflex helps to maintain an adequate blood pressure upon standing.

16.8 Measurement of Blood Pressure

Blood pressure is measured by means of sounds created by the turbulent flow of blood through a constricted artery.

A person's cardiovascular status can be assessed by taking pulse and blood pressure measurements. Most often, systemic arterial blood pressure is measured indirectly by the **auscultatory** (*aw-skul'tă-tor"e*) **method,** which is based on the correlation between blood pressure and arterial sounds. The auscultatory method involves the use of an apparatus called a **sphygmomanometer** (*sfig"mo-mă-nom'ĭ-ter*). An inflatable rubber bladder within a cloth cuff is wrapped around the upper arm, and a stethoscope is applied over the brachial artery (fig. 16.15). The artery is silent before inflation of the cuff because blood normally travels in a smooth *laminar flow* through the arteries. The term *laminar* means "layered"—blood in the central axial stream moves the fastest, and blood flowing closer to the artery wall moves more slowly. This smooth flow does not produce vibrations that create sounds.

When the artery is pinched, however, blood flow through the constriction becomes turbulent, causing the artery to vibrate and produce sounds. The ability of the cuff pressure to constrict the artery is opposed by the blood pressure. In order to constrict the artery, then, the cuff

auscultatory: L. *auscultare,* to listen to
sphygmomanometer: Gk. *sphygmos,* pulse; *manos,* thin; *metro,* measure

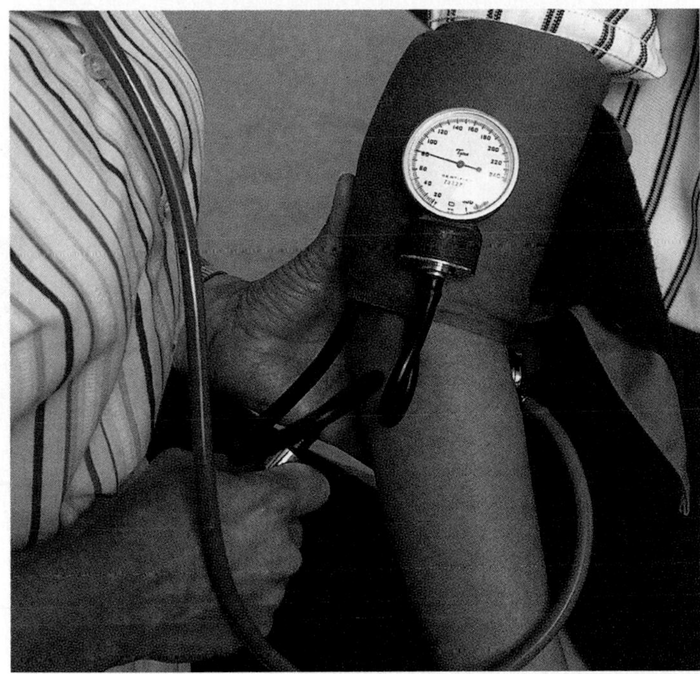

FIGURE 16.15
The use of a pressure cuff and sphygmomanometer to measure blood pressure.

pressure must be greater than the diastolic blood pressure. If the cuff pressure is also greater than the systolic blood pressure, the artery will be pinched off and silent. *Turbulent flow* and sounds produced by vibrations of the artery as a result of this flow occur only when the cuff pressure is greater than the diastolic pressure and less than the systolic pressure.

Suppose that a person has a systolic pressure of 120 mmHg and a diastolic pressure of 80 mmHg (the average normal values). When the cuff pressure is between 80 and 120 mmHg, the artery will be closed during diastole and open during systole. As the artery begins to open with every systole, turbulent flow of blood through the constriction will create vibrations that are heard as the **Korotkoff** (*kŏ-rot'kof*) **sounds,** as shown in figure 16.16. These are usually "tapping" sounds because the artery becomes constricted, blood flow stops, and silence resumes with every diastole. It should be understood that the Korotkoff sounds are not "lub-dub" sounds produced by closing of the heart valves. The latter sounds, discussed in chapter 15 (see module 15.2), can be heard only on the chest, not on the brachial artery.

Korotkoff sounds: from Nicolai S. Korotkoff, Russian physician (1874–1920)

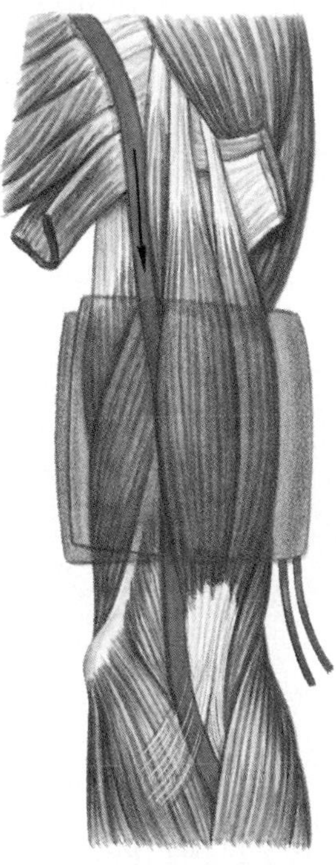

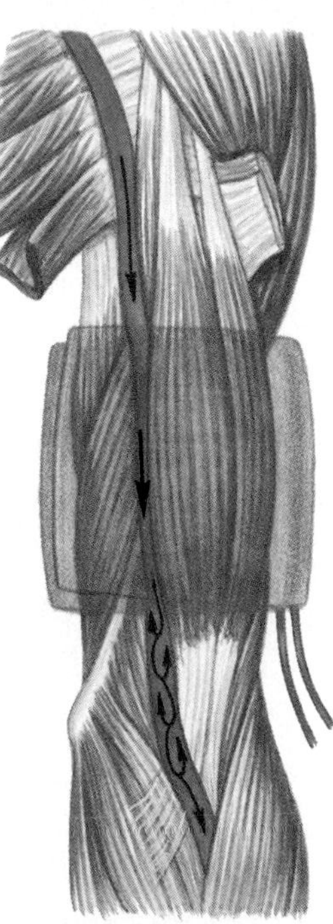

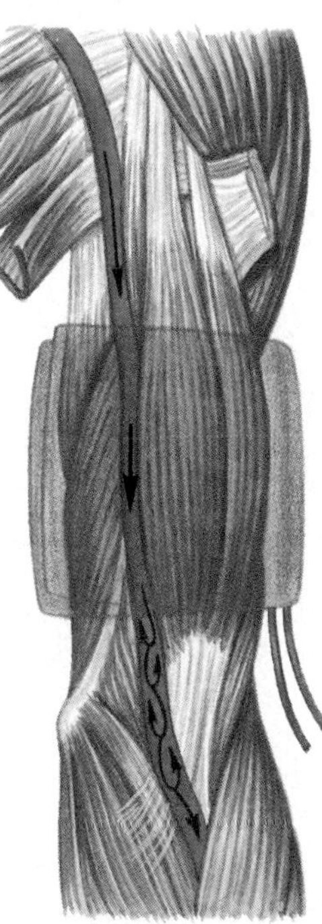

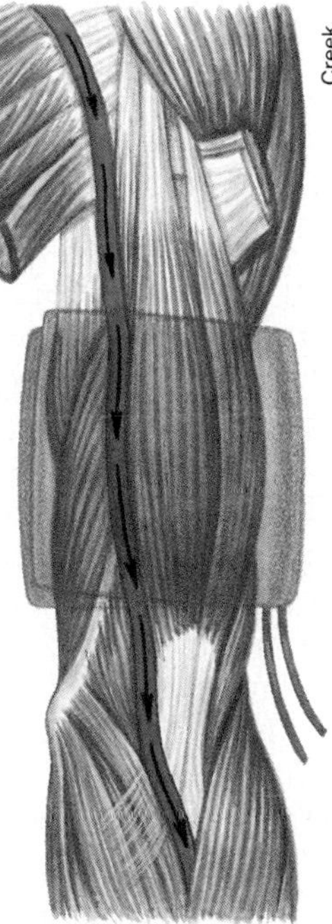

Blood pressure = 120/80

FIGURE 16.16
Korotkoff sounds are produced by the turbulent flow of blood through the partially constricted brachial artery. This occurs when the cuff pressure is greater than the diastolic pressure but less than the systolic pressure.

Initially, the cuff is inflated to produce a pressure greater than the systolic pressure so that the artery is pinched off and silent. The pressure in the cuff is read from the attached sphygmomanometer. A valve is then turned to allow slow release of air from the cuff, causing a gradual decrease in cuff pressure. When the cuff pressure is equal to the systolic pressure, the **first Korotkoff sound** is heard as blood flows turbulently through the constricted artery.

Korotkoff sounds will continue to be heard at every systole as long as the cuff pressure remains greater than

Clinical Considerations

Hypertension

Approximately 20% of all adults in the United States have *hypertension*—blood pressure in excess of the normal range for the person's age and sex. In general, a person has hypertension when the diastolic pressure is over 90 mmHg, and/or the systolic pressure is over 140 mmHg. Hypertension that is a result of ("secondary to") known disease processes is logically called **secondary hypertension.** Of the hypertensive population, secondary hypertension accounts for only about 10%. Hypertension that is the result of complex and poorly understood processes is not so logically called **primary,** or **essential, hypertension.**

Diseases of the kidneys and arteriosclerosis of the renal arteries can cause secondary hypertension because they tend to raise blood volume. More commonly, the reduction of renal blood flow can raise blood pressure by stimulating the secretion of vasoactive chemicals from the kidneys. Experiments in which the renal artery is pinched, for example, produce hypertension that is associated (at least initially) with elevated renin secretion.

Essential Hypertension The vast majority of people with hypertension have essential hypertension. An increased total peripheral resistance is a universal characteristic of this condition. Cardiac rate and the cardiac output are elevated in many, but not all, of these cases.

The secretion of renin, which is correlated with angiotensin II production and aldosterone secretion, is likewise variable. Although some people with essential hypertension have low renin secretion, most have either normal or elevated levels of renin secretion. Sustained high stress (acting via the sympathetic nervous system) and high salt intake appear to act synergistically in the development of hypertension.

The interactions between salt intake, sympathetic nerve activity, cardiovascular responses to sympathetic nerve activity, kidney function, and genetics make it difficult to sort out the cause-and-effect sequence that leads to essential hypertension. Many researchers have suggested that there is no single cause-and-effect sequence but rather a web of causes and effects. This view is currently controversial.

Dangers of Hypertension If other factors remain constant, blood flow increases as arterial blood pressure increases. The organs of people with hypertension are thus adequately perfused with blood until the excessively high pressure causes vascular damage. Because most patients have no symptoms until substantial vascular damage has occurred, hypertension is often referred to as a silent killer.

Hypertension is dangerous for the following reasons.

1. High arterial pressure makes it more difficult for the ventricles to eject blood. This increases the amount of work that the heart must perform and may result in pathological changes in heart structure and function, leading to congestive heart failure.
2. High arterial pressure may damage cerebral blood vessels, leading to cerebrovascular accident (stroke).
3. High arterial pressure contributes to the development of atherosclerosis, which can itself lead to heart disease and stroke.

Treatment of Hypertension Hypertension is usually treated by minimizing salt intake, stopping smoking, and losing weight, if applicable. Regular aerobic exercise may also be recommended. If lifestyle modifications alone are insufficient, various drugs may be prescribed. Most commonly, these drugs are diuretics that increase urine volume, thereby decreasing blood volume and pressure. Sympathetic-blocking drugs are also often used; drugs that block beta-adrenergic receptors (such as propranolol) decrease cardiac rate. Various vasodilators (see table 16.5) may also be used to decrease total peripheral resistance.

Inside Information—It is estimated that two-thirds of people over the age of 65 have elevated blood pressure. In fact, of all diseases, hypertension accounts for the largest number of outpatient prescriptions.

Circulatory Shock

Circulatory shock occurs when blood flow and/or oxygen utilization by the tissues is inadequate. Some of the signs of shock (table 16.6) are a result of inadequate tissue perfusion; other signs are produced by cardiovascular responses that help to compensate for the poor tissue perfusion (table 16.7). When these compensations are effective, they (together with emergency medical care) are able to reestablish adequate tissue perfusion. In some cases, however, and for reasons that are not clearly understood, the shock may progress to an irreversible stage, and death may result.

the diastolic pressure. When the cuff pressure becomes equal to or less than the diastolic pressure, the sounds disappear because the artery remains open, laminar flow is reestablished, and the vibrations stop. The **last Korotkoff sound** thus occurs when the cuff pressure is equal to the diastolic pressure.

In some people, the Korotkoff sounds do not disappear even when the cuff pressure is reduced to zero (zero pressure means that it is equal to atmospheric pressure). In these cases—and often routinely—the onset of muffling of the sounds is used rather than the onset of silence as an indication of diastolic pressure.

TABLE 16.5 Mechanisms of Action of Selected Antihypertensive Drugs

Category of Drugs	Examples	Mechanisms
Extracellular fluid volume depletors	Thiazide diuretics	Increase volume of urine excreted, and thus lower blood volume
Sympathoadrenal system inhibitors	Clonidine; alpha-methyldopa	Act to decrease sympathoadrenal stimulation by binding to α_2-adrenergic receptors in the brain
	Guanethidine; reserpine	Deplete norepinephrine from sympathetic nerve endings
	Propranolol; atenolol	Block beta-adrenergic receptors, decreasing cardiac output and/or renin secretion
	Phentolamine	Blocks alpha-adrenergic receptors, decreasing sympathetic vasoconstriction
Direct vasodilators	Hydralazine; sodium nitroprusside	Cause vasodilation by acting directly on vascular smooth muscle
Calcium channel blockers	Verapamil	Inhibits diffusion of Ca^{++} into vascular smooth muscle cells, causing vasodilation and reduced peripheral resistance
Angiotensin-converting enzyme (ACE) inhibitors	Captopril	Inhibits the conversion of angiotensin I into angiotensin II

TABLE 16.6 Signs of Shock

	Early Sign	Late Sign
Blood pressure	Decreased pulse pressure Increased diastolic pressure	Decreased systolic pressure
Urine	Decreased Na^+ concentration Increased osmolality	Decreased volume
Blood pH	Increased pH (alkalosis) due to hyperventilation	Decreased pH (acidosis) due to "metabolic" acids
Effects of poor tissue perfusion	Slight restlessness; occasionally warm, dry skin	Cold, clammy skin; "cloudy" senses

TABLE 16.7 Cardiovascular Reflexes That Help To Compensate for Circulatory Shock

Organ(s)	Compensatory Mechanisms
Heart	Sympathoadrenal stimulation produces increased cardiac rate and increased stroke volume due to stimulation of myocardial contractility
Digestive tract and skin	Decreased blood flow due to vasoconstriction as a result of sympathetic nerve stimulation (alpha-adrenergic effect)
Kidneys	Decreased urine production as a result of sympathetic-nerve-induced constriction of renal arterioles; increased salt and water retention due to increased aldosterone and antidiuretic hormone (ADH) secretion

SUMMARY

16.1 Cardiac Output and Cardiac Rate

1. The cardiac output is the amount of blood pumped per minute by each ventricle.
2. The cardiac rate is regulated by sympathetic and parasympathetic nerve stimulation.

16.2 Regulation of Stroke Volume

1. The stroke volume is directly proportional to the end-diastolic volume (EDV) and contractility; it is inversely proportional to the total peripheral resistance.
2. According to the Frank–Starling law of the heart, an increase in EDV causes the myocardium to become increasingly stretched, which, in turn, allows the myocardium to contract more forcefully.

16.3 Venous Return

1. The venous return is increased by the massaging of veins that occurs when surrounding skeletal muscles contract, acting as "skeletal muscle pumps."
2. Some blood plasma is filtered out of capillaries and becomes interstitial (tissue) fluid.

16.4 Exchange of Fluid across Capillaries

1. Fluid in the blood is filtered out of capillaries at their arteriolar end and returned to capillaries at their venular end.
2. The hydrostatic pressure and oncotic pressure forces that determine the direction of fluid movement are called Starling forces.
3. Excess interstitial (tissue) fluid is drained by the lymphatic system. An accumulation of excess tissue fluid is called edema.

16.5 Renal Regulation of Blood Volume

1. Urine formation begins in the kidneys as blood plasma is filtered through capillaries called glomeruli.
2. The blood volume is thus an inverse function of the urine volume, which is determined by the percentage of the glomerular filtrate reabsorbed.
3. Antidiuretic hormone (ADH) promotes the reabsorption of water into the blood, thus raising blood volume.
4. Aldosterone promotes the reabsorption of NaCl and water. Aldosterone secretion is stimulated by angiotensin II.

16.6 Regulation of Blood Flow

1. Blood flow is regulated extrinsically by sympathetic nerve stimulation, which promotes vasoconstriction in the viscera and skin and vasodilation in skeletal muscles.
2. Myogenic regulation occurs in the cerebral circulation, in which the vessels constrict or dilate in response to the blood pressure.
3. Metabolic vasodilation occurs in exercising skeletal muscles and in the heart. This intrinsic response is called active hyperemia.

16.7 Regulation of Blood Pressure

1. Blood pressure is directly related to the cardiac rate, blood volume, and total peripheral resistance.
2. Changes in blood pressure are detected by baroreceptors that relay the information to control centers in the medulla oblongata. These centers coordinate appropriate autonomic nerve responses.

16.8 Measurement of Blood Pressure

1. Korotkoff sounds are produced by turbulent blood flow through a constricted artery.
2. The first Korotkoff sound occurs at the systolic pressure; the last sound occurs at the diastolic pressure.

Review Activities

Objective Questions

1. Which of the following statements about the cardiac rate is *true?* (*module 16.1*)
  a. It is raised by increased sympathetic nerve activity.
  b. It is raised by decreased parasympathetic nerve activity.
  c. It is determined by the rate of spontaneous depolarization in the SA node.
  d. All of the above are true.

2. According to the Frank–Starling law of the heart, the strength of ventricular contraction is (*module 16.2*)
  a. directly proportional to the end-diastolic volume.
  b. inversely proportional to the end-diastolic volume.
  c. independent of the end-diastolic volume.

3. In the absence of compensations, the stroke volume will decrease when (*module 16.2*)
  a. blood volume increases.
  b. venous return increases.
  c. contractility increases.
  d. arterial blood pressure increases.

4. The average venous pressure is (*module 16.3*)
  a. 2 mmHg.
  b. 10 mmHg.
  c. 80 mmHg.
  d. 120 mmHg.

5. Which of the following statements about interstitial fluid is *false?* (*module 16.4*)
   a. It contains the same glucose and salt concentrations as blood plasma.
   b. It contains a lower protein concentration than blood plasma.
   c. Its oncotic pressure is greater than that of blood plasma.
   d. Its hydrostatic pressure is less than that of blood plasma.
6. An increase in blood volume will cause (*module 16.5*)
   a. a decrease in ADH secretion.
   b. a decrease in renin secretion.
   c. an increase in Na^+ excretion in the urine.
   d. all of the above.
7. Both ADH and aldosterone act to increase (*module 16.5*)
   a. urine volume.
   b. blood volume.
   c. total peripheral resistance.
   d. all of the above.
8. The greatest resistance to blood flow occurs in (*module 16.6*)
   a. large arteries.
   b. medium-sized arteries.
   c. arterioles.
   d. capillaries.
9. Vasodilation in the heart and skeletal muscles during exercise is primarily due to the effects of (*module 16.6*)
   a. alpha-adrenergic stimulation.
   b. beta-adrenergic stimulation.
   c. cholinergic stimulation.
   d. products released by the exercising muscle cells.
10. Blood pressure is lowest in (*module 16.7*)
    a. arteries.
    b. arterioles.
    c. capillaries.
    d. venules.
    e. veins.
11. The Korotkoff sounds are produced by (*module 16.8*)
    a. closing of the semilunar valves.
    b. closing of the AV valves.
    c. turbulent flow of blood through an artery.
    d. elastic recoil of the aorta.
12. If a person has a blood pressure of 130/86, (*module 16.8*)
    a. the first sound is heard at a pressure of 86 mmHg.
    b. the last sound is heard at a pressure of 130 mmHg.
    c. sounds are heard at every systole between 130 and 86 mmHg.
    d. all of the above apply.

Essay Questions

1. Describe how the cardiac output is affected by the cardiac rate and stroke volume. (*module 16.1*)
2. Explain how the cardiac rate is regulated. (*module 16.2*)
3. Using the Frank–Starling law, explain how the stroke volume is affected by (a) bradycardia and (b) a "missed beat." (*module 16.2*)
4. Which part of the circulatory system contains the most blood? Which part provides the greatest resistance to blood flow? Which part provides the greatest cross-sectional area? Explain. (*module 16.3*)
5. How is interstitial fluid formed? How is it returned to the vascular system? (*module 16.4*)
6. Explain how the kidneys regulate blood volume. (*module 16.5*)
7. A person who is dehydrated drinks more and urinates less. Explain the mechanisms involved. (*module 16.5*)
8. Describe the mechanisms that increase blood flow to skeletal muscles during exercise. (*module 16.6*)
9. With reference to the baroreceptor reflex, explain why a person who is dehydrated or who has lost a lot of blood has a rapid pulse and cold, clammy skin. (*module 16.7*)
10. Explain why blood flow through an artery is normally silent and why Korotkoff sounds are heard. (*module 16.8*)

Labeling Exercise

Label the structures indicated on the figure to the right.

1. ______________________________
2. ______________________________
3. ______________________________
4. ______________________________

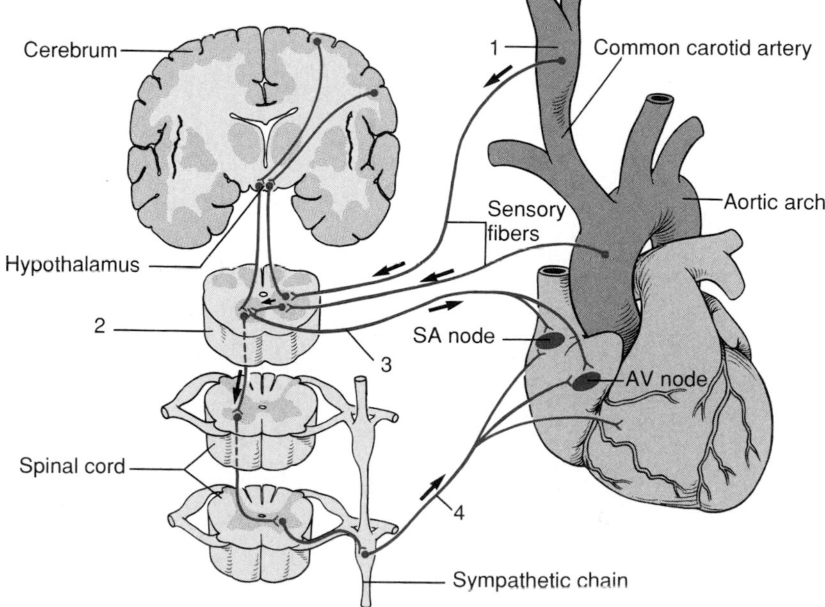

Critical Thinking Questions

1. You and your friends munch away happily on pretzels and salted popcorn during Superbowl Sunday. Could this temporarily affect your blood volume and pressure? If so, how? Why is this effect probably temporary? Explain the mechanisms involved.
2. You enter the Olympic stadium and the crowd goes wild! Thinking about your next gold medal, you notice that your palms are sweaty and your skin is cold. What mechanisms produce these effects? After your race, you notice that now your skin is hot and flushed. What produces these effects?
3. During your next Olympic event, the marathon race, you notice that people hand you drinks that contain sugar and salt. Why are these drinks better than pure water during a prolonged bout of exercise?
4. During your exercise, it suddenly occurs to you that your rapidly beating heart has less time to fill with blood between beats. You panic for a moment, but then your fear subsides when you remember that your stroke volume doesn't decrease, but actually increases, during normal exercise. What mechanisms produce this effect?
5. Pregnant women often develop swollen ankles. Why? (Hint: Think of the location of the fetus and its effect on maternal blood flow.)

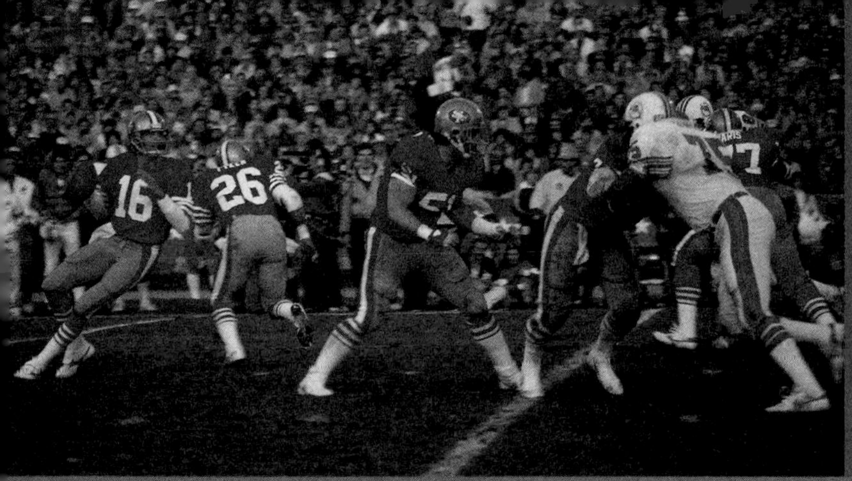

CHAPTER 17

"The body puts up an amazing variety of defenses to resist infection and harm from outside agents." *page 429*

Lymphatic System and Immunity

Chapter Outline

Terms to Remember

lymph, 426
lymph node, 426
spleen, 428
thymus (*thi'mus*), 428
nonspecific defense mechanisms, 429
specific immune response, 429
phagocytosis (*fag"ŏ-si-to'sis*), 430
Kupffer (*koop'fer*) cell, 430
diapedesis (*di"ă-pĕ-de'sis*), 430
interferon (*in"ter-fēr'on*), 431
antigen (*an'tĭ-jen*), 432
T lymphocyte, 432
B lymphocyte, 432
humoral immunity, 433
cell-mediated immunity, 433
plasma cell, 433
immunoglobulin (*im"yŭ-no-glob'yŭ-lin*), 433
antibody, 433
active immunity, 435
passive immunity, 436
lymphokine (*lim'fŏ-kīn*), 437
major histocompatibility complex (MHC), 439

Learning Objectives

After studying this chapter, you should be able to

- describe the structure and function of lymph vessels and lymph nodes, and identify the lymphatic organs.
- distinguish between nonspecific and specific defense mechanisms.
- identify the phagocytic cells of the immune system and explain the significance of interferons.
- explain what is meant by an antigen and identify the different types of lymphocytes.
- explain the function of B lymphocytes in immunity and describe the chemical nature of antibodies.
- discuss the clonal selection theory of active immunity and explain how passive immunity is produced.
- describe the functions of the subcategories of T lymphocytes and of different lymphokines.
- explain how T lymphocytes are activated and how killer T lymphocytes destroy a target cell.

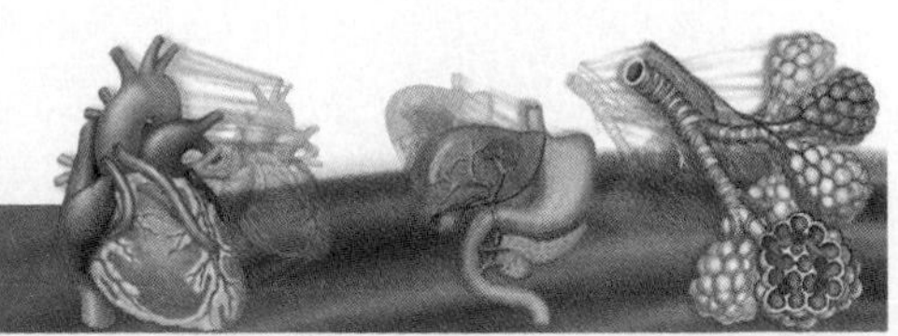

17.1 Lymphatic System

The lymphatic system, consisting of lymphatic vessels and lymph nodes, helps to maintain fluid balance in tissues and to absorb fats from the gastrointestinal tract. It also is part of the body's defense system against disease.

The lymphatic system is perhaps the least familiar of the body systems, yet without it the cardiovascular system would not have sufficient blood volume to operate properly. As discussed in the previous chapter (see module 16.4), a network of lymphatic vessels drains excess interstitial fluid (the approximate 15% that has not been returned directly to the capillaries) and returns it to the bloodstream in a one-way flow that moves slowly toward the heart. Additionally, the lymphatic system functions in fat absorption and in the body's defense against microorganisms and other foreign substances.

In short, the lymphatic system has three principal functions.

1. It transports interstitial (tissue) fluid, which was initially formed as a blood filtrate, back to the blood.
2. It serves as the route by which absorbed fats and some vitamins are transported from the small intestine to the blood.
3. Its cells—called **lymphocytes**—help to provide immunological defenses against disease-causing agents.

Lymph and Lymph Vessels

The lymphatic network of vessels begins with the microscopic **lymphatic capillaries** (see fig. 16.8). Lymphatic capillaries are closed-ended tubes that form vast networks in the intercellular spaces within most tissues. Within the villi of the small intestine, for example, lymphatic capillaries, called *lacteals* (*lak'te-alz*), transport the products of fat absorption away from the gastrointestinal tract (see fig. 20.19). Because the walls of lymphatic capillaries are composed of endothelial cells with porous junctions, interstitial fluid, proteins, microorganisms, and absorbed fat (in the small intestine) can easily enter. Once fluid enters the lymphatic capillaries, it is referred to as **lymph.**

From merging lymphatic capillaries, the lymph is carried into larger lymphatic vessels called **lymph ducts.** The walls of lymph ducts are much like those of veins. They have the same three layers and also contain valves to prevent backflow. Unlike blood, which has the pumping action of the heart to move it along, the pressure that keeps lymph moving comes from the massaging action produced by skeletal muscle contractions and intestinal movements, and from peristaltic contractions of some lymphatic vessels. The many valves keep lymph moving in one direction.

Interconnecting lymph ducts eventually empty into one of the two principal lymph vessels: the **thoracic duct** and the **right lymphatic duct** (fig. 17.1). These ultimately drain the lymph into the left and right subclavian veins, respectively, so that the lymphatic fluid is returned to the circulatory system.

The larger thoracic duct drains lymph from the lower extremities, abdomen, left thoracic region, left upper extremity, and left side of the head and neck. The main trunk of this vessel ascends along the spinal column and empties into the left subclavian vein. In the abdominal area is a saclike enlargement of the thoracic duct called the **cisterna chyli** (*sis-ter'nă ki'le*) (fig. 17.1). The smaller right lymphatic duct drains lymph vessels from the right upper extremity, right thoracic region, and right side of the head and neck. The right lymphatic duct empties into the right subclavian vein near the right internal jugular vein.

Lymph Nodes

Lymph filters through the reticular tissue of hundreds of lymph nodes clustered along the lymphatic vessels (fig. 17.2). The reticular tissue contains phagocytic cells that help to purify the fluid. Lymph nodes are small oval bodies enclosed within fibrous connective tissue capsules. Afferent lymphatic vessels deliver lymph into the node, where it is circulated through sinuses in the cortical tissue. Lymph leaves the node through the efferent lymphatic vessel. **Germinal centers** within the node are sites of lymphocyte production, and are thus important in the development of an immune response.

Clusters of lymph nodes usually occur in specific regions of the body. Some of the principal groups of lymph nodes are the **popliteal** and **inguinal nodes** of the lower

---

lacteal: L. *lacteus,* milk
lymph: L. *lympha,* clear water
cisterna chyli: L. *cisterna,* box; *chylos,* juice

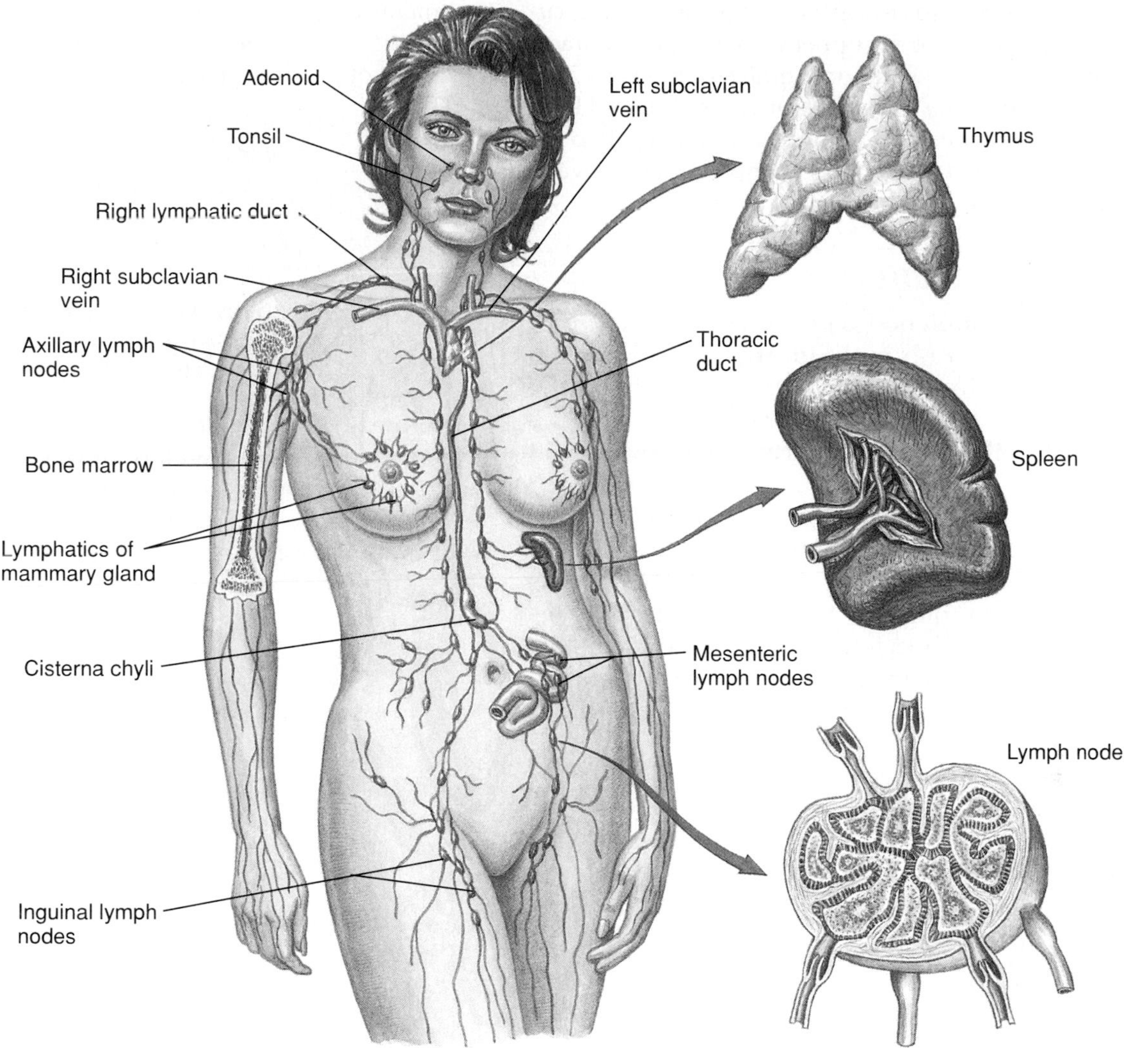

FIGURE 17.1
A diagram of the lymphatic system showing some of the principal lymph nodes and other lymphoid organs. Lymph from the upper right extremity, the right side of the head and neck, and the right thoracic region drains through the right lymphatic duct into the right subclavian vein. Lymph from the remainder of the body drains through the thoracic duct into the left subclavian vein.

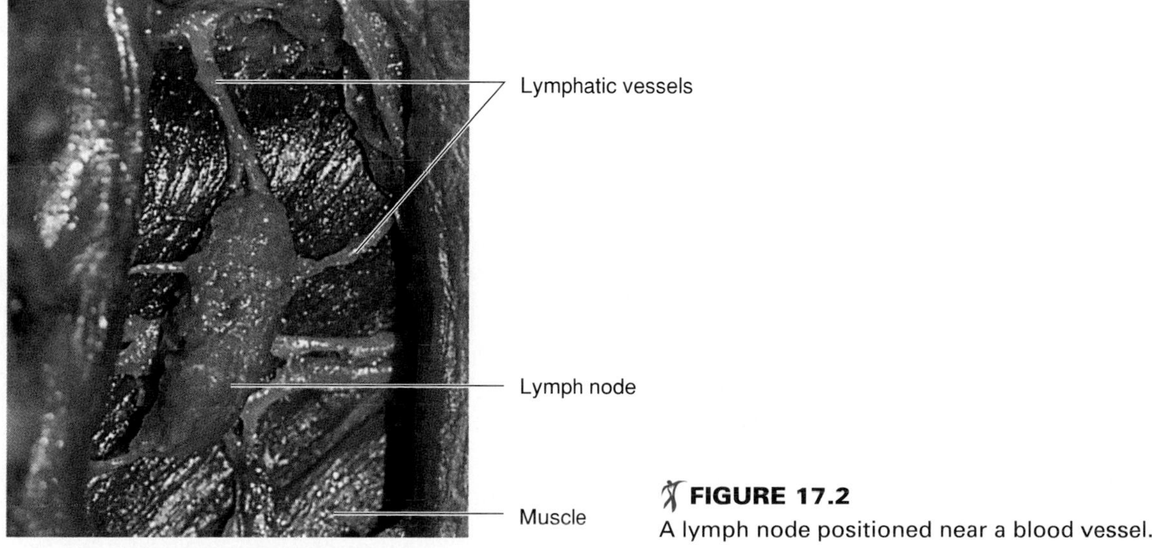

FIGURE 17.2
A lymph node positioned near a blood vessel.

extremity, the **lumbar nodes** of the pelvic region, the **cubital** and **axillary nodes** of the upper extremity, the **thoracic nodes** of the chest, and the **cervical nodes** of the neck. The wall of the small intestine contains numerous scattered lymphocytes and lymphatic nodules and larger aggregations of lymphatic tissue called **mesenteric lymph nodes** (Peyer's patches) (see fig. 17.1).

Other Lymphoid Organs

In addition to the lymph nodes just described, the *tonsils, spleen,* and *thymus* are lymphoid organs. The **tonsils** form a protective ring of lymphatic tissue around the openings between the nasal and oral cavities and the pharynx. They will be discussed further in chapter 18 (see module 18.1).

The **spleen** (fig. 17.3), the largest mass of lymphatic tissue, is located on the left side of the abdominal cavity, to the left of the stomach from which it is suspended. The spleen is not a vital organ in an adult, but it does assist other body organs in producing lymphocytes, filtering the blood, and destroying old erythrocytes. In an infant, the spleen is an important site for the production of erythrocytes. In an adult, it contains *red pulp,* which serves to destroy old erythrocytes, and *white pulp,* which contains germinal centers for the production of lymphocytes.

Of all of the abdominal organs, the spleen is the one most easily and frequently injured. Because it serves as a reservoir for blood, extensive—sometimes massive—hemorrhage occurs when the spleen is ruptured. To prevent death from loss of blood, a splenectomy *(removal of the spleen) is performed. Without immediate surgery for a ruptured spleen, the mortality rate is 90%.*

The **thymus** (*thi'mus*) (fig. 17.4) is located in the anterior thorax, deep to the manubrium of the sternum. Because it regresses in size during puberty, it is much larger in a fetus and child than in an adult. The thymus plays a key role in the immune system, as will be described in module 17.3.

The lymphoid organs of the body are summarized in table 17.1.

spleen: L. *splen,* low spirits (thought to cause melancholy)
thymus: Gk. *thymos,* thyme (compared to the flowers of this plant by Galen)

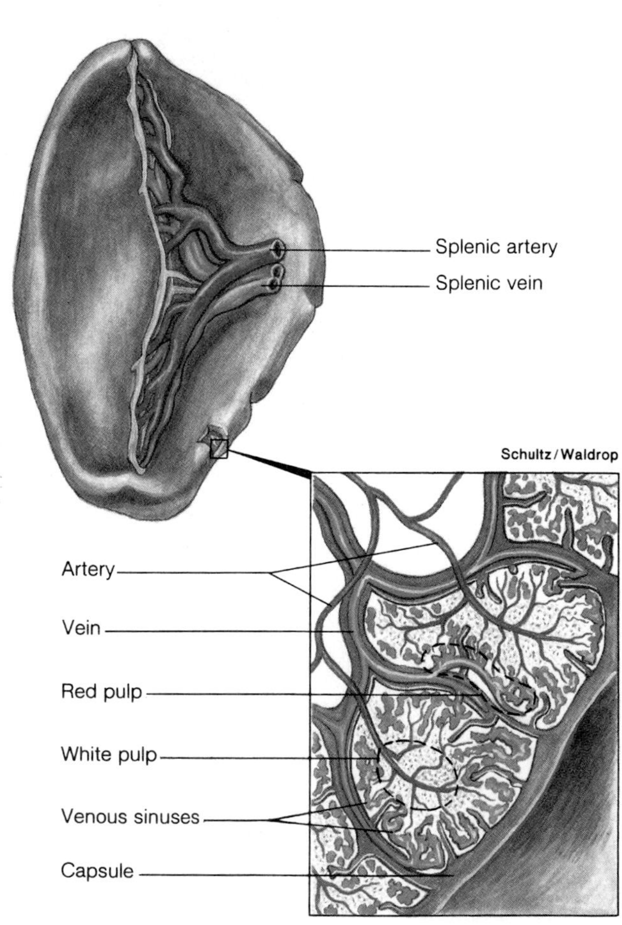

FIGURE 17.3
The structure of the spleen.

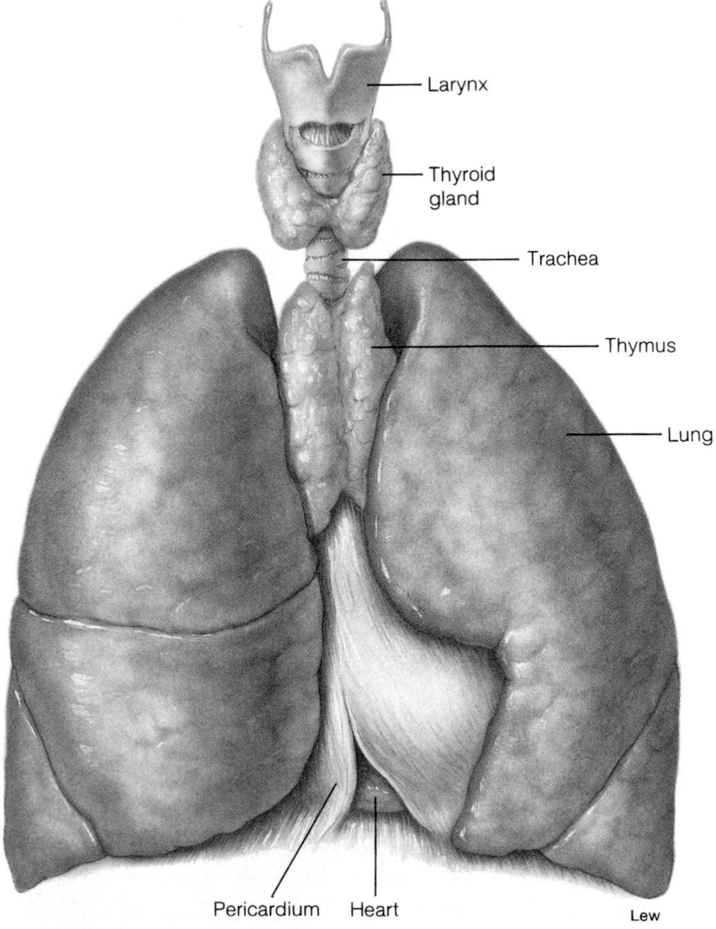

FIGURE 17.4
The location and structure of the thymus.

TABLE 17.1 **Lymphoid Organs**

Organ	Location	Function
Lymph nodes	In clusters or chains along the paths of larger lymphatic vessels	Sites of lymphocyte production; house T lymphocytes and B lymphocytes that are responsible for immunity; phagocytes filter foreign particles and cellular debris from lymph
Tonsils	In a ring at the junction of the oral cavity and pharynx	Protect against invasion of foreign substances that are inhaled or ingested
Spleen	In upper left portion of abdominal cavity, beneath the diaphragm and suspended from the stomach	Serves as blood reservoir; phagocytes filter foreign particles, cellular debris, and worn erythrocytes from the blood; houses lymphocytes
Thymus	Within the mediastinum, behind the manubrium	Important site of immunity in a child; houses lymphocytes; changes undifferentiated lymphocytes into T lymphocytes

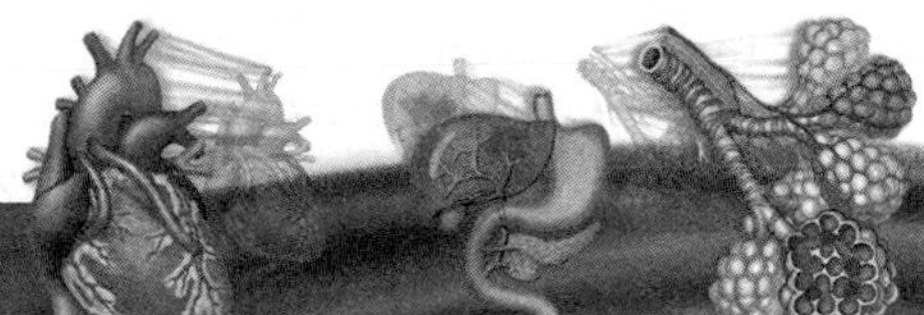

17.2 Defense Mechanisms

The defense mechanisms of the body guard against the disruption of homeostasis by disease-causing agents. They include nonspecific and specific mechanisms.

The body puts up an amazing variety of defenses to resist infection and harm from outside agents. These defenses may be grouped into two broad categories: *nonspecific* and *specific defense mechanisms.*

Nonspecific, or **innate, defense mechanisms** are inherited as part of the structure of each organism. Epithelial membranes that cover the body surfaces, for example, restrict infection by most harmful microorganisms. The strong acidity of gastric juice (pH 1–2) also helps to kill many microorganisms before they can invade the body. These external defenses are backed by internal defenses, such as phagocytosis, which function in both a specific and nonspecific manner (table 17.2).

Each individual can also acquire the ability to defend against specific pathogens (disease-causing agents) by a prior exposure to those pathogens. This **specific,** or **acquired, immune response** is a function of lymphocytes. Internal specific and nonspecific defense mechanisms work hand-in-hand to combat infection, with lymphocytes interacting in a coordinated effort with phagocytic cells.

TABLE 17.2 **Structures and Defense Mechanisms of Nonspecific Immunity**

	Structure	Mechanisms
External	Skin	Physical barrier to penetration by pathogens; secretions contain lysozyme (enzyme that destroys bacteria)
	Gastrointestinal tract	High acidity of stomach; protection by normal bacterial population of colon
	Respiratory tract	Secretion of mucus; movement of mucus by cilia; alveolar macrophages
	Urinary tract	Acidity of urine
	Female reproductive tract	Vaginal lactic acid
Internal	Phagocytic cells	Ingest and destroy bacteria, cellular debris, denatured proteins, and toxins
	Interferons	Inhibit replication of viruses
	Complement proteins	Promote destruction of bacteria and other effects of inflammation
	Endogenous pyrogen	Secreted by leukocytes and other cells; produces fever

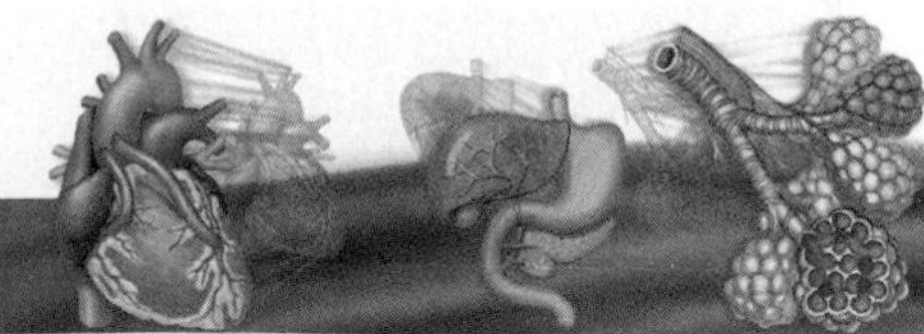

17.3 Nonspecific Defense Mechanisms

Nonspecific defense mechanisms include features of the skin and mucous membranes, phagocytosis, fever, and interferons.

External nonspecific defenses are the body's first line of defense against invading pathogens, such as bacteria. If the pathogens are successful in crossing epithelial barriers, however, they enter connective tissues. These invaders—or chemicals called *toxins* produced by them—may then enter blood or lymphatic capillaries and be carried to other areas of the body. To counter the invasion and spread of infection, nonspecific internal defenses are first employed. These include phagocytosis, fever, and interferon production. If these defenses are not sufficient to destroy the pathogens, lymphocytes may be recruited, and their specific actions used to reinforce the nonspecific defenses.

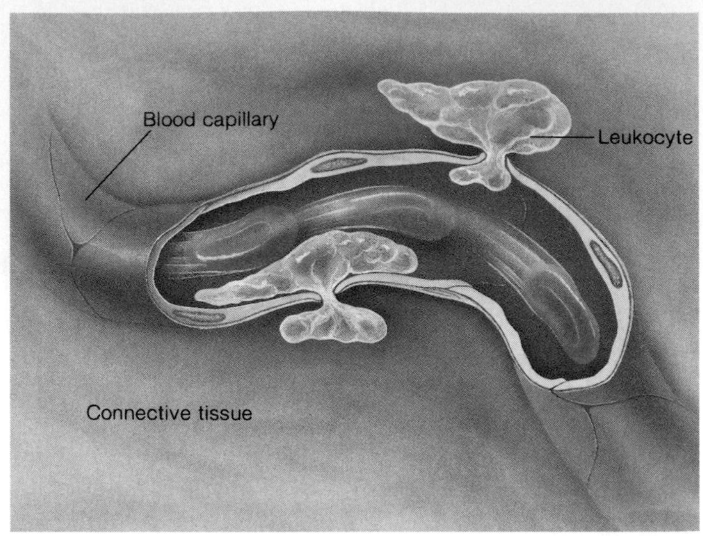

Collins

FIGURE 17.5
Diapedesis. White blood cells squeeze through openings between capillary endothelial cells to enter the connective tissue from the bloodstream.

Phagocytosis

Phagocytosis (*fag"ŏ-si-to'sis*) is the ingestion, killing, and digestion of bacteria and other small particles through the action of enzymes. The three major groups of phagocytic cells are

1. **neutrophils,** the most abundant type of leukocyte, constituting 54%–62% of the leukocytes in the blood;
2. the cells of the mononuclear phagocyte system, including **monocytes** in blood and **macrophages** (derived from monocytes) in other connective tissues; and
3. **organ-specific phagocytes** in the liver, spleen, lymph nodes, lungs, and brain.

The *Kupffer* (*koop'fer*) *cells* in the liver, together with phagocytic cells in the spleen and lymph nodes, are **fixed phagocytes**—so called because they are unable to move in the channels within these organs. As blood flows through the liver and spleen, and as lymph percolates through the lymph nodes, foreign chemicals and debris are removed by these stationary phagocytes and chemically inactivated within their cytoplasm.

In the connective tissues, a resident population of all leukocyte types scavenge for invaders and cellular debris. If the infection is sufficiently large, new phagocytic cells from the blood may join those already in the connective tissue. These new neutrophils and monocytes are able to squeeze through the tiny gaps between adjacent endothelial cells in the capillary wall and enter the connective tissues. This process, called **diapedesis** (*di"ă-pĕ-de'sis*), is illustrated in figure 17.5.

The neutrophils are mature phagocytes, ready to digest bacteria, viruses, and other foreign matter. The immature monocytes, on the other hand, are not very effective until they enter the connective tissues, at which point they begin to swell to four or five times their original size. To match their new stature, they are given a new name—*macrophage;* literally, "giant eater." Macrophages are considerably more phagocytic than the smaller neutrophils; in fact one of the jobs of macrophages is to help clear the infected area of neutrophils that have been destroyed.

Inside Information—Neutrophils can be considered the body's Kamikaze warriors. They do a thorough job of warding off invaders, but they poison themselves in the process. The more resilient macrophages release their digestive end-products by exocytosis, enabling them to survive (and to keep on eating) for months or even years.

As described in chapter 3 (see module 3.3), phagocytic cells engulf particles in much the same way as an amoeba eats. The particle becomes surrounded by cytoplasmic extensions called pseudopods, which ultimately fuse together (fig. 17.6). This forms a vacuole, which then fuses

Kupffer cell: from Karl W. Kupffer, German anatomist, 1829–1902
diapedesis: Gk. *dia,* through; *pedesis,* a leaping

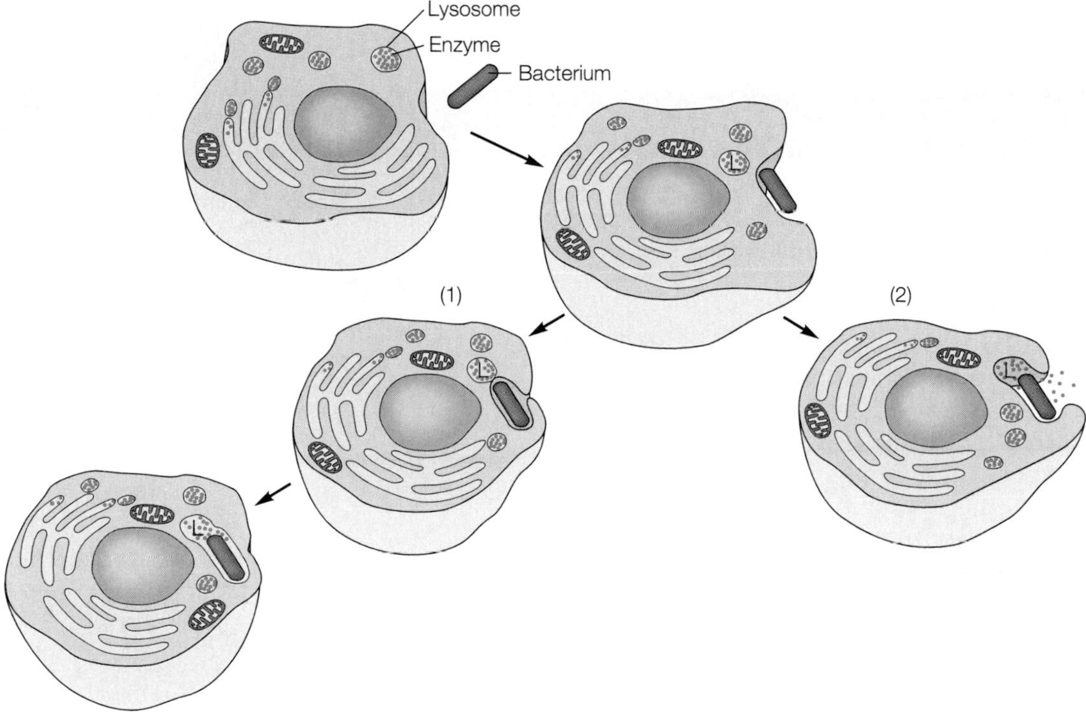

FIGURE 17.6

Phagocytosis by a neutrophil or macrophage. A phagocytic cell extends its pseudopods around the object to be engulfed (such as a bacterium). Blue dots represent lysosomal enzymes (*L* = lysosomes). (1) If the pseudopods fuse to form a complete food vacuole, lysosomal enzymes are restricted to the organelle formed by the lysosome and food vacuole. (2) If the lysosome fuses with the vacuole before fusion of the pseudopods is complete, lysosomal enzymes are released into the infected area of tissue.

with lysosomes. The release of lysosomal enzymes before the food vacuole has completely formed, however, may contribute to inflammation.

Fever

Fever may also be a component of the nonspecific defense system. Body temperature is regulated by the hypothalamus, which coordinates skeletal muscle shivering and the activity of the sympathoadrenal system to maintain body temperature at about 37°C. This thermostat is reset upwards in response to a chemical called **endogenous pyrogen,** secreted by monocytes. Endogenous pyrogen secretion is stimulated by a chemical called *endotoxin,* which is released by certain bacteria.

Although high fevers are definitely dangerous, many believe that a mild-to-moderate fever may be a beneficial response that aids recovery from bacterial infections. One theory holds that elevated body temperature interferes with the uptake of iron by some bacteria.

Interferons

Cells infected with a virus produce polypeptides that interfere with the ability of a second, unrelated strain of virus to infect other cells in the same culture. These **interferons** (*in″ter-fēr′onz*) produce a nonspecific, short-acting resistance to viral infection.

Leukocytes, fibroblasts, and probably many other cells produce their own characteristic types of interferons. Lymphocytes called *T lymphocytes* (described in module 17.7) release interferons in response to viral infections, and perhaps as part of their immunological surveillance against cancer. Interferons may destroy cancer cells directly and indirectly by the activation of T lymphocytes and *natural killer* (*NK*) *cells* (cells related to T lymphocytes).

Endotoxin, a lipopolysaccharide released by bacteria, and regulators secreted by cells of the immune system (e.g., gamma interferon) induce the synthesis of the enzyme nitric oxide synthetase *within macrophages. This enzyme catalyzes the production of* nitric oxide, *a gas used by macrophages to destroy toxins. Nitric oxide also causes relaxation of the smooth muscles in blood vessels, producing vasodilation. Too much nitric oxide produced in response to infection can thereby cause the dangerously low blood pressure of* septic shock. *Drugs that inhibit nitric oxide synthetase have recently been employed to treat the hypotension of septic shock.*

17.4 Specific Defense Mechanisms

Lymphocytes confer immunity against specific antigen molecules. In some cases, this is the result of the action of proteins called antibodies.

It was demonstrated in 1890 that exposure to a nonlethal dose of diphtheria toxin could protect a guinea pig against the disease, and that this protection, or **immunity,** could be transferred by molecules in the animal's blood serum to an unexposed guinea pig. The molecules in the serum that conferred this immunity were named **antibodies,** and they were found to bond in a specific fashion with molecules called **antigens** (an'tĭ-jenz).

Antigens

Antigens (*anti*body *gen*erators) are molecules that stimulate antibody production and combine with these specific antibodies. Most antigens are large molecules (such as proteins), with a molecular weight greater than about 10,000, and generally they are foreign to the blood and other body fluids. The ability of a molecule to function as an antigen depends not only on its size but also on the complexity of its structure. Proteins are more antigenic than polysaccharides, which have a simpler structure than do proteins.

A large, complex foreign molecule can have a number of different **antigenic determinant sites,** which are areas of the molecule that stimulate production of and combine with different antibodies. Most naturally occurring antigens have numerous antigenic determinant sites and stimulate the production of different antibodies with specificities for these sites.

When the antigen or antibody is attached to the surface of a cell or to particles of latex rubber (in commercial diagnostic tests), the antigen-antibody reaction becomes visible because the particles *agglutinate* (clump) as a result of antigen-antibody bonding (fig. 17.7). These agglutinated particles can be used to assay a variety of antigens, and tests that utilize this procedure are called *immunoassays* (*im"yŭ-no-as'āz*). Blood typing (see chapter 14, module 14.4) and pregnancy tests are examples of such immunoassays.

Lymphocytes

All blood cells, including lymphocytes, are ultimately derived from stem cells in red bone marrow. Lymphocytes produced in the bone marrow seed the thymus, spleen, and lymph nodes, producing self-replicating lymphocyte colonies in these organs.

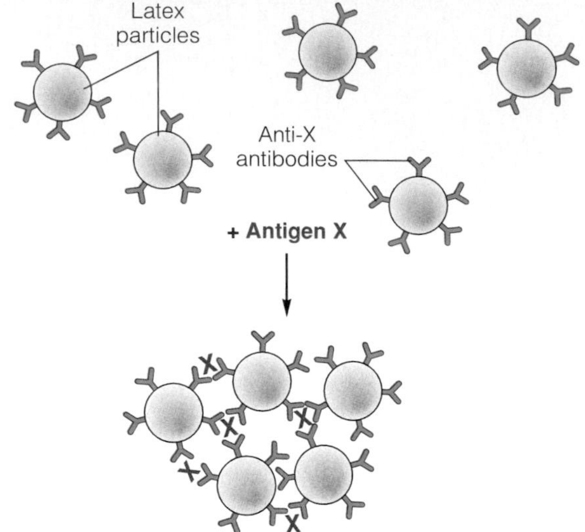

FIGURE 17.7
Immunoassay using the agglutination technique. Antibodies against a particular antigen are adsorbed to latex particles. When these are mixed with a solution that contains the appropriate antigen, the formation of the antigen-antibody complexes produces clumping (agglutination) that can be seen with the unaided eye.

The lymphocytes that become seeded in the thymus are processed into **T lymphocytes.** These cells have surface characteristics and an immunological function that differ from those of other lymphocytes. The thymus, in turn, seeds other organs: about 75% of the lymphocytes in blood and most of the lymphocytes in lymph nodes are T lymphocytes. T lymphocytes, therefore, come from or had an ancestor that came from the thymus.

Most of the lymphocytes that are not T lymphocytes are called **B lymphocytes.** They are so named because they were first identified in the *bursa of Fabricius,* a pouch of lymphatic tissue associated with the digestive tract in chickens. It is currently believed that B lymphocytes in mammals are processed in the bone marrow, which conveniently also begins with the letter *B.*

Both B lymphocytes and T lymphocytes function in specific immunity. The B lymphocytes combat bacterial and some viral infections by secreting antibodies into the blood and lymph. Since this immunity is transported in

bursa of Fabricius: from Hieronymus Fabricius, Italian anatomist, 1533–1619

TABLE 17.3 Comparison of B Lymphocytes and T Lymphocytes

Characteristic	B Lymphocytes	T Lymphocytes
Site where processed	Bone marrow	Thymus
Type of immunity	Humoral (secretes antibodies)	Cell-mediated
Subpopulations	Memory cells and plasma cells	Cytotoxic (killer) T cells, helper cells, suppressor cells
Presence of surface antibodies	Yes—IgM or IgD	Not detectable
Receptors for antigens	Present—are surface antibodies	Present—are related to immunoglobulins
Life span	Short	Long
Tissue distribution	High in spleen, low in blood	High in blood and lymph
Percentage of blood lymphocytes	10%–15%	75%–80%
Transformed by antigens to	Plasma cells	Small lymphocytes
Secretory product	Antibodies	Lymphokines
Immunity to viral infections	Enteroviruses, poliomyelitis	Most others
Immunity to bacterial infections	*Streptococcus, Staphylococcus,* many others	Tuberculosis, leprosy
Immunity to fungal infections	None known	Many
Immunity to parasitic infections	Trypanosomiasis; maybe to malaria	Most others

body fluids, or *humors,* the B lymphocytes are said to provide **humoral immunity,** although the term **antibody-mediated immunity** is also used. T lymphocytes attack host cells that have become infected with viruses or fungi, transplanted human cells, and cancerous cells. The T lymphocytes do not secrete antibodies; they must come into close proximity or have actual physical contact with the victim cell in order to destroy it. T lymphocytes are therefore said to provide **cell-mediated immunity.** B lymphocytes and T lymphocytes are compared in table 17.3.

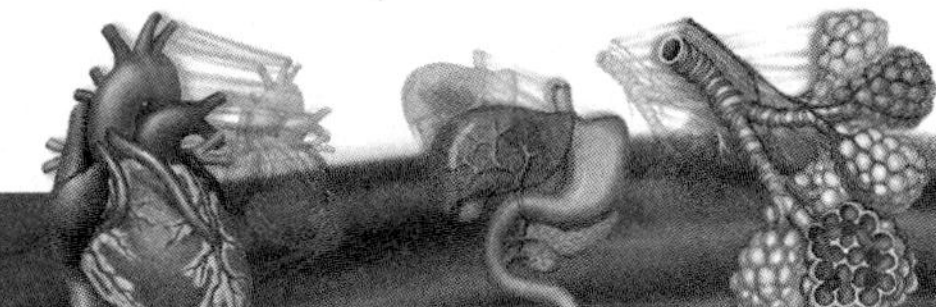

17.5 Functions of B Lymphocytes

B lymphocytes secrete protein antibodies that can bond in a specific fashion with antigens. This bonding activates other immune processes.

Exposure of a B lymphocyte to the appropriate antigen results in cell growth followed by many cell divisions. Some of the progeny become **memory cells,** which are indistinguishable from the original cell; others are transformed into **plasma cells** (fig. 17.8). Plasma cells are protein factories that produce about 2000 antibody proteins per second in their brief life span of 5 to 7 days.

The antibodies that are produced by plasma cells when B lymphocytes are exposed to a particular antigen react specifically with that antigen. Such antigens may be isolated molecules, or they may be molecules at the surface of an invading foreign cell. The combination of antibodies with antigens does not in itself produce destruction of the antigens or of the pathogenic organisms that contain these antigens. Rather, antibodies serve to identify the targets for immunological attack and to activate nonspecific immune processes that destroy the invaders. Bacteria that are buttered with antibodies, for example, are better targets for phagocytosis by neutrophils and macrophages.

Immune destruction of bacteria is also promoted by a group of serum proteins known as the **complement system.** This group of proteins circulates in the blood plasma and is activated by the binding of antibodies to antigens. When the antibodies bind to antigens on the surface of a bacterium, some of the complement proteins become inserted, or *fixed,* in the cell membrane of the bacterium, creating large pores in the membrane. These pores permit the osmotic influx of water, so that the victim cell swells and bursts.

Immunoglobulins

Antibody proteins are also known as immunoglobulins (*im"yŭ-no-glob'yŭ-linz*). These are found in the gamma globulin class of plasma proteins. Since antibodies are specific in their actions, it follows that different types of antibodies should have different structures. An antibody against smallpox, for example, does not confer immunity to poliomyelitis and, therefore, must have a slightly different structure than an antibody against polio. Despite these differences, antibodies are structurally related and form only a few classes.

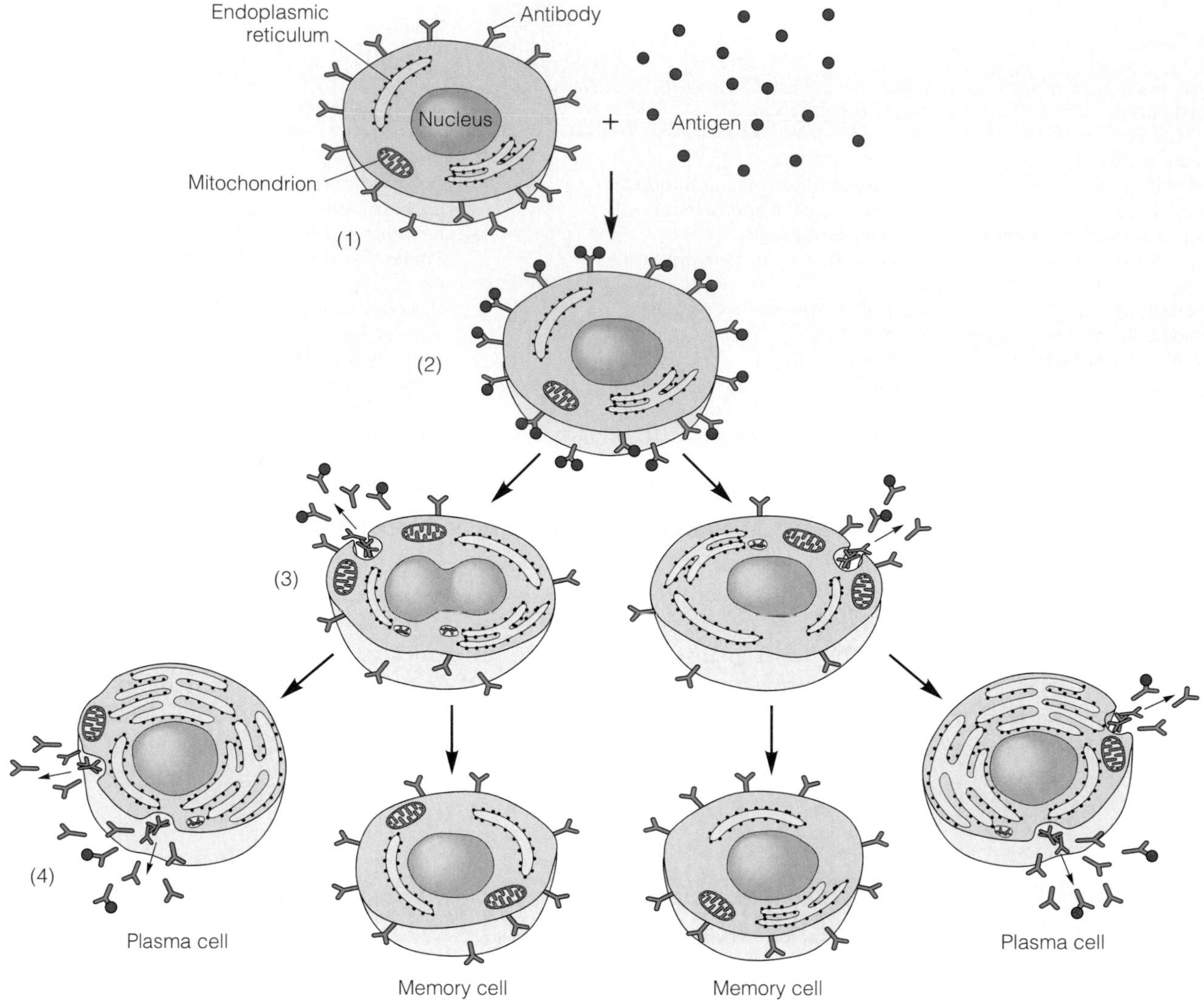

FIGURE 17.8

B lymphocytes have antibodies on their surface that function as receptors for specific antigens. The interaction of antigens and antibodies on the surface stimulates cell division and the maturation of the B cell progeny into memory cells and plasma cells. Plasma cells produce and secrete large amounts of the antibody. (Note the extensive rough endoplasmic reticulum in these cells.)

There are five classes of immunoglobulins (abbreviated Ig): *IgG, IgA, IgM, IgD,* and *IgE.* Most of the antibodies in serum are in the IgG class, whereas most of the antibodies in external secretions (saliva and milk) are IgA. Antibodies in the IgE class are involved in allergic reactions.

Antibody Structure

All antibody molecules consist of four interconnected polypeptide chains. Two long, "heavy" chains (the *H chains*) are joined to two short, "light" chains (*L chains*). These four chains are arranged in the form of a Y. The stalk of the Y has been called the "crystallizable fragment" (abbreviated F_c), whereas the tops of the Y are the "antigen-binding fragments" (F_{ab}). This configuration is shown in figure 17.9.

Analyses of these antibodies have shown that the F_c regions of different antibodies are the same (are constant), whereas the F_{ab} regions contain variable portions. Variability of the antigen-binding regions is required for the specificity of antibodies for antigens. Thus, it is the F_{ab} regions of an antibody that provides a specific site for bonding with a particular antigen.

B lymphocytes have antibodies on their cell membrane that serve as **receptors** for antigens. The combination of antigens with antibody receptors stimulates the B cell to divide and produce more of these antibodies, which are secreted. In this way, exposure to a given antigen results in increased amounts of the specific type of antibody that can attack that antigen. This provides active immunity, as described in the next module.

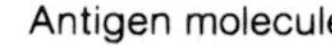

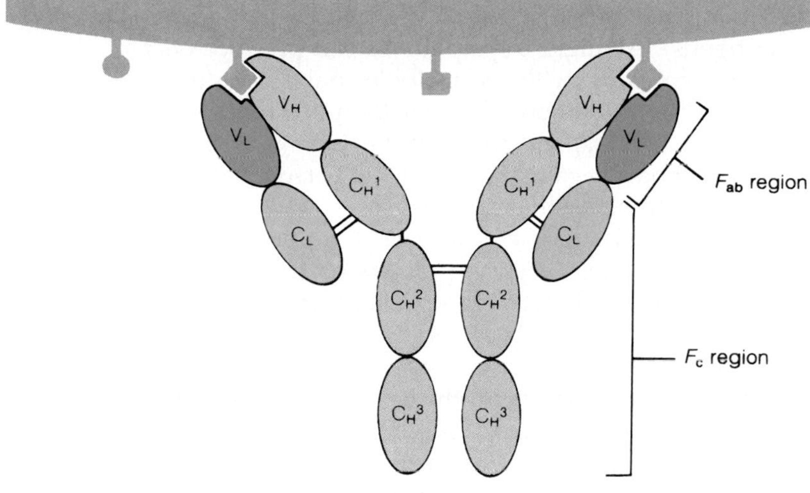

FIGURE 17.9
Antibodies are composed of four polypeptide chains—two are heavy (H) and two are light (L). The variable regions are abbreviated V, and the constant regions are abbreviated C. Antigens combine with the variable regions. Each antibody molecule is divided into two F_{ab} (antigen-binding) fragments and an F_c (crystallizable) fragment.

17.6 Active and Passive Immunity

Active immunity occurs as a result of prior exposure to an antigen, and is a response of the person's immune system. Passive immunity occurs when a person receives antibodies produced by another organism.

Active Immunity and the Clonal Selection Theory

When a person is exposed to a particular pathogen for the first time, there is a latent period of 5 to 10 days before measurable amounts of specific antibodies appear in the blood. This sluggish **primary response** may not be sufficient to protect the individual against the disease caused by the pathogen. Antibody concentrations in the blood during this primary response reach a plateau in a few days and decline after a few weeks.

A subsequent exposure of the same individual to the same antigen results in a **secondary response** (fig. 17.10). Compared to the primary response, antibody production during the secondary response is much more rapid. Maximum antibody concentrations in the blood are reached within a day's time and are maintained for a longer time than in the primary response. This rapid rise in antibody production is usually sufficient to prevent the disease.

Although the mechanisms by which secondary responses are produced are not completely understood, the **clonal selection theory** accounts for most of the evidence

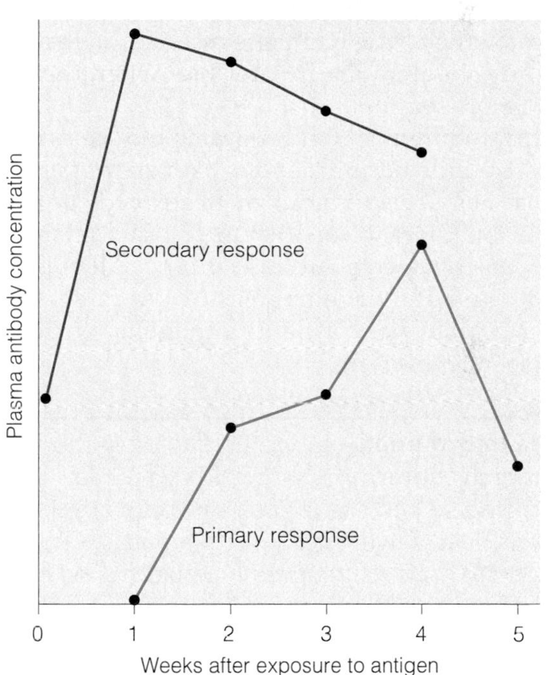

FIGURE 17.10
A comparison of antibody production in the primary response (upon first exposure to an antigen) and antibody production in the secondary response (upon subsequent exposure to the antigen). The greater production in the secondary response is believed to be due to the development of lymphocyte clones produced during the primary response.

and is widely accepted. According to this theory, B lymphocytes *inherit* the ability to produce particular antibodies (and T lymphocytes inherit the ability to respond to particular antigens). A single B lymphocyte can produce only one type of antibody, with specificity for only one antigen. Since this ability is inherited rather than acquired, some lymphocytes can respond to smallpox, for example, and produce antibodies against it even if the person has never been previously exposed to this disease.

The inherited specificity of each lymphocyte is reflected in the antigen receptor proteins on the surface of the lymphocyte's cell membrane. Exposure to smallpox antigens thus stimulates these specific lymphocytes to divide many times until a large population of genetically identical cells—a clone—is produced. Some of these cells become plasma cells that secrete antibodies for the primary response; others become memory cells that can be stimulated to secrete antibodies during later exposure and the secondary response (fig. 17.11).

The development of the active immunity derived from a secondary response requires prior exposure to the specific antigens, during which time a primary response is produced and the person may get sick. Before the development of vaccinations for measles, chickenpox and mumps, some parents would deliberately expose their children to others who had these diseases, so that the children would be immune in later life, when the diseases could be more serious.

Clinical immunization programs induce primary responses by inoculating people with pathogens whose virulence has been reduced or destroyed, or by using strains of microorganisms that are closely related to the disease-causing agents but less virulent. All of these procedures cause the development of lymphocyte clones that can combat the virulent pathogens by producing secondary responses.

Inside Information—It was first noted in the mid-eighteenth century that milkmaids who had contracted cowpox (a disease similar to smallpox but less deadly) were immune to smallpox—hence, their "milkmaids' complexion," unblemished by smallpox scars. Building on this observation, the English physician Edward Jenner developed a smallpox vaccine in 1798.

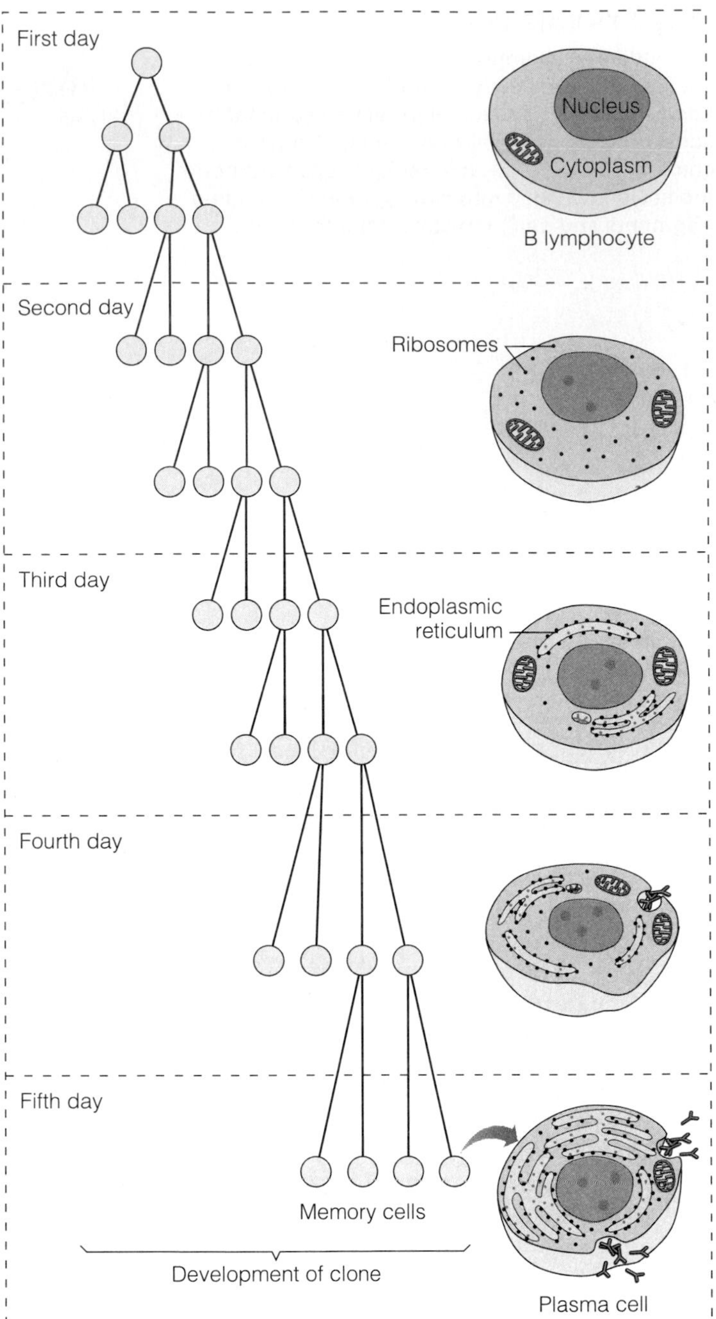

FIGURE 17.11
The clonal selection theory as applied to B lymphocytes. Most members of the B lymphocyte clone become memory cells, but some become antibody-secreting plasma cells.

Passive Immunity

The term passive immunity refers to the immune protection that can be produced by the transfer of antibodies to a recipient from another person or from an animal. The donor person or animal has been actively immunized, as explained by the clonal selection theory. The recipient of these ready-made antibodies is thus passively immunized to the same antigens. Passive immunity occurs naturally in the transfer of immunity from mother to fetus during pregnancy. It can also be artificially conferred by injecting antibodies into the individual requiring immunity.

The ability to mount an immune response—called **immunological competence**—does not develop until about a month after birth. The fetus, therefore, cannot immunologically reject its mother. The immune system of the mother is fully competent but does not usually respond to fetal antigens for reasons that are not completely understood. Some IgG antibodies from the mother do cross the placenta and enter the fetal circulation, however, and these serve to confer passive immunity to the fetus.

A fetus and newborn baby are immune to the same antigens as the mother. But since the newborn did not itself produce the lymphocyte clone needed to form these antibodies, such passive immunity largely disappears when the infant is about a month old. A breast-fed infant can receive additional antibodies of the IgA class in its mother's first milk.

Passive immunizations are used clinically to protect people who have been exposed to extremely virulent infections or toxins, such as snake venom or tetanus. In these cases the affected person is injected with *antiserum* (serum containing antibodies), also called *antitoxin,* from an animal that has been previously exposed to the pathogen. The animal develops the lymphocyte clones and active immunity, and thus has a high concentration of antibodies in its blood. Since the person who is injected with these antibodies does not develop active immunity, he or she must again be injected with antitoxin in the event of a subsequent exposure.

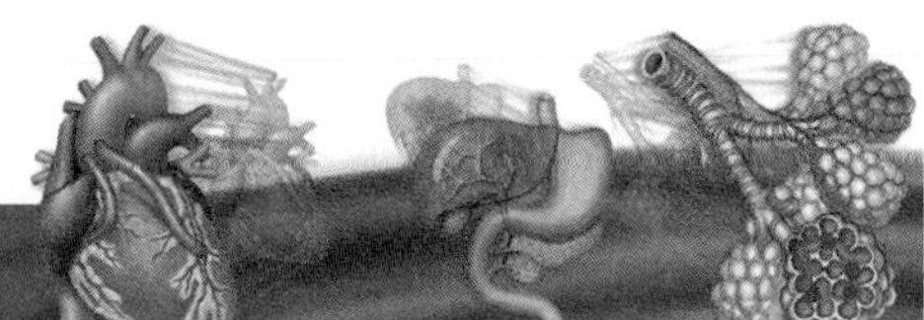

17.7 Types of T Lymphocytes

Three subcategories of T lymphocytes are recognized: killer, helper, and suppressor. Each type has specific immune functions.

Killer, or **cytotoxic, T lymphocytes** destroy specific victim cells that are identified by specific antigens on their surface. In order to effect this *cell-mediated destruction,* the T lymphocytes must be in actual contact with their victim cells (in contrast to B cells, which kill at a distance). Although the mechanisms by which killer lymphocytes destroy their victims are not completely understood, there is evidence that they accomplish this task by secreting certain molecules at the region of contact. Among these molecules, specific polypeptides called **perforins** have been identified. The long perforin molecules arrange themselves in a circular fashion, creating large "holes" in the cell membrane that cause the cell to rupture. Complement proteins use a similar mechanism, as described in module 17.5.

The killer T lymphocytes defend against viral and fungal infections, and are also responsible for transplant rejection reactions and for immunological surveillance against cancer. Although most bacterial infections are fought by B lymphocytes, some bacteria are the targets of cell-mediated attack by killer T lymphocytes. This is the case with the *tubercle bacilli* that cause tuberculosis. Injections of some of the bacteria under the skin produce inflammation after a latent period of 48 to 72 hours. This *delayed hypersensitivity reaction* is cell-mediated rather than humoral, as shown by the fact that it can be induced in an unexposed guinea pig by an infusion of lymphocytes, but not of serum, from an exposed animal.

The **helper T lymphocytes** and **suppressor T lymphocytes** indirectly participate in the specific immune response by regulating the responses of the B cells (fig. 17.12) and the killer T cells. The activity of B cells and killer T cells is increased by helper T lymphocytes and decreased by suppressor T lymphocytes. The amount of antibodies secreted in response to antigens is thus affected by the relative numbers of helper to suppressor T cells that develop in response to a given antigen.

Lymphokines

The T lymphocytes, as well as macrophages and some other cells, secrete a number of polypeptides that serve to regulate many aspects of the immune system. These products are called lymphokines (*lim'fŏ-kīnz*). When a lymphokine is first discovered, it is named according to its biological activity (e.g., *B cell–stimulating factor*). Such names can be misleading, however, since each lymphokine has many different actions (table 17.4). Scientists have thus agreed to use the name *interleukin,* followed by a number, to indicate a lymphokine once its amino acid sequence has been determined.

Interleukin-1, secreted by macrophages and other cells, can activate the T cell system. B cell–stimulating factor—now called *interleukin-4*—secreted by T lymphocytes, is required for proliferation and clone development of B cells. *Interleukin-2,* released by helper T lymphocytes, is required for activation of killer T lymphocytes. *Macrophage colony–stimulating factor,* a lymphokine secreted by helper T lymphocytes, promotes the activity of macrophages.

T Cell Receptors

Unlike B cells, T cells do not make antibodies, and thus do not have antibodies on their surfaces to serve as receptors for antigens. The T cell receptors differ from the antibody receptors on B cells in a very significant way: they *cannot bind to free antigens.* In order for a T lymphocyte to respond to a foreign antigen, the antigen must be presented to the T lymphocyte on the membrane of an **antigen-presenting cell.** The chief antigen-presenting cells are macrophages, as described in the next module.

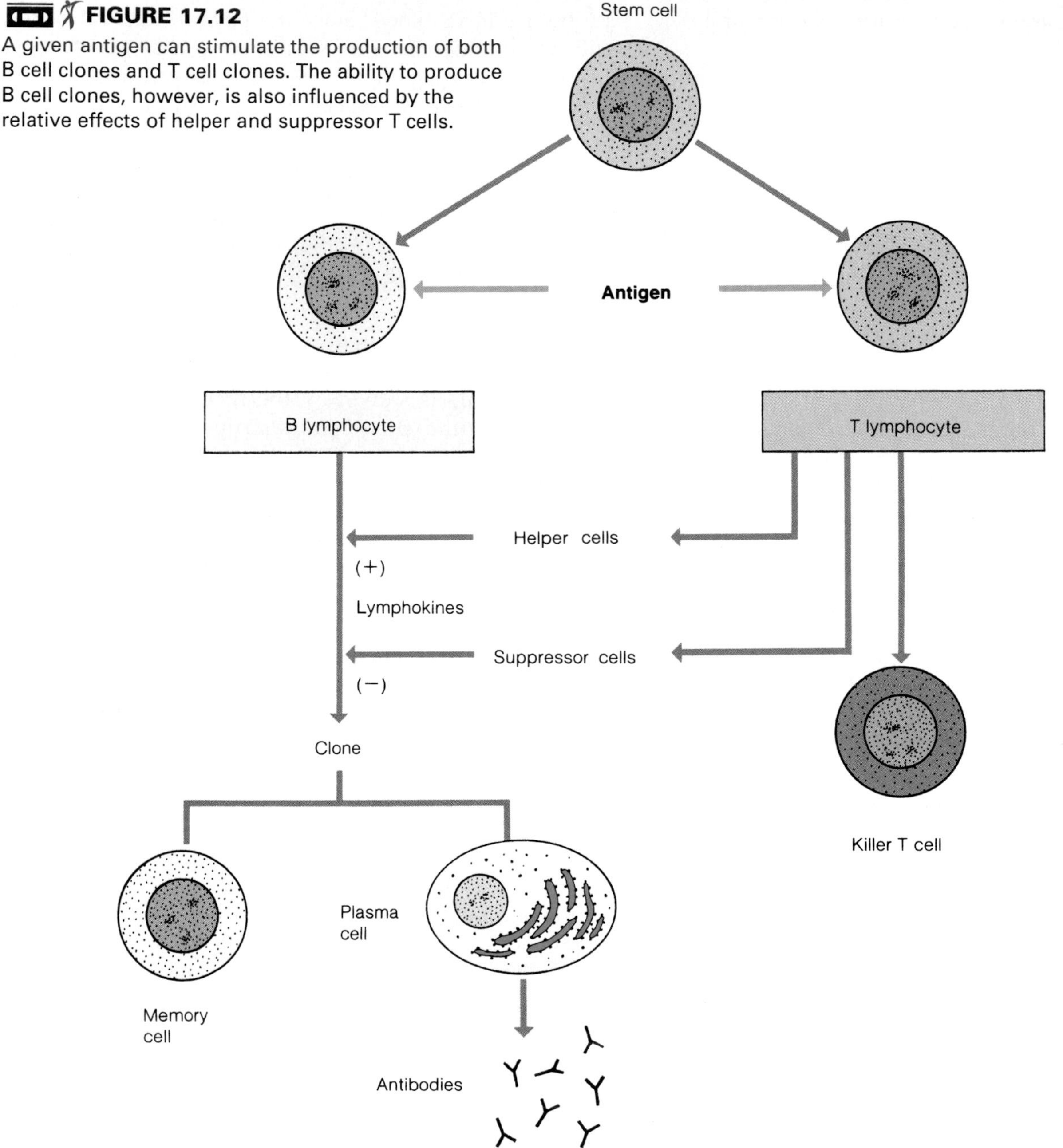

FIGURE 17.12
A given antigen can stimulate the production of both B cell clones and T cell clones. The ability to produce B cell clones, however, is also influenced by the relative effects of helper and suppressor T cells.

TABLE 17.4 Some Lymphokines That Regulate the Immune System

Lymphokine	Biological Functions	Secreted by
Interleukin-1	Activates resting B cells and T cells	Macrophages and others
Interleukin-2	Serves as growth factor for activated T cells; activates cytotoxic T cells	Helper T cells
Interleukin-3	Promotes the growth of bone marrow stem cells; serves as growth factor for mast cells	Helper T cells
Interleukin-4 (B cell–stimulating factor)	Promotes growth of activated B cells; promotes growth of resting T cells; enhances activity of cytotoxic T cells	Helper T cells
B cell–differentiating factor	Induces the conversion of activated B cells into antibody-secreting plasma cells	T cells and others
Colony-stimulating factors	Different colony-stimulating factors stimulate the proliferation of granulocytic leukocytes and of macrophages	T cells and others
Interferons	Activate macrophages; augment natural killer cell body activity; exhibit antiviral activity	T cells and others
Tumor necrosis factors	Exert direct cytotoxic effect on some tumor cells; stimulate production of other lymphokines	Macrophages and others

Adapted from C. A. Dinarello and J. W. Mier, *The New England Journal of Medicine,* 317:940.

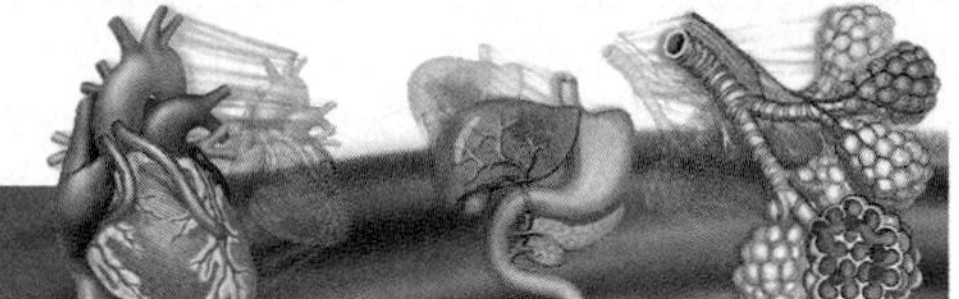

17.8 Activation of T Lymphocytes

Macrophages present foreign antigens to the T lymphocytes in conjunction with particular molecules produced by the person's own genes. Helper and killer T lymphocytes have different requirements for activation.

Macrophages present foreign antigens, together with other surface antigens called **histocompatibility antigens,** to the T lymphocytes. The histocompatibility antigens are proteins that are coded by a group of genes called the **major histocompatibility complex (MHC).** There are several hundred genes in this complex, all located on chromosome 6.

The major histocompatibility complex of genes produces two classes of MHC molecules, designated *class 1* and *class 2.* The class-1 MHC molecules are made by all cells in the body except red blood cells. The class-2 molecules, which are produced only by macrophages and B lymphocytes, promote the interactions between T cells and these other cells of the immune system.

Killer T lymphocytes can interact only with antigens presented with class-1 MHC molecules, whereas helper T lymphocytes can interact only with antigens presented with class-2 MHC molecules. These restrictions result from the presence of *coreceptors,* which are proteins associated with the T cell receptors. The coreceptor known as **CD8** is associated with the killer T lymphocyte receptor, and interacts with only the class-1 MHC molecules; the coreceptor known as **CD4** is associated with the helper T lymphocyte receptor, and interacts with only the class-2 MHC molecules. These structures are illustrated in figure 17.13.

When a foreign particle, such as a virus, infects the body, it is taken up by macrophages in the process of

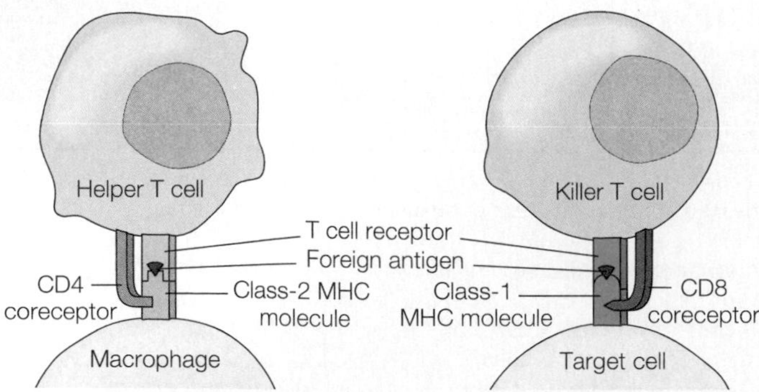

FIGURE 17.13
A foreign polypeptide antigen is presented to T lymphocytes in association with MHC molecules. The CD4 and CD8 coreceptors permit the T lymphocyte receptors to interact with only a specific class of MHC molecule.

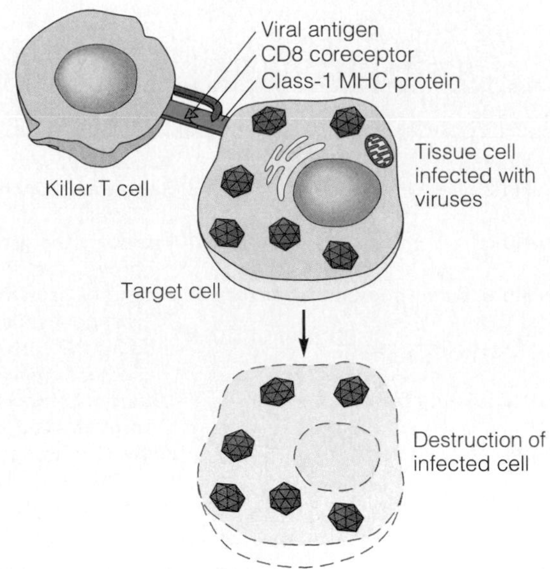

FIGURE 17.14
In order for a killer T cell to destroy a tissue cell infected with viruses, the T cell must interact with both the foreign antigen and the class-1 MHC molecule on the surface of the infected cell.

phagocytosis and partially digested. Within the macrophage, the partially digested virus particles provide foreign antigens that are moved to the surface of the cell membrane. At the membrane, these foreign antigens form a complex with the class-2 MHC molecules. This combination is required for interaction with the receptors on the surface of helper T cells. The macrophages thus "present" the antigens to the helper T cells, and in this way stimulate activation of the T cells (fig. 17.13).

At this point, the macrophage is stimulated to secrete the lymphokine known as interleukin-1. Interleukin-1 stimulates the cell division and proliferation of B and T lymphocytes. The activated helper T cells, in turn, secrete macrophage colony–stimulating factor and gamma interferon, which promote the activity of macrophages. In addition, interleukin-2 is secreted by the T lymphocytes and stimulates the macrophages to secrete *tumor necrosis factor,* which is particularly effective in killing cancer cells.

Killer T cells can destroy infected cells only if those cells display the foreign antigen together with their class-1 MHC molecules (fig. 17.14). Such interaction of killer T cells with the foreign antigen–MHC class-1 complex also stimulates the rapid increase of those killer T cells. This proliferation is supported by interleukin-2 secreted by the activated helper T cells (fig. 17.15).

The network of interaction among different cell types of the immune system now spreads outward. Activated helper T cells can also promote the humoral immune response of B cells. In order to do this, the membrane receptors on the surface of the helper T cells must interact with molecules on the surface of the B cells. This occurs when the foreign antigen attaches to the receptors on the B cells, so that the B cells can present this antigen together with its MHC class-2 molecules to the receptors on the helper T cells. This interaction stimulates proliferation of the B cells, their conversion to plasma cells, and their secretion of antibodies against the foreign antigens.

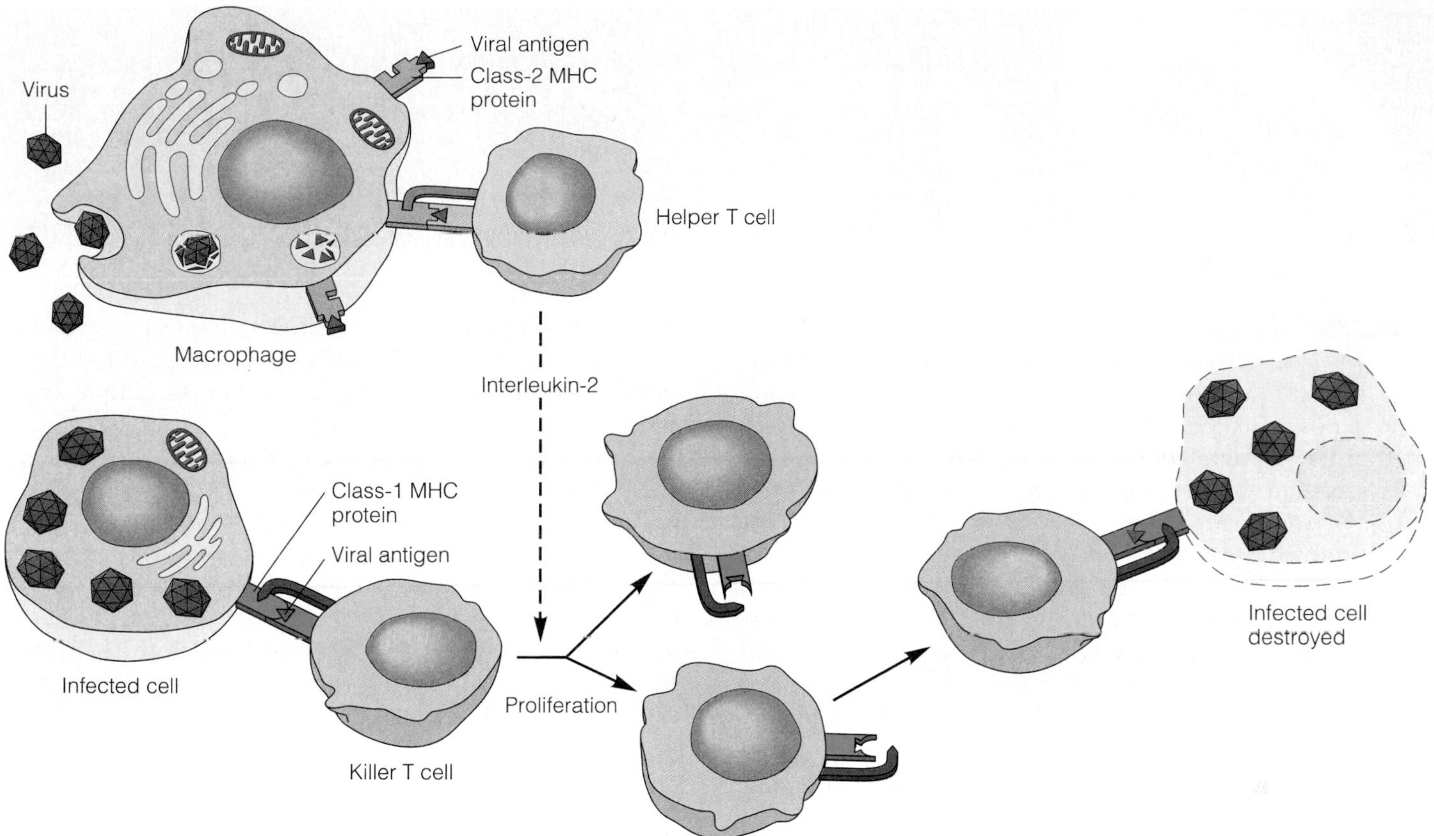

FIGURE 17.15
Interaction between macrophages, helper T lymphocytes, cytotoxic T lymphocytes, and infected cells in the immunological defense against viral infections.

Experimental cancer treatments include the use of different agents of the immune system. These immunotherapies *may involve different types of interferons and other lymphokines, such as interleukin-2* (IL-2) *and tumor necrosis factor* (TNF)*, to treat particular forms of cancer. In addition, lymphokine-activated killer cells* (LAKs) *and tumor-infiltrating lymphocytes* (TILs) *that are derived from the patient's own body have been employed. Thus far, these therapies show promise but do not present a general cure for cancer.*

Inside Information—Numerous studies indicate that prolonged stress, either physical or emotional, reduces the effectiveness of the immune system. For example, T cell functions are suppressed by glucocorticoids secreted by the adrenal cortex in response to stress. *Psychoneuroimmunology* is a new field of inquiry devoted to exploring the intimate relationships between the immune, endocrine, and nervous systems. Whether or not we can "think" ourselves sick or well is a question yet to be answered.

AIDS—A Status Report

There have been so many cases of AIDS in recent years that few of us have failed to think about the dangers it poses to ourselves or to someone we know. In 1995, the Centers for Disease Control in Atlanta, which keeps track of infectious diseases around the country, reported the 500,000th case of full-blown AIDS since 1981, the year the disease was first recognized. According to World Health Organization estimates, by the year 2000 some 40 million people worldwide may be infected with HIV (the virus that causes AIDS), and those who are infected but show no symptoms will continue to infect others. In light of what you now know about the strategies that enable our bodies to resist illness, it is appropriate to review current information about this plague of the twentieth century.

AIDS, short for **acquired immune deficiency syndrome,** is caused by a retrovirus identified as **human immunodeficiency virus (HIV).** A retrovirus is one of a group of RNA viruses that use an infected cell's DNA to reproduce, destroying the host cell's functioning in the process. For many years, retroviruses have been known to cause cancer in humans and animals. HIV is one of the first infectious retroviruses to be discovered. Once inside the body, it attacks and destroys the helper T lymphocytes, thereby severely impairing the individual's ability to combat infection. AIDS victims die from a host of opportunistic infections that would pose little threat to individuals with healthy immune systems.

HIV is transmitted through infected blood, semen, or vaginal secretions. A person infected with HIV may be asymptomatic or develop a brief flulike infection, including symptoms of fever, fatigue, headache, and swollen lymph nodes. These symptoms disappear rapidly, however, so that the person is unaware that he or she has contracted a deadly disease and is potentially dangerous to others.

From 2 weeks to a year or longer after the initial infection, the person's immune system reacts to HIV by producing antibodies that are detectable by blood tests. Antibody levels rise initially, but then fall as the immune system gradually begins to falter. For most people, the first sign that they are infected with the virus is the development of chronically enlarged lymph nodes, usually at several sites in the body. Severe fatigue and fever are other common symptoms.

As the helper T cell population continues to decline, the person moves on to the next level of HIV infection called **AIDS-related complex (ARC).** The most likely and dramatic visible expression is an extreme weight loss. Contributing to the weight loss is diarrhea, accompanied by a chronic low-grade fever, general fatigue, and a lack of energy. Hairy leukoplakia (white lesions in the mouth), thrush (mouth infection with candida), and night sweats are other symptoms associated with ARC. At this point, the helper T cells are declining rapidly and losing both their ability to secrete lymphokines and to activate some of the nonspecific defense mechanisms.

As the disease evolves into full-blown AIDS, the helper T cells are severely weakened and eventually completely destroyed, keeping them from stimulating production of killer T cells and B cells. Without the killer T cells, the system is vulnerable to other viral infections; without the B cells, the system cannot make memory cells or synthesize crucial antibodies. With no ability to synthesize specific antibodies, the entire immune system slips into total chaos. Symptoms at this stage are the same symptoms seen in ARC but intensified; added to them is a chronic cough, spots on the skin of the legs, fuzzy speech, and in the late stages, dementia.

Once the virus steps up its cellular attacks, the victim is extremely vulnerable to any infection that takes advantage of dysfunctional immune mechanisms. Such pathogens are called opportunistic. (The secondary infections that attack the weakened immune system are what first alerted the medical community to the presence of AIDS. When exotic diseases that were unusual in the general population became more prevalent, immunologists sought explanations.) Opportunistic infections include a previously rare form of cancer known as **Kaposi's sarcoma,** evidenced by coin-sized purple lesions of the skin. For about 60% of the AIDS patients, **Pneumocystis carnii pneumonia (PNP)** is the opportunistic infection that marks the progression of HIV infection to full-blown AIDS.

Although no cure for AIDS is currently known, treatment is available for many of the infections and cancers that afflict AIDS patients. The drug **AZT (zidovudine—formerly called azidothymidine)** blocks the action of reverse transcriptase, the enzyme needed by HIV to be incorporated in the host cell's DNA. Recent evidence suggests that AZT is effective in people with full-blown AIDS and may prolong the onset of AIDS symptoms. Additionally, AZT was shown to cut transmission of HIV from mothers to babies by about two-thirds in a study conducted in 1994. Unfortunately, HIV has developed strains that are resistant to AZT; moreover, AZT is costly and has severe side effects. Two other drugs—**ddI (dideoxyinosine)** and **ddC (dideoxycytidine)**—also inhibit viral replication and appear to have fewer side effects than AZT. Like AZT, however, they are effective in a patient only for a limited time.

Another class of drugs—**protease blockers**—inhibit other enzymes (proteases) that HIV needs in order to replicate. A combination of a protease inhibitor with two reverse transcriptase inhibitors has recently been demonstrated to be highly effective at inhibiting viral replication. If the viruses cannot replicate, the immune system might stand a chance to reduce or eliminate the HIV infection. Though exciting. this new drug regimen has severe side effects, is difficult to maintain, and is very expensive.

Many scientists believe that the development of a vaccine is the only way to bring AIDS under control. Most of the vaccines currently in use are made from live viruses that have been sufficiently weakened so that they no longer cause the disease yet stimulate the people who are vaccinated into producing their own antibodies. But with a virus as dangerous as the one that causes AIDS, using a live virus that has been weakened in the conventional way is like playing with a biological time bomb; it is possible that it could somehow regain strength and activate itself.

Advances in recombinant DNA technology have provided AIDS researchers with a new approach—the preparation of a subunit vaccine. Subunit vaccines present no danger of infection because they do not contain any genetic material from the virus; their content is limited to one or more viral proteins or pieces of proteins. To prepare such a vaccine, an attempt is made to identify the viral protein (antigen) most likely to cause a protective antibody response from the human immune system. This protein is usually part of the viral envelope—its outer coat. A hepatitis B vaccine based on this strategy was licensed in 1987. In addition, a subunit vaccine against feline leukemia virus is currently available for immunization of cats. Various subunit vaccines, utilizing different viral proteins, are currently under trial for use against AIDS.

Although progress is being made in the development of a vaccine, it has been hampered by the mutability of HIV. Investigations monitoring the evolution of HIV found as many as 150 different forms of the virus in one individual. Further compounding the problem are the logistics and expense of clinical trials and associated liability issues. For these reasons, mass immunizations against HIV should not be regarded as a reasonable option, at least for the near future.

At present, the only effective prevention of AIDS is to stop the transmission of HIV. Individuals can protect themselves by practicing sexual abstinence before marriage, by practicing "safe sex," and by avoiding multiple sex partners. Transmission through contaminated needles could be avoided if intravenous drug users were to use sterile needles only. Blood has been routinely screened for the presence of HIV since 1985. Although infection from donated blood is extremely rare, individuals who will need blood for upcoming operations are encouraged to donate blood ahead of time. Finally, any woman infected with HIV should avoid pregnancy. The normal immune system changes that occur in pregnancy may accelerate severe AIDS problems for a woman who has the virus, and the infection very likely will be transmitted to the baby.

If HIV infection trends persist, the threat of AIDS surely will be a rite of passage for many generations to come. As the search for a vaccine and more effective treatments continues, the best defense available to us is a stepped-up program of public education on how to protect ourselves.

Clinical Considerations

The ability of the normal immune system to tolerate self-antigens while it identifies and attacks foreign antigens provides a specific defense against invading pathogens. At times, however, this defense system against invaders is responsible for domestic offenses that may result in diseases ranging in severity from the sniffles to sudden death.

Autoimmunity

Autoimmune diseases result from failure of the immune system to recognize and tolerate self-antigens. As a consequence of this failure, autoreactive T cells are activated and autoantibodies are produced by B cells, causing inflammation and organ damage. Over 40 known or suspected autoimmune diseases affect 5% to 9% of the population. Some of the more well-known autoimmune diseases are rheumatoid arthritis, lymphocytic thyroiditis (Hashimoto's disease), Graves' disease, type I diabetes mellitus, heart damage in rheumatic fever, glomerulonephritis, systemic lupus erythematosus (SLE), and myasthenia gravis.

Allergy

The term *allergy,* usually used interchangeably with *hypersensitivity,* refers to particular types of abnormal immune response to antigens, which in these cases are called *allergens.* There are two major forms of allergy: (1) *immediate hypersensitivity,* which is due to an abnormal B lymphocyte response to an allergen that produces symptoms within seconds or minutes, and (2) *delayed hypersensitivity,* which is an abnormal T cell response that produces symptoms within about 48 hours after exposure to an allergen. These two types of hypersensitivity are compared in table 17.5.

Immediate Hypersensitivity Immediate hypersensitivity can produce symptoms that include allergic rhinitis (chronic runny or stuffy nose), conjunctivitis (red eyes), allergic asthma, and atopic dermatitis (hives), among others. These symptoms result from the production of antibodies of the IgE subclass rather than the normal IgG antibodies.

Unlike IgG antibodies, IgE antibodies do not circulate in the blood. Instead, they attach to tissue mast cells (which have membrane receptors for these antibodies). When the person is again exposed to the same allergen, the allergen bonds with the antibodies attached to the mast cells. This stimulates the mast cells to secrete various chemicals, including **histamine** (fig. 17.16). During this process, leukocytes may also secrete **prostaglandins** and related molecules called **leukotrienes,** both of which contribute to the symptoms of the allergic reactions. The itching, sneezing, tearing, and runny nose of people suffering from hay fever are produced largely by histamine and can be treated effectively by antihistamine drugs. Food allergies, causing diarrhea and colic, are mediated primarily by prostaglandins; these allergies can be treated with aspirin, which inhibits prostaglandin synthesis.

Allergens that provoke immediate hypersensitivity include various foods, bee stings, and pollen grains. The most common allergy of this type is seasonal hay fever, which may be provoked by ragweed (*Ambrosia*) pollen grains (fig. 17.17*a*). People with chronic allergic rhinitis and asthma due to an allergy to dust or feathers are usually allergic to a tiny mite (fig. 17.17*b*) that lives in dust and eats the scales of skin that are constantly shed from the body.

Inside Information—Pet cats, dogs, and birds are among the most common animal sources of allergens. Major allergenic proteins are present in the scales shed from skin (dander)—not in fur—so it makes little difference whether a cat or dog is long- or short-haired. Cats seem to present the most difficulty, and the danders of different breeds are closely related. Hypersensitivity to one breed of cat most likely indicates hypersensitivity to all breeds.

TABLE 17.5 Allergy: Comparison of Immediate and Delayed Hypersensitivity Reactions

Characteristic	Immediate Reaction	Delayed Reaction
Time for onset of symptoms	Within several minutes	Within 1 to 3 days
Lymphocytes involved	B cells	T cells
Immune effector	IgE antibodies	Cell-mediated immunity
Allergies most commonly produced	Hay fever, asthma, and most other allergic conditions	Contact dermatitis (as from poison ivy and poison oak)
Therapy	Antihistamines and adrenergic drugs	Corticosteroids (such as cortisone)

Delayed Hypersensitivity In delayed hypersensitivity, as the name implies, a longer time is required (hours to days) for symptoms to develop than in immediate hypersensitivity. This may be due to the fact that immediate hypersensitivity is mediated by antibodies, whereas delayed hypersensitivity is a cell-mediated T lymphocyte response. Since the symptoms are caused by the secretion of lymphokines, corticosteroids are the only drugs that can effectively treat delayed hypersensitivity. One of the best-known examples of delayed hypersensitivity is **contact dermatitis** caused by poison ivy, poison oak, and poison sumac.

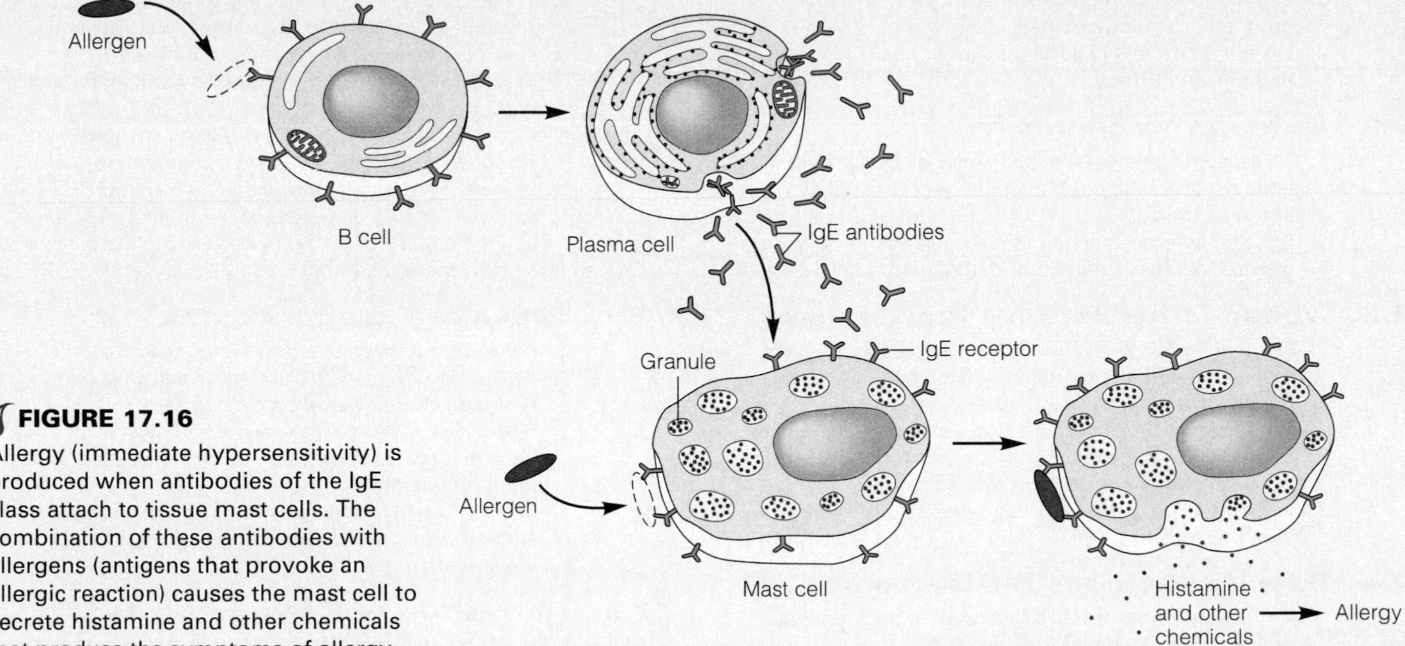

FIGURE 17.16
Allergy (immediate hypersensitivity) is produced when antibodies of the IgE class attach to tissue mast cells. The combination of these antibodies with allergens (antigens that provoke an allergic reaction) causes the mast cell to secrete histamine and other chemicals that produce the symptoms of allergy.

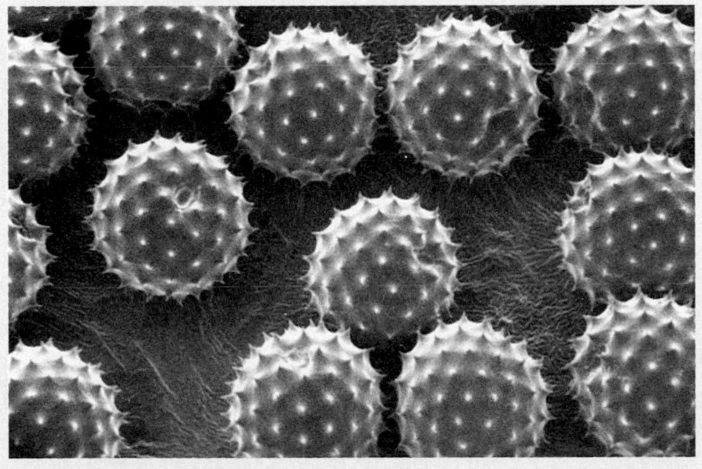

(a)

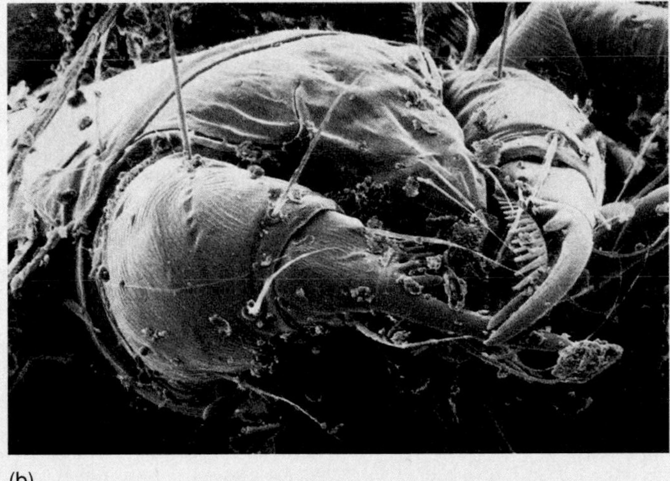

(b)

FIGURE 17.17
(*a*) A scanning electron micrograph of ragweed (*Ambrosia*), which is responsible for hay fever. (*b*) A scanning electron micrograph of a dust mite. Members of the spider family, dust mites are commonly found in house dust and are often responsible for chronic allergic rhinitis and asthma.
(*a*) From R. G. Kessel and C. Y. Shih, *Scanning Electron Microscopy in Biology*.

SUMMARY

17.1 Lymphatic System

1. Lymphatic capillaries are blind-ended lymphatic vessels that drain into lymph ducts.
2. The thoracic duct and right lymphatic duct drain lymph into the left and right subclavian veins, respectively.
3. Lymph nodes contain germinal centers for the production of lymphocytes.
4. Lymphoid organs include the lymph nodes, spleen, thymus, and tonsils.

17.2 Defense Mechanisms

1. Nonspecific defense mechanisms include barriers to penetration of the body by pathogens, as well as internal defenses.
2. Specific defense mechanisms involve responses directed against a particular pathogen or toxin.

17.3 Nonspecific Defense Mechanisms

1. Fixed phagocytes are found in the blood channels of the spleen and liver, and in the lymph channels of lymph nodes.
2. Connective tissues contain a population of phagocytic cells. New phagocytic cells from the blood can leave capillaries by the process of diapedesis.
3. Interferons provide nonspecific protection against viral infections.

17.4 Specific Defense Mechanisms

1. Antigens provoke the production of antibodies that react specifically with those antigens.
2. A large, complex molecule can have a number of antigenic determinant sites.
3. T lymphocytes, processed in the thymus, provide cell-mediated immunity.
4. B lymphocytes, processed in the bone marrow, provide humoral immunity by secreting antibodies.

17.5 Functions of B Lymphocytes

1. B lymphocytes are stimulated by specific antigens to divide and form memory cells and plasma cells, which are antibody factories.
2. Antibodies, or immunoglobulins, are divided into five classes: IgG, IgA, IgM, IgD, and IgE.
3. Each antibody molecule consists of two heavy and two light polypeptide chains that form constant and variable regions. There are two variable regions that can bind to two specific antigens.

17.6 Active and Passive Immunity

1. A person who is first exposed to a foreign antigen develops a primary response, which is sluggish.
2. Upon subsequent exposures to the same antigen, the immune system produces a secondary response in which more antibodies are secreted more quickly.
3. The secondary response is explained by the clonal selection theory; that is, an antigen interacts with specific receptors, which causes the lymphocyte capable of attacking that antigen to proliferate and produce a clone.
4. Clone development and a secondary immune response are elicited in active immunizations.
5. Passive immunity is produced when an individual receives antibodies produced by another organism; antibodies that cross the placenta from the mother to her fetus is an example.

17.7 Types of T Lymphocytes

1. Killer T lymphocytes perform cell-mediated destruction of infected or transformed cells by releasing chemicals called perforins.
2. Helper T lymphocytes release a variety of lymphokines that promote different aspects of the immune response.
3. In order for T lymphocytes to be stimulated by an antigen, a macrophage or B lymphocyte must present the antigen to them.

17.8 Activation of T Lymphocytes

1. Macrophages present foreign antigens in association with antigens produced by the person's own genes; these are the histocompatibility antigens, or MHC molecules.
2. Helper T lymphocytes interact only with antigen-presenting cells displaying class-2 MHC molecules. The activated helper T cells then secrete interleukin-2 and other lymphokines.
3. Killer T lymphocytes can attack only those cells that display the foreign antigen together with class-1 MHC molecules.

Review Activities

Objective Questions

1. Which of the following is *not* a function of the lymphatic system? (*module 17.1*)
   a. transport of interstitial fluid back to the blood
   b. transport of absorbed fats from the small intestine to the blood
   c. immunological defense
   d. filtration of metabolic wastes

2. Which of the following offers a nonspecific defense against a pathogen? (*module 17.2*)
   a. skin
   b. gastric juice
   c. Kupffer cells
   d. interferons
   e. all of the above

3. Which of the following is *not* a type of phagocytic cell? (*module 17.3*)
   a. neutrophil
   b. lymphocyte
   c. monocyte
   d. macrophage

4. B lymphocytes in mammals are processed within (*module 17.4*)
   a. the lymph nodes.
   b. the spleen.
   c. the thymus.
   d. the bone marrow.

5. Plasma cells secrete (*module 17.5*)
   a. antibodies.
   b. perforins.
   c. lysosomal enzymes.
   d. histamine.
   e. lymphokines.

6. Mast cell secretion during an immediate hypersensitivity reaction is stimulated when antigens combine with (*module 17.5*)
   a. IgG antibodies.
   b. IgE antibodies.
   c. IgM antibodies.
   d. IgA antibodies.

7. During a secondary immune response, (*module 17.6*)
   a. antibodies are made quickly and in great amounts.
   b. antibody production lasts longer than in a primary response.
   c. antibodies of the IgG class are produced.
   d. lymphocyte clones are believed to develop.
   e. all of the above apply.

8. Active immunity may be produced by (*module 17.6*)
   a. having a disease.
   b. receiving a vaccine.
   c. receiving gamma globulin injections.
   d. both a and b.
   e. both b and c.

9. Which of the following statements about T lymphocytes is *false?* (*module 17.7*)
   a. Some T cells promote the activity of B cells.
   b. Some T cells suppress the activity of B cells.
   c. some T cells secrete interferon.
   d. some T cells produce antibodies.

10. Which of the following cells aids the activation of lymphocytes by antigens? (*module 17.8*)
   a. macrophages
   b. neutrophils
   c. mast cells
   d. natural killer cells

11. Which of the following statements about class-2 MHC molecules is *false?* (*module 17.8*)
   a. They are found on the surface of B lymphocytes.
   b. They are found on the surface of macrophages.
   c. They are required for B cell activation by a foreign antigen.
   d. They are needed for interaction of helper and killer T cells.
   e. They are presented together with foreign antigens by macrophages.

Essay Questions

1. Describe the structure, function, and location of lymph nodes and the lymphoid organs. (*module 17.1*)
2. Give examples of nonspecific external and internal defense mechanisms. (*module 17.2*)
3. Identify the phagocytic cells of the immune system and explain the role of interferons in nonspecific immunity. (*module 17.3*)
4. Describe the characteristics of antigens. What is meant by an antigenic determinant site? (*module 17.4*)
5. Explain how antibodies help to destroy invading bacterial cells. (*module 17.5*)
6. With reference to the clonal selection theory, explain how active immunity is produced. (*module 17.6*)
7. Distinguish between the three classes of T lymphocytes. (*module 17.7*)
8. Explain the possible roles of helper and suppressor T lymphocytes in defense against infections. (*module 17.7*)
9. Explain how T lymphocytes interact with macrophages and the infected cells in fighting viral infections. (*module 17.8*)
10. Distinguish between class-1 and class-2 MHC molecules, and explain the function of the CD4 and CD8 coreceptors. (*module 17.8*)

Labeling Exercise

Label the structures indicated on the figure to the right.

1. ______________________________
2. ______________________________
3. ______________________________
4. ______________________________
5. ______________________________
6. ______________________________

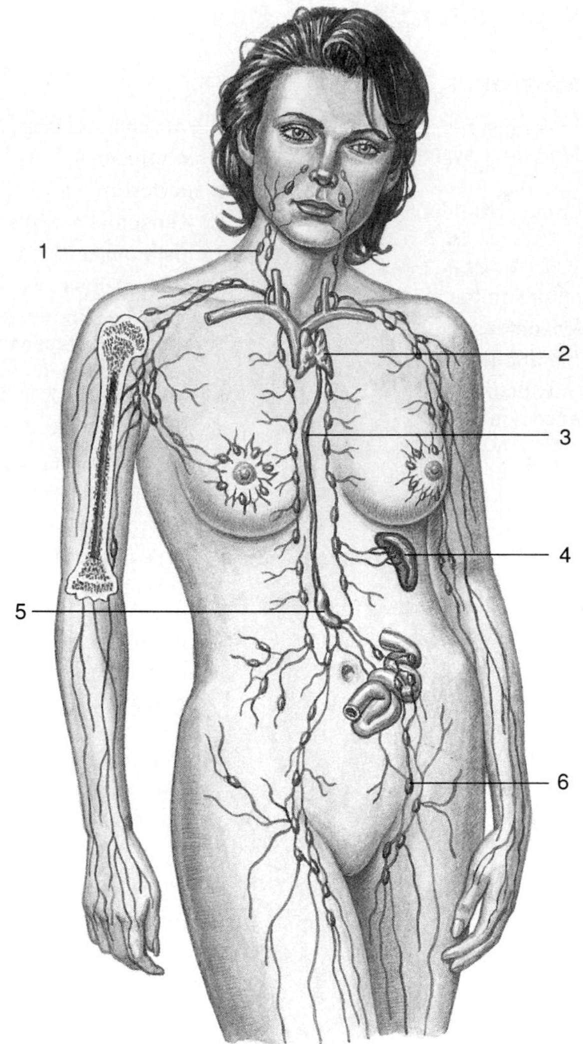

Critical Thinking Questions

1. People bitten by rattlesnakes may be passively immunized with injections of antitoxin. Propose a method by which this antitoxin might be prepared in sheep.
2. You get a flu shot each year to protect against flu. Why don't they just give "booster" injections? Why doesn't last year's flu shot protect you against this year's flu?
3. Laboratory tests made during the spring indicate that you have elevated levels of IgE antibodies. What might this indicate?
4. The clonal selection theory of immunity bears a striking resemblance to the concept of evolution by natural selection. Explain how these processes resemble each other.
5. If a child gets stung by a bee, an allergic reaction may occur quite rapidly. If a child crawls through some poison oak, the allergic reaction may not appear until the next day. What is different about these two allergic reactions? Which would respond better to antihistamines? Why?

CHAPTER 18

"[O]ur respiratory system processes a tremendous amount of air. . . over 8000 liters in a 24-hour period." *page 460*

Respiratory System

Chapter Outline

Terms to Remember

nasal cavity, 451
pharynx (*far' ingks*), 452
tonsil, 452
larynx (*lar' ingks*), 453
trachea (*tra' ke-ă*), 454
bronchus (*brong' kus*), 455
bronchiole (*brong' ke-ōl*), 455
alveolus (*al-ve' ō-lus*), 456
lung, 456
pleura (*ploor' ă*), 457
Boyle's law, 458
compliance, 458
law of Laplace, 459
tidal volume, 462
Dalton's law, 463
Henry's law, 463
hemoglobin, 467
Bohr effect, 468
acidosis, 469
alkalosis, 469

Learning Objectives

After studying this chapter, you should be able to

- discuss the functions of the respiratory system and distinguish between the conducting and respiratory division structures.
- describe the structure of the nose and identify the divisions of the pharynx.
- describe the structure of the larynx and the trachea.
- identify the components of the conductive and respiratory divisions of the lungs and describe the path of air through the airways.
- describe the structure and function of the lungs and pleural membranes, and explain how the intrapulmonary and intrapleural pressures change during inspiration and expiration.
- explain how the compliance, elasticity, and surface tension of the lungs affects breathing and discuss the nature and significance of pulmonary surfactant.
- explain how inspiration and expiration are accomplished and how pulmonary function is measured.
- define *partial pressure* and explain how measurements of the partial pressures of oxygen and carbon dioxide in the blood provide information regarding lung function.
- explain how breathing is regulated.
- describe how hemoglobin functions in oxygen transport and how carbon monoxide poisoning interferes with oxygen transport.
- describe how carbon dioxide is transported in the blood and explain how ventilation affects the blood pH.

18.1 Functions and Divisions of the Respiratory System

The respiratory system can be divided structurally into upper and lower divisions, and functionally into a conducting division and a respiratory division.

The term *respiration* refers to three separate but related functions: (1) **ventilation** (breathing); (2) **gas exchange,** which occurs between the air and blood in the lungs and between the blood and other tissues of the body; and (3) **oxygen utilization** by the tissues in the energy-liberating reactions of cell respiration. Ventilation and the exchange of gases (oxygen and carbon dioxide) between the air and blood are collectively called **external respiration.** Gas exchange between the blood and other tissues and oxygen utilization by the tissues are collectively known as **internal respiration.**

Ventilation is the mechanical process that moves air into and out of the lungs. Since air in the lungs has a higher oxygen concentration than the blood, oxygen diffuses from air to blood. Carbon dioxide, conversely, moves from the blood to the air within the lungs by diffusing down its concentration gradient. As a result of this gas exchange, the inspired air contains more oxygen and less carbon dioxide than the expired air. More importantly, blood leaving the lungs (in the pulmonary veins) has a higher oxygen concentration and a lower carbon dioxide concentration than the blood that entered the lungs in the pulmonary arteries. This is because the lungs function to bring the blood into gaseous equilibrium with the air.

The major passages and structures of the respiratory system (fig. 18.1) are the *nasal cavity, pharynx, larynx,* and *trachea,* and the *bronchi, bronchioles,* and *alveoli* within the *lungs.* The structures of the **upper respiratory system** include the nose, pharynx, and associated structures; the **lower respiratory system** includes the larynx, trachea, bronchial tree, alveoli, and lungs (table 18.1).

Based on general function, the respiratory system is frequently divided into a **conducting division** and a **respiratory division.** The conducting division includes all of the cavities and structures that transport gases to the respiratory division. The respiratory division consists of the structures involved in gas exchange between the air and blood.

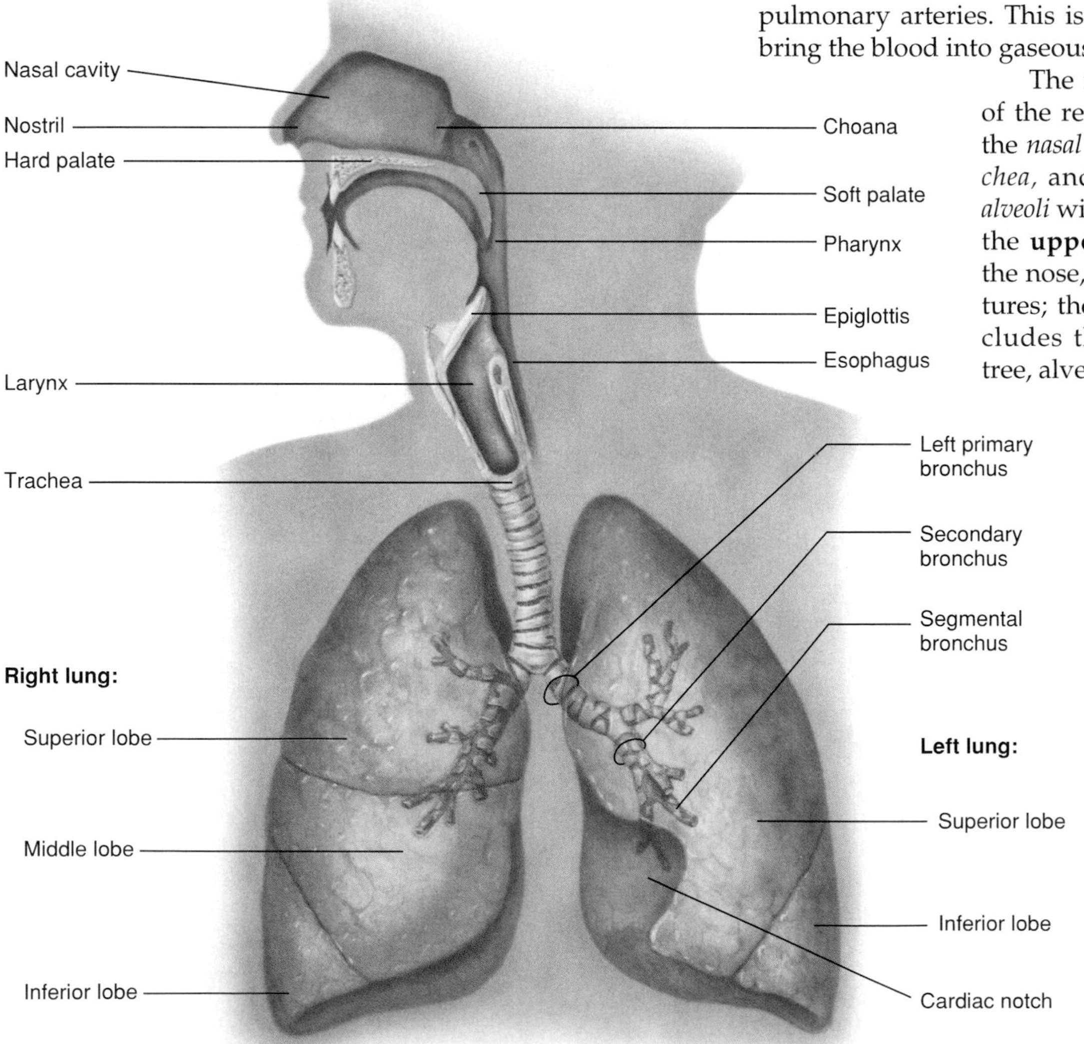

FIGURE 18.1
The basic anatomy of the respiratory system.

TABLE 18.1 **Major Structures of the Respiratory System**

Structure	Description	Function
Nose	Jutting external portion that is part of the face, plus internal nasal cavity	Warms, moistens, and filters inhaled air as it is conducted to the pharynx
Paranasal sinuses	Air spaces in the ethmoid, sphenoid, frontal, and maxillary bones	Produce mucus; provide sound resonance; lighten the skull
Pharynx	Chamber connecting oral and nasal cavities to the larynx	Serves as passageway for air into the larynx and for food into the esophagus
Larynx	Voice box; short passageway that connects the pharynx to the trachea	Serves as passageway for air; produces sound; prevents foreign materials from entering the trachea
Trachea	Flexible tubular connection between the larynx and the bronchial tree	Serves as passageway for air; pseudostratified ciliated columnar epithelium cleanses the air
Bronchial tree	Bronchi and branching bronchioles in the lung; tubular connection between the trachea and alveoli	Serves as passageway for air; continued cleansing of air
Alveoli	Microscopic membranous air sacs within the lungs	Functional units of respiration; site of gaseous exchange between the respiratory and circulatory systems
Lungs	Major organs of the respiratory system; located in the thoracic cavity	Contain bronchial trees, alveoli, and associated pulmonary vessels
Pleurae	Serous membranes covering the lungs and lining the thoracic cavity	Compartmentalize, protect, and lubricate the lungs

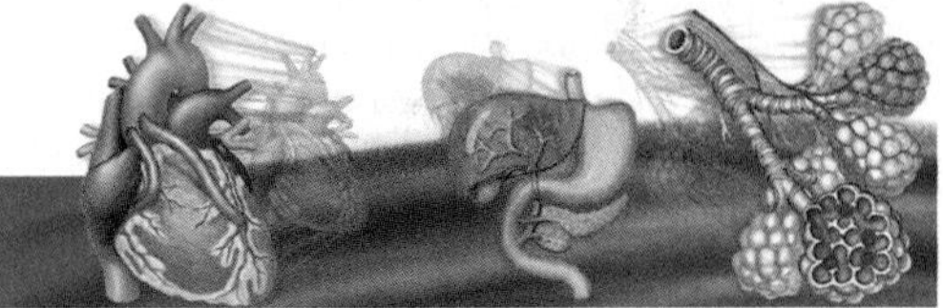

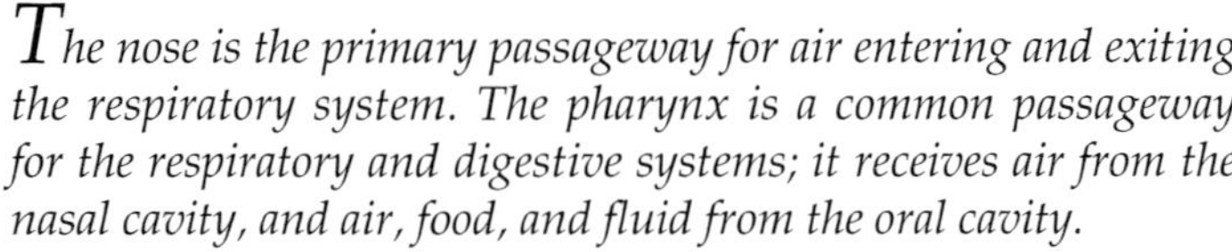

18.2 Nose and Pharynx

The nose is the primary passageway for air entering and exiting the respiratory system. The pharynx is a common passageway for the respiratory and digestive systems; it receives air from the nasal cavity, and air, food, and fluid from the oral cavity.

Nose

The nose includes an external portion that protrudes from the face and an internal **nasal cavity** for the passage of air (fig. 18.2). The nasal cavity is divided into right and left halves by a **nasal septum.** The bony part of the nasal septum consists of the *vomer* inferiorly and the perpendicular plate of the *ethmoid bone* superiorly. The anterior part of the nasal septum is formed by cartilage. Each half of the nasal cavity is referred to as a **nasal fossa.** Each nasal fossa opens anteriorly through the **nostril,** or *anterior nare* (*na're*), and communicates posteriorly with the **nasopharynx** through the **choana** (*ko-a'nă*), or *posterior nare* (fig. 18.2). On the lateral walls of the nasal cavity are three bony projections, the **superior, middle,** and **inferior conchae** (*kong'ke*). Air passages between the conchae are referred to as **meatuses.**

The nasal epithelium covering the conchae serves to warm, moisten, and cleanse the air. Nasal hairs called **vibrissae** (*vi-bris'e*) filter large particles that might otherwise be inhaled. Dust, pollen, smoke, and other fine particles are trapped along the moist mucous membrane lining the nasal cavity.

There are several drainage openings into the nasal cavity. The paranasal ducts drain mucus from the **paranasal sinuses** (see fig. 6.20) and the **nasolacrimal ducts** drain tears from the eyes (see fig. 12.9). The auditory tubes from the middle ear enter the upper respiratory tract posterior to the nasal cavity in the nasopharynx. These interconnections allow infections to spread easily from one chamber to another throughout the facial area.

---
nostril: O. E. *nosu,* nose; *thyrel,* hole
choana: Gk. *choane,* funnel
meatus: L. *meatus,* path

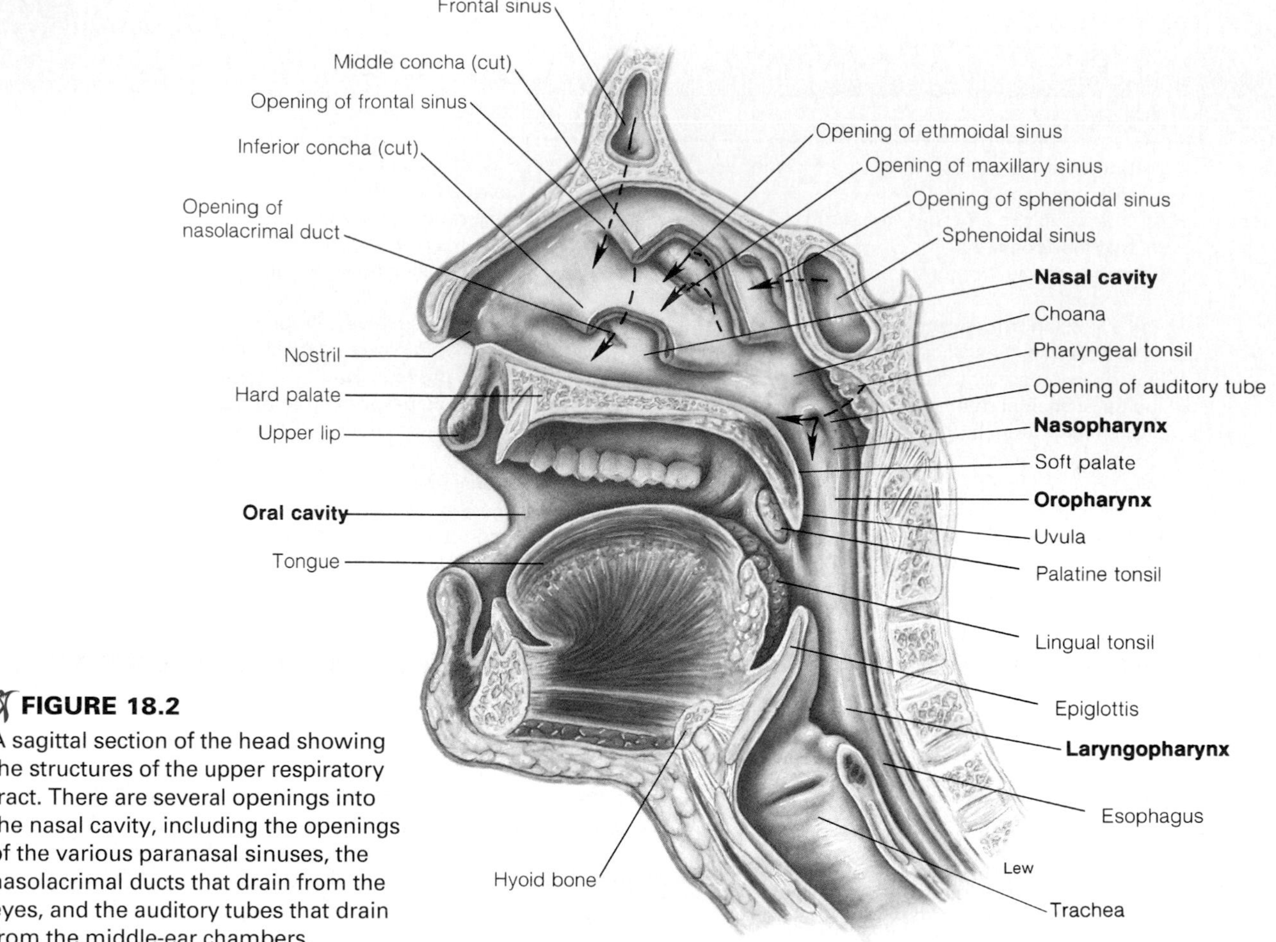

FIGURE 18.2
A sagittal section of the head showing the structures of the upper respiratory tract. There are several openings into the nasal cavity, including the openings of the various paranasal sinuses, the nasolacrimal ducts that drain from the eyes, and the auditory tubes that drain from the middle-ear chambers.

Pharynx

The pharynx (*far'ingks*) is a funnel-shaped passageway that connects the nasal and oral cavities to the larynx (fig. 18.2). On the basis of location and function, the pharynx is divided into three regions.

1. The **nasopharynx** serves only as a passageway for air, since it is located above the point of food entry into the body (the mouth). It is the uppermost portion of the pharynx, directly behind the nasal cavity and above the soft palate. A pendulous **uvula** hangs from the middle lower border of the soft palate. The paired **auditory** (eustachian) **tubes** connect the nasopharynx with the middle-ear cavities. The **pharyngeal tonsils,** or **adenoids,** are situated in the posterior wall of this cavity, just below the sphenoid sinus.
2. The **oropharynx** is the middle portion of the pharynx between the soft palate and the level of the hyoid bone. Both swallowed food and fluid and inhaled air pass through it. The base of the tongue forms the anterior wall of the oropharynx. Paired **palatine tonsils** are situated on the posterior lateral wall, and the **lingual tonsils** are found on the base of the tongue.
3. The **laryngopharynx** (*lă-ring"go-far'ingks*) is the lowermost portion of the pharynx. It extends inferiorly from the level of the hyoid bone to the larynx and opens into the esophagus and larynx. It is at the lower laryngopharynx that the respiratory and digestive systems become distinct. Swallowed food and fluid are directed into the esophagus, whereas inhaled air is directed anteriorly into the larynx.

During a routine physical examination, the physician will commonly depress your tongue and examine the condition of the palatine tonsils. Tonsils are lymphoid organs that tend to become swollen and inflamed by persistent infections. The tonsils may have to be surgically removed when they become so overrun with pathogens that they themselves become the source of infection. Removal of the palatine tonsils is called a tonsillectomy, *whereas removal of the pharyngeal tonsils is called an* adenoidectomy.

pharynx: L. *pharynx,* throat
uvula: L. *uvula,* small grape
adenoid: Gk. *adenoeides,* glandlike

18.3 Larynx and Trachea

The larynx is the structure between the pharynx and trachea that houses the vocal cords. The trachea is the airway leading from the larynx to the bronchi.

Larynx

The larynx (*lar'ingks*), or voice box, is a continuation of the conducting division that connects the laryngopharynx with the trachea. The primary function of the larynx is to prevent food or fluid from entering the trachea and lungs during swallowing and to permit passage of air while breathing. A secondary role is to produce sound.

The larynx is shaped like a triangular box (fig. 18.3). It is composed of a framework of nine cartilages: three are large unpaired structures, and six are smaller and paired. The largest of the unpaired cartilages is the anterior **thyroid cartilage.** The **laryngeal prominence** of the thyroid cartilage is commonly called the "Adam's apple." The thyroid cartilage is typically larger and more prominent in males than in females because of the effect of male sex hormones on the development of the larynx during puberty. The spoon-shaped **epiglottis** is located behind the root of the tongue; it covers the **glottis,** or laryngeal opening, during swallowing.

The entire larynx elevates during swallowing to close the glottis against the epiglottis. If it is not closed as it should be, food may become lodged within the glottis. In this case, the *abdominal thrust* (Heimlich) *maneuver* can be used to prevent suffocation (see "Clinical Considerations," p. 473).

The lower end of the larynx is formed by the ring-shaped **cricoid** (*kri'koid*) **cartilage.** This third unpaired cartilage connects the thyroid cartilage above and the trachea below. The paired **arytenoid** (*ar"ĭ-te'noid*) **cartilages,** located above the cricoid and behind the thyroid, furnish the attachments of the **vocal cords.** The other paired **cuneiform cartilages** and **corniculate** (*kor-nik'yŭ-lāt*) **cartilages** are small accessory cartilages that are closely associated with the arytenoid cartilages (fig. 18.3).

Two pairs of strong connective tissue bands stretch across the upper opening of the larynx from the thyroid cartilage anteriorly to the paired arytenoid cartilages posteriorly. These are the **true vocal cords** and the **false vocal cords** (fig. 18.4). The false vocal cords support the true vocal cords and are not used in sound production. The true vocal cords vibrate to produce sound. Mature males generally have thicker and longer vocal cords than females;

larynx: Gk. *larynx,* upper windpipe
thyroid: Gk. *thyreos,* shieldlike
cricoid: Gk. *krikos,* ring; *eidos,* form
arytenoid: Gk. *arytina,* ladle- or cup-shaped
cuneiform: L. *cuneus,* wedge-shaped
corniculate: L. *corniculum,* diminutive of *cornu,* horn

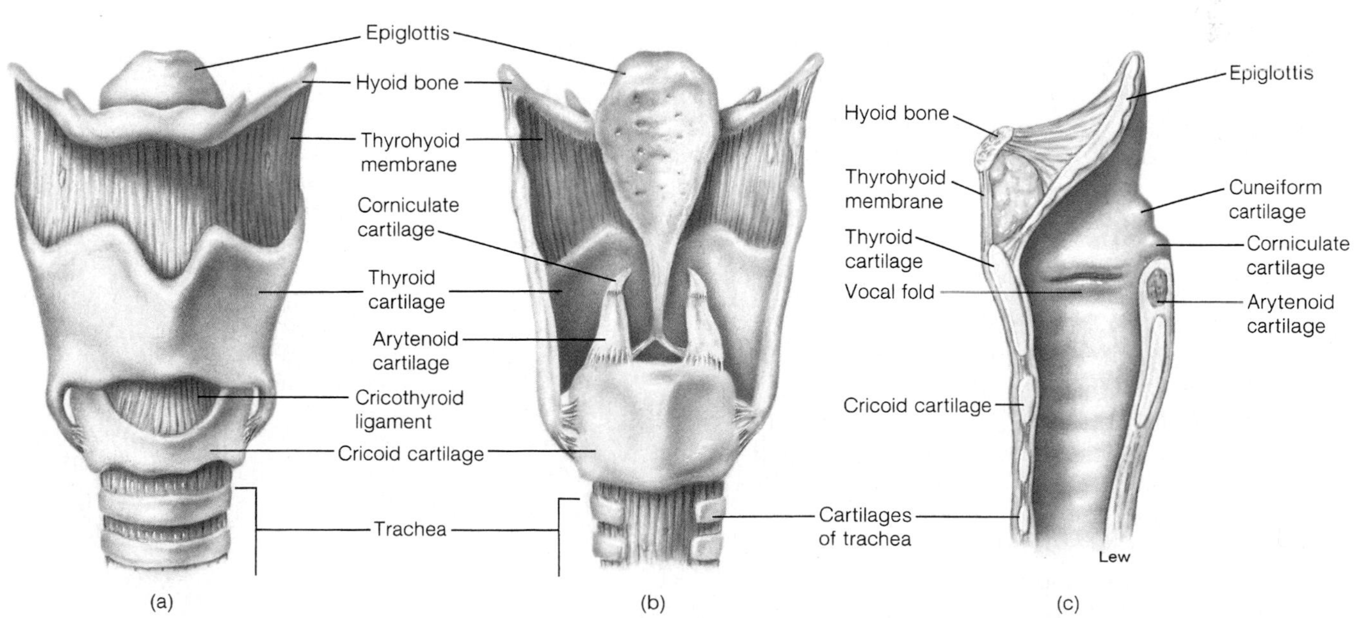

FIGURE 18.3
The structure of the larynx. (*a*) An anterior view, (*b*) a posterior view, and (*c*) a sagittal view.

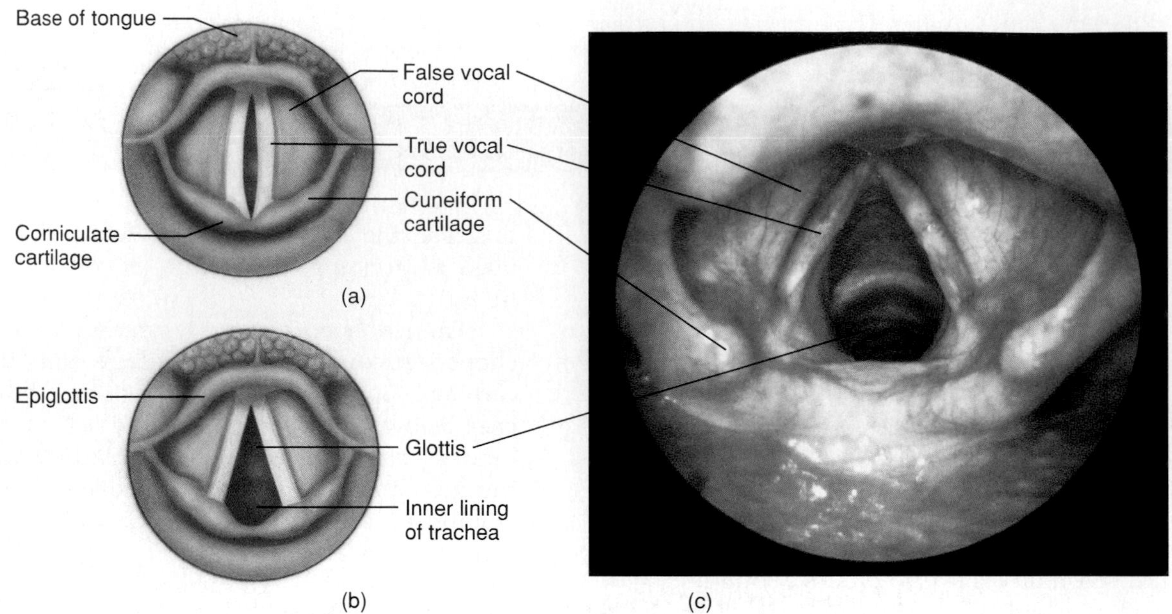

FIGURE 18.4
A superior view of the vocal cords. In (*a*) the vocal cords are taut; in (*b*) they are relaxed and the glottis is opened. (*c*) A photograph through a laryngoscope showing the true and false vocal cords and the glottis.

their vocal cords vibrate more slowly and produce lower pitches. The loudness of vocal sounds is determined by the force of air passed over the vocal cords and by the amount of vibration. The vocal cords do not vibrate when whispering.

Inflammation of the larynx and vocal cords, or **laryngitis**, *results in hoarseness or inability to speak above a whisper. When the laryngeal tissues become irritated—for example, from overuse of the voice, very dry air, harsh fumes, or bacterial infections—contraction of the cords may be impaired, or they may swell so that they cannot move freely. Acute laryngitis may cause serious breathing problems in children under the age of 5.*

Trachea

The trachea (*tra'ke-ă*), commonly called the windpipe, is a passageway, approximately 12 cm (4 in.) long and 2.5 cm (1 in.) in diameter, connecting the larynx to the primary bronchi (fig. 18.5). It is positioned in front of the esophagus as it extends into the thoracic cavity. A series of 16 to 20 C-shaped rings of hyaline cartilage form the supporting walls of the trachea. These cartilages ensure that the airway will always remain open. The open part of each cartilage ring faces the esophagus and permits it to expand slightly into the trachea during swallowing.

trachea: L. *trachia*, rough air vessel

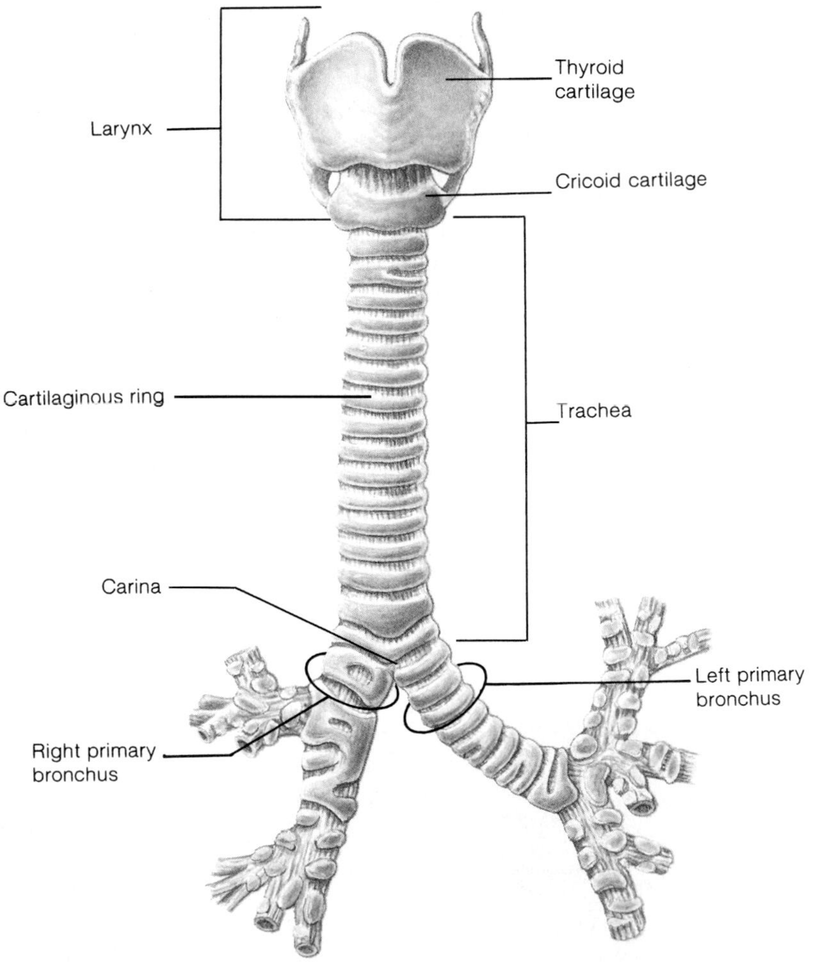

FIGURE 18.5
An anterior view of the larynx, trachea, and bronchi.

The mucosa (surface lining of the lumen) consists of pseudostratified ciliated columnar epithelium containing many mucus-secreting **goblet cells.** It provides the same protection against dust and other particles as the membrane lining the nasal cavity and larynx. Medial to the lungs, the trachea splits to form the right and left bronchi. This junction is reinforced by the **carina** (*kă-ri'nă*), a keel-like cartilage plate.

carina: L. *carina*, keel

If the trachea becomes blocked, as by aspiration of a foreign object, it may be necessary to create an emergency opening into this tube so that ventilation can still occur. A tracheotomy *is the process of surgically opening the trachea. A* tracheostomy *is the procedure of inserting a tube into the trachea to permit breathing and to keep the passageway open.*

18.4 Bronchial Tree, Alveoli, and Lungs

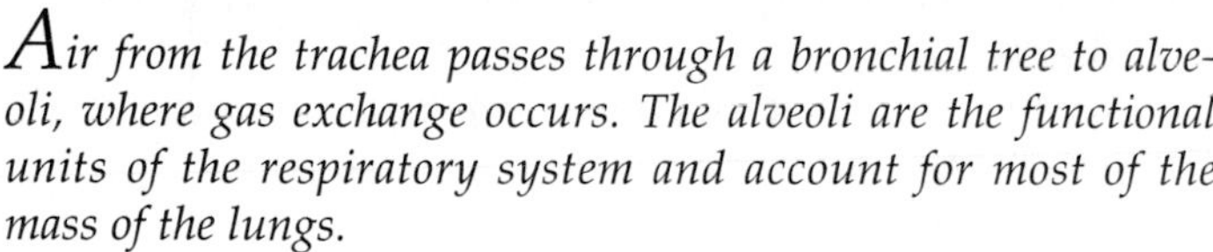

Air from the trachea passes through a bronchial tree to alveoli, where gas exchange occurs. The alveoli are the functional units of the respiratory system and account for most of the mass of the lungs.

Bronchial Tree

The bronchial tree is so named because it is composed of a series of respiratory tubes that branch into progressively narrower tubes as they extend into the lung. The trachea splits to form a right and a left **primary bronchus** (*brong'kus*) behind the manubrium of the sternum. Each bronchus has hyaline cartilage rings surrounding its lumen to keep it open as it extends into the lung.

The bronchus divides deeper in the lungs to form **secondary bronchi** and **segmental (tertiary) bronchi.** The bronchial tree continues to branch into even smaller tubules called **bronchioles** (*brong'ke-ōlz*) (fig. 18.6). There is no cartilage in the bronchioles, so the smooth muscle in their walls can constrict or dilate these airways. Bronchioles provide the greatest resistance to airflow in the conducting passages—a function analogous to that of

bronchus: L. *bronchus*, windpipe

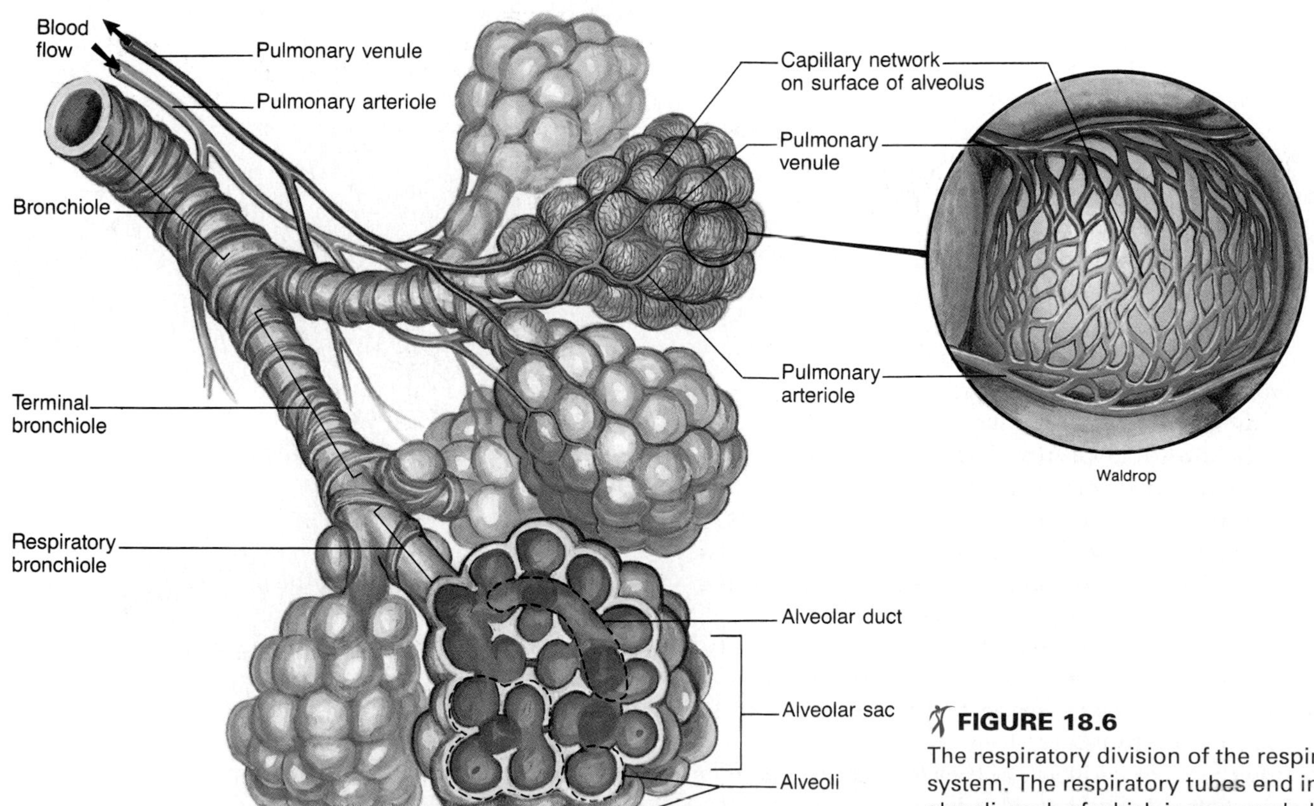

FIGURE 18.6
The respiratory division of the respiratory system. The respiratory tubes end in minute alveoli, each of which is surrounded by an extensive capillary network.

arterioles in the circulatory system. Numerous **terminal bronchioles** mark the end of the air-conducting pathway to the alveoli.

Alveoli

Air leaving terminal bronchioles enters the **alveolar ducts.** The alveolar ducts contain pulmonary alveoli (*al-ve'ŏ-li*) as outpouchings along their length, and open into clusters of alveoli called **alveolar sacs** at their ends (figs. 18.6 and 18.7). The alveolar ducts, alveoli, and alveolar sacs constitute the *respiratory division* of the lungs. Gas exchange occurs across the walls of the tiny alveoli; hence, these minute expansions are the functional units of the respiratory system. The vast number of alveoli (about 350 million per lung) provides a tremendous surface area for the diffusion of gases. The diffusion rate is further increased by the fact that the wall of each alveolus is only one cell layer thick, so that the total "air-blood barrier" is only one alveolar cell and one blood capillary cell across, or about 2 µm (fig. 18.8).

Type I alveolar cells form the major part of the alveolar walls and permit diffusion. Scattered among these squamous cells are cuboidal *type II alveolar cells* that secrete a lipoprotein substance called *surfactant.* Surfactant prevents alveolar collapse, as will be discussed in module 18.6.

Inside Information—The internal surface area of each lung is more than 40 times greater than the external surface area of the entire body.

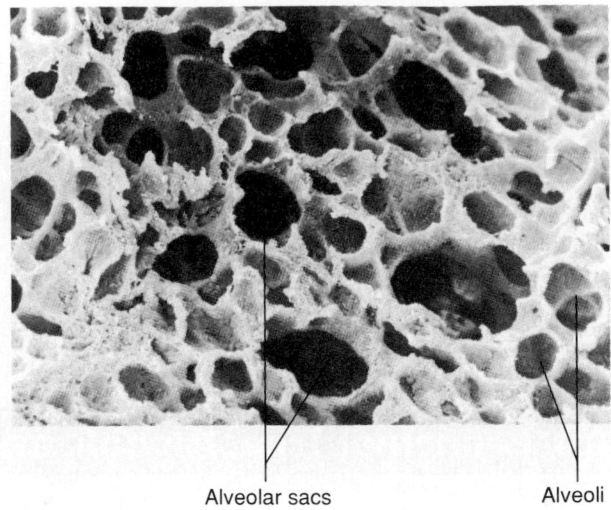

FIGURE 18.7
A scanning electron micrograph of lung tissue showing alveolar sacs and alveoli.

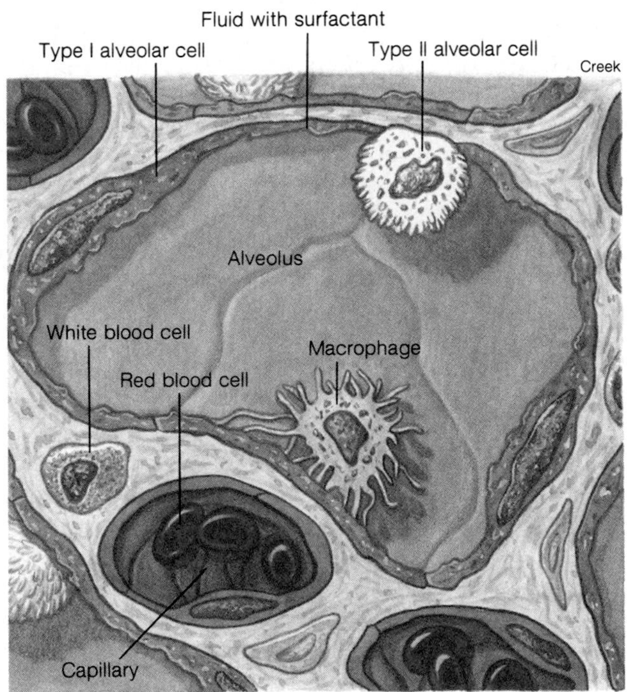

FIGURE 18.8
The relationship between lung alveoli and pulmonary capillaries.

Lungs

The lungs are spongy paired organs situated within the thoracic cavity. They extend from the diaphragm to a point just above the clavicles, and their surfaces are bordered by the ribs to the front and back. The lungs are separated from one another by the heart and other structures of the **mediastinum** (*me"de-ă-sti'num*), which is the area between the lungs. All structures of the respiratory system beyond the primary bronchi, including the bronchial tree and alveoli, are contained within the lungs.

Each lung presents four surfaces that match the contour of the thoracic cavity. The medial surface of each lung is slightly concave and contains a vertical slit, the **hilum,** through which pulmonary vessels, nerves, and bronchi pass. The inferior surface, called the **base of the lung,** is concave as it fits over the convex dome of the diaphragm. The superior surface, called the **apex of the lung,** extends above the level of the clavicle. Finally, the broad, rounded surface in contact with the membranes covering the ribs is called the **costal surface of the lung.**

Although the right and left lungs are basically similar, they are not identical. The left lung is somewhat smaller than the right and has a **cardiac notch** on its medial surface to accommodate the heart (see fig. 18.1). The left lung is subdivided into a *superior lobe* and an *inferior lobe* by a single fissure. The right lung is subdivided by two fissures into three lobes: *superior, middle,* and *inferior lobes.*

alveolus: L. diminutive of *alveus,* cavity
mediastinum: L. *mediastinus,* intermediate
hilum: L. *hilum,* a trifle (little significance)

18.5 Thoracic Compartments

The visceral and parietal pleural membranes enclose a cavity separate from the spaces within the lungs. Pressure differences between these compartments are produced during breathing.

Pleurae

Pleurae are serous membranes surrounding the lungs (fig. 18.9). The pleura (*ploor'ă*) of each lung is composed of both a visceral and parietal portion. The **visceral pleura** adheres to the outer surface of the lung, and the **parietal pleura** lines the thoracic walls and the thoracic surface of the diaphragm. A continuation of the parietal pleura around the heart and between the lungs forms the boundary of the mediastinum. Between the visceral and parietal pleurae is a moistened space called the **pleural cavity.**

The moistened serous membranes of the visceral and parietal pleurae are normally in contact with each other, so that the lungs are stuck to the thoracic wall. The pleural cavity (intrapleural space) between the two moistened membranes contains only a thin layer of lubricating serous fluid secreted by the pleural membranes. The pleural cavity in a healthy living organism is thus potential rather than real; it becomes real only in abnormal situations, when air enters the intrapleural space. Since the lungs normally remain stuck to the thoracic wall, they get larger and smaller together with the thoracic cavity during respiratory movements.

The thoracic cavity has four distinct compartments: a pleural cavity surrounds each lung; the pericardial cavity houses the heart; and the mediastinum contains the esophagus, thoracic duct, major vessels, various nerves, and portions of the respiratory tract. This *compartmentalization* has

pleura: Gk. *pleura,* side or rib

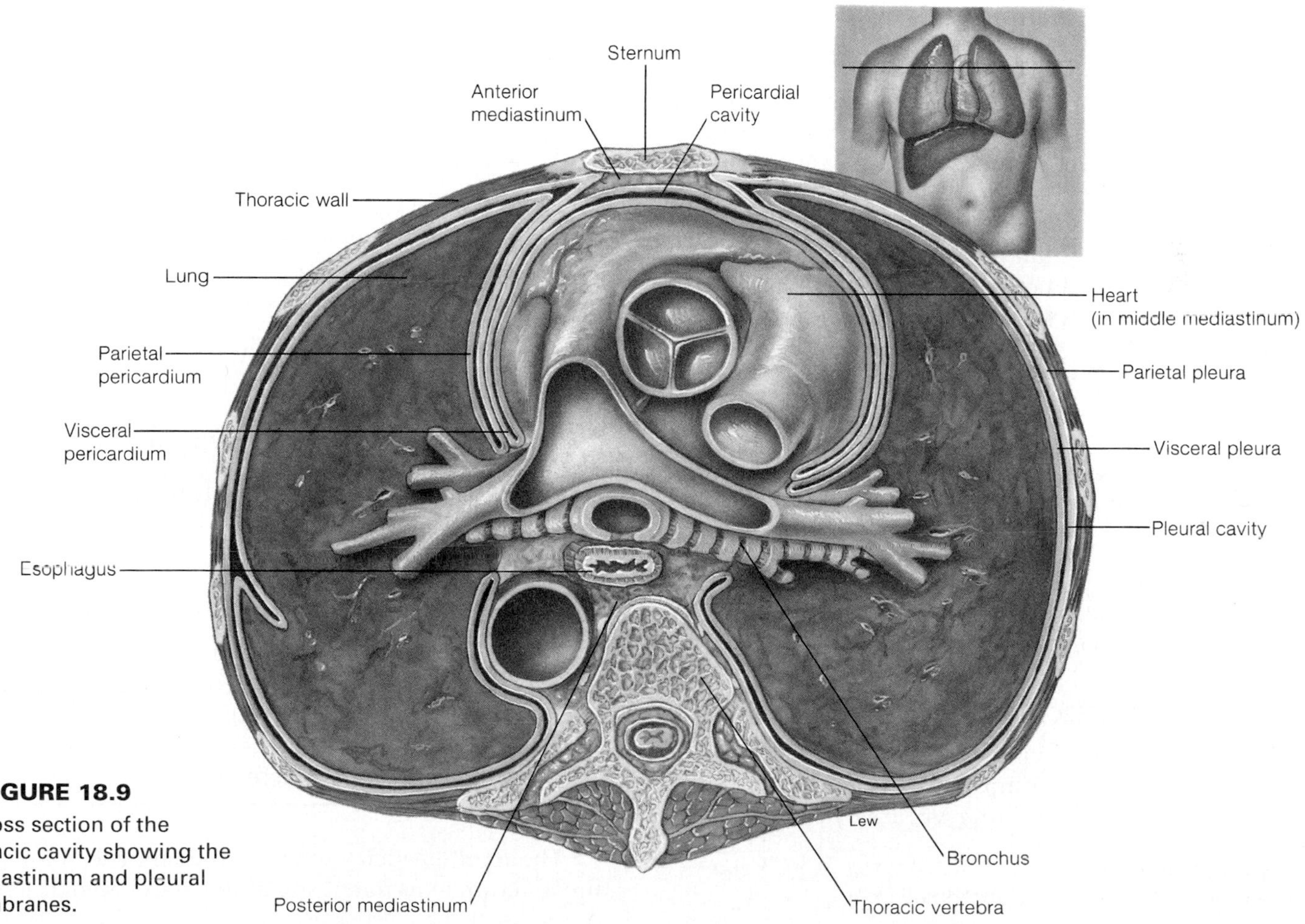

FIGURE 18.9
A cross section of the thoracic cavity showing the mediastinum and pleural membranes.

protective value in that infections are usually confined to one compartment. Also, damage to one organ usually will not involve another. For example, *pleurisy*, an inflamed pleura, is generally confined to one side; and a penetrating injury to the thoracic cavity, such as a knife wound, may cause collapse of one lung but not the other.

Intrapulmonary and Intrapleural Pressures

Air enters the lungs during inspiration because the atmospheric pressure is greater than the **intrapulmonary,** or **alveolar, pressure.** Since the atmospheric pressure does not usually change, the intrapulmonary pressure must fall below the atmospheric pressure to cause inspiration. A pressure below that of the atmosphere is called a *subatmospheric pressure,* or *negative pressure.* During quiet inspiration, for example, the intrapulmonary pressure may decrease to 3 mmHg below the pressure of the atmosphere. This subatmospheric pressure is commonly shown as -3 mmHg. Expiration, conversely, occurs when the intrapulmonary pressure is greater than the atmospheric pressure. During quiet expiration, for example, the intrapulmonary pressure may rise to +3 mmHg over the atmospheric pressure.

The lack of air in the intrapleural space produces a subatmospheric **intrapleural pressure,** which is lower than the intrapulmonary pressure (table 18.2). There is thus a pressure difference across the wall of the lung—called the **transpulmonary pressure**—which is the numerical difference between the intrapulmonary pressure and the intrapleural pressure. Since the pressure within the lungs (intrapulmonary pressure) is greater than that outside the lungs (intrapleural pressure), the difference in pressure (transpulmonary pressure) acts to expand the lungs as the thoracic volume expands during inspiration.

Changes in intrapulmonary pressure occur as a result of changes in lung volume. This follows from **Boyle's law,** which states that the pressure of a given quantity of gas is inversely proportional to its volume. An increase in lung volume during inspiration decreases intrapulmonary pressure to subatmospheric levels; air therefore goes in. A decrease in lung volume raises the intrapulmonary pressure above the atmospheric pressure, thus pushing air out. These changes will be described in module 18.7 on the mechanics of breathing.

TABLE 18.2 Intrapulmonary and Intrapleural Pressures in Normal Quiet Breathing, and the Transpulmonary Pressure Acting To Expand the Lungs

	Inspiration	Expiration
Intrapulmonary pressure (mmHg)	-3	+3
Intrapleural pressure (mmHg)	-6	-3
Transpulmonary pressure (mmHg)	+3	+6

Note: Pressures indicate mmHg below or above atmospheric pressure.

In summary, the arrangement of the pleurae serves three functions.

1. A small amount of serous fluid within the pleural cavities serves as a lubricant, allowing the lungs to slide along the chest wall during breathing.
2. The pressure in the pleural cavity is lower than the pressure in the lungs—a pressure difference that is needed for ventilation.
3. The pleurae effectively separate (compartmentalize) the thoracic organs, thus minimizing trauma and the spread of infections from one organ to another.

18.6 Physical Aspects of Ventilation

The physical properties of compliance, elasticity, and surface tension influence the behavior of the lungs. Surface tension is reduced by the action of a pulmonary surfactant.

Physical Properties of the Lungs

The ability of the lungs to inflate, or become distended or stretched, is known as **compliance.** Compliance is expressed as the change in lung volume per change in transpulmonary pressure. A given transpulmonary pressure, in other words, will cause greater or lesser distension, depending on the compliance of the lungs.

Inside Information—The lungs are remarkable in their stretching ability. Compared to a toy balloon, a lung is about 100 times more distensible.

The term **elasticity** refers to the tendency of a structure to return to its initial size after having been distended.

pulmonary: Gk. *pleumon,* lung
Boyle's law: from Robert Boyle, British physicist, 1627–91

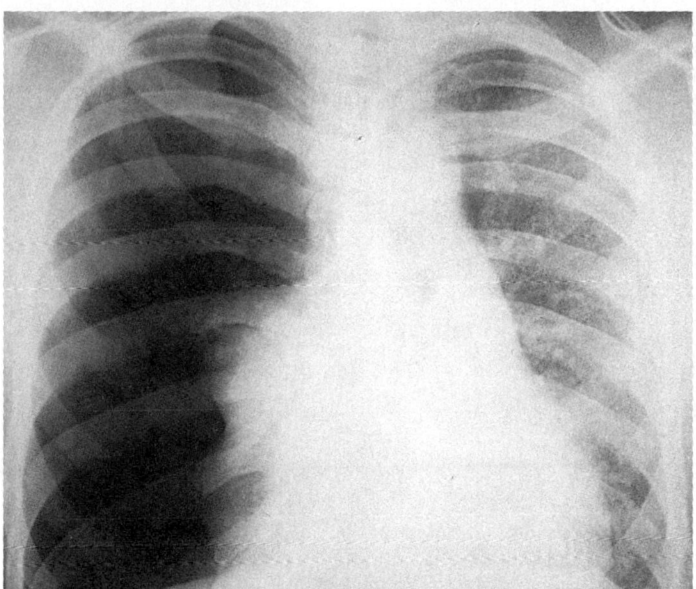

FIGURE 18.10

A pneumothorax of the right lung. The right side of the thorax appears uniformly dark because it is filled with air; the spaces between the ribs are also greater than those on the left, since the ribs are released from the elastic tension of the lungs. The left lung appears denser (less dark) because of shunting of blood from the right to the left lung.

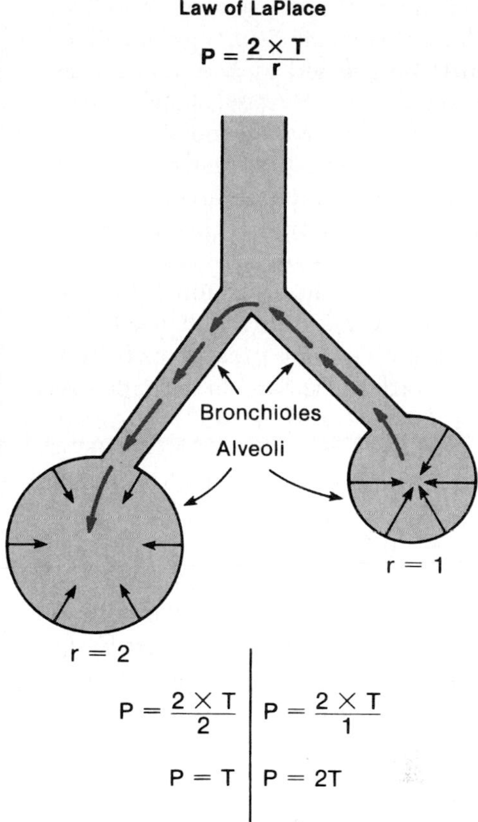

FIGURE 18.11

According to the law of Laplace, the pressure created by surface tension should be greater in the smaller alveolus (*right*) than in the larger alveolus (*left*). This implies that (without surfactant) smaller alveoli would collapse and empty their air into larger alveoli.

A high content of elastin proteins makes the lungs very elastic, so that they resist distension. Since the lungs are normally stuck to the chest wall, they are always in a state of elastic tension. This tension increases during inspiration when the lungs are stretched and is reduced by elastic recoil during expiration. The elasticity of the lungs and of other thoracic structures thus aids in pushing the air out during expiration.

The introduction of air into the intrapleural space, as can occur in the case of an open chest wound or a rupture of alveoli at the surface of the lung, is called a *pneumothorax* (*noo"mo-thor'aks*) (fig. 18.10). As the air enters, the intrapleural pressure rises until it is equal to the atmospheric pressure. When the intrapleural pressure is the same as the intrapulmonary pressure, the lung can no longer expand. In fact, it actually collapses away from the chest wall as a result of elastic recoil. Fortunately, a pneumothorax usually causes the collapse of only one lung; as described previously, each lung is enclosed within a separate pleural compartment.

In addition to elastic resistance, the **surface tension** exerted by the thin film of fluid in the alveoli acts to resist distension. The surface tension of an alveolus produces a force that is directed inward and, as a result, creates pressure within the alveolus. As described by the **law of Laplace,** this pressure is directly proportional to the surface tension and inversely proportional to the radius of the alveolus (fig. 18.11). Therefore, according to this law, the pressure in a smaller alveolus would be greater than that in a larger one if the surface tension were the same in both. The greater pressure of the smaller alveolus would then cause it to empty its air into the larger one. This does not normally occur, however, because of the action of a phospholipid in the alveolar fluid called **surfactant** (a contraction of the term *surface-active agent*) (fig. 18.12). The surfactant molecules interfere with the cohesiveness of water molecules, thus reducing the surface tension and preventing the alveoli from collapsing, as would be predicted by the law of Laplace.

Surfactant and Respiratory Distress Syndrome

The production of pulmonary surfactant by type II alveolar cells does not begin until late in fetal life—at about 8 months. For this reason, premature babies are sometimes born with lungs that lack sufficient surfactant—a condition

pneumothorax: Gk. *pneumon,* spirit (breath); L. *thorax,* chest
law of Laplace: From Pierre-Simon Laplace, French astronomer and mathematician, 1749–1827

called **respiratory distress syndrome (RDS).** When surfactant is insufficient, a great deal of energy must be exerted by the respiratory muscles to keep the lungs from collapsing. The condition is also called **hyaline membrane disease** because the high surface tension causes blood plasma to leak into the alveoli, giving the respiratory membranes a glassy appearance (*hyaline* means "resembling glass").

Even under normal conditions, the first breath of life is a difficult one because the newborn must overcome great surface tension forces in order to inflate its partially collapsed alveoli. The transpulmonary pressure required for the first breath is 15 to 20 times that required for subsequent breaths, and an infant with respiratory distress syndrome must duplicate this effort with every breath. Fortunately, many babies with this condition can be saved by mechanical ventilators that keep them alive long enough for their lungs to mature and manufacture sufficient surfactant. Pulmonary surfactant replacement therapy also may be used to treat these babies.

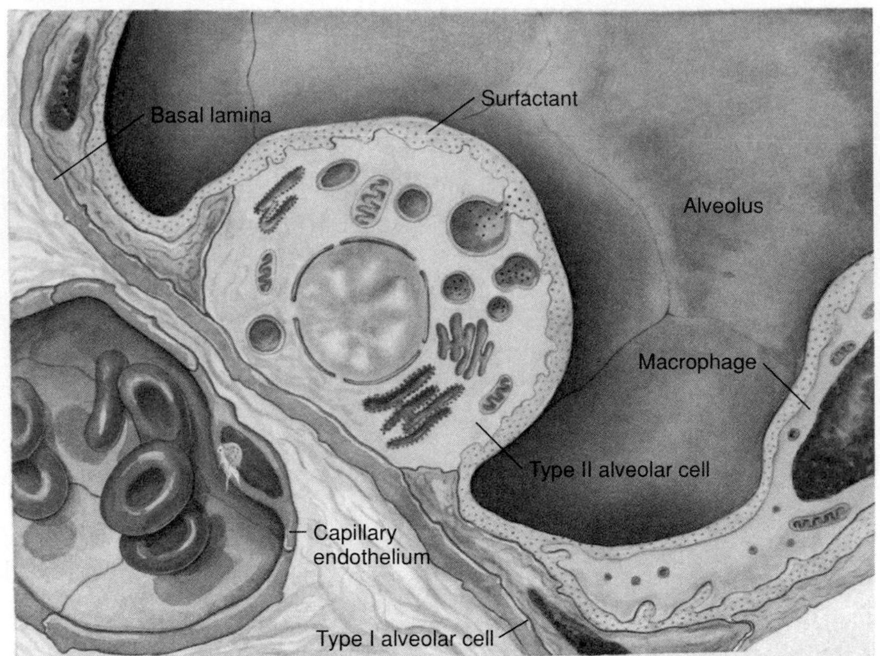

FIGURE 18.12
The production of pulmonary surfactant by type II alveolar cells. Surfactant appears to be composed of a derivative of lecithin combined with protein.

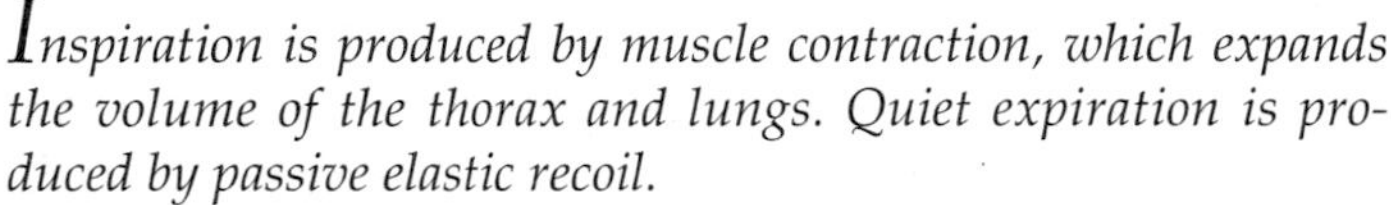

18.7 Mechanics of Breathing

Inspiration is produced by muscle contraction, which expands the volume of the thorax and lungs. Quiet expiration is produced by passive elastic recoil.

Breathing, or **pulmonary ventilation,** requires that the thorax be flexible in order to function as a bellows during the ventilation cycle. The structure of the rib cage and associated cartilages provide continuous elastic tension, so that when stretched by muscle contraction during inspiration, the rib cage can return passively to its resting dimensions when the muscles relax. This elastic recoil is greatly aided by the elasticity of the lungs.

Inspiration and Expiration

The thoracic cavity increases in size during *inspiration* (inhalation) in three directions: anteroposteriorly, laterally, and vertically. This is accomplished by contractions of the **diaphragm** and the **external intercostal muscles** (fig. 18.13). Other thoracic muscles become involved in *forced* (deep) *inspiration,* including the scalenes and sternocleidomastoid muscles of the neck and the pectoralis minor muscle of the chest.

Quiet expiration (exhalation) is a passive process. After becoming stretched by contractions of the diaphragm and thoracic muscles, the lungs recoil as a result of their elastic tension when the respiratory muscles relax. The decrease in lung volume raises the pressure within the alveoli above the atmospheric pressure and pushes the air out. During *forced expiration,* the **internal intercostal muscles** contract and depress the rib cage. The abdominal muscles may also aid expiration because, when they contract, they force abdominal organs up against the diaphragm and further decrease the volume of the thorax. The events that occur during inspiration and expiration are illustrated in figure 18.14.

Inside Information—We breathe an average of 15 times per minute during quiet breathing and more quickly during exertion or in the course of various illnesses. In the process, our respiratory system processes a tremendous amount of air. A normal individual exchanges approximately 6 liters of air in each minute of quiet breathing—over 8000 liters in a 24-hour period.

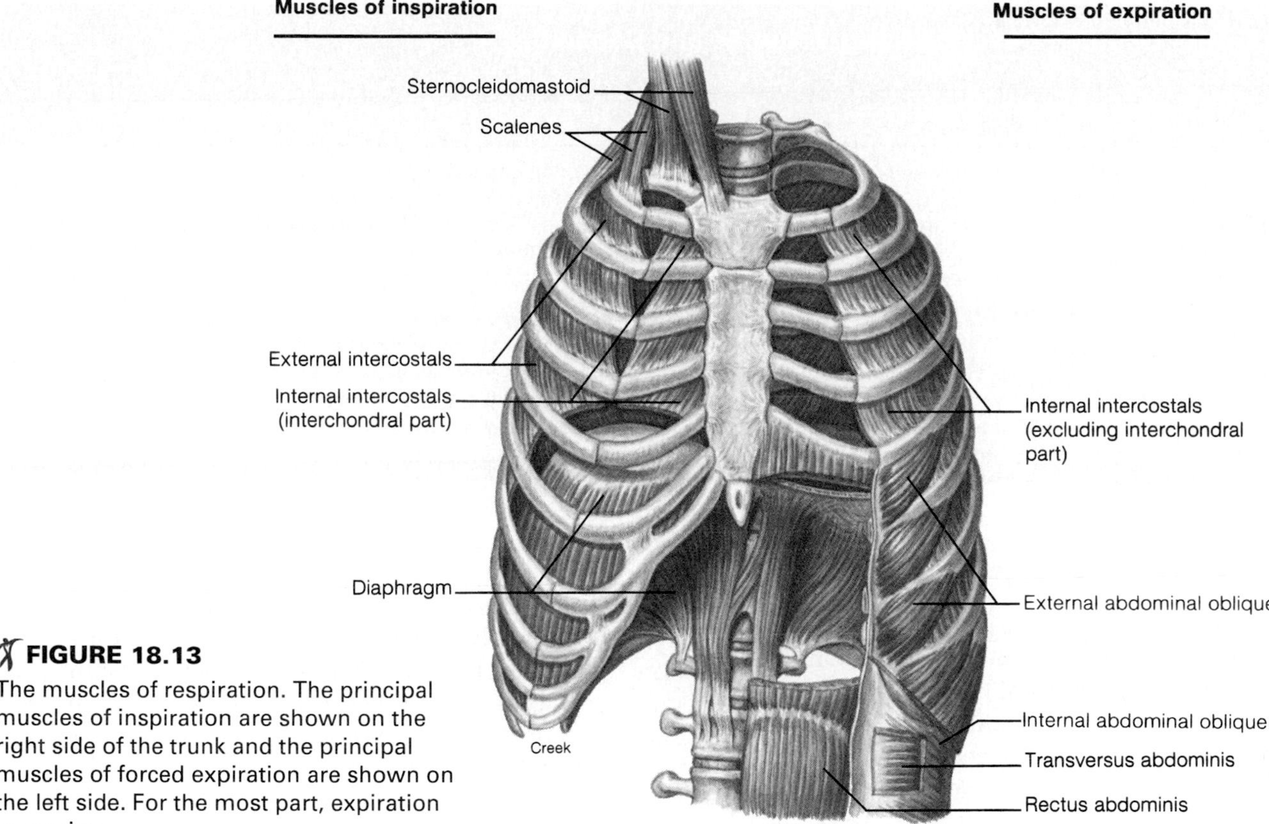

FIGURE 18.13
The muscles of respiration. The principal muscles of inspiration are shown on the right side of the trunk and the principal muscles of forced expiration are shown on the left side. For the most part, expiration is passive.

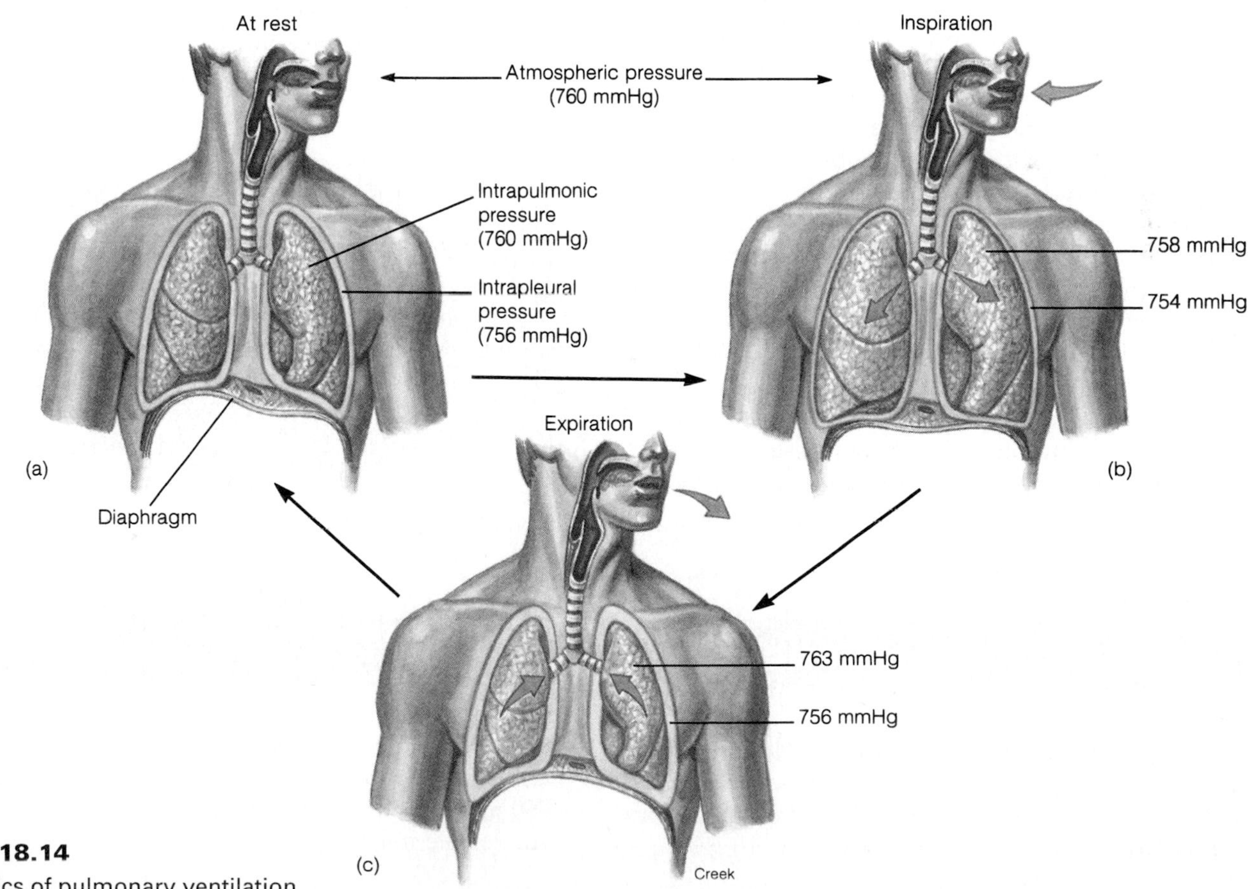

FIGURE 18.14
The mechanics of pulmonary ventilation.

TABLE 18.3 Terms Used To Describe Lung Volumes and Capacities

Term	Definition
Lung Volumes	The four nonoverlapping components of the total lung capacity
Tidal volume	The volume of gas inspired or expired in an unforced respiratory cycle
Inspiratory reserve volume	The maximum volume of gas that can be inspired during forced breathing in addition to tidal volume
Expiratory reserve volume	The maximum volume of gas that can be expired during forced breathing in addition to tidal volume
Residual volume	The volume of gas remaining in the lungs after a maximum expiration
Lung Capacities	Measurements that are the sum of two or more lung volumes
Total lung capacity	The total amount of gas in the lungs at the end of a maximum inspiration
Vital capacity	The maximum amount of gas that can be expired after a maximum inspiration
Inspiratory capacity	The maximum amount of gas that can be inspired at the end of a tidal expiration
Functional residual capacity	The amount of gas remaining in the lungs at the end of a tidal expiration

Lung Volumes and Capacities

Depending on the conditions of inspiration and expiration, our lungs exchange various amounts of air with the atmosphere; consequently, several different *respiratory volumes* can be described. Specific combinations, or sums, of these respiratory volumes—called *respiratory capacities*—are measured in assessing a person's respiratory status.

The various lung volumes and capacities are defined in table 18.3. During quiet breathing for example, the amount of air expired in each breath is the **tidal volume.** The maximum amount of air that can be forcefully exhaled after a maximum inhalation is called the **vital capacity,** and is equal to the sum of the **inspiratory reserve volume, tidal volume,** and **expiratory reserve volume.** Multiplying the tidal volume at rest by the number of breaths per minute yields a **total minute volume** of about 6 L per minute. During exercise, the tidal volume and the number of breaths per minute increase to produce a total minute volume as high as 100 to 200 L per minute. Because air in the conducting division has a lower oxygen concentration than ambient air, the higher tidal volumes produced during exercise bring a higher percentage of fresh air, and more oxygen, into the alveoli for gas exchange.

On the basis of pulmonary function tests, lung disorders can be classified as *restrictive* or *obstructive.* In restrictive disorders, such as pulmonary fibrosis, the vital capacity is reduced below normal. The rate at which the vital capacity can be forcibly exhaled, however, is normal. In disorders that are exclusively obstructive, such as asthma, the vital capacity is normal because lung tissue is not damaged. In asthma, the bronchioles constrict, and this bronchoconstriction increases the resistance to airflow. Although the vital capacity is normal, the increased airway resistance makes expiration more difficult, and it takes a longer time. Obstructive disorders are thus diagnosed by tests that measure the rate of expiration. One such test is the **forced expiratory volume (FEV),** in which the percentage of the vital capacity that can be exhaled in the first second ($FEV_{1.0}$) is measured.

Bronchoconstriction often occurs in response to inhaling noxious agents in the air (see the Synoptic Spotlight on page 470). The $FEV_{1.0}$ has therefore been used to determine the effects of inhaling various components of smog and of passive cigarette smoke on pulmonary function. Studies have shown that it is unhealthy to exercise on very smoggy days and that inhalation of smoke from other people's cigarettes in a closed environment can measurably affect pulmonary function.

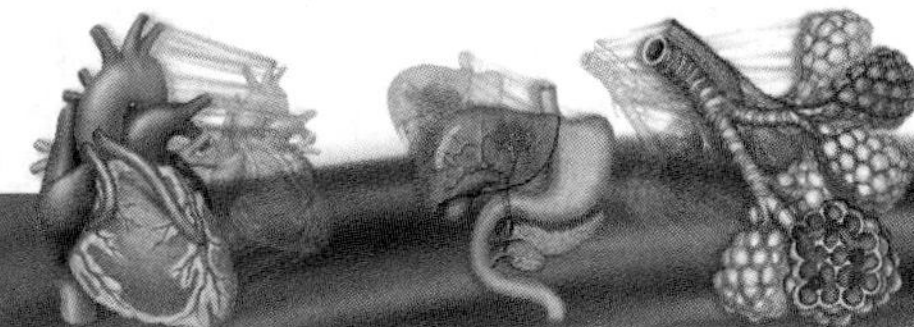

18.8 Gas Exchange in the Lungs

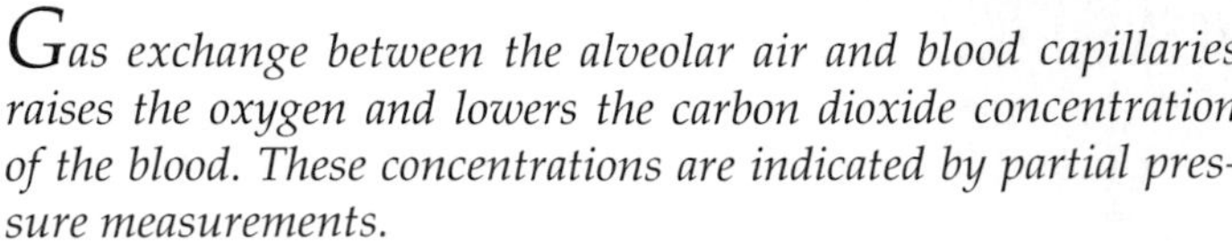

Gas exchange between the alveolar air and blood capillaries raises the oxygen and lowers the carbon dioxide concentration of the blood. These concentrations are indicated by partial pressure measurements.

The atmosphere is an ocean of gas that exerts pressure on all objects within it. At sea level, this pressure is equal to 760 mmHg (or 760 torr), which is also described as a pressure of *one atmosphere.*

TABLE 18.4 The Effect of Altitude on P_{O2}

Changes in P_{O_2} at Various Altitudes

Altitude (feet above sea level)	Atmospheric Pressure (mmHg)	P_{O_2} in air (mmHg)	P_{O_2} in alveoli (mmHg)	P_{O_2} in arterial blood (mmHg)
0	760	159	105	100
2000	707	148	97	92
4000	656	137	90	85
6000	609	127	84	79
8000	564	118	79	74
10,000	523	109	74	69
20,000	349	73	40	35
30,000	226	47	21	19

According to **Dalton's law,** the total pressure of a gas mixture (such as air) is equal to the sum of the pressures that each gas in the mixture would exert independently. The pressure that a particular gas in the mixture exerts independently is the **partial pressure** of that gas. The partial pressure of each gas in the mixture is determined by multiplying the percentage of the gas in the mixture by the total pressure. Since oxygen constitutes about 21% of the atmosphere, for example, its partial pressure (abbreviated P_{O_2}) is 21% of 760, or about 159 mmHg. Nitrogen constitutes about 78% of the atmosphere, so its partial pressure is equal to 78% of 760, or 593 mmHg. Oxygen and nitrogen thus contribute about 99% of the total pressure of 760 mmHg:

$$P_{\text{dry atmosphere}} = P_{N_2} + P_{O_2} + P_{CO_2} = 760 \text{ mmHg}$$

With increasing altitude, the total atmospheric pressure and the partial pressures of the constituent gases decrease, as indicated in table 18.4. Conversely, as one descends below sea level, as in ocean diving, the total pressure increases by one atmosphere for every 33 feet.

Air that enters the alveoli is saturated with water vapor. Water vapor, like the other constituent gases, contributes a partial pressure to the total atmospheric pressure. Since the total atmospheric pressure is constant (depending only on the height of the air mass), the water vapor "dilutes" the contribution of other gases to the total pressure:

$$P_{\text{wet atmosphere}} = P_{N_2} + P_{O_2} + P_{CO_2} + P_{H_2O}$$

As a result of gas exchange in the alveoli, the P_{O_2} of alveolar air is further diminished to about 105 mmHg. The partial pressures of the inspired air and the partial pressures of alveolar air are compared in figure 18.15.

When a fluid and a gas, such as blood and alveolar air, are at equilibrium, the amount of gas dissolved in the fluid reaches a maximum value. According to **Henry's law,** this value depends on the following factors:

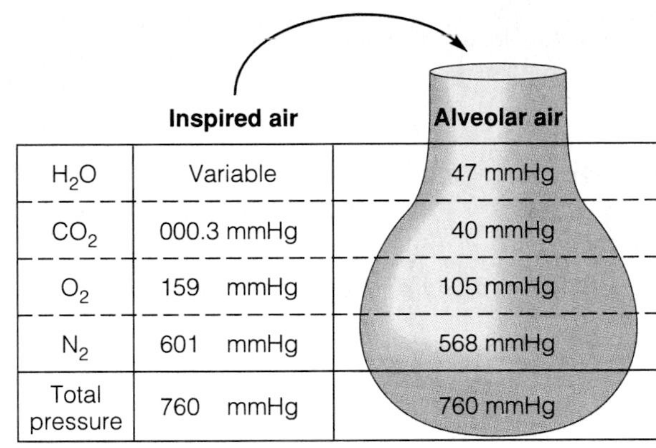

Inspired air
Alveolar air
H_2O
Variable
47 mmHg
CO_2
000.3 mmHg
40 mmHg
O_2
159 mmHg
105 mmHg
N_2
601 mmHg
568 mmHg
Total pressure
760 mmHg
760 mmHg

FIGURE 18.15
Partial pressures of gases in the inspired air and the alveolar air.

1. the *solubility of the gas in the fluid,* which is a physical constant;
2. the *temperature of the fluid*—more gas can be dissolved in cold water than in warm water; and
3. the *partial pressure of the gas.*

Since the temperature of the blood does not vary significantly, *the concentration of a gas dissolved in a fluid (such as plasma) depends directly on its partial pressure in the gas mixture.* If the P_{O_2} doubles, for example, the concentration of oxygen dissolved in the plasma will double. Measurements of blood P_{O_2} can thus be used as an indicator of the plasma O_2 concentration.

Dalton's law: from John Dalton, English chemist, 1766–1844
Henry's law: from William Henry, English chemist, 1775–1837

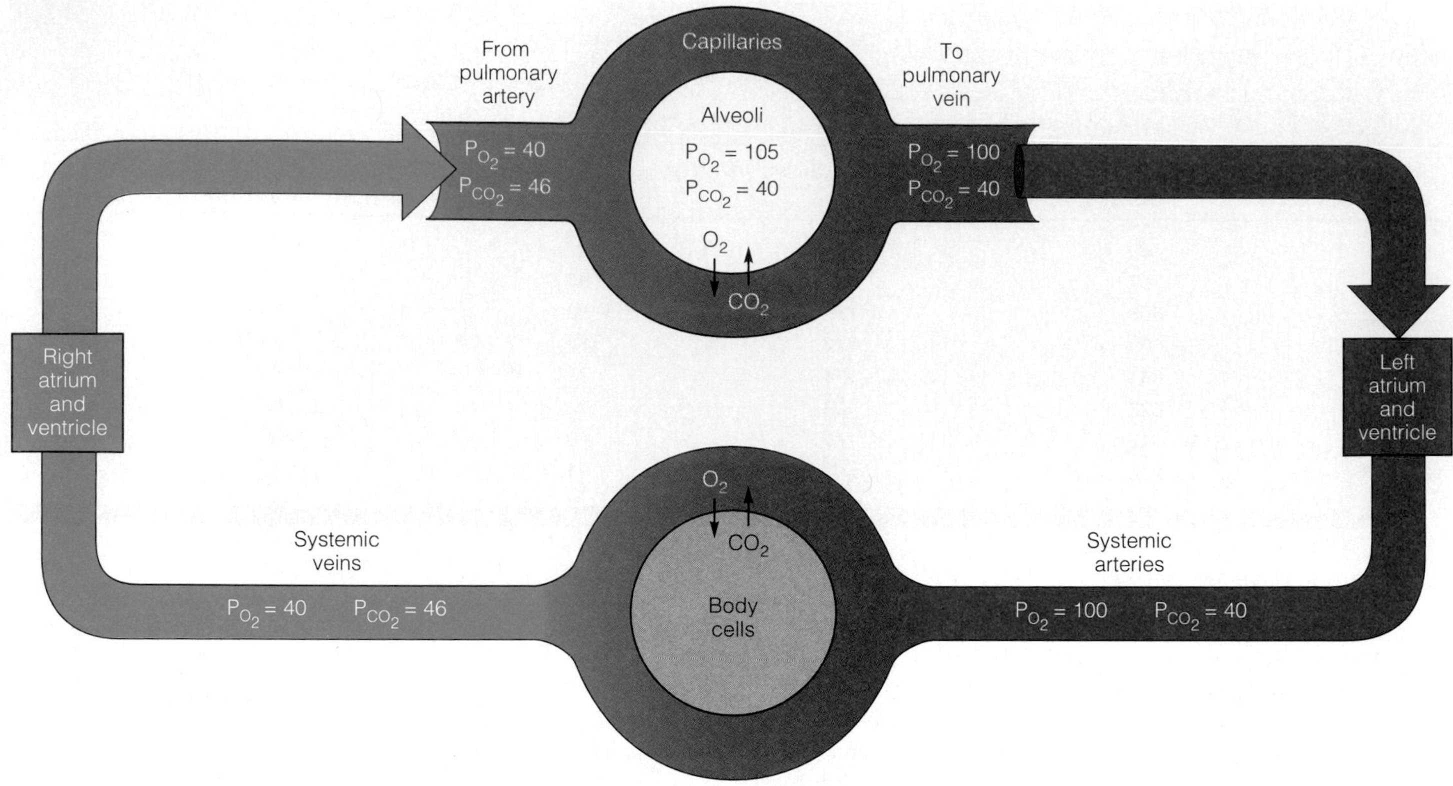

FIGURE 18.16
The P_{O_2} and P_{CO_2} of blood as a result of gas exchange in the lung alveoli and gas exchange between systemic capillaries and body cells.

Since blood P_{O_2} measurements are not directly affected by the oxygen in red blood cells, the P_{O_2} does not provide a measurement of the total oxygen content of whole blood. It does, however, provide a good index of lung function. If the inspired air has a normal P_{O_2} but the arterial P_{O_2} is below normal, for example, gas exchange in the lungs must be impaired. Measurements of arterial P_{O_2} thus provide valuable information in treating people with pulmonary disease, in performing surgery (when breathing may be depressed by anesthesia), and in caring for premature babies with respiratory distress syndrome.

Arterial P_{CO_2} is also measured in order to assess lung function. Blood in the pulmonary arteries usually has a P_{O_2} of 40 mmHg and a P_{CO_2} of 46 mmHg. After gas exchange in the alveoli of the lungs, blood in the pulmonary veins and systemic arteries has a P_{O_2} of about 105 mmHg and a P_{CO_2} of 40 mmHg (fig. 18.16).

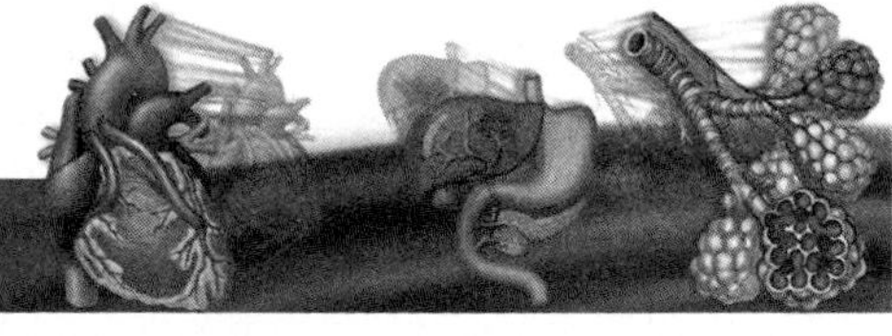

18.9 Regulation of Breathing

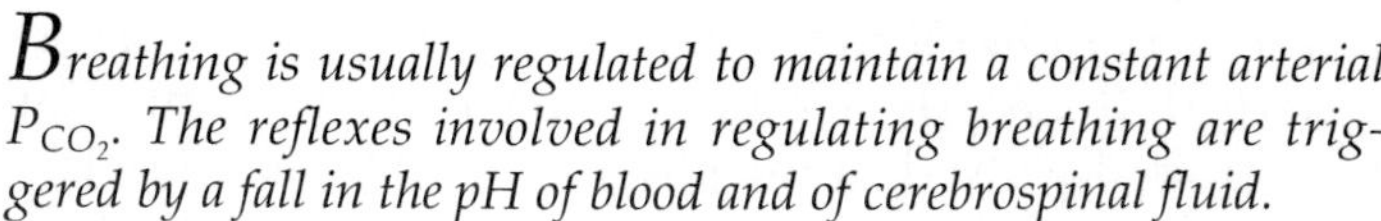

Breathing is usually regulated to maintain a constant arterial P_{CO_2}. The reflexes involved in regulating breathing are triggered by a fall in the pH of blood and of cerebrospinal fluid.

Brain Stem Respiratory Centers

A loose aggregation of neurons in the reticular formation of the *medulla oblongata* forms the **rhythmicity center** that controls automatic breathing. The rhythmicity center consists of interacting pools of neurons that respond either during inspiration or expiration. The activity of the medullary rhythmicity center is influenced by centers in the *pons*. One area—the **apneustic** (*ap-noo'stik*) **center**—appears to promote inspiration by stimulating the inspiratory neurons in the medulla oblongata. Another area in the pons—the **pneumotaxic** (*noo"mŏ-tak'sik*) **center**—seems to antagonize the apneustic center and inhibit inspiration (fig. 18.17).

The automatic control of breathing is also influenced by input from receptors sensitive to the chemical composition

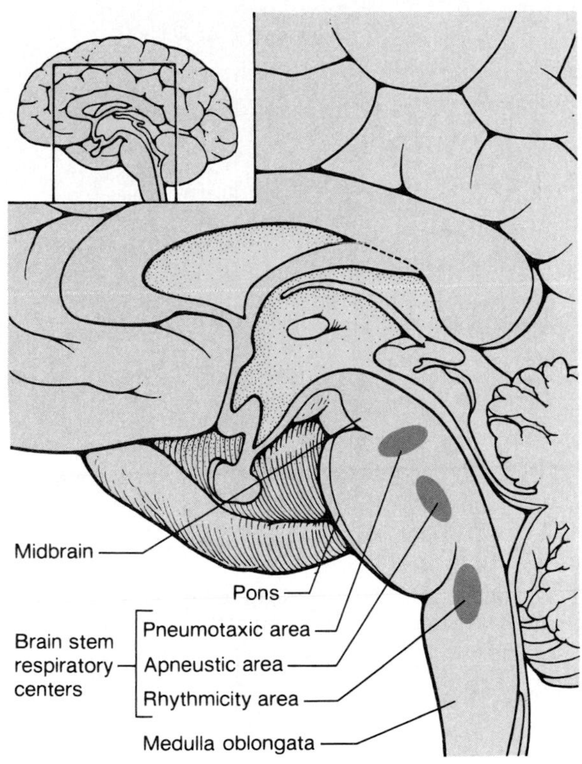

FIGURE 18.17
Approximate locations of the brain stem respiratory centers.

of the blood. There are two groups of *chemoreceptors* that respond to changes in blood P_{CO_2}, pH, and P_{O_2}. These are the **central chemoreceptors,** located in the medulla oblongata, and the **peripheral chemoreceptors,** contained within small nodules associated with the aorta and the carotid arteries. The peripheral chemoreceptors include the **aortic bodies,** located around the aortic arch, and the **carotid bodies,** located in each common carotid artery at the point just before it branches into the internal and external carotid arteries (fig. 18.18). The aortic and carotid bodies send sensory information to the medulla oblongata in the tenth and ninth cranial nerves, respectively.

Cessation of breathing during sleep, or sleep apnea, *is caused by a variety of disease processes.* Sudden infant death syndrome (SIDS) *is an especially tragic form of sleep apnea that claims the lives of about 10,000 babies annually in the United States. Victims of this condition are apparently healthy two- to five-month-old babies who die in their sleep without apparent reason—hence, the layperson's term,* crib death. *These deaths seem to be caused by failure of the respiratory control mechanisms in the brain stem and/or by failure of the carotid bodies to be stimulated by reduced arterial oxygen.*

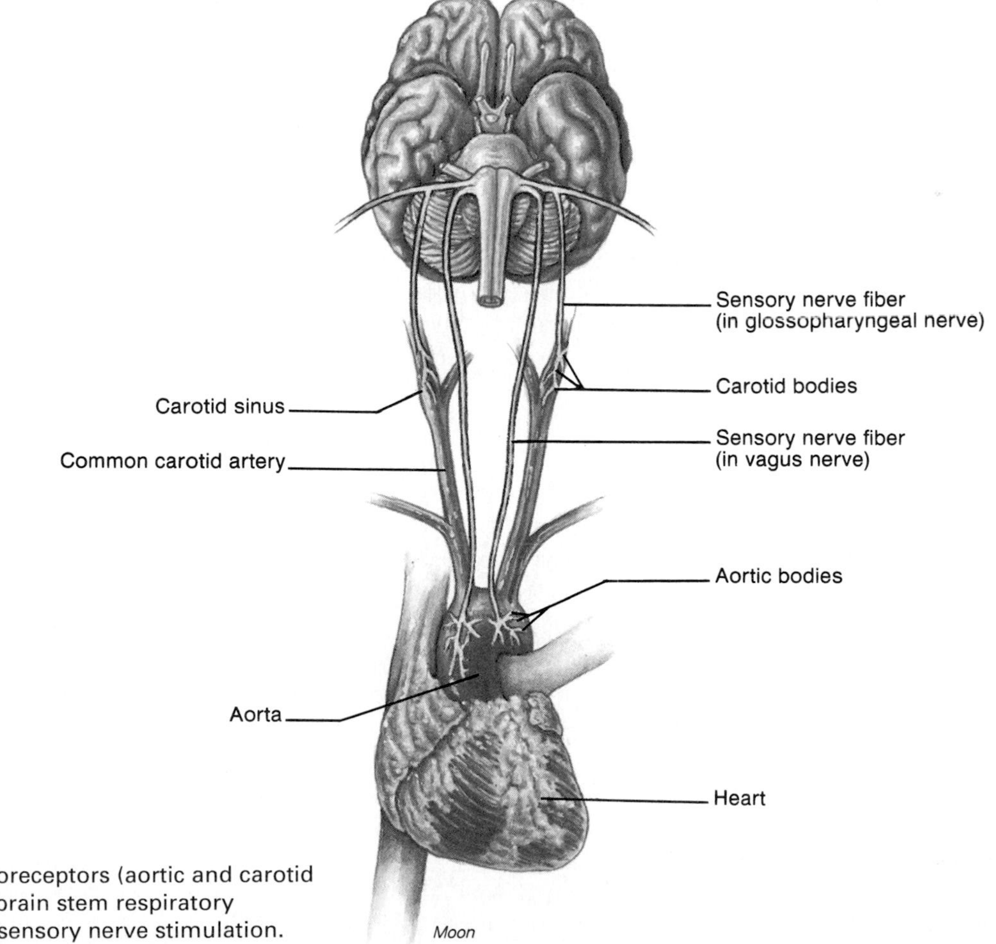

FIGURE 18.18
The peripheral chemoreceptors (aortic and carotid bodies) regulate the brain stem respiratory centers by means of sensory nerve stimulation.

Effects of Blood P_{CO_2} and pH on Ventilation

In the event of **hypoventilation,** or inadequate breathing, P_{CO_2} quickly rises and pH falls. The fall in pH is due to the fact that carbon dioxide can combine with water to form carbonic acid, which in turn can release H^+ to the solution. This is shown in the following equations:

$$CO_2 + H_2O \rightarrow H_2CO_3$$

$$H_2CO_3 \rightarrow HCO_3^- + H^+$$

During **hyperventilation,** or breathing that is excessive, blood P_{CO_2} quickly falls and pH rises due to the excessive elimination of carbonic acid. Hypoventilation causes a rise in P_{CO_2}—a condition called *hypercapnia* (*hi″per-kap′ne-ă*); hyperventilation results in *hypocapnia.* Ventilation is normally adjusted to maintain a constant arterial P_{CO_2}, and proper oxygenation of the blood occurs as a side product of this reflex control. The effects of arterial P_{CO_2} on ventilation are indirect, acting via associated changes in pH (fig. 18.19).

The chemoreceptors most sensitive to changes in the arterial P_{CO_2} are located in the medulla oblongata. Carbon dioxide in arterial blood can cross the blood-brain barrier and, through the formation of carbonic acid, can lower the pH of cerebrospinal fluid. This fall in cerebrospinal fluid pH directly stimulates the chemoreceptors in the medulla oblongata.

The aortic and carotid bodies also are not stimulated directly by blood CO_2. Rather, they are stimulated by a rise in the H^+ concentration (fall in pH) of arterial blood, which occurs when the levels of blood CO_2, and thus carbonic acid, are raised. The retention of CO_2 during hypoventilation stimulates the chemoreceptors in the medulla oblongata through a lowering of cerebrospinal fluid pH, and it stimulates the chemoreceptors in the aortic and carotid bodies through a lowering of blood pH.

Psychological stress or an anxiety attack can cause people to hyperventilate *involuntarily to the point where they become so dizzy that they faint. Breathing into a paper bag helps. By rebreathing their expired air enriched in CO_2, the hypocapnia that constricts cerebral blood vessels (thus causing dizziness) is relieved, and the hyperventilation stops.*

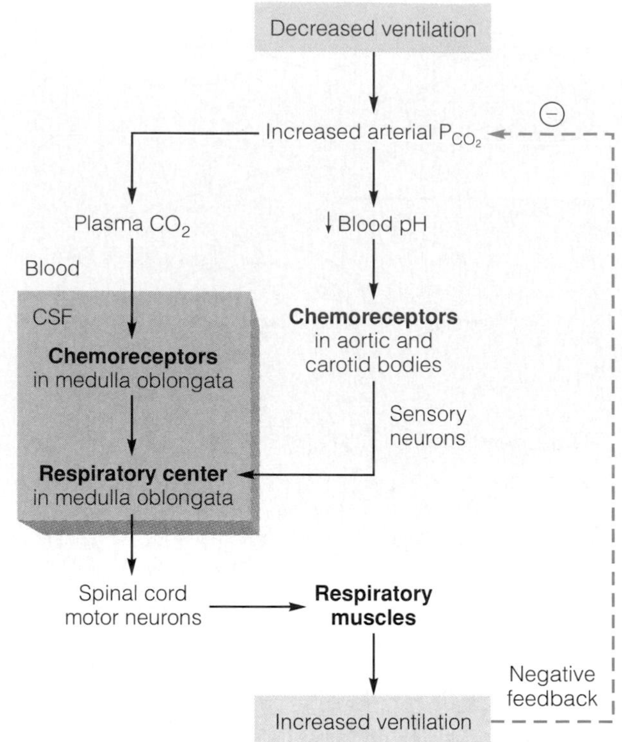

FIGURE 18.19

Negative feedback control of ventilation through changes in blood P_{CO_2} and pH. The orange box represents the blood-brain barrier, which allows CO_2 to pass into the cerebrospinal fluid while preventing the passage of H^+.

A deceased arterial P_{O_2}, as occurs when a person goes to a high altitude, can stimulate hyperventilation. This occurs first because the lowered P_{O_2} makes the chemoreceptors more sensitive to the arterial P_{CO_2}. When the arterial P_{O_2} falls to below 50 mmHg, however, it can directly stimulate the carotid bodies. This hyperventilation increases the proportion of fresh air brought into the alveoli, but additional adaptations (discussed in module 18.10) are needed to fully acclimatize to a high altitude.

Inside Information—It is impossible for a person to commit suicide by holding his or her breath. Rising CO_2 levels in the blood are a powerful respiratory stimulant; when they are high enough, breathing will resume automatically, despite conscious efforts to suppress it. Even if the person loses consciousness, automatic breathing will resume.

18.10 Hemoglobin and Oxygen Transport

Hemoglobin combines with oxygen in the lungs and unloads oxygen in the systemic capillaries. The extent of the loading and unloading reactions is adjusted by various mechanisms to ensure an adequate delivery of oxygen to the tissues.

Hemoglobin

Most of the oxygen in the blood is contained within the red blood cells, where it is chemically bound to hemoglobin. Each hemoglobin molecule consists of (1) a protein *globin* part, composed of four polypeptide chains, and (2) four nitrogen-containing, disc-shaped organic pigment molecules called *hemes* (fig. 18.20).

The protein part of hemoglobin is composed of two identical *alpha chains* and two identical *beta chains.* Each of the four polypeptide chains is combined with one heme group. In the center of each heme group is one atom of iron, which can combine with one molecule of oxygen. One hemoglobin molecule thus has the capacity to combine with four molecules of oxygen. Since each red blood cell contains about 280 million hemoglobin molecules, each can carry over a billion molecules of oxygen.

Normal heme contains iron in the reduced form (Fe^{++}, or ferrous iron). In this form, the iron can share electrons and bond with oxygen to form **oxyhemoglobin.** When oxyhemoglobin dissociates to release oxygen to the tissues, the heme iron is still in the reduced (Fe^{++}) form, and the hemoglobin is called **deoxyhemoglobin,** or **reduced hemoglobin.**

Carboxyhemoglobin is an abnormal form of hemoglobin in which the reduced heme is combined with *carbon monoxide* instead of oxygen. Since the bond with carbon monoxide is 210 times stronger than the bond with oxygen, carbon monoxide tends to displace oxygen in hemoglobin and remain attached to hemoglobin as the blood passes through systemic capillaries. The transport of oxygen to the tissues is thus reduced in carbon monoxide poisoning.

Inside Information—According to federal standards the percentage of carboxyhemoglobin in the blood of active nonsmokers should be no higher than 1.5%. However, concentrations of 3% in nonsmokers and 10% in smokers have been reported in some traffic-congested cities. Carbon monoxide is a major component both of automobile exhaust fumes and of cigarette smoke.

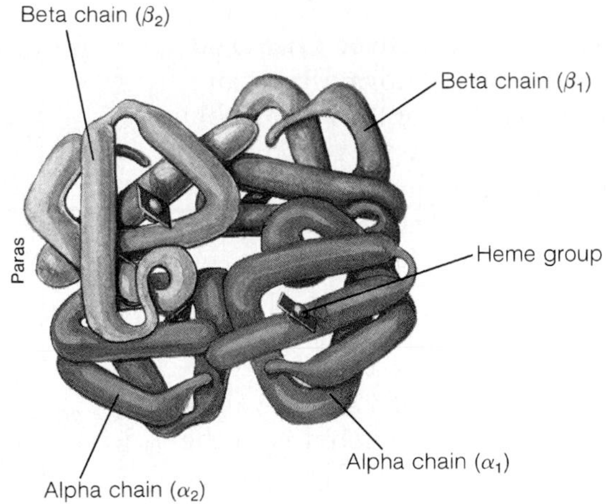

FIGURE 18.20
The three-dimensional structure of hemoglobin showing the two alpha and the two beta polypeptide chains. The heme groups are represented as flat structures with spheres in the center to indicate the iron atoms.

The production of hemoglobin and red blood cells in bone marrow is controlled by a hormone called **erythropoietin** (*e-rith"ro-poi'e-tin*), produced by the kidneys (see chapter 14, module 14.3). The production of erythropoietin is stimulated when the delivery of oxygen to the kidneys and other organs is lower than normal. This can occur, for example, as a response to life at a high altitude.

Loading and Unloading Reactions

Deoxyhemoglobin and oxygen combine to form oxyhemoglobin; this is called the **loading reaction.** Oxyhemoglobin, in turn, dissociates to yield deoxyhemoglobin and free oxygen molecules; this is the **unloading reaction.** The loading reaction occurs in the lungs, and the unloading reaction occurs in the systemic capillaries.

The extent to which the reaction will go in each direction depends on two factors: (1) the P_{O_2} of the environment and (2) the *affinity,* or bond strength, between hemoglobin and oxygen. High P_{O_2} favors the loading reaction in the pulmonary capillaries; low P_{O_2} favors the unloading reaction in the systemic capillaries.

The affinity between hemoglobin and oxygen also influences the loading and unloading reactions. A very strong bond would favor loading but inhibit unloading; a weak bond would hinder loading but improve unloading.

erythropoietin: Gk. *erythros,* red; *poiesis,* a making

TABLE 18.5 The Relationship between Percent Oxyhemoglobin Saturation and P_{O_2} (at pH = 7.40 and temperature = 37°C)

P_{O_2} (mmHg)	100	80	61	45	40	36	30	26	23	21	19
Percent Oxyhemoglobin Saturation	97	95	90	80	75	70	60	50	40	35	30
	Arterial Blood				Venous Blood						

The bond strength between hemoglobin and oxygen is normally strong enough so that 97% of the hemoglobin leaving the lungs is in the form of oxyhemoglobin, yet the bond is sufficiently weak so that adequate amounts of oxygen are unloaded to sustain aerobic respiration in the tissues.

Blood in the systemic arteries, at a P_{O_2} of 100 mmHg, has a *percent oxyhemoglobin saturation* of 97% (meaning that 97% of the hemoglobin is in the form of oxyhemoglobin). This blood is delivered to the systemic capillaries, where oxygen diffuses into the tissue cells and is consumed in aerobic respiration. Blood leaving in the systemic veins is thus reduced in oxygen; it has a P_{O_2} of about 40 mmHg and a percent saturation of about 75% (table 18.5). Therefore, 22% of the arterial oxygen is unloaded to the tissues. A graphic illustration of the percent oxyhemoglobin saturation at different values of P_{O_2} is called an **oxyhemoglobin dissociation curve** (fig. 18.21).

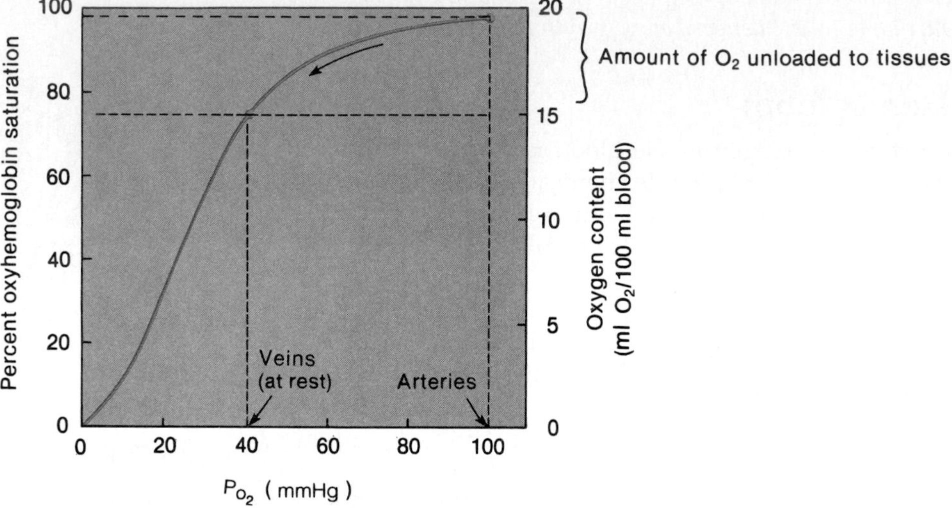

FIGURE 18.21
The percentage of oxyhemoglobin saturation and the blood oxygen content are shown at different values of P_{O_2}. Notice that there is about a 25% decrease in percent oxyhemoglobin as the blood passes through the tissue from arteries to veins, resulting in the unloading of approximately 5 ml O_2 per 100 ml to the tissues.

The affinity of hemoglobin for oxygen is decreased when the pH is lowered; this phenomenon is called the **Bohr effect.** The affinity is similarly decreased by a rise in temperature. There is slightly less loading of the blood with oxygen in the lungs but greater unloading of oxygen in the tissues under these conditions. The net effect of such changes is that active skeletal muscles, as a result of their higher metabolism, receive more oxygen from the blood than they do at rest.

Bohr effect: from Christian Bohr, Scandinavian physiologist, 1855–1911

Acclimatization to high altitude requires a number of physiological changes. First, a person hyperventilates due to the low P_{O_2} of the air. This brings a higher proportion of fresh air into the alveoli but can lead to dizziness and other problems. Next, changes within the red blood cells decrease the bond strength between hemoglobin and oxygen, so that more oxygen is unloaded to the tissues. This effect occurs after 2 to 3 days at the higher altitude. Finally, after a period of weeks, there is an increased production of red blood cells due to stimulation by erythropoietin, which results in an increased capacity of the blood to carry oxygen.

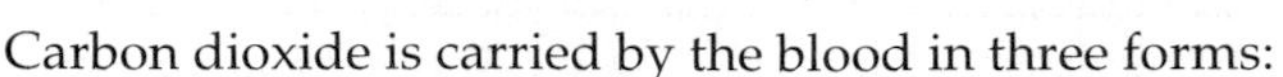

18.11 Carbon Dioxide Transport

Carbon dioxide forms carbonic acid and bicarbonate in the blood. As a result, the transport of carbon dioxide is related to the regulation of acid-base balance in the blood.

Carbon dioxide is carried by the blood in three forms:

1. **Dissolved CO_2**—about 10% of the total blood CO_2 is dissolved in blood plasma.

2. **Carbaminohemoglobin**—about 20% of the total blood CO_2 is carried attached to an amino acid in hemoglobin.
3. **Bicarbonate**—about 70% of the CO_2 is carried as HCO_3^- in the blood.

Carbon dioxide is able to combine with water to form carbonic acid. This reaction occurs spontaneously in the blood plasma at a slow rate but occurs more rapidly within the red blood cells due to the action of the enzyme **carbonic anhydrase.** Since this enzyme is confined to the red blood cells, most of the carbonic acid is produced there rather than in the blood plasma.

The Chloride Shift

As a result of the action of carbonic anhydrase, large amounts of carbonic acid are produced within the red blood cells. The carbonic acid then dissociates into hydrogen ions (protons, which contribute to the acidity of a solution) and HCO_3^- (bicarbonate).

The hydrogen ions released by the dissociation of carbonic acid are largely buffered by deoxyhemoglobin within the red blood cells. Although the unbuffered hydrogen ions are free to diffuse out of the red blood cells, more bicarbonate diffuses outward into the blood plasma than does H^+. As a result, the inside of the red blood cell gains a positive charge. This attracts chloride ions (Cl^-), which move into the red blood cells as HCO_3^- moves out. This exchange of anions as blood travels through the tissue capillaries is called the **chloride shift** (fig. 18.22).

A *reverse chloride shift* operates in the pulmonary capillaries to convert carbonic acid to H_2O and CO_2 gas, which is eliminated in the expired breath (fig. 18.23). The P_{CO_2}, carbonic acid, H^+, and bicarbonate concentrations in the systemic arteries are thus maintained relatively constant by normal ventilation.

Ventilation and Acid-Base Balance

The pH of body fluids may be adjusted by the rate and depth of breathing. *Hypoventilation* (see module 18.9), for example, produces an abnormally high arterial P_{CO_2}. Under these conditions, carbonic acid production is excessively high and **respiratory acidosis** occurs. In *hyperventilation,* conversely, the rate of ventilation is greater than the rate of CO_2 production. Arterial P_{CO_2} therefore decreases, causing less formation of carbonic acid than there would be under normal conditions. The depletion of carbonic acid raises the pH, and **respiratory alkalosis** occurs.

Disturbances in acid-base balance as a result of respiratory problems are discussed further in chapter 19 (see module 19.8). The metabolic component of acid-base balance is also considered in module 19.8.

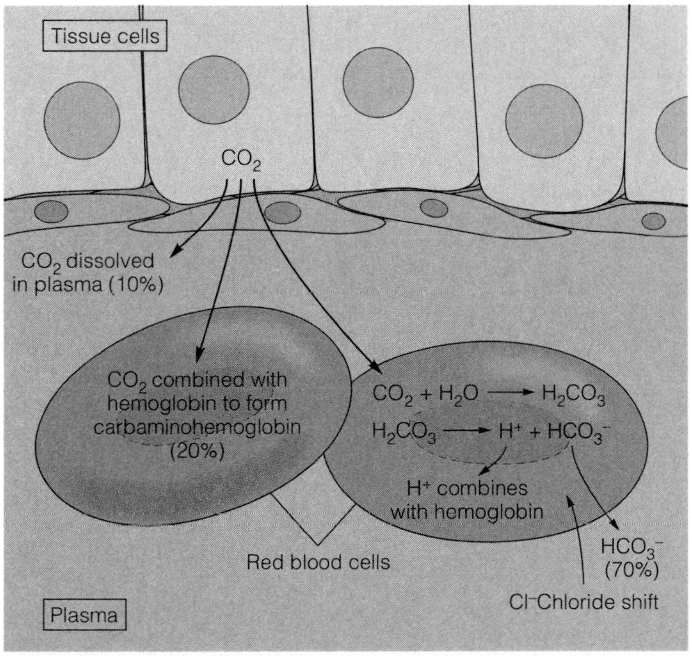

FIGURE 18.22
An illustration of carbon dioxide transport by the blood and the "chloride shift." Carbon dioxide is transported in three forms: as dissolved CO_2 gas, attached to hemoglobin as carbaminohemoglobin, and as carbonic acid and bicarbonate. Percentages indicate the proportion of CO_2 in each of the forms.

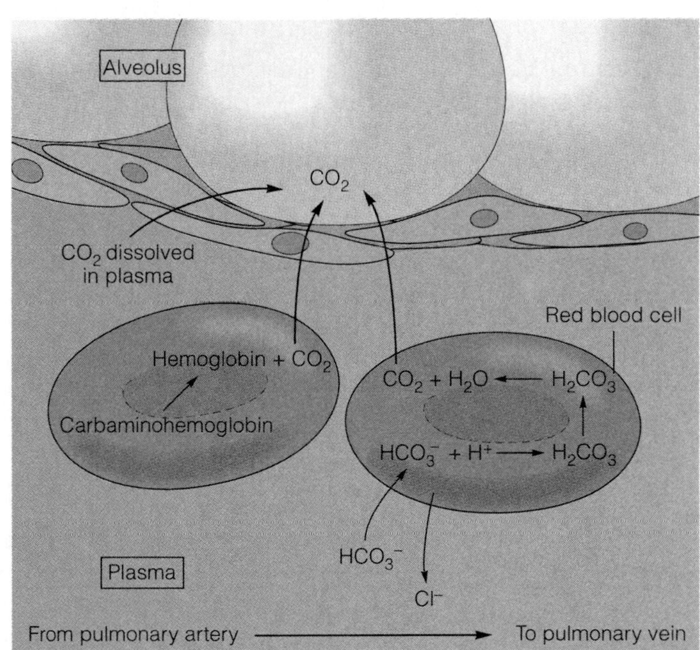

FIGURE 18.23
Carbon dioxide is released from the blood as it travels through the pulmonary capillaries. During this time a "reverse chloride shift" occurs and carbonic acid is transformed into CO_2 and H_2O.

SYNOPTIC SPOTLIGHT

An Atmosphere Worth Breathing?

Each year over a billion metric tons of pollutants enter the atmosphere as a result of human activity. *Primary pollutants,* which are released directly into the air, include particulate matter (small particles of solid or liquid substances), sulfur oxides, carbon monoxide, nitrogen oxides, and hydrocarbons. These substances make us vulnerable to a range of health problems from minor discomfort to life-threatening disease.

Nearly all industrial processes, as well as the burning of fossil fuels, release **particulate matter** into the atmosphere—much of it visible as smoke, soot, or dust. Included with the particulates are such materials as airborne asbestos particles and small particles of heavy metals, such as arsenic, copper, lead, and zinc. Among the most significant of the fine particulate pollutants are sulfates and nitrates, which are mainly *secondary pollutants;* that is, they are produced in the atmosphere through chemical reactions between the primary pollutants—sulfur oxides and nitrogen oxides—and normal atmospheric constituents.

Particulates that enter the lungs (see the figure to the right) may lodge there and have chronic effects on respiration. Children, with their small lungs and relatively rapid respiration, are the most vulnerable to damage—particularly from the effects of lead. Studies reveal that breathing of lead fumes leads to hyperactivity, stunted growth, and anemia. Of greatest concern is the growing body of evidence that suggests that lead exposure can also impair cognitive activity and cause mental retardation. Different studies indicate that the other metals found in air may contribute to the development of lung disease.

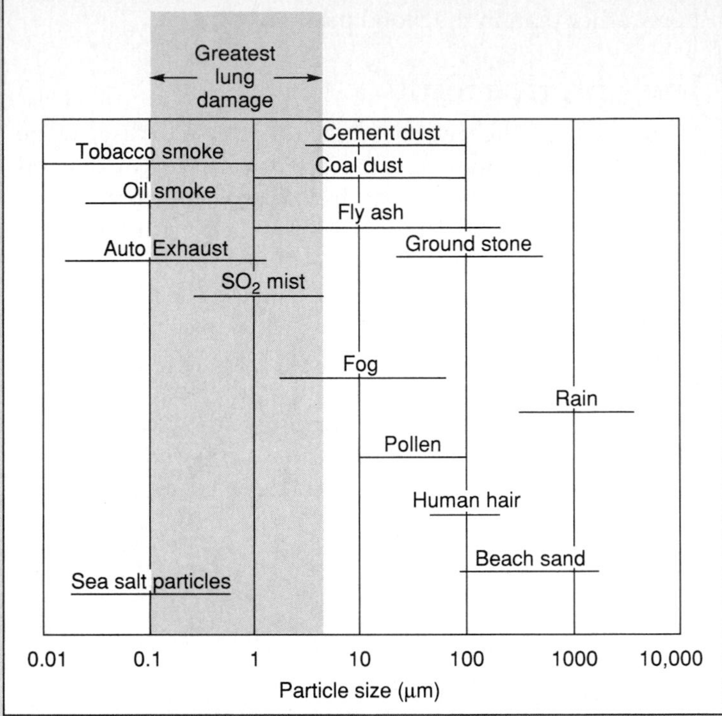

Sizes of selected particulates. The shaded area shows the size range that produces the greatest damage to the lungs.

Sources: Modified from Giddings, Chemistry, Man, & Environment Change: An Integrated Approach, *Canfield Press, 1973; and Hidy and Brock,* Topics in Current Aerosol Research, *Pergamon Press, 1971.*

Sulfur dioxide (SO_2) is a colorless, odorless gas, under normal conditions found at the earth's surface. Once emitted into the air through the burning of fossil fuels (mostly coal), it may be converted to fine particulate sulfate (SO_4). Sulfur dioxide, particularly in the sulfate form, can cause severe respiratory problems, including chronic bronchitis.

Another colorless, odorless gas is **carbon monoxide (CO),** a by-product of automobile combustion, certain industrial operations, and cigarette smoke. Hemoglobin in our blood will take up carbon monoxide more than 200 times more rapidly than it will oxygen. Therefore, when a person breathes in air containing high levels of CO, slowly the sites on the hemoglobin molecules that should carry oxygen are taken over by CO, and the individual suffers increasing respiratory insufficiency, general tissue hypoxia, and—in the extreme—death. Additional evidence suggests that long-term low exposure to CO may be involved in depressed neurological response and cardiac disease.

Burning of petroleum and coal releases another group of molecules into the air that we breathe: **nitrogen oxides.** Scientists have evidence that nitrogen dioxide is capable of causing serious lung impairment in people with asthma. There are other consequences of burning fossil fuels. In addition to the release of nitrogen oxides, a variety of **hydrocarbons** are added to the air each day. Power plants and automobiles are the two principal sources of the hydrocarbons that may end up in our lungs. This category of compounds, in the presence of sunlight, is capable of combining with nitrogen oxides to produce ozone and photochemical smog.

A colorless, unstable gas with a slightly sweet odor, **ozone (O_3)** forms a natural layer high in the earth's atmosphere (stratosphere), where it protects us from the harmful ultraviolet light of the sun. At the lower atmospheric levels where we live and breathe, ozone can irritate respiratory passageways and produce an inflammatory response. Overt symptoms range from coughing and chest pain to reduced vital capacity. Some studies even suggest that the bronchial tree may develop scar tissue as a result of exposure to ozone.

In nearly half the cities of the world, tailpipe exhaust is by far the single largest source of air pollution—surpassing wood fires, coal-burning power plants, and chemical manufacturing. In Mexico City, the pollution from the city's 3.2 million cars is so pervasive that young children reach for gray crayons when asked to draw the sky. The city is at an elevation of about 7400 feet in a natural basin surrounded by mountains, but the urban air is so smoggy, the mountains can hardly be seen. In the countries of the former Eastern Bloc, auto exhaust has already replaced much of the industrial smog of the Communist days. In China, as the recent surge of cars has pushed bikes off the streets, health costs are starting to take their toll. If China attains its dream of a car for every family, the resulting tailpipe emissions could increase carbon concentrations in the air to an extent that would affect the entire world.

Numerous efforts to control air pollution—whatever the source—have been undertaken in recent years. These methods vary from settling chambers for particulates to catalytic converters to remove carbon monoxide and hydrocarbons from automobile exhaust to scrubbers that use lime to remove sulfur before it enters the atmosphere. Although such efforts have achieved a measure of success, urban air quality is still very poor a good deal of the year in many areas of the United States and other countries throughout the world. For countless numbers of people—whether they are aware of it or not—this air pollution is undermining good health and the quality of their lives.

Clinical Considerations

Common Respiratory Disorders

A cough is the most common symptom of respiratory disorders. Acute problems may be accompanied by dyspnea, or wheezing. Respiratory or circulatory problems may cause **cyanosis**—a blue discoloration of the skin caused by blood with a low oxygen content.

As mentioned in module 18.7, pulmonary disorders are classified as **obstructive** when there is increased resistance to airflow in the bronchioles; they are classified as **restrictive** when alveolar tissue is damaged. Asthma and acute bronchitis are usually exclusively obstructive; chronic bronchitis and emphysema are both obstructive and restrictive. Pulmonary fibrosis, by contrast, is a purely restrictive disorder.

Asthma The obstruction of airflow through the bronchioles may occur as a result of excessive mucous secretion, inflammation, and/or contraction of the smooth muscles in the bronchioles. Asthma results from bronchiolar constriction, which increases airway resistance and makes breathing difficult. Constriction of the bronchiolar smooth muscles is stimulated by leukotrienes and, to a lesser degree, by histamine released by leukocytes and mast cells. This can be provoked by an allergic reaction or by the release of acetylcholine from parasympathetic nerve endings.

Emphysema Alveolar tissue is destroyed in emphysema, resulting in fewer but larger alveoli. The surface area for gas exchange is reduced, as is the ability of the bronchioles to remain open during expiration. Collapse of the bronchioles as a result of the compression of the lungs during expiration produces *air trapping,* which further decreases the efficiency of gas exchange in the alveoli.

There are different types of emphysema. The most common type occurs almost exclusively in people who have smoked cigarettes heavily over a period of years. A component of cigarette smoke apparently stimulates the macrophages and leukocytes to secrete proteolytic (protein-digesting) enzymes that destroy lung tissues. A less common type of emphysema results from the genetic inability to produce a plasma protein called alpha-1-antitrypsin. This protein normally inhibits proteolytic enzymes such as trypsin, and thus normally protects the lungs against the effects of enzymes that are released from alveolar macrophages.

Pulmonary Fibrosis Under certain conditions, for reasons that are poorly understood, lung damage leads to pulmonary fibrosis instead of emphysema. In this condition, the normal structure of the lungs is altered by the accumulation of fibrous connective tissue proteins. Fibrosis can result, for example, from the inhalation of particles less than 6 μm in size that are able to accumulate in the respiratory zone of the lungs. This type of fibrosis includes *anthracosis,* or black lung, which is produced by the inhalation of carbon particles from coal dust.

Other Respiratory Disorders The **common cold** is the most widespread of all respiratory diseases. Colds occur repeatedly because acquired immunity to one virus does not protect against other viruses that cause colds. Cold viruses cause acute inflammation of the respiratory mucosa, resulting in a flow of mucus, a fever, and often a headache.

Nearly all the structures and regions of the respiratory tract can become infected and inflamed. **Influenza** is a viral

cyanosis: Gk. *kyanosis,* dark blue color
asthrna: Gk. *asthma,* panting
emphysema: Gk. *emphysan,* to blow up, inflate
influenza: L. *influentia,* a flowing in

SUMMARY

18.1 Functions and Divisions of the Respiratory System

1. Respiration refers not only to breathing, but also to the exchange of gases between the atmosphere, the blood, and the individual cells, and to the use of oxygen by tissue cells (cellular respiration).
2. The respiratory system is divided into a respiratory division, in which gas exchange occurs, and a conducting division, which includes all the structures that conduct air to the respiratory division.

disease that causes inflammation of the upper respiratory tract. Influenza can become epidemic, but fortunately vaccines are available. **Sinusitis** is an inflammation of the paranasal sinuses. **Tonsillitis** may involve one or all of the tonsils, and frequently follows other lingering diseases of the oral or pharyngeal region. **Laryngitis** is inflammation of the larynx that often produces a hoarse voice and limits the ability to talk. **Tracheobronchitis** and **bronchitis** are infections of the regions for which they are named.

Diseases of the lungs are common and may be serious. **Pneumonia** is an acute infection and inflammation of lung tissue accompanied by exudation (the accumulation of fluid). It is usually caused by bacteria, most commonly the pneumococcus bacterium. **Tuberculosis (TB)** is an inflammatory disease of the lungs caused by the presence of tubercle bacilli. Tuberculosis softens lung tissues and leads to ulceration of the tissues. **Pleurisy** is an inflammation of the pleura that is usually secondary to some other respiratory disease. Inspiration may become painful, and fluid may collect within the pleural space.

A common form of lung cancer, **bronchiogenic carcinoma,** arises from the epithelium of the bronchial tree, usually as a result of constant irritation by inhaled cigarette smoke. Eventually, squamous cancer cells infiltrate surrounding tissues and may give rise to metastases that block the bronchial tubes. For people who regularly smoke a pack or more of cigarettes a day, the risk of developing bronchiogenic carcinoma is 20 times greater than for people who have never smoked.

Abdominal Thrust Maneuver

More than eight Americans choke to death each day on food lodged in their trachea. A simple technique called the **abdominal thrust** (Heimlich) **maneuver** can save the life of a person who is choking. The abdominal thrust maneuver is performed as follows:

If the victim is standing or sitting:

1. Stand behind the victim or the victim's chair and wrap your arms around his or her waist.
2. Grasp your fist with your other hand and place the fist against the victim's abdomen, slightly above the navel and below the rib cage.
3. Press your fist into the victim's abdomen with a quick upward thrust.
4. Repeat several times if necessary.

If the victim is lying down:

1. Position the victim on his or her back.
2. Face the victim, and kneel on his or her hips.
3. With one of your hands on top of the other, place the heel of your bottom hand on the abdomen, slightly above the navel and below the rib cage.
4. Press into the victim's abdomen with a quick upward thrust.
5. Repeat several times if necessary.

If you are alone and choking, use whatever is available to apply force just below your diaphragm. Press into a table or a sink, or use your own fist.

Reviving a Person Who Has Stopped Breathing

People saved from drowning and victims of shock frequently experience apnea (cessation of breathing). If breathing stops, a person will lose consciousness after 4 or 5 minutes. Brain damage is likely to occur after 7 to 8 minutes, and after 10 minutes the person will die.

18.2 Nose and Pharynx

1. The nose includes an external portion and an internal nasal cavity.
2. The pharynx is divided into a nasopharynx, oropharynx, and laryngopharynx.

18.3 Larynx and Trachea

1. The larynx, composed of nine cartilages, functions to direct air to the trachea. The vocal cords are contained within the larynx.
2. The trachea, or windpipe, contains C-shaped rings of cartilage.

18.4 Bronchial Tree, Alveoli, and Lungs

1. The bronchial tree consists of primary, secondary, and segmental bronchi, as well as microscopic bronchioles.
2. Bronchioles terminate in air sacs (composed of alveoli) that are thin walled for gas exchange between air and blood.
3. The lungs are spongy paired organs within the thoracic cavity. The left lung is divided into two lobes by a single fissure; the right lung is divided into three lobes by two fissures.

18.5 Thoracic Compartments

1. The pleurae are serous membranes surrounding the lungs that compartmentalize the thoracic organs.
2. A parietal pleura lines the thoracic wall, and a visceral pleura covers the lungs; between the two is a pleural cavity, or intrapleural space.

18.6 Physical Aspects of Ventilation

1. The lungs have compliance; thus, they can expand under pressure. They are elastic, so that they recoil after they are distended.
2. The action of pulmonary surfactant to reduce surface tension in alveoli prevents the alveoli from collapsing during expiration.

18.7 Mechanics of Breathing

1. Quiet inspiration is produced by expansion of the thorax and by contraction of the diaphragm and external intercostal muscles. This produces a subatmospheric intrapulmonary pressure.
2. Quiet expiration occurs when the lungs decrease in volume due to elastic recoil.

18.8 Gas Exchange in the Lungs

1. The amount of oxygen and carbon dioxide in the alveolar air is measured as their partial pressures.
2. The amount of oxygen and carbon dioxide dissolved in plasma depends on their partial pressures. These measurements are in P_{O_2} and P_{CO_2} units.
3. Arterial P_{O_2} and P_{CO_2} measurements are used clinically to assess lung function.

18.9 Regulation of Breathing

1. Breathing is controlled by centers in the medulla oblongata and is influenced by sensory information from chemoreceptors and by adjusting centers in the pons.
2. The plasma P_{CO_2} is the primary stimulus for chemoreceptor control of breathing.
3. Chemoreceptors in the aortic and carotid bodies are stimulated by a fall in arterial pH when the arterial P_{CO_2} increases.
4. Chemoreceptors in the medulla oblongata are stimulated by a decreased cerebrospinal fluid pH due to increased arterial P_{CO_2}.

18.10 Hemoglobin and Oxygen Transport

1. Hemoglobin contains four heme groups, each of which binds a molecule of oxygen.
2. Deoxyhemoglobin combines with oxygen in the lungs; oxyhemoglobin unloads oxygen in the systemic capillaries.
3. The affinity of hemoglobin for oxygen is decreased by a fall in pH or a rise in temperature.

18.11 Carbon Dioxide Transport

1. Most of the CO_2 is carried in the blood as bicarbonate, which is formed within red blood cells as part of the chloride shift.
2. A reverse chloride shift occurs in the pulmonary capillaries.
3. Hypoventilation and hyperventilation cause respiratory acidosis and alkalosis, respectively.

Review Activities

Objective Questions

1. The term *respiration* refers to (*module 18.1*)
   a. ventilation (breathing).
   b. gas exchange within the lungs.
   c. O_2 utilization within the cells.
   d. all of the above.

2. Which of the following is *not* a component of the nasal septum? (*module 18.2*)
   a. palatine bone
   b. vomer
   c. ethmoid bone
   d. septal cartilage

3. The vocal cords are attached to (*module 18.3*)
   a. the cricoid and thyroid cartilages.
   b. the cuneiform and cricoid cartilages.
   c. the corniculate and thyroid cartilages.
   d. the arytenoid and thyroid cartilages.

4. Which of the following is *not* characteristic of the left lung? (*module 18.4*)
   a. cardiac notch
   b. superior lobe
   c. single fissure
   d. inferior lobe
   e. middle lobe

5. Which of the following statements about intrapulmonary and intrapleural pressure is *true?* (*module 18.5*)
   a. The intrapulmonary pressure is always subatmospheric.
   b. The intrapleural pressure is always greater than the intrapulmonary pressure.
   c. The intrapulmonary pressure is greater than the intrapleural pressure.
   d. The intrapleural pressure equals the atmospheric pressure.

6. Collapse of the lungs in an infant with respiratory distress syndrome is caused by *(module 18.6)*
a. excessive lung elasticity.
b. excessive lung compliance.
c. lack of surface tension.
d. a deficiency in pulmonary surfactant.

7. The maximum amount of air that can be expired after a maximum inspiration is *(module 18.7)*
a. the tidal volume.
b. the forced expiratory volume.
c. the vital capacity.
d. the maximum expiratory flow rate.

8. If a person were to dive with scuba equipment to a depth of 66 ft, which of the following statements would be *false?* *(module 18.8)*
a. The arterial P_{O_2} would be three times its normal value.
b. The O_2 content of plasma would be three times its normal value.
c. The O_2 content of whole blood would be three times its normal value.

9. If a person with normal lung function were to hyperventilate for several seconds, there would be a significant *(module 18.9)*
a. increase in the arterial P_{O_2}.
b. decrease in the arterial P_{CO_2}.
c. increase in the arterial percent oxyhemoglobin saturation.
d. decrease in the arterial pH.

10. The chemoreceptors in the medulla oblongata are directly stimulated by *(module 18.9)*
a. CO_2 in the blood.
b. H^+ from the blood.
c. H^+ in cerebrospinal fluid, derived from blood CO_2.
d. decreased arterial P_{O_2}.

11. The rhythmic control of breathing is regulated by a center located in *(module 18.9)*
a. the medulla oblongata.
b. the cerebral cortex.
c. the apneustic center of the pons.
d. the pneumotaxic center of the pons.

12. Which of the following would be most affected by a decrease in the affinity of hemoglobin for oxygen? *(module 18.10)*
a. arterial P_{O_2}
b. arterial percent oxyhemoglobin saturation
c. venous oxyhemoglobin saturation
d. arterial P_{CO_2}

13. Erythropoietin is produced by *(module 18.10)*
a. the kidneys.
b. the liver.
c. the lungs.
d. the bone marrow.

14. Most of the carbon dioxide in the blood is carried in the form of *(module 18.11)*
a. dissolved CO_2.
b. carbaminohemoglobin.
c. bicarbonate.
d. carboxyhemoglobin.

Essay Questions

1. Explain how the respiratory system is organized on the basis of structure and function. *(module 18.1)*
2. Identify the three regions of the pharynx and state where the different tonsils are located. *(module 18.2)*
3. List the paired and unpaired cartilages of the larynx and describe the functions of the larynx. *(module 18.3)*
4. Describe the structure of the bronchial tree, alveoli, and lungs. Why are the alveoli called the functional units of the respiratory system? *(module 18.4)*
5. What is the significance of the compartmentalization of the thoracic cavity? Explain how the intrapulmonary and intrapleural pressures change during breathing. *(module 18.5)*
6. Explain the significance of pulmonary compliance, elasticity, and surface tension in the behavior of the lungs. *(module 18.6)*
7. Explain the law of Laplace and discuss the significance of surfactant. *(module 18.6)*
8. Using a flow diagram to show cause and effect, explain how contraction of the diaphragm produces inspiration. *(module 18.7)*
9. How does the partial pressure of oxygen in the arterial blood of someone at sea level compare with that of someone at a 5000-foot altitude? Explain. *(module 18.8)*
10. Using a flowchart, explain how a rise in blood P_{CO_2} stimulates breathing. Include both the central and peripheral chemoreceptors in your answer. *(module 18.9)*
11. A person with ketoacidosis may hyperventilate. Explain why this occurs and why it can be stopped by an intravenous fluid containing bicarbonate. *(module 18.9)*
12. Why is carbon monoxide poisonous? *(module 18.10)*
13. Explain how measurements of blood P_{CO_2} and pH are affected by hypoventilation and hyperventilation. *(module 18.11)*

Labeling Exercise

Label the structures indicated on the figure to the right.

1. ______________________________

2. ______________________________

3. ______________________________

4. ______________________________

5. ______________________________

6. ______________________________

Critical Thinking Questions

1. In the recovery room following the surgical removal of her ovaries (oophorectomy), a patient complains of a sore throat. If she didn't have a sore throat before the surgery, what is the most likely explanation for her sore throat at this time?
2. You are the winner of a skiing package to Mammoth Lakes, California. Nestled in the beautiful Sierra Nevada mountains, Mammoth has ski runs that go from 9000 to 12,000 feet in altitude. What causes your headache and difficulty breathing on the first day? What physiological changes occur that allow you to feel better a couple of days later?
3. Patients with pleurisy—inflammation of the pleural membranes—attempt to repress their breathing because it hurts when they take a breath. What physiological changes are produced by voluntary hypoventilation? How does the body respond to these changes?
4. How is a deep-sea diver like a bottle of champagne? (Hint: Think of what happens when a diver ascends too rapidly and when a champagne bottle is uncorked.)
5. A man who has been warming up his car in an enclosed garage for a period of time suddenly remembers that his lunch is still sitting on the kitchen counter. After getting out of his car and walking a few steps he collapses. Why did he pass out? Why did he pass out after walking, and not while he was sitting in his car?

CHAPTER

19

"Should our kidneys fail, we would be quickly poisoned by our own wastes." *page 478*

Urinary System

Chapter Outline

Terms to Remember

kidney, 478
renal cortex, 479
renal medulla, 479
renal pyramid, 479
renal pelvis, 479
nephron (*nef' ron*), 479
glomerulus (*glo-mer' yū" lus*), 480
glomerular capsule, 481
nephron loop, 481
collecting duct, 481
glomerular filtration rate (GFR), 482
countercurrent multiplier system, 486
antidiuretic (*an" te-di" yū-ret' ik*) hormone (ADH), 488
renal plasma clearance, 488
aldosterone (*al-dos' ter-ōn*), 491
renin, 492
atrial natriuretic (*na" trī-yoo-ret' ik*) factor, 492
ureter (*yoo-re' ter*), 495
urinary bladder, 495
urethra (*yoo-re' thră*), 496
micturition (*mik" tŭ-rish' un*), 497

Learning Objectives

After studying this chapter, you should be able to

- identify the components of the urinary system and describe the gross structure of the kidney.
- describe the anatomy of the nephron and explain how glomerular filtrate is formed.
- explain how salt and water are reabsorbed in the proximal convoluted tubule.
- explain how the descending and ascending limbs of the nephron loop interact to produce a countercurrent multiplier system.
- define the term *renal plasma clearance* and explain the renal clearance of inulin, urea, PAH, and glucose.
- describe the regulation of Na^+/K^+ balance by aldosterone and explain how aldosterone secretion is regulated.
- explain how the kidneys help to regulate the acid-base balance of the blood.
- describe the structure and location of the ureters, urinary bladder, and urethra.

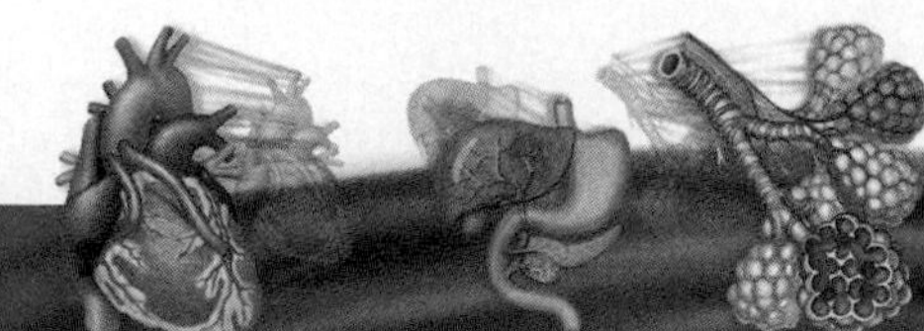

19.1 Urinary System and Kidney Structure

The two kidneys, each consisting of a renal cortex and a renal medulla, regulate the extracellular fluid of the body by forming urine. Urine drains into the renal pelvis and passes out of the kidney through the ureter.

The urinary system consists of two **kidneys,** two **ureters,** the **urinary bladder,** and the **urethra** (fig. 19.1). The primary function of the urinary system is to regulate the extracellular fluid (plasma and interstitial fluid) environment in the body. This function is accomplished through the formation of *urine* by the kidneys. The kidneys are the most important organs of the urinary system; the other organs serve simply as temporary reservoirs or as transport channels for the urine. Urine, a modified filtrate of blood plasma, passes out of the kidneys via the ureters to the urinary bladder, where it is stored for a time before being eliminated from the body through the urethra.

The kidney provides an excellent example of an organ whose primary function is homeostasis. In the process of urine formation, the kidneys regulate

1. the volume of blood plasma, and thus contribute significantly to the regulation of blood pressure;
2. the concentration of waste products in the blood;
3. the concentration of electrolytes (Na^+, K^+, HCO_3^-, and other ions) in the blood plasma; and
4. the pH of blood plasma.

These functions are vital to our survival. Should our kidneys fail, we would quickly be poisoned by our own wastes.

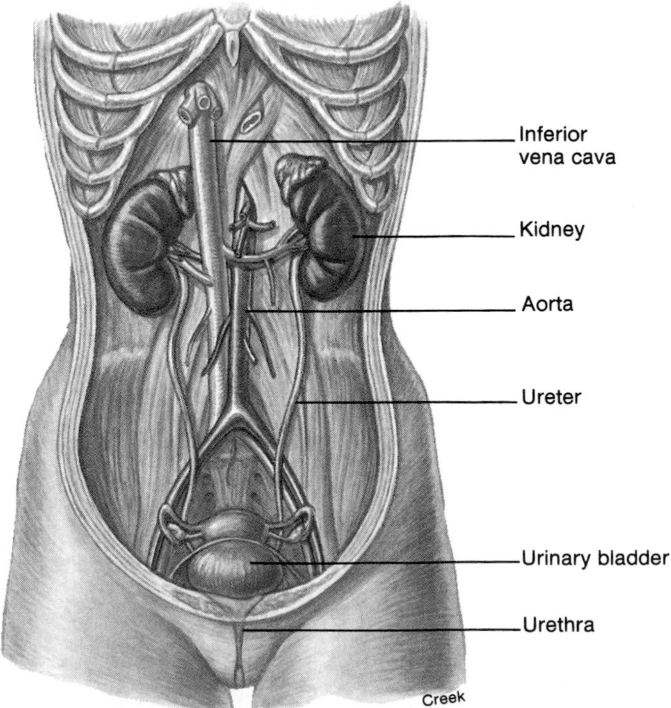

FIGURE 19.1
The organs of the urinary system are the two kidneys, two ureters, urinary bladder, and urethra.

Position and Structure of the Kidneys

The reddish-brown **kidneys** are positioned high in the abdominal cavity between the levels of the twelfth thoracic and third lumbar vertebrae (fig. 19.2). The kidneys are *retroperitoneal,* which means that they are located behind the parietal peritoneum. Each adult kidney is a lima-bean-shaped organ about 11.25 cm (4 in.) long, 5.5–7.7 cm (2–3 in.) wide, and 5.0 cm (2 in.) thick. The right kidney lies slightly lower than the left because it is crowded by the liver above it. The **hilum** of the kidney is the depression along the medial border through which the **renal artery** enters and the **renal vein** and **ureter** (*yoo-re'ter*) exit.

Each kidney is embedded in a fatty fibrous pouch consisting of three layers. The **renal capsule,** the innermost layer, is a strong, transparent fibrous membrane that serves as a barrier against trauma and infection. Surrounding the renal capsule is a firm protective mass of fatty tissue called

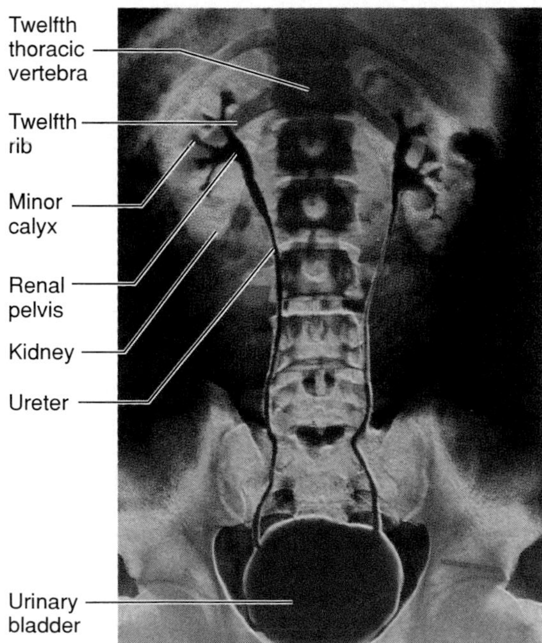

FIGURE 19.2
A color-enhanced radiograph of the calyces and renal pelvises of the kidneys, the ureters, and the urinary bladder. (Note the position of the kidneys relative to the vertebral column and ribs.)

the **adipose capsule.** The outermost layer, the **renal fascia,** is composed of dense irregular connective tissue. It is a supportive layer that anchors the kidney to the parietal peritoneum and the abdominal wall.

A coronal section of the kidney shows two distinct regions and a major cavity (fig. 19.3). The outer **renal cortex,** in contact with the renal capsule, is reddish brown and granular in appearance because of its many capillaries. The deeper **renal medulla** is darker in color, and the presence of microscopic tubules and blood vessels gives it a striped appearance. The renal medulla is composed of 8 to 15 conical **renal pyramids.** Portions of the renal cortex extend between the renal pyramid to form the **renal columns.** The apexes of the renal pyramids are known as the **renal papillae.** These nipplelike projections are directed toward the inner region of the kidney.

The cavity of the kidney collects and transports urine from the kidney to the ureter. It is divided into several portions. Each papilla of a pyramid projects into a small depression called the **minor calyx** (*ka'liks*)—in the plural, *calyces.* Several minor calyces unite to form a **major calyx.** In turn, the major calyces join to form the funnel-shaped **renal pelvis.** The renal pelvis collects urine from the calyces and transports it to the ureter.

The **nephron** (*nef'ron*) is the functional unit of the kidney that is responsible for the formation of urine. Each kidney contains more than a million nephrons. A nephron consists of a **renal tubule** (fig. 19.3*c*) and associated small

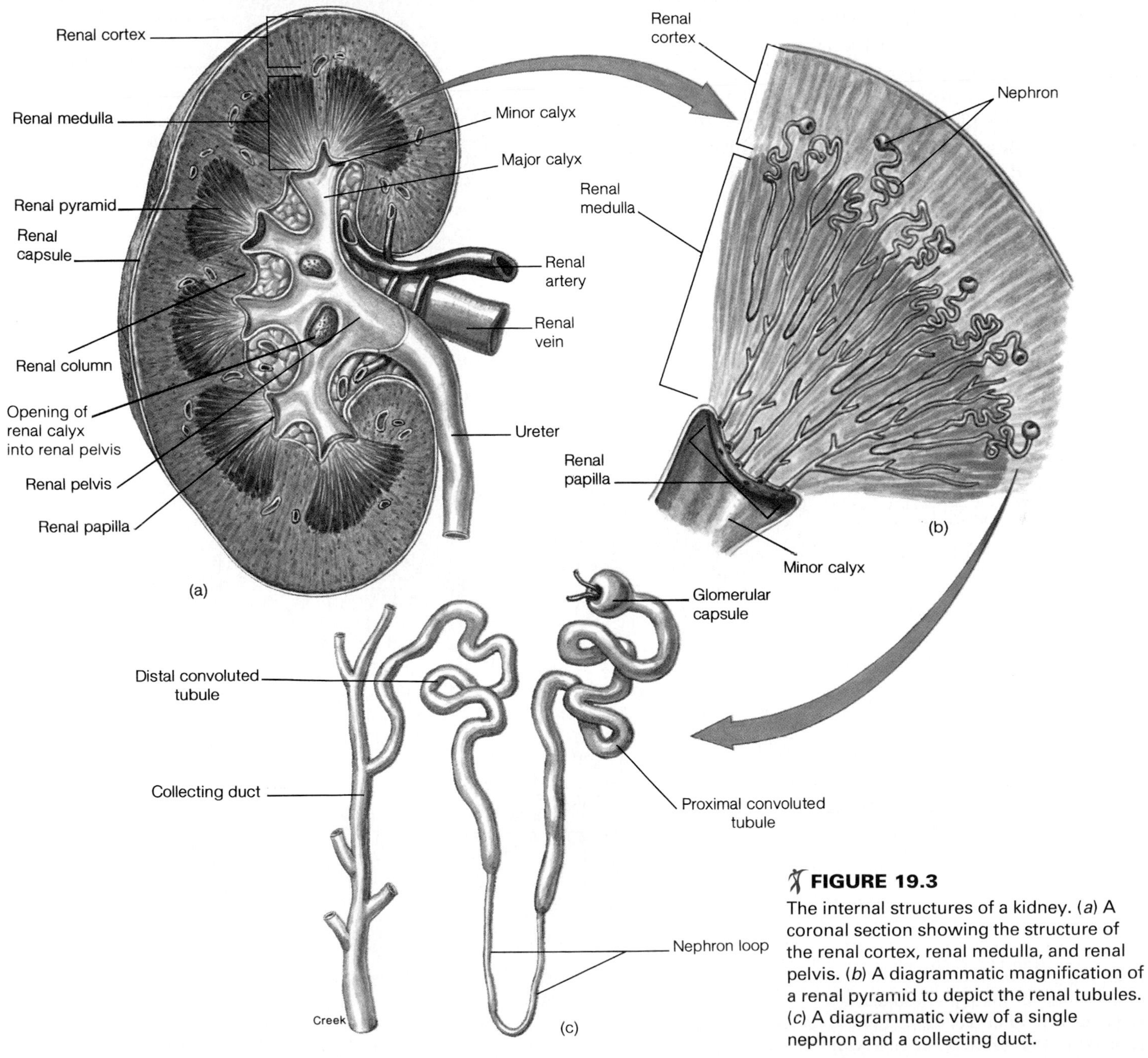

FIGURE 19.3

The internal structures of a kidney. (*a*) A coronal section showing the structure of the renal cortex, renal medulla, and renal pelvis. (*b*) A diagrammatic magnification of a renal pyramid to depict the renal tubules. (*c*) A diagrammatic view of a single nephron and a collecting duct.

blood vessels. Fluid formed by capillary filtration enters the renal tubule and is subsequently modified by transport processes. The resulting fluid that leaves the renal tubules is urine.

Although the kidneys are generally well protected by being encapsulated retroperitoneally, they may be injured by a hard blow to the lumbar region. Such an injury can produce blood in the urine, since the highly vascular kidneys are particularly susceptible to hemorrhage.

Renal Blood Vessels

The kidneys have an extensive circulatory network to allow for the continuous cleansing and modification of large volumes of blood (fig. 19.4). Arterial blood enters the kidney at the hilum through the **renal artery,** which divides into **interlobar** (*in"ter-lo'bar*) **arteries. Arcuate** (*ar'kyoo-āt*) **arteries** branch from the interlobar arteries at the boundary of the renal cortex and renal medulla. Small **interlobular arteries** radiate from the arcuate arteries and project into the renal cortex. Microscopic **afferent arterioles** arise from branches of the interlobular arteries. The afferent arterioles transport the blood into ball-shaped capillary networks, the **glomeruli** (*glo-mer'yŭ-li*), which produce a blood filtrate that enters the renal tubules. The blood remaining in the glomeruli leaves through **efferent arterioles** and is carried into another capillary network—the **peritubular capillaries** surrounding the renal tubules (fig. 19.5). This blood is drained by veins that parallel the course of the arteries. The renal vein collects the draining blood and delivers it to the inferior vena cava.

arcuate: L. *arcuare,* to bend
glomerulus: L. *glomerulus,* diminutive of *glomus,* ball

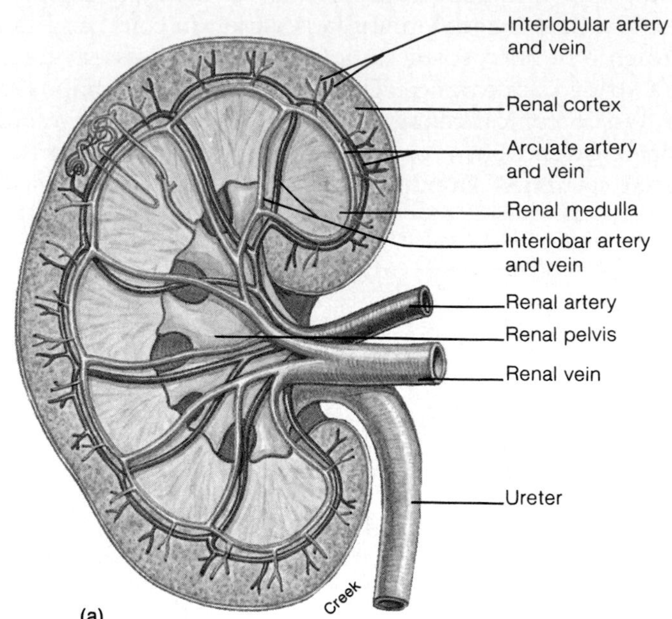

FIGURE 19.4
The principal arteries and veins of the kidney.

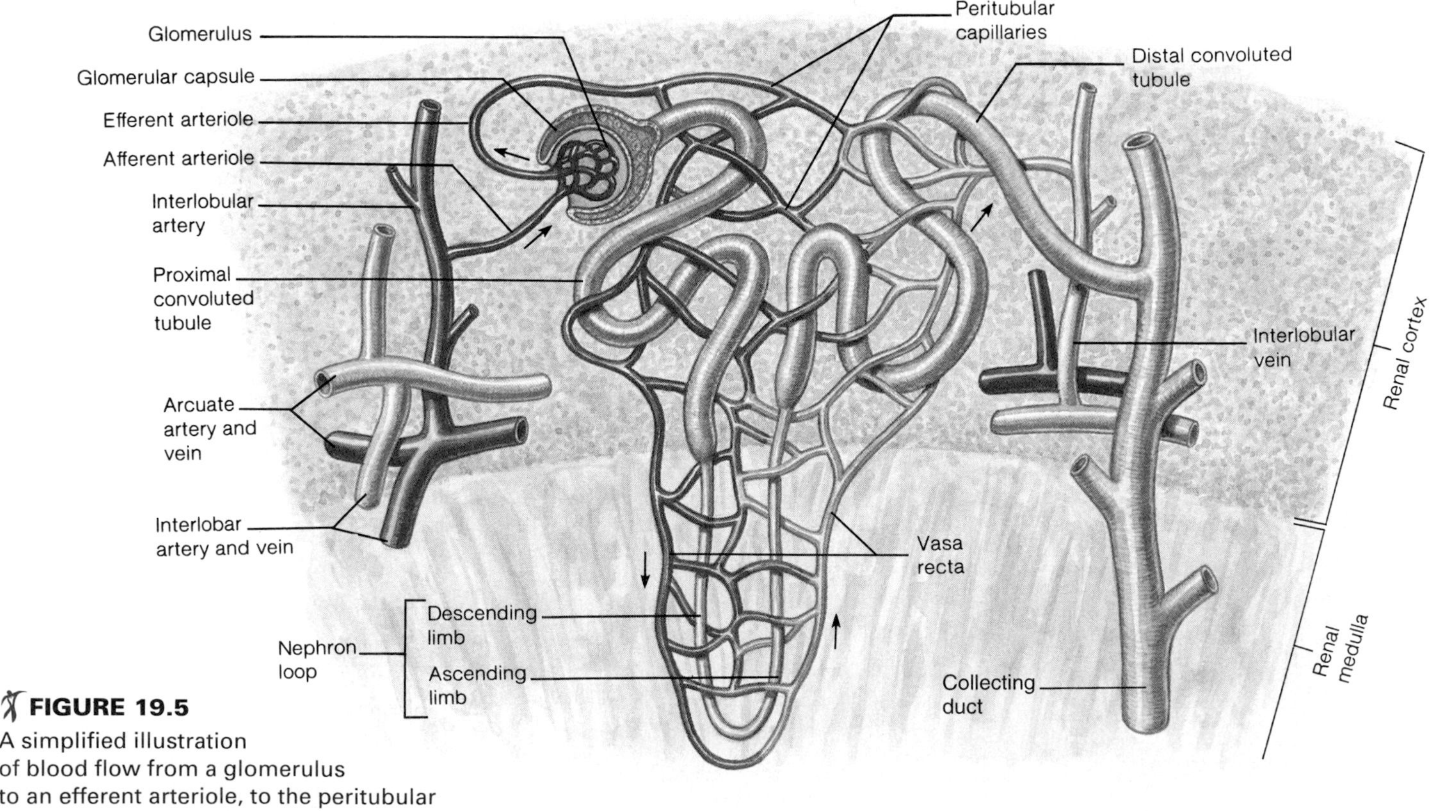

FIGURE 19.5
A simplified illustration of blood flow from a glomerulus to an efferent arteriole, to the peritubular capillaries, to the venous drainage of the kidney.

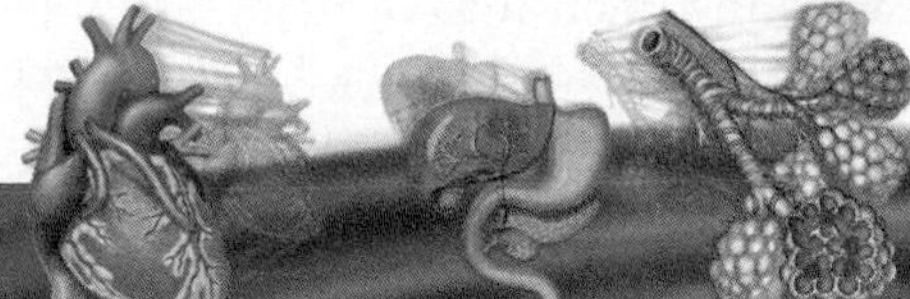

19.2 Nephron Tubules and Glomerular Filtration

The blood filtrate produced by the glomeruli passes through different regions of the nephron tubules. Most of this filtrate is returned to the circulatory system, and some eventually becomes urine.

Nephron Tubules

The nephron tubules are modified in different regions for different functions. The **glomerular** (Bowman's) **capsule,** the first part of the tubule, surrounds the glomerulus (fig. 19.5). Filtrate in the glomerular capsule passes into the lumen of the **proximal convoluted tubule.** The glomerulus, glomerular capsule, and proximal convoluted tubule are located in the renal cortex. Fluid passes from the proximal convoluted tubule to the **nephron loop** (loop of Henle). This fluid is carried into the renal medulla in the **descending limb** of the nephron loop and returns to the renal cortex in the **ascending limb** of the nephron loop (fig. 19.5). Back in the renal cortex, the tubule becomes coiled again, and is called the **distal convoluted tubule.**

The distal convoluted tubules of several nephrons drain into a **collecting duct** (fig. 19.6). Fluid is then drained from the renal cortex into the renal medulla as the collecting duct passes through a renal pyramid. This fluid, now called *urine,* passes out of a renal papilla into a minor

Bowman's capsule: from Sir William Bowman, English anatomist, 1916–92
loop of Henle: from Friedrich G. J. Henle, German anatomist, 1809–85

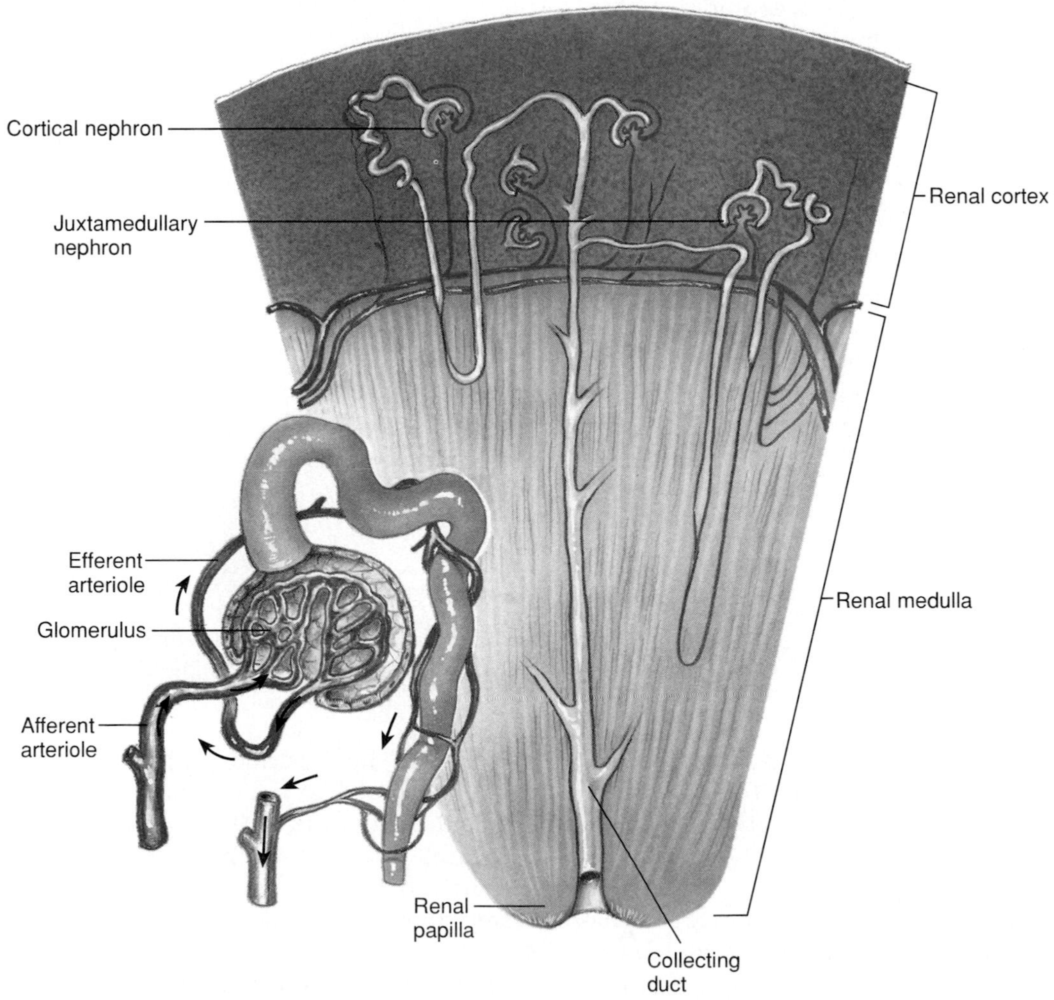

FIGURE 19.6
Cortical nephrons are located almost exclusively within the renal cortex. Juxtamedullary nephrons are located, for the most part, within the outer portion of the renal medulla. The direction of blood flow is indicated with arrows.

calyx and then into a major calyx (see fig. 19.3). The urine is then funneled through the renal pelvis and out of the kidney into the ureter.

Glomerular Filtration

The glomerular capillaries have large pores called fenestrae (figure 19.7), in their walls, and are thus said to be *fenestrated.* These large pores make the glomerular capillaries highly permeable to plasma water and dissolved solutes.

Before the filtrate can enter the interior of the glomerular capsule it must pass through the capillary pores, the basement membrane (a thin layer of glycoproteins), and the inner visceral layer of the glomerular capsule. The inner layer of the glomerular capsule is composed of unique cells, called **podocytes** (*pod'ŏ-sīts*), with numerous cytoplasmic extensions known as **pedicels** (*ped'ĭ-selz*), or "foot processes." These pedicels interdigitate, like the fingers of clasped hands, as they wrap around the glomerular capillaries. The narrow **filtration slits** between adjacent pedicels provide the passageways through which filtered molecules must pass in order to enter the interior of the glomerular capsule (fig. 19.7).

The fluid that enters the glomerular capsule is called *glomerular filtrate.* The **glomerular filtration rate (GFR)** is the volume of filtrate produced by both kidneys each minute. The GFR averages 115 ml per minute in women and 125 ml per minute in men. This volume is equivalent to 180 L per day (about 45 gallons)! Most of the filtrate must obviously be returned to the circulatory system immediately, or a person would literally urinate to death within minutes.

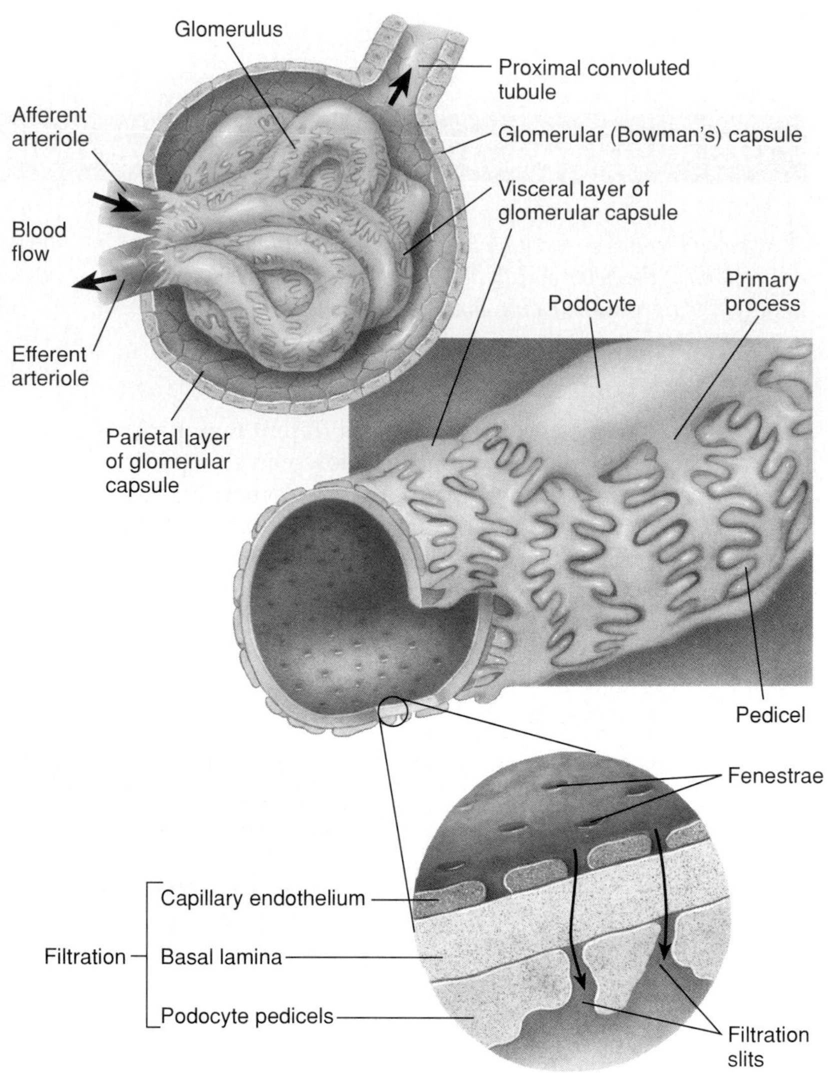

FIGURE 19.7
An illustration of the relationship between the glomerular capillaries and the inner layer of the glomerular (Bowman's) capsule.

19.3 Reabsorption in the Proximal Convoluted Tubule

Glomerular filtrate contains a large amount of water that must be reabsorbed to maintain blood volume. This is accomplished by the osmotic movement of water across the wall of the renal tubule, returning it to the blood.

About 180 L of glomerular filtrate are produced each day, but the kidneys normally excrete only 1 to 2 L of urine in a 24-hour period. Approximately 99% of the filtrate is returned to the circulatory system, while 1% is excreted in the urine. Although the urine volume varies according to the changing needs of the body, most of the filtered water must be returned to the circulatory system to maintain blood volume and pressure, regardless of the body's state of hydration. The return of filtered molecules from the renal tubules to the blood is called **reabsorption** (fig. 19.8).

The glomerular filtrate entering the proximal convoluted tubule is isotonic to blood plasma. As described in

fenestra: L. *fenestra,* window
podocyte: Gk. *pous,* foot; *kytos,* cell
pedicel: L. *peducekkus,* footplate

chapter 3, this means that water will not move by osmosis between these solutions because they have the same total solute concentration. This concentration is indicted in *osmolality* units. The exact definition of these units need not concern us; what is important to know at this point is that blood plasma has a total solute concentration of *300 milliosmolal* (300 mOsm). We can use this number as a reference. Any solution that is isotonic to blood must also have a concentration of 300 mOsm. Any solution with an osmolality less than 300 mOsm is hypotonic to (more dilute than) blood plasma; any solution with an osmolality greater than 300 mOsm is hypertonic to (more concentrated than) blood plasma. Later in this chapter, we will use osmolality units as reference numbers to follow the change in concentration of the urine.

Because osmosis cannot occur between isotonic solutions (see chapter 3, module 3.6), the plasma in the capillaries surrounding the tubules must be made more concentrated than the filtrate (hypertonic to the filtrate), so that water will move by osmosis from the filtrate to the surrounding capillaries. This is accomplished by the active transport of Na^+ from the filtrate to the blood in the surrounding peritubular capillaries.

The concentration of Na^+ in the fluid entering the proximal convoluted tubule is the same as in blood plasma. The epithelial cells of the renal tubule, however, have a much lower Na^+ concentration. This lower Na^+ concentration is partially due to the low permeability of the cell membrane to Na^+ and partially due to the active transport of Na^+ out of the cell by Na^+/K^+ pumps. In the cells of the proximal convoluted tubule, Na^+/K^+ pumps are located in the basal and lateral sides of the cell membrane (fig. 19.9), but not in the apical membrane (the part containing microvilli that faces the lumen of the tubule). As a result of these active transport pumps, a concentration gradient is created that favors the diffusion of Na^+ from the renal tubular fluid across the apical cell membranes and into the epithelial cells of the proximal convoluted tubule. The Na^+ is then extruded into the surrounding interstitial (tissue) fluid by the Na^+/K^+ pumps (fig. 19.9).

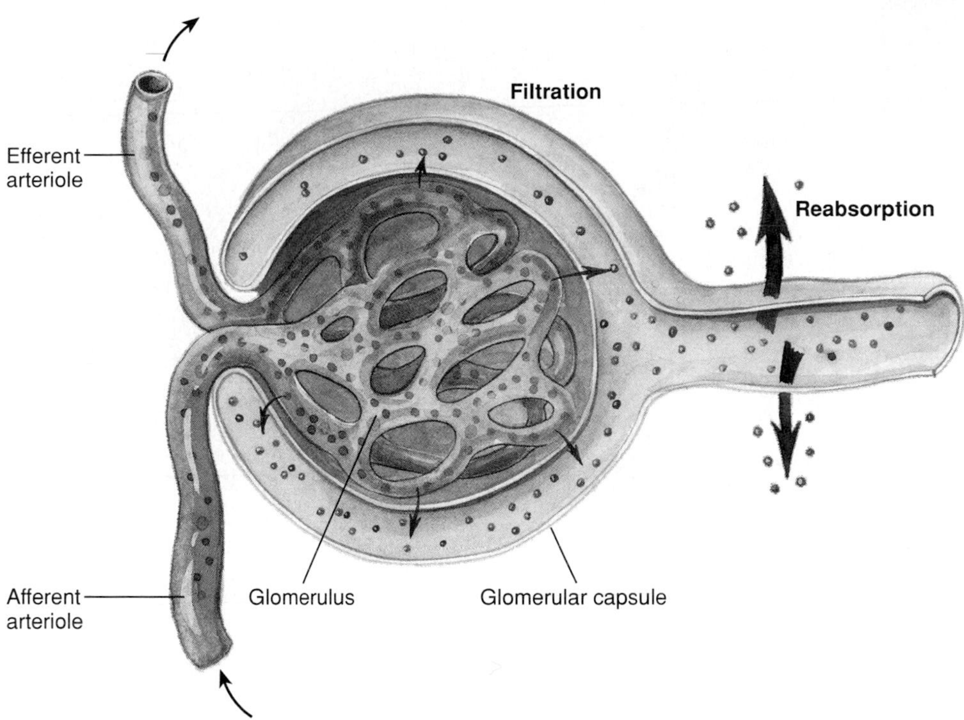

FIGURE 19.8
Water from blood plasma and its dissolved solutes (except proteins) enter the glomerular filtrate, but most of the filtered molecules are reabsorbed. The term *reabsorption* refers to the transport of molecules out of the renal tubular filtrate back into the blood.

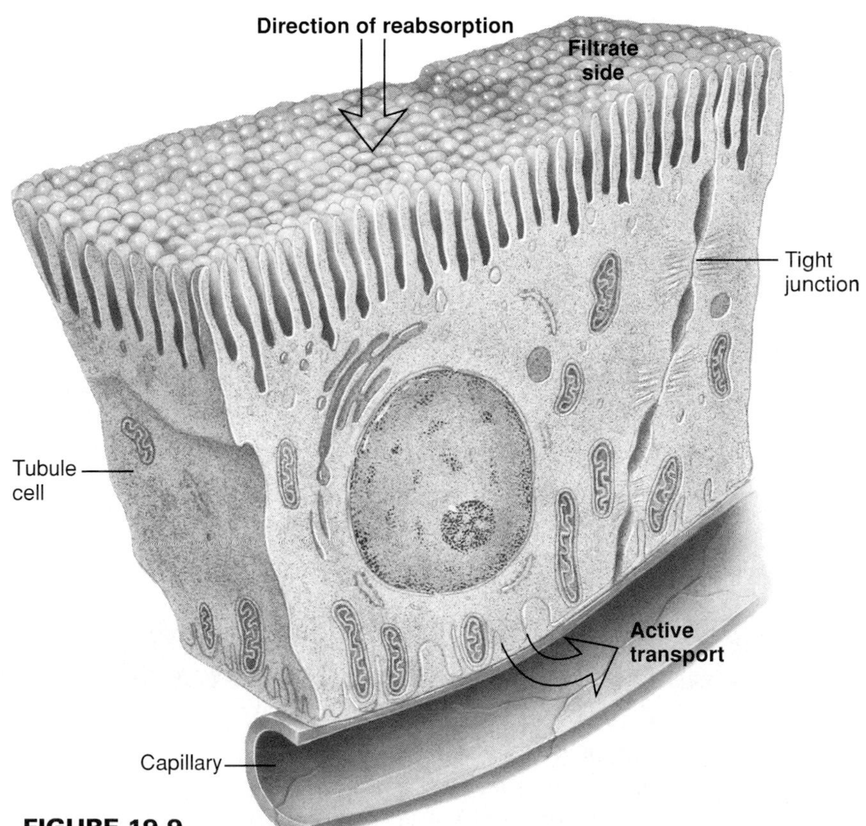

FIGURE 19.9
An illustration of the appearance of tubule cells in the electron microscope. Molecules that are reabsorbed pass through the apical membrane of the cells of a renal tubule to the endothelium of a capillary.

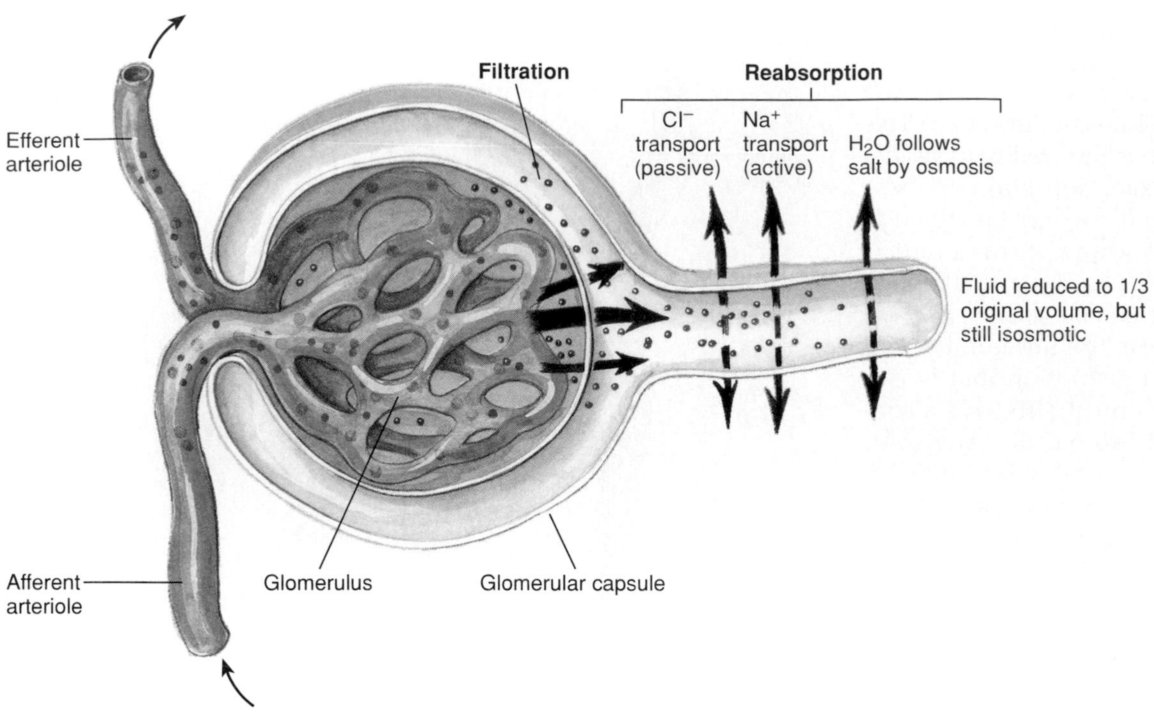

FIGURE 19.10

Mechanisms of salt and water reabsorption in the proximal convoluted tubule. Sodium is actively transported out of the filtrate, and chloride follows passively by electrical attraction. Water follows the salt out of the filtrate from the renal tubule into the peritubular capillaries by osmosis.

The transport of Na^+ from the tubular fluid to the interstitial fluid creates a potential difference across the wall of the renal tubule. Since this electrical gradient favors the passive transport of Cl^- toward the higher Na^+ concentration in the interstitial fluid, chloride ions passively follow sodium ions out of the filtrate into the interstitial fluid. As a result of the accumulation of NaCl (salt), the osmolality and osmotic pressure values of the interstitial fluid surrounding the epithelial cells are increased above those of the tubular fluid.

An osmotic gradient is thus created between the tubular fluid and the interstitial fluid surrounding the proximal convoluted tubule. Since the cells of the proximal convoluted tubule are permeable to water, water moves by osmosis from the tubular fluid into the epithelial cells and then across the basal and lateral sides of the epithelial cells into the interstitial fluid. The salt and water that were reabsorbed from the tubular fluid can then move into the surrounding peritubular capillaries and in this way be returned to the blood (fig. 19.10).

Approximately 65% of the salt and water in the original glomerular filtrate is reabsorbed across the proximal convoluted tubule in the manner just described and returned to the vascular system. The volume of tubular fluid remaining is reduced accordingly, but this fluid is still isotonic to the blood (has a concentration of 300 mOsm). This is because the cell membranes in the proximal convoluted tubule are freely permeable to water, so that water and salt are removed in proportionate amounts.

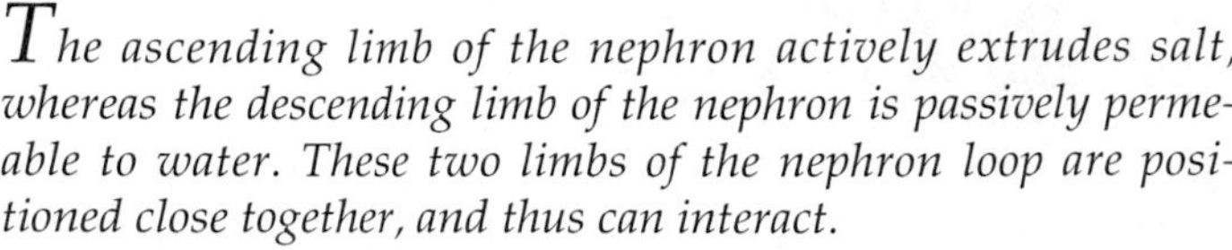

19.4 Reabsorption in the Nephron Loop

The ascending limb of the nephron actively extrudes salt, whereas the descending limb of the nephron is passively permeable to water. These two limbs of the nephron loop are positioned close together, and thus can interact.

The descending and ascending limbs of the nephron loop are close to one another, which allows them to interact. Since the ascending limb of the loop is the active partner in this interaction, we will first describe its properties before considering those of the descending limb.

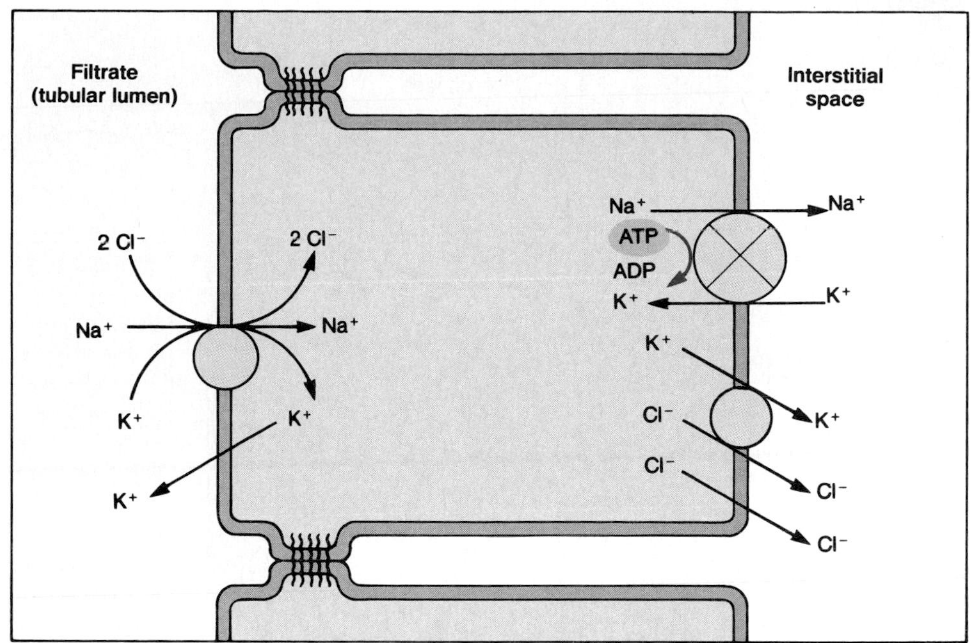

FIGURE 19.11
In the thick segment of the ascending limb of the nephron loop, Na^+ and K^+, together with two chloride ions, enter the cells of the renal tubule. Na^+ is then actively transported out into the interstitial space, and Cl^- follows passively. The K^+ diffuses back into the filtrate, and some also enters the interstitial space.

Ascending Limb of the Nephron Loop

Salt (NaCl) is actively extruded from the ascending limb of the nephron into the surrounding interstitial fluid. The way in which this is accomplished, however, is not the same as in the proximal convoluted tubule. Instead, Na^+, K^+, and Cl^- passively diffuse from the filtrate into the cells of the ascending limb of the nephron in a ratio of 1 Na^+ to 1 K^+ to 2 Cl^-. The Na^+ is then actively transported across the basolateral membrane to the interstitial fluid by the Na^+/K^+ pumps. Cl^- follows the Na^+ passively because of electrical attraction, and K^+ diffuses back into the filtrate (fig. 19.11).

The ascending limb of the nephron loop is structurally divisible into two regions: a *thin segment,* nearest to the bottom of the loop, and a *thick segment* of varying lengths, which carries the filtrate outward into the renal cortex and into the distal convoluted tubule. The thin segment is composed of squamous epithelium, whereas the thick segment is composed of a cuboidal epithelium. It is currently believed that only the cells of the thick segments of the ascending limb of the nephron are capable of actively transporting NaCl from the filtrate into the surrounding interstitial fluid.

Although the mechanism of NaCl transport is different in the ascending limb than in the proximal convoluted tubule, the net effect is the same. Salt (NaCl) is pumped out of the tubule and into the surrounding interstitial fluid. Unlike the epithelial walls of the proximal convoluted tubule, however, the walls of the ascending limb are *not permeable to water.* By pumping salt out, the fluid remaining in the tubule becomes increasingly dilute as it ascends toward the renal cortex, whereas the interstitial fluid around the nephron loops in the renal medulla becomes increasingly more concentrated. By means of these processes, the tubular fluid that enters the distal convoluted tubule in the renal cortex is made hypotonic (with a concentration of about 100 mOsm). By contrast, the interstitial fluid in the renal medulla is made hypertonic.

Descending Limb of the Nephron Loop

The interstitial fluid in the deeper regions of the renal medulla reaches a concentration of 1200 to 1400 mOsm. In order to reach this high a concentration, the salt pumped out of the ascending limb of the nephron must accumulate in the interstitial fluid of the renal medulla. This occurs as a result of the properties of the descending limb of the nephron, to be discussed next, and because blood vessels around the nephron loop do not carry back all of the extruded salt to the general circulation.

The descending limb does not actively transport salt, and indeed is believed to be impermeable to the passive diffusion of salt. It is, however, permeable to water. Since the surrounding interstitial fluid is hypertonic to the filtrate in the descending limb, water is drawn out of the descending limb by osmosis and enters blood capillaries. The concentration of tubular fluid is thus increased, and its volume is decreased, as it descends toward the bottom of the loop.

As a result of these passive transport processes in the descending limb, the fluid that "rounds the bend" at the nephron loop has the same osmolality as the surrounding interstitial fluid (1200–1400 mOsm). There is, therefore, a higher salt concentration arriving in the ascending limb than there would be if the descending limb simply delivered isotonic fluid. Salt transport by the ascending limb is increased accordingly, so that the "saltiness" of the interstitial fluid is multiplied (fig. 19.12).

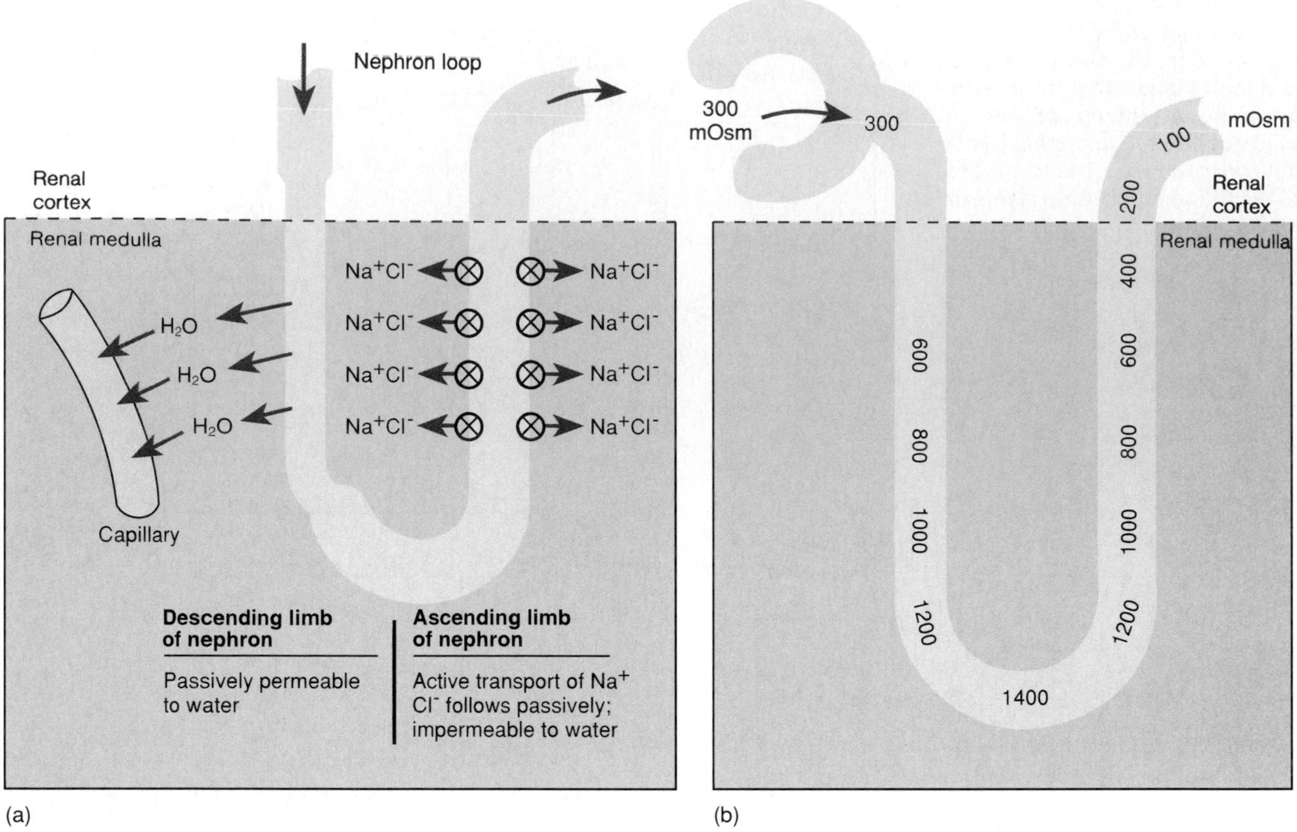

FIGURE 19.12
The countercurrent multiplier system. The extrusion of sodium chloride from the ascending limb of the nephron loop makes the surrounding interstitial fluid more concentrated. Multiplication of this concentration is due to the fact that the descending limb of the nephron loop is passively permeable, which causes its fluid to increase in concentration as the surrounding interstitial fluid becomes more concentrated. The transport properties of the nephron loop and their effect on the renal tubular fluid concentration are shown in (*a*). The values of these changes in osmolality, together with the effect on surrounding interstitial fluid concentration, are shown in (*b*).

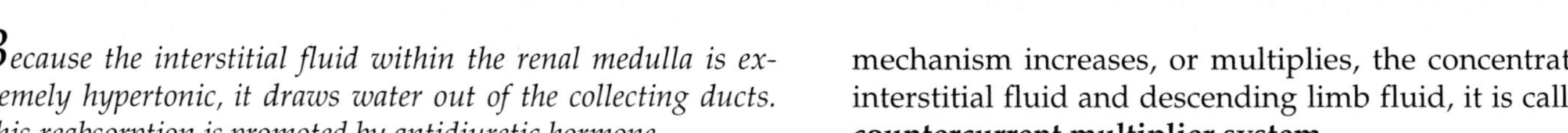

19.5 Countercurrent Multiplier System

Because the interstitial fluid within the renal medulla is extremely hypertonic, it draws water out of the collecting ducts. This reabsorption is promoted by antidiuretic hormone.

Countercurrent flow (flow in opposite directions) in the ascending and descending limbs of the nephrons and the close proximity of the two limbs allow for interaction. Since the concentration of the tubular fluid in the descending limb reflects the concentration of surrounding interstitial fluid, and since the concentration of this interstitial fluid is raised by the active extrusion of salt from the ascending limb, a *positive feedback mechanism* is created. The more salt the ascending limb or the nephron extrudes, the more concentrated the fluid that returns to it from the descending limb. Because this positive feedback mechanism increases, or multiplies, the concentration of interstitial fluid and descending limb fluid, it is called the **countercurrent multiplier system.**

The countercurrent multiplier system recirculates salt and thus traps some of the salt that enters the nephron loop in the interstitial fluid of the renal medulla. This system results in a gradually increasing concentration of renal interstitial fluid from the outer renal cortex to the inner renal medulla. In fact, the osmolality of interstitial fluid increases from 300 mOsm (isotonic) in the renal cortex to nearly 1400 mOsm in the deepest part of the renal medulla (fig. 19.13).

In order for the countercurrent multiplier system to be effective, most of the salt that is extruded from the ascending limbs of the nephrons must remain in the interstitial

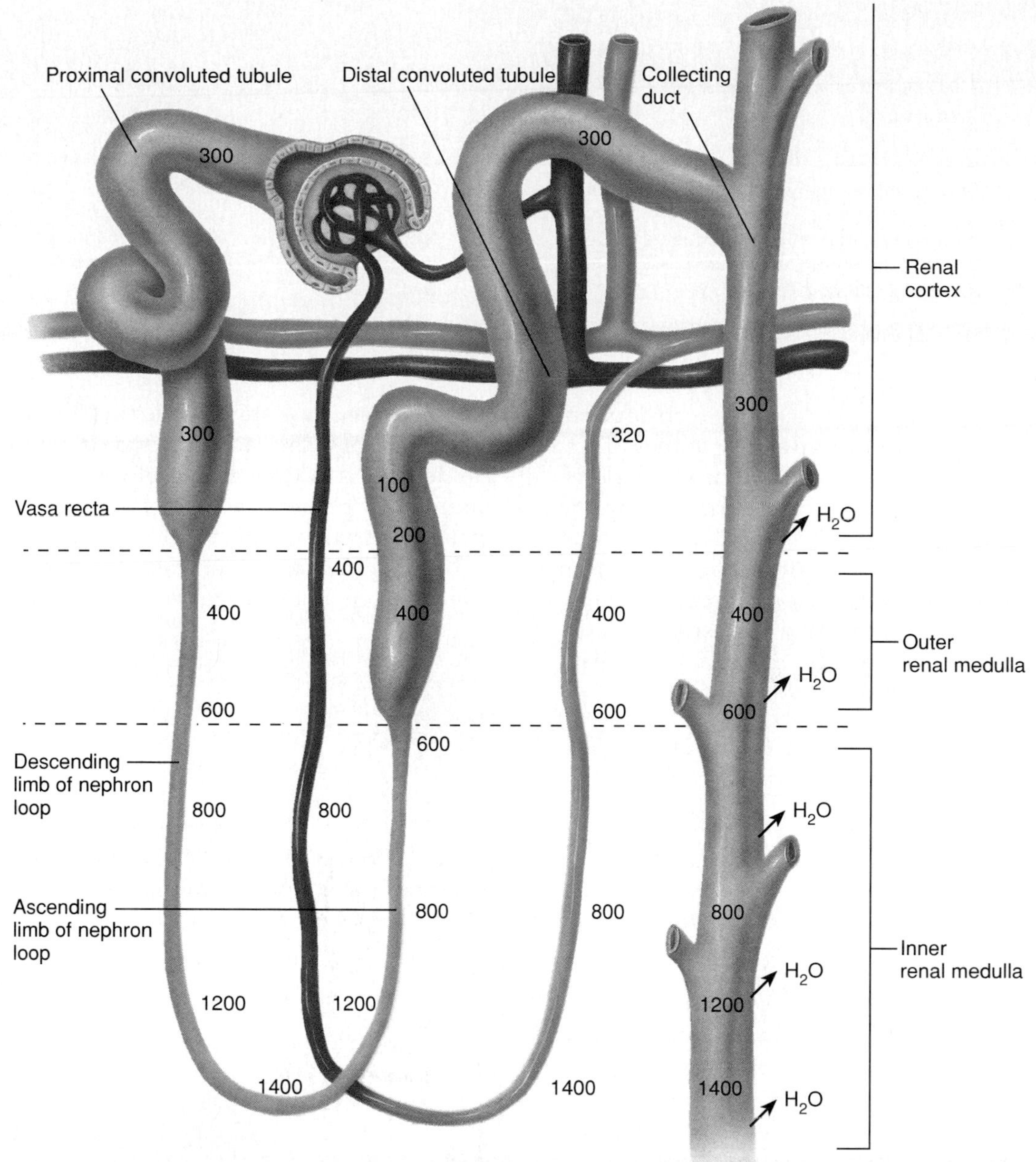

FIGURE 19.13
The countercurrent multiplier system in the nephron loop and countercurrent exchange in the vasa recta help to create a hypertonic renal medulla. Under the influence of antidiuretic hormone (ADH), the collecting duct becomes more permeable to water, so that water is drawn by osmosis out into the hypertonic renal medulla and into the peritubular capillaries.

fluid of the renal medulla, while most of the water that leaves the descending limbs of the nephrons must be removed by the blood. This is accomplished by thin-walled looping vessels, known as the **vasa recta,** which parallel the course of the long nephron loops (figs. 19.5 and 19.13). The vasa recta are freely permeable to water and salt. As a result, blood in the vasa recta makes passive exchanges with the surrounding interstitial fluid, achieving equilibrium all along its path. This mechanism, called **countercurrent exchange,** helps to maintain the osmotic gradient established by the cycling of salt.

The walls of the collecting ducts are *permeable to water but not to salt.* Since the surrounding interstitial fluid in the renal medulla is very hypertonic as a result of the countercurrent multiplier system, water is drawn out of the collecting ducts by osmosis. This water does not dilute the surrounding interstitial fluid because it is transported by capillaries to the general circulation. In this way, most of the water remaining in the filtrate is returned to the circulatory system.

The osmotic gradient created by the countercurrent multiplier system provides the force for water reabsorption

TABLE 19.1 **Antidiuretic Hormone Secretion and Action**

Stimulus	Receptors	Secretion of ADH	Effects on Urine Volume	Effects on Blood
↑ Osmolality (dehydration)	Osmoreceptors in hypothalamus	Increased	Decreased	Increased water retention; decreased blood osmolality
↓ Osmolality	Osmoreceptors in hypothalamus	Decreased	Increased	Water loss increases blood osmolality
↑ Blood volume	Stretch receptors in left atrium	Decreased	Increased	Decreased blood volume
↓ Blood volume	Stretch receptors in left atrium	Increased	Decreased	Increased blood volume

through the collecting ducts. The rate of this reabsorption, however, is determined by the permeability of the collecting duct cell membranes to water. The permeability of the collecting ducts to water, in turn, is determined by the concentration of **antidiuretic** (*an″te-di″yŭ-ret′ik*), **hormone (ADH)** in the blood. When more ADH is released from the posterior pituitary, the collecting ducts become more permeable to water, and more water is reabsorbed. A decrease in ADH release, conversely, results in less reabsorption of water and thus in the excretion of a larger volume of more dilute urine.

Inside Information—Both alcohol and caffeine interfere with the secretion of ADH—both are diuretics. This is why beer is a poor choice as a thirst-quencher, despite what the beer commercials would have you believe. And to the extent that a hangover is caused by dehydration, drinking coffee—a second diuretic—to "sober up" may be counterproductive in getting rid of a headache.

A person in a state of normal hydration excretes about 1.5 L of urine each day, indicating that 99.2% of the glomerular filtrate volume is reabsorbed. Even small changes in percent reabsorption translate into large changes in urine volume. Increasing the intake of water—and thus decreasing ADH secretion (table 19.1)—results in a correspondingly larger volume of urine excretion.

Diabetes insipidus *is a disease in which the posterior pituitary fails to secrete sufficient ADH or the renal tubules fail to respond to the presence of ADH. Therefore, the collecting ducts are not very permeable to water, and so a large volume (5–10 L per day) of dilute urine is produced. The dehydration that results causes intense thirst, but a person with this condition has difficulty drinking enough to compensate for the large volumes of water lost in the urine.*

19.6 Renal Plasma Clearance

The renal plasma clearance value for a substance indicates how the kidneys handle the excretion of that substance. The clearance value is affected by the processes of filtration, reabsorption, and secretion.

One of the major functions of the kidneys is the excretion of waste products. These molecules are filtered through the glomerulus into the glomerular capsule along with water, salt, and other plasma solutes. In addition, some waste products can gain access to the urine by a process called **secretion** (fig. 19.14). Secretion is the opposite of reabsorption. In secretion, molecules are transported out of the peritubular capillaries through the tubule cells and into the lumen of the renal tubule. In this way, molecules can still be excreted in the urine even though they were not initially filtered out of the blood in the glomerulus.

The **renal plasma clearance value** is the volume of plasma from which a substance is completely removed by excretion in the urine in 1 minute. If a substance is neither reabsorbed nor secreted by the renal tubules, the amount excreted in the urine each minute will be equal to the amount that is filtered out of the glomeruli. Such is the case with a plant product called *inulin*—a polymer of the monosaccharide fructose (fig. 19.15). The amount of inulin that enters the urine each minute is the same as the amount that was filtered each minute. The clearance value for inulin is thus equal to the glomerular filtration rate (GFR).

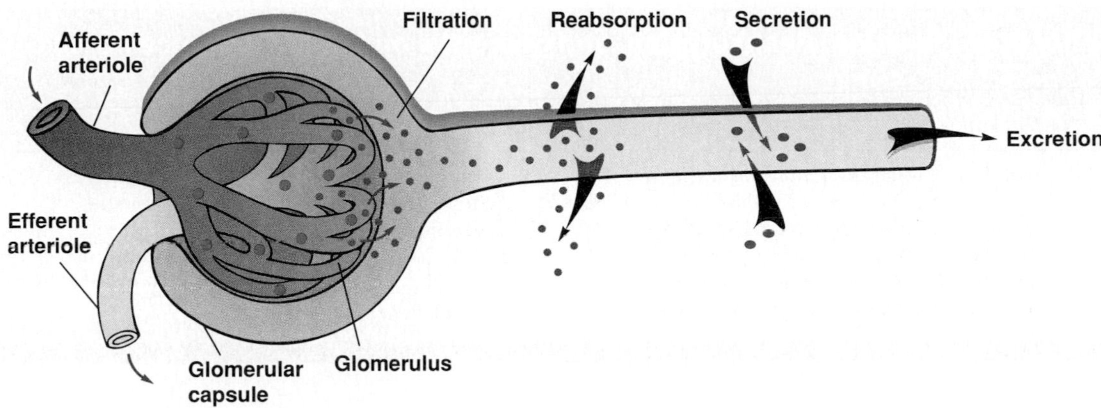

FIGURE 19.14
Secretion refers to the transport of substances from the peritubular capillaries into the fluid of the renal tubule. The direction of this transport is opposite to that of reabsorption.

Inulin

(a)

To peritubular capillaries

(b)

Renal artery– with inulin

Renal vein– inulin concentration less than in renal artery

Urine– containing all inulin that was filtered

(c)

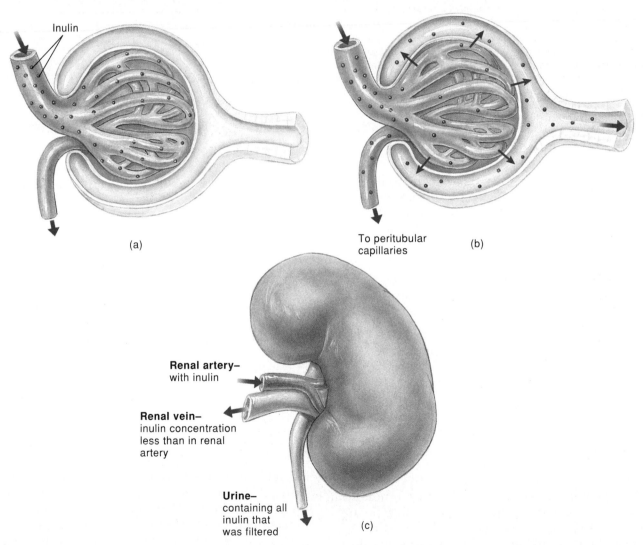

FIGURE 19.15
The renal clearance of inulin. (*a*) Inulin is present in the blood entering the glomeruli, and (*b*) some of this blood, together with its dissolved inulin, is filtered. All of this filtered inulin enters the urine, whereas most of the filtered water is returned to the circulatory system (is reabsorbed). (*c*) The blood leaving the kidneys in the renal vein, therefore, contains less inulin than the blood that entered the kidneys in the renal artery. Since inulin is filtered but neither reabsorbed nor secreted, the inulin clearance rate equals the glomerular filtration rate (GFR).

TABLE 19.2 **Effects of Filtration, Reabsorption, and Secretion on Renal Clearance Rates**

Term	Definition	Effect on Renal Clearance
Filtration	A substance enters the glomerular ultrafiltrate	Some or all of a filtered substance may enter the urine and be "cleared" from the blood
Reabsorption	A substance is transported from the filtrate, through tubular cells, and into the blood	Reabsorption decreases the rate at which a substance is cleared; the clearance rate is less than the glomerular filtration rate (GFR)
Secretion	A substance is transported from peritubular blood, through tubular cells, and into the filtrate	When a substance is secreted by the nephrons, its renal plasma clearance is greater than the GFR

TABLE 19.3 **Renal "Handling" of Different Plasma Molecules**

If Substance Is:	Example	Concentration in Renal Vein	Renal Clearance Rate
Not filtered	Proteins	Same as in renal artery	Zero (since not present in filtrate)
Filtered, not reabsorbed or secreted	Inulin	Less than in renal artery	Equal to GFR (115–125 ml/min)
Filtered, partially reabsorbed	Urea	Less than in renal artery	Less than GFR
Filtered, completely reabsorbed	Glucose	Same as in renal artery	Zero
Filtered and secreted	PAH	Less than in renal artery; approaches zero	Greater than GFR; up to total plasma flow rate (~625 ml/min)
Filtered, reabsorbed, and secreted	K^+	Variable	Variable

Measurements of the plasma concentration of creatinine (kre-at'n-ēn) *are often used clinically as an index of kidney function. Creatinine, produced as a waste product of muscle creatine, is secreted to a slight degree by the renal tubules, so that its excretion rate is a little above that of inulin. Since creatinine is released into the blood at a constant rate, an abnormal decrease in the GFR causes the plasma creatinine concentration to rise.*

The glomerular filtrate, however, also contains other solutes that may be reabsorbed to varying degrees. If a portion of a filtered solute is reabsorbed, the amount excreted in the urine will be less than the amount that was contained in the volume of plasma filtered. The renal plasma clearance for a substance that is reabsorbed must thus be less than the GFR (table 19.2).

Glucose, for example, is normally completely reabsorbed and thus absent from the urine (has a renal plasma clearance of zero). This is due to active transport by specific membrane carriers (see chapter 3, module 3.8). If the blood glucose concentration is sufficiently high, these carriers can become saturated; consequently, glucose will appear in the urine. This condition is called *glycosuria,* and the concentration of glucose needed to just saturate the carriers and produce glycosuria (about 180 mg/dl of blood) is called the **renal plasma threshold** for glucose. In *diabetes mellitus,* inadequate insulin causes fasting hyperglycemia that exceeds the renal plasma threshold, thereby producing glycosuria.

Inside Information—In performing experiments on dogs who had undergone surgical removal of their pancreas, researchers noticed that the urine from these dogs attracted ants. When the urine was tested, it was found to contain glucose, which explained why the ants were attracted to it. It also gave rise to the idea that the pancreas produces a substance that prevents glucose in the urine—a condition linked with diabetes mellitus. Thus, the stage was set for the discovery of insulin, which was finally isolated in 1921.

If a substance is not reabsorbed, all of the filtered amount will be cleared. If this substance is, in addition, secreted by active transport into the renal tubules from the peritubular blood, an additional amount of blood plasma can be cleared of that substance. The renal plasma clearance for a substance that is filtered and secreted is therefore greater than the GFR (table 19.3), which averages 115–125 ml/min. *Para-aminohippuric acid (PAH)*, for example, has a renal plasma clearance of about 625 ml/min. PAH is completely cleared in one pass through the kidneys, and its clearance is used to measure the total blood flow to the kidneys. Many other foreign molecules, including penicillin and other antibiotics, are quickly eliminated from the blood because they are secreted and thus have a high renal plasma clearance.

19.7 Renal Control of Electrolyte Balance: Na^+ and K^+

The renal excretion of Na^+ and K^+ is regulated primarily by aldosterone, which promotes Na^+ reabsorption and K^+ secretion into the renal tubules.

Effects of Aldosterone

The kidneys play a major role in controlling the electrolyte content of the blood. Two electrolytes regulated by the kidneys are sodium and potassium. **Aldosterone** (*al-dos'ter-ōn*) is a steroid hormone secreted by the adrenal cortex (see chapter 13, module 13.4). This hormone promotes sodium retention and potassium loss from the blood by stimulating the reabsorption of Na^+ and the secretion of K^+ across the wall of the last portions of the distal convoluted tubule and cortical portion of the collecting duct. Since aldosterone promotes the retention of Na^+, it contributes to an increased blood volume and pressure.

Without aldosterone, an average of 30 g of Na^+ is excreted in the urine each day. If this condition is not corrected, blood volume and pressure will continue to decrease until death occurs. Aldosterone stimulates the reabsorption of Na^+. When aldosterone is secreted in maximal amounts, all of the sodium delivered to the distal convoluted tubules is reabsorbed. Under these conditions, urine contains no Na^+ at all, although some will be lost in sweat.

About 90% of the filtered K^+ is reabsorbed in the early regions of the nephron (mainly in the proximal convoluted tubule). When aldosterone is absent, all of the filtered K^+ that remains is reabsorbed in the distal convoluted tubule. In the absence of aldosterone, therefore, no K^+ is excreted in the urine. The presence of aldosterone stimulates the secretion of K^+ from the peritubular blood into the last portions of the distal convoluted tubule and cortical portion of the collecting duct (fig. 19.16).

The body cannot rid itself of excess K^+ in the absence of aldosterone-stimulated secretion of K^+ into the distal convoluted tubules. Indeed, when both adrenal glands are removed from an experimental animal, the **hyperkalemia** (high blood K^+) that results can produce fatal cardiac arrhythmias. Abnormally low blood plasma K^+ concentrations, as might result from excessive aldosterone secretion, can also produce arrhythmias, as well as muscle weakness.

hyperkalemia: Gk. *hyper,* above; *kalium,* potassium

Regulation of Aldosterone Secretion

Since aldosterone promotes Na^+ retention and K^+ loss, one might assume (on the basis of negative feedback) that aldosterone secretion would be increased when there was

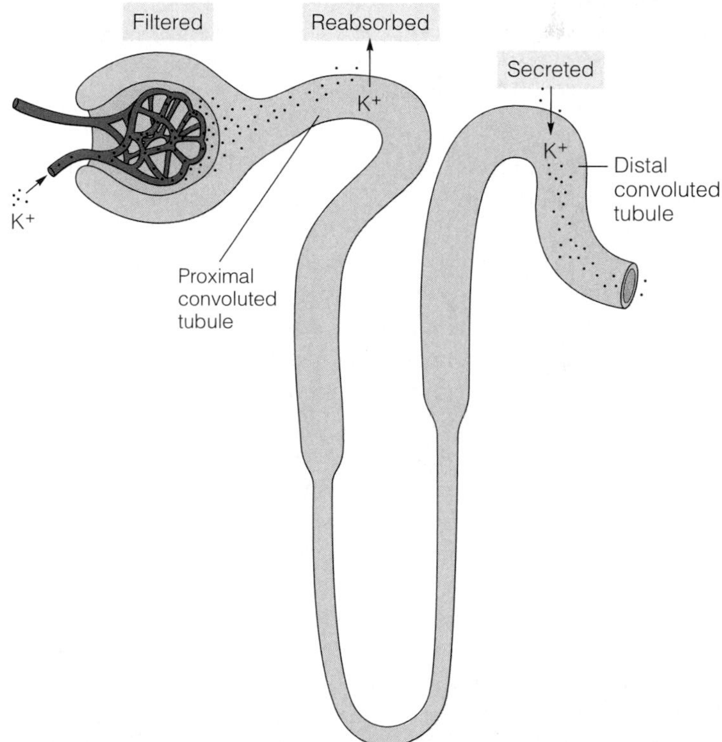

FIGURE 19.16
Potassium is almost completely reabsorbed in the proximal convoluted tubule, but under aldosterone stimulation it is secreted into the distal convoluted tubule. All of the K^+ in urine is derived from secretion rather than from filtration.

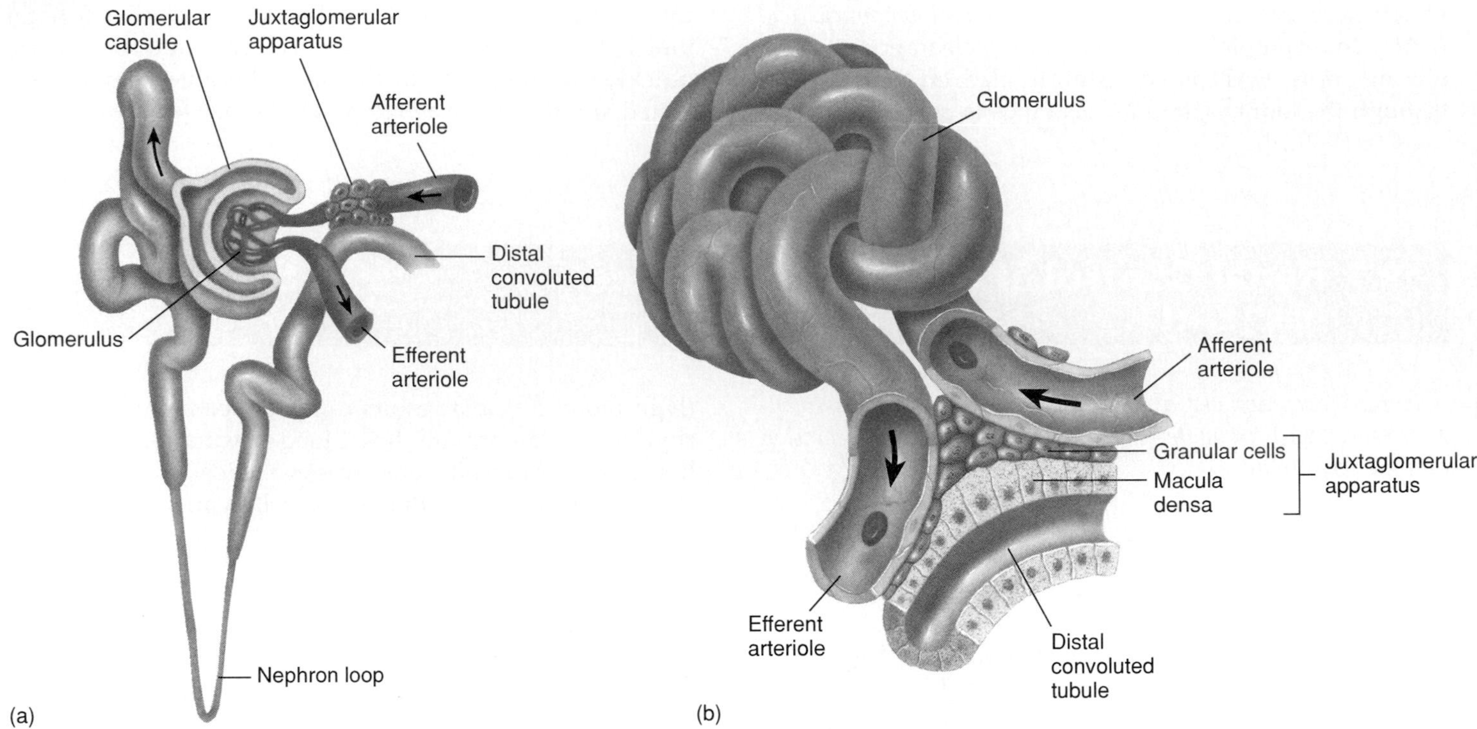

FIGURE 19.17

The juxtaglomerular apparatus (*a*) includes the region of contact of the afferent arteriole with the distal convoluted tubule. The afferent arterioles in this region contain granular cells that secrete renin; the distal convoluted cells in contact with the granular cells form an area called the macula densa (*b*).

a low Na^+ or a high K^+ concentration in the blood. This is indeed true. A rise in blood K^+ *directly* stimulates the secretion of aldosterone from the adrenal cortex. A decrease in plasma Na^+ concentration also promotes aldosterone secretion, but it does so indirectly, by means of a structure called the juxtaglomerular apparatus.

The **juxtaglomerular** (*juk″stă-glo-mer′yŭ-lar*) **apparatus** is the region in each nephron where the afferent arteriole and distal convoluted tubule come into contact (fig. 19.17). The microscopic appearance of the afferent arteriole and distal convoluted tubule in this small region differs from their appearances in other regions. *Granular cells* within the afferent arteriole secrete the enzyme **renin** into the blood; this enzyme catalyzes the conversion of *angiotensinogen* (a protein) into *angiotensin I* (a 10-amino-acid polypeptide).

Secretion of renin into the blood thus results in the formation of angiotensin I, which is then converted to **angiotensin II** by *angiotensin converting enzyme* (*ACE*) as blood passes through the lungs and other organs. Angiotensin II stimulates the adrenal cortex to secrete aldosterone, so that salt and water are retained by the kidneys. This **renin-angiotensin-aldosterone system** is activated by a fall in blood volume, as illustrated in figure 19.18.

Expansion of the blood volume, conversely, causes increased salt and water excretion in the urine. This is due in part to an inhibition of aldosterone secretion. There is also strong evidence that it is also partly due to the increased secretion of another substance with hormone properties. This substance, called **atrial natriuretic** (*na″trĭ-yoo-ret′ik*) **factor,** is secreted by the atria of the heart.

Drugs that act as inhibitors of angiotensin converting enzyme are known as ACE inhibitors. *These drugs reduce the production of angiotensin II, and thereby reduce the secretion of aldosterone. As a result, less salt and water are retained in the body and more are excreted in the urine. This lowers blood volume and blood pressure, and so ACE inhibitors are used to treat people with hypertension (high blood pressure) and heart disease. Angiotensin II is also a potent vasoconstrictor, so reducing its production by using an ACE inhibitor also helps to lower blood pressure through that means.*

natriuretic: L. *natrium,* sodium; *urina,* urine

FIGURE 19.18
The sequence of events by which a low sodium (salt) intake leads to increased sodium reabsorption by the kidneys. The dashed arrow and negative sign show the completion of the negative feedback loop.

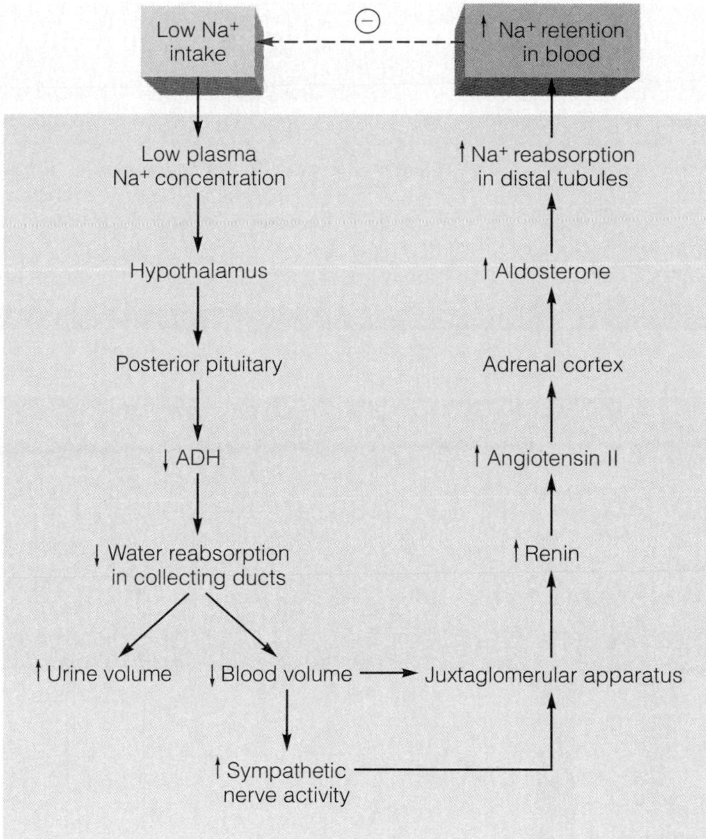

19.8 Renal Control of Acid-Base Balance: H^+ and HCO_3^-

The kidneys help to regulate the acid-base balance of the blood by excreting H^+ and reabsorbing HCO_3^-. Through these processes, the kidneys regulate the metabolic component of acid-base balance.

Acid-base balance involves both *respiratory* and *metabolic components.* As described in chapter 18 (see module 18.11), the respiratory component refers to the effect of ventilation on arterial P_{CO_2}, which in turn affects the production of carbonic acid (H_2CO_3). The metabolic component refers to the effect of nonvolatile metabolic acids—lactic acid, fatty acids, and ketone bodies—on blood pH. Since these acids are normally buffered by *bicarbonate* (HCO_3^-), the metabolic component can be described in terms of the free HCO_3^- concentration. An increase in metabolic acids is associated with a fall in plasma HCO_3^- concentrations, since the free bicarbonate is "used up" as it is converted to H_2CO_3. A decrease in metabolic acids, conversely, is associated with a rise in free HCO_3^-.

A normal blood pH (7.35–7.45) is maintained when there is a particular ratio of the plasma P_{CO_2} to the plasma HCO_3^- concentration. Respiratory acidosis or respiratory alkalosis occurs when the P_{CO_2} and the H_2CO_3 concentrations are altered. Metabolic acidosis or alkalosis occurs when the HCO_3^- concentration is abnormal. This classification is summarized in table 19.4.

Both vomiting *and* diarrhea *can lead to disturbances in acid-base balance. The loss of gastric acid from the body that occurs when vomiting is excessive produces a metabolic alkalosis. With diarrhea, there is a loss of bicarbonate, which is released into the intestine from the pancreas. Prolonged diarrhea can produce a metabolic acidosis.*

The kidneys help to regulate the blood pH by excreting H^+ in the urine and by reabsorbing bicarbonate.

TABLE 19.4 **Classification of Metabolic and Respiratory Components of Acidosis and Alkalosis**

P_{CO_2}	HCO_3^-	Condition	Causes
Normal	Low	Metabolic acidosis	Increased production of "nonvolatile" acids (lactic acid, ketone bodies, and others), or loss of HCO_3^- in diarrhea
Normal	High	Metabolic alkalosis	Vomiting of gastric acid; hypokalemia; excessive steroid administration
Low	Low	Respiratory alkalosis	Hyperventilation
High	High	Respiratory acidosis	Hypoventilation

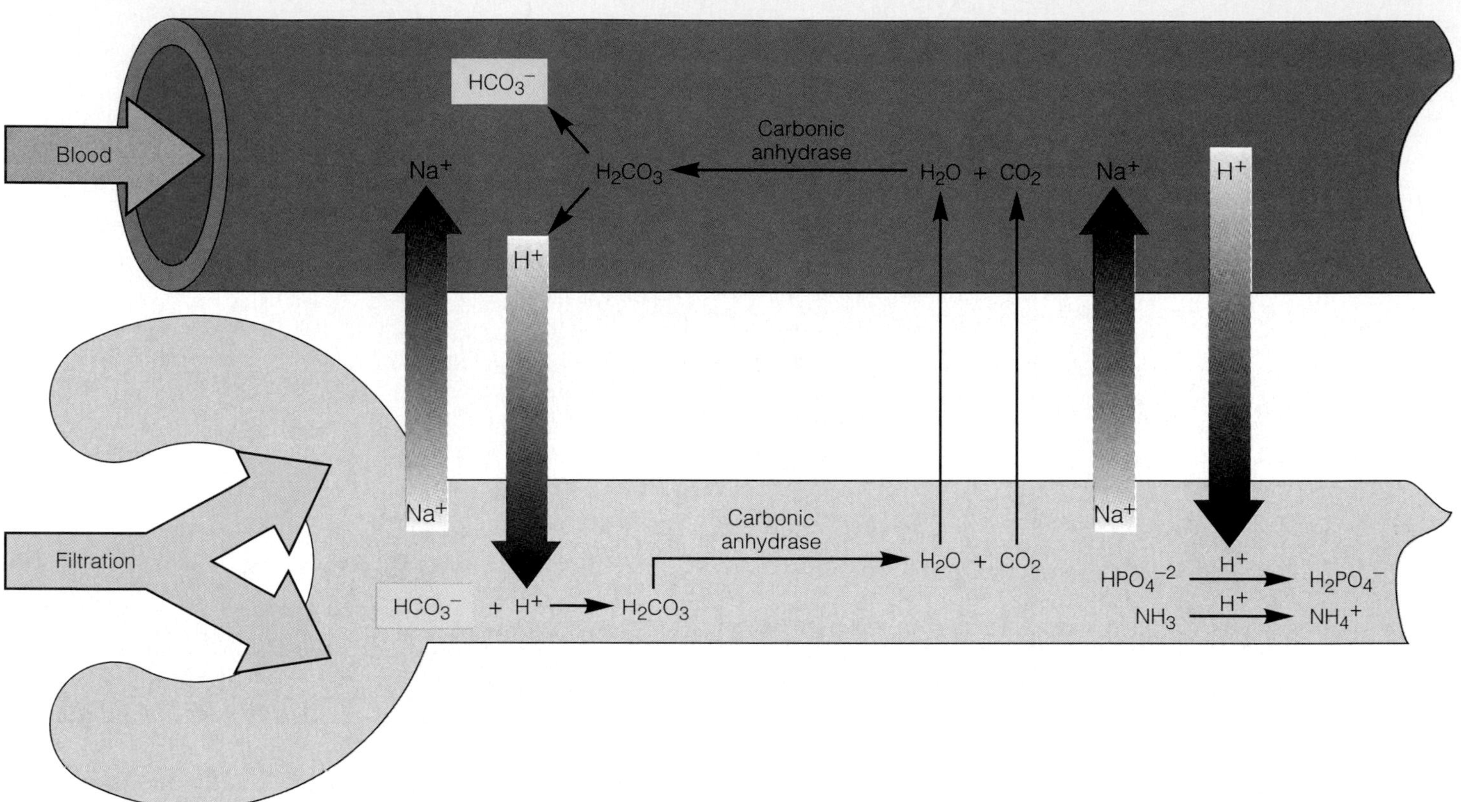

FIGURE 19.19
A diagram summarizing how the urine becomes acidified and how bicarbonate is reabsorbed from the filtrate.

Normal urine, therefore, is free of bicarbonate and is slightly acidic (with a pH range between 5 and 7).

The reabsorption of HCO_3^- (bicarbonate) from the filtrate to the blood is indirect. The HCO_3^- is first combined with H^+ in the filtrate to produce H_2CO_3. Then, in a reaction catalyzed by the enzyme *carbonic anhydrase,* the H_2CO_3 is changed into CO_2 and H_2O. These are reabsorbed into the blood, where they are changed back (by carbonic anhydrase within red blood cells) into H_2CO_3. This then dissociates into HCO_3^-, which travels in the blood plasma, and H^+. The H^+ may then be secreted back into the proximal tubule, producing a circular chain of events that results in the return of HCO_3^- to the blood without the return of a corresponding amount of H^+ (fig. 19.19).

Because they can reabsorb bicarbonate and excrete H^+, the kidneys are responsible for defending the body against metabolic acidosis. The kidneys also defend the body against metabolic alkalosis by excreting the excess bicarbonate under these conditions. As mentioned in chapter 18 (see module 18.11), the lungs defend the body

TABLE 19.5 Categories of Disturbances in Acid-Base Balance, Including Those That Involve Both Respiratory and Metabolic Components

P_{CO_2} (mmHg)	Bicarbonate (mEq/L)*		
	Less than 21	21–26	More than 26
More than 45	Combined metabolic and respiratory acidosis	Respiratory acidosis	Metabolic alkalosis and respiratory acidosis
35–45	Metabolic acidosis	Normal	Metabolic alkalosis
Less than 35	Metabolic acidosis and respiratory alkalosis	Respiratory alkalosis	Combined metabolic and respiratory alkalosis

*mEq/L = milliequivalents per liter. This is the millimolar concentration of HCO_3^- multiplied by its valence (×1).

against respiratory acidosis and alkalosis by regulating the concentration of plasma CO_2. When the kidneys or lungs fail in these defenses, the values for plasma HCO_3^- or plasma CO_2, or both, may become abnormal. Abnormal values, and the conditions of acidosis or alkalosis associated with them, are indicated in table 19.5.

When a person has a blood pH of less than 7.35 (acidosis), the urine pH almost always falls below 5.5. The nephron, however, cannot produce a urine pH significantly less than 4.5. In order for more H^+ to be excreted, the acid must be buffered. Actually, most of the H^+ excreted even in normal urine is in a buffered form. Bicarbonate cannot serve this function because it is normally completely reabsorbed. Instead, the buffering action of phosphates (mainly HPO_4^{-2}) and ammonia (NH_3) provide the means for excreting most of the H^+ in the urine (see fig. 19.19). Phosphate enters the urine by filtration. Ammonia (whose presence is strongly evident in a diaper pail or kitty litter box) is produced in the tubule cells from amino acids.

19.9 Ureters, Urinary Bladder, and Urethra

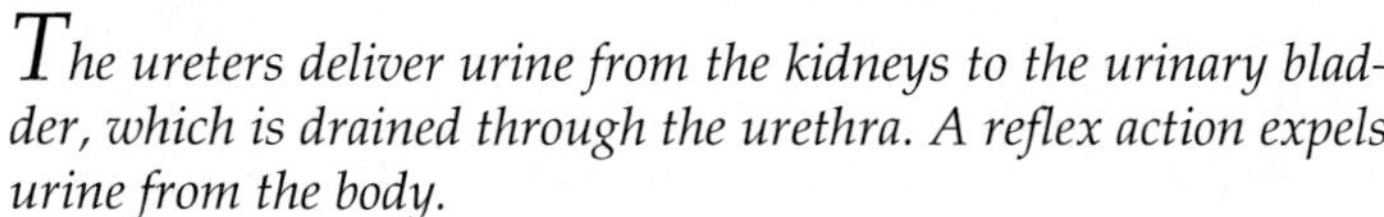

The ureters deliver urine from the kidneys to the urinary bladder, which is drained through the urethra. A reflex action expels urine from the body.

Ureters

The ureters (*yoo're'terz*), like the kidneys, are retroperitoneal. Each ureter is a tubular organ about 25 cm (10 in.) long that begins at the renal pelvis and courses interiorly to enter the urinary bladder.

The wall of the ureter consists of three layers, or tunics. The innermost layer, the **mucosa,** is composed of transitional epithelium (fig. 19.20). The middle layer, called the **muscularis,** consists of an inner longitudinal layer and an outer circular layer of smooth muscle. Muscular peristaltic waves, initiated by the presence of urine in the renal pelvis, move the urine through the ureter. The outermost layer of the ureter is called the **adventitia;** it is composed of loose fibrous connective tissue and protects the underlying layers.

A renal calculus, *or kidney stone (fig. 19.21), may obstruct the ureter and greatly increase the frequency of peristaltic waves in an attempt to pass through. The pain from a lodged calculus is extreme and extends throughout the pelvic area. A lodged calculus also causes a sympathetic reflex that constricts renal arterioles, reducing the production of urine in the kidney on the affected side.*

Urinary Bladder

The urinary bladder is a storage sac for urine. It lies just posterior to the symphysis pubis, anterior to the rectum. In females, the urinary bladder is in contact with the uterus and vagina. In males, the prostate is located below the urinary bladder (fig. 19.22).

calculus: L. *calculus,* small stone

The shape of the urinary bladder is determined by the volume of urine it contains. An empty urinary bladder is pyramidal; as it fills, it becomes ovoid and bulges upward into the abdominal cavity. The base of the urinary bladder receives the ureters, and the urethra exits at the inferior angle. The urethra is a tubular continuation of the neck of the urinary bladder.

The wall of the urinary bladder consists of four layers. The **mucosa** is the innermost layer, composed of transitional epithelium. The mucosa becomes thinner as the urinary bladder distends and the cells are stretched. Further distension is permitted by folds of the mucosa, called **rugae,** which can be seen when the urinary bladder is empty (fig. 19.22). The second layer of the urinary bladder, the **submucosa,** functions to support the mucosa. The **muscularis** consists of three interlaced smooth muscle layers and is referred to as the **detrusor muscle.** At the neck of the urinary bladder, the detrusor muscle is modified to form the upper of two muscular sphincters surrounding the urethra. The outer covering of the urinary bladder is the **serosa.** It appears only on the superior surface of the urinary bladder and is actually a continuation of the parietal peritoneum.

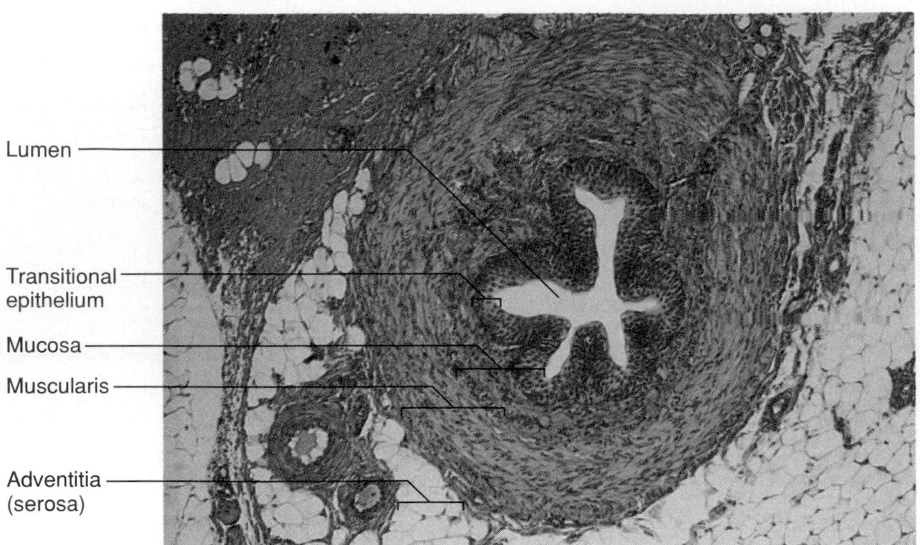

FIGURE 19.20
A photomicrograph of a ureter in transverse section. (Note the transitional epithelium of the mucosal layer.)

The urinary bladder becomes infected easily. Because a woman's urethra is so much shorter than a man's, women are particularly susceptible to these infections. A urinary bladder infection, called cystitis, *may easily ascend from the bladder to the ureters, since the mucous linings are continuous. An infection that involves the renal pelvis is called* pyelitis; *if it continues into the nephrons, it is known as* nephritis.

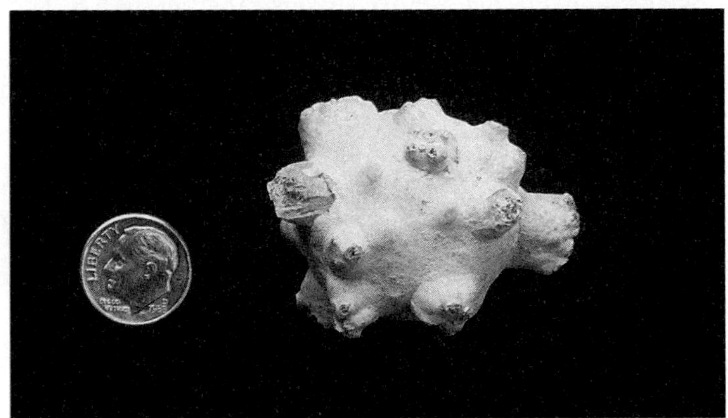

FIGURE 19.21
A calculus, or renal stone, positioned with a dime for size comparison. Factors contributing to stone formation may include the ingestion of excessive mineral salts, a decrease in water intake, and overactivity of the parathyroid glands. A renal stone generally consists of calcium oxalate, calcium phosphate, and uric acid crystals.

Urethra

The tubular urethra (*yoo-re'thră*) conveys urine from the urinary bladder to the outside of the body. Two muscular sphincters surround the urethra. The involuntary smooth muscle sphincter, the upper of the two, is the **internal urethral sphincter,** which is formed from the detrusor muscle of the urinary bladder. The lower sphincter, composed of voluntary skeletal muscle fibers, is called the **external urethral sphincter.**

The urethra of the female is a straight tube about 4 cm (1.5 in.) long that empties urine through the **urethral orifice** into the vestibule between the labia minora. The urethral orifice lies between the clitoris and the vaginal opening. It has a single function: to carry urine to the exterior.

The urethra of the male serves both the urinary and reproductive systems. It is about 20 cm (8 in.) long and S-shaped because of the shape of the penis. Three regions can be identified (fig. 19.22). The **prostatic urethra** is the proximal portion that passes through the prostate. The prostatic urethra receives drainage from small ducts of the prostate and two ejaculatory ducts of the reproductive system. The **membranous urethra** is the short portion of the urethra that passes through the urogenital diaphragm. The external urethral muscle is located in this region. The **spongy urethra** is the longest portion, extending from the outer edge of the urogenital diaphragm to the external urethral orifice on the glans penis.

detrusor: L. *detrudere,* thrust or forced down

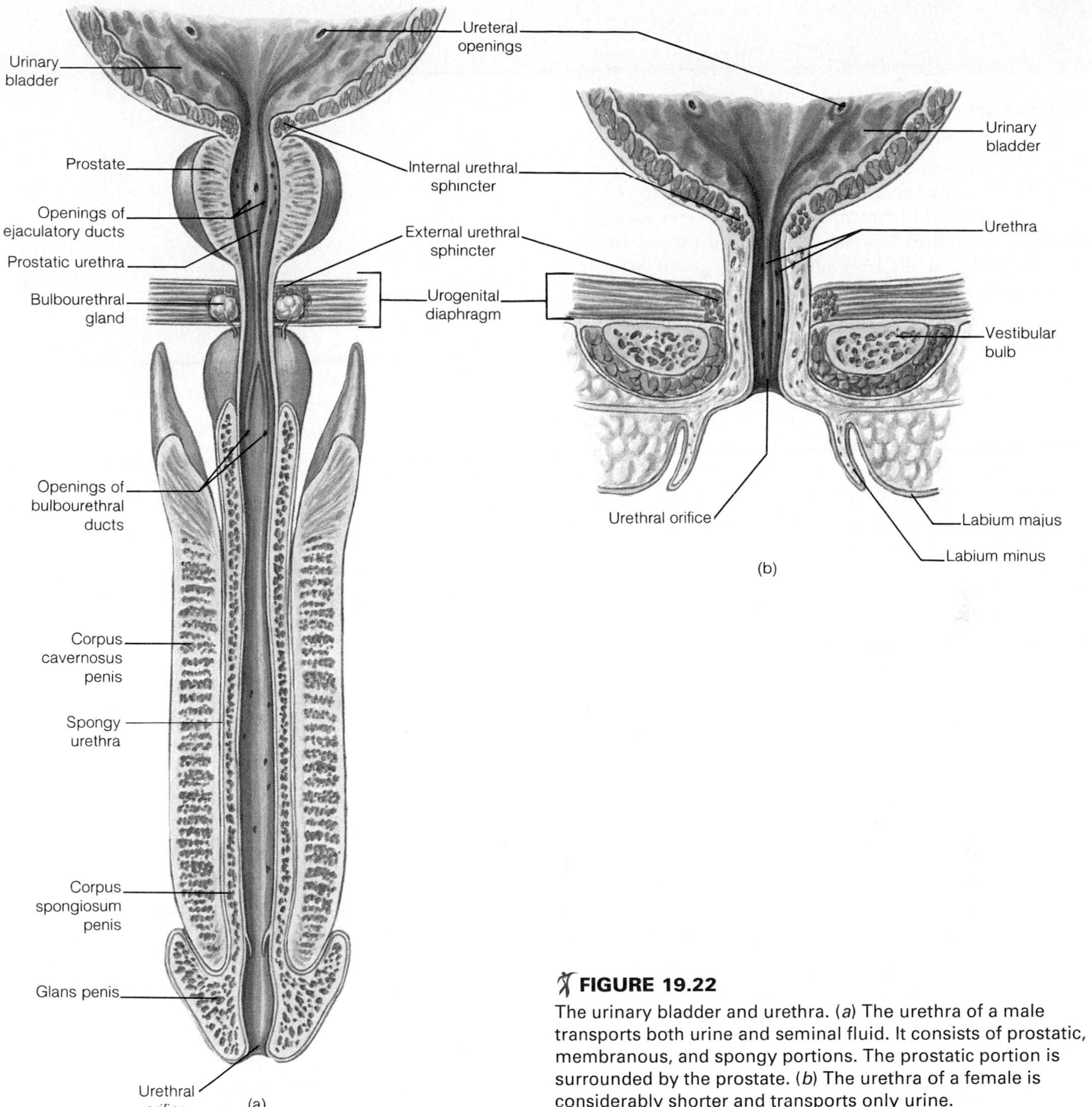

FIGURE 19.22
The urinary bladder and urethra. (*a*) The urethra of a male transports both urine and seminal fluid. It consists of prostatic, membranous, and spongy portions. The prostatic portion is surrounded by the prostate. (*b*) The urethra of a female is considerably shorter and transports only urine.

Micturition (*mik″tŭ-rish′un*), commonly called urination, is a reflex that expels urine from the urinary bladder. The average capacity of the urinary bladder is 700 to 800 ml, but a volume of 200 to 300 ml will distend the bladder enough to stimulate stretch receptors. The micturition reflex center is located in the second, third, and fourth sacral segments of the spinal cord. Following stimulation of this center by impulses arising from stretch receptors in the urinary bladder, parasympathetic nerves that stimulate the detrusor muscle and the internal urethral sphincter are activated. Stimulation of these muscles causes the urinary bladder wall to contract rhythmically and the internal sphincter to relax. At this point, a sensation of urgency is perceived in the brain, but there is still voluntary control over the external urethral sphincter. At the appropriate time, the conscious activity of the brain activates the motor nerve fibers to the external urethral sphincter, causing this sphincter to relax and urination to occur.

micturition: L. *micturire*, to urinate

Clinical Considerations

Diuretics

People who need to lower their blood volume because of hypertension, congestive heart failure, or edema take medications that increase the volume of urine excreted. Such medications are called **diuretics.** The various diuretics in clinical use act on the nephron in different ways. Based on their chemical structure or aspects of their actions, the commonly used diuretics are categorized as loop diuretics, thiazides, and potassium-sparing diuretics.

The most powerful diuretics act to inhibit active salt transport out of the ascending limb of the nephron loop. These loop diuretics include *furosemide* and *ethacrynic acid.* The thiazide diuretics, such as *hydrochlorothiazide,* inhibit salt and water reabsorption in the first segment of the distal convoluted tubule. These diuretics can result in excessive secretion of K^+ into the filtrate, and hence its excessive elimination in the urine. For this reason, potassium-sparing diuretics are sometimes used. *Spironolactones* are aldosterone antagonists that block the aldosterone stimulation of Na^+ reabsorption and K^+ secretion, preventing excessive K^+ loss.

Urinary Disorders

Normal micturition is painless. **Dysuria** (*dis-yur'e-ă*), or painful urination, is a sign of a bacterial infection or blockage in the urinary tract. **Hematuria** means blood in the urine, and is usually associated with trauma. **Oliguria** is an insufficient output of urine, whereas **polyuria** is an excessive output. Low blood pressure and kidney failure are two causes of oliguria. **Uremia** is a condition in which substances ordinarily excreted in the urine accumulate in the blood. **Enuresis** (*en"yŭ-re'sis*), or **incontinence,** is the inability to control micturition. It may be caused by psychological factors or by structural impairment.

An **intravenous pyelogram (IVP)** permits radiographs of the kidneys following the injection of radiopaque dye (see fig. 19.2). **Cystoscopy** (*sĭ-stos'kŏ-pe*) is the inspection of the inside of the urinary bladder by means of an instrument called a cystoscope (fig. 19.23). Using this technique, tissue samples can be obtained, as well as urine samples from each kidney prior to mixing in the urinary bladder.

Urinalysis is a simple but important laboratory aspect of a physical examination. The voided urine specimen is tested for color, specific gravity, chemical composition, and for the presence of microscopic bacteria, crystals, and casts.

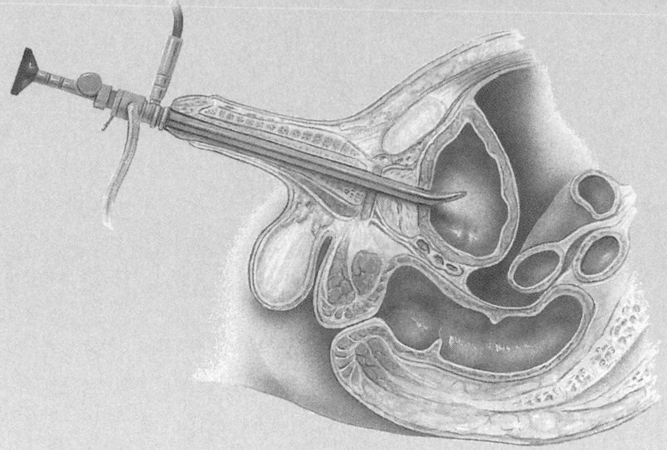

FIGURE 19.23
Cystoscopic examination of a male.

Inside Information—After urine is voided from the body, it is easily contaminated by bacteria. But at the moment it leaves a healthy body, it is sterile. In emergency situations, when no conventional disinfectant was available, fresh urine has been used as an antiseptic.

Infections of Urinary Organs

Urinary tract infections (UTIs) are a significant cause of illness and are also a major factor in the development of chronic renal failure. Females are more susceptible to urinary tract infections than are males because the urethra is shorter in females and the urethral and anal openings are closer together. The incidence of infection increases directly with sexual activity and aging in both sexes.

Infections of the urinary tract are named according to the infected organ. An infection of the urethra is called **urethritis,** and involvement of the urinary bladder is **cystitis.** Cystitis is frequently a secondary infection from some other part of the urinary tract. **Nephritis** is inflammation of the kidney tissue. **Glomerulonephritis** (*glo-mer"yŭ-lo-nĕ-fri'tis*) is inflammation of the glomeruli. Glomerulonephritis is frequently preceded by an upper respiratory tract infection because antibodies produced against streptococci bacteria can produce an autoimmune inflammation in the glomeruli. This inflammation may permanently change the glomeruli and figure significantly in the development of chronic renal diseases and renal failure.

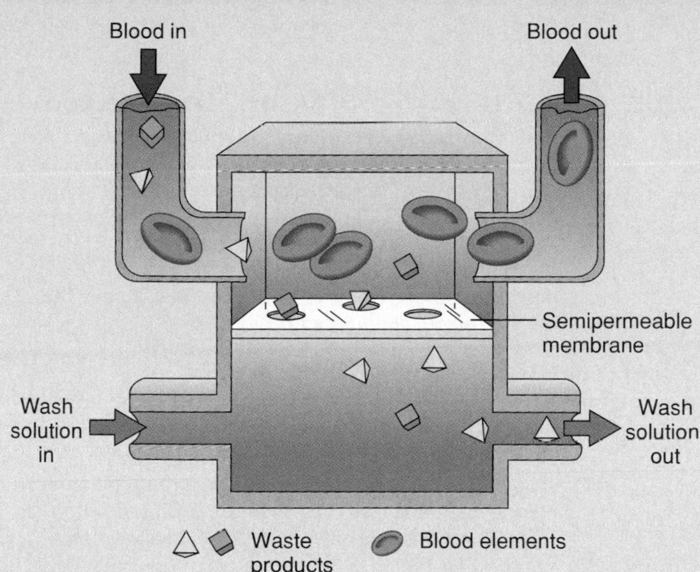

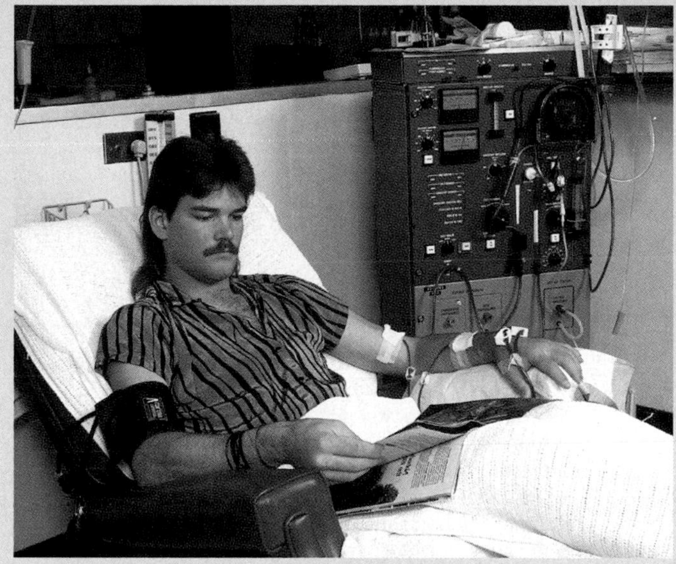

FIGURE 19.24
The hemodialysis process.

Any interference with the normal flow of urine, such as from a renal stone or an enlarged prostate in a male, causes stagnation of urine in the renal pelvis and the development of pyelitis. **Pyelitis** is an inflammation of the renal pelvis and its calyces. **Pyelonephritis** is a more extensive inflammation, involving the renal pelvis, the calyces, and the tubules of the nephron within one or both kidneys. Bacterial invasion from the blood or from the lower urinary tract is another cause of both pyelitis and pyelonephritis.

Renal Failure and Hemodialysis

An output of 50 to 60 cc of urine per hour is considered normal; an output of less than 30 cc per hour may indicate renal failure. *Renal failure* is the loss of the kidney's ability to maintain fluid and electrolyte balance and to excrete waste products. It can be either acute or chronic. **Acute renal failure** is the sudden loss of kidney function caused by shock and hemorrhage, thrombosis, or other physical trauma to the kidneys. The kidneys may sustain a 90% loss of their nephrons through tissue death and still not have an obvious loss of function. If a patient suffering acute renal failure is stabilized, the nephrons have an excellent capacity to regenerate.

A person with **chronic renal failure** cannot sustain life independently. Chronic renal failure is the end result of kidney disease in which the kidney tissue is progressively destroyed. As renal tissue continues to deteriorate, the only options for sustaining life are hemodialysis or kidney transplantation.

Hemodialysis equipment is designed to filter the wastes from the blood of a patient who has chronic renal failure. The patient's blood is pumped through a tube from the radial artery and passes through a machine, where it is cleansed and then returned to the body through a vein (fig. 19.24). The cleaning process involves pumping the blood past a semipermeable cellophane membrane that separates the blood from an isotonic solution containing molecules needed by the body (such as glucose). Waste products diffuse out of the blood through the membrane, while glucose and other molecules needed by the body remain in the blood.

More recent hemodialysis techniques include the use of the patient's own peritoneal membranes (which line the abdominal cavity) for filtering. Dialysis fluid is introduced into the peritoneal cavity, and then, after a period of time, discarded after wastes have accumulated. This procedure, called *continuous ambulatory peritoneal dialysis* (*CAPD*), can be performed several times a day by the patients themselves on an outpatient basis.

Patients with kidney failure who are on hemodialysis frequently suffer from anemia. This is due to lack of the hormone *erythropoietin*, secreted by normally functioning kidneys. Erythropoietin stimulates red blood cell production in the bone marrow (see chapter 14, module 14.3). Recombinant erythropoietin produced by genetic engineering techniques is administered to these patients.

SUMMARY

19.1 Urinary System and Kidney Structure

1. The urinary system consists of two kidneys, two ureters, the urinary bladder, and the urethra.
2. The urinary system regulates the extracellular environment of the body by forming urine.
3. The kidneys are retroperitoneal, embedded in a renal adipose capsule.
4. Each kidney consists of an outer vascular renal cortex and an inner renal medulla. The renal medulla is composed of renal pyramids.
5. The apex of each pyramid drains urine into calyces, which empty into the renal pelvis.
6. The functional units of the kidney are the nephrons, which consist of microscopic renal tubules and blood vessels.
7. Blood is delivered through afferent arterioles into capillary beds called glomeruli. The glomeruli produce a filtrate that enters the renal tubules.
8. Efferent arterioles carry blood from the glomeruli into peritubular capillaries that surround the renal tubules.

19.2 Nephron Tubules and Glomerular Filtration

1. The glomerular capsule receives the filtrate, which is passed into the proximal convoluted tubule, nephron loop, distal convoluted tubule, and collecting duct.
2. The glomerular filtrate contains all of the solutes in blood plasma, except the plasma proteins.
3. The glomerular filtration rate (GFR) is the volume of filtrate produced by both kidneys each minute.

19.3 Reabsorption in the Proximal Convoluted Tubule

1. Only about 1% of the glomerular filtrate is excreted as urine.
2. The proximal convoluted tubule actively transports Na^+ into the blood; Cl^- follows passively, and water follows the NaCl by osmosis.

19.4 Reabsorption in the Nephron Loop

1. The ascending limb of the nephron loop actively extrudes Na^+; Cl^- follows, but water remains within the renal tubule.
2. The descending limb of the nephron loop is permeable to water, so that water is drawn out by osmosis into the hypertonic renal medulla.
3. This makes the fluid within the descending limb more concentrated when it arrives at the bottom of the nephron loop.

19.5 Countercurrent Multiplier System

1. The interaction between the ascending and descending limbs of the nephron loops creates a very hypertonic renal medulla. This process is called the countercurrent multiplier system.
2. Antidiuretic hormone (ADH) increases the permeability of the collecting ducts to water, permitting more water to be reabsorbed across the wall of the collecting ducts.

19.6 Renal Plasma Clearance

1. Substances excreted in the urine are cleared from the blood. This clearance can occur by filtration and by secretion.
2. A substance that is reabsorbed has a renal plasma clearance that is less than the glomerular filtration rate. Glucose is normally completely reabsorbed.

19.7 Renal Control of Electrolyte Balance: Na^+ and K^+

1. Aldosterone stimulates the reabsorption of Na^+ in exchange for the secretion of K^+ in the distal convoluted tubule.
2. Aldosterone secretion is stimulated by angiotensin II, which is produced in response to the secretion of renin from the juxtaglomerular apparatus.

19.8 Renal Control of Acid-Base Balance: H^+ and HCO_3^-

1. The kidneys help to regulate acid-base balance by excreting H^+ and reabsorbing HCO_3^-.
2. Bicarbonate is reabsorbed indirectly via the carbonic anhydrase reaction. H^+ is excreted with ammonium and phosphate buffers.

19.9 Ureters, Urinary Bladder, and Urethra

1. The ureters drain the kidneys. The walls of the ureters consist of three layers: the mucosa, muscularis, and adventitia.
2. The urinary bladder stores urine. It is lined by a transitional epithelium that is folded into rugae to permit distension.
3. The male urethra conducts urine during urination and seminal fluid during ejaculation. The shorter female urethra conducts only urine.
4. Micturition is controlled by a parasympathetic reflex.

Review Activities

Objective Questions

1. Oxygenated blood enters the kidney through (*module 19.1*)
a. the hilum artery.
b. the afferent arterioles.
c. the renal artery.
d. the renal portal system.

2. Urine flowing from the papillary ducts enters directly into (*module 19.2*)
a. the renal calyces.
b. the ureter.
c. the renal pelvis.
d. the distal convoluted tubules.

3. About 65% of the glomerular filtrate is reabsorbed in (*module 19.3*)
a. the proximal convoluted tubule.
b. the distal convoluted tubule.
c. the nephron loop.
d. the collecting duct.

4. Which part of the nephron tubule is permeable to water and does not actively transport salt? (*module 19.4*)
a. proximal convoluted tubule
b. descending limb of the nephron
c. ascending limb of the nephron
d. distal convoluted tubule
e. collecting duct

5. Which part of the nephron tubule actively transports salt and is impermeable to water? (*module 19.4*)
a. proximal convoluted tubule
b. descending limb of the nephron
c. ascending limb of the nephron
d. distal convoluted tubule
e. collecting duct

6. Antidiuretic hormone promotes the retention of water by increasing (*module 19.5*)
a. the rate of active transport of water.
b. the rate of active transport of chloride.
c. the rate of active transport of sodium.
d. the permeability of the collecting duct to water.

7. Substance X has a clearance greater than zero but less than that of inulin. What can be concluded about substance X? (*module 19.6*)
a. It is not filtered.
b. It is filtered, but neither reabsorbed nor secreted.
c. It is filtered and partially reabsorbed.
d. It is filtered and secreted.

8. Aldosterone stimulates sodium reabsorption and potassium secretion in (*module 19.7*)
a. the proximal convoluted tubule.
b. the descending limb of the nephron.
c. the ascending limb of the nephron.
d. the distal convoluted tubule.
e. the collecting duct.

9. If a drug inhibits the kidneys' ability to reabsorb bicarbonate so that bicarbonate is excreted in the urine, the person will develop (*module 19.8*)
a. respiratory acidosis.
b. respiratory alkalosis.
c. metabolic acidosis.
d. metabolic alkalosis.

10. The detrusor muscle is located in (*module 19.9*)
a. the kidneys.
b. the ureters.
c. the urinary bladder.
d. the urethra.

Essay Questions

1. Describe the location of the kidneys and trace the flow of urine from the kidneys out of the body. (*module 19.1*)
2. Explain how glomerular filtrate is produced and why it has a low protein concentration. (*module 19.2*)
3. Explain how the proximal convoluted tubule reabsorbs salt and water. (*module 19.3*)
4. How do the descending and ascending limbs of the nephron loop interact with each other? What is the net result of this interaction? (*module 19.4*)
5. Explain how the countercurrent multiplier system works and discuss its functional significance. (*module 19.5*)
6. Explain why a person does not normally have glycosuria but may develop this condition by consuming too much sugar. (*module 19.6*)
7. Explain the roles of ADH and aldosterone in the renal regulation of fluid and electrolyte balance. (*module 19.7*)
8. Distinguish between the respiratory and metabolic components of acid-base balance. (*module 19.8*)
9. Describe the structure and location of the ureters and urinary bladder. How does the structure of the urethra differ in males and females? (*module 19.9*)
10. Describe the micturition reflex. (*module 19.9*)

Labeling Exercise

Label the structures indicated on the figure to the right.

1. ______________________
2. ______________________
3. ______________________
4. ______________________
5. ______________________
6. ______________________

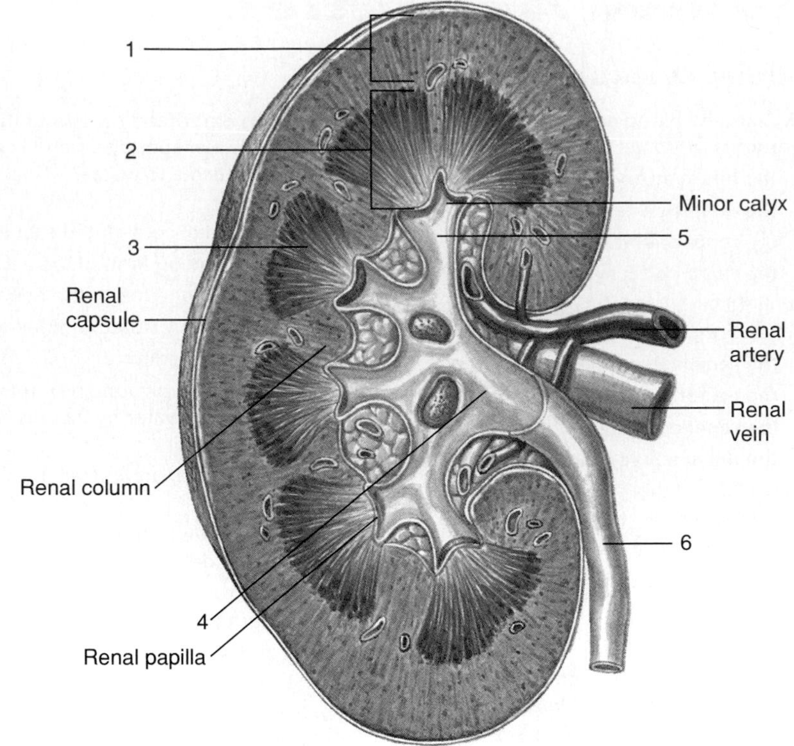

Critical Thinking Questions

1. Suppose you suddenly develop a wild craving for bananas and eat a half-dozen. Since bananas are rich in potassium, this will raise your blood potassium concentration. How does the body get rid of the extra potassium? Why are people who take diuretic drugs for their hypertension advised to eat bananas or take potassium supplements?
2. A person with kidney disease may develop hypertension. What are two different mechanisms by which abnormal kidney function might produce hypertension? What are two different ways that drugs could act via the kidneys to reduce hypertension?
3. Young women who are anorexic may become so thin that their kidneys droop, a condition called *renal ptosis*. What is the anatomical basis of this condition?
4. A friend complains that he is always thirsty, and is producing large quantities of very dilute urine. How would you explain these symptoms?
5. Two people with the same degree of body hydration each reach for a drink on a hot day. One drinks a can of lemonade, the other a can of beer. How will each drink affect the body's state of hydration? Which person will be better hydrated 2 hours later? Explain.

CHAPTER 20

"Food requires about 24 hours to be processed as it travels the length of the GI tract." *page 504*

Digestive System

Chapter Outline

Terms to Remember

gastrointestinal tract, 504
peritoneum (*per"ĭ-tŏ-ne'um*), 507
mucosa (*myoo-ko'să*), 509
teeth, 513
salivary glands, 514
pharynx (*far'ingks*), 514
esophagus (*ĕ-sof'ă-gus*), 515
stomach, 515
chyme (*kīm*), 516
gastric juice, 517
duodenum (*doo"ŏ-de'num*), 519
jejunum (*jĕ-joo'num*), 519
ileum (*il'e-um*), 519
villi (*vil'e*), 519
peristalsis (*per"ĭ-stal'sis*), 521
cecum (*se'kum*), 522
colon (*ko'lon*), 523
liver, 524
bile, 525
gallbladder, 527
pancreas, 528
pancreatic juice, 529
carbohydrate, 531
enzyme, 531
protein, 531
lipid, 532
emulsification, 532

Learning Objectives

After studying this chapter, you should be able to

- describe the activities of the digestive system and list its structures and regions.
- explain how the serous membranes are arranged within the abdominal cavity.
- describe the generalized structure of the four tunics (layers) that compose the wall of the GI tract.
- describe the anatomy of the oral cavity and the structure of a typical tooth.
- describe the structure of the stomach and list the cell types of the gastric mucosa, including their secretions.
- describe the structure and function of intestinal villi and crypts, and explain the nature and significance of brush border enzymes.
- describe the structure of the liver and gallbladder, and discuss their various functions.
- describe the structure of the pancreas and explain the role of pancreatic juice in digestion.
- explain how carbohydrates, proteins, and lipids are digested.
- describe the mechanisms that regulate stomach function and the secretion of pancreatic juice and bile.

20.1 Introduction to the Digestive System

The organs of the digestive system are specialized for the digestion and absorption of food. The digestive system consists of a tubular gastrointestinal tract and accessory digestive organs.

Food is basic to our survival. It provides us with the essential nutrients the body cannot produce for itself. These nutrients are required for chemical reactions involving synthesis of enzymes, cellular division and growth, repair, and the production of heat energy. Without an adequate food supply, the body will use its own tissues as fuel to keep itself going. In periods of prolonged fasting, it will even start breaking down muscle in the heart to feed itself.

But even if we do eat three square meals a day, most of the food is not yet suitable for use by the cells. First, it must be mechanically and chemically reduced to forms that can be absorbed through the intestinal wall and transported to the cells by the blood. Technically, ingested food is not inside the body until it is absorbed; and, in fact, a large portion of consumed food remains undigested and passes through the body as waste material.

The principal function of the digestive system—preparing food for cellular utilization—involves the following activities:

1. **Ingestion**—taking food into the mouth.
2. **Mastication**—chewing the food to pulverize it and mix it with saliva.
3. **Deglutition**—swallowing food to move it from the mouth to the pharynx and into the esophagus.
4. **Peristalsis**—rhythmic, wavelike contractions that propel food through the gastrointestinal tract.
5. **Digestion**—the mechanical and chemical breakdown of food material to prepare it for absorption.
6. **Absorption**—the passage of molecules of food through the mucous membrane of the small intestine and into the blood or lymph for distribution to cells.
7. **Defecation**—the discharge of indigestible wastes, called feces, from the gastrointestinal tract.

Anatomically and functionally, the digestive system can be divided into a tubular **gastrointestinal tract (GI tract),** or *digestive tract,* and **accessory digestive organs.** The GI tract, which extends from mouth to anus, is a continuous tube approximately 9 m (30 ft) long. Piercing through the diaphragm, the GI tract traverses both the thoracic and abdominopelvic cavities.

The organs of the GI tract include the **oral cavity** (*mouth*), **pharynx, esophagus, stomach, small intestine,** and **large intestine** (fig. 20.1). The accessory digestive organs include the **teeth, tongue, salivary glands, liver, gallbladder,** and **pancreas.** The term *viscera* is frequently used to refer to the abdominal organs of digestion, but it can be broadly applied in reference to any of the organs of the thoracic and abdominal cavities. *Gut* is the anatomical term for the developing stomach and intestines in the embryo.

Food usually requires between 24 hours to 48 hours to be processed as it travels the length of the GI tract. Most of the organic molecules of food are similar to the molecules that form the structure of human tissue. These are generally large molecules, called *polymers,* which are composed of subunits, called *monomers.* Within the GI tract, polymers are digested into their monomers by *hydrolysis reactions* (fig. 20.2). Once formed, the monomers are absorbed across the innermost layer of the small intestine into the blood and lymph. Each region of the GI tract has specific functions in preparing food for utilization (table 20.1).

Although there is an abundance of food in the United States, so many people are malnourished that eating patterns have become a critical public health concern. Obesity *is a major health problem. Grossly overweight people are at greater risk for cardiovascular disease, hypertension, osteoarthritis, and diabetes mellitus. People with good nutritional habits are better able to withstand trauma, are less likely to get sick, and are usually less seriously ill when they do become sick.*

Inside Information—According to the National Center for Health Statistics, at the current rate of increase, all Americans will be overweight by the year 2059.

ingestion: L. *ingerere,* carry in
mastication: Gk. *mastichan,* gnash the teeth
deglutition: L. *deglutire,* swallow down
peristalsis: Gk. *peri,* around; *stellein,* compress
polymer: Gk. *poly,* many; *meros,* part
monomer: Gk. *monos,* single; *meros,* part
hydrolysis: Gk. *hydor,* water; *lysis,* break

TABLE 20.1 Regions of the GI Tract and Basic Functions

Region	Function
Oral cavity	Ingests food; receives saliva; grinds food and mixes it with saliva (mastication); initiates digestion of carbohydrates; forms and swallows soft mass of chewed food called bolus (deglutition)
Pharynx	Receives bolus from oral cavity; autonomically continues deglutition of bolus to esophagus
Esophagus	Transports bolus to stomach by peristalsis; lower esophageal sphincter restricts backflow of food
Stomach	Receives bolus from esophagus; churns bolus with gastric juice; initiates digestion of proteins; carries out limited absorption; moves mixture of partly digested food and secretions (chyme) into duodenum and prohibits backflow of chyme; regurgitates when necessary
Small intestine	Receives chyme from stomach and secretions from liver and pancreas; chemically and mechanically breaks down chyme; absorbs nutrients; transports wastes through peristalsis to large intestine; prohibits backflow of intestinal wastes from large intestine
Large intestine	Receives undigested wastes from small intestine; absorbs water and electrolytes; forms, stores, and expels feces through defecation reflex

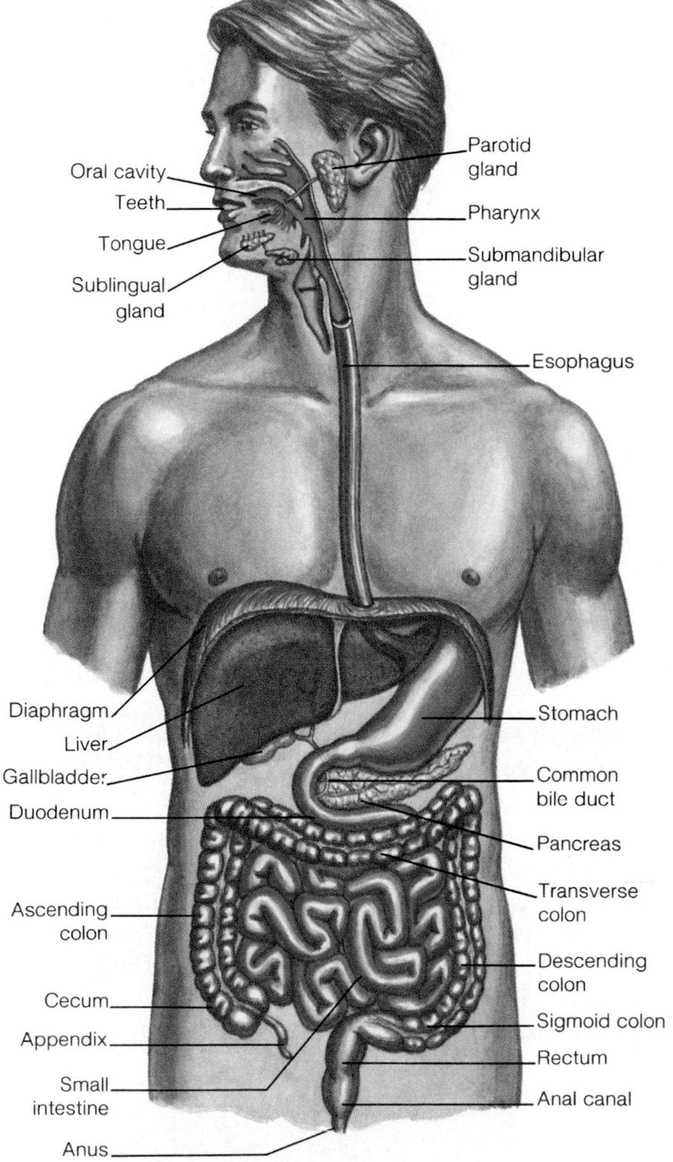

FIGURE 20.1
The digestive system, including the gastrointestinal tract and accessory digestive organs.

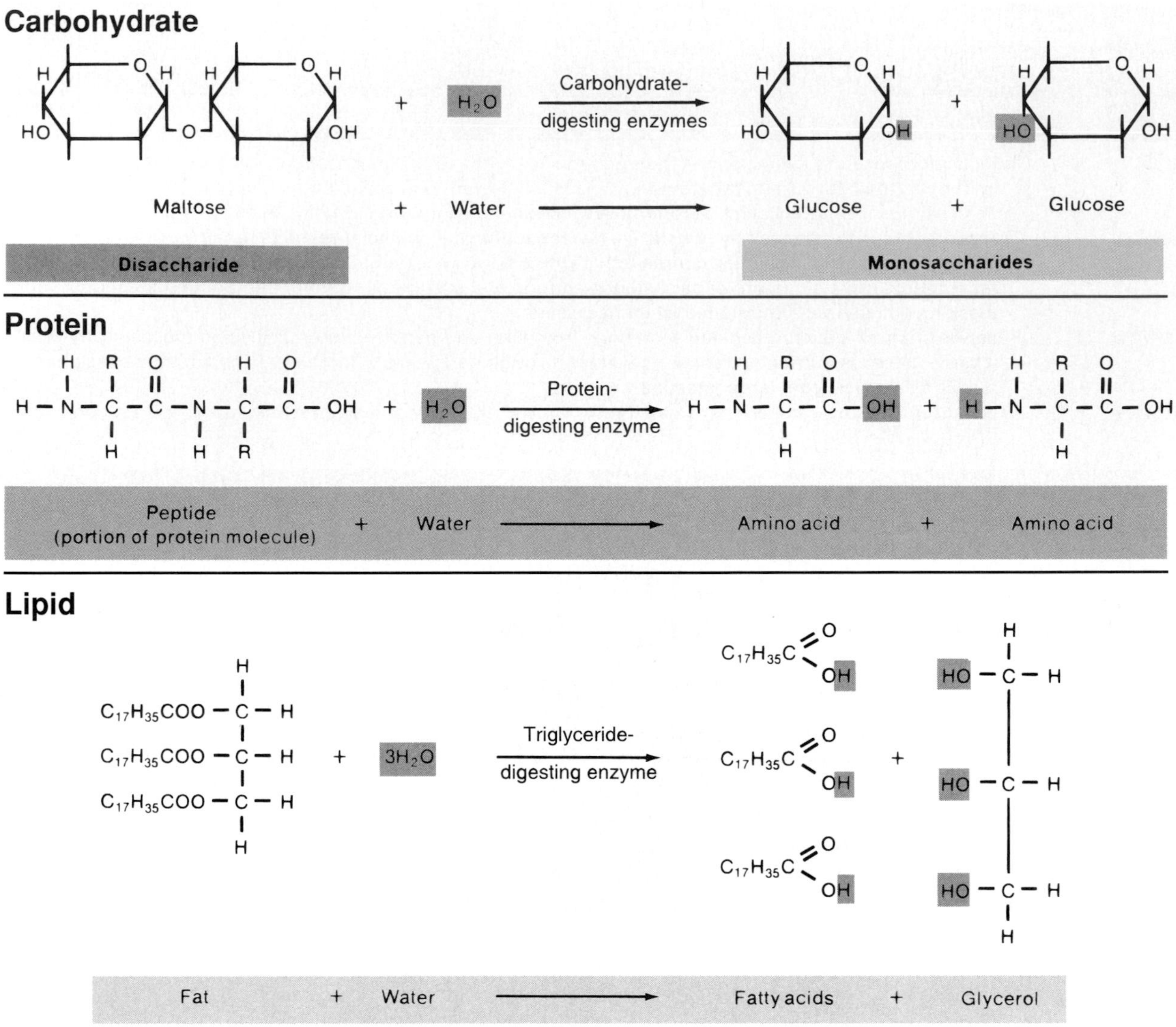

Carbohydrate

Maltose + Water $\xrightarrow{\text{Carbohydrate-digesting enzymes}}$ Glucose + Glucose

Disaccharide → Monosaccharides

Protein

$$H-N-\underset{H}{\overset{H\ R}{C}}-\overset{O}{\overset{\|}{C}}-\underset{H}{N}-\underset{R}{\overset{H}{C}}-\overset{O}{\overset{\|}{C}}-OH + H_2O \xrightarrow{\text{Protein-digesting enzyme}} H-N-\underset{H}{\overset{H\ R}{C}}-\overset{O}{\overset{\|}{C}}-OH + H-N-\underset{H}{\overset{H\ R}{C}}-\overset{O}{\overset{\|}{C}}-OH$$

Peptide (portion of protein molecule) + Water → Amino acid + Amino acid

Lipid

$$\begin{array}{l} C_{17}H_{35}COO-CH_2 \\ C_{17}H_{35}COO-CH \\ C_{17}H_{35}COO-CH_2 \end{array} + 3H_2O \xrightarrow{\text{Triglyceride-digesting enzyme}} 3\ C_{17}H_{35}COOH + \begin{array}{l} HO-CH_2 \\ HO-CH \\ HO-CH_2 \end{array}$$

Fat + Water → Fatty acids + Glycerol

FIGURE 20.2
The digestion of food molecules occurs by means of hydrolysis reactions.

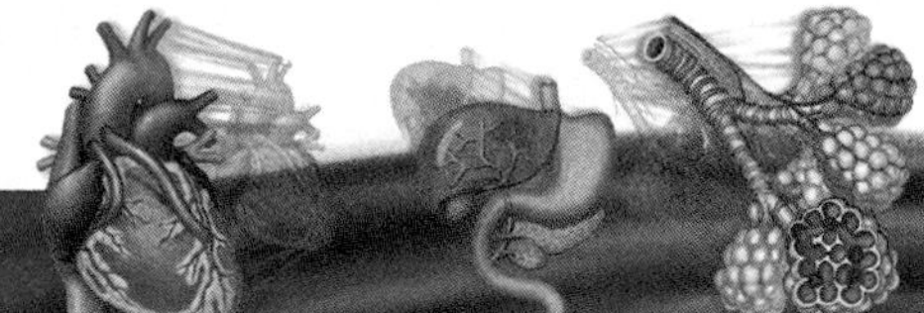

20.2 Serous Membranes of the GI Tract

Protective and lubricating serous membranes line the abdominal cavity and cover the visceral organs. The mesentery is a specialized serous membrane that supports the GI tract and provides a structure through which nerves and vessels pass.

Most of the visceral organs are positioned within the abdominopelvic cavity. These organs are supported and covered by **serous membranes** that line body cavities and cover the organs that lie within these cavities (fig. 20.3). Serous membranes secrete a lubricating serous fluid that continuously moistens the cavity containing the abdominal viscera. The *parietal portion* of the serous membrane lines the body wall, and the *visceral portion* covers the internal organs (fig. 20.4). The serous membranes of the

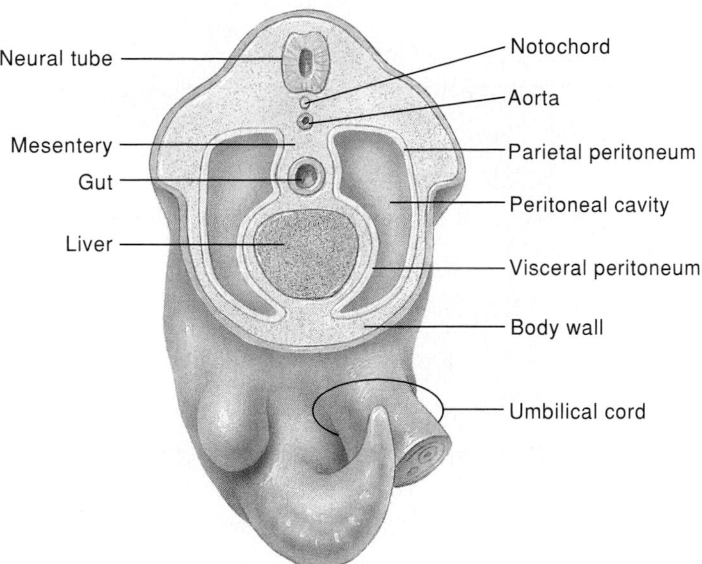

FIGURE 20.3
A diagram of the abdominal serous membranes.

abdominal cavity are called **peritoneal membranes,** or **peritoneum** (*per″ĭ-tŏ-ne′um*). The peritoneum is composed of simple squamous epithelium, portions of which are reinforced with connective tissue.

The **parietal peritoneum** lines the wall of the abdominal cavity. Along the posterior abdominal cavity, the sheets of parietal peritoneum come together to form a double-layered peritoneal fold called the **mesentery** (see fig. 20.3). The mesentery supports the GI tract, at the same time allowing the small intestine freedom for peristaltic movement. It also provides a structure for the passage of intestinal nerves and vessels. The **mesocolon** is the specific portion of the mesentery that supports the large intestine.

The peritoneal covering extends around the intestinal viscera as the **visceral peritoneum.** The **peritoneal cavity** is the space between the parietal and visceral portions of the peritoneum.

Peritonitis *is a bacterial inflammation of the peritoneum. It may be caused by trauma, rupture of a visceral organ, ectopic pregnancy (see fig. 24.20), or postoperative complications. Peritonitis is usually extremely painful and serious. Treatment usually involves the injection of massive doses of antibiotics and perhaps peritoneal intubation (insertion of a tube) to permit drainage.*

Extensions of the parietal peritoneum into the peritoneal cavity serve other functions. The **falciform** (*fal′sĭ-form*) **ligament,** a serous membrane reinforced with connective tissue, attaches the liver to the diaphragm and anterior abdominal wall. The **greater omentum** (fig. 20.4*a*) extends from the greater curvature of the stomach to the transverse colon, forming an apronlike structure over most of the small intestine. Functions of the greater omentum include storing fat, cushioning visceral organs, supporting lymph nodes, and protecting against the spread of infections. In cases of localized inflammation, such as appendicitis, the greater omentum may actually compartmentalize the inflamed area, sealing it off from the rest of the peritoneal cavity. The **lesser omentum** (fig. 20.4*b*) passes from the lesser curvature of the stomach and the upper duodenum to the inferior surface of the liver.

In short, there are two cavities within the abdomen. The *abdominopelvic cavity* is the space inside the muscular body wall. Most, but not all, of the visceral organs in the abdominopelvic cavity project into the *peritoneal cavity,* which is the space to the inside of the parietal peritoneum. Because the parietal peritoneum is attached to the inside of the body wall, it would seem as though the abdominopelvic cavity and the peritoneal cavity were one and the same. They are defined as separate cavities because a few of the abdominal viscera lie posterior to the parietal peritoneum. These *retroperitoneal organs* are thus within the abdominopelvic cavity but not within the peritoneal cavity. They include most of the pancreas, the kidneys, the adrenal glands, a portion of the duodenum, and the abdominal aorta.

The peritoneal cavity provides a warm, moist, normally aseptic environment for the abdominal viscera. In a male, the peritoneal cavity is totally closed off from the outside body environment. In a female, however, it is not isolated from the outside, which presents the potential for contamination through the entry of microorganisms. A fairly common gynecological condition is pelvic inflammatory disease (PID), *which results from the entry of pathogens into the peritoneal cavity at the sites of the open-ended uterine (fallopian) tubes.*

peritoneum: Gk. *peritonaion,* stretched over
mesentery: Gk. *mesos,* middle; *enteron,* intestine
falciform: L. *falcis,* sickle; *forma,* form
omentum: L. *omentum,* apron

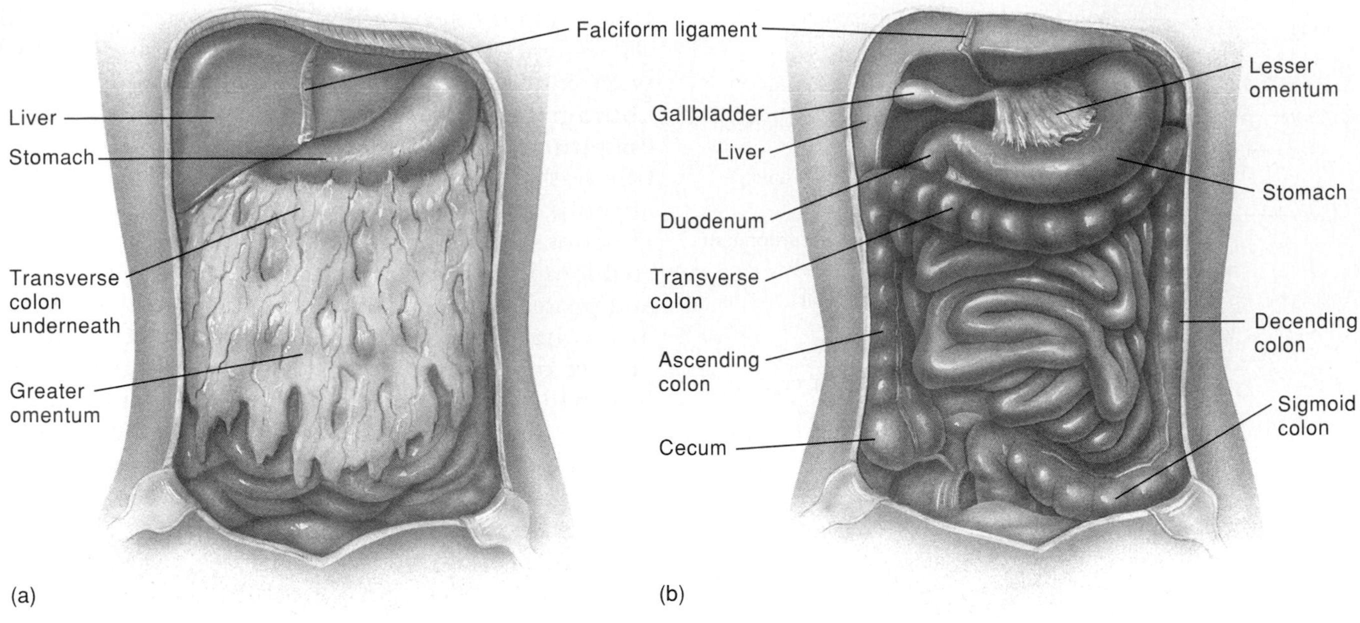

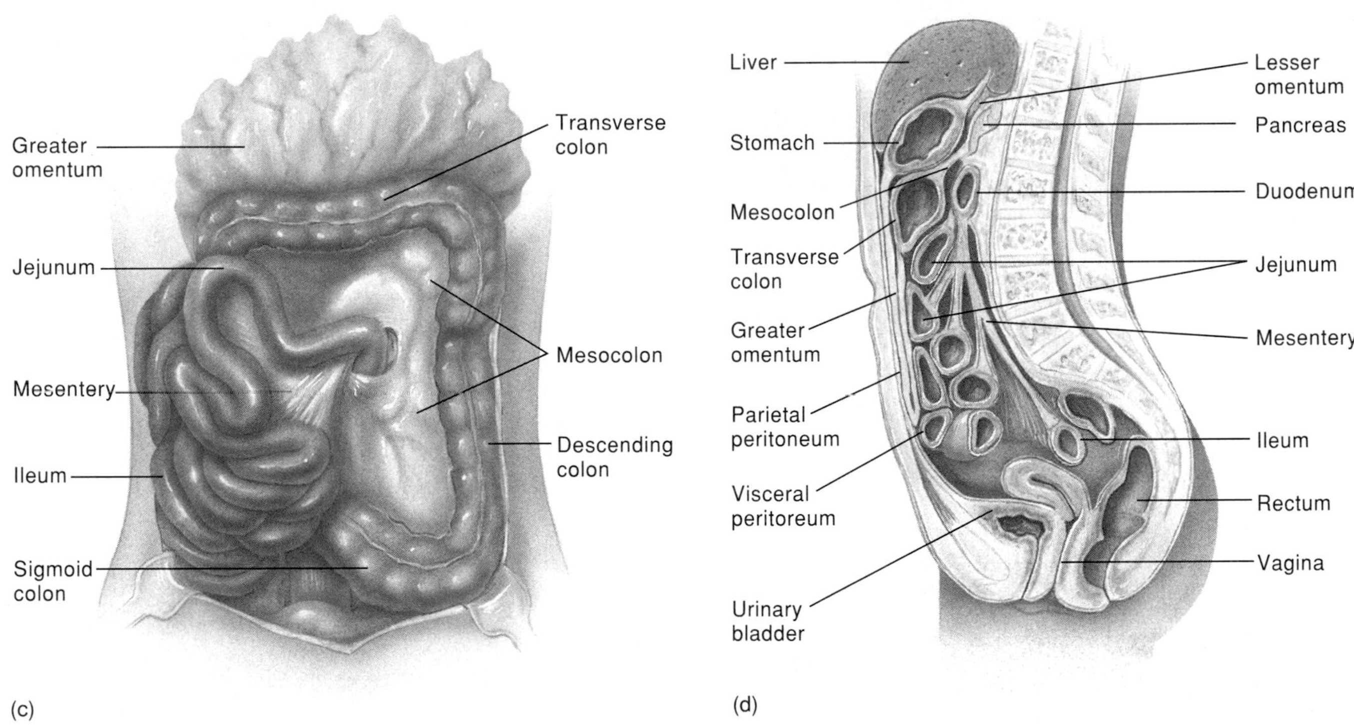

FIGURE 20.4

The structural arrangement of the abdominal organs and peritoneal membranes within the peritoneal cavity. (*a*) The greater omentum, (*b*) the lesser omentum with the liver lifted, (*c*) the mesentery with the greater omentum lifted, and (*d*) the relationship of the peritoneal membranes to the visceral organs as shown in a sagittal view.

20.3 Layers of the GI Tract

Each of the four tunics (mucosa, submucosa, muscularis, and serosa) of the GI tract contains a dominant tissue type and performs specific functions in the digestive process.

The GI tract from the esophagus to the anal canal is composed of four layers, or **tunics.** From the inside out, these are the mucosa (*myoo-ko'să*), submucosa, muscularis, and serosa (fig. 20.5).

Mucosa

The mucosa, which lines the lumen of the GI tract, is both an absorptive and a secretory layer. It consists of a simple columnar epithelium supported by the **lamina propria** (fig. 20.6), a thin layer of connective tissue. The lamina propria contains numerous lymph nodules, which are important in protecting against disease. External to the lamina propria is a thin smooth muscle layer called the **muscularis mucosa.** This is the muscle layer that throws portions of the GI tract into numerous small folds, which greatly increases the absorptive surface area. Specialized **goblet cells** in the mucosa throughout most of the GI tract secrete mucus (fig. 20.7).

Submucosa

The relatively thick submucosa is a highly vascular layer of connective tissue serving the mucosa. Absorbed molecules that pass through the columnar epithelial cells of the mucosa enter into blood vessels of the submucosa. In addition to blood vessels, the submucosa contains glands and nerve plexuses. The *submucosal plexus* (*Meissner's plexus*) (see fig. 20.5*b*) provides an autonomic nerve supply to the muscularis mucosa.

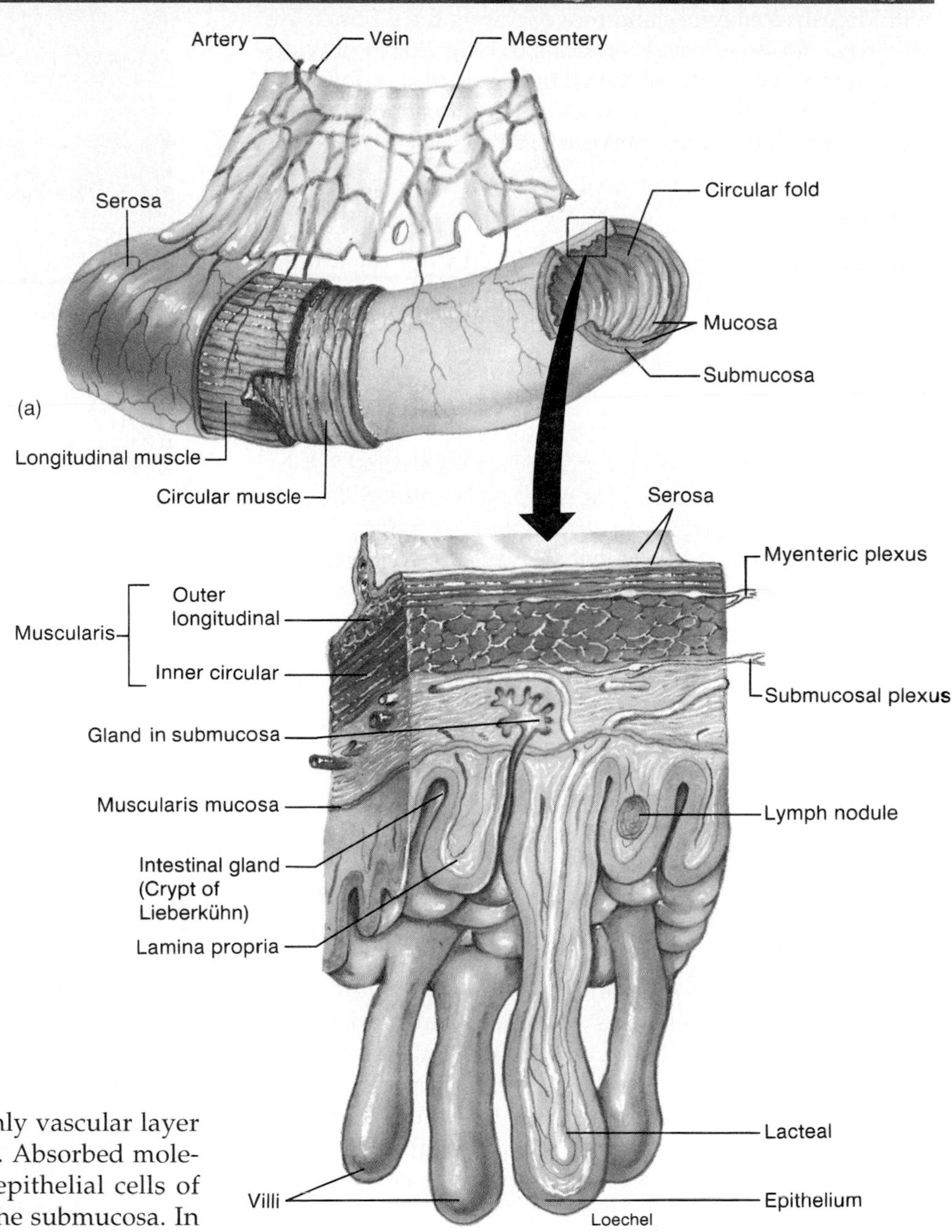

FIGURE 20.5
The tunics (layers) of the GI tract. (*a*) A section of the small intestine with each of the four tunics exposed, and (*b*) a section showing the detailed structure and relative thickness of each tunic. (Note the location of the exocrine gland and the innervation of the small intestine.)

---
tunic: L. *tunica*, covering or coat

Meissner's plexus: from Georg Meissner, German histologist, 1829–1905

Muscularis

The muscularis is responsible for the intestinal contractions and peristaltic movement through the GI tract. The muscularis has an inner circular and an outer longitudinal layer of smooth muscle. Contractions of these layers churn the food with digestive enzymes and move it along the GI tract. The *myenteric plexus* (*Auerbach's plexus*), (see fig. 20.5*b*) provides the major nerve supply to the GI tract; it includes fibers and ganglia from both the sympathetic and parasympathetic divisions of the autonomic nervous system.

Serosa

The outer serosa completes the wall of the GI tract. It is a binding and protective layer consisting of loose connective tissue covered with a layer of simple squamous epithelium. The simple squamous epithelium is actually the visceral peritoneum.

The body has several defenses against ingested material that may be harmful if absorbed. The acidic environment of the stomach and the lymphatic system kill many harmful bacteria. The mucous lining throughout the GI tract serves as a protective layer. Vomiting, and in certain cases diarrhea, are reactions to substances that irritate the GI tract. Vomiting is a reflexive response to many toxic chemicals, and as such can be beneficial even though unpleasant.

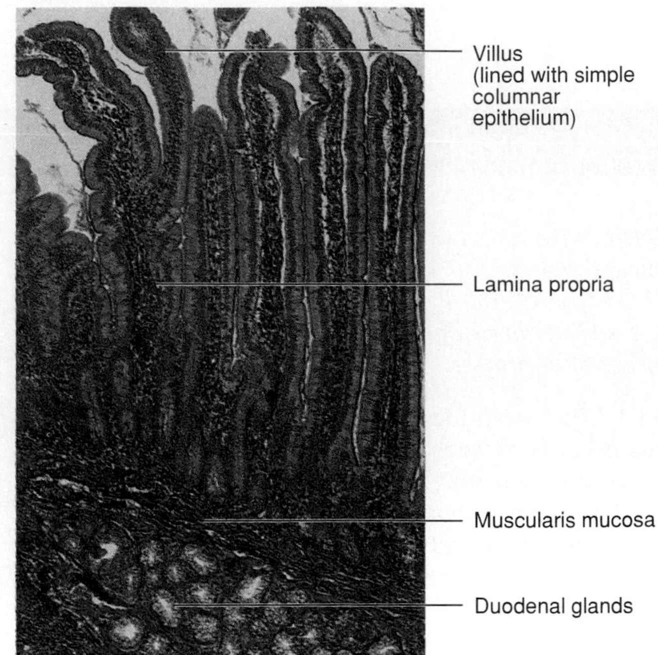

FIGURE 20.6
Histology of the duodenum of the small intestine.

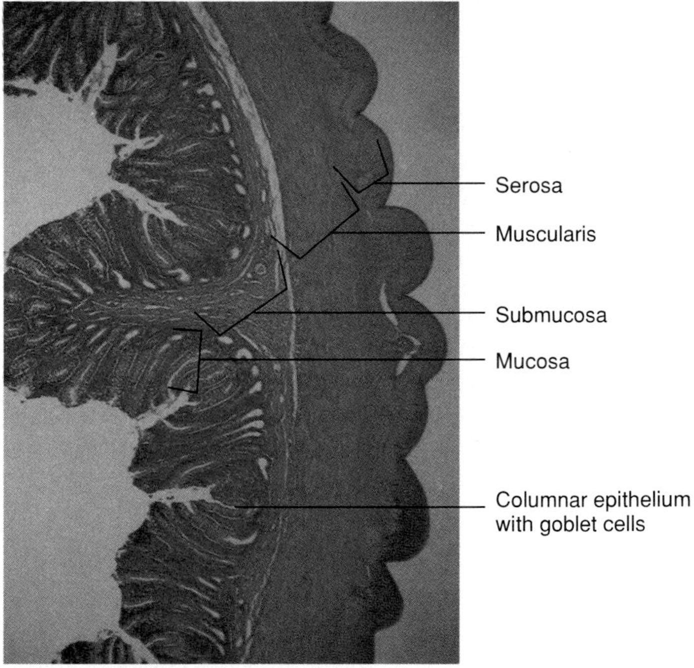

FIGURE 20.7
The histology of the large intestine.

Innervation of the GI Tract

The GI tract is innervated by the sympathetic and parasympathetic divisions of the autonomic nervous system. Impulses through the paired vagus nerves are the source of parasympathetic activity in the esophagus, stomach, pancreas, gallbladder, small intestine, and upper portion of the large intestine. The lower portion of the large intestine receives parasympathetic innervation from spinal nerves in the sacral region. The submucosal plexus and myenteric plexus are the sites where preganglionic fibers synapse with postganglionic fibers that innervate the smooth muscle of the GI tract. Stimulation of the parasympathetic fibers promotes peristalsis and increases secretory activity in the GI tract.

Postganglionic sympathetic fibers pass through the submucosal and myenteric plexuses and innervate the GI tract. The effects of sympathetic nerve stimulation are antagonistic to those of parasympathetic nerve stimulation. Sympathetic nerve effects inhibit peristalsis, reduce secretions, and constrict sphincters along the GI tract.

Auerbach's plexus: from Leopold Auerbach, German anatomist, 1828–97

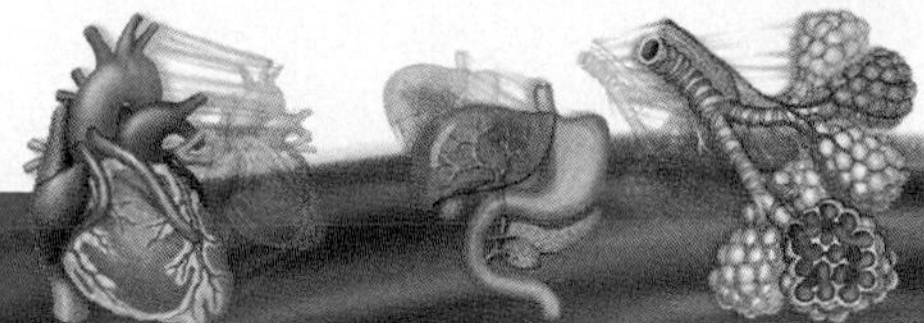

20.4 Mouth, Pharynx, and Associated Structures

Ingested food is changed into a bolus by the mechanical action of teeth and by the chemical activity of saliva. The bolus is swallowed in the process of deglutition.

The functions of the **mouth** and associated structures are to form a receptacle for food, to initiate mechanical digestion through mastication, to swallow food, and to form words in speech. The mouth can also serve the respiratory system in breathing air. The **pharynx,** or throat, which is posterior to the mouth, serves as a common passageway for both the respiratory and digestive systems. Both the mouth and pharynx are lined with nonkeratinized stratified squamous epithelium, which is constantly moistened by the secretion of saliva. The mouth is referred to as the **oral cavity** (fig. 20.8). It is formed by the **cheeks, lips, hard palate** and **soft palate,** and **tongue.** The **vestibule of the oral cavity** is the depression between the cheeks and lips externally and the gums and teeth internally (fig. 20.9). The opening of the oral cavity is referred to as the **oral orifice,** and the opening between the oral cavity and the pharynx is called the **fauces.**

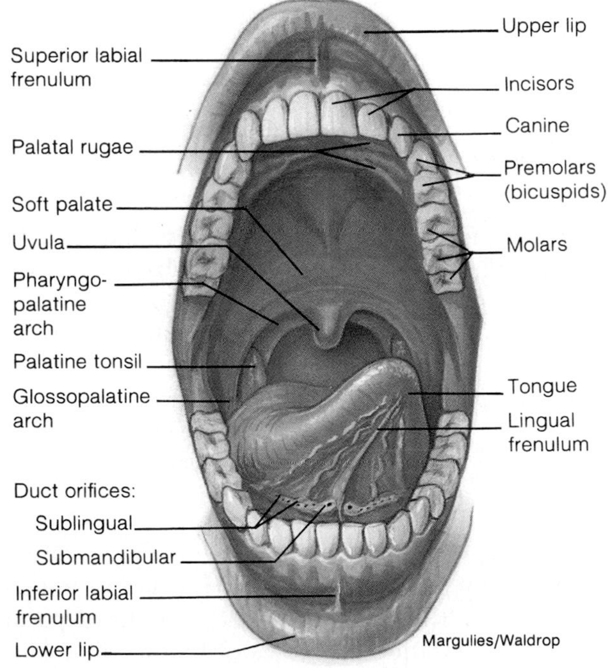

FIGURE 20.8
The superficial structures of the oral cavity.

Cheeks and Lips

The **cheeks** form the lateral walls of the oral cavity. They consist of outer layers of skin, subcutaneous fat, facial muscles that assist in manipulating food, and inner linings of moistened stratified squamous epithelium. The anterior portion of the cheeks terminates in the superior and inferior lips that surround the oral orifice.

The **lips** are fleshy, highly mobile organs whose principal function in humans is associated with speech. Lips also serve for suckling and manipulating food. Each lip is attached from its inner surface to the gum by a midline fold of mucous membrane called the **labial frenulum** (see fig. 20.8). The lips are formed from the orbicularis oris muscle (see fig. 8.19*a*) and associated connective tissue, and they are covered with soft, pliable skin. Between the outer skin and the mucous membrane of the oral cavity is a highly vascular, reddish-brown transition zone called the **vermilion** (fig. 20.9).

Tongue

As a digestive organ, the tongue functions to move food around in the mouth during mastication and to assist in swallowing food. It is also essential in producing speech. The tongue is a mass of skeletal muscle covered with a mucous membrane. Extrinsic tongue muscles (those that insert upon the tongue) move the tongue from side to side and in and out. Only the anterior two-thirds of the tongue lies in the oral cavity; the remaining one-third lies in the pharynx (fig. 20.9) and is attached to the hyoid bone. Rounded masses of **lingual tonsils** are located on the surface of the base of the tongue (fig. 20.10). Anteriorly, the undersurface of the tongue is connected along the midline to the floor of the mouth by the vertically positioned **lingual frenulum.**

On the surface of the tongue are numerous small elevations called **papillae** (see chapter 12, module 12.2). The papillae give the tongue a distinct roughened surface that aids the handling of food. Some of them also contain **taste buds** that respond to sweet, salty, sour, and bitter chemical stimuli. Three types of papillae are present on the surface

pharynx: L. *pharynx,* throat
fauces: L. *fauces,* throat
vermilion: O.E. *vermeylion,* red colored
papilla: L. *papula,* little nipple

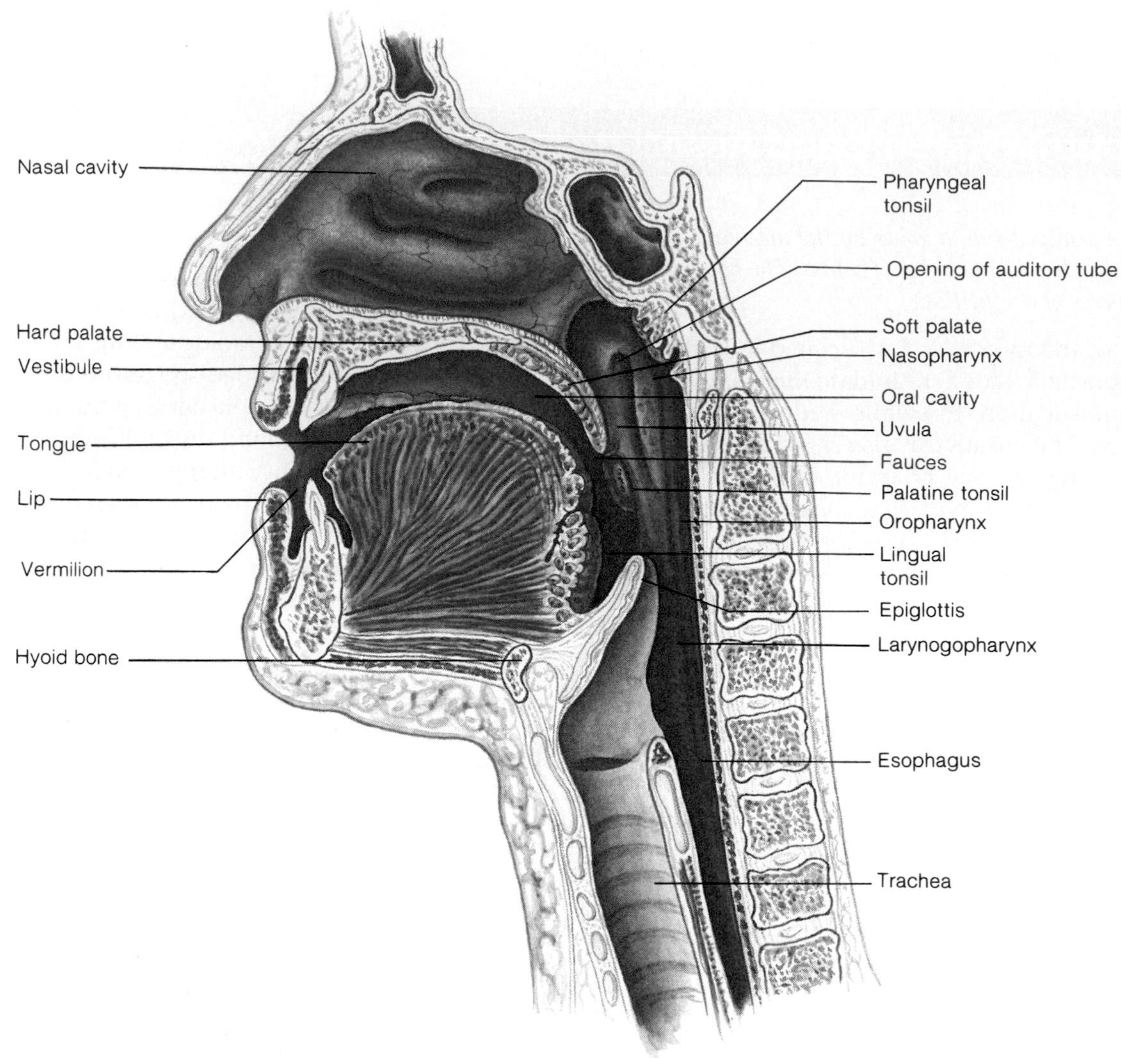

FIGURE 20.9
A sagittal section of the facial region showing the oral cavity, nasal cavity, and pharynx.

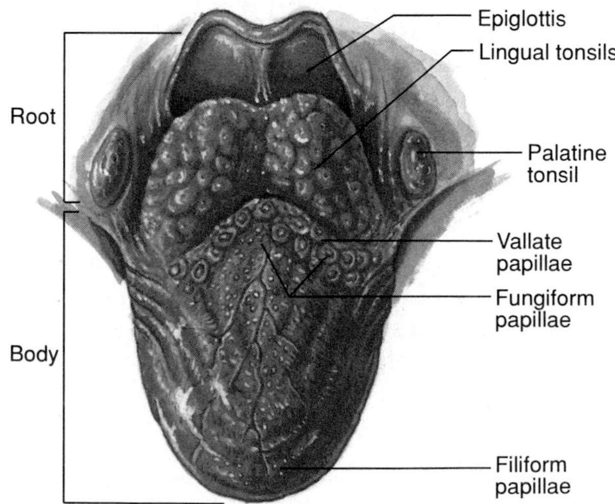

FIGURE 20.10
The surface of the tongue.

of the tongue: **filiform, fungiform,** and **vallate** (fig. 20.10). Filiform papillae are sensitive to touch, have tapered tips, and are by far the most numerous. These papillae lack taste buds and are not involved in the perception of taste. The larger, rounded fungiform papillae are scattered among the filiform type. The few vallate papillae are arranged in a V-shape on the posterior surface of the tongue.

Palate

The palate, which forms the roof of the oral cavity, consists of the bony hard palate anteriorly and the soft palate posteriorly (see figs. 20.8 and 20.9). The **hard palate** is formed by the palatine processes of the maxillae and the

---

filiform: L. *filum,* thread; *forma,* form
fungiform: L. *fungus,* fungus; *forma,* form
vallate: L. *vallatus,* surrounded with a rampart

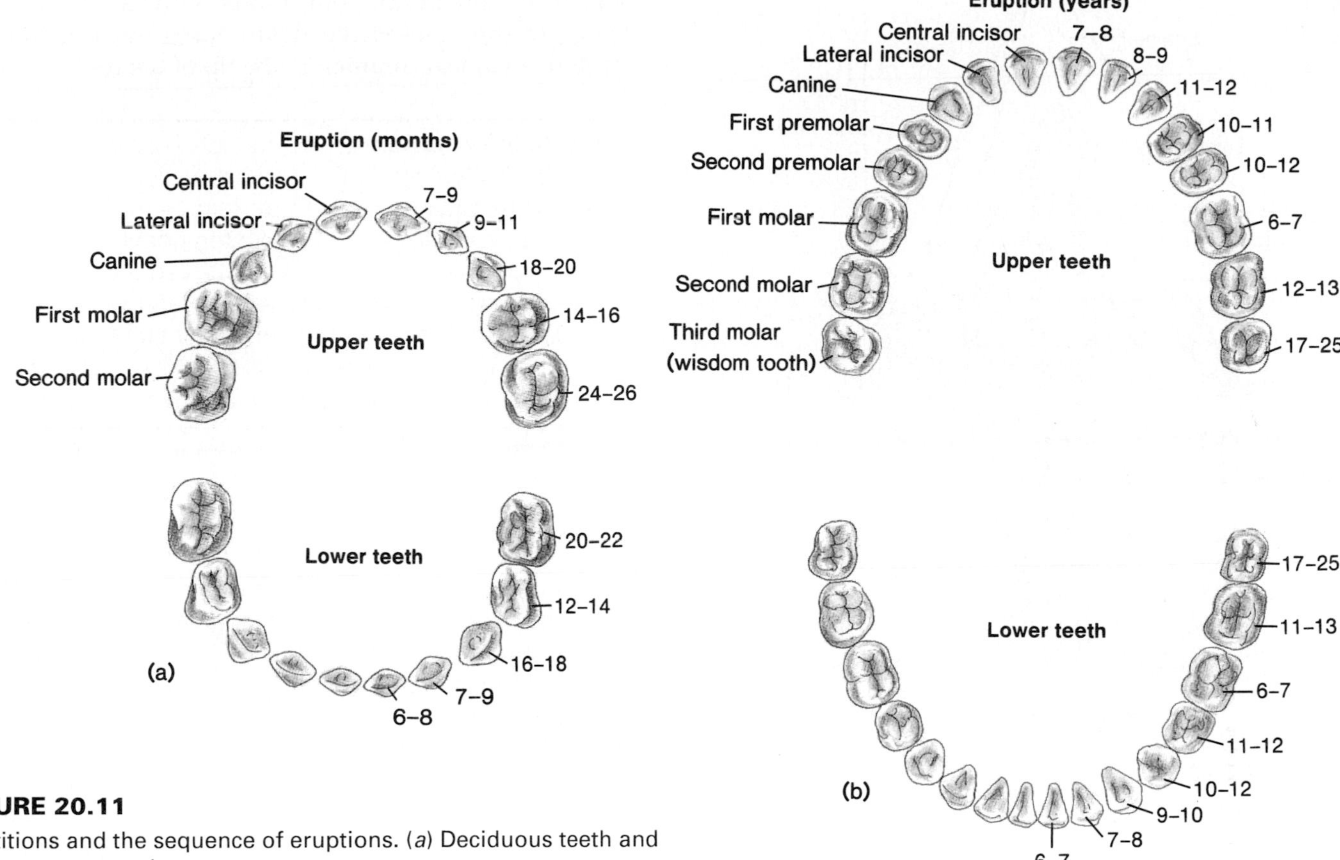

FIGURE 20.11
Dentitions and the sequence of eruptions. (*a*) Deciduous teeth and (*b*) permanent teeth.

horizontal plates of the palatine bones and is covered with a mucous membrane. Transverse ridges called **palatal rugae** (*roo'je*) are located along the mucous membrane of the hard palate. These structures serve as friction ridges against which the tongue is placed during swallowing. The **soft palate** is a muscular arch covered with mucous membrane; it is continuous anteriorly with the hard palate. Suspended from the middle lower border of the soft palate is a cone-shaped projection called the **uvula.** During swallowing, the soft palate and uvula are drawn upward, closing the nasopharynx and preventing food and fluid from entering the nasal cavity.

Two muscular folds extend downward from both sides of the base of the uvula (see fig. 20.9). The anterior fold is called the **glossopalatine arch,** and the posterior fold is the **pharyngopalatine arch.** Between these two arches is the **palatine tonsil.**

Teeth

Humans and other mammals have various types of teeth that are structurally adapted to handle food in particular ways (fig. 20.11). The four pairs (upper and lower jaws) of anteriormost teeth are the **incisors** (*in-si'sorz*). The chisel-shaped incisors are adapted for cutting and shearing food. The two pairs of cone-shaped **canines,** or **cuspids,** located at the anterior corners of the mouth, are adapted for holding and tearing. Incisors and canines are further characterized by a single root on each tooth. Located behind the canines are the **premolars,** or **bicuspids,** and the **molars.** These teeth have two or three roots, and they have somewhat rounded, irregular surfaces, called **dental cusps,** for crushing and grinding food.

Two sets of teeth develop in a person's lifetime. Twenty **deciduous (milk) teeth** begin to erupt at about 6 months of age, beginning with the incisors. All of the deciduous teeth should have erupted by the age of 2½. Thirty-two **permanent teeth** replace the deciduous teeth in a predictable sequence. This process begins at about age 6 and continues until about age 17. The **third molars,** or **wisdom teeth,** are the last to erupt. There may not be room in the jaw to accommodate the wisdom teeth, however, in which case they may grow sideways and become impacted, or emerge only partially. If they do erupt at all, it

uvula: L. *uvula,* small grape
incisor: L. *incidere,* to cut
canine: L. *canis,* dog
molar: L. *mola,* millstone
deciduous: L. *deciduus,* to fall away

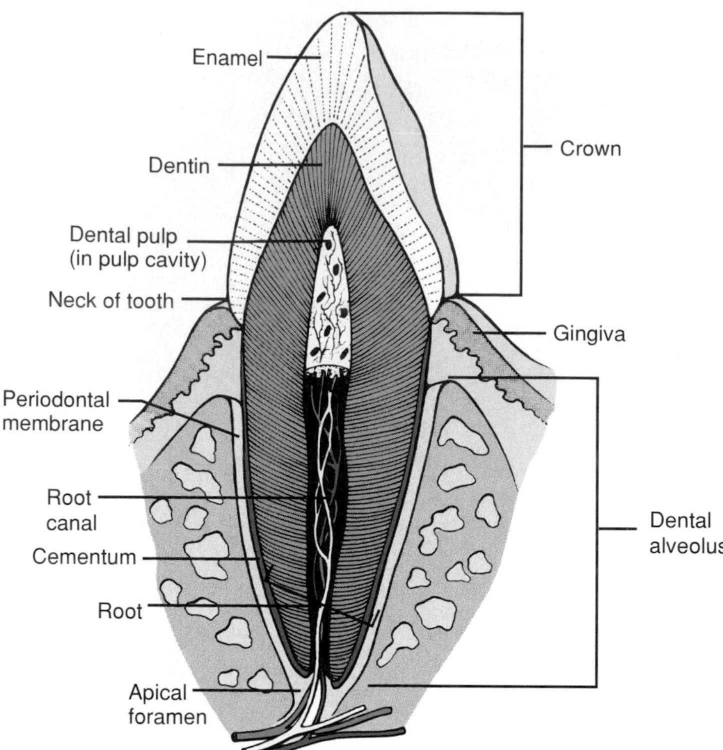

FIGURE 20.12
The structure of a tooth shown in a vertical section through a canine.

is usually between the ages of 17 and 25. Presumably, a person has acquired some wisdom by then—hence, the popular name for the third molars.

The cusps of the upper and lower premolars and molars occlude for chewing food, whereas the upper incisors normally form an overbite with the incisors of the lower jaw. An overbite of the upper incisors creates a shearing action as these teeth slide past one another. Masticated food is mixed with saliva, which initiates chemical digestion and aids swallowing. The soft, flexible mass of food that is swallowed is called a *bolus.*

A tooth consists of an exposed **crown,** supported by a **neck,** anchored firmly into the jaw by one or more **roots** (fig. 20.12). The roots of teeth fit into sockets called **dental alveoli** in the alveolar processes of the mandible and maxillae. Each socket is lined with a **periodontal membrane.** The root of a tooth is covered with a bonelike material, the **cementum,** that fastens the tooth in its socket. The **gingiva** (*gum*) is the mucous membrane surrounding the alveolar process in the oral cavity.

The bulk of a tooth consists of **dentin,** a substance similar to bone but harder. Covering the dentin on the outside and forming the crown is a tough, durable layer of **enamel.** Enamel is composed primarily of calcium phosphate and is the hardest substance in the body. The central region of the tooth contains the **pulp cavity.** The pulp cavity contains the **pulp,** which is composed of connective tissue with blood vessels, lymph vessels, and nerves. A **root canal,** continuous with the pulp cavity, opens to the connective tissue surrounding the root through an **apical foramen** at the tip of the root.

Inside Information—Even though enamel is the hardest substance in the body, it could not withstand the nitric acid treatment used by medieval barbers and blacksmiths to whiten peoples' teeth. The corrosive liquid actually destroyed the enamel and caused massive tooth decay. Refluxed stomach acids also destroy tooth enamel; constant vomiting, as in the eating disorder bulimia nervosa, contributes to the development of dental caries.

Salivary Glands

The salivary glands (fig. 20.13) are accessory digestive glands that produce a fluid secretion called *saliva.* Saliva functions as a solvent in cleansing the teeth and in dissolving food molecules so that they can be tasted. Saliva also contains starch-digesting enzymes and lubricating mucus, which aids swallowing. The amount of saliva secreted daily ranges from 1.0 to 1.5 L.

The **parotid gland** is the largest of three paired salivary glands. It is positioned below and in front of the ear, between the skin and the masseter muscle. The parotid gland drains through the **parotid duct** into the oral cavity opposite the second upper molar. It is the parotid gland that becomes infected and swollen with the mumps. The **submandibular gland** is inferior to the body of the mandible. It empties through the **submandibular duct** into the floor of the mouth on either side of the lingual frenulum. The **sublingual gland** lies under the mucosa in the floor of the mouth, where it drains through a series of **sublingual ducts** that empty into the mouth under the tongue.

Pharynx

The funnel shaped pharynx (*far'ingks*) is a passageway approximately 13 cm (5 in.) long connecting the oral and nasal cavities to the esophagus and trachea. The pharynx has both digestive and respiratory functions. The supporting walls of the pharynx are composed of skeletal muscle, and the lumen is lined with a mucous membrane of stratified squamous epithelium. The pharynx is divided into three regions: the *nasopharynx,* posterior to the nasal cavity; the *oropharynx,* posterior to the oral cavity; and the *laryngopharynx,* at the level of the larynx (see fig. 18.2).

bolus: Gk. *bolos,* lump
gingiva: L. *gingiva,* gum
dentin: L. *dens,* tooth
parotid: Gk. *para,* beside; *otos,* ear

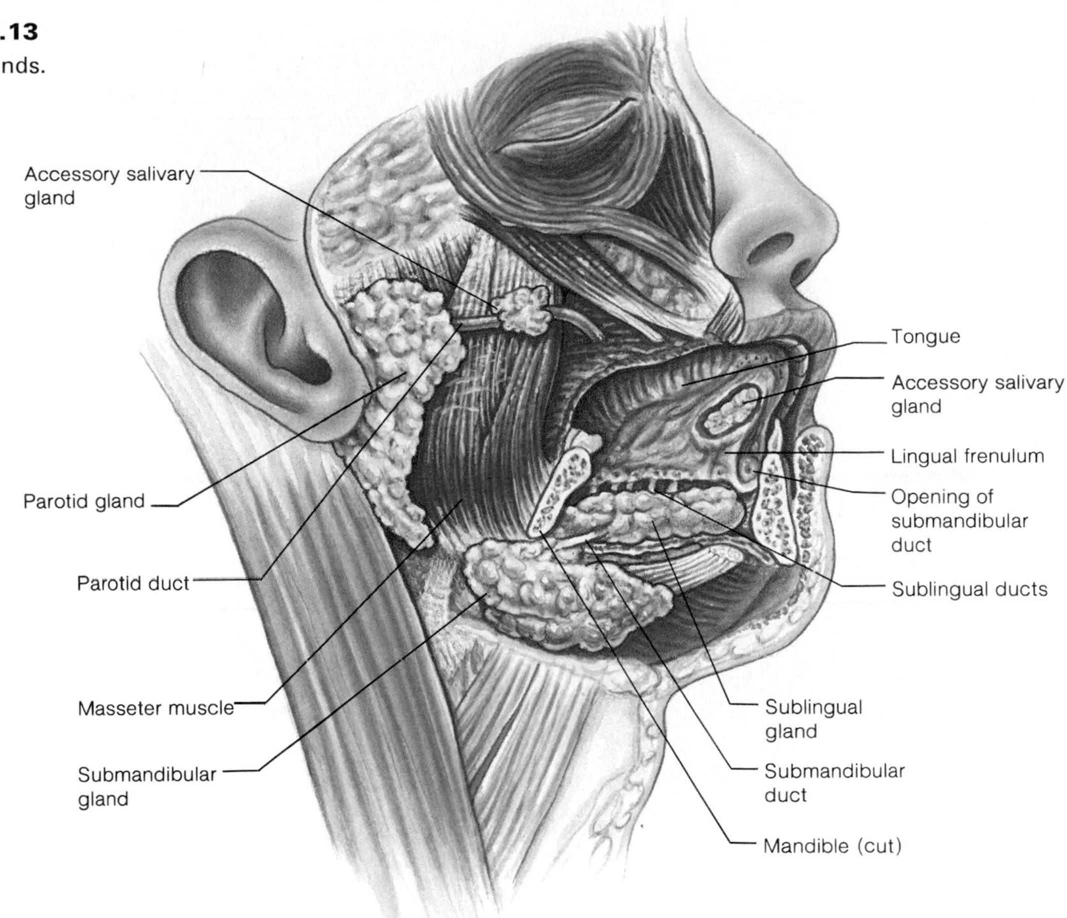

FIGURE 20.13
The salivary glands.

20.5 Esophagus and Stomach

A bolus is passed from the esophagus to the stomach, where it is mixed with gastric secretions to become chyme. Chyme is sent on past the pyloric sphincter to enter the duodenum.

Esophagus

The esophagus (*ĕ-sof'ă-gus*) is the portion of the GI tract that connects the pharynx to the stomach (figs. 20.9 and 20.14). It is a collapsible muscular tube approximately 25 cm (10 in.) long, located behind the trachea within the mediastinum of the thorax. The esophagus is lined with a stratified squamous epithelium. The upper third of the esophageal wall contains skeletal muscle; the middle third, a mixture of skeletal and smooth muscle; and the terminal portion, smooth muscle only. The **lower esophageal sphincter** is a thickening of the circular muscle fibers at the junction of the esophagus and the stomach. After food or fluid passes into the stomach, this sphincter constricts to prevent the stomach contents from regurgitating into the esophagus. There is a normal tendency for this to occur because the thoracic pressure is lower (due to the air-filled lungs) than the abdominal pressure. The "heartburn" people commonly experience is due to a backflow (reflux) of the acidic gastric juices through the lower esophageal sphincter up into the esophagus.

Stomach

The stomach—the most distensible part of the GI tract—is located higher up in the body than most people think. This J-shaped pouch is positioned in the upper left portion of the peritoneal cavity, immediately below the diaphragm. It is continuous with the esophagus superiorly,

esophagus: Gk. *oisein*, to carry; *phagema*, food

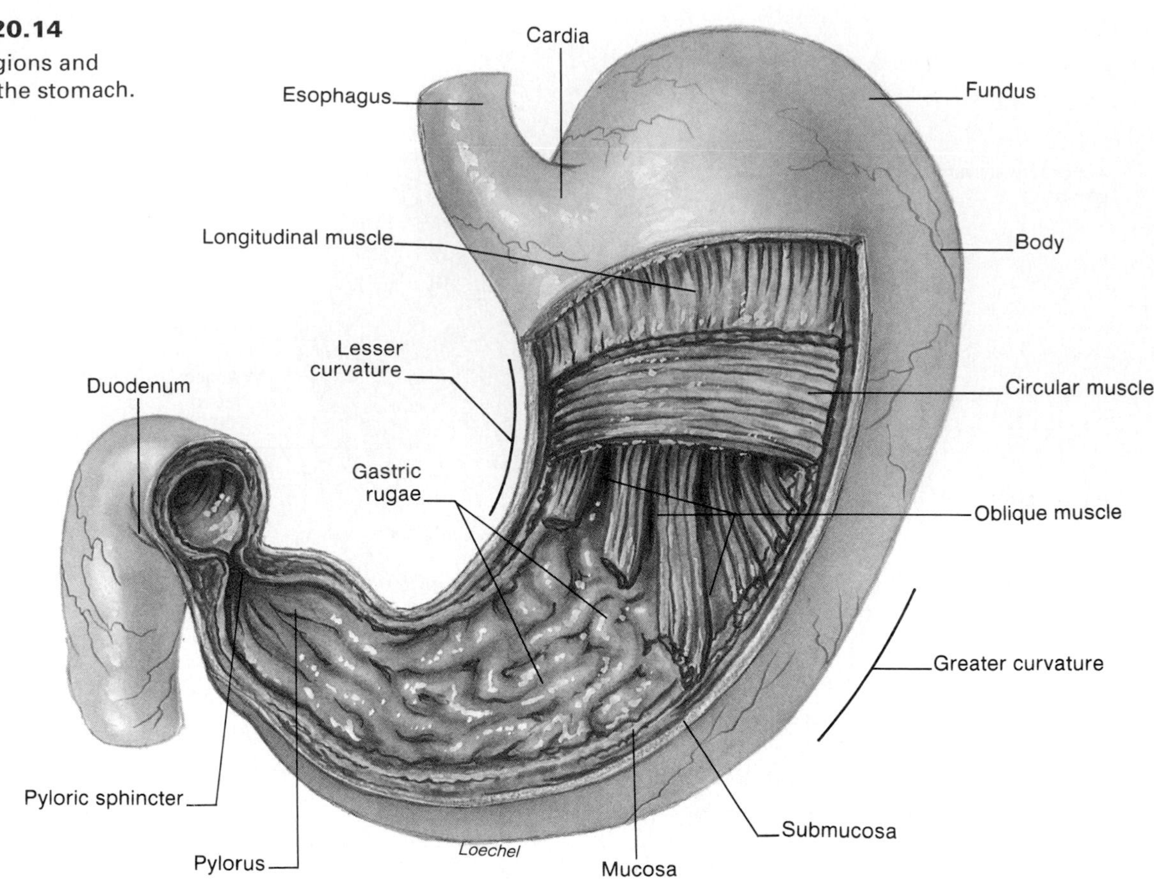

FIGURE 20.14
The major regions and structures of the stomach.

emptying into the duodenal portion of the small intestine inferiorly. In the stomach, which serves as a "holding area" for ingested food, the food is mechanically churned with gastric secretions to form a pasty material called *chyme* (kīm). Once formed, chyme is moved from the stomach to the small intestine.

The **cardia** is the upper narrow region of the stomach (fig. 20.14), immediately below the lower esophageal sphincter. The **fundus** is the dome-shaped portion in direct contact with the diaphragm. The **body** is the large central portion, and the **pylorus** is the funnel-shaped terminal portion. The **pyloric sphincter** is the modified circular muscle at the end of the pylorus. It regulates the movement of chyme into the duodenum and prevents backflow.

The stomach has two surfaces and two borders. The broadly rounded surfaces are referred to as the **anterior** and **posterior surfaces.** The medial concave border is the **lesser curvature** (fig. 20.14), and the lateral convex border is the **greater curvature.**

The wall of the stomach is composed of the same four tunics found in other regions of the GI tract, with two principal modifications: (1) an extra *oblique muscle layer* is present in the muscularis, and (2) the mucosa is thrown into numerous longitudinal folds, called **gastric rugae,** which permit stomach distension. The mucosa is further characterized by the presence of microscopic **gastric pits** and **gastric glands** (fig. 20.15).

There are five types of cells in the gastric glands that secrete specific products.

1. **Goblet cells** secrete protective mucus.
2. **Parietal cells** secrete hydrochloric acid (HCl).
3. **Chief cells** secrete pepsinogen, an inactive form of the protein-digesting enzyme pepsin.
4. **Argentaffin cells** secrete serotonin and histamine and autocrine regulators.
5. **G cells** secrete the hormone gastrin into the blood.

In addition to these products, the gastric mucosa (probably the parietal cells) secretes a polypeptide called *intrinsic factor,* which is required for the absorption of vitamin B_{12} in the small intestine.

chyme: L. *chymus,* juice
cardia: Gk. *karida,* heart (upper portion, nearer the heart)
fundus: L. *fundus,* bottom
pylorus: Gk. *pyloros,* gatekeeper
argentaffin: L. *argentum,* silver; *affinis,* attraction (become colored with silver stain)

Gastric Juice

The secretions of the goblet cells, parietal cells, and chief cells are collectively called *gastric juice.* The HCl contributed by the parietal cells makes gastric juice very acidic, with a pH less than 2. This strong acidity serves three functions: (1) ingested proteins are denatured at a low pH—that is, their structure is altered so that they become more digestible; (2) under acidic conditions, weak pepsinogen enzymes partially digest each other—this frees the active pepsin enzyme as small peptide fragments are removed (fig. 20.16); and (3) pepsin is more active under acidic conditions. The peptide bonds of ingested protein are broken through hydrolysis by pepsin under acidic conditions; the HCl itself does not directly digest proteins.

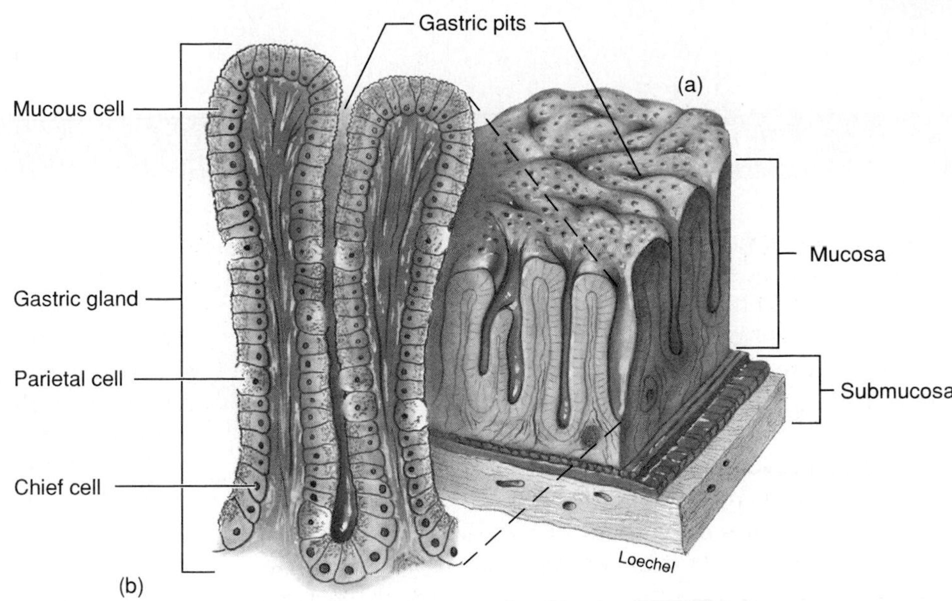

FIGURE 20.15
Gastric pits and gastric glands of the mucosa. (*a*) Gastric pits are the openings of the gastric glands. (*b*) Gastric glands consist of mucous cells, chief cells, and parietal cells, each type producing a specific secretion.

Digestion and Absorption in the Stomach

Proteins are only partially digested in the stomach by the action of pepsin. Carbohydrates and fats are not digested at all in the stomach. The complete digestion of food molecules occurs later, when chyme enters the small intestine. Patients with partial gastric resections, therefore, and even those with complete gastrectomies (stomach removal), can still adequately digest and absorb their food.

Little absorption takes place in the stomach; almost all of the products of digestion are absorbed through the wall of the small intestine. The only commonly ingested substances that can be absorbed across the stomach wall are alcohol and aspirin. This occurs as a result of the lipid solubility of these molecules. The passage of aspirin through the gastric mucosa has been shown to cause bleeding, which may be significant if large amounts of aspirin are taken.

Inside Information—In a series of experiments that lasted 8 years, Dr. William Beaumont (1785–1853) monitored the digestive process by direct observation of a patient's stomach. The patient, Alexis St. Martin, had sustained a massive gunshot wound to his side. After the wound healed, a small opening into his stomach remained, permitting its interior to be exposed to view. St. Martin's wound led to the discovery of HCl in gastric juice and served as a basis for modern studies of digestion.

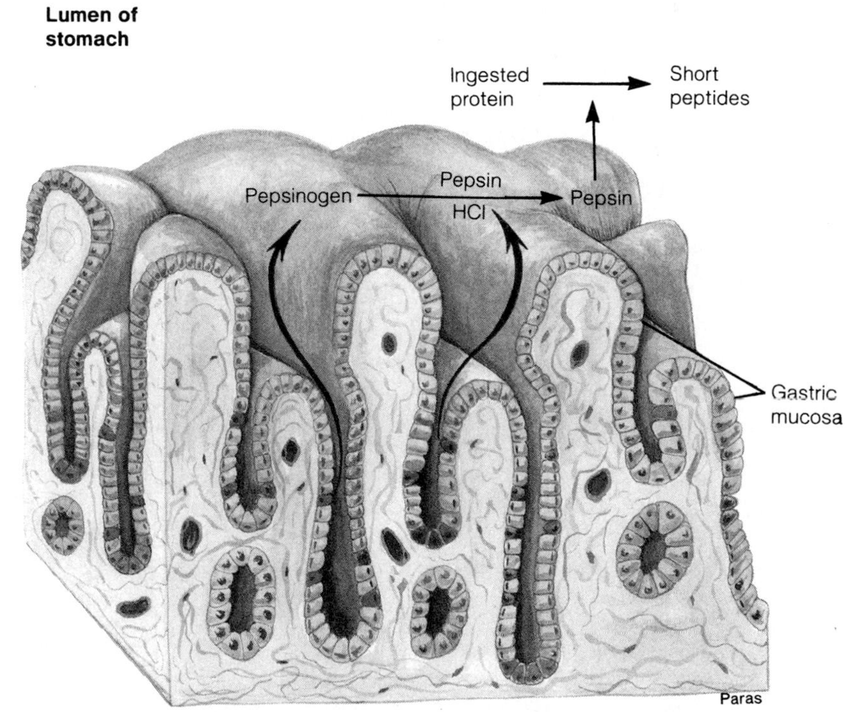

FIGURE 20.16
The gastric mucosa secretes the inactive enzyme pepsinogen and hydrochloric acid (HCl). In the presence of HCl, the active enzyme pepsin is produced. Pepsin digests proteins into shorter polypeptides.

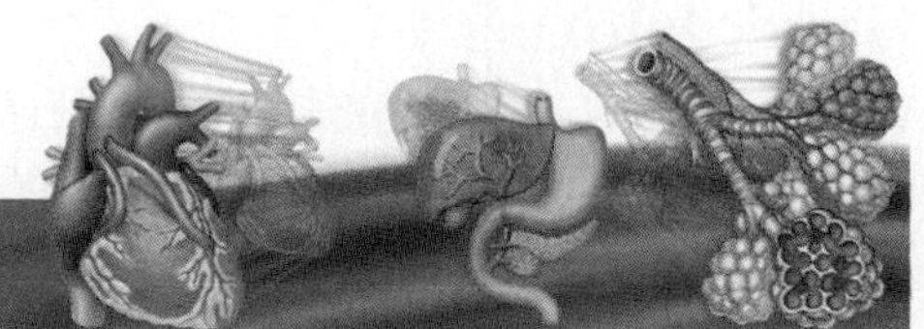

20.6 Small Intestine

The small intestine is adapted to provide a large surface area for the exposure of chyme to brush border enzymes and for absorption of nutrients.

The **small intestine** is the portion of the GI tract between the pyloric sphincter of the stomach and the ileocecal valve opening into the large intestine. It is positioned in the central and lower portions of the peritoneal cavity and is supported by **mesentery** (fig. 20.17). The fan-shaped mesentery permits movement of the small intestine but leaves little chance for it to become twisted or kinked. Enclosed within the mesentery are blood vessels, nerves, and lymphatic vessels that supply the intestinal wall.

The small intestine is about 3 m (12 ft) long and 2.4 cm (1 in.) wide in a living person, but it will measure nearly twice this length in a cadaver when the muscle wall is relaxed. It is called the "small" intestine because of its relatively small diameter compared to that of the large intestine. The small intestine serves as the major site of digestion and absorption in the GI tract. It is innervated by the *superior mesenteric plexus.* The branches of the plexus contain sensory fibers, postganglionic sympathetic fibers, and preganglionic parasympathetic fibers. The arterial blood supply to the small intestine is through the *superior mesenteric artery* and branches from the *celiac* and *inferior mesenteric arteries.* The venous drainage is through the *superior mesenteric vein.* This vein unites with the splenic vein to form the *hepatic portal vein,* which carries nutrient-rich blood to the liver (see fig. 15.22).

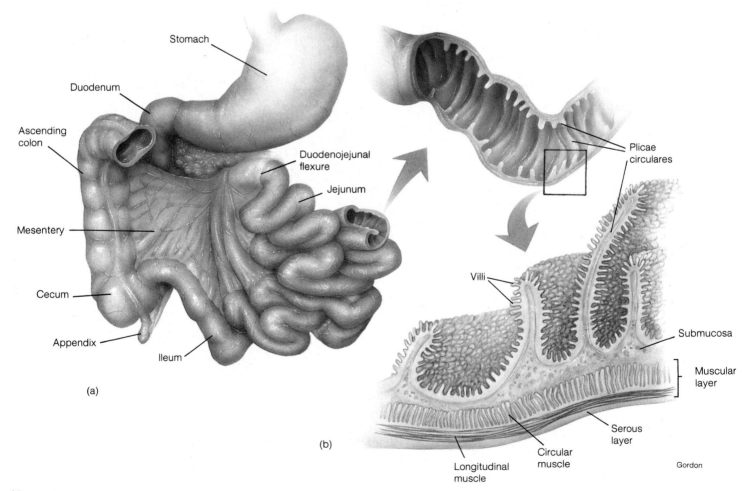

FIGURE 20.17
The small intestine. (*a*) The regions and the mesenteric attachment. (*b*) A section of the intestinal wall showing the mucosa and submucosa folded into structures called plicae circulares.

Regions of the Small Intestine

Based on function and histological structure, the small intestine is divided into three regions:

1. The **duodenum** (*doo"ŏ-de'num*) is a C-shaped tube measuring approximately 25 cm (10 in.) from the pyloric sphincter of the stomach to the *duodenojejunal flexure.* The concave surface of the duodenum faces to the left, where it receives bile secretions through the **common bile duct** from the liver and gallbladder, and pancreatic secretions through the **pancreatic duct** of the pancreas (fig. 20.18). These two ducts merge to form a common entry into the duodenum, called the **hepatopancreatic ampulla** (ampulla of Vater), which pierces the duodenal wall and drains into the duodenum from an elevation called the **duodenal papilla.** It is here that bile and pancreatic juice (described in modules 20.8 and 20.9, respectively) enter the small intestine. The papilla can be opened or closed by the action of the **sphincter of ampulla** (of Oddi). The duodenum differs histologically from the rest of the small intestine by the presence of mucus-secreting **duodenal** (Brunner's) **glands,** located in the submucosa (see fig. 20.6).
2. The **jejunum** (*jĕ-joo'num*) is approximately 1 m (3 ft) long and extends from the duodenum to the ileum. The jejunum has a slightly larger lumen and more internal folds than does the ileum, but it has a similar histological structure.
3. The **ileum** (*il'e-um*)—not to be confused with the ilium, or hipbone—makes up the remaining 2 m (6–7 ft) of the small intestine. The terminal portion of the ileum empties into the medial side of the cecum through the **ileocecal valve.** The walls of the ileum have an abundance of lymph nodules, called **mesenteric** (Peyer's) **patches.**

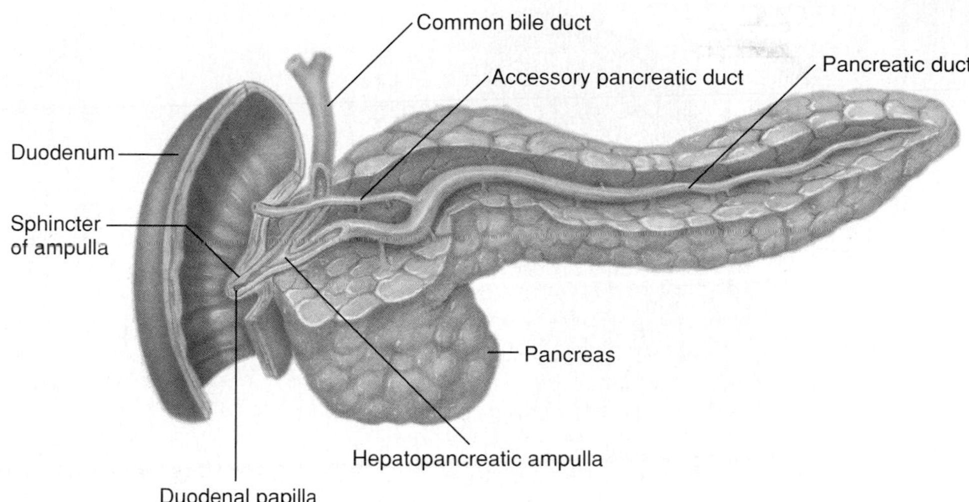

FIGURE 20.18
The duodenum and associated structures.

Structural Adaptations of the Wall of the Small Intestine

The products of digestion are absorbed across the epithelial lining of the intestinal mucosa. Absorption occurs primarily in the jejunum, although some also occurs in the duodenum and ileum. Absorption occurs at a rapid rate as a result of three specializations that increase the intestinal surface area.

1. The **plicae** (*pli'se*) **circulares** are large macroscopic folds of mucosa (see fig. 20.17).
2. The **villi** (*vil'e*) are fingerlike macroscopic folds of the mucosa that project into the lumen of the small intestine.
3. The **microvilli** are microscopic projections formed by the folding of each epithelial cell membrane. In a light microscope, the microvilli display a somewhat vague brush border on the edges of the columnar epithelium. The terms *brush border* and *microvilli* are often used interchangeably in describing the small intestine.

The villi are covered with columnar epithelial cells, among which are interspersed the mucus-secreting goblet cells. The lamina propria forms the connective core of each villus and contains numerous lymphocytes, blood capillaries, and a lymphatic vessel called the **lacteal** (fig. 20.19). Absorbed monosaccharides and amino acids enter the blood capillaries; absorbed fat enters the lacteals.

---

duodenum: L. *duodeni,* twelve each (distance of twelve fingers' breadth)
ampulla of Vater: from Abraham Vater, German anatomist, 1684–1751
Brunner's glands: from Johann C. Brunner, Swiss anatomist, 1653–1727
Peyer's patches: from Johann K. Peyer, Swiss anatomist, 1653–1712
plica: L. *plicatus,* folded
villus: L. *vilosus,* shaggy
lacteal: L. *lacteus,* milk

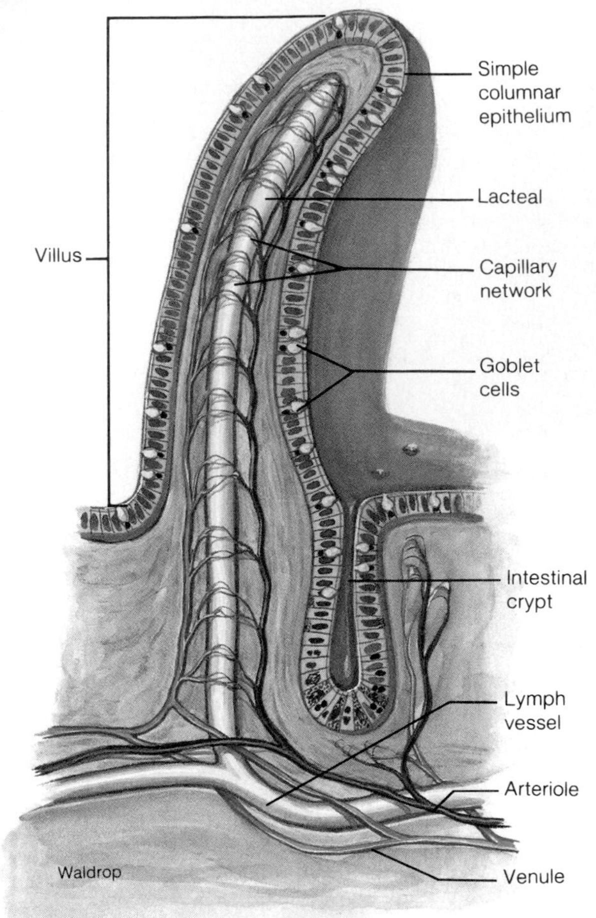

FIGURE 20.19
The structure of an intestinal villus.

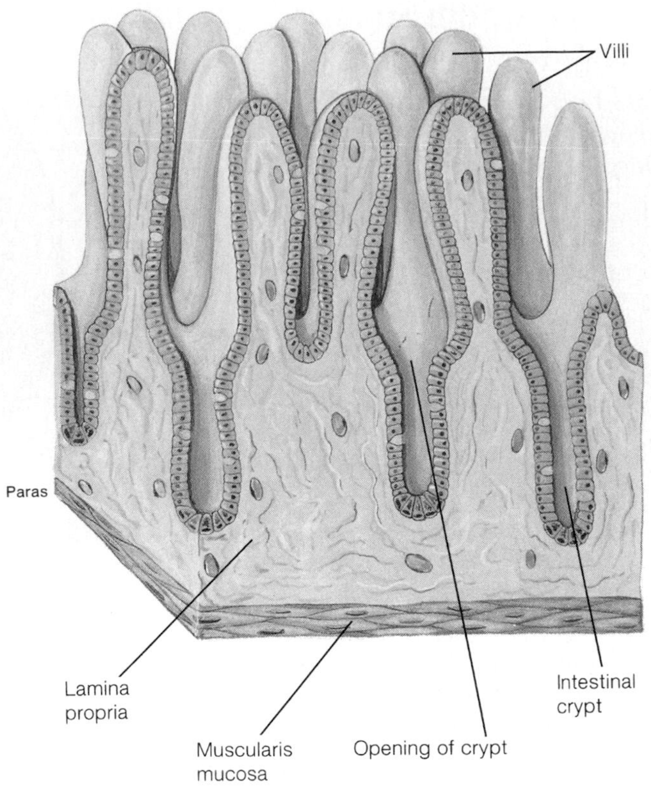

Villi
Paras
Lamina
propria
Muscularis
mucosa
Opening of crypt
Intestinal
crypt

FIGURE 20.20
Villi and intestinal crypts are specializations of the mucosa of the small intestine.

Epithelial cells at the tips of the villi are continuously shed and are replaced by cells that are pushed up from the bases of the villi. The epithelium at the base of the villi invaginates downward at various points to form narrow pouches that open through pores to the intestinal lumen. These structures are called the **intestinal crypts** (crypts of Lieberkühn) (fig. 20.20).

Inside Information—The absorption area of the small intestine is three times the entire surface area of the human body.

Intestinal Enzymes

In addition to providing a large surface area for absorption, the cell membranes of the microvilli contain digestive enzymes. These enzymes are not secreted into the lumen, but instead remain attached to the cell membrane with their active sites exposed to the chyme. These **brush border enzymes** hydrolyze disaccharides, polypeptides, and other molecules (table 20.2). One brush border enzyme, *enterokinase,* is required for activation of the protein-digesting enzyme *trypsin,* which enters the small intestine in pancreatic juice.

The ability to digest milk sugar, or lactose, depends on the presence of a brush border enzyme called lactase. *This enzyme is present in children under the age of 6 but becomes inactive to some degree in most adults. A deficiency of lactase can result in* lactose intolerance. *The presence of large amounts of undigested lactose in the intestine causes diarrhea, gas, cramps, and other unpleasant symptoms. Yogurt is better tolerated than milk because it contains lactase produced by the yogurt bacteria. This lactase becomes activated in the duodenum and digests lactose.*

crypts of Lieberkühn, from Johann N. Lieberkühn, German anatomist, 1711–56

TABLE 20.2 **Brush Border Enzymes Attached to the Cell Membrane of Microvilli in the Small Intestine**

Category	Enzyme	Comments
Disaccharidase	Sucrase	Digests sucrose to glucose and fructose; deficiency produces gastrointestinal disturbances
	Maltase	Digests maltose to glucose
	Lactase	Digests lactose to glucose and galactose; deficiency produces gastrointestinal disturbances (lactose intolerance)
Peptidase	Aminopeptidase	Produces free amino acids, dipeptides, and tripeptides
	Enterokinase	Activates trypsin (and indirectly other pancreatic juice enzymes); deficiency results in protein malnutrition
Phosphatase	Ca^{++}, Mg^{++}—ATPase	Needed for absorption of dietary calcium; enzyme activity regulated by vitamin D
	Alkaline phosphatase	Removes phosphate groups from organic molecules; enzyme activity may be regulated by vitamin D

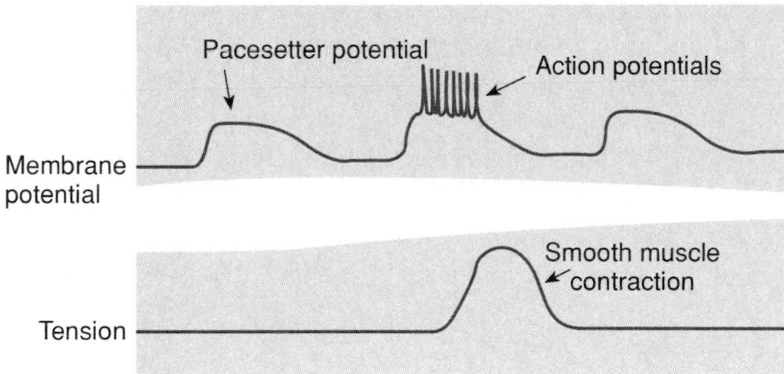

FIGURE 20.21
The smooth muscle of the gastrointestinal tract produces and conducts spontaneous pacesetter potentials. As these potential changes reach a threshold level of depolarization, they stimulate the production of action potentials, which in turn stimulate smooth muscle contraction.

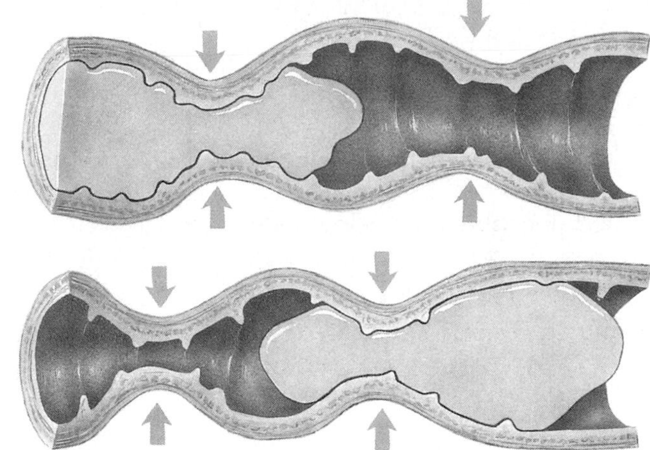

FIGURE 20.22
Segmentation of the small intestine. Simultaneous contractions of numerous segments of the small intestine help to mix the chyme with digestive enzymes and mucus.

Intestinal Contractions and Motility

Like cardiac muscle, intestinal smooth muscle is capable of spontaneous electrical activity and automatic rhythmic contractions. Spontaneous depolarizations begin in the longitudinal smooth muscle and are conducted to the circular smooth muscle layer across *nexuses,* which are electrical synapses between smooth muscle cells. The spontaneous depolarizations, called *pacesetter potentials,* decrease in amplitude as they are conducted from one muscle cell to another. Pacesetter potentials stimulate the production of action potentials in the smooth muscle cells through which they are conducted (fig. 20.21). When action potentials are produced, they stimulate smooth muscle contraction. The rate at which this automatic activity occurs is influenced by nerve impulses through autonomic nerves. Contraction is stimulated by parasympathetic innervation and inhibited by sympathetic nerve activity.

Two principal types of contractions occur in the small intestine: peristalsis and segmentation. **Peristalsis** (*per″ĭ-stal′sis*), the series of wavelike contractions that propels food through the GI tract, is much weaker in the small intestine than in the esophagus and stomach. **Segmentation** (fig. 20.22), the major contractile activity of the small intestine, refers to the muscular constrictions of the lumen that occur simultaneously at different intestinal segments. This back-and-forth movement mixes the chyme more thoroughly.

20.7 Large Intestine

The large intestine receives undigested food from the small intestine, absorbs water and electrolytes from chyme, and passes feces out of the GI tract.

Structure of the Large Intestine

The large intestine averages 1.5 m (5 ft) in length and 6.5 cm (2.5 in.) in diameter. It begins at the end of the ileum in the lower right portion of the peritoneal cavity (fig. 20.23). From there, it leads superiorly on the right side to just below the liver, where it crosses to the left, descends into the pelvis, and terminates at the anus. A specialized portion of the mesentery, called the **mesocolon,** supports the large intestine along the posterior abdominal wall.

The large intestine is structurally divided into the *cecum, colon, rectum,* and *anal canal.* The **cecum** (*se'kum*) is a dilated pouch that lies below the ileocecal valve. A finger-like projection called the **appendix** is attached to the inferior medial margin of the cecum. The 8-cm (3-in.) appendix contains an abundance of lymphatic tissue, which may serve to resist infection. Although the appendix serves no digestive function, it may be a vestigial remnant of an organ that was functional in human ancestors.

A common disorder of the large intestine is inflammation of the appendix, or appendicitis. *Wastes that collect in the appendix cannot be moved easily by peristalsis, since the appendix has only one opening. Although the symptoms of appendicitis are quite variable, they often include a high white blood cell count, localized pain in the lower right quadrant, and vomiting. Rupture of the appendix (a "burst appendix") spreads infectious material throughout the peritoneal cavity, resulting in* peritonitis.

cecum: L. *caecum,* blind pouch
appendix: L. *appendix,* attachment

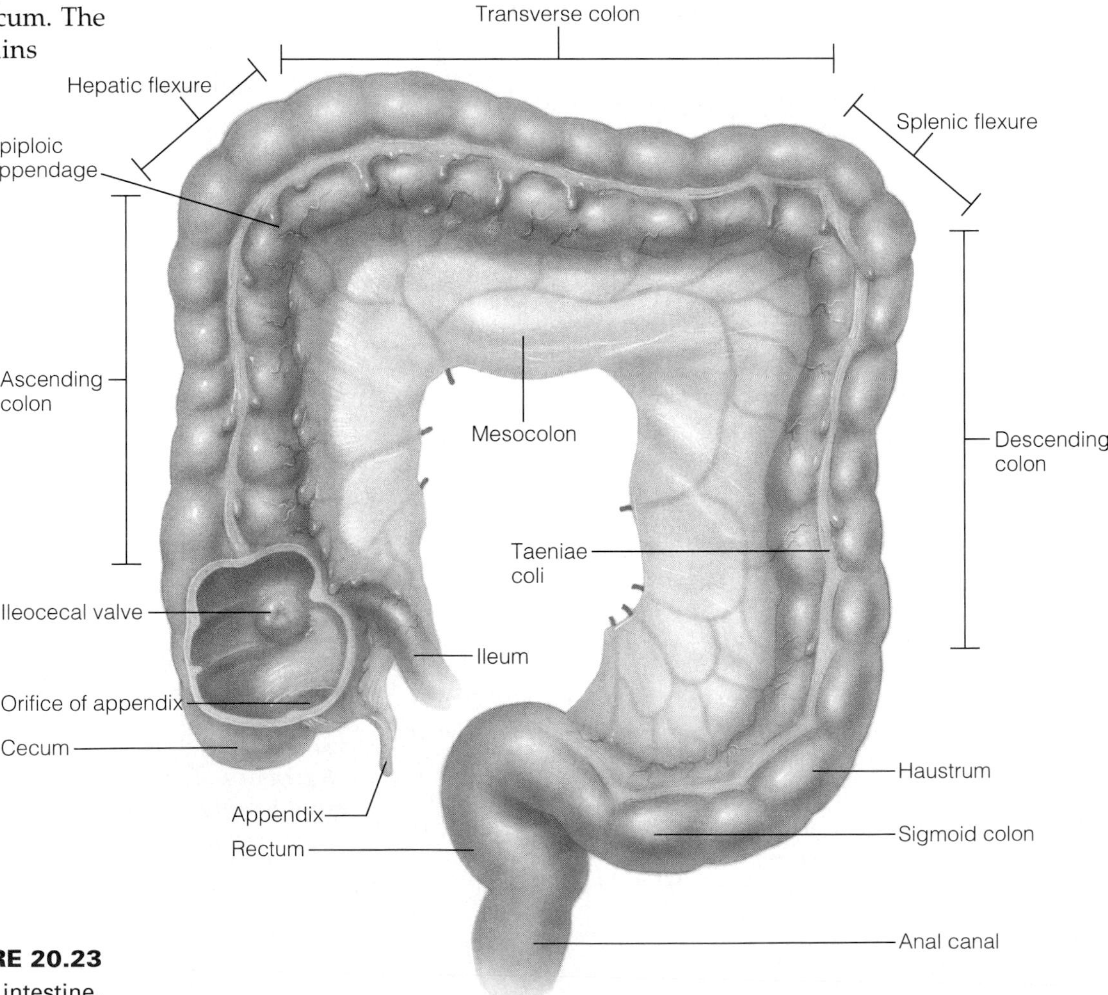

FIGURE 20.23
The large intestine.

The superior portion of the cecum is continuous with the *colon* (*ko'lon*), which consists of *ascending, transverse, descending,* and *sigmoid* portions. The **ascending colon** extends superiorly from the cecum along the right abdominal wall to the liver. Here the colon bends sharply to the left at the **hepatic flexure** and crosses the upper peritoneal cavity as the **transverse colon.** At the left abdominal wall, another right-angle bend called the **splenic flexure** marks the beginning of the **descending colon.**

The terminal 20 cm (7.5 in.) of the GI tract is the **rectum,** and the last 2 to 3 cm is the **anal canal,** which opens to the outside through the **anus** (fig. 20.24). Two sphincter muscles guard the anal opening: the **internal anal sphincter,** composed of smooth muscle, and the **external anal sphincter,** composed of skeletal muscle. Varicose veins in the anal area are known as *hemorrhoids.*

Although the large intestine consists of the same tunics as the small intestine, there are some structural differences. The large intestine lacks villi but does have numerous goblet cells in the mucosal layer. The longitudinal muscle layer of the muscularis externa forms three muscle bands, called **taeniae coli,** which run the length of the large intestine. A series of bulges in the walls of the large intestine form sacculations, or **haustra,** along its entire length (figs. 20.23 and 20.25). Numerous small fat-filled pouches called **epiploic appendages** are attached superficially to the taeniae coli (see fig. 20.23).

The *sympathetic innervation* of the large intestine arises from superior and inferior mesenteric plexuses as well as from the celiac plexus. The *parasympathetic innervation* arises from the paired pelvic splanchnic and vagus nerves. Sensory fibers from the large intestine respond to bowel pressure and signal the need to defecate. Blood is supplied to the large intestine by branches from the *superior* and *inferior mesenteric arteries.* Venous blood is returned through the *superior* and *inferior mesenteric veins,* which in turn drain into the *hepatic portal vein* into the liver.

Fluid and Electrolyte Absorption

The large intestine does not contribute significantly to digestion, since it produces no enzymes. Its major functions are to form, store, and expel feces from the body. In addition, it absorbs some water and electrolytes, although most of the fluid and electrolytes in the GI tract are absorbed by the small intestine. A person may drink only about 1.5 L of water per day, but the small intestine receives 7 to 9 L per day as a result of the fluid secreted into the GI tract by the salivary glands, stomach, pancreas, and liver. The small intestine absorbs most of this fluid and passes 1.5 to 2.0 L of fluid per day to the large intestine. The large intestine absorbs about 90% of this remaining volume, leaving less than 200 ml of fluid to be excreted in the feces.

The epithelial cells of the intestinal mucosa contain Na^+/K^+ pumps in the basolateral membrane. Absorption of water in the large intestine occurs passively as a result of the osmotic gradient created by the active transport of ions. Both salt and water absorption seem to be influenced by *aldosterone,* similar to the mechanism in the renal tubules of the kidneys.

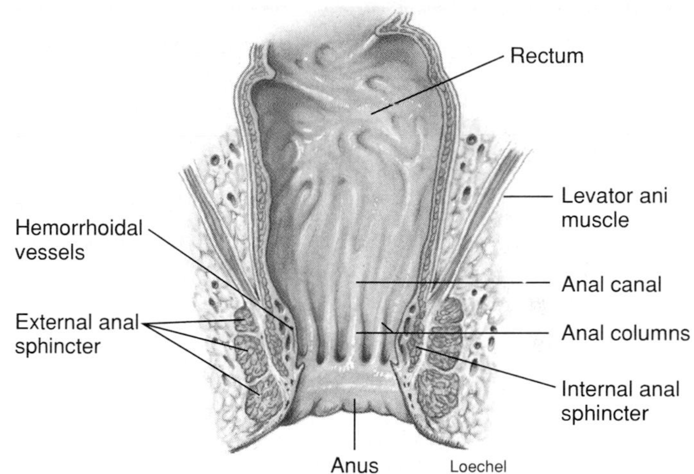

FIGURE 20.24
The anal canal.

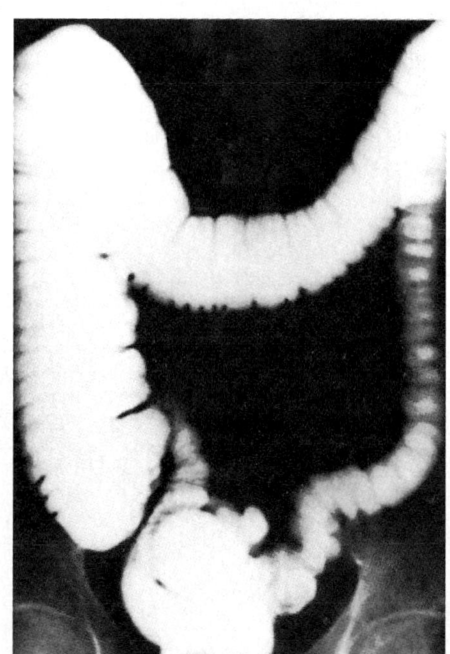

FIGURE 20.25
A radiograph after a barium enema, showing the regions, flexures, and the haustra of the large intestine.

colon: Gk. *kolon,* member of the whole
sigmoid: Gk. *sigmoeides,* shaped like a sigma, Σ
rectum: L. *rectum,* straight tube
anus: L. *anus,* ring
taenia: L. *tainia,* a ribbon
haustrum: L. *haustrum,* bucket or scoop
epiploic: Gk. *epiplein,* to float on

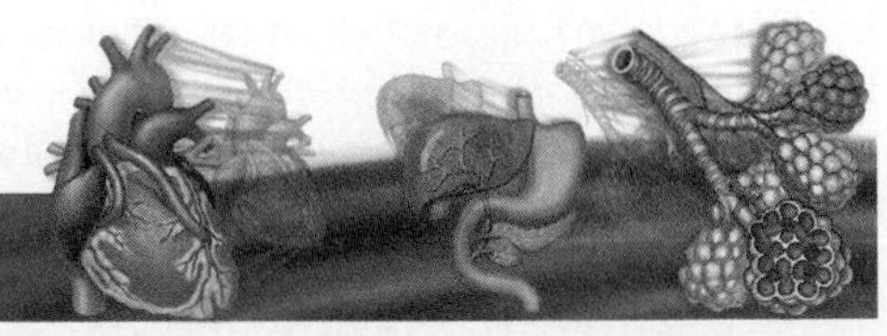

20.8 Liver and Gallbladder

The liver regulates the chemical composition of blood and performs many vital metabolic functions. Its principal role in digestion is the production and secretion of bile, which is stored and concentrated in the gallbladder prior to its discharge into the duodenum.

Structure of the Liver

The liver is the largest internal organ of the body, weighing about 1.7 kg (3.5 to 4.0 lb) in an adult. It is positioned immediately beneath the diaphragm in the right side of the peritoneal cavity. Its reddish-brown color is due to its great vascularity.

The liver has two major lobes and two minor lobes. Anteriorly, the **right lobe** is separated from the smaller **left lobe** by the **falciform ligament** (fig. 20.26). Inferiorly, the **caudate lobe** is positioned near the inferior vena cava, and the **quadrate lobe** is adjacent to the gallbladder. The falciform ligament attaches the liver to the abdominal wall and the diaphragm. The **ligamentum teres** extends from the falciform ligament to the umbilicus. This ligament is the remnant of the fetal umbilical vein.

Liver Lobules

The lobes of the liver are made up of numerous functional subunits (fig. 20.27). A liver lobule consists of **hepatic plates** that are only one or two cells thick. The principal cells of the hepatic plates are known as **hepatocytes,** and there are approximately 300 billion of them in the liver. The hepatic plates are separated from each other by large capillary spaces called **sinusoids.** Phagocytic **Kupffer cells** line the sinusoids.

In the center of each liver lobule is a **central vein,** and at the periphery are branches of the hepatic portal vein (discussed below) and of the hepatic artery. These vessels open into the sinusoids. Because of the plate structure of the liver lobules, each hepatocyte is in direct contact with the blood.

Hepatic Portal System

As described in chapter 15 (see module 15.6), the products of digestion that are absorbed into blood capillaries in the GI tract do not directly enter the general circulation. Instead, this blood is delivered first to the liver. Capillaries in the stomach, small intestine, and large intestine drain venous blood into veins that converge to form the *hepatic portal vein,* which carries blood to capillaries in the liver. It is not until the blood has passed through this second capillary bed that it enters the general circulation through the two *hepatic veins* that drain the liver. The term **hepatic portal system** (see fig. 15.22)

hepatic: Gk. *hepatos,* liver

Kupffer cells: from Karl Wilhelm von Kupffer, Bavarian anatomist and embryologist, 1829–1902

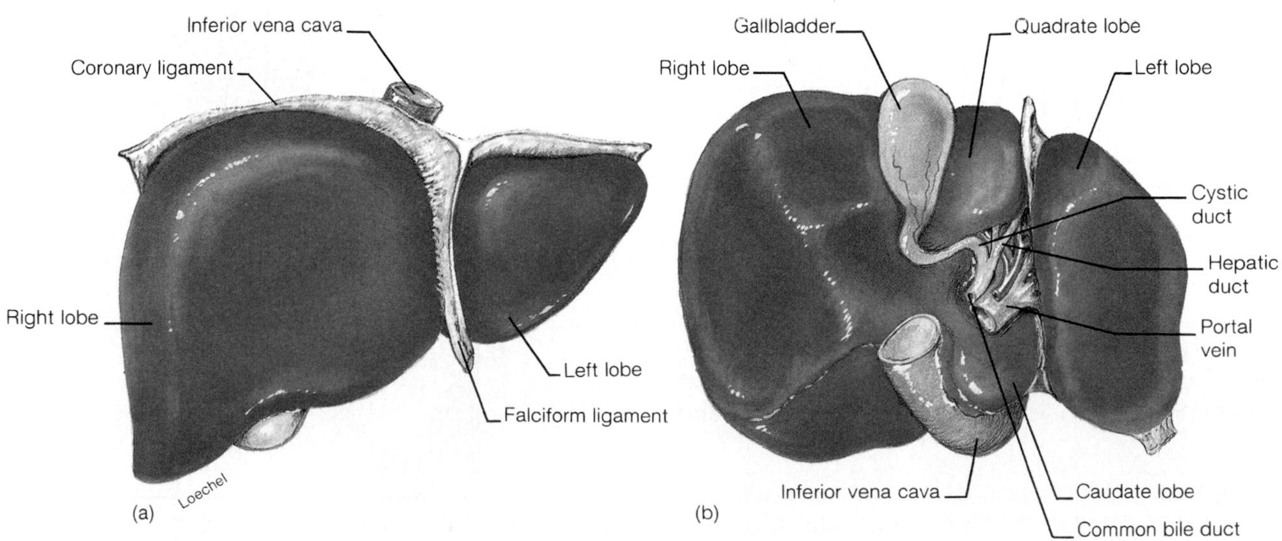

FIGURE 20.26
The liver. (*a*) An anterior view and (*b*) an inferior view showing the lobes of the liver, gallbladder, hepatic vessels, and ducts.

is used to describe this unique pattern of circulation: capillaries → vein → capillaries → vein. In addition to receiving venous blood from the GI tract, the liver also receives arterial blood from the *hepatic artery.*

Portal venous blood, laden with nutrients from the digestive organs, mixes with arterial blood as the blood flows within the sinusoids from the periphery of the lobule to the central vein. The central vein of the lobule converges with the central veins of other lobules to form the two hepatic veins, which carry blood from the liver to the inferior vena cava.

Functions of the Liver

Because of its unique structure and diverse enzymatic content, and because it receives food-laden blood from the GI tract, the liver performs more functions than any other organ. So many and varied are its functions, that it is frequently likened to a chemical processing plant.

Bile Production and Secretion

Bile is produced by the hepatocytes and secreted into thin channels called **bile canaliculi** (kan"ă-lik'yoo-li) located within each hepatic plate. The bile drains toward the periphery of the hepatic plates to the **bile ducts,** which in turn drain into **hepatic ducts** that carry bile away from the liver. As shown in figure 20.28, bile travels in the hepatic plates toward the periphery and blood travels in the opposite direction through the sinusoids toward the central vein. This arrangement ensures that bile and blood will not mix in the liver lobules.

The liver produces and secretes 250 to 1500 ml of bile per day. The major constituents of bile include bile salts, bile pigment (bilirubin), phospholipids (mainly lecithin), cholesterol, and inorganic ions. Bile pigment, or *bilirubin* (*bil"ĭ-roo'bin*) is produced in the spleen, liver, and bone marrow from heme groups (minus the iron) derived from hemoglobin. Without the protein

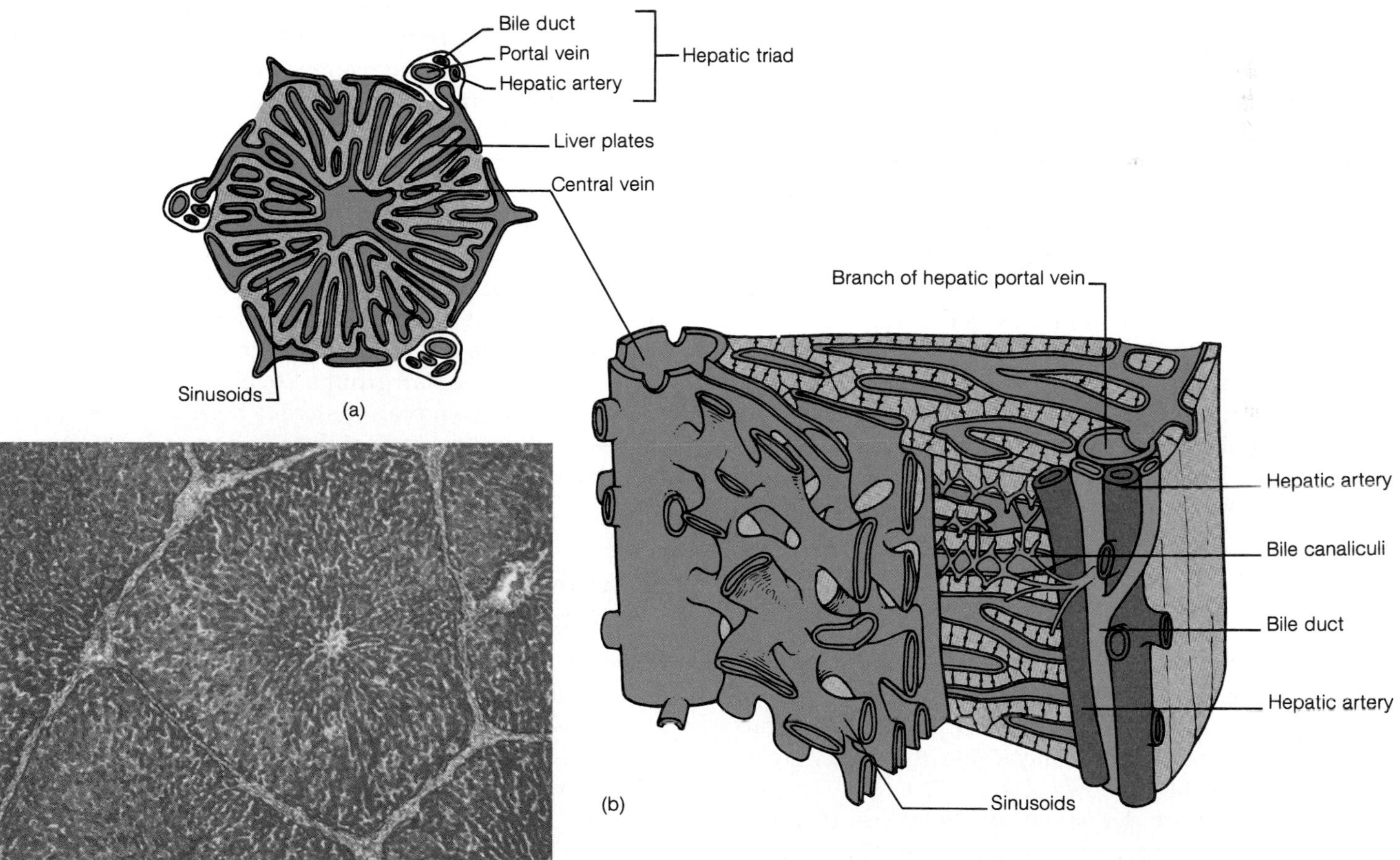

FIGURE 20.27
A liver lobule and the histology of the liver. (*a*) A cross section of a liver lobule and (*b*) a longitudinal section. Blood enters a liver lobule through the vessels in a hepatic triad, passes through hepatic sinusoids, and leaves the lobule through a central vein. The central veins converge to form hepatic veins that transport venous blood from the liver. (*c*) A photomicrograph of a liver lobule in cross section.

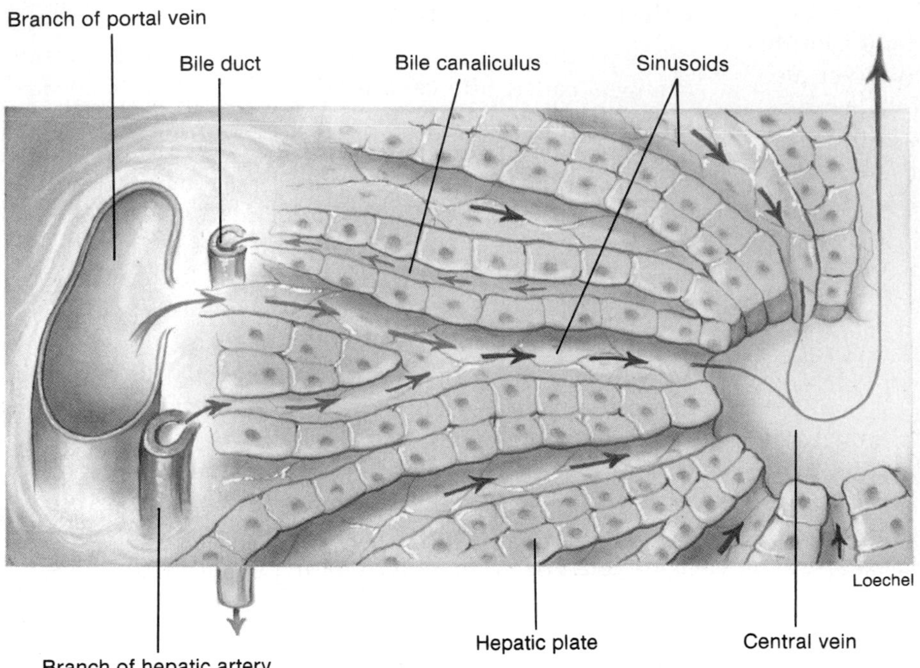

FIGURE 20.28

The flow of blood and bile in a liver lobule. Blood flows within sinusoids from a portal vein to the central vein (from the periphery to the center of a lobule). Bile flows within hepatic plates from the center of a lobule to bile ducts at the periphery.

part of hemoglobin, the **free bilirubin** is not very water soluble and is, therefore, carried in the blood attached to albumin proteins. This protein-bound bilirubin can neither be filtered by the kidneys into the urine nor directly excreted into the bile.

The liver can take some of the free bilirubin out of the blood and conjugate (combine) it with glucuronic acid. The **conjugated bilirubin** is water soluble and can be secreted into the bile. Once it enters the small intestine, it is converted by bacteria into another pigment, **urobilinogen,** which is partially responsible for the color of the feces. About 30% to 50% of the urobilinogen, however, is absorbed by the small intestine and enters the hepatic portal blood. Some of this is returned to the small intestine in an *enterohepatic circulation;* the rest enters the general circulation (fig. 20.29). The urobilinogen in plasma, unlike free bilirubin, is not attached to albumin; therefore, it is easily filtered by the kidneys into the urine, giving urine its characteristic yellow color.

The **bile salts** are derivatives of cholesterol that have two to four polar groups on each molecule. The principal bile salts are *cholic acid* and *chenodeoxycholic acid.* In aqueous solutions, these molecules form aggregates, or clusters, known as **micelles** (*mi-selz'*). The nonpolar parts are located in the central region of the micelle (away from water), whereas the polar groups face water around the periphery of the micelle. Lecithin, cholesterol, and other lipids enter these micelles in a process that aids the digestion and absorption of fats.

Detoxification of the Blood

The liver can remove hormones, drugs, and other biologically active molecules from the blood by (1) excretion of these compounds in the bile, (2) phagocytosis by Kupffer cells, or (3) chemical alteration of these molecules by hepatocytes. Ammonia, for example, is a toxic molecule produced by bacteria in the large intestine. Upon receiving the ammonia through the hepatic portal vein, liver enzymes convert ammonia into less toxic *urea* molecules. Urea molecules leave the liver through the draining venous blood and are eventually excreted by the kidneys in the urine. Similarly, the liver converts metabolic toxins into *bilirubin* and *uric acid.*

Steroid hormones and other nonpolar compounds, including many drugs, are inactivated in their passage through the liver by alteration of their chemical structures. The liver has enzymes that convert these molecules into forms that are more water soluble by *hydroxylation* (the addition of OH groups) or by *conjugation* with highly polar groups such as sulfate and glucuronic acid. Derivatives of steroid hormones and drugs are less active than the original forms and, because of their increased water solubility, are more easily excreted by the kidneys.

Secretion of Glucose, Triglycerides, and Ketone Bodies

The liver helps to regulate the blood glucose concentration by either removing glucose from or adding glucose to the blood, according to the needs of the body. After a carbohydrate-rich meal, the liver can remove some glucose from the hepatic portal blood and convert it into glycogen through **glycogenesis.** It can also convert glucose into triglycerides through **lipogenesis.** During fasting, the liver secretes glucose into the blood. This glucose can be derived from the breakdown of stored glycogen in a process called **glycogenolysis,** or it can be

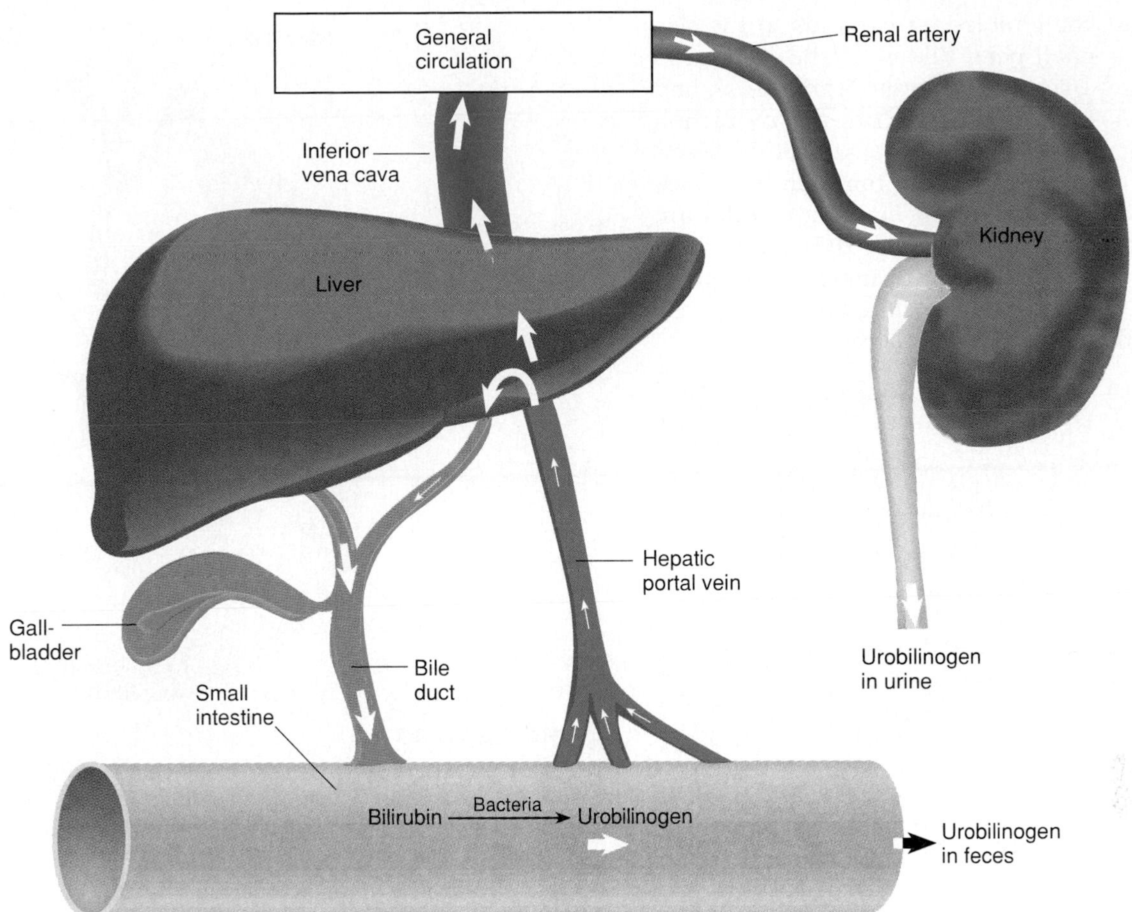

FIGURE 20.29
Enterohepatic circulation. Substances secreted in the bile may be absorbed by the intestinal epithelium and recycled to the liver via the hepatic portal vein.

produced by the conversion of noncarbohydrate molecules (such as amino acids) into glucose in a process called **gluconeogenesis.** The liver also contains the enzymes required to convert free fatty acids into ketone bodies in a process called **ketogenesis.** Ketone bodies are secreted into the blood in large amounts during fasting.

Production of Blood Plasma Proteins

Blood plasma albumin and most of the blood globulins (not immunoglobulins) are produced by the liver. Albumin accounts for about 70% of the total blood plasma protein and contributes most to the oncotic pressure of the blood. The globulins produced by the liver transport cholesterol, triglycerides, and steroid and thyroid hormones. In addition, globulins inhibit trypsin activity and assist blood clotting.

Inside Information—In ancient Greece and Rome, the liver was regarded as the seat of life. Before going into battle, Greek and Roman warriors would sacrifice animals to the gods. A healthy liver and bright red blood meant victory was at hand; a diseased liver predicted defeat.

Gallbladder

The gallbladder is a saclike organ attached to the inferior surface of the liver. This organ stores and concentrates bile, which drains to it from the liver by way of the **hepatic ducts, bile duct,** and **cystic duct.** A sphincter valve at the neck of the gallbladder allows a storage capacity of about 35 to 100 ml. The inner mucosal layer of the gallbladder is arranged in rugae similar to those of the stomach. When filled with bile, the gallbladder expands to the size and shape of a small pear. Bile is a yellowish-green fluid containing bile salts, bilirubin, cholesterol, and other compounds previously discussed. Contraction of the muscularis in the wall of the gallbladder ejects bile from the cystic duct into the **common bile duct,** through which it is conveyed into the duodenum when it is needed for digestion (fig. 20.30). When the small intestine is empty, the **sphincter of ampulla** (sphincter of Oddi) closes, and bile backs up in the cystic duct to the gallbladder for storage.

> *Approximately 20 million Americans have* gallstones*—small, hard mineral deposits that can obstruct the cystic or common bile ducts, producing severe pain. Gallstones are formed from cholesterol crystals that become hardened by the precipitation of inorganic salts. Gallstones may sometimes be dissolved by oral ingestion of bile acids, or they may have to be removed surgically. In a relatively new technique called* lithotripsy, *gallstones are fragmented by ultrasound vibrations.*

sphincter of Oddi: from Ruggero Oddi, nineteenth-century Italian physician

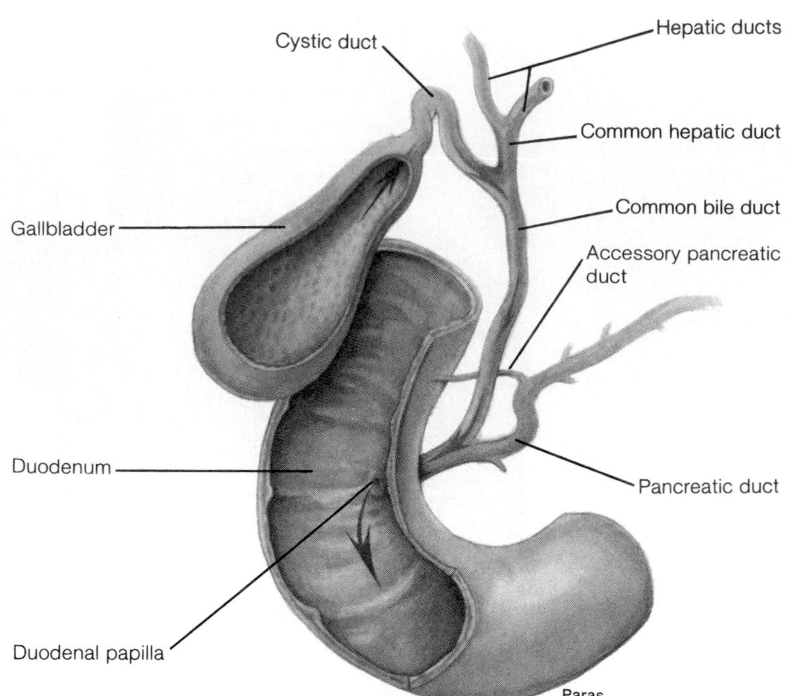

FIGURE 20.30
The pancreatic duct joins the common bile duct to empty its secretions through the duodenal papilla into the duodenum. The release of bile and pancreatic juice into the duodenum is controlled by the sphincter of ampulla (see fig. 20.18).

20.9 Pancreas

In addition to secreting the hormones insulin and glucagon, the pancreas produces an exocrine secretion containing important digestive enzymes.

Structure of the Pancreas

The soft, lobulated pancreas is known as a *mixed gland* because it has both the exocrine and endocrine functions. The endocrine function (see chapter 13, module 13.6) is performed by clusters of cells called the **pancreatic islets** (islets of Langerhans). The islet cells secrete the hormones *insulin* and *glucagon* into the blood. As an exocrine gland, the pancreas secretes *pancreatic juice* into the **pancreatic duct,** which empties into the duodenum (fig. 20.31).

The pancreas is positioned horizontally along the posterior abdominal wall, adjacent to the greater curvature of the stomach. It is about 12.5 cm (6 in.) long and 2.4 cm (1 in.) thick and consists of an expanded **head** near the duodenum, a centrally located **body,** and a tapering **tail.** All but a portion of the head is retroperitoneal. Within the lobules of the pancreas are the exocrine secretory units, called **acini.** Each acinus (fig. 20.31) consists of a single layer of acinar cells surrounding a lumen into which the constituents of pancreatic juice are secreted.

The pancreas is innervated by branches of the *celiac plexus.* The glandular portion of the pancreas receives parasympathetic innervation, whereas the pancreatic blood vessels receive sympathetic innervation. The pancreas is supplied with blood by the *pancreatic branch of the splenic artery* arising from the celiac artery and by the *pancreatoduodenal branches* from the superior mesenteric artery. Venous blood is returned through the *splenic* and *superior mesenteric veins* into the *hepatic portal vein.*

pancreas: Gk. *pan,* all; *kreas,* flesh
islets of Langerhans: from Paul Langerhans, German anatomist, 1847–88
acinus: L. *acinus,* grape

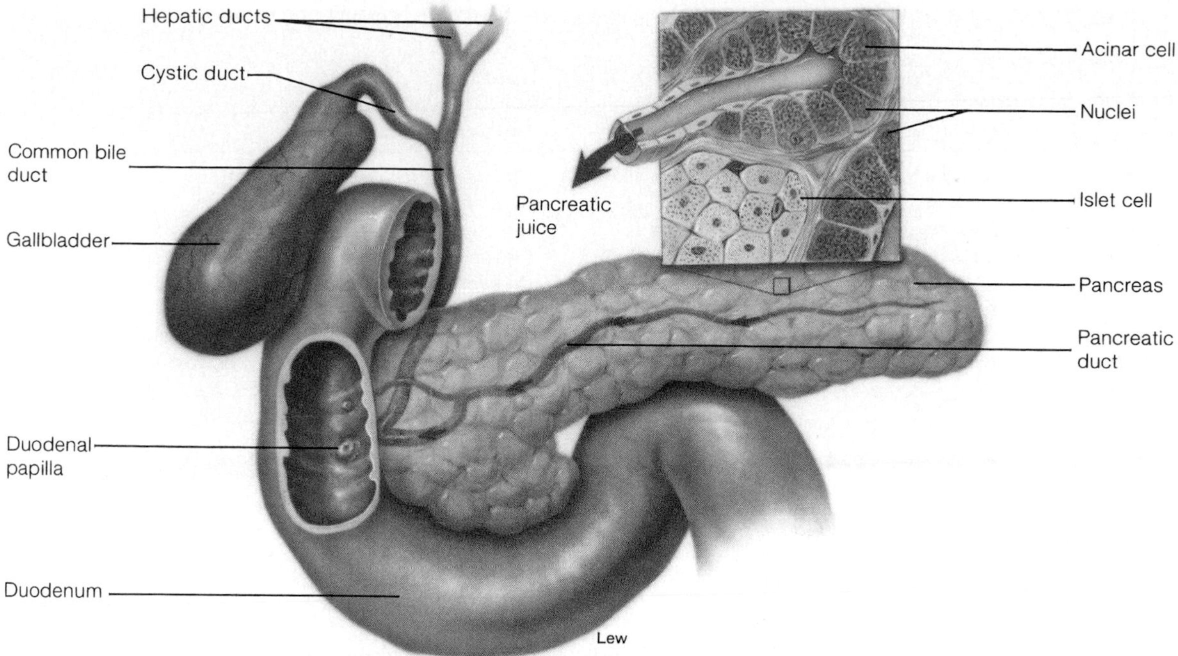

FIGURE 20.31
The pancreas is both an exocrine and an endocrine gland. Pancreatic juice—the exocrine product—is secreted by acinar cells into the pancreatic duct. Scattered islands of cells, called pancreatic islets (islets of Langerhans), secrete the hormones insulin and glucagon into the blood.

Pancreatic cancer *has the worst prognosis of all types of cancer. A variety of symptoms that can also be signals of diseases other than pancreatic cancer makes diagnosis difficult. For this reason, pancreatic cancer is often advanced before it is detected. Pancreatic surgery presents a problem because the soft, spongy tissue is difficult to suture.*

Pancreatic Juice

Pancreatic juice contains water, bicarbonate, and a wide variety of digestive enzymes that are delivered to the duodenum. These enzymes include (1) **amylase,** which digests starches; (2) **trypsin,** which digests proteins; and (3) **lipase,** which digests triglycerides. Other pancreatic enzymes are indicated in table 20.3. It should be noted that the complete digestion of food molecules in the small intestine requires the action of both pancreatic enzymes and brush border enzymes.

Most pancreatic enzymes are produced as inactive molecules, or *zymogens,* which reduces the risk of self-digestion within the pancreas. The inactive form of trypsin, called *trypsinogen,* is activated within the small intestine by the brush border enzyme *enterokinase.* Enterokinase converts trypsinogen to active trypsin. Trypsin, in turn, activates the other zymogens of pancreatic juice (fig. 20.32) by cleaving off polypeptide sequences that inhibit the activity of these enzymes.

The activation of trypsin is, therefore, the triggering event for the activation of other pancreatic enzymes. Actually, the pancreas does produce small amounts of active trypsin, yet the other enzymes do not become active until pancreatic juice has entered the duodenum. This is because pancreatic juice also contains a small protein called *pancreatic trypsin inhibitor* that attaches to trypsin and inactivates it in the pancreas.

Inflammation of the pancreas occurs when the various safeguards against self-digestion fail. Acute pancreatitis *is believed to be caused by the reflux of pancreatic juice and bile from the duodenum into the pancreatic duct. The leakage of trypsin into the blood also occurs, but trypsin is inactive in the blood because of the inhibitory action of various plasma proteins. Pancreatic amylase may also leak into the blood, but it is not active because its substrate (glycogen) is not present in blood. Pancreatic amylase activity can be measured in vitro, however, and these measurements are commonly performed to assess the health of the pancreas.*

TABLE 20.3 Enzymes Contained in Pancreatic Juice

Enzyme	Zymogen	Activator	Action
Trypsin	Trypsinogen	Enterokinase	Cleaves internal peptide bonds
Chymotrypsin	Chymotrypsinogen	Trypsin	Cleaves internal peptide bonds
Elastase	Proelastase	Trypsin	Cleaves internal peptide bonds
Carboxypeptidase	Procarboxypeptidase	Trypsin	Cleaves last amino acid from carboxyl-terminal end of polypeptide
Phospholipase	Prophospholipase	Trypsin	Cleaves fatty acids from phospholipids such as lecithin
Lipase	None	None	Cleaves fatty acids from glycerol
Amylase	None	None	Digests starch to maltose and short chains of glucose molecules
Cholesterolesterase	None	None	Releases cholesterol from its bonds with other molecules
Ribonuclease	None	None	Cleaves RNA to form short chains
Deoxyribonuclease	None	None	Cleaves DNA to form short chains

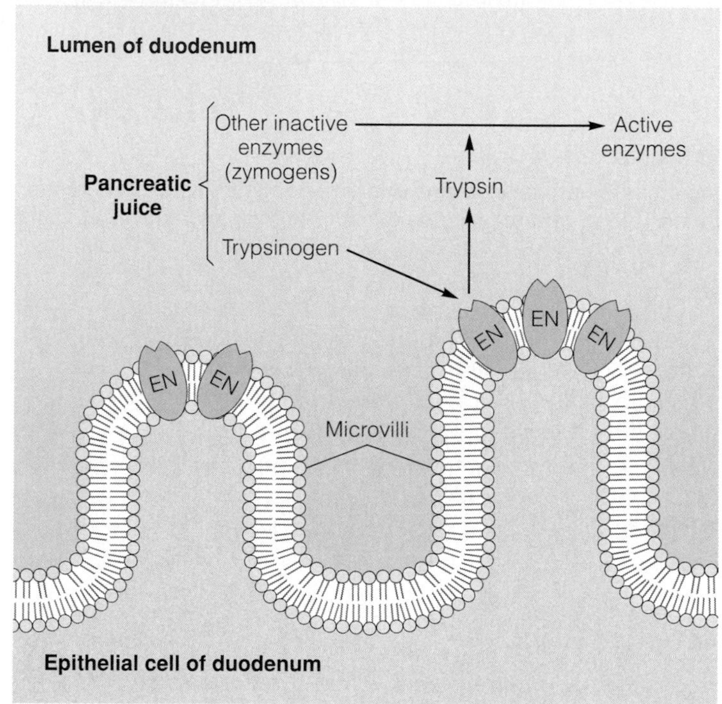

FIGURE 20.32
The pancreatic protein-digesting enzyme trypsin is secreted in an inactive form known as trypsinogen. This inactive enzyme is activated by a brush border enzyme, enterokinase (EN), located in the cell membrane of microvilli. Active trypsin in turn activates other zymogens (inactive enzymes) in pancreatic juice.

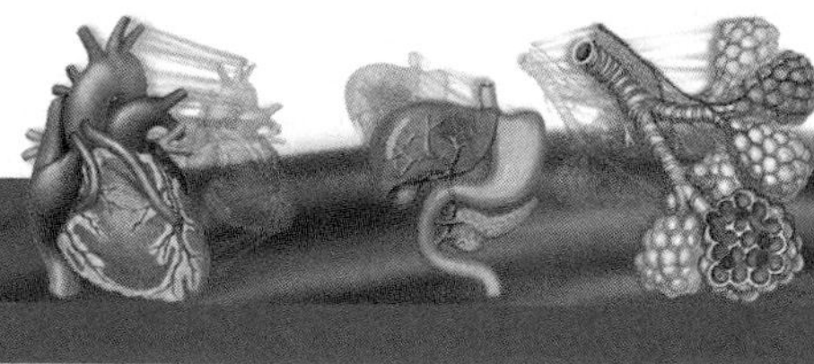

20.10 Digestion and Absorption of Carbohydrates, Proteins, and Lipids

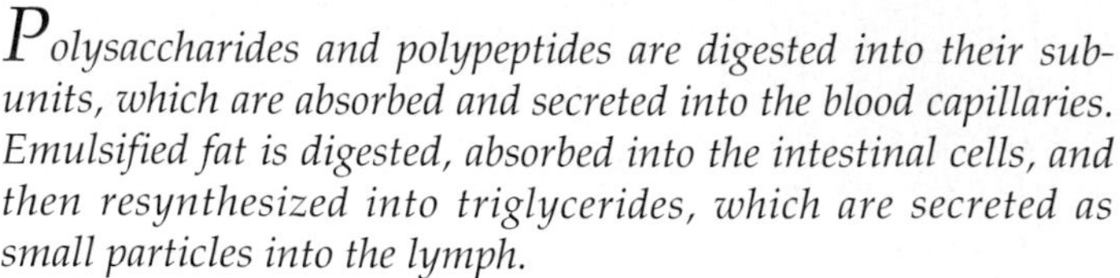

Polysaccharides and polypeptides are digested into their subunits, which are absorbed and secreted into the blood capillaries. Emulsified fat is digested, absorbed into the intestinal cells, and then resynthesized into triglycerides, which are secreted as small particles into the lymph.

The caloric (energy) value of food is derived predominantly from its content of carbohydrates, lipids, and proteins. In the average American diet, carbohydrates account for approximately 50% of the total calories, protein accounts for 11% to 14%, and lipids make up the balance. These food molecules consist primarily of long combinations of subunits (monomers), which must be digested by *hydrolysis reactions* into *free monomers* before absorption can occur. The characteristics of the major digestive enzymes are summarized in table 20.4.

TABLE 20.4 **Characteristics of the Major Digestive Enzymes**

Enzymes	Site of Production	Source	Substrate	Optimum pH	Products
Salivary amylase	Mouth	Saliva	Starch	6.7	Maltose
Pepsin	Stomach	Gastric glands	Protein	1.6–2.4	Shorter polypeptides
Pancreatic amylase	Duodenum	Pancreatic juice	Starch	6.7–7.0	Maltose, maltriose, and oligosaccharides
Trypsin, chymotrypsin, carboxypeptidase			Polypeptides	8.0	Amino acids, dipeptides, and tripeptides
Pancreatic lipase			Triglycerides	8.0	Fatty acids and monoglycerides
				5.0–7.0	
Maltase		Epithelial membranes	Maltose	5.0–7.0	Glucose
Sucrase			Sucrose	5.8–6.2	Glucose + fructose
Lactase			Lactose	8.0	Glucose + galactose
Aminopeptidase			Polypeptides		Amino acids, dipeptides, and tripeptides

Digestion and Absorption of Carbohydrates

Most of the carbohydrates we consume are in the form of starch, which is a long polysaccharide of glucose monomers in straight chains with occasional branchings. The most commonly ingested sugars are the disaccharides sucrose (table sugar, consisting of glucose and fructose) and lactose (milk sugar, consisting of glucose and galactose).

The digestion of starch begins in the mouth with the action of **salivary amylase,** or **ptyalin** (*ti'ă-lin*). This enzyme cleaves some of the bonds between adjacent glucose molecules. However, most people do not chew their food long enough for sufficient digestion to occur in the mouth. The digestive action of salivary amylase stops when the bolus enters the stomach because the low pH of gastric juice inactivates the enzyme.

The digestion of starch, therefore, occurs mainly in the duodenum as a result of the action of **pancreatic amylase.** This enzyme cleaves the straight chains of starch to produce the disaccharide *maltose* and the trisaccharide *maltriose.* Pancreatic amylase, however, cannot hydrolyze the bonds between glucose molecules at the branch points in the starch. As a result, short branched chains of glucose molecules, called *oligosaccharides,* are released together with maltose and maltriose by the activity of this enzyme (fig. 20.33).

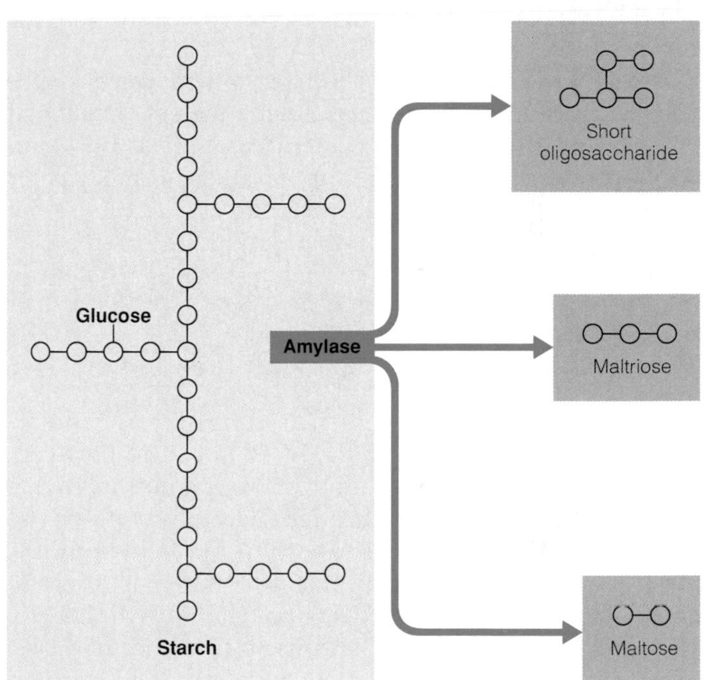

FIGURE 20.33
Pancreatic amylase digests starch into maltose, maltriose, and short oligosaccharides containing branch points in the chain of glucose molecules.

Maltose, maltriose, and oligosaccharides released from partially digested starch, together with the disaccharides sucrose and lactose, are hydrolyzed to their monosaccharides by brush border enzymes. These enzymes are located on the microvilli of the epithelial cells in the mucosa of the small intestine. The absorption of monosaccharides across the membrane of the microvilli occurs by means of *coupled transport.* In this process, glucose binds to the same carrier as Na^+ and enters the epithelial cell as Na^+ diffuses down its electrochemical gradient. This is a type of active transport, since energy from ATP is needed to maintain the Na^+ gradient. Glucose is then secreted from the epithelial cells into capillaries within the villi.

Digestion and Absorption of Proteins

Protein digestion begins in the stomach with the action of *pepsin.* Pepsin activity liberates some amino acids, but it produces mainly short-chain polypeptides. While this activity helps to produce a more homogenous chyme, it is not essential for the complete digestion of protein that occurs in the small intestine.

Most protein digestion occurs in the duodenum and jejunum. The pancreatic juice enzymes **trypsin, chymotrypsin** (*ki"mŏ-trip'sin*), and **elastase** cleave peptide bonds within the interior of the polypeptide chains. These enzymes are thus classified as *endopeptridases.* Enzymes that remove amino acids from the ends of polypeptide chains, by contrast, are *exopeptidases.* These include the pancreatic juice enzyme **carboxypeptidase,** which removes amino acids from the carboxyl-terminal end of polypeptide chains, and the brush border enzyme **aminopeptidase.** Aminopeptidase cleaves amino acids from the amino-terminal end of polypeptide chains.

As a result of the action of these enzymes, polypeptide chains are digested into free amino acids, dipeptides, and tripeptides. The free amino acids are absorbed by cotransport with Na^+ into the epithelial cells and secreted into blood capillaries. The dipeptides and tripeptides enter epithelial cells by a different carrier system, but they are then digested within these cells into amino acids, which are secreted into the blood (fig. 20.34).

Newborns appear to be capable of absorbing a substantial amount of undigested proteins (hence, they can absorb antibodies from their mother's first milk); in adults, however, only the free amino acids normally enter the hepatic portal vein. An interesting exception is the protein toxin that causes botulism, produced by the bacterium *Clostridium botulinum.* This protein is resistant to digestion and is absorbed into the blood.

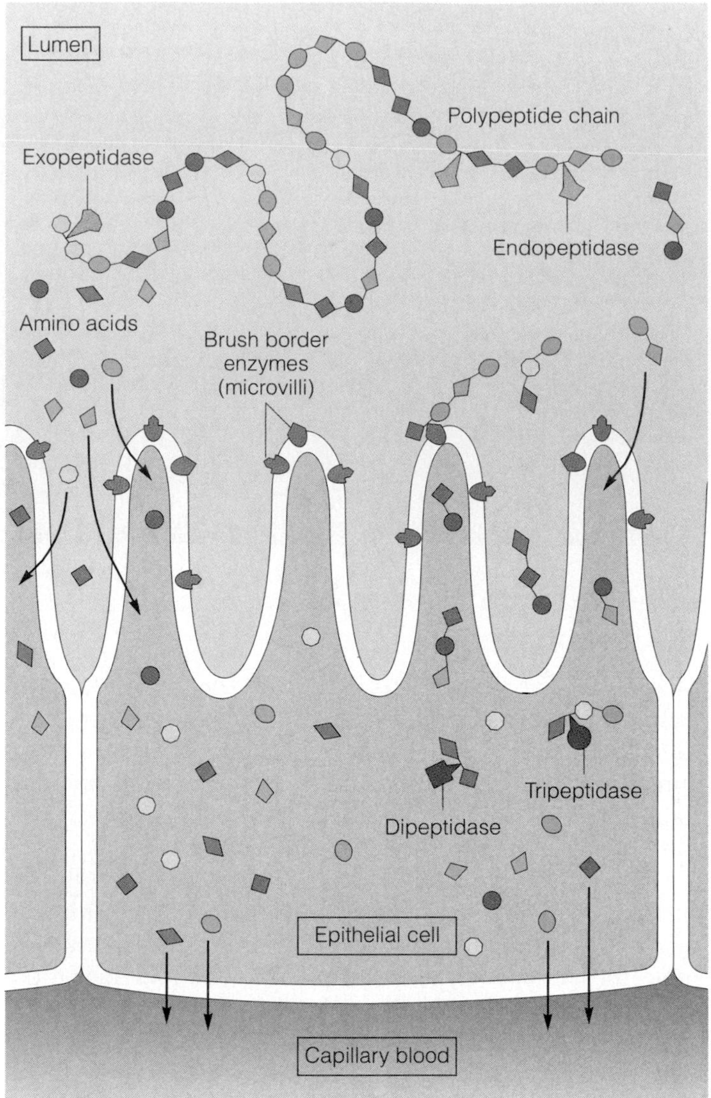

FIGURE 20.34
Polypeptide chains are digested into free amino acids, dipeptides, and tripeptides by the action of pancreatic juice enzymes and brush border enzymes. The amino acids, dipeptides, and tripeptides enter duodenal epithelial cells. Dipeptides and tripeptides are hydrolyzed into free amino acids within the epithelial cells, and these products are secreted into capillaries that carry them to the hepatic portal vein.

Digestion and Absorption of Lipids

The salivary glands and stomach of newborns produce lipid-digesting enzymes called *lipases.* In adults, however, little fat digestion occurs until the fat globules in chyme arrive in the duodenum. Through mechanisms to be described shortly, the arrival of fat in the duodenum is a stimulus for the secretion of bile. Mixed micelles of bile salts, lecithin, and cholesterol are secreted into the duodenum and break up fat droplets into much finer particles. This process, called *emulsification,* results in the formation of tiny *emulsification droplets* of triglycerides.

Digestion of Lipids

The emulsification of fat aids digestion because the smaller and more numerous emulsification droplets present a much greater surface area than that of the unemulsified fat droplets that entered the duodenum originally. Fat digestion occurs at the surface of the droplets through the enzymatic action of **pancreatic lipase.** This enzyme is aided in its action by a pancreatic protein called *colipase* that coats the emulsification droplets and anchors lipase to the droplets. Through hydrolysis, lipase removes two of the three fatty acids from each triglyceride molecule, liberating *free fatty acids* and *monoglycerides* (fig. 20.35). **Phospholipase** likewise digests phospholipids (such as lecithin) into fatty acids and lysolecithin (the remainder of the lecithin molecule after two fatty acids have been removed).

Free fatty acids, monoglycerides, and lysolecithin are more water soluble than the undigested lipids and quickly become associated with micelles of bile salts, lecithin, and cholesterol (fig. 20.36). These micelles then move to the brush border of the intestinal epithelium, where absorption occurs.

Absorption of Lipids

Free fatty acids, monoglycerides, and lysolecithin leave the micelles and pass through the membrane of the microvilli

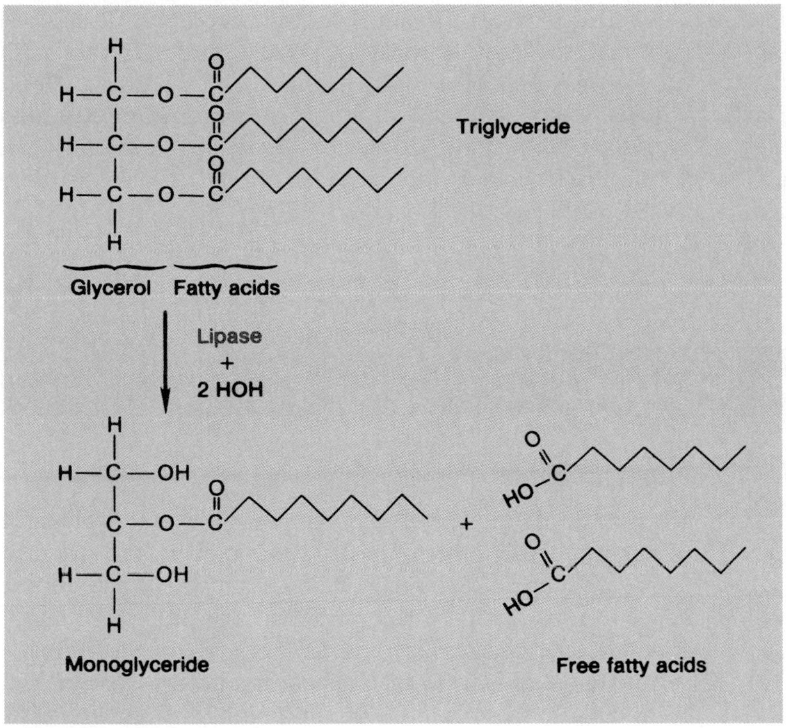

Triglyceride

Glycerol Fatty acids

Lipase + 2 HOH

Monoglyceride

Free fatty acids

FIGURE 20.35

Pancreatic lipase digests fat (triglycerides) by cleaving off the first and third fatty acids. This produces free fatty acids and monoglycerides. Sawtooth structures indicate hydrocarbon chains in the fatty acids.

to enter the epithelial cells of the small intestine. There is also some evidence that the micelles may be transported intact into the epithelial cells, and that the lipid digestion products may be removed intracellularly from the micelles. These products are used to *resynthesize* triglycerides and phospholipids within the epithelial cells. This is different from the absorption of amino acids and monosaccharides, which pass through the epithelial cells without being altered.

Triglycerides, phospholipids, and cholesterol are then combined with protein inside the epithelial cells to form small particles called **chylomicrons** (*ki"lŏ-mi'kronz*). These tiny lipid and protein combinations pass into the lymphatic capillaries of the intestinal villi. Absorbed lipids pass through the lymphatic system, eventually entering the venous blood by way of the thoracic duct (see fig. 17.1). Amino acids and monosaccharides are absorbed into the bloodstream and eventually enter the hepatic portal vein.

Transport of Lipids in Blood

Once the chylomicrons are in the blood, their triglyceride content is removed by the enzyme **lipoprotein lipase,** which is attached to the endothelium of blood vessels. This enzyme provides free fatty acids and glycerol for use by the tissue cells by hydrolyzing

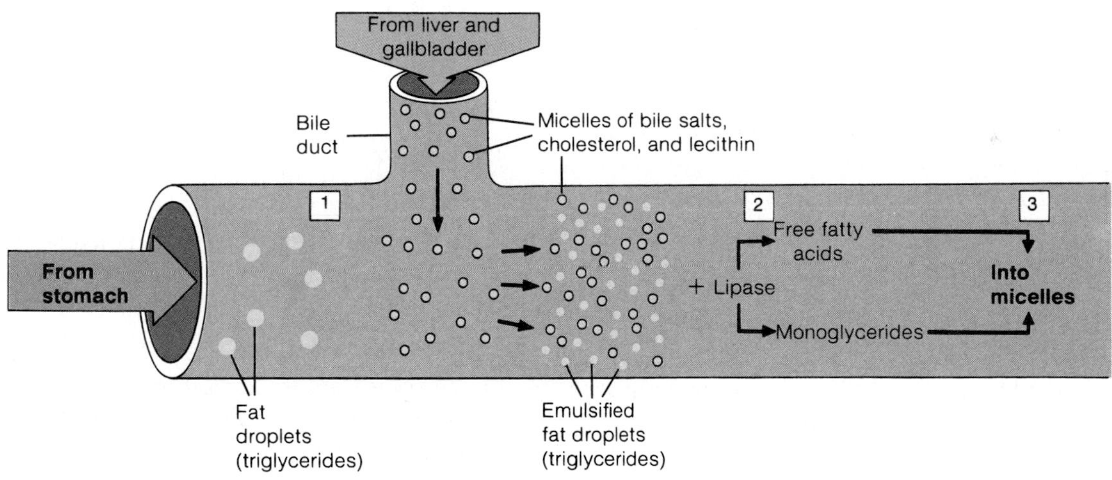

FIGURE 20.36

Steps in the digestion of fat (triglycerides) and the entry of fat digestion products (fatty acids and monoglycerides) into micelles of bile salts secreted by the liver into the duodenum.

triglycerides. The remaining *remnant particles,* containing cholesterol, are taken up by the liver by the process of *endocytosis.*

Cholesterol and triglycerides produced by liver cells are combined with other proteins and secreted into the blood as **very low-density lipoproteins (VLDLs),** which serve to deliver these triglycerides to different organs. Once the triglycerides have been removed, the VLDLs are converted to **low-density lipoproteins (LDLs),** which transport cholesterol to various organs, including blood vessels. Excess cholesterol is returned from these organs to the liver attached to **high-density lipoproteins (HDLs).** (For a discussion of "good" versus "bad" cholesterol, see the Synoptic Spotlight in chapter 15, p. 397.)

20.11 Regulation of the Digestive System

Neural and hormonal mechanisms coordinate the activities of different regions of the GI tract and regulate the actions of the liver, gallbladder, and pancreas.

The motility and glandular secretions of the GI tract are, to a large degree, regulated intrinsically. Neural and endocrine control mechanisms, however, can stimulate or inhibit these intrinsic mechanisms to help coordinate the different stages of digestion. A hormonal discussion of the digestive system is particularly interesting because many of the organs of the GI tract are not only target sites for the action of various hormones but themselves function as endocrine glands. The hormones produced by organs of the GI tract include *gastrin, secretin, cholecystokinin,* and *gastric inhibitory peptide* (table 20.5).

Regulation of Stomach Function

Hormones and autonomic innervation (parasympathetic and sympathetic) extrinsically control stomach contraction. To some extent, these activities are also intrinsic. Peristaltic waves that move chyme through the pyloric sphincter, for example, are initiated spontaneously by pacesetter cells in the body of the stomach. The secretion of HCl and pepsinogen, likewise, can be stimulated in the absence of neural and hormonal influences by the presence of cooked or partially digested protein in the stomach.

The extrinsic control of stomach function is divided into three phases: *cephalic, gastric,* and *intestinal.*

Cephalic Phase

This phase of stomach regulation refers to parasympathetic stimulation through the vagus nerves. Activation of the vagus nerves can stimulate HCl and pepsinogen secretion by two mechanisms: (1) direct vagal stimulation of the gastric parietal and chief cells (the primary mechanism) and (2) vagal stimulation of gastrin secretion by the G cells, which in turn stimulates the parietal and chief cells to secrete HCl and pepsinogen, respectively. The cephalic phase continues into the first 30 minutes of a meal, but gradually declines as the process moves into the gastric phase.

Gastric Phase

The arrival of food into the stomach stimulates this phase of regulation. Gastric secretion is stimulated by the stomach being stretched and by the chemical nature of the chyme being formed. While intact proteins have little stimulatory effect, short polypeptides and amino acids in the stomach stimulate the G cells to secrete gastrin and the parietal and chief cells to secrete HCl and pepsinogen, respectively. Since gastrin also stimulates HCl and pepsinogen secretion, a *positive feedback mechanism* develops (fig. 20.37). Glucose in the chyme, by contrast, has no effect on gastric secretion, and the presence of fat actually inhibits acid secretion.

Secretion of HCl during the gastric phase is also regulated by a *negative feedback mechanism.* As the pH of gastric juice drops, so does the secretion of gastrin—at a pH of 2.4, gastrin secretion is reduced, and at a pH of 1.0 gastrin secretion shuts off entirely. The secretion of HCl declines accordingly. The presence of proteins and polypeptides in the stomach helps to buffer the acid and prevent a rapid fall in gastric pH; as a result, more acid can be secreted when proteins are present than when they are absent. As the stomach is emptied, the protein buffers leave, the pH falls, and the secretion of gastrin and HCl is accordingly inhibited.

Intestinal Phase

This phase refers to the inhibition of gastric activity as chyme enters the duodenum, which is due to both a neural reflex originating from the duodenum as it is stretched and to the activity of a hormone secreted from the duodenum. This hormone, known as *enterogastrone,* is secreted in response to the stimulus of fat in the chyme. As a result, fatty meals take longer to digest and they remain in the stomach longer than meals that are low in fat.

Regulation of Pancreatic Juice and Bile Secretion

The arrival of chyme into the duodenum stimulates the secretion of pancreatic juice and bile by both neural reflexes and by the secretion of the duodenal hormones *cholecystokinin* (*CCK*) and *secretin* (table 20.6). The release of

TABLE 20.5 **Physiological Effects of Gastrointestinal Hormones**

Secreted by	Hormone	Effects
Stomach	Gastrin	Stimulates parietal cells to secrete HCl Stimulates chief cells to secrete pepsinogen Maintains structure of gastric mucosa
Small intestine	Secretin	Stimulates water and bicarbonate secretion in pancreatic juice Potentiates actions of cholecystokinin on pancreas
Small intestine	Cholecystokinin (CCK)	Stimulates contraction of gallbladder Stimulates secretion of pancreatic juice enzymes Potentiates action of secretin on pancreas Maintains structure of exocrine pancreas (acini)
Small intestine	Gastric inhibitory peptide (GIP)	Inhibits gastric emptying (?) Inhibits gastric acid secretion (?) Stimulates secretion of insulin from endocrine pancreas (islets of Langerhans)

TABLE 20.6 **Regulation of Pancreatic Juice and Bile Secretion by the Hormones Secretin and Cholecystokinin (CCK) and by the Neurotransmitter Acetylcholine, Released from Parasympathetic Nerve Endings**

Feature	Secretin	CCK	Acetylcholine (Vagus Nerve)
Stimulus for release	Decrease in duodenal pH below 4.5 due to acidity of chyme	Fat and protein in chyme	Sight, smell of food; distension of stomach
Second messenger	Cyclic AMP	Ca^{++}	Ca^{++}
Effect on pancreatic juice	Stimulates water and bicarbonate secretion; potentiates action of CCK	Stimulates enzyme secretion; potentiates action of secretin	Stimulates enzyme secretion
Effect on bile	Stimulates secretion	Potentiates action of secretin; stimulates contraction of gallbladder	Stimulates contraction of gallbladder

secretin from specialized duodenal endocrine cells occurs when the duodenal pH drops below 4.5. This fall in pH is temporary, however, because the acidic chyme is rapidly neutralized by alkaline pancreatic juice. The secretion of CCK occurs in response to the fat content of incoming chyme. Secretin also stimulates the production of bicarbonate by the pancreas. Since bicarbonate neutralizes the acidic chyme and since secretin is released in response to the low pH of chyme, this completes a negative feedback loop in which the effects of secretin inhibit its secretion. Cholecystokinin, by contrast, stimulates the production of pancreatic enzymes such as trypsin, lipase, and amylase.

The liver secretes bile continuously, but this secretion increases following a meal due to the release of secretin and CCK from the duodenum. Secretin is the major stimulator of bile secretion by the liver, and CCK enhances this effect. The arrival of chyme in the duodenum also causes the gallbladder to contract and eject bile. Gallbladder contraction occurs in response to neural reflexes from the duodenum and in response to stimulation by CCK, but not in response to secretin.

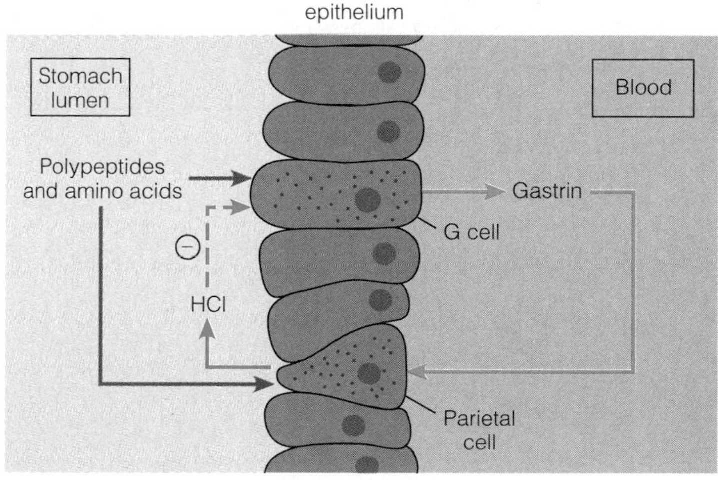

FIGURE 20.37
Stimulation of gastric acid (HCl) secretion by the presence of proteins in the stomach lumen and by the hormone gastrin. The secretin of gastrin is inhibited by gastric acidity, forming a negative feedback loop.

The Inimitable Liver

Although every body organ serves some purpose, many can be artificially supported or even replaced in the event of failure or damage. Kidney function can be replicated temporarily or permanently by hemodialysis. A failing heart can be supported by an intra-aortic balloon pump or a left ventricular assist device. The heart can be replaced by a mechanical version until a suitable donor heart becomes available for transplant. The function of the pancreas can be simulated by oral enzyme replacement and insulin injections. Poorly functioning lungs can be supported by mechanical ventilation almost indefinitely. Limbs can be replaced with prostheses; skin grafts made of biological textiles can temporarily cover burns and wounds; and metallic joints can serve as sturdy replacements for worn-out knees and hips. Through increasingly sophisticated biomedical engineering, sight can be restored to the blind, hearing to the deaf, and movement to the paralyzed.

Some organs, however, have functions so complex that modern technology has been unable to replicate the basic services they provide. The brain is an obvious example. Could its function be replaced without compromising the fundamental element that defines you as human? And with someone else's brain in your skull, if such a transplantation were possible, would you still be you?

Second only to the brain in organ complexity, and not as obviously so to most people, is the liver. Although it was revered in ancient times as the center of the soul, we tend to immediately associate "liver" with the homely food item served up in roadside diners under a pile of onions. Even if we had the opportunity to look at a specimen of human liver, we would probably see little beyond a mass of homogeneous dark red tissue.

Despite its less than impressive appearance, no one has been able to devise a replacement part for the liver. Why this should be so can be best appreciated by examining the physiological ramifications of liver failure—caused for example, by long-standing *cirrhosis,* in which the liver is irreversibly scarred and all of its functions eventually depressed.

- When old or damaged red blood cells are broken down, free bilirubin (part of the hemoglobin molecule) is a by-product that cannot be eliminated without chemical modification by the liver. The most visible sign of liver failure is *jaundice,* or yellow skin, which is caused by an excess of bilirubin in the blood and skin. Elevated bilirubin levels (often as high as 40 times normal) also cause yellowing of the sclera of the eye and insatiable itching of the skin.
- End-stage liver failure is characterized by low protein levels in the blood, since the liver is the

chief manufacturer of serum proteins. The clinical picture of malnutrition occurs despite normal dietary intake. As the oncotic pressure exerted by proteins to hold fluid in the vessels drops, interstitial fluid collects in the soft tissues of the extremities. This results in edema, primarily in the legs. There also may be abnormal pooling of fluid in the abdomen—a condition known as *ascites*. The main dangers of ascites are spontaneous infection and physical restriction of the diaphragm as a result of fluid mass.

- Glucose is stored in the liver in the form of glycogen, and then converted back to glucose as dictated by the body's metabolic demands. When cirrhosis replaces the cells responsible for storage and reconstitution of usable glucose with fibrous tissue, fatigue and weakness are the obvious outcomes.
- One of the most striking difficulties encountered in treating patients with alcoholic cirrhosis is the complication of severe bleeding. This is due to a deficit of certain blood clotting factors produced by the healthy liver. Any surgical procedure or traumatic injury, if accompanied by excessive bleeding, can be life threatening.
- Liver cirrhosis dramatically elevates the pressure in the hepatic portal system. The congested blood causes the vessels surrounding the stomach and esophagus to dilate and possibly break open and bleed into the upper GI tract. This condition, added to the clotting deficiencies described above, can trigger massive bleeding—a common cause of rapid demise in patients with long-standing cirrhosis.
- The huge array of molecules of varying size and structure that can be chemically metabolized by the liver is nearly incomprehensible. Potent drugs and potentially dangerous chemicals can be rendered inert by one pass through a healthy liver. Cirrhotic livers lose the ability to detoxify these substances; consequently, they remain active in the body for prolonged periods of time. Physicians must be cautious in administering certain drugs to patients with cirrhosis to avoid overdosage.

Considering the consequences of cirrhosis, it is clear that an artificial liver could not be just a pump or filter, as for the heart and kidneys. It would have to store energy molecules, chemically alter bile products, produce proteins and clotting factors, and clear the blood of toxins, drugs, and old or damaged red blood cells. No wonder, then, that the only option for a failed liver is a transplant—a healthy living organ replacing a dying one. The liver is truly an inimitable organ. Its functions simply cannot be duplicated, even by the most elaborately engineered spare parts.

Clinical Considerations

Pathogens and Poisons

The GI tract presents a suitable environment for an array of microorganisms. Many of these are beneficial, but some bacteria and protozoa can cause diseases. **Dysentery** is an inflammation of the intestinal mucosa, characterized by the discharge of loose stools that contain mucus, pus, and blood. The most common dysentery is *amoebic dysentery*, caused by the intestinal parasite *Entamoeba histolytica*. Cysts from this organism are ingested in contaminated food, and after the protective coat has been removed by HCl in the stomach, the active form invades the mucosal walls of the ileum and colon.

Food poisoning is caused by consuming the toxins produced by pathogenic bacteria. *Salmonella* is a bacterium that commonly infects food. **Botulism** is the most serious type of food poisoning, and is caused by consuming food contaminated with the toxin produced by the bacterium *Clostridium botulinum*. This organism is widely distributed in nature, and the spores it produces are frequently found on food being processed by canning. For this reason, food must be heated to 120°C (248°F) before it is canned. It is the toxins produced by the bacterium growing in the food that are pathogenic rather than the organisms themselves. The poison is a neurotoxin that is readily absorbed into the blood, at which point it can affect the nervous system.

Clinical Problems of the Teeth

Dental caries, or tooth decay, is the gradual decalcification of tooth enamel (fig. 20.38) and underlying dentin, caused by the acid products of bacteria. These bacteria thrive between teeth where food particles accumulate. They eventually form part of a thin layer of debris called *plaque* that covers teeth. The development of dental caries can be reduced by brushing at least once a day and by flossing between the teeth at regular intervals.

People over the age of 35 are particularly susceptible to **periodontal disease,** which involves inflammation and deterioration of the gums, alveolar sockets, periodontal membranes, and the cementum that covers the roots of teeth. Some of the symptoms are loosening of the teeth, bad breath, bleeding gums when brushing, and some edema. Periodontal disease may result from impacted plaque, cigarette smoking, crooked teeth, and poor diet. It accounts for 80% to 90% of tooth loss in adults.

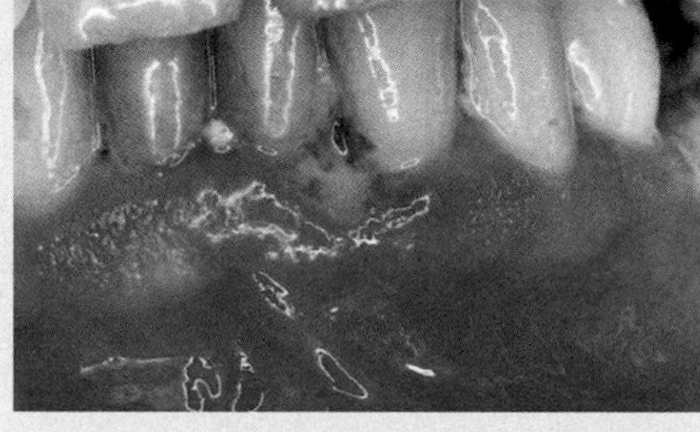

(a)

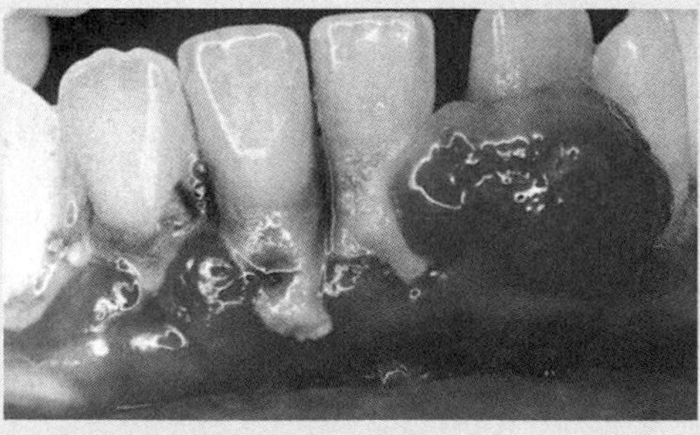

(b)

FIGURE 20.38
Clinical problems of the teeth. (*a*) Trench mouth and dental caries, and (*b*) pyrogenic granuloma and dental caries.

dysentery: Gk. *dys*, bad; *entera*, intestine

Disorders of the Liver and Gallbladder

The liver is a remarkable organ that has the ability to regenerate itself even if up to 80% has been removed. The most serious diseases of the liver (*hepatitis, cirrhosis,* and *hepatomas*) damage the liver throughout, so that it cannot repair itself. **Hepatitis** is inflammation of the liver. Certain chemicals may cause hepatitis, but generally it is caused by infectious viral agents. *Hepatitis A* (infectious hepatitis) is a viral disease transmitted through contaminated foods and liquids. *Hepatitis B* (serum hepatitis) is also caused by a virus and is transmitted in blood plasma during transfusions or by improperly sterilized needles and syringes. Other types of viral hepatitis are designated as *hepatitis C, D, E,* and *G.*

In **cirrhosis,** the liver becomes infused with fibrous tissue. This causes the liver tissue to break down and become filled with fat. Eventually, all functions of the liver are compromised (see the Synoptic Spotlight on page 536). Cirrhosis is most often the result of long-term alcohol abuse, but it can also result from malnutrition, hepatitis, or other infections.

Jaundice is a yellow staining of tissues produced by high blood concentrations of either free or conjugated bilirubin. Since free bilirubin is derived from hemoglobin, abnormally high concentrations of this pigment may result from an unusually high rate of red blood cell destruction. This can occur, for example, as a result of *Rh disease* (*erythroblastosis fetalis*) in an Rh positive baby born to a sensitized Rh negative mother (see chapter 14, module 14.4). Jaundice may also occur in otherwise healthy infants because excess red blood cells are normally destroyed at about the time of birth. This condition is called *physiological jaundice of the newborn,* and is not indicative of disease. In adults, jaundice due to high levels of conjugated bilirubin in the blood is commonly produced when the excretion of bile is blocked by gallstones (fig. 20.39).

Disorders of the GI Tract

Peptic ulcers are erosions of the mucous membranes of the stomach (fig. 20.40) or duodenum produced by the action of HCl. Exposure to agents that weaken the mucosal lining of the stomach, such as alcohol and aspirin, and abnormally high secretions of HCl increase the likelihood of developing peptic ulcers. Many people subject to chronic stress produce too much gastric acid and, as a result, develop a peptic ulcer. A relatively recent finding is that most people who have peptic ulcers are infected with a bacterium known as *Helicobacter pylori,* which resides in the GI tract. Clinical trials have demonstrated that antibiotics that eliminate this bacterium appear to help in the treatment of the peptic ulcers. It is now thought that *H. pylori* does not itself cause the ulcer, but rather contributes to the weakening of the mucosal barriers to gastric acid damage.

Enteritis, an inflammation of the mucosa of the small intestine, is frequently referred to as intestinal flu. Causes of enteritis include bacterial or viral infections, irritating foods or fluids (including alcohol), and emotional stress. The symptoms include abdominal pain, nausea, and diarrhea. *Diarrhea* is the passage of watery, unformed stools. This condition is symptomatic of inflammation, stress, and other body dysfunctions.

Peritonitis is inflammation of the peritoneum lining the abdominal cavity and covering the viscera. The causes of peritonitis include bacterial contamination of the peritoneal cavity through accidental or surgical wounds in the abdominal wall or perforation of the intestinal wall (as with a ruptured appendix).

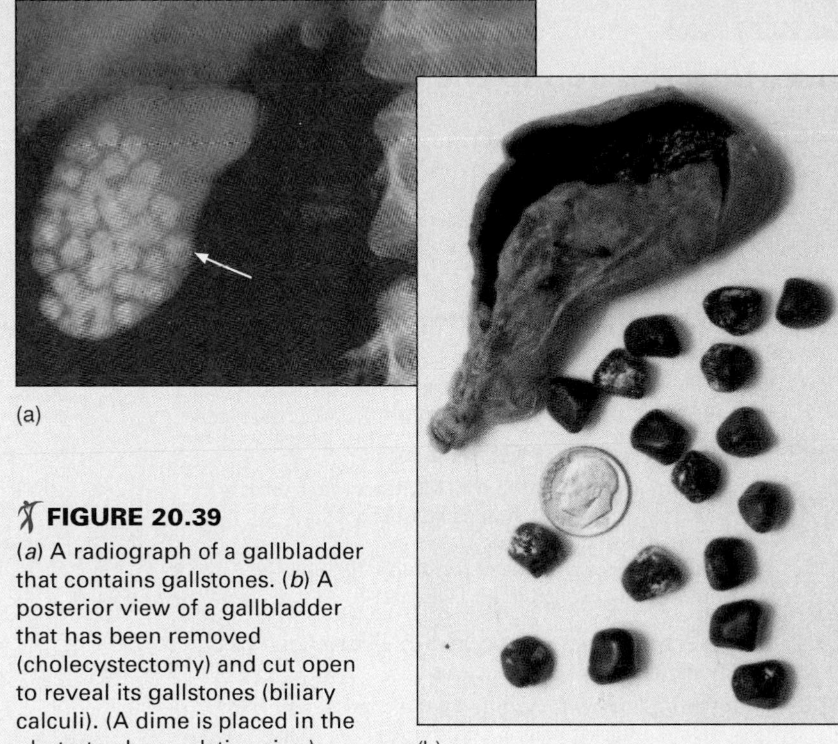

FIGURE 20.39
(*a*) A radiograph of a gallbladder that contains gallstones. (*b*) A posterior view of a gallbladder that has been removed (cholecystectomy) and cut open to reveal its gallstones (biliary calculi). (A dime is placed in the photo to show relative size.)

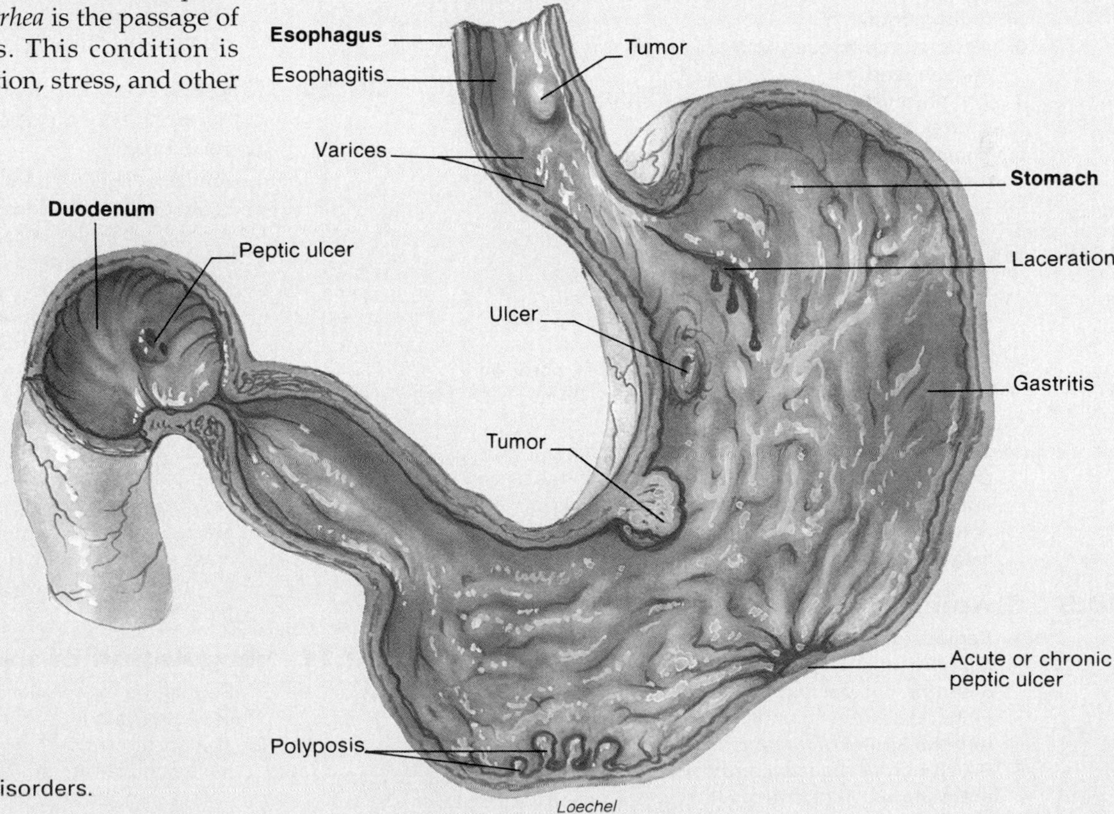

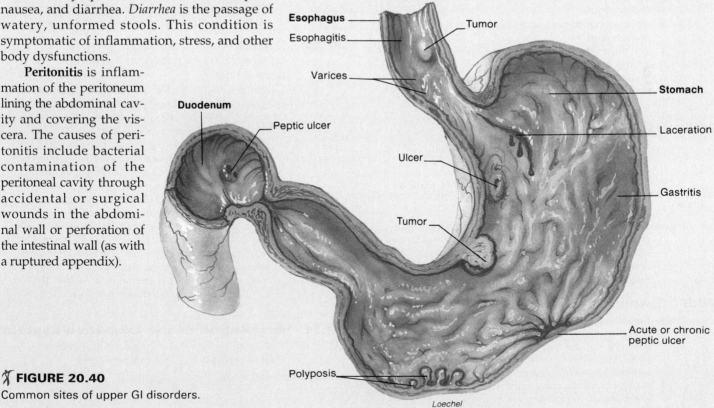

FIGURE 20.40
Common sites of upper GI disorders.

SUMMARY

20.1 Introduction to the Digestive System

1. The digestive system breaks down food into its component monomers, which are absorbed into the blood or lymph.
2. The digestive system consists of a GI tract and accessory digestive organs.

20.2 Serous Membranes of the GI Tract

1. Peritoneal membranes line the abdominal wall and cover the visceral organs.
2. The GI tract is supported by a double layer of peritoneum called the mesentery.

20.3 Layers of the GI Tract

1. The tunics of the abdominal GI tract are, from the inside outward, the mucosa, submucosa, muscularis, and serosa.
2. Composed of simple columnar epithelium, the mucosa is absorptive and secretory.

20.4 Mouth, Pharynx, and Associated Structures

1. The oral cavity is formed by the cheeks, lips, hard palate and soft palate, and tongue.
2. The most anterior teeth are the incisors and canines, which have one root each. The most lateral teeth are the premolars (bicuspids) and molars, which have two or three roots.
3. A tooth is composed primarily of enamel, dentin, and cementum.
4. The parotid, submandibular, and sublingual glands secrete saliva.
5. The muscular pharynx is a passageway connecting the oral and nasal cavities to the esophagus and trachea.

20.5 Esophagus and Stomach

1. Peristaltic waves of contraction push a bolus of food through the lower esophageal sphincter into the stomach.
2. The stomach consists of a cardia, fundus, body, and pylorus; the pylorus terminates at the pyloric sphincter.
3. The parietal cells of the gastric glands secrete HCl, and the chief cells secrete pepsinogen. In the acidic environment of gastric juice, pepsinogen is converted into the active protein-digesting enzyme called pepsin.

20.6 Small Intestine

1. Regions of the small intestine include the duodenum, jejunum, and ileum. The duodenum receives the common bile duct and the pancreatic duct.
2. Fingerlike extensions of mucosa, called villi, project into the lumen of the small intestine.
3. Brush border digestive enzymes are located in the membranes of the microvilli.

20.7 Large Intestine

1. The large intestine is divided into the cecum, colon, rectum, and anal canal. The appendix is attached to the inferior medial margin of the cecum.
2. The colon consists of ascending, transverse, descending, and sigmoid portions.
3. The large intestine functions mainly to form, store, and expel feces from the body. It also absorbs water and some electrolytes.

20.8 Liver and Gallbladder

1. The liver is divided into four lobes. Each lobe contains hepatic lobules composed of hepatocytes separated by capillary sinusoids.
2. Blood flows from small branches of the hepatic portal vein through capillary sinusoids to the central vein.
3. Bile flows within the hepatocyte plates, in canaliculi, to the bile ducts.
4. The liver detoxifies the blood and modifies the blood plasma concentration of glucose, triglycerides, ketone bodies, and proteins.
5. The gallbladder stores and concentrates bile, and releases it through the cystic duct and common bile duct to the duodenum.

20.9 Pancreas

1. The pancreatic islets secrete the hormones insulin and glucagon.
2. The exocrine acini produce pancreatic juice, which contains digestive enzymes and bicarbonate.

20.10 Digestion and Absorption of Carbohydrates, Proteins, and Lipids

1. Amylase digests starch into disaccharides and short-chain oligosaccharides. Complete digestion into monosaccharides is accomplished by brush border enzymes in the intestinal mucosa.
2. Protein-digestion begins in the stomach with the action of pepsin. It continues in the small intestine with the action of pancreatic enzymes, including trypsin. Amino acids, such as monosaccharides, are absorbed and secreted into the capillary blood entering the hepatic portal vein.
3. Lipids are digested in the small intestine after being emulsified by bile salts. Once within the epithelial cells, triglycerides, phospholipids, and cholesterol are resynthesized into fat droplets called chylomicrons that migrate to the lacteals of the villi. Lymph in the lacteals carries the droplets to the bloodstream.

20.11 Regulation of the Digestive System

1. Regulation of stomach functions occurs during cephalic, gastric, and intestinal phases.
2. The hormones secretin and CCK regulate pancreatic juice and bile secretions.

Review Activities

Objective Questions

1. Which of the following organs is *not* considered a part of the digestive system? (*module 20.1*)
a. pancreas
b. spleen
c. tongue
d. gallbladder

2. The double layer of peritoneum that supports the GI tract is called (*module 20.2*)
a. the visceral peritoneum.
b. the mesentery.
c. the greater omentum.
d. the lesser omentum.

3. An incision into the lumen of the GI tract would cut the layers, or tunics, in the following sequence: (*module 20.3*)
a. mucosa, submucosa, muscularis, serosa
b. serosa, submucosa, muscularis, mucosa
c. serosa, muscularis, submucosa, mucosa
d. muscularis, submucosa, mucosa, serosa

4. Which of the following types of teeth are found in the permanent but *not* in the deciduous dentition? (*module 20.4*)
a. incisors
b. canines
c. premolars
d. molars

5. The secretion of HCl by the stomach mucosa is inhibited by (*module 20.5*)
a. neural reflexes from the duodenum.
b. the secretion of gastric inhibitory peptide from the duodenum.
c. the lowering of gastric pH.
d. all of the above.

6. Which of the following statements about gastric secretion of HCl is *false?* (*module 20.5*)
a. HCl is secreted by parietal cells.
b. HCl hydrolyzes peptide bonds.
c. HCl is needed for the conversion of pepsinogen to pepsin.
d. HCl is needed for maximum activity of pepsin.

7. Intestinal enzymes such as lactase are (*module 20.6*)
a. secreted by the small intestine into the chyme.
b. produced by the intestinal crypts.
c. produced by the pancreas.
d. attached to the cell membrane of microvilli in the epithelial cells of the mucosa.

8. The splenic flexure is located between (*module 20.7*)
a. the ascending colon and transverse colon.
b. the cecum and ascending colon.
c. the descending colon and sigmoid colon.
d. the transverse colon and descending colon.

9. The first organ to receive the bloodborne products of digestion is (*module 20.8*)
a. the liver.
b. the pancreas.
c. the heart.
d. the brain.

10. Which of the following is *not* a digestive enzyme in pancreatic juice? (*module 20.9*)
a. trypsin
b. lipase
c. enterokinase
d. amylase
e. nuclease

11. Which of the following statements about fat digestion and absorption is *false?* (*module 20.10*)
a. Emulsification by bile salts increases the rate of fat digestion.
b. Triglycerides are hydrolyzed by the action of pancreatic lipase.
c. Triglycerides are resynthesized from monoglycerides and fatty acids in the intestinal epithelial cells.
d. Triglycerides, as particles called chylomicrons, are absorbed into blood capillaries within the villi.

12. During the gastric phase, the secretion of HCl and pepsinogen is stimulated by (*module 20.11*)
a. vagus nerve stimulation that originates in the brain.
b. polypeptides in the gastric lumen and by gastrin secretion.
c. secretin and cholecystokinin from the duodenum.
d. all of the above.

Essay Questions

1. Explain what is meant by digestion. Differentiate between the mechanical and chemical aspects of digestion. (*module 20.1*)
2. List in order the regions of the GI tract through which ingested food would pass from the mouth to the anus. (*module 20.1*)
3. Define the term *serous membrane.* How are the serous membranes of the abdominal cavity classified and what are their functions? (*module 20.2*)
4. Why are there two autonomic innervations to the GI tract? Identify the specific sites of autonomic stimulation in the tunics. (*module 20.3*)
5. Describe the different types of teeth and explain their functions. Sketch a tooth and label its parts. (*module 20.4*)
6. What are the specializations of the mucosa of the stomach that permit distension and secretion? (*module 20.5*)
7. Describe the structural modifications of the small intestine that increase the surface area for absorption. (*module 20.6*)
8. What are the regions of the large intestine? In what portion of the abdominal cavity and pelvic cavity is each region located? (*module 20.7*)
9. Describe how the gallbladder is filled with and emptied of bile fluid. What is the function of bile? (*module 20.8*)
10. Describe how pancreatic enzymes become activated in the lumen of the small intestine. Why are these mechanisms necessary? (*module 20.9*)
11. Explain the steps involved in the digestion and absorption of fat. (*module 20.10*)
12. Explain how the gastric secretion of HCl and pepsin is regulated during the cephalic, gastric, and intestinal phases. (*module 20.11*)

Labeling Exercise

Label the structures indicated on the figure to the right.

1. ______________________
2. ______________________
3. ______________________
4. ______________________
5. ______________________
6. ______________________

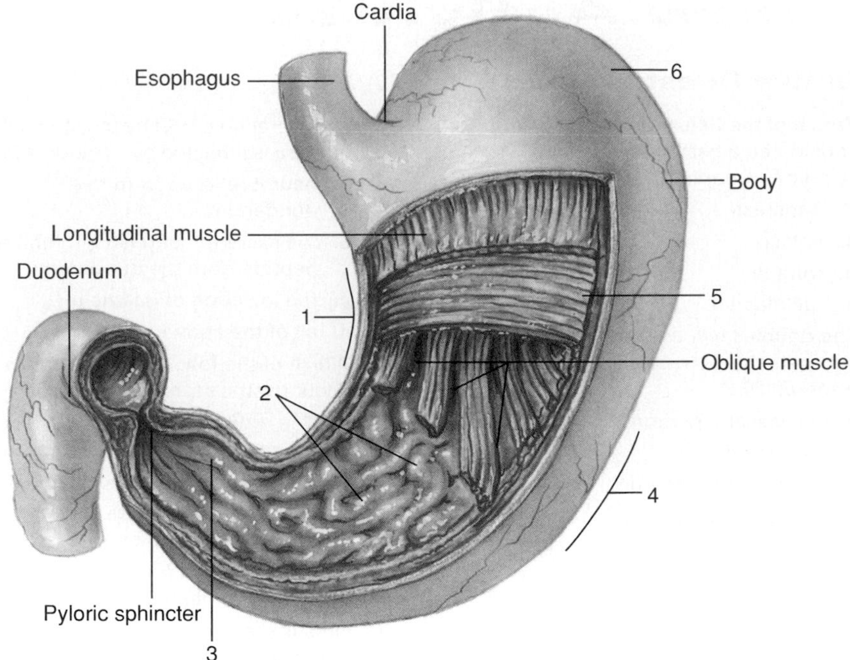

Critical Thinking Questions

1. Technically speaking, ingested food is not in the body. Neither are feces excreted from within the body (except bile residue). Explain these statements. Why would this information be important to a drug company interested in preparing a new oral medication?
2. What is the function of bicarbonate in the pancreatic juice? Explain why stress ulcers are more likely to be located in the duodenum than in the stomach.
3. A treatment that was once popular for morbid obesity was intestinal bypass surgery, a procedure that removed fairly long portions of the jejunum and ileum. Can you explain why this procedure would help someone lose weight?
4. Your aerobics instructor tells you that exercise helps to prevent constipation. Is the instructor correct? Can you think of any plausible explanation?
5. You go to an out-of-control party and see that some of the partygoers have gotten so inebriated that they are vomiting. Why would drinking to excess have this effect on the body?

CHAPTER 21

"Catabolic reactions release energy, usually by the breakdown of larger organic molecules into smaller molecules." *page 544*

Regulation of Metabolism

Chapter Outline

Terms to Remember

metabolism (*mĕ-tab'ŏ-liz"em*), 544
anabolism (*ă-nab'ŏ-liz"em*), 544
catabolism (*kă-tab'ŏ-liz"em*), 544
amino acid, 544
vitamin, 544
trace element, 545
glycolysis (*gli"kol'ĭ sis*), 546
anaerobic (*an-ă-ro'bik*) respiration, 548
gluconeogenesis (*gloo"kŏ-ne"ŏ-jen'ĭ-sis*), 548
aerobic (*ă-ro'bik*) respiration, 549
Krebs cycle, 549
electron-transport system, 550
glycogenesis (*gli"kŏ-jen'ĭ-sis*), 551
glycogenolysis (*gli"kŏ-jĕ-nol'ĭ-sis*), 551
lipogenesis (*lip"ŏ-jen'ĕ-sis*), 553
lipolysis (*lĭ-pol'ĭ-sis*), 553
ketone body, 553
kilocalorie (*kil'ŏ-kal"ŏ-re*), 555
basal metabolic rate, 556
insulin, (*in'-sŭ-lin*), 557
glucagon (*gloo'kă-gon*), 557

Learning Objectives

After studying this chapter, you should be able to

- explain the body's nutritional need for energy and identify the essential amino acids and essential fatty acids.
- list the common vitamins and minerals, and explain why they are needed for a balanced diet.
- describe the pathway of glycolysis in general terms and identify the end products.
- describe the pathway of anaerobic respiration and explain the significance of this pathway; also describe the process of gluconeogenesis and explain its significance.
- explain what happens to pyruvic acid when it enters into a mitochondrion and identify the products of the Krebs cycle.
- describe the function of the electron-transport system and explain how oxygen is used in aerobic respiration.
- describe how carbohydrates can be converted into fat, and how fat and protein can be used for energy.
- identify the sources of glucose, fatty acids, and amino acids in the body, and explain how the metabolic rate is regulated.
- state the functions of insulin and glucagon, and explain how the secretion of these hormones is regulated.

21.1 Nutritional Requirements

The body's energy requirements must be met by the caloric value of food. Additionally, the essential amino acids and fatty acids are needed to replace molecules in the body that are continuously degraded, and vitamins and minerals obtained in the diet are needed for diverse enzymatic reactions.

Organisms maintain their highly ordered structure and life-sustaining activities by constantly using up energy obtained ultimately from the environment. Central to life processes are biochemical reactions that are coupled, so that energy released by one reaction is incorporated into the products of another reaction. The sum of all the biochemical reactions that occur within an organism to make energy available for use is termed **metabolism** (*mĕ-tab'ŏ-liz"em*). Metabolism may be divided into *anabolism* (*ă-nab'ŏ-liz"em*) and *catabolism* (*kă-tab'ŏ-liz"em*).

1. **Anabolic reactions** include the synthesis of large energy-storage molecules such as glycogen, fat, and protein; these synthesis reactions require the input of energy.
2. **Catabolic reactions** release energy, usually by the breakdown of larger organic molecules into smaller molecules.

Food provides molecules that are delivered by the blood to the body cells. The catabolism of food molecules yields energy for the body. Food also supplies the raw materials for the synthesis reactions of anabolism that occur constantly within the cells of the body.

Anabolic Requirements

A wide variety of synthetic reactions are part of our daily metabolic "housekeeping." Anabolic reactions include those that synthesize DNA and RNA, proteins, glycogen, and triglycerides. The general purpose of anabolic reactions is to constantly replace those molecules that are hydrolyzed (broken down—see chapter 2, module 2.4) into their component subunits of glucose, fatty acids, and amino acids.

Exercise and fasting, acting through changes in hormonal secretion, cause an increase in the catabolism of stored glycogen, fat, and body protein. These molecules are also broken down at a certain rate in a person who is neither exercising nor fasting. Some of the subunits that are formed (amino acids, glucose, and fatty acids) are used in cell respiration for energy. For this reason, new subunits must be obtained from food to prevent a continual decline in the amount of protein, glycogen, and fat in the body.

Of the naturally occurring amino acids, nine cannot be synthesized by the body; thus they must be part of our diet. These **essential amino acids** are lysine, tryptophan, phenylalanine, threonine, valine, methionine, leucine, isoleucine, and histidine (see table 21.3). In addition, the **essential fatty acids**—linoleic acid and linolenic acid—must be obtained in the diet. Fortunately, most vegetable oils contain both of these fatty acids.

Vitamins and Minerals

Vitamins are small organic molecules, most of which cannot be synthesized by the body. They serve as coenzymes, or "helpers" of enzymes, in metabolic reactions (table 21.1). There are two groups of vitamins: fat soluble and water soluble. The *fat-soluble vitamins* include vitamins A, D, E, and K. The *water-soluble vitamins* include thiamine (B_1), riboflavin (B_2), niacin (B_3), pyridoxine (B_6), pantothenic acid, biotin, folic acid, vitamin B_{12}, and vitamin C (ascorbic acid). Recommended daily allowances for these vitamins are listed in table 21.2.

Many of the water-soluble vitamins serve as coenzymes in the metabolism of carbohydrates, lipids, and proteins. Thiamine, for example, is needed for the activity of the enzyme that converts pyruvic acid to acetyl coenzyme A (as will be discussed in module 21.4). Riboflavin and niacin are needed for the production of **FAD** (flavine adenine dinucleotide) and **NAD** (nicotinamide adenine dinucleotide), respectively; these latter compounds serve as coenzymes that transfer hydrogens during cell respiration. Deficiencies of the water-soluble vitamins can thus have widespread effects in the body.

---
metabolism: Gk. *metabole,* change
anabolism: Gk. *anabole,* a raising up
catabolism: Gk. *katabole,* a casting down

TABLE 21.1 **The Major Vitamins**

Vitamin	Function	Deficiency Symptoms/Diseases	Sources
A	Constituent of visual pigment; strengthens epithelial membranes	Night blindness; dry skin	Yellow vegetables and fruit
B_1 (thiamine)	Cofactor for enzymes that catalyze decarboxylation	Beriberi; neuritis	Liver, unrefined cereal grains
B_2 (riboflavin)	Part of flavoproteins (such as FAD)	Cracked lips; inflammation of tongue	Liver, milk
B_6 (pyridoxine)	Coenzyme for decarboxylase and transaminase enzymes	Convulsions	Liver, corn, wheat, and yeast
B_{12} (cyanocobalamin)	Coenzyme for amino acid metabolism; needed for erythropoiesis	Pernicious anemia	Liver, meat, eggs, milk
Biotin	Needed for fatty acid synthesis	Dermatitis; enteritis	Egg yolk, liver, tomatoes
C	Needed for collagen synthesis in connective tissue	Scurvy	Citrus fruits, green leafy vegetables
D	Needed for intestinal absorption of calcium and phosphate	Rickets; osteomalacia	Fish, liver
E	Antioxidant	Muscular dystrophy	Milk, eggs, meat, leafy vegetables
Folates	Needed for reactions that transfer one carbon	Sprue; anemia	Green leafy vegetables
K	Promotes reactions needed for function of clotting factors	Hemorrhage; inability to form clot	Green leafy vegetables
Niacin	Part of NAD and NADP	Pellagra	Liver, meat, yeast
Pantothenic acid	Part of coenzyme A	Dermatitis; enteritis; adrenal insufficiency	Liver, eggs, yeast

Many fat-soluble vitamins have highly specialized functions. Vitamin K, for example, is required for the production of several clotting factors in the blood plasma. The visual pigments in the rods and cones of the retina are derived from vitamin A. Vitamin A and related compounds, called *retinoids,* also affect genetic expression in epithelial cells; these compounds are now used clinically in the treatment of some skin conditions.

Inside Information—The way you prepare and cook your food has a great deal to do with your actual vitamin intake from it. The skins of some plant foods are vitamin-rich and should not be discarded. Overcooking may destroy the chemical structure of some vitamins, and boiling vegetables can leach the water-soluble vitamins out into the broth.

Minerals, or **elements,** are needed as cofactors for specific enzymes and for a wide variety of other critical functions. Elements that are required in relatively large amounts each day include sodium, potassium, magnesium, calcium, phosphorus, and chlorine (table 21.2). **Trace elements** are those required in amounts of less than 100 mg per day. Among these are iron, zinc, manganese, fluorine, copper, molybdenum (*mŏ-lib'dĕ-num*), chromium, and selenium.

Inside Information—Minerals make up only a small portion of your body weight—only 4%. Calcium accounts for 2%, and phosphorus for 1%. The remaining 1% comes from all the other minerals combined.

TABLE 21.2 **Recommended Daily Allowances for Vitamins and Minerals**

Category or condition	Age (years)	Weight		Height		Protein			Fat-soluble Vitamins	
		(kg)	(lb)	(cm)	(in.)	(g)	Vitamin A (μg RE)[1]	Vitamin D (μg)[2]	Vitamin E (mg α-TE)[3]	Vitamin K (μg)
Infants	0.0–0.5	6	13	60	24	13	375	7.5	3	5
	0.5–1.0	9	20	71	28	14	375	10	4	10
Children	1–3	13	29	90	35	16	400	10	6	15
	4–6	20	44	112	44	24	500	10	7	20
	7–10	28	62	132	52	28	700	10	7	30
Males	11–14	45	99	157	62	45	1000	10	10	45
	15–18	66	145	176	69	59	1000	10	10	65
	19–24	72	160	177	70	58	1000	10	10	70
	25–50	79	174	176	70	63	1000	5	10	80
	51+	77	170	173	68	63	1000	5	10	80
Females	11–14	46	101	157	62	46	800	10	8	45
	15–18	55	120	163	64	44	800	10	8	55
	19–24	58	128	164	65	46	800	10	8	60
	25–50	63	138	163	64	50	800	5	8	65
	51+	65	143	160	63	50	800	5	8	65
Pregnant						60	800	10	10	65
Lactating	1st 6 months					65	1300	10	12	65
	2nd 6 months					62	1200	10	11	65

[1]Retinol equivalent (1 RE = 1 μg retinol or 6 μg β-carotene)

[2]As cholecalciferol (10 μg cholecalciferol = 400 W of Vitamin D)

[3]α-tocopherol equivalents (1 mg α-tocopherol = 1 α-TE)

Reprinted with permission from *Recommended Dietary Allowances:* 10th Edition. Copyright 1989 by the National Academy of Sciences. Courtesy of the National Academy Press, Washington, DC.

21.2 Glycolysis

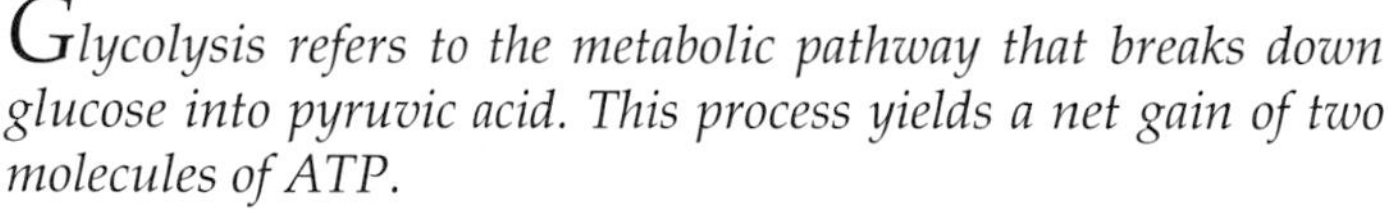

Glycolysis refers to the metabolic pathway that breaks down glucose into pyruvic acid. This process yields a net gain of two molecules of ATP.

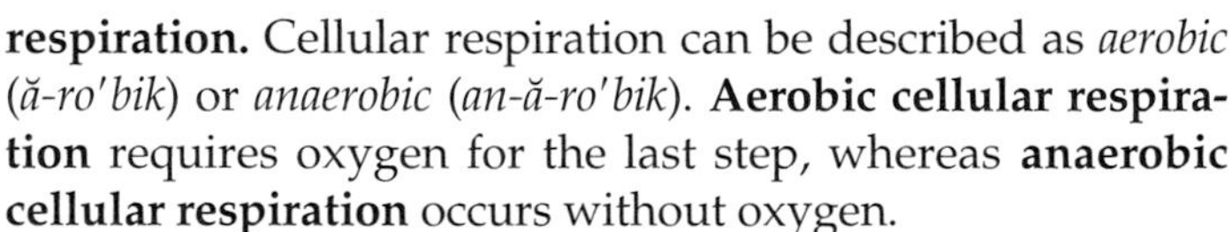

The catabolic reactions that break down glucose and other food molecules for energy do not directly power the cell. Instead, the energy liberated by the breakdown of food molecules is used to drive the reaction ADP + P_i → ATP. As discussed in chapter 2 (see module 2.7), some of the food energy is captured in the chemical bond formed in this reaction. Therefore, when the reaction is reversed (ATP → ADP + P_i), energy is released. Since all energy that cells use for their activities is directly obtained by the breakdown of ATP, ATP is called the *universal energy carrier.*

The reactions that break down glucose, fatty acids, and amino acids yield energy for the synthesis of ATP. These metabolic pathways are known collectively as **cellular respiration.** Cellular respiration can be described as *aerobic* (*ă-ro'bik*) or *anaerobic* (*an-ă-ro'bik*). **Aerobic cellular respiration** requires oxygen for the last step, whereas **anaerobic cellular respiration** occurs without oxygen.

The cell respiration of glucose, whether aerobic or anaerobic, begins with a multistep metabolic pathway known as **glycolysis** (*gli"kol'ĭ-sis*) (fig. 21.1). This is the metabolic pathway by which glucose is converted into two molecules of *pyruvic* (*pi-roo'vik*) *acid.*

The breakdown of glucose into two molecules of pyruvic acid releases energy, and the energy is used to form 4 ATP from 4 ADP and 4 P_i. However, 2 ATP were broken down (into 2 ADP and 2 P_i) at the beginning of glycolysis (the "up staircase" in fig. 21.1). This is akin to striking a match; energy is put in (as friction on the match

anaerobic: Gk. *an,* without; *aer,* air; *bios,* life

glycolysis: Gk. *glyco,* sugar; *lysis,* breaking

Water-soluble Vitamins							Minerals						
Vitamin C (mg)	Thiamin (mg)	Riboflavin (mg)	Niacin (mg NE)[4]	Vitamin B_6 (mg)	Folate (µg)	Vitamin B_{12} (µg)	Calcium (mg)	Phosphorus (mg)	Magnesium (mg)	Iron (mg)	Zinc (mg)	Iodine (µg)	Selenium (µg)
30	0.3	0.4	5	0.3	25	0.3	400	300	40	6	5	40	10
35	0.4	0.5	6	0.6	35	0.5	600	500	60	10	5	50	15
40	0.7	0.8	9	1.0	50	0.7	800	800	80	10	10	70	20
45	0.9	1.1	12	1.1	75	1.0	800	800	120	10	10	90	20
45	1.0	1.2	13	1.4	100	1.4	800	800	170	10	10	120	30
50	1.3	1.5	17	1.7	150	2.0	1200	1200	270	12	15	150	40
60	1.5	1.8	20	2.0	200	2.0	1200	1200	400	12	15	150	50
60	1.5	1.7	19	2.0	200	2.0	1200	1200	350	10	15	150	70
60	1.5	1.7	19	2.0	200	2.0	800	800	350	10	15	150	70
60	1.2	1.4	15	2.0	200	2.0	800	800	350	10	15	150	70
50	1.1	1.3	15	1.4	150	2.0	1200	1200	280	15	12	150	45
60	1.1	1.3	15	1.5	180	2.0	1200	1200	300	15	12	150	50
60	1.1	1.3	15	1.6	180	2.0	1200	1200	280	15	12	150	55
60	1.1	1.3	15	1.6	180	2.0	800	800	280	15	12	150	55
60	1.0	1.2	13	1.6	180	2.0	800	800	280	10	12	150	55
70	1.5	1.6	17	2.2	400	2.2	1200	1200	300	30	15	175	65
95	1.6	1.8	20	2.1	280	2.6	1200	1200	355	15	19	200	75
90	1.6	1.7	20	2.1	260	2.6	1200	1200	340	15	16	200	75

[4]Niacin equivalent (1 NE = 1 mg of niacin or 60 mg of dietary tryptophan)

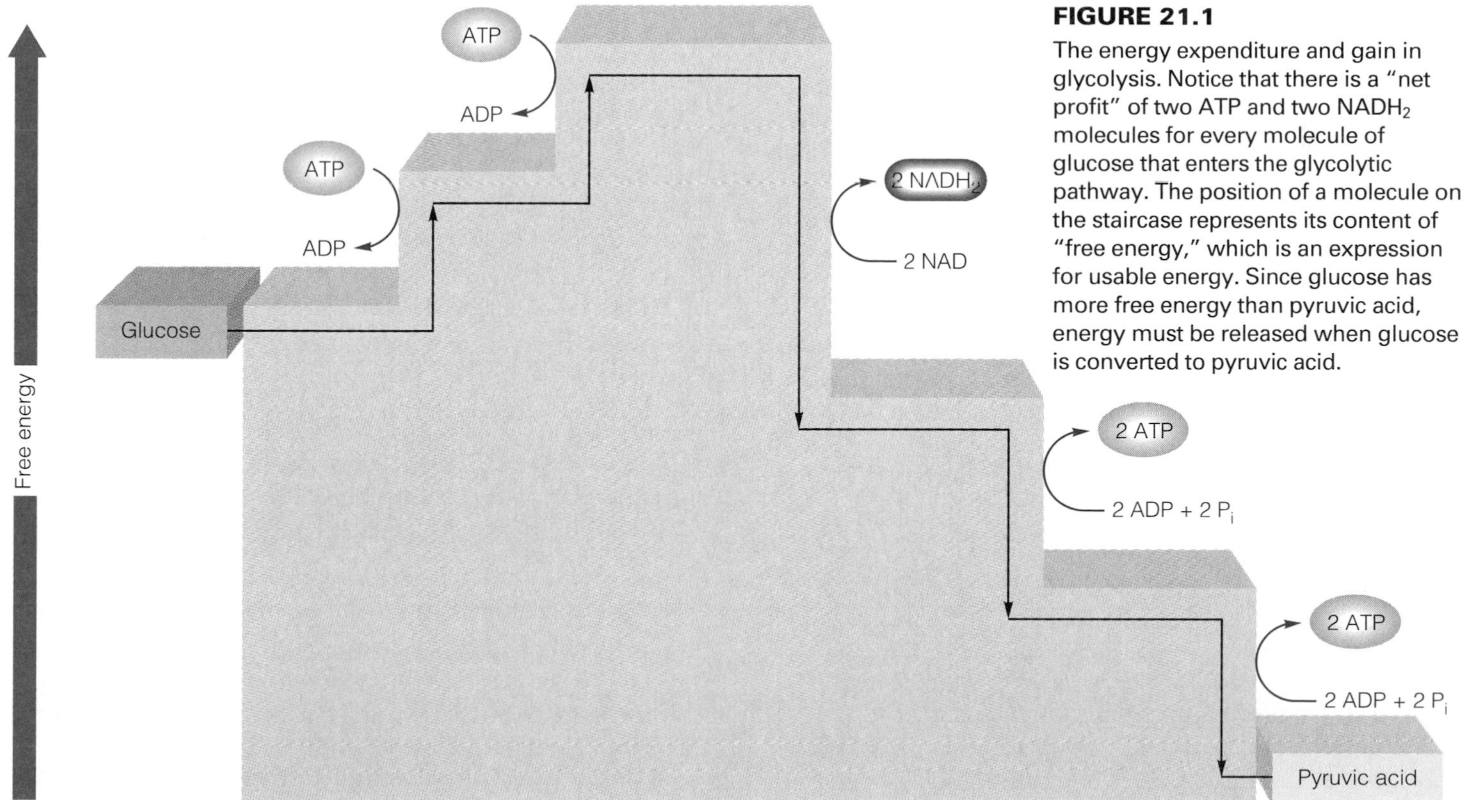

FIGURE 21.1

The energy expenditure and gain in glycolysis. Notice that there is a "net profit" of two ATP and two $NADH_2$ molecules for every molecule of glucose that enters the glycolytic pathway. The position of a molecule on the staircase represents its content of "free energy," which is an expression for usable energy. Since glucose has more free energy than pyruvic acid, energy must be released when glucose is converted to pyruvic acid.

head) in order to get more energy out (as the heat of the flame). In glycolysis, therefore, there is a net gain of 4 ATP – 2 ATP = 2 ATP per glucose converted into pyruvic acid.

Each of the pyruvic acid molecules formed from glucose has three carbons and three oxygens, so the two together account for the six carbons and six oxygens in the glucose molecule. Glucose, however, has 12 hydrogens—its formula is $C_6H_{12}O_6$—while each pyruvic acid molecule only has 4 hydrogens, for a total of 8. The other 4 hydrogens from glucose are added as pairs of hydrogen to two molecules of the coenzyme NAD. For simplicity, we can show the combination as $NADH_2$, although it is chemically more accurate to call it NADH + H^+. The $NADH_2$ will then release its hydrogens in other reactions, as described in the next two modules. Therefore, you can think of NAD as a "hydrogen shuttle."

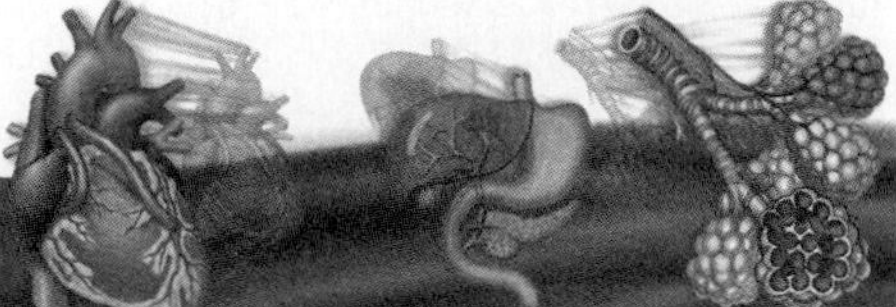

21.3 Anaerobic Respiration and Gluconeogenesis

Skeletal muscles can obtain sufficient ATP for a limited time under anaerobic conditions. This involves the conversion of pyruvic acid to lactic acid.

Anaerobic Respiration

In order for glycolysis to continue, adequate amounts of NAD must be available to accept hydrogen atoms. Therefore, the $NADH_2$ that is produced in glycolysis must donate its hydrogens to another molecule. In aerobic respiration, this other molecule is located in the mitochondria of the cell and the hydrogens in $NADH_2$ ultimately pass to oxygen.

When oxygen is not available in sufficient amounts, the $NADH_2$ produced in glycolysis is reconverted in the cytoplasm by donating its hydrogens to pyruvic acid. This results in the re-formation of NAD, which can act as a hydrogen acceptor and allow glycolysis to continue. The addition of two hydrogen atoms to pyruvic acid reduces it and forms a new molecule: *lactic acid* (fig. 21.2). This metabolic pathway, by which glucose is converted through pyruvic acid to lactic acid, is called **anaerobic** (*an-ă-ro'bik*) **respiration.**

Anaerobic respiration yields a net gain of two ATP (produced by glycolysis) per glucose molecule. Skeletal muscles respire anaerobically and produce lactic acid normally for limited periods of time during exercise. Cardiac muscle, however, respires anaerobically only under abnormal conditions when it is deprived of oxygen, and then can survive only for a few minutes under these conditions. The brain, by contrast, is exquisitely sensitive to any shortage of oxygen and will be damaged by that shortage sooner than any other tissue.

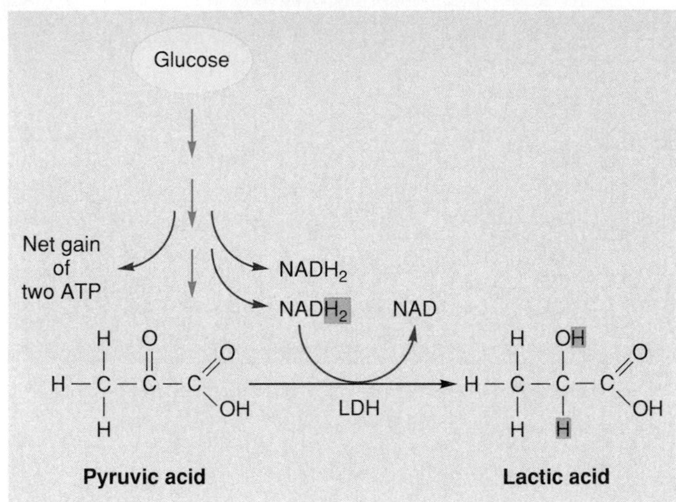

FIGURE 21.2
Anaerobic respiration. The addition of two hydrogen atoms (colored boxes) from reduced NAD to pyruvic acid produces lactic acid and oxidized NAD. This reaction is catalyzed by lactic acid dehydrogenase (LDH).

Ischemia (ĭ-ske'me-ă) *refers to inadequate blood flow to an organ, such that the rate of oxygen delivery is insufficient to maintain aerobic respiration. Inadequate blood flow to the heart, or* myocardial ischemia, *may occur if the coronary blood flow is occluded by atherosclerosis, a blood clot, or by an artery spasm. People with myocardial ischemia often experience* angina pectoris*—severe pain in the chest and left (or sometimes, right) arm area. This pain is associated with increased blood levels of lactic acid, which are produced by anaerobic respiration of the heart muscle. The degree of ischemia and angina can be decreased by vasodilator drugs, such as nitroglycerin and amyl nitrite. These drugs improve blood flow to the deprived heart and also decrease the work of the heart by dilating peripheral blood vessels.*

Gluconeogenesis

Some of the lactic acid produced by exercising skeletal muscles is delivered by the blood to the liver. An enzyme within liver cells is then able to convert the lactic acid back into pyruvic acid. Unlike most other organs, the liver contains the enzymes needed to then take two pyruvic acid molecules and convert them back to glucose. This process is essentially the reverse of glycolysis.

The conversion of noncarbohydrate molecules (lactic acid, amino acids, and glycerol) into pyruvic acid and then into glucose is an extremely important process called **gluconeogenesis** (*gloo"kŏ-ne"ŏ-jen'ĭ-sis*). In a state of starvation and during prolonged exercise, when glycogen stores are depleted, gluconeogenesis in the liver is the only way that adequate blood glucose levels can be maintained to prevent brain death.

21.4 Aerobic Respiration and the Krebs Cycle

In aerobic respiration, pyruvic acid is converted into acetyl coenzyme A. This begins a cyclic pathway known as the Krebs cycle.

Just as with anaerobic respiration, the aerobic respiration of glucose begins with glycolysis. In both anaerobic and aerobic respiration, glycolysis results in the production of two molecules of pyruvic acid, two molecules of ATP, and two molecules of $NADH_2$ per glucose molecule. In anaerobic respiration, the $NADH_2$ is changed to NAD during the conversion of pyruvic acid to lactic acid in the cytoplasm of the cell.

In aerobic respiration, pyruvic acid leaves the cell cytoplasm and enters the interior (the matrix) of mitochondria. Once pyruvic acid is inside a mitochondrion, carbon dioxide is enzymatically removed from each three-carbon-long pyruvic acid to form a two-carbon-long organic acid—acetic acid. The enzyme that catalyzes this reaction combines the acetic acid with a coenzyme (derived from the vitamin pantothenic acid) called coenzyme A. The molecule thus produced is called **acetyl** (*as'ĕ-tl*) **coenzyme A,** abbreviated **acetyl CoA.**

Remember that glycolysis converts one glucose molecule into two molecules of pyruvic acid. Since each pyruvic acid molecule is converted into one acetyl CoA and one CO_2, two molecules of acetyl CoA and two molecules of CO_2 are derived from each glucose. The acetyl CoA molecules serve as substrates for mitochondrial enzymes in the aerobic pathway; the carbon dioxide—a waste product in this process—is carried by the blood to the lungs for elimination. It should be noted that the oxygen in CO_2 is derived from pyruvic acid, not from oxygen gas.

Once acetyl CoA is formed, the acetic acid subunit (two carbons long) is combined with oxaloacetic acid (four carbons long) to form a molecule of citric acid (six carbons long). Coenzyme A acts only as a transporter of acetic acid from one enzyme to another (similar to the transport of hydrogen by NAD).

The formation of citric acid begins a cyclic metabolic pathway known as the **citric acid cycle,** or **TCA cycle** (for tricarboxylic acid; citric acid has three carboxylic acid groups). More often, however, this cyclic pathway is called the **Krebs cycle,** after its principal discoverer, Sir Hans Krebs. Through a series of reactions involving the elimination of two carbons and four oxygens (as two CO_2 molecules) and the removal of hydrogens, citric acid is eventually converted to oxaloacetic acid, which completes the cyclic metabolic pathway (fig. 21.3).

Krebs cycle: from Hans A. Krebs, German biochemist, 1900–1981

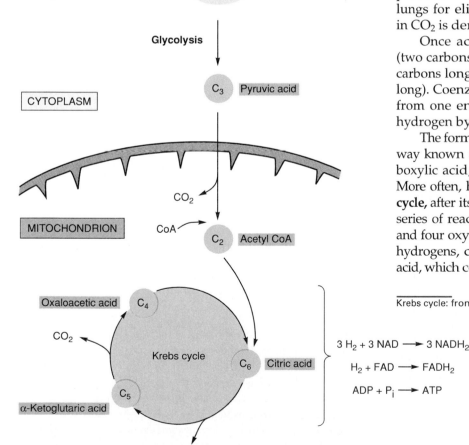

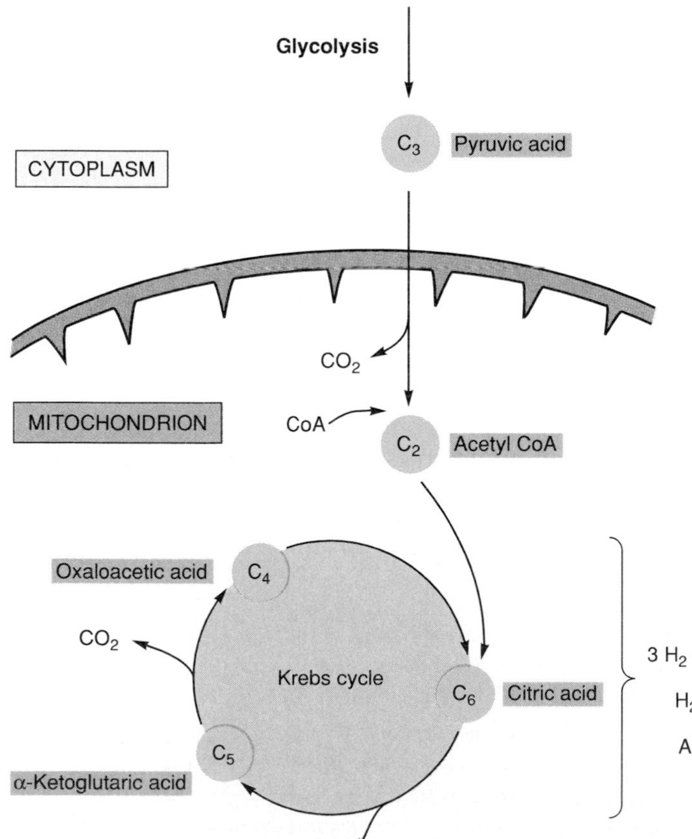

FIGURE 21.3
A simplified diagram of the Krebs cycle showing how the original four-carbon-long oxaloacetic acid is regenerated at the end of the cyclic pathway. Only the numbers of carbon atoms in the Krebs cycle intermediates are shown; the numbers of hydrogens and oxygens are not accounted for in this simplified scheme.

Since this metabolic pathway is cyclic, it might seem pointless; it ends up where it begins. This view is misleading, however, because hydrogen atoms are removed at several steps during the Krebs cycle and added to the coenzymes NAD and FAD. Therefore, "as the cycle turns," molecules of $NADH_2$ and $FADH_2$ are produced. These are important high-energy products of the Krebs cycle. It should also be noted that one ATP is produced directly by each turn of the Krebs cycle.

The hydrogens from $NADH_2$ and $FADH_2$ will, after going through an intermediate process described in the next module, combine with oxygen to form water (H_2O). This water formation is extremely significant because the intermediate steps leading up to it constitute the major energy-generating process in the cell.

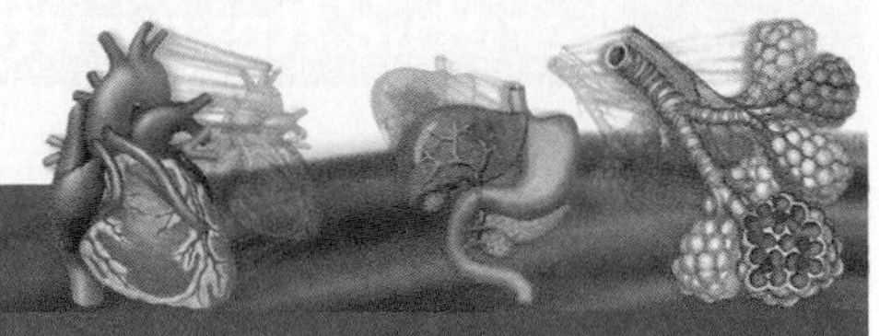

21.5 Electron Transport and Oxidative Phosphorylation

Electrons from the hydrogen atoms in NADH and $FADH_2$ are transported by a system of molecules in the mitochondria and passed to oxygen. This process is used to make ATP from ADP and P_i.

Electron-Transport System

Built into the foldings of the inner membrane of the mitochondria are molecules that function as an **electron-transport system.** Each of these molecules is fixed in position in the membrane, so that they occur in a definite order. The first member of the electron-transport system picks up electrons from the $NADH_2$ produced by the Krebs cycle. Removing an electron from a hydrogen atom leaves a proton (H^+). From one $NADH_2$, two electrons will be passed to the first member of the electron-transport chain and two H^+ will be left behind. The $FADH_2$ will pass its electrons to the electron-transport system at a later step, but otherwise its story is similar to that of $NADH_2$.

Once the $NADH_2$ or $FADH_2$ has lost its hydrogens, it is free to go back to the Krebs cycle and pick up new pairs of hydrogens. In this way, hydrogens (or rather the electrons from the hydrogens) are shuttled continuously from the Krebs cycle to the electron-transport system.

The pairs of electrons are passed, bucket-brigade fashion, from the first member of the electron-transport system to the second member, and from the second member to the third. This process continues until the electrons are picked up by the last member of the electron-transport system. (Their fate will be discussed a little later.) As the electrons move along the electron-transport system, energy is released. Just as the flow of electricity in a wire can drive a motor, the movement of electrons along the electron-transport system drives the production of ATP (fig. 21.4).

In total, 34 ATP per glucose are produced by the electron-transport system. To this number, we can add the one ATP made per turn of the Krebs cycle. Since we get two pyruvic acids per glucose, we get two turns of the Krebs cycle, and therefore we get two more ATP. Finally, we can add the net gain of two ATP per glucose made during glycolysis. This gives us a grand total of 38 ATP made from the aerobic respiration of glucose. This number is a potential total; in actual practice, a cell obtains less than this number (about 30 ATP), because of inefficiencies that need not concern us here.

Function of Oxygen

At this point, you might be wondering where oxygen fits into the story; after all, this is a discussion of aerobic respiration, and oxygen hasn't yet played a part! You might also be curious about what happens to the two electrons after they reach the last member of the electron-transport system. They can't stay there; if they did, the electron-transport system would stop. And while we're listing loose ends, what about those two H^+ left over from the hydrogens (in $NADH_2$ and $FADH_2$) that gave up their electrons to the electron-transport system?

Oxygen atoms, obtained from the oxygen molecules in the air we breathe, are required for the last step of aerobic respiration. The two electrons from the last member of the electron transport system are passed to an atom of oxygen. This would give the oxygen two negative charges, but it quickly binds to the two H^+ to form H_2O. To summarize: oxygen is the **final electron acceptor** of the electron-transport system, and water is formed at the last step from oxygen and the two hydrogens originally contributed by $NADH_2$ or $FADH_2$.

Cyanide *is a fast-acting lethal poison that produces symptoms of rapid heart rate, hypotension, coma, and death in the absence of quick treatment. The reason cyanide is so deadly is that it has one very specific action: it blocks the transfer of electrons from cytochrome a_3 to oxygen. The effects are the same as would occur if oxygen were completely removed: aerobic respiration and the production of ATP by oxidative phosphorylation comes to a halt.*

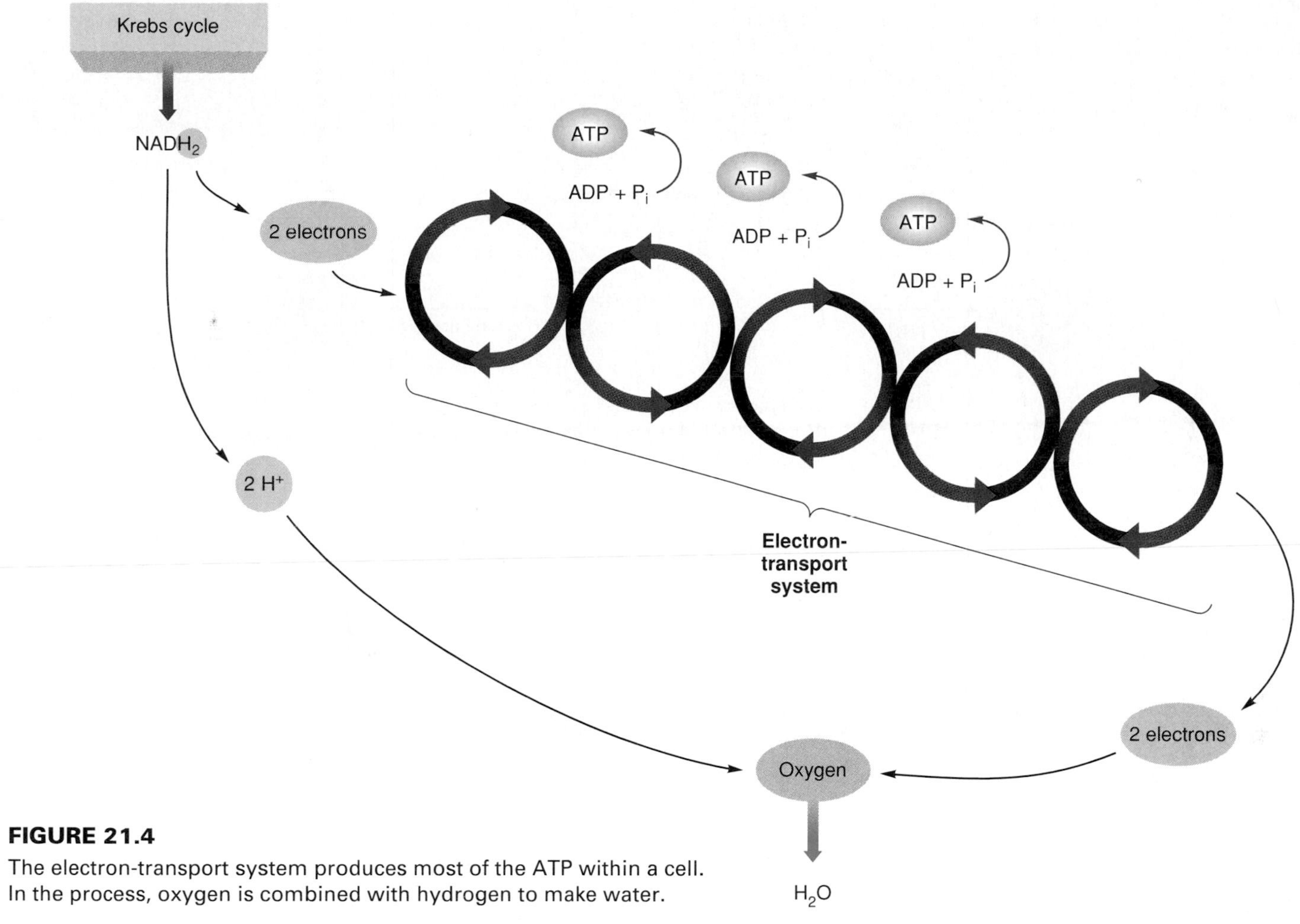

FIGURE 21.4
The electron-transport system produces most of the ATP within a cell. In the process, oxygen is combined with hydrogen to make water.

21.6 Metabolism of Carbohydrates, Lipids, and Proteins

Carbohydrates, lipids, and proteins can be used for energy in cell respiration, or they can be interconverted and stored as energy reserves.

Metabolism of Carbohydrates

The carbohydrates present in food are digested into their component monosaccharides (primarily glucose) and absorbed through the small intestine into the blood. Many organs use this blood glucose as an important energy source. The brain and heart obtain energy (ATP) from glucose via the reactions of aerobic cell respiration. Exercising skeletal muscles also can obtain energy from glucose in the process of anaerobic cell respiration.

When more glucose is available than is immediately needed for ATP synthesis (such as following a meal), the liver, skeletal muscles, and heart store carbohydrates in the form of glycogen. The formation of glycogen from glucose is called **glycogenesis** (*gli"kŏ-jen'ĭ-sis*). The reverse process, in which glycogen is converted into glucose, is called **glycogenolysis** (*gli"kŏ-jĕ-nol'ĭ-sis*).

The glucose involved in these reactions contains a phosphate group attached to the first or sixth carbon (fig. 21.5). This is important, because organic molecules with phosphate groups cannot cross membranes. The glucose

glycogenesis: Gk. *glyco,* sugar; *genesis,* production

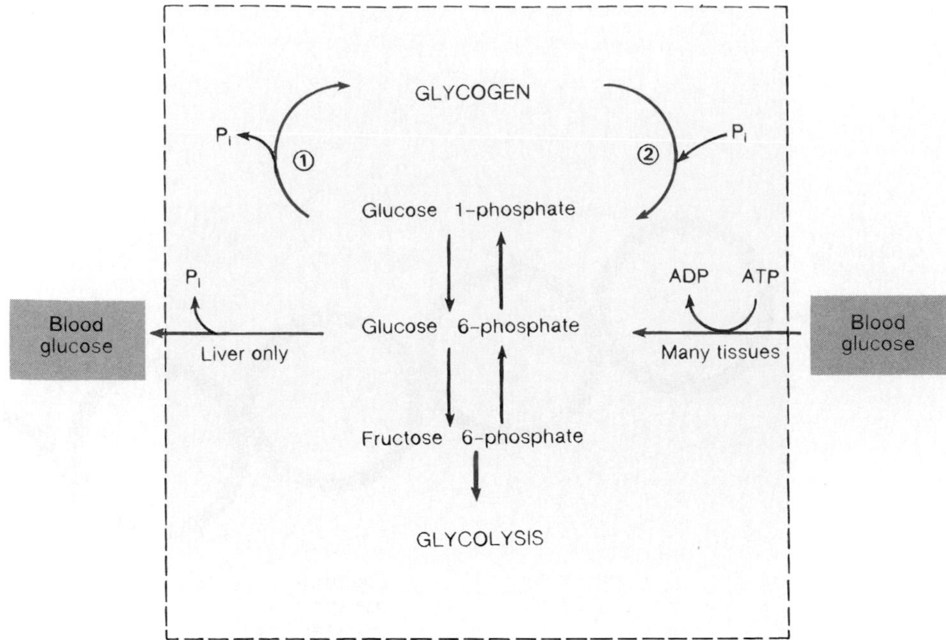

FIGURE 21.5

Blood glucose that enters tissue cells is rapidly converted to glucose 6-phosphate. This intermediate can be metabolized for energy in glycolysis, or it can be converted to glycogen (1), a process called *glycogenesis.* Glycogen represents a storage form of carbohydrates, which can be used as a source for new glucose 6-phosphate (2) in a process called *glycogenolysis.* The liver contains an enzyme that can remove the phosphate from glucose 6-phosphate; liver glycogen thus serves as a source for new blood glucose.

derived from glycogen thus cannot leak out of the cell. Similarly, glucose that enters the cell from the blood is "trapped" within the cell when it is converted to glucose 6-phosphate. Skeletal muscles, which store large amounts of glycogen, can generate glucose 6-phosphate for their own needs but cannot release glucose into the blood because they lack the ability to remove the phosphate group.

Unlike skeletal muscles, the liver contains an enzyme—known as *glucose 6-phosphatase*—that can remove the phosphate groups and produce free glucose (fig. 21.5). This free glucose can then be transported through the cell membrane. The liver, therefore, can secrete glucose into the blood, whereas skeletal muscles cannot. Liver glycogen can thus supply blood glucose for use by other organs, including exercising skeletal muscles that may have depleted much of their own stored glycogen during exercise.

Metabolism of Lipids

The body can convert excess carbohydrates into fat through the metabolic pathway shown in figure 21.6. Fat (triglyceride) is the most efficient form of energy storage—for each gram of fat, the body obtains 9 kilocalories; for each gram of carbohydrate or protein, it

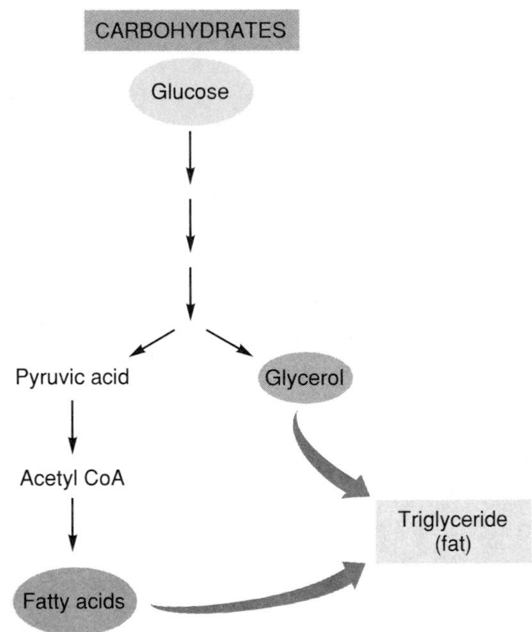

FIGURE 21.6

The formation of triglyceride (fat) from glucose. This occurs primarily within adipose cells.

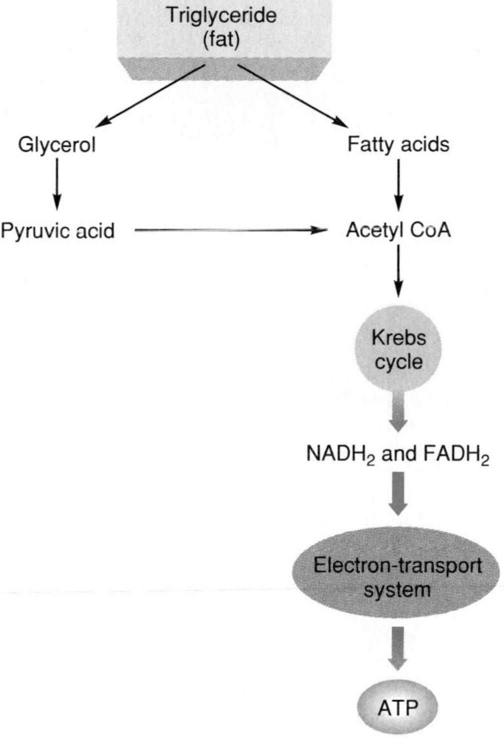

FIGURE 21.7
The metabolic pathways by which fat is broken down for energy. The number of ATP molecules made per fat molecule is variable (depending on the particular fatty acids), but can range in the hundreds.

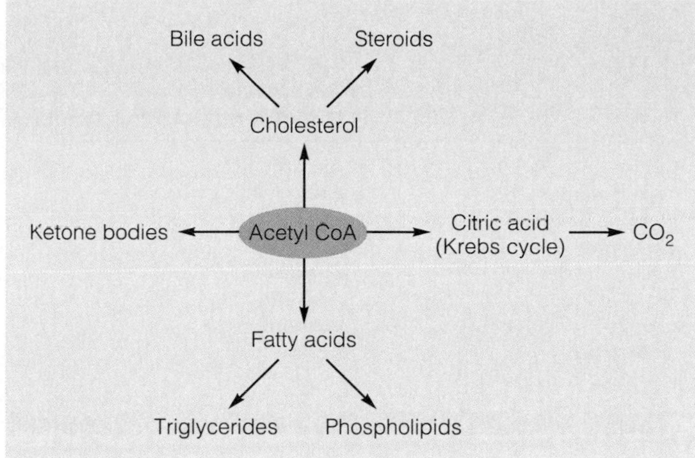

FIGURE 21.8
Divergent metabolic pathways for acetyl coenzyme A.

obtains only 4 kilocalories. Triglycerides provide energy primarily through the conversion of fatty acids into acetyl CoA. Acetyl CoA then enters into Krebs cycles and generates ATP through the electron-transport system (fig. 21.7).

Formation of Fat (Lipogenesis)

When glucose is going to be converted into fat, glycolysis occurs, and pyruvic acid is converted into acetyl CoA. Instead of entering Krebs cycles, the two-carbon acetic acid subunits of these acetyl CoA molecules can be used to produce a variety of lipids. These include steroids, such as cholesterol, as well as ketone bodies and fatty acids (fig. 21.8).

In the formation of fatty acids, a number of acetic acid (two-carbon) subunits are joined to one another to form the fatty acid chain. Six acetyl CoA molecules, for example, will produce a fatty acid that is 12 carbons long. When three of these fatty acids combine with one glycerol, a triglyceride molecule is formed. The formation of fat, or **lipogenesis** (*lip"ŏ-jen'ĕ-sis*), occurs primarily in adipose tissue and in the liver when the concentration of blood glucose is elevated following a meal.

Breakdown of Fat (Lipolysis)

When fat that is stored in adipose tissue is going to be used as an energy source, lipase enzymes hydrolyze triglycerides into *glycerol* and *free fatty acids* in a process called **lipolysis** (*lĭ-pol'ĭ-sis*). These molecules (primarily the free fatty acids) serve as energy sources that can be used by the liver, skeletal muscles, and other organs for aerobic respiration.

Most fatty acids consist of a long hydrocarbon chain with a carboxylic acid group (COOH) at one end. In a process known as **β-oxidation** (β is the Greek letter *beta*), enzymes remove two-carbon acetic acid molecules from the acid end of a fatty acid. This results in the formation of acetyl CoA, as the third carbon from the end becomes a new carboxylic acid group. The fatty acid chain is thus decreased in length by two carbons. The process of β-oxidation continues until the entire fatty acid molecule is converted to acetyl CoA. These acetyl CoA molecules can then be aerobically respired to yield a large number of ATP molecules.

Ketone Bodies

Some of the acetyl CoA derived from fatty acids is channeled within the liver into an alternate pathway. This pathway involves the conversion of two molecules of acetyl CoA into four-carbon-long acidic derivatives,

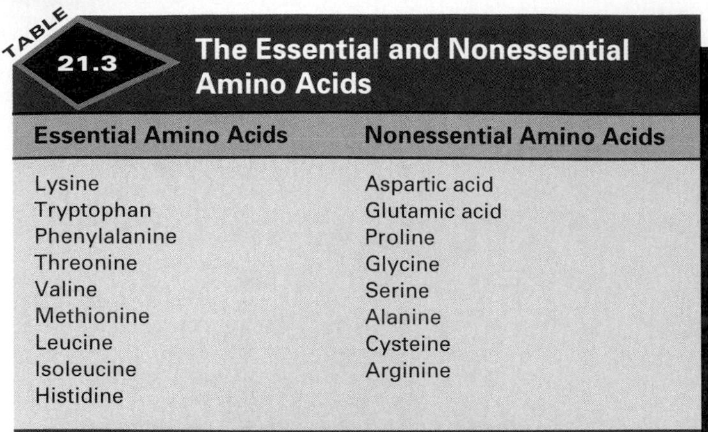

TABLE 21.3 The Essential and Nonessential Amino Acids

Essential Amino Acids	Nonessential Amino Acids
Lysine	Aspartic acid
Tryptophan	Glutamic acid
Phenylalanine	Proline
Threonine	Glycine
Valine	Serine
Methionine	Alanine
Leucine	Cysteine
Isoleucine	Arginine
Histidine	

acetoacetic acid and *β-hydroxybutyric acid.* Together with *acetone,* which is a three-carbon-long derivative of acetoacetic acid, these products are known as **ketone bodies.**

Under conditions of fasting or of diabetes mellitus, *the increased liberation of free fatty acids from adipose tissue results in an elevated production of ketone bodies by the liver. The secretion of abnormally high amounts of ketone bodies into the blood produces* ketosis, *which is one of the signs of fasting or an uncontrolled diabetic state. A person in this condition may also have a sweet-smelling breath due to the presence of acetone, which leaves the blood in the exhaled air.*

Metabolism of Proteins

An adequate amount of all 20 amino acids is required to build proteins for growth and to replace the proteins that are turned over. As mentioned in module 21.1, nine cannot be produced by the body and so must be obtained in the diet; these are the *essential amino acids* (table 21.3). The remaining amino acids are "nonessential" only in the sense that the body can produce them if provided with the essential amino acids and with carbohydrates in sufficient amounts.

If there are more amino acids in the body than are needed for protein synthesis, the amine group may be removed and excreted as *urea* in the urine (fig. 21.9). The metabolic pathway that removes amine groups from amino acids is known as **oxidative deamination** (*de-am"ĭ-na'shun*). The molecules formed by this process may then be used in the Krebs cycle as a source of energy.

Depending on which amino acid is deaminated, the molecule left over may be either pyruvic acid or one of the Krebs cycle acids. These can be respired for energy, converted to fat, or converted to glucose. In the last case, the amino acids are eventually changed to pyruvic acid, which is used to form glucose. As mentioned previously (see module 21.3), the formation of new glucose from amino acids (or other noncarbohydrate molecules), is called **gluconeogenesis.** Gluconeogenesis from amino acids occurs primarily in the liver. It is an extremely important process in maintaining blood glucose levels during prolonged fasting and exercise.

The possible interrelationships between amino acids, carbohydrates, and fat are illustrated in figure 21.10.

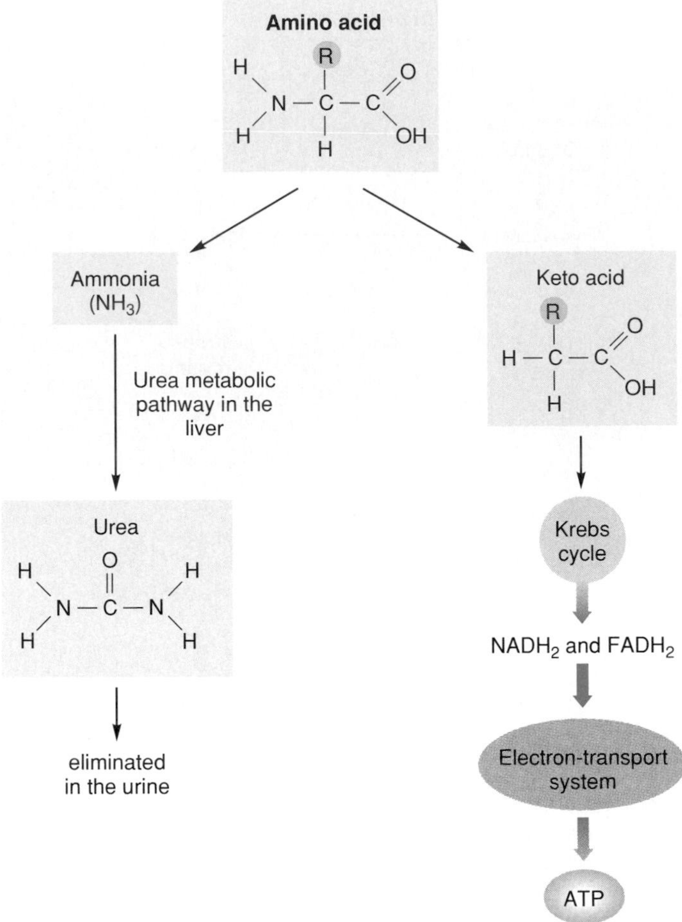

FIGURE 21.9
The formation of urea by the oxidative deamination of amino acids. The organic acid left over, called a keto acid, can enter the Krebs cycle and be used for energy.

Inside Information—Starving people have extremely thin arms and legs. This is because their muscle protein is being broken down for conversion to glucose. The brain requires blood glucose for its energy, so muscle is sacrificed to keep the brain alive. Cardiac muscle is no exception. In starvation, therefore, people die of heart failure.

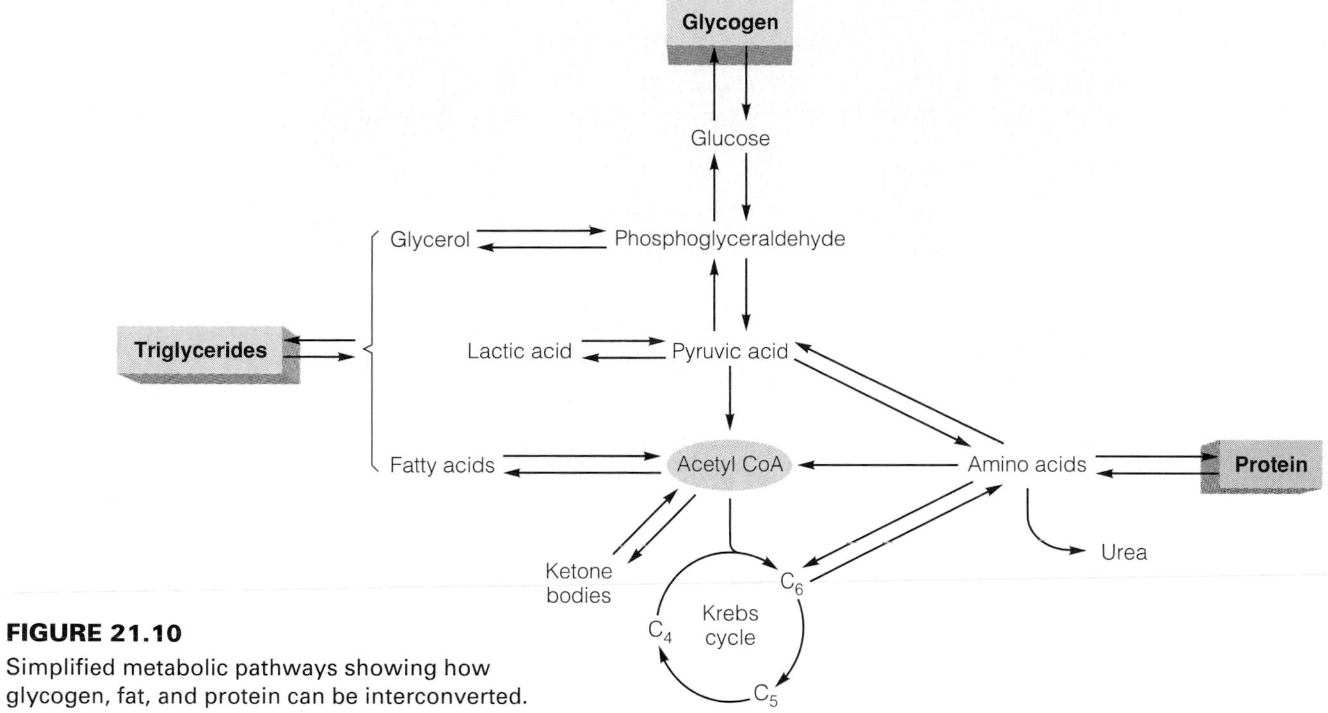

FIGURE 21.10
Simplified metabolic pathways showing how glycogen, fat, and protein can be interconverted.

21.7 Metabolic Balance

The molecules needed for cell respiration are obtained from food and from energy storage molecules within the body. The rate of energy consumption determines the metabolic rate.

Energy Reserves and Energy Substrates

The molecules that can be oxidized for energy by the processes of cell respiration may be derived from energy reserves of glycogen, fat, or protein. Alternatively, the molecules used for cell respiration can be derived from the products of digestion that are absorbed through the small intestine. Since these molecules—glucose, fatty acids, amino acids, and others—are carried by the blood to the tissue cells for use in cell respiration, they can be called **circulating energy substrates** (fig. 21.11).

Because of differences in their cellular enzyme content, organs differ in their *preferred energy sources.* The brain, for example, has an almost absolute requirement for blood glucose as its energy source. Resting skeletal muscles, by contrast, use fatty acids as their preferred energy source. Similarly, ketone bodies (derived from fatty acids), lactic acid, and amino acids can be used as energy sources by various organs. The blood plasma normally contains adequate concentrations of all of these circulating energy substrates to meet the energy needs of the body.

Following meals, the absorption of circulating energy substrates from the small intestine rises to high levels. This is called the **absorptive state.** Between meals—in the **postabsorptive,** or **fasting, state**—molecules are no longer being absorbed from the small intestine. Despite this, the plasma concentration of glucose and other energy substrates does not normally fall below a certain level during these periods. During the absorption of digestion products from the small intestine, energy substrates are removed from the blood and deposited as energy reserves. The energy reserves are like a savings account from which withdrawals can be made during times of fasting (fig. 21.12). This ensures that there will be an adequate plasma concentration of energy substrates to sustain tissue metabolism at all times.

Metabolic Rate

The energy obtained or consumed by the body is commonly measured in **kilocalories** (*kil′ŏ-kal″ŏ-rēz*); these

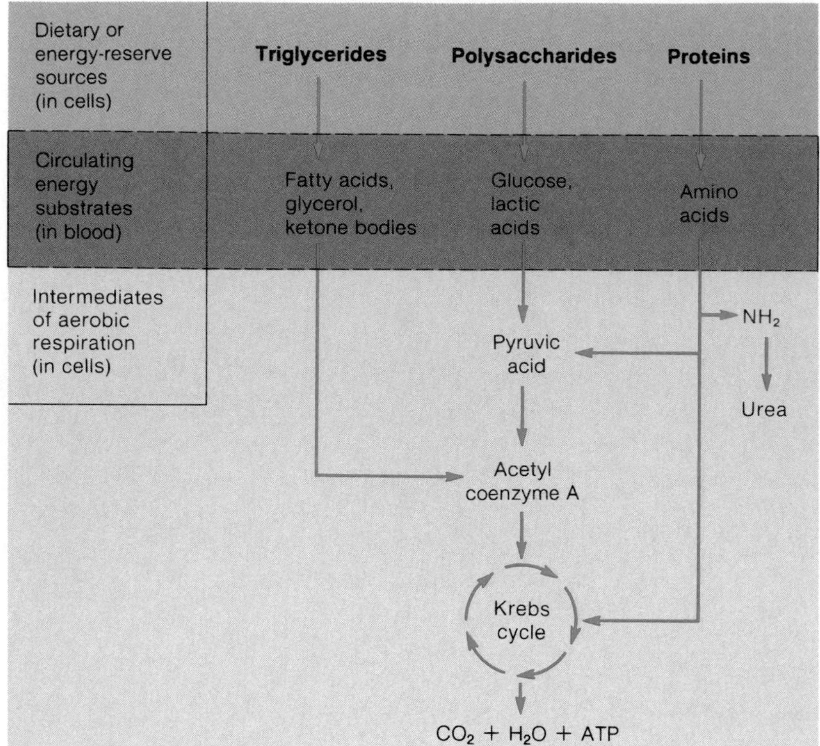

FIGURE 21.11
A schematic flowchart of energy pathways in the body.

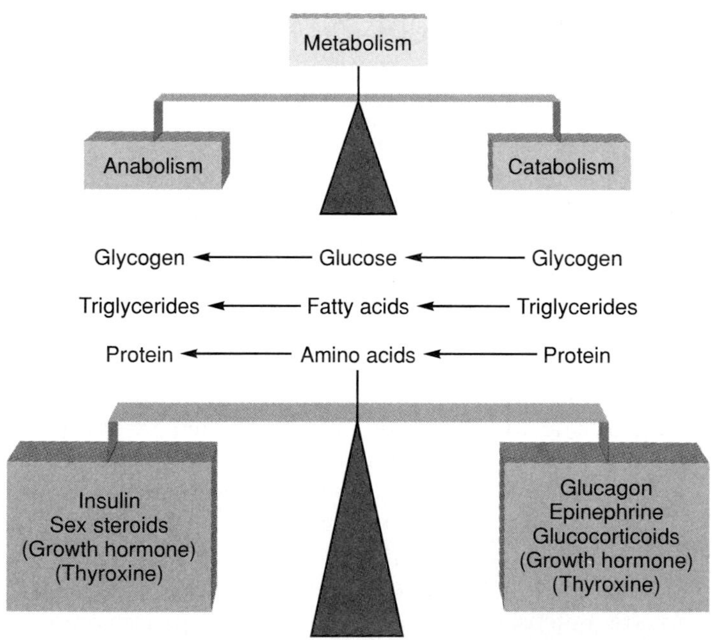

FIGURE 21.12
The balance of metabolism can be tilted toward anabolism (synthesis of energy reserves) or catabolism (utilization of energy reserves) by the combined actions of various hormones. Growth hormone and thyroxine have both anabolic and catabolic effects.

units are also called "big calories" and spelled with a capital letter (Calories). One kilocalorie is equal to 1000 calories; one calorie is defined as the amount of heat required to raise the temperature of one cubic centimeter of water from 15° to 16°C. When this energy is released by cell respiration, some is transferred to the high-energy bonds of ATP and some is lost as heat.

The total amount of energy used by the body per unit of time is called the *metabolic rate.* Because this rate is influenced by a variety of factors, it is usually measured under standardized conditions. The metabolic rate of an awake, relaxed person 12–14 hours after eating and at a comfortable temperature is known as the **basal metabolic rate (BMR).** The BMR is determined primarily by a person's age, sex, and body surface area, but it is also strongly influenced by the level of thyroid secretion. A person with hyperthyroidism has an abnormally high BMR, and a person with hypothyroidism has a low BMR.

As we know from experience, when our caloric intake exceeds our energy expenditures, we store the excess calories primarily as fat. This is true regardless of the

calorie: L. *calor,* heat

source of the calories—carbohydrates, protein, or fat—because these molecules can be converted to fat by the metabolic pathways previously described. We gain weight when our caloric intake is greater than our caloric expenditure over a period of time. We lose weight when the caloric value of the food we consume is less than the amount required in cell respiration over a period of time.

Inside Information—The most common cause of obesity is underactivity, not overeating. According to the Centers for Disease Control, only 22% of Americans are physically active to levels recommended for good health.

21.8 Metabolic Regulation by Insulin and Glucagon

The secretion of insulin is increased during absorption and decreased during the postabsorptive state. Glucagon secretion varies inversely, and insulin and glucagon have antagonistic effects on metabolism.

The rate of deposit and withdrawal of energy substrates into and from the energy reserves and the conversion of one type of energy substrate into another are regulated by the actions of hormones. The balance between anabolism and catabolism is determined by the antagonistic effects of such hormones as insulin, glucagon, growth hormone, and thyroxine. The metabolic effects of these hormones are illustrated in figure 21.13.

Insulin (*in'-sŭ-lin*) and **glucagon** (*gloo'kă-gon*) are secreted by the pancreatic islets (of Langerhans), as described in chapter 13 (see module 13.6). Insulin is secreted by the *beta cells* and glucagon by the *alpha cells* of the pancreatic islets. The rate of insulin and glucagon secretion is largely regulated by the blood glucose levels.

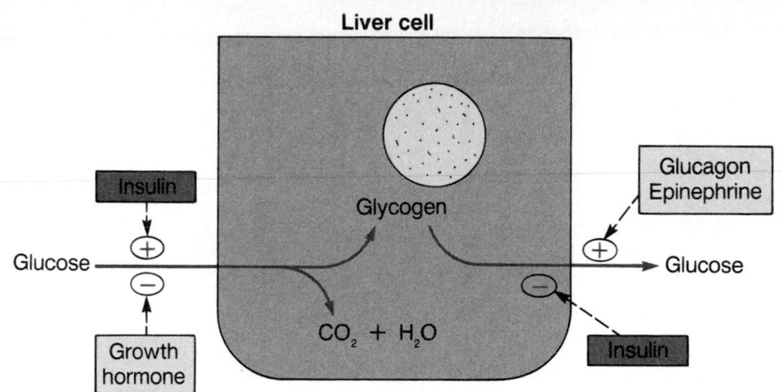

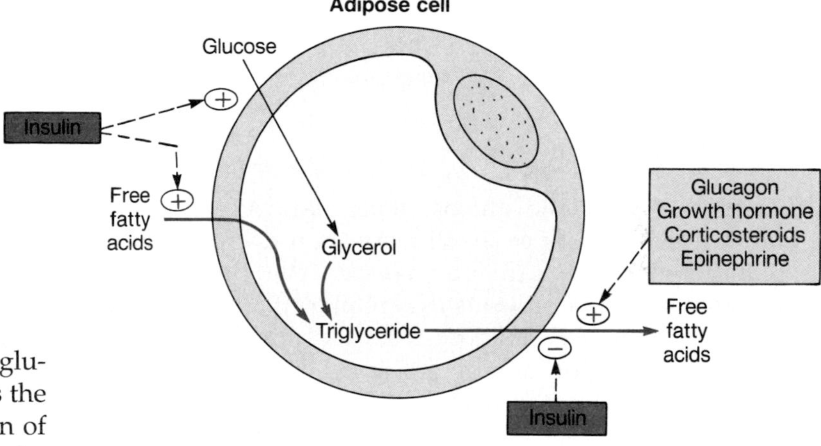

FIGURE 21.13

Different hormones participate both synergistically and antagonistically in the regulation of metabolism. A ⊕means the hormone stimulates a process; a ⊖means the hormone inhibits a process.

State of Food Absorption

During the absorption of a carbohydrate meal, blood glucose levels rise. This rise in blood glucose (1) stimulates the beta cells to secrete insulin and (2) inhibits the secretion of glucagon from the alpha cells. Insulin acts to stimulate the cellular uptake of blood glucose into the muscles, liver, and adipose tissue. A rise in the rate of insulin secretion therefore lowers the blood glucose concentration. The action of glucagon is antagonistic to that of insulin: glucagon raises the blood glucose concentration (by stimulating the breakdown of liver glycogen into glucose). Since glucagon secretion is inhibited when insulin secretion is stimulated during the absorption of food, the blood glucose levels are prevented from remaining in an elevated state.

During absorption of the products of digestion, the increased rate of insulin secretion promotes the cellular uptake of blood glucose, so that the cells can utilize the glucose. Some of the glucose will be metabolized for energy, and some may be stored as an energy reserve for later use. Insulin therefore promotes the synthesis of glycogen and fat (fig. 21.14).

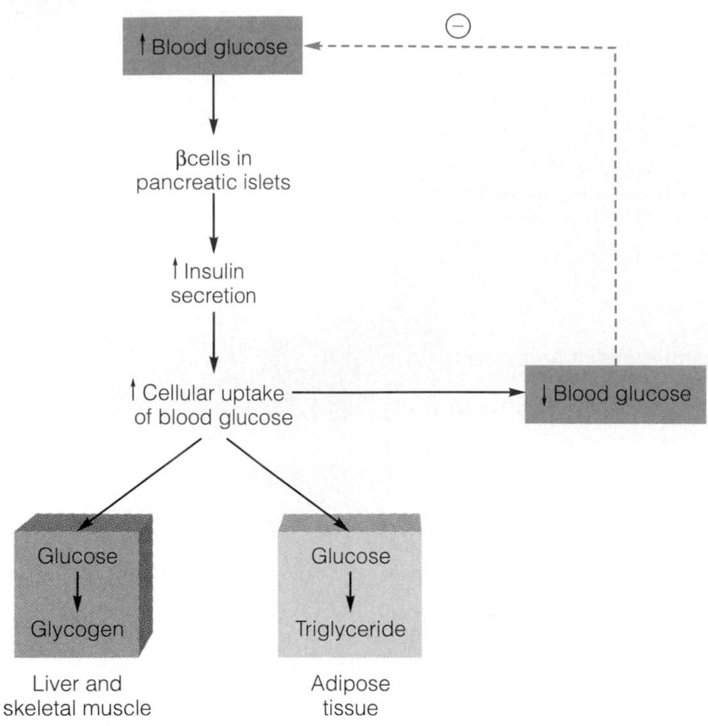

FIGURE 21.14
A rise in blood glucose concentration stimulates insulin secretion. Insulin promotes a fall in blood glucose by stimulating the cellular uptake of glucose and the conversion of glucose to glycogen and fat.

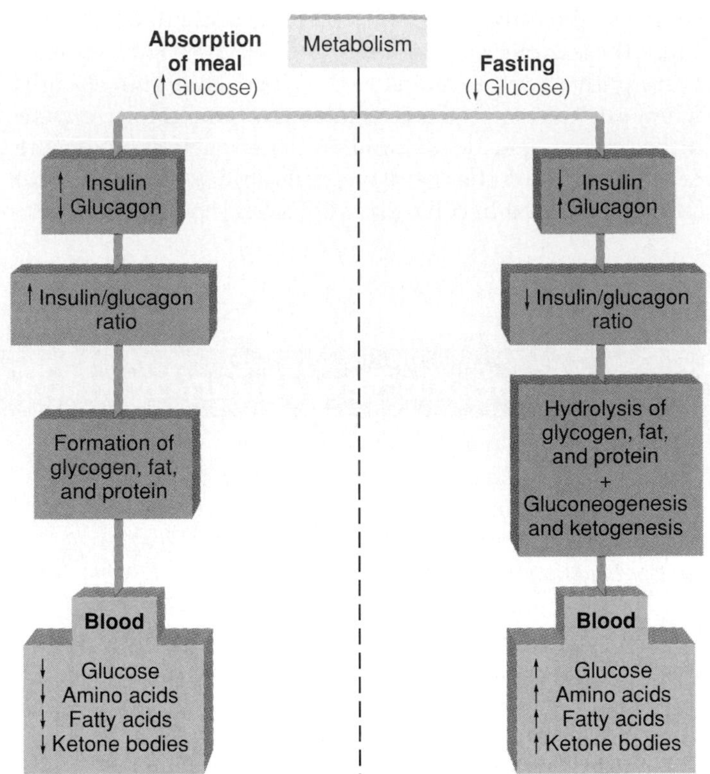

FIGURE 21.15
The inverse relationship between insulin and glucagon secretion during the absorption of a meal and during fasting. Changes in the ratio of insulin to glucagon tilt metabolism toward anabolism during the absorption of food and toward catabolism during fasting.

State of Fasting

During fasting, blood glucose levels fall. At this time, therefore (1) the rate of insulin secretion decreases and (2) that of glucagon secretion increases. The reduced rate of insulin secretion prevents the cellular uptake of blood glucose into the liver, adipose cells, and muscles. The increased rate of glucagon secretion promotes the breakdown of liver glycogen and the release of glucose from the liver. A negative feedback loop is therefore completed, which helps to retard the fall in blood glucose levels that occur during fasting.

Since there are only about 100 g of stored glycogen in the liver, adequate blood glucose levels could not be maintained for very long during fasting using this source alone. The low rate of insulin secretion during fasting, together with an elevated rate of glucagon secretion, however, promotes gluconeogenesis (production of glucose from non-carbohydrate molecules). Low insulin levels allow the release of amino acids from skeletal muscle proteins, while an increase in glucagon and cortisol (discussed later) stimulates the production of enzymes in the liver that convert amino acids to pyruvic acid, which is used to form glucose. As mentioned previously, during prolonged fasting and exercise, the conversion of amino acids into glucose by the liver may be the only source of blood glucose.

Glucagon also stimulates the hydrolysis of stored fat and the release of free fatty acids and glycerol into the blood. Some fatty acids are converted within the liver into ketone bodies, which are also released into the blood. Several organs in the body can use ketone bodies, like fatty acids, as a source of acetyl CoA in aerobic respiration. The antagonistic action of insulin and glucagon (fig. 21.15) thus promotes appropriate metabolic responses during periods of fasting and absorption.

Clinical Considerations

Chronic high blood glucose, or *hyperglycemia,* is the hallmark of the disease **diabetes mellitus.** The name of this disease is derived from the fact that glucose "spills over" into the urine when the blood glucose concentration is too high (*mellitus* is derived from the Latin word meaning "honeyed" or "sweet"). Diabetes mellitus results from the inadequate secretion or action of insulin.

There are two forms of diabetes mellitus. In **type I,** or **insulin-dependent, diabetes,** the beta cells are destroyed and secrete little or no insulin. This form of the disease accounts for only about 10% of the cases of diabetes in the United States. About 90% of the people who have diabetes have **type II,** or **non-insulin-dependent, diabetes.** These two forms of diabetes mellitus are compared in table 21.4.

Type I Diabetes Mellitus

Type I diabetes mellitus results when the beta cells of the pancreatic islets are destroyed by the effects of a virus or some other environmental agent. This environmental agent may act directly, or it may provoke an autoimmune reaction in which the body's own antibodies attack beta cells. Removal of the insulin-secreting cells in this way causes hyperglycemia and the appearance of glucose in the urine.

A rapid breakdown of fat occurs in a person with uncontrolled type I diabetes, and many of the fatty acids released from adipose cells are converted into ketone bodies in the liver. This may result in an elevated concentration of ketone bodies in the blood (ketosis); moreover, if the buffer reserve of bicarbonate is neutralized, it may also result in *ketoacidosis.* During this time, the glucose and excess ketone bodies that are excreted in the urine exert an osmotic force that causes the excessive excretion of water in the urine. This can produce severe dehydration, which, together with ketoacidosis and associated disturbances in electrolyte balance, may result in coma and death (fig. 21.16).

Before the development of recombinant DNA technology (genetic engineering), people with insulin-dependent diabetes had to inject themselves with insulin derived from the pancreases of pigs or cows. Now they have the option of using human insulin, which is available in large quantities from recombinant bacteria that have incorporated the human insulin genes.

People with type I diabetes mellitus are dependent upon insulin injections to prevent hyperglycemia and ketoacidosis. If inadequate insulin is injected, the person may enter a coma as a result of the ketoacidosis, electrolyte imbalance, and dehydration that develop. An overdose of insulin, however, can also produce a coma as a result of the hypoglycemia (abnormally low blood glucose levels) produced. The physical signs and symptoms of diabetic and hypoglycemic coma are sufficiently different (table 21.5) to allow hospital personnel to distinguish between these two types.

Type II Diabetes Mellitus

Tissue responsiveness to insulin varies under normal conditions. Exercise increases insulin sensitivity and obesity decreases insulin sensitivity of the target tissues. Therefore, the pancreatic islets of a nondiabetic obese person must secrete high amounts of insulin to maintain the blood glucose concentration within the normal range.

Type II diabetes is hereditary. It is usually slow to develop and occurs most often in people who are overweight. Most people who have type II diabetes have normal or even elevated levels of insulin in their blood, but their tissue sensitivity to insulin is abnormally low. This is true even if the person is not obese, but the problem is compounded by the decreased tissue sensitivity

Table 21.4 Comparison of Insulin-Dependent and Non-Insulin-Dependent Diabetes Mellitus

	Insulin-Dependent (Type I)	Non-Insulin-Dependent (Type II)
Usual Age at Onset	Under 20 years	Over 40 years
Development of Symptoms	Rapid	Slow
Percentage of Diabetic Population	About 10%	About 90%
Development of Ketoacidosis	Common	Rare
Association with Obesity	Rare	Common
Beta Cells of Islets (at Onset of Disease)	Destroyed	Not destroyed
Insulin Secretion	Decreased	Normal or increased
Autoantibodies to Islet Cells	Present	Absent
Treatment	Insulin injections	Diet; oral stimulators of insulin secretion

Clinical Considerations

that accompanies obesity. There is also evidence, however, that the beta cells are not functioning correctly. Whatever amount of insulin they secrete is inadequate to the task.

Type II diabetes can usually be controlled by increasing tissue sensitivity to insulin through diet and exercise. If this is not sufficient, oral drugs (generically known as the *sulfonylureas*) may be administered to stimulate insulin secretion.

Inside Information—Among certain Native American populations, type II diabetes is a virtual epidemic. The Pima Indians of Arizona, for example, have one of the highest rates of diabetes in the world—more than 40% of their adult population suffers from this condition. In addition, a disproportionately large segment of the adult population is also obese.

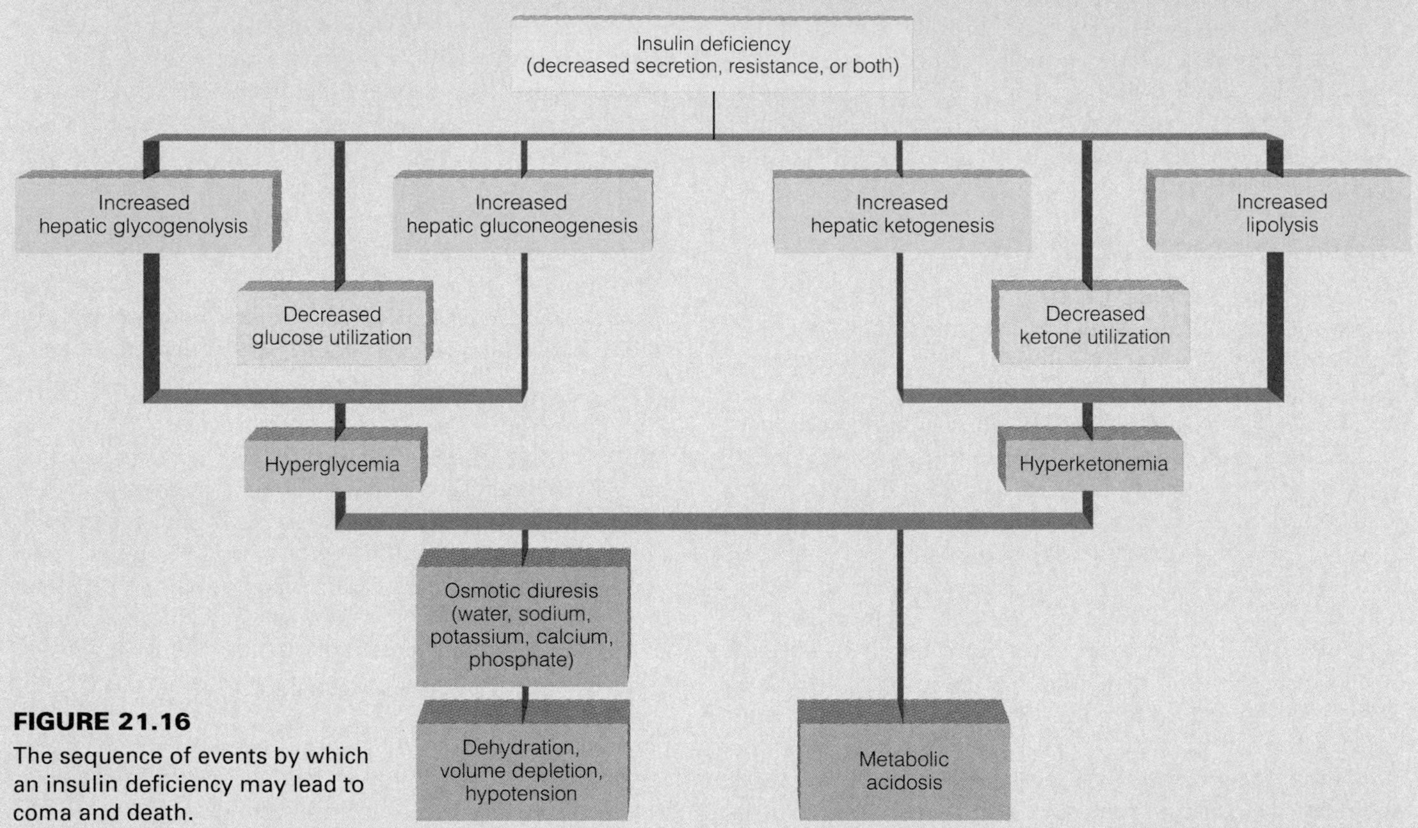

FIGURE 21.16
The sequence of events by which an insulin deficiency may lead to coma and death.

TABLE 21.5 Comparison of Diabetic and Hypoglycemic Coma

	Diabetic Ketoacidosis	Hypoglycemia
Onset	Hours to days	Minutes
Causes	Insufficient insulin; other diseases	Excess insulin; insufficient food; excessive exercise
Symptoms	Excessive urination and thirst; headache, nausea, and vomiting	Hunger, headache, confusion, stupor
Physical findings	Deep, labored breathing; acetone odor on breath; blood pressure decreased; pulse weak; skin dry	Pulse, blood pressure, and respiration normal; skin pale and moist
Laboratory findings	Urine: glucose present, ketone bodies increased Plasma: glucose and ketone bodies increased, bicarbonate decreased	Urine: normal Plasma: glucose concentration low, bicarbonate normal

SUMMARY

21.1 Nutritional Requirements

1. The diet must provide adequate calories and sufficient amounts of the essential amino acids and fatty acids.
2. Water-soluble and fat-soluble vitamins, as well as various elements, also must be obtained in the diet.
3. NAD is a coenzyme derived from niacin; FAD is a coenzyme derived from riboflavin.

21.2 Glycolysis

1. The conversion of glucose to pyruvic acid is known as glycolysis.
2. Glycolysis produces a net gain of 2 ATP and 2 $NADH_2$.

21.3 Anaerobic Respiration and Gluconeogenesis

1. Anaerobic respiration occurs primarily in exercising skeletal muscles.
2. The conversion of lactic acid and amino acids to pyruvic acid and then to glucose is called gluconeogenesis.

21.4 Aerobic Respiration and the Krebs Cycle

1. Pyruvic acid enters a mitochondrion and is converted into acetyl coenzyme A. This begins a cyclic metabolic pathway called the Krebs cycle.
2. When they accept electrons from hydrogen atoms, NAD and FAD are reduced to $NADH_2$ and $FADH_2$, respectively. Molecules become oxidized when they donate electrons.
3. The Krebs cycle produces 1 ATP, 3 $NADH_2$, and 1 $FADH_2$.

21.5 Electron Transport and Oxidative Phosphorylation

1. Electrons from $NADH_2$ and $FADH_2$ are passed to an electron-transport system of molecules in the cristae of mitochondria.
2. Electrons are ultimately passed to oxygen, which then binds protons (H^+) to become H_2O.
3. Production of ATP by electron transport is called oxidative phosphorylation.
4. Beginning with glucose, a total of 38 ATP molecules can be produced by aerobic respiration.

21.6 Metabolism of Carbohydrates, Lipids, and Proteins

1. Glycogenesis occurs in the liver and skeletal muscles. In the liver, glycogenolysis can release free glucose into the blood.
2. Fatty acids are hydrolyzed through beta-oxidation into acetyl CoA.
3. Amino acids can be converted into pyruvic acid or Krebs cycle acids through oxidative deamination; they can be interconverted through transanimation.
4. Amino acids can be converted into glucose in the process of gluconeogenesis.

21.7 Metabolic Balance

1. Anabolism is favored in the absorptive state, and catabolism occurs during the postabsorptive (fasting) state.
2. Energy reserves of glycogen, fat, and protein can provide circulating energy substrates of glucose, fatty acids, ketone bodies, and amino acids for use in cell respiration.

21.8 Metabolic Regulation by Insulin and Glucagon

1. A rise in blood glucose levels stimulates insulin and inhibits glucagon secretion.
2. Insulin acts to lower the blood glucose by promoting the cellular uptake of glucose and its conversion into glycogen and fat.
3. During fasting, the rate of insulin secretion decreases and the rate of glucagon secretion increases. These changes promote the hydrolysis of stored glycogen and fat.

Review Activities

Objective Questions

1. Which of the following is *not* a fat-soluble vitamin? (*module 21.1*)
   a. vitamin A
   b. vitamin C
   c. vitamin E
   d. vitamin K

2. Which of the following statements about glycolysis is *true?* (*module 21.2*)
   a. Glucose is converted into two pyruvic acid molecules.
   b. Two pairs of hydrogens are removed from intermediates of glycolysis.
   c. Two ATP are converted into ADP + P_i at the start of glycolysis.
   d. Two $NADH_2$ are produced.
   e. All of the above are true.

3. In anaerobic respiration, the molecule that removes hydrogen from NADH is (*module 21.3*)
   a. pyruvic acid.
   b. lactic acid.
   c. citric acid.
   d. oxygen.

4. The formation of glucose from pyruvic acid derived from lactic acid or amino acids is called (*module 21.3*)
a. glycogenesis.
b. glycogenolysis.
c. glycolysis.
d. gluconeogenesis.

5. The net gain of ATP per glucose in anaerobic respiration is _________; the potential net gain in aerobic respiration is _________. (*modules 21.4 and 21.5*)
a. 2; 4
b. 2; 38
c. 38; 2
d. 24; 30

6. Which of the following statements about the oxygen we breathe is *true?* (*module 21.5*)
a. It functions as the final electron acceptor of the electron-transport system.
b. It combines with hydrogen to form water.
c. It combines with carbon to form CO_2.
d. Both a and b are true.
e. Both a and c are true.

7. A lowering of blood glucose levels results in (*module 21.6*)
a. decreased lipogenesis.
b. increased lipolysis.
c. increased glycogenolysis.
d. all of the above.

8. Glucose can be secreted into the blood by (*module 21.6*)
a. the liver.
b. the muscles.
c. the liver and muscles.
d. the liver, muscles, and brain.

9. The basal metabolic rate is determined primarily by (*module 21.7*)
a. hydrocortisone.
b. insulin.
c. growth hormone.
d. thyroxine.

10. The absorption of a meal is accompanied by (*module 21.8*)
a. a rise in insulin and a rise in glucagon.
b. a fall in insulin and a rise in glucagon.
c. a rise in insulin and a fall in glucagon.
d. a fall in insulin and a fall in glucagon.

11. Which of the following is *not* promoted by the action of insulin? (*module 21.8*)
a. breakdown of liver glycogen
b. synthesis of fat
c. synthesis of protein
d. lowering of the blood glucose concentration

Essay Questions

1. Distinguish between the water-soluble and fat-soluble vitamins and state some of the functions of vitamins in the body. (*module 21.1*)
2. Define the terms *anabolism, catabolism,* and *cell respiration.* (*module 21.1*)
3. Explain why the two ATPs formed in glycolysis represent a net gain. What are the other products of glycolysis? (*module 21.2*)
4. Define *gluconeogenesis.* Why is this process so important? (*module 21.3*)
5. What purpose is served by the formation of lactic acid during anaerobic respiration? How is this purpose achieved during aerobic respiration? (*modules 21.3 and 21.4*)
6. Identify the products of the Krebs cycle and explain the significance of this cyclic metabolic pathway. (*module 21.4*)
7. Cyanide blocks the transfer of electrons from the last cytochrome to oxygen. What are the consequences of this action on oxidative phosphorylation and the Krebs cycle? (*module 21.5*)
8. Using the metabolic pathways involved, explain how a meal rich in carbohydrates can result in an increased synthesis of fat. (*module 21.6*)
9. Explain how proteins are used for energy and how glucose can be formed from amino acids. (*module 21.6*)
10. Define *basal metabolic rate.* Why does a person with hypothyroidism have a tendency to gain weight and less tolerance for cold? (*module 21.7*)
11. How does insulin lower blood glucose levels? Does insulin promote anabolism or catabolism? Explain. (*module 21.8*)
12. How does fasting affect the secretion of insulin and glucagon? How do these changes help to maintain homeostasis? (*module 21.8*)

Labeling Exercise

Label the products indicated on the diagram to the right.

1. ______________________________
2. ______________________________
3. ______________________________
4. ______________________________
5. ______________________________

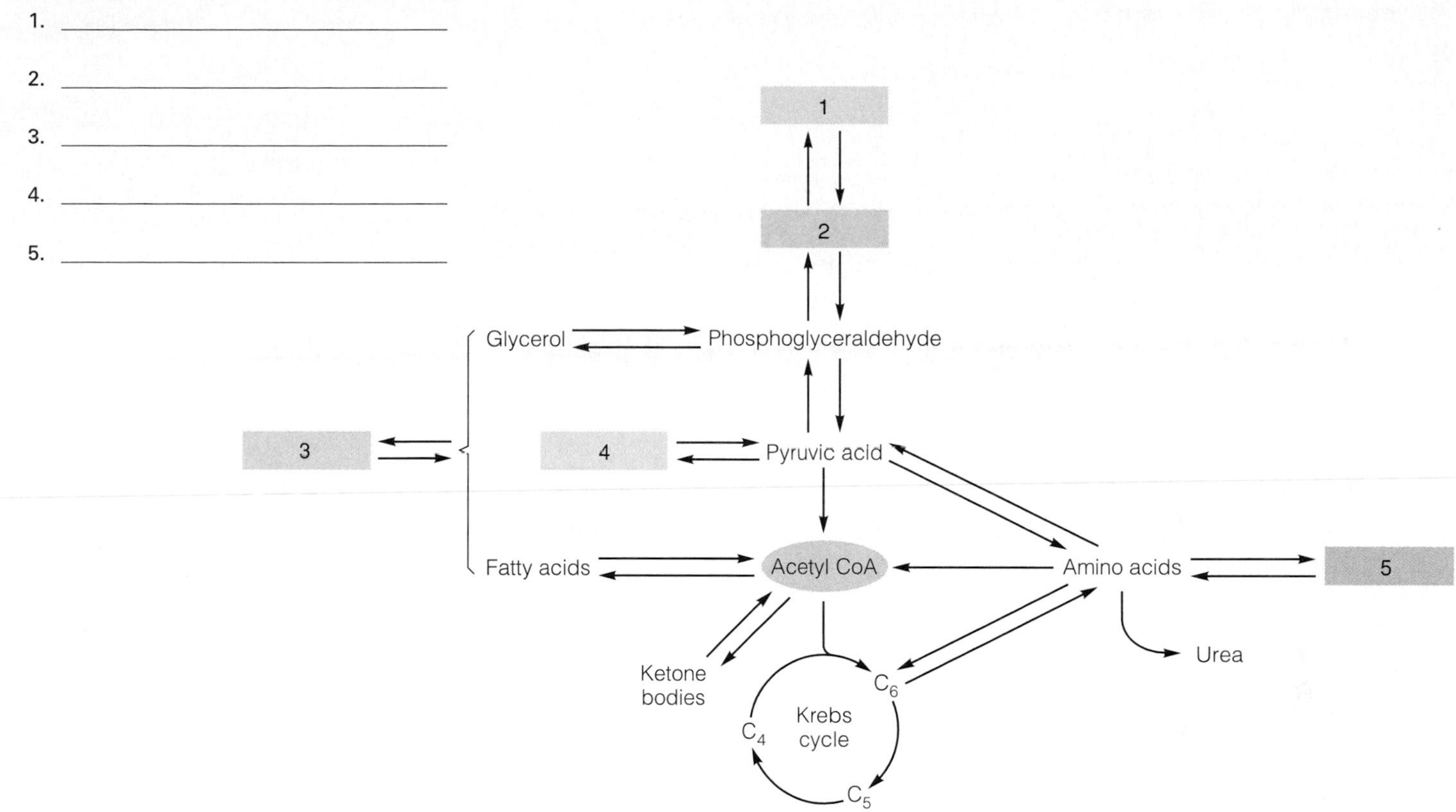

Critical Thinking Questions

1. You may have noticed that, gram for gram, potato chips have many more calories than does a baked potato. What produces this greater number of calories? By what metabolic pathway does the body derive energy for ATP synthesis from the additional calories in potato chips compared to a baked potato?
2. Your friend tells you that she is taking extra niacin to "burn calories." In light of what you know about niacin's role in metabolism, is there any sense in that statement? Would taking excessive doses of niacin drive metabolism faster and faster? Explain.
3. You may have heard that competitive runners often "carbo-load" (eat extraordinary amounts of carbohydrates) before a race. How do the liver and skeletal muscles handle this carbohydrate load? Do all of these carbohydrate calories end up in the liver and muscles? What other fate might the excess glucose molecules have?
4. The arms and legs of a person who is starving eventually become very skinny. What happens to the muscle protein, and how does this aid the body during starvation? A starving person may also develop a protruding belly, due to edema. How does protein malnutrition lead to edema? (Hint: Review osmotic pressure and the distribution of fluid across blood capillaries in chapter 16.)
5. You decide to go on an extremely strict diet and cut out fat and oil entirely. Some time later, you notice that your night vision isn't what it used to be. How might this symptom be related to your new diet?

CHAPTER

22

"The functions of the male reproductive system are to produce the male gametes, or sperm cells, and to transfer them to the female. . . ." *page 565*

Male Reproductive System

Chapter Outline

Terms to Remember

testosterone (*tes-tos' tĕ-rōn*), 567
puberty (*pyoo' ber-te*), 570
perineum (*per"ĭ-ne' um*), 572
scrotum (*skro' tum*), 572
testis (*tes' tis*), 574
spermatogenesis (*sper-mat"ŏ-jen' ĕ-sis*), 576
spermatogonia (*sper-mat"ŏ-go' ne-ă*), 576
meiosis (*mi-o' sis*), 576
spermiogenesis (*sper"me-ŏ-jen' ĕ-sis*), 577
spermatozoon (*sper-mat"ŏ-zo' on*), 578
epididymis (*ep"ĭ-did' ĭ-mis*), 578
ductus deferens (*duk' tus def' er-enz*), 578
spermatic cord, 578
seminal (*sem' ĭ-nal*) vesicle, 578
prostate (*pros' tāt*), 580
bulbourethral (*bul"bo-yoo-re' thral*) gland, 580
penis (*pe' nis*), 580
ejaculation (*ĕ-jak"yū-la' shun*), 582

Learning Objectives

After studying this chapter, you should be able to

- explain why sexual reproduction is biologically advantageous.
- distinguish between primary and secondary sex organs.
- explain what determines whether testes or ovaries will form in the embryo and describe the function of the embryonic testes.
- discuss the mechanisms that regulate the onset of puberty and describe the changes that occur in young males during this phase of development.
- state the location of the perineum and describe the structure and function of the scrotum.
- describe the location, structure, and functions of the testes.
- describe the hormonal interactions between the hypothalamus, anterior pituitary, and testes.
- discuss the process of spermatogenesis and explain how this process is regulated.
- describe the location, structure, and functions of the seminal vesicles, prostate, and bulbourethral glands.
- describe the structure and function of the penis.
- describe the chief phases of the male sexual response.

22.1 Introduction to the Reproductive System

The organs of the male and female reproductive systems produce and unite gametes that contain specific genes. A random combination of the genes during sexual reproduction produces individuals with genetic differences. Both the male and female reproductive systems consist of primary and secondary sex organs; in addition, secondary sex characteristics serve as sexual attractants.

Unlike the other body systems, the reproductive system is not essential for the survival of the individual; it is, however, required for the survival of the species. It is through reproduction that new individuals of a species are produced and the genetic code passed from one generation to the next. This can be accomplished by either asexual or sexual reproduction. But sexual reproduction, in which genes from two individuals are combined in random ways with each new generation, offers the overwhelming advantage of introducing great variability into a population. This variability of genetic constitution helps to ensure that some members of a population will survive changes in the environment over evolutionary time.

The reproductive system is unique in two other respects. First, the fact that it does not become functional until it is "turned on" at puberty by the actions of sex hormones sets the reproductive system apart. By contrast, all of the other body systems are functional at birth, or shortly thereafter. Second, while the other organ systems of the body exhibit slight sexual differences, no other system approaches the level of dissimilarity of the reproductive system. Because sexual reproduction requires the production of two types of **gametes** (*gam'ēts*), or sex cells, we have a male individual and a female individual, each with a unique reproductive system. The male and female reproductive systems complement each other in their common purpose of producing offspring.

The functions of the male reproductive system are to produce the male gametes, or **sperm cells,** and to transfer them to the female through *coitus* (*sexual intercourse*), or *copulation.* The female not only produces her own gametes (called **oocytes,** or **ova**) and receives the sperm cells from the male, but her reproductive organs are specialized to provide sites for fertilization, implantation of the developing embryonic mass (the blastocyst), pregnancy, and delivery of a baby. The more complex reproductive system of the female also provides a means for nourishing the baby through the secretion of milk from the mammary glands.

In this chapter we will consider the anatomy and physiology of the male reproductive system; the female reproductive system is the focus of chapter 23.

Categories of Reproductive Structures

The structures of the male reproductive system can be grouped into three categories based on function.

1. **Primary sex organs.** The primary sex organs are called **gonads;** specifically, the *testes* in the male. Gonads produce the sperm cells and produce and secrete sex hormones. The secretion of male sex hormones at the appropriate times and in sufficient quantities causes the development of secondary sex organs and the expression of secondary sex characteristics.
2. **Secondary sex organs.** Secondary sex organs are those structures that are essential in caring for and transporting sperm cells. The three categories of male secondary sex organs are the sperm-transporting ducts, the accessory glands, and the copulatory organ. The ducts that transport sperm cells include the *epididymides, ductus deferentia, ejaculatory ducts,* and *urethra.* The accessory glands are the *seminal vesicles,* the *prostate,* and the *bulbourethral glands.* The *penis,* which contains erectile tissue, is the copulatory organ. The *scrotum* is a pouch of skin that encloses and protects the testes. These structures will be described in detail in subsequent modules.
3. **Secondary sex characteristics.** Secondary sex characteristics are features that are not essential for the reproductive process but that are generally considered sexual attractants. In the male, they include body physique, body hair, and voice pitch.

The organs of the male reproductive system are shown in figure 22.1.

---

gamete: Gk. *gameta,* husband or wife

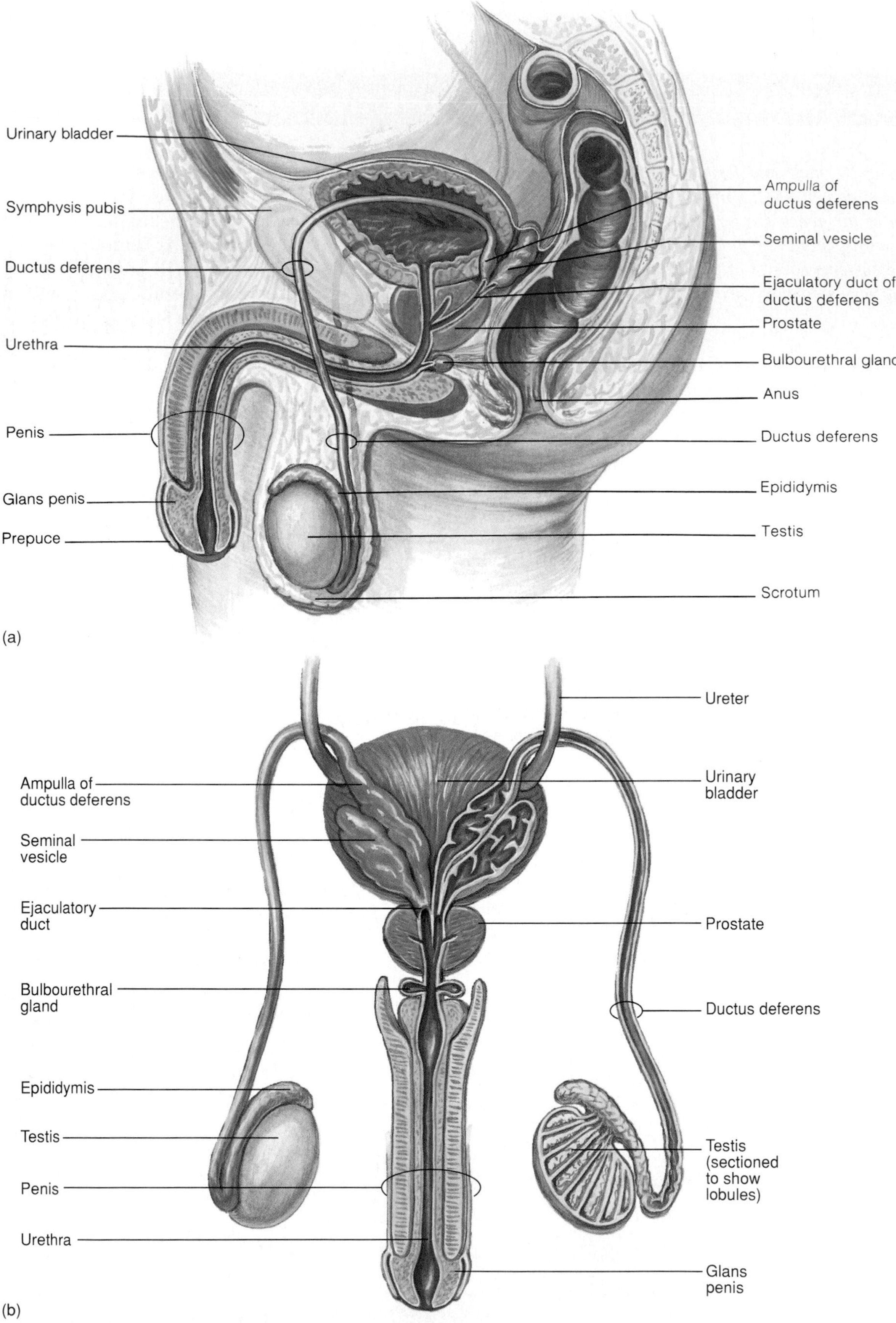

FIGURE 22.1
Organs of the male reproductive system. (*a*) A sagittal view and (*b*) a posterior view.

22.2 Sex Determination and the Influence of Hormones on Reproductive Development

A person's gender (male or female) is determined upon fertilization of an ovum by a sperm cell containing an X or a Y chromosome. Subsequent masculinization of embryonic reproductive structures occurs as a result of testosterone secreted by the embryonic testes.

Sex Determination

Sexual identity is initiated at conception when the *genetic sex* of the zygote (fertilized egg) is determined. The ovum is fertilized by a sperm cell containing either an X or a Y chromosome. If the sperm cell contains an X chromosome, it will pair with the X chromosome of the ovum, and a female child will develop (fig. 22.2). A sperm cell carrying a Y chromosome produces an XY combination, and a male child will develop.

The presence or absence of a Y chromosome determines whether the gonads will be testes or ovaries. This is because the Y chromosome contains a gene that causes the formation of testes. In the absence of this gene, ovaries form. In the event that testes form, they will produce and secrete male sex hormones during late embryonic and early fetal development and cause the secondary sex organs of the male to develop. In the event that ovaries form, the lack of hormonal secretion from the testes will cause the secondary sex organs of the female to develop.

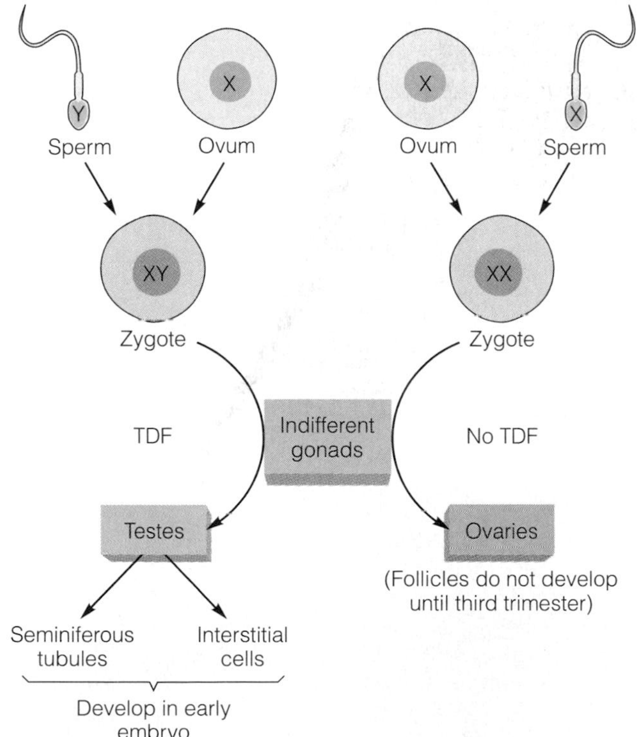

FIGURE 22.2
The formation of the chromosomal sex of the embryo and the development of the gonads. The very early embryo has "indifferent gonads" that can develop into either testes or ovaries. The testis-determining factor (TDF) is a gene located on the Y chromosome. In the absence of TDF, ovaries will develop.

A person born with Klinefelter's syndrome *has 47 instead of 46 chromosomes because of the presence of an extra X chromosome. These people, with XXY genotypes, develop testes despite the presence of two X chromosomes. People with* Turner's syndrome, *who have the genotype XO (and therefore only 45 chromosomes), develop ovaries.*

Hormonal Influence on Reproductive Development

Once the testes have formed, they begin to produce hormones. Large amounts of male sex hormones, called *androgens,* are secreted from interstitial cells (cells of Leydig). The major androgen secreted by these cells is **testosterone** (*tes-tos'tĕ-rōn*). Testosterone secretion begins as early as 8 to 10 weeks after conception, reaches a peak at 12 to 14 weeks, and thereafter declines so that very low levels are being secreted by the end of the second trimester (about 21 weeks). High levels of testosterone will not occur again until the onset of puberty.

The seminiferous tubules, which will eventually produce sperm cells within the testes, appear about 45 to 50 days following conception. Although the sperm-producing tissue is formed during embryonic life, it remains inactive until puberty.

Other embryonic reproductive structures (the *urogenital sinus, genital tubercle, urethral folds,* and *labioscrotal swellings*) are also masculinized by secretions of the testes (see the Synoptic Spotlight on page 584). The prostate derives from the urogenital sinus, and the other embryonic structures differentiate into the external genitalia (fig. 22.3). In the absence of androgens, the female genitalia are formed.

In summary, the genetic sex is determined by whether a Y-bearing or an X-bearing sperm cell fertilizes the ovum. The presence or absence of a Y chromosome determines

Klinefelter's syndrome: from Harry F. Klinefelter, Jr., American physician, b. 1912
Turner's syndrome: from Henry H. Turner, American endocrinologist, 1892–1970
androgen: Gk. *andros,* male-producing

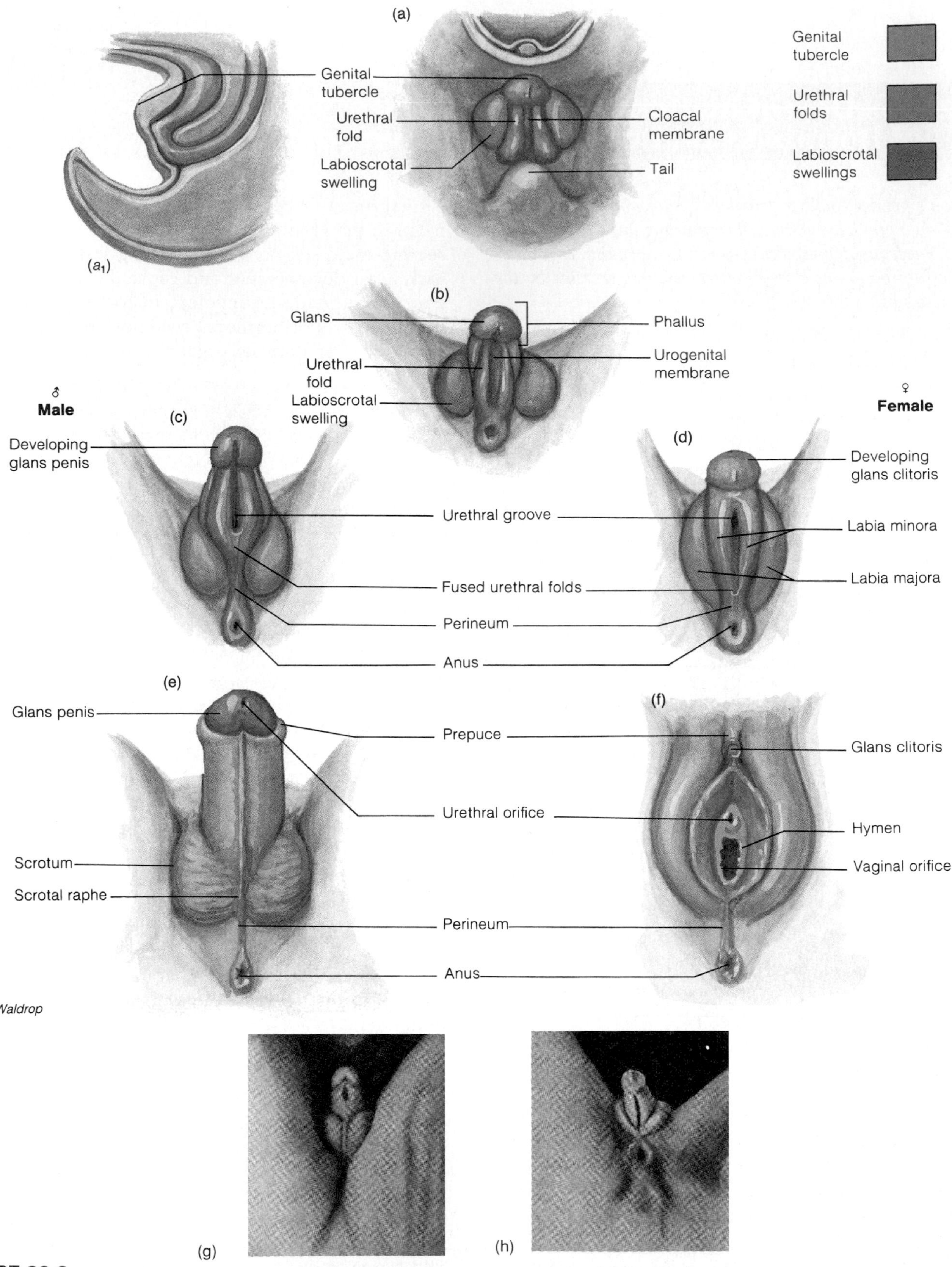

FIGURE 22.3
(*a*) Differentiation of the external genitalia in the male and female. (*a*$_1$) A sagittal view. At 6 weeks, the urethral fold and labioscrotal swelling are differentiated from the genital tubercle. (*b*) At 8 weeks, a distinct phallus is present during the indifferent stage. From the tenth through the twelfth week, the genitalia become distinctly male (*c*) or female (*d*), being derived from homologous structures. (*e, f*) At 16 weeks, the genitalia are completely formed. (*g, h*) Photographs at week 10 of male and female genitalia, respectively.

whether the gonads of the embryo will be testes or ovaries, and the presence or absence of testes (which secrete androgens) determines whether the accessory sex organs and external genitalia will be male or female. This mechanism of sex determination makes sense in the light of the fact that both male and female embryos develop within an environment high in estrogen, which is produced by the mother's ovaries and the placenta. If estrogen determined the sex, all embryos would be feminized.

Inside Information—Because an embryo has the potential to become either a male or a female, developmental errors can result in various degrees of intermediate sex, or *hermaphroditism.* A person with undifferentiated or ambiguous external genitalia is called a *hermaphrodite* (*her-maf' rŏ-dīt*).

True hermaphroditism—in which both ovarian and testicular tissue is present in the body—is rare. True hermaphrodites usually have a 46 XX chromosome constitution. *Male pseudohermaphroditism* (*pseudo* = false) occurs more commonly and generally results from hormonal influences during early fetal development. This condition is caused either by inadequate amounts of androgenic hormones being secreted or by the delayed development of the reproductive organs after the period of tissue sensitivity has passed. These individuals have a 46 XY chromosome constitution and male gonads, but genitalia that are indistinguishable as male or female.

The treatment of hermaphroditism varies, depending on the extent of ambiguity of the reproductive organs. Although people with this condition are sterile, they may engage in normal sexual relations following hormonal therapy and plastic surgery.

Development of the Male Genitalia and Descent of the Testes

The external genitalia of the male can be distinguished by the tenth week (fig. 22.3). Prior to that, a swelling called the **genital tubercle** elongates in both sexes, and becomes known as the **phallus.** A **urethral groove** forms on the ventral surface of the phallus.

The differentiation of the external genitalia of a male from the indifferent stage to discernible organs is caused by the androgens produced and secreted by the testes. These changes include the differentiation of the phallus into a **penis,** the fusion of the **urethral folds** surrounding the urethral groove along the ventral surface of the penis, and a midventral fusion of the **labioscrotal swellings** to form the wall of the **scrotum.** The external genitalia of a male are completely formed by the end of the twelfth week.

A significant fact of embryonic development is that the reproductive organs of both sexes are derived from

TABLE 22.1 Summary of Homologous Structures

Indifferent Stage	Male	Female
Gonads	Testes	Ovaries
Urogenital groove	Membranous urethra	Vestibule
Genital tubercle	Glans penis	Clitoris
Urethral folds	Spongy urethra	Labia minora
Labioscrotal swellings	Scrotum	Labia majora
	Bulbourethral glands	Vestibular glands

the same embryonic tissue; thus, they are considered *homologous structures.* For example, as noted in table 22.1, the genital tubercle eventually forms the glans penis in the male; in the female, it gives rise to the clitoris. Similarly, the labioscrotal swellings form the male scrotum or the female labia majora.

The descent of the testes from the site of development begins between the sixth and tenth week. Descent into the scrotal sac, however, does not occur until about week 28, when paired inguinal canals form in the abdominal wall to provide openings from the pelvic cavity into the scrotal sac. The process by which a testis descends is not well understood, but it seems to be due to the shortening of the **gubernaculum.** This strong fibromuscular cord is attached to the testis and extends through the inguinal canal to the wall of the scrotum (fig. 22.4). As the testis descends, it passes to the side of the urinary bladder, in front of the symphysis pubis. It carries with it the ductus deferens, the testicular vessels and nerve, a portion of the abdominal muscle, and lymph vessels. All of these structures remain attached to the testis and form the **spermatic cord** (see fig. 22.7). By the time the testis is in the scrotal sac, the gubernaculum is no more than a remnant of scarlike tissue.

During the physical examination of a neonatal male child, the physician will palpate the scrotum to determine if the testes are in position. Cryptorchidism, *meaning "hidden testis," occurs in about 3% of male infants and is characterized by the failure of one or both testes to descend into the scrotum. It may be possible to induce descent by administering certain hormones. If this procedure fails, surgery is necessary. The surgery is generally performed before the age of 5. Failure to correct the situation may result in sterility, and possibly the development of a tumorous testis.*

hermaphrodite: Gk. (mythology) *Hermaphroditos,* son of Hermes (Mercury)
phallus: Gk. *phallus,* penis
homologous: Gk. *homos,* the same
gubernaculum: L. *gubernaculum,* helm
cryptorchidism: Gk. *crypto,* hidden; *orchis,* testis

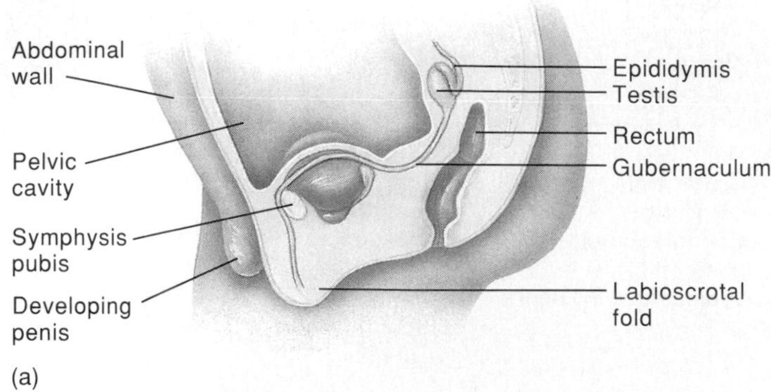

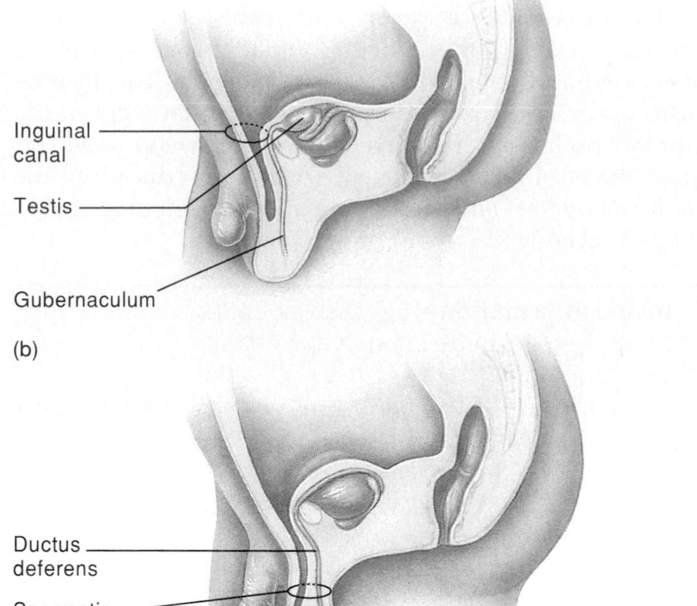

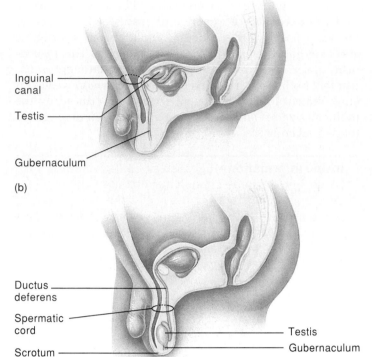

FIGURE 22.4
The descent of the testes (*a*) at 10 weeks, (*b*) at 18 weeks, and (*c*) at 28 weeks. During development, each testis descends through an inguinal canal in front of the symphysis pubis and enters the scrotum.

22.3 Puberty

Gonadotropic hormones from the anterior pituitary and maturational changes in the central nervous system stimulate the secretion of sex hormones from the testes. Sex hormones, in turn, stimulate a growth spurt and sexual maturation during puberty.

Puberty (*pyoo'ber-te*) is the period of growth and development when the sex organs become functional. In a male, puberty is characterized by a deepening of the voice and increased sebum secretion and hair growth—especially on the face, axillae (underarms), genital region, and chest. A boy generally goes through puberty between the ages of 12 and 14. For girls, it is generally about 6 months to a year earlier.

From the time of birth up until the onset of puberty, there are equal (low) concentrations of *sex steroids*—androgens and estrogens—in both males and females. Apparently, this is not due to deficiencies in the ability of the gonads to produce these hormones, but rather to lack of sufficient stimulation. During puberty, the gonads secrete increased amounts of sex steroid hormones as a result of increased stimulation by **gonadotropic hormones** from the anterior pituitary.

puberty: L. *puberty*, grown up

Interactions between the Hypothalamus, Pituitary, and Gonads

The anterior pituitary produces and secretes **follicle-stimulating hormone (FSH)** and **luteinizing hormone (LH).** Although these two hormones are named for their role in the female, the same hormones are secreted by the male's pituitary. (In the male, LH is also called **interstitial cell-stimulating hormone—ICSH.**) The gonadotropic hormones of both sexes have three primary effects on the gonads:

1. stimulation of spermatogenesis or oogenesis (to form sperm cells or ova);
2. stimulation of gonadal hormone secretion; and
3. maintenance of the structure of the gonads.

TABLE 22.2 Development of Secondary Sex Characteristics and Other Changes That Occur during Puberty in Boys

Characteristic	Age of First Appearance	Hormonal Stimulation
Growth of testes	10–14	Testosterone, FSH, growth hormone
Pubic hair	10–15	Testosterone
Body growth	11–16	Testosterone, growth hormone
Growth of penis	11–15	Testosterone
Growth of larynx (voice lowers)	Same time as growth of penis	Testosterone
Facial and axillary (underarm) hair	About 2 years after the appearance of pubic hair	Testosterone
Eccrine sweat glands and sebaceous glands; acne (from blocked sebaceous glands)	About the same time as facial and axillary hair growth	Testosterone

With regard to gonadotropin action, the testes are strictly compartmentalized. Cellular receptor proteins for FSH are located exclusively in the seminiferous tubules. LH receptor proteins are located exclusively in the interstitial cells. Thus, secretion of testosterone by the interstitial cells is stimulated by LH but not by FSH. Spermatogenesis in the tubules is stimulated only by FSH. As described in chapter 13 (see module 13.3), the secretion of both LH and FSH from the anterior pituitary is stimulated by a hormone produced by the hypothalamus and released into the hypothalamo-hypophyseal portal vessels. This releasing hormone is called *gonadotropin–releasing hormone* (*GnRH*).

Similarities in the gonadal regulation of males and females are illustrated in figure 22.5. Important differences in hypothalamus-pituitary-gonad interactions exist, however, between males and females. The secretion of gonadotropins and sex steroids is more or less constant in adult males. The secretion of gonadotropins and sex steroids in adult females, by contrast, shows cyclic variations (during the menstrual cycle). Also, during one point of the normal female cycle, estrogen exerts a stimulatory effect on LH secretion.

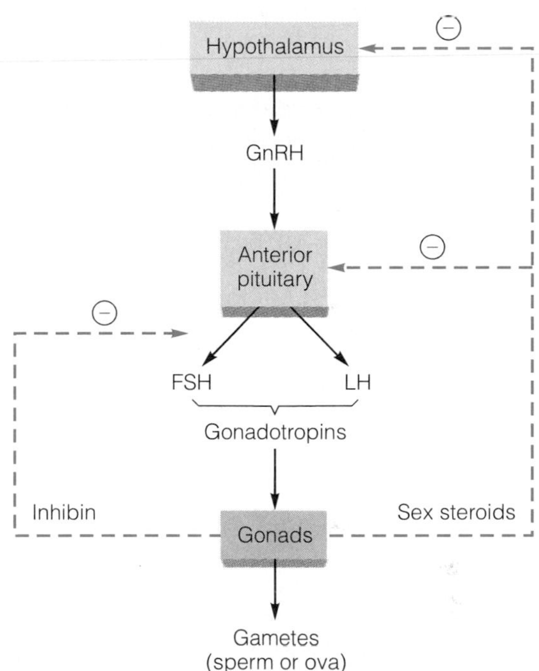

FIGURE 22.5
Interactions between the hypothalamus, anterior pituitary, and gonads. Sex steroids secreted by the gonads have a negative feedback effect on the secretion of GnRH (gonadotropin-releasing hormone) and on the secretion of gonadotropins. The gonads may also secrete a polypeptide hormone called *inhibin* that exerts negative feedback control of FSH secretion.

The Onset of Puberty

The secretion of FSH and LH is high in the newborn but falls to very low levels a few weeks after birth. These secretions remain low until the beginning of puberty, which is marked by rising levels of FSH followed by LH secretion.

During late puberty, FSH and LH secretion increases during periods of sleep and decreases during periods of wakefulness. These pulses of increased gonadotropin secretion during puberty stimulate a rise in hormonal secretion from the gonads. The increased secretion of testosterone from the testes during puberty causes development of the male *secondary sex characteristics* (table 22.2).

Inside Information—Even in the last century, castration (removal of the testes) was practiced to maintain the high-pitched voices of prepubescent males. The papal choir of the Sistine Chapel was filled with eunuchs (castrated men) until 1878, when Pope Leo XIII ended the tradition.

22.4 Perineum and Scrotum

The perineum is the specific portion of the pelvic region that contains the external genitalia and the anus. The scrotum, a pouch that supports the testes, is divided into two internal compartments by a connective tissue septum.

Perineum

The perineum (*per″ĭ-ne′um*) is a diamond-shaped region between the symphysis pubis and the coccyx (fig. 22.6). It is the muscular region of the outlet of the pelvis (see fig. 8.26). The perineum is divided into a *urogenital triangle* in front and an *anal triangle* in back. In the male perineum, the penis and scrotum are attached at the anterior portion of the urogenital triangle; the anus is located within the posterior portion of the anal triangle.

Scrotum

The saclike scrotum (*skro′tum*) is suspended immediately behind the base of the penis (fig. 22.7). The functions of the scrotum are to support and protect the testes and to regulate their position relative to the pelvic region of the body. The soft, textured skin of the scrotum is covered with sparse hair in mature males and is darker in color than most of the other skin of the body. It also contains numerous sebaceous glands.

The external appearance of the scrotum varies at different times in the same individual as a result of the contraction and relaxation of the scrotal muscles. The **dartos** (*dar′tos*) is a layer of smooth muscle fibers in the subcutaneous tissue of the scrotum, and the **cremaster** (*krĕ-mas′ter*) is a thin strand of skeletal muscle extending through the spermatic cord (fig. 22.7). Both muscles involuntarily contract in response to low temperatures to move the testes closer to the heat of the body in the pelvic region. The cremaster muscle is a continuation of the internal abdominal oblique muscle of the abdominal wall, which is derived as the testes descend into the scrotum. Because it is a skeletal muscle, it can contract voluntarily as well. When these muscles are contracted, the scrotum appears tightly wrinkled as the testes are pulled closer to the warmth of the body wall. High temperatures cause the dartos and cremaster muscles to relax and the testes to be suspended lower in the relaxed scrotum. The temperature of the testes is maintained at about 35°C (95°F, or about 3.6°F below normal body temperature) by the contraction or relaxation of the scrotal muscles. This temperature is ideal for the production and storage of sperm cells.

Inside Information—Although uncommon, *male infertility* may result from an excessively high temperature of the testes over an extended period of time. Frequent hot baths or saunas may destroy sperm cells to the extent that the sperm count will be too low to enable fertilization.

The scrotum is subdivided into two longitudinal compartments by a fibrous **scrotal septum** (fig. 22.7*b*). The purpose of the scrotal septum is to compartmentalize each testis so that trauma to one will not affect the other. In addition, the left testis is generally suspended lower in the scrotum than the right so that the two are not as likely to be compressed forcefully together. The site of the scrotal septum is apparent on the surface of the scrotum along a median longitudinal ridge called the **perineal raphe** (*ra′fe*). The perineal raphe extends forward to the undersurface of the penis and backward to the anal opening (see fig. 22.6).

The blood supply and innervation of the scrotum are extensive. The arteries that serve the scrotum are the internal pudendal branch of the internal iliac artery, the external pudendal branch of the femoral artery, and the cremasteric branch of the inferior epigastric artery. The venous drainage follows a pattern similar to the arteries. The scrotal nerves are primarily sensory; they include the pudendal nerves, ilioinguinal nerves, and posterior cutaneous nerves of the thigh.

dartos: Gk. *dartos,* skinned or flayed
cremaster: Gk. *cremaster,* a suspender, to hang
septum: L. *septum,* a partition
raphe: Gk. *raphe,* a seam
pudendal: L. *pudeo,* to feel ashamed

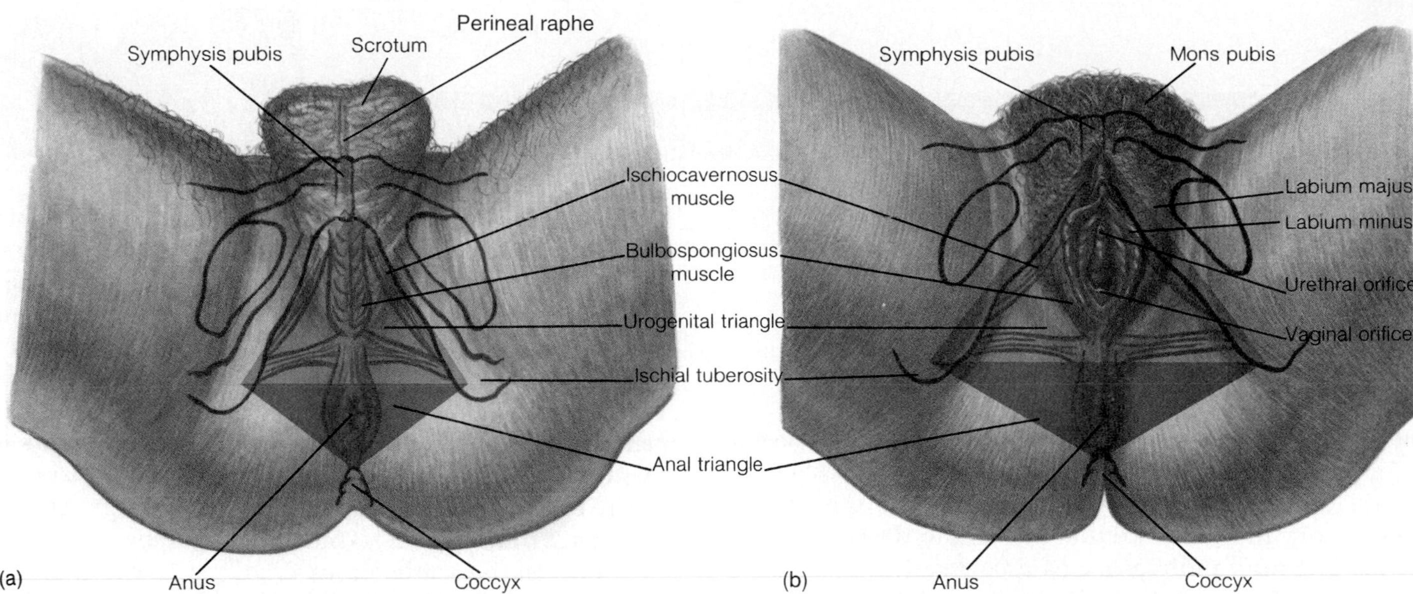

FIGURE 22.6
A superficial view of (*a*) the male perineum and (*b*) the female perineum. Divided into urogenital and anal triangles, the perineum is the region between the symphysis pubis and the coccyx.

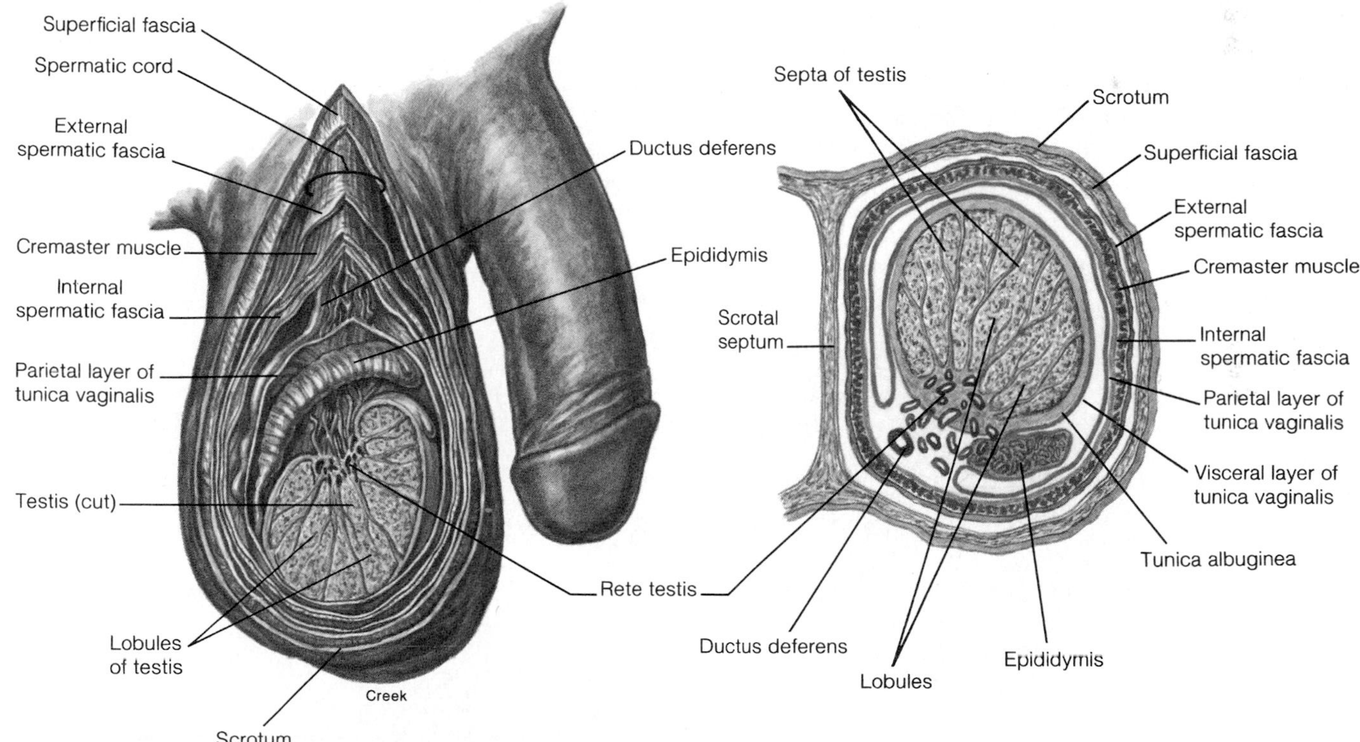

FIGURE 22.7
Structural features of a testis within the scrotum. (*a*) A longitudinal view and (*b*) a transverse view.

22.5 Structure and Function of the Testes

Located within the scrotum, the testes produce sperm cells and androgens. Androgens regulate spermatogenesis and the development and functioning of the secondary sex organs.

Structure of the Testes

The **testes** (*tes'tēz*—singular, *testis*) are paired, whitish, ovoid organs, each about 4 cm (1.5 in.) long and 2.5 cm (1 in.) in diameter. Two tissue layers, or tunicas, cover the testes. The outer **tunica vaginalis** is a thin serous sac derived from the peritoneum during the descent of the testes. The **tunica albuginea** (*al"byoo-jin'e-ă*) is a tough fibrous membrane that directly encapsulates each testis (fig. 22.7). Fibrous strands of the tunica albuginea divide the testis into 250 to 300 wedge-shaped **lobules.**

Each lobule of the testis contains tightly convoluted **seminiferous tubules** that may exceed 70 cm (28 in.) in length if uncoiled. The seminiferous tubules are the functional units of the testis because it is here that *spermatogenesis* (*sper-mat"ŏ-jen'ĕ-sis*), the production of sperm cells, occurs. Sperm cells are produced at the rate of thousands per second—more than 100 million per day—throughout the life of a healthy, sexually mature male.

Various stages of spermatogenesis can be observed in a section of a seminiferous tubule (fig. 22.8). The process begins as specialized germinal cells, called *spermatogonia,* undergo meiosis to produce, in order of advancing maturity, the *primary spermatocytes, secondary spermatocytes,* and *spermatids* (see fig. 22.10). Forming the walls of the seminiferous tubules are **sustentacular** (Sertoli) **cells** (also called *nurse cells*) that produce and secrete nutrients for the developing sperm cells embedded between them. The sperm cells are formed, but not fully matured, by the time they reach the lumen of a seminiferous tubule.

Between the seminiferous tubules are specialized endocrine cells called **interstitial cells** (of Leydig). The function of these cells is to produce and secrete the male sex hormones. The testes are thus considered mixed exocrine and endocrine glands because they produce both sperm cells and androgens.

tunica: L. *tunica,* a coat
vaginalis: L. *vagina,* a sheath
albuginea: L. *albus,* white
Sertoli cells: from Enrico Sertoli, Italian histologist, 1842–1910

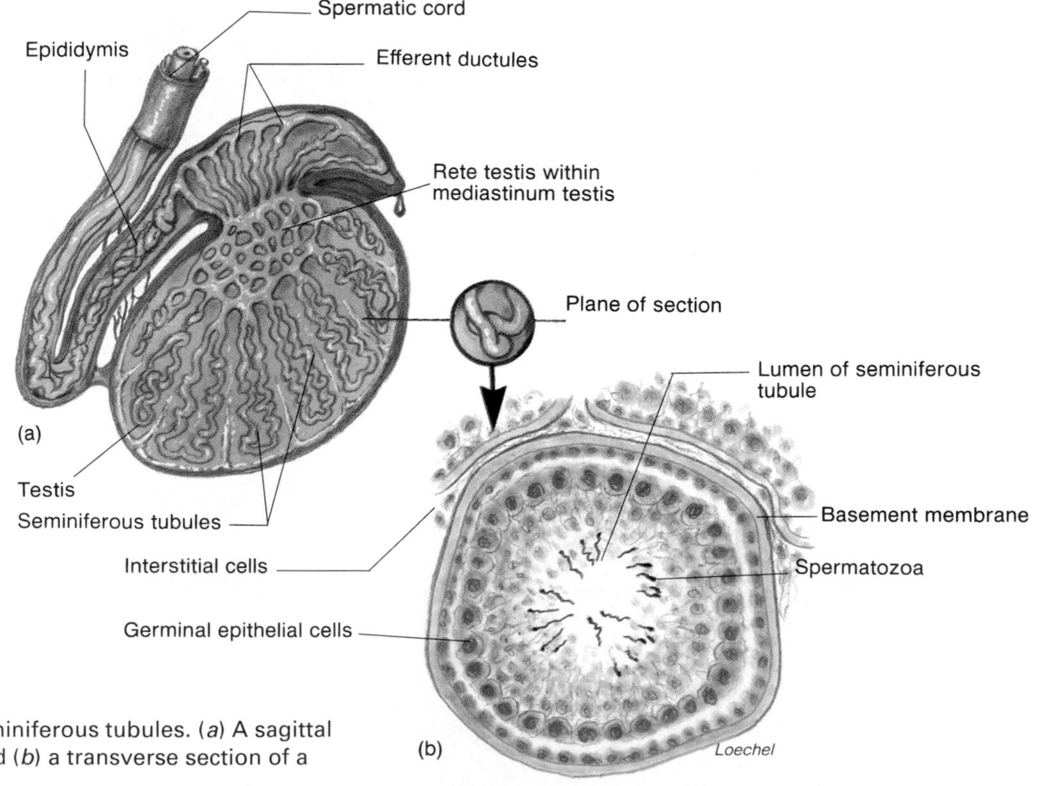

FIGURE 22.8
A diagram of the seminiferous tubules. (*a*) A sagittal section of a testis and (*b*) a transverse section of a seminiferous tubule.

Once the sperm cells are produced, they move through the seminiferous tubules and enter a tubular network called the **rete testis** for further maturation. Cilia are located on some of the cells of the rete testis, presumably for moving sperm cells. The sperm cells are transported out of the testis and into the epididymis through a series of **efferent ductules.**

Inside Information—In all, it takes from 8 to 10 weeks for a spermatogonium to become a mature sperm cell. In the spermatic ducts, sperm cells can remain fertile for several months, in a state of suspended animation. If they are not ejaculated, they eventually degenerate and are absorbed by the body.

The testes receive blood through the testicular arteries, which arise from the abdominal aorta immediately below the origin of the renal arteries. The testicular veins drain the testes. The testicular vein of the right side enters directly into the inferior vena cava, whereas the testicular vein of the left side drains into the left renal vein. Testicular nerves innervate the testes with both motor and sensory fibers.

The most common cause of male infertility is a condition called varicocele (var'ĭ-ko-sēl). *Varicocele occurs when one or both of the testicular veins draining from the testes becomes swollen, resulting in poor circulation in the testes. A varicocele generally occurs on the left side, because the left spermatic vein drains into the renal vein, where the blood pressure is higher than in the inferior vena cava, into which the right testicular vein empties.*

Endocrine Functions of the Testes

Testosterone is by far the major androgen secreted by the adult testes. Androgens are sometimes called *anabolic steroids* because they stimulate the growth of muscles and other structures (table 22.3). Increased testosterone secretion during puberty is also required for the growth of the accessory sex organs, primarily the seminal vesicles and the prostate. Androgens stimulate growth of the larynx (causing a deepening of the voice), hemoglobin synthesis (so that males have higher hemoglobin levels than females), and bone growth. At puberty, androgens are required, along with FSH, to stimulate spermatogenesis. Within the adult testes, spermatogenesis is maintained by androgens alone, in the absence of FSH.

rete: L. *rete,* a net
efferent ductules: L. *efferre,* to bring out: *ducere,* to lead
varicocele: L. *varico,* a dilated vein; Gk. *kele,* tumor or hernia

Although androgens are by far the predominant product of the testes, the testes do produce and secrete small amounts of *estradiol,* the major estrogen, or female sex steroid (fig. 22.9). The source of estradiol in the testes is not known, and its possible physiological functions are not well understood.

TABLE 22.3 Effects of Androgens in the Male

Category	Effect
Sex determination	Growth and development of mesonephric ducts into epididymides, ductus deferentia, seminal vesicles, and ejaculatory ducts Development of urogenital sinus and tubercle into prostate Development of male external genitalia (penis and scrotum)
Spermatogenesis	At puberty: completion of meiotic division and early maturation of spermatids After puberty; maintenance of spermatogenesis
Secondary sex characteristics	Growth and maintenance of accessory sex organs Growth of penis Growth of facial and axillary hair Body growth
Anabolic effects	Protein synthesis and muscle growth Growth of bones Growth of other organs (including larynx) Erythropoiesis (red blood cell formation)

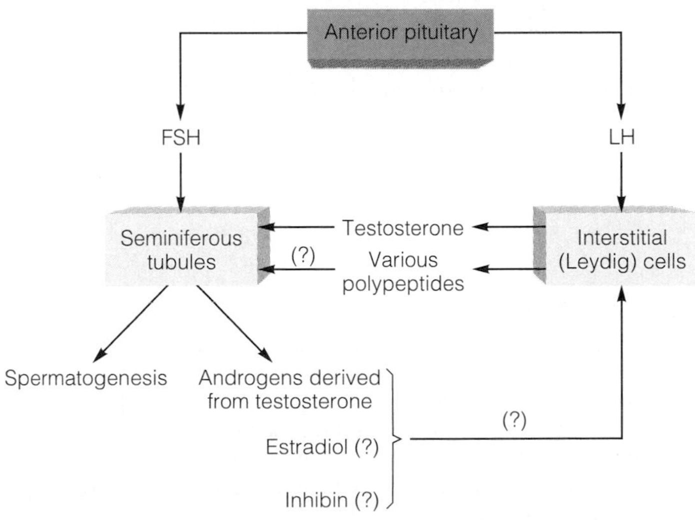

FIGURE 22.9
Interactions between the two compartments of the testes. Testosterone secreted by the interstitial cells stimulates spermatogenesis in the tubules. Secretions of the tubules may affect the sensitivity of the interstitial cells to LH stimulation.

22.6 Spermatogenesis and the Structure of a Spermatozoon

Spermatozoa are formed through the process of meiosis during spermatogenesis. As they develop, they are nurtured by sustentacular cells until they become mature gametes.

Spermatogenesis

The embryonic cells that migrate from the yolk sac to the testes during early development become stem cells, or **spermatogonia** (*sper-mat"ŏ-go'ne-ă*), within the outer region of the seminiferous tubules. Spermatogonia are diploid cells (with 46 chromosomes) that give rise to mature haploid gametes through the process of *meiosis* (*mi-o'sis*).

Actually, only about 1500 stem cells migrate from the yolk sac into the embryonic testes. In order to produce billions of sperm cells throughout adult life, spermatogonia cells duplicate themselves by mitotic division, and only one of the two cells—now called a **primary spermatocyte**—undergoes meiotic division (fig. 22.10). In this way, spermatogenesis can occur continuously without exhausting the number of spermatogonia.

When a diploid primary spermatocyte completes the first meiotic division (at telophase I), the two haploid daughter cells produced are called **secondary spermatocytes.** At the end of the second meiotic division, each of the two secondary spermatocytes produces two haploid **spermatids.** One primary spermatocyte therefore produces four spermatids.

As mentioned in the previous module, the sequence of events in spermatogenesis is reflected in the wall of the seminiferous tubule. The epithelial wall of the tubule,

spermatogonia: Gk. *sperma,* seed; *gone,* generation

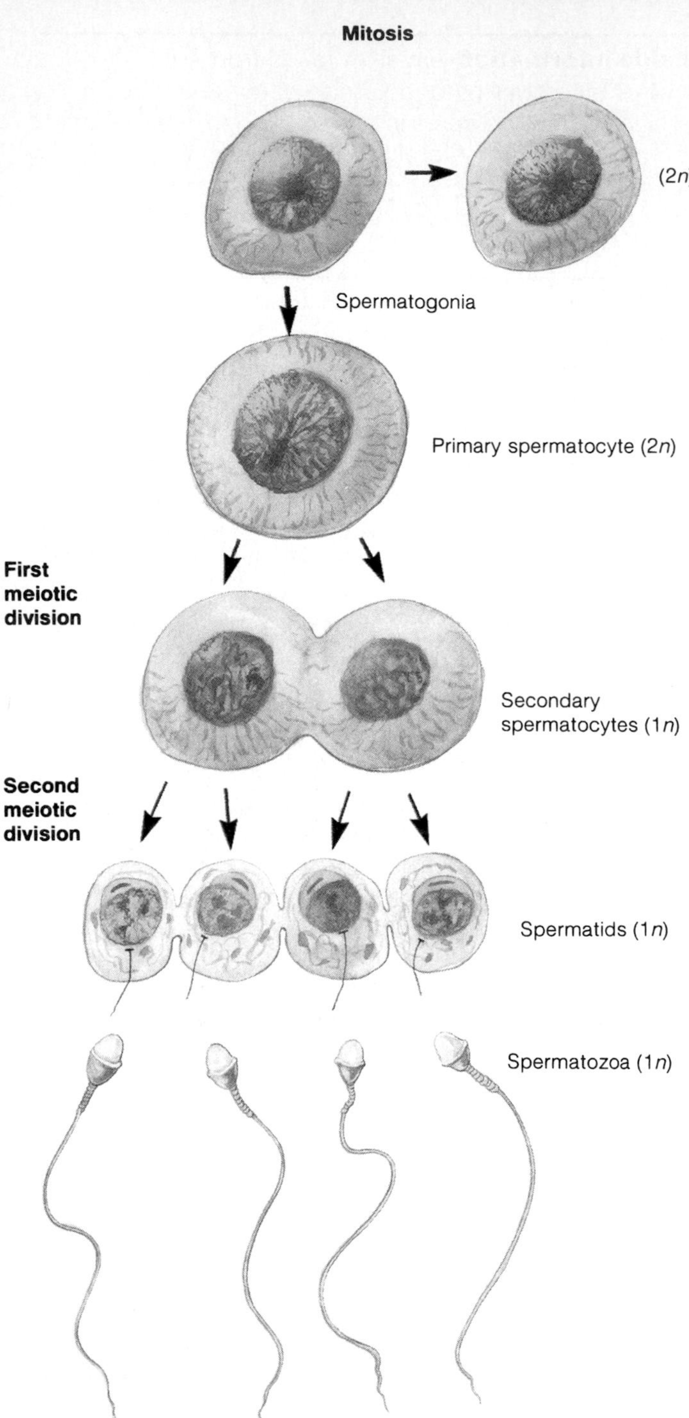

FIGURE 22.10
Spermatogonia undergo mitotic division to replace themselves and produce a daughter cell that will undergo meiotic division. This cell is called a primary spermatocyte. Upon completion of the first meiotic division, the daughter cells are called secondary spermatocytes. Each of them completes a second meiotic division to form spermatids. Notice that the four spermatids produced by the meiosis of a primary spermatocyte are interconnected. Each spermatid forms a mature spermatozoon.

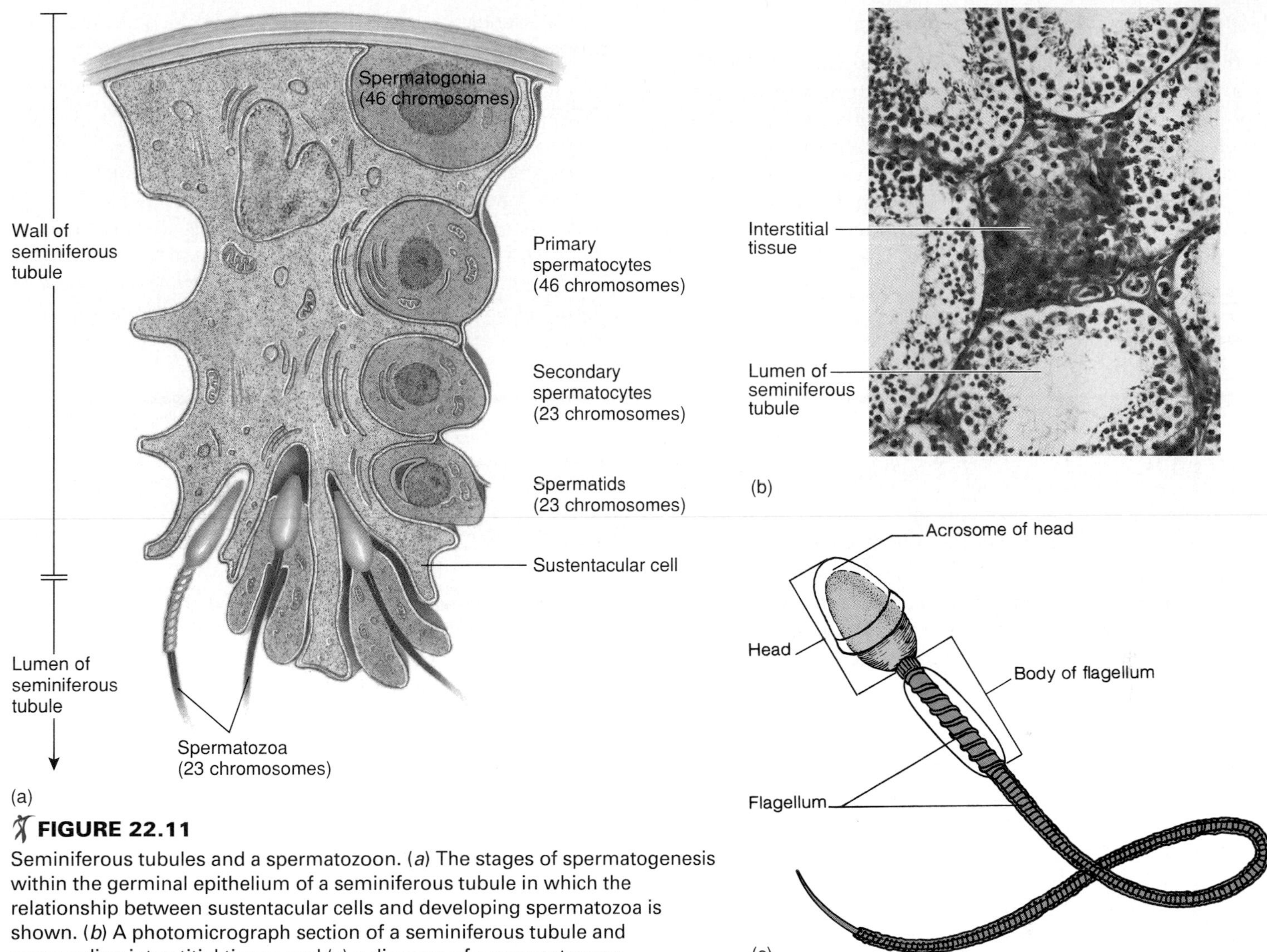

FIGURE 22.11

Seminiferous tubules and a spermatozoon. (*a*) The stages of spermatogenesis within the germinal epithelium of a seminiferous tubule in which the relationship between sustentacular cells and developing spermatozoa is shown. (*b*) A photomicrograph section of a seminiferous tubule and surrounding interstitial tissue, and (*c*) a diagram of a spermatozoon.

called the **germinal epithelium,** consists of cells in different stages of spermatogenesis. The spermatogonia and primary spermatocytes are located toward the outer side of the tubule, whereas the spermatids and mature sperm cells are located on the side facing the lumen.

At the end of the second meiotic division, the four spermatids produced by meiosis of one primary spermatocyte are interconnected—their cytoplasm does not completely pinch off at the end of each division. The development of these interconnected spermatids into separate, mature sperm cells occurs during the process called **spermiogenesis** (*sper″me-ŏ-jen′ĕ-sis*). This process requires the participation of another type of cell in the tubules—the sustentacular (nurse) cells, or Sertoli cells.

Sustentacular cells (fig. 22.11), connected by tight junctions, form a continuous layer around the circumference of each tubule. In this way, the sustentacular cells form a *blood-testis barrier*. The molecules from the blood must pass through the cytoplasm of sustentacular cells before entering germinal cells.

During the initial stage of spermatogenesis, meiosis progresses to early prophase I where it is arrested until puberty. Although testosterone secretion is not required for the initiation of spermatogenesis, it is required for the completion of meiotic division (see fig. 22.10) and for the early stages of spermatid maturation. Following puberty, both androgens and FSH are needed for the completion of spermatogenesis.

Structure of a Spermatozoon

A mature sperm cell, or **spermatozoon** (*sper-mat"ŏ-zo'on*), is a microscopic tadpole-shaped structure approximately 0.06 mm long (fig. 22.11*c*). It consists of an oval **head** and an elongated **flagellum.** The head contains a nucleus with 23 chromosomes. The tip of the head, called the *acrosome,* contains enzymes that help the spermatozoon penetrate the ovum. The flagellum contains numerous mitochondria spiraled around a filamentous core. The mitochondria provide the energy necessary for locomotion. The flagellum propels the spermatozoon with a lashing movement.

spermatozoon: Gk. *sperma,* seed; *zoon,* animal
acrosome: Gk. *akros,* extremity; *soma,* body

Recent findings indicate that ejaculated spermatozoa (plural of spermatozoon) can survive up to 5 days at body temperature—longer than previously thought. In the average fertile man, up to 20% of these sperm cells are defective, however, and are of no value. It is not uncommon for them to have enlarged heads, dwarfed and misshapen heads, two flagella, or a flagellum that is bent. Spermatozoa such as these are unable to propel themselves adequately.

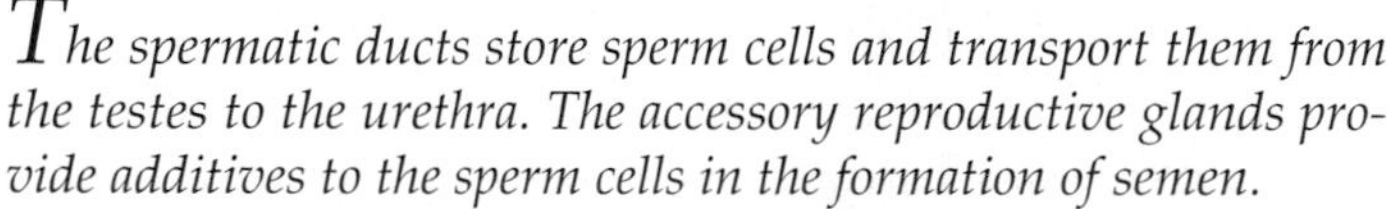
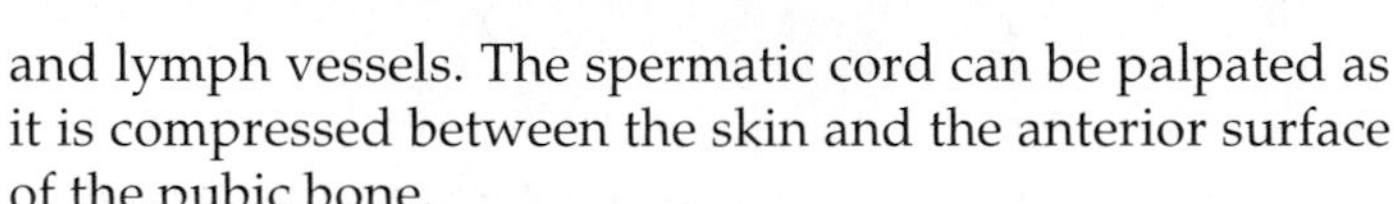

22.7 Spermatic Ducts and Accessory Reproductive Glands

The spermatic ducts store sperm cells and transport them from the testes to the urethra. The accessory reproductive glands provide additives to the sperm cells in the formation of semen.

Spermatic Ducts

After the sperm cells leave the testes through the efferent ductules, they pass through a series of ducts leading to the outside of the body. The duct system includes the *epididymis,* the *ductus deferentia,* the *ejaculatory ducts,* and finally the *urethra,* which opens at the tip of the penis. The histology of the spermatic ducts and the accessory reproductive glands is depicted in figure 22.12.

The **epididymis** (*ep"ĭ-did'ĭ-mis*—plural, *epididymides*) is a long, flattened organ attached to the posterior surface of the testis (see figs. 22.1, 22.7, and 22.13). The highly coiled, tubular *tail portion* contains sperm cells in their final stages of maturation. The upper expanded portion is the *head,* and the tapering middle section is the *body.* The tail is continuous with the beginning portion of the ductus deferens; both store sperm cells to be discharged during *ejaculation.*

The **ductus deferens** (*duk'tus def'er-enz*—plural, *ductus deferentia*) is a fibromuscular tube that conveys sperm cells from the epididymis to the ejaculatory duct. Also called the *vas deferens* (plural, *vasa deferentia*), it exits the scrotum as it ascends along the posterior border of the testis (see figs. 22.1 and 22.7). From here, it enters the pelvic cavity and passes to the side of the urinary bladder, medial to the ureter. The *ampulla* of the ductus deferens is the terminal portion that joins the ejaculatory duct.

Much of the ductus deferens is located within the **spermatic cord** (see fig. 22.7). The spermatic cord extends from the testis to the inguinal ring and consists of the ductus deferens, spermatic vessels, nerves, cremaster muscle, and lymph vessels. The spermatic cord can be palpated as it is compressed between the skin and the anterior surface of the pubic bone.

The **ejaculatory duct** is formed by the union of the ampulla of the ductus deferens and the duct of the seminal vesicle. Both ejaculatory ducts receive secretions from the seminal vesicles, and then eject the sperm cells and additives into the urethra to be mixed with secretions from the prostate.

The **urethra** of the male serves as a common tube for both the urinary and reproductive systems (see fig. 19.22*a*). Urine and semen cannot simultaneously pass through the urethra, however, because the nervous reflex during ejaculation automatically inhibits urination.

Accessory Reproductive Glands

The accessory reproductive glands of the male include the *seminal vesicles,* the *prostate,* and the *bulbourethral glands* (see fig. 22.1). The contents of the seminal vesicles and the prostate are mixed with the spermatozoa during ejaculation to form *semen* (*seminal fluid*). The fluid from the bulbourethral glands is released in response to sexual stimulation prior to ejaculation.

The **seminal** (*sem'ĭ-nal*) **vesicles** are club-shaped glands that lie at the base of the urinary bladder, in front of the rectum (see fig. 22.1). They secrete a sticky, slightly alkaline, yellowish substance that contributes to the motility and viability of the sperm cells. The secretion from the seminal vesicles contains a variety of nutrients, including fructose, that provide an energy source for the sperm cells. It also contains citric acid and prostaglandins.

deferens: L. *deferens,* conducting away
ampulla: L. *ampulla,* a two-handled bottle

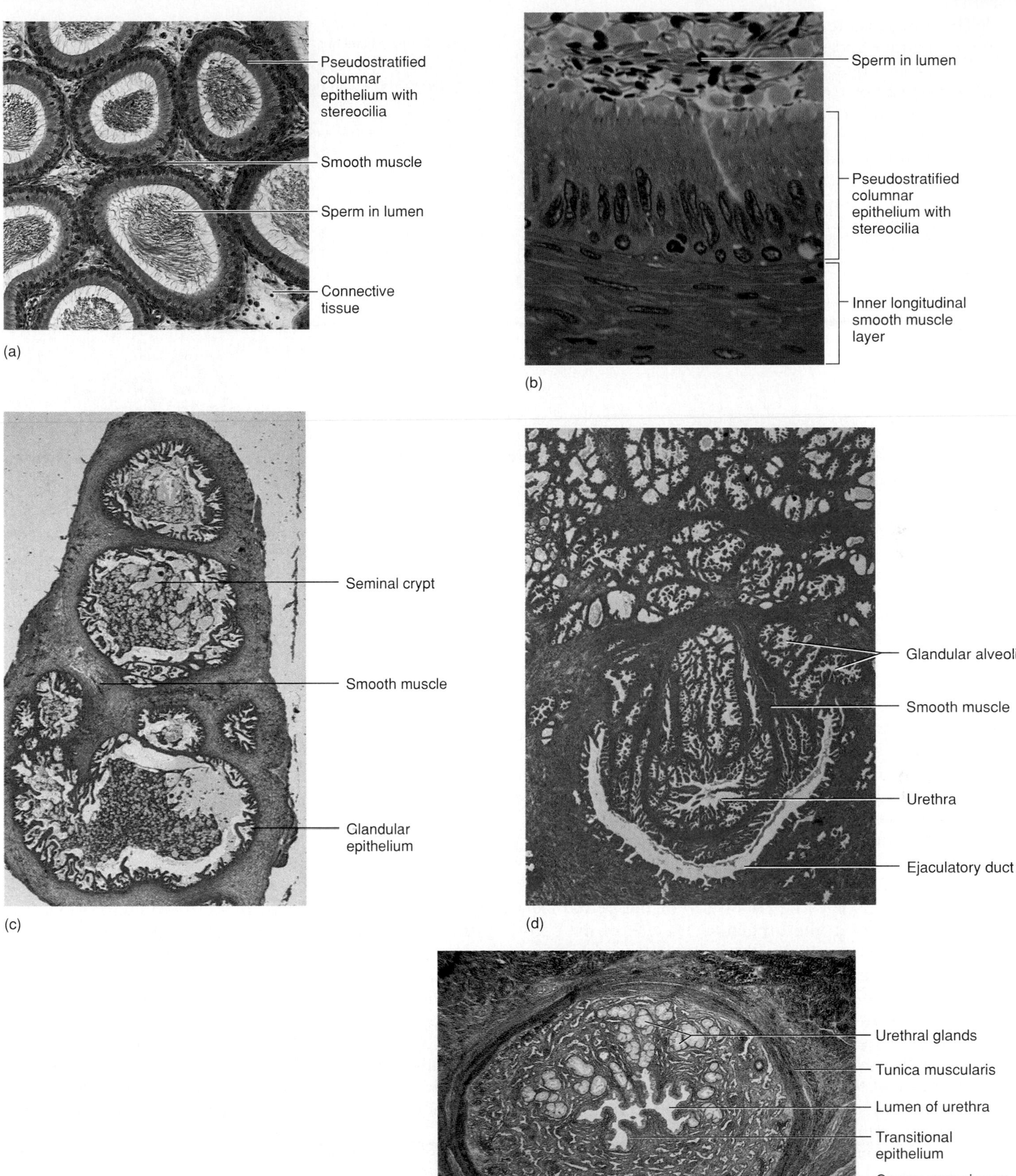

FIGURE 22.12

Histology of the male reproductive organs and glands. (*a*) Epididymis, (*b*) ductus deferens, (*c*) seminal vesicle, (*d*) prostate, and (*e*) urethra.

The firm **prostate** (*pros'tāt*) lies immediately below the urinary bladder and surrounds the beginning portion of the urethra (see figs. 19.22*a* and 22.1). It is about the size and shape of a horse chestnut. The prostate is enclosed by a fibrous capsule and divided into lobules by the urethra and by the ejaculatory ducts that extend through the gland. The ducts from the lobules open into the urethra. The thin, milky-colored prostatic secretion assists sperm cell motility as a liquefying agent. The prostate also secretes the enzyme *acid phosphatase.* This enzyme is often measured clinically to assess prostate function.

A routine physical examination of the male includes rectal palpation of the prostate. Enlargement or overgrowth of the prostate, called benign prostatic hypertrophy, *is relatively common in older men. It constricts the urethra, causing difficult urination. This condition may require surgery.*

The paired, pea-sized **bulbourethral** (*bul"bo-yoo-re'thral*) **glands** (Cowper's glands) are located beneath the prostate (see fig. 22.1), where their secretions drain into the urethra upon sexual excitement prior to ejaculation. The mucoid secretion from these glands coats the lining of the urethra to neutralize the pH of the urine residue. It also lubricates the tip of the penis in preparation for coitus.

prostate: Gk. *prostate,* one standing before
Cowper's gland: from William Cowper, English anatomist, 1666–1709

22.8 The Penis

The penis, which contains the distal portion of the urethra, is specialized with three columns of erectile tissue to become engorged with blood for insertion into the vagina during sexual intercourse.

The **penis** (*pe'nis*), when distended, serves as the copulatory organ of the male reproductive system. The penis and scrotum (see module 22.4), which hang suspended from the perineum, constitute the **external genitalia** of the male. Under the influence of sexual stimulation, the penis becomes engorged with blood. This engorgement results from the filling of intricate blood sinuses, or spaces, in the *erectile tissue* of the penis (as will be described more fully in the next module). The penis is divided into a *root,* which is attached to the pubic arch; an elongated tubular *body,* or *shaft;* and a distal cone-shaped *glans penis* (fig. 22.13).

The **root of the penis** expands posteriorly to form the **bulb of the penis** and the **crus** (*krus*) **of the penis.** The bulb is positioned in the urogenital triangle of the perineum, where it is attached to the undersurface of the perineal membrane and surrounded by the bulbocavernosus muscle (see fig. 8.26). The crus, in turn, attaches the root of the penis to the pubic arch and to the perineal membrane. The crus is positioned above the bulb and is enveloped by the ischiocavernosus muscle.

The **body of the penis** is composed of three cylindrical columns of erectile tissue that are bound together by fibrous tissue and covered with skin. The paired masses are called the **corpora cavernosa penis.** The fibrous tissue between the two corpora forms a **median septum.** The smaller **corpus spongiosum penis** surrounds the urethra. The penis is flaccid and relaxed when the sponge-like tissue is not engorged with blood. In the sexually aroused male, the penis becomes firm and erect as the columns of erectile tissue fill with blood.

Trauma to the penis, testes, and scrotum is common because of their pendent (hanging) position. Because the penis and testes are extremely sensitive to pain, a male will respond reflexively to protect the groin area. Urethral injuries are more common in men than in women because of the position of the urethra in the penis. "Straddle injuries" may be caused, for example, when a man walking along a raised beam slips and compresses his urethra and penis between the hard surface and his pubic arch, rupturing the urethra.

The **glans penis** is the cone-shaped terminal portion of the penis, which is formed from the expanded corpus spongiosum. The opening of the urethra at the tip of the glans is called the **external urinary meatus.** The **corona glandis** is the prominent posterior ridge of the glans penis. On the undersurface of the glans penis, a vertical fold of tissue called the **frenulum** (*fren'yŭ-lum*) attaches the skin covering the penis to the glans penis.

The skin covering the penis is hairless, lacks fat cells, and generally is more darkly pigmented than the rest of the body skin. The skin of the shaft is loosely attached and is continuous over the glans as a protective retractable sheath called the **prepuce,** or **foreskin.** The prepuce is commonly removed from an infant on the third or fourth

crus: L. *crus,* leg, resembling a leg
cavernosa: L. *cavus,* hollow
glans: L. *glans,* acorn
corona: L. *corona,* garland or crown
frenulum: L. diminutive of *frenum,* a bridle
prepuce: L. *prae,* before; *putium,* penis

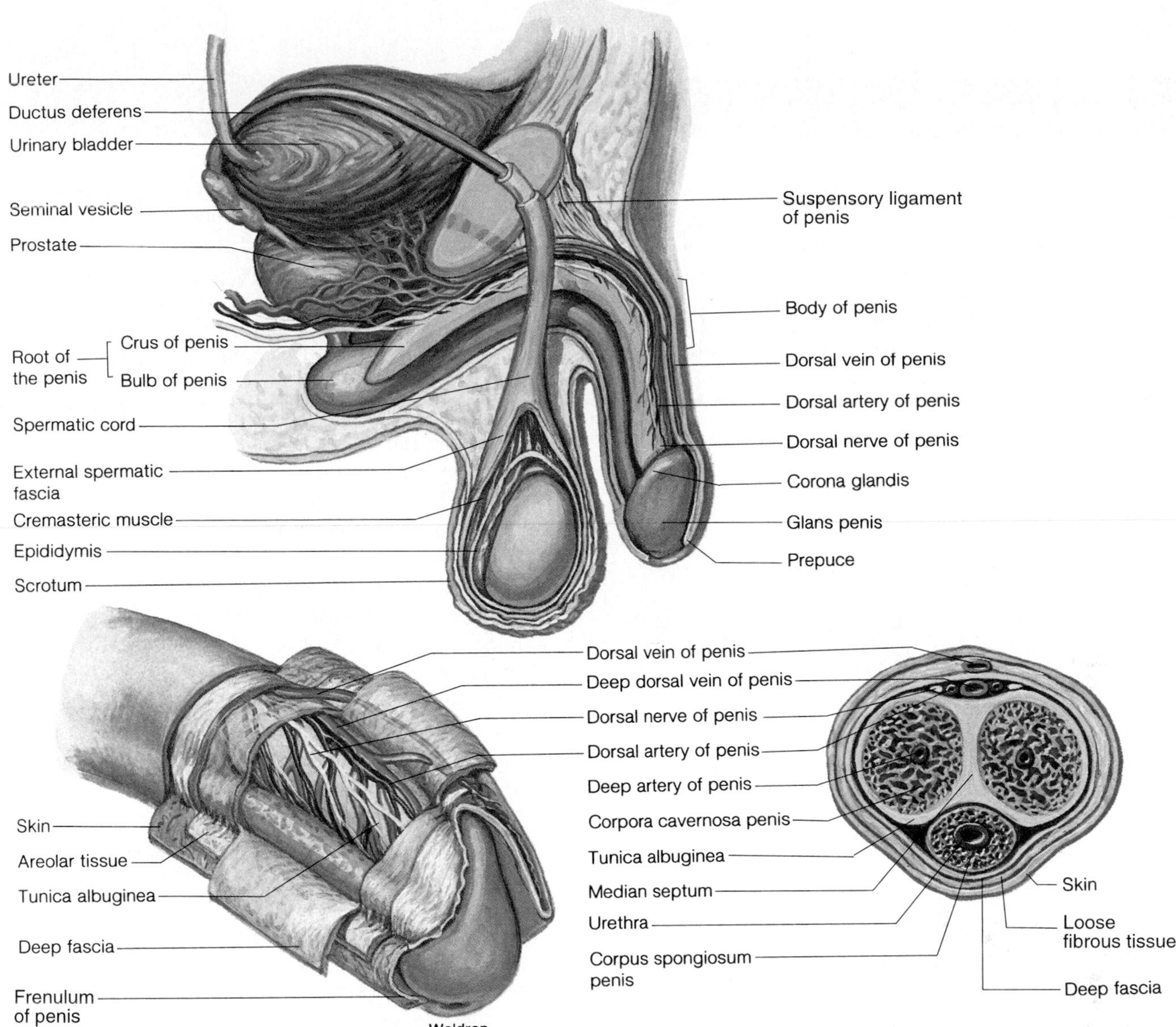

FIGURE 22.13
The structure of the penis showing the attachment, blood and nerve supply, and the arrangement of the erectile tissue.

day after birth, or on the eighth day as part of a Jewish religious rite. This procedure is called a *circumcision.*

A circumcision *is generally performed for hygienic purposes because the glans is easier to clean if exposed. A sebaceous secretion from the glans, called* smegma, *will accumulate along the border of the corona glandis if good hygiene is not practiced. Smegma can foster bacteria that may cause infections, and therefore should be removed through washing. Cleaning the glans of an uncircumcised male requires retraction of the prepuce. Occasionally, a child is born with a prepuce that is too tight to permit retraction. This condition, called* phimosis, *necessitates circumcision.*

The penis is supplied with blood on each side through a pudendal branch of the femoral artery. The venous return is through a superficial median dorsal vein that drains into the great saphenous vein in the thigh and through the deep median vein that drains into the prostatic plexus.

The penis has many sensory tactile receptors, especially in the glans penis, that make it a highly sensitive organ. In addition, the penis has extensive motor innervation from both parasympathetic and sympathetic fibers.

phimosis: Gk. *phimosis,* a muzzling

22.9 Mechanisms of Erection, Emission, and Ejaculation

Erection of the penis results from parasympathetic-induced vasodilation of arteries within the penis. Emission and ejaculation are stimulated by sympathetic impulses, resulting in the forceful expulsion of semen from the penis.

Erection, emission, and ejaculation are a series of interrelated events by which semen from the male is deposited into the female vagina. *Erection* usually occurs as a male becomes sexually aroused and the erectile tissue of the penis becomes engorged with blood. *Emission* is the movement of spermatozoa from the epididymides to the ejaculatory ducts. *Ejaculation* (*ĕ-jak"yŭ-la'shun*) is the forceful expulsion of the ejaculate, or *semen,* from the penis.

Erection of the Penis

Erection depends on the volume of blood that enters the arteries of the penis as compared to the volume that exits through venous drainage. During sexual excitement, parasympathetic impulses cause marked draining (fig. 22.14). At the same time, there may be slight vasoconstriction of the dorsal vein of the penis and an increase in cardiac output. These combined events cause the spongy erectile tissue to become distended with blood and the penis to become wider, longer, and firmer. In this condition, the penis can be inserted into the vagina and function as a copulatory organ to discharge semen.

Erection is controlled by the hypothalamus and the sacral portion of the spinal cord. The hypothalamus responds to conscious sexual thoughts originating in the cerebrum. Impulses from the hypothalamus elicit parasympathetic responses from the sacral region. These parasympathetic responses cause vasodilation of the arterioles within the penis. Conscious thought is not required for an erection, however. Stimulation of the penis can cause an erection because of a reflex response in the spinal cord. This reflexive action makes possible an erection in a sleeping male or in an infant—perhaps from the stimulus of a diaper.

Inside Information—The penis of many mammals contains a bone called an *os penis,* or *baculum,* of a highly variable shape. The human male, having no such bone, relies exclusively on blood-filled erectile tissue to give rigidity to the penis. On average, the erect penis in an adult is 15 cm (6 in.) long and 4 cm (1.5 in.) in diameter.

Emission and Ejaculation of Semen

Emission Continued sexual stimulation following erection of the penis causes emission (fig. 22.15). Emission is the movement of sperm cells from the epididymides to the ejaculatory ducts and the secretions of the accessory glands into the ejaculatory ducts and urethra in the formation of semen. Emission occurs as sympathetic impulses from the pelvic nerve plexus cause a rhythmic contraction of the smooth muscle layers of the testes, epididymides, ductus deferentia, ejaculatory ducts, seminal vesicles, and prostate.

Ejaculation Ejaculation immediately follows emission and is accompanied by *orgasm*—the climax of the sex act. Ejaculation occurs in a series of spurts of semen from the urethra. This takes place as parasympathetic impulses traveling through the pudendal nerves stimulate the bulbocavernosus muscles at the base of the penis and cause them to contract rhythmically. In addition, the smooth muscles in the urethral wall contract in response to sympathetic stimulation, helping to eject the semen.

Sexual function in the male thus requires the synergistic action (rather than antagonistic action) of the parasympathetic and sympathetic systems. The mechanism of emission and ejaculation is summarized in figure 22.15.

Immediately following ejaculation or cessation of a sexual stimulus, sympathetic impulses cause vasoconstriction of the arterioles within the penis, reducing the inflow of blood. At the same time, cardiac output returns to normal, as does venous return of blood from the penis. With the normal flow of blood through the penis, it returns to its flaccid condition.

Adolescent males commonly experience erection of the penis and spontaneous emission and ejaculation during sleep. These nocturnal emissions, *sometimes called "wet dreams," are triggered by psychic stimuli associated with dreaming. They are thought to be caused by changes in hormonal concentrations that accompany adolescent development.*

Semen Semen, also called *seminal fluid,* is the substance discharged during ejaculation. Generally, between 1.5 and 5.0 ml of semen are ejected during ejaculation. The bulk of the fluid (about 60%) is produced by the seminal vesicles, and the remainder (about 40%) is contributed by the prostate. Spermatozoa account for less than 1% of the volume. There are usually between 60 and 150 million sperm cells per milliliter of ejaculate. In the condition of *oligospermia,* the male ejaculates fewer than 10 million sperm cells per milliliter and is likely to have fertility problems.

Sexual excitement; stimulation
Inhibition of sympathetic impulses to penis
Increase in parasympathetic impulses causes vasodilation of arterioles within penis
Reduction of venous flow from penis; increased cardiac output through sympathetic stimulation to heart
Accumulation of blood within erectile tissue of penis
Penis becomes turgid and erect

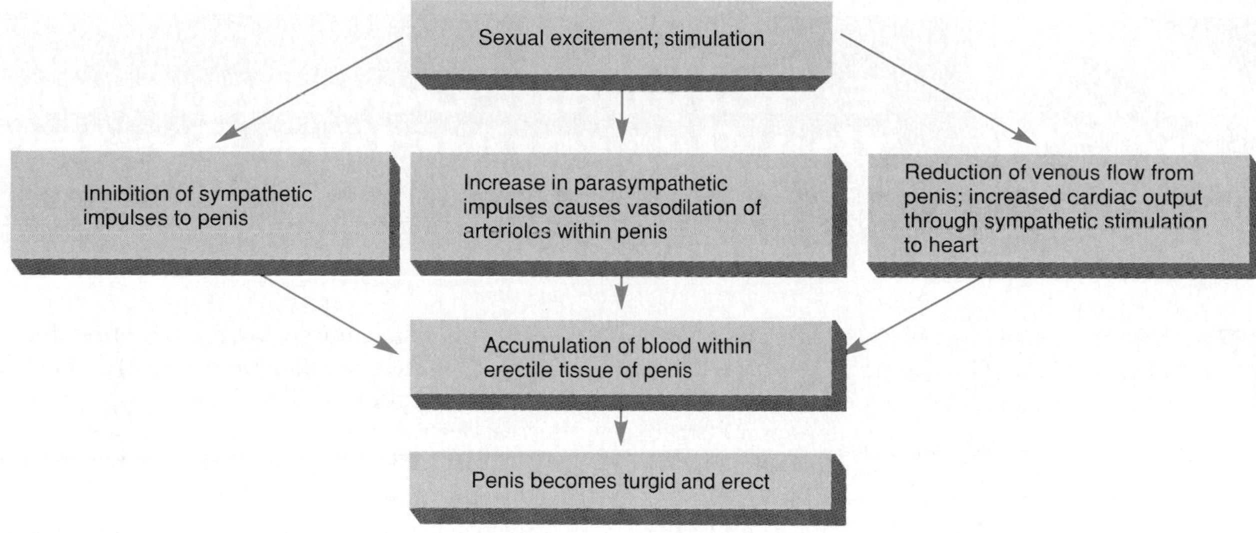

FIGURE 22.14
The mechanism of erection of the penis.

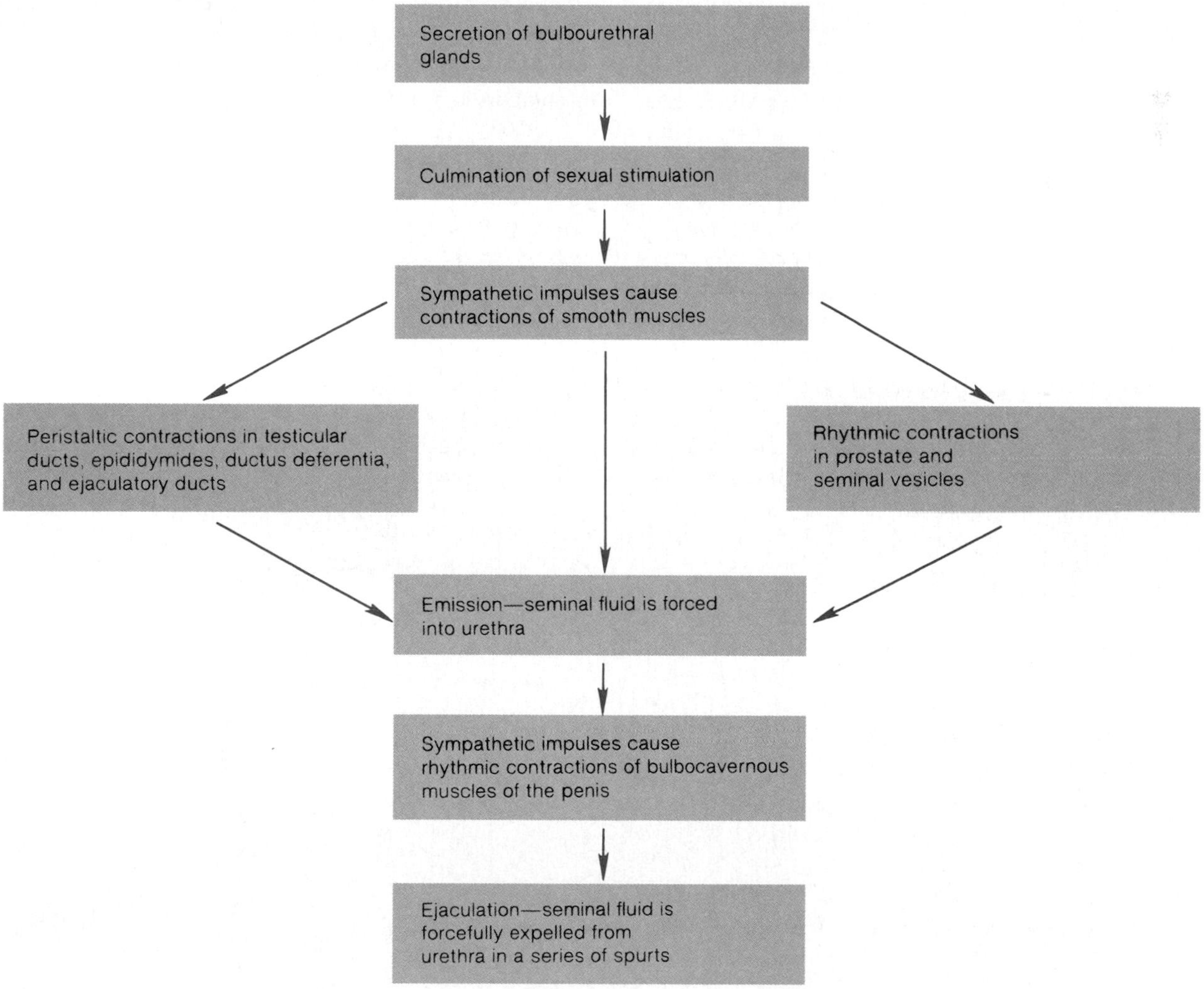

Secretion of bulbourethral glands
Culmination of sexual stimulation
Sympathetic impulses cause contractions of smooth muscles
Peristaltic contractions in testicular ducts, epididymides, ductus deferentia, and ejaculatory ducts
Rhythmic contractions in prostate and seminal vesicles
Emission—seminal fluid is forced into urethra
Sympathetic impulses cause rhythmic contractions of bulbocavernous muscles of the penis
Ejaculation—seminal fluid is forcefully expelled from urethra in a series of spurts

FIGURE 22.15
The mechanism of emission and ejaculation.

Embryonic Development of the Reproductive System

The male and female reproductive systems follow a similar pattern of development, with sexual distinction coming about as a result of the influence of hormones. As mentioned previously, the reproductive organs of both sexes derive from the same developmental tissue and are considered *homologous* (see table 22.1).

The first sign of development of reproductive organs (male or female) occurs during the fifth week with the formation of the **gonadal ridge** (see the figure to the right). The gonadal ridge continues to grow behind the developing peritoneal membrane lining the abdominal cavity. By the sixth week, stringlike masses called **primary sex cords** form within the enlarging gonadal ridge. The primary sex cords in the male will eventually mature to become the sperm-nurturing seminiferous tubules. In the female, the primary sex cords will contribute to nurturing tissue of developing ova.

Each gonad develops near a **mesonephric duct** and a **paramesonephric duct.** In the male embryo, each testis connects through a series of tubules to the mesonephric duct. During further development, the connecting tubules become the seminiferous tubules, and the mesonephric duct becomes the *efferent ductules, epididymis, ductus deferens, ejaculatory duct,* and the *seminal vesicle.* The paramesonephric duct in the male degenerates without contributing any functional structures to the reproductive system.

In the female embryo, the mesonephric duct degenerates, and the paramesonephric duct contributes greatly to structures of the female reproductive system. The distal ends of the paired paramesonephric ducts fuse to form the *vagina* and *uterus.* The proximal unfused portions become the *uterine tubes.*

By the sixth week, a swelling called the **genital tubercle** is apparent in the groin of the embryo. The mesonephric and paramesonephric ducts open to the outside through the genital tubercle. The genital tubercle consists of a *glans,* a *urethral groove,* paired *urethral folds,* and paired *labioscrotal swellings* (see fig. 22.3). As the glans portion of the genital tubercle enlarges, it becomes known as the *phallus.*

Early in fetal development (the tenth through the twelfth week), sexual distinction of the external genitalia becomes apparent. In the male, the phallus enlarges and develops into the *glans of the penis.* The urethral folds fuse around the urethra and become the *erectile tissue* that forms the *body of the penis.* The labioscrotal swellings fuse to form the *scrotum,* into which the testes will descend. In the female, the phallus gives rise to the *clitoris,* the urethral folds remain separated as the *labia minora,* and the urethral groove is retained as a longitudinal cleft known as the *vestibule* (see chapter 23, module 23.1).

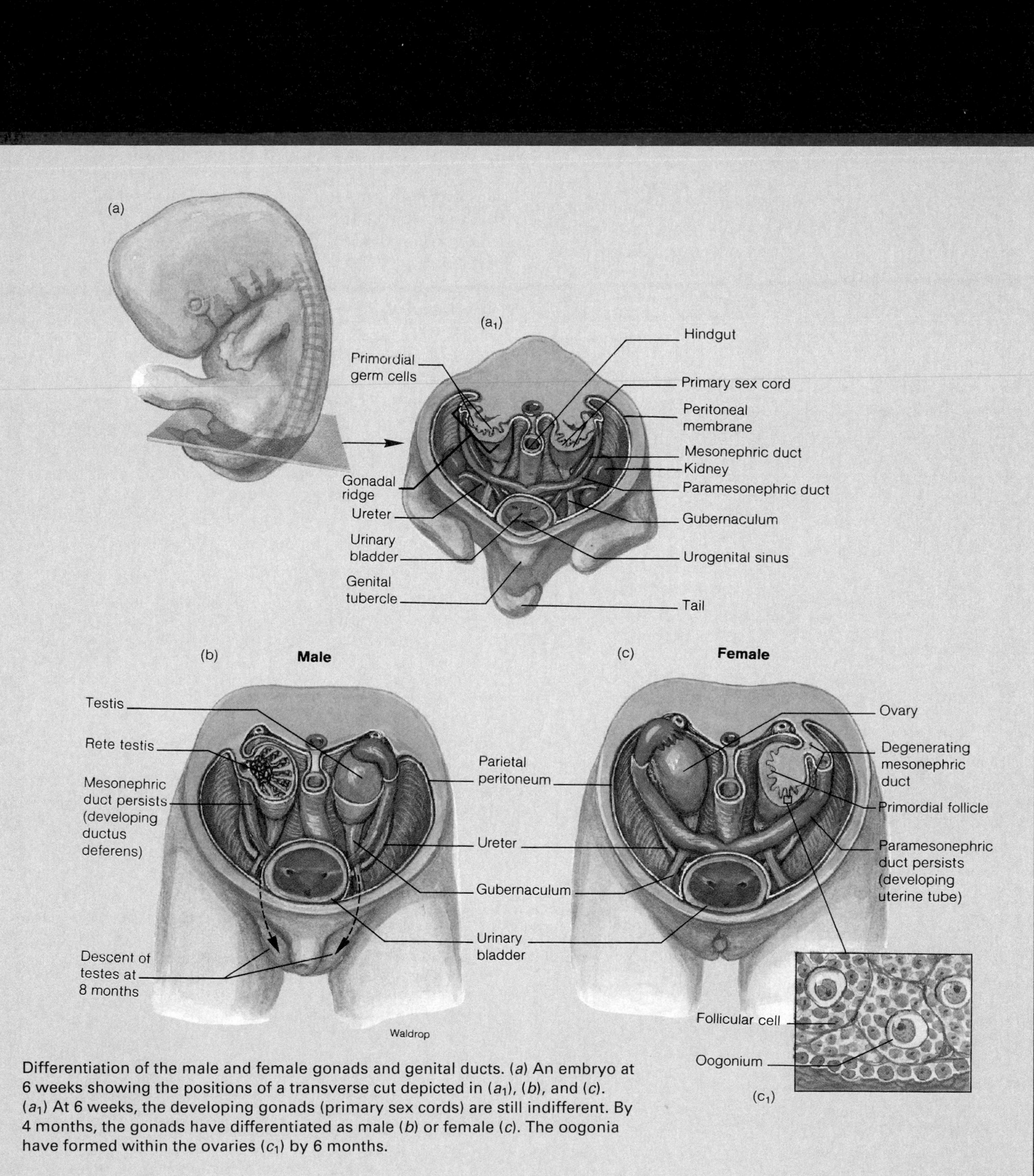

Differentiation of the male and female gonads and genital ducts. (*a*) An embryo at 6 weeks showing the positions of a transverse cut depicted in (a_1), (*b*), and (*c*). (a_1) At 6 weeks, the developing gonads (primary sex cords) are still indifferent. By 4 months, the gonads have differentiated as male (*b*) or female (*c*). The oogonia have formed within the ovaries (c_1) by 6 months.

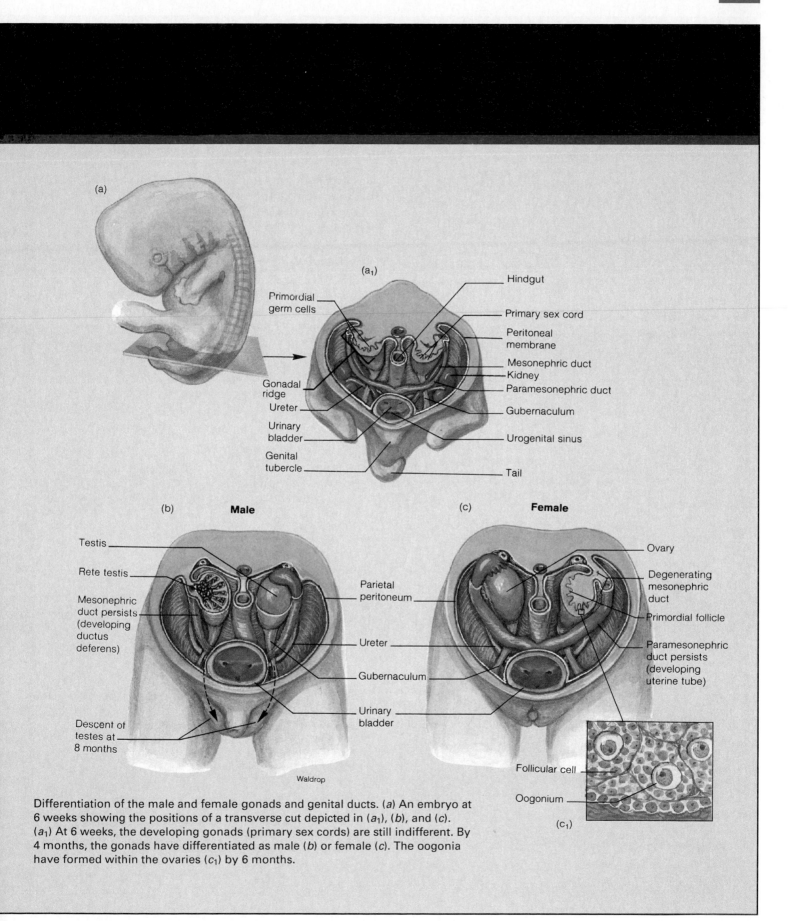

Clinical Considerations

Functional Reproductive Disorders

Functional disorders of the male reproductive system include *impotence, infertility,* and *sterility.* These disorders make it impossible for a couple to produce offspring through sexual intercourse.

Impotence is the inability of a sexually mature male to achieve and maintain penile erection and/or the inability to achieve ejaculation. The causes of impotence may be physical (such as abnormalities of the penis, vascular irregularities, or neurological disorders), or they may be the result of diseases. Generally, however, the cause of impotence is psychological, and the patient will require skilled counseling by a sex therapist.

Infertility is the inability of the sperm to fertilize the ovum and may involve the male or female, or both. The most common cause of male infertility is inadequate production of viable sperm. This may be due to alcoholism, dietary deficiencies, local injury, varicocele, excessive heat, hormonal imbalance, or excessive exposure to X rays. Many of the causes of infertility can be treated through proper nutrition, gonadotropic hormone treatment, or microsurgery.

Sterility is similar to infertility except that sterility is a permanent condition. Sterility may be genetically caused, or it may be the result of degenerative changes in the seminiferous tubules (for example, mumps in a mature male may secondarily infect the testes and cause irreversible tissue damage).

Voluntary sterilization of the male by means of a **vasectomy** is a common technique of birth control (see the Synoptic Spotlight on page 608 in chapter 23). In this procedure, a small section of each ductus deferens near the epididymis is surgically removed, and the cut ends of the ducts are tied (fig. 22.16). A vasectomy prevents transport

impotence: L. *im,* not: *potens,* potent
sterility: L. *sterilis,* barren
vasectomy: L. *vas,* vessel; Gk. *ektome,* excision

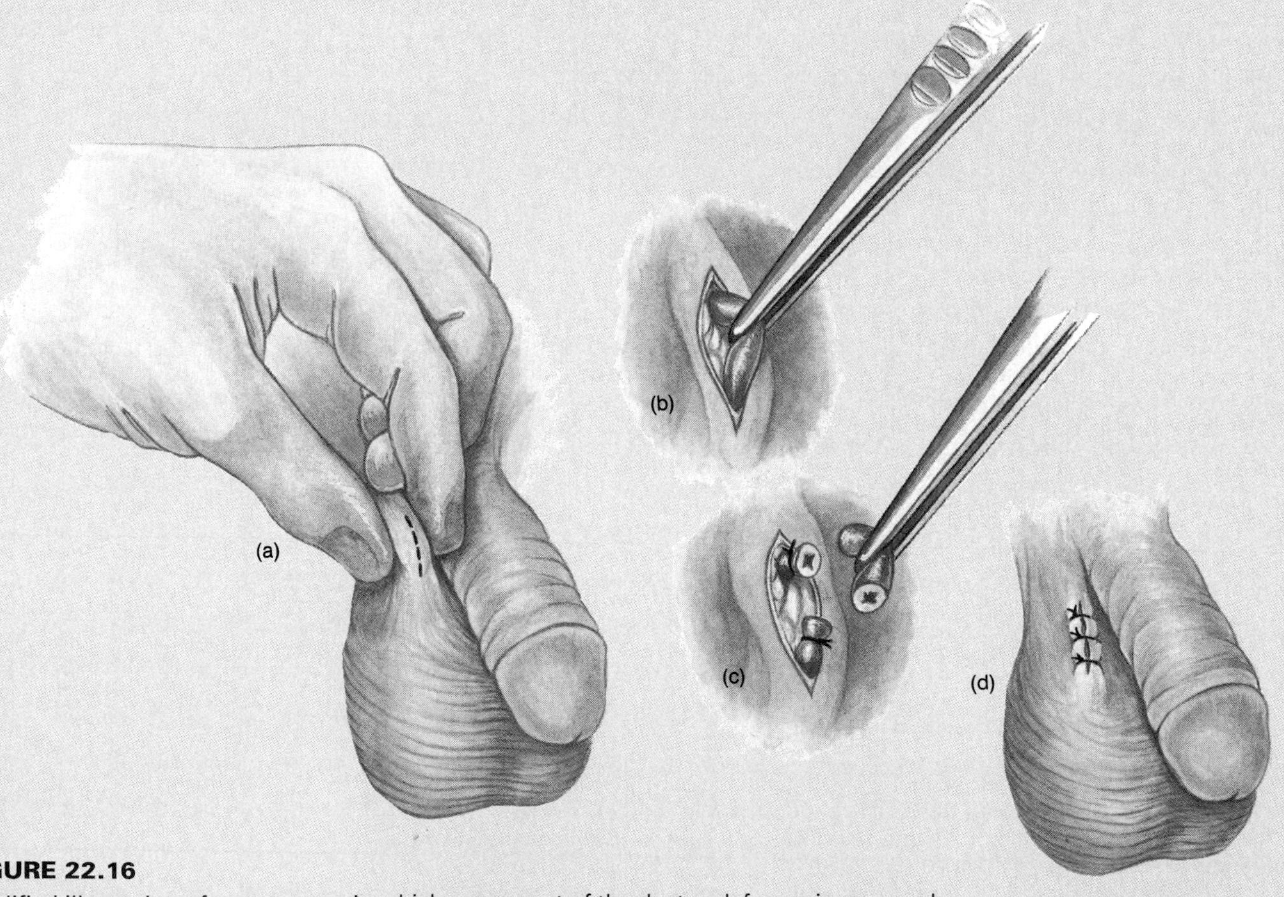

FIGURE 22.16
A simplified illustration of a vasectomy in which a segment of the ductus deferens is removed through an incision in the scrotum. The procedure is then repeated on the opposite side.

of spermatozoa but does not directly affect the secretion of androgens, the sex drive, or ejaculation. Since sperm cells make up less than 1% of an ejaculate, even the volume is not noticeably affected.

Inside Information—According to the National Center for Health Statistics, there are currently 5.3 million childless infertile couples in the United States. In 40% of the cases, it is the woman's infertility that prevents conception; in 40% it is the man's. In the remaining 20%, either both partners have some abnormality, or the cause of the problem is unknown. Of those couples who seek help for infertility, only 21% are eventually successful in producing a child.

Diseases of the Male Reproductive System

Sexually transmitted diseases (STDs) are contagious diseases that affect the reproductive systems of both the male and the female (table 22.4). STDs are transmitted during sexual activity, and their frequency of occurrence in the United States is regarded by health authorities as epidemic. Commonly called *venereal diseases,* STDs have not been eradicated mainly because humans cannot develop immunity to them and because increased sexual activity increases the likelihood of infection and reinfection. Furthermore, many of the causative organisms can mutate so fast that available drug treatments are no longer effective.

Gonorrhea, commonly called "clap," is caused by the bacterium gonococcus, or *Neisseria gonorrhoeae.* Males with this disease suffer inflammation of the urethra, accompanied by painful urination and frequently the discharge of pus. In females, the condition is usually asymptomatic (without symptoms); therefore, many women may be unsuspecting carriers of the disease.

venereal: L. (mythology) from *Venus,* the goddess of love
gonorrhea: L. *gonos,* seed; *rhoia,* a flow

continued

TABLE 22.4 Sexually Transmitted Diseases

Name	Organism	Resulting Condition	Treatment
Gonorrhea	*Gonococcus* (bacterium)	Adult: sterility due to scarring of epididymides and tubes (in rare cases, blood poisoning); newborn: blindness	Penicillin injections; tetracycline tablets; eyedrops (silver nitrate or penicillin) in newborns as preventative
Syphilis	*Treponema pallidum* (bacterium)	Adult: gummy tumors, cardiovascular neurosyphilis; newborn: congenital syphilis (abnormalities, blindness)	Penicillin injections; tetracycline tablets
Chancroid (soft chancre)	*Hemophilus ducreyi* (bacterium)	Chancres, buboes	Tetracycline; sulfa drugs
Urethritis in men	Various microorganisms	Clear discharge	Tetracycline
Vaginitis	*Trichomonas* (protozoan)	Frothy white or yellow discharge	Metronidazole
	Candida albicans (yeast)	Thick, white, curdy discharge (moniliasis)	Nystatin
Acquired immune deficiency syndrome (AIDS)	Human immunodeficiency virus (HIV)	Early symptoms include extreme fatigue, weight loss, fever, diarrhea; extreme susceptibility to pneumonia, rare infections, and cancer	Azidothymidine (AZT), dideoxyinosine (ddlI), and dideoxycytidine (ddC); new drugs being developed
Chlamydia	*Chlamydia trachomatis* (bacterium)	Whitish discharge from penis or vagina; pain during urination	Tetracycline and sulfonamides
Venereal warts	Virus	Warts	Podophyllin; cautery, cryosurgery, or laser treatment
Genital herpes	Herpes simplex virus	Sores	Palliative treatment
Crabs	Arthropod	Itching	Gamma benzene hexachloride

Clinical Considerations

Syphilis is caused by the bacterium *Treponema pallidum.* Syphilis is less common than gonorrhea but is the more serious of the two diseases. During the *primary stage,* a lesion called a *chancre* develops at the point where contact was made with a similar sore on an infected person. The chancre persists for 10 days to 3 months before the disease enters the *secondary stage,* which is characterized by lesions or a skin rash accompanied by fever. This stage lasts from 2 weeks to 6 months, and the symptoms disappear of their own accord. The *tertiary stage* of untreated syphilis may occur 10 to 20 years following the primary infection. This stage is characterized by degenerative changes in various systems of the body that may lead to blindness, insanity, and death.

Acquired immune deficiency syndrome (AIDS) is a viral disease that is transmitted primarily through intimate sexual contact and through drug abuse (by sharing contaminated syringe needles). This disease, for which there is currently no cure, is discussed in detail in chapter 17 (see the Synoptic Spotlight on page 584).

chancre: Fr. *chancre,* indirectly from L. *cancer,* a crab

The prostate and the testes are subject to several disorders, most of which are common in older men. **Prostatis,** inflammation of the prostate, is usually caused by a bacterial infection. **Prostatic hyperplasia** is an enlarged prostate from unknown causes. This condition occurs in nearly one-third of all males over the age of 60, and is characterized by painful and difficult urination. **Prostatic carcinoma,** cancer of the prostate, is the second leading cause of death from cancer in males in the United States. It, too, is common in males over 60 and accounts for about 25,000 deaths annually. The metastases of this cancer to the spinal column and brain are generally what kills the patient. Advanced prostatic carcinoma is treated by removal of the prostate (*prostatectomy*) and frequently by removal of the testes (*orchiectomy*). An orchiectomy inhibits metastases by eliminating testosterone secretion.

An infection in the testes is called **orchitis.** Orchitis may develop as the result of a primary bacterial infection or as a secondary complication of mumps contracted after puberty. If orchitis from mumps involves both testes, it usually causes sterility.

SUMMARY

22.1 Introduction to the Reproductive System

1. The common purpose of the male and female reproductive systems is to produce offspring.
2. The functions of the male reproductive system are to produce sperm cells, secrete androgens, and transfer sperm cells to the reproductive system of the female.
3. Features of the male reproductive system include primary and secondary sex organs and secondary sex characteristics.

22.2 Sex Determination and the Influence of Hormones on Reproductive Development

1. An XY chromosome combination produces a male; an XX chromosome combination produces a female.
2. The testes secrete testosterone, which stimulates the development of male accessory sex organs. Lack of hormonal secretion from the testes causes the development of female accessory sex organs.
3. The penis and scrotum are formed by the end of the twelfth week of prenatal development, and the descent of the testes occurs during the twenty-eighth week.

22.3 Puberty

1. Puberty is the time of development when the sex organs become functional.
2. At the time of puberty, increased levels of gonadotropic hormones from the anterior pituitary stimulate the secretion of sex steroid hormones from the gonads. The sex steroids, in turn, stimulate the development of secondary sex characteristics.

22.4 Perineum and Scrotum

1. The saclike scrotum within the urogenital portion of the perineum supports and protects the testes. It also regulates their position relative to the pelvic region of the body.
2. Each testis is contained within its own scrotal compartment and is separated from the other by the scrotal septum.

22.5 Structure and Function of the Testes

1. The testes are partitioned into lobules. The lobules are composed of seminiferous tubules, which produce sperm cells, and of interstitial cells, which secrete androgens.
2. Luteinizing hormone (LH) stimulates the interstitial cells to secrete androgens; follicle-stimulating hormone (FSH) stimulates the seminiferous tubules to produce sperm cells.

22.6 Spermatogenesis and the Structure of a Spermatozoon

1. Spermatogenesis is the process of producing sperm cells.
2. Spermatogenesis occurs by meiotic division of the cells that line the seminiferous tubules. The conversion of spermatids to spermatozoa is called spermiogenesis.
3. A mature spermatozoon (sperm cell) consists of an oval head and an elongated flagellum. Within the head are the nucleus and an enzyme-containing acrosome. The body of the flagellum contains mitochondria that provide the energy for locomotion.

22.7 Spermatic Ducts and Accessory Reproductive Glands

1. The epididymides, ductus deferentia, and ejaculatory ducts comprise the duct system of the male reproductive system.
2. The seminal vesicles and prostate provide additives to the sperm cells in the formation of semen.

22.8 The Penis

1. Composed of three columns of erectile tissue, the penis is specialized to become rigid for insertion into the vagina during sexual intercourse.
2. The glans penis is the terminal end of the penis; in an uncircumcised male, it is covered with a prepuce, or foreskin.

22.9 Mechanisms of Erection, Emission, and Ejaculation

1. Erection of the penis occurs as the erectile tissue becomes engorged with blood. Emission is the movement of the spermatozoa from the epididymides to the ejaculatory ducts; ejaculation is the forceful expulsion of semen from the ejaculatory ducts and urethra of the penis.
2. Parasympathetic stimuli to arteries in the penis cause the erectile tissue to fill with blood as arteriole flow increases and venous drainage decreases.
3. Ejaculation results from sympathetic reflexes in the smooth muscles of the reproductive organs.

Review Activities

Objective Questions

1. Which of the following is the primary sex organ of the male? (*module 22.1*)
 a. penis
 b. scrotum
 c. testis
 d. prostate

2. An embryo with the genotype XY develops male accessory sex organs because of (*module 22.2*)
 a. androgens.
 b. estrogens.
 c. the absence of androgens.
 d. the absence of estrogens.

3. The two principal hormones that influence the onset of male puberty are (*module 22.3*)
 a. FSH and LH.
 b. testosterone and GH.
 c. FSH and testosterone.
 d. testosterone and estradiol.

4. The scrotum is (*module 22.4*)
 a. part of the primary sex organ.
 b. important in regulating the temperature of the testes.
 c. poorly innervated.
 d. an extension of the spermatic cord.

5. Testosterone is produced and secreted by (*module 22.5*)
 a. spermatogonia.
 b. sustentacular cells.
 c. seminiferous tubules.
 d. interstitial cells.

6. Which of the following is a *false* statement? (*module 22.6*)
 a. One primary spermatocyte produces four spermatids.
 b. Testosterone is required for the completion of meiotic division and spermatid maturation.
 c. The head of a sperm cell contains a nucleus with 23 chromosomes.
 d. Spermatogenesis eventually exhausts the number of spermatogonia.

7. Which of the following is *not* a spermatic duct? (*module 22.7*)
 a. epididymis
 b. spermatic cord
 c. ejaculatory duct
 d. ductus deferens

8. In an uncircumcised male, the prepuce covers (*module 22.8*)
 a. the glans penis.
 b. the bulb of the penis.
 c. the crus of the penis.
 d. the corpus spongiosum.

9. Which statement regarding erection is *false?* (*module 22.9*)
 a. It is a parasympathetic response.
 b. It may be both a voluntary and an involuntary response.
 c. It has to be followed by emission and ejaculation.
 d. It is controlled by the hypothalamus of the brain and the sacral portion of the spinal cord.

Essay Questions

1. List the organs of the male reproductive system and indicate whether they are primary or secondary sex organs. (*module 22.1*)
2. Define *gubernaculum* and describe how the testes descend into the scrotum. (*module 22.2*)
3. Define *puberty* and describe how it is hormonally controlled. (*module 22.3*)
4. Explain how the temperature of the testes is maintained. Why is this important? (*module 22.4*)
5. Describe the internal structure of a testis. Discuss the function of the sustentacular cells, interstitial cells, rete testis, and efferent ductules. (*module 22.5*)
6. Diagram a spermatozoon and an adjacent sustentacular cell. Briefly describe the process of spermatogenesis. (*module 22.6*)
7. Compare the seminal vesicles and the prostate in terms of location, structure, and function. (*module 22.7*)
8. Describe the external structure of the penis and the internal arrangement of the erectile tissue within the penis. (*module 22.8*)
9. Explain the mechanisms by which erection, emission, and ejaculation are achieved. (*module 22.9*)
10. How much semen is normally ejected during an ejaculation? What are the constituents of semen and where is each produced? (*module 22.9*)

Labeling Exercise

Label the structures indicated on the figure to the right.

1. ____________
2. ____________
3. ____________
4. ____________
5. ____________
6. ____________

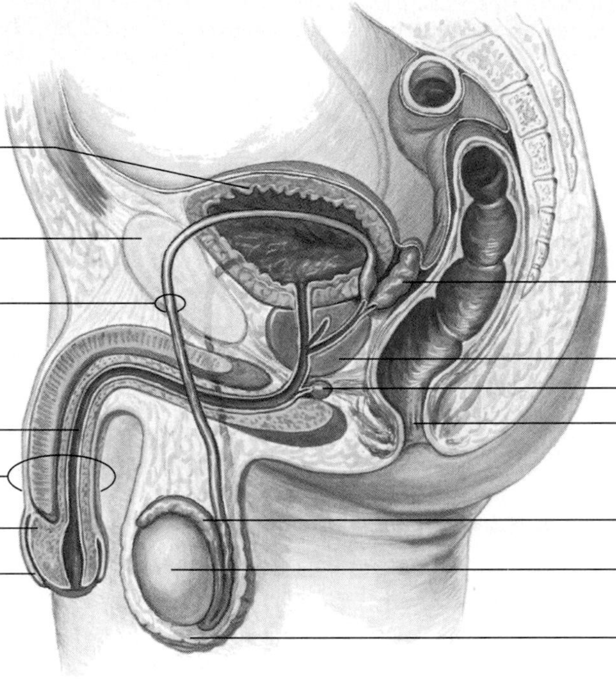

Critical Thinking Questions

1. Explain what is meant by the latent development of the reproductive organs during puberty. What hormonal factors bring about puberty in a male?
2. The reproductive system has been described as a "nonessential" body system because it is not necessary for the survival of the individual. Upon maturation, however, the reproductive system does produce hormones that maintain adult features as well as bone structure. These hormones also influence behavior, blood composition, and metabolism. Discuss your thoughts as to whether or not the reproductive system is nonessential.
3. Explain what effect, if any, an enlargement of the prostate would have on reproductive functions.
4. As a safeguard against having more children, a 40-year-old man is considering a vasectomy. This procedure involves surgically removing a short section of each ductus (vas) deferens within the spermatic cord. Will the vasectomy affect his sexual performance? Will his sexual desire be altered? What will be the components and volume of his ejaculate? Will he be sterile or infertile?
5. A number of pathogenic (disease-causing) organisms are transmitted through sexual activity. From the standpoint of the pathogen, why does coitus provide an ideal means of propagation?

CHAPTER

23

"Under proper hormonal stimulation, [the graafian follicle] will rupture. . . and expel its oocyte (ovulate) into the uterine tube." *page 599*

Female Reproductive System

Chapter Outline

Terms to Remember

ovary (*o' vă-re*), 592
uterine (*yoo' ter-in*) tube, 593
uterus (*yoo' ter-us*), 594
endometrium (*en"do-me' tre-um*), 594
vagina (*vă-ji' nă*), 595
vulva (*vul' vă*), 596
mons pubis (*pyoo' bis*), 596
labia majora (*la' be-ă mă-jor' ă*), 596
labia minora (*mī-nor' ă*), 596
clitoris (*klit' or-is, kli-tor' is*), 596
oogenesis (*o"ŏ-jen' ĕ-sis*), 597
primary oocyte (*o' ŏ-sīt*), 597
ovarian follicle, 597
ovulation (*ov-yŭ-la' shun*), 599
menstruation (*men"stroo-a' shun*), 601
menopause (*men' ŏ-pawz*), 605
mammary gland, 605
lactation (*lak-ta' shun*), 606

Learning Objectives

After studying this chapter, you should be able to

- describe the structure and functions of the organs of the female reproductive system and discuss the body changes associated with puberty in the female.
- describe oogenesis and explain why meiosis of one primary oocyte results in the formation of only one mature ovum.
- explain how hormones regulate the time of ovulation and describe the pathway taken by the ovulated oocyte.
- describe the changes that occur in the ovary in the course of the menstrual cycle.
- describe the changes that occur in the endometrium of the uterus in the course of the menstrual cycle.
- describe the structure of the mammary glands and explain how milk production and the milk-ejection reflex are hormonally regulated.

23.1 Anatomy of the Female Reproductive System

The ovaries are the primary sex organs of the female reproductive system; the uterine tubes, uterus, vagina, and mammary glands are the secondary sex organs. Secondary sex characteristics develop in response to elevated levels of ovarian steroid hormones.

Primary Sex Organs

The differences between the reproductive systems of the female and male are based on the specific functions of each in sexual reproduction and on the cyclic events characteristic of the female. In the female, the *ovaries* (*o'vă-rēz*) (fig. 23.1) are the gonads, or **primary sex organs.** The *gametes* (sex cells) of the ovaries are the *egg cells,* also called *oocytes* (*o'ŏ-sīts*) or *ova.*

A male does not produce any sperm cells until puberty but is then capable of producing viable sperm cells throughout his life if he remains healthy. By contrast, all of the oocytes of a female are completely formed, but not totally matured, during fetal development of the ovaries. In fact, the ovaries of a newborn girl contain about 2 million oocytes—all she will ever have.

Oocytes are discharged, or *ovulated,* from the ovaries (usually one at a time) in a cyclic pattern throughout the reproductive period of the female, which extends from puberty to menopause. In girls, puberty occurs between the ages of 10 and 14—generally 6 months to a year earlier than in boys—and is heralded by *menarche* (*mĕ-nar'ke*), which is the first menstrual period. *Menstruation* is the discharge of *menses* (blood and solid tissue) from the uterus at the end of each ovulation cycle. *Menopause* is the period marked by the termination of ovulation and menstruation. The reproductive period in females generally extends from about age 12 to about age 50. The cyclic reproductive pattern of ovulation and the span of a woman's reproductive years are determined by the actions of hormones, as will be discussed in subsequent modules.

menses: L. *menses,* plural of *mensis,* monthly

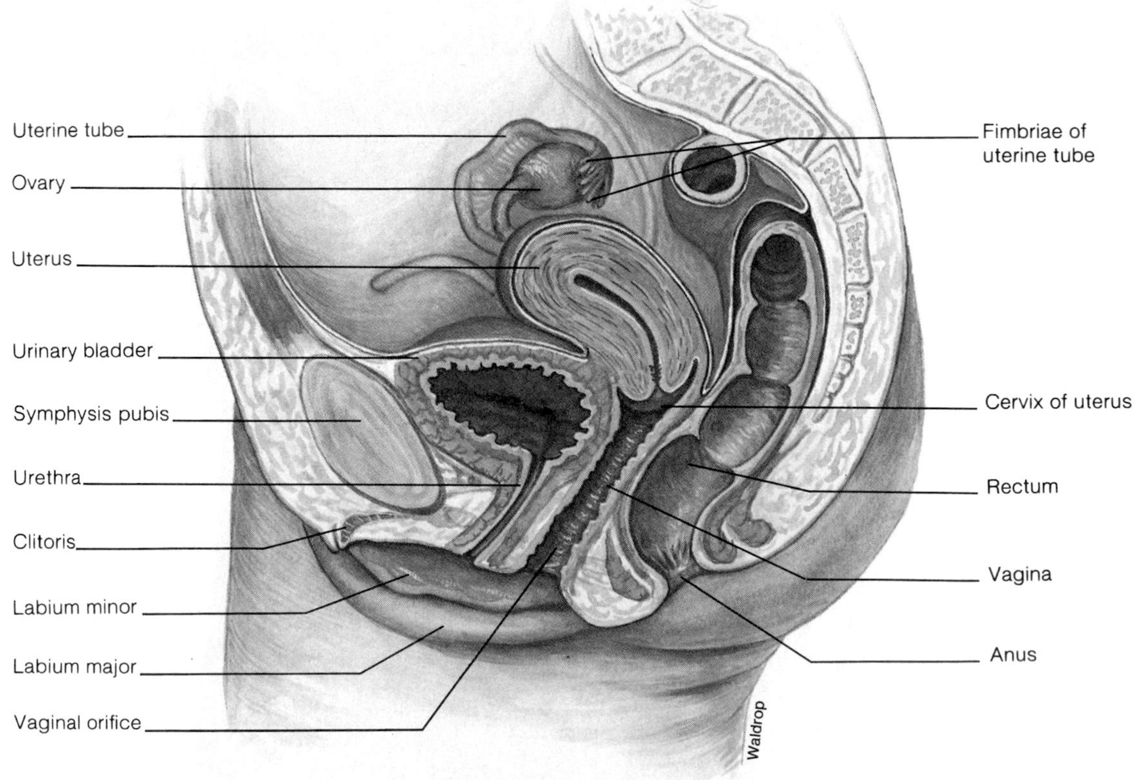

FIGURE 23.1
Organs of the female reproductive system seen in sagittal section.

TABLE 23.1 The Female Reproductive Organs

Organ	Function/Description
Ovaries	Produce female gametes (oocytes) and female sex hormones
Uterine tubes	Convey oocytes toward uterus; site of fertilization; convey developing embryo to uterus
Uterus	Site of implantation; protects and sustains embryo and fetus during pregnancy; plays active role in parturition (childbirth)
Vagina	Conveys uterine secretions to outside of body; receives erect penis and semen during coitus and ejaculation; serves as passageway for fetus during parturition
Labia majora	Form margins of pudendal cleft; enclose labia minora
Labia minora	Form margins of vestibule; protect openings of vagina and urethra
Clitoris	Glans of the clitoris is richly supplied with sensory nerve endings associated with feeling of pleasure during sexual stimulation
Pudendal cleft	Cleft between labia majora within which labia minora and clitoris are located.
Vaginal vestibule	Cleft between labia minora within which vaginal and urethral openings are located
Vestibular glands	Secrete fluid that moistens and lubricates the vestibule and vaginal opening during coitus
Mammary glands	Produce and secrete milk for nourishment of an infant

Inside Information—The average age of menarche—the first menstrual flow—is later (age 15) in girls who are very active physically than in the general population (age 12.6). This appears to be due to a requirement for a minimum percentage of body fat for menstruation to begin, and may represent a mechanism favored by natural selection to ensure the ability to successfully complete a pregnancy and nurse the baby.

Secondary Sex Organs

The secondary sex organs (fig. 23.1 and table 23.1) include the *uterine tubes,* through which an oocyte travels toward the uterus after ovulation; the *uterus,* where implantation and development occur; the *vagina,* which receives the penis and ejaculated semen during sexual intercourse and through which the baby passes during delivery; the *external genitalia* (*vulva*); and the *mammary glands.* The mammary glands within the *breasts* are considered to be secondary sex organs because the milk they secrete after childbirth provides nourishment to the baby. (The structure and function of mammary glands and breasts are discussed in module 23.6.)

Uterine Tubes

The paired uterine (*yoo'ter-in*) tubes, also known as the *fallopian tubes* or *oviducts,* transport oocytes from the ovaries to the uterus. Each uterine tube is approximately 10 cm (4 in.) long and 0.7 cm (0.3 in.) in diameter, and is positioned between the folds of the broad ligament of the uterus. The funnel-shaped, open-ended portion of the uterine tube, the **infundibulum,** lies close to the ovary but is not attached. A number of fringed, fingerlike processes called **fimbriae** (*fim'bre-e*) project from the margins of the infundibulum over the lateral surface of the ovary. Wave-like movements of the fimbriae sweep the ovarian surface. The beating cilia on the fimbriae create currents that carry an ovulated oocyte into the lumen of the uterine tube. Fertilization of an oocyte usually occurs in the region of the uterine tube near the infundibulum. From the infundibulum, the uterine tube extends medially and inferiorly to open into the cavity of the uterus (see fig. 23.4). The lumen of the uterine tube is lined with a ciliated mucous membrane. If fertilization is successful, the movements of the cilia transport the developing embryo toward the cavity of the uterus.

Two frequently occurring gynecological problems that involve the uterine tubes are salpingitis and ectopic pregnancies. Salpingitis *is an inflammation or infection of one or both uterine tubes. Infection of the uterine tubes is generally caused by a sexually transmitted disease (see table 22.4), although secondary bacterial infections from the vagina may also cause salpingitis. Sterility may result if the uterine tubes become blocked.*

An ectopic pregnancy *results from implantation of the embryo in a location other than the body of the uterus. The most frequent ectopic site is in the uterine tube, where an implanted embryo causes what is commonly called a* tubal pregnancy. *If this condition is not diagnosed and treated early, the uterine tube will rupture and subsequently hemorrhage, possibly endangering the life of the would-be mother. A tubal pregnancy is generally terminated by removing the affected uterine tube.*

fallopian tubes: from Gabriele Fallopius, Italian anatomist, 1523–62
fimbriae: L. *fimbria,* fringe
salpingitis: L. *salpinx,* trumpet
ectopic: Gk. *ex,* out; *topos,* place

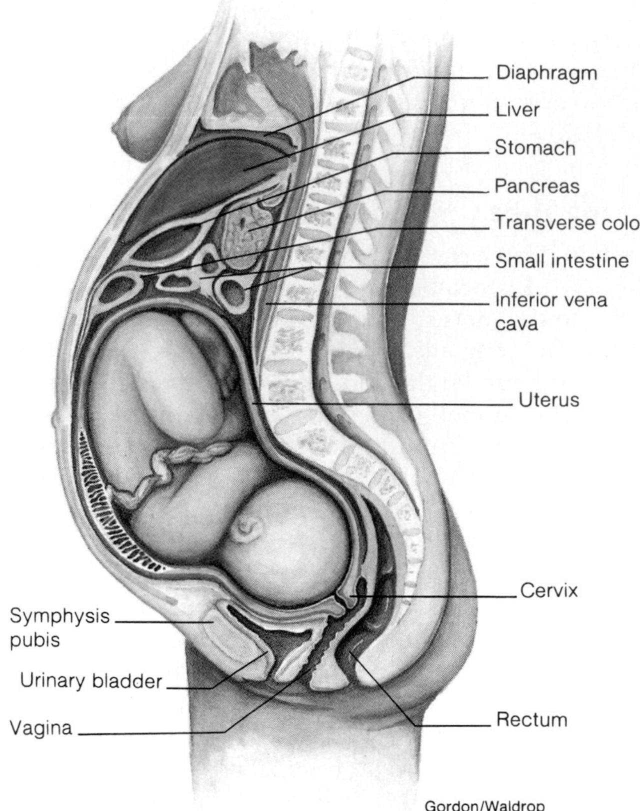

FIGURE 23.2
The size and position of the uterus in a full-term pregnant woman in sagittal section.

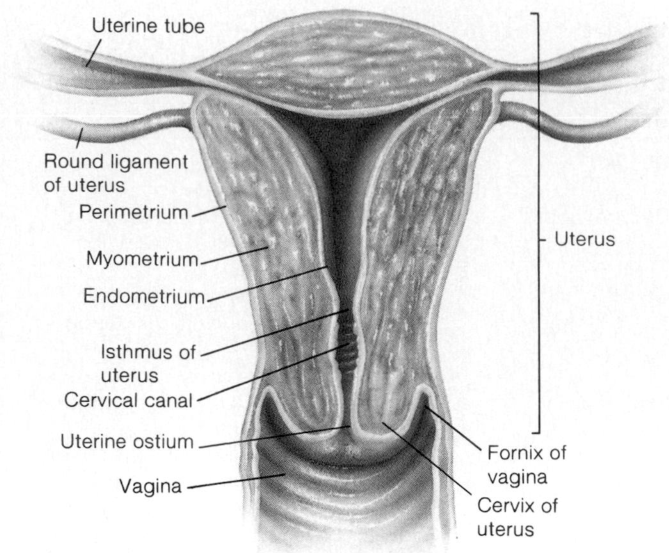

FIGURE 23.3
Layers of the uterine wall.

Since the infundibulum of the uterine tube is open-ended, it provides a potential pathway for pathogens to enter the abdominopelvic cavity. The mucosa of the uterine tube is continuous with that of the uterus and vagina. It is therefore possible for infectious agents to enter the vagina and cause infections that ultimately spread to the peritoneal linings, resulting in pelvic inflammatory disease (PID)*.*

Uterus

The uterus (*yoo'ter-us*) is the normal site of implantation of the embryo (see chapter 24, module 24.2). Prenatal development continues within the uterus until gestation (the duration of pregnancy) is completed, at which time the uterus plays an active role in childbirth.

The uterus is a hollow, thick-walled, muscular organ with the shape of an inverted pear. Although the shape and position of the uterus changes dramatically during pregnancy (fig. 23.2), in its nonpregnant state it is about 7 cm (2.8 in.) long, 5 cm (2 in.) wide (through its broadest region), and 2.5 cm (1 in.) in diameter. The anatomical regions of the uterus include the uppermost dome-shaped portion above the entrance of the uterine tubes, called the **fundus of the uterus;** the enlarged main portion, called the **body of the uterus;** and the inferior constricted portion opening into the vagina, called the **cervix of the uterus** (figs. 23.3 and 23.4). The cervix of the uterus projects posteriorly and inferiorly, joining the vagina at nearly a right angle (see fig. 23.1).

fundus: L. *fundus,* bottom
cervix: L. *cervix,* neck

The **uterine cavity** is the space within the fundus and body regions of the uterus. The narrow **cervical canal** extends through the cervix and opens into the lumen of the vagina (see fig. 23.3). The junction of the uterine cavity with the cervical canal is called the **isthmus of uterus,** and the opening of the cervical canal into the vagina is called the **uterine ostium.**

The wall of the uterus is composed of three layers (fig. 23.4).

1. The **perimetrium,** the outermost serous layer, consists of the thin visceral peritoneum.
2. The **myometrium** is composed of three thick, poorly defined layers of smooth muscle, arranged in longitudinal, circular, and spiral patterns.
3. The **endometrium** (*en"do-me'tre-um*), the inner mucosal lining of the uterus, is composed of two distinct layers. The superficial *stratum functionale,* consisting of columnar epithelium and containing excretory glands, is shed as *menses* during menstruation and built up again under the stimulation of ovarian steroid hormones. The deeper *stratum basale* is highly vascular and serves to regenerate the stratum functionale after each menstruation.

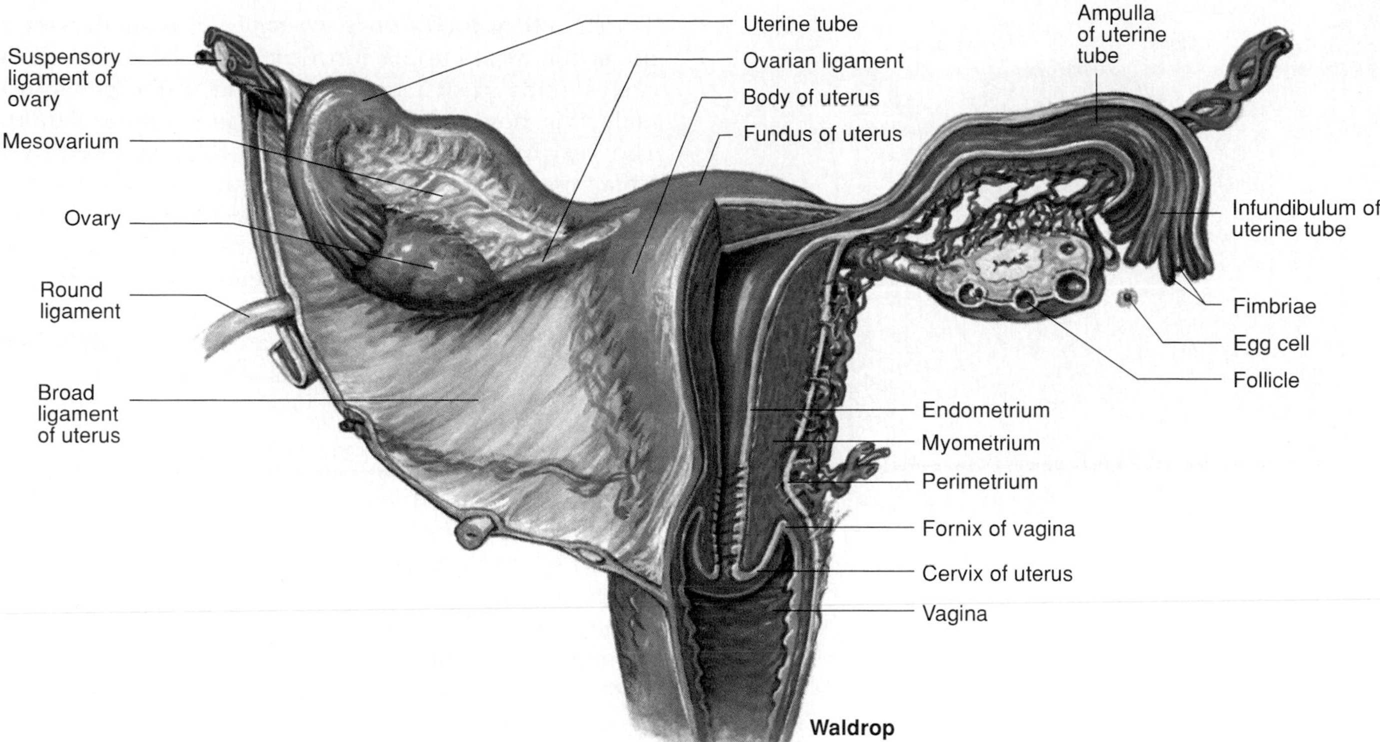

FIGURE 23.4
An anterior view of the female reproductive organs showing the relationship of the ovaries, uterine tubes, uterus, cervix, and vagina.

The internal reproductive organs are supported by paired **broad ligaments** and **round ligaments** (fig. 23.4). In addition, each ovary and uterine tube is supported by the **ovarian ligament,** the **suspensory ligament of the ovary,** and the **mesovarium.** Both the suspensory ligament and the mesovarium are part of the broad ligament. The fibrous ovarian ligament is enclosed within the broad ligament.

Inside Information—The extent to which the uterus enlarges during pregnancy is nothing short of remarkable. From a fist-sized organ within the pelvis, it grows to occupy the bulk of the abdominal cavity, becoming about 16 times heavier than it was before conception. After parturition (childbirth), the uterus rapidly shrinks, but it may remain somewhat enlarged until menopause, at which time there is marked atrophy.

Vagina

The vagina (*vă-ji'nă*) is a tubular organ about 9 cm (3.6 in.) in length that extends from the cervix of the uterus to the outside of the body. The deep recess surrounding the projection of the cervix into the vagina is called the **fornix.** The exterior opening of the vagina, at its lower end, is called the **vaginal orifice.** A thick fold of mucous membrane, called the **hymen,** may partially cover the vaginal opening.

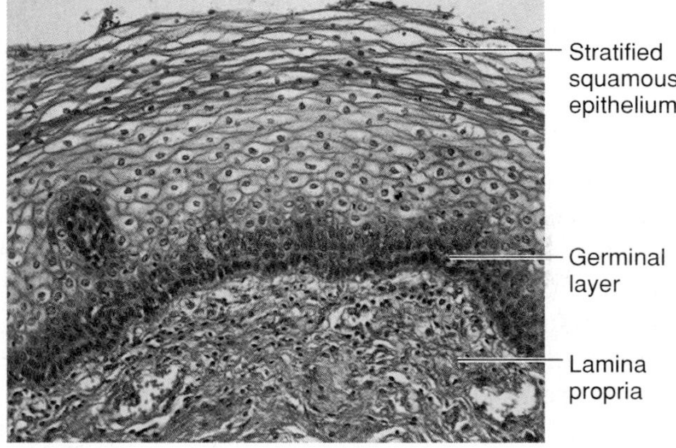

FIGURE 23.5
The histology of a vaginal ruga.

The vaginal wall is composed of three layers: an inner *mucosa,* a middle *muscularis,* and an outer *fibrous layer.* The mucosal layer consists of stratified squamous epithelium, which is folded in a series of transverse ridges called **vaginal rugae** (fig. 23.5). The vaginal rugae

vagina: L. *vagina,* sheath or scabbard

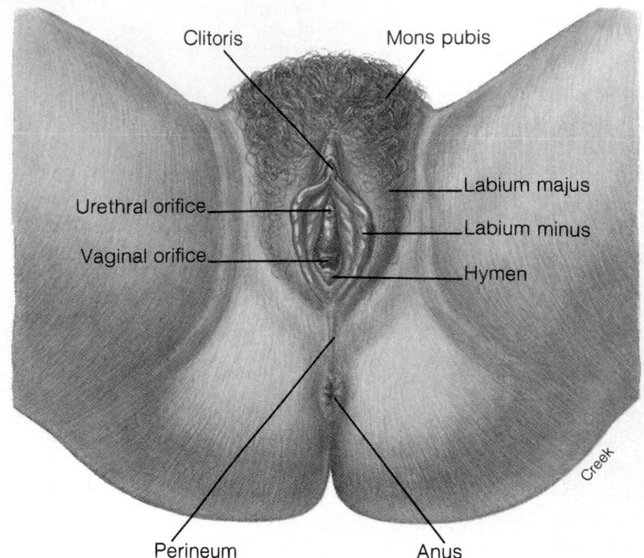

FIGURE 23.6
The external female genitalia.

provide friction for stimulation of the erect penis during sexual intercourse. The underlying *lamina propria* is thick and elastic; it contains small blood vessels, nerves, and lymph nodes. The muscular layer is composed of smooth muscle that can stretch considerably to accommodate the penis during intercourse and allow for birth of a baby.

Vulva

The external genitalia of the female are referred to collectively as the vulva (*vul'vă*) (fig. 23.6). The structures of the vulva surround the vaginal orifice in the region of the perineum (see also fig. 22.6). They include the *mons pubis, labia majora, labia minora, clitoris, vestibule, vestibular glands,* and *vestibular bulbs.*

The **mon pubis** (*pyoo'bis*) is the subcutaneous pad of adipose connective tissue covering the symphysis pubis. The **labia majora** (*la'be-ă mă-jor'ă*—singular, *labium majus*) are two thickened longitudinal folds of skin that contain loose fibrous connective tissue and adipose tissue, as well as some smooth muscle. After puberty, the surface of the mons pubis and the lateral surfaces of the labia majora are covered with pubic hair. Medial to the labia majora are two smaller, hair-free longitudinal folds called the **labia minora** (*la'be-ă mĭ-nor'ă*—singular, *labium minus*). They are covered with a mucous membrane, richly supplied with oil and sweat glands.

The **clitoris** (*klit'or-is, kli-tor'is*) is a small rounded projection at the upper junction of the labia. The clitoris corresponds in structure and origin to the penis in the male; it is, however, much smaller and without a urethra. The **vestibule** is the longitudinal cleft enclosed by the labia minora. The openings for the urethra and vagina are located in the vestibule. The vagina is lubricated by secretions from paired **vestibular glands** (Bartholin's glands) located within the wall of the region immediately inside the vaginal orifice. Bodies of vascular erectile tissue, called **vestibular bulbs,** are located immediately below the skin forming the lateral walls of the vestibule.

Secondary Sex Characteristics

As you will recall from the previous chapter (see module 22.3), puberty is the period during which sexual maturity is reached and reproduction becomes possible. In girls, this means that the ovaries begin their cycles of egg cell maturation and sex steroid secretion. It is in response to changing levels of female sex hormones (*estrogens* and *progesterone*) in the bloodstream that the first episode of menstrual bleeding, or menarche, occurs in girls. In addition, ovarian estrogens are responsible for certain body changes associated with female puberty. These are the female **secondary sex characteristics.**

Estrogen-induced secondary sex characteristics of females include growth of the breasts; increased deposit of subcutaneous fat, especially in the hips, thighs, and breasts; widening of the pelvis; and the appearance of axillary (underarm) and pubic hair. Although these features are not essential for the reproductive process, they distinguish the female body from that of the male and are considered sexual attractants.

Inside Information—The fatty tissue deposited at puberty determines the size and shape of the breasts, but it is the glandular tissue that produces the milk. The amount of milk the mammary glands secrete is therefore not related to breast size. Large breasts, apparently, serve primarily as sexual attractants.

vulva: L. *volvere,* to roll, wrapper
mons pubis: L. *mons,* mountain; *pubis,* genital area
vestibule: L. *vestibule,* an entrance, court
Bartholin's glands: from Casper Bartholin Jr., Danish anatomist, 1655–1738

23.2 Ovaries and Ovarian Follicles

Egg cells, or oocytes, develop within hollow structures called follicles within the ovaries. The follicles mature as the oocytes go through different phases of meiosis.

Ovaries

The color and texture of the ovaries vary according to the age and reproductive stage of the female. The ovaries of a young girl are smooth and pinkish. Following puberty, they are pinkish-gray and their surface is irregular because of the scarring caused by ovulation. The ovaries of a sexually mature female are solid ovoid structures about 3.5 cm (1.4 in.) long, 2 cm (0.8 in.) wide, and 1 cm (0.4 in.) thick. The lateral portion of the ovary is positioned near the open end of the uterine tube (see fig. 23.4).

Ovarian tumors, *which occur most often in women over the age of 60, can grow to be massive. Ovarian tumors as heavy as 5 kg (14 lb) are not uncommon, and some weighing as much as 110 kg (300 lb) have been reported. Some ovarian tumors produce estrogen and thus cause feminization in elderly women, including the resumption of menstrual periods. The prognosis for women with ovarian tumors varies depending on the type of tumor, whether or not it is malignant, and if it is, the stage of the cancer.*

Oogenesis

Meiosis, the specialized cell division that occurs in the testes to produce sperm cells, also takes place in the ovaries. In this case, female sex cells are produced, and the process is known as **oogenesis** (*o"ŏ-jen'ĕ-sis*).

The cells that will later become oocytes (egg cells) migrate into the ovaries during early embryonic development and multiply rapidly. By about 5 months of gestation, the ovaries contain approximately 6 to 7 million primordial oocytes, called **oogonia.** The production of new oogonia stops at this point and never resumes. Toward the end of gestation, the oogonia begin meiosis, at which time they are called **primary oocytes.** Meiosis is arrested at prophase I of the first meiotic division, and therefore the primary oocytes are still diploid (have 46 chromosomes). Although the ovaries of a newborn girl contain about 2 million oocytes, this number declines to 300,000 to 400,000 by the time she enters puberty. On average, 400 oocytes are ovulated during a woman's reproductive lifetime.

Primary oocytes that are not stimulated to complete the first meiotic division are contained within tiny clusters of cells called **primary follicles** (fig. 23.7*a*). Immature primary follicles consist of only a single layer of *granulosa cells.* These are the cells that secrete estrogen, and their name refers to their "grainy" appearance as compared with the much larger egg cell. In response to FSH stimulation, some of the oocytes and follicles begin to grow, and the granulosa cells divide and secrete estrogen. Some primary follicles will be stimulated to grow still more and develop a number of fluid-filled cavities called *vesicles.* At this point, they are called **secondary follicles** (fig. 23.7*a*). With the continued growth of one of these follicles, its vesicles will fuse to form a single fluid-filled cavity called an *antrum.* At this stage, the follicle is known as a **graafian follicle** (fig. 23.7*b*).

As the graafian follicle develops, the primary oocyte completes its first meiotic division, yielding two cells of unequal size. One cell—the **secondary oocyte**—gets all of the cytoplasm, whereas the other becomes a small *polar body* (fig. 23.8) that eventually fragments and disappears. This unequal division of cytoplasm ensures that the ovum will have enough nutrient reserves to become a viable embryo should fertilization later occur. The secondary oocyte then begins the second meiotic division, but stops at metaphase II. Meiosis is arrested at metaphase II while the ovum is still in the ovary; the second meiotic division is completed only by an ovum that has been ovulated and fertilized.

ovary: L. *ovum,* egg
oogonium: Gk. *oion,* egg; *gonos,* generation
follicle: L. *folliculus,* little bag, diminutive of *follis,* bellows
graafian follicle: from Regnier de Graaf, Dutch anatomist and physician, 1641–73

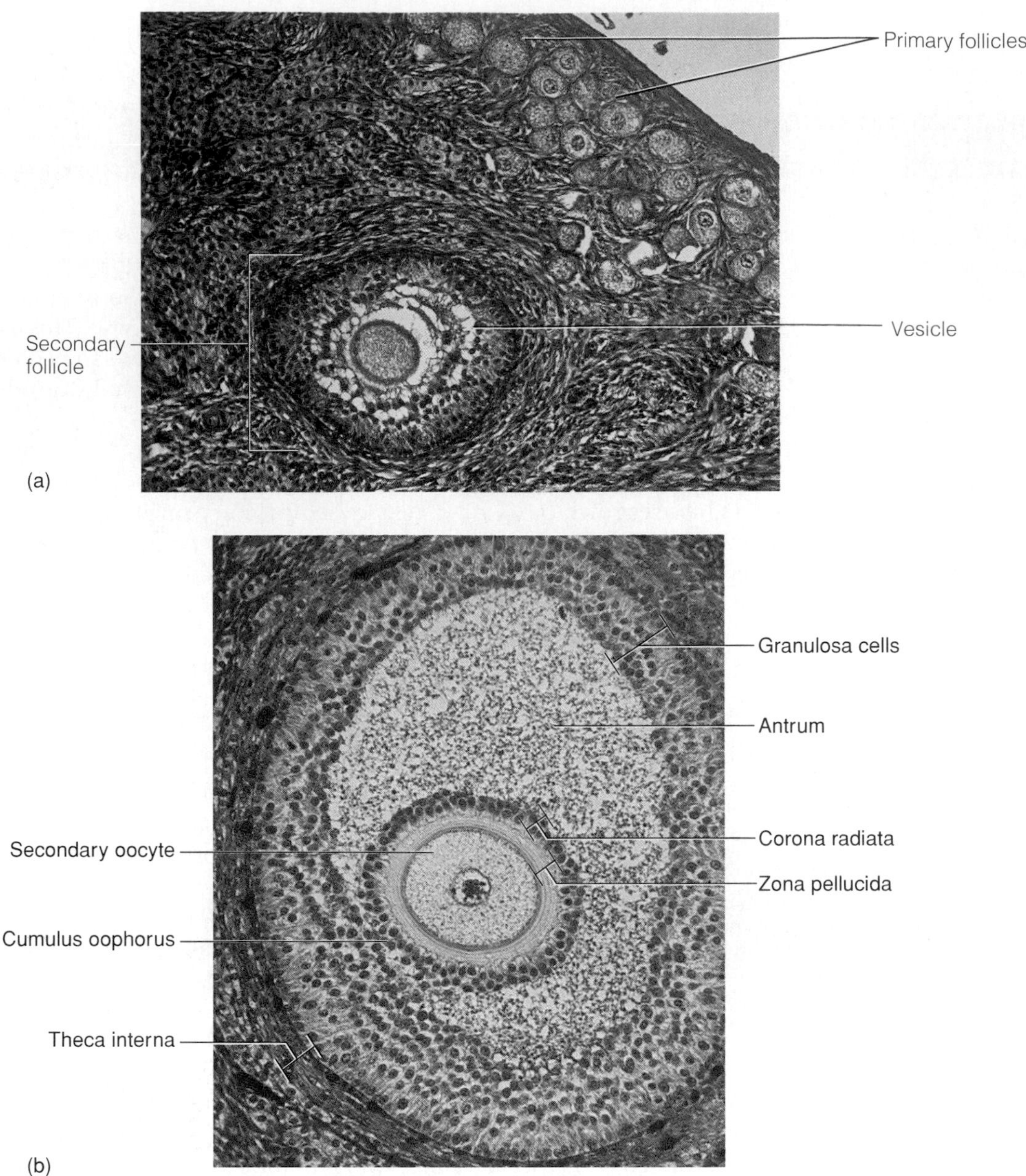

FIGURE 23.7
Photomicrographs of (*a*) developing follicles and (*b*) a graafian follicle.

A graafian follicle therefore contains a secondary oocyte, arrested at metaphase II. The granulosa cells of this follicle form a ring around the circumference of the follicle, called the *corona radiata,* and a mound that supports the ovum. The mound is called the *cumulus oophorus.* Between the oocyte and the corona radiata is a thin gel-like layer of proteins and polysaccharides called the *zona pellucida.* When stimulated by FSH from the anterior pituitary, the granulosa cells of the ovarian follicles secrete increasing amounts of estrogen as the follicles grow. Interestingly, the granulosa cells synthesize estrogen from its precursor testosterone, which is supplied by a layer of cells immediately outside the follicle called the *theca interna* (see fig. 23.7*b*).

---

corona radiata: Gk. *korone,* crown; L. *radiata,* radiate
cumulus oophorus: L. *cumulus,* a mound; Gk. *oophoros,* egg-bearing
zona pellucida: L. *zone,* girdle; L. *pellis,* skin
theca interna: Gk. *theke,* a box; L. *internus,* interior

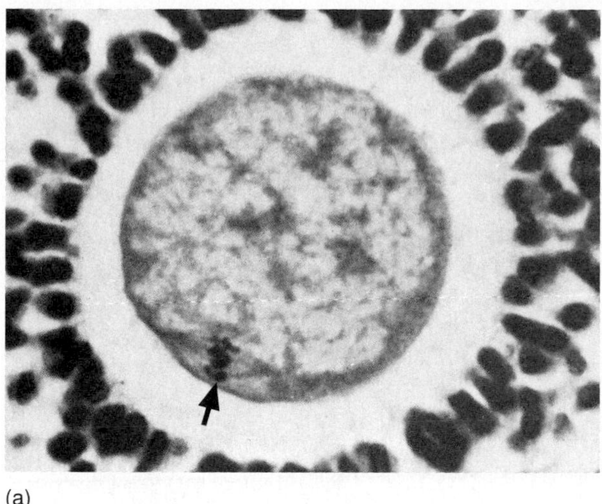

(a)

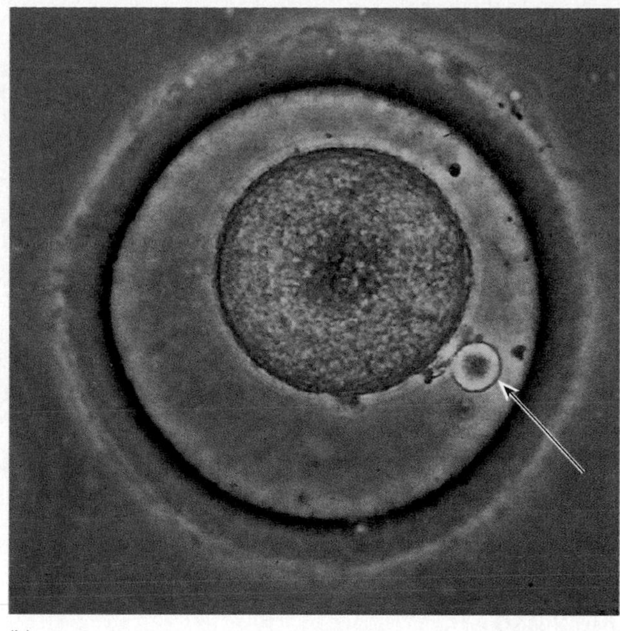

(b)

FIGURE 23.8
(*a*) A primary oocyte at metaphase I of meiosis. (Note the alignment of chromosomes [arrow].) (*b*) A human secondary oocyte formed at the end of the first meiotic division, and the first polar body (arrow).

23.3 Ovulation

Ovulation occurs when a graafian follicle ruptures and expels its secondary oocyte. The now-empty follicle becomes a corpus luteum.

Changes in the Ovary

Only one follicle per cycle continues its growth to become a fully mature graafian follicle. Other secondary follicles during that cycle regress and undergo degeneration, or *atresia.* The graafian follicle becomes so large that it forms a bulge on the external surface of the ovary. Under proper hormonal stimulation, this follicle will rupture—much like the popping of a blister—and expel its oocyte (ovulate) into the uterine tube (fig. 23.9).

The released cell is a secondary oocyte, surrounded by a zona pellucida and corona radiata. If it is not fertilized, it disintegrates in a couple of days. If a sperm cell passes through the corona radiata and zona pellucida and enters the cytoplasm of the secondary oocyte, the oocyte will then complete the second meiotic division. In this process, the cytoplasm is again not divided equally; most

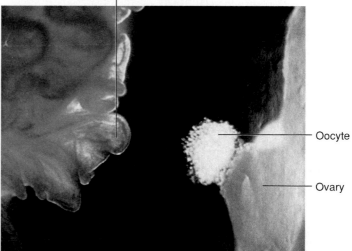

FIGURE 23.9
The release of the secondary oocyte from a human ovary. An ovulated oocyte is free in the peritoneal cavity until it enters the lumen of the uterine tube.

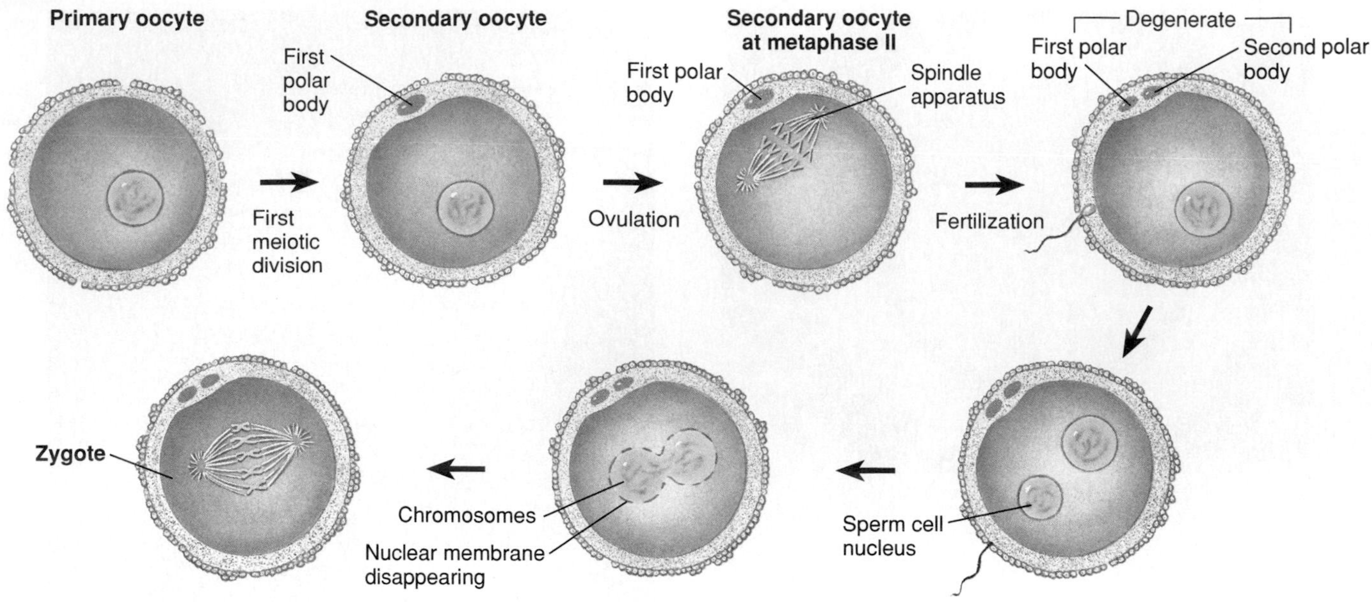

FIGURE 23.10
A schematic diagram of the process of oogenesis. During meiosis, each primary oocyte produces a single haploid gamete. If the secondary oocyte is fertilized, it forms a second polar body and becomes a zygote.

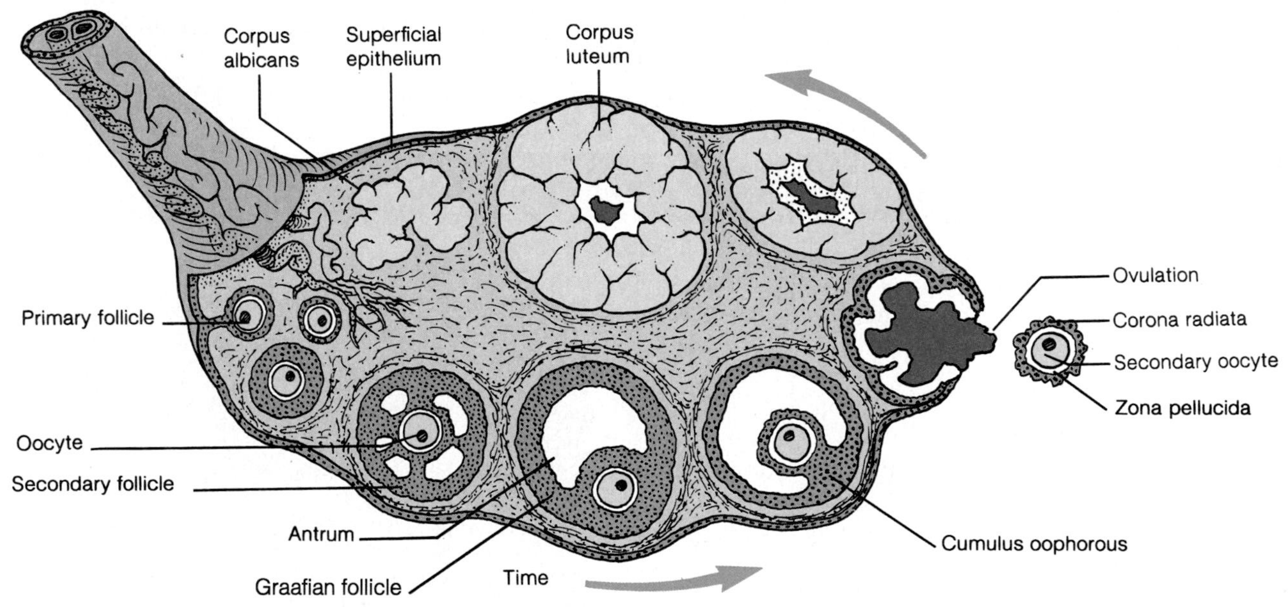

FIGURE 23.11
A schematic diagram of an ovary, showing the various stages of ovum and follicle development.

of the cytoplasm remains in the *zygote* (fertilized egg), leaving another small polar body that, like the first, disintegrates (fig. 23.10).

Changes continue in the ovary following ovulation. The empty follicle, under the influence of luteinizing hormone from the anterior pituitary, undergoes structural and biochemical changes to become a **corpus luteum** ("yellow body"). Unlike the ovarian follicles, which secrete only estrogen, the corpus luteum secretes two sex steroid hormones: estrogen and progesterone. Toward the end of a nonfertile cycle, the corpus luteum regresses and is changed into a nonfunctional mass of scar tissue called the **corpus albicans** ("white body"). These cyclic changes in the ovary are summarized in figure 23.11.

corpus luteum: L. *corpus,* body; *luteum,* yellow
albicans: L. *albicare,* to whiten

Changes in Hormone Secretion

The term *pituitary-ovarian axis* refers to the hormonal interactions between the anterior pituitary and the ovaries. The anterior pituitary secretes two gonadotropic hormones—*follicle-stimulating hormone* (*FSH*) and *luteinizing hormone* (*LH*)—that promote cyclic changes in the structure and function of the ovaries. As mentioned in the previous chapter (see module 22.3), the secretion of both gonadotropic hormones is controlled by a single releasing hormone from the hypothalamus: *gonadotropin-releasing hormone* (*GnRH*). Secretion of FSH and LH is also regulated by feedback effects (see chapter 13, module 13.3) from the ovarian hormones. The nature of these interactions will be described in detail in the next module.

Since one releasing hormone can stimulate the secretion of both FSH and LH, one might expect always to see parallel changes in the secretion of these gonadotropins. This, however, is not the case. During an early phase of the menstrual cycle, the rate of FSH secretion slightly exceeds that of LH secretion, and just prior to ovulation LH secretion greatly exceeds FSH secretion. These differences are believed to be a result of the feedback effects of ovarian sex steroids that change the amount of GnRH secreted and the response of the anterior pituitary to GnRH. These complex interactions result in a pattern of hormone secretion that is responsible for each phase of the menstrual cycle.

23.4 Menstrual Cycle: Pituitary Gland and Ovary

Rising levels of estrogen secretion from the developing follicles stimulate a surge in LH secretion, and this LH surge triggers ovulation. The empty follicle becomes a corpus luteum, which secretes progesterone as well as estrogen.

Human females have cycles of ovarian activity that repeat at approximately 1-month intervals; hence the name menstrual cycle (*menstru* = monthly). The term *menstruation* is used to indicate the most obvious event of the cycle—the regular shedding of the stratum functionale of the endometrium. The average menstrual cycle is about 28 days long, the first day of menstruation being "day 1." The monthly cycle is divided into three phases.

Follicular Phase

Menstruation lasts from day 1 to day 4 or 5 of the average cycle. During this time, the secretions of ovarian steroid hormones are at their lowest ebb, and the ovaries contain only primary follicles. During the **follicular phase** of the ovaries, which lasts from day 1 to about day 13 of the cycle, some of the primary follicles grow, form an antrum, and become secondary follicles. Toward the end of the follicular phase, one follicle in one ovary reaches maturity and becomes a graafian follicle. As follicles grow, the granulosa cells secrete an increasing amount of **estradiol** (the principal estrogen), which reaches its highest concentration in the blood at about day 12 of the cycle (2 days before ovulation).

The rapid rise in estradiol secretion during the late follicular phase stimulates the hypothalamus to secrete GnRH, which, in turn, stimulates the anterior pituitary to secrete LH. This stimulatory effect of estradiol on the anterior pituitary is referred to as a *positive feedback effect*. As a result of this positive feedback, there is an enormous burst of LH release, called an **LH surge,** near the end of the follicular phase (fig. 23.12). The LH surge begins about 23 hours before ovulation and reaches its peak about 16 hours before ovulation. It is this surge that triggers ovulation.

Ovulation

The surge of LH causes the wall of the graafian follicle to rupture at about day 14 and release its secondary oocyte in the process of **ovulation** (*ov-yŭ-la'shun*). Ovulation occurs, therefore, as a result of the sequential effects of FSH and LH. First, FSH is needed to make the follicles grow and secrete increasing amounts of estradiol. Then, the ovary "informs" the pituitary that it contains a mature graafian follicle by secreting rising levels of estradiol. Lastly, the LH surge that is stimulated by the estradiol causes the follicle to rupture and ovulate.

Inside Information—Most women are quite unaware that one follicle, approximately 2.5 cm (1 in.) in diameter, has ruptured and released its solitary egg. A substantial number of women, however (approximately 30%), experience a sharp, cramplike pain at the time of ovulation. Not infrequently, this *mittelschmerz* (German for "middle pain") is confused with appendicitis.

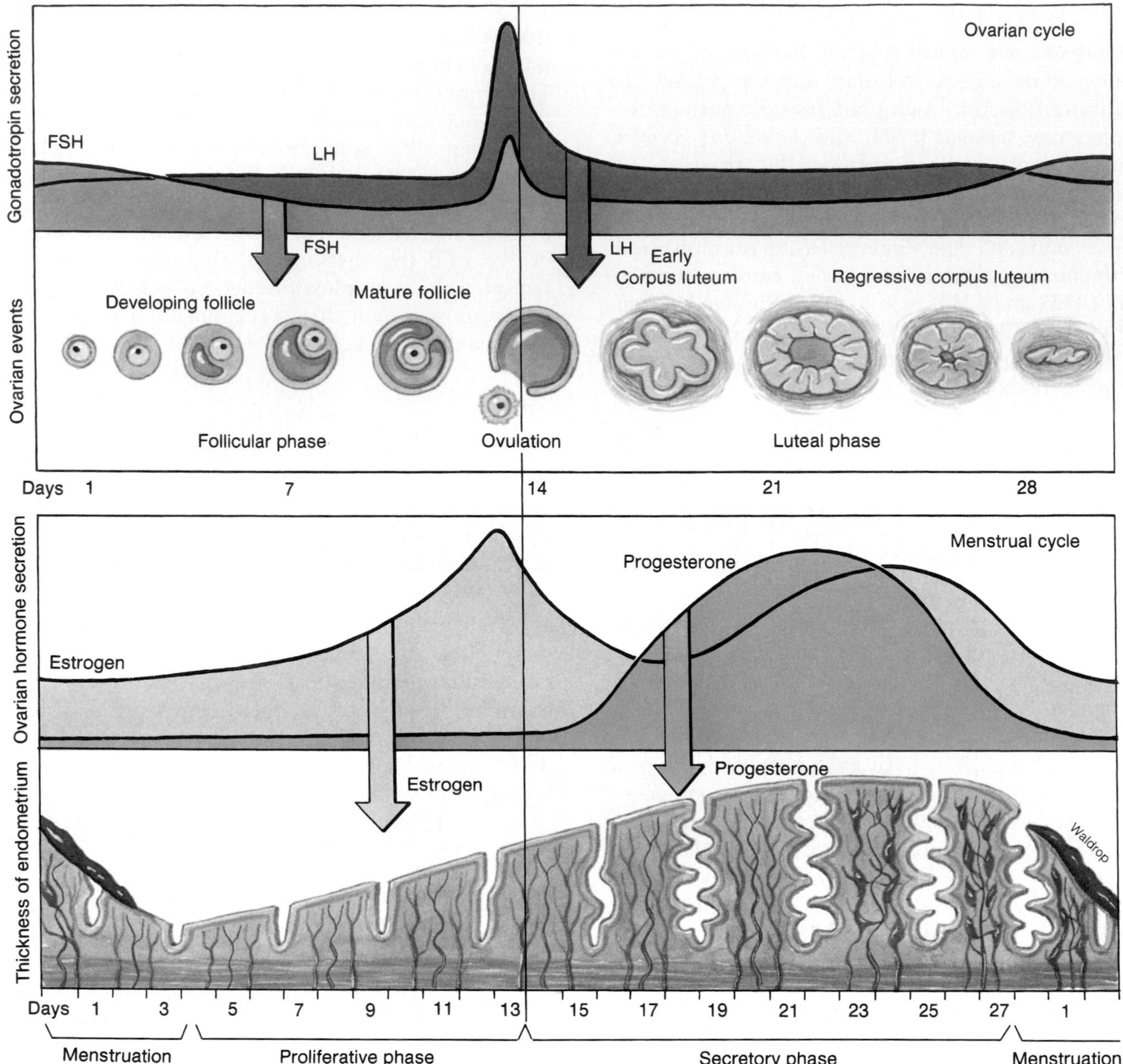

FIGURE 23.12
The cycle of ovulation and menstruation.

Luteal Phase

After ovulation, the ovaries are in the **luteal phase** until the first day of menstruation. During this phase, the empty follicle is stimulated by LH to become a new structure, the **corpus luteum** (fig. 23.13). Whereas the developing follicles secrete only estradiol, the corpus luteum secretes both estradiol and **progesterone.** Progesterone levels in the blood reach a peak approximately 1 week after ovulation (see fig. 23.12).

The combined high levels of estradiol and progesterone during the luteal phase exert a *negative feedback inhibition* of FSH and LH secretion. This serves to retard development of new follicles, so that further ovulation does not normally occur during that cycle. In this way, multiple ovulations (and possible pregnancies) on succeeding days of the cycle are prevented.

High levels of estrogen and progesterone during the nonfertile cycle do not persist for very long, however. Estrogen and progesterone levels decline during the late luteal phase (starting about day 22) because the corpus luteum regresses and stops functioning.

With the declining function of the corpus luteum, estrogen and progesterone fall to very low levels by day 28 of the cycle. It is this fall in estradiol and progesterone that triggers menstruation, thus permitting a new cycle of ovarian follicle development to progress.

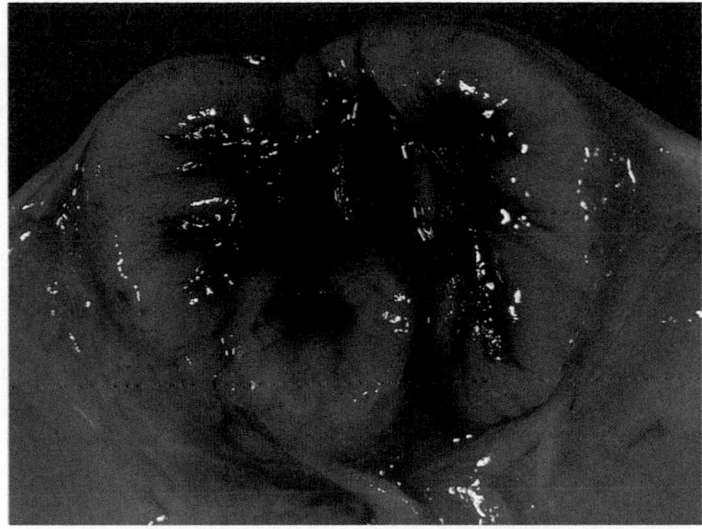

FIGURE 23.13
A corpus luteum in a human ovary.

23.5 Menstrual Cycle: Endometrium

The endometrium becomes thicker, spongier, and more vascular as a result of the action of ovarian steroid hormones. Withdrawal of these hormones causes menstruation.

Cyclic Changes

Just as the ovaries undergo change during the menstrual cycle, so do the uterus and other reproductive organs. Three phases of the menstrual cycle (fig. 23.12 *[bottom]* and fig. 23.14) can be identified on the basis of cyclic changes in the endometrium.

1. The **proliferative phase** of the endometrium occurs while the ovary is in its follicular phase. The increasing amounts of estradiol secreted by the developing follicles stimulate growth (proliferation) of the stratum functionale of the endometrium. In humans and other primates, spiral, or coiled, arteries develop in the endometrium during this phase, enriching the blood supply.
2. The **secretory phase** of the endometrium occurs when the ovary is in its luteal phase. In this phase, increased progesterone secretion stimulates the production of glycogen and the development of mucous glands. As a result of the combined actions of estradiol and progesterone, the endometrium becomes thick, vascular, and "spongy" in appearance during the time of the cycle following ovulation. It is therefore well prepared to accept and nourish an embryo should fertilization occur.
3. The **menstrual phase** or **menstruation** (*men"stroo-a'shun*) occurs as a result of the fall in the ovarian secretion of estradiol and progesterone during the late luteal phase. Necrosis (cellular death) and sloughing of the stratum functionale of the endometrium is caused by constriction of the spiral arteries when ovarian hormone secretion declines.

The cyclic changes in ovarian secretion cause other cyclic changes in the female reproductive ducts. Early in the cycle, high levels of estradiol, for example, cause the production of a thin, watery cervical mucus that can easily be penetrated by sperm cells. During the later luteal phase of the cycle, by contrast, high levels of progesterone cause the cervical mucus to become thick and sticky after ovulation has occurred. The thick mucus probably serves as a barrier—another mechanism to prevent possible successive pregnancies.

Cyclic changes in ovarian hormone secretion also cause cyclic changes in *basal body temperature*—the temperature of the body taken upon awakening, before engaging in any other activity. In the *rhythm method* of birth control, a woman measures her oral basal body temperature to determine when ovulation has occurred. On the day of the LH peak, when estradiol secretion begins to decline, there is a slight drop in basal body temperature. Starting about 1 day after the LH peak, the basal body temperature sharply rises as a result of progesterone secretion and remains elevated throughout the luteal phase of the cycle

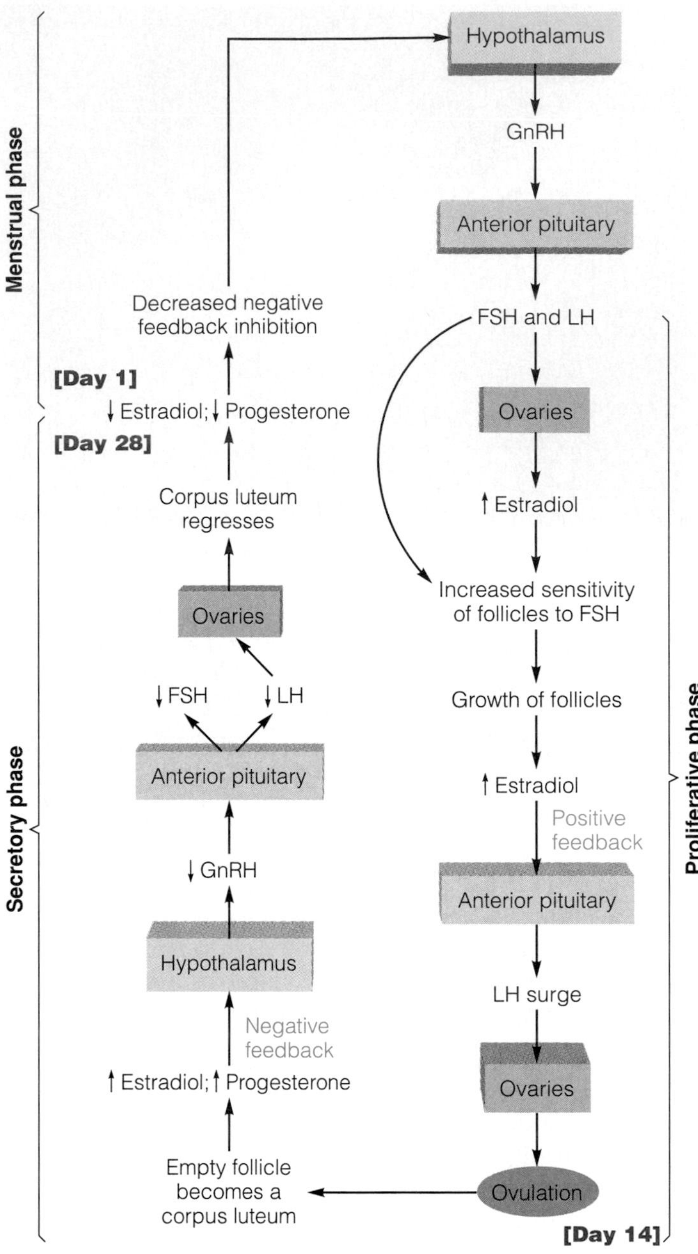

FIGURE 23.14

The sequence of events in the endocrine control of the ovarian cycle in context of the phases of the endometrium during the menstrual cycle.

(fig. 23.15). The day of ovulation can be accurately determined by this method, making the method useful if conception is desired. Since the day of the cycle on which ovulation occurs is quite variable in many women, however, the rhythm method is not very reliable for preventing conception by predicting when the next ovulation will occur. In short, ovulation can be determined only when it has occurred; an accurate test to predict the time of ovulation does not exist.

Inside Information—No woman is absolutely regular in her menstrual cycle. Cycles as long as 36 days or as short as 21 days are considered within the normal range. Menstrual bleeding generally lasts 3 to 7 days and totals 4 to 6 tablespoons of fluid, one-third of which is blood.

Amenorrhea (a-men"ŏ-re'ă) *is the absence of menstruation; it can be categorized as normal, primary, or secondary.* Normal amenorrhea *follows menopause, occurs during pregnancy, and in some women may occur during lactation.* Primary amenorrhea *is the failure to have menstruated by the age at which menstruation normally begins. Primary amenorrhea is generally accompanied by failure of the secondary sex characteristics to develop. Endocrine disorders may cause primary amenorrhea and abnormal development of the ovaries and uterus.* Secondary amenorrhea *is the cessation of menstruation in women who previously have had normal menstrual periods and who are not pregnant and have not gone through menopause. Various endocrine disturbances, as well as psychological factors, may cause secondary amenorrhea. It is not uncommon, for example, for young women who are in the process of making major changes or adjustments in their lives to miss menstrual periods. Secondary amenorrhea is also frequent in women athletes during periods of intense training. A low percentage of body fat may be a contributing factor. Sickness, fatigue, poor nutrition, or emotional stress also may cause secondary amenorrhea.*

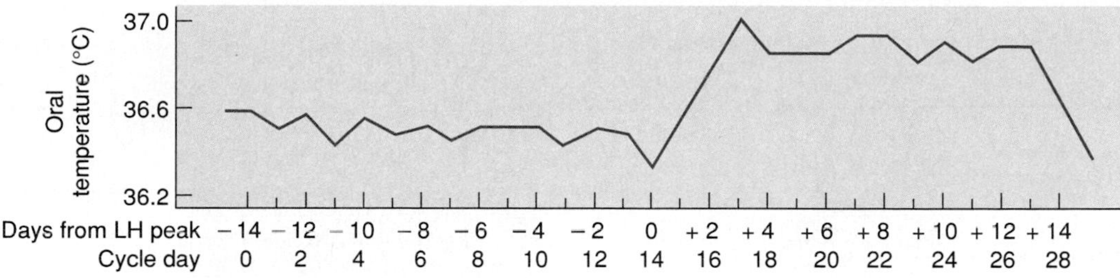

FIGURE 23.15
Changes in basal body temperature during the menstrual cycle.

Menopause

The term menopause (*men'ŏ-pawz*) means literally "pause in the menses" and refers to the cessation of ovarian activity that occurs at about the age of 50. During the postmenopausal years, which account for about a third of a woman's life span, no new ovarian follicles develop and the ovaries stop secreting estradiol. Like prepubertal boys and girls, the only estrogen found in the blood of postmenopausal women is that formed by conversion of the weak androgen androstenedione, secreted principally by the adrenal cortex, into a weak estrogen called *estrone.*

It is the lack of estradiol secretion from the ovaries that is most responsible for the many symptoms of menopause. These include vasomotor disturbances and urogenital atrophy. Vasomotor disturbances produce the "hot flashes" of menopause, where a fall in core body temperature is followed by feelings of heat and profuse perspiration. Atrophy (shrinkage) of the urethra, vaginal wall, and vaginal glands occur, with loss of lubrication. There is also an increased risk of cardiovascular disease and a faster progression of osteoporosis (see chapter 6, "Clinical Considerations"). These changes can be reversed by hormone replacement therapy (HRT) to a significant degree.

Inside Information—Among mammals, only human females lose their ability to reproduce. Most animals in the wild never live to an advanced age, falling victim to enemies or diminished hunting skills. But in zoos females continue to breed in old age, even if they do so less frequently.

23.6 Mammary Glands

At puberty, estrogen from the ovaries stimulates growth of the mammary glands and the deposition of fatty tissue in the breasts. The production of milk and its release from the mammary glands depend on the actions of prolactin and oxytocin.

Structure of the Breasts and Mammary Glands

The size and shape of the **breasts** vary widely from person to person because of differences in genetic makeup, age, and percentage of body fat. Each breast overlies the pectoralis major muscle and portions of the serratus anterior and external abdominal oblique muscles (fig. 23.16). The **axillary process of the breast** extends upward and laterally toward the axilla (underarm). This region of the breast is clinically significant because of the high incidence of breast cancer within the lymphatic drainage of the axillary process.

Located within the breast, the **mammary gland** is composed of 15 to 20 **lobes,** separated by adipose tissue. Each lobe possesses a single duct that opens independently to the outside of the body. Within each lobe are smaller **lobules,** in which glandular **mammary alveoli** are found. The mammary alveoli are the structures that

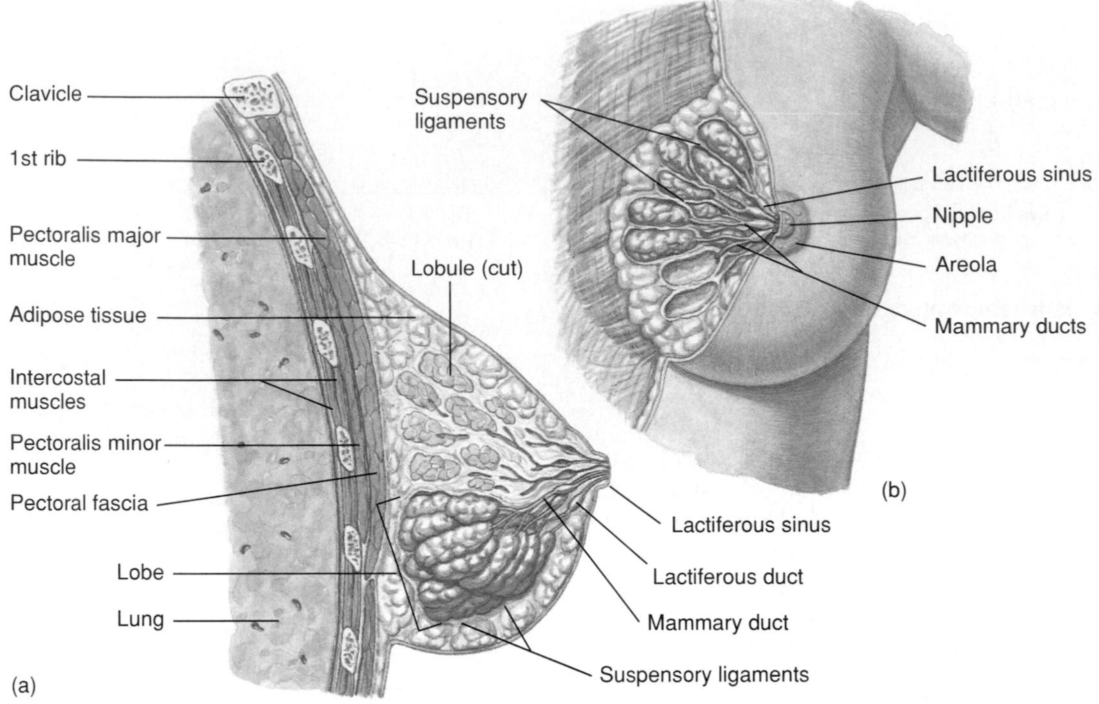

FIGURE 23.16
The structure of the breast and mammary glands. (*a*) A sagittal section and (*b*) an anterior view partially sectioned.

produce the milk. **Suspensory ligaments** (of Cooper) between the lobules extend from the skin to the deep fascia overlying the pectoralis major muscle. These ligaments support the breasts. The clustered mammary alveoli secrete milk into a series of **mammary ducts** that converge to form **lactiferous** (*lak-tif'er-us*) **ducts.** The lumen of each lactiferous duct expands near the nipple to form a **lactiferous sinus.** Milk is stored in the lactiferous sinuses before draining at the tip of the nipple.

The **nipple** is a cylindrical projection of the breast that contains some erectile tissue. A circular pigmented **areola** (*ă-re'o-lă*) surrounds the nipple. The surface of the areola may appear rough because of the sebaceous **areolar glands** close to the surface. The secretions of these glands keep the nipple pliable. The color of the areola and nipple varies with the complexion of the woman. During pregnancy, the areola becomes darker in most women, and enlarges somewhat, presumably to become more conspicuous to a nursing infant.

ligaments of Cooper: from Sir Astley P. Cooper, English anatomist and surgeon, 1768–1841

Lactation

Lactation (*lak-ta'shun*) is the production of milk by the hormone-prepared mammary glands. The high levels of estrogen that are secreted by the placenta during pregnancy stimulate the development of the mammary alveoli. Similarly, high levels of progesterone from the placenta stimulate proliferation of the tubules and ducts (fig. 23.17).

The production of milk proteins is stimulated after the birth of the baby by **prolactin,** a hormone secreted by the anterior pituitary. The secretion of prolactin is controlled by *prolactin-inhibiting hormone* (*PIH*), produced by the hypothalamus, which inhibits prolactin secretion from the anterior pituitary. The reason a pregnant woman doesn't lactate (produce milk) is because the secretion of

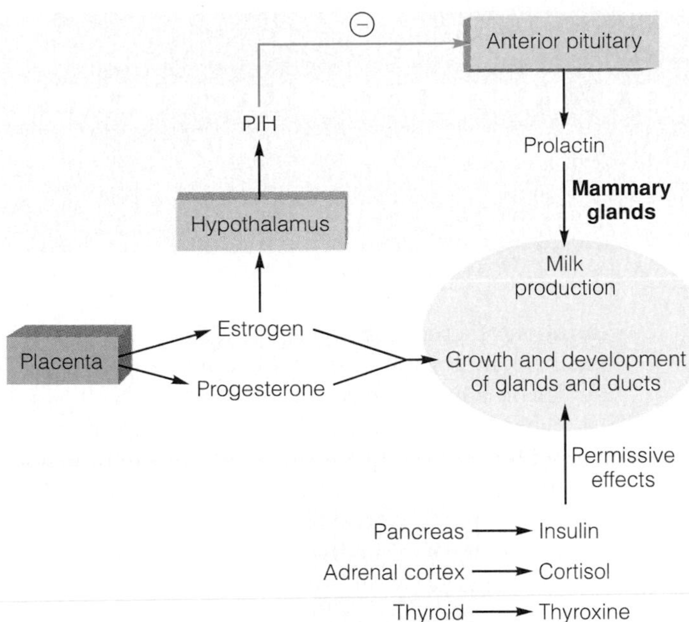

FIGURE 23.17

The hormonal control of mammary gland development during pregnancy and lactation. Notice that milk production is prevented during pregnancy by estrogen inhibition of prolactin secretion. This inhibition is accomplished by the stimulation of PIH (prolactin-inhibiting hormone) secretion from the hypothalamus.

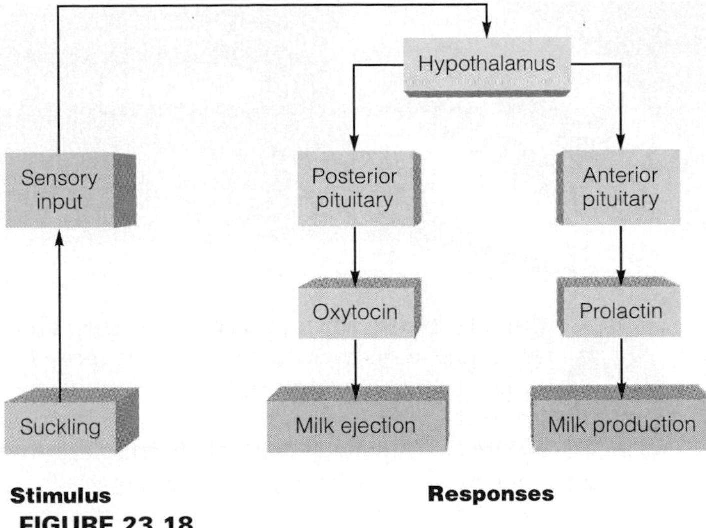

FIGURE 23.18

Lactation occurs in two stages: milk production (stimulated by prolactin) and milk ejection (stimulated by oxytocin). The stimulus of suckling triggers a neuroendocrine reflex that results in increased secretion of oxytocin and prolactin.

PIH is stimulated by the high levels of estrogen that are present during pregnancy. In addition, high levels of estrogen act directly on the mammary glands to block their stimulation by prolactin. During pregnancy, consequently, the high levels of estrogen prepare the breasts for lactation but prevent the secretion and action of prolactin.

The act of nursing helps to maintain high levels of prolactin secretion via a *neuroendocrine reflex* (fig. 23.18). Sensory endings in the breast, activated by the stimulus of suckling, relay impulses to the hypothalamus and inhibit the secretion of PIH. With less PIH, the anterior pituitary can secrete more prolactin. Suckling thus results in the reflex secretion of prolactin, which promotes the secretion of milk from the mammary alveoli into the ducts. In order for the baby to get the milk, however, the action of another hormone is needed.

The stimulus of suckling also results in the reflex secretion of oxytocin from the posterior pituitary. As described in chapter 13 (see module 13.2), this hormone is produced in the hypothalamus and stored in the posterior pituitary. Oxytocin secretion in a lactating woman stimulates the **milk-ejection reflex,** or milk letdown, as a result of the contractions of the lactiferous ducts. Oxytocin also stimulates contractions of the uterus. This is why women who breast-feed regain uterine muscle tone faster than women who bottle-feed.

Milk letdown can become a conditioned reflex in response to visual or auditory cues; the crying of a baby can elicit oxytocin secretion and the milk-ejection reflex. On the other hand, this reflex can be suppressed by the adrenergic effects produced in the flight-or-fight reaction. Thus, if a woman becomes anxious while breast feeding, she will produce milk but it will not flow (there will be no milk letdown). This can cause increased pressure, intensifying her anxiety and frustration, and further inhibiting the milk-ejection reflex. A quiet and calm environment is conducive to successful breast feeding.

Control of Reproduction

Although contraceptive potions, devices, and techniques date back to antiquity, as recently as the early part of this century it was illegal virtually all over the world to provide, print, or disseminate information about birth control. It was largely through the efforts of one woman, Margaret Sanger, that public access to birth control was finally achieved. Sanger, an obstetrical nurse who witnessed the horrors of botched abortions, relentlessly advocated women's rights to control their own destinies. She is considered the founder of the modern birth-control movement.

Today, a wide variety of birth-control methods is readily available. Your task, if you wish to separate sexual intercourse from conception, is to discover which of them comes closest to meeting your own needs. Safety and reliability are major concerns, of course, but many other factors come into play in making a decision about birth control. These include your health status, your lifestyle, and your attitudes about sex.

Natural methods of contraception that have been practiced for centuries include *coitus interruptus,* or withdrawal of the penis from the vagina before ejaculation; breast feeding; and natural family planning (formerly called the *rhythm method*), which requires abstinence on the days when a woman is believed to be fertile. All of these methods are unreliable. Modern methods are based on three general approaches: (1) hormones that interrupt ovulation, (2) long-term mechanical interruption of fertilization or implantation, and (3) barriers that keep semen from moving into a female's reproductive tract. Practical applications of these approaches are illustrated in the figure on page 610. The more you know about how the contraceptives work, the better equipped you will be to decide among them.

Hormones

- **Oral Contraceptives.** Oral contraceptives ("the Pill") are currently the most popular reversible contraceptive in the United States, used by almost 30% of all women. They usually consist of a synthetic estrogen combined with a synthetic progesterone in the form of pills that are taken daily for 21 days after the last day of a menstrual period. This procedure causes an immediate increase in blood levels of ovarian steroids (from the pill), which is maintained for the normal duration of a monthly cycle. As a result of negative feedback inhibition of gonadotropin secretion, ovulation never occurs.

  Since the contraceptive pills contain ovarian steroid hormones, the uterine lining proliferates and becomes secretory, just as it does in a normal cycle. In order to prevent an abnormal growth of the lining, women stop taking the pills after 3 weeks (placebo pills are taken during the fourth week). This causes estrogen and progesterone levels to fall, and menstruation follows.

  The side effects of earlier versions of the birth-control pill have been reduced by decreasing the content of estrogen and by using newer generations of progestogens (analogues of progesterone). The newer pills are very effective (with only about a 2% failure rate) and have a number of beneficial side effects, including a reduced risk of endometrial and ovarian cancer, a reduced risk of heart disease, and a reduction in osteoporosis. However, the risk for breast cancer, and possibly cervical cancer, may be increased with oral contraceptives.
- **Norplant.** The Norplant implant, introduced in the United States in 1990, consists of six flexible capsules filled with a progestin, each about the size of a matchstick. The capsules are usually implanted in the hypodermis of the inner arm, below the armpit, and offer protection for 5 years. Although their failure rate is less than 1%, up to 7% of women discontinue use because of persistent bleeding problems. Removal of the implant may be difficult, giving Norplant recent negative publicity.
- **Depo-Provera.** An injectable progesterone contraceptive, Depo-Provera has been in use worldwide for over 30 years, but it was not approved for the U. S. market until 1992. A few studies had shown a relationship to cancer in animals, but the millions of women using the drug had no such problem; in fact, they actually showed a decrease in endometrial cancer as a result of taking the drug. Depo-Provera suppresses ovulation for 3 months after it is injected. It has an effectiveness rate of about 99% and carries few side effects, although a small number of women report irregular bleeding, amenorrhea, and weight gain.

- **The "Morning-After Pill" and RU486.** The theory behind the *morning-after pill* is that extremely high doses of estrogen over a period of days will prevent implantation of a fertilized ovum. In the United States, the morning-after pill is usually reserved as an emergency measure after rape or contraceptive failure. The side effects of its high estrogen content can be severe, including constant vomiting and water retention.

  Another "emergency contraceptive" is *mifepristone* (*RU486*). Developed in France, this so-called abortion pill causes the loss of an implanted embryo by blocking the progesterone receptors of the cells of the uterine lining. If progesterone is prohibited from acting, the lining sloughs off, carrying the embryo with it. When taken in conjunction with a small amount of a prostaglandin to stimulate uterine contraction, about 96% of the women undergoing the procedure have a successful abortion. RU486 can be taken up to 5 weeks after conception. The drug's side effects include cramps and nausea. About one in a thousand women experiences bleeding severe enough to require a transfusion. Vocal pro-life forces in the United States have taken a strong stand against RU486, but many people believe that it will be approved for use soon.

Long-Term Mechanical Interruption

- **IUDs.** An *i*ntra*u*terine *d*evice is a small plastic or metal object, often molded into a loop, spiral, or T shape. It is inserted into the uterus by a physician, where it can remain in place for at least a year. Only about 3% of the women who use IUDs become pregnant; in most cases the device has an effect like that of the morning-after pill—that is, it causes the expulsion of the fertilized ovum from the uterus. An IUD can present serious problems, however. As a foreign body in the uterus, it creates irritation and inflammation and is associated with an increased incidence of pelvic inflammatory disease and subsequent infertility. Also, perforation of the uterine wall by an IUD (although this rarely occurs) is a life-threatening situation.
- **Sterilization.** *Tubal ligation* and *vasectomy* (cutting or cauterizing the uterine tubes or ductus deferentia, respectively) are surgical procedures that can be performed on an outpatient basis. They have failure rates of less than 1% and in no way diminish the sex drive. Patients are advised to consider these procedures irreversible, however. Although a successful reconnection can sometimes be achieved by microsurgical techniques, fertility is usually reduced substantially.

Barrier Methods

- **Spermicides, Diaphragms, and Cervical Caps.** *Spermicides* are made of two basic components: a spermicidal (sperm-killing) chemical and a harmless bulky base. The base is heavy enough to block the cervix, so even if some sperm cells aren't killed by the chemical, they are prevented from entering the cervical canal. Foams are the most effective spermicides—they spread quickly and coat the cervix evenly. Creams and jellies are more likely to fail because they don't spread as evenly throughout the cervix. *Sponges*—polyurethane discs soaked in spermicide—were once fairly popular, but they are no longer on the market. Spermicides are not highly effective when used alone; they are significantly more effective when used in conjunction with diaphragms, cervical caps, and condoms.

  A *diaphragm* is a rubber or plastic dome with a flexible rim that fits over the cervix. It must be fitted by a physician and should be used with a spermicidal jelly or cream for maximum protection. A diaphragm should be left in place for at least 6 hours after sexual relations. *Cervical caps,* smaller versions of the diaphragm, fit over the very end of the cervix. When used together with spermicides, they are as effective as full-sized diaphragms.
- **Condoms.** A condom is a thin sheath made of rubber or animal membrane that is rolled onto the erect penis to receive the ejaculate, thus preventing spermatozoa from entering the vagina. Used with a spermicide, a condom is about 90% effective. An added benefit is the protection it affords against AIDS and other sexually transmitted diseases. A female version is also available; it fits into the vagina and may also offer some protection against STDs.

continued

A number of other methods of birth control are currently under investigation. These include reversible vaccines targeted against specific hormones involved in conception, such as hCG (human chorionic gonadotropin), LH, and FSH. Vaccines that target proteins on the surface of the sperm cell, or on the zona pellucida of the ovum, are also being studied. In addition, researchers are exploring the efficacy of agents that will inhibit sperm production in males, or render sperm cells incapable of fertilizing an ovum. At present, however, the only completely safe, 100% effective method continues to be *abstinence*—refraining from sexual intercourse. This form of birth control, particularly appropriate for many young people, warrants careful consideration.

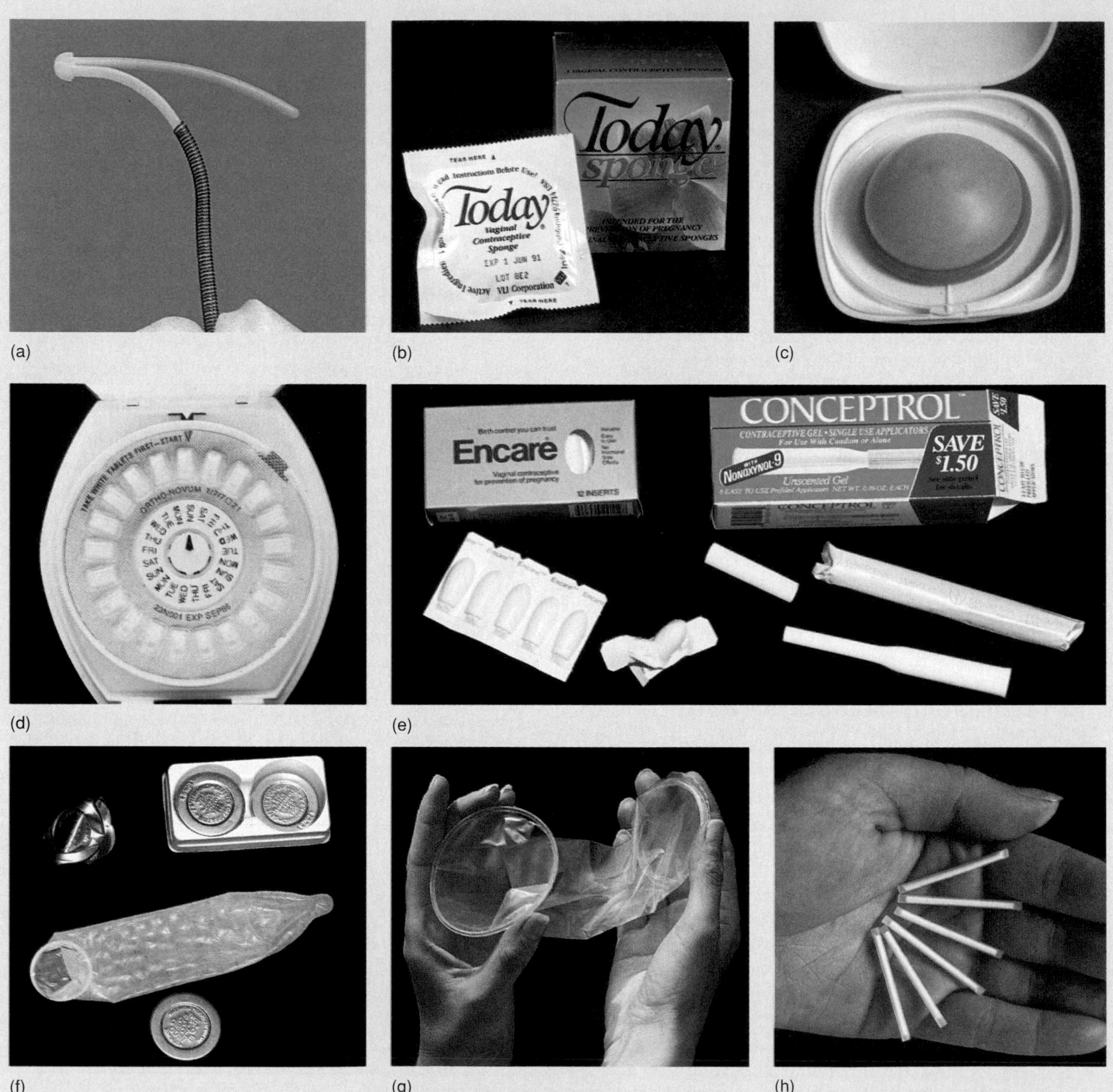

(a) (b) (c) (d) (e) (f) (g) (h)

Various types of birth-control devices. (*a*) IUD, (*b*) contraceptive sponge, (*c*) diaphragm, (*d*) birth-control pills, (*e*) vaginal spermicide, (*f*) condom, (*g*) female condom, and (*h*) Norplant.

Clinical Considerations

Diseases and Dysfunctions of the Female Reproductive System

A comprehensive discussion of the numerous clinical aspects of the female reproductive system is beyond the scope of this text. *Gynecology* is a medical specialty devoted entirely to the administration of health care to women, especially the diagnosis and treatment of disorders affecting the female reproductive organs. *Obstetrics* is the specialty dealing with pregnancy and childbirth. Frequently, a physician will specialize in both obstetrics and gynecology (OBGYN). In the following paragraphs, we will briefly discuss two important diagnostic procedures and only a few of the more important clinical conditions.

One diagnostic procedure that should be routinely performed by a woman is **breast self-examination (BSE).** The importance of a BSE is not to prevent diseases of the breast but to detect any problems before they become serious. A BSE should be performed monthly, 1 week after the cessation of menstruation so that the breasts will not be swollen or especially tender.

Another important diagnostic procedure is a **Papanicolaou (Pap) smear.** The Pap smear permits microscopic examination of cells covering the tip of the cervix. Samples of cells are obtained by gently scraping the surface of the cervix with a specially designed wooden spatula. Women should have Pap smears on a regular basis for the early detection of cervical cancer.

Infertility, or the inability to conceive, is a clinical problem that can involve either the male or female reproductive system, or both. Generally, when a male is infertile, it is because of inadequate sperm counts. Female infertility is frequently caused by an obstruction of the uterine tubes or by abnormal ovulation. **Sterility** is also the inability to conceive but generally refers to surgical alterations of the reproductive tract. *Tubal ligation* (fig. 23.19) is a common birth-control technique that produces female sterility (see the preceding Synoptic Spotlight). Likewise, surgical removal of the uterus in a procedure called a *hysterectomy* (*his"tĕ-rek'tŏ-me*) produces female sterility.

Endometriosis is a condition characterized by the presence of endometrial tissues at sites other than the inner lining of the uterus (called *ectopic sites*). Ectopic endometrial cells are often found on the ovaries, on the outer layer of the uterus, on the abdominal wall, and on the urinary bladder. Although it is not certain how endometrial cells become established outside the uterus, it is speculated that some discharged endometrial tissue might be flushed backward from the uterus and through the uterine tubes during menstruation. Women with endometriosis will bleed internally with each menstrual period because the ectopic endometrial cells respond to ovarian hormones, just as the endometrium does. The most common symptoms of endometriosis are severe *dysmenorrhea* (menstrual pain) and a feeling of fullness during each menstrual period. Endometriosis can cause infertility. It is most often treated by suppressing the endometrial tissues with oral contraceptive pills or by surgery. A hysterectomy, and possibly an *oophorectomy* (removal of the ovaries), may be necessary in extreme cases.

Ovarian neoplasms are usually nonmalignant, fluid-filled cysts. These tumors often can be palpated during a gynecological examination. If they exceed 4 cm in diameter, they may have to be removed surgically. They are generally removed as a precaution because it is impossible to determine by palpation whether the mass is malignant or benign.

Uterine neoplasms are an extremely common problem of the female reproductive tract. These neoplasms, most of which are benign, include cysts, polyps, and smooth muscle tumors. Any of them may provoke irregular menstrual periods. If they grow to be massive, infertility may result.

Cancer of the uterus is the most common malignancy of the female reproductive tract. The most common site of uterine cancer is the cervix. Cervical cancer, which is second only to breast cancer in frequency of occurrence, generally strikes women between the ages of 30 and 50. If detected early enough, through regular Pap smears, the disease can be cured before it metastasizes. The treatment of cervical cancer depends on the stage of the malignancy and the age and general health of the woman. In the case of women for whom future fertility is not an issue, a hysterectomy is usually performed.

Pap smear: from George N. Papanicolaou, American anatomist and physician, 1883–1962

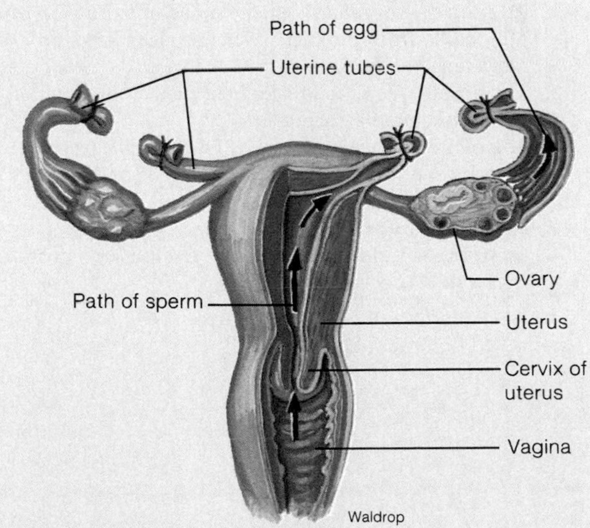

FIGURE 23.19
Tubal ligation involves the removal of a portion of each uterine tube. In actual practice, cautery, clips, or rings are used for tubal closure more often than ligation (tying) with suture thread.

SUMMARY

23.1 Anatomy of the Female Reproductive System

1. The female gonads, the ovaries, produce the female gametes—the ova, or oocytes.
2. The secondary sex organs of the female include the vagina, uterus, uterine tubes, external genitalia (vulva), and mammary glands.
3. Fingerlike processes called fimbriae extend from the uterine tubes to the surface of the ovaries.
4. The uterus consists of a fundus, body, and cervix. The cervical canal opens into the vagina at the uterine ostium.
5. The inner epithelium of the uterus is the endometrium.
6. Female secondary sex characteristics develop in response to increased levels of ovarian hormones.

23.2 Ovaries and Ovarian Follicles

1. Ova, or egg cells, are contained within hollow structures called follicles.
2. Primary oocytes have begun meiosis but meiotic division is arrested at prophase I.
3. Unstimulated primary oocytes are contained within primary follicles.
4. In response to FSH stimulation, some primary follicles mature each month to become secondary follicles. Secondary follicles contain granulosa cells and vesicles.
5. Continued stimulation results in a graafian follicle that contains an antrum, or cavity.
6. A primary oocyte may be stimulated to become a secondary oocyte, but meiosis is arrested at metaphase II; the other product of the first meiotic division is a tiny, nonfunctional polar body.

23.3 Ovulation

1. Rupture of a graafian follicle results in expulsion of the oocyte from the ovary. This process is called ovulation.
2. The empty ovulated follicle becomes a new ovarian structure called a corpus luteum.
3. The ovaries are stimulated by FSH and LH from the anterior pituitary; these gonadotropins are in turn controlled by a hypothalamic-releasing hormone known as GnRH.
4. Ovarian steroid hormones exert feedback control of FSH and LH secretion.

23.4 Menstrual Cycle: Pituitary Gland and Ovary

1. During the follicular phase, follicles grow under FSH stimulation to become secondary follicles. By day 13, one graafian follicle is usually produced.
2. The growing follicles secrete an increasing amount of estradiol; this stimulates an LH surge from the pituitary just before ovulation.
3. The LH surge is a positive feedback effect of estradiol that stimulates ovulation.
4. The corpus luteum secretes progesterone as well as estradiol. These hormones inhibit FSH and LH secretion during the luteal phase of the cycle.

23.5 Menstrual Cycle: Endometrium

1. During the follicular phase, rising levels of estradiol stimulate a thickening of the endometrium.
2. The actions of progesterone and estradiol during the luteal phase cause the endometrium to enter its secretory phase.
3. At the end of the cycle, the corpus luteum dies; the consequent fall in estradiol and progesterone stimulates menstruation.

23.6 Mammary Glands

1. Each mammary gland is subdivided into lobules that contain the glandular mammary alveoli.
2. High levels of estrogen during pregnancy cause growth of the mammary glands but block the stimulatory effects of prolactin.
3. After the birth of the baby, prolactin from the anterior pituitary stimulates milk production.
4. The stimulus of the baby suckling causes reflex secretion of oxytocin from the posterior pituitary; this produces the milk-ejection reflex.

Review Activities

Objective Questions

1. Secondary sex characteristics are (*module 23.1*)
a. the only body features that distinguish the sexes.
b. not essential for the reproductive process.
c. those structures that develop only after puberty.
d. the only reproductive structures that are affected by hormones.

2. Which of the following is the correct sequence of structures through which an ovulated egg would pass as it moves into the uterine cavity? (*module 23.1*)
a. infundibulum, uterine tube, fimbriae, uterine cavity
b. fimbriae, infundibulum, uterine tube, uterine cavity
c. uterine tube, fimbriae, infundibulum, uterine cavity
d. fimbriae, uterine tube, infundibulum, uterine cavity

3. The cervix is a portion of (*module 23.1*)
a. the vulva.
b. the vagina.
c. the uterus.
d. the uterine tubes.

4. Which of the following statements about oogenesis is *true?* (*module 23.2*)
a. Oogonia form continuously in postnatal life.
b. Primary oocytes are haploid.
c. Meiosis is completed prior to ovulation.
d. A secondary oocyte is released at ovulation.

5. Following ovulation, the corpus luteum develops under the influence of (*module 23.3*)
a. progesterone.
b. FSH.
c. LH.
d. estradiol.

6. The follicular phase is characterized by (*module 23.4*)
a. high levels of estrogen and progesterone.
b. low levels of estrogen and progesterone.
c. increasing estrogen and little or no progesterone.
d. high levels of LH.

7. The luteal phase is characterized by (*module 23.4*)
a. high levels of estrogen and progesterone.
b. low levels of estrogen and progesterone.
c. increasing estrogen and little or no progesterone.
d. high levels of LH.

8. The secretory phase of the endometrium corresponds to which of the following ovarian phases? (*module 23.5*)
a. follicular phase
b. ovulation
c. luteal phase
d. menstrual phase

9. Contractions of the mammary ducts are stimulated by (*module 23.6*)
a. prolactin.
b. oxytocin.
c. estrogen.
d. progesterone.

10. Uterine contractions are stimulated by (*module 23.6*)
a. oxytocin.
b. vasopressin.
c. prolactin.
d. luteinizing hormone.

Essay Questions

1. Describe the structure and position of the uterus and explain the significance of the strata functionale and basale of the endometrium. (*module 23.1*)
2. Describe the process of oogenesis and explain why cleavage (cell division) during meiosis is unequal. (*module 23.2*)
3. Explain the hormonal interactions that control ovulation and make it occur at the proper time. (*module 23.3*)
4. Describe the events of the ovarian cycle. What role do FSH and LH play and how is their secretion regulated? (*module 23.4*)
5. Describe the structural changes that occur in the endometrium during the cycles of ovulation and menstruation, and explain how these changes are controlled by hormones. (*module 23.5*)
6. Describe the structure and position of the mammary glands in the breasts and explain the hormonal mechanisms that control lactation. (*module 23.6*)

Labeling Exercise

Label the structures indicated on the figure to the right.

1. ______________________
2. ______________________
3. ______________________
4. ______________________
5. ______________________
6. ______________________
7. ______________________

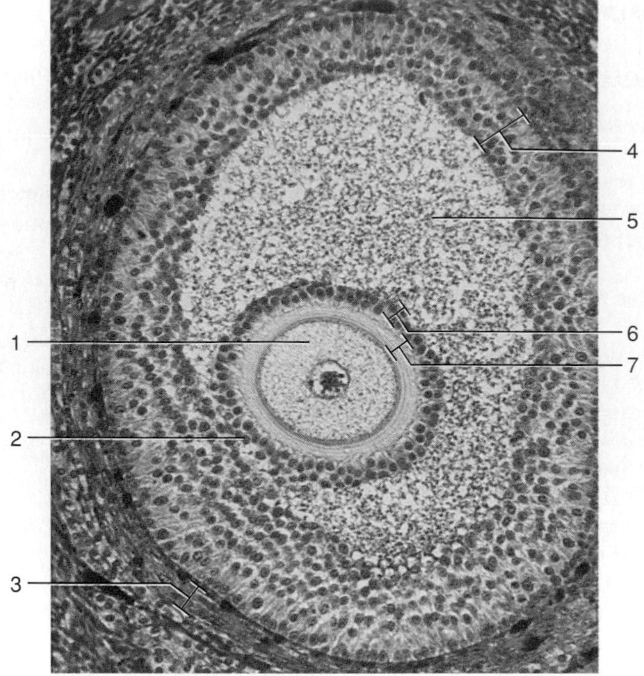

Critical Thinking Questions

1. One of the changes that occurs during the menstrual cycle is a tremendous increase in blood flow to the uterus. Explain how this might represent a biological advantage.
2. One way to describe the action of contraceptive pills is that they "trick the brain into thinking you're pregnant." In what way does the Pill mimic pregnancy?
3. Adult women who routinely engage in strenuous activity may experience amenorrhea. Should they seek medical advice? Explain and justify your response.
4. For a female to produce testosterone would appear to be a gender contradiction. How can you explain this statement?
5. When a woman is nursing, why is it that milk letdown occurs in both breasts, not just the one being suckled?

CHAPTER

24

"Gregor Mendel discovered the universal laws of inheritance by studying particular traits in pea plants that are passed from one generation to the next." *page 634*

Prenatal Development and Inheritance

Chapter Outline

Terms to Remember

fertilization, 616
zygote (*zi' gōt*), 618
morphogenesis (*mor" fo-jen' ĕ-sis*), 618
cleavage, 618
morula (*mor' yŭ-lă*), 618
blastocyst (*blas' tŏ-sist*), 618
implantation, 618
ectoderm (*ek' tŏ-derm*), 620
endoderm (*en' dŏ-derm*), 620
mesoderm (*mes' ŏ-derm*), 620
embryo, 621
conceptus, 621
amnion (*am' ne-on*), 621
allantois (*ă-lan' to-is*), 622
chorion (*kor' e-on*), 622
placenta (*plă-sen' tă*), 622
umbilical (*um-bĭ' lĭ-kal*) cord, 624
fetus, 625
parturition (*par" tyoo-rish' un*), 629
gene, 632
allele (*ă-lēl'*), 632
genotype, 632
phenotype, 632

Learning Objectives

After studying this chapter, you should be able to

- describe the changes that occur in the sperm cell and ovum prior to, during, and immediately following fertilization.
- describe the events of the pre-embryonic period that result in the formation of a blastocyst.
- give the names of the primary germ layers and list the principal structures that form from each layer.
- define *embryo* and describe the major events of the embryonic period of development.
- describe the structure and function of each of the extraembryonic membranes.
- define *fetus* and describe the major events of the fetal period of development.
- describe some of the techniques that are used to examine a fetus or monitor fetal activity.
- explain how the parturition (delivery) date is determined.
- explain the hormonal action that controls labor and parturition.
- describe the three stages of labor.
- explain how probability is involved in predicting inheritance and describe the mechanism of sex-linked inheritance.

24.1 Fertilization

The fertilization of an egg cell (oocyte) by a sperm cell stimulates the egg cell to complete its second meiotic division. The product of fertilization is known as the zygote.

Fertilization refers to the penetration of a secondary oocyte (see chapter 23, module 23.2) by a sperm cell, with the subsequent union of their genetic material. It is this event that determines a person's biological inheritance. Fertilization cannot occur, however, unless certain conditions are met. First of all, a secondary oocyte must be present in the uterine tube, waiting to be fertilized. It can wait for at most 24 hours before it becomes incapable of undergoing fertilization. Second, very large quantities of sperm cells must be available to encounter this oocyte. Although a recent study has shown that sperm cells can remain viable 5 days after ejaculation, this still leaves a "window of fertility" of only 6 days each month—the day of ovulation and the 5 days leading up to it.

As described in chapter 23, a woman usually ovulates one egg cell (ovum) a month, totaling about 400 during her reproductive years. Each ovulation releases an immature secondary oocyte arrested at metaphase of the second meiotic division (fig. 24.1*a*). As the oocyte enters the uterine tube, it is surrounded by a thin layer of protein and polysaccharides, called the *zona pellucida,* and a layer of granulosa cells, called the *corona radiata.* These layers form a protective shield around the oocyte.

zona pellucida: L. *zone,* girdle; *pellis,* skin
corona radiata: Gk. *korone,* crown; *radiata,* radiate

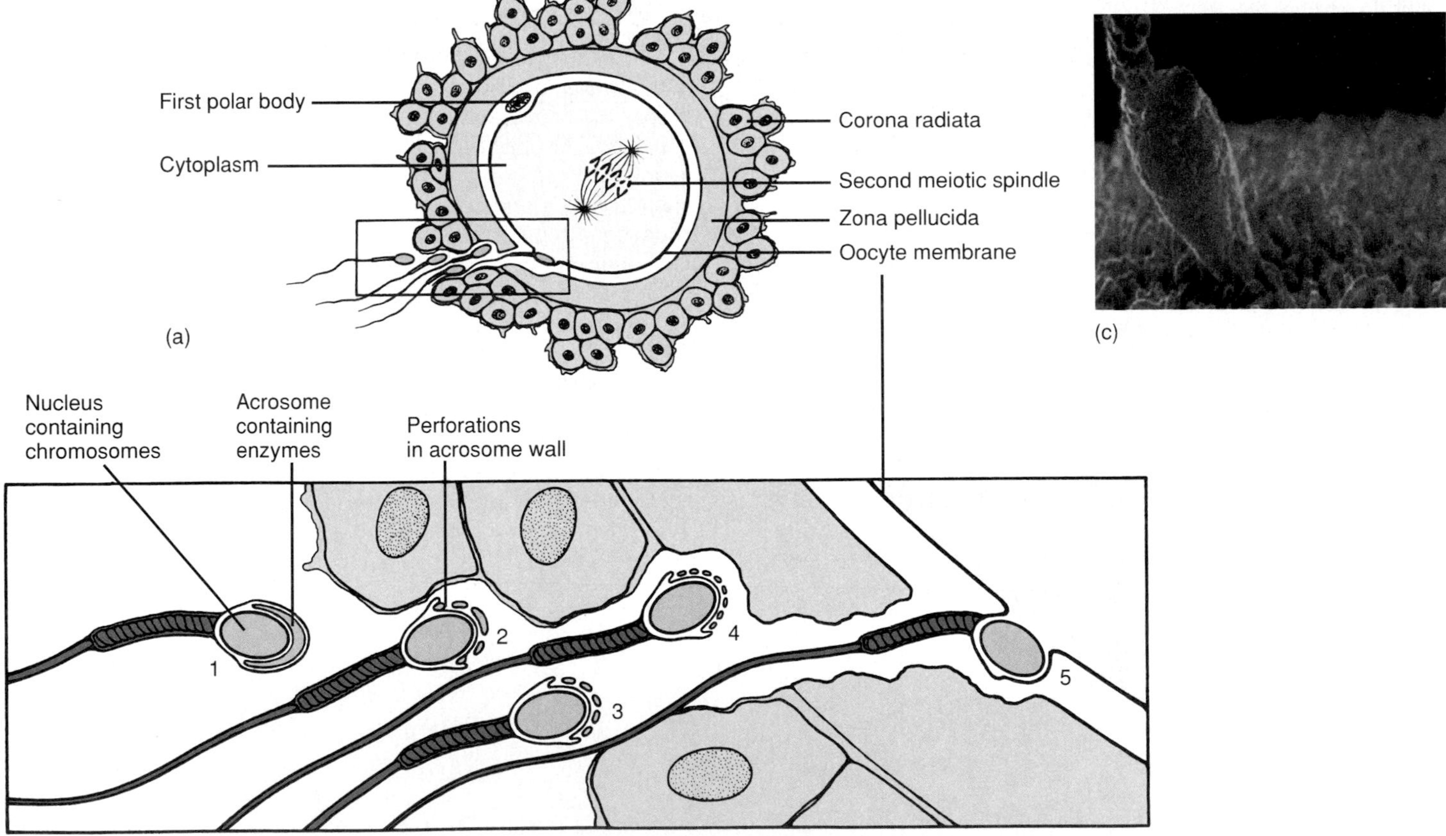

FIGURE 24.1
The process of fertilization. (*a, b*) As the head of the sperm cell encounters the corona radiata of the oocyte (2), digestive enzymes are released from the acrosome (3, 4), clearing a path to the oocyte membrane. When the membrane of the sperm cell contacts the oocyte membrane (5), the membranes become continuous, and the nucleus and other contents of the sperm cell move into the cytoplasm of the oocyte. (*c*) A scanning electron micrograph of a sperm cell bound to the surface of an oocyte.

During sexual intercourse, a male ejaculates between 100 million and 500 million sperm cells into the female's vagina. This tremendous number is needed because of the high fatality rate—only about 100 sperm cells survive to encounter the oocyte in the uterine tube. In addition to deformed sperm cells—up to 20% in the average fertile male—another quarter will perish as soon as they reach the vagina's acid environment. Others fall victim to the woman's immune cells, which try to defend against the invaders. Still others mistake vaginal cells for egg cells and vainly try to unite with them. And then there are some that become disoriented and swim in the wrong direction. Even if they manage to reach the cervix, many sperm cells will get stuck there, and they too drop out of the race.

The final challenge of a sperm cell, if it at last encounters an oocyte, is to break through the corona radiata and zona pellucida that make up the egg's "shell." To do this, each sperm cell is equipped with a cap of enzymes called an *acrosome* (fig. 24.1*b*). The acrosome contains protein-digesting enzymes and an enzyme called *hyaluronidase* (*hi"ă-loo-ron'ĭ-dās*). Hyaluronidase digests hyaluronic acid, an important constituent of connective tissue. When a sperm cell meets an oocyte in the uterine tube, an *acrosomal reaction*—the release of the acrosome's digestive enzymes—allows the sperm cell to penetrate. A sperm cell that comes along relatively late—after many others have undergone acrosomal reactions to expose the oocyte membrane—is more likely to be *the* sperm cell that finally achieves penetration.

Experiments confirm that freshly ejaculated sperm cells must remain in the female reproductive tract for at least 7 hours before they can fertilize an oocyte. Their membranes must become fragile enough to permit the release of the acrosomal enzymes—a process called capacitation. *During in vitro fertilization (see "Clinical Considerations"), capacitation is induced artificially by treating the ejaculate with a solution of gamma globulin, serum, follicular fluid, dextran, serum dialysate, and adrenal gland extract.*

As soon as the "chosen" sperm cell penetrates the zona pellucida, a chemical change in the zona prevents other sperm cells from attaching to it. Therefore, only one sperm cell is allowed to fertilize an oocyte. With the entry of the single sperm cell through its cell membrane, the oocyte is stimulated to complete its second meiotic division (fig. 24.2). Like the first meiotic division, the second produces one cell that contains all of the cytoplasm and one *polar body*. The healthy cell is the mature ovum, and the second polar body, like the first, ultimately fragments and disintegrates.

At fertilization, the entire sperm cell enters the cytoplasm of the much larger ovum. Within 12 hours, the nuclear membrane in the ovum disappears, and the *haploid number* of chromosomes (23) in the ovum is joined by the

acrosome: Gk. *akron,* extremity; *soma,* body
capacitation: L. *capacitas,* capable of
haploid: Gk. *haplous,* single; L. *ploideus,* multiple in form

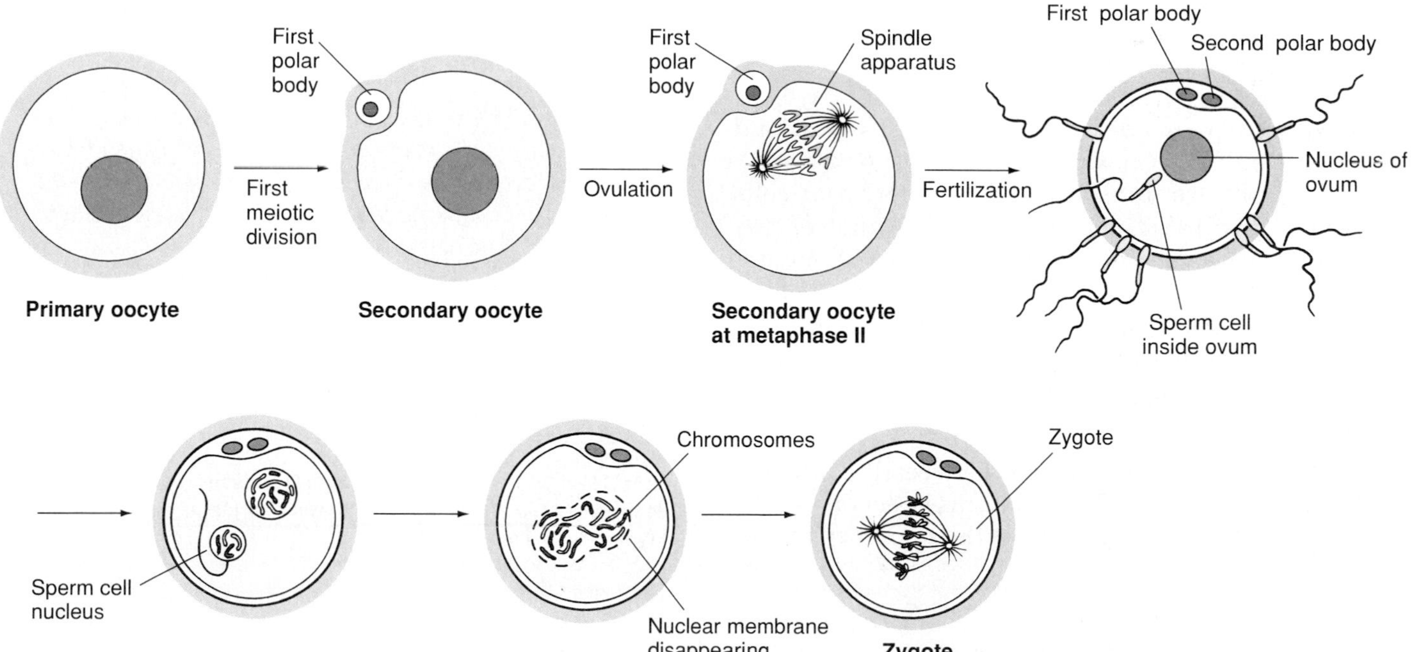

FIGURE 24.2
A secondary oocyte, arrested at metaphase II of meiosis, is released at ovulation. If this cell is fertilized, it will become a mature ovum, complete its second meiotic division, and produce a second polar body. The fertilized ovum is known as a zygote.

haploid number of chromosomes from the sperm cell. A fertilized egg, or **zygote** (*zi'gōt*), containing the *diploid number* of chromosomes (46), is thus formed.

Within hours after conception, the structure of the body begins to form from this single fertilized egg, culminating some 38 weeks later with the birth of a baby. The transformation involved in the growth and differentiation of cells and tissues is known as **morphogenesis** (*mor"fo-jen'ĕ-sis*), and it is through this awesome process that the organs and systems of the body are established in a functional relationship. Moreover, there are sensitive stages of morphogenesis for each organ and system, during which genetic or environmental factors may affect the normal development of the baby.

Prenatal development can be divided into a *pre-embryonic period,* which is initiated by the fertilization of an egg cell; an *embryonic period,* during which the body's organ systems are formed; and a *fetal period,* which culminates in parturition (childbirth).

zygote: Gk. *zygotos,* yolked, joined
diploid: Gk. *diplous,* double; L. *ploideus,* multiple in form
morphogenesis: Gk. *morphe,* form; *genesis,* beginning

24.2 Pre-Embryonic Period

The events of the 2-week pre-embryonic period include fertilization, transportation of the zygote through the uterine tube, mitotic divisions, implantation, and the formation of the primordial embryonic tissue.

Cleavage and Formation of the Blastocyst

Within 30 hours following fertilization, the zygote undergoes a mitotic division called **cleavage.** This first division results in the formation of two identical daughter cells called *blastomeres.* Additional cleavages occur as the structure passes down the uterine tube and enters the uterus on about the third day (fig. 24.3). It is now composed of a ball of cells called a **morula** (*mor'yŭ-lă*). The morula floats freely in the uterine cavity for about 3 days. During this time, the center of the morula fills with fluid passing in from the uterine cavity. As the fluid-filled space develops within the morula, two distinct groups of cells form, and the structure becomes known as a **blastocyst** (*blas'tŏ-sist*). The blastocyst is composed of an outer layer of cells known as the *trophoblast,* and an inner aggregation of cells called the *embryoblast.* With further development, the trophoblast differentiates into a structure called the *chorion,* which will later become a portion of the placenta. The embryoblast will eventually become the embryo.

Implantation

The process of implantation begins between the fifth and seventh day following fertilization. This is the process by which the blastocyst burrows into the endometrium of the uterine wall. Implantation is made possible by the secretion of *proteolytic enzymes* by the trophoblast, which digest a portion of the endometrium. The blastocyst sinks into the depression, and endometrial cells move back to cover the defect in the wall. Implantation is usually completed at the end of 2 weeks, as the primary germ layers develop (fig. 24.4).

The blastocyst saves itself from being aborted by secreting the hormone **human chorionic gonadotropin (hCG)** that indirectly prevents menstruation. This hormone is identical to LH in its effects, and therefore is able to maintain the corpus luteum past the time when it would otherwise regress. The secretion of estrogen and progesterone is maintained, and menstruation is normally prevented (fig. 24.5).

All pregnancy tests *assay for the presence of hCG in the blood or urine because this hormone is secreted only by the blastocyst. Since there is no other source of hCG, the presence of this hormone confirms a pregnancy. Modern pregnancy tests detect the presence of hCG by use of antibodies against hCG or by the use of cellular receptor proteins for hCG.*

morula: Gk. *morus,* mulberry
trophoblast: Gk. *trophe,* nourishment; *blastos,* germ
embryoblast: Gk. *embryon,* to be full, swell; *blastos,* germ
implantation: L. *im,* in; *planto,* to plant

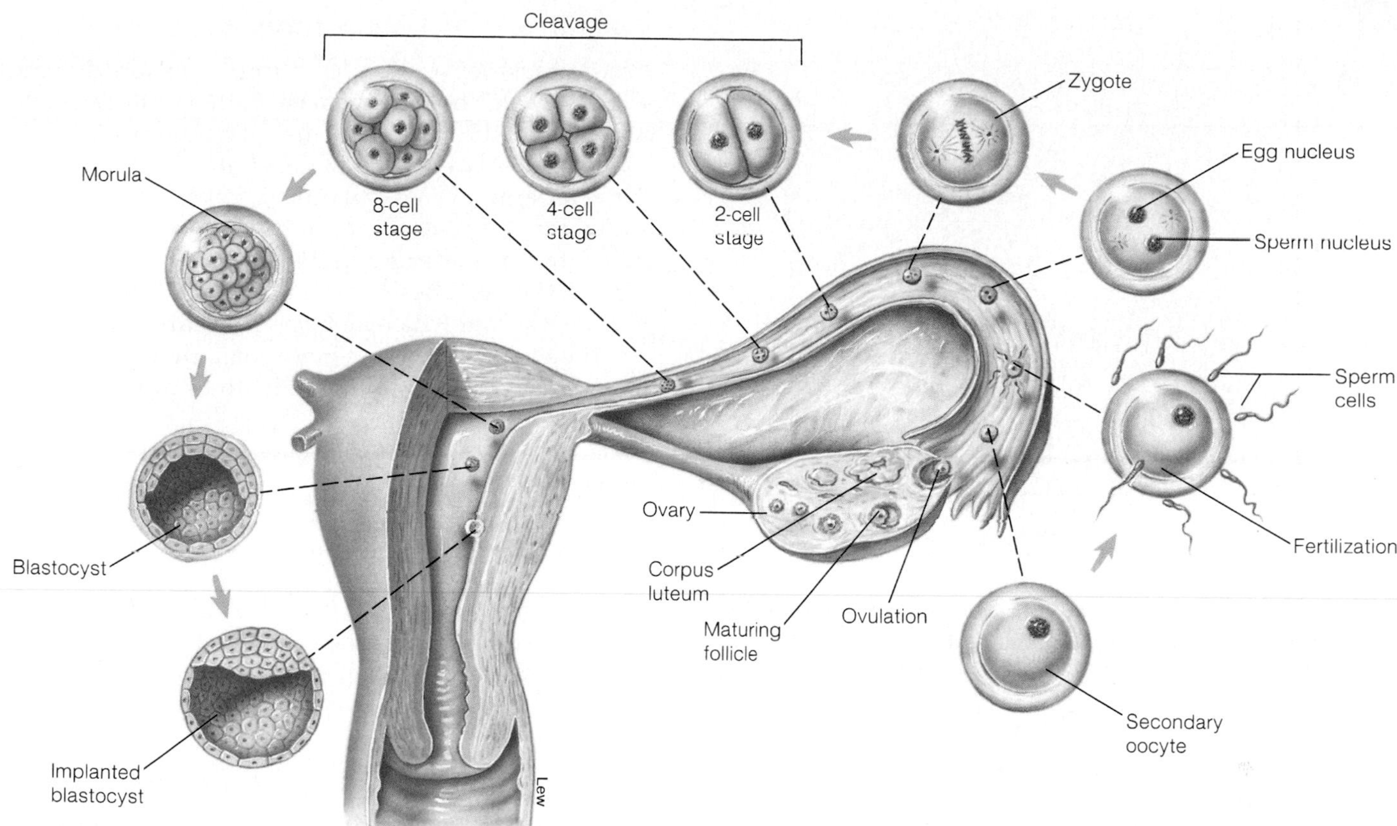

FIGURE 24.3
A diagram of the ovarian cycle, fertilization, and the events of the first week. Implantation of the blastocyst begins between the fifth and seventh day and is generally completed by the tenth day.

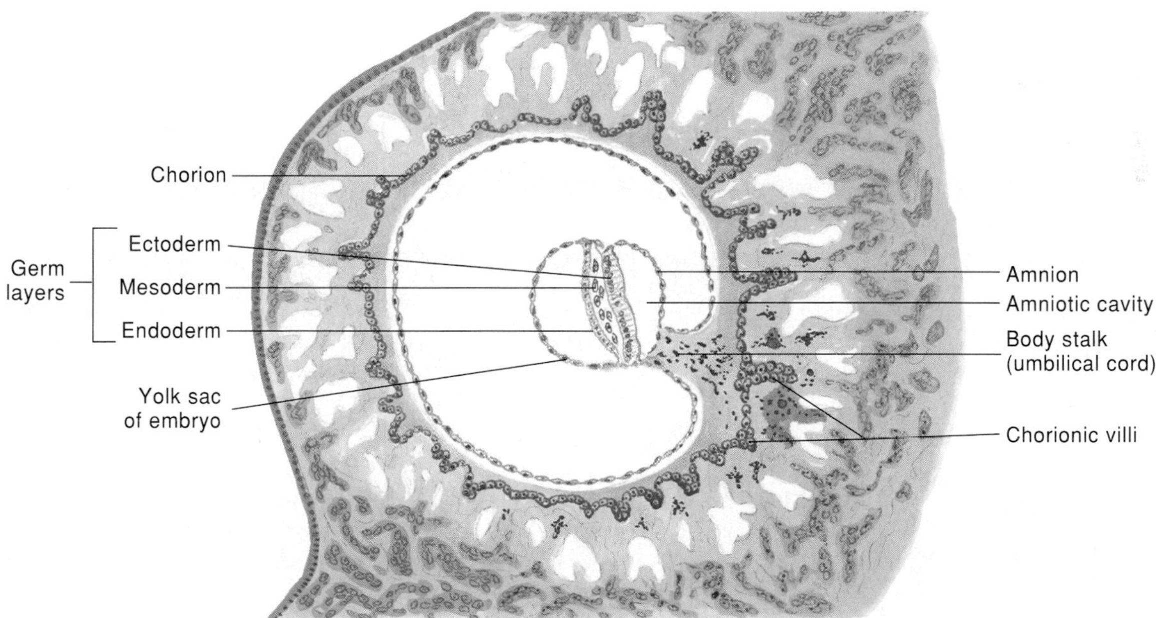

FIGURE 24.4
The completion of implantation occurs as the primary germ layers develop at the end of the second week.

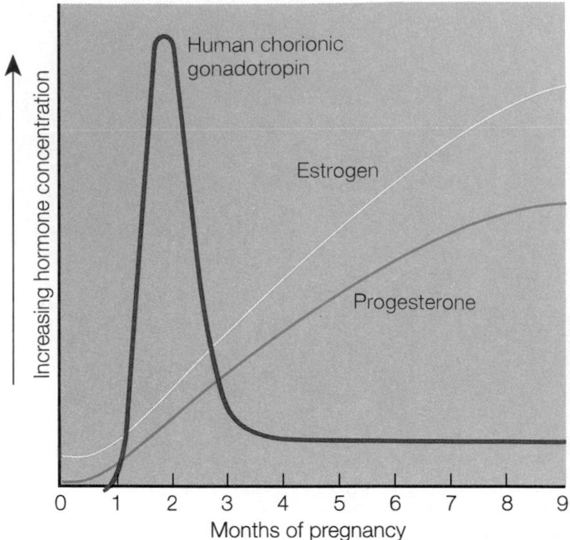

FIGURE 24.5
Human chorionic gonadotropin (hCG) is secreted by cells of the blastocyst during the first trimester of pregnancy. This pituitary-like hormone maintains the mother's corpus luteum for the first 5 1/2 weeks of pregnancy. The placenta then assumes the role of estrogen and progesterone production, and the corpus luteum degenerates.

Formation of Germ Layers

As the blastocyst completes implantation during the second week, the embryoblast flattens into a cluster of cells called the *embryonic disc.* The embryonic disc is composed of two layers of cells: an upper layer, called the **ectoderm** (*ek'tŏ-derm*), and a lower layer, called the **endoderm** (*en'dŏ-derm*). A short time later, a third layer called the **mesoderm** (*mes'ŏ-derm*) forms between the first two layers. These three layers constitute the *primary germ layers* from which all body structures form.

The events of the pre-embryonic period are summarized in table 24.1, and the derivatives of the three germ layers are listed in table 24.2.

ectoderm: Gk. *ecto,* outside; *derm,* skin
endoderm: Gk. *endo,* within; *derm,* skin
mesoderm: Gk. *meso,* middle; *derm,* skin

TABLE 24.1 Morphogenic Stages and Principal Events of Pre-embryonic Development

Stage	Time Period	Principal Events
Zygote	24 to 30 hours following ovulation	Egg is fertilized; zygote has 23 pairs of chromosomes (diploid) from haploid sperm and haploid egg, and is genetically unique
Cleavage	30 hours to third day	Mitotic divisions produce increased number of cells
Morula	Third to fourth day	Hollow ball-like structure forms, a single layer thick
Blastocyst	Fifth day to end of second week	Embryoblast and trophoblast form; implantation occurs; embryonic disc forms, followed by primary germ layers

TABLE 24.2 Derivatives of Germ Layers

Ectoderm	Mesoderm	Endoderm
Epidermis of skin and epidermal derivatives: hair, nails, glands of the skin; linings of oral, nasal, anal, and vaginal cavities	Muscle: smooth, cardiac, and skeletal	Epithelium of pharynx, auditory canal, tonsils, thyroid, parathyroid, thymus, larynx, trachea, lungs, GI tract, urinary bladder and urethra, and vagina
Nervous tissue; sense organs	Connective tissue: embryonic, connective tissue proper, cartilage, bone, blood	Liver and pancreas
Lens of eye; enamel of teeth	Dermis of skin; dentin of teeth	
Pituitary gland	Epithelium of blood vessels, lymphatic vessels, body cavities, and joint cavities	
Adrenal medulla	Internal reproductive organs	
	Kidneys and ureters	
	Adrenal cortex	

24.3 Embryonic Period

The events of the 6-week embryonic period include the differentiation of the germ layers into specific body organs and the formation of the placenta, the umbilical cord, and the extraembryonic membranes. Through these morphogenic events, the needs of the embryo are met.

During the embryonic period—from the beginning of the third week to the end of the eighth week—the developing organism is correctly called an **embryo.** It is at this stage that all of the body organs form, as well as the placenta, umbilical cord, and extraembryonic membranes. The term *conceptus* refers to the embryo, or to the fetus later on, and all of the extraembryonic structures—the products of conception.

Extraembryonic Membranes

At the same time that the internal organs of the embryo are being formed, a complex system of extraembryonic membranes is also developing (figs. 24.6 and 24.7). The **extraembryonic membranes** are the *amnion, yolk sac, allantois,* and *chorion.* These membranes are responsible for the protection, respiration, excretion, and nutrition of the embryo and subsequent fetus. At parturition, the baby and the extraembryonic membranes are expelled from the uterus as the *afterbirth.*

Amnion

The amnion (*am'ne-on*) is a thin membrane, derived from ectoderm and mesoderm. It loosely envelops the embryo, forming an **amniotic sac** that is filled with *amniotic fluid.* The buoyant amniotic fluid performs four functions for the embryo and subsequent fetus.

1. It ensures symmetrical structural development and growth.
2. It cushions and protects by absorbing jolts that the mother may receive.
3. It helps to maintain consistent pressure and temperature.
4. It permits freedom of fetal movement, which is important for musculoskeletal development and blood flow.

Amniotic fluid is formed initially as an isotonic fluid absorbed from the maternal blood in the endometrium. Later, the volume is increased and the concentration changed by urine excreted from the fetus into the amniotic sac. Amniotic fluid also contains cells that are sloughed off from the fetus, placenta, and amniotic sac. Since all of these cells are derived from the same fertilized egg, all have the same genetic composition. Many genetic abnormalities can be detected by aspirating this fluid and examining the cells obtained in a procedure called *amniocentesis.*

Amniocentesis *is usually performed during the fourteenth or fifteenth week of pregnancy, when the amniotic sac contains 175–225 ml of fluid. Genetic diseases such as* Down syndrome *(in which there are three instead of two number 21 chromosomes) can be detected by examining chromosomes; diseases such as* Tay–Sachs disease, *in which there is a defective enzyme involved in formation of myelin sheaths, can be detected by biochemical techniques.*

amniocentesis: Gk. *amnion,* lamb (fetal membrane); *kentesis,* puncture
Down syndrome: from John L. H. Down, English physician, 1828–96

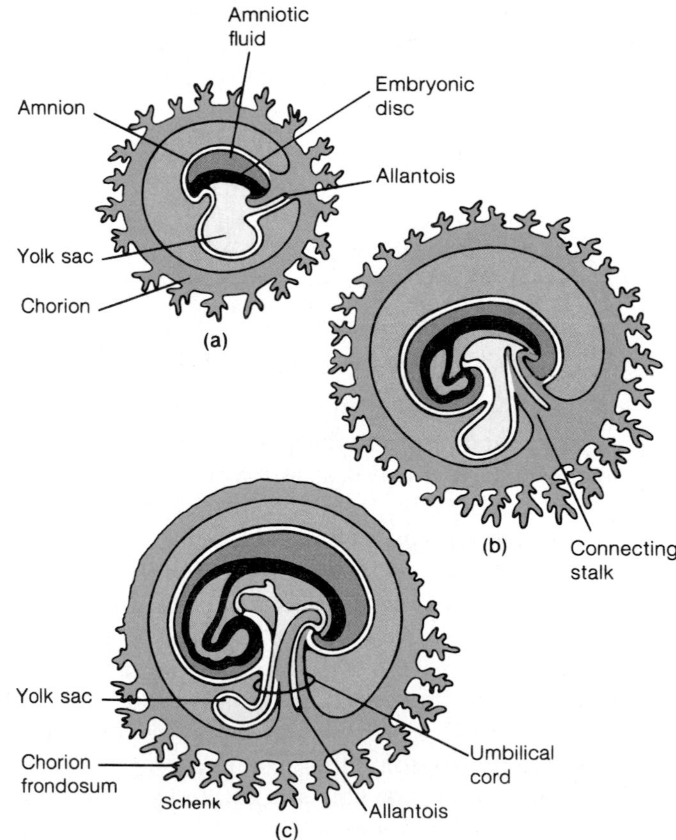

FIGURE 24.6
The formation of the extraembryonic membranes during a single week of embryonic development. (*a*) At 3 weeks, (*b*) at 3½ weeks, and (*c*) at 4 weeks.

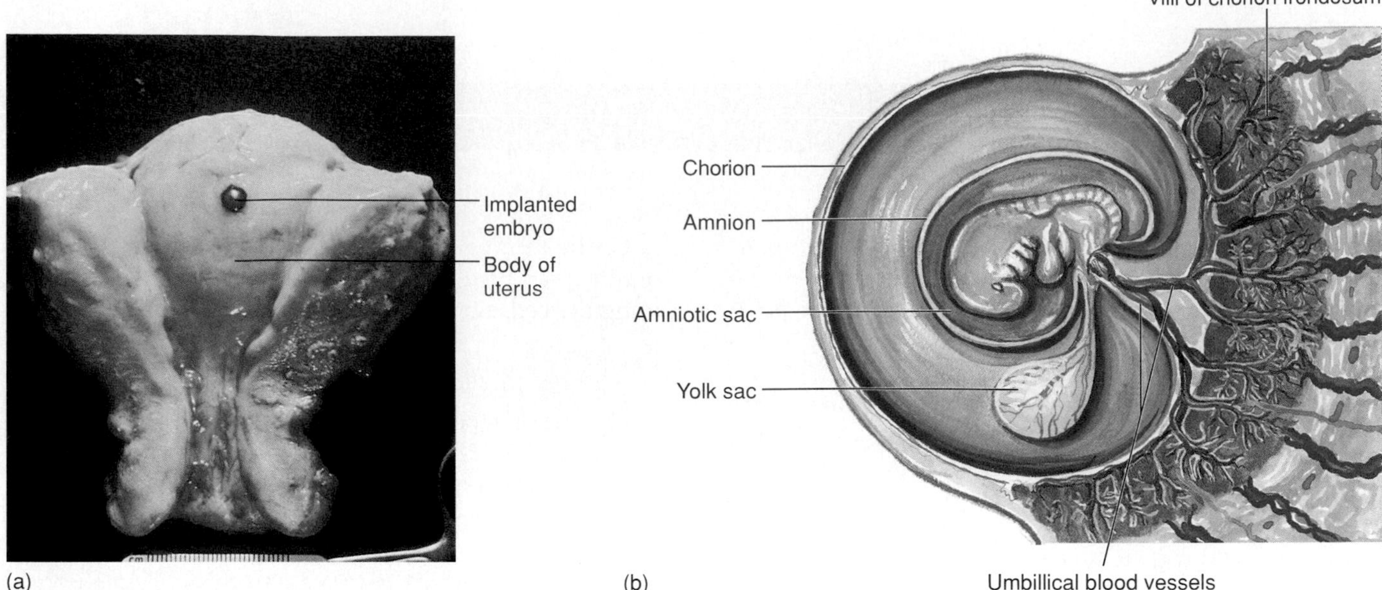

FIGURE 24.7
An implanted embryo at approximately 4 1/2 weeks. (*a*) The interior of a uterus showing the implantation site. (*b*) The developing embryo, extraembryonic membranes, and placenta.

Amniotic fluid is normally swallowed by the fetus and absorbed in the GI tract. Prior to delivery, the amnion is naturally or surgically ruptured, releasing the fluid—commonly described as breaking the "bag of waters."

Inside Information—As the fetus grows, the amount of amniotic fluid increases. It is also continually absorbed and renewed. For the near-term baby, almost 8 liters of fluid are completely replaced each day.

Yolk Sac

The yolk sac is established during the end of the second week as cells from the trophoblast form a thin *exocoelomic membrane.* Unlike the yolk sac of many vertebrates, the human yolk sac contains no nutritive yolk but is an essential structure during early embryonic development. Attached to the underside of the embryonic disc (see fig. 24.6), it produces blood for the embryo until the liver forms during the sixth week. A portion of the yolk sac is also involved in the formation of the primitive gut. In addition, primordial germ cells form in the wall of the yolk sac. During the fourth week, they migrate to the developing gonads, where they become the primitive germ cells. Following the sixth week, the yolk sac gradually shrinks and serves no additional function.

Allantois

The allantois (*ă-lan'to-is*) forms during the third week as a small outpouching of the endoderm near the base of the yolk sac (see fig. 24.6). It remains small but is involved in the formation of blood cells and gives rise to the fetal umbilical arteries and vein. It also contributes to the development of the urinary bladder.

Chorion

The chorion (*kor'e-on*) is the outermost extraembryonic membrane. It contributes to the formation of the placenta as small fingerlike extensions, called *villi,* penetrate deeply into the uterine tissue (fig. 24.7*b*). The **villous chorion** is the vascular portion in contact with the uterine wall.

Placenta

The placenta (*plă-sen'tă*) is a vascular structure by which an unborn child is attached to its mother's uterine wall and through which metabolic exchange occurs (fig. 24.8). The placenta is formed in part from maternal tissue and in part from embryonic tissue. The embryonic portion consists

allantois: Gk. *allanto,* sausage; *iodos,* resemblance
chorion: Gk. *chorion,* external fetal membrane
villous: L. *villus,* tuft of hair
placenta: L. *placenta,* a flat cake

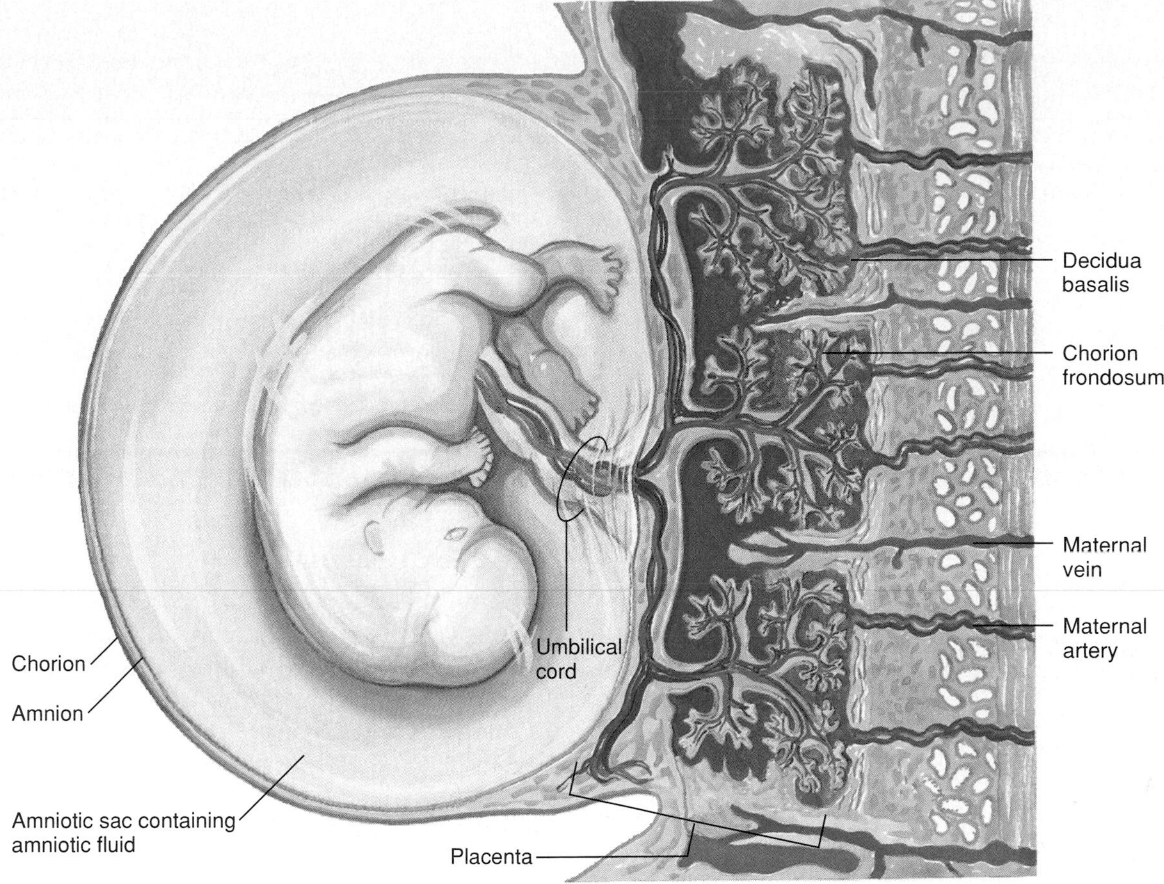

FIGURE 24.8
The embryo, extraembryonic membranes, and placenta at approximately 7 weeks of development. At this time, the adjacent amnion and chorion are frequently referred to as the amniochorionic membrane. Blood from the embryo is carried to and from the chorion frondosum by the umbilical arteries and vein. The maternal tissue between the chorionic villi is known as the decidua basalis; this tissue, together with the villi, form the functioning placenta.

of the villi of the **chorion frondosum.** The maternal portion is composed of an area of the uterine wall called the **decidua basalis.** Blood does not flow directly between these two portions, but because their membranes are in close proximity, certain substances diffuse readily.

Exchange of Molecules across the Placenta

When fully formed, the placenta is a reddish-brown oval disc that weighs 500–600 g (17.5–21 oz)—about one-sixth the weight of the fetus. The two umbilical arteries deliver fetal blood to vessels within the chorion frondosum. This blood circulates within the villi and returns to the fetus via the umbilical vein. Maternal blood is delivered to and drained from the cavities within the decidua basalis. In this way, maternal and fetal blood are brought close together but never mix within the placenta.

The placenta serves as a site for the exchange of gases and other molecules between the maternal and fetal blood. Oxygen diffuses from the mother to fetus, and carbon dioxide diffuses in the opposite direction. Nutrient molecules and waste products likewise pass between maternal and fetal blood.

The placenta has a high metabolic rate. It utilizes about one-third of all the oxygen and glucose supplied by the maternal blood. In fact, the rate of protein synthesis is actually higher in the placenta than it is in the fetal liver. Like the liver, the placenta produces a great variety of enzymes capable of converting biologically active molecules (such as hormones and drugs) into less active, more water-soluble forms. In this way, potentially dangerous molecules in the maternal blood are often prevented from harming the fetus.

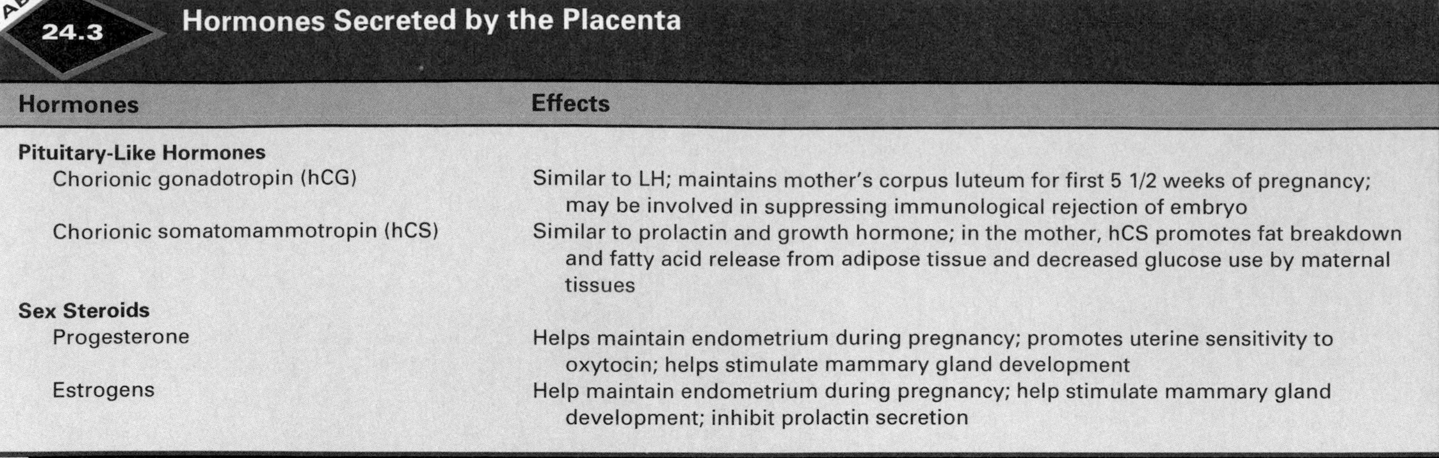

TABLE 24.3 Hormones Secreted by the Placenta

Hormones	Effects
Pituitary-Like Hormones	
Chorionic gonadotropin (hCG)	Similar to LH; maintains mother's corpus luteum for first 5 1/2 weeks of pregnancy; may be involved in suppressing immunological rejection of embryo
Chorionic somatomammotropin (hCS)	Similar to prolactin and growth hormone; in the mother, hCS promotes fat breakdown and fatty acid release from adipose tissue and decreased glucose use by maternal tissues
Sex Steroids	
Progesterone	Helps maintain endometrium during pregnancy; promotes uterine sensitivity to oxytocin; helps stimulate mammary gland development
Estrogens	Help maintain endometrium during pregnancy; help stimulate mammary gland development; inhibit prolactin secretion

Some substances ingested by a pregnant woman are able to pass through the placenta readily, to the detriment of the fetus. These include nicotine, heroin, and certain antidepressant drugs. Excessive nicotine will stunt the growth of the fetus; heroin can lead to fetal drug addiction; *and certain antidepressants can cause respiratory problems.*

Endocrine Functions of the Placenta

The placenta secretes protein hormones that have actions similar to those of some anterior pituitary hormones. In addition, as the corpus luteum regresses, the placenta becomes the major sex steroid–producing gland (table 24.3). The placenta, however, is an incomplete endocrine gland because it cannot produce estrogen and progesterone without the aid of precursors supplied to it by both the mother and the fetus. In terms of steroid hormone production, the fetus and placenta form a single functioning system called the *fetal-placental unit.*

The ability of the placenta to convert androgens into estrogen protects a female embryo from becoming masculinized by the androgens secreted from the mother's adrenal glands. In addition to forming estradiol, the placenta secretes large amounts of a weak estrogen called *estriol.*

The production of estriol increases tenfold during pregnancy. By the third trimester, estriol accounts for about 90% of the estrogens excreted in the mother's urine. Since almost all of the estriol comes from the placenta, measurements of urinary estriol can be used clinically to assess the health of the placenta.

Umbilical Cord

The umbilical (*um-bĭ'lĭ-kal*) cord forms as the yolk sac shrinks and the amnion expands to envelop the tissues on the underside of the embryo (fig. 24.8). When fully formed, the umbilical cord is about 1 to 2 cm (0.5 to 1 in.) in diameter and 55 cm (2 ft) in length. On average, the umbilical cords of boys are a couple of inches longer than those of girls. The umbilical vessels are surrounded by embryonic connective tissue called *mucoid connective tissue* (*Wharton's jelly*).

The umbilical cord has a helical, or screwlike, form that keeps it from kinking. The spiral shape of the umbilical cord forms as the umbilical vein grows faster and longer than the umbilical arteries. In about one-fifth of all deliveries, the cord is looped around the baby's neck. If drawn tightly, the cord may cause death or serious perinatal problems.

Structural Changes in the Embryo by Weeks

Third Week

During the eventful third week following conception, many embryonic tissues are formed and positioned to become distinct organs in the developing body. The first evidence of change comes early in the third week with the appearance of a thick linear band on the dorsal midline of the embryonic disc. Formed from mesoderm, this band is known as the **primitive streak.** As the primitive

Wharton's jelly: from Thomas Wharton, English anatomist, 1614–73

streak elongates, a thickening known as the **primitive knot** appears at its cranial end. The primitive knot later gives rise to the mesodermal structures of the head and to a rod of mesodermal cells called the **notochord.** The notochord supports the embryo and forms a midline axis that is the basis of the embryonic skeleton. The primitive streak also gives rise to embryonic connective tissue called **mesenchyme.** Mesenchyme differentiates into all the various kinds of connective tissue found in the adult (see chapter 4, module 4.3).

Inside Information—One of the earliest formed organs is the skin, which develops during the third week to support and maintain homeostasis within the embryo.

A tremendous amount of change and specialization occurs during the embryonic stage. The factors that cause precise, sequential change from one cell or tissue type to another are not fully understood. It is known, however, that the potential for change is programmed into the genetics of each cell, and that under the right signals this change takes place. The process of developmental change is referred to as **induction.** Induction occurs when one tissue, called the *inductor tissue,* has a marked effect on an adjacent tissue, causing it to become *induced tissue* and stimulating it to differentiate.

Fourth Week

During the fourth week of development, the embryo increases about 4 mm (0.16 in.) in length. A **connecting stalk,** which is later involved in the formation of the umbilical cord, is established between the body of the embryo and the developing placenta. By this time, the heart is already pumping blood to all parts of the embryo through simple arteries and veins. The head and jaws are apparent, and the primordial tissue that will form the eyes, brain, spinal cord, lungs, and digestive organs has developed. The four **limb buds** are recognizable as small swellings on the lateral body walls.

Fifth Week

Embryonic changes during the fifth week are not as extensive as those during the fourth week. The head enlarges, and the developing eyes, ears, and nasal pit become obvious. The appendages have formed from the limb buds, and paddle-shaped hand plates develop ridges called **digital rays.**

Sixth Week

During the sixth week, the embryo measures 16 to 24 mm (0.64 to 0.96 in.), half of which is the head—a share that holds constant until the end of the eighth week. At this time, the brain is undergoing marked differentiation. The limbs are lengthened and slightly flexed, and notches appear between the digital rays. Many of the internal organs are developing rapidly, and it is at this period of development that they are most vulnerable. In addition, the gonads are beginning to produce hormones that will influence the development of the external genitalia.

Seventh and Eighth Weeks

During the last 2 weeks of the embryonic period, the embryo, which is now 28 to 40 mm (1.12 to 1.6 in.) long, has distinctly human features. The body organs are formed, and the nervous system is beginning to coordinate body activity. The eyes are well developed, but the lids are stuck together to protect against probing fingers as the arms wave about. The nostrils are developed but are plugged with mucus. The external genitalia are forming but cannot yet be distinguished as male or female (see fig. 23.5). The major body systems are developed by the end of the eighth week, and from this time on the embryo is called a **fetus.**

Inside Information—The most precarious time of prenatal development is the embryonic period—yet, well into this period many women are still unaware that they are pregnant. For this reason, a woman should abstain from taking certain drugs (including some antibiotics) if there is even a remote chance that she is pregnant or might become pregnant in the near future.

Some of the structural changes that occur during the embryonic period are illustrated in figure 24.9.

induction: L. *inductus,* to lead in

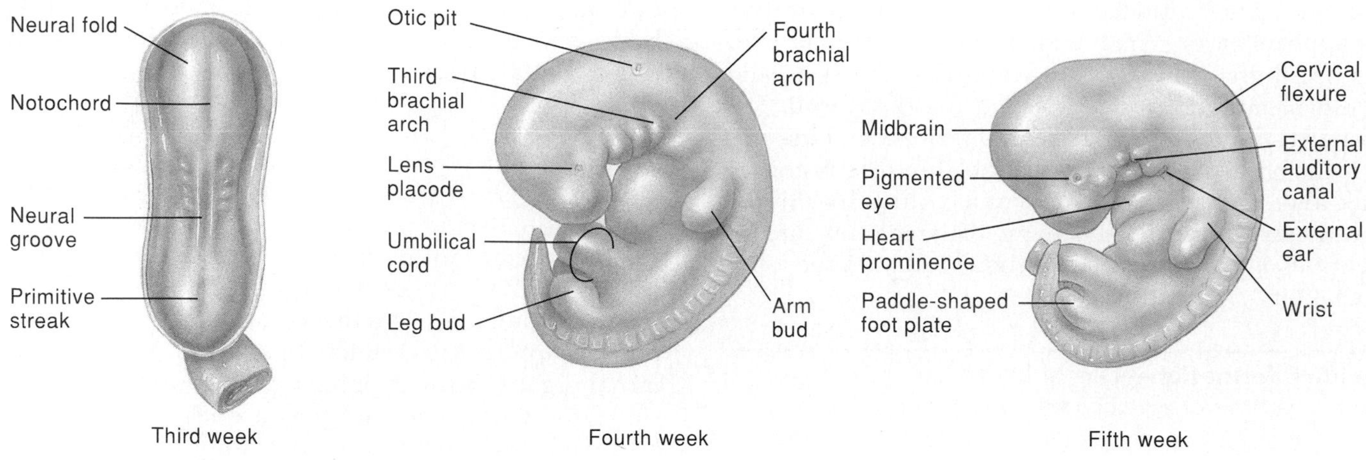

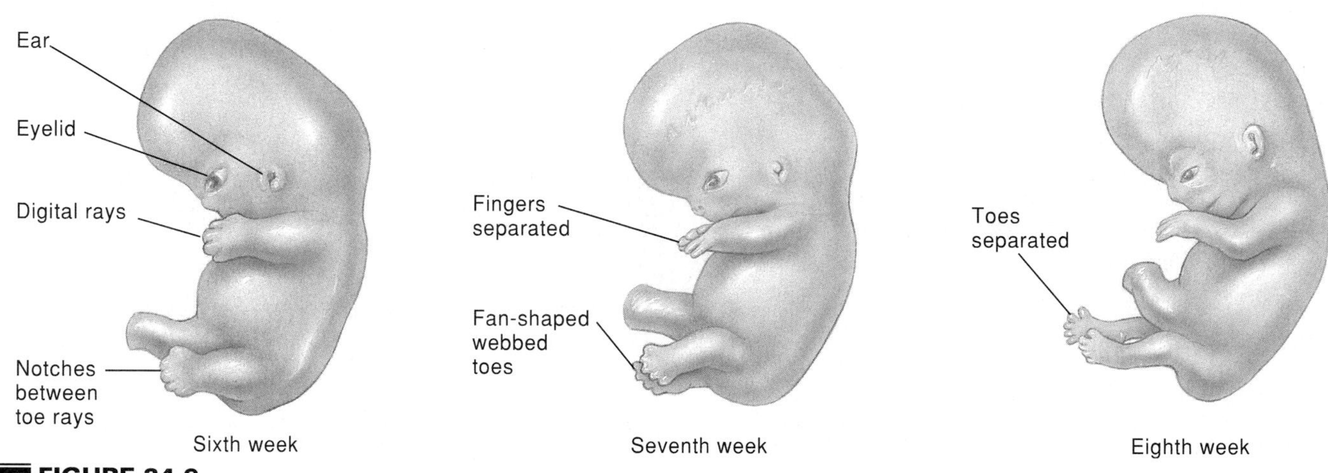

FIGURE 24.9
Structural changes in the embryo by weeks.

24.4 Fetal Period

The fetal period, beginning at week 9 and culminating at birth, is characterized by tremendous growth and the specialization of body structures.

Since most of the tissues and organs of the body form during the embryonic period, the fetus is recognizable as a human being at 9 weeks. The fetus is far less vulnerable than the embryo to the deforming effects of viruses, drugs, and radiation. Tissue differentiation and organ development continue during the fetal stage, but to a lesser degree than before. For the most part, fetal development is limited to sheer body growth (fig. 24.10).

Nine to Twelve Weeks

At the beginning of the ninth week, the head of the fetus is as large as the rest of the body. Head growth slows during the next 3 weeks, whereas growth in body length

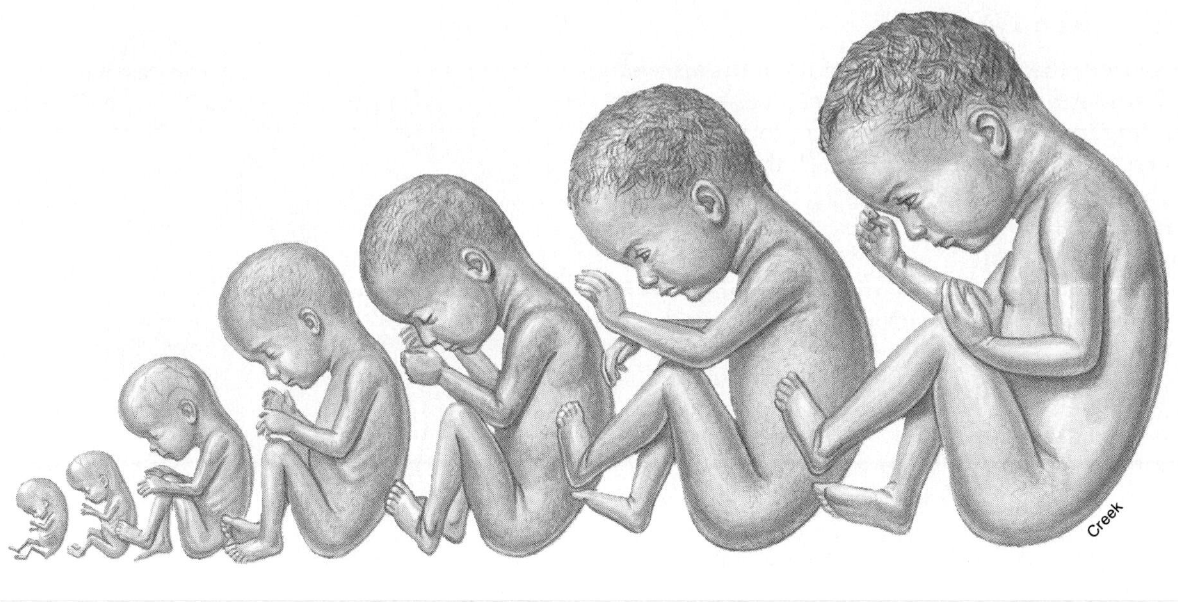

FIGURE 24.10
Changes in the external appearance of the fetus from week 9 through week 38.

accelerates. Ossification centers appear in most bones during the ninth week. Differentiation of the external genitalia begins at the end of the ninth week but not until the twelfth week can the external genitalia be distinguished as male or female (see fig. 22.3). By the end of the twelfth week, the fetus is 86 mm (3.5 in.) long and weighs about 45 g (1.6 oz). It can swallow, digest the fluid that passes through its digestive tract, and defecate and urinate into the amniotic fluid. Its nervous system has developed sufficiently to enable it to withdraw its leg if prodded. It begins inhaling through its nose but can take in only amniotic fluid.

Major structural abnormalities, which may not be predictable from genetic analysis, can often be detected by ultrasonography *(fig. 24.11). In this procedure, organs are bombarded with sound waves that reflect back in a certain pattern determined by tissue densities. For example, sound waves bouncing off amniotic fluid will produce an image much different from that produced by sound waves bouncing off the placenta or the mother's uterus. Ultrasonography is so sensitive that it can detect a fetal heartbeat several weeks before it can be heard with a stethoscope.*

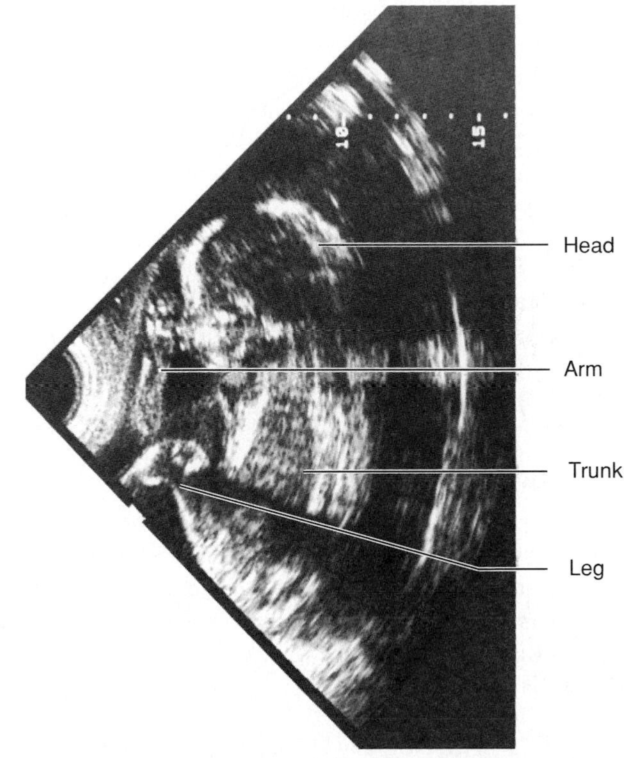

FIGURE 24.11
Structures of the fetus as seen through an ultrasound scan.

Thirteen to Sixteen Weeks

During this period, the facial features of the fetus are well formed, and epidermal structures such as eyelashes, hair on the head, fingernails, and nipples begin to appear. The limbs lengthen, and by the sixteenth week the skeleton is sufficiently developed so that it shows up clearly on X-ray films. During the sixteenth week, the fetal heartbeat can be heard by applying a stethoscope to the mother's abdomen. By the end of the sixteenth week, the fetus is 140 mm (5.6 in.) long and weighs 200 g (7 oz).

After the sixteenth week, the reported length of a fetus is generally derived from a straight-line measurement taken from the crown of the head to the developing ischium (crown-rump length). Measurements made on an embryo prior to the fetal period, however, are not reported as crown-rump measurements, but as total length.

Seventeen to Twenty Weeks

During this period, the legs reach their final relative proportions, and fetal movements, known as *quickening*, are commonly felt by the mother. The skin of the fetus is covered with a white, cheeselike material known as **vernix caseosa.** It consists of fatty secretions from the sebaceous glands and dead epidermal cells. Vernix caseosa protects the fetus while it is bathed in amniotic fluid. Twenty-week-old fetuses also have fine, silklike fetal hair, called **lanugo,** covering the skin. Lanugo is thought to hold the vernix caseosa in place on the skin. In addition, its ciliary motion moves the amniotic fluid. A 20-week-old fetus is about 190 mm (7.6 in.) long, and it weighs about 460 g (16 oz). Because of cramped space, it develops a marked spinal flexure and is in what is commonly called the *fetal position,* with the head bent down, in contact with the flexed knees.

Twenty-one to Twenty-five Weeks

During the period from 21 to 25 weeks, the fetus increases its weight substantially to about 900 g (32 oz). Its body length increases only moderately, however, so the weight is evenly distributed. Its translucent skin is quite wrinkled and is pinkish in color because the blood flowing in the capillaries is now visible.

Twenty-six to Twenty-nine Weeks

Toward the end of this period, the fetus will be about 275 mm (11 in.) long and will weigh about 1,300 g (45.5 oz). If born during this period, a fetus might survive if it is put in an incubator with a respirator to maintain its breathing. Its eyes are open, and if it is a male, the testes should have descended into the scrotum. As the time of birth approaches, the fetus rotates to a *vertex position,* with its head positioned against the cervix of the uterus (fig. 24.12).

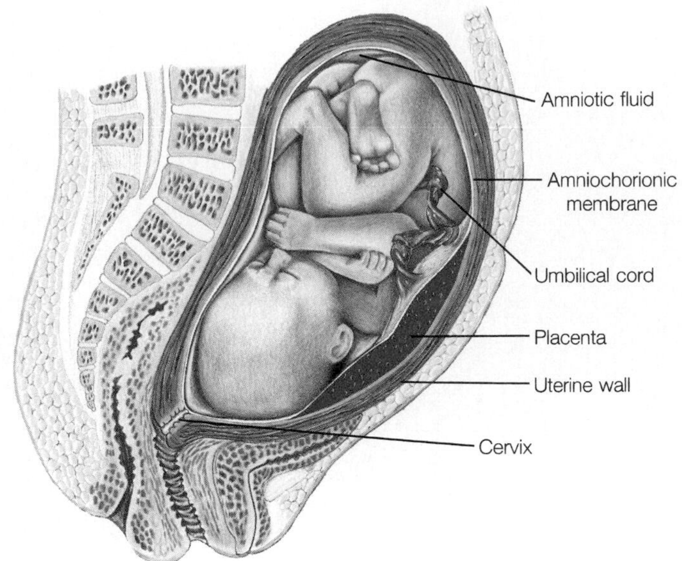

FIGURE 24.12
A fetus in vertex position. Toward the end of most pregnancies, the weight of the fetal head causes the body to rotate, positioning the head against the cervix of the uterus.

Thirty to Thirty-eight Weeks

By the end of 38 weeks, the fetus is considered full-term. It has reached a crown-rump length of 360 mm (14 in.) and weighs 3,400 g (7.4 lb). Its skin is smooth and pinkish blue, even when parents are dark-skinned, because melanocytes do not produce melanin until the skin is exposed to sunlight. Its chest is prominent, and the mammary area protrudes in both sexes.

The period of prenatal development is referred to as gestation. *Normal gestation for humans is 9 months. Knowing this and the pattern of menstruation makes it possible to determine the delivery date of a baby. In a typical reproductive cycle, a woman ovulates 14 days prior to the onset of the next menstruation and is fertile for approximately 20 to 24 hours following ovulation. Adding 9 months, or 38 weeks, to the time of ovulation gives one the estimated delivery date.*

A photographic summary of prenatal development through the fourteenth week is presented in figure 24.13.

vernix caseosa: L. *vernix,* varnish; L. *caseus,* cheese
lanugo: L. *lana,* wool
vertex: L. *vertex,* summit
gestation: L. *gestatus,* to bear

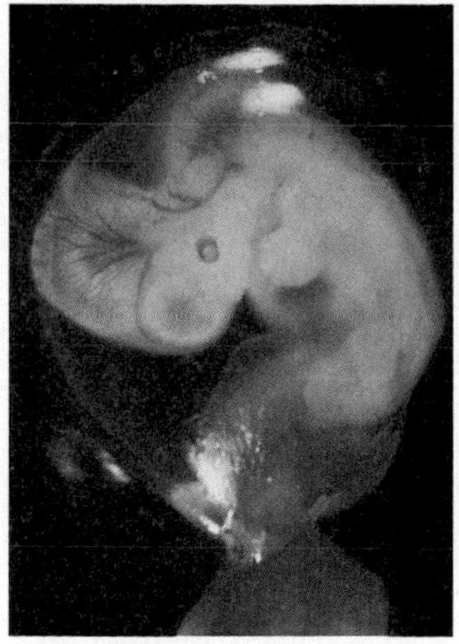
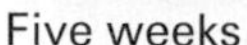
Five weeks

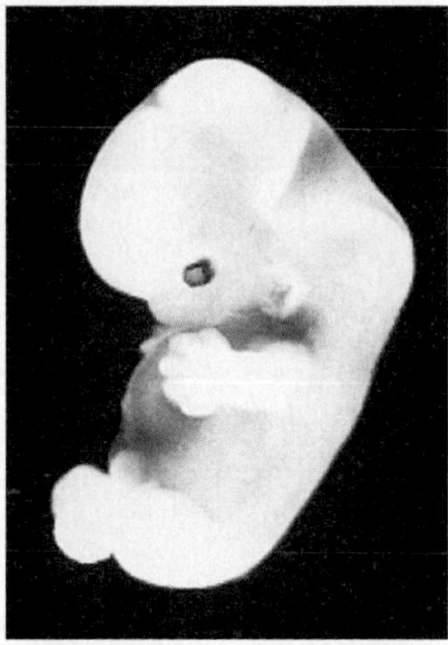
Six weeks

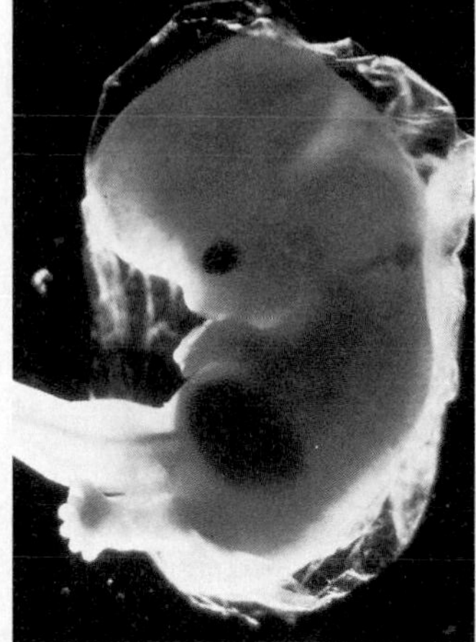
Seven weeks

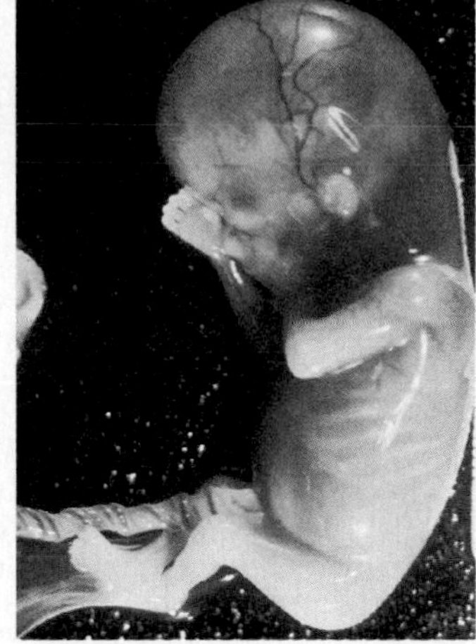
Eight weeks

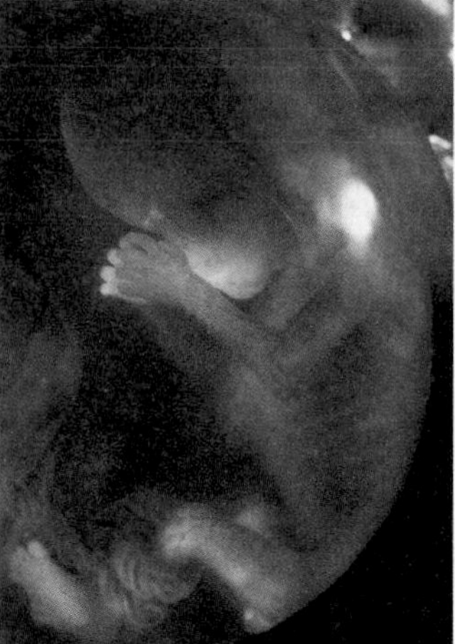
Ten weeks

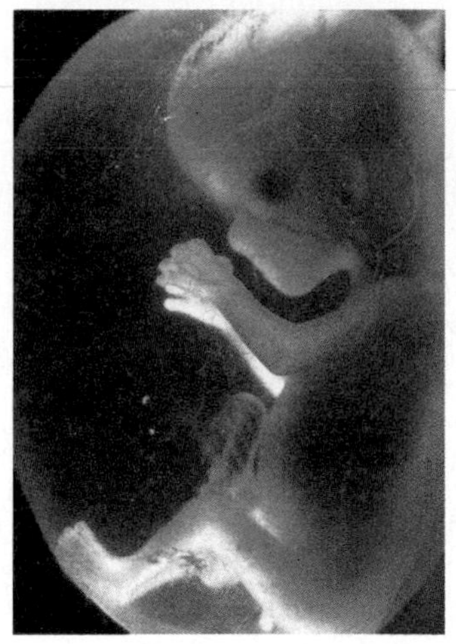
Twelve weeks

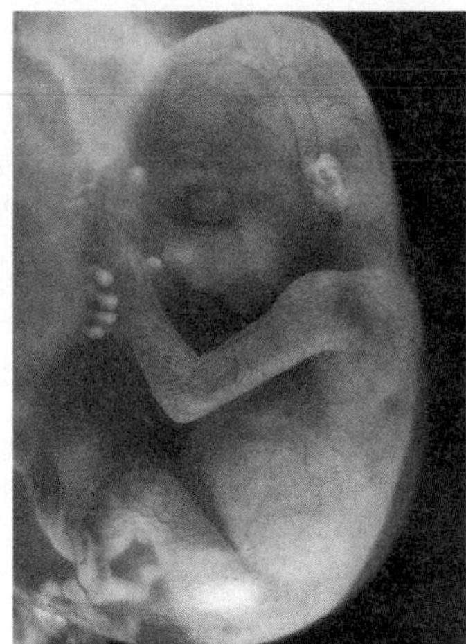
Fourteen weeks

FIGURE 24.13
A photographic summary of embryonic and fetal development.

24.5 Labor and Parturition

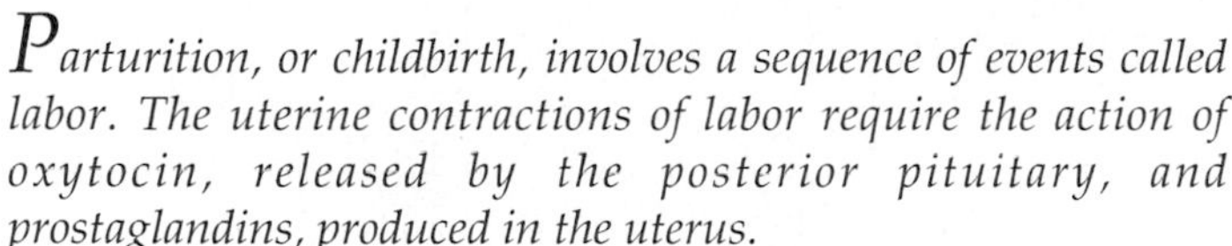

Parturition, or childbirth, involves a sequence of events called labor. The uterine contractions of labor require the action of oxytocin, released by the posterior pituitary, and prostaglandins, produced in the uterus.

Parturition (*par"tyoo-rish'un*), or childbirth, is the culmination of pregnancy. The sequence of physiological and physical events that expel the infant from the uterus through the vagina are collectively referred to as **labor.**

The *onset of labor* is indicated by rhythmic and forceful contractions of the myometrial layer of the uterus (table 24.4). In *true labor*, the pains from uterine contractions occur at regular intervals and intensify as the time between contractions shortens. A reliable evidence of true labor is dilation of the cervix and a "show," or vaginal discharge, of blood-containing mucus from the uterus. In *false labor*, abdominal pain is felt at irregular intervals, and cervical dilation and cervical show are absent.

The uterine contractions of labor are stimulated by two agents: (1) **oxytocin** (*ok"sĭ-to'sin*), a polypeptide hormone produced in the hypothalamus and released by the posterior pituitary, and (2) **prostaglandins** (*pros"tă-glan'dinz*), a class of fatty acids produced within the uterus itself. Labor can be induced artificially by injections of oxytocin or by the insertion of prostaglandins into the vagina as a suppository.

TABLE 24.4 **Events Preceding Labor**

Sequence	Event
1	High estrogen secretion from the placenta stimulates the production of oxytocin receptors in the uterus
2	Uterine muscle (myometrium) becomes increasingly sensitive to oxytocin
3	Oxytocin stimulates the production of prostaglandins in the uterus
4	Prostaglandins stimulate uterine contractions
5	Contractions of the uterus stimulate the release of oxytocin from the posterior pituitary
6	Increased oxytocin levels further stimulate uterine contractions, resulting in labor

The hormone relaxin, *produced by the corpus luteum, is involved in labor and parturition. Relaxin softens the symphysis pubis in preparation for parturition and is also thought to soften the cervix in preparation for dilation. It may be, however, that relaxin does not affect the uterus, but rather progesterone and estradiol may be responsible for this effect. Further research is necessary to completely understand the physiological effect of these hormones.*

As illustrated in figure 24.14, there are three stages of labor and parturition.

1. **Dilation stage.** In this period, the cervix dilates to a diameter of approximately 10 cm. Contractions are regular, and the amniotic sac generally ruptures. If the amniotic sac does not rupture spontaneously, it is

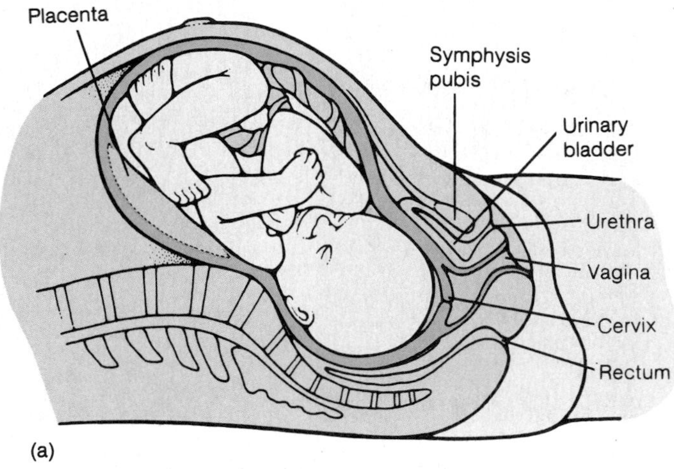

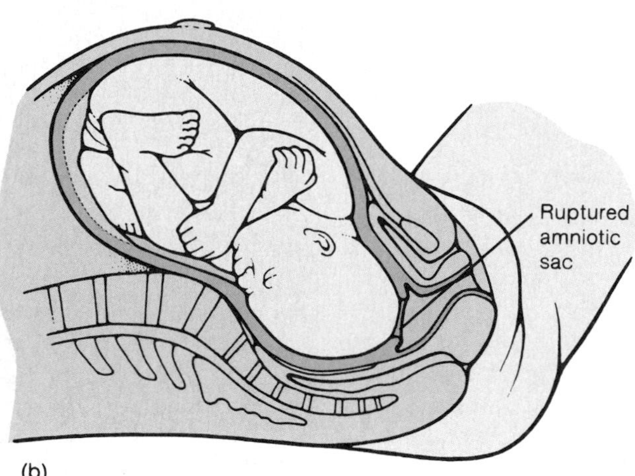

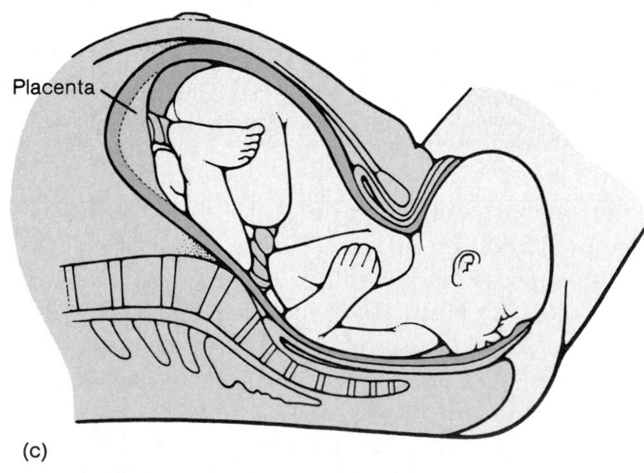

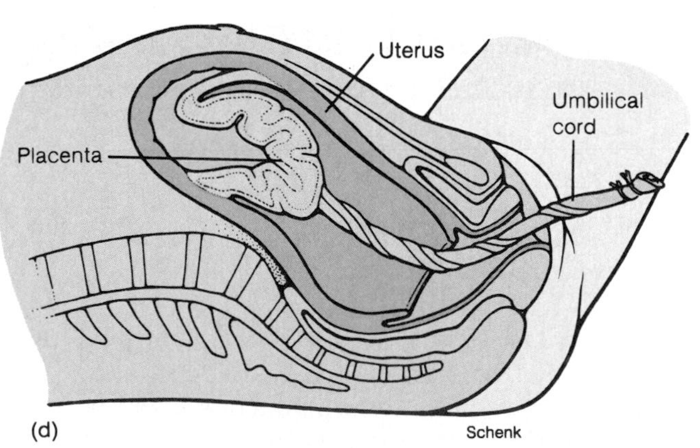

FIGURE 24.14

The stages of labor and parturition. (*a*) The position of the fetus prior to labor. (*b*) The ruptured amniotic sac during the early *dilation stage* of the cervix. (*c*) The *expulsion stage,* or period of parturition. (*d*) The *placental stage,* as the afterbirth is being expelled.

broken surgically. The dilation stage may last 8 to 24 hours, depending on whether it is occurring in the first or a subsequent pregnancy.

2. **Expulsion stage.** This stage involves actual childbirth. It consists of forceful uterine contractions and abdominal compressions that expel the fetus from the uterus through the vagina. This stage may require 30 minutes in a first pregnancy, but only a few minutes in subsequent pregnancies.
3. **Placental stage.** Generally within 15 minutes after parturition, the placenta separates from the uterine wall and is expelled as the *afterbirth.* Forceful uterine contractions characterize this stage, constricting uterine blood vessels to prevent hemorrhage. In a normal delivery, blood loss does not exceed 350 ml (10.5 oz); to be classified as a hemorrhage, it would have to reach 500 ml (15 oz).

A *pudendal nerve block* may be administered during the early part of the expulsion stage to ease the trauma of delivery for the mother and to permit an *episiotomy.* An episiotomy is a surgical incision into the perineum and vagina to prevent traumatic tearing during parturition.

Five percent of newborns are born breech—*buttocks first. The principal concern of a breech birth is the increased time and difficulty of the expulsion stage of labor. Attempts to rotate the fetus through the use of forceps may injure the infant. If an infant cannot be delivered breech, a* cesarean section *(C-section) must be performed. A cesarean section is delivery through a surgical incision made through the abdominal and uterine walls. Other reasons for a C-section include bleeding from placental abnormalities, fetal distress (such as from the umbilical cord wrapped around the neck), a baby too large to fit through the pelvis, and toxemia of pregnancy (metabolic disturbances in the pregnant woman).*

24.6 Inheritance

Inheritance is the acquisition of characteristics or qualities by transmission from parent to offspring. Hereditary information is transmitted by genes.

Genetics is the branch of biology that deals with inheritance. Genetics and inheritance are important in anatomy and physiology because of the numerous developmental and functional disorders that have a genetic basis. Genetic counseling is the practical application of knowing which disorders and diseases are inherited.

As described earlier (see module 24.1), each zygote inherits 23 chromosomes from its mother and 23 chromosomes from its father. This does not produce 46 different chromosomes; rather, it produces 23 pairs of *homologous chromosomes.* With the important exception of the sex chromosomes, the members of a homologous pair appear to be structurally identical and contain similar genes (such as those coding for eye color, height, and so on). These homologous pairs of chromosomes can be *karyotyped* (photographed or illustrated) and identified (fig. 24.15). Each cell that contains 46 chromosomes (that is *diploid*) has two number 1 chromosomes, two number 2 chromosomes, and so on through chromosomes number 22. The first 22 pairs of chromosomes are called **autosomal chromosomes.** The twenty-third pair are the **sex chromosomes,** and these may look different and may carry different genes. In a female they consist of two X chromosomes, whereas in a male there is one X chromosome and one Y chromosome.

(a)

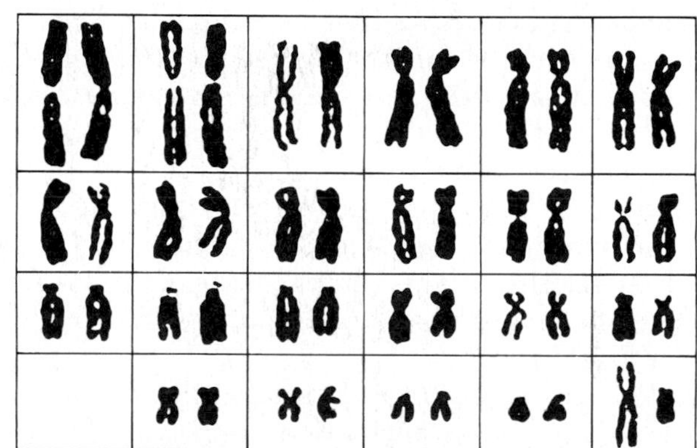

(b)

FIGURE 24.15

A karyotype of homologous pairs of chromosomes obtained from a human diploid cell. The first 22 pairs of chromosomes are called the autosomal chromosomes. The sex chromosomes are (*a*) XY for a male and (*b*) XX for a female.

Genes and Alleles

A **gene** is the portion of the DNA of a chromosome that contains the information needed to synthesize a particular protein molecule. Although each diploid cell has a pair of gene locations for each characteristic, the genes may be present in variant forms. Those alternative forms of a gene that affect the same characteristic but that produce different expressions of that characteristic are called **alleles** (*ă-lēlz'*). One allele of each pair originates from the female parent and the other from the male. The shape of a person's ears, for example, is determined by the kind of allele received from each parent and the interaction of the alleles with one another. Alleles are always located on the same spot, or **locus,** on homologous chromosomes (fig. 24.16). If the alleles are identical, the person is said to be **homozygous** (*ho"mo-zi'gus*) for that particular characteristic. But if the two alleles are different, the person is **heterozygous** (*het"er-o-zi'gus*) for that particular trait.

Genotype and Phenotype

A person's DNA contains a catalog of genes known as the **genotype** of that person. The expression of those genes results in observable features referred to as the **phenotype.** If the alleles for a particular trait are homozygous, the characteristic expresses itself in a specific manner (two alleles for attached earlobes, for example, results in a person with attached earlobes). If the alleles for a particular trait are heterozygous, however, the allele that expresses itself and the way in which the genes for that trait interact will determine the phenotype. The allele that expresses itself is called the **dominant allele;** the one that does not is the **recessive allele.** The various combinations of dominant and recessive alleles are responsible for a person's physical traits (table 24.5).

In describing genotypes, it is traditional to use letter symbols to refer to the alleles of an organism. The dominant alleles are symbolized by uppercase letters and the recessive alleles are symbolized by lowercase. Thus, the *genotype* of a person who is homozygous for unattached (free) earlobes due to a dominant allele is symbolized *EE;* a heterozygous pair is symbolized *Ee.* In both of these instances, the *phenotypes* of the individuals would be free earlobes, since a dominant allele is present in each genotype. A person who inherited two recessive alleles for earlobes would have the genotype *ee* and would have attached earlobes.

Thus, three genotypes are possible when gene pairing involves dominant and recessive alleles. They are *homozygous dominant* (*EE*), *heterozygous* (*Ee*), and *homozygous recessive* (*ee*). Only two phenotypes are possible, however, since the dominant allele is expressed in both the homozygous dominant (*EE*) and heterozygous (*Ee*) individuals. The recessive allele is expressed only in the homozygous recessive (*ee*) condition. Refer to figure 24.17 for an illustration of how a homozygous recessive trait may be expressed in a child of parents who are heterozygous.

Probability

A **Punnett square** is a convenient way to express probabilities (the "odds") of allele combinations for a particular inheritable trait. In constructing a Punnett square, the male gametes (spermatozoa) carrying a particular trait are

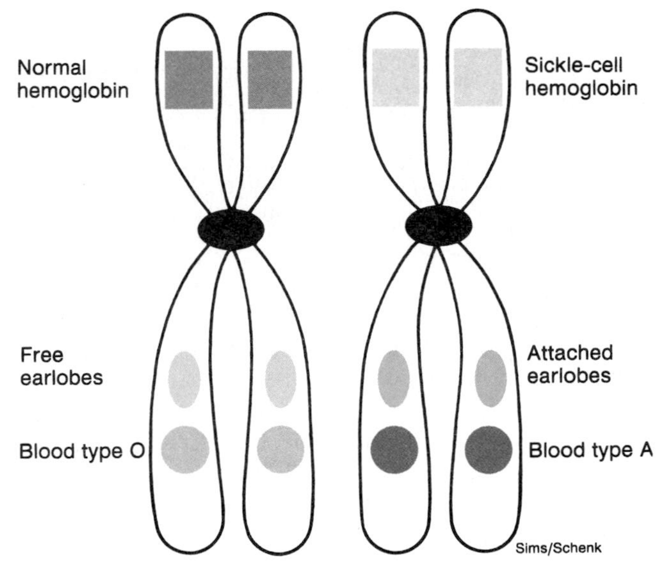

FIGURE 24.16
A pair of homologous chromosomes. Homologous chromosomes contain genes for the same characteristic at the same locus.

TABLE 24.5 Hereditary Traits in Humans Determined by Single Pairs of Dominant and Recessive Alleles

Dominant	Recessive	Dominant	Recessive
Free earlobes	Attached earlobes	Color vision	Color blindness
Dark brown hair	All other colors	Broad lips	Thin lips
Curly hair	Straight hair	Ability to roll tongue	Lack of this ability
Pattern baldness (♂♂)	Baldness (♀♀)	Arched feet	Flat feet
Pigmented skin	Albinism	A or B blood factor	O blood factor
Brown eyes	Blue or green eyes	Rh blood factor	No Rh blood factor

placed at the side of the chart, and the female gametes (ova) at the top (as in fig. 24.18). The four spaces on the chart represent the possible combinations of male and female gametes that could form zygotes. The probability of an offspring having a particular genotype is 1 in 4 for homozygous dominant and homozygous recessive, and 1 in 2 for heterozygous.

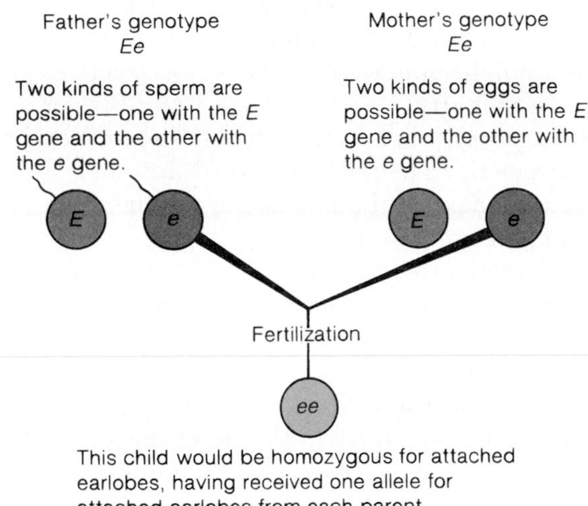

FIGURE 24.17

Inheritance of earlobe characteristics. Two parents with unattached (free) earlobes can have a child with attached earlobes.

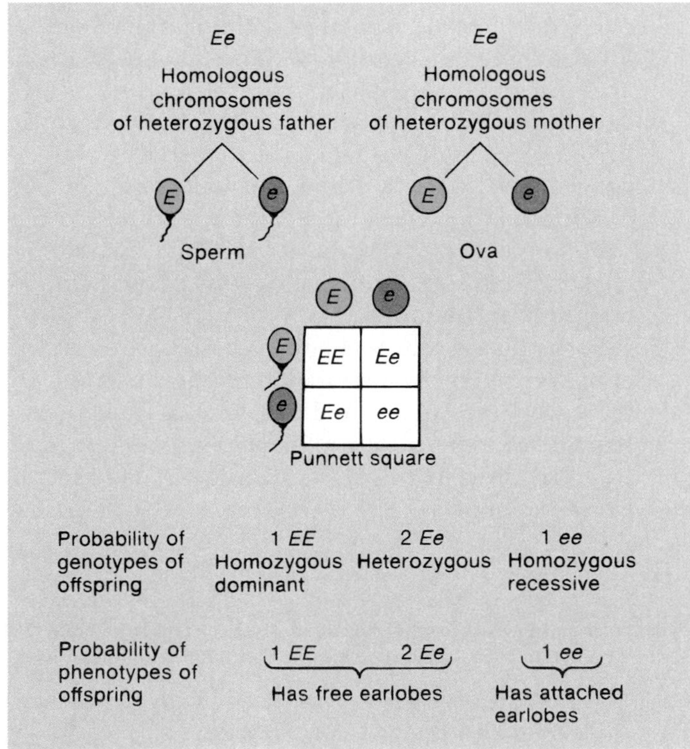

FIGURE 24.18

Use of a Punnett square to determine genotypes and phenotypes that could result from the mating of two heterozygous parents.

A genetic study in which a single characteristic (e.g., ear shape) is followed from parent to offspring is referred to as a **monohybrid cross.** A genetic study in which two characteristics are followed from parents to offspring is referred to as a **dihybrid cross** (fig. 24.19). The term *hybrid* refers to an offspring descended from parents that have different genotypes.

Sex-Linked Inheritance

Certain inherited traits are determined by genes on the sex chromosomes, and are thus called **sex-linked characteristics.** *Red-green color blindness,* for example, is determined by a recessive allele (designated *c*) found on the X

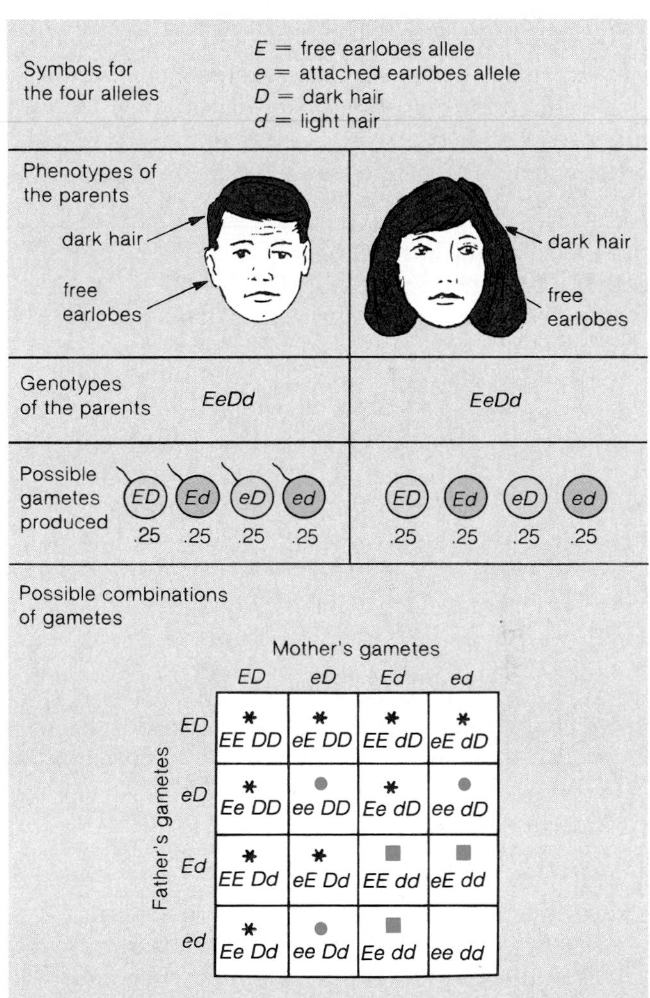

FIGURE 24.19

In dihybrid cross, two pairs of traits are followed simultaneously. Any of the combinations of genes that have a *D* and an *E* (nine possibilities) will have free earlobes and dark hair. These are indicated with an asterisk (*). Three of the possible combinations have two alleles for attached earlobes (*ee*) and at least one allele for dark hair. They are indicated with a dot (●). Three of the combinations have free earlobes and light hair. These are indicated with a square (■). The remaining possibility has the genotype *eedd* for attached earlobes and light hair.

chromosome but not on the Y chromosome. Normal color vision (designated *C*) dominates. The ability to discriminate red from green, therefore, depends entirely on the X chromosomes. The genotype possibilities are as follows:

X^CY Normal male

X^cY Colorblind male

X^CX^C Normal female

X^CX^c Normal female carrying the recessive allele

X^cX^c Colorblind female

In order for a female to be red-green colorblind, she must have the recessive allele on both of her X chromosomes. Her father would have to be red-green colorblind, and her mother would have to be a carrier for this condition. A male with only one such allele on his X chromosome, however, will show the characteristic. Since a male receives his X chromosome from his mother, the inheritance of sex-linked characteristics usually passes from mother to son.

Inside Information—Born in Austria in 1822, an Augustinian monk named Gregor Mendel discovered the universal laws of inheritance by studying particular traits in pea plants that are passed from one generation to the next. Mendel had never heard of genes, chromosomes, or the position of genes on chromosomes, but he seemed to comprehend the nature of inheritance even without knowing the mechanics. How he came upon his remarkable insights will always remain a mystery, since the monk who became head abbot of the monastery after Mendel's death in 1884 destroyed all of his notebooks. Although Mendel is now considered the founder of genetics, the scientific community totally ignored his findings for almost 35 years.

Common Genetic Disorders

cretinism An autosomal recessive disorder characterized by a lack of thyroid secretion, due to a defect in the iodine transport mechanism. Untreated children are dwarfed, sterile, and may be mentally retarded.

cystic fibrosis An autosomal recessive disorder characterized by the formation of thick mucus in the lungs and pancreas that interferes with normal breathing and digestion.

galactosemia An autosomal recessive disorder characterized by an inability to metabolize galactose, a component of milk sugar. Patients with this disorder have cataracts, damaged livers, and mental retardation.

gout An autosomal dominant disorder characterized by an accumulation of uric acid in the blood and tissue due to an abnormal metabolism of purines.

hemophilia A sex-linked condition caused by a recessive allele. The blood of a person with hemophilia either fails to clot or clots very slowly after an injury.

hepatic porphyria An autosomal dominant condition characterized by painful GI disorders and neurologic disturbances due to an abnormal metabolism of porphyrins.

Huntington's chorea An autosomal dominant disorder characterized by uncontrolled twitching of skeletal muscles and the deterioration of mental capacities.

Marfan's syndrome An autosomal dominant disorder characterized by tremendous growth of the extremities, looseness of the joints, dislocation of the lenses of the eyes, and congenital cardiovascular defects.

phenylketonuria (*fen"il-kēt"n-oor'e-ă*) **(PKU)** An autosomal recessive disorder characterized by an inability to metabolize the amino acid phenylalanine. It is accompanied by brain and nerve damage and mental retardation. (Newborns are routinely tested for PKU; those that are affected are placed on a diet low in phenylalanine.)

pseudohypertrophic muscular dystrophy A sex-linked recessive disorder characterized by progressive muscle atrophy. It usually begins during childhood and causes death in adolescence.

retinitis pigmentosa A sex-linked recessive disorder characterized by progressive atrophy of the retina and eventual blindness.

Tay–Sachs disease An autosomal recessive disorder characterized by a deterioration of physical and mental abilities, early blindness, and early death. It has a disproportionately high incidence in Jews of Eastern European origin (see chapter 9, "Clinical Considerations").

Huntington's chorea: from George Huntington, American physician, 1850–1916

Marfan's syndrome: from Antoine Bernard-Jean Marfan, French physician, 1858–1942

Tay–Sachs disease: from Warren Tay, English physician, 1843–1927, and Bernard Sachs, American neurologist, 1858–1944

Clinical Considerations

In clinical terms, gestation is frequently divided into three phases, or **trimesters,** each lasting three calendar months. By the end of the *first trimester,* all of the major body systems are formed, the fetal heart can be detected, and the external genitalia are developed. The fetus is about the width of the palm of an adult's hand. During the *second trimester,* fetal quickening can be detected, epidermal features are formed, and although the vital body systems are functioning, the fetus would be unlikely to survive if birth were to occur. At the end of the second trimester, fetal length is about equal to the length of an adult's hand. The fetus experiences tremendous growth and refinement in system functioning during the *third trimester.* A fetus of this age may survive if born prematurely, and of course, the chances of survival improve as the natural delivery date draws closer.

Abnormal Implantation Sites

In an **ectopic pregnancy** (see chapter 23, module 23.1) the blastocyst implants outside the uterus or in an abnormal site within the uterus (fig. 24.20). The most common ectopic location is within the uterine tube, and is referred to as a *tubal pregnancy.* Occasionally, implantation occurs near the cervix, where development of the placenta blocks the cervical opening. This condition, called *placenta previa,* causes serious bleeding. Ectopic pregnancies do not develop normally; the fetus seldom survives beyond the first trimester.

Tubal pregnancies are terminated through medical intervention. If a tubal pregnancy is permitted to progress, the uterine tube generally ruptures, followed by hemorrhaging. Depending on the location and stage of development of a tubal pregnancy, it may or may not be life-threatening to the woman.

In Vitro Fertilization and Artificial Implantation

In the increasingly common technique of **in vitro fertilization,** several human ova are fertilized *in vitro* (outside the body). To obtain the ova, a specialized laparoscope (fig. 24.21) is used to draw out preovulatory eggs from graafian follicles. The eggs are then placed in a suitable culture medium, where they are fertilized with sperm cells and allowed to develop to the blastocyst stage. At this stage of development, the blastocysts are artificially deposited in the uterine cavity in the hope that at least one will implant. If implantation is successful, normal full-term development and delivery usually follows. In vitro fertilization with artificial implantation is a means of overcoming infertility problems due to blocked uterine tubes in females or low sperm counts in males.

continued

previa: L. *previa,* appearing before or in front of

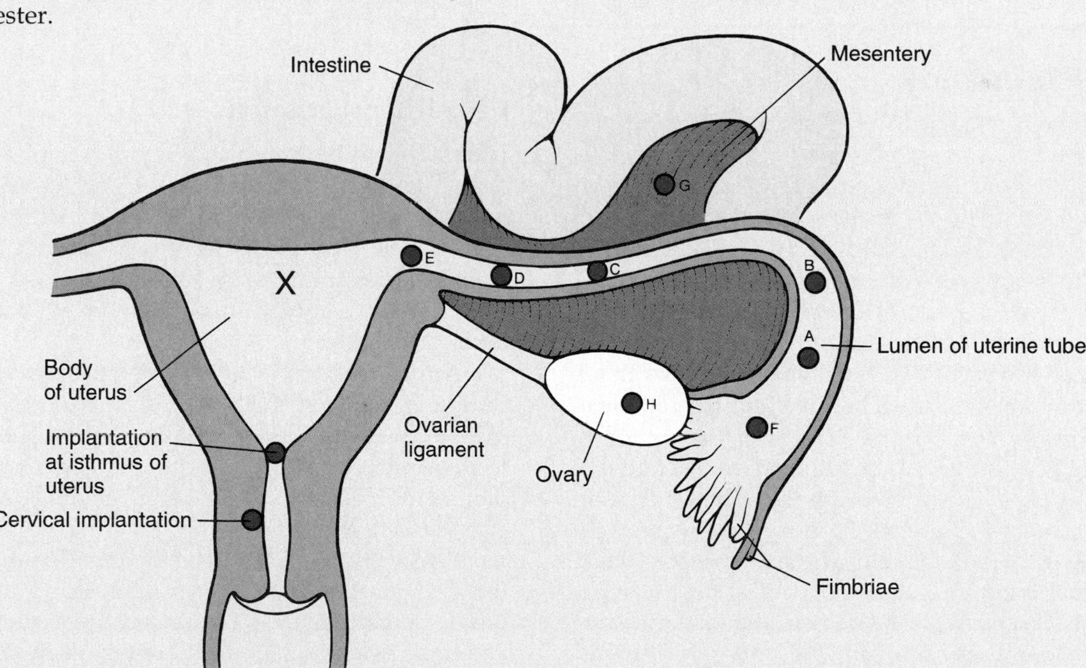

FIGURE 24.20
Sites of ectopic pregnancies. The normal implantation site is indicated by an X, and the abnormal sites are indicated by letters in order of frequency of occurrence.

Clinical Considerations

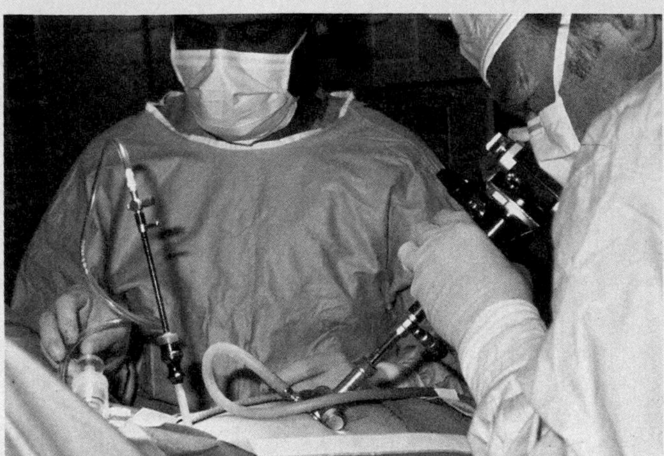

FIGURE 24.21
A laparoscope is used for various abdominal operations, including the extraction of a preovulatory ovum.

Inside Information—In 1978, Robert Edwards, a Cambridge University physiologist, and Patrick Steptoe, an English gynecologist, announced that a woman was finally pregnant as a result of a technique they had been perfecting for years—*in vitro fertilization.* The woman, 30-year-old Lesley Brown, became an instant celebrity on July 25, 1978, when she gave birth to Louise Joy Brown—the first "test-tube" baby.

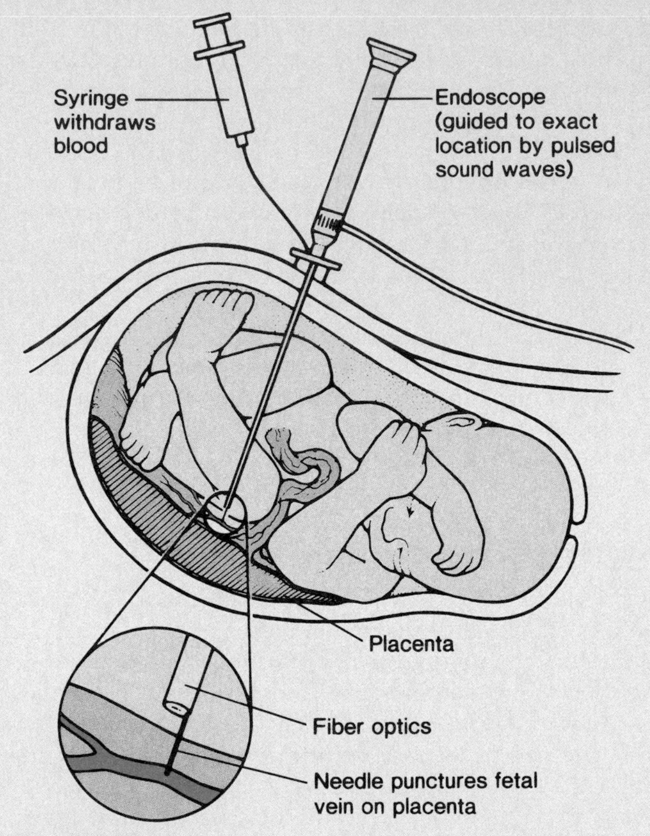

FIGURE 24.22
Fetoscopy.

Multiple Pregnancy

Twins occur about once in 85 pregnancies. They can develop in two ways. *Dizygotic (fraternal) twins* develop from two zygotes produced by the fertilization of two oocytes by two spermatozoa in the same ovulatory cycle. *Monozygotic (identical) twins* form from a single zygote. About one-third of all twins are monozygotic.

Dizygotic twins may be of the same sex or different sexes and are no more alike than brothers or sisters born at different times. Dizygotic twins always have two chorions and two amnions, but the chorions and the placentas may be fused.

Monozygotic twins are of the same sex and are genetically identical. They have two amnions but only one chorion and a common placenta. Monozygotic twinning is usually initiated toward the end of the first week, when the embryoblast divides. If the embryoblast fails to separate completely, *conjoined twins* (Siamese twins) may form.

Triplets occur about once in 7,600 pregnancies and may be (1) all from the same ovum and identical, (2) two identical and the third from another ovum, or (3) three zygotes from three different ova. Similar combinations occur in quadruplets, quintuplets, and so on.

Fetal Monitoring

Ultrasonography, produced by a mechanical vibration of high frequency, produces a safe, high-resolution (sharp) image of fetal structure (see fig. 24.11). Ultrasonic imaging is a reliable way to determine pregnancy as early as 6 weeks after ovulation. It can also be used to determine fetal weight, length, and position, as well as to identify multiple fetuses.

Amniocentesis is a technique used to obtain a small sample of amniotic fluid with a syringe, so that the fluid can be checked. This technique is most often performed to determine fetal maturity, but it can also help to predict serious disorders like *Down syndrome* and *Gaucher's disease* (see module 24.3).

Fetoscopy (fig. 24.22) allows direct examination of the fetus. Using fetoscopy, physicians scan the uterus with pulsed sound waves to locate and observe fetal structures and take tissue samples. Fetoscopy is also used to diagnose several diseases, including hemophilia, thalassemia, and the 40% of sickle-cell anemia cases missed by amniocentesis.

fetoscopy: L. *fetus,* offspring; *skopein,* to view

SUMMARY

24.1 Fertilization

1. Upon fertilization of a secondary oocyte by a sperm cell in the uterine tube, the oocyte completes its second meiotic division and a diploid zygote is formed.
2. The sequential formation of body structures during the prenatal period of human life is known as morphogenesis.
3. Prenatal development lasts 38 weeks and is divided into pre-embryonic, embryonic, and fetal periods.

24.2 Pre-Embryonic Period

1. Cleavage of the zygote is initiated within 30 hours and continues until a morula forms. The morula enters the uterine cavity on about the third day.
2. The morula hollows out and fills with fluid, at which point it is called a blastocyst.
3. Implantation of the blastocyst occurs between the fifth and seventh day and is enabled by the secretion of enzymes that digest a portion of the endometrium. During implantation, the trophoblast secretes hCG. This hormone prevents the breakdown of the endometrium and subsequent menstruation.
4. The embryoblast of the implanted blastocyst flattens into the embryonic disc, from which the primary germ layers of the embryo develop.

24.3 Embryonic Period

1. The events of the 6-week embryonic period include the differentiation of the germ layers into specific body organs and the formation of the placenta, the umbilical cord, and the extraembryonic membranes that sustain and protect the embryo.
2. The extraembryonic membranes include the amnion, which contains the protective amniotic fluid; the yolk sac, which produces blood for the embryo; the allantois, which also produces blood and which gives rise to the umbilical vessels; and the chorion, which participates in the formation of the placenta.
3. The umbilical cord, containing two umbilical arteries and one umbilical vein, is formed as the amnion envelops the tissues on the underside of the embryo.
4. From the third to the eighth week, the structure of all the body organs, except the genitalia, becomes apparent. By the end of the fourth week, the heart is beating. By the end of the fifth week, the sensory organs have formed in the enlarged head and the appendages have developed, with digital rays present. The brain is well developed by the end of the sixth week, the limbs are elongated, and the digits are separated.

24.4 Fetal Period

1. A small amount of tissue differentiation and organ development occurs during the fetal period, but for the most part fetal development is limited to body growth.
2. Between weeks 9 and 12, ossification centers appear; the genitalia are formed; and the digestive, urinary, respiratory, and muscle systems begin to function.
3. Between weeks 17 and 20, quickening can be felt by the mother, and vernix caseosa and lanugo cover the skin of the fetus.
4. Toward the end of the period between weeks 26 and 29, the eyes have opened, the gonads have descended in a male, and the fetus has developed to the point where it may survive if born prematurely.
5. By 38 weeks, the fetus is considered full-term. Normal gestation for humans is 266 days.

24.5 Labor and Parturition

1. Labor and parturition are the culmination of gestation. The uterine contractions of labor require the action of oxytocin, released by the posterior pituitary, and prostaglandins, produced in the uterus.
2. Labor is divided into dilation, expulsion, and placental stages.

24.6 Inheritance

1. Inheritance is the acquisition of hereditary traits by transmission from parent to offspring. The branch of biology that studies inheritance is called genetics.
2. Each zygote contains 22 pairs of autosomal chromosomes and 1 pair of sex chromosomes—XX in a female and XY in a male.
3. A gene is the portion of a DNA molecule that contains information for the production of one kind of protein molecule. Alleles are different forms of genes that occupy corresponding positions on homologous chromosomes.
4. The combination of genes in a person's cells constitutes his or her genotype. The observable expression of the genotype is the person's phenotype.
5. A Punnett square is a simple diagram used to express possible genetic combinations. For a cross of two heterozygous genotypes, the probability of a particular genotype is 1 in 4 for homozygous dominant and homozygous recessive, and 1 in 2 for heterozygous. A single trait is studied in a monohybrid cross, and two traits are studied in a dihybrid cross.
6. Traits determined by genes on the X and Y chromosomes are said to be sex-linked characteristics.

Review Activities

Objective Questions

1. An ovulated egg is surrounded by a thin layer of protein and polysaccharides called (*module 24.1*)
   a. hyaluronidase.
   b. the corona radiata.
   c. the translucent membrane.
   d. the zona pellucida.

2. The pre-embryonic period is completed when (*module 24.2*)
   a. the blastocyst implants.
   b. the placenta forms.
   c. the blastocyst reaches the uterus.
   d. the primary germ layers form.

3. The yolk sac produces blood for the embryo until (*module 24.3*)
   a. the heart becomes functional.
   b. the kidneys become functional.
   c. the liver becomes functional.
   d. the baby is delivered.

4. Which of the following is a function of the placenta? (*module 24.3*)
   a. production of steroids and hormones
   b. diffusion of nutrients and oxygen
   c. production of enzymes
   d. all of the above

5. Which of the following could diffuse across the placenta? (*module 24.3*)
   a. nicotine
   b. alcohol
   c. heroin
   d. all of the above

6. As the delivery date nears, the fetus rotates into (*module 24.4*)
   a. the vernix position.
   b. the fetal position.
   c. the vertex position.
   d. the retroflex position.

7. The uterine contractions of labor are stimulated by (*module 24.5*)
   a. estrogen.
   b. oxytocin.
   c. prostaglandins.
   d. both b and c.

8. Which pairing of symbols and genotype descriptions is *correct?* (*module 24.6*)
   a. *Bb*-homozygous recessive
   b. *bb*-heterozygous
   c. *BB*-homozygous dominant

9. If the genotypes of the parents are *Aa* and *Aa,* the offspring probably will be (*module 24.6*)
   a. 1/2 *AA* and 1/2 *aa.*
   b. all *Aa.*
   c. 1/4 *AA,* 1/2 *Aa,* and 1/4 *aa.*
   d. 3/4 *AA,* and 1/4 *aa.*

Essay Questions

1. Discuss the changes that occur in a spermatozoon from the time of ejaculation to the time of fertilization. What changes occur in an ovulated egg at the time of fertilization? (*module 24.1*)
2. Discuss the implantation of the trophoblast in the uterine wall and its involvement in the formation of the placenta. (*module 24.2*)
3. Explain how the primary germ layers form. What major structures does each germ layer give rise to? (*module 24.2*)
4. Explain why development during the embryonic period is so critical. Why is proper maternal health care important during this period of gestation? (*module 24.3*)
5. State the approximate time period (in weeks) for the following occurrences: (*modules 24.3 and 24.4*)
   a. appearance of the arm and leg buds
   b. differentiation of the external genitalia
   c. perception of quickening by the mother
   d. functioning of the embryonic heart
   e. initiation of bone ossification
   f. appearance of lanugo and vernix caseosa
   g. survival of the fetus, if born prematurely
   h. formation of all major body organs completed
6. Distinguish between labor and parturition. (*module 24.5*)
7. Define *genetics, genotype, phenotype, allele, dominant, recessive, homozygous,* and *heterozygous.* (*module 24.6*)

Labeling Exercise

Label the structures indicated on the figure to the right.

1. ______________________________
2. ______________________________
3. ______________________________
4. ______________________________
5. ______________________________
6. ______________________________

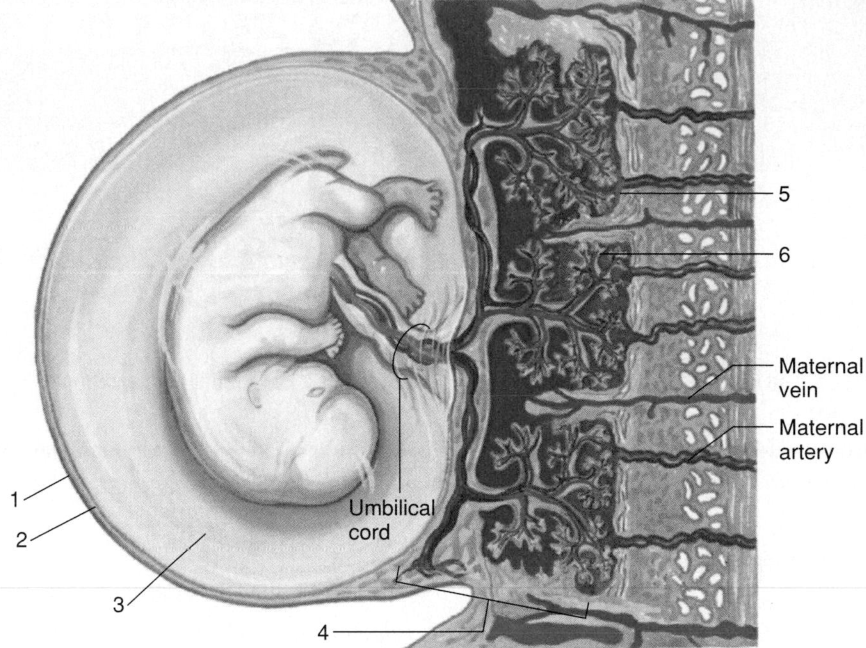

Critical Thinking Questions

1. Write a short paragraph about pregnancy that includes the terms *ovum, blastocyst, implantation, embryo, fetus, gestation,* and *parturition.*
2. If a pregnant woman had a pack-a-day cigarette habit, what health predictions could you make about her baby? Justify your responses.
3. The sedative thalidomide was used by thousands of pregnant women in the 1960s to alleviate their morning sickness. This drug inhibited normal limb development and resulted in tragically deformed infants with flipperlike arms and legs. At what period of prenatal development did such abnormalities originate?
4. In which stage of labor is the fetus most vulnerable to damage? Justify your answer.
5. Given an X^CY male and an X^CX^c female, construct a Punnett square to show the possible genotypes for color blindness.

APPENDIX A

Answers to Objective Questions

Chapter 1 The Human Body: Organizing Principles

1. (d) 4. (c) 7. (a)
2. (a) 5. (d) 8. (a)
3. (d) 6. (c) 9. (c)

Chapter 2 Chemical Composition of the Body

1. (e) 5. (c) 8. (c)
2. (b) 6. (c) 9. (b)
3. (d) 7. (d) 10. (c)
4. (d)

Chapter 3 Cell Structure and Function

1. (a) 7. (b) 12. (b)
2. (c) 8. (d) 13. (a)
3. (d) 9. (b) 14. (a)
4. (a) 10. (d) 15. (b)
5. (a) 11. (d) 16. (c)
6. (a)

Chapter 4 Histology

1. (b) 4. (a) 7. (b)
2. (b) 5. (c) 8. (b)
3. (d) 6. (a) 9. (b)

Chapter 5 Integumentary System

1. (d) 4. (c) 7. (c)
2. (b) 5. (d) 8. (a)
3. (a) 6. (a) 9. (b)

Chapter 6 Skeletal System

1. (b) 5. (d) 9. (d)
2. (d) 6. (b) 10. (b)
3. (d) 7. (d) 11. (d)
4. (a) 8. (c) 12. (a)

Chapter 7 Articulations

1. (b) 4. (d) 7. (c)
2. (a) 5. (d) 8. (b)
3. (c) 6. (d) 9. (d)

Chapter 8 Muscular System

1. (c) 6. (b) 11. (d)
2. (c) 7. (c) 12. (c)
3. (a) 8. (c) 13. (a)
4. (b) 9. (a) 14. (e)
5. (c) 10. (b)

Chapter 9 Functional Organization of Nervous Tissue

1. (d) 5. (c) 8. (c)
2. (a) 6. (d) 9. (a)
3. (c) 7. (d) 10. (c)
4. (a)

Chapter 10 Nervous System

1. (c) 5. (d) 9. (b)
2. (a) 6. (d) 10. (d)
3. (c) 7. (c) 11. (c)
4. (c) 8. (b)

Chapter 11 Autonomic Nervous System

1. (d) 4. (b) 7. (c)
2. (d) 5. (c) 8. (c)
3. (a) 6. (b) 9. (c)

Chapter 12 The Senses

1. (b) 5. (c) 9. (b)
2. (a) 6. (d) 10. (d)
3. (c) 7. (c) 11. (a)
4. (c) 8. (c)

Chapter 13 Endocrine System

1. (d) 5. (d) 8. (e)
2. (b) 6. (d) 9. (d)
3. (d) 7. (a) 10. (d)
4. (e)

Chapter 14 Circulatory System: Blood

1. (d) 5. (c) 8. (c)
2. (d) 6. (b) 9. (d)
3. (a) 7. (d) 10. (c)
4. (a)

Chapter 15 Circulatory System: Heart and Blood Vessels

1. (d) 5. (a) 8. (b)
2. (b) 6. (c) 9. (a)
3. (c) 7. (b) 10. (e)
4. (e)

Chapter 16 Cardiac Output and Blood Flow

1. (d) 5. (c) 9. (d)
2. (a) 6. (d) 10. (e)
3. (d) 7. (b) 11. (c)
4. (a) 8. (c) 12. (c)

Chapter 17 Lymphatic System and Immunity

1. (d) 5. (a) 9. (d)
2. (e) 6. (b) 10. (a)
3. (b) 7. (e) 11. (c)
4. (d) 8. (d)

Chapter 18 Respiratory System

1. (d) 6. (d) 11. (a)
2. (a) 7. (c) 12. (c)
3. (d) 8. (c) 13. (a)
4. (e) 9. (b) 14. (c)
5. (c) 10. (c)

Chapter 19 Urinary System

1. (c) 5. (c) 8. (d)
2. (a) 6. (d) 9. (d)
3. (a) 7. (d) 10. (c)
4. (b)

Chapter 20 Digestive System

1. (b) 5. (d) 9. (a)
2. (b) 6. (b) 10. (c)
3. (c) 7. (d) 11. (c)
4. (c) 8. (d) 12. (b)

Chapter 21 Regulation of Metabolism

1. (b) 5. (b) 9. (d)
2. (e) 6. (d) 10. (c)
3. (a) 7. (d) 11. (c)
4. (d) 8. (a)

Chapter 22 Male Reproductive System

1. (c) 4. (b) 7. (b)
2. (a) 5. (d) 8. (a)
3. (a) 6. (d) 9. (c)

Chapter 23 Female Reproductive System

1. (b) 5. (c) 8. (c)
2. (b) 6. (b) 9. (b)
3. (c) 7. (a) 10. (d)
4. (d)

Chapter 24 Prenatal Development and Inheritance

1. (d) 4. (d) 7. (d)
2. (d) 5. (d) 8. (c)
3. (c) 6. (c) 9. (c)

APPENDIX B

Reference Plates

FIGURE 1
The surface anatomy of the facial region. (Also refer to figure 8.18 for the underlying muscles.)

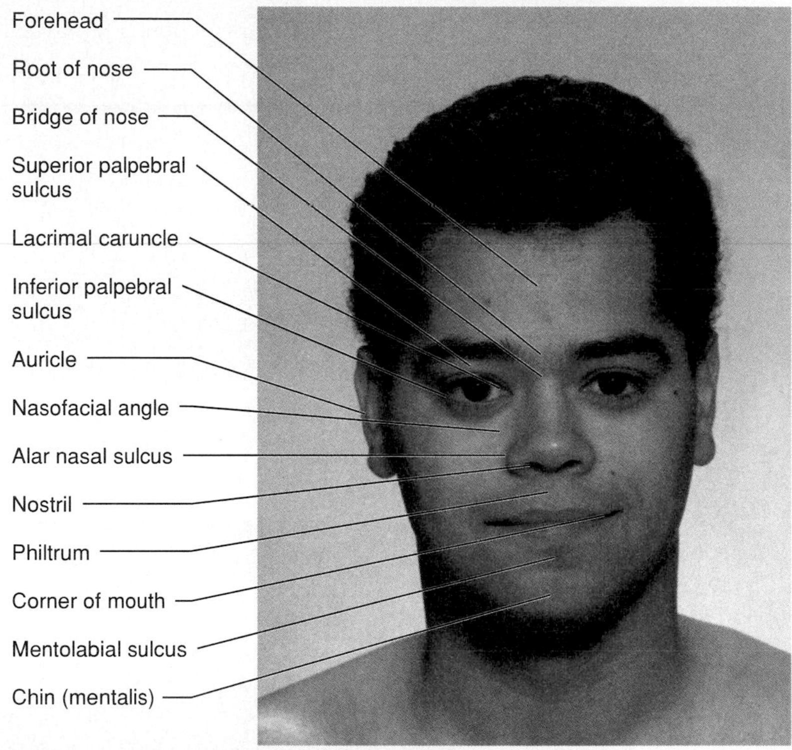

FIGURE 2
An anterolateral view of the neck. (Also refer to figure 8.23 for the underlying muscles.)

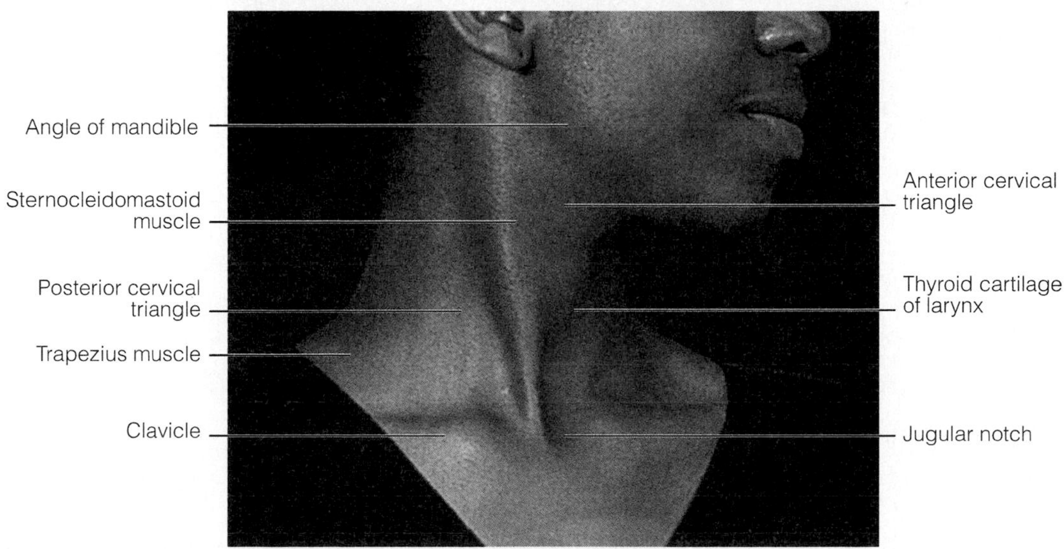

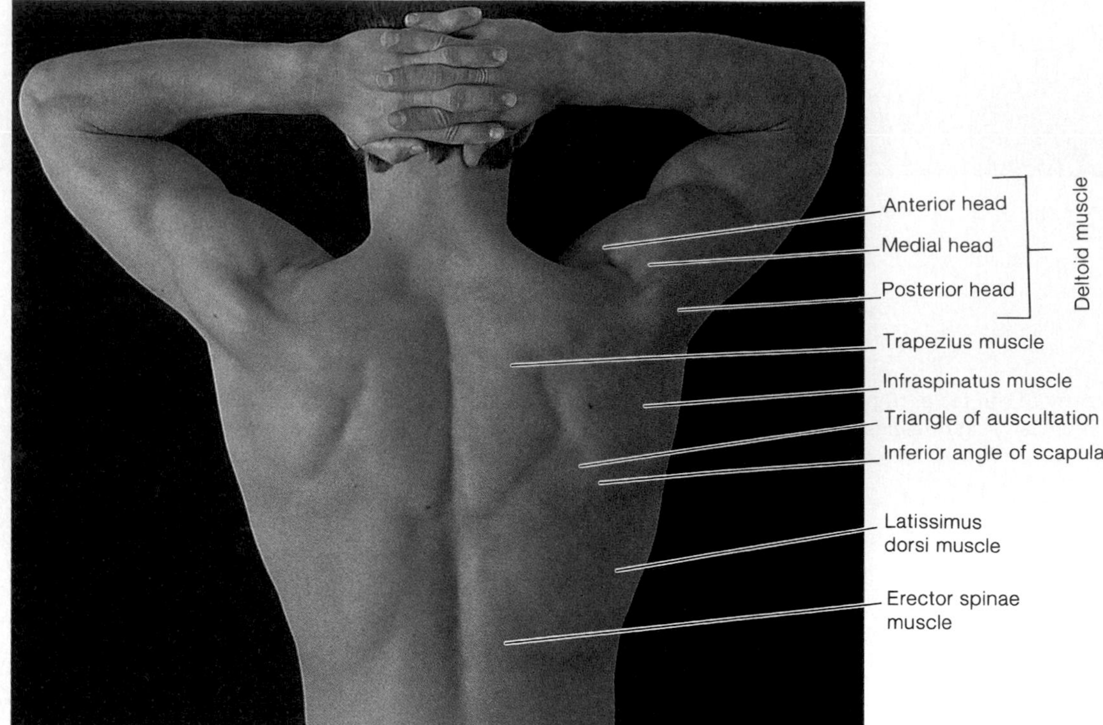

FIGURE 3
The surface anatomy of the back. (Also refer to figure 8.29 for the underlying muscles.)

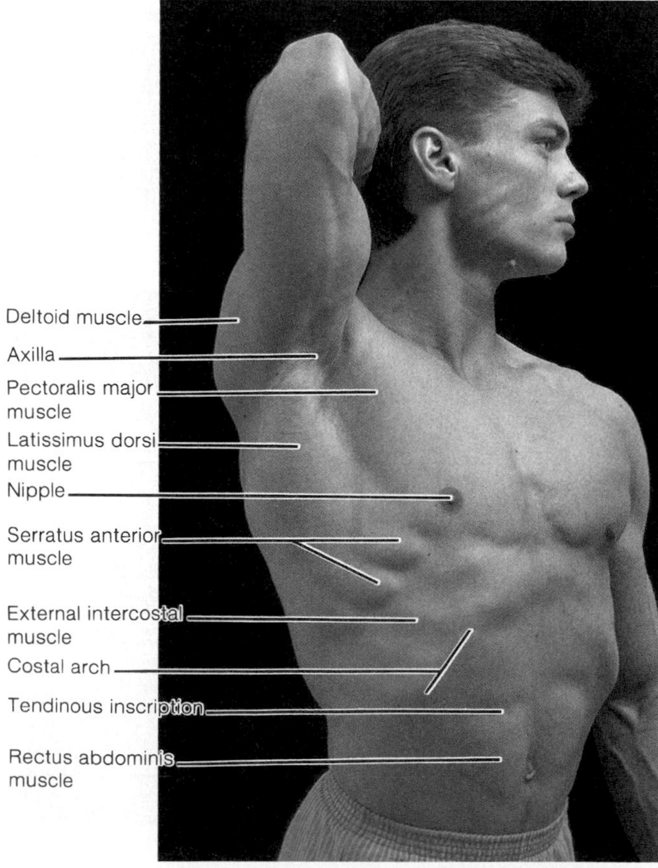

FIGURE 4
An anterolateral view of the trunk and axilla. (Also refer to figures 8.25, 8.30, and 8.31 for the underlying muscles.)

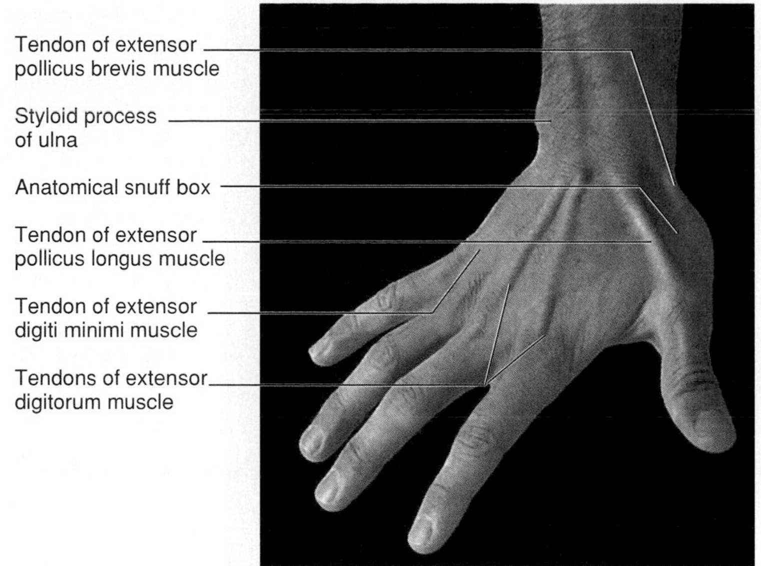

FIGURE 5

The anatomical snuffbox. (Also refer to figure 8.32*b* for the underlying muscles and tendons.)

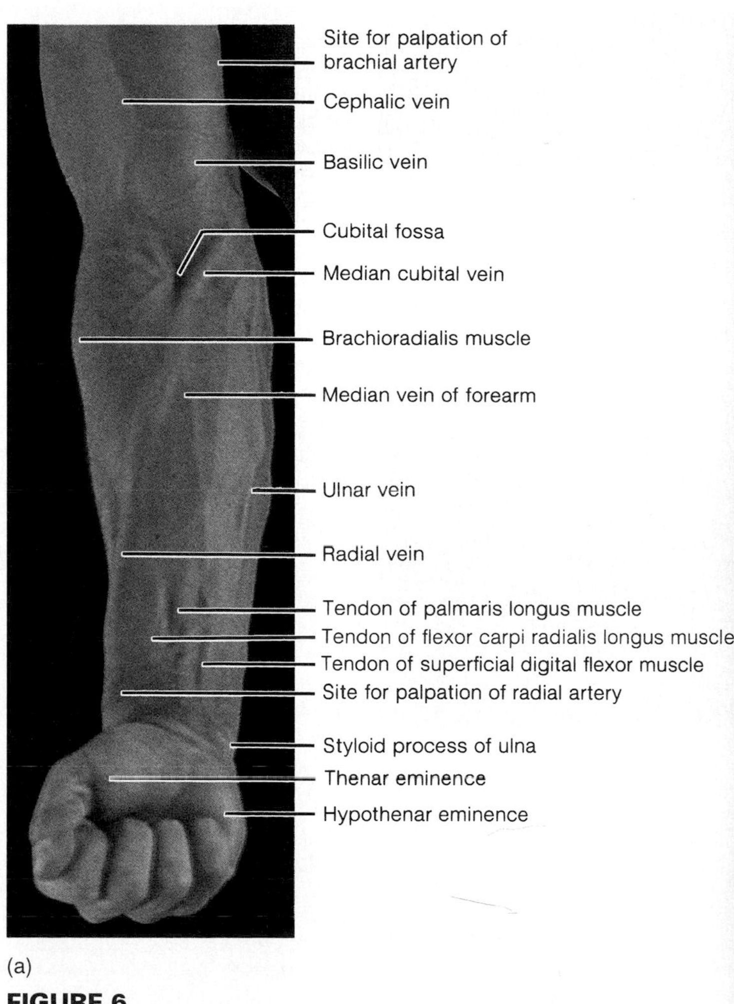

(a)

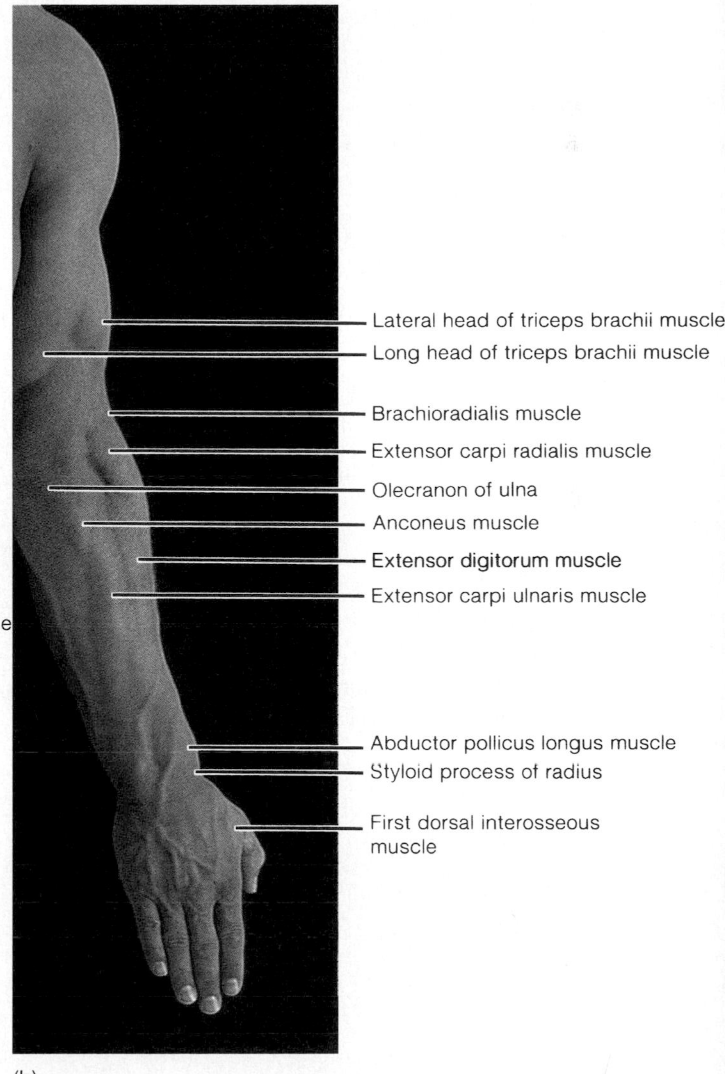

(b)

FIGURE 6

The forearm and hand. (*a*) An anterior view and (*b*) a posterior view. (Also refer to figure 8.32 for the underlying muscles and tendons.)

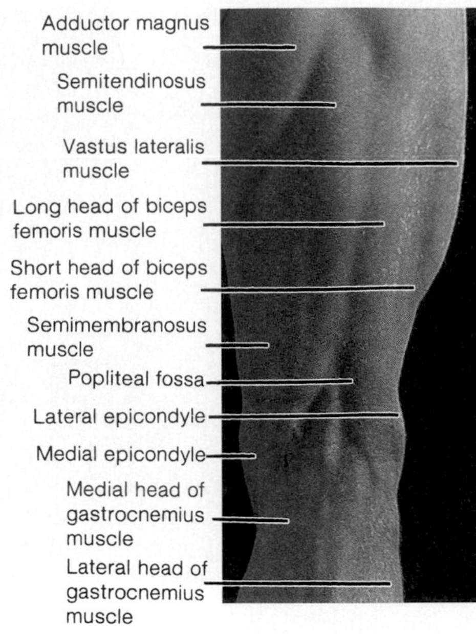

(a)

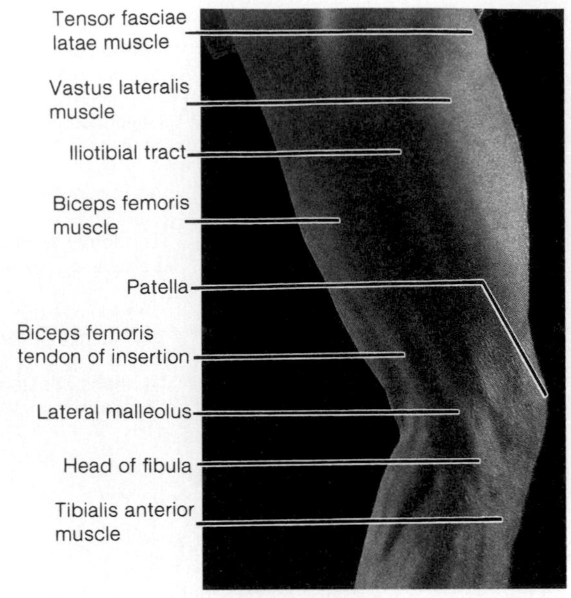

(b)

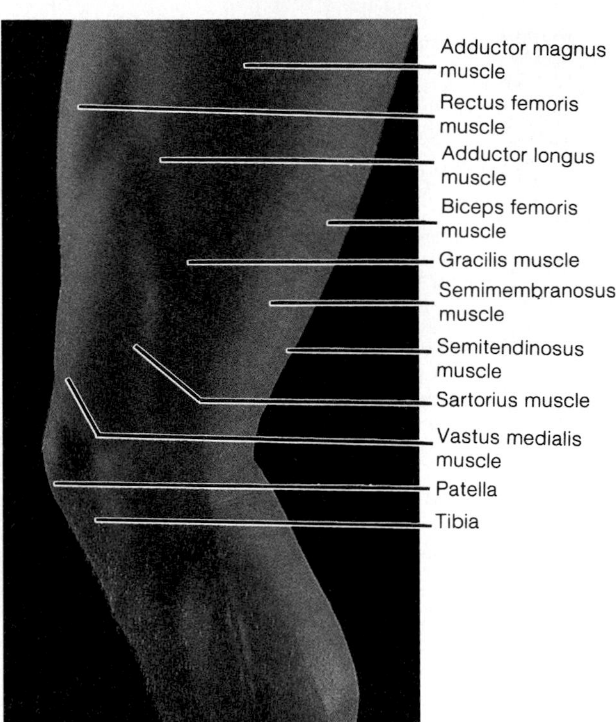

(c)

FIGURE 7

The (*a*) lateral, (*b*) posterior, and (*c*) medial surfaces of the lower limb. (Also refer to figures 8.39 and 8.40 for the underlying muscles.)

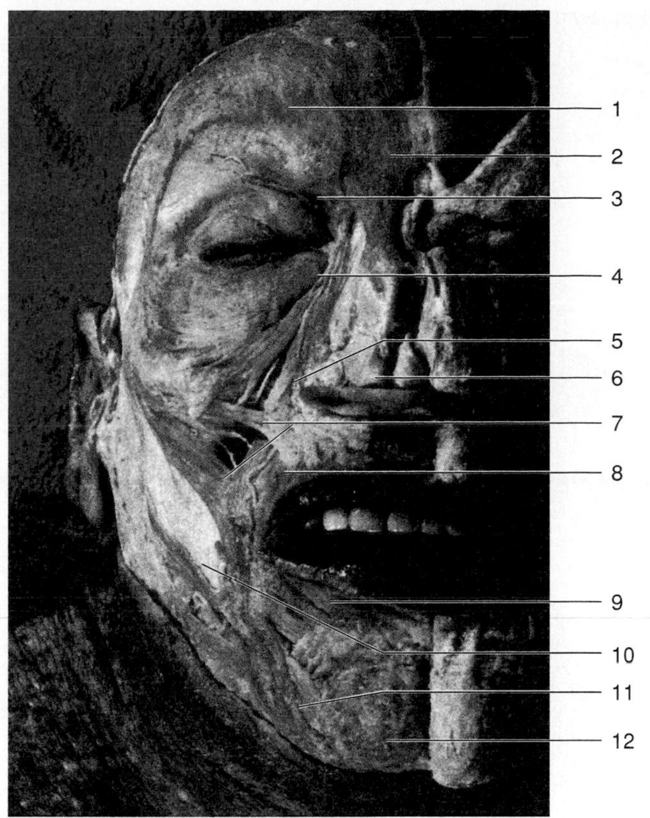

1	Frontalis m.	7	Zygomaticus mm.
2	Supratrochlear a.	8	Facial a.
3	Corrugator m.	9	Orbicularis oris m.
4	Orbicularis oculi m.	10	Risorius m.
5	Levator labii superioris m.	11	Depressor angularis oris m.
6	Alar cartilage	12	Mentalis m.

FIGURE 8

An anterior view of the muscles of the head. (Also refer to figure 8.18.)

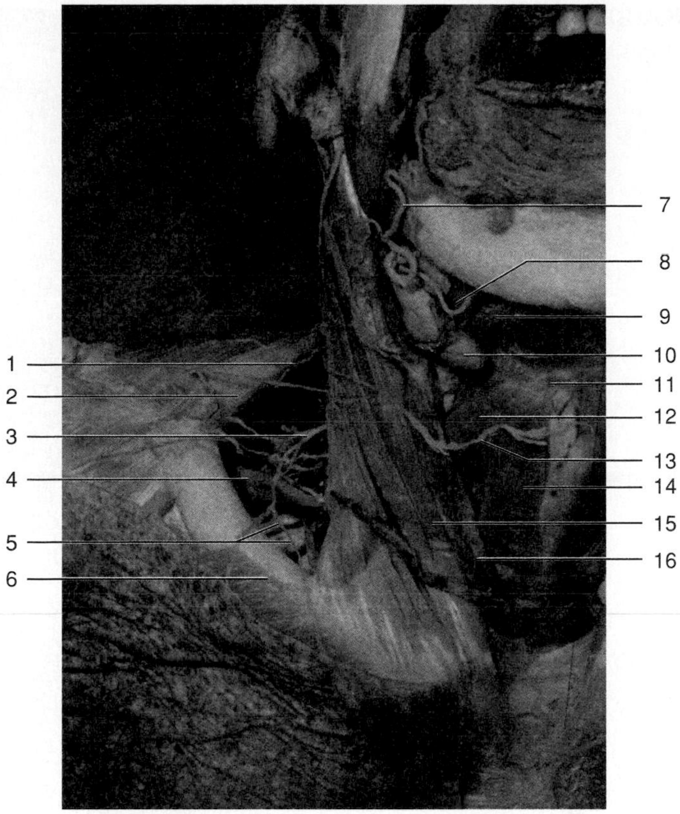

1	Accessory nerve	9	Digastric m.
2	Trapezius m.	10	Submandibular gland
3	Supraclavicular nerve	11	Hyoid bone
4	Omohyoid m.	12	Omohyoid m.
5	Brachial plexus	13	Transverse cervical nerve
6	Clavicle	14	Sternohyoid m.
7	Facial artery	15	Sternocleidomastoid m.
8	Mylohyoid m.	16	External jugular vein

FIGURE 9

An anterior view of the right cervical region. (Also refer to figure 8.23.)

FIGURE 10
A sagittal section of the head and neck.

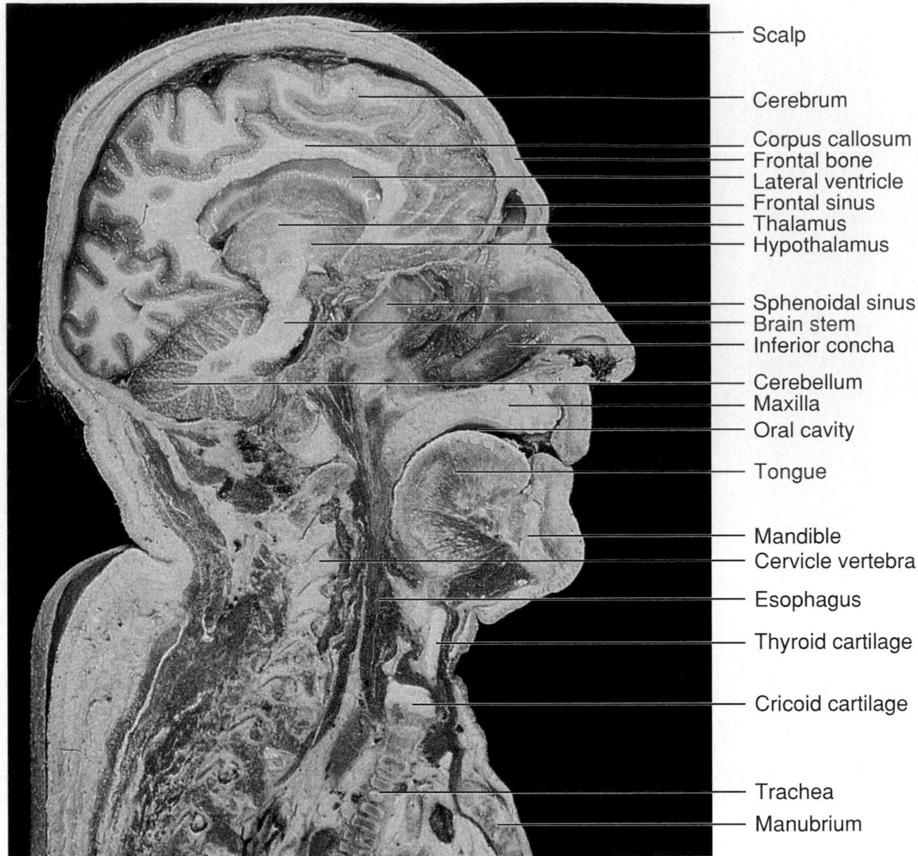

FIGURE 11
An anterior view of the right thorax, shoulder, and brachium. (Also refer to figures 8.28 and 8.30.)

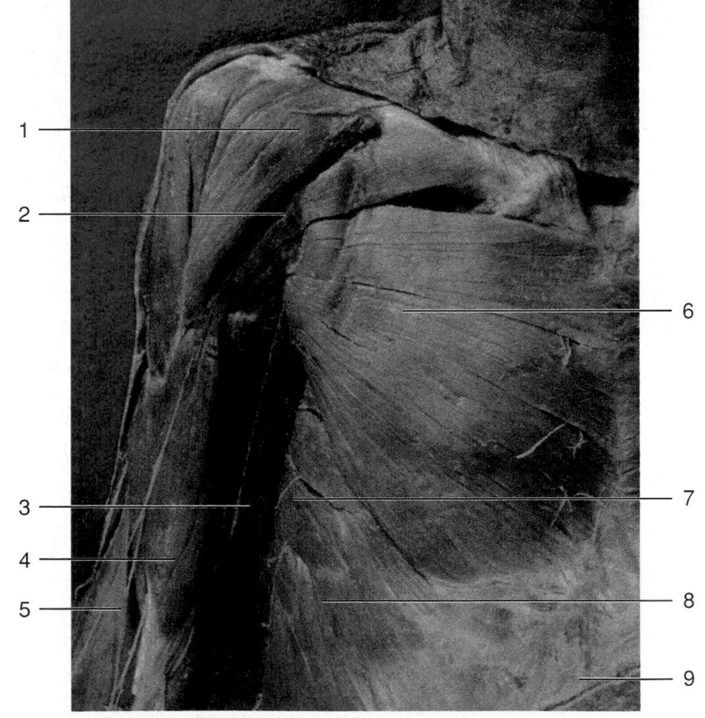

1 Deltoid m.
2 Cephalic vein
3 Latissimus dorsi m.
4 Biceps brachii m.
5 Brachioradialis m.
6 Pectoralis major m.
7 Serratus anterior m.
8 External abdominal oblique m.
9 Rectus sheath

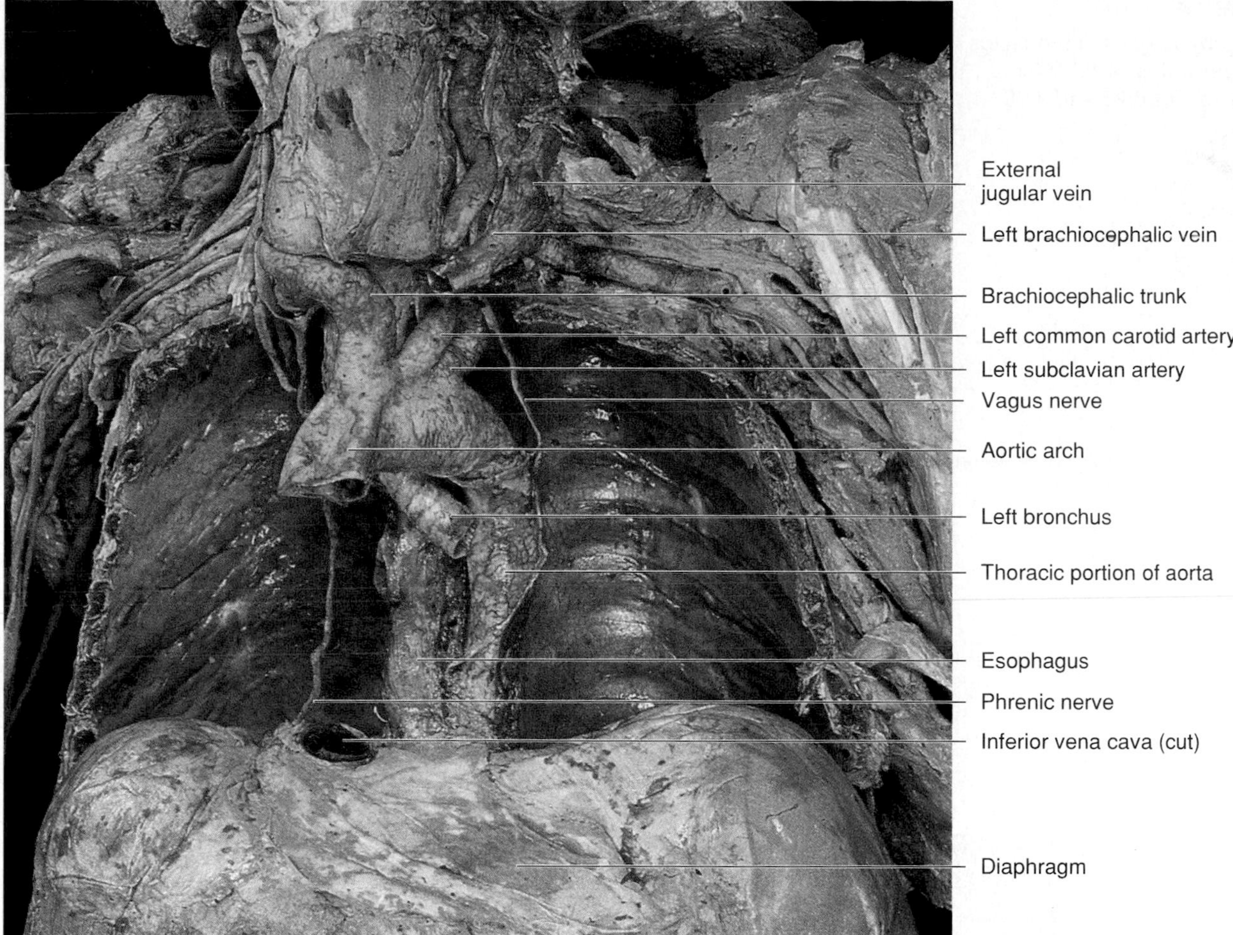

FIGURE 12

The thoracic cavity with the heart and lungs removed.

FIGURE 13

A posterior view of the right thorax and neck. (Also refer to figure 8.29.)

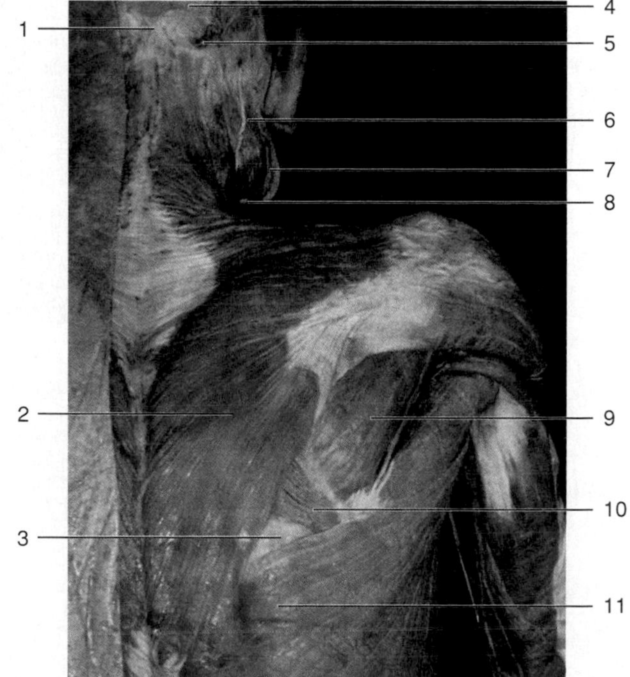

1 External occipital protuberance
2 Trapezius m.
3 Triangle of auscultation
4 Occipital artery
5 Greater occipital n.
6 Lesser occipital nerve
7 Sternocleidomastoid m.
8 Great auricular nerve
9 Infraspinatus m.
10 Rhomboideus major m.
11 Latissimus dorsi m.

FIGURE 14

An anterior view of the muscles of the abdominal wall. (Also refer to figures 8.25 and 8.28.)

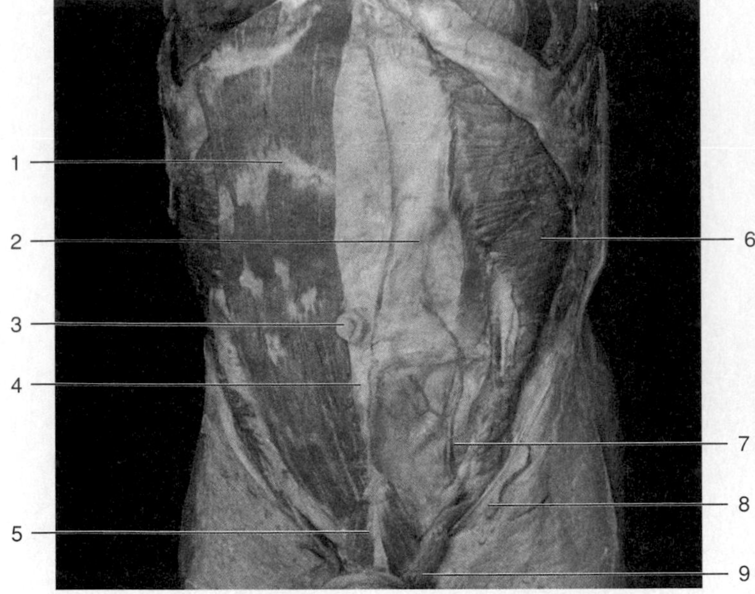

1 Rectus abdominis m.
2 Rectus sheath
3 Umbilicus
4 Linea alba
5 Pyramidalis m.
6 Transverse abdominis m.
7 Inferior epigastric artery
8 Inguinal ligament
9 Spermatic cord

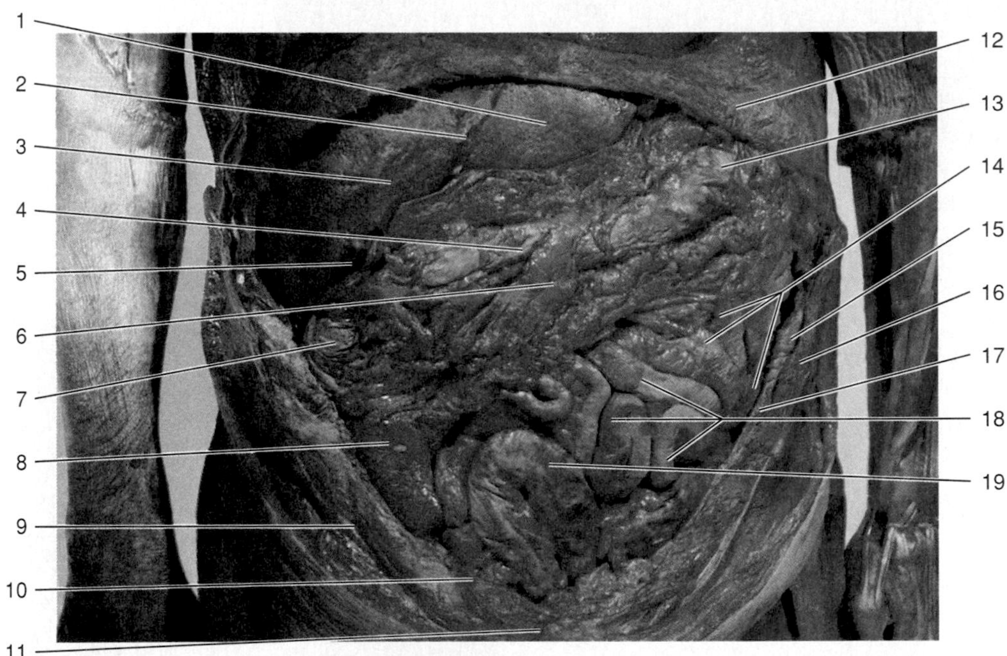

1 Left lobe of liver
2 Falciform ligament
3 Right lobe of liver
4 Transverse colon
5 Gallbladder
6 Greater omentum
7 Heptic (right colic) flexure
8 Fat deposit within greater omentum
9 Aponeurosis of internal abdominal oblique m.
10 Rectus abdominis m. (cut)
11 Rectus sheath (cut)
12 Diaphragm
13 Splenic (left colic) flexure
14 Jejunum
15 Transversus abdominis m. (cut)
16 Internal and external abdominal oblique mm. (cut)
17 Parietal peritoneum (cut)
18 Ileum
19 Sigmoid colon

FIGURE 15

Viscera of the abdomen.

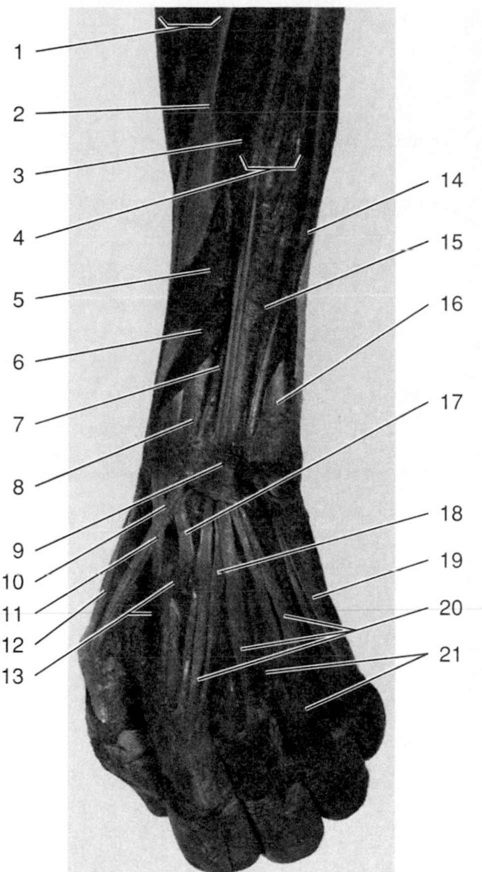

1 Brachioradialis m.
2 Extensor carpi radialis longus tendon
3 Extensor carpi radialis brevis m.
4 Extensor digitorum communis m.
5 Abductor pollicis longus m.
6 Extensor pollicis brevis m.
7 Extensor pollicis longus m.
8 Radius
9 Extensor retinaculum
10 Tendon of extensor carpi radialis longus m.
11 Tendon of extensor pollicis longus m.
12 Tendon of extensor pollicis brevis m.
13 First dorsal interosseous m.
14 Extensor carpi ulnaris m.
15 Extensor digiti minimi m.
16 Ulna
17 Tendon of extensor carpi radialis brevis m.
18 Tendon of extensor indicis m.
19 Tendon of extensor digiti minimi m.
20 Tendons of extensor digitorum m.
21 Intertendinous connections

FIGURE 16

A posterior view of the left forearm and hand. (Also refer to figures 8.32 and 8.33 for views of the right forearm.)

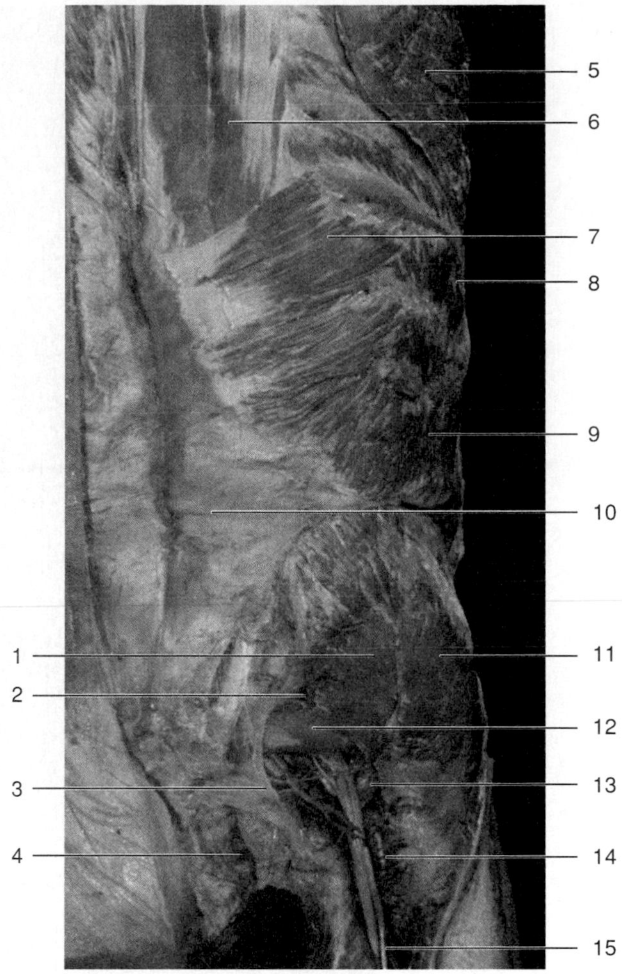

1 Superior gluteal vessels
2 Inferior gluteal vessels
3 Sacrotuberous ligament
4 Levator ani m.
5 Serratus anterior m.
6 Erector spinae m.
7 Serratus posterior m.
8 External intercostal m.
9 Internal abdominal oblique m.
10 Lumbar aponeurosis
11 Gluteus medius m.
12 Piriformis m.
13 Obturator internus m.
14 Quadratus femoris m.
15 Sciatic nerve

FIGURE 17

A posterior view of the deep muscles of the right abdominal and gluteal regions. (Also refer to figures 8.27, 8.28, and 8.29.)

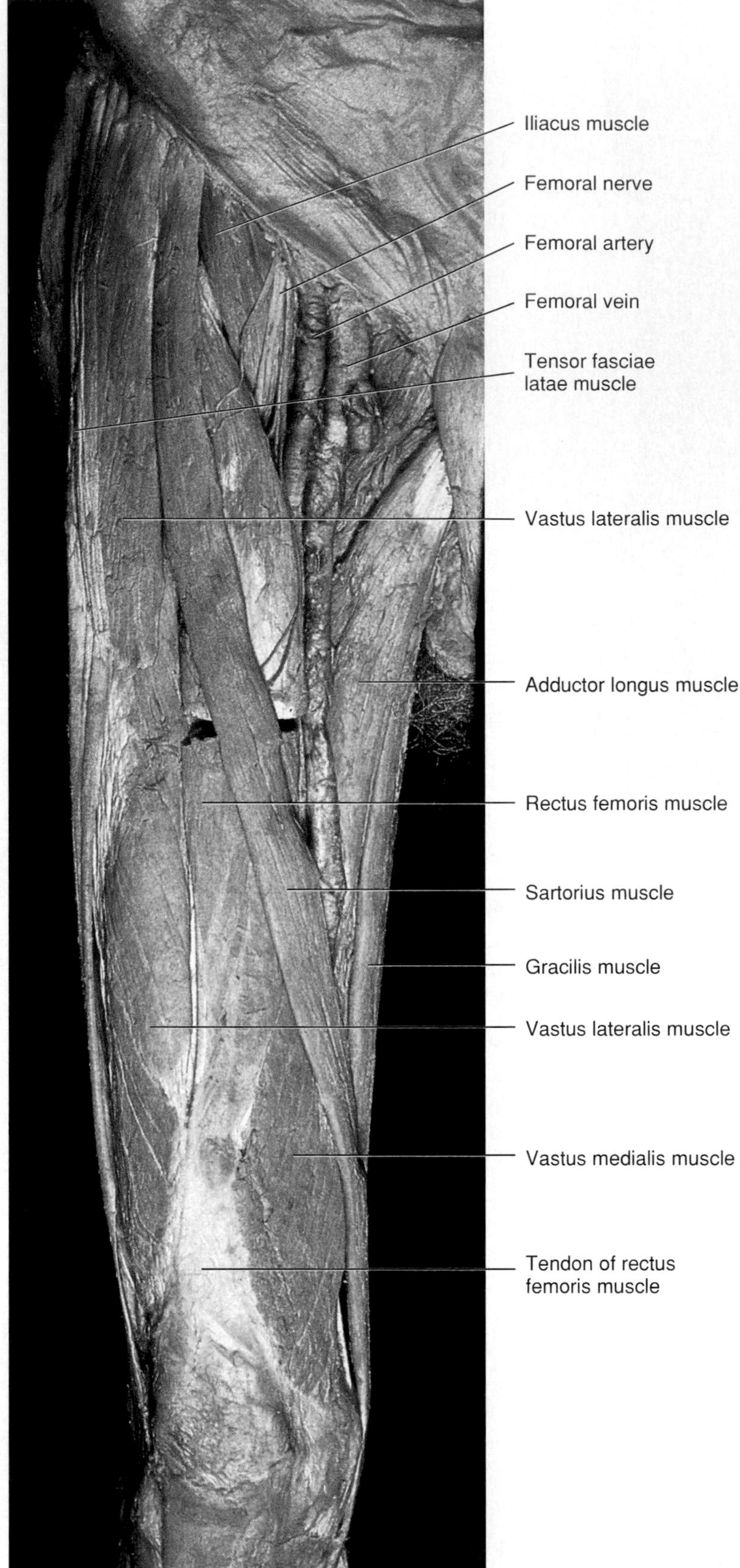

FIGURE 18
An anterior view of the right thigh. (Also refer to figures 8.36 and 8.37.)

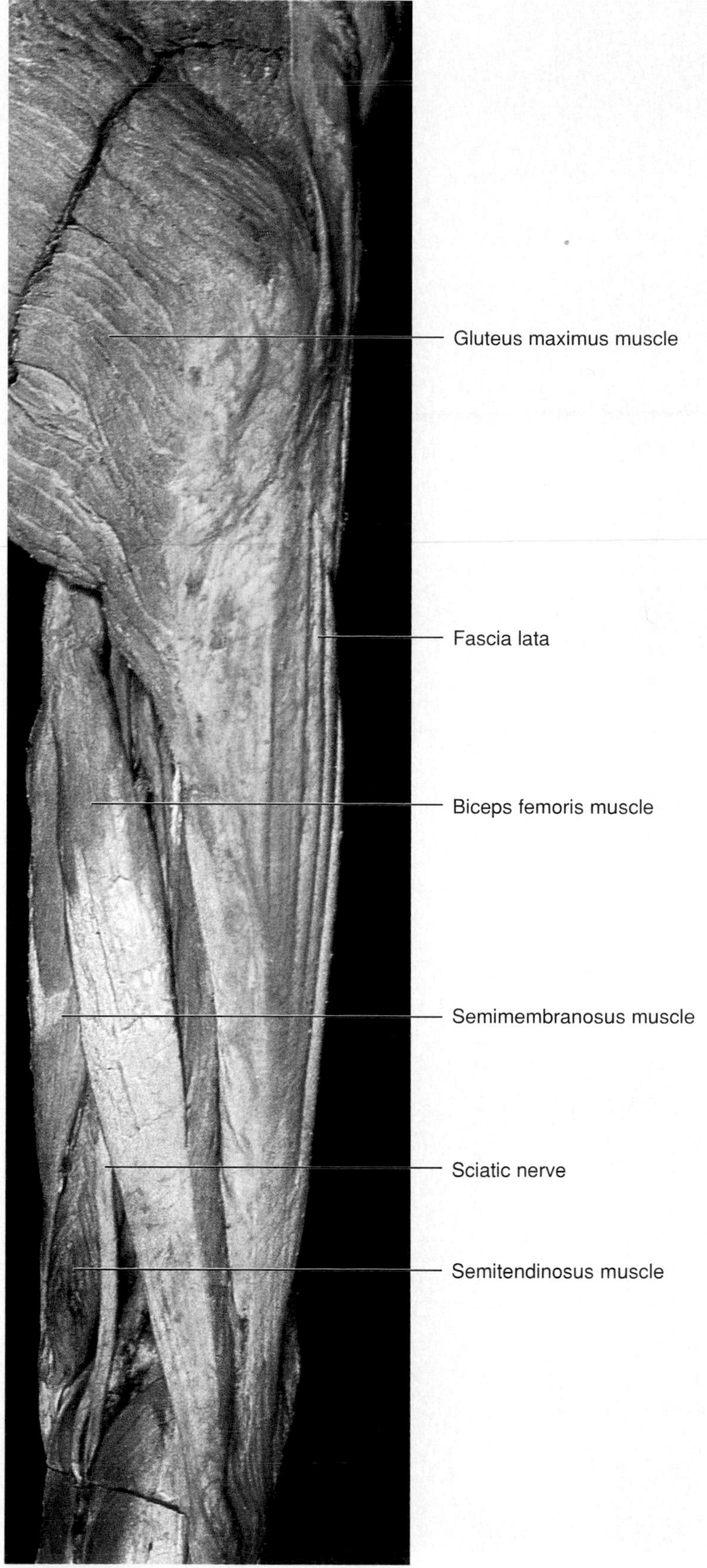

FIGURE 19
A posterior view of the right hip and thigh. (Also refer to figures 8.35 and 8.38.)

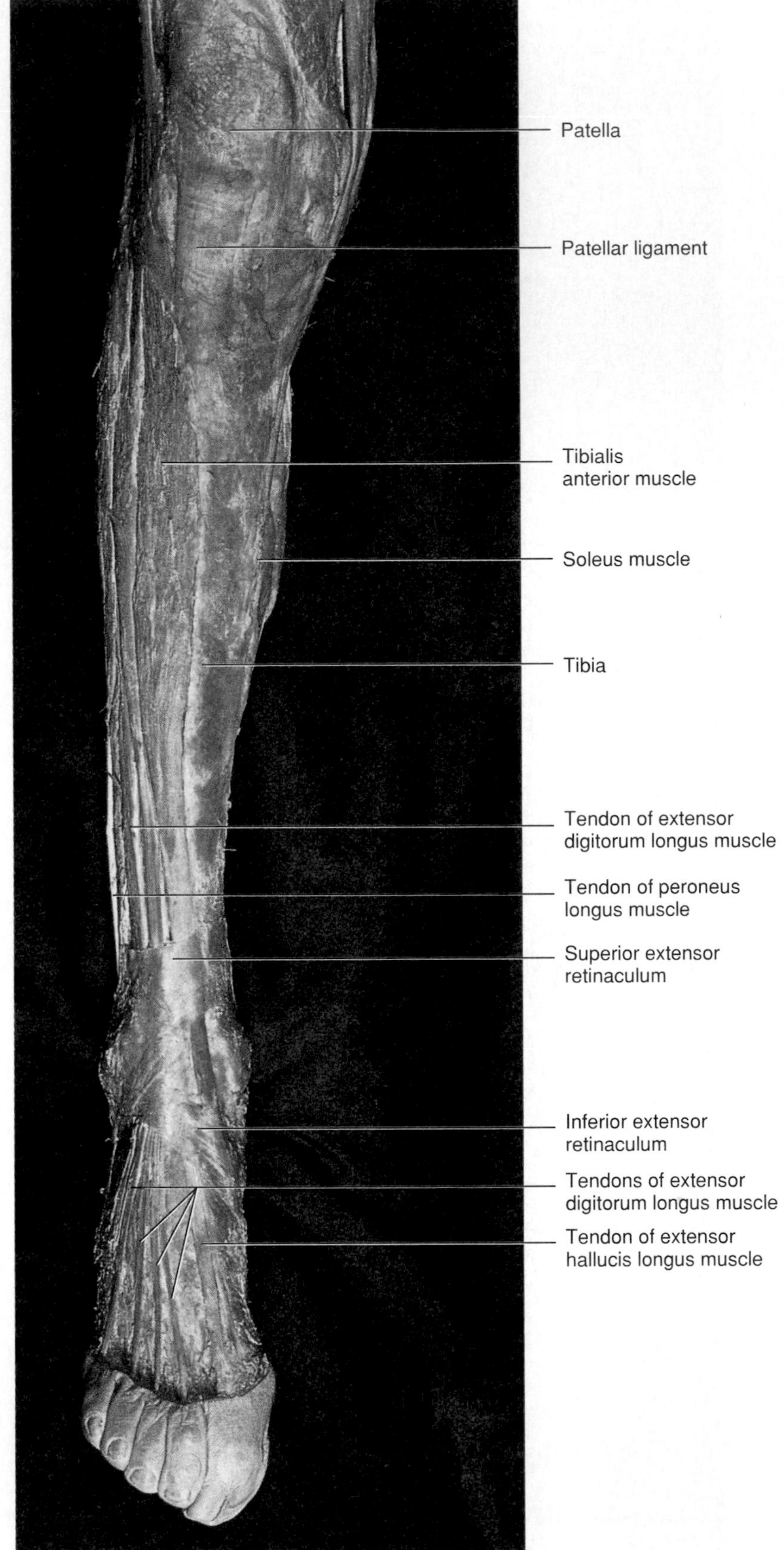

FIGURE 20
An anterior view of right leg. (Also refer to figure 8.39*a*.)

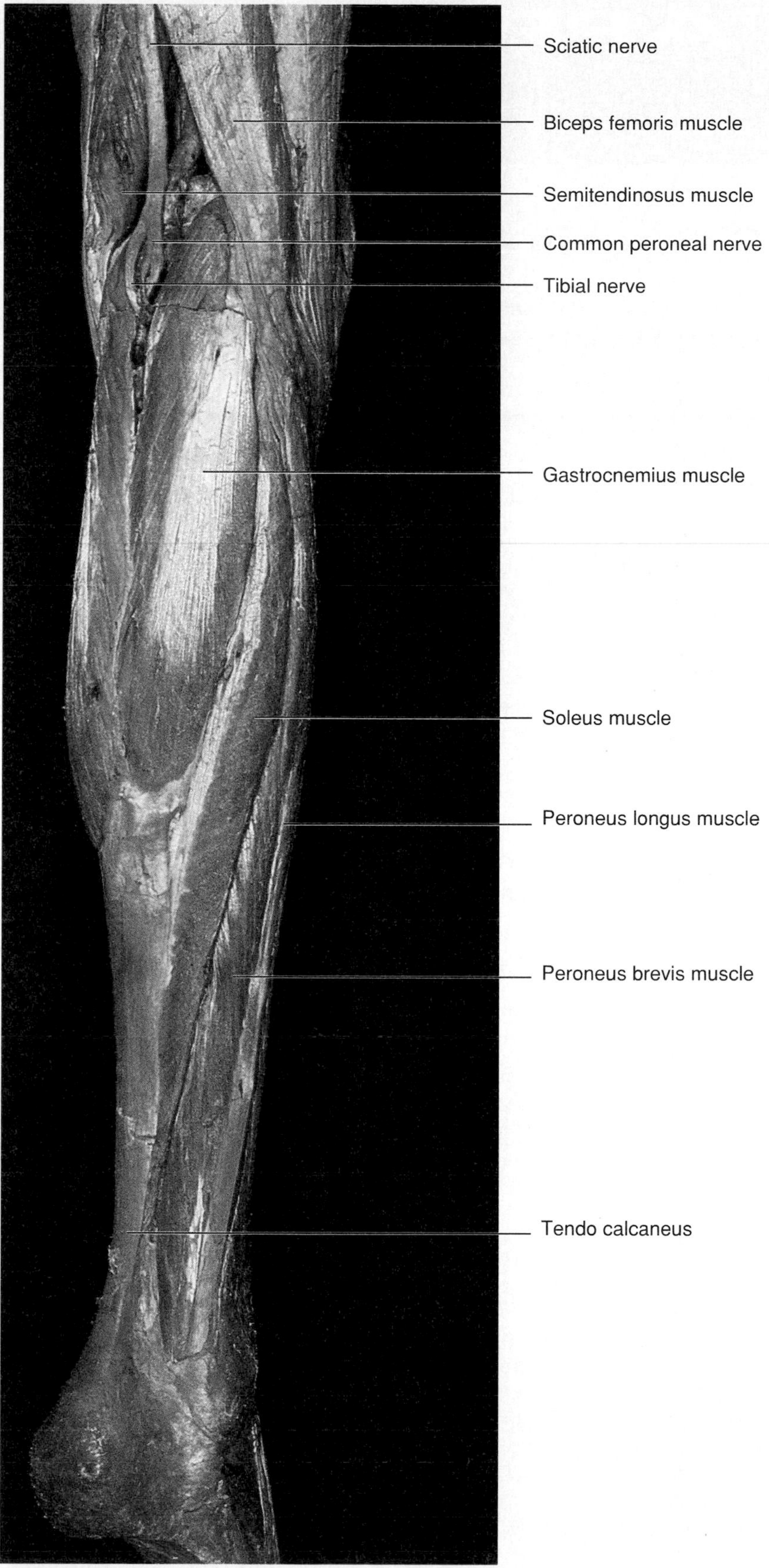

FIGURE 21
A posterior view of the right leg. (Also refer to figure 8.40.)

APPENDIX C

Prefixes and Suffixes in Anatomical and Medical Terminology

Element	Definition and Example
a-	absent, deficient, lack of: *atrophy*
ab-	off, away from: *abduct*
abdomin	relating to the abdomen: *abdominal*
-able	capable of: *viable*
ac-	toward, to: *actin*
acou-	hear: *acoustic*
ad-	toward, to: *adduct*
af-	movement toward a central point: *afferent artery*
alb-	white: *corpus albicans*
-algia	pain: *neuralgia*
ambi-	both: *ambidextrous*
angi-	pertaining to the vessels: *angiology*
ante-	before, in front of: *antebrachium*
anti-	against: *anticoagulant*
aque-	water: *aqueous*
arch-	beginning, origin: *archenteron*
arthr-	joint: *arthritis*
-asis	condition or state of: *homeostasis*
aud-	hearing, sound: *auditory*
auto-	self: *autolysis*
bi-	two: *bipedal*
bio-	life: *biopsy*
blast-	generative or germ bud: *blastocyst*
brachi-	arm: *brachialis*
brachy-	short: *brachydont*
brady-	slow: *bradycardia*
bucc-	cheek: *buccal cavity*
cac-	bad, ill: *cachexia*
calc-	stone: *calculus*
capit-	head: *capitis*
carcin-	cancer: *carcinogenic*
cardi-	heart: *cardiac*
cata-	lower, under, against: *catabolism*
caud-	tail: *cauda equina*
cephal-	head: *cephalic*
cerebro-	brain: *cerebrospinal fluid*
chol-	bile: *cholic*
chondr-	cartilage: *chondrocyte*
chrom-	color: *chromocyte*
-cide	destroy: *germicide*
circum-	around: *circumduct*
co-	together: *copulation*
coel-	hollow cavity: *coelom*
-coele	swelling, an enlarged space or cavity: *blastocoele*
con-	with, together: *congenital*
contra-	against, opposite: *contraception*
corn-	denoting hardness: *cornified*
corp-	body: *corpus*
crypt-	hidden: *cryptorchidism*

Element	Definition and Example
cyan-	blue: *cyanosis*
cyst-	sac or bladder: *cystoscope*
cyto-	cell: *cytology*
de-	down, from: *descent*
derm-	skin: *dermatology*
di-	two: *diarthrotic*
dipl-	double: *diploid*
dis-	apart, away from: *disarticulate*
duct-	lead, conduct: *ductus deferens*
dur-	hard: *dura mater*
dys-	bad, difficult, painful: *dysentery*
e-	out, from: *eccrine*
ec-	outside, outer, external: *ectoderm*
-ectomy	surgical removal: *tonsillectomy*
ede-	swelling: *edema*
-emia	pertaining to a condition of the blood: *lipemia*
en-	within: *endoderm*
enter-	intestine: *enteritis*
epi-	upon, over: *epidermis*
erythro-	red: *erythrocyte*
ex-	out of: *excise*
exo-	outside: *exocrine*
extra-	outside of, beyond, in addition: *extracellular*
fasci-	band: *fascia*
febr-	fever: *febrile*
-ferent	bear, carry: *efferent*
fiss-	split: *fissure*
for-	opening: *foramen*
-form	shape: *fusiform*
gastro-	relating to the stomach: *gastrointestinal*
-gen	an agent that produces or originates: *pathogen*
-genic	produced from, producing: *carcinogenic*
gloss-	tongue: *glossopharyngeal*
glyco-	sugar: *glycosuria*
-gram	a record, recording: *electroencephalogram*
gran-	grain, particle: *granulosa cells*
-graph	instrument for recording: *electrocardiograph*
gravi-	heavy: *gravid*
gyn-	female sex: *gynecology*
haplo-	simple or single: *haploid*
hem(at)-	blood: *hematology*
hemi-	half: *hemiplagia*
hepat-	liver: *hepatic portal*
hetero-	other, different: *heterosexual*
histo-	web, tissue: *histology*
holo-	whole, entire: *holocrine*
homo-	same, alike: *homologous*
hydro-	water: *hydrocoel*
hyper-	beyond, above, excessive: *hypertension*
hypo-	under, below: *hypoglycemia*

Element	Definition and Example
-ia	abnormal state or condition: *hypoglycemia*
-iatrics	medical specialties: *pediatrics*
idio-	self, separate, distinct: *idiopathic*
ilio-	ilium: *iliosacral*
infra-	beneath: *infraspinatus*
inter-	among, between: *interosseous*
intra-	inside, within: *intracellular*
-ion	process: *acromion*
-ism	condition or state: *dimorphism*
iso-	equal, like: *isotonic*
-itis	inflammation: *meningitis*
labi-	lip: *labium majus*
lacri-	tears: *lacrimal apparatus*
later-	side: *lateral*
leuk-	white: *leukocyte*
lip-	fat: *lipid*
-logy	science of: *morphology*
-lysis	solution, dissolve: *hemolysis*
macro-	large, great: *macrophage*
mal-	bad, abnormal, disorder: *malignant*
medi-	middle: *medial*
mega-	great, large: *megakaryocyte*
meso-	middle or moderate: *mesoderm*
meta-	after, beyond: *metatarsal*
micro-	small: *microtome*
mito-	thread: *mitochondrion*
mono-	alone, one, single: *monocyte*
morph-	form, shape: *morphology*
multi-	many, much: *multinuclear*
myo-	muscle: *myology*
narc-	numbness, stupor: *narcotic*
necro-	corpse, dead: *necrosis*
neo-	new, young: *neonatal*
nephr-	kidney: *nephritis*
neuro-	nerve: *neurolemma*
noto-	back: *notochord*
ob-	against, toward, in front of: *obturator*
oc-	against: *occlusion*
-oid	resembling, likeness: *sigmoid*
oligo-	few, small: *oligodendrocyte*
-oma	tumor: *lymphoma*
oo-	egg: *oocyte*
or-	mouth: *oral*
orchi-	testis: *orchiectomy*
ortho-	straight, normal: *orthopnea*
-ory	pertaining to: *sensory*
-ose	full of: *adipose*
osteo-	bone: *osteoblast*
oto-	ear: *otolith*
ovo-	egg: *ovum*
par-	give birth to, bear: *parturition*
para-	near, beyond, beside: *paranasal*
path-	disease, that which undergoes sickness: *pathology*
-pathy	abnormality, disease: *neuropathy*
ped-	children: *pediatrician*
pen-	need, lack: *penicillin*
-penia	deficiency: *thrombocytopenia*
per-	through: *percutaneous*
peri-	near, around: *pericardium*
phag-	to eat: *phagocyte*
-phil	have an affinity for: *neutrophil*

Element	Definition and Example
phleb-	vein: *phlebitis*
-phobia	abnormal fear, dread: *hydrophobia*
-plasty	reconstruction of: *rhinoplasty*
platy-	flat, side: *platysma*
-plegia	stroke, paralysis: *paraplegia*
-pnea	to breathe: *apnea*
pneumo(n)-	lung: *pneumonia*
pod-	foot: *podiatry*
-poiesis	formation of: *hemopoiesis*
poly-	many, much: *polyploid*
post-	after, behind: *postnatal*
pre-	before in time or place: *prenatal*
pro-	before in time or place: *prophase*
proct-	anus: *proctology*
pseudo-	false: *pseudostratified*
psycho-	mental: *psychology*
pyo-	pus: *pyorrhea*
quad-	fourfold: *quadriceps femoris*
re-	back, again: *repolarization*
rect-	straight: *rectus abdominis*
ren-	kidney: *renal*
rete-	network: *rete testis*
retro-	backward, behind: *retroperitoneal*
rhin-	nose: *rhinitis*
-rrhagia	excessive flow: *menorrhagia*
-rrhea	flow or discharge: *diarrhea*
sanguin-	blood: *sanguineous*
sarc-	flesh: *sarcoma*
-scope	instrument for examining a part: *stethoscope*
-sect	cut: *dissect*
semi-	half: *semilunar*
-sis	process or action: *dialysis*
steno-	narrow: *stenosis*
-stomy	surgical opening: *tracheostomy*
sub-	under, beneath, below: *subcutaneous*
super-	above, beyond, upper: *superficial*
supra-	above, over: *suprarenal*
syn- (sym-)	together, joined, with: *synapse*
tachy-	swift, rapid: *tachycardia*
tele-	far: *telencephalon*
tens-	stretch: *tensor tympani*
tetra-	four: *tetrad*
therm-	heat: *thermogram*
thorac-	chest: *thoracic cavity*
thrombo-	lump, clot: *thrombocyte*
-tomy	cut: *appendectomy*
tox-	poison: *toxemia*
tract-	draw, drag: *traction*
trans-	across, over: *transfuse*
tri-	three: *trigone*
trich-	hair: *trichology*
-trophy	a state relating to nutrition: *hypertrophy*
-tropic	turning toward, changing: *gonadotropic*
ultra-	beyond, excess: *ultrasonic*
uni-	one: *unicellular*
-uria	urine: *polyuria*
uro-	urine, urinary organs or tract: *uroscope*
vas-	vessel: *vasoconstriction*
viscer-	organ: *visceral*
vit-	life: *vitamin*
zoo-	animal: *zoology*
zygo-	union, join: *zygote*

GLOSSARY

Pronunciation Key

Most of the words in this glossary are followed by a phonetic spelling that serves as a guide to pronunciation. The phonetic spellings reflect standard scientific usage and can be easily interpreted following a few basic rules.

1. Any unmarked vowel that ends a syllable or that stands alone as a syllable has the long sound. For example, *ba, ma,* and *na* rhyme with *fay; be, de,* and *we* rhyme with *fee; bi, di,* and *pi* rhyme with *sigh; bo, do,* and *mo* rhyme with *go.* Any unmarked vowel that is followed by a consonant has the short sound (for example, the vowel sounds in *hat, met, pit, not,* and *but*).
2. If a long vowel appears in the middle of a syllable (followed by a consonant), it is marked with a macron (¯). Similarly, if a vowel stands alone or ends a syllable but should have a short sound, it is marked with a breve (˘) .
3. Syllables that are emphasized are indicated by stress marks. A single stress mark (′) indicates the primary emphasis; a secondary emphasis is indicated by a double stress mark (″).

Page numbers indicate where entries may be found in the text.

A

abdomen (ab′dŏ-men) The portion of the trunk between the diaphragm and pelvis. 13

abduction A movement of a body part away from the main axis of the body. 164

ABO system The major class of red blood cell antigens. On the basis of antigens on the red blood cell surface, individuals can be type A, type B, type AB, or type O. 362

absorption The transport of molecules across epithelial membranes into the body fluids. 504

acetone (as′ĕ-tōn) A ketone body produced as a result of the oxidation of fats. 554

acetyl (as′ĕ-tl, ă-set′l) **coenzyme A (acetyl CoA)** A coenzyme that contributes two-carbon-long substrates to the Krebs cycle. 549

acetylcholine (ă-set″l-ko′lēn) **(ACh)** A substance that functions as a neurotransmitter in the CNS, somatic motor nerve fibers, and parasympathetic nerve fibers. 239

acetylcholinesterase (ă-set″l-ko″lĭ-nes′tĕ-rās) An enzyme in the membrane of postsynaptic cells that catalyzes the conversion of ACh into choline and acetic acid. This enzymatic reaction inactivates the neurotransmitter. 240

acid A substance that releases hydrogen ions when ionized in water. 27

acidosis An abnormal increase in the H^+ concentration of the blood that lowers the arterial pH to below 7.35. 469

actin A protein in muscle fibers that, together with myosin, is responsible for contraction. 178

action potential An all-or-none electrical event in an axon or muscle fiber in which the polarity of the membrane potential is rapidly reversed and reestablished. 236

active immunity Specific immunity involving prior sensitization. A rapid immune response is produced upon subsequent exposure to the antigen. 435

active transport The movement of molecules or ions across the cell membranes of epithelial cells by membrane carriers. An expenditure of cellular energy (ATP) is required. 53

adduction A movement of a body part toward the main axis of the body. 164

adenohypophysis (ad″n-o-hi-pof′ĭ-sis) The anterior glandular lobe of the pituitary gland that secretes FSH (follicle-stimulating hormone), LH (luteinizing hormone), ACTH (adrenocorticotropic hormone), TSH (thyroid-stimulating hormone), GH (growth hormone), and prolactin. Secretions of the adenohypophysis are controlled by hormones produced by the hypothalamus. 336

adenoids (ad′ĕ-noidz) The tonsils located in the nasopharynx; also called *pharyngeal tonsils.* 452

adenylate cyclase (ă-den′l-it si′klās) An enzyme found in cell membranes that catalyzes the conversion of ATP to cyclic AMP and pyrophosphate (PP_i). This enzyme is activated by an interaction between a specific hormone and its membrane receptor protein. 349

ADH Antidiuretic hormone; the hormone produced by the hypothalamus and released by the posterior pituitary that acts on the kidneys to promote water reabsorption; also called *vasopressin.* 412, 488

adipocyte (ad′ĭ-po-sīt) A fat cell found within adipose tissue. 83

ADP Adenosine diphosphate; a molecule that together with inorganic phosphate is used to make ATP (adenosine triphosphate). 37

adrenal cortex (ă-dre′nal kor′teks) The outer part of the adrenal gland that secretes corticosteroid hormones (such as aldosterone and hydrocortisone). 341

adrenal gland One of two endrocrine glands that cap the superior borders of the kidneys. 341

adrenal medulla (mĕ-dul′ă) The inner part of the adrenal gland that secretes catecholamine hormones—epinephrine and (to a lesser degree) norepinephrine. 341

adrenergic (ad″rĕ-ner′jik) A term used to describe the actions of epinephrine, norepinephrine, or other molecules with similar activity (as in *adrenergic receptor* and *adrenergic stimulation*). 293

aerobic (ă-ro′bik) **respiration** The complete breakdown of organic substrates to carbon dioxide and water. This process yields large amounts of ATP and requires mitochondria and oxygen. 549

afferent (af′er-ent) Conveying or transmitting to. 232

afferent arteriole (ar-tir'e-ōl) A blood vessel within the kidney that supplies blood to the glomerulus. 480

afferent neuron (noor'on) *See* sensory neuron. 232

agglutinate (ă-gloot'n-āt) A clump of cells (usually erythrocytes) formed as a result of specific chemical interaction between surface antigens and antibodies. 364

agranular leukocytes (a-gran'yŭ-lar loo'kŏ-sīts) White blood cells (leukocytes) that do not contain cytoplasmic granules; specifically, lymphocytes and monocytes. 359

albumin (al-byoo'min) A water-soluble protein produced in the liver; the major component of the plasma proteins. 358

aldosterone (al-dos'ter-ōn) The principal corticosteroid hormone involved in the regulation of electrolyte balance (a mineralocorticoid). 413, 491

alkalosis An abnormal decrease in the H^+ concentration of the blood that raises the arterial pH to above 7.45. 469

allantois (ă-lan'to-is) An extraembryonic membrane. It gives rise to the umbilical blood vessels and contributes to the formation of the urinary bladder. 622

alleles (ă-lēlz') Alternative forms of a gene that affect the same characteristic but that produce different forms of that characteristic. 632

allergens (al'er-jenz) Antigens that evoke an allergic response rather than a normal immune response. 444

allergy A state of hypersensitivity caused by exposure to allergens. It results in the liberation of histamine and other molecules with histamine-like effects. 444

all-or-none principle The statement that muscle fibers of a motor unit contract to their maximum extent when exposed to a stimulus of threshold strength. 236

alveolus (al-ve'ŏ-lus) 1. An individual air capsule within the lung. The respiratory alveoli are the basic functional units of respiration. 2. The socket that secures a tooth (tooth socket). 456

amino (ă-me'no) **acid** Any of a class of organic acids that comprise the building blocks for proteins. 544

amnion (am'ne-on) The innermost fetal membrane—a thin sac that holds the fetus suspended in amniotic fluid; also called the *bag of waters.* 621

ampulla of Vater (fă'ter) *See* hepatopancreatic ampulla. 519

anabolic steroids (an"ă-bol'ik ster'oidz) Steroids with androgen-like stimulatory effects on protein synthesis. 218

anabolism (ă-nab'ŏ-liz"em) A phase of metabolism involving chemical reactions within cells that result in the production of larger molecules from smaller ones; specifically, the synthesis of protein, glycogen, and fat. 544

anaerobic (an-ă-ro'bik) **respiration** A form of cell respiration involving the conversion of glucose to lactic acid in which energy is obtained without the use of molecular oxygen; also called *lactic acid fermentation.* 548

anal (a'nal) **canal** The terminal tubular portion of the large intestine that opens through the anus of the GI tract. 523

anatomical position An erect body stance with the eyes directed forward, the arms at the sides, the palms of the hands facing forward, and the fingers pointing straight down. 10

anatomy The branch of science concerned with the structure of the body and the relationship of its organs. 2

androgens (an'drŏ-jenz) Steroids containing 18 carbons that have masculinizing effects; primarily testosterone secreted by the testes, although weaker androgens are also secreted by the adrenal cortex. 567

anemia (ă-ne'me-ă) An abnormal reduction in the red blood cell count, hemoglobin concentration, hematocrit, or any combination of these measurements. This condition is associated with a decreased ability of the blood to carry oxygen. 359

angiotensin II (an"je-o-ten'sin) An 8-amino-acid polypeptide formed from angiotensin I (a 10-amino-acid precursor), which in turn is formed from cleavage of a protein (angiotensinogen) by the action of renin (an enzyme secreted by the kidneys). Angiotensin II is a powerful vasoconstrictor and a stimulator of aldosterone secretion from the adrenal cortex. 413

anion (an'i-on) A negatively charged ion, such as chloride, bicarbonate, or phosphate. 26

antagonist A muscle that acts in opposition to another muscle. 189

anterior Toward the front; also called *ventral.* 10

anterior pituitary (pĭ-too'ĭ-ter-e) *See* adenohypophysis. 336

anterior root The anterior projection of the spinal cord, composed of axons of motor neurons. 275

antibodies Immunoglobuin proteins secreted by B lymphocytes that have transformed into plasma cells. Antibodies are responsible for humoral immunity. Their synthesis is induced by specific antigens, and they combine with these antigens but not with unrelated antigens. 433

anticodon (an"tĭ-ko'don) A base triplet provided by three nucleotides within a loop of transfer RNA that is complementary in its base-pairing properties to a triplet (the codon) in mRNA. The matching of codon to anticodon provides the mechanism for translating the genetic code into a specific sequence of amino acids. 63

antigen (an'tĭ-jen) A molecule that can induce the production of antibodies and react in a specific manner with antibodies. 432

antiserum (an'tĭ-sir"um) A serum that contains specific antibodies. 437

anus (a'nus) The terminal opening of the GI tract. 523

aorta (a-or'tă) The major systemic vessel of the arterial system of the body, emerging from the left ventricle. 383

apneustic (ap-noo'stik) **center** A collection of nuclei (nerve cell bodies) in the brain stem that participates in the rhythmic control of breathing. 464

apocrine (ap'ŏ-krin) **gland** A type of sweat gland that functions in evaporative cooling. It may respond during periods of emotional stress. 107

appendix A short pouch that attaches to the cecum. 522

aqueous (a'kwe-us) **humor** The watery fluid that fills the anterior and posterior chambers of the eye. 314

arachnoid (ă-rak'noid) **mater** The weblike middle covering (meninx) of the central nervous system. 265

arbor vitae (ar'bor vi'te) The branching arrangement of white matter within the cerebellum. 263

arteriole (ar-tir'e-ōl) A minute arterial vessel. 357

artery A blood vessel that carries blood away from the heart. 357, 381

articular cartilage (ar-tik'yŭ-lar kar'tl-ij) A hyaline cartilaginous covering over the articulating surface of the bones of synovial joints. 122

articulation A joint. 155

ascending colon (ko'lon) The portion of the large intestine between the cecum and the hepatic flexure. 523

association neuron (noor'on) A nerve cell located completely within the central nervous system. It conveys impulses in an arc from sensory to motor neurons; also called an *interneuron* or *internuncial neuron.* 233

atom The smallest particle of matter that characterizes an element. 24

atomic number The number of protons in the nucleus of an atom. 24

ATP Adenosine triphosphate; the universal energy donor of the cell. 37

atrial natriuretic (na"trĭ-yoo-ret'ik) **factor** A substance secreted by the atria of the heart that promotes the urinary excretion of sodium, thereby reducing blood volume and pressure. 492

atrioventricular (a"tre-o-ven-trik'yŭ-lar) **bundle** A group of specialized cardiac fibers that conduct impulses from the atrioventricular node to the ventricular muscles of the heart; also called the *bundle of His* or *AV bundle.* 377

atrioventricular node The microscopic aggregation of specialized cardiac fibers located in the interatrial septum of the heart that is part of the conduction system of the heart; AV node. 377

atrioventricular valve A cardiac valve located between an atrium and a ventricle of the heart; also called an *AV valve.* 373

atrium (a′tre-um) Either of the two superior chambers of the heart that receive venous blood. 373

atrophy (at′rŏ-fe) A gradual wasting away or decrease in the size of a tissue or an organ. 92

auditory Pertaining to the structures of the ear associated with hearing.

auditory ossicles Three tiny bones of the middle ear: the malleus, incus, and stapes. 319

auditory tube The narrow canal that connects the middle ear chamber to the pharynx; also called the *eustachian canal.* 319

auricle (or′ĭ-kul) 1. The fleshy pinna of the ear. 318 2. An ear-shaped appendage of each atrium of the heart.

autoantibodies Antibodies formed in response to, and that react with, molecules that are part of one's own body. 444

autonomic nervous system The sympathetic and parasympathetic portions of the nervous system that function to control the actions of the visceral organs and skin; ANS. 288

autoregulation The ability of an organ to regulate the rate of its own blood flow. 414

autosomal chromosomes (aw″to-so′mal kro′mŏ-sōmz) The paired chromosomes; those other than the sex chromosomes. 631

axilla (ak-sil′ă) The depressed hollow commonly called the armpit. 12

axon (ak′son) The elongated process of a nerve cell that transmits an impulse away from the cell body of a neuron. 232

B

baroreceptor (bar″o-re-sep′tor) A cluster of neuroreceptors stimulated by blood pressure changes. 416

basal metabolic (met″ă-bol′ik) **rate (BMR)** The rate of metabolism (expressed as oxygen consumption or heat production) under resting or basal conditions (14 to 18 hours after eating). 556

basal nucleus (noo′kle-us) A mass of nerve cell bodies located deep within a cerebral hemisphere of the brain; also called *basal ganglion.* 256

base A chemical substance that ionizes in water to release hydroxyl ions (OH^-) or other ions that combine with hydrogen ions. 27

basement membrane A thin sheet of extracellular substance to which the basal surfaces of membranous epithelial cells are attached; also called the *basal lamina.* 75

B cell lymphocytes (lim′fŏ-sīts) Lymphocytes that can be transformed by antigens into plasma cells that secrete antibodies (and are thus responsible for humoral immunity). The *B* stands for *bursa equivalent.* 432

benign (bĭ-nīn′) Not malignant.

bile A liver secretion that is stored and concentrated in the gallbladder and released through the common bile duct into the duodenum. It is essential for the absorption of fats. 525

bilirubin (bil″ĭ-roo′bin) Bile pigment derived from the breakdown of the heme portion of hemoglobin. 525

blastocyst (blas′tŏ-sist) An early stage of embryonic development consisting of a hollow ball of cells with an inner cell mass and an outer layer called the trophoblast. 618

blood The fluid connective tissue that circulates through the cardiovascular system to transport substances throughout the body. 86, 357

blood-brain barrier A specialized mechanism that inhibits the passage of certain materials from the blood into brain tissue and cerebrospinal fluid. 267

Bohr effect A weakening of the hemoglobin-oxygen bond with declining pH. 468

bolus (bo′lus) A moistened mass of food that is swallowed from the oral cavity into the pharynx. 505

bone A solid, rigid, ossified connective tissue that forms an organ of the skeletal system. 85, 122

Bowman's (bo′manz) **capsule** *See* glomerular capsule. 481

brachial plexus (bra′ke-al plek′sus) A network of nerve fibers arising from spinal nerves C5–C8 and T1. Nerves arising from the brachial plexuses supply the upper extremities. 276

bradycardia (brad″ĭ-kar′de-ă) A slow cardiac rate; fewer than 60 beats per minute. 405

bradykinins (brad″ĭ-ki′ninz) Short polypeptides that stimulate vasodilation and other cardiovascular changes. 415

brain The enlarged superior portion of the central nervous system located in the cranial cavity of the skull. 248

brain stem The portion of the brain consisting of the midbrain, pons, and medulla oblongata. 260

bronchial (brong′ke-al) **tree** The bronchi and their branching bronchioles. 455

bronchiole (brong′ke-ōl) A small division of a bronchus within the lung. 455

bronchus (brong′kus) A branch of the trachea that leads to a lung. 455

buccal (buk′al) **cavity** The mouth, or oral cavity. 511

buffer A system of molecules and ions that serves to prevent large changes in pH by either combining with H^+ or by releasing H^+ into solution. 28

bulbourethral (bul″bo-yoo-re′thral) **glands** A pair of glands that secrete a viscous fluid into the male urethra during sexual excitement; also called *Cowper's glands.* 580

bundle of His *See* atrioventricular bundle. 377

bursa A saclike structure filled with synovial fluid. Bursae are located at friction points, as around joints, over which tendons can slide without contacting bone. 160

C

calcitonin (kal″si-to′nin) A polypeptide hormone produced by the parafollicular cells of the thyroid and secreted in response to hypercalcemia. It acts to lower blood calcium and phosphate concentrations and may serve as an antagonist of parathyroid hormones; also called *thyrocalcitonin.* 344

calorie A unit of heat equal to the amount needed to raise the temperature of one gram of water by 1 C°. 556

calyx (ka′liks) A cup-shaped portion of the renal pelvis that encircles a renal papilla. 479

cAMP *See* cyclic AMP. 349

canal of Schlemm (shlem) *See* scleral venous sinus. 314

capillary A microscopic blood vessel that connects an arteriole and a venule; the functional unit of the circulatory system. 357, 381

carbohydrate Any of the group of organic molecules composed of carbon, hydrogen, and oxygen, including sugars and starches; usually has the formula CH_2O. 29, 531

carbonic anhydrase (kar-bon′ik an-hi′drāse) An enzyme that catalyzes the formation or breakdown of carbonic acid. When carbon dioxide concentrations are relatively high, this enzyme catalyzes the formation of carbonic acid from CO_2 and H_2O. When carbon dioxide concentrations are low, the breakdown of carbonic acid to CO_2 and H_2O is catalyzed. These reactions aid the transport of carbon dioxide from tissues to alveolar air. 469

cardiac (kar′de-ak) **muscle** Muscle of the heart, consisting of striated muscle cells. These cells are interconnected into a mass called the myocardium. 88

cardiac output The volume of blood pumped each minute by either the right or left ventricle. 405

carotid(kă-rot′id) **sinus** An expanded portion of the internal carotid artery located immediately above the point of branching from the external carotid artery. The carotid sinus contains baroreceptors that monitor blood pressure. 383

carpus (kar′pus) The proximal portion of the hand that contains the eight carpal bones. 142

carrier-mediated transport The transport of molecules or ions across a cell membrane by means of specific protein carriers. It includes both facilitated diffusion and active transport. 52, 53

cartilage (kar′tl-ij) A type of connective tissue with a solid organic matrix. 84

cartilaginous (kar″tĭ-laj′ĭ-nus) **joint** A joint that lacks a joint cavity, permitting little movement between the bones held together by cartilage. 158

catabolism (kă-tab′ŏ-liz″em) The metabolic breakdown of complex molecules into simpler ones, often resulting in a release of energy. 544

catecholamine (kat″ĕ-kol′ă-mēn) Any of a group of molecules, including epinephrine, norepinephrine, L-dopa, and related molecules, with effects similar to those produced by activation of the sympathetic nervous system. 335

cation (kat′i-on) A positively charged ion, such as sodium, potassium, calcium, or magnesium. 26

cauda equina (kaw′dă e-kwi′nă) The lower end of the spinal cord, where the roots of spinal nerves have a tail-like appearance. 272

cecum (se′kum) The pouchlike portion of the large intestine to which the ileum of the small intestine is attached. 522

cell The structural and functional unit of an organism; the smallest structure capable of performing all the functions necessary for life. 6

cell-mediated immunity Immunological defense provided by T cell lymphocytes that come within close proximity of their victim cells (as opposed to humoral immunity provided by the secretion of antibodies by plasma cells). 433

cellular respiration (res″pĭ-ra′shun) The energy-releasing metabolic pathways in a cell that oxidize organic molecules such as glucose and fatty acids. 546

central nervous system The part of the nervous system consisting of the brain and the spinal cord; CNS. 248

centrioles (sen′trĭ-ōlz) Cell organelles that form the spindle apparatus during cell division. 68

centromere (sen′trŏ-mēr) The central region of a chromosome to which the chromosomal arms are attached. 68

centrosome (sen′trŏ-sōm) A dense body near the nucleus of a cell that contains a pair of centrioles. 67

cerebellum (ser″ĕ-bel′um) The portion of the brain concerned with the coordination of skeletal muscle contraction. It consists of two hemispheres and a central vermis. 262

cerebral (ser′ĕ-bral) **arterial circle** An arterial vessel that encircles the pituitary gland. It provides alternate routes for blood to reach the brain should a carotid or vertebral artery become occluded; also called the *circle of Willis.* 386

cerebrospinal (ser″ĕ-bro-spi′nal) **fluid** A fluid produced by the choroid plexus of the ventricles of the brain. It fills the ventricles and surrounds the central nervous system in association with the meninges. 269

cerebrum (ser′ĕ-brum) The largest portion of the brain, composed of the right and left hemispheres. 250

ceruminous (sĕ-roo′mĭ-nus) **gland** A specialized gland of the external auditory canal that secretes cerumen, or earwax. 107

cervical plexus (plek′sus) A network of spinal nerves formed by the anterior branches of the first four cervical nerves. 276

cervix (ser′viks) 1. The narrow necklike portion of an organ. 2. The inferior end of the uterus that adjoins the vagina (cervix of the uterus). 594

chemoreceptor (ke″mo-re-sep′tor) A neuroreceptor that is stimulated by the presence of chemical molecules. 305

choane (ko-a′ne) The two posterior openings from the nasal cavity into the nasal pharynx; also called the *internal nares.* 451

cholesterol (kŏ-les′ter-ol) A 27-carbon steroid that serves as the precursor of steroid hormones. 32

cholinergic (ko″lĭ-ner′jik) Denoting nerve endings that liberate acetylcholine as a neurotransmitter, such as those of the parasympathetic system. 297

chondrocyte (kon′dro-sīt) A cartilage-forming cell. 84

chordae tendineae (kor′de ten-din′e-e) Chordlike tendinous bands that connect papillary muscles to the cusps of the atrioventricular valves within the ventricles of the heart. 375

chorion (kor′e-on) An extraembryonic membrane that participates in the formation of the placenta. 622

choroid (kor′oid) The vascular, pigmented middle layer of the wall of the eye. 310

choroid plexus (plek′sus) A mass of vascular capillaries from which cerebrospinal fluid is secreted into the ventricles of the brain. 267

chromatids (kro′mă-tidz) Duplicated chromosomes, joined together at the centromere, that separate during cell division. 67

chromatin (kro′mă-tin) Threadlike structures in the cell nucleus consisting primarily of DNA and protein. They represent the extended form of chromosomes during interphase. 66

chromatophilic (kro″mă-to-fil′ik) **substances** Clumps of rough endoplasmic reticulum in the cell bodies of neurons; also called *Nissl bodies.* 232

chromosomes (kro′mŏ-sōmz) Structures in the nucleus that contain the genes for genetic expression. 66

chyme (kīm) The mass of partially digested food that passes from the pylorus of the stomach into the duodenum of the small intestine. 516

cilia (cil′e-ă) Microscopic hairlike processes that move in a wavelike manner on the exposed surfaces of certain epithelial cells. 45

ciliary (sil-e-er″e) **body** A portion of the choroid layer of the eye that secretes aqueous humor. It contains the ciliary muscle. 310

circle of Willis *See* cerebral arterial circle. 386

circumduction A circular movement of a body part in which a cone-shaped airspace is traced. 165

cleavage Rapid cell divisions of the zygote approximately 36 hours after fertilization of the secondary oocyte. 618

clitoris (klit′or-is, kli′tor-is) A small erectile structure in the vulva of the female, homologous to the glans penis in the male. 596

CNS *See* central nervous system. 248

cochlea (kok′le-ă) The organ of hearing in the inner ear where nerve impulses are generated in response to sound waves. 321

cochlear window A membrane-covered opening between the middle and inner ear, directly below the vestibular window; also called the *round window.* 319

codon (ko′don) The sequence of three nucleotide bases in mRNA that specifies a given amino acid and determines the position of that amino acid in a polypeptide chain through complementary base pairing with an anticodon in tRNA. 63

coenzyme (ko′en′zīm) An organic molecule, usually derived from a water-soluble vitamin, that combines with and activates specific enzyme proteins. 544

collateral ganglia Sympathetic ganglia found at the origin of large abdominal arteries; include the celiac, superior mesenteric, and inferior mesenteric ganglia. 290

colon (ko′lon) The first portion of the large intestine. 523

common bile duct A tube formed by the union of the hepatic duct and cystic duct that transports bile to the duodenum. 519

compact bone Tightly packed bone that is superficial to spongy bone and covered by the periosteum; also called *dense bone.* 122

compliance A measure of the ease with which a structure such as the lung expands under pressure; a measure of the change in volume as a function of pressure changes. 408, 458

conceptus The product of conception at any point between fertilization and birth. It includes the embryo or fetus, as well as the extraembryonic structures. 621

conduction myofibers Specialized large-diameter cardiac muscle fibers that conduct electrical impulses from the AV bundle into the ventricular walls; also called *Purkinje fibers.* 377

condyle (kon′dīl) A rounded prominence at the end of a bone, most often for articulation with another bone. 121

cone A color receptor cell in the retina of the eye. 312

congenital (kon-jen′ĭ-tal) Present at the time of birth.

conjunctiva (kon″jungk-ti′vă) The thin membrane covering the anterior surface of the eyeball and lining the eyelids. 308

connective tissue One of the four basic tissue types within the body. It is a binding and supportive tissue with abundant matrix. 81

convolution (kon-vŏ-loo′shun) An elevation on the surface of a structure and an infolding of the tissue upon itself. 250

cornea (kor′ne-ă) The transparent anterior portion of the outer layer of the eyeball. 310

cornification The drying and flattening of the outer keratinized cells of the epidermis. 98

coronal (kor′ŏ-nal, kŏ-ro′nal) **plane** A plane that divides the body into anterior and posterior portions; also called a *frontal plane.* 11

coronary (kor′ŏ-nar″e) **circulation** The arterial and venous blood circulation to the wall of the heart. 376

coronary sinus A large venous channel on the posterior surface of the heart into which the cardiac veins drain. 375

corpora quadrigemina (kor′por-ă kwad″rĭ-jem′ĭ-nă) Four superior lobes of the midbrain concerned with visual and auditory functions. 261

corpus callosum (kor′pus kă-lo′sum) A large tract of white matter within the brain that connects the right and left cerebral hemispheres. 250

corpuscle (kor′pus′l) **of touch** A touch sensory receptor found in the dermis of the skin; also called *Meissner's corpuscle.* 302

cortex (kor′teks) 1. The outer layer of an internal organ or body structure, as of the kidney or adrenal gland. 341
2. The convoluted layer of gray matter that covers the surface of each cerebral hemisphere. 250

corticosteroids (kor″tĭ-ko-ster′oidz) Steroid hormones of the adrenal cortex, consisting of glucocorticoids (such as hydrocortisone) and mineralocorticoids (such as aldosterone). 32

countercurrent multiplier system The interaction between the ascending and descending limbs of the nephron loops that creates a very hypertonic renal medulla. 486

covalent bond A chemical bond formed by the sharing of one or more electrons, especially pairs of electrons, between atoms. 25

cranial nerves One of 12 pairs of nerves that arise from the brain. 269

cranium (kra′ne-um) The bones of the skull that enclose or support the brain and the organs of sight, hearing, and balance. 115

creatine phosphate (kre′ă-tin fos′fāt) An organic phosphate molecule in muscle cells that serves as a source of high-energy phosphate for the synthesis of ATP; also called *phosphocreatine.* 186

cyclic AMP Cyclic adenosine monophosphate; a molecule formed from ATP by the action of the enzyme adenylate cyclase. It serves as a second messenger in mediating the action of some hormones. 349

cystic (sis′tik) **duct** The tube that transports bile from the gallbladder to the common bile duct. 527

cytokinesis (si″to-kĭ-ne′sis) The division of the cytoplasm that occurs in mitosis and meiosis, when a parent cell divides to produce two daughter cells. 67

cytoplasm (si′tŏ-plaz″em) In a cell, the material between the nucleus and the cell membrane. 43

D

decussation (dek″uh-sa′shun) A crossing of nerve fibers from one side of the CNS to the other. 273

delayed hypersensitivity An allergic response in which the onset of symptoms may not occur until 2 or 3 days after exposure to an antigen. Produced by T cells, it is a type of cell-mediated immunity. 444

dendrite (den′drīt) A nerve cell process that transmits impulses toward a neuron cell body. 232

dentin (den′tin) The main substance of a tooth, covered by enamel over the crown of the tooth and by cementum on the root. 514

depolarization The loss of membrane polarity in which the inside of the cell membrane becomes less negative in comparison to the outside of the membrane. The term is also used to indicate the reversal of membrane polarity that occurs during the production of action potentials in nerve and muscle cells. 236

dermis (der′mis) The second, or deep, layer of skin beneath the epidermis. 99

descending colon (ko′lon) The segment of the large intestine that descends on the left side from the level of the spleen to the level of the left iliac crest. 523

diapedesis (di″ă-pĕ-de′sis) The passage of blood cells through intact vessel walls to and from surrounding connective tissues. 430

diaphragm (di′ă-fram) A sheetlike dome of muscle and connective tissue that separates the thoracic and abdominal cavities. 196

diaphysis (di-af′ĭ-sis) The shaft of a long bone. 121

diastole (di-as′tŏ-le) The sequence of the cardiac cycle during which a heart chamber wall is relaxed. 377

diencephalon (di″en-sef′ă-lon) A major region of the brain that includes the third ventricle, thalamus, hypothalamus, and pituitary gland. 258

diffusion The net movement of molecules or ions from regions of higher to regions of lower concentration. 47

digestion The process by which larger molecules of food substance are broken down mechanically and chemically into smaller molecules that can be absorbed. 504

diploid (dip′loid) Denoting cells having two of each chromosome or twice the number of chromosomes that are present in sperm cells or ova. 631

disaccharide (di-sak′ă-rīd) Any of a class of double sugars; carbohydrates that yield two simple sugars, or monosaccharides, upon hydrolysis. 29

diuretic (di″yŭ-ret′ik) An agent that promotes the excretion of urine, thereby lowering blood volume and pressure. 498

DNA Deoxyribonucleic acid; composed of nucleotide bases and deoxyribose sugar. It is found in all living cells and contains the genetic code. 60

dopamine (do′pă-mēn) A type of neurotransmitter in the central nervous system; also is the precursor of norepinephrine, another neurotransmitter molecule. 243

dorsal (dor′sal) Pertaining to the back or posterior portion of a body part; the opposite of ventral; also called *posterior.* 10

dorsal root ganglion *See* posterior root ganglion. 275

ductus arteriosus (duk′tus ar-tir″e-o′sus) The blood vessel that connects the pulmonary trunk and the aorta in a fetus. 396

ductus deferens (def′er-enz); pl., *ductus deferentia* A tube that carries spermatozoa from the epididymis to the ejaculatory duct; also called the *vas deferens.* 578

duodenum (doo″ŏ-de′num, doo-od′ĕ-num) The first portion of the small intestine that leads from the pylorus of the stomach to the jejunum. 519

dura mater (door′ă ma′ter) The outermost meninx. 263

E

eccrine (ek′rin) **gland** A sweat gland that functions in thermoregulation. 107

ECG *See* electrocardiogram. 379

ectoderm The outermost of the three primary germ layers of the developing embryo. 620

ectopic (ek-top′ik) **focus** An area of the heart other than the SA node that assumes pacemaker activity. 379

ectopic pregnancy Embryonic development that occurs anywhere other than in the uterus (as in the uterine tubes or body cavity). 513

edema (ĕ-de′ma) Abnormal accumulation of fluid in body parts or tissues. 411

EEG *See* electroencephalogram. 255

effector An organ, such as a gland or muscle, that responds to a motor stimulation. 17

efferent (ef′er-ent) Conveying away from the center of an organ or structure. 232

efferent arteriole (ar-tir′e-ōl) An arteriole of the renal vascular system that conducts blood away from the glomerulus of a nephron. 480

efferent ductules (duk'toolz) A series of coiled tubules through which spermatozoa are transported from the rete testis to the epididymis. 575

efferent neuron (noor'on) *See* motor neuron. 232

ejaculation (ĕ-jak"yŭ-la'shun) The discharge of semen from the male urethra that accompanies orgasm. 582

electrocardiogram A recording of the electrical activity that accompanies the cardiac cycle; ECG or EKG. 379

electroencephalogram A recording of the brain-wave patterns or electrical impulses of the brain from electrodes placed on the scalp; EEG. 255

electrolyte (ĕ-lek'tro-līt) An ion in solution that is capable of carrying an electric current. The most common electrolytes in the plasma are Na^+, HCO_3^-, and K^+. 478

electromyogram (ĕ-lek"tro-mi'ŏ-gram) A recording of the electrical impulses or activity of skeletal muscles using surface electrodes; EMG. 219

electron-transport system A phase of aerobic respiration in which molecules in the inner mitochondrial membrane receive electrons from NADH and $FADH_2$, and use them in the formation of ATP and water. 550

embryo The developing organism from the beginning of the third week to the end of the eighth week of prenatal development. 621

EMG *See* electromyogram. 219

emulsification The process of producing an emulsion or fine suspension. In the small intestine, fat globules are emulsified by the detergent action of bile. 532

enamel (ĕ-nam'el) The outer dense substance covering the crown of a tooth. 514

endocardium (en"do-kar'de-um) The endothelial lining of the heart chambers and valves. 373

endochondral (en"dŏ-kon'dral) **bone** Denoting bones that develop as hyaline cartilage models first, and that are then ossified. 118

endocrine (en'dŏ-krin) **gland** A ductless, hormone-producing gland that is part of the endocrine system. 334

endocytosis (en"do-si-to'sis) A general term for the cellular uptake of particles that are too large to cross the cell membrane. *See also* phagocytosis and pinocytosis. 46

endoderm The innermost of the three primary germ layers of the developing embryo. 620

endogenous (en-doj'ĕ-nus) Denoting a product or process arising from within the body (as opposed to exogenous products or influences from external sources).

endolymph A fluid within the membranous labyrinth and cochlear duct of the inner ear that aids in the conduction of vibrations involved in hearing and the maintenance of equilibrium. 320

endometrium (en"do-me'tre-um) The inner lining of the uterus. 594

endomysium (en"do-mis'e-um) The connective tissue sheath that surrounds each skeletal muscle fiber, separating the muscle cells from one another. 176

endoplasmic reticulum (en-do-plaz'mik rĕ-tik'yŭ-lum) A cytoplasmic organelle composed of a network of canals running through the cytoplasm of a cell. 57

endorphin (en-dor'fin) Any of a group of endogenous opiate molecules that may act as natural analgesics. 243

endothelium (en"do-the'le-um) The layer of epithelial tissue that forms the thin inner lining of blood vessels and heart chambers. 381

endotoxin (en"do-tok'sin) A toxin found within certain types of bacteria that is able to stimulate the release of endogenous pyrogen and produce a fever. 431

enkephalins (en-kef'ă-linz) Short polypeptides, containing five amino acids, that have analgesic effects and that may function as neurotransmitters in the brain. The two known enkephalins (which differ in only one amino acid) are endorphins. 268

enzyme (en'zīm) A protein catalyst that increases the rate of specific chemical reactions. 35, 531

epicardium (ep"ĭ-kar'de-um) A thin outer layer of the heart; also called the *visceral pericardium.* 372

epidermis (ep"ĭ-der'mis) The outermost layer of the skin, composed of stratified squamous epithelium. 97

epididymis (ep"ĭ-did'ĭ-mus); pl., *epididymides* (ep"ĭ-dĭ-dim'ĭ-dēz) A highly coiled tube located along the posterior border of the testis. It stores spermatozoa and transports them from the seminiferous tubules of the testis to the ductus deferens. 578

epidural (ep"ĭ-door'al) **space** A space between the spinal dura mater and the bone of the vertebral canal. 265

epiglottis (ep"ĭ-glot'is) A leaflike structure positioned on top of the larynx. It covers the glottis during swallowing. 453

epinephrine (ep"ĭ-nef'rin) A hormone secreted from the adrenal medulla resulting in actions similar to those resulting from sympathetic nervous system stimulation; also called *adrenaline.* 292

epiphyseal (ep"ĭ-fiz'e-al) **plate** A hyaline cartilaginous layer located between the epiphysis and diaphysis of a long bone. It functions as a longitudinal growing region. 122

epiphysis (ĕ-pif'ĭ-sis) The end segment of a long bone, separated from the diaphysis early in life by an epiphyseal plate but later becoming part of the larger bone. 121

episiotomy (ĕ-pe"ze-ot'ŏ-me) An incision of the perineum at the end of the second stage of labor to facilitate delivery and to avoid tearing the perineum. 631

epithelial (ep"ĭ-the'le-al) **tissue** One of the four basic tissue types; the type of tissue that covers or lines all exposed body surfaces; also called *epithelium.* 75

EPSP Excitatory postsynaptic potential; a graded depolarization of a postsynaptic membrane in response to stimulation by a neurotransmitter chemical. EPSPs can be summated but can be transmitted only over short distances. They can stimulate the production of action potentials when a threshold level of depolarization has been attained. 240

erythroblastosis fetalis (ĕ-rith"ro-blas-to'sis fĭ-tal'is) Hemolytic anemia in an Rh positive newborn caused by maternal antibodies against the Rh factor that have crossed the placenta. 365

erythrocyte (ĕ-rith'rŏ-sīt) A red blood cell. 86, 359

erythropoietin (ĕ-rith"ro-poi-e'tin) A hormone secreted by the kidneys that stimulates the production of red blood cells. 362

esophagus (ĕ-sof'ă-gus) A tubular portion of the GI tract that leads from the pharynx to the stomach as it passes through the thoracic cavity. 515

essential amino acids Those amino acids that cannot be made by the human body; therefore, they must be obtained in the diet. 544

estrogens (es'trŏ-jenz) Any of several female sex hormones secreted from the ovarian (graafian) follicle. 596

eustachian (yoo-sta'ke-an) **canal** *See* auditory tube. 319

exocrine (ek'sŏ-krin) **gland** A gland that secretes its product to an epithelial surface, either directly or through ducts. 78

exocytosis (ek"so-si-to'sis) The process of cellular secretion in which the secretory products are contained within a membrane-enclosed vesicle. The vesicle fuses with the cell membrane so that the lumen of the vesicle is open to the extracellular environment. 47

expiration The process of expelling air from the lungs through breathing out; also called *exhalation.* 460

extension A movement that increases the angle between parts of a joint. 164

external acoustic meatus (ă-koo'stik me-a'tus) An opening through the temporal bone that connects with the tympanum and the middle-ear chamber and through which sound vibrations pass; also called the *external auditory meatus.* 318

exteroceptor (ek"stĕ-ro-sep'tor) A specialized sensory nerve cell located near the surface of the body that responds to stimuli from the external environment. 301

extraocular (ek"stră-ok'yŭ-lar) **muscles** The muscles that insert onto the sclera of the eye and act to change the position of the eye in its orbit (as opposed to the intraocular muscles, such as those of the iris and ciliary body within the eye). 192

F

face 1. The anterior aspect of the head not supporting or covering the brain. 12
2. The exposed surface of a structure.

facet (fas′et) A flattened, shallow articulating surface on a bone. 121

facilitated diffusion The carrier-mediated transport of molecules through the cell membrane along the direction of their concentration gradients. It does not require the expenditure of metabolic energy. 53

FAD Flavin adenine dinucleotide; a coenzyme derived from riboflavin that participates in electron transport within the mitochondria. 544

fallopian (fă-lo′pe-an) **tube** *See* uterine tube. 593

false vocal cords The supporting folds of tissue for the true vocal cords within the larynx. 453

fascia (fash′e-ă) A tough sheet of fibrous tissue binding the skin to underlying muscles or supporting and separating muscles. 176

fasciculus (fă-sik′yŭ-lus) A small bundle of muscle or nerve fibers. 176

feces (fe′sēz) Material expelled from the GI tract during defecation, composed of undigested food residue, bacteria, and secretions; also called *stool.* 504

fertilization The fusion of an ovum and spermatozoon. 616

fetus A prenatal human after 8 weeks of development. 625

fibrinogen (fi-brin′ŏ-jen) A clotting factor in the blood plasma that is converted to fibrin by the action of thrombin. 358

fibroblast (fi′bro-blast) An elongated connective tissue cell with cytoplasmic extensions that is capable of forming collagenous fibers or elastic fibers. 82

fibrous joint A type of articulation bound by fibrous connective tissue that allows little or no movement (e.g., a syndesmosis). 155

filiform papillae (fil′ĭ-form pă-pil′e) Numerous small projections on the surface of the tongue in which taste buds are absent. 306

fimbriae (fim′bre-e) Fringelike extensions from the borders of the open end of the uterine tube. 593

fissure (fish′ur) A groove or narrow cleft that separates two parts, such as the cerebral hemispheres of the brain. 250

flagellum (flă-jel′um) A whiplike structure that provides motility for spermatozoa. 45

flavoprotein (fla″vo-pro′te-in) A conjugated protein containing a flavin pigment that is involved in electron transport within the mitochondria. 545

flexion A movement that decreases the angle between parts of a joint. 163

fontanel (fon″tă-nel′) A membranous-covered region on the skull of a fetus or baby where ossification has not yet occurred; commonly called a *soft spot.* 119

foot The terminal portion of the lower extremity, consisting of the tarsal bones, metatarsal bones, and phalanges. 14

foramen (fŏ-ra′men); pl., *foramina* (fŏ-ram′ĭ-nă) An opening in an anatomical structure, usually in a bone, for the passage of a blood vessel or a nerve. 121

foramen ovale (o-val′e) An opening through the interatrial septum of the fetal heart. 396

formed elements The cellular portion of blood. 357

fossa (fos′ă) A depression in or on a bone. 121

fourth ventricle (ven′trĭ-k′l) A cavity within the brain, between the cerebellum and the medulla oblongata and the pons, containing cerebrospinal fluid. 265

fovea centralis (fo′ve-ă sen-tra′ lis) A depression on the macula lutea of the eye, where only cones are located; the area of keenest vision. 313

frontal 1. Pertaining to the region of the forehead. 2. A plane through the body, dividing the body into anterior and posterior portions; also called the *coronal plane.* 11

FSH Follicle-stimulating hormone; one of the two gonadotropic hormones secreted from the anterior pituitary. In females, FSH stimulates the development of the ovarian follicles; in males, it stimulates the production of sperm cells in the seminiferous tubules. 338

fungiform papillae (fun′jĭ-form pă-pil′e) Flattened, mushroom-shaped projections on the surface of the tongue that contain taste buds. 306

G

GABA Gamma-aminobutyric acid; believed to function as an inhibitory neurotransmitter in the central nervous system. 243

gallbladder A pouchlike organ attached to the underside of the liver in which bile secreted by the liver is stored and concentrated. 527

gamete (gam′ēt) A haploid sex cell; either an egg cell or a sperm cell. 565

ganglion (gang′gle-on) An aggregation of nerve cell bodies occurring outside the central nervous system. 232

gastric (gas′trik) **intrinsic factor** A glycoprotein secreted by the stomach that is needed for the absorption of vitamin B_{12}. 516

gastric juice The colorless, watery, acidic digestive fluid composed of the secretions of the gastric glands of the stomach. It consists chiefly of pepsinogen, HCl, mucus, and gastrin. 517

gastrin (gas′trin) A hormone secreted by the stomach that stimulates the gastric secretion of hydrochloric acid and pepsin. 534

gastrointestinal tract (GI tract) A continuous tube through the anterior (ventral) body cavity that extends from the mouth to the anus; also called the *digestive tract.* 504

gates Structures composed of one or more protein molecules that regulate the passage of ions through channels within the cell membrane. Gates may be chemically regulated (by neurotransmitters) or voltage regulated (in which case they open in response to a threshold level of depolarization). 236

gene The portion of the DNA of a chromosome that contains the information needed to synthesize a particular protein molecule. 632

genetic transcription The process by which RNA is produced with a sequence of nucleotide bases that is complementary to a region of DNA. 62

genetic translation The process by which proteins are produced with amino acid sequences specified by the sequence of codons in mRNA. 63

genotype The genetic makeup of an individual. 632

gland An organ that produces a specific substance or secretion. 78

glans penis (glanz pe′nis) The cone-shaped terminal portion of the penis formed from the expanded corpus spongiosum. 580

globulin (glob′yoo-lin) Any of a class of proteins found extensively in blood plasma; grouped into three subtypes: alpha, beta, and gamma globulins. 358

glomerular (glo-mer′yŭ-lar) **capsule** The double-walled proximal portion of a renal tubule that encloses the glomerulus of a nephron; also called *Bowman's capsule.* 481

glomerular filtration rate (GFR) The volume of filtrate produced each minute by both kidneys. 482

glomerulus (glo-mer′yŭ-lus) A coiled tuft of capillaries surrounded by the glomerular capsule that filtrates urine from the blood. 480

glottis (glot′is) A slitlike opening into the larynx, positioned between the true vocal cords. 453

glucagon (gloo′kă-gon) A polypeptide hormone secreted by the alpha cells of the pancreatic islets. It acts primarily on the liver to promote glycogenolysis and raise blood glucose levels. 557

glucocorticoids Steroid hormones secreted by the adrenal cortex (corticosteroids). They affect the metabolism of glucose, protein, and fat and also have anti-inflammatory and immunosuppressive effects. The major glucocorticoid in humans is hydrocortisone (cortisol). 341

gluconeogenesis (gloo″kŏ-neŏ-jen′ĭ-sis) The formation of glucose from noncarbohydrate molecules, such as amino acids and lactic acid. 548

glycerol (glis'ĕ-rol) A 3-carbon alcohol that serves as a building block of fats. 31

glycogen (gli'kŏ-jen) A polysaccharide of glucose—also called animal starch—produced primarily in the liver and skeletal muscles. Similar to plant starch in composition, glycogen contains more highly branched chains of glucose subunits than does plant starch. 29

glycogenesis (gli"kŏ-jen'ĭ-sis) The formation of glycogen from glucose. 551

glycogenolysis (gli"kŏ-jĕ-nol'ĭ-sis) The hydrolysis of glycogen to glucose 1-phosphate, which can be converted to glucose 6-phosphate, which then may be oxidized via glycolysis or (in the liver) converted to free glucose. 551

glycolysis (gli"kol'ĭ-sis) The metabolic pathway that converts glucose to pyruvic acid; the final products are two molecules of pyruvic acid and two molecules of $NADH_2$, with a net gain of two ATP molecules. In anaerobic respiration, the $NADH_2$ is oxidized by the conversion of pyruvic acid to lactic acid. In aerobic respiration, pyruvic acid enters the Krebs cycle in mitochondria, and $NADH_2$ is ultimately oxidized by oxygen, yielding water. 546

glycosuria (gli"kŏ-soor'e-ă) The excretion of an abnormal amount of glucose in the urine (urine normally only contains trace amounts of glucose). 490

goblet cell A unicellular mucus-secreting gland associated with columnar epithelia; also called a *mucous cell.* 76

Golgi (gol'je) **apparatus** A network of stacked, flattened membranous sacs within the cytoplasm of cells. Its major function is to concentrate and package proteins for secretion from the cell. 58

Golgi tendon organ A sensory receptor found near the junction of tendons and muscles. 305

gomphosis (gom-fo'sis) A fibrous joint between the root of a tooth and the periodontal ligament of the tooth socket. 158

gonad A reproductive organ, testis or ovary, that produces gametes and sex hormones. 347

gonadotropic (go-nad"ŏ-tro'pik) **hormones** Hormones of the anterior pituitary that stimulate gonadal function—the formation of gametes and secretion of sex steroids. The two gonadotropins are FSH (follicle-stimulating hormone) and LH (luteinizing hormone), each of which is essentially the same in males and females. 338, 601

graafian (graf'e-an) **follicle** A mature ovarian follicle containing a single fluid-filled cavity, with the ovum located toward one side of the follicle and positioned on a cluster of granulosa cells. 517

granular leukocytes (loo'kŏ-sīts) Leukocytes with granules in the cytoplasm; on the basis of the staining properties of the granules, these cells are classified as neutrophils, eosinophils, or basophils. 359

gray matter The region of the central nervous system composed of nonmyelinated nerve tissue. 249

greater omentum (o-men'tum) A double-layered peritoneal membrane that originates on the greater curvature of the stomach. It hangs inferiorly like an apron over the contents of the abdominal cavity. 507

growth hormone A hormone secreted by the anterior pituitary that stimulates growth of the skeleton and soft tissues during the growing years and that influences the metabolism of protein, carbohydrate, and fat throughout life. 338

gustation The sense of taste. 305

gyrus A convoluted elevation or ridge. 252

H

hair A threadlike appendage of the epidermis consisting of keratinized dead cells that have been pushed up from a dividing basal layer. 103

hair cells Specialized receptor nerve endings for detecting sensations, such as in the spiral organ (organ of Corti). 305, 322

hair follicle A tubular depression in the dermis of the skin in which a hair develops. 104

hand The terminal portion of the upper extremity, containing the carpal bones, metacarpal bones, and phalanges. 142

haploid (hap'loid) A cell that has one of each chromosome type and therefore half the number of chromosomes present in most other body cells. Only the gametes (sperm and ova) are haploid. 617

hard palate (pal'it) The bony partition between the oral and nasal cavities, formed by the maxillae and palatine bones and lined by mucous membrane. 512

haustra (haws'tră) Sacculations or pouches of the colon. 523

haversian (hă-ver'zhun) **system** *See* osteon.

heart A four-chambered muscular pumping organ positioned in the thoracic cavity, slightly to the left of midline. 357

helper T cells A subpopulation of T cells (lymphocytes) that helps to stimulate the antibody production of B lymphocytes by antigens. 437

hematocrit (hĭ-mat'ŏ-krit) The ratio of packed red blood cells to total blood volume in a centrifuged sample of blood, expressed as a percentage. 357

heme (hēm) The iron-containing red pigment that, together with the protein globin, forms hemoglobin. 467

hemoglobin (he'mŏ-glo"bin) The pigment of red blood cells that transports oxygen; some carbon dioxide is also transported by hemoglobin. 467

hemopoiesis (hem"ŏ-poi-e'sis) The production of red blood cells. 117

heparin (hep'ar-in) A mucopolysaccharide found in many tissues, but most abundantly in the lungs and liver, that is used medically as an anticoagulant. 367

hepatic (hĕ-pat'ik) **duct** A duct formed from the fusion of several bile ducts that drain bile from the liver. The hepatic duct merges with the cystic duct from the gallbladder to form the common bile duct. 525

hepatic portal circulation The return of venous blood from the digestive organs and spleen through a capillary network within the liver before draining into the heart. 393

hepatopancreatic (hep"ă-to-pan"kre-at'ik) **ampulla** A small ductule within the wall of the duodenum where the pancreatic and common bile ducts join to form a common entry into the duodenum; also called the *ampulla of Vater.* 519

high-density lipoproteins (lip"o-pro'te-inz) **(HDLs)** Combinations of lipids and proteins that migrate rapidly to the bottom of a test tube during centrifugation. HDLs are carrier proteins for cholesterol and other lipids, and appear to offer some protection from atherosclerosis. 397

hilum (hi'lum) A concave or depressed area where vessels or nerves enter or exit an organ; also called a *hilus.* 456

hinge joint A type of synovial articulation in which the convex surface of one bone fits into the concave surface of another, thus confining movement to one plane, as in the knee or interphalangeal joint. 160

histamine (his'tă-mēn) A compound secreted by tissue mast cells and other connective tissue cells that stimulates vasodilation and increases capillary permeability. It is responsible for many of the symptoms of inflammation and allergy. 444

histology Microscopic anatomy of the structure and function of tissues. 75

homeostasis (ho"me-o-sta'sis) The dynamic constancy of the internal environment, the maintenance of which is the principal function of physiological regulatory mechanisms. The concept of homeostasis provides a framework for understanding most physiological processes. 17

homologous (hŏ-mol'ŏ-gus) **chromosomes** The matching pairs of chromosomes in a diploid cell. 631

horizontal (transverse) plane A directional plane that divides the body, organ, or appendage into superior and inferior or proximal and distal portions. 11

hormone A chemical substance produced in an endocrine gland and secreted into the bloodstream to cause effects in specific target organs. 334

humoral immunity The form of acquired immunity in which antibody molecules are secreted in response to antigenic stimulation (as opposed to cell-mediated immunity); also called *antibody-mediated immunity.* 433

hyaline (hi'ă-līn) **cartilage** A cartilage with a homogeneous matrix. It is the most common type, occurring at the articular ends of bones, in the trachea, and within the nose. Most of the bones in the body are formed from hyaline cartilage. 84

hydrocortisone (hi″drŏ-kor′tĭ-sōn) The principal corticosteroid hormone secreted by the adrenal cortex, with glucocorticoid action; also called *cortisol.* 342
hydrolysis (hi-drol′ĭ-sis) Breakdown of a molecule into its subunits by reaction with water. 30
hydrophilic (hi″drŏ-fil′ik) Denoting a substance that readily absorbs water; literally, "water loving." 26
hydrophobic (hi″drŏ-fo′bik) Denoting a substance that repels water and that is repelled by water; "water fearing." 26
hyperemia (hi″per-e′me-ă) An excess of blood in part of the body caused by increased blood flow. Skin in the area usually becomes reddened and warm. 414
hyperglycemia (hi″per-gli-se′me-ă) An abnormally increased concentration of glucose in the blood. 53
hyperkalemia (hi″per-kă-le′me-ă) An abnormally high concentration of potassium in the blood. 491
hypertension Blood pressure in excess of the normal range for the person's age and sex. 420
hypertonic (hi″per-ton′ik) Denoting a solution with a greater solute concentration and thus a greater osmotic pressure than plasma. 52
hyperventilation A high rate and depth of breathing that results in a decrease in the blood carbon dioxide concentration to below normal. 466
hypodermis (hi″pŏ-der′mis) Subcutaneous tissue that binds the dermis to underlying organs. 100
hypothalamic (hi″po-thă-lam′ik) **hormones** Hormones produced by the hypothalamus. These include antidiuretic hormone and oxytocin, which are secreted by the posterior pituitary, and both releasing and inhibiting hormones that regulate the secretions of the anterior pituitary. 339
hypothalamo-hypophyseal (hi″pŏ-fiz′e-al) **portal system** A vascular system that transports releasing and inhibiting hormones from the hypothalamus to the anterior pituitary. 339
hypothalamo-hypophyseal tract The tract of nerve fibers (axons) that transports antidiuretic hormone and oxytocin from the hypothalamus to the posterior pituitary. 339
hypothalamus (hi″po-thal′ă-mus) A portion of the forebrain within the diencephalon that lies below the thalamus, where it functions as an autonomic nerve center and regulates the pituitary gland. 258

I

ileocecal (il″e-ŏ-se′kal) **valve** A modification of the mucosa at the junction of the small intestine and large intestine that forms a one-way passage and prevents the backflow of food materials. 519
ileum (il′e-um) The terminal portion of the small intestine between the jejunum and cecum. 519
immediate hypersensitivity Hypersensitivity (allergy) mediated by antibodies of the IgE class that results in the release of histamine and related compounds from tissue cells. 444
immunization The process of increasing one's resistance to pathogens. In active immunity a person is injected with antigens that stimulate the development of clones of specific B or T lymphocytes; in passive immunity a person is injected with antibodies produced by another organism. 436
immunoassay (im″yŭ-no-as′a) Any of a number of laboratory or clinical techniques that employ the specific bonding of an antigen with its antibody in order to identify and quantify a substance in a sample. 432
immunoglobulins (im″yŭ-no-glob′yŭ-linz) Classes of the gamma globulin fraction of plasma proteins that have antibody functions, providing humoral immunity. 433
immunosurveillance (im″yŭ-no-ser-va′lens) The concept that the immune system recognizes and attacks malignant cells that produce antigens not recognized as "self." This function is believed to be cell mediated rather than humoral. 437
implantation The process by which a blastocyst attaches itself to and penetrates into the endometrium of the uterus. 618
incus (ing′kus) The middle of three auditory ossicles within the middle-ear chamber; commonly called the *anvil.* 319
inferior vena cava (ve′nă ka-vă) A major vein that collects blood from parts of the body inferior to the heart and empties into the right atrium. 389
infundibulum (in″fun-dib′yŭ-lum) The stalk that attaches the pituitary gland to the hypothalamus of the brain. 337
ingestion (in-jes′chun) The process of taking food or liquid into the body by way of the oral cavity. 504
inhibin (in-hib′in) A polypeptide hormone secreted by the testes that is believed to specifically exert negative feedback inhibition of FSH secretion from the anterior pituitary. 571
inner ear The innermost portion or chamber of the ear, containing the cochlea and the vestibular organs. 320
insertion The more movable attachment of a muscle, usually more distal. 175
inspiration The act of breathing air into the alveoli of the lungs; also called *inhalation.* 460
insula (in′sŭ-lă) A fifth lobe of the cerebrum that lies deep to the lateral sulcus. It cannot be seen in an external view. 255
insulin (in′sŭ-lin) A polypeptide hormone secreted by the beta cells of the pancreatic islets that promotes the anabolism of carbohydrates, fat, and protein. Insulin acts to promote the cellular uptake of blood glucose and, therefore, to lower the blood glucose concentration; insulin deficiency results in hyperglycemia and diabetes mellitus. 557
integument (in-teg′yoo-ment) The skin; the largest organ of the body. 96
intercalated (in-ter′kă-lāt-ed) **disc** A thickened portion of the sarcolemma that extends across a cardiac muscle fiber, indicating the boundary between cells. 88
intercellular substance The matrix or material between cells that largely determines tissue types. 75
interferon (in″ter-fēr-on) Any of a group of small proteins that inhibit the multiplication of viruses inside host cells and that also have antitumor properties. 431
interphase The interval between successive cell divisions, during which time the chromosomes are in an extended state and are active in directing RNA synthesis. 68
interstitial (in″ter-stish′al) **cells** Cells located in the interstitial tissue between adjacent convolutions of the seminiferous tubules of the testes. They secrete androgens (mainly testosterone); also called *cells of Leydig.* 574
interstitial fluid Fluid between the cells; also called *tissue fluid.* 409
intervertebral (in″ter-ver′tĕ-bral) **disc** A pad of fibrocartilage located between the bodies of adjacent vertebrae. 135
intestinal crypt A simple tubular digestive gland opening onto the surface of the intestinal mucosa that secretes digestive enzymes; also called the *crypt of Lieberkühn.* 520
intrapleural (in″tră-ploor′al) **space** An actual or potential space between the visceral pleural membrane covering the lungs and the pleural membrane lining the thoracic wall. 457
intrinsic (in-trin′zik) Situated within or pertaining to internal origin.
inulin (in′yŭ-lin) A polysaccharide of fructose, produced by certain plants, that is filtered by the human kidneys but neither reabsorbed nor secreted. The clearance rate of injected inulin is thus used to measure the glomerular filtration rate. 488
in vitro (in ve′tro) Occurring outside the body, in a test tube or other artificial environment. 180
in vivo (in ve′vo) Occurring within the body. 180
ion (i′on) An atom or group of atoms that has either lost or gained electrons and thus has a net positive or a net negative charge. 26
ionization (i-on-ĭ-za′shun) The dissociation of a solute to form ions. 26
IPSP Inhibitory postsynaptic potential; hyperpolarization of the postsynaptic membrane in response to a particular neurotransmitter chemical, which makes it more difficult for the postsynaptic cell to attain a threshold level of depolarization required to produce action potentials. It is responsible for postsynaptic inhibition. 240

iris The pigmented portion of the vascular tunic of the eye that surrounds the pupil and regulates its diameter. 312

ischemia (ĭ-ske′me-ă) A rate of blood flow to an organ that is inadequate to supply sufficient oxygen and maintain aerobic respiration in that organ. 398

islets of Langerhans (i′letz of lang′er-hanz) *See* pancreatic islets. 345

isometric contraction Muscle contraction in which there is no appreciable shortening of the muscle. 181

isotonic contraction Muscle contraction in which the muscle shortens in length and maintains approximately the same amount of tension throughout the shortening process. 181

isotonic solution A solution having the same osmotic pressure as the solution with which it is compared, so that osmosis will not occur between the two if they are separated by a semipermeable membrane; a solution with the same osmotic pressure as plasma. 52

isotope (i′sŏ-tōp) One of two or more atoms having the same atomic number but different mass numbers. Isotopes vary only in number of neutrons. 25

J

jaundice (jawn′dis) A condition characterized by high blood bilirubin levels and staining of the tissues with bilirubin, which imparts a yellow color to the skin and mucous membranes. 536

jejunum (jĕ-joo′num) The middle portion of the small intestine, located between the duodenum and the ileum. 519

joint capsule The fibrous tissue that encloses the joint cavity of a synovial joint. 159

K

keratin (ker′ă-tin) An insoluble protein present in the epidermis and in epidermal derivatives, such as hair and nails. 97

ketoacidosis (ke″to-ă-sĭ-do′sis) A type of metabolic acidosis resulting from the excessive production of ketone bodies, as in diabetes mellitus. 599

ketogenesis (ke″to-jen′ĭ-sis) The production of ketone bodies. 527

ketone bodies The substances derived from fatty acids via acetyl coenzyme A in the liver; namely, acetone, acetoacetic acid, and β-hydroxybutyric acid. Ketone bodies are oxidized by skeletal muscles for energy. 553

ketosis (ke-to′sis) An abnormal elevation in the blood concentration of ketone bodies. 554

kidney One of a pair of organs of the urinary system that, among other functions, filters wastes from the blood in the formation of urine. 478

kilocalorie (kil′ŏ-kal″ŏ-re) A unit of measurement equal to 1000 calories, which are units of heat (a kilocalorie is the amount of heat required to raise the temperature of 1 kilogram of water by 1 C°). In nutrition, the kilocalorie is called a big calorie (Calorie). 555

Korotkoff (kŏ-rot′kof) **sounds** Sounds produced by turbulent blood flow through a constricted artery. They are used in measuring blood pressure. 419

Krebs cycle A cyclic metabolic pathway in the matrix of mitochondria by which the acetic acid part of acetyl CoA is oxidized and substrates provided for reactions that are coupled to the formation of ATP. 549

Kupffer (koop′fer) **cells** Phagocytic cells that line the sinusoids of the liver. They are part of the body immunity system. 430

L

labia majora (la′be-ă mă-jor′ă); sing., *labium majus* A portion of the external genitalia of a female consisting of two longitudinal folds of skin extending downward and backward from the mons pubis. 596

labia minora (mĭ-nor′ă), sing., *labium minus* Two small folds of skin, devoid of hair and sweat glands, lying between the labia majora of the external genitalia of a female. 596

labyrinth (lab′ĭ-rinth) An intricate structure consisting of interconnecting passages (e.g., the bony and membranous labyrinths of the inner ear). 320

lacrimal (lak′rĭ-mal) **gland** A tear-secreting gland located on the superior lateral portion of the eyeball, underneath the upper eyelid. 308

lactation (lak-ta′shun) The production and secretion of milk by the mammary glands. 606

lacteal (lak′te-al) A small lymphatic duct associated with a villus of the small intestine. 519

lactose (lak′tōs) Milk sugar; a disaccharide of glucose and galactose. 29

lactose intolerance A disorder resulting in the inability to digest lactose because of an enzyme, lactase, deficiency. Symptoms include bloating, intestinal gas, nausea, diarrhea, and cramps. 520

lacuna (lă-kyoo′nă) A small hollow chamber that houses an osteocyte in mature bone tissue or a chondrocyte in cartilage tissue. 123

lambdoid (lam′doid) **suture** The immovable joint in the skull between the parietal bones and the occipital bone. 120

lamella (lă-mel′ă) A concentric ring of matrix surrounding the central canal in an osteon of mature bone tissue. 123

lamellated (lam′ĕ-la-ted) **corpuscle** A sensory receptor for pressure, found in tendons, around joints, and in visceral organs; also called a *pacinian corpuscle.* 302

lanugo (lă-noo′go) Short, silky fetal hair, which may be present for a short time on a premature infant. 104

large intestine The last major portion of the GI tract, consisting of the cecum, colon, rectum, and anal canal. 522

laryngopharynx (lă-ring″go-far′ingks) The inferior or lower portion of the pharynx in contact with the larynx. 482

larynx (lar′ingks) The structure located between the pharynx and trachea that houses the vocal cords; commonly called the *voice box.* 453

lateral Pertaining to the side; farther from the midplane. 10

lateral ventricle (ven′trĭ-k′l) A cavity within the cerebral hemisphere of the brain that is filled with cerebrospinal fluid. 265

L-Dopa Levodopa; a derivative of the amino acid tyrosine. It serves as the precursor for the neurotransmitter molecule dopamine and is given to patients with Parkinson's disease to stimulate dopamine production. 243, 267

lens A transparent refractive organ of the eye lying posterior to the pupil and iris. 310

lesion (le′zhun) A wounded or damaged area.

lesser omentum (o-men′tum) A peritoneal fold of tissue extending from the lesser curvature of the stomach to the liver. 507

leukocyte (loo′kŏ-sīt) A white blood cell; variant spelling, leucocyte. 86, 359

ligament A tough cord or fibrous band of connective tissue that binds bone to bone to strengthen and provide flexibility to a joint. It also may support viscera. 82

limbic system A portion of the brain concerned with emotions and autonomic activity. 260

linea alba (lin′e-ă al′bă) A vertical fibrous band extending down the anterior medial portion of the abdominal wall. 198

lipid (lip′id) Any of a group of organic molecules, including fats, phospholipids, and steroids, that are generally insoluble in water. 31, 532

lipogenesis (lip″ŏ-jen′ĕ-sis) The formation of fat, or triglycerides. 553

lipolysis (lĭ-pol′ĭ-sis) The hydrolysis of triglycerides into free fatty acids and glycerol. 553

liver A large visceral organ inferior to the diaphragm in the right hypochondriac region. The liver detoxifies the blood and modifies the blood plasma concentration of glucose, triglycerides, ketone bodies, and proteins. 524

loop of Henle *See* nephron loop. 481

lumbar plexus (plek′sus) A network of nerves formed by the anterior branches of spinal nerves L1 through L4. 276

lumen (loo′men); pl. lumina (loo′mĭ-nă) The space within an artery, vein, or other tubular structure, or within a hollow organ. 75

lung One of the two major organs of respiration positioned within the thoracic cavity on either side of the mediastinum. 456

lung surfactant (sur-fak'tant) A mixture of lipoproteins (containing phospholipids) secreted by type II alveolar cells into the alveoli of the lungs. It lowers surface tension and prevents collapse of the lungs as occurs in hyaline membrane disease, in which surfactant is absent. 459

luteinizing (loo'te-ĭ-ni"zing) **hormone (LH)** A hormone secreted by the adenohypophysis (anterior lobe) of the pituitary gland that stimulates ovulation and the secretion of progesterone by the corpus luteum. It also influences mammary gland milk secretion in females and stimulates testosterone secretion by the testes in males. 338

lymph A clear, plasmalike fluid that flows through lymphatic vessels. 426

lymphatic (lim-fat'ik) **system** The lymphatic vessels and lymph nodes. 426

lymph node A small, ovoid mass of reticular tissue located along the course of lymph vessels. 426

lymphocyte (lim'fŏ-sīt) A type of white blood cell characterized by agranular cytoplasm. Lymphocytes usually constitute about 20% to 25% of the white blood cell count. 432

lymphoid tissue A type of connective tissue dominated by lymphocytes. 361

lymphokine (lim'fŏ-kīn) Any of a group of chemicals released from T cells that contribute to cell-mediated immunity. 437

lysosome (li'sŏ-sōm) A membrane-bound organelle containing digestive enzymes. 57

M

macromolecule (mak"ro-mol'ĭ-kyool) A large molecule; a term that usually refers to protein, RNA, and DNA. 33

macrophage (mak'rŏ-faj) A wandering phagocytic cell. 430

macula lutea (mak'yŭ-lă loo'te-ă) A yellowish depression in the retina of the eye that contains the fovea centralis, the area of keenest vision. 313

major histocompatibility complex (MHC) A group of genes that controls the production of histocompatibility antigens, or MCH molecules. The MCH molecules are used by antigen-presenting cells to present antigens to lymphocytes. 439

malleus (mal'e-us) The first of three auditory ossicles that attaches to the tympanum; commonly called the *hammer.* 319

mammary gland The gland of the female breast responsible for lactation and nourishment of the young. 107, 605

marrow The soft connective tissue found within the inner cavity of certain bones that produces red blood cells. 121

mast cell A type of connective tissue cell that produces and secretes histamine and heparin and promotes local inflammation. 445

matrix (ma'triks) The intercellular substance of a tissue. 75

maximal oxygen uptake The maximum amount of oxygen that can be consumed by the body per unit time during heavy exercise. 187

mechanoreceptor (mek"ă-no-re-sep'tor) A sensory receptor that responds to a mechanical stimulus. 301

medial Toward or closer to the midplane of the body. 10

mediastinum (me"de-ă-sti'num) The partition in the center of the thorax between the two pleural cavities. 456

medulla (mĕ-dul'ă) The center portion of an organ. 341

medulla oblongata (ob"long-gă'tă) A portion of the brain stem located between the spinal cord and the pons. 261

medullary (med'l-er"e) **(marrow) cavity** The hollow core of the diaphysis of a long bone in which marrow is found. 121

megakaryocyte (meg"ă-kar'e-o-sīt) A bone marrow cell that gives rise to blood platelets. 360

meiosis (mi-o'sis) A specialized type of cell division by which gametes or haploid sex cells are formed. 576

Meissner's (mīs'nerz) **corpuscle** *See* corpuscle of touch. 302

melanin (mel'ă-nin) A dark pigment found within the epidermis or epidermal derivatives of the skin. 98

melanocyte (mel'ă-no-sīt) A specialized melanin-producing cell found in the deepest layer of the epidermis. 98

melanoma (mel"ă-no'mă) A malignant tumor of the skin that frequently forms in moles. 108

melatonin (mel"ă-to'nin) A hormone secreted by the pineal gland that produces lightening of the skin in lower vertebrates and that may contribute to the regulation of gonadal function in mammals. Secretion follows a circadian rhythm and peaks at night. 346

membrane potential The potential difference or voltage that exists between the inner and outer sides of a cell membrane. It exists in all cells but is capable of being changed by excitable cells (neurons and muscle cells). 236

menarche (mĕ-nar'ke) The first menstrual discharge. 592

meninges (mĕ-nin'jēz); sing., *meninx* (me'ningks) A group of three fibrous membranes covering the central nervous system, composed of the dura mater, arachnoid mater, and pia mater. 263

meniscus (mĕ-nis'kus) pl., *menisci* (mĕ-nis'i, mĕ-nis'ki) A wedge-shaped pad of fibrocartilage in certain synovial joints. 160

menopause (men'ŏ-pawz) The period marked by the cessation of menstrual periods in the human female. 605

menstrual (men'stroo-al) **cycle** The rhythmic female reproductive cycle characterized by changes in hormone levels and physical changes in the uterine lining. 601

menstruation (men"stroo-a'shun) The discharge of blood and tissue from the uterus at the end of the menstrual cycle. 601

mesencephalic aqueduct (mez"en-sĕ-fal'ik ak'wĕ-dukt) The channel that connects the third and fourth ventricles of the brain; also called the *aqueduct of Sylvius.* 265

mesencephalon (mes"en-sef'ă-lon) The midbrain, which contains the corpora quadrigemina and the cerebral peduncles. 249

mesenchyme (mez'en-kīm) An embryonic connective tissue that can migrate, and from which all connective tissues arise. 81

mesenteric (mes"en-ter'ik) **patches** Clusters of lymph nodes on the walls of the small intestine; also called *Peyer's patches.* 428

mesentery (mes'en-ter"e) A fold of peritoneal membrane that attaches an abdominal organ to the abdominal wall. 17

mesoderm (mes'ŏ-derm) The middle layer of the three primary germ layers of the developing embryo. 620

messenger RNA (mRNA) A type of RNA that contains a base sequence complementary to a part of the DNA that specifies the synthesis of a particular protein. 62

metabolism (mĕ-tab'ŏ-liz"em) The sum total of the chemical changes that occur within a cell. 544

metatarsus (met"ă-tar'sus) The region of the foot between the ankle and the phalanges that includes the five metatarsal bones. 146

metencephalon (met"en-sef'ă-lon) The most superior portion of the hindbrain that contains the cerebellum and the pons. 249

micelles (mi-selz') Colloidal particles formed by the aggregation of many molecules. 526

microglia (mi-krog'le-ă) Small phagocytic cells found in the central nervous system. 234

microvilli (mi"kro-vil'i) Microscopic hairlike projections of cell membranes on certain epithelial cells. 234

micturition (mik"tŭ-rish'un) The process of voiding urine; also called *urination.* 497

midbrain The portion of the brain between the pons and the forebrain. 261

middle ear The middle of the three portions of the ear that contains the three auditory ossicles. 319

midsagittal (mid-saj'ĭ-tal) **plane** A plane that divides the body into equal right and left halves; also called the *median plane* or *midplane.* 11

mineralocorticoids (min"er-al-o-kor'tĭ-koidz) Steroid hormones of the adrenal cortex (corticosteroids) that regulate electrolyte balance. 341

mitochondrion (mi"tŏ-kon'dre-on); pl., *mitochondria* A cytoplasmic organelle that serves as a site for the production of most of the energy of a cell; the so-called powerhouse of the cell. 57

mitosis (mi-to'sis) The process of cell division that results in two identical daughter cells, containing the same number of chromosomes. 67

mitral (mi'tral) **valve** The left atrioventricular heart valve; also called the *bicuspid valve.* 375

mixed nerve A nerve that contains both motor and sensory nerve fibers. 250

monocyte (mon'o-sit) A phagocytic type of white blood cell, normally constituting about 3% to 8% of the white blood cell count. 360

monomer (mon'ŏ-mer) A single molecular unit of a longer, more complex molecule. Monomers are joined together to form dimers, trimers, and polymers; the hydrolysis of polymers eventually yields separate monomers. 504

monosaccharide (mon"ŏ-sak'ă-rīd) The monomer of the more complex carbohydrates, examples of which include glucose, fructose, and galactose; also called a *simple sugar.* 29

mons pubis (pyoo'bis) A fatty tissue pad covering the symphysis pubis and covered by pubic hair in the female. 596

morphogenesis (mor"fo-jen'ĕ-sis) During prenatal development, the transformation involved in the growth and differentiation of cells and tissues. 618

morula (mor'yŭ-lă) An early stage of embryonic development characterized by a solid ball of cells. 618

motor area A region of the cerebral cortex from which motor impulses to muscles or glands originate. 254

motor nerve A nerve composed of motor nerve fibers. 233

motor neuron (noor'on) A nerve cell that conducts action potentials away from the central nervous system and innervates effector organs (muscle and glands). It forms the anterior roots of the spinal nerves; also called an *efferent neuron.* 232

motor unit A single motor neuron and the muscle fibers it innervates. 181

mucosa (myoo-ko'să) A mucous membrane that lines cavities and tracts opening to the exterior. 77, 509

mucous (myoo'kus) **cell** *See* goblet cell. 76

mucous membrane A thin sheet consisting of layers of visceral organs that include the lining epithelium, submucosal connective tissue, and (in some cases) a thin layer of smooth muscle (the muscularis mucosa). 16

multipolar neuron A nerve cell with many processes originating from the cell body. 233

muscle A major type of tissue adapted to contract. The three kinds of muscle are cardiac, smooth, and skeletal. 87

muscle spindles Sensory organs within skeletal muscles composed of intrafusal fibers. They are sensitive to muscle stretch and provide a length detector within muscles. 304

muscularis (mus"kyŭ-lar'is) A muscular layer or tunic of an organ, composed of smooth muscle tissue. 509

myelencephalon (mi"ĕ-len-sef'ă-lon) The posterior portion of the hindbrain that contains the medulla oblongata. 249

myelin (mi'ĕ-lin) A lipoprotein material that forms a sheathlike covering around nerve fibers. 234

myelin sheath A sheath surrounding axons formed by successive wrappings of a neuroglial cell membrane. Myelin sheaths are formed by neurolemmocytes in the peripheral nervous system and by oligodendrocytes within the central nervous system. 234

myeloid tissue The red bone marrow in which blood cells are produced. 361

myenteric plexus (mi"en-ter'ik plek'sus) A network of sympathetic and parasympathetic nerve fibers located in the muscularis tunic of the small intestine; also called the *plexus of Auerbach.* 509

myocardial infarction (mi"ŏ-kar'de-al in-fark'shun) An area of necrotic tissue in the myocardium that is filled in by scar (connective) tissue. 399

myocardium (mi"ŏ-kar'de-um) The cardiac muscle layer of the heart. 372

myofibril A bundle of contractile fibers within muscle cells. 178

myofilament The filament that constitutes myofibrils. It is composed of either actin or myosin. 178

myogenic (mi"ŏ-jen'ik) Originating within muscle cells; used to describe self-excitation by cardiac and smooth muscle cells.

myoglobin (mi"ŏ-glo'bin) A molecule composed of globin protein and heme pigment. It is related to hemoglobin but contains only one subunit (instead of the four in hemoglobin) and is found in skeletal and cardiac muscle cells where it serves to store oxygen. 187

myogram (mi'ŏ-gram) A recording of electrical activity within a muscle. 219

myometrium (mi"o-me'tre-um) The layer or tunic of smooth muscle within the uterine wall. 594

myoneural (mi"ŏ-noor'al) **junction** The site of contact between an axon of a motor neuron and a muscle fiber. 239

myosin A thick myofilament protein that, together with actin, causes muscle contraction. 178

N

NAD Nicotinamide adenine dinucleotide; a coenzyme derived from niacin that helps to transport electrons from the Krebs cycle to the electron-transport chain within mitochondria. 544

nail A hardened, keratinized plate that develops from the epidermis and forms a protective covering on the surface of the distal phalanges of fingers and toes. 105

nasal cavity A mucosa-lined space above the oral cavity, divided by a nasal septum. It is the first chamber of the respiratory system. 451

nasal concha (kong'kă); pl., *conchae* (kong'ke) A scroll-like bone extending medially from the lateral wall of the nasal cavity; also called a *turbinate.* 129

nasal septum (sep'tum) A bony and cartilaginous partition that separates the nasal cavity into two portions. 451

nasopharynx (na"zo-far'ingks) The first or uppermost chamber of the pharynx, lying posterior to the nasal cavity and extending down to the soft palate. 452

negative feedback A mechanism in the body for maintaining a state of internal constancy, or homeostasis; effectors are activated by changes in the internal environment, and the actions of the effectors serve to counteract these changes and maintain a state of balance. 17

neonatal (ne"o-na'tal) The stage of life from birth to the end of 4 weeks.

nephron (nef'ron) The functional unit of the kidney, consisting of a glomerulus, convoluted tubules, and a nephron loop. 479

nephron loop The U-shaped part of the nephron, consisting of descending and ascending limbs; also called the *loop of Henle.* 481

nerve A bundle of nerve fibers outside the central nervous system. 233

nerve plexus One of several interlacing nerve networks that occur in the cervical, brachial, lumbar, and sacral regions of the vertebral column. 276

neurofibril node A gap in the myelin sheath of a nerve fiber; also called a *node of Ranvier.* 234

neuroglia (noo-rog'le-ă) Specialized supportive cells of the central nervous system; also called *glial cells,* or *glia.* 89, 234

neurohypophysis (noor"o-hi-pof'ĭ-sis) The posterior lobe of the pituitary gland derived from the brain. Its major secretions include antidiuretic hormone (ADH), also called vasopressin, and oxytocin, produced in the hypothalamus. 336

neurolemmocyte (noor"ŏ-lem'ŏ-sīt) A specialized neuroglial cell that surrounds an axon fiber of a peripheral nerve and forms the neurilemmal sheath; also called a *Schwann cell.* 234

neuron (noor'on) The structural and functional unit of the nervous system, composed of a cell body, dendrites, and an axon; also called a *nerve cell.* 88, 232

neurotransmitter A chemical contained in synaptic vesicles in nerve endings that is released into the synaptic cleft, where it stimulates the production of either excitatory or inhibitory postsynaptic potentials. 239

neutrons (noo'tronz) Electrically neutral particles that exist together with positively charged protons in the nucleus of atoms. 24

nipple A darkly pigmented, rounded projection at the tip of the breast. 606

node of Ranvier (rahn-ve-a'; ran'vēr) *See* neurofibril node. 234

nonspecific defense mechanisms Barriers to penetration of the body by pathogens, as well as internal defenses. 429

norepinephrine (nor"ep-ĭ-nef'rin) A catecholamine released as a neurotransmitter from postganglionic sympathetic nerve endings and as a hormone (together with epinephrine) from the adrenal medulla. 292

notochord (no'tŏ-kord) A flexible rod of tissue that extends the length of the back of an embryo. 625

nucleolus (noo-kle'ŏ-lus) A dark-staining area within a cell nucleus; the site where ribosomal RNA is produced. 43

nucleoplasm (noo'kle-ŏ-plaz"em) The protoplasmic contents of the nucleus of a cell. 59

nucleotide (noo'kle-ŏ-tīd) The subunit of DNA and RNA macromolecules. Each nucleotide is composed of a nitrogenous base (adenine, guanine, cytosine, and thymine or uracil); a sugar (deoxyribose or ribose); and a phosphate group. 60

nucleus (noo'kle-us) A spheroid body within a cell that contains the genetic factors of the cell. 43

nystagmus (nĭ-stag'mus) Involuntary oscillary movements of the eye. 326

O

olfaction The sense of smell. 305

olfactory bulb An aggregation of sensory neurons of an olfactory nerve, lying inferior to the frontal lobe of the cerebrum on either lateral side of the crista galli of the ethmoid bone. 305

olfactory tract The olfactory sensory tract of axons that conveys impulses from the olfactory bulb to the olfactory portion of the cerebral cortex. 305

oligodendrocyte (ol"ĭ-go-den'drŏ-sīt) A type of neuroglial cell concerned with the formation of the myelin of nerve fibers within the central nervous system. 234

oncotic (on-kot'ik) **pressure** The osmotic pressure of solutions produced by proteins. In plasma, it serves to counterbalance the outward filtration of fluid from capillaries due to hydrostatic pressure. 410

oocyte (o'ŏ-sīt) A developing egg cell. 597

oogenesis (o"ŏ-jen'ĕ-sis) The process of female gamete formation. 597

optic (op'tik) Pertaining to the eye.

optic chiasma (ki-az'mă) An X-shaped structure on the inferior aspect of the brain, anterior to the pituitary gland, where there is a partial crossing over of fibers in the optic nerves; also called the *optic chiasm.* 316

optic disc A small region of the retina where the fibers of the ganglion neurons exit from the eyeball to form the optic nerve; also called the *blind spot.* 313

optic tract A bundle of sensory axons located between the optic chiasma and the thalamus that functions to convey visual impulses from the photoreceptors within the eye. 317

oral Pertaining to the mouth.

organ A structure consisting of two or more tissues that performs a specific function. 6

organelle (or"gă-nel') A minute cellular structure that performs a specific function for the cell as a whole. 43

organism An individual living creature. 7

organ of Corti (kor'te) *See* spiral organ. 322

origin The place of muscle attachment—usually the more stationary point or the proximal bone; opposite the insertion. 175

oropharynx (o"ro-far'ingks) The second portion of the pharynx, lying posterior to the oral cavity and extending from the soft palate to the hyoid bone. 452

osmolality (oz"mŏ-lal'ĭ-te) A measure of the total concentration of a solution; the number of moles of solute per kilogram of solvent. 51

osmoreceptors (oz"mŏ-re-cep'torz) Sensory neurons that respond to changes in the osmotic pressure of the surrounding fluid. 51

osmosis (oz-mo'sis) The passage of solvent (water) from a more dilute to a more concentrated solution through a membrane that is more permeable to water than to the solute. 50

osmotic (oz-mot'ik) **pressure** A measure of the tendency of a solution to gain water by osmosis when separated by a membrane from pure water. Directly related to the osmolality of the solution, it is the pressure required to just prevent osmosis. 51

ossicle (os'ĭ-kul) One of the three bones of the middle ear; also called the *auditory ossicle.* 319

ossification (os"ĭ-fĭ-ka'shun) The process of bone tissue formation. 118

osteoblast (os'te-ŏ-blast) A bone-forming cell. 122

osteoclast (os'te-ŏ-klast) A cell that causes erosion and resorption of bone tissue. 122

osteocyte (os'te-ŏ-sīt) A mature bone cell. 85

osteon (os'te-on) A group of osteocytes and concentric lamellae surrounding a central canal, constituting the basic unit of structure in osseous tissue; also called a *haversian system.* 86, 123

outer ear The outer portion of the ear, consisting of the auricle and the external auditory canal. 318

oval window *See* vestibular window. 319

ovarian (o-var'e-an) **follicle** A developing ovum and its surrounding epithelial cells. 597

ovary (o'vă-re) The female gonad in which ova and certain sexual hormones are produced. 592

ovulation (ov-yŭ-la'shun) The rupture of an ovarian (graafian) follicle with the release of an ovum. 599

ovum (o'vum) A secondary oocyte capable of developing into a new individual when fertilized by a spermatozoon. 592

oxidative phosphorylation (ok"sĭ-da'tiv fos"for-ĭ-la'shun) The formation of ATP using energy derived from electron transport to oxygen. It occurs in the mitochondria. 550

oxidizing agent An atom that accepts electrons in an oxidation-reduction reaction.

oxygen debt The added oxygen taken into the body after exercise, over and above the resting oxygen consumption, for restorative purposes. 187

oxyhemoglobin (ok"se-he"mŏ-glo'bin) A compound formed by the bonding of molecular oxygen to hemoglobin. 467

oxyhemoglobin saturation The ratio, expressed as a percentage, of the amount of oxyhemoglobin relative to the total amount of hemoglobin in blood. 468

oxytocin (ok"sĭ-to'sin) One of the two hormones produced in the hypothalamus and secreted by the posterior pituitary (the other hormone is vasopressin). Oxytocin stimulates the contraction of uterine smooth muscles and promotes milk ejection in females. 338, 607

P

pacemaker *See* sinoatrial node. 337

pacinian (pă-sin'e-an) **corpuscle** *See* lamellated corpuscle. 302

PAH Para-aminohippuric acid; a substance used to measure total renal plasma flow because its clearance rate is equal to the total rate of plasma flow to the kidneys. PAH is filtered and secreted but not reabsorbed by the renal nephrons. 491

palate (pal'at) The roof of the oral cavity. 512

palpebra (pal'pĕ-bră) An eyelid. 307

pancreas A mixed organ in the abdominal cavity that secretes pancreatic juices into the GI tract and insulin and glucagon into the blood. 528

pancreatic duct A drainage tube that carries pancreatic juice from the pancreas into the duodenum of the hepatopancreatic ampulla. 528

pancreatic islet A cluster of cells within the pancreas that forms the endocrine portion and secretes insulin and glucagon; also called *islet of Langerhans.* 345

pancreatic juice The exocrine secretion of the pancreas, consisting of water, bicarbonate, and a wide variety of digestive enzymes. 529

papillae (pă-pil'e) Small, nipplelike projections. 306

papillary (pap′ĭ-ler″e) **muscles** Muscular projections from the ventricular walls of the heart to which the chordea tendineae are attached. 375

paranasal (par″ă-naz′al) **sinus** An air chamber lined with a mucous membrane that communicates with the nasal cavity. 451

parasympathetic division Pertaining to the division of the autonomic nervous system concerned with activities that, in general, inhibit or oppose the physiological effects of the sympathetic nervous system. 289

parathyroid gland One of four small glands embedded in the posterior surfaces of the lateral lobes of the thyroid gland. 344

parathyroid hormone (PTH) A polypeptide hormone secreted by the parathyroid glands. PTH acts to raise the blood Ca^{++} levels primarily by stimulating reabsorption of bone. 345

paravertebral ganglia A series of ganglia that lie in a vertical row on either side of the verebral column; also called *sympathetic trunk ganglia.* 290

parietal (pă-ri′ĕ-tal) Pertaining to a wall of an organ or cavity. 16

parotid (pă-rot′id) **gland** One of the paired salivary glands located on the side of the face over the masseter muscle just anterior to the ear and connected to the oral cavity through a salivary duct. 514

parturition (par″tyoo-rish′un) The process of giving birth; childbirth. 629

passive immunity Specific immunity granted by the administration of antibodies made by another organism. 436

pathogen (path′ŏ-jen) Any disease-producing microorganism or substance. 538

pectoral girdle The portion of the skeleton that supports the upper extremities. 139

pelvic Pertaining to the pelvis.

pelvic girdle The portion of the skeleton to which the lower extremities are attached. 143

pelvis A basinlike bony structure formed by the sacrum and ossa coxae. 143

penis (pe′nis) The male organ of copulation, used to introduce sperm cells into the female vagina and through which urine passes during urination. 580

pepsin (pep′sin) The protein-digesting enzyme secreted in gastric juice. 531

peptic ulcer An injury to the mucosa of the esophagus, stomach, or small intestine due to the action of acidic gastric juice. 539

pericardium (per″ĭ-kar′de-um) The protective serous membrane that surrounds the heart. 16, 372

perilymph A fluid of the inner ear that provides a liquid-conducting medium for the vibrations involved in hearing and the maintenance of equilibrium. 320

perineum (per″ĭ-ne′um) The floor of the pelvis, which is the region between the anus and the symphysis pubis; the region that contains the external genitalia. 572

periosteum (per″e-os′te-um) A fibrous connective tissue covering the outer surface of bone. 122

peripheral nervous system The nerves and ganglia of the nervous system that lie outside of the brain and spinal cord; PNS. 248

peristalsis (per″ĭ-stal′sis) Rhythmic contractions of smooth muscle in the walls of various tubular organs by which the contents are forced onward. 521

peritoneum (per″ĭ-tŏ-ne′um) The serous membrane that lines the abdominal cavity and covers the abdominal visceral organs. 507

Peyer′s (pi′erz) **patches** *See* mesenteric patches. 428

pH A measure of the relative acidity or alkalinity of a solution, numerically equal to 7 for neutral solutions. The pH scale in common use ranges from 0 to 14. Solutions with a pH lower than 7 are acidic, and those with a higher pH are basic. 27

phagocytosis (fag″ŏ-si-to′sis) Cellular eating; the ability of some cells (such as white blood cells) to engulf large particles (such as bacteria) and digest these particles by merging the food vacuole in which they are contained with a lysosome containing digestive enzymes. 45, 430

phalanx (fa′langks); pl., *phalanges* (fă-lan′jēz) A bone of a finger or toe. 143, 146

pharynx (far′ingks) The organ of the digestive system and respiratory system located at the back of the oral and nasal cavities that extends to the larynx anteriorly and to the esophagus posteriorly; also called the *throat.* 452, 514

phenotype Observable features in an individual that result from expression of the genotype. 632

phosphocreatine (fos″fo-kre′ă-tēn) *See* creatine phosphate. 186

photoreceptor A sensory nerve ending that responds to the stimulation of light. 313

physiology The science that deals with the study of body functions. 2

pia mater (pi′ă ma′ter) The innermost meninx that is in direct contact with the brain and spinal cord. 265

pineal (pin′e-al) **gland** A small cone-shaped gland located in the roof of the third ventricle. 345

pinocytosis (pin″ŏ-si-to′sis) Cell drinking; invagination of the cell membrane forming narrow channels that pinch off into vacuoles. This allows for cellular intake of extracellular fluid and dissolved molecules. 45

pituitary (pĭ-too′ĭ-ter-e) **gland** A small, pea-shaped endocrine gland situated on the inferior surface of the brain, consisting of anterior and posterior lobes; also called the *hypophysis.* 259, 336

placenta (plă-sen′tă) The organ of metabolic exchange between the mother and the fetus. 347, 395, 622

plasma (plaz′mă) The fluid extracellular portion of circulating blood. 358

plasma cells Cells derived from B lymphocytes that produce and secrete large amounts of antibodies. They are responsible for humoral immunity. 433

platelets Small fragments of specific bone marrow cells that function in blood coagulation; also called *thrombocytes.* 360

pleura (ploor′ă); pl., *pleurae* (ploor′e) The serous membrane associated with the lungs. 16, 457

pleural cavity The potential space between the visceral pleura and parietal pleura. 457

pleural membranes Serous membranes that surround the lungs and provide protection and compartmentalization. 457

plexus (plek′sus) A network of interlaced nerves or vessels. 276

plexus of Auerbach (ow′er-bak) *See* myenteric plexus. 506

plexus of Meissner (mīs′ner) *See* submucosal plexus. 509

plicae circulares (pli′se sur-kyŭ-lar′ēz) Deep folds within the wall of the small intestine that increase the absorptive surface area. 519

pneumotaxic (noo″mŏ-tak′sik) **area** The region of the respiratory control center located in the pons of the brain. 464

polar body A small daughter cell formed by meiosis that degenerates in the process of oocyte production. 597

polar molecule A molecule in which the shared electrons are not evenly distributed, so that one side of the molecule is negatively (or positively) charged in comparison with the other side. Polar molecules are soluble in polar solvents, such as water. 26

polymer (pol′ĕ-mer) A large molecule formed by the combination of smaller subunits, or monomers. 504

polypeptide A chain of amino acids connected by covalent bonds called peptide bonds. A very large polypeptide is called a protein. 335

polysaccharide (pol″e-sak′ă-rīd) A carbohydrate formed by covalent bonding of numerous monosaccharides. Examples include glycogen and starch. 29

polyuria (pol″e-yoor′e-ă) Excretion of an excessively large volume of urine in a given period. 498

pons The portion of the brain stem just above the medulla oblongata and anterior to the cerebellum. 261

popliteal (pop″lĭ-te′al, pop-lit′e-al) Pertaining to the concave region on the posterior aspect of the knee. 14

posterior Toward the back; also called *dorsal.* 10

posterior pituitary (pĭ-too′ĭ-ter-e) *See* neurohypophysis. 336

posterior root An aggregation of sensory neuron fibers lying between a spinal nerve and the posterolateral aspect of the spinal cord; also called the *dorsal root* or *sensory root.* 275

posterior root ganglion (gang′gle-on) A cluster of cell bodies of sensory neurons located along the posterior root of a spinal nerve. 275

postganglionic (pōst″gang-gle-on′ik) **neuron** The second neuron in an autonomic motor pathway. Its cell body is outside the central nervous system, and it terminates at an effector organ. 289

postsynaptic (pōst″sĭ-nap′tik) **inhibition** The inhibition of a postsynaptic neuron by axon endings that release a neurotransmitter that induces hyperpolarization (inhibitory postsynaptic potentials). 240

preganglionic (pre″gang-gle-on′ik) **neuron** The first neuron in an autonomic motor pathway. Its cell body is inside the central nervous system, and it terminates on a postganglionic neuron. 289

pregnancy A condition in which a female is carrying a developing offspring within the body. 611

prepuce (pre′pyoos) A fold of loose, retractable skin covering the glans of the penis or clitoris; also called the *foreskin.* 580

prolactin (pro-lak′tin) A hormone secreted by the anterior pituitary that, in conjunction with other hormones, stimulates lactation in the postpartum female. 338, 606

proprioceptor (pro″pre-o-cep′tor) A sensory nerve ending that responds to changes in tension in a muscle or tendon. 302

prostaglandin (pros″tă-glan′din) Any of a family of fatty acids that have numerous autocrine regulatory functions, including the stimulation of uterine contractions and gastric acid secretion, and the promotion of inflammation. 629

prostate (pros′tāt) A walnut-shaped gland surrounding the male urethra just below the urinary bladder that secretes an additive to seminal fluid during ejaculation. 580

prosthesis (pros-the′sis) An artificial device to replace a diseased or worn body part. 171

protein Any of a group of large organic molecules made up of amino acid subunits linked by peptide bonds. 33, 531

proton (pro′ton) A unit of positive charge in the nucleus of atoms. 24

protraction The movement of part of the body forward, on a plane parallel to the ground. 165

proximal (prok′sĭ-mal) Closer to the midplane of the body or to the origin of an appendage; the opposite of distal. 10

pseudopods (soo′dŏ-podz) Footlike extensions of the cytoplasm that enable some cells (with amoeboid motion) to move across a substrate. Pseudopods are also used to surround food particles in the process of phagocytosis. 45

ptyalin (ti′ă-lin) An enzyme in saliva that catalyzes the hydrolysis of starch into smaller molecules; also called *salivary amylase.* 531

puberty (pyoo′ber-te) The period of development in which the reproductive organs become functional. 570

pulmonary (pul′mŏ-ner″e) Pertaining to the lungs.

pulmonary circulation The system of blood vessels from the right ventricle of the heart to the lungs that transports deoxygenated blood and returns oxygenated blood from the lungs to the left atrium of the heart. 375

pupil The opening through the iris that permits light to enter the posterior cavity of the eyeball and be refracted by the lens through the vitreous chamber. 312

Purkinje (pur-kin′je) **fibers** *See* conduction myofibers. 377

pyloric sphincter (pi-lor′ik sfingk′ter) A modification of the muscularis tunic between the stomach and the duodenum that functions to regulate the food material leaving the stomach. 516

pyramid Any of several structures that have a pyramidal shape (e.g., the renal pyramids in the kidney and the medullary pyramids on the anterior surface of the brain). 479

pyrogen (pi′rŏ-jen) A fever-producing substance. 431

Q

QRS complex The principal deflection of an electrocardiogram that is produced by depolarization of the ventricles. 380

R

receptor A sense organ or a specialized distal end of a sensory neuron that receives stimuli from the environment. 301

rectum (rek′tum) The terminal portion of the GI tract, between the sigmoid colon and the anal canal. 523

red marrow (mar′o) A tissue that forms blood cells, located in the medullary cavity of certain bones. 121

reduced hemoglobin (he′mŏ-glo″bin) Hemoglobin with iron in the reduced ferrous state. It is able to bond with oxygen but is not combined with oxygen. Also called *deoxyhemoglobin.* 467

reflex A rapid involuntary response to a stimulus. 279

reflex arc The basic conduction pathway through the nervous system, consisting of a sensory neuron, an association neuron, and a motor neuron. 277

refractory period The period of time during which a region of axon or muscle cell membrane is incapable of responding to further stimulation. 237

renal (re′nal) Pertaining to the kidney.

renal cortex The outer portion of the kidney, containing the glomeruli and the proximal and distal tubules of the nephrons. 479

renal medulla (mĕ-dul′ă) The inner portion of the kidney, including the renal pyramids and renal columns. 479

renal pelvis The inner cavity of the kidney formed by the expanded ureter and into which the calyces open. 479

renal plasma clearance rate The milliliters of plasma cleared of a particular solute each minute by the excretion of that solute in the urine. If there is no reabsorption or secretion of that solute by the nephron tubules, the plasma clearance rate is equal to the glomerular filtration rate. 488

renal pyramid A triangular structure within the renal medulla composed of nephron loops and the collecting ducts. 479

renin An enzyme secreted by the juxtaglomerular apparatus of the kidneys. It catalyzes the conversion of angiotensinogen into angiotensin I. 492

repolarization The reestablishment of the resting membrane potential after depolarization has occurred. 236

respiration (res″pĭ-ra′shun) The exchange of gases between the external environment and the cells of an organism. 450

respiratory acidosis (as″ĭ-do′sis) A lowering of the blood pH to below 7.35 due to accumulation of CO_2 as a result of hypoventilation. 469

respiratory alkalosis (al″kă-lo′sis) A rise in blood pH to above 7.45 due to excessive elimination of blood CO_2 as a result of hyperventilation. 469

respiratory center The structure or portion of the brain stem that regulates the depth and rate of breathing. 464

respiratory membrane A thin, moistened membrane within the lungs, composed of an alveolar portion and a capillary portion, through which gaseous exchange occurs. 456

resting membrane potential The potential difference (voltage) across the cell membrane when the cell is in an unstimulated state. 236

rete testis (re′te tes′tis) A network of ducts in the center of the testis associated with the production of spermatozoa. 575

reticular formation A network of nervous tissue fibers in the brain stem that arouses the higher brain centers. 262

retina (ret′ĭ-nă) The principal portion of the internal tunic of the eyeball that contains the photoreceptors. 312

retraction The pulling back of a protracted part of the body on a plane parallel to the ground. 165

Rh factor An inherited agglutinogen (antigen) on the surface of red blood cells. 364

rhythmicity (rith-mis′ĭ-te) **area** A portion of the respiratory control center located in the medulla oblongata that controls inspiratory and expiratory phases. 464

ribosome (ri'bŏ-sōm) A cytoplasmic organelle composed of protein and RNA in which protein synthesis occurs. 57

right lymphatic (lim-fat'ik) **duct** A major vessel of the lymphatic system that drains lymph from the upper right portion of the body into the right subclavian vein. 426

RNA Ribonucleic acid; a nucleic acid consisting of the nitrogenous bases adenine, guanine, cytosine, and uracil; the sugar ribose; and phosphate groups. There are three types of RNA found in cytoplasm: messenger RNA (mRNA), transfer RNA (tRNA), and ribosomal RNA (rRNA). 61

rod A photoreceptor in the retina of the eye that is specialized for colorless, dim-light vision. 312

rotation The movement of a bone around its own axis. 164

round window *See* cochlear window. 319

rugae (roo'je) Folds or ridges of the mucosa of an organ, such as those of the stomach or urinary bladder. 496, 516

S

saccule (sak'yool) A saclike cavity in the membranous labyrinth inside the vestibule of the inner ear that contains a vestibular organ for equilibrium. 325

sacral plexus (plek'sus) A network of nerve fibers that arises from spinal nerves L4 through S3. Nerves arising from the sacral plexus merge with those from the lumbar plexus to form the lumbosacral plexus and supply the lower extremity. 276

sagittal (saj'ĭ-tal) **plane** A vertical plane, running parallel to the midsagittal plane, that divides the body into unequal right and left portions. 11

salivary gland An accessory digestive gland that secretes saliva into the oral cavity. 514

saltatory (sal'tă-to"re) **conduction** The rapid passage of action potentials from one neurofibril node to another in myelinated axons. 239

sarcolemma (sar"kŏ-lem'ă) The cell membrane of a muscle fiber. 178

sarcomere (sar"kŏ-mēr) The portion of a skeletal muscle fiber between the two adjacent Z lines that is considered the functional unit of a myofibril. 179

sarcoplasm (sar'kŏ-plaz"em) The cytoplasm within a muscle fiber. 178

sarcoplasmic reticulum (rĕ-tik'yŭ-lum) The smooth or agranular endoplasmic reticulum of skeletal muscle cells. It surrounds each myofibril and stores Ca^{++} when the muscle is at rest. 178

scala tympani (ska'lă tim'pă-ne) The lower channel of the cochlea that is filled with perilymph. 321

scala vestibuli (vĕ-stib'yŭ-le) The upper channel of the cochlea that is filled with perilymph. 321

Schwann cell *See* neurolemmocyte. 234

scientific method The general method of scientific investigation involving the statement and testing of a hypothesis, followed by a conclusion that validates or modifies the hypothesis. 4

sclera (skler'ă) The outer white layer of fibrous connective tissue that forms the protective covering of the eyeball. 310

scleral venous (ve'nus) **sinus** A circular venous drainage for the aqueous humor from the anterior cavity of the eye; located at the junction of the sclera and the cornea; also called the *canal of Schlemm.* 314

scrotum (skro'tum) A pouch of skin that contains the testes and their accessory organs. 572

sebaceous (sĕ-ba'shus) **gland** An exocrine gland of the skin that secretes sebum. 106

sebum (se'bum) An oily, waterproofing secretion of the sebaceous glands. 106

second messenger A molecule or ion whose concentration within a target cell is increased by the action of a regulatory compound (e.g., a hormone or neurotransmitter) and which stimulates the metabolism of that target cell in a way that mediates the intracellular effects of that regulatory compound. 349

secretin (sĕ-kre'tin) A polypeptide hormone secreted by the small intestine in response to acidity of the intestinal lumen. Along with cholecystokinin, secretin stimulates the secretion of pancreatic juice into the small intestine. 535

semen (se'men) The thick, whitish secretion of the reproductive organs of the male, consisting of spermatozoa and additives from the prostate and seminal vesicles. 582

semicircular canals Tubular channels within the inner ear that contain receptors for equilibrium. 321

semilunar valve Crescent- or half-moon-shaped heart valves positioned at the entrances to the aorta and the pulmonary trunk. 373

seminal (sem'ĭ-nal) **vesicles** A pair of male accessory reproductive organs lying posterior and inferior to the urinary bladder that secrete additives to spermatozoa into the ejaculatory ducts. 578

seminiferous (sem"ĭ-nif'er-us) **tubules** Numerous small ducts in the testes, where spermatozoa are produced. 574

semipermeable membrane A membrane with pores of a size that permits the passage of solvent and some solute molecules while restricting the passage of other solute molecules. 42

sensory area A region of the cerebral cortex that receives and interprets sensory nerve impulses. 254

sensory neuron (noor'on) A nerve cell that conducts an impulse from a receptor organ to the central nervous system; also called an *afferent neuron.* 232

serous (ser'us) **membrane** An epithelial and connective tissue membrane that lines body cavities and covers visceral organs within these cavities; also called *serosa.* 16

Sertoli (ser-to'le) **cells** *See* sustentacular cells. 574

serum Blood plasma with the clotting elements removed. 358

sesamoid (ses'ă-moid) **bone** A membranous bone formed in a tendon in response to joint stress (e.g., the patella). 119

sex chromosomes The X and Y chromosomes; the unequal pairs of chromosomes involved in sex determination (which is based on the presence or absence of a Y chromosome). Females lack a Y chromosome and normally have the genotype XX; males have a Y chromosome and normally have the genotype XY. 631

sigmoid colon (sig'moid ko'lon) The S-shaped portion of the large intestine between the descending colon and the rectum. 523

sinoatrial (sin"no-a'tre-al) **node** A mass of specialized cardiac tissue in the wall of the right atrium that initiates the cardiac cycle; the SA node; also called the *pacemaker.* 337

sinusoid (si'nŭ-soid) A small, blood-filled space in certain organs, such as the spleen or liver. 524

skeletal muscle A specialized type of multinucleated muscle tissue that occurs in bundles, has crossbands of proteins, and contracts in either a voluntary or involuntary fashion. 88

sliding filament theory The theory that the thin filaments of a myofibril slide past the thick ones during muscle contraction in a way that causes the sarcomeres to shorten. 182

small intestine The portion of the GI tract between the stomach and the cecum whose function is the absorption of food nutrients. 518

smooth muscle A specialized type of nonstriated muscle tissue composed of fusiform single-nucleated fibers. It contracts in an involuntary rhythmic fashion within the walls of visceral organs. 87

sodium/potassium pump An active transport carrier with ATPase enzymatic activity that acts to accumulate K^+ within cells and extrude Na^+ from cells, thus maintaining gradients for these ions across the cell membrane. 54

somatic (so-mat'ik) Pertaining to the nonvisceral parts of the body.

somatotropic (sŏ-mat'ŏ-trop'ik, sŏ"mă-tŏ-trop'ik) **hormone** Growth hormone; an anabolic hormone secreted by the anterior pituitary that stimulates skeletal growth and protein synthesis in many organs. 338

specific immune response The defense of the body against specific pathogens as a result of prior exposure to those pathogens. 432

spermatic cord The structure of the male reproductive system composed of the ductus deferens, spermatic vessels, nerves, cremaster muscle, and connective tissue. The spermatic cord extends from a testis to the inguinal ring. 578

spermatogenesis (sper-mat″ŏ-jen′ĕ-sis) The production of male sex gametes, or spermatozoa. 576

spermatogonia (sper-mat″ŏ-go′ne-ă) Sperm stem cells within the seminiferous tubules of the testes; the progenitors of spermatocytes. 576

spermatozoon (sper-mat″ŏ-zo′on); pl., *spermatozoa* or, loosely, *sperm* A mature male sperm cell, or gamete. 578

spermiogenesis (sper″me-ŏ-jen′ĕ-sis) The maturational changes that transform spermatids into spermatozoa. 577

sphincter (sfingk′ter) A circular muscle that functions to constrict a body opening or the lumen of a tubular structure. 189

sphincter of ampulla The muscular constriction at the opening of the common bile and pancreatic ducts; also called the *sphincter of Oddi.* 528

sphincter of Oddi (o′de) *See* sphincter of ampulla. 528

sphygmomanometer (sfig″mo-mă-nom′ĭ-ter) A manometer (pressure transducer) used to measure the blood pressure. 418

spinal cord The portion of the central nervous system that extends downward from the brain stem through the vertebral canal. 272

spinal ganglion A cluster of nerve cell bodies on the posterior root of a spinal nerve. 275

spinal nerves One of the 31 pairs of nerves that arise from the spinal cord. 275

spindle fibers Filaments that extend from the poles of a cell to its equator and attach to the chromosomes during the metaphase stage of cell division. Contraction of the spindle fibers pulls the chromosomes to opposite poles of the cell. 67

spiral organ The functional unit of hearing, consisting of a basilar membrane that supports receptor hair cells, the tips of which are embedded in a tectorial membrane within the endolymph of the cochlear duct; also known as the *organ of Corti.* 322

spleen A large, blood-filled, glandular organ located in the upper left quadrant of the abdomen and attached by mesenteries to the stomach. 428

spongy bone Bone tissue with a latticelike structure; also called *cancellous bone.* 122

squamous (skwa′mus) Flat or scalelike. 76

stapes (sta′pēz) The innermost of the auditory ossicles that fits against the vestibular window of the inner ear; also called the *stirrup.* 319

Starling forces The hydrostatic pressure and oncotic pressure forces that determine the direction of fluid movement across capillaries. 410

steroid (ster′oid) Any of a large family of lipids, derived from cholesterol, that has three 6-sided carbon rings and one 5-sided carbon ring. Steroids form the sex hormones of the gonads and the corticosteroids of the adrenal cortex. 32, 335

stomach A pouchlike digestive organ located between the esophagus and the duodenum. 515

stratified (strat′ĭ-fīd) Arranged in layers, or strata. 76

stratum basale (stra′tum bă-să′le) The deepest epidermal layer, where mitotic activity occurs. 76, 98

stratum corneum (kor′ne-um) The outer, cornified layer of the epidermis of the skin. 98

stroke volume The amount of blood ejected from each ventricle at each heartbeat. 405

subarachnoid (sub″ă-rak′noid) **space** The space within the meninges between the arachnoid mater and pia mater, where cerebrospinal fluid flows. 265

sublingual (sub-ling′gwal) **gland** One of the three pairs of salivary glands. 514

submandibular (sub″man-dib′yŭ-lar) **gland** One of the three pairs of salivary glands. 514

submucosa (sub″myoo-ko′sa) A layer of supportive connective tissue that underlies a mucous membrane. 514

submucosal plexus (plek′sus) A network of sympathetic and parasympathetic nerve fibers located in the submucosa tunic of the small intestine; also called the *plexus of Meissner.* 509

sudoriferous (soo″dor-if′er-us) **gland** An exocrine gland that excretes perspiration, or sweat, onto the surface of the skin. 106

sulcus (sul′kus) A shallow impression or groove. 252

summation In neural physiology, the additive effects of graded synaptic potentials. In muscle physiology, the additive effects of contractions of different muscle fibers. 180

superior vena cava (ve′nă ka′vă) A major vein that collects blood from parts of the body superior to the heart and empties into the right atrium. 389

suppressor T cell A subpopulation of T lymphocytes that acts to inhibit the production of antibodies against specific antigens by B lymphocytes. 437

surfactant (sur-fak′tant) A substance produced by the lungs that decreases the surface tension within the alveoli. 459

suspensory (sŭ-spen′sŏ-re) **ligament**
1. A portion of the peritoneum that extends laterally from the surface of the ovary to the wall of the pelvic cavity. 595
2. A ligament that supports an organ or body part, such as that supporting the lens of the eye. 310

sustentacular (sus-ten-tak′yŭ-lar) **cells** Specialized cells within the testes that supply nutrients to developing spermatozoa; also called *Sertoli cells* or *nurse cells.* 574

suture A type of fibrous joint found between bones of the skull. 157

sweat gland A skin gland that secretes a fluid substance for evaporative cooling. 78, 106

sympathetic division Pertaining to the division of the autonomic nervous system concerned with activities that, in general, arouse the body for physical activity; also called the *thoracolumbar division.* 289

sympathoadrenal system A term used in reference to the secretion of epinephine and norepinephrine by the adrenal medulla in concert with the activity of the sympathetic division of the ANS. 292

symphysis (sim′ fĭ-sis) A cartilaginous joint in which the adjoining bones are separated by a pad of fibrocartilage. 158

synapse (sin′aps) A minute space between the axon terminal of a presynaptic neuron and a dendrite of a postsynaptic neuron. 239

synchondrosis (sin″kon-dro′sis) A cartilaginous joint in which the articulating bones are separated by hyaline cartilage. 158

syndesmosis (sin″des-mo′sis) A type of fibrous joint in which two bones are united by an interosseous ligament. 157

synergist (sin′er-jist) A muscle that assists the action of the prime mover. 189

synovial (sĭ-no′ve-al) **cavity** The fluid-filled space between the two bones of a synovial joint. 156

synovial fluid The lubricating fluid that fills the synovial cavity of a synovial joint. 159

synovial joint A freely movable joint in which there is a synovial cavity between the articulating bones. 159

synovial membrane The inner membrane of a synovial capsule that secretes synovial fluid into the joint cavity. 159

system A group of body organs that function together. 7

systemic (sis-tem′ik) Relating to the entire organism rather than to individual parts.

systemic circulation The portion of the circulatory system concerned with blood flow from the left ventricle of the heart to the entire body and back to the heart via the right atrium (in contrast to the pulmonary system, which involves the lungs). 375

systole (sis′tŏ-le) The muscular contraction of a heart chamber during the cardiac cycle. 377

systolic pressure (sis-tol′ik) Arterial blood pressure during the ventricular systolic phase of the cardiac cycle. 419

T

tachycardia (tak″ĭ-kar′de-ă) An excessively rapid heart rate, usually in excess of 100 beats per minute (in contrast to bradycardia, in which the heart rate is very slow).

tactile (tak′til) Pertaining to the sense of touch.

taeniae coli (te′ne-e ko′li) The three longitudinal bands of muscle in the wall of the large intestine; variant spelling, teniae coli. 523

target organ The specific body organ that a particular hormone affects. 334

taste bud An organ containing the chemoreceptors associated with the sense of taste. 305

T cell A type of lymphocyte that provides cell-mediated immunity (in contrast to B lymphocytes, which provide humoral immunity through the secretion of antibodies). There are three subpopulations of T cells: cytotoxic, helper, and suppressor. 432

tectorial (tek-to′re-al) **membrane** A gelatinous membrane positioned over the hair cells of the spiral organ in the cochlea. 322

teeth Accessory structures of digestion adapted to cut, shred, crush, and grind food. 513

telencephalon (tel″en-sef′ă-lon) The anterior portion of the forebrain, constituting the cerebral hemispheres and related parts. 249

tendon A band of dense regular connective tissue that attaches muscle to bone. 175

terminal ganglia Parasympathetic ganglia that are located next to, or actually within, the organs they innervate. 294

testis (tes′tis) The primary reproductive organ of a male that produces spermatozoa and male sex hormones. 574

testosterone (tes-tos′tĕ-rōn) The major androgenic steroid secreted by the interstitial (Leydig) cells of the testes after puberty. 567

tetanus (tet′n-us) A smooth contraction of a muscle (as opposed to muscle twitching). 180

thalamus (thal′ă-mus) An oval mass of gray matter within the diencephalon that serves as a sensory relay center. 258

third ventricle (ven′trĭ-k′l) A narrow cavity between the right and left halves of the thalamus and between the lateral ventricles that contains cerebrospinal fluid. 265

thoracic (thor′ă-sik) **duct** The major lymphatic vessel of the body that drains lymph from the entire body, except for the upper right quadrant, and returns it to the left subclavian vein. 426

thorax (thor′aks) The chest. 12

threshold stimulus The weakest stimulus capable of producing an action potential in an excitable cell. 236

thrombocyte (throm′bŏ-sīt) A blood platelet formed from a fragmented megakaryocyte. 87, 360

thymus (thi′mus) A bilobed lymphoid organ positioned in the upper thorax, posterior to the sternum and between the lungs. 347, 428

thyroid cartilage The largest cartilage in the larynx that supports and protects the vocal cords; commonly called *Adam's apple.* 453

thyroid gland An endocrine gland located just below the larynx, in front of the trachea, consisting of two lobes connected by a narrow band of tissue called the isthmus. 343

thyroxine (thi-rok′sin) Also called tetraiodothyronine, or T_4. The major hormone secreted by the thyroid gland, which regulates the basal metabolic rate and stimulates protein synthesis in many organs. A deficiency of this hormone in early childhood produces cretinism. 344

tidal volume During quiet breathing, the amount of air expired in each breath. 462

tissue An aggregation of similar cells and their binding intercellular substance, joined to perform a specific function. 6

tongue A protrusible muscular organ on the floor of the oral cavity. 511

tonsil A node of lymphoid tissue located in the mucous membrane of the pharynx. 452

trace elements Chemical elements present in minute concentrations in the body, many of which have functions essential to life. 545

trachea (tra′ke-ă) The airway leading from the larynx to the bronchi, composed of cartilaginous rings and a ciliated mucosal lining of the lumen; commonly called the *windpipe.* 454

tract A bundle of nerve fibers within the central nervous system. 273

transpulmonary (trans″pul′mŏ-ner″e) **pressure** The pressure difference across the wall of the lung, equal to the difference between intrapulmonary pressure and intrapleural pressure. 458

transverse colon (ko′lon) A portion of the large intestine that extends from right to left across the abdomen between the hepatic and splenic flexures. 523

transverse plane A plane that divides the body into superior and inferior portions; also called a *horizontal,* or *cross-sectional, plane.* 11

treppe (trep′e) The gradual increase in the strength of a muscle contraction caused by repetitive stimuli of the same strength. 180

tricuspid (tri-kus′pid) **valve** The heart valve located between the right atrium and the right ventricle. 375

triiodothyronine (tri″i-o″do-thi′rŏ-nēn) Abbreviated T_3; a hormone secreted in small amounts by the thyroid; the active hormone in target cells, formed from thyroxine. 344

trophic hormones The hormones secreted by the anterior pituitary. High amounts of these hormones stimulate hypertrophy in their target organs. 337

tropomyosin (tro″pŏ-mi′ŏ-sin) A filamentous protein, that attaches to actin in the thin myofilaments and that acts, together with another protein, to inhibit and regulate the attachment of myosin cross bridges to actin. 185

troponin (tro′pŏ-nin) The protein that works with tropomyosin in regulating muscle contraction. 185

true vocal cords Folds of the mucous membrane in the larynx that produce sound as they are pulled taut and vibrated. 453

trunk The thorax and abdomen together. 12

trypsin (trip′sin) A protein-digesting enzyme in pancreatic juice that is released into the small intestine. 529

tunica albuginea (too′nĭ-kă al″byoo-jin′e-ă) A tough, fibrous tissue surrounding the testis. 574

twitch A single rapid contraction of a muscle in response to a stimulus. 180

tympanic membrane The membranous eardrum positioned between the external and middle ear. 319

U

umbilical (um-bĭ′lĭ-kal) **artery** One of two arteries in the umbilical cord by which blood passes from the fetus to the placenta. 396

umbilical cord A cordlike structure containing the umbilical arteries and vein and connecting the fetus with the placenta. 624

umbilical vein The vein in the umbilical cord by which the fetus receives nourishment from the maternal system. 395

umbilicus (um-bĭ-lĭ-kus) The site where the umbilical cord was attached to the fetus; commonly called the *navel.* 13

unipolar neuron (yoo′nĭ-po-lar noor′on) A nerve cell that has a single nerve fiber extending from its cell body. 233

universal donor A person with blood type O who is able to donate blood to people with other blood types in emergency blood transfusions. 364

universal recipient A person with blood type AB who can receive blood of any type in emergency transfusions. 364

urea (yoo-re′ă) The chief nitrogenous waste product of protein catabolism in the urine, formed in the liver from amino acids. 490

uremia (yoo-re′me-ă) The retention of urea and other products of protein catabolism as a result of inadequate kidney function. 498

ureter (yoo-re′ter) A tube that transports urine from the kidney to the urinary bladder. 495

urethra (yoo-re′thră) A tube that transports urine from the urinary bladder to the outside of the body. 496

urinary bladder A distensible sac that stores urine, situated in the pelvic cavity posterior to the symphysis pubis. 495

urobilinogen (yoo″rŏ-bi-lin′ŏ-jen) A compound formed from bilirubin in the small intestine; some is excreted in the feces, and some is absorbed and enters the enterohepatic circulation, where it may be excreted either in the bile or in the urine. 526

uterine (yoo′ter-in) **tube** The tube through which the ovum is transported to the uterus and the site of fertilization; also called the *oviduct* or *fallopian tube.* 593

uterus (yoo′ter-us) A hollow, muscular organ in which a fetus develops. It is located within the female pelvis between the urinary bladder and the rectum; commonly called the *womb.* 594

utricle (yoo′trĭ-k′l) An enlarged portion of the membranous labyrinth, located within the vestibule of the inner ear. 325

uvula (yoo′vyŭ-lă) A fleshy, pendulous portion of the soft palate that blocks the nasopharynx during swallowing. 513

V

vacuole (vak-yoo′ōl) A small space or cavity within the cytoplasm of a cell. 45

vagina (vă-ji′nă) A tubular organ leading from the uterus to the vestibule of the female reproductive tract that receives the male penis during coitus. 595

vallate papillae (val′āt pă-pil′e) The largest papillae on the surface of the tongue. They are arranged in an inverted V-shaped pattern at the posterior portion of the tongue. 306

vasectomy (vă-sek′tŏ-me, va-zek′tŏ-me) Surgical removal of portions of the ductus deferentia to induce infertility. 586

vasoconstriction (va″zo-kon-strik′shun) Narrowing of the lumina of blood vessels due to contraction of the smooth muscles in their walls. 414

vasodilation (va″zo-di-la′shun) Widening of the lumina of blood vessels due to relaxation of the smooth muscles in their walls. 414

vasomotor (va″zo-mo′tor) **center** A cluster of nerve cell bodies in the medulla oblongata that controls the diameter of blood vessels. It is therefore important in regulating blood pressure. 417

vein A blood vessel that conveys blood toward the heart. 357, 381

vena cava (ve′nă ka′vă) One of two large vessels that return deoxygenated blood to the right atrium of the heart. 389

ventilation Breathing; the process of moving air into and out of the lungs. 450

ventral (ven′tral) Toward the front or facing surface; the opposite of dorsal; also called *anterior.* 10

ventricle (ven′trĭ-k′l) A cavity within an organ; especially those cavities in the brain that contain cerebrospinal fluid and those in the heart that contain blood to be pumped from the heart. 373

venule (ven′yool) A small vessel that carries venous blood from capillaries to a vein. 381

vermis (ver′mis) The coiled middle lobular structure that separates the two cerebellar hemispheres. 262

vertebral (ver′tĕ-bral) **canal** The tubelike cavity extending through the vertebral column that contains the spinal cord; also called the *spinal canal.* 273

vestibular window An oval opening in the bony wall between the middle and inner ear, into which the footplate of the stapes fits; also called the *oval window.* 319

vestibule (ves′tĭ-byool) A space or cavity at the entrance to a canal, especially that of the nose, inner ear, or vagina. 321, 596

villus (vil′us); pl., *villi* A minute projection that extends outward into the lumen from the mucosal layer of the small intestine. 519

viscera (vis′er-ă) The organs within the abdominal or thoracic cavities. 14

visceral (vis′er-al) Pertaining to the membranous covering of the viscera. 16

visceral peritoneum (per″ĭ-tŏ-ne′um) A serous membrane that covers the surfaces of abdominal viscera. 17

visceral pleura (ploor′ă) A serous membrane that covers the surfaces of the lungs. 16

visceroceptor (vis″er-ŏ-sep′tor) A sensory receptor located within body organs that responds to information concerning the internal environment. 302

vitamin Any of various organic substances that are required in diverse functions throughout the body. Most vitamins cannot be synthesized by the body, and so they must be obtained in the diet. 544

vitreous (vit′re-us) **humor** The transparent gel that occupies the space between the lens and retina of the eyeball. 314

vulva (vul′vă) The external genitalia of the female that surround the opening of the vagina. 596

W

white matter Bundles of myelinated axons located in the central nervous system. 249

yellow marrow (mar′o) Specialized lipid storage tissue within bone cavities. 121

Z

zygote (zi′gōt) A fertilized egg cell formed by the union of a sperm cell and an ovum. 618

zymogen (zi′mŏ-jen) An inactive enzyme that becomes active when part of its structure is removed by the action of another enzyme or by some other means; also called a *proenzyme.* 529

CREDITS

Photos

Table of Contents

Unit 1: © James Amos/Corbis; **Unit 2:** © Kennan Ward/Corbis; **Unit 3:** © Hermann Eisenbeiss/Photo Researchers, Inc.; **Unit 4:** © Michael P. Gadomski/Photo Researchers, Inc.; **Unit 5:** © Glyn Kirk/Tony Stone Images

Chapter 1

Opener: © James Amos/Corbis; **1.1:** Courtesy of the New York Academy of Medicine Library; **1.2:** © Stock Montage; **1.5a:** © Lester V. Bergman & Associates, Inc.; **1.5b, 1.5c:** © Hank Morgan/Photo Researchers, Inc.; **1.3a, 1.3b:** Kent M. Van De Graaff; **1.7a:** © Dr. Sheril D. Burton

Chapter 2

Opener: © Chromosohm Media, Inc./Corbis

Chapter 3

Opener: Herbert L. Mirels PhD, Photographer; **3.3a:** © Keith R. Porter; **3.3b:** © Richard Chao; **3.4a, 3.4b:** Courtesy of Kwang W. Jeon; **3.5a, 3.5b, 3.5c, 3.5d:** M. M. Perry & A. B. Gilbert, *Journal of Cell Science* 39:257–72, 1979/Company of Biologists Ltd.; **3.14, 3.21:** © Richard Chao; **3.22a, 3.23a:** © Keith R. Porter; **3.24a:** © David M. Phillips/Visuals Unlimited; **3.33a-e:** © Ed Reschke

Chapter 4

Opener: © Kees Van Den Berg/Photo Researchers, Inc.; **4.1a:** © Ray Simons/Photo Researchers, Inc.; **4.1b, 4.1c, 4.1d, 4.1e, 4.2a, 4.2b, 4.2c,** © Ed Reschke; **4.5b:** © Biophoto/Science Source/Photo Researchers, Inc.; **4.6a:** © Ed Reschke; **4.6b:** © Biology Media/Robert Knauft, Photographer/Photo Researchers, Inc.; **4.6c:** © Ed Reschke/Peter Arnold, Inc.; **4.6d:** Ed Reschke; **4.6e:** © Ed Reschke; **4.6f:** © Ed Reschke/Peter Arnold, Inc.; **4.7a, 4.7b, 4.7c:** © Ed Reschke/Peter Arnold, Inc.; **4.8b:** © Ed Reschke; **4.10a:** © Ed Reschke/Peter Arnold, Inc.; **4.10b, 4.10c:** © Ed Reschke; **4.11a:** © Manfred Kage/Peter Arnold, Inc.; **4.12a:** © Larry Mulvehill/Photo Researchers, Inc.

Chapter 5

Opener: © Kennan Ward/Corbis; **5.3:** © Bruno P. Zehnder/Peter Arnold, Inc.; **5.6:** © Dr. Sheril D. Burton; **5.7:** © James M. Clayton; **5.9a:** World Health Organization; **5.9b:** George P. Bogumill, M.D.; **5.10a, 5.10b:** © Dr. Sheril D. Burton; **5.11a:** © Michael Abbey/Photo Researchers, Inc.; **5.11b:** Dr. Kerry L. Openshaw; **5.12b:** © John D. Cunningham/Visuals Unlimited; **p. 109 top:** © Zeva Oelbaum/Peter Arnold, Inc.; **p. 109 middle:** © Dr. P. Marazzi/SPL/Photo Researchers, Inc.; **p. 109 bottom:** © James Stevenson/SPL/Photo Researchers, Inc.; **5.15:** © John Radcliffe/SPL/Photo Researchers, Inc.

Chapter 6

Opener: © Werner H. Muller/Peter Arnold, Inc.; **6.4b:** Ted Conde; **6.8:** © Biophoto Associates/Photo Researchers, Inc.; **6.9a:** From R. G. Kessel and R. H. Kardon: *Tissues and Organs: A Text-Atlas of Scanning Electron Microscopy,* W.H. Freeman and Company © 1979; **6.9b, 6.12:** © Ed Reschke; **6.13, 6.20a, 6.20b, 6.24a:** Courtesy of Utah Valley Regional Medical Center, Dept. of Radiology; **6.31:** Kent M. Van De Graaff; **6.44e:** Reprinted courtesy Eastman Kodak Company

Chapter 7

Opener: © Michael Rosenfeld/Tony Stone Images; **7.1:** © Paolo Koch/Photo Researchers, Inc.; **7.2:** © Bruce Curtis/Peter Arnold, Inc.; **7.8:** Courtesy of Utah Valley Regional Medical Center, Department of Radiology; **7.19a, 7.19b, 7.19c, 7.19d, 7.19e, 7.19f, 7.19g, 7.19h:** © Dr. Sheril D. Burton; **7.20a:** Kent M. Van De Graaff; **7.20b, 7.20c:** © Dr. Sheril D. Burton; **7.20d:** Kent M. Van De Graaff; **7.20e, 7.20f:** © Dr. Sheril D. Burton; **7.21a, 7.21b:** Kent M. Van De Graaff; **7.22:** © Lester V. Bergman & Associates, Inc.; **7.23, 7.24a, 7.24b:** SIU, School of Medicine

Chapter 8

Opener: © Craig Lovell/Corbis; **8.3b:** © Ed Reschke; **8.6a:** International Bio-Medical, Inc.; **8.6b:** Stuart Ira Fox; **8.8a, 8.8b:** Kent M. Van De Graaff; **8.9a:** © Dr. H. E. Huxley; **8.16:** Hans Hoppler, *Respiratory Physiology* 44:94 (1981)

Chapter 9

Opener: © Hermann Eisenbeiss/Photo Researchers, Inc.; **9.2:** © Ed Reschke; **9.13:** © John Heuser, Washington University, School of Medicine, St. Louis, MO

Chapter 10

Opener: © U.S. Dept. of Defense/Corbis; **10.3a, 10.3b:** Kent M. Van De Graaff; **10.20:** © Monte S. Buchsbaum, M.D.; p. 286: Kent Van De Graaff

Chapter 11

Opener: © Chromosohm Media, Inc./Corbis

Chapter 12

Opener: © Arvind Garg/Corbis; **12.6:** © Dr. Sheril D. Burton; **12.13:** © Thomas Sims; **12.18:** Kent Van De Graaff; **12.28:** © Penny Tweedie/Tony Stone Images; **12.29:** © Emil Muench/Photo Researchers, Inc.; **12.30:** Courtesy Dr. Stephen Clark

Chapter 13

Opener: © Morton Beebe Photography/Corbis; **13.14:** © Fred Hossler/Visuals Unlimited; **13.18:** © Ed Reschke; **13.23:** © Lester V. Bergman & Associates, Inc.; **13.24:** © Lester V. Bergman & Associates, Inc.

Chapter 14

Opener: © Michael P. Gadomski/Photo Researchers, Inc.; **14.2b:** © Bill Longcore/Science Source/Photo Researchers, Inc.; **14.5:** Stuart Ira Fox

Chapter 15

Opener: © Michael T. Sedam/Corbis; **15.26a, 15.26b:** Donald S. Bain from Hurst et al: *The Heart*, 5/E, Fig. 47-4, pg. 1165. Reproduced with the Permission of The McGraw Hill Companies.; **15.27a:** American Lung Association; **15.27b:** © Lewis Lainey

Chapter 16

Opener: © Roger Wood/Corbis; **16.9:** From E. K. Markell and M. Vogue, *Medical Parasitology*, , 5th ed, W.B. Saunders; **16.15:** © Blair Seitz/Photo Researchers, Inc.

Chapter 17

Opener: © Morton Beebe Photography/Corbis; **17.2:** Kent M. Van De Graaff; **17.17a:** From R. G. Kessel and C. Y. Shih, *Scanning Electron Microscopy in Biology*; **17.17b:** © Dr. Jeremy Burgess/SPL/Photo Researchers, Inc.

Chapter 18

Opener: © F. Stuart Westmorland/Corbis; **18.4c:** © CNRI/Phototake, Inc.; **18.7:** © David

M. Phillips/Visuals Unlimited; **18.10:** Edward C. Vasquez R.T. C.R.T./Dept. of Radiologic Technology, Los Angeles City College

Chapter 19

Opener: © Adam Woolfitt/Corbis; **19.2:** © SPL/Photo Researchers, Inc.; **19.20:** © Per H. Kjeldsen, University of Michigan, Ann Arbor; **19.21:** Kent M. Van De Graaff; **19.24b:** © SIU/Peter Arnold, Inc.

Chapter 20

Opener: © David R. Frazier/Photo Researchers, Inc.; **20.6:** © Manfred Kage/Peter Arnold, Inc.; **20.7:** © Ed Reschke; **20.25:** From W.A. Sodeman and Sodeman Sodeman's *Pathologic Physiology*, 6th ed., pg. 861, 1979 W.B. Saunders Co.; **20.27c:** © Victor B. Eichler; **20.38a, 20.38b:** John Van De Graaff, DDS & Larry S. Pierce, DDS both at Northwestern University; **20.39a:** © Carroll Weiss/Camera M.D. Studios; **20.39b:** © Dr. Sheril D. Burton

Chapter 21

Opener: © Eunice Harris/Photo Researchers, Inc.

Chapter 22

Opener: © Glyn Kirk/Tony Stone Images; **22.3g, 22.3h:** © Dr. Landrum Shettles; **22.11b:** © Biophoto Associates/Photo Reseachers, Inc.; **22.12a, 22.12b:** © Ed Reschke; **22.12c, 22.12d:** © Manfred Kage/Peter Arnold, Inc.; **22.12e:** © Ed Reschke

Chapter 23

Opener: © Kim Westerskov/Tony Stone Images; **23.5:** © Ed Reschke; **23.7a, 23.7b:** © Ed Reschke; **23.8a, 23.8b:** From R. J. Blandau *A Textbook of Histology*, 10th ed. Chapman & Hall; **23.9:** © Dr. Landrum Shettles; **23.13:** © Martin M. Rotker; **p. 610: a:** Courtesy Gyno Pharma, Inc., **b-g:** © Times Mirror Higher Education Group, Inc./Bob Coyle, photographer; **h:** © Hank Morgan/ Science Source/Photo Researchers, Inc.; **p. 614 :** © Ed Reschke

Chapter 24

Opener: © Michael Viard/Peter Arnold, Inc.; **24.7a:** © Donald Yeager/Camera M.D. Studios; **24.11:** Kent Van De Graaff; **24.13a:** © Donald Yeager/Camera M.D. Studios; **24.13b:** © Dr. Landrum Shettles; **24.13c:** © Petit Format/ Nestle/Science Source/Photo Researchers, Inc.; **24.13d, 24.13e, 24.13f:** © Donald Yeager/Camera M.D. Studios; **24.13g:** © Petit Format/Nestle /Science Source/Photo Researchers, Inc.; **24.15:** Babu, A. & Hirschhom, K. *A Guide to Human Chromosomes Reflects* 3e White Plains, NY: March of Dimes Birth Defects Foundation, BD:OAS, 28(2), 1992. With permission of copyright holder.; **24.21:** © Lester V. Bergman & Associates, Inc.

Appendix B Reference Plates

Figure 1-5, 6a-b, 7a-c: Kent M. Van De Graaff; **8:** From B. Vidic and F. R. Suarez, *Photographic Atlas of the Human Body* (Pl. 27, Pg. 34), 1984, St. Louis, Mosby; **9:** From B. Vidic and F. R. Suarez, *Photographic Atlas of the Human Body* (Pl. 34, Pg. 48), 1984, St. Louis, Mosby; **10:** © Times Mirror Higher Education Group, Inc./Karl Rubin, photographer; **11:** From B. Vidic and F. R. Suarez, *Photographic Atlas of the Human Body* (Pl. 113, Pg. 182), 1984, St. Louis, Mosby; **12:** © Times Mirror Higher Education Group, Inc./Karl Rubin, photographer; **13:** From B. Vidic and F.R. Suaraz, *Photographic Atlas of the Human Body* (Pl. 39, Pg. 58), 1984, St. Louis, Mosby; **14:** From B. Vidic and F. R. Suarez, *Photographic Atlas of the Human Body* (Pl. 148, Pg. 240), 1984, St. Louis, Mosby; **15,16:** Kent M. Van De Graaff; **17:** From B. Vidic and F.R. Suarez, *Photographic Atlas of the Human Body* (Pl. 206, Pg. 338), 1984, St. Louis, Mosby; **18,19,20,21:** © Times Mirror Higher Education Group, Inc./Karl Rubin, photographer

Line Art

Chapter 3

3.22b: From David Shier, et al., *Hole's Human Anatomy & Physiology*, 7th ed. Copyright © 1996 Times Mirror Higher Education Group, Inc., Dubuque, Iowa. All Rights Reserved. Reprinted by permission.

Chapter 8

8.2: From John W. Hole, Jr., *Human Anatomy and Physiology*, 6th ed. Copyright © 1993 Times Mirror Higher Education Group, Inc., Dubuque, Iowa. All Rights Reserved. Reprinted by permission.

Chapter 10

10.6: From David Shier, et al., *Hole's Human Anatomy & Physiology*, 7th ed. Copyright © 1996 Times Mirror Higher Education Group, Inc., Dubuque, Iowa. All Rights Reserved. Reprinted by permission; **10.16:** From Joan G. Creager, *Human Anatomy and Physiology*, 2d ed. Copyright © 1992 Joan G. Creager. Reprinted by permission of Times Mirror Higher Education Group, Inc., Dubuque, Iowa. All Rights Reserved; **10.19:** From David Shier, et al., *Hole's Human Anatomy & Physiology*, 7th ed. Copyright © 1996 Times Mirror Higher Education Group, Inc., Dubuque, Iowa. All Rights Reserved. Reprinted by permission; **10.23:** From Joan G. Creager, *Human Anatomy and Physiology*, 2d ed. Copyright © 1992 Joan G. Creager. Reprinted by permission of Times Mirror Higher Education Group, Inc., Dubuque, Iowa. All Rights Reserved.

Chapter 12

12.7, 12.19: From David Shier, et al., *Hole's Human Anatomy & Physiology*, 7th ed. Copyright © 1996 Times Mirror Higher Education Group, Inc., Dubuque, Iowa. All Rights Reserved. Reprinted by permission.

Chapter 13

13.10: From David Shier, et al., *Hole's Human Anatomy & Physiology*, 7th ed. Copyright © 1996 Times Mirror Higher Education Group, Inc., Dubuque, Iowa. All Rights Reserved. Reprinted by permission.

Chapter 14

14.4: From John W. Hole, Jr., *Human Anatomy and Physiology*, 6th ed. Copyright © 1993 Times Mirror Higher Education Group, Inc., Dubuque, Iowa. All Rights Reserved. Reprinted by permission.

Chapter 16

16.12: From E. O. Feigel, "Physics In the Cardiovascular System" in *Physiology and Biophysics*, Vol. II, 20th ed., by T. C. Ruch and H. D. Patton. Copyright © 1974 W. B. Saunders. Reprinted by permission.

Chapter 20

20.2: From John W. Hole, Jr., *Human Anatomy and Physiology*, 5th ed. Copyright © 1990 Times Mirror Higher Education Group, Inc., Dubuque, Iowa. All Rights Reserved. Reprinted by permission; **20.9, 20.15, 20.26:** From David Shier, et al., *Hole's Human Anatomy & Physiology*, 7th ed. Copyright © 1996 Times Mirror Higher Education Group, Inc., Dubuque, Iowa. All Rights Reserved. Reprinted by permission.

Chapter 22

22.1: From David Shier, et al., *Hole's Human Anatomy & Physiology*, 7th ed. Copyright © 1996 Times Mirror Higher Education Group, Inc., Dubuque, Iowa. All Rights Reserved. Reprinted by permission; **22.11a:** From Stuart Ira Fox, *Human Physiology*, 5th ed. Copyright © 1996 Times Mirror Higher Education Group, Inc., Dubuque, Iowa. All Rights Reserved. Reprinted by permission; **22.11c:** From David Shier, et al., *Hole's Human Anatomy & Physiology*, 7th ed. Copyright © 1996 Times Mirror Higher Education Group, Inc., Dubuque, Iowa. All Rights Reserved. Reprinted by permission; **page 590:** From David Shier, et al., *Hole's Human Anatomy & Physiology*, 7th ed. Copyright © 1996 Times Mirror Higher Education Group, Inc., Dubuque, Iowa. All Rights Reserved. Reprinted by permission.

Chapter 23

23.1, 23.12: From David Shier, et al., *Hole's Human Anatomy & Physiology*, 7th ed. Copyright © 1996 Times Mirror Higher Education Group, Inc., Dubuque, Iowa. All Rights Reserved. Reprinted by permission.

Illustrators

Ernest Beck

15.24, 24.7B, 24.8, page 639

Sam Collins

17.5, 22.16

Chris Creek

1.3, 1.4, 1.9, page 18, 4.3, 4.4, 4.8A & C, 5.1, 5.2, 5.5, 5.8, 5.13, 5.14, 6.1, 6.5, 6.6, 6.7, 6.10, 6.11, 6.14, 6.15, 6.16, 6.17, 6.18, 6.19, 6.21, 6.22, 6.23,

6.24B & C, 6.25, 6.26, 6.27, 6.28, 6.29, 6.30, 6.32, 6.33, 6.34, 6.35, 6.36, 6.37, 6.38, 6.39, 6.40, 6.41, 6.42, 6.43, 6.44A–D, page 153, 7.4, 7.5, 7.7, 7.9, 7.10, 7.11, 7.12, 7.13, 7.14, 7.15, 7.16, 7.17, 7.18, page 173, 8.7, Table 8.3, 8.18, 8.19, 8.20, 8.21, 8.22, 8.23, 8.24, 8.25, 8.26, 8.27, 8.28, 8.29, 8.30, 8.31, 8.32, 8.33, 8.34, 8.35, 8.36, 8.37, 8.38, 8.39A–C, 8.40A–B, page 222, Plate 1, Plate 2, Plate 3, Plate 4, Plate 5, Plate 6, Plate 7, 10.4, 10.5, 10.8, 10.12, 10.13, 10.14, 10.16, 10.21, 10.22, 10.24, 10.25, Table 10.5, 10.27, 10.28, page 286, Table 12.1, 12.1, 12.4, 13.1, 15.9, 16.16, 17.1, page 448, 18.8, 18.13, 18.14, 19.1, 19.3, 19.4, 19.5, 19.22, page 502, 22.7, 23.6, 23.16, 24.10

FineLine

3.6, 22.15, 24.17, 24.18, 24.19

Rob Gordon

8.5, 10.7, 10.15, 13.13, 14.3, 14.4, page 370, 20.17, 23.3

Rob Gordon/Tom Waldrop

10.1, 12.20, 23.2

Illustrious, Inc.

1.16, 1.17, 2.1, 2.2, 2.3, 2.4, 2.5, 2.6, 2.11, 2.15C, 2.16, 2.17, 3.10, 3.11, 3.12, 3.13, 3.18, 3.19, 3.20, 3.22B, 3.26, 3.28, 3.29, 3.30, 3.31, 9.10, 9.11, 9.12, 11.6, 12.24, 13.8, 13.9, 13.12, 13.15, 13.20, 13.21, 13.22, 14.6, 16.3, 16.5, 16.7, 16.10, 16.11, 16.14, 17.6, 17.7, 17.8, 17.10, 17.11, 17.13, 17.14, 17.15, 17.16, 18.15, 18.16, 18.19, 18.22, 18.23, 19.16, 19.18, 20.32, 20.33, 20.34, 20.37, 21.1, 21.8, 21.10, 21.12, 21.14, page 563, 21.15, 21.16, 22.2, 22.5, 22.9, 23.14, 23.17, 23.18

J&R Art Services

16.1

Ruth Krabach

3.23B, 15.25, 24.12

Rictor Lew

1.6, 1.13, 3.1, 4.1, 4.2, 4.6, 4.7, 4.10, 4.11, 4.12, page 94, 6.2, 6.3, 9.5, 9.6, 9.7, 9.14, page 242, page 246, 10.9, 10.18, 11.5, 11.7, 12.8, 12.9, 12.10, 12.14, 12.21, 12.22, 12.23, page 332, 13.7, 13.16, 13.17, 13.19, 14.1, 15.6, 15.20, 15.22, 17.4, 18.1, 18.2, 18.3, 18.4A&B, 18.9, 19.7, 19.9, 19.13, 19.15, 19.17, 19.23, 20.3, 20.4, 20.18, 20.22, 20.23, 20.31, 22.4, 22.11A, 23.10, 24.3, 24.9

Rictor Lew/Diphrent Strokes

page 476

Bill Loechel

13.10, page 355, 19.6, 20.5, 20.14, 20.15, 20.24, 20.28, 20.40, page 542, 22.8

Bill Loechel/Tom Waldrop

20.26

Rob Margulies/Tom Waldrop

8.1, 15.10, 15.13, 15.14, 15.15, 15.16, 15.17, 15.19, 15.21, 15.23, 20.8

Nancy Marshburn/Tom Waldrop

5.11C, 5.16, page 113

Steve Moon

11.3, 12.5, 12.12, 15.1, 15.11, 16.8, 18.5, 18.18, 20.9, 20.10, 20.13

Diane Nelson

Table 15.2, 15.2A&B, 15.3, 15.12, 15.18, 22.11C, 24.1A&B

Mark Nero

3.9, 3.16, 3.17, 8.11, 8.12, 8.13, 9.8, 9.15, 16.4, 19.14, 19.19

Felecia Paras

1.8A–C, 1.11, 1.12, 1.14, page 22, 3.24B, 3.32, 3.33A–E, 4.9, 9.3, 9.4, 10.23, 10.26, 11.1, 11.2, 11.4, page 299, 12.3, 12.11, 12.25, 12.26, Table 15.1, 18.12, 18.20, 19.8, 19.10, 20.16, 20.20, 20.30, 22.10

Precision Graphics

12.27, 23.11

Mike Schenk

5.12A, 7.6, 8.9B, 12.15, 12.17, 13.6, 20.12, 20.27, 24.2, 24.6, 24.14, 24.16, 24.20, 24.22

Schultz/Waldrop

17.3

Tom Sims

16.13, page 423, 18.17

Tom Waldrop

6.4A, 7.3, 8.2, 8.3A, 8.4, 10.6, 10.10, 10.11, 13.4, 13.5, 18.6, 20.19, 22.1A&B, 22.3A–F, 22.13, page 585, page 590, 23.1, 23.4, 23.12, 23.19

Tom Waldrop/Rictor Lew

12.2

John Walters & Assoc.

20.11

INDEX

B

D

E

F

G

I

J

M

N

O

P

Q

R

S

T

U

V

W

Z